AF721559

Stress-Resilient Crops

CABI BIOTECHNOLOGY SERIES

Biotechnology, in particular the use of transgenic organisms, has a wide range of applications including agriculture, forestry, food and health. There is evidence that it could make a major impact in producing plants and animals that are able to resist stresses and diseases, thereby increasing food security. There is also potential to produce pharmaceuticals in plants through biotechnology, and provide foods that are nutritionally enhanced. Genetically modified organisms can also be used in cleaning up pollution and contamination. However, the application of biotechnology has raised concerns about biosafety, and it is vital to ensure that genetically modified organisms do not pose new risks to the environment or health. To understand the full potential of biotechnology and the issues that relate to it, scientists need access to information that not only provides an overview of and background to the field, but also keeps them up to date with the latest research findings.

This series, which extends the scope of CABI's successful "Biotechnology in Agriculture" series, addresses all topics relating to biotechnology including transgenic organisms, molecular analysis techniques, molecular pharming, in vitro culture, public opinion, economics, development and biosafety. Aimed at researchers, upper-level students and policy makers, titles in the series provide international coverage of topics related to biotechnology, including both a synthesis of facts and discussions of future research perspectives and possible solutions.

Titles Available

1. Animal Nutrition with Transgenic Plants
   Edited by G. Flachowsky
2. Plant-derived Pharmaceuticals: Principles and Applications for Developing Countries
   Edited by K.L. Hefferon
3. Transgenic Insects: Techniques and Applications
   Edited by M.Q. Benedict
4. Bt Resistance: Characterization and Strategies for GM Crops Producing Bacillus thuringiensis Toxins
   Edited by Mario Soberón, Yulin Gao and Alejandra Bravo
5. Plant Gene Silencing: Mechanisms and Applications
   Edited by Tamas Dalmay
6. Ethical Tensions from New Technology: The Case of Agricultural Biotechnology
   Edited by Harvey James
7. GM Food Systems and their Economic Impact
   Tatjana Brankov and Koviljko Lovre
8. Endophyte Biotechnology: Potential for Agriculture and Pharmacology
   Edited by Alexander Schouten
9. Forest Genomics and Biotechnology
   Richard Meilan and Matias Kirst
10. Transgenic Insects: Techniques and Applications, 2nd Edition
    Edited by M.Q. Benedict and Maxwell J. Scott
11. Plant Omics: Advances in Big Data Biology
    Edited by Hajime Ohyanagi, Eiji Yamamoto, Ai Kitazumi and Kentaro Yano
12. Next-generation Sequencing and Agriculture
    Edited by Phillip Bayer and Dave Edwards
13. Aquaculture and Fisheries Biotechnology: Genetic Approaches
    Rex A Dunham
14. Algal Biotechnology
    Edited by Qiang Wang
15. Genetic Engineering of Vegetable Crops
    Edited by Pritam Kalia
16. Plant Breeding Technology: Future Trends and Challenges
    Edited by Zeba Khan
17. Stress-Resilient Crops: Coordinated Omics-CRISPR-Nanotechnology Strategies
    Edited by Jen-Tsung Chen

Stress-Resilient Crops

Coordinated Omics-CRISPR-Nanotechnology Strategies

Edited by

Jen-Tsung Chen

Department of Life Sciences, National University of Kaohsiung, Taiwan

CABI

CABI is a trading name of CAB International

CABI
Nosworthy Way
Wallingford
Oxfordshire OX10 8DE
UK

Tel: +44 (0)1491 832111
E-mail: info@cabi.org
Website: www.cabi.org

CABI
200 Portland Street
Boston
MA 02114
USA

T: +1 (617)682-9015
E-mail: cabi-nao@cabi.org

© CAB International 2025. All rights, including for text and data mining, AI training, and similar technologies, are reserved. No part of this publication may be reproduced in any form or by any means, electronically, mechanically, by photocopying, recording or otherwise, without the prior permission of the copyright owners.

The views expressed in this publication are those of the author(s) and do not necessarily represent those of, and should not be attributed to, CAB International (CABI). Any images, figures and tables not otherwise attributed are the author(s)' own. References to internet websites (URLs) were accurate at the time of writing.

CAB International and, where different, the copyright owner shall not be liable for technical or other errors or omissions contained herein. The information is supplied without obligation and on the understanding that any person who acts upon it, or otherwise changes their position in reliance thereon, does so entirely at their own risk. Information supplied is neither intended nor implied to be a substitute for professional advice. The reader/user accepts all risks and responsibility for losses, damages, costs and other consequences resulting directly or indirectly from using this information.

CABI's Terms and Conditions, including its full disclaimer, may be found at https://www.cabidigitallibrary.org/terms-and-conditions.

A catalogue record for this book is available from the British Library, London, UK.

ISBN-13: 9781800627284 (hardback)
9781800627291 (ePDF)
9781800627307 (ePub)

DOI: 10.1079/9781800627307.0000

Commissioning Editor: Ward Cooper
Editorial Assistant: Emma McCann
Production Editor: James Bishop

Typeset by Exeter Premedia Services Pvt Ltd, Chennai, India
Printed and bound in the UK by CPI Group (UK) Ltd, Croydon, CR0 4YY

Contents

List of Contributors

Banzeer Ahsan Abbasi, Department of Botany, Rawalpindi Women University, 6th Road, Satellite Town, Rawalpindi 46300, Pakistan. Email: benazirahsanabbasi786@gmail.com

Naglaa A. Abdallah, Department of Genetics Faculty of Agriculture, Cairo University, Cairo, Egypt and National Biotechnology Network of Expertise, ASRT, Egypt. Email: naglaa.abdallah@agr.cu.edu.eg

Sarfraz Ahmad, Department of Genetics and Plant Breeding, College of Agriculture Mandawa, Swami Keshewanand Rajasthan Agriculture University, Bikaner, Rajasthan, India. Email: sarfraz_firoz@rediffmail.com

Shumaila Ahmad, Department of Plant Sciences, Faculty of Biological Sciences, Quaid-i-Azam University Islamabad, 45320, Pakistan

SK MD Ajaharuddin, Department of Entomology, Bidhan Chandra Krishi Viswavidyalaya, Mohanpur, Nadia, West Bengal, India. Email: iamajaharuddin@gmail.com

Mohd. Sayeed Akhtar, Department of Botany, Gandhi Faiz-e-Aam College, Shahjahanpur 242001, Uttar Pradesh, India. Email: sayeedbot@gmail.com

Sayed Abdul Akher, Department of Genetics and Plant Breeding, Tobacco Research Institute, Graduate School of Chinese Academy of Agricultural Science, China. Email: akher.hstu090@gmail.com

Jameel M Al-Khayri, Department of Agricultural Biotechnology, College of Agriculture and Food Sciences, King Faisal University, Al-Ahsa 31982, Saudi Arabia. Email: jkhayri@kfu.edu.sa

Zunnorain Ali, Department of Plant Sciences, Faculty of Biological Sciences, Quaid-i-Azam University Islamabad, 45320, Pakistan. Email: zunnorainali5@gmail.com

Arumugam Vijaya Anand, Department of Human Genetics and Molecular Biology, Bharathiar University, Coimbatore 641046, Tamil Nadu, India.

Anshu, Department of Medical Oncology, All India Institute of Medical Sciences, New Delhi, India. Email: anshu08021998@gmail.com

Boro Arthi, Department of Human Genetics and Molecular Biology, Bharathiar University, Coimbatore 641046, Tamil Nadu, India.

Muthukrishnan Arun, Department of Biotechnology, Bharathiar University, Coimbatore 641046, Tamil Nadu, India

Zaman Ashraf, Department of Chemistry, Rawalpindi Women University, 46300, Rawalpindi, Punjab, Pakistan. Email: zaman.ashraf@f.rwu.edu.pk

Sunali Bandral, Nematology Laboratory, Department of Zoology, University of Jammu, Jammu (J&K) India-180006. Email: sunalibandral005@gmail.com

Akshatha Banadka, Department of Biotechnology and Genetics, School of Sciences, Jain (Deemed to be University), Bengaluru 560 027, Karnataka, India and Department of Life Sciences, School of Sciences, CHRIST (Deemed to be University), Bengaluru 560 029, Karnataka, India. Email: b.akshatha@jainuniversity.ac.in

Michael Baum, Department of Biotechnology, International Centre for Agricultural Research in the Dry Areas (ICARDA), Giza, Egypt. Email: m.baum@cgiar.org

Ahamed Kabeer Umera Begam, Department of Biochemistry, Holy Cross College, Affiliated to Bharathidasan University, Tiruchirappalli 620002, Tamil Nadu, India. Email: umera150176@gmail.com

Sirajudeen Benazir Begum, PG and Research Department of Zoology, V. O. Chidambaram College, Thoothukudi 628008, Tamil Nadu, India.

Tajudheen Nargis Begum, Department of Biotechnology, Jamal Mohamed College, (Autonomous), Tiruchirapalli 620020, Tamil Nadu, India.

Vikas Belwal, Department of Tree Improvement, College of Forestry Ranichauri, VCSG Uttarakhand University of Horticulture and Forestry, Bharsar, Uttarakhand, India. Email: vikasbelwal@hotmail.com

Khadija Benamar, Department of Plant Protection, Phytopathology Unit, Ecole Nationale d'Agriculture de Meknès, Km10, Rte Haj Kaddour, BP S/40, Meknès 50001, Morocco and Microbial Biotechnology and Bioactive Molecules Laboratory, Sciences and Technologies Faculty, Sidi Mohamed Ben Abdellah University, P.O. Box 2202, Imouzzer Road, Fez, Morocco.

Rashi Bhavsar, Department of Biotechnology, The Oxford College of Science, HSR layout, Bangalore, 560102, Karnataka, India. Email: bhavsar.rashi@gmail.com

Jitul Bora, Department of Agricultural Biotechnology, Assam Agricultural University, Jorhat, Assam, India. Email: jitulbora01@gmail.com

Sudipta Sankar Bora, Multi-disciplinary Research Unit, Jorhat Medical College and Hospital, Jorhat, Assam, India. Email: sudip.asm@gmail.com

Sayyada Bushra, Department of Botany, Aligarh Muslim University, Aligarh 202002, Uttar Pradesh, India. Email: sbushra012@gmail.com

Priyadarshini Deka, Hereditary Biosciences, Guwahati, Assam, India. Email: dekapriyadarshini@gmail.com

Vandana Chendanda Devaiah, Department of Biotechnology and Genetics, School of Sciences, Jain (Deemed to be University), Bengaluru 560 027, Karnataka, India. Email: vandana.cd@jainuniversity.ac.in

Sana Choudhary, Department of Botany, Aligarh Muslim University, Aligarh, U.P. 202002, India. Email: sana.12amu@gmail.com

Urvashi Dhiman, Natural Products and Medicinal Chemistry Division, CSIR-Indian Institute of Integrative Medicine, Jammu (J&K) India-180001 and Academy of Scientific and Innovative Research (AcSIR), Ghaziabad (Uttar Pradesh), India-201002. Email: Urvashi.iiim22a@acsir.res.in

Farzana Mustafa Era, Department of Genetics and Plant Breeding, Faculty of Agriculture, Gazipur Agricultural University, Gazipur 1706, Bangladesh. Email: farzana@bsmrau.edu.bd

Rachid Ezzouggari, Department of Plant Protection, Phytopathology Unit, Ecole Nationale d'Agriculture de Meknès, Km10, Rte Haj Kaddour, BP S/40, Meknès 50001, Morocco and Laboratory of Biotechnology, Conservation, and Valorization of Natural Resources (LBCVNR), Faculty of Sciences Dhar El Mehraz, Sidi Mohamed Ben Abdallah University, Fez 30000, Morocco.

Shahla Faizan, Department of Botany, Aligarh Muslim University, Aligarh 202002, Uttar Pradesh, India. Email: sfaizan.bt.@amu.ac.in

Amin Fathi, Department of Agronomy, Ayatollah Amoli Branch, Islamic Azad University, Amol, Iran. Email: dr.aminfathi@gmail.com

Nourhan Fouad, Department of Biotechnology, International Centre for Agricultural Research in the Dry Areas (ICARDA), Giza, Egypt. Email: N.Mahmoud@cgiar.org

Rahul Gogoi, Department of Agricultural Biotechnology, Assam Agricultural University, Jorhat, Assam, India. Email: rahulgogoi.biotech@gmail.com

Pemula Gowtham, Department of Human Genetics and Molecular Biology, Bharathiar University, Coimbatore 641046, Tamil Nadu, India.

Prasoon Gupta, Natural Products and Medicinal Chemistry Division, CSIR-Indian Institute of Integrative Medicine, Jammu (J&K) India-180001 and Academy of Scientific and Innovative Research (AcSIR), Ghaziabad (Uttar Pradesh) India-201002. Email: guptap@iiim.ac.in

Ali Haider, Department of Chemistry, Quaid-I-Azam University, Islamabad, Pakistan. Email: ahaider@qau.edu.pk

Aladdin Hamwieh, Department of Biotechnology, International Centre for Agricultural Research in the Dry Areas (ICARDA), Giza, Egypt. Email: a.hamwieh@cgiar.org

Habeebuheen Fathima Seyed Mafia Haniff, Department of Zoology, Jamal Mohamed College, (Autonomous), Tiruchirapalli 620020, Tamil Nadu, India.

Daniel Heltberg, School of Biological Sciences, University of Aberdeen, Cruickshank Building, St Machar Dr, Old Aberdeen, Aberdeen, AB24 3UL, UK. Email: d.heltberg.22@abdn.ac.uk

Farin Tasnova Hia, Institute of Biotechnology and Genetic Engineering, Gazipur Agricultural University, Gazipur, Bangladesh. Email: farintasnova1998@gmail.com

MD Saddam Hosen, Department of Agroforestry, Bangladesh Agricultural University, Bangladesh. Email: saddamhosen1022@gmail.com

Akbar Hossain, Soil Science Division, Bangladesh Wheat and Maize Research Institute, Dinajpur 5200, Bangladesh. Email: akarhossainwrc@gmail.com

Alisha Hussain, Department of Botany, Aligarh Muslim University, Aligarh 202002, Uttar Pradesh, India. Email: alishahussain302@gmail.com

Muhammad Awais Ibrahim, Department of Plant Sciences, Faculty of Biological Sciences, Quaid-i-Azam University Islamabad, 45320, Pakistan. Email: awaisse1@gmail.com

Naila Ijaz, Nano composite lab, Department of Physics, University of Agriculture Faisalabad, Faisalabad 38000, Pakistan. Email: Nailaijaz1999@gmail.com

Shumaila Ijaz, Department of Burn and Plastic Surgery, Shenzhen Institute of Translational Medicine, The First Affiliated Hospital of Shenzhen University, School of Biomedical Engineering, Shenzhen University Medical School, Shenzhen University, Shenzhen 518035, China. Email: shumailaijaz1996@yahoo.com

Javed Iqbal, Department of Botany, Bacha Khan University, Charsadda, 24420, Khyber Pakhtunkhwa, Pakistan. Email: javed89qau@gmail.com

Rashid Iqbal, Department of Agronomy, Faculty of Agriculture and Environment, The Islamia University of Ba-hawalpur, Bahawalpur 63100, Pakistan and Department of Life Sciences, Western Caspian University, Baku, Azerbaijan. Email: rashid.iqbal@iub.edu.pk

A.K.M. Aminul Islam, Department of Genetics and Plant Breeding, Faculty of Agriculture, Gazipur Agricultural University, Gazipur 1706, Bangladesh. Email: aminulgpb@bsmrau.edu.bd

Md Aminul Islam, Department of Botany, Majuli College, Majuli, Assam, India-785106. Email: aminul.islam102@gmail.com

Sharmin Islam, Institute of Biotechnology and Genetic Engineering, Gazipur Agricultural University, Gazipur, Bangladesh. Email: sharminislam39983@gmail.com

Tofazzal Islam, Institute of Biotechnology and Genetic Engineering, Gazipur Agricultural University, Gazipur, Bangladesh. Email: tofazzalislam@bsmrau.edu.bd

Shivam Jasrotia, Department of Biosciences, Chandigarh University, Mohali (Punjab) India-140413. Email: shivamzoology@gmail.com

Nimitha K, Department of Genetics and Plant Breeding, Bidhan Chandra Krishi Viswavidyalaya, Mohanpur, Nadia, West Bengal-741252, India. Email: nimithak007@gmail.com

Sobia Kanwal, Department of Biology and Environmental Sciences, Allama Iqbal Open University, Islamabad-45320, Pakistan. Email: sobia.kanwal@aiou.edu.pk

Nitika Kapoor, Department of Botanical and Environmental Sciences, Guru Nanak Dev University, Amritsar, Punjab, India-143005. Email: nitikaarora8@gmail.com

Aritra Karmakar, Dept. of Biotechnology, Techno India University, EM-4, Sector V, Bidhannagar, Kolkata – 700091. Email: karmakara07@gmail.com

Swagata Karmakar, Department of Environmental Studies, Ram Lal Anand College, University of Delhi, Delhi, India-110021. Email: sway.swagata@gmail.com

Md. Omar Kayess, Institute of Biotechnology and Genetic Engineering, Gazipur Agricultural University, Gazipur, Bangladesh. Email: kayess.1405036@std.hstu.ac.bd

Mohsin Kazi, Department of Pharmaceutics, College of Pharmacy, POBOX-2457, King Saud University, Riyadh 11451, Saudi Arabia. Email: mkazi@ksu.edu.sa

Kashaf Khan, Department of Plant Sciences, Faculty of Biological Sciences, Quaid-i-Azam University Islamabad, 45320, Pakistan.

Shoaib Khan, Department of Chemistry, Abbottabad University of Science and Technology (AUST) Abbottabad 22500, Pakistan. Email: shoaibkhanswati@gmail.com

Marium Khatun, Department of Genetics and Plant Breeding, Faculty of Agriculture, Gazipur Agricultural University, Gazipur 1706, Bangladesh. Email: mariumgpb@bsmrau.edu.bd

Sandeep Kour, Department of Zoology, Guru Nanak Dev University, Amritsar, Punjab, India-143005. Email: sandeepkour.ch@gmail.com

Santanu Kundu, Division of Agronomy, ICAR-Indian Agricultural Research Institute, New Delhi, India. Email: santanukundu120@gmail.com

Deepak Kumar, Department of Zoology, Guru Nanak Dev University, Amritsar, Punjab, India-143005. Email: deepakdhaka415@gmail.com

Pankaj Kumar, Department of Crop Improvement, College of Forestry Ranichauri, VCSG Uttarakhand University of Horticulture and Forestry, Bharsar, Uttarakhand, India. Email: vaidya.pankal4@gmail.com

Prerna Kumari, University Department of Botany, Ranchi University, Ranchi, India-834008. Email: prekmri1999@gmail.com

Rinkee Kumari, Department of Botany, Aligarh Muslim University, Aligarh 202002, Uttar Pradesh, India. Email: krinki252@gmail.com

Ritika Kumari, University Department of Botany, Ranchi University, Ranchi, India-834008. Email: 8797900573ritikasi@gmail.com

Rachid Lahlali, Department of Plant Protection, Phytopathology Unit, Ecole Nationale d'Agriculture de Meknès, Km10, Rte Haj Kaddour, BP S/40, Meknès 50001, Morocco. Email: rlahlali@enameknes.ac.ma

Sameer Gunjan Lakra, Department of Botany, J. N. College, Dhurwa, Ranchi, India-834002. Email: gunjansameer5@gmail.com

Tariq Mahmood, Department of Plant Sciences, Faculty of Biological Sciences, Quaid-i-Azam University Islamabad, 45320, Pakistan. Email: tmahmood@qau.edu.pk

Arockiya Anita Margret, Department of Biotechnology and Bioinformatics, Bishop Heber College, (Autonomous), Tiruchirappalli 6200177, Tamil Nadu, India.

Binod Kumar Mahto, University Department of Botany, Ranchi University, Ranchi, India-834008. Email: krvinod09@gmail.com; binodkrmahto@ranchiuniversity.ac.in

Lamya Muazzeda Medha, Institute of Biotechnology and Genetic Engineering, Gazipur Agricultural University, Gazipur, Bangladesh. Email: lamyamedha98@gmail.com

Johiruddin Molla, Department of Botany, Ghatal Rabindra Satabarsiki Mahavidyalaya, Ghatal, Paschim Medinipur, West Bengal 721212, India. Email: johiruddinmolla@gmail.com

Kutubuddin Molla, Crop Improvement Division, ICAR-NRRI, Cuttack, Odisha-753006, India. Email: kutubuddin.molla@icar.gov.in

Akhtar Munir, Department of Chemistry, Quaid-I-Azam University, Islamabad, Pakistan. Email: akhtarmunir991@gmail.com

Faiza Munir, Department of Agricultural Sciences and Technology, Atta-ur-Rahman School of Applied Biosciences, National University of Sciences and Technology, Islamabad, Pakistan. Email: faiza.munir@asab.nust.edu.pk

Sk Rehan Murshed, State Aided College, Ghatal Rabindra Satabarsiki, Mahavidyalaya, Ghatal, Paschim Medinipur, West Bengal 721212, India. Email: rehan.keshpur@gmail.com

Ghulam Murtaza, School of Agriculture, Yunnan University, Kunming, Yunnan 650504, China. Email: murtazabotanist@gmail.com

Praveen Nagella, Department of Life Sciences, School of Sciences, Christ University, Hosur Road, Bengaluru-560029, India. Email: praveen.n@christuniversity.in

Puja Ohri, Department of Zoology, Guru Nanak Dev University, Amritsar, Punjab, India-143005. Email: ohri_puja_11@gmail.com

Ekta Pandey, Department of Botany, Aligarh Muslim University, Aligarh, U.P. 202002, India. Email: epandey251@gmail.com

Ripu Daman Parihar, Nematology Laboratory, Department of Zoology, University of Jammu, Jammu (J&K) India-180006. Email: ripuparihar@jammuuniversity.ac.in

Parkirti, Department of Zoology, Guru Nanak Dev University, Amritsar, Punjab, India-143005. Email: parkirti209@gmaiul.com

Damien J. Platen, International Rice Research Institute, Los Baños, Laguna 4031, Philippines. Email: J.Platten@irri.org

Anbazhagan Priya, PG and Research Department of Zoology, V. O. Chidambaram College, Thoothukudi 628008, Tamil Nadu, India.

Shampa Purkaystha, Department of Genetics and Plant Breeding, Centurion University of Technology and Management, Paralekhamundi, Odisha-761211, India. Email: shampa.rimi3@gmail.com

Saira Qadir, Department of Botany, Rawalpindi Women University, 6th Road, Satellite Town, Rawalpindi 46300, Pakistan. Email: saira.qadir@f.rwu.edu.pk

Khaled Radwan, National Biotechnology Network of Expertise, ASRT, Egypt and Department of Biotechnology, Agricultural Genetic Engineering Research Institute (AGERI), ARC, Giza, Egypt. Email: khradwan@gmail.com

Mohammad Ghani Raghi, Department of Horticulture, Faculty of Agriculture, Nangarhar University, Afghanistan. Email: mohammadghaniraghib@gmail.com

Manoharan Rajesh, Department of Biochemistry, Bharathiar University, Coimbatore 641046, Tamil Nadu, India.

Asirvatham Alwin Robert, Department of Endocrinology and Diabetes, Prince Sultan Military Medical City, Ministry of Defense, Riyadh 11159, Saudi Arabia.

Arokiaraj Sherlin Rosita, Department of Biotechnology and Bioinformatics, Bishop Heber College, (Autonomous), Tiruchirappalli 6200177, Tamil Nadu, India.

Sanjay S, Department of Biotechnology, School of Biological Sciences and Biotechnology, Goa University, Goa 403 206, India. Email: sanjaykarthikeyan253@gmail.com

Shoumik Saha, Division of Agriculture, Rural and Tribal Development (ARTD), Ramakrishna Mission Vivekananda Educational and Research Institute (RKMVERI), Morabadi, Ranchi-834008, Jharkhand, India. Email: shoumiksaha92@gmail.com

Shobana Sampath, Department of Biotechnology, Vel Tech Rangarajan Dr. Sagunthala R&D Institute of Science and Technology, Avadi-600072, Chennai, Tamil Nadu, India. Email: swtshoba@gmail.com

Palanisamy Sampathkumar, Department of Chemistry and Biosciences, SASTRA Deemed to be University, Srinivasa Ramanujan Centre, Kumbakonam 612 001, Tamil Nadu, India.

Saniya, Department of Horticulture, ICAR-National Research Centre for Grapes, Pune, Maharashtra, India. Email: saniyacg.17@gmail.com

Sumi Sarkar, Department of Genetics and Plant Breeding, Faculty of Agriculture, Gazipur Agricultural University, Gazipur 1706, Bangladesh. Email: sumisarkar@bsmrau.edu.bd

Gunjan Kumar Saurav, Insect Biology Laboratory, Department of Zoology, Rajiv Gandhi University, Doimukh, Arunachal Pradesh, India-791112. Email: gunjan.saurav@rgu.ac.in; gunjanshukla24@gmail.com

Shreya Sen, Hiyaa Lab, National Centre for Biological Sciences, Bengaluru-560097, Karnataka, India. Email: sshreya2k18@gmail.com

Navid Hussain Shah, Optoelectronic Research Center, School of Science, Minzu University of China, Beijing 100081, China. Email: navidshah742@gmail.com

Maksud Hasan Shaha, Department of Agronomy, Bidhan Chandra Krishi Viswavidyalaya, Mohanpur, Nadia, West Bengal, India. Email: maksudhasanshah@gmail.com

Tayyaba Shahzadi, Department of Chemistry, Government College for Women, University Sialkot, Sialkot 51310, Pakistan. Email: tayyaba.shehzadi@gcwus.edu.pk

Akash Shaikh, Department of Life Sciences, School of Sciences, Christ University, Hosur Road, Bengaluru-560029, India. Email: akash.shaikh@res.christuniversity.in

Alisha Sharma, Department of Zoology, Guru Nanak Dev University, Amritsar, Punjab, India-143005. Email: sharmaalisha405@gmail.com

Madhurjya Ranjan Sharma, Department of Agricultural Biotechnology, Assam Agricultural University, Jorhat, Assam, India. Email: madhurjyaranjansharma@gmail.com

Nidhi Sharma, Department of Botany, Aligarh Muslim University, Aligarh, U.P. 202002, India. Email: nicky.sharma165@gmail.com

Roohi Sharma, Department of Zoology, Guru Nanak Dev University, Amritsar, Punjab, India-143005. Email: sharmaroohi77.rs@gmail.com

Sonali Sharma, Nematology Laboratory, Department of Zoology, University of Jammu, Jammu (J&K) India-180006. Email: sharma03.sonali@gmail.com

Vaishali Sharma, Nematology Laboratory, Department of Zoology, University of Jammu, Jammu (J&K) India-180006. Email: vaishalisharma7468@gmail.com

Anchita Singh, Department of Zoology, Guru Nanak Dev University, Amritsar, Punjab, India-143005. Email: anchitasinghthakur@gmail.com

Nimrita Singh, Nematology Laboratory, Department of Zoology, University of Jammu, Jammu (J&K) India-180006. Email: namratarajput5610@gmail.com

Seweta Srivastava, Department of Plant Pathology, School of Agriculture, Lovely Professional University, Phagwara-144411, Punjab, India. Email: seweta.21896@lpu.co.ib

Kalpana Suman, University Department of Botany, Ranchi University, Ranchi, India-834008. Email: sumankalpana582@gmail.com

Gunna Sureshbabu Suruthi, Department of Human Genetics and Molecular Biology, Bharathiar University, Coimbatore 641046, Tamil Nadu, India.

Mohammed Taoussi, Department of Plant Protection, Phytopathology Unit, Ecole Nationale d'Agriculture de Meknès, Km10, Rte Haj Kaddour, BP S/40, Meknès 50001, Morocco and Environment and Valorization of Microbial and Plant Resources Unit, Faculty of Sciences, Moulay Ismail University, Meknes, Morocco.

Varsha Toppo, Department of Life Sciences, School of Sciences, Christ University, Hosur Road, Bengaluru-560029, India. Email: varsha.toppo@res.christuniversity.in

Zakir Ullah, Department of Plant Sciences, Faculty of Biological Sciences, Quaid-i-Azam University Islamabad, 45320, Pakistan. Email: zakirullah@bs.qau.edu.pk

Parul Verma, Department of Botany, Aligarh Muslim University, Aligarh 202002, Uttar Pradesh, India. Email: vermaparul9955@gmail.com

Vikram, Department of Zoology, Guru Nanak Dev University, Amritsar, Punjab, India-143005. Email: vikramhomosapiens29@gmail.com

Tabassum Yaseen, Department of Botany, Bacha Khan University, Charsadda, 24420, Khyber Pakhtunkhwa, Pakistan. Email: tabassumyaseen@bkuc.edu.pk

Lune Yirang, Insect Biology Laboratory, Department of Zoology, Rajiv Gandhi University, Doimukh, Arunachal Pradesh, India-791112. Email: lune.yirang@rgu.ac.in

Preface

Over the past decades, the world has faced critical crises chiefly due to the expanding human population, resulting in environmental pollution, and shortage of food and natural resources. The situation is made worse by climate change, mostly caused by rising global temperatures. To secure sufficient agricultural production, scientists are expected to solve problems by exploring knowledge of plant sciences, particularly highlighting stress-tolerant traits and underlying machinery. This can advance agricultural biotechnology and crop breeding focusing on crop improvement with climate-resilient capacities, which might ensure nutrition and food security amid the challenge of global climate change.

Nowadays, the demand to organize crop breeding programmes for developing climate-resilient crops is increasing. To support such demand, modern plant molecular biotechnology has been upgraded to achieve high-throughput, high-resolution, smart and precision manners such as multiple omics, CRISPR/Cas9-mediated genome editing, and nanotechnology. In the post-genomics era, the approaches of multiple omics continue to mine huge amounts of genetic resources and accumulate interesting key genes and, definitely, there is the need for efficient ways of genetic engineering to benefit crop breeding and agricultural production. In the past decade, an emerging genetic engineering technology, i.e. genome editing technology, has released the potential of plant genomes through the precise delivery of tolerant genes for crop improvement. The system of genome editing has evolved quickly, and now CRISPR/Cas9 has become the dominant technology. Thus, the integration and co-ordination of multiple omics and CRISPR/Cas9 for organizing omics-CRISPR breeding strategies inevitably become the next wave of important tasks in plant science. In addition, in recent years, nanotechnology has been introduced into diverse fields of plant science and has proved its great potential in mitigating plant stress responses as well as being a smart carrier in some ways to advance genomics and CRISPR approaches. The role of nanotechnology in the management of CRISPR-edited crop production has become an important topic. In the future, achieving climate-resilient agriculture needs the co-ordination of three mainstream technologies: multiple omics, CRISPR/Cas9, and nanotechnology.

In this book, the ways for vertical integration and horizontal co-ordination of these three crucial technologies is comprehensively described to advance crop breeding programmes towards stress-resilient agriculture. Strategies for combating diverse stressors and complicated or multiple stress conditions are proposed, covering stressors of salinity, temperature, drought, heavy metals, pests, pathogens etc. Also, the book proposes strategies to develop stress-tolerant crops with high-yield and high-quality traits by integrating or co-ordinating the three mainstream technologies listed above.

This is an ideal reference work to integrate the emerging field of multiple omics, CRISPR/Cas9, and nanotechnology by sharing crucial aspects of methods, applications and future directions, and it opens doors for students and researchers to efficiently overview these critical subtopics of plant science and technology. Hopefully, it will inspire ideas for future experiments and expand knowledge, and lead to better development of crops by scientists, plant biologists and crop breeders in the future. The book supports, through knowledge of stress-resilient crop-breeding technologies, the Sustainable Development Goals (SDGs) of the United Nations and the Paris Agreement, and the vision of sustainable agriculture.

Acknowledgement

The Editor, Dr Jen-Tsung Chen, appreciates the work of all chapter authors and the staff of CABI for their important contributions and assistance.

1 Developing Nutritionally Enhanced Crops through Approaches of Functional Genomic and CRISPR/Cas Technologies

Ritika Kumari[1#], Gunjan Kumar Saurav[2#], Kalpana Suman[1], Sameer Gunjan Lakra[3], Swagata Karmakar[4], Prerna Kumari[1], Md Aminul Islam[5], Lune Yirang[2] and Binod Kumar Mahto[1*]

[1]University Department of Botany, Ranchi University, Ranchi, India; [2]Insect Biology Laboratory, Department of Zoology, Rajiv Gandhi University, Doimukh, Arunachal Pradesh, India; [3]Department of Botany, J.N. College, Dhurwa, Ranchi, India; [4]Department of Environmental Studies, Ram Lal Anand College, University of Delhi, India; [5]Department of Botany, Majuli College, Majuli, Assam, India

Abstract

As the world's population continues to rapidly increase, guaranteeing food security and tackling nutritional gaps have become pressing issues of the utmost importance. In response, cutting-edge biotechnological approaches such as functional genomics and CRISPR/Cas technologies have emerged as powerful tools to revolutionize crop improvement. Functional genomics, with its ability to decipher gene functions, has played a pivotal role in identifying key genes associated with traits such as yield, stress tolerance, and nutrient content. Leveraging this knowledge, CRISPR/Cas technology has enabled precise and targeted genome editing, allowing for the modification of specific genes to enhance desired traits. This synergy between functional genomics and CRISPR/Cas has paved the way for the development of crops with heightened nutritional value, offering solutions to combat malnutrition and enhance overall human health. This chapter delves into the convergence of these technologies in creating nutritionally enhanced crops with improved yields, resilience, and nutritional profiles. Also, it contextualizes specific examples of successful applications of these technologies in developing nutritionally enhanced crops.

Keywords: functional genomics, CRISPR/Cas, nutritional value, nutritionally enhanced crop, stress tolerance

1.1 Introduction

The global challenge of ensuring food security and adequate nutrition for a rapidly growing population is increasingly complex due to the impacts of climate change, soil degradation, and diminishing arable land. According to the United Nations, the world's population is estimated to reach 9.7 billion by 2050, which necessitates a dramatic increase in both food production and

*Corresponding author: krvinod09@gmail.com

[#]These authors have contributed equally to this work and share first authorship.

© CAB International 2025. *Stress-Resilient Crops: Coordinated omics-CRISPR-Nanotechnology Strategies* (ed. Jen-Tsung Chen)
DOI: 10.1079/9781800627307.0001

nutritional quality to meet global demand (Fritz *et al.*, 2019). Traditional breeding methods have been instrumental in improving crops; however, these methods are often constrained by the time required for development and genetic variability within crop species. In contrast, the advent of advanced biotechnological tools, particularly functional genomics and CRISPR/Cas-based genome editing, has revolutionized plant science by enabling precise and efficient enhancement of crop nutritional profiles.

Functional genomics involves the comprehensive study of gene functions and interactions, which provides valuable insights into the biological processes that govern plant traits, including nutritional content. High-throughput technologies such as genomics, transcriptomics, proteomics, metabolomics, and phenomics have been instrumental in the systematic identification and characterization of genes and metabolic pathways associated with nutrient biosynthesis, accumulation, and regulation in crops (Kim *et al.*, 2021). These approaches have enabled researchers to gain a deeper understanding of the complex networks that control nutrient levels in plants. For instance, functional genomics has identified key regulatory networks and metabolic pathways involved in the biosynthesis of essential nutrients in staple crops. One important application is the development of Golden Rice, which is enriched with micronutrient β-carotene, a precursor to vitamin A. The identification of genes responsible for provitamin A synthesis, such as phytoene synthase (PSY) and phytoene desaturase (PDS), enabled the creation of this biofortified variety (Paine *et al.*, 2005). Similarly, functional genomics has elucidated iron and zinc transport mechanisms in crops, leading to the development of varieties with enhanced micronutrient content (Johnson *et al.*, 2011).

Recent advances in functional genomics have leveraged RNA sequencing (RNA-seq) to uncover candidate genes associated with increased accumulation of essential minerals. For example, a study by Sun *et al.* (2022) utilized transcriptomic analysis to identify genes related to iron and zinc accumulation in rice grains. This research provided targets for genetic improvement to enhance the mineral content of rice. Similarly, Zhang *et al.* (2021) identified genes involved in lysine and tryptophan biosynthesis in maize through transcriptomic analysis, addressing the deficiency of these essential amino acids in cereal crops. Metabolomic profiling has also played a crucial role in understanding and improving nutritional traits. Li *et al.* (2020) used metabolomics to enhance the understanding of flavonoid biosynthesis in maize, leading to the development of varieties with increased antioxidant content. Additionally, metabolomic studies in wheat have uncovered pathways associated with vitamin E biosynthesis, facilitating biofortification efforts aimed at increasing vitamin E content (Li *et al.*, 2018). Integrating multi-omics approaches, such as combining transcriptomics and metabolomics, has further enhanced our understanding of the interplay between genes, transcripts, proteins, and metabolites in determining nutritional traits (Pandohee *et al.*, 2023).

CRISPR/Cas technology has revolutionized plant biotechnology by enabling precise and efficient genome editing. Unlike traditional breeding or earlier genetic modification methods, CRISPR/Cas allows for the direct modification of specific genes associated with desirable traits without the introduction of foreign DNA (Zhang *et al.*, 2020). This precision makes CRISPR/Cas a powerful tool for enhancing the nutritional content of crops while preserving their genetic integrity. Applications of CRISPR/Cas in crop improvement have led to significant advancements in developing nutritionally enhanced varieties. For example, CRISPR/Cas has been used to increase the content of essential amino acids in rice by editing genes involved in lysine biosynthesis. This approach addresses the common deficiency of lysine in cereal-based diets (Liu *et al.*, 2024a). Additionally, CRISPR/Cas has been employed to reduce anti-nutritional factors, such as phytic acid in maize, thereby improving the bioavailability of essential minerals like iron and zinc (Sahu *et al.*, 2024). Furthermore, CRISPR/Cas technology is utilized to enhance crop resistance to environmental stressors, indirectly contributing to nutritional enhancement by ensuring stable yields under adverse conditions. For instance, CRISPR/Cas-mediated editing of genes related to drought and heat tolerance in crops such as wheat and maize has shown promise in maintaining nutritional quality despite climate stress (Wang *et al.*, 2022b).

Recent developments in CRISPR/Cas applications include the creation of rice varieties with increased levels of essential amino acids, like lysine and tryptophan, by editing genes in their biosynthetic pathways (Ku and Ha, 2020). CRISPR/Cas has also been used to enhance provitamin A content in cassava, a staple crop in many developing countries, by targeting genes regulating carotenoid biosynthesis (Dwivedi *et al.*, 2023). Additionally, CRISPR/Cas has been applied to reduce anti-nutritional factors like phytate in maize, which enhance the bioavailability of iron and zinc (Yoon *et al.*, 2016; Kumar *et al.*, 2023). Another notable advancement is the use of CRISPR/Cas to improve iron content in beans by editing genes involved in iron transport and storage, which addresses iron deficiency anaemia in populations reliant on beans as a protein source (Jha *et al.*, 2022). Multiplexed CRISPR/Cas systems, which allow for the simultaneous editing of multiple genes, have further expanded the potential for nutritional enhancement. A recent study demonstrated the simultaneous enhancement of vitamin C, iron, and zinc content in tomato plants using multiplexed CRISPR/Cas-mediated editing (Adeyinka *et al.*, 2023). Recent research efforts have increasingly focussed on integrating functional genomics with CRISPR/Cas technologies to develop crops with tailored nutritional profiles. For example, functional genomics has identified key genes involved in the biosynthesis of essential fatty acids in soybean, and subsequent CRISPR/Cas editing has been used to enhance omega-3 fatty acid content, improving the nutritional value of soybean oil (Verma *et al.*, 2023). Similarly, in maize, the integration of genome-wide association studies (GWAS) with CRISPR/Cas has enabled the identification and editing of genes associated with increased zinc content, contributing to the development of zinc-enriched maize varieties (Zulfiqar *et al.*, 2024). Barupal and Kumar (2024) used metabolomic data to identify genes involved in vitamin E synthesis in soybean and employed CRISPR/Cas12a to elevate tocopherol levels, improving both the nutritional quality and oxidative stability of soybean oil. Wang *et al.* (2022a) combined transcriptomic analysis with CRISPR/Cas9 editing to enhance folate content in maize by upregulating key enzymes in the folate biosynthesis pathway, resulting in biofortified maize varieties with potential to alleviate folate deficiency. The integration of functional genomics and CRISPR/Cas technologies holds immense potential for addressing global nutritional challenges. These advancements enable precise, efficient, and targeted improvements in crop nutritional content, offering promising solutions to enhance food security and nutrition. Continued research and collaboration will be crucial in harnessing these technologies to meet the growing demands of a rapidly expanding global population. So, in this chapter, we assess the convergence of these technologies in creating nutritionally enhanced crops with improved yields, resilience, and nutritional profiles. We also contextualize specific examples of successful applications of these technologies in developing nutritionally enhanced crops.

1.2 Functional Genomics

Functional genomics is the study of genes, transcripts, proteins, and metabolites, and their interactions on a genome-wide scale to understand biological processes precisely (Caudai *et al.*, 2021). Functional genomics can be studied at various levels: DNA level (genomics and epigenomics): RNA level (transcriptomics): protein (proteomics): and metabolites (metabolomics) (Caudai *et al.*, 2021). Functional genomics at various levels have been widely used for a comprehensive understanding of the plant genomic structure and evolution, the expression profile of various genes, protein and metabolites (Fig. 1.1). Thus, our overall understanding of plant molecular biology and metabolism is enhanced.

1.2.1 Genomics

Genomics uses advanced tools, such as next-generation sequencing techniques (Illumina sequencing, Shotgun sequencing, Ion Torrent, and SOLiD sequencing, etc. for the purpose of whole-genome sequencing (Qin, 2019). Plant genomic study started with genome sequencing of *Arabidopsis thaliana*. Currently, from 1482 plant species, 4604 plant genomes have been sequenced

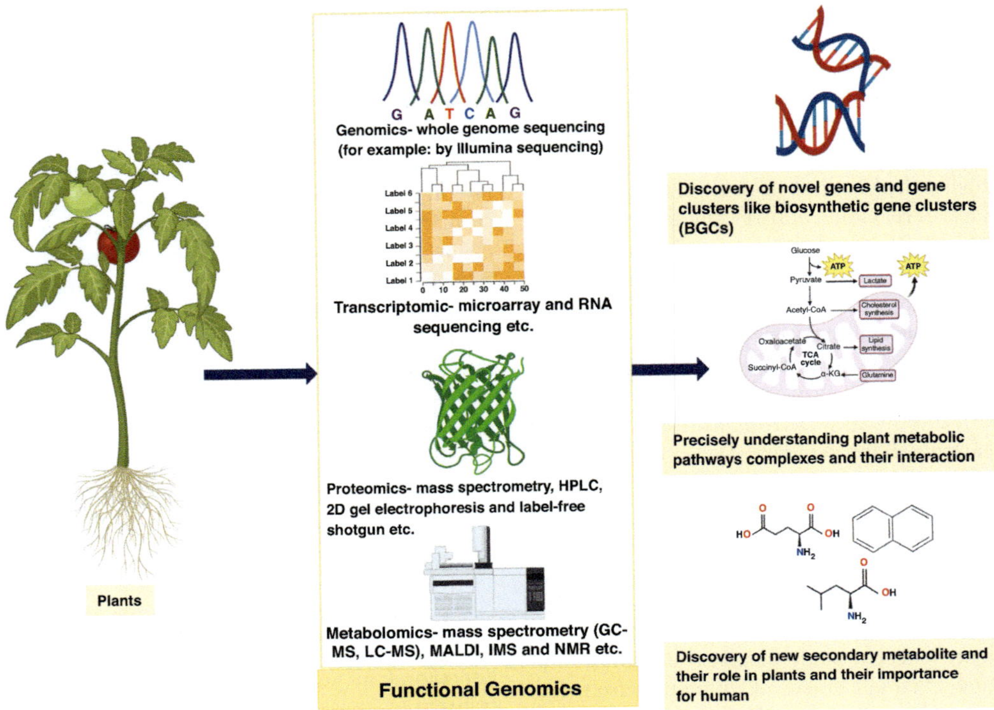

Fig. 1.1. Different types of functional genomic studies and their significance. (This figure was made by using BioRender (https://www.biorender.com.)

from a variety of plants due to its affordability and extensive research (The Arabidopsis Genome Initiative, 2000; 1001 Genomes Consortium, 2016; Bernal-Gallardo and de Folter, 2024). To date, only about 0.3% of the plant genome has been sequenced. The study revealed that the mean of the protein-coding gene is 34,071 and large genome sizes are mainly due to the activity of transposable elements and polyploidy activates (Bernal-Gallardo and de Folter, 2024). Plant genome projects such as Earth BioGenome Project affiliated to Darwin Tree of Life Project, 10KP Initiative, African BioGenome Project, African Orphan Crops Consortium, and Genomics for Australian Plants Consortium, aim to achieve genome sequencing of many plant species across the globe (Genomics for Australian Plants Initiative, 2018; Cheng *et al.*, 2018; Hendre *et al.*, 2019; Darwin Tree of Life Project Consortium, 2022; Ebenezer *et al.*, 2022; Lewin *et al.*, 2022; McLay *et al.*, 2022) (Table 1.1). The study of plant genomes leads to a precise understanding of the functioning of plants, their evolutionary histories such as adaptation, speciation, and organismal dependency, etc. (Blaxter *et al.*, 2022).

Plant genomics studies revealed that gene duplication and whole-genome duplication (WGD) in plants play a major role in plant metabolic diversification and evolution (Clark and Donoghue, 2018; Qiao *et al.*, 2022). WGD is the process of formation of polyploid organisms and is considered one of the causes of macroevolution, mainly in land plants. WGD is reported in both monocot and dicot plants of clade angiosperms (Clark and Donoghue, 2018; Zhou and Liu, 2022). The reference genome of soybean indicated that it had undertaken two WGD in the last 60 million years. WGD genes show less frequency of alternate splicing due to reduced transcriptional level, exon numbers, and intron length. WGD genes with more connectivity are also

Table 1.1. Various genomic projects with their aims/outputs.

Projects	Aim/outputs	References
Arabidopsis Genome Initiative, 2000	From 1482 plant species, 4604 plant genomes have been sequenced.	Bernal-Gallardo and de Folter, 2024
International Rice Genome Sequencing Project (IRGSP)	It aims for the whole-genome sequence of rice.	Sasaki and Burr, 2000
Earth BioGenome Project affiliated Darwin Tree of Life Project	It aims for whole-genome sequencings of 70,000 plants from Ireland and Britain.	Lewin *et al.*, 2022
10KP (10,000 Plants) Initiative	It aims for whole-genome sequencing of 10,000 species from embryophytes, green algae, and Protist.	Cheng *et al.*, 2018
African BioGenome Project	It aims for whole-genome sequencing of 1,05,000 plant species.	Ebenezer *et al.*, 2022
African Orphan Crops Consortium	It aims for whole-genome sequencing of 101 orphan crops/ trees of Africa. Six genomes are completed.	Hendre *et al.*, 2019; Ghazal *et al.*, 2021
Genomics for Australian Plants Consortium	It aims for whole-genome sequencing of Australian plant species.	McLay *et al.*, 2022

more conserved than less-connected WGD genes (Zhang *et al.*, 2022). A genomic study of tomato revealed that two WGD events had occurred in the last 140–205 million years (Song *et al.*, 2012). Another genomic study of tomato indicated that the domesticated tomato (*Solanum lycopersicum*) genome shows 6% divergence from a wild relative, *Solanum pimpinellifolium*, and more than 8% divergence from the genome of potato along with several inversions (Tomato Genome Consortium, 2012). A large number of genome analyses revealed that WGD occurred frequently throughout the evolutionary history of plants (Qiao *et al.*, 2022; Zhou *et al.*, 2022).

We know that plant genes are even found in the clusters: biosynthetic gene clusters (BGCs) that are collectively responsible for particular metabolic pathways. Such BGCs have led the path to the identification of earlier unknown pathways, enzymes and their chemistries (Nutzmann *et al.*, 2016; Zhou *et al.*, 2022). Combining distinctive features of BCG DNA using plant genomes along with transcriptomic and metabolomic data empowers us to precisely understand the metabolic diversity of plants (Zhou and Liu, 2022). Further combining genomics with RNA interference (RNAi) and artificial microRNAs (amiRNAs) helps in the identification of many gene functions, which can be further utilized for crop improvement (*Ossowski et al., 2008; Saurabh et al., 2014*; Muhammad *et al.*, 2019). Thus, it can be inferred that genomic is a valuable tool to understand plant genome, gene organization, and discovery of new genes and gene clusters (Table 1.2).

1.2.2 Transcriptomic

Transcriptomic using microarray and RNA sequencing has become a prominent tool of functional genomics. This technique helps us to study gene expression profiles and differential gene expression of plants on the high-throughput scale in a cost-effective manner, which can be utilized to understand

Table 1.2. Significance/applications of the function genomics.

Functional genomics	Significance/applications	References
Genomics	It has led to whole-genome sequencing of 4604 plant genomes from 1482 plant species, and many more are underway.	Bernal-Gallardo and de Folter, 2024
	Studies of gene duplication, whole-genome duplication (WGD), and inversion are possible through plant genomic. Also it was revealed that these processes are responsible for plant diversification and macroevolution.	Clark and Donoghue, 2018; Qiao *et al.*, 2022; Zhou and Liu, 2022
	The discovery of biosynthetic gene clusters (BGCs) that are collectively responsible for particular metabolic pathways led to the discovery of many novel genes of many metabolic pathways.	Nützmann *et al.*, 2016; Zhou and Liu, 2022
Transcriptomics	Microarray helped to identify subsets of genes responsible for the formation of a thick protective cover of periderm (a defensive tissue) during heat stress in potatoes.	Ginzberg *et al.*, 2009
	Microarray analysis along with various downstream experiments proved that *OsMYB3R-2* gene is the master switch of tolerance during stress.	Dai *et al.*, 2007; Ganie *et al.*, 2019
	RNA sequencing in combination with other omics studies revealed that chalcone synthase (CHS), like enzyme, take part in the formation of the styrylpyrone backbone of kavalactone in the kava plant.	Pluskal *et al.*, 2019
	RNA sequencing study led to the discovery of genes that can be used in heterosis breeding of plants.	Howlader *et al.*, 2020
	RNA sequencing study led to the identification of the main gene expression in alfalfa plant in response to interaction with its different rhizobial strains.	Kang *et al.*, 2020
Proteomics	Iron and manganese deficiency causes extensive cell wall remodeling in tomato plants.	Ceballos-Laita *et al.*, 2022
	Millimeter-wave irradiation in wheat could result in tolerance in flooding situations, an abiotic stress.	Komatsu *et al.*, 2022
	The study of post-translational modification (PTMs) is possible through proteomics and advanced computational methods.	Ramazi and Zahiri, 2021
	Plant signaling pathway analysis is performed through mapping protein–protein interaction. For example, the identification of components of MPK complexes.	Leissing *et al.*, 2021; Yan *et al.*, 2022

Continued

Table 1.2. Continued

Functional genomics	Significance/applications	References
Metabolomics	Understanding the diversity of plants' primary and secondary metabolites.	Zhou and Liu, 2022
	Quantification and assignment of metabolic profiling of plant species.	Patel *et al*., 2021
	Identification of plants with a rich source of antioxidants that can be used as human food sources.	Patel *et al*., 2016
	It was identified that membrane lipid oxidation can be used as an indicator of salt stress in plants.	Yu *et al*., 2020
	It led to the identification of compounds, which provide protection against biotic stress	Niculaes *et al*., 2018; Florean *et al*., 2023

the role of various genes and discover new genes of specific metabolic pathways (Upton *et al.*, 2023; You, 2023; Table 1.2). Other transcriptomic techniques like Nanopore direct RNA sequencing and Iso-Seq are used to analyze splicing events and transcript isoforms (Schaarschmidt *et al.*, 2020; Jain *et al.*, 2022). Plant phenotypic plasticity study using transcriptomic study complements the understanding of the plant response to various types of stress (Monforte, 2021). Transcriptome profiling through microarray helped to identify subsets of genes responsible for the formation of a thick protective cover of periderm (a defensive tissue) during heat stress in potato (*Solanum tuberosum* L.) (Ginzberg *et al.*, 2009). A differential gene expression study using a microarray of rice plants during cold stress indicated that many genes are overexpressed; out of which *OsMYB3R-2* gene was proven to be overexpressed during salt, cold, and drought stress, in transgenic *Arabidopsis thaliana* plants, indicating that it is the master switch of tolerance during stress (Dai *et al.*, 2007; Ganie *et al.*, 2019). Thus, it can be inferred that transcriptomic study along with others can be useful to understand the role of various genes during stress, which can be utilized to tackle climate change effects on various agricultural plants and to secure food safety (Ganie *et al.*, 2019; Monforte, 2021).

High-throughput RNA sequencing was used for differential gene expression between hybrid and parent Easter lily. This study resulted in the identification of the main genes, which can be used for heterosis breeding of plants (Howlader *et al.*, 2020). Transcriptomics study using RNA sequencing and combining it with other omics studies revealed that chalcone synthase (CHS), like enzyme, takes part in the formation of styrylpyrone backbone of kavalactone in the kava plant (Pluskal *et al.*, 2019). Another RNA sequencing study helped to understand molecular interaction between alfalfa plants and rhizobial symbionts at the biotype or strain level. It indicated alfalfa expresses genes for noduline, Medsa002106, NBS-LRR family proteins, and NCR peptide in response to interaction with its different rhizobial strains (Kang *et al.*, 2020).

1.2.3 Proteomics

Proteomics deals with large-scale study of proteins (proteome) for functional genomics and can be integrated with other omics to precisely understand plant biology. It provides valuable information on protein profiles and their subcellular localization and post-translational modification (PTM)(Komatsu and Colgrave, 2022; Yan *et al.*, 2023). Proteomics uses 2D gel electrophoresis, various mass spectrometry (MS) techniques (gas chromatography (GC)-MS, nano-liquid chromatography (LC)-MS), and label-free shotgun to analyze various molecular aspects of plants (Yan *et al.*, 2022) (Table 1.2). A proteomic study utilizing 2D gel electrophoresis and shotgun analysis revealed iron and manganese deficiency causes extensive cell-wall remodeling in tomato plants (Ceballos-Laita *et al.*, 2022). LC-MS analysis along with other biochemical analyses indicated that millimeter-wave irradiation in wheat could result in tolerance in flooding situations and abiotic stress (Komatsu *et al.*, 2022).

PTMs are important events to increase protein diversity, comprised mainly of acylation, phosphorylation, glycosylation, ubiquitylation, etc. High-throughput proteomic study with advance computational methods is useful for the study of such PTMs (Ramazi and Zahiri, 2021). Signaling pathway analyses in plants are performed through mapping of protein–protein interactions through various proteomics tools (Yan *et al.*, 2022). XL–TAP–MS analysis in *Arabidopsis thaliana* helped in the identification of candidate components of mitogen-activated protein kinases (MPKs) complexes (Leissing *et al.*, 2021).

1.2.4 Metabolomics

The plant produces a large number of diverse molecules, some of which are required for growth and reproduction, known as primary metabolites (PMs) because they play a significant role in the biochemical and physiological processes of plants (Patel *et al.*, 2021). The majority of others are known as secondary metabolites (SMs), which play various roles: plant defense against abiotic and biotic stress; shaping of a healthy bacterial community of rhizosphere; modulation of root growth; metal transporting agents; symbiotic agent; micronutrient uptake; and component of innate immunity (Huang *et al.*, 2019; Erb *et al.*, 2020; Bai *et al.*, 2021; Zhou and Liu, 2022). SMs are low-molecular-weight compounds with overwhelming diversity, having importance to plants and humans alike. SMs are categorized into terpenoids, phenylpropanoids, and alkaloids (Zhou and Liu, 2022). Different advanced analytical methods, like GC-MS, LC-MS, Fourier transform ion cyclotron resonance-mass spectrometry (FTICR-MS), capillary electrophoresis-mass spectrometry (CE-MS), ion mobility spectrometry (IMS), matrix-assisted laser desorption/ionization (MALDI), ion mobility spectrometry (IMS), and nuclear magnetic resonance (NMR), significantly speed up and improve the plant metabolic profiling (Patel *et al.*, 2021). A metabolomic study reveals that psyllium (*Plantago ovata*) leaves are a rich source of nutraceuticals (containing a variety of natural antioxidants) and can be eaten by humans as a green salad for health benefits (Patel *et al.*, 2016). It was observed that the salinity stress causes the oxidation of membrane lipids in the barley roots that can be utilized as an indicator of salt stress in plants (Yu *et al.*, 2020). Another metabolomic study revealed that nitric oxide (NO) enhances the seed germination process by controlling reactive oxygen species (ROS) accumulation and respiration (Pandey *et al.*, 2019).

Studies of secondary metabolites indicated that benzoxazinoids (BXs), specialized secondary metabolites found in many flowering plants, provide protection against biotic stress in maize (Niculaes *et al.*, 2018; Florean *et al.*, 2023). Thus, it can be concluded that plant metabolomics aims at a comprehensive understanding of metabolite characteristics, role, and significance in plant biology (Table 1.2). This knowledge can be utilized for crop improvements for human benefits in various aspects (Manickam *et al.*, 2023).

Multi-omics for metabolic pathway analysis of plants

Plant metabolism, its regulation and control are very complex, where different components interact with each other to build a functioning system (Araújo *et al.*, 2012). The study of plant metabolic diversity is challenging due to its size, complexity, and presence of huge repetitive sequences in its genome (Zhou and Liu, 2022). Integration of plant genomic information to molecular profiling (transcriptomics, proteomics, and metabolite profiles) has enhanced our ability to precisely understand plant metabolism (Araújo *et al.*, 2012). Multi-omics, along with tools of metabolic pathway mapping and enzyme annotation, helps in the discovery of new metabolic pathways in various plant species. Enzyme annotation public recourses (UniProt, Rhea and BRENDA), and metabolic pathway-mapping infrastructures like Kyoto Encyclopedia of Genes and Genomes (KEGG), MetaCyc, Plant Metabolic Network (PMN), and PlantReactome, facilitate understanding and discovery of new plant metabolic pathways (Wang *et al.*, 2022a). Assigning enzyme genes into metabolic pathways also requires understanding of substrate, product, and substrate-product reaction (Wang *et al.*, 2022a). For example, multi-omics along with quantitative real-time reverse-transcription PCR (qRT-PCR) and gene ontology annotations study on *Arabidopsis thaliana* revealed that *alternative oxidase1a* gene plays a role in normal redox balance in cells during stress response (Giraud *et al.*, 2008). Recent multi-omics, along with metabolic pathway mapping by KEGG and weighted gene co-expression network analysis (WGCNA) study on the rose plant for salt tolerance, revealed that under salt stress, the phenylpropane pathway, mainly flavonoid pathway, is greatly enhanced (Ren *et al.*, 2024).

1.2.5 Role of functional genomics in nutritionally enhanced crops

In a time of climate change and growing demand for food, crop improvement is an urgent need. It aims at higher yields with the nutritional enhancement of crops and vigorous tolerance against abiotic and biotic stresses in plants (Ku and Ha, 2020). Humans require 40 different nutrients categorized as macronutrients, micronutrients, and minerals for a healthy life. Micronutrient deficiency can be found in over 2 billion of the global population (mainly in Africa and Asia). One of the major causes of this deficiency is that commercially cultivated crops are deficient in important macronutrients due to a variety of abiotic and biotic stresses (Gaikwad *et al.*, 2020). Biofortification is a process of enhancing vitamin and mineral contents in crops through agronomic and genetic means in a cost-effective and sustainable manner for the long term. Various approaches, like plant agronomy practice, molecular breeding, and transgenic techniques have been utilized for this (Garg *et al.*, 2018). Understanding plant biology at the genomic and molecular level via multi-omics of functional genomics study becomes very important for developing nutritionally enhanced crops (Fig. 1.2). In the above sections, we have already discussed how the various levels of functional genomics along with metabolic pathway mapping and enzyme annotation help us to understand plant metabolism very precisely.

Functional genomics can be utilized for the identification of gene/quantitative trait loci (QTLs) related to essential nutrients. Further QTLs are used for specific breeding programs for the nutritional enhancement of crops (Gaikwad *et al.*, 2020). Knowledge from functional genomics can be utilized in developing genetically modified (GM) crops with nutritional enhancement through transgenic crop development and genome-editing methods. Genome-editing methods have been utilized: first-generation methods – like Zinc Finger Nucleases (ZFNs) and Transcription Activator-Like Effector Nucleases (TALENs) – and second-generation and advanced methods; Clustered Regularly Interspaced Short Palindromic Repeats (CRISPR)/Cas system (CRISPR/Cas) in successively developing many nutritionally enhanced crops (Abdul Aziz *et al.*, 2022).

1.3 CRISPR/Cas Technologies for Crop Enhancement

CRISPR/Cas technologies have rapidly become a cornerstone of modern plant biotechnology, offering unprecedented precision and flexibility in genome editing for crop improvement. By enabling targeted modifications to specific genes, CRISPR/Cas allows researchers to augment desirable traits such as nutritional content, yield, and stress tolerance, while minimizing the introduction of undesired traits (Rasheed *et al.*, 2021). The technology's ability to perform site-specific edits, including gene knockouts, insertions, and base editing, has revolutionized the process of crop enhancement, making it faster and more efficient than traditional breeding methods (Tabassum *et al.*, 2021).

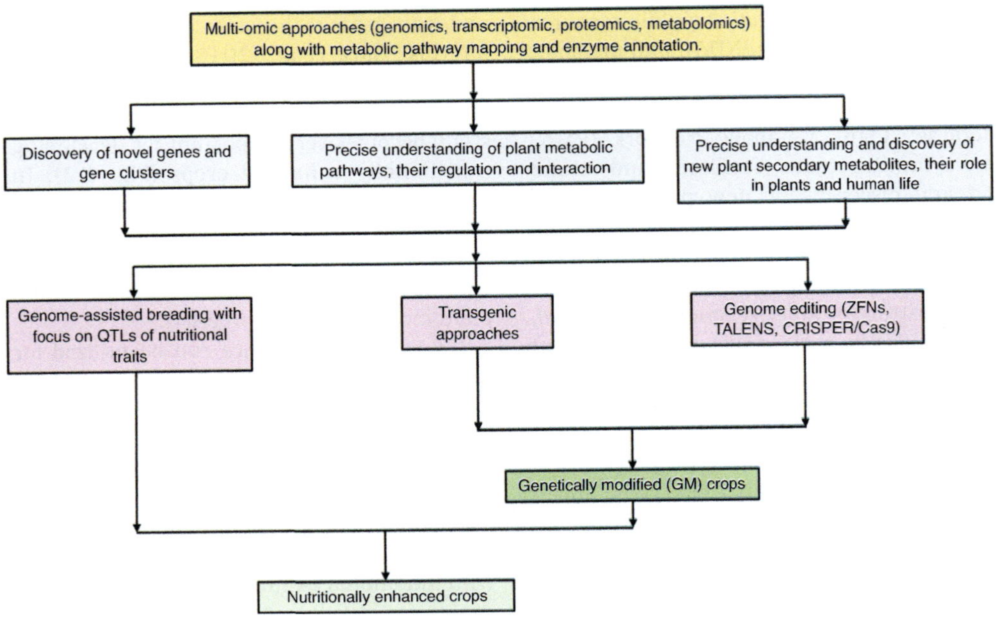

Fig. 1.2. Flowchart of application of multi-omics in the nutritional enhancement of crops (gene/quantitative trait loci (QTL), Zinc Finger Nucleases (ZFNs), Transcription Activator-Like Effector Nucleases (TALENs), Clustered Regularly Interspaced Short Palindromic Repeats (CRISPR/Cas)).

Recent advancements in CRISPR/Cas systems, such as CRISPR/Cas9, CRISPR/Cas12a, and prime editing, have further expanded the scope of gene editing, facilitating the development of crops with increased nutritional profiles, improved resistance to biotic and abiotic stresses, and better adaptation to changing environmental conditions (Rasheed *et al.*, 2021; Yadav *et al.*, 2023). These innovations are paving the way for the next generation of biofortified crops, with the potential to address global food security and nutritional challenges (Dhaliwal *et al.*, 2022) (Table 1.3).

1.3.1 CRISPR/Cas9

CRISPR/Cas9 technology has revolutionized the development of nutritionally enhanced crops by enabling precise genetic modifications that directly improve the nutritional content of food plants. This tool allows scientists to target and edit specific genes responsible for the biosynthesis of vitamins, minerals, and essential amino acids, resulting in crops with significantly higher nutritional value. Kumar et al. (2022) successfully used CRISPR/Cas9 to increase the lysine content in rice by editing genes in the aspartate pathway, a critical pathway for the synthesis of essential amino acids. This improvement addresses the common deficiency of lysine in cereal-based diets, particularly in developing countries. In addition, CRISPR/Cas technology has been applied to reduce anti-nutritional factors in crops. Wei *et al.* (2023) demonstrated the use of CRISPR/Cas9 to reduce phytic acid levels in maize, enhancing the bioavailability of important micronutrients such as iron and zinc. This is particularly important for populations that rely heavily on maize as a dietary staple, as it can help combat micronutrient deficiencies that are prevalent in many regions. The flexibility of CRISPR/Cas systems, including the ability to perform multiplexed gene editing, allows for the simultaneous enhancement of multiple nutritional traits within a single crop. This capability has been illustrated in research aimed at increasing both β-carotene and vitamin E content in tomatoes, which was achieved through the targeted editing of multiple genes

Table 1.3. CRISPR/Cas9 technology is used for the nutritional enhancement of crops. Author's own table.

Crops	Targeted genes	Trait improvement	Reference
Oryza sativa (rice)	DEP1	Used to tweak genes associated with starch synthesis that boosted their incidence	Liu *et al.*, 2015; Klap *et al.*, 2017
	GW2, GW5 and TGW6	Used to create GW2 knockout mutants in the Indica rice cultivar MTU1010	Achary and Reddy, 2021
	GW5	Producing knockout mutants. When compared to rice of the wild type, these mutants have different grain widths.	Yang *et al.*, 2024
	SBE and SBE11b	By modifying the SBE and SBE11b genes using CRISPR/Cas9, scientists have altered the production of starch.	Marín-Guirao *et al.*, 2017
	OsHAK-1	CRISPR/Cas9 knockout of the *OsHAK-1* gene in rice leads to reduced cadmium uptake and accumulation, highlighting its role in cadmium transport and potential biofortification.	Nieves-Cordones *et al.*, 2014
	OsNramp5	CRISPR/Cas9 knockout of the *OsNramp5* gene in rice creates mutants with reduced cadmium uptake, significantly lowering cadmium accumulation in grains, enhancing crop safety and consumer health.	Chang *et al.*, 2022
	OsNRAMP2	Gene in rice produces mutants with altered nutrient transport, disrupting the remobilization of iron and cadmium, and affecting their distribution within the plant.	Chang *et al.*, 2022
	OSFAD2-1	Produces mutants with altered fatty acid composition, achieving a high oleic and low linoleic acid profile in rice bran oil.	Gao *et al.*, 2018
	GS9	Rice grain form is mostly controlled by the GS9 gene. It specifically impacts the change from spherical to slender grains.	Lin *et al.*, 2018

Continued

Table 1.3. Continued

Crops	Targeted genes	Trait improvement	Reference
	GS3 and Gn1a	CRISPR/Cas9-mediated knockout of the GS3 and Gn1a genes in rice produces mutants with altered grain size and length, which could potentially increase crop yield and improve overall productivity.	Roorkiwal *et al.*, 2018
	GL2/OsGRFA and OsGRF3	These modifications aim to enhance grain size by promoting growth and development.	Gerona *et al.*, 2019
	OsGS3, OsGW2, OsGn1a	Enhance grain size, width, and weight by modulating the expression of these specific genes	Koochak *et al.*, 2019
	OsPLDalpha1	Reduced phytic acid levels. This reduction enhances the bioavailability of essential minerals, specifically iron (Fe) and zinc (Zn).	Syu *et al.*, 2019
	GS3 and GL3.1	Improves grain size and length by modulating the expression of these genes.	Virdi *et al.*, 2020
	OsAAP6 and OsAAP10	Reduced phytic acid levels, resulting in increased bioavailability of essential minerals, including iron (Fe) and zinc (Zn).	Gangurde *et al.*, 2020
	OsSGBSS1	Increasing amylose content in rice grains through the alteration of this gene's expression	Huang *et al.*, 2020
	OsBADH2	Introduce aromatic properties into the nonaromatic rice variety ASD16 by modifying the gene's expression	Wu *et al.*, 2020
	OsGAD3	Increasing GABA content in rice grains through the regulation of this gene's expression	Bai, 2020
	GW2	Enhance grain architecture by modulating the expression of this gene	Havlin and Schlegel, 2021
	OsSPL16/qGWB	Improving grain architecture through the modulation of this gene's expression	Tabassum *et al.*, 2021

Continued

Table 1.3. Continued

Crops	Targeted genes	Trait improvement	Reference
	OsNRAMP2	Improved iron (Fe) uptake efficiency in rice enhances grain Fe content, addressing iron deficiency anaemia in staple-food regions. Reduced cadmium (Cd) levels in grains ensure food safety and prevent Cd toxicity.	Yang *et al.*, 2023
	OsBADH2	Increasing the production of 2-acetyl-1-pyrroline (2AP), the primary aromatic compound in fragrant rice	Mendiondo *et al.*, 2016
	Crt1 and PSY	Insertion of carotenoid-production gene	Mendiondo *et al.*, 2016
Glycine max (soybean)	FAD2-1A and FAD2-1B	Reduced phytic acid levels to enhance mineral absorption	Byrne and Murphy, 2022; Awasthi *et al.*, 2024
	FAD2-1A and FAD2-1B	Resulting in increased oleic acid content and decreased linoleic acid content in soybean oil	Mendiondo *et al.*, 2016
	GlymBd28k and GlymBd30k	Reduction in allergenic proteins in soybeans	Matsuoka and Sugano, 2022
Vitis vinifera (grape)	IdnDH	Reduced tartaric acid levels within the organism	Wu *et al.*, 2016
Lycopersicon esculentum (tomato)	Fasciated, SlPti1	Improved fruit sweetness, prolong shelf life, and ripen fruit	Li and Weng, 2017
	ALC	Extending the shelf life of the product	Li and Weng, 2017
	SlGAD2 and SlGAD3	Leading to increased GABA production in tomatoes	Zujur *et al.*, 2017
	RIN	Reduced production of volatile organic compounds (VOCs) in fruits	Brodie *et al.*, 2017
	SlAN2	Investigated biosynthesis and regulation of anthocyanins in purple tomatoes	Yerramsetty *et al.*, 2017
	OVATE, FAS, FW2.2, CLV3	Leading to enhanced yield and productivity in tomato plants	Lu *et al.*, 2015
	SGR1, LCY-E, Blc, LCY-B1and LCY-B2	Higher lycopene levels in tomato fruit	Du *et al.*, 2020
	PL, PG2a and TBGA	Extending the shelf life of products	Dai *et al.*, 2018
	STAYGREEN (SGR) gene	Negative regulator of carotenoid biosynthesis	Yin and Qiu, 2019

Continued

Table 1.3. Continued

Crops	Targeted genes	Trait improvement	Reference
Triticum aestivum (wheat)	Alpha-gliadin genes	Increase the amounts of proteins	Rosati *et al.*, 2018
	TaGW2	Increased grain width, grain length, and thousand grain weight (TGW) in both tetraploid and hexaploidy wheat	Zheng *et al.*, 2021
	TaGW7	Increased grain width and weight in wheat	Koenigshofer and Loeppert, 2019
Hordeum vulgare (barley)	GBSS1and protein targeting to starch 1 (PTST1)	Enhancing GABA levels	Tsutsui *et al.*, 2020
Solanum tuberosum (potato)	Steroid 16 alpha hydroxylase (st16DOX)	Enhance the quality of starch	Bennewitz *et al.*, 2018
	StGBSS	Enhancing amylase biosynthesis	Molla and Yang, 2019
	Sterol side chain reductase 2	Reduced levels of toxic SGAs in potatoes	Molla and Yang, 2019
	Starch branching enzyme (Sbe)	Increasing the proportion of amylose (a type of starch) and promoting longer amylopectin chains in crops	Yang *et al.*, 2021
Sorghum bicolor (L.) (jowar)	k1C gene family	Various trait-specific climate-resilient lines developed	Dai *et al.*, 2018
Daucus carota (carrot)	F3H	Editing the F3H gene enhances carotenoid accumulation, especially beta-carotene, by modulating the flavonoid biosynthetic pathway in carrots	Sertse *et al.*, 2019
Lactuca sativa L. (lettuce)	LsGGP2	Editing LsGGP2, a glucosinolate biosynthesis gene, can increase antioxidant levels in lettuce.	Abu-Abied *et al.*, 2018
Monihot esculenta Crantz (cassava)	GBSSI	GBSSI is essential for amylose synthesis in starch granules; its editing can modify the amylose-to-amylopectin ratio.	Bräutigam and Cronk, 2018
Malus pumila (apple)	IdnDH	Fine-tuned qualitative traits for better consumer apple	Wada *et al.*, 2020

Continued

Table 1.3. Continued

Crops	Targeted genes	Trait improvement	Reference
Brassica napus (rapeseed)	FAD2	Increased oleic acid content for improved oil quality	Matsumoto and Takenaka, 2018
	SFAR4 and SEARS	Increased oleic acid content and decreased levels of linoleic and linolenic acids	Fernando and Gamage, 2021
	BnTT8	Altered oil content and changes in the types of fatty acids present in oilseed crops	Du *et al*., 2020
	BnITPK	Reduced phytic acid levels in oilseed crops	Du *et al*., 2020
Ipomoea batatas (sweet potato)	GBSSI	GBSSI is crucial for amylose synthesis in starch; editing it adjusts the amylose-to-amylopectin ratio.	Shanmugabalaji and Kessler, 2019
Arachis hypogaea (peanut)	FAD2A and FAD2B	FAD2A and FAD2B are key for polyunsaturated fatty acid synthesis; editing increases oleic acid, reduces linoleic acid.	Neelakandan *et al*., 2022
Punica granatum (pomegranate)	PgUGT84A23 and PgUGT84A24	Regulate flavonoid metabolism; editing enhances antioxidants	Neelakandan *et al*., 2022
Solanum melongena (eggplant)	PP04, PPOS and PP06	Improves fruit quality, yield, and disease resistance.	Sertse *et al*., 2019
Commelina benghalensis (willow leaved dayflower)	CsFAD2	Enhance specific traits, such as disease resistance or flower color	Ariel and Manavella, 2021
Musa paradisiaca Linn. (banana)	MaACO1	MaACO1 is involved in ethylene biosynthesis, which accelerates fruit ripening. Editing this gene can delay ripening.	Verma *et al*., 2023

involved in carotenoid biosynthesis (Nie *et al.*, 2024).

These advancements demonstrate the significant potential of CRISPR/Cas technologies to improve the nutritional quality of crops, offering a sustainable and scalable solution to global malnutrition. As research in this area continues to progress, the development of nutritionally enhanced crops through CRISPR/Cas technologies is expected to play a crucial role in addressing food security and improving public health worldwide.

1.3.2 CRISPR/Cas12

Like CRISPR/Cas9, Cas12 enables precise genetic modifications, but it offers unique advantages, such as the ability to target multiple genes simultaneously and improved specificity in certain plant genomes. This makes Cas12 particularly effective for complex trait improvements, such as increasing the content of essential nutrients in crops. For instance, Zhang *et al.* (2021) demonstrated the use of CRISPR/Cas12a to simultaneously edit multiple genes in wheat, leading to a significant reduction in phytic acid content and an increase in the bioavailability of essential minerals such as iron and zinc. This capability makes CRISPR/Cas12 particularly valuable for improving the nutritional quality of staple crops, providing a more comprehensive approach to addressing global micronutrient deficiencies. Additionally, Cas12's improved specificity in certain plant genomes reduces off-target effects, making it a more reliable tool for developing crops with enhanced levels of essential nutrients like vitamins and fatty acids (Li *et al.*, 2022).

1.3.3 CRISPR/Cas13

CRISPR/Cas13, an RNA-targeting system distinct from the DNA-targeting CRISPR/Cas9 and CRISPR/Cas12, has emerged as a powerful tool for post-transcriptional gene regulation in plants. Unlike other CRISPR systems that primarily introduce mutations at the DNA level, Cas13 specifically targets and degrades RNA, offering a unique approach for gene silencing without altering the genome. This RNA-targeting capability opens new possibilities for enhancing the nutritional quality of crops by precisely controlling the expression of genes involved in nutrient biosynthesis, anti-nutritional factors, and stress responses.

CRISPR/Cas13 systems utilize a guide RNA (gRNA) to direct the Cas13 protein to complementary RNA sequences, leading to RNA degradation or interference with translation. Cas13 can be programmed to target and cleave specific messenger RNAs (mRNAs) that encode proteins responsible for undesired traits, such as low nutrient content or high levels of anti-nutritional factors. By degrading these mRNAs, CRISPR/Cas13 can reduce the abundance of specific proteins, thereby enhancing the nutritional profile of the crop without permanently altering its genetic code. The ability to target RNA also allows for the dynamic and reversible control of gene expression, making Cas13 particularly useful for fine-tuning metabolic pathways that influence nutrient composition. This flexibility is crucial for optimizing nutrient levels without compromising the plant's growth, yield, or resistance to environmental stresses.

CRISPR/Cas13 offers a novel approach by targeting and degrading mRNAs that encode enzymes or regulatory proteins that divert metabolic flux away from the synthesis of essential amino acids. For instance, in maize, Zhang *et al.* (2021) demonstrated the use of CRISPR/Cas13 to selectively degrade mRNAs encoding the enzyme lysine-ketoglutarate reductase, which catalyzes the first step in lysine catabolism. By reducing the expression of this enzyme, the researchers were able to increase lysine content in maize kernels without adversely affecting plant growth or yield. This approach highlights the potential of CRISPR/Cas13 to enhance the nutritional quality of crops by manipulating amino acid biosynthesis pathways. In rice, Hamdan *et al.* (2022) employed CRISPR/Cas13 to degrade mRNAs encoding the enzyme carotenase, which converts β-carotene (provitamin A) into other carotenoids. By reducing the activity of this enzyme, the researchers successfully increased β-carotene levels in rice grains, offering a potential strategy to address vitamin A deficiency, particularly in regions where rice is a dietary staple. This application demonstrates the versatility of CRISPR/Cas13 in enhancing vitamin content in crops by modulating the expression of key enzymes at the RNA level. In soybean, a major source of plant-based protein, high levels of phytic acid can reduce the bioavailability of minerals such as iron and zinc. Tong *et al.* (2023) used CRISPR/Cas13 to degrade mRNAs encoding inositol-3-phosphate synthase, a key enzyme in the phytic acid biosynthesis pathway. The resulting soybean plants showed a significant reduction in phytic acid content, leading to improved mineral bioavailability and nutritional quality. This example illustrates how CRISPR/Cas13 can be applied to reduce anti-nutritional factors in crops, enhancing their overall nutritional value.

1.3.4 Prime editing

Prime editing is an innovative genome-editing technology that offers unparalleled precision and versatility in modifying plant genomes, making it particularly suitable for developing nutritionally enhanced crops. Unlike traditional CRISPR/Cas9 or CRISPR/Cas12 systems, which rely on creating double-strand breaks (DSBs) in DNA, prime editing uses a modified Cas9 enzyme fused to a reverse transcriptase. This complex, guided by a prime editing guide RNA (pegRNA), introduces specific edits, such as base substitutions, small insertions, or deletions, directly at the target site without inducing DSBs. This approach reduces the risk of off-target effects and increases the accuracy of genome modifications, making prime editing a powerful tool for improving the nutritional quality of crops (Alharbi *et al.*, 2024).

Prime editing has been successfully applied to enhance various nutritional traits in staple crops, addressing deficiencies in essential vitamins, minerals, and amino acids. For example, prime editing has been used to precisely modify genes involved in the biosynthesis of β-carotene, which is a precursor to vitamin A, in rice. In a study by Palaniswamy *et al.* (2024), prime editing was employed to introduce specific nucleotide changes in the OsCRTISO gene, which plays a crucial role in the carotenoid biosynthesis pathway. The resulting rice plants exhibited significantly increased β-carotene content in their grains, providing a potential solution to vitamin A deficiency which is a major public health issue in many developing countries. In addition to enhancing vitamin content, prime editing has been applied to improve the amino acid profile of crops. Tiwari *et al.* (2021) used prime editing to modify the Gln1 gene in soybean, which is involved in the synthesis of glutamine, an essential amino acid. The edited soybeans showed elevated levels of glutamine, contributing to improved protein quality in the crop. This development is particularly important for regions where soybean serves as a primary source of dietary protein. Another significant application of prime editing is in the reduction of anti-nutritional factors, which can inhibit the absorption of essential nutrients. For instance, Liu *et al.* (2024b) utilized prime editing to introduce specific mutations in the TaMATE1B gene in wheat, reducing the accumulation of aluminum ions, which are known to interfere with the uptake of phosphorus, a critical nutrient for plant growth and human nutrition. The edited wheat plants demonstrated improved phosphorus bioavailability, enhancing the nutritional value of the grain.

1.3.5 CRISPRi

CRISPR interference (CRISPRi) is a powerful tool in the CRISPR/Cas system that allows for the precise, reversible, and tuneable downregulation of gene expression without altering the DNA sequence. Unlike traditional CRISPR/Cas9, which creates double-strand breaks to edit the genome, CRISPRi utilizes a catalytically dead Cas9 (dCas9) that is fused to repressor domains or functions on its own to block transcription by binding to target gene promoters or coding regions. This technology has significant potential for developing nutritionally enhanced crops by selectively silencing or reducing the expression of genes that negatively impact nutrient composition, biosynthesis, or bioavailability (Banerjee *et al.*, 2023).

CRISPRi works by guiding the dCas9 protein to specific genomic loci using a single-guide RNA (sgRNA). When dCas9 is directed to the promoter or coding region of a target gene, it physically obstructs the RNA polymerase from initiating or elongating transcription, effectively reducing gene expression. This targeted gene repression is highly specific and can be finely tuned by adjusting the concentration of dCas9 or the sgRNA, making CRISPRi an ideal tool for modulating metabolic pathways involved in nutrient synthesis. One of the key applications of CRISPRi in nutritionally enhanced crops is the downregulation of anti-nutritional factors. Anti-nutritional compounds such as phytic acid in cereals and legumes can bind essential minerals like iron and zinc, reducing their bioavailability. By using CRISPRi to silence genes involved in phytic acid biosynthesis, researchers can decrease its levels in crops, thus enhancing the nutritional quality. For example, Mishra and Zhao (2018) demonstrated the use of CRISPRi to target and repress the expression of the IPK1 gene in maize, which plays a central role in phytic acid synthesis. The resultant maize lines showed reduced phytic acid content and improved mineral bioavailability, which is crucial for addressing micronutrient deficiencies in populations that rely heavily on maize as a dietary staple. Another application is in the modulation of metabolic pathways to increase the production of health-promoting compounds. For instance, CRISPRi has been employed to downregulate genes that compete with carotenoid biosynthesis pathways in crops like tomatoes, leading to an increase in the accumulation of provitamin A (β-carotene). By repressing genes such as DET1, which negatively regulates carotenoid accumulation, Sinn *et al.* (2021) successfully enhanced β-carotene levels in tomatoes, offering a potential strategy to combat vitamin A deficiency. Moreover, CRISPRi can be used to manipulate pathways that control the levels of essential amino acids in crops. For example, downregulating genes that limit the flow of carbon into essential amino acid synthesis pathways can lead to higher concentrations of these nutrients. In rice, Zhong *et al.* (2020) applied CRISPRi to reduce the expression of genes in the lysine catabolism pathway, thereby increasing the lysine content, a critical amino acid often lacking in cereal grains.

1.3.6 CRISPR for reducing anti-nutritional factors in crops

Anti-nutritional factors are naturally occurring compounds in crops that interfere with the absorption and utilization of essential nutrients in the human diet. These compounds, such as phytic acid, tannins, and glucosinolates, can significantly diminish the nutritional value of food crops by binding to vitamins and minerals, making them unavailable for absorption. The application of CRISPR technology in reducing these anti-nutritional factors has opened new avenues for enhancing the nutritional quality of staple crops, thereby addressing global health issues related to micronutrient deficiencies.

CRISPR/Cas systems, particularly CRISPR/Cas9, allow for precise genome editing by creating targeted double-strand breaks at specific loci in the DNA, followed by repair processes that can lead to gene disruption, deletion, or modification. In the context of reducing anti-nutritional factors, CRISPR/Cas9 can be used to knock out or modify the genes involved in the biosynthesis or accumulation of these compounds. This process reduces or eliminates the presence of anti-nutritional factors in the edible parts of the crop, thereby enhancing nutrient bioavailability.

Phytic acid reduction

Phytic acid (myo-inositol hexakisphosphate) is a major storage form of phosphorus in seeds and grains, but it is poorly digested by humans and monogastric animals. It chelates essential minerals, like iron, zinc, and calcium, forming insoluble complexes that are excreted rather than absorbed. This leads to deficiencies in these crucial micronutrients, particularly in populations that rely heavily on cereals and legumes as staple foods. CRISPR/Cas9 has been successfully employed to reduce phytic acid levels in several crops. In maize, Gao *et al.* (2015) used CRISPR/Cas9 to knock out the IPK1 gene, which encodes inositol pentakisphosphate 2-kinase, an enzyme involved in the final step of phytic acid biosynthesis. The resulting maize lines exhibited significantly lower phytic acid content and improved bioavailability of phosphorus and other essential minerals. This advancement is crucial for enhancing the nutritional quality of maize, which is a staple food for millions of people worldwide. Similarly, Wang *et al.* (2016) applied CRISPR/Cas9 to reduce phytic acid levels in rice by targeting and disrupting the OsMRP5 gene, which encodes a multi-drug resistance-associated protein involved in phytic acid transport. The modified rice plants showed a marked decrease in phytic acid accumulation in the grains, leading to enhanced iron and zinc bioavailability. These developments highlight the potential of CRISPR technology to address micronutrient deficiencies through targeted reduction of anti-nutritional factors in staple crops.

Reduction of tannins

Tannins are polyphenolic compounds found in various plant species, including legumes, cereals, and fruits. While tannins have some beneficial antioxidant properties, they are also known to inhibit the absorption of proteins and minerals by binding to them and forming insoluble complexes. High tannin levels in food can lead to reduced nutritional intake and are particularly problematic in regions where tannin-rich crops form a significant part of the diet. CRISPR/Cas9 technology has been utilized to reduce tannin content in sorghum, a staple crop in many arid regions of Africa and Asia. Li and Holding

(2021) employed CRISPR/Cas9 to knock out the Tannin1 (TAN1) gene in sorghum, which is responsible for tannin biosynthesis in the seed coat. The edited sorghum plants exhibited a significant reduction in tannin levels, leading to improved digestibility and nutrient absorption. This modification not only enhances the nutritional value of sorghum but also makes it a more attractive crop for broader agricultural use.

Glucosinolate reduction

Glucosinolates are sulfur-containing compounds found in cruciferous vegetables like broccoli, cabbage, and canola. While some glucosinolates have health-promoting properties, others can have goitrogenic effects, interfering with thyroid function by disrupting iodine uptake. Moreover, high levels of glucosinolates can impart a bitter taste to food, reducing its palatability. In canola (*Brassica napus*), which is widely cultivated for its oil-rich seeds, CRISPR/Cas9 has been applied to reduce glucosinolate content, thereby improving both the nutritional quality and taste of the crop. Jiang *et al.* (2020) targeted and knocked out the BnAOP2 gene, which encodes a key enzyme in the glucosinolate biosynthetic pathway. The edited canola plants showed a significant reduction in glucosinolate levels in the seeds, leading to the production of oil with improved flavor and nutritional properties. This approach not only enhances the suitability of canola oil for human consumption but also increases the market value of the crop.

1.4 Future Prospects

The convergence of functional genomics and CRISPR/Cas technologies marks a pivotal advancement in crop science, offering the potential to revolutionize agricultural practices and nutritional outcomes worldwide. These technologies are poised to significantly enhance the nutritional quality of crops by enabling precise gene editing with unprecedented accuracy. This includes the simultaneous editing of multiple genes to improve complex traits such as protein quality, vitamin content, and mineral bioavailability. For instance, editing genes involved in both iron and zinc biosynthesis in rice could produce varieties that address multiple micronutrient deficiencies in a single crop, particularly in regions where rice is a staple food. Moreover, with the growing challenges posed by climate change, there is an urgent need for crops that are both nutritionally enhanced and resilient to environmental stresses. Integrating functional genomics with CRISPR/Cas technology could lead to the development of crops that maintain or even improve their nutritional quality under conditions like drought, salinity, and extreme temperatures. This could be achieved by editing genes linked to both stress tolerance and nutrient biosynthesis, creating crops that are both nutritious and adaptable. Furthermore, as our understanding of human nutrition becomes more personalized, there is potential to tailor crops to meet the specific nutritional needs of different populations. CRISPR/Cas technology could be used to develop biofortified crops with nutrients tailored to particular dietary requirements, helping to prevent diet-related diseases and improving public health. The broader adoption of CRISPR/Cas technologies will require addressing ethical, regulatory, and societal concerns. Public perception, safety, and acceptance of genetically edited crops are critical for their widespread implementation. Additionally, global collaboration and knowledge-sharing among research institutions, governments, and the private sector will be essential to ensure these innovations benefit regions and populations most in need, particularly in developing countries where food insecurity is prevalent.

1.5 Conclusion

The integration of various tools of functional genomics and CRISPR/Cas technologies represents a powerful and transformative approach to the development of nutritionally enhanced crops. These technologies offer unprecedented opportunities to improve the nutritional quality of staple foods, address micronutrient deficiencies, and contribute to global food security. By enabling precise modifications to plant genomes, researchers can develop crops that are not only richer in essential nutrients but also more resilient to environmental stresses, adaptable to changing dietary needs, and aligned with the principles of sustainable agriculture.

As we look to the future, the continued advancement of these technologies will require a multifaceted approach that includes rigorous scientific research, ethical and regulatory oversight, and proactive engagement with society. The success of nutritionally enhanced crops will depend on the ability to balance innovation with responsibility, ensuring that these crops are developed and deployed in ways that are safe, equitable, and beneficial to all. By fostering global collaboration and embracing the potential of functional genomics and CRISPR/Cas technologies, we can pave the way for a new era of agricultural innovation that meets the nutritional needs of a growing world population.

Acknowledgments

The authors acknowledge all the contributors to the field Developing 'Nutritionally-enhanced crop through the approaches of functional genomic and CRISPR/Cas technologies' that could not be included in this chapter due to space limitations. GKS is supported by a Minor Research Project (memo no. RGU/UGC/SM-415/2019/108, dated 23 May 2022) granted by Rajiv Gandhi University, Rono Hill, Doimukh, Arunachal Pradesh-791112. SK is supported by a Research Project (project code no. CRG/RLA/208/2023; ref no. RLAC/2023/277, dated 12 June 2023), funded by Ram Lal Anand College, University of Delhi.

Conflict of Interest

The authors declare that there is no conflict of interest.

References

1001 Genomes Consortium (2016) 1,135 genomes reveal the global pattern of polymorphism in *Arabidopsis thaliana*. *Cell* 166(2), 481–491.

Abdul Aziz, M., Brini, F., Rouached, H. and Masmoudi, K. (2022) Genetically engineered crops for sustainably enhanced food production systems. *Frontiers in Plant Science* 13, 1027828.

Abu-Abied, M., Belausov, E., Hagay, S., Peremyslov, V., Dolja, V. *et al.* (2018) Myosin XI-K is involved in root organogenesis, polar auxin transport, and cell division. *Journal of Experimental Botany* 69(12), 2869–2881.

Achary, V.M.M. and Reddy, M.K. (2021) CRISPR-Cas9 mediated mutation in GRAIN WIDTH and WEIGHT2 (GW2) locus improves aleurone layer and GRAIN nutritional quality in rice. *Scientific Reports* 11(1), 21941.

Adeyinka, O.S., Tabassum, B., Koloko, B.L. and Ogungbe, I.V. (2023) Enhancing the quality of staple food crops through CRISPR/Cas-mediated site-directed mutagenesis. *Planta* 257(4), 78.

Alharbi, S., Al-Dakhil, M. and Alotaibi, F.S. (2024) Precise genome editing of plants through base and prime editor. In: Faiz, S. and Prakash, C.S. (eds) *OMICs-Based Techniques for Global Food Security*. Wiley, pp. 269–281.

Arabidopsis Genome Initiative (2000) Analysis of the genome sequence of the flowering plant *Arabidopsis thaliana*. *Nature* 408(6814), 796–815.

Araújo, W.L., Nunes-Nesi, A. and Williams, T.C. (2012) Functional genomics tools applied to plant metabolism: A survey on plant respiration, its connections and the annotation of complex gene functions. *Frontiers in Plant Science* 3, 210.

Ariel, F.D. and Manavella, P.A. (2021) When junk DNA turns functional: Transposon-derived non-coding RNAs in plants. *Journal of Experimental Botany* 72(11), 4132–4143.

Awasthi, S., Kaushik, N., Plaha, N.S., Kaur, V. and Kumar, A. (2024) Exploring lipid health indices and protein quality in ninety Indian linseed varieties by comprehensive analysis of fatty acid composition, lignan content, and amino acid composition. *Industrial Crops and Products* 212, 118366.

Bai, S.-N. (2020) Are unisexual flowers an appropriate model to study plant sex determination? *Journal of Experimental Botany* 71(16), 4625–4628.

Banerjee, S., Roy, P., Nandi, S. and Roy, S. (2023) Advanced biotechnological strategies towards the development of crops with enhanced micronutrient content. *Plant Growth Regulation* 100(2), 355–371.

Barupal, M. and Kumar, A. (2024) CRISPR/Cas systems for enhancing photosynthesis: Climate resilience and food production. In: Kumar, A., Arora, S., Ogita, S., Yau, Y.-Y. and Mukherjee, K. (eds) *Gene Editing in Plants: CRISPR-Cas and Its Applications*. Springer, pp. 477–519.

Bennewitz, S., Bergau, N. and Tissier, A. (2018) QTL mapping of the shape of type VI glandular trichomes in tomato. *Frontiers in Plant Science* 9, 1421.

Bernal-Gallardo, J.J. and de Folter, S. (2024) Plant genome information facilitates plant functional genomics. *Planta* 259(5), 117.

Blaxter, M., Archibald, J.M., Childers, A.K., Coddington, J.A., Crandall, K.A. *et al.* (2022) Why sequence all eukaryotes? *Proceedings of the National Academy of Sciences* 119(4), e2115636118.

Bräutigam, K. and Cronk, Q. (2018) DNA methylation and the evolution of developmental complexity in plants. *Frontiers in Plant Science* 9, 1447.

Brodie, J., Chan, C.X., Clerck, O., Cock, J.M., Coelho, S.M. *et al.* (2017) The algal revolution. *Trends in Plant Science* 22(8), 726–738.

Byrne, L. and Murphy, R.A. (2022) Relative bioavailability of trace minerals in production animal nutrition: A review. *Animals* 12(15), 1981.

Caudai, C., Galizia, A., Geraci, F., Le Pera, L., Morea, V. *et al.* (2021) AI applications in functional genomics. *Computational and Structural Biotechnology Journal* 19, 5762–5790.

Ceballos-Laita, L., Takahashi, D., Uemura, M., Abadía, J., López-Millán, A.F. *et al.* (2022) Effects of fe and mn deficiencies on the root protein profiles of tomato (*Solanum lycopersicum*) using two-dimensional electrophoresis and label-free shotgun analyses. *International Journal of Molecular Sciences* 23(7), 3719.

Chang, J.-D., Gao, W., Wang, P. and Zhao, F.-J. (2022) OsNRAMP5 is a major transporter for lead uptake in rice. *Environmental Science & Technology* 56(23), 17481–17490.

Cheng, S., Melkonian, M., Smith, S.A., Brockington, S., Archibald, J.M. *et al.* (2018) 10KP: A phylodiverse genome sequencing plan. *Gigascience* 7(3), giy013.

Clark, J.W. and Donoghue, P.C. (2018) Whole-genome duplication and plant macroevolution. *Trends in Plant Science* 23(10), 933–945.

Dai, F., Wang, X., Zhang, X.Q., Chen, Z., Nevo, E. *et al.* (2018) Assembly and analysis of a qingke reference genome demonstrate its close genetic relation to modern cultivated barley. *Plant Biotechnology Journal* 16(3), 760–770.

Dai, X., Xu, Y., Ma, Q., Xu, W., Wang, T. *et al.* (2007) Overexpression of an R1R2R3 MYB gene, osMYB3R-2, increases tolerance to freezing, drought, and salt stress in transgenic *Arabidopsis*. *Plant Physiology* 143(4), 1739–1751.

Darwin Tree of Life Project Consortium (2022) Sequence locally, think globally: The darwin tree of life project. *Proceedings of the National Academy of Sciences of the United States of America* 119(4), e2115642118.

Dhaliwal, S.S., Sharma, V., Shukla, A.K., Verma, V., Kaur, M. *et al.* (2022) Biofortification: A frontier novel approach to enrich micronutrients in field crops to encounter the nutritional security. *Molecules* 27(4), 1340.

Du, Q., Jiao, X., Song, X., Zhang, J., Bai, P. *et al.* (2020) The response of water dynamics to long-term high vapor pressure deficit is mediated by anatomical adaptations in plants. *Frontiers in Plant Science* 11, 758.

Dwivedi, S.L., Garcia-Oliveira, A.L., Govindaraj, M. and Ortiz, R. (2023) Biofortification to avoid malnutrition in humans in a changing climate: Enhancing micronutrient bioavailability in seed, tuber, and storage roots. *Frontiers in Plant Science* 14, 1119148.

Ebenezer, T.E., Muigai, A.W., Nouala, S., Badaoui, B., Blaxter, M. *et al.* (2022) Africa: Sequence 100,000 species to safeguard biodiversity. *Nature* 603(7901), 388–392.

Fernando, K. and Gamage, J. (2021) Biotechnology – The panacea when everything else fails.

Florean, M., Luck, K., Hong, B., Nakamura, Y., O'Connor, S.E. *et al.* (2023) Reinventing metabolic pathways: Independent evolution of benzoxazinoids in flowering plants. *Proceedings of the National Academy of Sciences* 120(42), e2307981120.

Fritz, S., See, L., Carlson, T., Haklay, M., Oliver, J.L. *et al.* (2019) Citizen science and the United Nations sustainable development goals. *Nature Sustainability* 2(10), 922–930.

Gaikwad, K.B., Rani, S., Kumar, M., Gupta, V., Babu, P.H. *et al.* (2020) Enhancing the nutritional quality of major food crops through conventional and genomics-assisted breeding. *Frontiers in Nutrition* 7, 533453.

Gangurde, S.S., Wang, H., Yaduru, S., Pandey, M.K., Fountain, J.C. *et al.* (2020) Nested-association mapping (NAM)-based genetic dissection uncovers candidate genes for seed and pod weights in peanut (*Arachis hypogaea*). *Plant Biotechnology Journal* 18(6), 1457–1471.

Ganie, S.A., Molla, K.A., Henry, R.J., Bhat, K. and Mondal, T.K. (2019) Advances in understanding salt tolerance in rice. *Theoretical and Applied Genetics* 132, 851–870.

Gao, H., Wen, L., Wu, Y., Yan, X., Li, J. *et al.* (2018) Sensitive and facile electrochemiluminescent immunoassay for detecting genetically modified rapeseed based on novel carbon nanoparticles. *Journal of Agricultural and Food Chemistry* 66(20), 5247–5253.

Gao, J., Wang, G., Ma, S., Xie, X., Wu, X. *et al.* (2015) CRISPR/Cas9-mediated targeted mutagenesis in *Nicotiana tabacum*. *Plant Molecular Biology* 87, 99–110.

Garg, M., Sharma, N., Sharma, S., Kapoor, P., Kumar, A. *et al.* (2018) Biofortified crops generated by breeding, agronomy, and transgenic approaches are improving lives of millions of people around the world. *Frontiers in Nutrition* 5, 12.

Genomics for Australian Plants Initiative (2018). Available at: https://doi.org/10.25953/3108-3v82 (accessed 12 May 2025).

Gerona, M.E.B., Deocampo, M.P., Egdane, J.A., Ismail, A.M. and Dionisio-Sese, M.L. (2019) Physiological responses of contrasting rice genotypes to salt stress at reproductive stage. *Rice Science* 26(4), 207–219.

Ghazal, H., Adam, Y., Idrissi Azami, A., Sehli, S., Nyarko, H.N. *et al.* (2021) Plant genomics in Africa: Present and prospects. *The Plant Journal* 107(1), 21–36.

Ginzberg, I., Barel, G., Ophir, R., Tzin, E., Tanami, Z. *et al.* (2009) Transcriptomic profiling of heat-stress response in potato periderm. *Journal of Experimental Botany* 60(15), 4411–4421.

Giraud, E., Ho, L.H., Clifton, R., Carroll, A., Estavillo, G. *et al.* (2008) The absence of ALTERNATIVE OXIDASE1a in *Arabidopsis* results in acute sensitivity to combined light and drought stress. *Plant Physiology* 147(2), 595–610.

Havlin, J.L. and Schlegel, A.J. (2021) Review of phosphite as a plant nutrient and fungicide. *Soil Systems* 5(3), 52.

Hendre, P.S., Muthemba, S., Kariba, R., Muchugi, A., Fu, Y. *et al.* (2019) African orphan crops consortium (AOCC): Status of developing genomic resources for African orphan crops. *Planta* 250, 989–1003.

Howlader, J., Robin, A.H.K., Natarajan, S., Biswas, M.K., Sumi, K.R. *et al.* (2020) Transcriptome analysis by rna-seq reveals genes related to plant height in two sets of parent-hybrid combinations in easter lily (*Lilium longiflorum*). *Scientific Reports* 10(1), 9082.

Huang, L., Li, Q., Zhang, C., Chu, R., Gu, Z. *et al.* (2020) Creating novel wx alleles with fine-tuned amylose levels and improved grain quality in rice by promoter editing using CRISPR/Cas9 system. *Plant Biotechnology Journal* 18(11), 2164.

Jha, R., Yadav, H.K., Raiya, R., Singh, R.K., Jha, U.C. *et al.* (2022) Integrated breeding approaches to enhance the nutritional quality of food legumes. *Frontiers in Plant Science* 13, 984700.

Jiang, F.-N., Liang, Y.-X., Wei, W., Zou, C.-Y., Chen, G.-X. *et al.* (2020) Functional classification of prostate cancer-associated miRNAs through CRISPR/Cas9-mediated gene knockout. *Molecular Medicine Reports* 22(5), 3777–3784.

Johnson, A.A., Kyriacou, B., Callahan, D.L., Carruthers, L., Stangoulis, J. *et al.* (2011) Constitutive overexpression of the OsNAS gene family reveals single-gene strategies for effective iron-and zinc-biofortification of rice endosperm. *PLoS One* 6(9), e24476.

Kang, W., Jiang, Z., Chen, Y., Wu, F., Liu, C. *et al.* (2020) Plant transcriptome analysis reveals specific molecular interactions between alfalfa and its rhizobial symbionts below the species level. *BMC Plant Biology* 20, 1–19.

Kim, S., Chen, J., Cheng, T., Gindulyte, A., He, J. *et al.* (2021) PubChem in 2021: New data content and improved web interfaces. *Nucleic Acids Research* 49(D1), D1388–D1395.

Klap, C., Yeshayahou, E., Bolger, A.M., Arazi, T., Gupta, S.K. *et al.* (2017) Tomato facultative parthenocarpy results from Sl AGAMOUS-LIKE 6 loss of function. *Plant Biotechnology Journal* 15(5), 634–647.

Koenigshofer, H. and Loeppert, H.-G. (eds) (2019) The up-regulation of proline synthesis in the meristematic tissues of wheat seedlings upon short-term exposure to osmotic stress. *Journal of Plant Physiology* 237, 21–29.

Komatsu, S. and Colgrave, M.L. (2022) Plant proteomic research 5.0: From data to insights. *International Hournal of Molecular Sciences* 24(1), 258.

Komatsu, S., Tsutsui, Y., Furuya, T., Yamaguchi, H., Hitachi, K. *et al.* (2022) Proteomic and biochemical approaches elucidate the role of millimeter-wave irradiation in wheat growth under flooding stress. *International Journal of Molecular Sciences* 23(18), 10360.

Koochak, H., Puthiyaveetil, S., Mullendore, D., Li, M. and Kirchhoff, H. (2019) The structural and functional domains of plant thylakoid membranes. *Plant Journal* 100(1), 213.

Ku, H.-K. and Ha, S.-H. (2020) Improving nutritional and functional quality by genome editing of crops: Status and perspectives. *Frontiers in Plant Science* 11, 577313.

Kumar, D., Yadav, A., Ahmad, R., Dwivedi, U.N. and Yadav, K. (2022) CRISPR-based genome editing for nutrient enrichment in crops: A promising approach toward global food security. *Frontiers in Genetics* 13, 932859.

Kumar, M., Prusty, M.R., Pandey, M.K., Singh, P.K., Bohra, A. *et al.* (2023) Application of CRISPR/Cas9-mediated gene editing for abiotic stress management in crop plants. *Frontiers in Plant Science* 14, 1157678.

Leissing, F., Misch, N.V., Wang, X., Werner, L., Huang, L. *et al.* (2021) Purification of MAP–kinase protein complexes and identification of candidate components by XL–TAP–MS. *Plant Physiology* 187(4), 2381–2392.

Lewin, H.A., Richards, S., Lieberman Aiden, E., Allende, M.L., Archibald, J.M. *et al.* (2022) The earth biogenome project 2020: Starting the clock. In: *Proceedings of the National Academy of Science USA*, 119, e2115635118.

Li, F.-S. and Weng, J.-K. (2017) Demystifying traditional herbal medicine with modern approach. *Nature Plants* 3(8), 1–7.

Li, A. and Holding, D.R. (2021) Genome editing in sorghum. In: Willmann, M.R. (ed.) *Genome Editing for Precision Crop Breeding*. Burleigh Dodds Science Publishing, Sawston, UK, pp. 377–402.

Li, C., Zong, Y., Wang, Y., Jin, S., Zhang, D. *et al.* (2018) Expanded base editing in rice and wheat using a Cas9-adenosine deaminase fusion. *Genome Biology* 19, 1–9.

Li, Y., Zhang, B. and Yu, H. (2022) Molecular genetic insights into orchid reproductive development. *Journal of Experimental Botany* 73(7), 1841–1852.

Lin, S., Yue, X., Miao, Y., Yu, Y., Dong, H. *et al.* (2018) The distinct functions of two classical arabinogalactan proteins BcMF8 and BcMF18 during pollen wall development in *Brassica campestris*. *The Plant Journal* 94(1), 60–76.

Liu, F., Baye, W., Zhao, K., Tang, S., Xie, Q. *et al.* (2024a) Unravelling sorghum functional genomics and molecular breeding: Past achievements and future prospects. *Journal of Genetics and Genomics* [in press].

Liu, N., Zeng, F., Wang, J., Wang, T.J., Zhang, P. *et al.* (2024b) Large scale massively parallel numerical simulation for blast effects on structures. *Physica Scripta* 99, 10.

Liu, Z.-Q., Qiu, A.-L., Shi, L.-P., Cai, J.-S., Huang, X.-Y. *et al.* (2015) SRC2-1 is required in PcINF1-induced pepper immunity by acting as an interacting partner of PcINF1. *Journal of Experimental Botany* 66(13), 3683–3698.

Lu, J., Luo, C., Bali, K.K., Xie, R.-G., Mains, R.E. *et al.* (2015) A role for Kalirin-7 in nociceptive sensitization via activity-dependent modulation of spinal synapses. *Nature Communications* 6(1), 6820.

Marín-Guirao, L., Entrambasaguas, L., Dattolo, E., Ruiz, J.M. and Procaccini, G. (2017) Molecular mechanisms behind the physiological resistance to intense transient warming in an iconic marine plant. *Frontiers in Plant Science* 8, 1142.

Matsumoto, T. and Takenaka, A. (2018) Recent developments of Ge. In: Lee, T.-R., Becker, P.R., Fuchimoto, D.-I., Matsumoto, T. and Kess, P. (eds) *Case Studies: Insights on Agriculture Innovation 2016*. IAAS Series.

Matsuoka, R. and Sugano, M. (2022) Health functions of egg protein. *Foods* 11(15), 2309.

McLay, T.G., Murphy, D.J., Holmes, G.D., Mathews, S., Brown, G.K. *et al.* (2022) A genome resource for Acacia, Australia's largest plant genus. *PLoS One* 17(10), e0274267.

Mendiondo, G.M., Gibbs, D.J., Szurman-Zubrzycka, M., Korn, A., Marquez, J. *et al.* (2016) Enhanced waterlogging tolerance in barley by manipulation of expression of the n-end rule pathway E3 ligase PROTEOLYSIS 6. *Plant Biotechnology Journal* 14(1), 40–50.

Mishra, R. and Zhao, K. (2018) Genome editing technologies and their applications in crop improvement. *Plant Biotechnology Reports* 12, 57–68.

Molla, K.A. and Yang, Y. (2019) CRISPR/Cas-mediated base editing: Technical considerations and practical applications. *Trends in Biotechnology* 37(10), 1121–1142.

Monforte, A.J. (2021) Identification of tomato accessions as source of new genes for improving heat tolerance: From controlled experiments to field. *BMC Plant Biology* 21.

Neelakandan, A.K., Wright, D.A., Traore, S.M., Chen, X., Spalding, M.H. *et al.* (2022) CRISPR/Cas9 based site-specific modification of FAD2 cis-regulatory motifs in peanut (*Arachis hypogaea* L.). *Frontiers in Genetics* 13, 849961.

Niculaes, C., Abramov, A., Hannemann, L. and Frey, M. (2018) Plant protection by benzoxazinoids: Recent insights into biosynthesis and function. *Agronomy* 8(8), 143.

Nie, H., Yang, X., Zheng, S. and Hou, L. (2024) Gene-based developments in improving quality of tomato: Focus on firmness, shelf life, and pre- and post-harvest stress adaptations. *Horticulturae* 10(6), 641.

Nieves-Cordones, M., Alemán, F., Martínez, V. and Rubio, F. (2014) K+ uptake in plant roots: The systems involved, their regulation and parallels in other organisms. *Journal of Plant Physiology* 171(9), 688–695.

Nützmann, H.W., Huang, A. and Osbourn, A. (2016) Plant metabolic clusters – From genetics to genomics. *New Phytologist* 211(3), 771–789.

Ossowski, S., Schneeberger, K., Clark, R.M., Lanz, C., Warthmann, N. *et al.* (2008) Sequencing of natural strains of *Arabidopsis thaliana* with short reads. *Genome Research* 18(12), 2024–2033.

Paine, J.A., Shipton, C.A., Chaggar, S., Howells, R.M., Kennedy, M.J. *et al.* (2005) Improving the nutritional value of golden rice through increased pro-vitamin A content. *Nature Biotechnology* 23(4), 482–487.

Palaniswamy, R., Kambale, R., Mohanavel, V., Rajagopalan, V.R., Manickam, S. *et al.* (2024) Identifying molecular targets for modulating carotenoid accumulation in rice grains. *Biochemistry and Biophysics Reports* 40, 101815.

Pandey, S., Kumari, A., Shree, M., Kumar, V., Singh, P. *et al.* (2019) Nitric oxide accelerates germination via the regulation of respiration in chickpea. *Journal of Experimental Botany* 70(17), 4539–4555.

Pandohee, J., Basu, R., Dasgupta, S., Sundarrajan, P., Shaikh, N. *et al.* (2023) Applications of multi-omics approaches for food and nutritional security. In: Li, Q. and Yan, J. (eds) *Sustainable Agriculture in the Era of the OMICs Revolution*. Springer, pp. 103–118.

Patel, M.K., Mishra, A. and Jha, B. (2016) Non-targeted metabolite profiling and scavenging activity unveil the nutraceutical potential of psyllium (*Plantago ovata* Forsk). *Frontiers in Plant Science* 7, 431.

Patel, M.K., Pandey, S., Kumar, M., Haque, M.I., Pal, S. *et al.* (2021) Plants metabolome study: Emerging tools and techniques. *Plants* 10(11), 2409.

Pluskal, T., Torrens-Spence, M.P., Fallon, T.R., Abreu, A., Shi, C.H. *et al.* (2019) The biosynthetic origin of psychoactive kavalactones in Kava. *Nature Plants* 5(8), 867–878.

Qiao, X., Zhang, S. and Paterson, A.H. (2022) Pervasive genome duplications across the plant tree of life and their links to major evolutionary innovations and transitions. *Computational and Structural Biotechnology Journal* 20, 3248–3256.

Ramazi, S. and Zahiri, J. (2021) Post-translational modifications in proteins: Resources, tools and prediction methods. *Database* 2021, baab012.

Rasheed, A., Gill, R.A., Hassan, M.U., Mahmood, A., Qari, S. *et al.* (2021) A critical review: Recent advancements in the use of CRISPR/Cas9 technology to enhance crops and alleviate global food crises. *Current Issues in Molecular Biology* 43(3), 1950–1976.

Ren, H., Yang, W., Jing, W., Shahid, M.O., Liu, Y. *et al.* (2024) Multi-omics analysis reveals key regulatory defense pathways and genes involved in salt tolerance of rose plants. *Horticulture Research* 11(5), uhae068.

Roorkiwal, M., Jain, A., Kale, S.M., Doddamani, D., Chitikineni, A. *et al.* (2018) Development and evaluation of high-density Axiom® Cicer SNP array for high-resolution genetic mapping and breeding applications in chickpea. *Plant Biotechnology Journal* 16(4), 890–901.

Rosati, R.G., Lario, L.D., Hourcade, M.E., Cervigni, G.D., Luque, A.G. *et al.* (2018) Primary metabolism changes triggered in soybean leaves by *Fusarium tucumaniae* infection. *Plant Science* 274, 91–100.

Sahu, A., Verma, R., Gupta, U., Kashyap, S. and Sanyal, I. (2024) An overview of targeted genome editing strategies for reducing the biosynthesis of phytic acid: An anti-nutrient in crop plants. *Molecular Biotechnology* 66(1), 11–25.

Sasaki, T. and Burr, B. (2000) International rice genome sequencing project: The effort to completely sequence the rice genome. *Current Opinion in Plant Biology* 3(2), 138–142.

Saurabh, S., Vidyarthi, A.S. and Prasad, D. (2014) RNA interference: Concept to reality in crop improvement. *Planta* 239(3), 543–564.

Sertse, D., You, F.M., Ravichandran, S. and Cloutier, S. (2019) The complex genetic architecture of early root and shoot traits in flax revealed by genome-wide association analyses. *Frontiers in Plant Science* 10, 1483.

Shanmugabalaji, V. and Kessler, F. (2019) CHLORAD: Eradicating translocon components from the outer membrane of the chloroplast. *Molecular Plant* 12(4), 467–469.

Sinn, J.P., Held, J.B., Vosburg, C., Klee, S.M., Orbovic, V. *et al.* (2021) Flowering Locus T chimeric protein induces floral precocity in edible citrus. *Plant Biotechnology Journal* 19(2), 215.

Sun, C., Liang, X., Gong, X., Chen, H., Liu, X. *et al.* (2022) Comparative transcriptomics provide new insights into the mechanisms by which foliar silicon alleviates the effects of cadmium exposure in rice. *Journal of Environmental Sciences* 115, 294–307.

Syu, C.-H., Wu, P.-R., Lee, C.-H., Juang, K.-W. and Lee, D.-Y. (2019) Arsenic phytotoxicity and accumulation in rice seedlings grown in arsenic-contaminated soils as influenced by the characteristics of organic matter amendments and soils. *Journal of Plant Nutrition and Soil Science* 182(1), 60–71.

Tabassum, J., Ahmad, S., Hussain, B., Mawia, A.M., Zeb, A. *et al.* (2021) Applications and potential of genome-editing systems in rice improvement: Current and future perspectives. *Agronomy* 11(7), 1359.

Tiwari, M., Kumar Trivedi, P. and Pandey, A. (2021) Emerging tools and paradigm shift of gene editing in cereals, fruits, and horticultural crops for enhancing nutritional value and food security. *Food and Energy Security* 10(1), e258.

Tomato Genome Consortium. (2012) The tomato genome sequence provides insights into fleshy fruit evolution. *Nature* 485(7400), 635–641.

Tong, H., Huang, J., Xiao, Q., He, B., Dong, X. *et al.* (2023) High-fidelity Cas13 variants for targeted RNA degradation with minimal collateral effects. *Nature Biotechnology* 41(1), 108–119.

Tsutsui, H., Yanagisawa, N., Kawakatsu, Y., Ikematsu, S., Sawai, Y. *et al.* (2020) Micrografting device for testing systemic signaling in *Arabidopsis*. *The Plant Journal* 103(2), 918–929.

Verma, V., Kumar, A., Partap, M., Thakur, M. and Bhargava, B. (2023) CRISPR-Cas: A robust technology for enhancing consumer-preferred commercial traits in crops. *Frontiers in Plant Science* 14, 1122940.

Virdi, K.S., Spencer, M., Stec, A.O., Xiong, Y., Merry, R. *et al.* (2020) Similar seed composition phenotypes are observed from CRISPR-generated in-frame and knockout alleles of a soybean KASI ortholog. *Frontiers in Plant Science* 11, 1005.

Wada, N., Ueta, R., Osakabe, Y. and Osakabe, K. (2020) Precision genome editing in plants: State-of-the-art in CRISPR/Cas9-based genome engineering. *BMC Plant Biology* 20, 1–12.

Wang, F., Wang, C., Liu, P., Lei, C., Hao, W. *et al.* (2016) Enhanced rice blast resistance by CRISPR/Cas9-targeted mutagenesis of the ERF transcription factor gene OsERF922. *PLoS One* 11(4), e0154027.

Wang, P., Schumacher, A.M. and Shiu, S.-H. (2022a) Computational prediction of plant metabolic pathways. *Current Opinion in Plant Biology* 66, 102171.

Wang, Y., Zafar, N., Ali, Q., Manghwar, H., Wang, G. *et al.* (2022b) CRISPR/Cas genome editing technologies for plant improvement against biotic and abiotic stresses: Advances, limitations, and future perspectives. *Cells* 11(23), 3928.

Wei, X., Long, Y., Yi, C., Pu, A., Hou, Q. *et al.* (2023) Bibliometric analysis of functional crops and nutritional quality: Identification of gene resources to improve crop nutritional quality through gene editing technology. *Nutrients* 15(2), 373.

Wu, J., Chen, C., Xian, G., Liu, D., Lin, L. *et al.* (2020) Engineering herbicide-resistant oilseed rape by CRISPR/Cas9-mediated cytosine base-editing. *Plant Biotechnology Journal* 18(9), 1857.

Wu, W., Liu, S., Ruwe, H., Zhang, D., Melonek, J. *et al.* (2016) SOT 1, a pentatricopeptide repeat protein with a small MUTs-related domain, is required for correct processing of plastid 23S–4.5 S rRNA precursors in *Arabidopsis thaliana*. *The Plant Journal* 85(5), 607–621.

Yadav, R.K., Tripathi, M.K., Tiwari, S., Tripathi, N., Asati, R. *et al.* (2023) Genome editing and improvement of abiotic stress tolerance in crop plants. *Life* 13(7), 1456.

Yan, S., Bhawal, R., Yin, Z., Thannhauser, T.W. and Zhang, S. (2022) Recent advances in proteomics and metabolomics in plants. *Molecular Horticulture* 2(1), 17.

Yang, Q.Q., Yu, W.H., Wu, H.Y., Zhang, C.Q., Sun, S.S.M. *et al.* (2021) Lysine biofortification in rice by modulating feedback inhibition of aspartate kinase and dihydrodipicolinate synthase. *Plant Biotechnology Journal* 19(3), 490–501.

Yang, T., Dong, J., Xiong, X., Zhang, L., Wang, J. *et al.* (2024) A novel function of GW5 on controlling the early growth vigor and its haplotype effect on shoot dry weight and grain size in rice (*Oryza sativa* L.). *Rice* 17(1), 1–16.

Yang, W., Yang, Y., Wang, L., Lv, X., Li, J. *et al.* (2023) Comparative characterization of flavor precursors and volatiles of Taihe black-boned silky fowl and Hy-line brown yolks using multiomics and GC-O-MS-based volatilomics. *Food Research International* 172, 113168.

Yerramsetty, P., Agar, E.M., Yim, W.C., Cushman, J.C. and Berry, J.O. (2017) An rbc L mRNA-binding protein is associated with C3 to C4 evolution and light-induced production of Rubisco in Flaveria. *Journal of Experimental Botany* 68(16), 4635–4649.

Yin, K. and Qiu, J.-L. (2019) Genome editing for plant disease resistance: Applications and perspectives. *Philosophical Transactions of the Royal Society B* 374(1767), 20180322.

Yoon, E.K., Dhar, S., Lee, M.-H., Song, J.H., Lee, S.A. *et al.* (2016) Conservation and diversification of the SHR-SCR-SCL23 regulatory network in the development of the functional endodermis in *Arabidopsis* shoots. Molecular Plant 9(8), 1197–1209.

Yu, D., Boughton, B.A., Hill, C.B., Feussner, I., Roessner, U. *et al.* (2020) Insights into oxidized lipid modification in barley roots as an adaptation mechanism to salinity stress. *Frontiers in Plant Science* 11, 1.

Zhang, D., Hussain, A., Manghwar, H., Xie, K., Xie, S. *et al.* (2020) Genome editing with the CRISPR-Cas system: An art, ethics and global regulatory perspective. *Plant Biotechnology Journal* 18(8), 1651–1669.

Zhang, D., Zhang, Z., Unver, T. and Zhang, B. (2021) CRISPR/Cas: A powerful tool for gene function study and crop improvement. *Journal of Advanced Research* 29, 207–221.

Zhang, M., Liu, S., Wang, Z., Yuan, Y. and Zhang, Z. (2022) Progress in soybean functional genomics over the past decade. *Plant Biotechnology Journal* 20(2), 256–282.

Zhao, K. and Rhee, S.Y. (2022) Omics-guided metabolic pathway discovery in plants: Resources, approaches, and opportunities. *Current Opinion in Plant Biology* 67, 102222.

Zheng, Y., Li, Q., Ye, M., Cehn, A. and Wang, H. (2021) Applications of CRISPR/Cas9-based genome editing in the plant biology. *Turkish Journal of Botany* 45(4), 253–268.

Zhong, Z., Liu, S., Liu, X., Liu, B., Tang, X. *et al.* (2020) Intron-based single transcript unit CRISPR systems for plant genome editing. *Rice* 13, 1–13.

Zhou, X. and Liu, Z. (2022) Unlocking plant metabolic diversity: A (pan)-genomic view. *Plant Communications* 3(2).

Zhou, J., Liu, Y., Guo, X., Birchler, J.A., Han, F. *et al.* (2022) Centromeres: From chromosome biology to biotechnology applications and synthetic genomes in plants. *Plant Biotechnology Journal* 20(11), 2051–2063.

Zujur, D., Kanke, K., Lichtler, A.C., Hojo, H., Chung, U.-I. *et al.* (2017) Three-dimensional system enabling the maintenance and directed differentiation of pluripotent stem cells under defined conditions. *Science Advances* 3(5), e1602875.

Zulfiqar, U., Khokhar, A., Maqsood, M.F., Shahbaz, M., Naz, N. *et al.* (2024) Genetic biofortification: Advancing crop nutrition to tackle hidden hunger. *Functional & Integrative Genomics* 24(2), 34.

2 High-Yield Crops Developed by Climate-Resilient Strategies

Johiruddin Molla[1]*, Sk Rehan Murshed[1] and Kutubuddin Molla[2]
[1]Ghatal Rabindra Satabarsiki Mahavidyalaya Ghatal, Paschim Medinipur, West Bengal, India; [2]Crop Improvement Division, ICAR-NRRI, Cuttack, Odisha, India

Abstract

It is estimated that the global population is projected to reach 9.7 billion by 2050, requiring a 70–90% increase in food production. Rising temperatures, changing precipitation patterns, and extreme weather conditions threaten agricultural productivity, potentially reducing yields by up to 18% by 2050. To cope with the increasing population and rising temperature, it seems urgent to develop high-yield climate-resilient crops to ensure food security. In this context, scientists put effort tirelessly into developing high-yield climate-resilient crops through different strategies like traditional breeding, marker-assisted breeding, next-generation breeding (speed breeding), agroecological practices, genetic engineering, and gene editing (CRISPR/Cas9). Existing crops like wheat, corn, soybean, rice, potato, cotton, maize, cassava, and sorghum are cultivated throughout the world to diminish hunger. Now there is a need to improve those crops in ways that increase their optimum yield, make them resistant to different biotic and abiotic stresses, and meet future demand.

Keywords: CRISPR/Cas9, genome editing, plant stress tolerance, stress-resilient crops

2.1 Introduction

It is a challenge to our scientists to develop high-yielding, climate-resilient crops for several reasons: a global population set to reach 9.7 billion by 2050 according to the Food and Agriculture Organization of the United Nations (FAO, 2017); extreme weather changes (rising temperature, change in precipitation patterns), which may lead to reduced crop yields of 18% by 2050 (PIB Delhi, 2023); and shrinking of arable land and freshwater resources (UN Water, 2020). Therefore, adaptation of conventional farming and advanced breeding strategies may help to develop climate-resilient crops, resistant to different biotic and abiotic factors, and meet the demands of future generations. Here the different breeding strategies that are significant in developing high-yield crops are described.

2.1.1 Traditional breeding

Through traditional breeding, wheat varieties like Borlaug 100 were developed for high yield and disease resistance (Rajaram and Braun, 2008). One such disease that threatens most wheat varieties is stem rust race Ug99. To combat this disease, scientists developed two wheat varieties, Sr26 and SrR, which show significant resistance to the Ug99 race (Singh *et al.*, 2011).

*Corresponding author: johiruddinmolla@gmail.com

© CAB International 2025. *Stress-Resilient Crops: Coordinated omics-CRISPR-Nanotechnology Strategies* (ed. Jen-Tsung Chen)
DOI: 10.1079/9781800627307.0002

Drought is a major consequence of environmental or climate change and one which directly affects the yield of different crops worldwide. One such crop is maize. In 1999, International Maize and Wheat Improvement Centre (CIMMYT) developed two drought-tolerant varieties, namely ZM309 and ZM523 (Edmeades *et al.*, 1999). Another challenge for underdeveloped countries is the lack of nutritious food. Biofortification is one way to enhance nutrition in crops. An example is orange maize, which is a provitamin A biofortified crop (Nestel *et al.*, 2006).

For soybean, traditional breeding produced two varieties named Peking (Arelli and Liu, 2005) and Williams 82 (Specht *et al.*, 1999), which show resistance to soybean cyst nematode and high yield, respectively.

Rice is the second-largest crop in the world and needs to be improved to meet the world's demand through different strategies. Breeding is one such strategy used to develop Swarna-sub1, which is submergence-tolerant. In field trials, Swarna-sub1 demonstrated yield improvements ranging from 30% to 45% compared to non-submergence-tolerant varieties when subjected to flooding (Xu *et al.*, 2006). IRRI146 is an iron- and zinc-fortified variety developed in 2010 to enhance its nutritional value (Bouis and Welch, 2010).

Renowned Chinese agronomist Yuan Longping, known as the 'father of hybrid rice,' has dedicated his research to rice and has recently achieved a significant breakthrough by successfully developing salt-tolerant hybrid rice varieties using traditional breeding techniques. This was achieved by overexpressing a gene from a selected wild rice species that shows greater tolerance to salinity. Through a series of experiments conducted in coastal areas of China, these hybrid varieties have demonstrated remarkable yields of up to 7.5 tons per hectare (Qian *et al.*, 2021).

Potato varieties like Victoria (Gebhardt and Valkonen, 2001) and Kufri Jyoti (Bradshaw and Ramsay, 2009) were developed for late blight resistance and high yield, respectively.

Cotton biotic-resistant variety CIM-616 was developed to increase yield (Riaz Marral *et al.*, 2020). The majority of cotton is grown for high-quality fiber. In this regard, high-quality fiber-yielding varieties like FiberMax832 have been developed through conventional breeding (BASF company: https://agriculture.basf.us/crop-protection/products/seeds/fibermax.html?utm_source=chatgpt.com [accessed 13/5/25]).

In 2017, over 26 million hectares of land were dedicated to cassava cultivation globally (https://www.fao.org/faostat/ [accessed 22/3/25]). Breeding methods have enhanced its yield capacity by developing Narocass1 (Legg and Thresh, 2000) and Yellow Cassava varieties (Nassar and Ortiz, 2010), which are resistant to cassava mosaic disease and fortified with vitamin A.

For sorghum, the drought-tolerant variety Parbhani shakti was developed in 2000 by Borrell (Borrell *et al.*, 2014).

2.1.2 Molecular breeding/marker-assisted breeding

The International Rice Research Institute (IRRI), a member organization of the Consortium of International Agricultural Research Centers (CGIAR), is utilizing marker-assisted breeding, also known as molecular breeding, to cultivate rice varieties that have the ability to endure drought and flooding. Several drought-tolerant rice varieties have been developed, released, and cultivated by farmers. Examples of these varieties include Sahbhagi Dhan in India, Sahod Ulan in the Philippines, and Sukha Dhan in Nepal. These varieties have shown an average yield increase of 0.8–1.2 t/ha under drought conditions. The flash-flood tolerance gene (SUB1A) was introduced into Swarna (India, Bangladesh and Nepal); Samba Mahsuri, IR64, and CR1009 in India; and BR11 in Bangladesh, using marker-assisted backcrossing. The Sub1 varieties, especially Swarna-Sub1 in India, Bangladesh, and Nepal, and BR11-Sub1 in Bangladesh, exhibited a yield advantage of up to 1–3 t/ha (Singh *et al.*, 2013).

Marker-assisted breeding has successfully developed wheat lines that possess long-lasting resistance to major diseases like stem rust and yellow rust. One prominent example is the Sr2/Yr30-linked wheat line, which offers substantial yield advantages in regions susceptible to rust. The utilization of such lines can result in a yield boost of 10–15%, underscoring the significance of marker-assisted breeding in contemporary wheat breeding initiatives (Bajgain *et al.*, 2016).

Soybean lines containing the Rhg1 and Rhg4 genes have been found to display resistance to the soybean cyst nematode (SCN). In addition, these lines have shown an increase in yield of up to 10–30% in fields infested with SCN (Cook *et al.*, 2012).

2.1.3 Speed breeding (Fig. 2.1)

Development of new varieties through conventional breeding requires 7–12 years, which includes crossing, fixing the trait genetically stable, and field trial. Another drawback of conventional breeding is the generation times of crops. There are some crops which grow only once or twice a year. To overcome this, constrained single-seed descendent speed breeding is the emerging choice for the scientific community. In this regard, Watson and his colleagues showed that prolonged photoperiods can be used to accelerate the developmental rate of plants (Sysoeva *et al.*, 2010). Also, they harvest single immature seeds and germinate to reduce the generation time. Through the above method, they successfully developed up to six generations per year for spring wheat (*Triticum aestivum*), durum wheat (*T. durum*), barley (*Hordeum vulgare*), chickpea (*Cicer arietinum*) and pea (*Pisum sativum*), and four generations for canola (*Brassica napus*), instead of 2–3 under normal glasshouse conditions (Watson *et al.*, 2018). Speed-breeding methods have been shown to effectively reduce generation time for various crop species. For instance, sunflower (*Helianthus annuus*), pepper (*Capsicum annuum*), and radish (*Raphanus sativus*) have all demonstrated positive responses to extended photoperiods (Watson *et al.*, 2018). Additionally, speed-breeding techniques have already been successfully applied to expedite breeding goals for amaranth (*Amaranthus* spp.) (Stetter *et al.*, 2016), peanut (*Arachis hypogaea*) (O'Connor *et al.*, 2013), and pea (*P. sativum*) (Mobini and Warkentin, 2016). While it is unlikely that the speed-breeding protocols can be directly applied to short-day species such as maize or rice and yield successful results, there is potential for optimization in order to achieve faster generation times in these species. This strategy can also be used to refine gene stacking, transgenic pipeline, genomic selection (GS), and express editing for crop development (Watson *et al.*, 2018).

2.1.4 Genetic engineering

Extensive crossbreeding is not required in the development of genetically modified (GM) crops. Additionally, the development of GM crops is more controlled compared to traditional breeding. We have the ability to control the timing and location of gene expression, determining which organ or tissue it will be expressed in. Nowadays, the stringency of food safety considerations is greatly impacted by the origin of the DNA utilized in the development of the GM crop. Commercially used GM crops primarily target two traits: herbicide resistance and insect resistance. Through mutagenesis and selection, several variants of the acetohydroxyacid synthase (AHAS) gene, which confers herbicide imidazolinone tolerance, have been discovered in plants. Most of the herbicide-resistant crops currently on the market were developed using conventional breeding techniques by Clearfield* crops between 1992 and the present. These crops include imidazolinone-tolerant maize (*Zea mays* L.), wheat (*Triticum aestivum* L.), rice (*Oryza sativa* L.), oilseed rape (*Brassica napus* L.), and sunflower (*Helianthus annuus* L.) (Tan *et al.*, 2005).

Insect-resistant crops, particularly those engineered to produce Bt (*Bacillus thuringiensis*) toxins, have significantly transformed pest control in agriculture. Bt crops have achieved remarkable success and are extensively cultivated worldwide. Prominent examples of insect-resistant crops are Bt cotton (James, 2013), Bt corn (maize) (Tabashnik *et al.*, 2013), and Bt soybean (Romeis *et al.*, 2006).

The report finds that, globally, GM crops have led to a 22% increase in crop yields. From 1996 to 2018, GM technology resulted in an additional 231 million tonnes of soybean, 498 million tonnes of corn, 32.6 million tonnes of cotton lint, and 14 million tonnes of canola (Brookes and Barfoot, 2020). Qaim discusses the substantial increase in crop yields observed across different GM crops. On average, GM maize

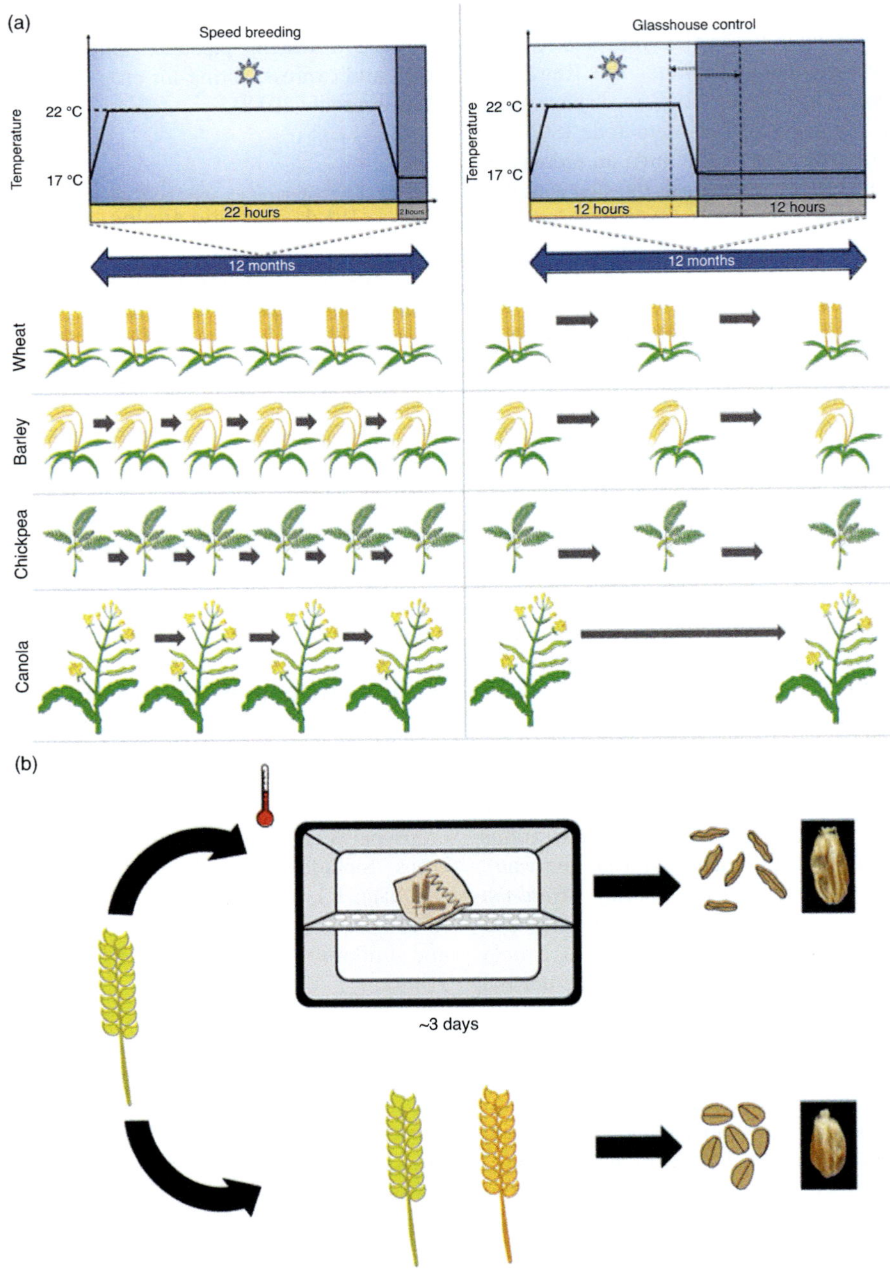

Fig. 2.1. Examples of plant speed breeding: (a) Compared to a glasshouse with a natural variable photoperiod of 10–16 hours, where only 2–3 generations of wheat, barley, chickpea, and canola can be achieved per year (*right*), speed breeding enables 4–6 generations of these crops to be grown in a year (*left*). These values represent the relatively rapid cycling cultivars of each crop; (b) Harvesting immature spikes and drying them in an oven/dehydrator for approximately 3 days allows for a quicker seed-to-seed cycle compared to the natural ripening process, which takes about 15 days. However, this method results in a loss of grain weight.

has shown a yield increase of 25%, while GM cotton has demonstrated an impressive 30% boost in yield (Qaim, 2020).

2.1.5 CRISPR/Cas9

Clustered Regularly Interspaced Short Palindromic Repeats (CRISPR)/CRISPR-associated (Cas) 9 (CRISPR/Cas9) have been previously widely used in precision crop breeding (Feng *et al.*, 2013; Baltes and Voytas, 2015; Weeks *et al.*, 2016), including for salinity tolerance. For instance, CRISPR/Cas9-mediated mutagenesis of the *OsRR22* gene has been used to create salinity-tolerant Japonica rice mutants (Zhang *et al.*, 2019). Rice OsSPL14 gene modification improved grain number and yield by up to 30% (Jiao *et al.*, 2010).

In wheat, the knockout of the negative yield regulator TaGW2 gene resulted in an increase in grain width and weight. This genetic modification has been found to contribute to a yield increase of up to 20% in grain weight (Wang *et al.*, 2018).

In maize, shutting down ARGOS8 can lead to an up to 15% increase in yield under drought conditions (Shi *et al.*, 2017).

In the case of soybean, there is a potential for a 10–20% increase in yield through genome editing of flowering genes, namely GmFT2a and GmFT5a (Cai *et al.*, 2018).

Editing the granule-bound starch synthase (GBSS) gene in potatoes resulted in a higher amylopectin content and improved yield. This led to an approximate 20% increase in starch yield (Andersson *et al.*, 2017). Similarly, editing the GhMYB25-like gene in cotton enhanced fiber length and yield. This could increase fiber yield by up to 15% (Qin *et al.*, 2007).

In cassava, editing PTST1 resulted in an increase in starch content and overall yield. Specifically, there was an approximately 30% increase in starch yield (Bull *et al.*, 2018). Likewise, in sorghum, editing the Dw1 gene led to an increase in plant height and biomass yield. In fact, there was an up to 30% increase in biomass yield (Che *et al.*, 2018).

2.1.6 Agroecological practices

Agroecological practices focus on developing crop varieties that are resilient, sustainable, and well-adapted to specific local environments. Below are some of the varieties that have been developed using this practice.

Emmer Wheat (*Triticum dicoccum*) is an ancient variety of wheat that can thrive in poor soil conditions and requires minimal inputs (Zaharieva and Monneveux, 2014a). Einkorn Wheat (*Triticum monococcum*) is another ancient variety that is highly adaptable to harsh conditions with low fertility (Zaharieva and Monneveux, 2014b).

In the case of Oaxacan Green Dent Maize, it is a traditional variety originating from Oaxaca, Mexico (Bellon and Brush, 1994). This particular maize is well known for its adaptability to different environmental conditions. Another variety worth mentioning is Chapalote, which is an ancient maize variety native to northern Mexico (Vázquez-Carrillo *et al.*, 2019). It possesses a natural resistance to pests and is specifically adapted to low-input farming systems.

SRI (System of Rice Intensification) is a method that focuses on soil health and plant spacing to enhance yield and sustainability (Stoop *et al.*, 2002). It is used with specific rice varieties adapted to this method. Another traditional variety from India is known as Khaddar (Pandey *et al.*, 2007). This variety is resilient to local pests and diseases, making it well suited for low-input farming practices.

Papa Chola, a traditional variety from the Andes, is well known for its resistance to pests and diseases, as well as its ability to thrive in challenging climates (Zimmerer and Douches, 1991). Another variety, Sarpo Mira, is highly regarded for its resistance to late blight, a significant potato disease, and its adaptability to low-input farming systems (Bradshaw, 2009).

Desi Cotton is a type of cotton that is native to India (Bajaj, 2012). It is known for its ability to withstand pests and diseases, and it thrives in rainfed conditions. On the other hand, Pima Cotton, or *Gossypium barbadense*, is famous for its resistance to pests and its production of high-quality fiber (Ulloa *et al.*, 2009). It is often grown using organic farming methods.

TME 419 is a cassava variety developed with a specific focus on resistance to cassava mosaic disease and adaptability to low-input farming systems (Okogbenin *et al.*, 2013). Bitter Cassava, on the other hand, is a traditional variety naturally resistant to pests and diseases, making it an ideal choice for agroecological farming (Elias *et al.*, 2001).

2.1.7 5G breeding strategy (Fig. 2.2)

It has been anticipated that, on average, every degree Celsius increase in global mean temperature may reduce global yields of major crops like wheat by 6.0%, rice by 3.2%, maize by 7.4%, and soybean by 3.1% (Zhao *et al.*, 2017); so in this context, scientists are facing difficulties in producing improved crop varieties with single strategies of breeding to feed the growing population. To cope with the increasing temperature, Varshney group proposes a 5G breeding strategy to dramatically accelerate crop genetic improvement. The first G refers to the genome assembly for each crop species. The second G represents germplasm characterized at genomic and agronomic levels. The third G pertains to gene function identification. The fourth G relates to genomic breeding methodologies. Lastly, the fifth G denotes gene-editing technologies. (Varshney *et al.*, 2020).

Genome assembly or genome sequences have been completed for most of the important crops like rice 389 Mb genome (Sasaki, 2005), maize 2.3 Gb genome (Schnable *et al.*, 2009), bread wheat 16 Gb, durum wheat 12 Gb, wild emmer wheat 10 Gb (Appels *et al.*, 2018), soybean 1.1 Gb (Schmutz *et al.*, 2010), barley 5.1 Gb (Mascher *et al.*, 2017), tomato 900 Mb (Tomato Genome Consortium X, 2012) and Potato 840 Mb (Xu *et al.*, 2011). In the same way, germplasm characterized for different crops like more than 20,000 rice accessions have been extensively characterized, particularly through projects such as the International Rice Genebank Platform (IRGC) and the Rice Diversity Project. Over 30,000 maize lines have also been characterized, with extensive genomic and phenotypic data available through databases like MaizeGDB.

Thousands of soybean lines have undergone genomic and agronomic characterization, with significant data available from projects like the Soybean Genomics and Improvement Network (SGIN). Furthermore, about 15,000 barley accessions have been characterized, and resources such as the International Barley Sequencing Consortium (IBSC) have contributed to the knowledge base. Additionally, numerous tomato germplasm accessions have been characterized, particularly through resources like the Tomato Genome Sequencing Consortium and various breeding programs.

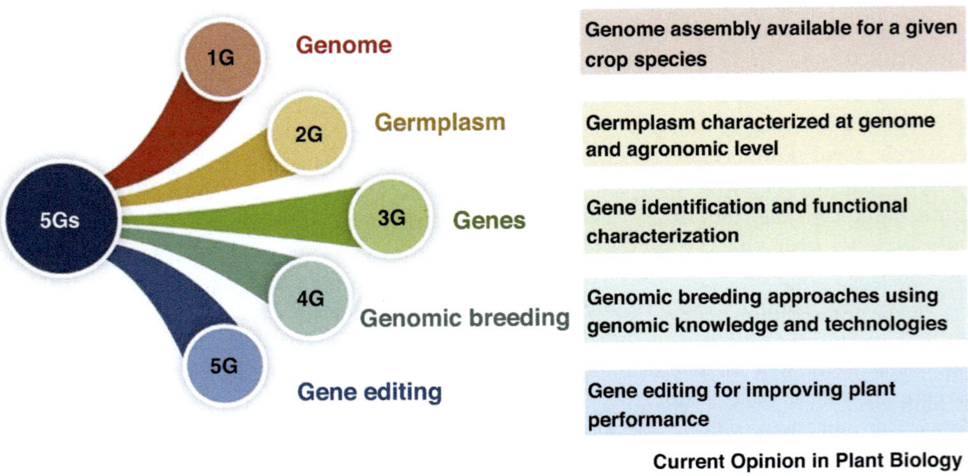

Fig. 2.2. Illustration of 5G breeding strategy.

Lastly, there are several thousand potato accessions that have been characterized genomically and agronomically, and significant resources are available from projects like the Potato Genome Sequencing Consortium.

The third G includes genes available on different platforms; wheat, for example, has approximately 90,000–100,000 genes, while rice has about 35,000 genes that can be found in databases such as the Rice Annotation Project (RAP) and the International Rice Genome Sequencing Project (IRGSP). Corn has approximately 32,000–39,000 genes, which are available from MaizeGDB and other resources. Soybean has around 46,000 genes that can be accessed through resources like the SoyBase database. Barley has approximately 30,000–35,000 genes that can be found in databases like the International Barley Genome Sequencing Consortium (IBSC). Tomato has about 35,000 genes, which are available from the Solanaceae Genomics Network and other sources. Lastly, the potato has approximately 39,000–41,000 genes that can be accessed through databases like Solanaceae Genomics Network.

The fourth G, genome breeding methodologies, includes marker-assisted selection (MAS), genomic selection (GS), genome-wide association studies (GWAS), genomic prediction, Genomic Selection Index (GSI), genome editing, high-throughput phenotyping (HTP), and integrative approaches available that have made important contribution in the recent past for different crops.

The fifth G, gene-editing technologies, includes CRISPR-associated protein 9 (CRISPR-Cas9), as well as TALENs (Transcription Activator-like Effector Nucleases) and ZFNs (Zinc Finger Nucleases), which have all played crucial roles in the development of various agriculturally important crop varieties. In particular, the CRISPR-Cas9 strategies have potential to flourish in the near future for different crops because of having fewer ethical issues.

All the information related to the five Gs may play an important role in shaping agronomically important crops for climate resistance and high yield. For instance, Asian soybean has been facing difficulties in developing a resistant variety for a long time that can withstand *Phakopsora pachyrhizi*, one of the disease-causing agents of Asian soybean rust. In this context, the 1G of pigeon pea genome sequence played a role in developing resistant varieties for Asian soybean rust. The Kawashima Group inserted a gene, *Cajanus cajan*, resistant against *Phakopsora pachyrhizi* 1 (CcRpp1), from pigeon pea (*Cajanus cajan*) into soybean. As a result, the developed soybean showed that CcRpp1 confers full resistance to *P. pachyrhizi* (Kawashima *et al.*, 2016).

2.2 Conclusion

Traditional breeding has long been the bedrock of agricultural innovation, crucially contributing to the creation of crop varieties boasting superior yields, robust disease resistance, and enriched nutritional profiles. Landmark achievements in this arena include wheat strains impervious to the devastating stem rust race Ug99, drought-resilient maize, and biofortified crops such as orange maize and iron-rich rice. Furthermore, the advent of salt-tolerant hybrid rice, disease-resistant potatoes, and premium-quality cotton and cassava exemplifies the indispensable role traditional breeding plays in surmounting global agricultural challenges and bolstering food security.

Marker-assisted breeding, also dubbed molecular breeding, has emerged as a pivotal technique in crafting crop varieties that are adept at withstanding environmental stresses and diseases. Triumphs in this domain include drought-resilient rice varieties like Sahbhagi Dhan and Swarna-Sub1; wheat lines fortified against diseases such as Sr2/Yr30; and soybean strains that fend off soybean cyst nematodes. These breakthroughs have ushered in remarkable yield surges, underscoring the significance of marker-assisted breeding in the evolution of contemporary agriculture.

Conventional breeding, however, is a time-intensive endeavor, often spanning 7–12 years to develop new crop varieties, with progress hampered by the protracted generation times of certain crops. Enter speed breeding, a game-changing approach that accelerates plant growth by extending photoperiods and hastening seed harvesting, thereby slashing generation times. This method has dramatically shortened breeding cycles for a range of crops, although fine-tuning may be necessary for short-day species like maize or rice. Moreover, speed breeding amplifies the efficacy of advanced techniques such as gene stacking and genomic selection in crop development.

The development of GM crops offers a more controlled and efficient alternative to traditional breeding, allowing for precise manipulation of gene expression in specific tissues or organs. GM crops predominantly focus on conferring herbicide and insect resistance, with notable successes in elevating global crop yields. For instance, Bt crops have revolutionized pest management, while GM technology has significantly bolstered the production of vital crops like maize, cotton, soybean, and canola. The overall impact of GM crops has been profound, driving substantial yield increases and enhancing agricultural productivity on a global scale.

CRISPR/Cas9 technology has rapidly emerged as a formidable tool in precision crop breeding, facilitating the development of crops endowed with enhanced traits such as salinity tolerance, increased yield, and superior quality. By precisely targeting specific genes, researchers have engineered substantial yield enhancements across various crops, including a 30% rise in grain yield in rice, a 20% uptick in grain weight in wheat, and a 15% boost in drought-resistant maize. Additionally, CRISPR/Cas9 has catalyzed notable improvements in yield and quality in crops like soybean, potato, cotton, cassava, and sorghum, heralding a new era in agricultural innovation.

Agroecological practices pivot towards cultivating crop varieties that are not only resilient and sustainable but also finely attuned to local environmental conditions, often thriving with minimal inputs. Exemplary cases include ancient grains like Emmer and Einkorn wheat, which flourish in nutrient-poor soils, and traditional maize varieties from Mexico, such as Oaxacan Green Dent and Chapalote, renowned for their adaptability and pest resistance. Similarly, crops like Papa Chola and Sarpo Mira potatoes, Desi and Pima cotton, and cassava varieties like TME 419 and Bitter Cassava have been specifically bred or selected for their resilience against local pests, diseases, and harsh climates, making them ideal for low-input, sustainable farming systems.

As the planet warms, the yields of major staple crops like wheat, rice, maize, and soybean are projected to plummet. To counter this looming threat, the Varshney Group has put forth a visionary 5G breeding strategy aimed at turbocharging crop improvement. This multifaceted strategy encompasses genome assembly, germplasm characterization, gene function identification, genomic breeding methodologies, and gene editing technologies. These cutting-edge approaches, with CRISPR-Cas9 at the forefront, hold immense promise for creating climate-resilient, high-yield crops. The seamless integration of these 5G elements is poised to play a pivotal role in boosting crop resilience and productivity, as exemplified by the successful incorporation of a pigeon pea gene into soybean to combat Asian soybean rust.

2.2.1 Prospects

To combat food insecurity in a growing population, it is time to integrate all strategies –traditional, molecular, transgenic, CRISPR/Cas9, 5G, and agronomical practices – to produce climate-resistant and high-yielding crop varieties. For instance, the pigeon pea genome sequence has helped develop rust-resistant soybeans. Similarly, all available information related to crop development can assist scientists in achieving fruitful outcomes.

References

Andersson, M., Turesson, H., Nicolia, A., Fält, A.S., Samuelsson, M. *et al.* (2017) Efficient targeted multiallelic mutagenesis in tetraploid potato (*Solanum tuberosum*) by transient CRISPR-Cas9 expression in protoplasts. *Plant Cell Reports* 36(1), 117–128.

Appels, R., Eversole, K., Stein, N., Feuillet, C., Keller, B. *et al.* (2018) Shifting the limits in wheat research and breeding using a fully annotated reference genome. *Science* 361(6403), eaar7191.

Arelli, P.R. and Liu, S. (2005) Genetics and breeding for resistance to soybean cyst nematode in soybean. *Plant Breeding Reviews* 25, 305–314.

Bajaj, Y.P. (2012) *Transgenic Medicinal Plants*. Springer Science & Business Media.

Bajgain, P., Rouse, M.N., Tsilo, T.J., Macharia, G.K., Bhavani, S. *et al*. (2016) Nested association mapping of stem rust resistance in wheat using genotyping by sequencing. *PLoS One* 11(5), e0155760.

Baltes, N.J. and Voytas, D.F. (2015) Enabling plant synthetic biology through genome engineering. *Trends in Biotechnology* 33(2), 120–131.

Bellon, M.R. and Brush, S.B. (1994) Keepers of maize in Chiapas, Mexico. *Economic Botany* 48(2), 196–209.

Borrell, A.K., Mullet, J.E., George-Jaeggli, B., Oosterom, E.J., Hammer, G.L. *et al*. (2014) Drought adaptation of stay-green sorghum is associated with canopy development, leaf anatomy, root growth, and water uptake. *Journal of Experimental Botany* 65(21), 6251–6263.

Bouis, H.E. and Welch, R.M. (2010) Biofortification: A sustainable agricultural strategy for reducing micronutrient malnutrition in the global south. *Crop Science* 50(Supplement1), S–20.

Bradshaw, J.E. (2009) Potato breeding at the Scottish plant breeding station and the Scottish crop research institute: 1920–2008. *Potato Research* 52(2), 141–172.

Bradshaw, J.E. and Ramsay, G. (2009) Potato origin and production. In: *Advances in Potato Chemistry and Technology*. Academic Press, pp. 1–26.

Brookes, G. and Barfoot, P. (2020) *GM Crops: Global Socio-economic and Environmental Impacts 1996–2018*. PG Economics Ltd, Dorchester, UK.

Bull, S.E., Seung, D., Chanez, C., Mehta, D., Kuon, J.E. *et al*. (2018) Accelerated *ex situ* breeding of GBSS- and PTST1-edited cassava for modified starch. *Science Advances* 4(9), eaat6086.

Cai, Y., Chen, L., Liu, X., Guo, C., Sun, S. *et al*. (2018) CRISPR/Cas9-mediated targeted mutagenesis of GmFT2a delays flowering time in soyabean. *Plant Biotechnology Journal* 16(1), 176–185.

Che, P., Anand, A., Wu, E., Sander, J.D., Simon, M.K. *et al*. (2018) Developing a flexible, high-efficiency agrobacterium-mediated sorghum transformation system with broad application. *Plant Biotechnology Journal* 16(7), 1388–1395.

Cook, D.E., Lee, T.G., Guo, X., Melito, S., Wang, K. *et al*. (2012) Copy number variation of multiple genes at Rhg1 mediates nematode resistance in soybean. *Science* 338(6111), 1206–1209.

Edmeades, G.O., Bolaños, J., Chapman, S.C., Lafitte, H.R. and Bänziger, M. (1999) Selection improves drought tolerance in tropical maize populations. I. Gains in biomass, grain yield, and harvest index. *Crop Science* 39(5), 1306–1315.

Elias, M., McKey, D., Panaud, O., Anstett, M.C. and Robert, T. (2001) Traditional management of cassava morphological and genetic diversity by the *Makushi amerindians* (Guyana, South America): Perspectives for on-farm conservation of crop genetic resources. *Euphytica* 120(1), 143–157.

FAO (2017) The future of food and agriculture: Trends and challenges. Available at: https://openknowledge.fao.org/server/api/core/bitstreams/2e90c833-8e84-46f2-a675-ea2d7afa4e24/content (accessed 22 March 2025).

Feng, Z., Zhang, B., Ding, W., Liu, X., Yang, D.-L. *et al*. (2013) Efficient genome editing in plants using a CRISPR/Cas system. *Cell Research* 23(10), 1229–1232.

Gebhardt, C. and Valkonen, J.P.T. (2001) Organization of genes controlling disease resistance in the potato genome. *Annual Review of Phytopathology* 39(1), 79–102.

James, C. (2013) *Global status of commercialized biotech/GM crops: 2013*. ISAAA Brief No. 46, ISAAA, Ithaca, New York.

Jiao, Y., Wang, Y., Xue, D., Wang, J., Yan, M. *et al*. (2010) Regulation of OsSPL14 by OsmiR156 defines ideal plant architecture in rice. *Nature Genetics* 42(6), 541–544.

Kawashima, C.G., Guimarães, G.A., Nogueira, S.R., MacLean, D., Cook, D.R. *et al*. (2016) A pigeonpea gene confers resistance to Asian soybean rust in soybean. *Nature Biotechnology* 34(6), 661–665.

Legg, J.P. and Thresh, J.M. (2000) Cassava mosaic virus disease in East Africa: A dynamic disease in a changing environment. *Virus Research* 71(1–2), 135–149.

Mascher, M., Gundlach, H., Himmelbach, A., Beier, S., Twardziok, S.O. *et al*. (2017) A chromosome conformation capture ordered sequence of the barley genome. *Nature* 544(7651), 427–433.

Mobini, S.H. and Warkentin, T.D. (2016) A simple and efficient method of *in vivo* rapid generation technology in pea (*Pisum sativum* L.). *In Vitro Cellular & Developmental Biology – Plant* 52(5), 530–536.

Nassar, N.M.A. and Ortiz, R. (2010) Breeding cassava to feed the poor. *Scientific American* 302(5), 78–84.

Nestel, P., Bouis, H.E., Meenakshi, J.V. and Pfeiffer, W. (2006) Biofortification of staple food crops. *The Journal of Nutrition* 136(4), 1064–1067.

O'Connor, D.J., Wright, G.C., Dieters, M.J., George, D.L., Hunter, M.N. *et al.* (2013) Development and application of speed breeding technologies in a commercial peanut breeding program. *Peanut Science* 40(2), 107–114.

Okogbenin, E., Moreno, I., Tomkins, J., Fauquet, C.M., Mkamilo, G. *et al.* (2013) Marker-assisted breeding for cassava mosaic disease resistance. In: Varshney, R.K. and Tuberosa, R. (eds) *Translational Genomics for Crop Breeding: Biotic Stress*. Wiley Blackwell, pp. 291–325.

Pandey, S., Bhandari, H., Sharan, R., Naik, D., Hirway, I. *et al.* (2007) Economic costs of drought and rice farmers' drought-coping mechanisms in Eastern India. In: Pandey, S., Bhandari, H.S. and Hardy, B. (eds) *Economic Costs of Drought and Rice Farmers' Coping Mechanisms: A Cross-Country Comparative Analysis*. International Rice Research Institute.

PIB Delhi (2023) Impact of climate change on agriculture. Available at: https://pib.gov.in/PressReleaseIframePage.aspx?PRID=1909206#:~:text=Climate%20change%20is%20projected%20to%20reduce%20the%20kharif%20maize%20yields,and%20its%20impact%20on%20farmers (accessed 22 March 2025).

Qaim, M. (2020) *Genetically Modified Crops and Agricultural Development*. Palgrave Macmillan.

Qian, Q., Zhang, F. and Xin, Y. (2021) Yuan longping and hybrid rice research. *Rice* 14(1), 101.

Qin, Y.M., Hu, C.Y., Pang, Y., Kastaniotis, A.J., Hiltunen, J.K. *et al.* (2007) Saturated very-long-chain fatty acids promote cotton fiber and *Arabidopsis* cell elongation by activating ethylene biosynthesis. *The Plant Cell* 19(11), 3692–3704.

Rajaram, S. and Braun, H.J. (2008) Wheat yield potential. In: *International Symposium on Wheat Yield Potential: Challenges to International Wheat Breeding*, CIMMYT.

Riaz Marral, M.W., Khan, M.B., Ahmad, F., Farooq, S. and Hussain, M. (2020) The influence of transgenic (Bt) and non-transgenic (non-Bt) cotton mulches on weed dynamics, soil properties and productivity of different winter crops. *PLoS One* 15(9), e0238716.

Romeis, J., Meissle, M. and Bigler, F. (2006) Transgenic crops expressing bacillus thuringiensis toxins and biological control. *Nature Biotechnology* 24(1), 63–71.

Sasaki, T. (2005) The map-based sequence of the rice genome. *Nature* 436(7052), 793–800.

Schmutz, J., Cannon, S.B., Schlueter, J., Ma, J., Mitros, T. *et al.* (2010) Genome sequence of the palaeopolyploid soybean. *Nature* 463(7278), 178–183.

Schnable, P.S., Ware, D., Fulton, R.S., Stein, J.C., Wei, F. *et al.* (2009) The B73 maize genome: Complexity, diversity, and dynamics. *Science* 326(5956), 1112–1115.

Shi, J., Habben, J.E., Archibald, R.L., Drummond, B.J., Chamberlin, M.A. *et al.* (2017) Overexpression of ARGOS8 improves maize grain yield under field drought stress conditions. *Plant Biotechnology Journal* 15(2), 207–216.

Singh, R.P., Hodson, D.P., Huerta-Espino, J., Jin, Y., Bhavani, S. *et al.* (2011) The emergence of Ug99 races of the stem rust fungus is a threat to world wheat production. *Annual Review of Phytopathology* 49, 465–481.

Singh, U.S., Dar, M.H., SINGH, S., Zaidi, N.W., Bari, M.A. *et al.* (2013) Field performance, dissemination, impact and tracking of submergence tolerant (sub1) rice varieties in South Asia. *SABRAO Journal of Breeding & Genetics* 45(1).

Specht, J.E., Hume, D.J. and Kumudini, S.V. (1999) Soybean yield potential: A genetic and physiological perspective. *Crop Science* 39(6), 1560–1570.

Stetter, M.G., Zeitler, L., Steinhaus, A., Kroener, K., Biljecki, M. *et al.* (2016) Crossing methods and cultivation conditions for rapid production of segregating populations in three grain *Amaranth* species. *Frontiers in Plant Science* 7, 816.

Stoop, W.A., Uphoff, N. and Kassam, A. (2002) A review of agricultural research issues raised by the system of rice intensification (SRI) from Madagascar: Opportunities for improving farming systems for resource-poor farmers. *Agricultural Systems* 71(3), 249–274.

Sysoeva, M.I., Markovskaya, E.F. and Shibaeva, T.G. (2010) Plants under continuous light: A review. *Plant Stress* 4, 5–17.

Tabashnik, B.E., Brévault, T. and Carrière, Y. (2013) Insect resistance to Bt crops: Lessons from the first billion acres. *Nature Biotechnology* 31(6), 510–521.

Tan, S., Evans, R.R., Dahmer, M.L., Singh, B.K. and Shaner, D.L. (2005) Imidazolinone-tolerant crops: History, current status and future. *Pest Management Science: Formerly Pesticide Science* 61(3), 246–257.

Tomato Genome Consortium, X (2012) The tomato genome sequence provides insights into fleshy fruit evolution. *Nature* 485(7400), 635–641.

Ulloa, M., Percy, R., Zhang, J., Hutmacher, R.B., Wright, S.D. *et al.* (2009) Registration of four pima cotton germplasm lines having good levels of fusarium wilt race 4 resistance with moderate yields and good fibers. *Journal of Plant Registrations* 3(2), 198–202.

UN Water (2020) World Water Development Report 2020 - Water and Climate Change. Available at: https://reliefweb.int/report/world/world-water-development-report-2020-water-and-climate-change?gad_source=1&gclid=CjwKCAjw1920BhA3EiwAJT3ISQo9J4B5OFBQWbLVkT57Te5Vfkd1N-f895QGbm7Q6mg6RQeN4g3qrBoCsNEQAvD_BwE (accessed 22 March 2025).

Varshney, R.K., Sinha, P., Singh, V.K., Kumar, A., Zhang, Q. *et al.* (2020) 5Gs for crop genetic improvement. *Current Opinion in Plant Biology* 56, 190–196.

Vázquez-Carrillo, M.G., Santiago-Ramos, D. and Figueroa-Cárdenas, J. de D. (2019) Kernel properties and popping potential of Chapalote, a Mexican ancient native maize. *Journal of Cereal Science* 86, 69–76.

Wang, W., Pan, Q., He, F., Akhunova, A., Chao, S. *et al.* (2018) Transgenerational CRISPR-Cas9 activity facilitates multiplex gene editing in allopolyploid wheat. *The CRISPR Journal* 1(1), 65–74.

Watson, A., Ghosh, S., Williams, M.J., Cuddy, W.S., Simmonds, J. *et al.* (2018) Speed breeding is a powerful tool to accelerate crop research and breeding. *Nature Plants* 4(1), 23–29.

Weeks, D.P., Spalding, M.H. and Yang, B. (2016) Use of designer nucleases for targeted gene and genome editing in plants. *Plant Biotechnology Journal* 14(2), 483–495.

Xu, K., Xu, X., Fukao, T., Canlas, P., Maghirang-Rodriguez, R. *et al.* (2006) Sub1A is an ethylene-response-factor-like gene that confers submergence tolerance to rice. *Nature* 442(7103), 705–708.

Xu, X., Pan, P., Cheng, S., Zhang, B., Mu, D. *et al.* (2011) Genome sequence and analysis of the tuber crop potato.

Zaharieva, M. and Monneveux, P. (2014a) Cultivated emmer wheat (*Triticum dicoccon schrank*), an old crop with promising future: A review. *Genetic Resources and Crop Evolution* 61(4), 929–943.

Zaharieva, M. and Monneveux, P. (2014b) Cultivated einkorn wheat (*Triticum monococcum* L. subsp. monococcum): The long life of a founder crop of agriculture. *Genetic Resources and Crop Evolution* 61, 677–706.

Zhang, A., Liu, Y., Wang, F., Li, T., Chen, Z. *et al.* (2019) Enhanced rice salinity tolerance via CRISPR/Cas9-targeted mutagenesis of the *OsRR22* gene. *Molecular Breeding* 39, 47.

Zhao, C., Liu, B., Piao, S., Wang, X., Lobell, D.B. *et al.* (2017) Temperature increase reduces global yields of major crops in four independent estimates. *Proceedings of the National Academy of Sciences of the United States of America* 114(35), 9326–9331.

Zimmerer, K.S. and Douches, DS. (1991) Geographical approaches to crop conservation: The partitioning of genetic diversity in Andean potatoes. *Economic Botany* 45(2), 176–189.

3 Disease-resistant Crops: Plant Protection through Nanotechnology and CRISPR-enhanced Plant Immunity

Aritra Karmakar*

Department of Biotechnology, Techno India University, Bidhannagar, Kolkata, India

Abstract

CRISPR/Cas9 tool and nanotechnology are two versatile techniques that have brought a paradigm shift in crop improvement research, both by enhancing disease resilience and improving crop quality and yield. Despite the crop production management system has advanced by leaps and bounds, two serious challenges before agricultural scientists are mitigating the problem of disease resistance and production of crops without the burden of toxic pollutants in the environment. Using transgene-free CRISPR/Cas9 technology and bolstering nanotechnology-based protocols has enabled agricultural scientists to overcome universal food scarcity. The emergence of CRISPR/Cas9 technology has helped to avert the jab on the commercialization of genetically modified (GM) crops. Using CRISPR/Cas9 technology-mediated genome-editing precise alteration of the target nucleotide sequence has been achieved and the desirable trait is obtained in a much shorter duration than the conventional breeding program. It has been found that the strategic use of nanotechnology has induced disease tolerance in crops and has significantly impacted in enhancement of crop yield. Due to the small size, large surface area, and high reactivity, the nanoparticles can be used as biosensors for plant disease diagnostics, as well as microfertilizers without augmenting pollutants in the environment. Due to the simple preparation process, cost-effectiveness, and nontoxic nature, different forms of nanoparticles such as metalloids, metallic oxides, nonmetals, and carbon nanomaterials have been used as nanocarriers, as vectors of genetic material to cells. This chapter presents an overview of applications of CRISPR/Cas9 technology and engineered nanoparticles in developing disease-resistant crops. It aims to give an insight into the limitations and challenges of CRISPR/Cas9 technology in the genome editing of crops. It also analyzes the safe use and social acceptance of nanoparticles as an optimal tool in the modern era of genetic engineering and suggests their risks in the environmental food chain.

Keywords: disease resistance, genetically modified crops, genome editing, nanocarriers, nanoparticles

3.1 Introduction

Ensuring global food nutrition is a critical concern that is placing agricultural scientists on the brink of danger due to rising food demand and there is a consistent reduction in crop yield due to climate change, degradation of soil quality, and enhanced disease susceptibility of crops (Ahmar *et al.*, 2021). There is an alarming spike in the world population,

*Corresponding author: karmakara07@gmail.com

© CAB International 2025. *Stress-Resilient Crops: Coordinated omics-CRISPR-Nanotechnology Strategies* (ed. Jen-Tsung Chen)
DOI: 10.1079/9781800627307.0003

which may reach around 9 billion by 2050. As a result, the daunting task that lies before the farmers is that they ought to produce a significantly enhanced quantity of crops on declining arable land, with less water, and on low-quality soil. Additionally, fertilizer and pesticides must be optimally used to minimize the burden of harmful chemicals in the soil (Karmakar *et al.*, 2022a).

The use of traditional technologies in agricultural crop production at present is capped as they are cumbersome and time-consuming. Sustainable crop improvement involves enhancing crop productivity, with minimal environmental impact, as well as preserving natural resources. Adhering to techniques like conventional breeding, targeted genetic modification and gene editing for better yield, disease resistance, and environmental resilience are commonly practiced in modern agriculture. One of the main challenges of crossbreeding is that it takes several years to incorporate the desired agronomic characters into indigenous cultivars by this technique. Natural breeding processes often result in the fixation of large segments of crop genomes, which reduce the chances of incorporating genetic variability and limit the potential for continuous directed evolution. Conventional breeding processes, also referred to as 'variation breeding', have incorporated genetic variation by means of chemical mutagens, physical irradiation, or by using enzymes, and subsequently breeding with other cultivars (Chen *et al.*, 2019b). However, these untargeted breeding programs are plagued by the appearance of chimeras and a reduction in survival percentage in future progenies. They are also time-consuming and are unable to meet the demands for enhanced crop production. Transgenic breeding has made it possible to obtain desired traits by transferring foreign genes into elite cultivar lines. Transgenic technology has broken the barrier of inter-species reproductive constraints. However, regulatory challenges and a lack of public awareness are significantly restricting the commercialization of genetically modified (GM) crops (Chen *et al.*, 2019b).

With the advancement of genome editing tools, the limitations associated with traditional breeding protocol and transgenic techniques have been effectively averted. Genome-editing technology is a new breeding technique (NBT) that has brought a paradigm shift in biological sciences and genome engineering for the future development of crops. One of the main advantages of genome-edited (GE) varieties over GM crops is that GE crops do not carry any foreign DNA and hence do not face any constraint related to commercialization in most countries across the world. The other boon of using genome-editing technology is that it enables us to carry out target-specific single-gene mutations or single nucleotide mutations and thereby obtain the desired traits in the crop genome. It enables genetic engineers to design and program the target nucleotide sequence and edit it with greater precision to obtain the desired genotype and phenotype in crops. Miao *et al.* (2018) developed a mutated rice line by CRISPR/Cas9 tool, which has up to 31% enhanced yield than conventional breeding.

Probably, the first attempt at targeted genome engineering was carried out in 1988 in tobacco (*N.tabacum*) protoplasts by Paszkowski *et al.* (1988) and since then the GE has evolved into three generations. Meganucleases (MegaN) and zinc finger nuclesases (ZFN) belong to first-generation genome-editing tool. Meganucleases are endodeoxy ribocluases that are found in a large number of organisms like bacteria, fungi, phages, etc. Meganucleases characteristically have large recognition sites of 12–40 bp, which occur only once in any genome (Silva *et al.*, 2011).

Zinc finger nucleases (ZFNs) are sequence-specific nucleases, which can recognize 3–6 nucleotide triplets and have been widely used for genome editing in both plant and animal kingdoms (Mushtaq and Molla, 2021; Karmakar *et al.*, 2022a). Transcription activators like effector nucleases (TALENs), which belonged to the second-generation GE tools, are restriction enzymes that were engineered to cut specific nucleotide sequences and are superior in efficiency to ZFNs. Both ZFN and TALEN follow the same DNA–protein interaction mechanism to perform the genome-editing function.

In 2012, Jinek *et al.* (2012) reported the usage of the CRISPR/Cas9 mechanism, where a small guide RNA has been used to induce double-strand break (DSB) in the target DNA. In contrast to the ZFN and TALEN mechanism of action, the CRISPR/Cas9 tool performs the genome-editing function by DNA–RNA interaction. Subsequently, more advanced genome-editing techniques like base editing and prime editing were used for precise modifications of the target genome sequence and thereby helped to overcome the constraints of conventional CRISPR/Cas9 technology (Molla and Yang, 2019; Molla *et al.* 2020, 2021). Hence CRISPR/Cas9 system, base editing, and prime editing technology all belonged to the third-generation GE tools (Ahmar *et al.*, 2021; Karmakar *et al.*, 2022b).

The first- and second-generation GE tools faced few constraints either in design or in their applications. The design of ZFNs is cumbersome as there is a complex interaction of ZFNs among themselves and it also involves the danger of off-target mutation (Sander *et al.*, 2011). The design of TALEN was relatively easier but repetitive sequence in TALEN increased the rate of homologous recombination. The shortcomings of first- and second-generation techniques were overcome to a large extent with the advent of third-generation GE techniques (Nekrasov *et al.*, 2013). The CRISPR/Cas9 system consists of only two major components: the Cas9 protein and a single guide RNA (sgRNA). The Cas9.sgRNA complex requires a 20bp DNA sequence known as a protospacer and binds by Watson-Crick base pairing. The target site should be located immediately upstream of a sequence known as the protospacer adjacent motif (PAM) and is essential for target localization and function of SpCas9 (Molla and Yang, 2020). This design mediates Cas9 to bind precisely to the target domains and produce DSB in the genome. In this manner, multiple sgRNA can be designed with several target sequences to carry out multiple editing of the target site. Hence CRISPR/Cas9 tool offers higher manipulation options and has much better accuracy in inducing target mutation than ZFN and TALENs (Zhang *et al.*, 2017b; Karmakar *et al.*, 2022b).

Several transformation methods have been developed to transfer biomolecules into the plant genomes like DNA, RNA such as electroporation, biolistics, and *Agrobacterium*-mediated genetic transformation. The conventional biomolecule delivery methods face a few critical challenges like low efficiency of gene transformation, narrow species range, limitation in cargo molecule delivery, and high risk of tissue damage.

Additionally, generating genetically modified plants is a time-consuming process and the transformation protocols that have been developed are effective only for a limited range of plant species. Biolistic transformation technology, which delivers biomolecules via high mechanical force inflicts damage to the target tissue, thereby resulting in low levels of gene expression. The *Agrobacterium*-mediated genetic transformation (AMT) is popularly used but its application is limited to a few plant species. Also, AMT leads to random transgene integration disrupting important genes and thereby unstable gene expression. Thus conventional transformation methods throw a few serious challenges for GE's capacity. Scientists have found that nanotechnology-based delivery of biomolecules is a key transformation protocol to address delivery challenges and enhance plant transformation and regeneration efficiency (Cunningham *et al.*, 2018).

Nanoparticles (NPs) have been used in agricultural crop biotechnology as nanosensors, nanopesticides, nanofertilizers, and nanocarriers, all of which led to considerable advancements in modern crop production. Owing to a small size (1–100 nm) of NPs, they served as effective carrier vectors to transport cargos that easily bypassed the barrier of the cell wall and plasma membrane, thereby leading to stable genetic transformation in various plant species. The key advantages that make NP-mediated gene transformation the preferred choice in modern plant genetic transformation are the easy and cost-effective operation of NPs, a short cycle, higher expression and transformation efficiency, and biosafety. The emergence of nanomaterial (NM) engineering led to the progress of sustainable crop production systems to cutting-edge technology. NM has shown promising impetus in reducing stress on food and energy production while optimally managing limited resources like water and nutrients (Lv *et al.*, 2020; Wu *et al.*, 2023).

3.2 Importance of Genome Editing in Agriculture

Genome editing in agriculture emerged as a significant promise for crop protection and enhancing crop productivity amidst serious global issues such as exponential population growth, drastic climatic changes due to global warming, and diminishing resources. CRISPR/Cas9 tool has enabled researchers to make precise modifications in plant genomes, improving traits like yield and quality, and enhancing resilience against biotic and abiotic stresses (Waghmode *et al.*, 2024).

Researchers have successfully harnessed CRISPR/Cas9 technology to manipulate flowering time, fruit size, and grain yield in crops like rice, wheat, and maize. Targeted delaying in flowering time has enabled the crops to accumulate more biomass and nutrients before the commencement of the reproductive stage. Thus, the overall productivity of grains has enhanced significantly under changing environmental conditions (Xu *et al.*, 2020). In a research study, targeted gene manipulation of *GmFT2a* was carried out by the CRISPR/Cas9 tool which resulted in delayed flowering in the GE plants. *GmFT2a* is an important regulator gene in the flowering pathway in soybean. Homozygous *GmFT2a* mutants exhibited delayed flowering under both short-day and long-day conditions (Cai *et al.*, 2018).

By means of genome editing, scientists were also able to alter the genes involved in cell division, fruit development and hormone signaling pathways, which led to an increase in size and production of crop grains. Usman *et al.* (2020) reported enhanced grain yield by targeted editing of *OsSPL16* gene in rice by CRISPR/Cas9 technology. Further genetic analysis revealed that the programmed mutation increased grain yield. Similarly, researchers have used genome editing to target genes involved in stress response pathways, osmotic regulation, and heat-shock protein to develop crops resistant to drought-, salinity- and temperature-stress conditions. Knockout of *FLN2* gene by CRISPR/Cas9 technology resulted in enhanced salinity tolerance in mutant rice lines. The levels of enzymes like sucrose synthase, and acid invertase activity were lower in mutant plants, which has a significant influence on rice plants' response to salinity stress (Chen *et al.*, 2019a). In a recent research, lodging resistant tef (*Eragrostis tef*) has been developed by CRISPR/ Cas9-mediated mutation of *SEMIDWARF-1 (SD-1)* gene. Tef is a popular staple crop for people in Ethiopia. The genome-edited plants showed significantly higher resistance to lodging during the heading stage as compared to control plants (Beyene *et al.*, 2022).

Genome editing offers a plausible pathway for biofortification of crops. By CRISPR/Cas9 tool it has been possible to enhance the levels of Vit A content in rice (Dong *et al.*, 2020), Vit E content in *H.vulgare* (Zeng *et al.*, 2020), iron and zinc content in wheat and maize (Ibrahim *et al.*, 2022), and omega-3-fatty acids in soybean (Al Amin *et al.*, 2019). Likewise, several researchers have used genome-editing technology to alter crucial genes to produce genome-edited crops through targeted engineering to achieve sustainable agriculture.

3.3 Applications of CRISPR/Cas9 Tool in Developing Disease-resistant Crops

3.3.1 Developing fungal disease resistance by CRISPR/Cas9 mechanism

Fungi are a potent threat to plants because they destroy plant tissues directly or by producing plant toxins which lead to the death of the host plant. Fungal pathogens are responsible for a wide range of serious diseases in crops like rot, smut, rust, mildew, etc. Several mechanisms have been harnessed by scientists to combat fungal pathogenesis in plant species based on the current understanding of plant–pathogen interaction. It has been observed in plant–pathogen interaction that the effector-triggered immunity (ETI) gets activated following the interaction of pathogen effectors with plant resistance (R) protein, thereby finally upregulating the plant defense system (Ahmad *et al.*, 2020; Tyagi *et al.*, 2021). The usual method of incorporating the R gene into crop plants failed to establish a promising approach as pathogens tend to infect only a particular species of crop. Also, fungi tend to avoid such resistance by deliberately mutating the respective *avr* gene, thereby producing resistant pathogens. Instead, targeting genes like host susceptible genes, and disease development receptors proved to be effective approaches to develop resistance in crop plants. The CRISPR/Cas9 mechanism-mediated mutation in host S genes has proved to be a successful approach in conferring disease resistance against fungal pathogens.

In several research studies, the CRISPR-Cas9 tool has been used to modify plant genomes to develop resistance against diverse groups of fungal pathogens (Karmakar *et al.*, 2022b). The mildew-resistant locus O (MLO) is a host susceptible gene (S gene) that plays an essential role in the plant defense system and earlier studies proved that mutations in the MLO locus conferred resistance to mildew diseases in several cases. Using CRISPR/Cas9 technology, MLO gene was modified in tomato, wheat, and grapevine and developed resistance against the powdery mildew disease. In wheat, MILDEW-RESISTANCE LOCUS (MLO) encodes proteins that are involved in repressing resistance against powdery mildew disease. Transgenic wheat has been developed by mutating *TaMLO-A1* allele through CRISPR/Cas9 technology and showed the feasibility of manipulation of *TaMLO* locus to incorporate beneficial heritable traits into the genome (Wang *et al.*, 2014b). Similarly, Nekrasov *et al.* (2017) reported the development of a nontransgenic variety of tomato called Tomelo, resistant to mildew fungal pathogens by means of CRISPR/Cas9 technology. In this study, they developed a nontransgenic mutant of mildew-resistant locus O (Mlo). The mutant *SlMlo1* tomato variety showed improved resistance against the fungal pathogen *O. neolycopersici*. In grapevine, *VvMLO7* gene was also mutated using the CRISPR/Cas9-ribonucleoprotein (RNP) complex. The mutated plants developed enhanced fungal resistance in protoplast culture (Malnoy *et al.*, 2016).

It has been found that in powdery mildew disease, caused by the fungus *Erysiphe cichoracearum*, the enhanced disease resistant1 gene (*EDR1*) acts as a negative regulatory gene. In *Arabidopsis*, *a* mutation in the *EDR1* mitogen-activated protein kinase (MAPKKK) exhibited enhanced resistance against powdery mildew infection (Frye *et al.*, 2001). In another study, simultaneous modifications of three homologs of the wheat *EDR1 gene* have been carried out by CRISPR/Cas9 technology. The *Taedr1* mutant plants did not exhibit any off-target mutations and showed higher resistance to powdery mildew disease (Zhang *et al.*, 2017a).

Fungi cause pathogenesis by secreting wall-degrading enzymes to enable fungi to penetrate inside the host tissue. To counter fungal invasion, plants secrete inhibitory substances against cell-wall degrading enzymes and strengthen the defense of the cell wall by secreting callose (Imam *et al.*, 2016). GE has been used to target the inhibitors of cell-wall degrading enzymes or genes involved in callose deposition to confer resistance against the fungal infection. It was found that the ortholog of Powdery Mildew Resistance4 (PMR4) gene *SlPMR4* is responsible for callose deposition. However, the mutation of *PMR4* gene resulted in enhanced salicylic acid levels and stimulated hypersensitive response (HR) during pathogen invasion. Knockout mutants of tomato *PMR4* gene generated using CRISPR/Cas9 showed significantly enhanced resistance compared to RNAi-mutated plants (Santillán Martínez *et al.*, 2020).

Plant ethylene response factors (ERF) are a sub-family of APETELLA2/ethylene response transcription factor (AP2/ERF) super family in plants, which are involved in various biotic and abiotic stress tolerance; upregulation of expression of *OsERF* genes like *OsBIERF1*, *OsBIERF3*, and *OsBIERF4* are observed due to induction of abiotic stresses (such as salt, cold, drought), as well as due to *M. oryzae* infection. In tobacco, upregulation of *BrERF11* resulted in significant disease resistance against the pathogen *Ralstonia solanacearum*. However, silencing of *StERF3* in potato developed enhanced resistance to *Phytophthora infestans*. Similarly, enhanced resistance against *M. oryzae* in rice was obtained by RNAi-mediated knockdown of *OsERF922*. These results showed that *OsERF922* acts as a negative regulator in blast disease resistance in rice (Wang *et al.*, 2016). In rice, Wang *et al.* (2016) reported a significant enhancement in blast resistance by CRISPR/Cas9-mediated knockout of *OsERF922* without any significant change in agronomic traits of the mutated rice plants. Similarly, the CRISPR/Cas9 tool was used to mutate *OsSEC3A* gene which resulted in enhancement in salicylic acid content and disease resistance against the fungal pathogen *Magnaporthe oryzae* (Ma *et al.*, 2018).

It is well acknowledged that WRKY transcription factors contribute to PAMP-triggered immunity (PTI) and ETI and regulate defense responses (Tyagi *et al.*, 2021). The WRKY transcription factor in grapes was mutated by CRISPR/Cas9 technology. Mutation in *VvWRKY52* resulted in increased disease resistance against *Botrytis cinerea* and the mutant plants showed no difference in phenotype as compared to control plants (Wang *et al.*, 2018).

Theobroma cacao (cocoa tree) suffers a significant economic loss due to pathogens like *Phytophthora tropicalis*. Using CRISPR/Cas9 technology, the cacao nonexpressor of the pathogenesis-related 3 (*TcNPR3*) gene was mutated to develop mutant *TcNPR3* plants which showed enhanced resistance against *Phytophthora tropicalis* (Fister *et al.*, 2018). In a recent study in soybean, novel chromosome rearrangement has been reported using CRISPR/Cas9 technique. In this study, two nucleotide-binding site-leucine rich-repeat (NBS-LRR) gene clusters called Rpp1-like and Rps1 were targeted and the mutated plants conferred resistance against *Phakopsora pachyrhizi* and *P. sjae*. This study showed that enhanced plant resistance can be conferred by simultaneous mutagenesis of single to multiple CRISPR/Cas9 targets (Nagy *et al.*, 2021).

The above studies exploited the potential of S genes to develop resistance against a broad spectrum of pathogens. The potential of the host susceptibility (S) gene over the resistance (R) gene in developing resistance against pathogens was also revealed. Hence there is a plethora of opportunities to identify various S genes that can be exploited shortly. The S gene is tightly linked with growth and development in various crops. Hence, the mutation of the S gene and its subsequent negative implications on agronomic properties should be carefully analyzed. Table 3.1 illustrates a few examples of the development of resistance against fungal pathogens using CRISPR/Cas9 technology.

3.3.2 Developing bacterial disease resistance by CRISPR/Cas9 mechanism

Various pathogenic bacteria cause several diseases to crops resulting in a huge loss of economy as well as crop yield. Due to the high multiplication rate, it is extremely difficult to control bacterial infestation resulting in the cause of several types of diseases such as bacterial blight, spots, mosaic disease, and rots (Karmakar *et al.*, 2022b). Depending on host specificity, phytopathogenic bacteria can be classified into three types: crop-specific species; polyphagus species; and kingdom crosser. When a specific type of bacteria produces typical disease symptoms in a specific host, it is known as crop-specific bacteria, e.g. *Pseudomonas syringae* pv. *Glycinea* causes disease on growing soybeans resulting in bacterial blight. On the contrary, polyphagus species bacteria cause multiple disease symptoms and infect both the monocot as well as dicot plants, e.g. *Ralstonia solanacearum* causes the widespread bacterial wilt disease in banana, potato, and eggplants; whereas kingdom crosser pathogens are opportunistic organisms and cause diseases across different kingdoms such as animals and plants, e.g. *Dickeya dedantii* causes soft rot disease in ornamental plants, tuber crops like potato, and other plants of the Solanaceae family. These pathogens inflict different responses in host plant species and many host plant genes, such as S genes, participate in the host–pathogen interaction process. Constant genetic mutation and horizontal gene transfer enable bacteria to develop resistance to antibacterial agents and facilitate the spread of infections (Borrelli *et al.*, 2018).

There are a few studies that reported harnessing CRISPR/Cas9 technology to develop resistance against bacterial pathogens. The citrus canker disease is caused by pathogenic bacteria *Xanthomonas citri*, and results in significant economic losses to citrus plants worldwide. Citrus canker disease resistance has been developed in grapefruit by modulating PthA4 effector binding element (EBE), driven by *CsLOB1* promoter ($EBEP_{th}A4$-*CsLOBP*) of the *CsLOB1* (*Citrus sinensis Lateral Organ Boundaries*) gene. The mutation of the promoter resulted in a loss in the ability to recognition and response to bacterial effectors and consequently showed significant resistance against citrus canker disease. Using CRISPR/cas9 tool, Jia *et al.* (2016) carried out targeted modification of *EBEPthA4* element of *CsLOB1* gene. They reported a significant reduction in disease symptoms without producing any off-target mutation. Further, Peng *et al.* (2017) exhibited that the removal of *EBEPthA4* sequence from *CsLOB1* enhanced resistance against *X. citri* without affecting phenotypic characteristics. This study

Table 3.1. Genome-edited plants to develop resistance against fungal pathogens.

Disease	Plant affected	Target S-gene	System	Causal fungal pathogen	Result	References
Powdery mildew	Wheat	*TaMLO*	CRISPR/Cas9	*Blumeria gramini* f.sp. *tritici*	Broad spectrum heritable resistance to powdery mildew	Wang *et al.*, 2014b
Powdery mildew	Wheat	*TaEDR1*	CRISPR/Cas9	*Blumeria gramini* f.sp. *tritici*	Enhanced resistance to powdery mildew	Zhang *et al.*, 2017a
Powdery mildew	Tomato	*SlMlo1*	CRISPR/Cas9	*Oidium neolycopersici*	Mutant plants were resistant to the pathogen	Nekrasov *et al.*, 2017
Powdery mildew	Tomato	PMR4	CRISPR/Cas9	*Oidium neolycopersici*	Development of partial resistance causal pathogen	Santillán Martínez *et al.*, 2020
Fusarium wilt	Tomato	*Solyc08g075770*	CRISPR/Cas9	*Fusarium oxysporium* f.sp *lycopersici*	Mutant plants showed *Fusarium* wilt tolerance	Prihatna *et al.*, 2018
Gray mold	Tomato	*SlMAPK3*	CRISPR/Cas9	*Botrytis cinerea*	Mutant plants showed enhanced resistance to the pathogen and regulated reactive oxygen species (ROS) accumulation salicylic acid (SA), and jasmonic acid (JA) defense signaling pathway	Zhang *et al.*, 2018b
Powdery mildew	Grape and apple	*MLO-7*	CRISPR/Cas9	*Erysiphe nector*	Mutant plants showed enhanced resistance to the pathogen	Malnoy *et al.*, 2016
Rice blast	Rice	*OsERF922*	CRISPR/Cas9	*Magnaporthe oryzae*	Mutated plants exhibited enhanced disease resistance	Wang *et al.*, 2016
Rice blast and bacterial blight	Rice	*TMS5,Pi21, Xa13*	CRISPR/Cas9	*Magnaporthe oryzae*	Mutant plants showed enhanced resistance to disease	Li *et al.*, 2019
Not tested	Rice	*OsMPK5*	CRISPR/Cas9	*Magnaporthe oryzae*	Disease resistance not confirmed	Xie and Yang, 2013
Gray mold	Grape plant	*VvWRKY52*	CRISPR/Cas9	*Botrytis cinerea*	Mutant plants exhibited enhanced resistance to *Botrytis cinerea*	Wang *et al.*, 2018
NA	Cacao	*TcNPR3*	CRISPR/Cas9	*Phytophthora tropicalis*	Enhanced resistance against *P. tropicalis*	Fister *et al.*, 2018
NA	Papaya plant	*AlEPIC8*	CRISPR/Cas9	*Phpalmivoraytophthora*	Enhanced resistance against *P. palmivora*	Tripathi *et al.*, 2019
NA	NA	*Rpp1,Rps1*	CRISPR/Cas9	*Phakopsora pachyrhizi, P. sojae*	Targeted mutagenesis conferred enhanced resistance	Nagy *et al.*, 2021

NA: information not available

showed that CRISPR/Cas9-mediated targeted mutation of *CsLOB1* promoter can prove to be an efficient strategy to develop resistance against citrus canker in citrus cultivars. In another study, deletion of SlDMR6-1 (tomato ortholog) by CRISPR/Cas9 tool enhanced resistance against several pathogens like *P. syringae* pv. tomato, *Phytophthora capsici* and *Xanthomonas* species, etc. (Thomazella *et al.*, 2016). Similarly, *DMR6* gene was mutated using CRISPR/Cas9 tool in *A.thaliana*. The targeted mutation not only enhanced the level of salicylic acid but, concomitantly, provided broad-spectrum pathogen resistance (Zeilmaker *et al.*, 2015). Zhou *et al.* (2014) reported a major targeted genomic modification in four rice *SWEET* genes and successfully generated large chromosome deletion in separate clusters of *SWEET* genes. *OsSWEET* gene is an important host susceptible gene, essentially carrying out sucrose transportation during pathogenesis. They also reported Cas9/sg-RNA induced large deletion in three segments of genes in rice protoplasts and generated genome-edited rice plants. In a further study, Zhou *et al.* (2015) carried out a targeted mutation of *OsSWEET-13* gene in rice. *OsSWEET-13* is a disease-susceptible host gene that is involved in sucrose transportation during *X. oryzae* pathogenesis. The mutant *OsSWEET-13* plants exhibited significantly enhanced resistance against *X. oryzae*.

A tomato variety was developed by CRISPR/Cas9 technology, which is resistant to the bacterial speck disease caused by pathogen *P. syringae* pv. tomato. When *P. syringae* encounters the host plant, it secretes the chemical coronatine (COR), which is sensed by co-receptor *AtJAZ2*. This response results in stomatal opening, facilitating initiation of bacterial infection. Targeted CRISPR/Cas9 mutation in JAZ2 generated truncated JAZ2. This genome-edited *SlJAZ2* mutant fully checked stomatal opening by COR, thus providing resistance against bacterial speck disease (Ortigosa *et al.*, 2019). *SlJAZ2* mutant also showed resistance against the fungal pathogen *Botrytis cinerea*, responsible for grey mold disease in tomato. This study exhibited the way of uncoupling the jasmonate (JA) and salicylate (SA) defense pathways and proposed a novel CRISPR/Cas9-based strategy to develop resistance against a broad spectrum of pathogens.

Fire blight, a devastating disease in apple, caused by the pathogen *Erwinia amylovora*, produces an effector protein designated as DspE during a pathogen attack. The effector protein interacts with serine/threonine kinase proteins produced by DspE interacting protein of Malus (DIPM) genes – *DIPM 1,2,3,4*. Genome-editing tool has been used to target *DIPM 1,2,3,4* genes, which resulted in the development of resistance against *E. amylovora* (Malnoy *et al.*, 2016). This experiment successfully introduced the sgRNA/Cas9 ribonucleoprotein (sgRNA/Cas9) complex into plant protoplasts leading to effective targeted genome editing and minimized off-target activity (Malnoy *et al.*, 2016; Borrelli *et al.*, 2018).

CRISPR/Cas9 genome editing has been an effective tool in the enhancement of resistance against bacterial diseases. In another study, Luu *et al.* (2019) reported a novel rice immune receptor XA21 which is responsible for the development of immunity designated as RaxX, produced by *X. oryzae* pv. *oryzae* (*Xoo*). RaxX is an unclassified eukaryotic tyrosine sulfated ribosomally synthesized peptide (RiPPs) which has not been studied in detail. Genome-editing technology can be used to alter such unclassified receptors like XA21 to detect *Xanthomonas* strains and ligands. Emerging editing technologies can be leveraged to discover additional tools for targeting various bacterial pathogens and enhancing resistance in crop plants. Table 3.2 highlights several examples of how CRISPR/Cas9 technology has been employed to engineer bacterial pathogen resistance in crop plants.

3.3.3 Developing viral disease resistance by CRISPR/Cas9 mechanism

Viruses are parasites that depend entirely on host organisms for their replication and survival. Viral pathogenesis in crop plants can be detected only via the symptoms they cause. Viral infection is almost impossible to control by the application of pesticides or other chemicals. As the viral genome is mutating at a fast pace, viral

Table 3.2. Genome-edited plants to develop resistance against bacterial pathogens.

Bacterial disease	Plant affected	Target S-gene	System	Causal bacterial pathogen	Result	References
Citrus canker	Grape fruit	*CsLOB1* cds region	CRISPR/Cas9	*X. citri* subsp. *citri*	Enhanced resistance against citrus canker disease	Jia *et al*., 2016
Citrus canker	Grape fruit	*CsLOB1* promoter	CRISPR/Cas9	*X. citri* subsp. *citri*	Enhanced resistance against citrus canker disease	Peng *et al*., 2017
Bacterial speck	Tomato	*SlDMR6-1*	CRISPR/Cas9	*P.syringae*	Enhanced disease resistance	Thomazella *et al*., 2016
Phytophthora blight	Tomato	*SlDMR6-1*	CRISPR/Cas9	*Phytophthora capsici*	Enhanced disease resistance	Thomazella *et al*., 2016
Bacterial spot	Tomato	*SlDMR6-1*	CRISPR/Cas9	*Xanthomonas* sp.	Enhanced disease resistance	Thomazella *et al*., 2016
Bacterial speck	Tomato	*SlJAZ2*	CRISPR/Cas9	*P.syringae* pv. *tomato*	Enhanced disease resistance	Ortigosa *et al*., 2019
Tomato grey mold	Tomato	*SlJAZ2*	CRISPR/Cas9	*Botrytis cinerea*	Enhanced disease resistance	Ortigosa *et al*., 2019
Bacterial blight	Rice	Promoter of *SWEET11*, *SWEET 13* and *SWEET14*	CRISPR/Cas9	*Xanthomonas oryzae* pv. *oryzae*	Enhanced disease resistance	Oliva *et al*., 2019
Bacterial blight	Rice	*Os8N3* promoter	CRISPR/Cas9	*Xanthomonas oryzae* pv. *oryzae*	Enhanced disease resistance	Kim *et al*., 2019
Bacterial blight	Rice	Os*SWEET11* and Os*SWEET14* promoter region	CRISPR/Cas9	*Xanthomonas oryzae* pv. *oryzae*	Enhanced disease resistance	Xu *et al*., 2019
Bacterial blight	Rice	*Xa13*	CRISPR/Cas9	*Xanthomonas oryzae* pv. *oryzae*	Enhanced disease resistance	Li *et al*., 2019
Bacterial blight	Rice	Os*SWEET11* and Os*SWEET14* promoter region	CRISPR/Cas9	*Xanthomonas oryzae* pv. *oryzae*	Not defined	Jiang *et al*., 2013
Bacterial blight	Rice	Os*SWEET13 cds* region	CRISPR/Cas9	*Xanthomonas oryzae* pv. *oryzae*	Enhanced disease resistance	Zhou *et al*., 2015

diseases present a significant threat to sustainable agriculture, incurring billions of losses every year. Due to the devastating economic losses caused by viruses, their detection, mode of transmission between plants, and prophylactic control measures have been studied in a large number of scientific investigations (Valkonen, 2014; Tyagi *et al.*, 2021).

Viruses may contain two types of genomes: DNA (either double strand or single strand) or RNA. RNA genomes are of four types: double-stranded (dsRNA), negative single-stranded (-ssRNA), positive single-stranded (+ssRNA), and retroviruses. The geminiviridae family, which contains single-stranded DNA, causes substantial yield losses to a large number of economically important crops. Viruses lack their own translational machinery and instead rely on the host plant's replication system to propagate. Therefore, effective resistance can be attained by targeting genes related to pathogenicity or by altering the host susceptibility genes (Ali *et al.*, 2020).

Over the past two decades, various CRISPR/Cas9-based tools utilizing Cas9 or Cas13a nucleases have been employed to target viral genomes or modify the host plant genome, enabling the development of resistant crop lines against disease-causing viruses. In 2015, a few research groups used CRISPR/Cas9 technology to engineer resistance against beet severe curly top virus (BSCTV), tomato leaf curl virus (TYLCV), and bean yellow dwarf virus (BeYDV) in plants like *A. thaliana* and *N. benthamiana* (Baltes *et al.*, 2015; Ji *et al.*, 2015). Baltes *et al.* (2015) used 11 sgRNA to target six regions within the BeYDV genome-ReP binding sites, hairpin, nonnucleotide sequence and 3 ReP motifs which were essential for its replication. They reported a significant reduction in viral infection in *Nicotiana benthamiana*. Ji *et al.* (2015) targeted 43 sites and designed sgRNAs to target the genome sequences of BSCTV. They reported substantial inhibition of virus accumulation in the inoculated leaves of *Nicotiana* and *Arabidopsis* plants, where the Cas9 and sgRNA expression was highest.

Gemini virus, a plant DNA virus poses a serious threat to food security by causing substantial crop production losses worldwide. Ali *et al.* (2015, 2016) utilized the CRISPR/Cas9 tool to develop resistance in plants against begomoviruses, which are members of the Geminiviridae family. In both the studies, CRISPR/Cas9 system was used within the host cell nucleus to target and cleave the viral genome during its replication. sgRNAs were designed and delivered via tobacco rattle virus (TRV) vector into *N. benthamiana* plants to target various sequences of TYLCV. The sgRNA targeting the invariant stem-loop sequence within the IR region demonstrated a notable decrease in viral load. Likewise, CRISPR/Cas9 was deployed to target beet curly top virus (BCTV) and Merremia mosaic virus (MeMV) as they both feature an identical stem-loop sequence in their intergenic regions (IR). A marked decrease in disease symptoms was observed for both viruses, showing that a single sgRNA designed to target a conserved sequence can confer immunity to mixed viral infections (Ali *et al.*, 2016). The study by Ali *et al.* (2016) also showed that mutations induced by CRISPR/Cas9 in the open reading frames (ORF) of Gemini viruses led to the generation of mutated viruses with retained replication and systemic movement abilities. CRISPR/Cas9 was utilized to edit both coding and noncoding sequences of cotton-leaf curl kokhran virus (CLCuKoV), MeMV, and several strains of TYLCV. It was observed that the sgRNA/Cas9 tool successfully edited the coding regions across all the viruses. The muted viruses eventually escaped the CRISPR/Cas9 machinery and thereby replicated successfully. On the other hand, when sgRNA was targeted against the noncoding IR of CLCuKoV and TYLCSV, the mutated viruses were unable to replicate which gave a better strategy to interfere with the viral genome.

Achieving viral resistance is problematic for agricultural scientists when using SpCas9 from *Streptococcus pyogens*, as it exclusively targets double-stranded DNA. Zhang *et al.* (2018c) pointed out that FnCas9 from *Francisella novicida* and LwaCas913a from *Leptotrichia wadei* can bind and cleave RNA. Targeting the RNA of cucumber mosaic virus (CMV) and tobacco mosaic virus (TMV) with sgRNA and FnCas9 in *N. benthamiana* and *Arabidopsis* resulted in mutated plants that had 40–80% lower virus accumulation vs the control plants.

The expression of sgRNA-FnCas9 was found to be stable till the T_6 generation. FnCas9 offers another advantage of not requiring a nuclear localization signal, thereby minimizing interference with RNA viruses located in the cytoplasm.

In their 2013 study, Larson *et al.* (2013) presented CRISPR interference (CRISPRi), an RNA-driven approach for targeted suppression of transcription in both bacterial and human cell systems. The CRISPRi system is adapted from the CRISPR pathway of *Streptococcus pyogens* wherein the Cas9-sgRNA complex binds to the complementary DNA and the sgRNA restricting the transcription elongation process by RNA polymerase, thereby repressing the target gene. The work by Aman *et al.* (2018) involved the application of CRISPR/Cas13a to alter the RNA genome of turnip mosaic virus (TuMV). This tool, a Class 2 Type VI-A ribonuclease, is capable of targeting and cutting ssRNA within the phage genome. The design focussed on four distinct regions of the viral genome: two regions associated with green fluorescent protein (GFP), one corresponding to the helper component proteinase (HC-Pro) silencing suppressor, and one coding for the coat protein. Targeting the HC-Pro and the GFP2 genes with CRISPR RNA successfully disrupted the viral activity, which was accompanied by decreased replication and spread of TuMV in tobacco leaves.

To achieve viral disease resistance, an alternative strategy modifies the host plants' disease-response genes, enabling them to develop resistant traits and produce nontransgenic mutants for agricultural field release. Plant host genes like eukaryotic translational initiation factors like eIF4E,eIF(iso)4E, eIF4G are utilized by plant RNA viruses to maintain their life cycle (Chandrasekaran *et al.*, 2016). Using CRISPR/Cas9, researchers recently targeted the eIF(iso)4E locus in *Arabidopsis thaliana* to create resistance to TuMV. Both 1bp insertion and deletions in the eIF(iso)4E locus enhanced resistance significantly to the ssRNA TuMV, and no off-target mutations were detected among the mutant plants (Pyott *et al.*, 2016).

In a recent report, Macovei *et al.* (2018) introduced mutation in the eIF4G alleles in rice plants by CRISPR/Cas9 system. For the mutated eIF4G plants that were developed, only those in-frame mutations that took place in SVLPNLAGKS residues, adjacent to the YVV residues, conferred resistance against the viral pathogen attack. The rice tungro spherical virus (RTSV)-resistant mutant plants showed no significant alteration in agronomic parameters as compared to nonmutant plants. Another study revealed that targeting host susceptibility genes for knockout in affected plants is a more effective strategy. This approach leverages recessive resistance, which is often more desirable than dominant R genes as it exerts less selective pressure on viruses, thus slowing their evolution and their ability to overcome host defenses (Abdul-Razzak *et al.*, 2009). Using CRISPR/Cas9 and multiplex genome editing, a recent study developed mutant *N. benthamiana* plants with targeted resistance against chickpea chlorotic dwarf virus (CpCDV). Targeted gDNA constructs aimed at conserved CpCDV genomic regions were utilized, both individually and in combination, to effectively counter the virus. The results showed that multiplexing of three sgRNA exhibited a significant reduction of viral infection (Malik *et al.*, 2023). Table 3.3 outlines a few examples of the application of CRISPR/Cas9 tool in developing resistance to viral pathogens.

3.4 Challenges in CRISPR/Cas9 Genome-Editing Technology

CRISPR/Cas9 genome editing has shown great promise by precisely modifying genome sequences to achieve the desired crop trait. It has become a powerful tool for developing nontransgenic crops to ensure the food security of ever-increasing worldwide populations. Despite these remarkable advantages, there are a few challenges which are faced by scientists in the CRISPR/Cas9 genome-editing tool:

1. **Off-target effects**. This tool can occasionally cause unintended changes in the genome. Such effects occur on two occasions: firstly, when improperly designed sgRNA binds to nontarget sequence; due to the complexity of the crop genome, sometimes sgRNA binds to multiple similar

Table 3.3. Examples of genome-edited plants to develop resistance against viral pathogens in crop plants.

Disease	Plant affected	Target S-gene	System	Causal viral pathogen	Result	References
Gemini virus infection	Tobacco, Arabid opsis	CP, ReP, and IR	CRISPR/Cas9	BeYDV	Significant inhibition of viral accumulation in the mutant plants	Ji *et al.*, 2015
Gemini virus infection	*N.tabacum*	Six regions within the BeYDV genome- ReP binding sites, hairpin, nonnucleotide sequence and 3 ReP motifs	CRISPR/Cas9	BeYDV	Up to 87% reduction of virus accumulation in the mutant plants	Baltes *et al.*, 2015
TuMV infection	*N.tabacum*	GFP infection, HC-Pro and coat protein	CRISPR/Cas13a	TuMV	Successful viral interference was observed.	Aman *et al.*, 2018
Potyvirus infection	*A.thaliana*	eIF(iso)4E locus of *Arabidopsis thaliana*	CRISPR/Cas9	TuMV	Mutant plants exhibited significant resistance against TuMV.	Chandrasekaran *et al.*, 2016
Potyvirus infection	*A.thaliana*	1bp insertion and 1bp deletion in eIF(iso)4E locus of *Arabidopsis thaliana*	CRISPR/Cas9	TuMV	Significant resistance conferred against TuMV and no off-target mutation was observed among the mutant plants.	Pyott *et al.*, 2016
Rice tungro disease	*O.sativa*	*eIF4G* alleles of rice plants	CRISPR/Cas9	RTSV	Mutant plants were resistant to RTSV infection.	Macovei *et al.*, 2018
Mastrevirus infection	*N. benthamiana*	gDNA constructs targeting conserved genomic regions	CRISPR/Cas9	CpCDV	Plants harboring multiplex constructs showed significant reduction of CpCDV infection.	Malik *et al.*, 2023

sequences, resulting in off-target effects; and secondly, sometimes Cas9 detects nonstandard PAM in addition to standard PAM, which finally results in off-target effects. Careful designing of sgRNA and rigorous validation can substantially minimize the possibility of off-target events (Liu *et al.*, 2022). Some researchers demonstrated that modifying certain regions of the SpCas9 enzyme can limit off-target effects (Slaymaker *et al.*, 2016).

2. **Limitation related to delivery methods**. Delivering CRISPR components into target cells poses a challenge, due to the requirement of an efficient and stable genetic transformation system. The leading techniques for crop gene transfer include *Agrobacterium*-mediated transformation and particle bombardment. Due to the dependence on cell culture regeneration mechanisms, these transformation systems are not suitable for all crops and tissues. Other methods like viral vectors, liposomes, and electroporation have their advantages and limitations (Borrelli *et al.*, 2018).
3. **Regulatory issues**. Different countries have different regulations concerning genome editing. Researchers must adhere to local and international guidelines and regulations to ensure legal and ethical adherence. The editing of crop genomes involves heritable changes in the genome of future generations, which raises profound ethical concerns (Ruikar *et al.*, 2024).
4. **Immunogenicity issues**. The plant immune system may recognize and trigger counter-responses against Cas9 protein or other components, which may limit its effectiveness and cause adverse reactions (Saxena *et al.*, 2024).

3.5 Definition and Properties of Nanoparticles

3.5.1 Definition

Nanoparticles (NPs) are defined as small molecules or particulate materials with sizes typically ranging from 1 to 100 nm (or more specifically 0.2–100 nm). The term 'nanotechnology' was first scientifically described by Norio Taniguchi of Japan, who defined it as the study of matter at the atomic or molecular level. A range of nanoparticles is currently utilized in agriculture and farming systems to enhance sustainability and improve the development of crops resistant to diseases. The classification of nanoparticles is based on multiple parameters, including surface morphology, chemical composition, physicochemical properties, size, source and origin, magnetic properties, crystallinity, and their intended functions (Vargas-Hernandez *et al.*, 2020; Banerjee *et al.*, 2021). Natural NPs are found in several forms, such as volcanic dust and salt aerosols from the ocean. Many viral and viroid particles are characterized as nanoparticles. The disadvantages of natural NPs are their irregular size and variable shape, whereas uniformly shaped and sized chemically synthesized or engineered NPs include spherical particles, sheets, rods, multi-walled tubes (like carbon nanotubes), and dendrimers with tree-like branching structures. Since engineered NPs can be designed and manufactured both chemically and biosynthetically as per requirements, they can be utilized in various fields (Elmer and White, 2018).

3.5.2 Properties of nanoparticles

Nanoparticles, with their small size and higher surface area-to-volume ratio, are inherently more reactive than their larger counterparts. With high efficiency, they can bind to and transport small molecule drugs, DNA, RNA, proteins, and probes. Due to the differences in size and shape NPs have markedly different properties as compared to their bulk equivalents. For example, gold (Au) in larger form is inert and golden in color but is more reactive and red at the nanoscale. Although TiO and ZnO are normally white, they exhibit a colorless appearance at the nanoscale. NPs have lower melting points and exhibit greater reactivity compared to their larger counterparts. Due to these marked differences in properties, NPs are currently being aggressively used in many industries and sustainable agriculture (Elmer and White, 2018).

3.6 Synthesis of Nanoparticles

Various methods, including physical, chemical, and biological processes can be used to synthesize NPs. To synthesize NPs of desired shape, size, and function, two fundamental approaches are generally followed: (i) the top-down approach; and (ii) the bottom-up approach. The top-down approach reduces large-size materials to nanoscale particles (1–100 nm) through physical and chemical methods. This process incorporates techniques such as mechanical milling, etching, sputtering, electro-explosion, and laser ablation to break down bulk materials into nano-sized particles. The bottom-up method focuses on constructing NPs by assembling small atoms. This approach uses techniques such as chemical vapor deposition, laser pyrolysis, and atomic condensation. In this approach, generally green nanoparticles are synthesized using biological organisms and plants. The physical properties of NPs can be tailored by optimizing factors such as reactant concentration, reaction duration, temperature, and pH (Banerjee *et al.*, 2021). However, most of the physical and chemical methods of the NP synthesis involve complex protocols, extremely high-temperature conditions, and the use of toxic chemicals as reducing agents, which incur high operational costs and are harmful to the environment (Hebbalalu *et al.*, 2013).

3.6.1 Green synthesis on nanoparticles

Green synthesis of NPs is an environmentally friendly method that facilitates the preparation of NPs while minimizing toxicity to the environment. By employing green synthesis methods, the production of NPs reduces environmental contaminants, minimizes toxic by-products, and relies on sustainable raw materials.

The synthesis of small-scale inorganic materials is naturally present in various biological systems. NPs are synthesized through green methods from various natural sources such as bacteria, algae, fungi, and plants, supporting large-scale, environmentally safe production with lower contamination levels. In green NP synthesis, capping agents or stabilizers are commonly used to manage particle size and avoid agglomeration, with the addition of plant extracts, yeast, or bacteria (Singh *et al.*, 2015). A few methods of green synthesis of biological NPs are explained below.

3.6.1.1 Green synthesis of nanoparticles using plant extracts

Plant extracts are rich in valuable bioactive compounds such as phenols, flavonoids, ascorbic acid, ketones, aldehydes, and carboxylic acids, and are increasingly used for the synthesis of metal and metal oxide NPs. These compounds play a dual role as reducing agents and stabilizers in green nanomaterial synthesis (Ahmad *et al.*, 2019).

Plants have the potential to accumulate a certain quantity of heavy metals in their parts. Hence, plant extracts have been used by scientists as a simple, efficient, cost-effective, and feasible approach, as well as an excellent alternative to conventional preparation methods for the synthesis of NPs (Singh *et al.*, 2018). In this technique, metal precursors are treated with leaf extracts, and factors like pH, temperature, pressure, and solvent are optimized to produce NPs. The phytochemicals in the leaf extracts reduce metal ions, thereby acting as both reducing and stabilizing agents in the green synthesis of NPs. For example, flavonoids acted as reducing agents during the green synthesis of AgNPs using *Ocimum basilicum* leaf extract. Glucose and fructose, found abundantly in various plant extracts, are promising reducing agents for the synthesis of metal NPs (Singh *et al.*, 2018; Banerjee *et al.*, 2021).

Initially, gold (Au) and silver (Ag) NPs were used for plant extract-based synthesis of NPs. Various plants like Aloe vera (*Aloe barbadensis*), oats (*Avena sativa*), alfalfa (*Medicago sativa*), tulsi (*Ocimum sanctum*), lemon (*Citus lemon*), neem (*Azadirachta indica*), among others, have been extensively utilized to synthesize Au and AgNPs. Table 3.4 lists

Table 3.4. Green synthesis of nanoparticles from plant origin.

Sr. no.	Plant origin	Type of NP	Size (nm)	Morphology	Application	References
1.	*Ocimum sanctum* (tulsi)	AgNPs	9.5 nm	Spherical	Medical field, pharmaceutical field	Mallikarjuna *et al.*, 2011
2.	*Pterocarpus santalinus* (sandalwood)	AgNPs	20 nm	Spherical	Antimicrobial activity	Gopinath *et al.*, 2013
3.	*Acacia nilotica* (babul)	AgNPs	20–30 nm	Distorted spherical	Antimicrobial activity	Edison and Sethuraman, 2013
4.	*Camelia sinensis* (tea)	AgNPs	2–10 nm	Spherical	Cosmetic, food, medicine	Loo *et al.*, 2012
5.	*Citrus sinensis* (orange)	AgNPs	30 nm	Spherical	Strong antibacterial activity	Kaviya *et al.*, 2011
6.	*Coriandrum sativum* (coriander)	AgNPs	26 nm	Spherical	Nonlinear optical and optical limiting properties	Sathyavathi *et al.*, 2010
7.	*Aloe vera*	AgNPs	15–20 nm	Spherical	Antimicrobial activity	Chandran *et al.*, 2006
8.	*Hibiscus rosa sinensis* (China rose)	AgNPs	13 nm	Spherical	Anti-aging	Philip, 2010
9.	*Cinnamomum camphora* (rose mallow)	AgNPs	64.8 nm	Quasi spherical	Pharmacological anticancer property	Huang *et al.*, 2007
10.	*Piper longum* (Indian long pepper)	AgNPs	46 nm	Spherical	Medicine, drug delivery, anticancer property	Reddy *et al.*, 2014
11.	*Moringa oleifera* (drum stick)	AgNPs	40 nm	Spherical, pentagonal	Anticancerous property	Vasanth *et al.*, 2014
12.	*Origanum vulgare* (oregano)	AgNPs	63–85 nm	Spherical	Antibacterial, anticancer activity	Sankar *et al.*, 2013
13.	*Vitex negundo* (Chinese chaste tree)	AgNPs	5–47 nm	Spherical	Anticancer property	Prabhu *et al.*, 2013
14.	*Altenanthera dentate*	AgNPs	50–100 nm	Spherical	Potent antimicrobial property	Kumar *et al.*, 2014
15.	*Dalbergia spinosa* (liana)	AgNPs	18–22 nm	Spherical	Antibacterial, antiinflammatory, antioxidant, catalytic activity	Muniyappan and Nagarajan, 2014

16.	*Milingtonia hortensis* (Indian cork tree)	AgNPs	2–8 nm	NA	Cardioprotection effect	Savitha and Saraswathi, 2016
17.	*Syzygium cumini* (jamun)	AgNPs	40–100 nm	Irregular spherical contour	Therapeutic application and antidiabetic	Atale *et al.*, 2017
18.	*Nelumbo nucifera* (lotus)	AgNPs	25–80 nm	Spherical (SEM) and triangular (TEM)	Larvicidal activity against malaria and filariasis vectors	Santhoshkumar *et al.*, 2011
19.	*Polyalthia longifolia* (ashoka tree)	AgNPs	13.9 nm	Spherical	Antioxidant activity	Kumar *et al.*, 2016b
20.	*Azadirachta indica* (neem)	AgNPs	NA	Spherical	Antimicrobial activity	Gohil *et al.*, 2024
	Magnifera indica (mango)	AgNPs	42–201 nm	Spherical	Antimicrobial activity	
	Eucalyptus fenugreek	AgNPs	68–106 nm	Spherical	Antimicrobial activity	
21.	*Aloe fleurentinorum* (Yemeni brown aloe)	AgNPs	26.7 nm	Tetrahedral	Antibacterial activity	Jamil *et al.*, 2024
22.	*Phyllanthus emblica* (Indian gooseberry, amla)	AgNPs	15.3–20.1 nm	Spherical	Antibacterial activity	Hossain *et al.*, 2024
23.	*Aconitum violaceum* (violet monkshood)	AuNPs	< 100 nm	Spherical, triangular	Antioxidant, photocatalytic activity	Ahmad *et al.*, 2024
		AgNPs	< 100 nm	Spherical	Antioxidant, photocatalytic activity	
24.	*P. benghalensis* (Bengal shrub mint)	AuNP	10–100 nm	Spherical triangular	Antimicrobial, antifungal, anticancer activity	Jafarizad *et al.*, 2015
25.	*Pelargonium*	AuNP	10–100 nm	Spherical	Antimicrobial, antifungal, anticancer activity	Jafarizad *et al.*, 2015
26.	*Garcinia mangostana* (purple mangostee)	AuNP	28–38 nm	Spherical triangular	Potential biomedical application	Xin Lee *et al.*, 2016

Continued

Table 3.4. Continued

Sr. no.	Plant origin	Type of NP	Size (nm)	Morphology	Application	References
27.	*Mimusops elengi* (Spanish cherry, malsari)	AuNP	9–14 nm	Spherical	Colloidal NP used for reduction of 3-nitrophenol as well as 4-nitrophenol in aqueous medium as a reducing agent	Majumdar *et al.*, 2016
28.	*Parkia raxburghii* (tree bean)	Au & AgNPs	5–25 nm	Spherical	Significant phytocatalytic and antibacterial activity	Paul *et al.*, 2016
29.	*Genipa Americana*	AuNP	15–40 nm	Spherical	Antioxidant and cytotoxic activity against human cancer cell lines	Kumar *et al.*, 2016a
30.	*Couronpita guianensis* leaf (cannol ball tree)	AgNP	10–45 nm	Spherical	Mosquito repellent	Vimala *et al.*, 2015
	Couronpita guianensis fruit (cannol ball tree)	AgNP	5–15 nm	Spherical	Mosquito repellent	
31.	*Alysicarpus monilifer* (alyce clover)	AgNP	5–45 nm	Spherical, hexagonal	Antibacterial activity against MRSA and CONS isolates in HIV patients	Kasithevar *et al.*, 2017
32.	*Capparis spinosa* (caper bush)	AgNP	5–30 nm	Spherical	Antibacterial property, pharmaceutical applications	Benakashani *et al.*, 2016
33.	*Chlorella vulgaris*	PdNP	5–20 nm	Crystalline	NA	Arsiya *et al.*, 2017
34.	*Lithodora hispidula* (rock garden plant)	PdNP	21.60 nm	Rod-shaped	Electrolytic reduction of H_2O_2	Turunc *et al.*, 2017
35.	*Filicium decipiens*	PdNP	2–22 nm	Spherical	Antibacterial activity, biomedical application	Sharmila *et al.*, 2017
36.	*Camellia sinensis* (black tea)	PdNP	5–8 nm	Spherical	Reducing and stabilizing agent	Lebaschi *et al.*, 2017
37.	*Boswellia serrata* (gum olibanum)	PdNP	2.5–8.8 nm	Spherical polydisperse	Antibacterial and antioxidant activity, biocatalyst in environmental remediation of polluted waste water	Kora and Rastogi, 2016

38.	*Gloriosa suparba* (flame lily)	CuONP	5–10 nm	Spherical	Antibacterial activity	Naika *et al.*, 2015
39.	*Thymus vulgaris*	CuONP	23–94 nm	Spherical	Used for reduction of organic dyes	Issaabadi *et al.*, 2017
40.	*Glycine max* (black soya bean)	CuONP	26.6 nm	Spherical	Anticancer activity against human carcinoma cells	Nagajyothi *et al.*, 2017
41.	*Malva sylvestris* (common mallow)	CuONP	5–30 nm	Monoclinic	Antibacterial activity	Kuppusamy *et al.*, 2016
42.	*Urtica dioica* (stinging nettle)	CuONP	9–31 nm	Nanoclusters	Detoxification, wastewater treatment	Ebrahiminezhad *et al.*, 2017
43.	*Mognolia kobus*	CuNP	40–100 nm	Spherical	Antibacterial activity	Lee *et al.*, 2011
44.	*Zingiber officinal* (ginger)	Ag-doped SnO_2 NPs	10–17 nm	Tetragonal rutile	Antibacterial activity	Tiki *et al.*, 2024
45.	*Spirodela polyrrhiza* (greater duckweed plant)	CuONPs	100 nm	Spherical	Significant antimicrobial, antifungal, nematicidal effect	Khan *et al.*, 2023
46.	*Matricaria chamomilla* (chamomile)	Ag-doped ZnO NPs	NA	NA	Antimicrobial activity	Mortezagholi *et al.*, 2022

NA: information not available

various plant materials employed in the green synthesis of NPs.

3.6.1.2 Use of bacteria in the synthesis of nanoparticles

Besides plant extracts, microorganisms like bacteria have been studied in detail for decades by scientists for the green synthesis of metallic NPs. A few advantages of using bacteria for the synthesis of NPs are: their rapid cultivation, ease of NP recovery from culture solutions, and exceptional ability to adapt to challenging environmental conditions. Hence, cultures of various bacterial strains like *Pseudomonas meridian*, *Arthrobacter gangotriensis*, and *Arthrobacter kerguelensis* have been used for the synthesis of AgNPs. Despite the advantages, bacteria-mediated NP synthesis comes with drawbacks, including the risk of culture contamination, lengthy procedural times, and difficulty in controlling NP size (Mageswari *et al.*, 2016).

The synthesis of NPs by microbes generally involves the following steps: metal capture, enzymatic reduction, and capping. Metal ions initially adhere to the surface and enter the microbial cells. Enzymes inside the cells then catalyze the reduction of these ions, resulting in the formation of NPs. Microbes impact mineral formation through two distinct mechanisms: by modifying the solution's composition to become super-saturated, or by generating organic polymers that stabilize mineral seeds, thereby affecting the nucleation process (Singh *et al.*, 2015). The initial publication on the synthesis of AgNPs mediated by the bacterium was by Klaus *et al.* (1999), who revealed the presence of these NPs inside the cells of *Pseudomonas stutzeri* AG259, a bacterium sourced from a silver mine. After isolating the bacterium from a silver-rich environment, subsequent findings revealed that several gram-positive and gram-negative bacteria also harbored AgNPs. The synthesis of NPs within the bacterial cells takes place either in the intracellular or extracellular regions (Singh *et al.*, 2015).

EXTRACELLULAR SYNTHESIS OF AGNPS. Extracellular synthesis of NP occurs outside bacterial cells and yields NP of various shapes such as spherical, disk, cuboidal, hexagonal, and triangular. This extracellular synthesis can utilize cells, culture supernatants, or aqueous cell-free extracts (Srivastava and Constanti, 2012). In this approach, the cell culture is centrifuged, and NPs are collected in the form of a pellet, which can be redispersed in a solvent. The main advantage of extracellular synthesis is the ease with which the NPs can be recovered from the culture solution (Singh *et al.*, 2015).

INTRACELLULAR SYNTHESIS OF AGNPS. In the intracellular mechanism, the bacteria reduce toxic metal ions to nontoxic forms through biochemical mechanisms. The synthesis of AgNPs generally takes place outside the cytoplasmic membrane, which may be due to metal ion efflux and metal ion binding. Deposition of triangular and hexagonal-shaped AgNPs in the periplasmic space of *Pseudomonas stutzeri* AG259 has been reported by Klaus *et al.* (1999). In the intracellular synthesis of AgNPs, bacterial cells are grown in a nutrient medium supplemented with silver salt and incubated under suitable conditions. The resulting NP size is influenced by the specific culture medium used (Srivastava *et al.*, 2013). To extract the NPs, techniques such as ultrasonication, heat treatment, and the application of detergents and salts are typically used to lyse the cells (Kalishwaralal *et al.*, 2010; Fesharaki *et al.*, 2010).

However, bacterial strains like *Lactobacillus* sp. (Nair and Pradeep, 2002), and *Vibrio alginolyticus* sp. (Rajeshkumar *et al.*, 2013) have been reported to secrete NPs in the extracellular environment in addition to accumulating them intracellularly. The actual reasons and mechanisms for this dual mode of NP synthesis are not properly understood. Several reports have been published regarding the synthesis of AgNPs by various bacterial strains. The synthesis of metal NP, including gold, silver, titanium, platinum and zinc can be synthesized by bacterial strains such as *Escherichia coli*, *Geobacillus* sp., *Lactobacillus* sp., *Bacillus* sp., *Shewanella* algae and *Rhodobacter sphaeroides* (Sweet *et al.*, 2012; Mageswari *et al.*, 2016; Shao *et al.*, 2023). Studies have shown that a single microorganism can generate NP of different metals based on its physiological traits.

Table 3.5. Bacteria-mediated synthesis of nanoparticles.

Sr no.	Bacterial strain	Type	Size	Shape	Gram nature	Application	Reference
1.	*Paenibacillus* Sp. *MAHUQ-63*	AgNPs	15–55 nm	Spherical	Gram positive	Significant antibacterial and antifungal activity against drug-resistant pathogens *S. enteritidis* and *C. albicans*	Huq *et al.*, 2023
2.	*Bacillus subtilis*	AgNPs	62 nm	Spherical	Gram positive	Significant antifungal activity against *A. flavus* fungus	Bharose *et al.*, 2024
3.	*Planococcus maritimus MBP-2* (Marine bacteria)	AgNPs	24.9 nm	Spherical, cubic	Gram positive	Antimicrobial, antifungal, cytotoxic effect	Pallath *et al.*, 2024
4.	*Burkholderia* sp. *EIKU21*	ZnO-NPs	55.08 ± 2.28 nm	NA	Gram negative	Antibacterial activity, antibacterial nanocoating in cotton fabric	Roy *et al.*, 2024
5.	*Priestia megaterium BASMA 2022* (OP672246)	ZnO-NPs	5.77–13.9 nm	Semi-spherical	Gram positive	Pharmaceutical and biomedical application	Ashour and Abd-Elhalim, 2024
6.	*Lactobacillus fermentum*	ZnO-NPs	90–100 nm	Spherical	Gram positive	Antibacterial activity	Shanmugam *et al.*, 2023
7.	*Streptomyces monashensis* MSK03	AuNPs	7.1–40 nm	Cubic lattice	Gram positive	Significant antibacterial activity	Kerdtoob *et al.*, 2024
8.	*Stenotrophomonas* sp. BS95	CuONPs	18.24–21.3 nm	Roughly spherical	Gram negative	Antibacterial, antioxidant, anticancer properties	Talebian *et al.*, 2023
9.	*Klebsiella* sp.	AgNPs	Varied depending on bacterial strain and conditions for synthesis	Varied depending on bacterial strain and conditions for synthesis	Gram negative	Antimicrobial activity	Malik *et al.*, 2023
	Citrobacter freundii	AgNPs			Gram negative	Antimicrobial activity	
	Bacillus cereus sp.	AgNPs			Gram positive	Antimicrobial activity	
10.	*E. coli*	AgNPs	5–50 nm	Spherical	Gram negative	Potent antimicrobial agent	Saeed *et al.*, 2020
	Exiguobacterium aurantiacumm	AgNPs	5–50 nm	Spherical	Gram positive	Potent antimicrobial agent	
	Breveendimonas diminuta	AgNPs	5–50 nm	Spherical	Gram negative	Potent antimicrobial agent	
11.	*B. cereus*	ZnNPs	NA	Irregular	Gram positive	Antibacterial activity	Iqtedar et al. 2020
12.	*S.nematodiphila*	ZnNPs	15–30 nm	Roughly spherical	Gram negative	Antibacterial, photocatalytic	Jain *et al.*, 2020
13.	*B. cereus*	ZnNPs	21–35 nm	Spherical	Gram negative	Antibacterial	Ahmed *et al.*, 2021
14.	*Lactobacillus* sp.	ZnNPs	32 nm	Spherical	Gram positive	Antimicrobial, anticancerous activity	Suba *et al.*, 2021

Table 3.5 highlights a few examples of AgNP synthesis using various bacterial strains.

3.6.1.3 Use of fungus in the synthesis of nanoparticles

Recently, the focus has shifted to employing fungi for synthesizing various NPs, thanks to their abundant availability, economic viability, and capacity for large-scale production. They are both easy to manage and simple to cultivate. There are also a few additional advantages that make fungi the material of choice for the biosynthesis of NPs. Fungi are known for their superior tolerance, and ability to accumulate metals, outperforming other microorganisms in these aspects. In the synthesis of gold NPs, fungus-mediated biosynthesis is advantageous due to its high efficiency, low energy consumption, no requirement for capping and stabilizing agents, and simple operation, isolation, and purification (Xu *et al.*, 2024). Table 3.6 illustrates some examples of fungi used for the synthesis of various NPs.

3.6.1.4 Use of algae in the synthesis of nanoparticles

A wide range of microorganisms such as algae have been harnessed for the synthesis of NPs. Microalgae have some distinguishing properties that make them a potential alternative source for the synthesis of NPs such as fast growth rate, effective CO_2 sequestration, heavy metal hyperaccumulation, minimal production of toxic byproducts, and low energy consumption. Algal extracts also act as natural reducing and capping agents, making them ideal living cell factories for the green synthesis of NPs. The natural physiological and biological mechanisms of algae allow them to absorb heavy metals and use these elements as nutrients and convert them into various types of NPs. Table 3.7 illustrates a few examples of the synthesis of NPs from algae.

3.7 Types, Mode of Action and Characterization of Nanoparticles

3.7.1 Types of nanoparticles

Based on composition, NPs can be categorized into the following types:

3.7.1.1 Metallic nanoparticles

1. **Silver NPs (AgNPs)**. AgNPs are the most studied, well-characterized, and popular NPs due to their significant antimicrobial activities. Numerous reports have been published on the physical, chemical, and biological synthesis of AgNPs. Due to their effectiveness in killing bacteria, fungi, and viruses, they are valuable in medical devices, wound dressings, and antimicrobial coatings (Banerjee *et al.*, 2021). AgNPs can also be employed as catalysts in chemical reactions, for the reduction of organic compounds, and in fuel cells (Erikson *et al.*, 2019). Furthermore, AgNPs are employed in electronic and optical devices owing to their superior electrical conductivity and distinctive optical characteristics (Algethami *et al.*, 2022). AgNPs are utilized in water purification, drug delivery, and various biomedical applications (Ivanova *et al.*, 2018). The application of AgNPs has shown promising results in suppressing diseases caused by soilborne pathogens such as *Phytophthora parasitica* and *Fusarium species* (Elmer and White, 2018).
2. **Gold NPs (AuNPs)**. AuNPs typically have sizes between 1 nm and 100 nm, and are known for their light absorption and scattering, which is attributed to surface plasmon resonance (SPR). This property makes them useful in optical applications and sensing (Amendola *et al.*, 2017). AuNPs are employed in imaging techniques like CT scans and MRI, and they are also used for the detection of biomarkers in diagnostic tests (Luo *et al.*, 2021a). Additionally, AuNPs are harnessed for

Table 3.6. Fungus-mediated synthesis of nanoparticles.

Sr No.	Fungal strain	Type of NPs	Size	Shape	Application	References
1.	*Rhizoctonia solani Cladosporium Cladosporioides*	AgNPs	80–100 nm	Asymmetrical, spherical when aggregated	Antigungal activity against fungus sp. like *A. flavus, Pecitrinum, Fusarium oxysporum* etc.	Malik *et al*., 2024
2.	*Aspergillus niger*	CuNPs	NA	NA	Anticancer, antidiabetic and antimicrobial property	Noor *et al*., 2020
3.	*F.oxysporum*	Chitos an NPs	273.20 nm	Spherical	Antifungal property	Boruah and Dutta, 2021
	Metarhizium anisopliae	Chitos an NPs	172.50 nm	Spherical	Antifungal property	
	Beauveria bassiana	Chitos an NPs	78.36 nm	Spherical	Antifungal property	
	T.viridae	Chitos an NPs	89.03 nm	Spherical	Antifungal property	
4.	*Fusarium scirpi*.	AgNPs	2–20 nm	Quasi-spherical	Antimicrobial property	Rodríguez-Serrano *et al*., 2020
5.	*F.semitectum*	SeNPs	32.80–103.82 nm	Spherical	Antimicrobial, anticancer property	Abbas and Abou Baker, 2020
6.	*Fusarium solani*	$Zr0_2$ NPs	NA	Spherical	NA	Kavitha *et al*., 2020
7.	*Fusarium solani*	AuNPs	40–45 nm	Needle flower-like structure	Anticancer, biomedical application	Clarance *et al*., 2020
8.	*Fusarium solani* KJ623702	AgNPs	7.65–18.89 nm	Spherical	Biocidal effect against *P. aeruginosa*	El Sayed and El-Sayed, 2020
	Fusarium solani KJ623702	CuNPs	9.97–19.49 nm	Spherical	NA	
	Fusarium solani KJ623702	ZnNPs	8.55–21.76 nm	Spherical	Biocidal effect against *F.oxysporum*	
9.	*Penicillium oxalicum strain* LA-1	AgNPs	52.26 nm	Cubic crystalline structure	Antibacterial activity	Seetharaman *et al*., 2021
10.	*Agaricus bisporus*	AgNPs	50.44 nm	Uniform cubic structure	Antibacterial activity	Jameel *et al*., 2022

Table 3.7. Algae-mediated synthesis of nanpoparticles.

Sr. no.	Algae	Type of NPs	Size	Shape	Application	References
1.	*Cystoseira* extract (brown algae)	ZnO NPs	Less than 100 nm	Hexagonal crystal structure	Solar cell application	Jalali *et al.*, 2024
2.	*Sargassum plagiophyllum*	AgCl NPs	18–42 nm	Spherical	NA	González-Ballesteros and Rodríguez-Argüelles, 2020
3.	*Ulva fasciata*	ZnO NPs	77.81 nm	Spherical	Shrimp biopreservation	Alsaggaf *et al.*, 2021
4.	*Jania rubens*	Fe_3O_4 NPs	22.22–33.33 nm	Spherical	Antibiofilm activity	Salem *et al.*, 2020
	Sargassum vulgare	Fe_3O_4 NPs	17.05–34.09 nm	Spherical	Antibiofilm activity	
	Ulva fasciata	Fe_3O_4 NPs	22.73–39.77 nm	Spherical	Antibiofilm activity	
5.	*Portieria hornemannii*	AgNPs	70–75 nm	Spherical	Antibacterial activity against pathogenic *Vibrio* species	Fatima *et al.*, 2020
6.	*Padina pavonica* (marine algae)	FeNPs	10.53–20.55 nm	Cubic crystal	Photocatalytic degradation of Congo red dye, antioxidant activity	Caf, 2022
7.	*Spirulina platensis*	TiO_2 NPs	300 nm	Spherical	Environmental application, semiconductor application	Vasanth *et al.*, 2022
8.	*Petalonia fascia*	FeO NPs	NA	NA	Wastewater bioremediation	El-Sheekh *et al.*, 2021
	Colpomenia sinuosa	FeO NPs	NA	NA	Wastewater bioremediation	
	Padina pavonica	FeO NPs	NA	NA	Wastewater bioremediation	
9.	*Chlorella vulgaris*	Fe_30_4 NPs	109 nm	Cylindrical	Crystal violet photodegradation activity	Kiew *et al.*, 2023
10.	*Coelastrella aeroterrestrica* strain BA_chlo4	AgNPs	14.5 ± 0.5 nm	Hexagonal	Anticancer, antibacterial, antioxidant activity	Hamida *et al.*, 2022

targeted drug delivery and act as promising agents for improving disease diagnosis and cancer therapy (Ning *et al.*, 2017).

3. **Platinum NPs (PtNPs)**. PtNPs exhibit exceptional catalytic properties enabling them to carry out various types of chemical reactions such as hydrogenation, oxidation, and reduction. They are also widely used in electrochemical reactions including those occurring in fuel cells and batteries. Due to their stability and resistance to corrosion, PtNPs enhance the longevity and durability of instruments (Peng and Yang, 2009) of instruments. Moreover, PtNPs are used as biosensors, in targeted drug delivery, and medical imaging techniques like MRI (Pedone *et al.*, 2017).

3.7.1.2 Metalloid nanoparticles

Metalloids are chemical elements whose properties are intermediate between metals and nonmetals. NPs have been prepared from metalloids such as silicon, arsenic, boron, and germanium. Metalloid NPs are well known for their semiconductor properties and are widely used in electronics, optoelectronics and biosensors (Yazdi *et al.*, 2016).

3.7.1.3 Metallic oxide nanoparticles

When metals combine with oxygen they form metal oxides such as TiO_2, ZnO, and Fe_2O_3. These NPs are used in catalysis, bioremediation, energy storage, and biomedical applications (Yazdi *et al.*, 2016).

3.7.1.4 Carbon nanoparticles

Carbon NPs are tiny particles primarily composed of carbon atoms. They vary in size and shape and possess unique properties that make them useful in a number of applications. There are primarily five types of carbon NPs: (i) carbon nanotubes (CNT); (ii) fullerenes; (iii) graphene; (iv) carbon quantum dots (CQD); and (v) carbon nanofibers (Elmer and White, 2018).

1. CNTs are cylindrical nanostructures with exceptional strength and electrical conductivity. Their potential antimicrobial activity has also been explored (Wang *et al.* 2013, 2014a).
2. Fullerenes are a novel carbon allotrope with a polygonal structure made of 60 carbon atoms. They have extraordinary mechanical properties. Various fullerene-based materials have been developed and their antiviral, antioxidant, and chemotactic properties have been reported (Fiorito *et al.*, 2006).
3. Graphene is a material made up of a single layer of carbon atoms structured in a hexagonal grid. It has high strength, conductivity, and flexibility. Graphene oxide has exhibited significant antimicrobial properties (Chen *et al.*, 2013). It has been proposed that the sharp edges of graphene oxide directly interact with the cell walls of bacteria and fungi potentially leading to inactivation, due to the reactive oxygen functionalities of graphene oxide (Liu *et al.*, 2009; Elmer and White, 2018).
4. CQDs are nanoscale particles of carbon with quantum mechanical properties. They are used in imaging, sensing, and as fluorescent probes in biosensors. The use of CQDs as biosensors is under extensive research (Rad *et al.*, 2012).
5. Carbon nanofibers are similar to CNTs but have a different structural arrangement. They can be used as biosensors, in batteries, and as supercapacitors (Feng *et al.*, 2014).

3.7.1.5 Chitosan nanoparticles

They are a type of biodegradable NPs made from chitosan, a natural polymer derived from chitin. The recent focus on chitosan NPs stems from their low toxicity, biodegradability, and affordability. Evidence from recent studies suggests that these NPs can promote disease resistance in plants (Narasimhamurthy *et al.*, 2022). Research also indicates that chitosan NPs boost plant immune responses, underscoring their potential as effective tools for managing plant diseases.

3.7.1.6 Dendrimers

Dendrimers are a type of synthetic macromolecule characterized by their highly branched, tree-like structure. Due to their branched design, they possess large surface area relative to their volume, enhancing their effectiveness in various applications such as drug delivery and catalysis. Given their effectiveness as carrier molecules for transporting materials to tissues, dendrimers are currently being explored for their potential to address plant pathogenesis (Folliero *et al.*, 2021).

3.7.1.7 Liposomes

Liposomes consist of spherical lipid bilayers capable of encapsulating various substances, such as nucleotides, drugs, or nutrients. Since liposomes are stable in an aqueous medium, they are increasingly being explored in plant disease management to enhance the delivery of agrochemicals, improve the stability of active ingredients, and thereby reduce environmental impact (Taylor *et al.*, 2005).

3.7.2 Mode of action of nanoparticles against plant pathogens

Many NPs have been found to exhibit strong biocidal activity against phytopathogens; however, the precise mechanisms behind their effectiveness are still not clearly elucidated. A few mechanisms have been reported by which NPs combat phytopathogens such as direct toxicity, protein dysfunction, surface interaction and membrane damage, genotoxicity (e.g. dsDNA damage), and essential enzyme dysfunction (Ali *et al.*, 2020).

NPs, like AgNPs and CuNPs, exert their effects by producing reactive oxygen species (ROS), which directly damage the cell membranes and DNA of phytopathogens, leading to their death. Zhang *et al.* (2018a) observed that AgNPs compromised cell membrane integrity and elevated ROS levels leading to cell damage and reduced growth of *Azotobacter vinelandii*. Interaction of NPs has also been reported to induce protein dysfunction such as destruction of Fe-S clusters, oxidation of cysteine in Fe-binding sites, and exchange of catalytic metals. Several studies have demonstrated that interaction with metal NPs leads to genotoxicity, specifically lethal double-strand breaks in DNA. Exposure to toxic amounts of Cu and Pb has been shown to boost mutation rates in the genomes of pathogenic bacteria (Lemire *et al.*, 2013).

Another mechanism of action is the surface interaction and adhesion of NPs with the negatively charged cell membrane. The electrostatic attraction between NPs and the cell membrane leads to the disorganization of cell membrane components, causing disruption of membrane permeability and respiratory functions, ultimately leading to cell death. Several studies have indicated that the AgNPs exhibit antifungal properties by distorting fungal hyphae and damaging cell walls as confirmed by SEM and TEM studies (Ibrahim *et al.*, 2020). Certain NPs also inhibit key enzyme systems that are essential for pathogen metabolism and reproduction, leading to cell death.

3.7.3 Characterization techniques of nanoparticles

Characterization of NPs involves a variety of techniques and methods to determine their physical, chemical, and structural properties. Before using NPs for various applications, their characterization is an essential step to ensure efficacy. The following highlights key aspects and methods for characterizing NPs:

3.7.3.1 Size and shape analysis of nanoparticles

1. **Transmission electron microscopy**. This is a high-resolution imaging technique, used to determine the internal structure and morphology of individual NPs by transmitting a beam of high-energy electrons through them.
2. **Scanning electron microscopy**. This technique projects and scans a focussed beam of electrons over the surface of NPs, and collects different signals produced using various high-end detectors. It provides a detailed image of the surface morphology of the NPs and information on their size.
3. **Atomic force microscopy (AFM)**. This cutting-edge microscopy technique offers exceptional imaging capabilities for a wide

range of surfaces. It offers precise insights into surface topology and NP dimension enabling detailed analysis at the nanoscale.

4. **Dynamic light scattering (DLS).** This technique measures the size of NPs in suspension by analyzing the scattering of light. It determines the Brownian motion of NPs in colloidal suspension.

3.7.3.2 Structural analysis of nanoparticles

1. **X-ray diffraction (XRD).** This method analyzes the crystalline structure and phase composition of NPs, providing detailed insights into their material properties.
2. **Fourier transform infrared spectroscopy (FTIR).** Employing infrared light, this method analyzes NPs to reveal their functional groups and molecular bonding, enabling comprehensive analysis of their chemical properties.
3. **Raman spectroscopy.** This technique delivers a nondestructive chemical analysis, revealing intricate details about the chemical structure, phase and polymorphy, crystallinity, and molecular interactions of NPs, through the interaction of light with the chemical bonds.

3.7.3.3 Surface properties

1. **Zeta potential measurement.** It determines the amount of effective charge on the NPs' surface. The surface charge of NPs influences their stability in suspension and their interaction with surrounding biological systems.

3.7.3.4 Brunauer-Emmett-Teller (BET) surface area analysis

It measures the specific surface area and the porosity of the NPs using gas adsorption analysis, where an inert gas, such as nitrogen, is either passed over the sample or used in a controlled volume.

3.7.3.5 Chemical composition of nanoparticles

1. **Energy dispersive X-ray spectroscopy (EDX or EDS).** It measures the structure and components of magnetic NPs by detecting the characteristic X-ray emitted from the NPs.
2. **Inductively coupled plasma mass spectroscopy (ICP-MS).** In this method, the atomic elements are delivered through a plasma source, where they are ionized, and the trace elements in NPs are quantified, providing detailed information on the composition of NPs. It measures low concentrations (parts per billion) or ultra-low concentrations of elements (parts per trillion).

3.7.3.6 Optical properties

1. **UV-Vis spectroscopy.** It studies the UV or visible light absorption and band gap energy levels of NPs.
2. **Photoluminescence spectroscopy.** It is a nondestructive and contactless optical method that determines the electronic properties of materials. It evaluates the emission properties and energy levels of NPs.

3.7.3.7 Magnetic properties

1. **Vibrating sample magnetometry (VSM).** This method evaluates how NPs generate magnetic fields and their degree of responsiveness to external magnetic influences.
2. **Superconducting quantum interference device (SQUID) magnetometry.** This highly sensitive technique gauges extremely weak magnetic fields and delivers comprehensive magnetic analysis at low temperatures.

3.7.3.8 Thermal properties of nanoparticles

1. **Differential scanning calorimetry (DSC).** It helps to analyze and understand the fusion and crystallization events in NPs. It also determines the thermal transitions and stability of NPs.
2. **Thermogravimetric analysis (TGA).** It is a method that measures weight changes with respect to time and temperature to assess thermal stability and composition.

Table 3.8. Characterization parameters of nanoparticles and the related instruments. (Adapted and modified from Mourdikoudis *et al*., 2018)

Sr. no.	Parameters	Related instruments
1.	Size and shape	TEM, SEM, AFM, DLS,
2.	Structural analysis	XRD, FTIR, raman spectroscopy, energy-dispersive x-ray spectroscopy (EDX), high-angle annular dark-field imaging (HAADF), x-ray photoelectron spectroscopy (XPS)
3.	Surface properties	Zeta potential measurement, BET surface area analysis, electrophoretic mobility (EPM)
4.	Crystal structure	XRD, x-ray absorption fine structure (EXAFS), electron diffraction, scanning transmission electron microscope (STEM)
5.	3D structure	3D tomography, atomic force microscopy (AFM), SEM
6.	Chemical composition	EDX or EDS, ICPMS, XPS
7.	Optical properties	UV-Vis spectroscopy, photoluminescence spectroscopy
8.	Magnetic properties	VSM, SQUID, magnetic force microscopy (MFM), ferrormagnetic resonance microscopy (FMR)
9.	Thermal properties	DSC, TGA
10.	Structural defects	High-resolution TEM (HRTEM), electron backscatter diffraction (EBSD)
11.	Detection of NPs	TEM, SEM, STEM, EBSD, magnetic susceptibility
12.	Functionalization and biocompatibility	Cytotoxic assays, drug loading and release studies

3.7.3.9 Functionalization and biocompatibility

1. **Cytotoxicity assays.** This test measures the ability of NPs to cause cell damage or cell death, thereby evaluating the biocompatibility and toxic effects of NPs on biological systems.
2. **Drug loading and release studies.** It determines the capacity of NPs for drug delivery applications.

This combination of techniques gives a quick insight into the properties of NPs, their design, and their practical applications. Effective characterization of NPs is crucial for determining their application in various fields such as medicine, electronics and environmental science. Table 3.8 gives a brief overview of the different characterization parameters and the instruments used to measure these parameters (Mourdikoudis *et al.*, 2018).

3.8 Interaction of Nanoparticles with Plants

The mechanism of interaction between NPs and plants is a topic of keen interest among scientists in nanotechnology and agriculture. NPs can be absorbed into the plant either through the roots or the leaves. They may be absorbed via endocytosis or through transport pathways such as stomata, plasmodesmata, or root hairs.

1. **Stomatal uptake.** The pore of the stomata is wide enough to allow convenient entry of NPs into the leaves. Small NPs can enter through the open stomata by several mechanisms including diffusion, endocytosis, and translocation. They easily diffuse through the stomatal openings, subsequently traverse the stomatal pore, and finally reach the intercellular spaces of the leaves (Hu *et al.*, 2020). Once inside the leaf tissue, cells take up NPs via endocytosis. The cell membrane

engulfs the NPs forming vesicles and transports them into the cytoplasm for further processing (Dong *et al.*, 2023).

2. **Entry through cuticle**. Certain NPs gain access into the leaves by passing through the waxy cuticle that coats the leaf surface. NPs with specific surface charges or ligands can enter the mesophyll cells through the cuticle layer, and through intricate internal transport pathways get distributed throughout the plant body (Francis *et al.*, 2024).
3. **Entry through the root surface**. NPs enter the root surface through the root hairs. Root hairs excrete mucus and organic acids which impart a negative charge to the root surface. NPs with a positive charge on their outer surface accumulate and are subsequently internalized into the root surface (Mondal *et al.*, 2023).

Metal NPs reach the plant's vascular transport system (xylem and phloem) primarily through two routes: the apoplast and the symplast routes. The apoplast pathway involves passage through the extracellular spaces, the gap widths between cell walls, and intercellular spaces, allowing NPs to navigate and reach the root or stem xylem tissue system. In the symplastic pathway, the NPs move through the intercellular channels, known as plasmodesmata, which range in diameter from 3 nm to 20 nm. NPs pass through this route to accumulate within the cytoplasm. They then traverse the plant's endothelial layer and Casparian strips to finally reach the vascular transport system. Studies have shown that NPs with particle sizes below 50 nm predominantly follow the symplast route while larger NPs, ranging from 50 nm to 200 nm, mainly follow the apoplastic pathway (Wang *et al.*, 2023).

NPs absorption through the plant surface is affected by various factors, including the size of the NPs, the chemical composition of the surface, and their surface charge (Francis *et al.*, 2024). The absorption of NPs through the cell-wall pores also depends on factors like osmotic pressure, passive diffusion through the cell wall, and the interplay of capillary action forces (Kurczyńska *et al.*, 2021). It has been observed that the size of the NPs is a critical factor in their absorption. NPs with a diameter below 50 nm enter the leaves through the stomatal pathway. ZnO NPs with a 30 nm particle size enter the leaves through the stomatal pathway, whereas larger NPs are absorbed via the cuticular pathway.

Electrostatic attraction between positively charged NPs and the negatively charged plant cell wall significantly impacts their absorption. Furthermore, environmental factors such as temperature, humidity, and sunlight also play a major role in NP absorption. Hence, the absorption of NPs is significantly influenced by the interplay between their properties, the nature of the absorbing surfaces, and environmental factors (Hong *et al.*, 2021).

3.9 Use of NPs for Virus Resistance

The serious diseases caused by plant viruses make them a significant threat to crop production in agriculture. The emergence of nanotechnology has led to the use of various NPs for designing and managing plant viruses. A new term has emerged for the application of nanotechnology in plant virology: 'nanophytovirology.' The range of applications includes viral disease diagnosis, drug delivery, genetic engineering, plant defense activation, and biostimulation. The use of NPs in viral disease management is still in its nascent stage and opens up various areas of research interest, particularly regarding the interaction between viruses and NPs at the molecular level, and the application of NPs in viral disease management to achieve sustainable agriculture (Dutta *et al.*, 2022).

3.9.1 Nanobiosensors for detection of plant viruses

Nanobiosensors, based on nanotechnology, are analytical devices that identify pathogen infestation in plants by tracking signal changes in plant pathways and metabolism without disrupting the plant's metabolic functions. The biological signal is transformed into an electrical signal, which is subsequently modulated and presented as an output display.

The signals from the sensors provide real-time analysis of metabolic changes due to biotic or abiotic stress affecting the plant (Cheng *et al.*, 2009).

Initially, the detection of viruses was based on the identification of intact viruses. With advancements in optics, microfabrication, and nanotechnology, viruses can now be detected and quantified with greater precision. Currently, quantum dots (QDs) and NPs labeled with detection tags, such as dyes or antiviral antibodies, have been developed for the efficient identification and labeling of viruses. NPs have been conjugated with probes like nucleic acids and antibodies for the detection of virus-based molecules (Dougan *et al.*, 2007; Wang *et al.*, 2009). Safarpour *et al.* (2012) introduced a nanobiosensor utilizing antibodies conjugated with cadmium-telluride QDs (CQDs) and glutathione-S-transferase protein (GST) protein. This FRET-based biosensor was engineered to detect *Polymyxa betae*, the vector of beet necrotic yellow vein virus (BNYVV), which is responsible for the severe rhizomania disease in sugarcane. In 2021, Lavanya and Arun (2021) created a visual detection technique for begomovirus based on gold NP (AuNP) assay with functionalization. Using the clccpi1 probe, the AuNP assay successfully identified begomoviral infections in plants such as chili, tomato, common bean, gram, and black gram plants. It proved to be highly sensitive and provided faster results compared to conventional methods like PCR and ELISA. The Quartz-crystal microbalance (QCM) immunosensor is another effective mass-sensor technique. It operates on the principle that the resonance frequency of a quartz crystal shifts in response to the adsorption of substances on its surface. A QCM-based nanosensor has been created to detect plant viruses, including cymbidium mosaic potexvirus (CymMV) and Odontoglossum ringspot tobamovirus (ORSV). The device was treated with antiviral antibodies designed to capture specific plant viruses. When the targeted viruses adhere to the antibody-coated surface, they increase the mass at the QCM interface. This increased mass leads to a decrease in the crystal resonance frequency, providing a direct measurement of the virus concentration. QCM was highly sensitive and provided faster detection than ELISA (Eun *et al.*, 2002).

In addition to chemical, electrochemical, and optical methods of identification, there are many biosensors based on mechanical and electrical detection techniques that detect a broad range of species without the need for labeling. Examples of these biosensors include microcantilever, photoacoustic, micropillar, and piezoelectric sensors. Although many of these sensors are highly sensitive, label-free, and commercially available, they have some disadvantages, including complex preparation procedures, high costs, functionalization issues, and complex image processing (Chalklen *et al.*, 2020). Another immunosensor-based assay, the colloidal gold immunochromatographic assay (GICA), has proven to be a popular method for the rapid detection of plant viruses. This assay uses antibodies labeled with colloidal gold to detect specific viruses. This laboratory-based method is cost-effective, reliable, highly precise, and efficient, making it a dependable option for plant virology diagnostics and plant disease management (Ghorbani *et al.*, 2023).

Most nanobiosensors are quantum dot (QD) sensors. QDs exhibit photoluminescence properties, making them ideal for imaging and labeling. QDs are resistant to photobleaching and can be labeled with nucleic acids, antibodies, and oligonucleotides. Abubakar *et al.* (2018) engineered an NP-based particle bombardment system using a gene gun for the detection and assessment of viral infectivity. This approach has been further employed to analyze both transient and stable gene expression in citrus leaf cells at the cellular scale (Acanda *et al.*, 2019). Further research is needed to develop biosensors with higher precision, selectivity, and efficiency.

3.9.2 Nanoparticles for inhibition of phytoviruses

A range of NPs has displayed viricidal activity against numerous viruses and has been successful in impeding viral growth. Silver nanoparticles (AgNPs), specifically, have demonstrated significant effectiveness in controlling multiple viral strains (Dutta *et al.*, 2022). In faba bean plants, AgNPs have been successfully used to combat bean yellow mosaic virus (BYMV). AgNPs were found to significantly inhibit the virus by binding to its particles, thereby reducing both replication and accumulation in plant tissues. AgNPs also contributed to the upregulation of the pathogenesis-related gene PR-1 and the activation of defense-related oxidizing enzymes in plants affected by the virus (El Gamal *et al.*, 2022).

In a related study, zinc oxide nanoparticles (ZnONPs) were synthesized through a green method using the aqueous extract from *Mentha spicata* leaves. These NPs demonstrated significant antiviral efficacy against tobacco mosaic virus (TMV), leading to a 90.21% decrease in viral accumulation when administered 24 hours before and after TMV inoculation. Furthermore, ZnONPs-treated plants showed upregulated transcription of phenylalanine ammonia lyase (PAL), pathogenesis-related protein (PR-1), chalcone synthase (CHS), and peroxidase (POD), indicating a strengthened defense response (Abdelkhalek and Al-Askar, 2020). The study by Adeel *et al.* (2021) revealed that carbon nanotubes (CNT) can successfully inhibit TMV replication and infection. Moreover, CNTs induced an increase in abscisic acid and salicylic acid levels and stimulated the expression of genes linked to the biosynthesis of phytohormones. Bean yellow mosaic virus (BYMV) causes significant damage to faba bean plants (*Vicia faba*). *Rhizobium* leguminosarum-mediated AgNPs have been biosynthesized, and AgNPs successfully reduced BYMV-mediated plant disease. AgNP-treated plants showed higher expression of PR-1 and HQT than control plants (Abdelkhalek *et al.*, 2022).

In a recent report, AgNPs were synthesized using *Punica granatum* peel extract. AgNPs were found to markedly enhance the plant growth, and inhibit the replication of TMV. They also led to the higher expression of pathogenesis-related (PR) genes – PR-1 and PR-2 – as well as increased accumulation of polyphenolic compounds, relative to control plants (Al-Askar *et al.*, 2023). The research by Mahfouze *et al.* (2020) revealed that AgNPs effectively combat banana bunchy top virus (BBTV) in banana plants, demonstrating strong viricidal activity. AgNP treatment enhanced chlorophyll and carotenoid content and significantly elevated phenol and proline content in banana plants.

The evidence from these studies establishes that the metal and metal oxide NPs can disrupt the viral replication cycle, and promote systemic acquired resistance (SAR) in plants. To advance the management of viral diseases in crops, additional research should focus on evaluating the antiviral effects of metal and metal oxide NPs, along with organic polymer-based NPs. Table 3.9 lists examples of NPs that have been employed to suppress plant viruses in agricultural crop plants.

3.10 Nanoparticles for Inhibition of Plant Pathogenic Bacteria

Bacterial pathogenesis results in significant damage to crops and also poses a serious threat due to its tendency to develop antibiotic resistance. Several NPs have been developed, and their antibacterial properties have been studied to address the problem of antibiotic resistance. Green-synthesized NPs offer superior performance and are more eco-friendly than chemically synthesized NPs (Gul *et al.*, 2024).

Dickeya dadantii, the bacterium responsible for stem and root rot disease in sweet potatoes, causes considerable damage. Silver nanoparticles (AgNPs) derived from the culture supernatant of the bacterium *Pseudomonas rhodesiae*, with their activity enhancing as AgNP concentration increased. These environmentally friendly AgNPs have been utilized to control soft rot disease in sweet potato seed tubers (Hossain *et al.*, 2019). In their 2024 study, Chowdhury *et al.* examined the effectiveness of Cu and Ag nanocomposites in inhibiting *Xanthomonas oryzae*, the bacterium responsible for the destructive bacterial leaf blight disease in rice plants. Although AgNPs have reported better antibacterial activities against bacterial pathogens, they are expensive and rarely used in agriculture. Hence, the antibacterial efficacy of Ag and CuNPs was studied. The disease-causing protein of *X. oryzae*, peptide deformylase, formed a complex with the composite nanoparticles and showed antibacterial activity against *X. oryzae*. Hence, the CuNPs and AgNPs composite proved effective in mitigating leaf blight disease caused by *X. oryzae*.

In a recent study, AgNPs were synthesized from extracts of *Arctium lappa* fruit (Al-NP), *Solanum melongena* leaves (Sm-NP), and *Taraxacum mongolicum* leaves (Tm-NP). The use of green-synthesized NPs proved effective in

Table 3.9. Antiviral properties of nanoparticles for inhibition of plant viruses.

Sr. no.	Type of NPs	Size	Shape	Target Virus	Viricidal activity	References
1.	AgNPs	NA	NA	Bean Yellow mosaic virus (BYMV)	Potent viricidal activity against *BYMV*. Activated PR-1 genes and induced defense- related oxydizing enzymes	El Gamal *et al.*, 2022
2.	ZnONPs	11–88 nm	Crystalline	Tobacco mosaic virus (TMV)	90.21% reduced TMV accumulation	Abdelkhalek and Al-Askar, 2020
3.	Carbon nanotubes (CNT)	NA	NA	TMV	Inhibition of TMV replication and induced resistance	Adeel *et al.*, 2021
4.	*R.leguminoserum*-mediated AgNPs	13.7–40 nm	Spherical	BYMV	Inhibition of BYMV	Abdelkhalek *et al.*, 2022
5.	ZnONPs	18 nm	NA	TMV	Suppressed invasion of TMV	Cai *et al.*, 2019
6.	Fe_3O_4 NPs	0.19 nm	NA	TMV	Inhibited TMV infection, activated antioxidants against TMV	Cai *et al.*, 2019
7.	AgNPs	NA	NA	TMV, Potato virus Y(PVY)	AgNPs conferred resistance against TMV, PVY	Noha *et al.*, 2018
8.	AgNPs	12.6 nm	Spherical	Tomato spotted wilt virus (TSWV)	AgNP reduced TSWV infectivity	Shafie *et al.*, 2018
9.	AgNPs mediated by *Punica granatum*	33.37 ± 12.7 nm	Round shape	TMV	Inhibited TMV pathogens and enhanced PR genes	Al-Askar *et al.*, 2023
10.	AgNPs	15 nm	Spherical	Banana bunchy top virus (BBTV)	AgNPs induced viricidal effect against BBTV	Mahfouze *et al.*, 2020

NA: information not available

controlling the growth of *X. oryzae* pv. *oryzae* (*Xoo*), the pathogen that causes bacterial leaf blight in rice. However, Al-NP showed better antibacterial activity against *X. oryzae* (Tian *et al.*, 2022). Khan *et al.* (2023) described the green synthesis of CuONPs derived from the greater duckweed plant, *Spirodela polyrhiza*. The synthesized NPs were tested for antibacterial, antifungal, and nematicidal effectiveness against plant pests and pathogens. CuONPs showed potent antifungal activity against *Pythium vexans*, inhibiting 50% of fungal mycelium. The NPs were effective against both the pathogenic bacterium *Pectobacterium carotovorum* and the nematode *Meloidogyne incognita*, showing significant inhibitory effects.

A less explored field involves the synthesis of AgNPs using the endophytic fungus *Talaromyces purpureogenus*. These NPs showed potent antimicrobial activity against *Listeria monocytogenes*, *Escherichia coli*, *Shigella dysenteriae*, and *Salmonella typhi*. The AgNPs displayed antioxidant properties, as evidenced by their ability to scavenge free radicals, including 2,2'-diphenyl-1-picrylhydrazyl (Sharma *et al.*, 2022). A similar approach has been taken using the endophytic bacterium *Bacillus zanthoxyli*, with the antimicrobial properties of AgNPs being explored. The AgNPs were effective in inhibiting the growth of multiple bacterial strains including *Staphylococcus aureus*, *Salmonella typhi*, *E. coli* DH5α, *Bacillus subtilis*, and *Pseudomonas aeruginosa* (Li *et al.*, 2024). Table 3.10 illustrates a few examples of antibacterial properties of NPs for inhibition of pathogenic bacteria.

3.11 Nanoparticles-mediated Induction of Fungal Resistance in Crops

Different types of fungal pathogens, such as *Fusarium* sp., Rhizoctonia sp., *Aspergillus* sp., Alternaria sp., and *Phytophthora* sp., infect cereal crops like rice, barley, and maize, causing substantial economic losses to farmers. Mycotoxins are a substantial hazard to the wellbeing and safety of both humans and animals. The first choice for protecting crops against fungal pathogens has traditionally been chemical fungicides; however, these have raised concerns regarding fungicide resistance, human health, the environment, and the ecosystem. Therefore, alternative strategies need to be explored to mitigate fungal pathogenesis (Dong *et al.*, 2023).

AgNPs were green-synthesized with the help of *Pseudomonas poae*, an endophytic bacterium derived from garlic plants. These NPs exhibited strong antifungal activity against *Fusarium graminearum*, inhibiting mycelial growth, spore germination, germ tube extension, and the production of Mycotoxins. Hence, AgNPs can potentially be used to protect wheat from fungal infections (Ibrahim *et al.*, 2020). Das and Dutta (2021) demonstrated that biologically synthesized AgNPs and AuNPs effectively inhibited *Rhizoctonia solani*, the pathogen responsible for sheath blight in rice. AgNPs were more effective AuNPs in inhibiting fungal mycelium, causing greater distortion of fungal cells. Furthermore, AgNPs significantly boosted the production of secondary metabolites such as phenols, flavonoids, and terpenoids.

Hoang *et al.* (2022) reported the evaluation of metal NPs (Fe^0, Cu^0, Co^0), ZnONPs, and chitosan-stabilized AgNPs for the inhibition of *Helminthosporium teres*. The highest inhibition was observed with Co^0 and chitosan-AgNPs. The application of NPs enhanced chlorophyll and carotenoid contents in both infected and noninfected plants. A recent study focussed on the biosynthesis of AgNPs employing the fungi *R. solani* and *Cladosporium cladosporioides*. The AgNPs exhibited significant zones of inhibition against various fungal pathogens, including *A. flavus*, *P. citrinum*, *F. oxysporum*, *F. metavorans*, and *A. aflatoxiformans*. AgNPs hold significant potential in biomedical applications, food processing and packaging industries for reducing contamination and extending shelf life (Malik *et al.*, 2024). In another study, CuONPs were synthesized from the *Spirodela polyrrhiza* plant. The synthesized CuONPs showed significant antifungal activity against *Phytophthora vexans*, which causes substantial destruction of important crop plants (Khan *et al.*, 2023).

Fusarium oxysporum is a highly destructive fungal pathogen. In a recent study, bimetallic ZnO-CuONPs were biosynthesized with the help of *Aspergillus fumigates* to address and manage the pathogenicity of *F. oxysporum*. Assessment of the fungal activity of ZnO-CuONPs exhibited potent antifungal activity towards *F. oxysporum*.

Table 3.10. Antibacterial properties of nanoparticles for inhibition of plant pathogenic bacteria.

Sr no.	Target bacterial pathogen	Type of NPs	Size	Shape	Application	References
1.	*Dickeya dadantii*	AgNPs	20–100 nm	Uniform spherical shape	Potent antimicrobial activity	Hossain *et al.*, 2019
2.	*X.oryzae*	AgNP +CuNP composite	306–375 nm	Spherical	Antibacterial activity against *X. oryzae*	Chowdhury *et al.*, 2024
3.	*X.oryzae*	AgNPs produced using rice leaf extract	16.5 nm	Cubic crystal shape	Effective in inhibiting growth of *X. oryzae*	Namburi *et al.*, 2021
4.	Bacterial leaf blight pathogen *X. oryzae* pv. *oryzae*	AgNPs	20–40 nm	Closely packed uniform shape	AgNPs inhibited growth of *X. oryzae*	Tian *et al.*, 2022
5.	*Rhizotonia solani*	CuNPs	80 nm	Spherical to circular	CuNPs exhibited strong antibacterial property	Kashyap *et al.*, 2023
6.	*X. oryzae*	FeONPs	86 m	Spherical	Significant antibacterial activity	Ahmed *et al.*, 2022
7.	Bacterial leaf blight, tomato bacterial wilt pathogens *Ralstonia solanacearum, X. oryzae*	Cu-AgNPs	NA	NA	Showed significant broad-spectrum antibacterial acitivity	Ning *et al.*, 2024
8.	*Pectobacterium carotovorum*	CuONPs from *Spirodela polyrhiza plant* (greater duckweed plant)	100 nm	Spherical	Potent antibacterial activity	Khan *et al.*, 2023
9.	*P. aeruginosa* and *S. aureus*	AgNPs (synthesized from *Lactiplantibacillus*sp. strain C1)	129 ± 12.31 nm	Spherical	Significantly inhibited biofilm and virulence properties of *P. aeruginosa* and *S. aureus*	Kang *et al.*, 2023
		AuNPs (synthesized from *Lactiplantibacillus*sp. strain C1)	100.54 ± 14.07 nm	Spherical		
10.	*Ralstonia solanacearum* (cause bacterial wilt disease in tomato)	ZnONPs	NA	NA	ZnONPs exhibited strong antibacterial activity against *Ralstonia solanacearum*	Guo *et al.*, 2022
11.	*S. aureus ATCC6538P, Salmonella typhi enteric ATCC25922, E.coli DH5α, Bacillus subtilis WB800, Pseudomonas aeruginosa ATCC27853*	AgNPs synthesized using endophytic bacterium *Bacillus zanthoxyli GBE11*	3.68–31.60 nm	Crystalline	Exhibited strong antibacterial activity	Li *et al.*, 2024
12.	*Listeria monocytogenes, Escherichia coli, Shigella dysenteriae, Salmonella typhi*	AgNPs synthesized from utilizing endophytic fungus *Talaromyces purpureogenus*	30–60 nm	Spherical	AgNPs showed antibacterial and antioxidant properties	Sharma *et al.*, 2022

NA: Information not available

The TEM results indicated that ZnO-CuONPs caused significant disintegration and destruction of the cell wall, as well as damage to nuclear materials. Thus, ZnO-CuONPs emerged as a promising antifungal agent for combating *F. oxysporum* (Gaber *et al.*, 2023).

Paraguay-Delgado *et al.* (2022) reported an innovative approach involving the hydrothermal synthesis of ZnO NPs, ZnO/Cu$2_{\%}$ NPs, and CuONPs, followed by an evaluation of their antifungal properties. The antifungal efficacy was evaluated against four pathogenic fungi: *Neofusicoccum arbuti, Alternaria alternata, Fusarium solani, and Colletotrichum gloeosporioides*. ZnO/Cu$2_{\%}$ NPs showed enhanced photocatalytic and antifungal activities relative to pure oxide and composite NPs. These results highlight the potential of ZnO/Cu$2_{\%}$ NPs in controlling plant pathogenic fungi in agriculture and forestry applications.

The use of metal and metal oxide NPs provides numerous benefits compared to conventional fungicides for managing fungal infections such as effective antifungal activity and multi-target action that disrupts multiple cellular processes, thereby reducing the chances of resistance development. NPs also ensure targeted delivery to the desired fungal cells, and their longer shelf lives provide greater stability than conventional fungicides. Adopting green synthesis for NPs ensures a sustainable and eco-friendly approach and reports suggest that NPs can induce plant defense mechanisms, thus improving resistance against fungal infections. Their affordability will further promote the development of a more diverse array of metallic and metal oxide NPs, offering a promising alternative to conventional methods with greater safety and environmental benefits (Tortella *et al.*, 2023). Table 3.11 outlines a few examples of how NPs contribute to resistance against pathogenic fungi.

3.12 New Strategies in Nanotechnology in the Genetic Engineering of Plants

Agricultural scientists have turned to genetic engineering to expedite the production capabilities of plants in response to climate change and population growth pressures. Current genetic manipulation methods are labor-intensive and time-consuming, which hinders advancements in agriculture and molecular farming. Therefore, new approaches and tools are essential to overcome these limitations. Innovations in nanotechnology have recently tackled these issues and advanced the field of plant genetic engineering (Yan *et al.*, 2022). Some key strategies include the following.

3.12.1 NPs-mediated gene delivery

NPs such as AuNPs, AgNPs, and CuNPs are used as nanocarriers to deliver genetic material into plant cells more efficiently. NPs have been engineered to improve uptake and minimize damage to plant cells. Additionally, NPs are capable of entering cells independently, in contrast to traditional transformation methods (Wang *et al.*, 2019). This ability has allowed NPs to be employed for delivering molecular cargo and therapeutic agents with high precision to specific subcellular locations, such as nuclei or organelles (Wu *et al.*, 2023).

3.12.2 Quantum dots (QDs) for gene tracking

Semiconductor NPs such as quantum dots (QDs) are valuable tools for tracking gene expression and assessing gene editing effectiveness in plants. The application of high-resolution imaging techniques facilitates real-time analysis of gene dynamics. Recent research utilized carbon quantum dots (CQDs) to explore transport mechanisms in *Arabidopsis thaliana*, aiming to understand their influence on the inheritance of the Argonaute family of genes and their role in micro-RNA (miRNA) mediated gene regulation. Findings revealed that the CQDs interacted with miRNA, resulting in the repression of post-transcriptional regulation of target genes in the cytoplasm. Microscopic observation also validated the rapid localization of CQDs on cell membranes and nuclei, with enhanced nuclear entry. Moreover, the influence of CQDs on the expression of plant genes was found to be heritable (Lin *et al.*, 2023). In a separate study, CdSe/

Table 3.11. Antifungal properties of nanoparticles for inhibition of plant pathogenic fungus.

Sr. no.	Target fungal pathogen	Target crop	Type of NPs	Size	Shape	Application	References
1.	Head blight pathogen *F. graminearum*	Wheat	AgNPs	19.8–44.9 nm	Spherical	Showed strong inhibition of fungal pathogen	Ibrahim *et al.*, 2020
2.	*Rhizotonia solani*	Rice	AgNPs, AuNPs	NA	NA	AgNPs showed greater inhibition of *R. solani* than AuNPs	Das and Dutta, 2021
3.	*H. teres*	Wheat, barley	Metal NPs (Fe^0, Cu^0, Co^0), ZnONPs and Chitosan stabilized AgNPs	50 nm	NA	CuONPs and chitosan stabilized AgNPs showed better inhibition against *H.teres*	Hoang *et al.*, 2022
4.	*A. flavus, P. citrinum, F. oxysporium, F. metavorans A. aflatoxiformans*	NA	Green synthesis of AgNPs from fungi *R. solani, Cladosporium cladosporioides*	80–100 nm	Spherical, sub- spherical	Strong antifungal property of AgNPs and controls various pathogenic diseases caused by fungi	Malik *et al.*, 2024
5.	*P. vexans*	NA	CuONPs (synthesized from CeO_2NPs)	100 nm	Spherical	Growth of *P. vexans* significantly reduced	Khan *et al.*, 2023
6.	*Ustaligo tritici*	Wheat	CeO_2NPs	7–10 nm	Round shape	Potent antifungal tioactivity against *U.tritici*	Alotaibi *et al.*, 2023
7.	*Alternaria brassicae*	Oil seed crops	CuONPs	22 nm	NA	CuONPs were efficient antifungal agent against *Alternaria* blight	Gaba *et al.*, 2022
8.	*F. oxysporum*	NA	Bimetallic ZnO-CuONPs	54.18 ± 1.9 nm	Semispherical	ZnO-CuONPs exhibited significant antifungal activity against *F.oxysporum*	Gaber *et al.*, 2023
9.	*Fusarium solani* (causative agent of root rot disease in cucumber)	Cucumber	Cu_2ONPs	25 nm	Distinct crystal shape	Cu_2ONPs showed potent inhibitory effect against *F.solani* and reduced disease incidence in cucumber roots	Kamel *et al.*, 2022
10.	*Neofusicoccum arbuti, Alternaria alternata, Fusarium solani* and *Colletotrichum gloeosporioides*	NA	ZnONPs, ZnO/$Cu2_{\%}$ and CuONPs	NA	NA	The ZnO/$Cu2_{\%}$ exhibited adequate photocatalytic and antifungal activity	Paraguay-Delgado *et al.*, 2022

NA: information not available

ZnS QDs were utilized for real-time tracking of essential nutrients in *S. cerevisiae*. The cysteine-conjugated CdSe/ZnS QDs were less toxic than CdQDs and remained on the cell surfaces, rather than penetrating the cells, indicating that CQDs may be a better option for molecular tracking in fungi (Färkkilä *et al.*, 2023).

3.12.3 nanocarriers for RNA interference (RNAi)

Nanotechnology-based carriers have been developed to deliver RNAi molecules precisely to target cells. This technique helps avoid unwanted gene silencing and enables precise alteration of the gene of interest. A recent research highlighted the use of QDs to oversee RNAi delivery and enhance gene silencing. Their outstanding photostability and versatile optical characteristics make them highly selectable for detailed selection processes, making them ideal for monitoring and sorting cells treated with different siRNA/QD complexes (Chen *et al.*, 2005).

Spray-induced gene silencing (SIGS) involves applying dsRNA directly onto plant surfaces to induce RNAi specifically into fungal cells. The dsRNA is absorbed by fungal cells, where it is processed by the fungal RNAi system, leading to specific gene silencing and reduced fungal growth. Several studies have shown that exogenously applied dsRNA/siRNAs significantly inhibit the growth of various phytopathogenic fungi, such as *Fusarium species, Aspergillus niger,* and *R. solani*, through the SIGS mechanism (Ray *et al.*, 2022).

3.12.4 Nanostructured materials for improved transformation

Advances in nanostructured materials enhanced the efficiency of plant tissue transformation. One key obstacle in traditional transformation methods is the plant cell wall. Novel nanostructured materials have been developed to enhance the transformation efficiency of plant tissues. One of the major barriers in conventional plant cell transformation methods is the plant cell wall. NP-mediated gene transformation can overcome this challenge and accurately deliver DNA or RNA into plant cells, resulting in stable or transient transformations. Conventional methods are hindered by their laborious nature, potential for tissue damage, and limited efficiency. NP-mediated approaches offer superior efficiency, versatility across species, and targeted delivery of genetic material (Wu *et al.*, 2023. Moreover, NPs have a unique advantage in that they can easily reach and localize in target sub-organelles such as chloroplasts, mitochondria, and the nucleus (Yan *et al.*, 2022).

Carbon nanotubes (CNTs) have been synthesized with nucleic acids anchored to the side walls of CNTs using covalently grafted polymer chains such as polyethylene glycol (PEG), polyallylamine (PAA), and chitosan to enhance their water solubility. The nucleic acids are strongly adhered to the positively charged CNTs by electrostatic attraction, while protein molecules are attached by Van der Waals forces and XVI-XVI stacking (Zheng *et al.*, 2003). In a research study, a low dose of single-walled CNTs (SWCNTs) entered the *Arabidopsis* mesophyll cells, cell walls, and cell membranes, and subsequently penetrated the nucleus, chloroplast, vacuole, and mitochondria through a non-energy-dependent endocytosis pathway. The study found that low levels of SWCNTs encouraged cell growth, but high levels triggered ROS generation and resulted in necrosis or apoptosis (Yuan *et al.*, 2011). Research showed that fluorescein isothiocyanate (FITC)-labeled SWCNTs penetrated the plant cell wall of *N. tabacum* BY-2 suspension cells and facilitated the delivery of ssDNA into the cells (Liu *et al.*, 2009). Several studies also highlighted the use of graphene, magnetic NPs, and SiNPs in plant cell transformation. Combining genome editing, with NP-mediated gene transformation and advanced regeneration techniques could lead to improved crop protection (Wu *et al.*, 2023).

3.13 Interaction of Nanoparticles and Induction of Defense Mechanisms in Plants

By interacting with phytopathogens, various NPs provoke plant defense responses, activating signal transduction pathways through increased expression of pathogenesis-related genes and generation of reactive oxygen species (ROS). The study revealed that MgO NPs effectively reduced bacterial wilt disease incidence in tomato plant roots. The application of MgO NPs resulted in ROS generation, including the O^{2-} radical, and activated genes in tomato roots such as *PR1* (salicylic acid-inducible), *LoxA* (jasmonic acid-inducible), *Osm* (ethylene-inducible) and *GluA* (systemic resistance-related) (Imada *et al.*, 2016). In another study, when plants were treated with Ce NPs, there was upregulation of ROS and Ca2+ mediated signaling genes. Both ROS and Ca2+ significantly affected plant stress responses (Yu *et al.*, 2020).

Studies indicated that NPs can modulate the production of ROS depending on plant requirements. When viral particle interactions trigger a rapid burst of ROS that impacts host cells, NPs activate the antioxidant system to mitigate this oxidative surge. Notably, nanoparticles such as CuO, CeO_2, ZnO, and Fe_3O_4 NPs have been shown to stimulate the antioxidant defense system and mitigate oxidative stress in various crops (Soares *et al.*, 2018). In another similar study, chitosan NPs (CNP) significantly suppressed the sheath blight pathogen *R. solani* and subsequently activated defense-related enzymes like peroxidase, phenylalanine lyase (PAL), and chitinase enzyme (Divya *et al.*, 2020). In their 2015 study, Chandra *et al.* demonstrated that CNP activated the plant's innate immune response by enhancing the activity of defense enzymes, including many antioxidant enzymes, and total phenolic compounds. The study also revealed that CNP treatment increased nitric oxide (NO) levels, a vital signaling molecule in plant defense.

External stimuli such as pathogen infections trigger plant defense responses through signaling molecules like salicylic acid (SA), jasmonic acid (JA), and ethylene. The interaction between these defense-related signals and growth-regulatory hormones – such as gibberellic acid (GA), cytokinin, indole acetic acid (IAA), and abscisic acid (ABA) – influences both plant growth and physiological processes (Rastogi *et al.*, 2017). Another research showed that the interaction of CuONPs enhanced the levels of SA and JA in *A.thaliana*, as well as metalloprotein-like genes involved in plant defense responses (Landa *et al.*, 2017). Cai *et al.* (2020) reported that foliar exposure to Fe_3O_4 NPs on *N. benthamiana* elicited defense responses against the tobacco mosaic virus (TMV). Fe_3O_4 NPs triggered plant antioxidant responses, elevated the synthesis of salicylic acid (SA), and activated the expression of SA-responsive PR genes (PR1 and PR2). These results demonstrate that FeNPs enhance plant resistance to TMV by inducing SA. According to Ali *et al.* (2018), phytopathogens and defense-related signaling molecules induce the expression of pathogenesis-related (PR) proteins and antimicrobial proteins (AMPs). These PR genes act as markers for the plant defense signaling pathway. The increased expression of PR-1, PR-2, and PR-5 genes signifies the activation of the SA signaling pathway. Several other reports also demonstrated that the interaction of NPs activated phytohormones and PR proteins, thereby inducing plant defense mechanisms.

3.14 Risks of Using Nanoparticles and Impact on the Environment

NPs have potential benefits in crop disease management and improvement programs. However, alongside these advantages, NPs pose certain risks to ecosystems and nontarget organisms. It has been reported that silver nanoparticles (AgNPs) and copper nanoparticles (CuNPs) may be toxic to plants, beneficial microorganisms, and even humans if they accumulate in the environment or food chain. Consequently, NPs can disrupt the ecological balance and negatively impact environmental ecosystems. AgNPs, CuNPs, and titanium dioxide nanoparticles (TiONPs) have been shown to significantly influence the enzymatic activities of crucial soil enzymes, thereby harming the soil bacterial community (Parada *et al.*, 2019).

However, the long-term environmental effects of NPs are not fully understood. They may persist in soil and water, potentially disrupting ecosystems. Thus, assessing the possible risks of consuming food treated with NPs is crucial. Studies have shown that metal NPs undergo chemical transformations and interactions with soil components, affecting soil properties in the environment (Klaine *et al.*, 2008). Metal NPs also affect aquatic organisms, thereby disrupting aquatic ecosystems. It has been reported that gold nanoparticles (AuNPs) at a concentration of 1 mg/L caused oxidative stress in zebrafish embryos and disturbed normal developmental processes (Strungaru *et al.*, 2018). Li *et al.* (2023) documented the abundance of titanium-based NPs in both seawater and sediments and assessed AgNPs and TiNPs as high risks to marine ecosystems.

Iron oxide nanoparticles (IONPs) have potential applications in crop improvement, but their negative impact on soil quality and microorganisms should be considered. The impact of high concentrations of IONPs on soil microorganisms, human health, and ecological safety should be thoroughly studied (Tao *et al.*, 2023). Research indicated that NPs can stimulate the production ROS and oxidative stress while boosting the synthesis of glutathione, protein, lipid, and molting protein. These effects led to adverse effects on the growth, reproductive health, and immune responses of aquatic organisms (Liu *et al.*, 2021). Therefore, the application dosage of NPs to crops should be optimized, and their long-term impact on ecosystems should be assessed. Careful optimization of NPs' concentration in various environmental compartments will mitigate their potential risks to a significant extent.

3.15 Conclusion and Future Prospectives of CRISPR/Cas9 and Nanotechnology in Developing Disease-resistant Crops

Both CRISPR and nanotechnology provide breakthrough advancements in agricultural biotechnology, particularly in developing disease-resistant crops. The advent of CRISPR/Cas9 genome editing has enabled researchers to accurately alter plant genomes, thereby enhancing their resistance to diseases, pests, and environmental stresses. Unlike traditional breeding, which can be time-consuming, CRISPR/Cas9 allowed for the swift incorporation of disease-resistant traits (Ali *et al.*, 2022). These technologies have empowered agricultural scientists to create crops with superior traits, like enhanced disease resistance and higher yields, thus ultimately reducing the need for harmful chemical pesticides and fertilizers (Akanmu *et al.*, 2024). As CRISPR/Cas9 technology continues to evolve, using cutting-edge CRISPR techniques such as base editing and prime editing, scientists can develop disease-resistant crops with inheritable traits, free from transgene (Liu *et al.*, 2022). Base editors provide a way to change individual bases in crop genomes without inducing double-strand breaks or engaging non-homologous end-joining (NHEJ) or homology-directed repair (HDR). This allowed the introduction of multiple amino acid substitutions, creating a variety of disease-resistant crops (Zhang *et al.*, 2019). By eliminating foreign DNA, genome-edited crops have a distinct commercialization edge over crops that are either conventionally mutated or genetically engineered (Lassoued *et al.*, 2019). Regulatory approaches will play a crucial role in the commercialization of genome-edited (GE) crops. The United States Department of Agriculture (USDA) does not regulate these crops, whereas in the European Union, GE crops are regulated as genetically modified organisms and face stringent regulatory hurdles (Jaganathan *et al.*, 2018; Callaway, 2018).

Multiplex genome editing is a new CRISPR advancement that permits the concurrent targeting of several genes through a single construct (Wei *et al.*, 2024). Luo *et al.* (2021b) reported the development of a genome-edited wheat variety by multiplex genome editing tool. Their work demonstrated the ability to simultaneously target and edit up to 15 genomic loci in hexaploid wheat using the rice Pol II promoter Actin to express the tandem repeats of tRNA-gRNA. Another study utilized Cas12a orthologs to perform multiplex genome editing at 16 different sites in rice. CRISPR-12a is a Class II, Type V endonuclease, and is the second most popular CRISPR system used for genome editing. This study successfully explores the effectiveness of CRISPR-12a orthologs for achieving multiplex genome editing in crop engineering (Zhang *et al.*, 2021). Several other studies have confirmed the high specificity and low frequency of off-target mutation associated with this technique. In a recent advancement, Transposase B (TnpB) has been explored which represents a next-generation genome editor and is much smaller than Cas9 and Cas12a. The TnpB has recently been successfully employed for plant genome editing with high efficiency (Karmakar *et al.*, 2024).

Nanotechnology, on the other hand, has shown enormous promise in the agricultural sector and offers novel approaches to enhance crop protection and productivity. NPs facilitate the efficient delivery of various micro- and macro-molecules-like nucleic acids, proteins, pesticides, and fertilizers – directly to targets, whereas nanosensors offer real-time insights into plant health and conditions (Singh *et al.*, 2024). Further investigation is needed to enhance and perfect the application of nano-technology in advanced genome editing. For more precise and efficient genetic engineering and targeted treatment, the refinement of NP transformation techniques, including gene gun applications, electromagnetic field approaches and nano-material-mediated delivery systems, is necessary. To optimize the use of NPs, it is crucial to gain deeper insights into their interaction with protoplasts and cells, and to elucidate the mechanisms of NPs absorption, transport and intracellular accumulation (Wu *et al.*, 2023). Thus, the synergistic development of CRISPR and nanotechnology could be crucial in addressing global food security issues, as it offers the potential to enhance crop productivity and resilience against climate variability and population growth.

References

Abbas, H. and Abou Baker, D. (2020) Biological evaluation of selenium nanoparticles biosynthesized by *Fusarium semitectum* as antimicrobial and anticancer agents. *Egyptian Journal of Chemistry* 63(4), 1119–1133.

Abdelkhalek, A. and Al-Askar, A.A. (2020) Green synthesized ZnO nanoparticles mediated by *Mentha spicata* extract induce plant systemic resistance against Tobacco mosaic virus. *Applied Sciences* 10(15), 5054.

Abdelkhalek, A., Yassin, Y., Abdel-Megeed, A., Abd-Elsalam, K.A., Moawad, H. *et al.* (2022) *Rhizobium leguminosarum* bv. *viciae*-mediated silver nanoparticles for controlling bean yellow mosaic virus (BYMV) infection in faba bean plants. *Plants* 12(1), 45.

Abdul-razzak, A.N.A.S., Guiraud, T., Peypelut, M., Walter, J., Houvenaghel, M. *et al.* (2009) Involvement of the cylindrical inclusion (CI) protein in the overcoming of an eIF4E-mediated resistance against lettuce mosaic potyvirus. *Molecular Plant Pathology* 10(1), 109–113.

Abubakar, A.L., Abarshi, M.M. and Maruthi, M.N. (2018) Testing the infectivity of a begomovirus by particle Bombardment method using a gene gun. *Nigerian Journal of Biotechnology* 35(2), 58.

Acanda, Y., Wang, C. and Levy, A. (2019) Gene expression in citrus plant cells using Helios® gene gun system for particle bombardment. In: Catara, A.F., Bar-Joseph, M. and Licciardello, G. (eds) *Citrus Tristeza Virus: Methods and Protocols*. Springer, pp. 219–228.

Adeel, M., Farooq, T., White, J.C., Hao, Y., He, Z. *et al.* (2021) Carbon-based nanomaterials suppress tobacco mosaic virus (TMV) infection and induce resistance in *Nicotiana Benthamiana*. *Journal of Hazardous Materials* 404, 124167.

Ahmad, S., Munir, S., Zeb, N., Ullah, A., Khan, B. *et al.* (2019) Green nanotechnology: A review on green synthesis of silver nanoparticles—an ecofriendly approach. *International Journal of Nanomedicine* 5087–5107.

Ahmad, S., Wei, X., Sheng, Z., Hu, P. and Tang, S. (2020) CRISPR/Cas9 for development of disease resistance in plants: Recent progress, limitations and future prospects. *Briefings in Functional Genomics* 19(1), 26–39.

Ahmad, S., Ahmad, S., Xu, Q., Khan, I., Cao, X. *et al.* (2024) Green synthesis of gold and silver nanoparticles using crude extract of *Aconitum violaceum* and evaluation of their antibacterial, antioxidant and photocatalytic activities. *Frontiers in Bioengineering and Biotechnology* 11, 1320739.

Ahmar, S., Mahmood, T., Fiaz, S., Mora-Poblete, F., Shafique, M.S. *et al.* (2021) Advantage of nanotechnology-based genome editing system and its application in crop improvement. *Frontiers in Plant Science* 12, 663849.

Ahmed, T., Wu, Z., Jiang, H., Luo, J., Noman, M. *et al.* (2021) Bioinspired green synthesis of zinc oxide nanoparticles from a native *Bacillus cereus* strain RNT6: Characterization and antibacterial activity against rice panicle blight pathogens *Burkholderia glumae* and *B. gladioli*. *Nanomaterials* 11(4), 884.

Ahmed, T., Noman, M., Jiang, H., Shahid, M., Ma, C. *et al.* (2022) Bioengineered chitosan-iron nanocomposite controls bacterial leaf blight disease by modulating plant defense response and nutritional status of rice (*Oryza sativa* L.). *Nano Today* 45, 101547.

Akanmu, A.O., Asemoloye, M.D., Marchisio, M.A. and Babalola, O.O. (2024) Adoption of CRISPR-Cas for crop production: Present status and future prospects. *PeerJ* 12, e17402.

Al Amin, N., Ahmad, N., Wu, N., Pu, X., Ma, T. *et al.* (2019) CRISPR-cas9 mediated targeted disruption of FAD2–2 microsomal omega-6 desaturase in soybean (*Glycine max*.L). *BMC Biotechnology* 19(1), 1–10.

Al-Askar, A.A., Aseel, D.G., El-Gendi, H., Sobhy, S., Samy, M.A. *et al.* (2023) Antiviral activity of biosynthesized silver nanoparticles from pomegranate (*Punica granatum* L.) peel extract against tobacco mosaic virus. *Plants* 12(11), 2103.

Algethami, N., Rajeh, A., Ragab, H.M., Tarabiah, A.E. and Gami, F. (2022) Characterization, optical, and electrical properties of chitosan/polyacrylamide blend doped silver nanoparticles. *Journal of Materials Science* 33(13), 10645–10656.

Ali, Z., Abulfaraj, A., Idris, A., Ali, S., Tashkandi, M. *et al.* (2015) CRISPR/Cas9-mediated viral interference in plants. *Genome Biology* 16, 1–11.

Ali, Z., Ali, S., Tashkandi, M., Zaidi, S.S.E.A. and Mahfouz, M.M. (2016) CRISPR/Cas9-mediated immunity to geminiviruses: Differential interference and evasion. *Scientific Reports* 6(1), 26912.

Ali, S., Ganai, B.A., Kamili, A.N., Bhat, A.A., Mir, Z.A. *et al.* (2018) Pathogenesis-related proteins and peptides as promising tools for engineering plants with multiple stress tolerance. *Microbiological Research* 212, 29–37.

Ali, M.A., Ahmed, T., Wu, W., Hossain, A., Hafeez, R. *et al.* (2020) Advancements in plant and microbe-based synthesis of metallic nanoparticles and their antimicrobial activity against plant pathogens. *Nanomaterials* 10(6), 1146.

Ali, Q., Yu, C., Hussain, A., Ali, M., Ahmar, S. *et al.* (2022) Genome engineering technology for durable disease resistance: Recent progress and future outlooks for sustainable agriculture. *Frontiers in Plant Science* 13, 860281.

Alotaibi, M.O., Alotaibi, N.M., Ghoneim, A.M., Ain, N.U., Irshad, M.A. *et al.* (2023) Effect of green synthesized cerium oxide nanoparticles on fungal disease of wheat plants: A field study. *Chemosphere* 339, 139731.

Alsaggaf, M.S., Diab, A.M., El Saied, B.E., Tayel, A.A. and Moussa, S.H. (2021) Application of ZnO nanoparticles phycosynthesized with *Ulva fasciata* extract for preserving peeled shrimp quality. *Nanomaterials* 11(2), 385.

Aman, R., Ali, Z., Butt, H., Mahas, A., Aljedaani, F. *et al.* (2018) RNA virus interference via CRISPR/Cas13a system in plants. *Genome Biology* 19(1), 1–9.

Amendola, V., Pilot, R., Frasconi, M., Maragò, O.M. and Iatì, M.A. (2017) Surface plasmon resonance in gold nanoparticles: A review. *Journal of Physics. Condensed Matter* 29(20), 203002.

Arsiya, F., Sayadi, M.H. and Sobhani, S. (2017) Green synthesis of palladium nanoparticles using *Chlorella vulgaris*. *Materials Letters* 186, 113–115.

Ashour, M.A. and Abd-Elhalim, B.T. (2024) Biosynthesis and biocompatibility evaluation of zinc oxide nanoparticles prepared using *Priestia megaterium* bacteria. *Scientific Reports* 14(1), 4147.

Atale, N., Saxena, S., Nirmala, J.G., Narendhirakannan, R.T., Mohanty, S. *et al.* (2017) Synthesis and characterization of *Sygyzium cumini* nanoparticles for its protective potential in high glucose-induced cardiac stress: A green approach. *Applied Biochemistry and Biotechnology* 181(3), 1140–1154.

Baltes, N.J., Hummel, A.W., Konecna, E., Cegan, R., Bruns, A.N. *et al.* (2015) Conferring resistance to geminiviruses with the CRISPR-Cas prokaryotic immune system. *Nature Plants* 1(10), 1–4.

Banerjee, A., Sarkar, A., Acharya, K. and Chakraborty, N. (2021) Nanotechnology: An emerging hope in crop improvement. *Letters in Applied NanoBioScience* 10(4), 2784–2803.

Benakashani, F., Allafchian, A.R. and Jalali, S.A.H. (2016) Biosynthesis of silver nanoparticles using *Capparis spinosa* L. leaf extract and their antibacterial activity. *Karbala International Journal of Modern Science* 2(4), 251–258.

Beyene, G., Chauhan, R.D., Villmer, J., Husic, N., Wang, N. *et al.* (2022) CRISPR/Cas9-mediated tetra-allelic mutation of the 'Green Revolution' SEMIDWARF-1 (SD-1) gene confers lodging resistance in tef (*Eragrostis tef*). *Plant Biotechnology Journal* 20(9), 1716–1729.

Bharose, A.A., Hajare, S.T., HP, G., Soni, M., Prajapati, K.K. *et al.* (2024) Bacteria-mediated green synthesis of silver nanoparticles and their antifungal potentials against *Aspergillus flavus*. *PLoS One* 19(3), e0297870.

Borrelli, V.M., Brambilla, V., Rogowsky, P., Marocco, A. and Lanubile, A. (2018) The enhancement of plant disease resistance using CRISPR/Cas9 technology. *Frontiers in Plant Science* 9, 407423.

Boruah, S. and Dutta, P. (2021) Fungus mediated biogenic synthesis and characterization of chitosan nanoparticles and its combine effect with *Trichoderma asperellum* against *Fusarium oxysporum, Sclerotium rolfsii* and *Rhizoctonia Solani*. *Indian Phytopathology* 74(1), 81–93.

Caf, F. (2022) Biogenic synthesis of iron oxide nanoparticle using *Padina pavonica* extract: Application for photocatalytic degradation of congo red dye, neurotoxicity and antioxidant activity. *Turkish Journal of Fisheries and Aquatic Sciences* 23(2).

Cai, Y., Chen, L., Liu, X., Guo, C., Sun, S. *et al.* (2018) CRISPR/Cas9-mediated targeted mutagenesis of *GmFT2a* delays flowering time in soya bean. *Plant Biotechnology Journal* 16(1), 176–185.

Cai, L., Cai, L., Jia, H., Liu, C., Wang, D. *et al.* (2020) Foliar exposure of Fe_3O_4 nanoparticles on Nicotiana benthamiana: Evidence for nanoparticles uptake, plant growth promoter and defense response elicitor against plant virus. *Journal of Hazardous Materials* 393(12), 122415.

Cai, L., Liu, C., Fan, G., Liu, C. and Sun, X. (2019) Preventing viral disease by ZnONPs through directly deactivating TMV and activating plant immunity in *Nicotiana benthamiana*. *Environmental Science: Nano* 6(12), 3653–3669.

Callaway, E. (2018) CRISPR plants now subject to tough GM laws in European Union. *Nature* 560(7716), 16–17.

Chalklen, T., Jing, Q. and Kar-Narayan, S. (2020) Biosensors based on mechanical and electrical detection techniques. *Sensors* 20(19), 5605.

Chandran, S.P., Chaudhary, M., Pasricha, R., Ahmad, A. and Sastry, M. (2006) Synthesis of gold nanotriangles and silver nanoparticles using *Aloe vera* plant extract. *Biotechnology Progress* 22(2), 577–583.

Chandrasekaran, J., Brumin, M., Wolf, D., Leibman, D., Klap, C. *et al.* (2016) Development of broad virus resistance in non-transgenic cucumber using CRISPR/Cas9 technology. *Molecular Plant Pathology* 17(7), 1140–1153.

Chen, A.A., Derfus, A.M., Khetani, S.R. and Bhatia, S.N. (2005) Quantum dots to monitor RNAi delivery and improve gene silencing. *Nucleic Acids Research* 33(22), e190–e190.

Chen, J., Wang, X. and Han, H. (2013) A new function of graphene oxide emerges: Inactivating phytopathogenic bacterium *Xanthomonas oryzae* pv. *oryzae*. *Journal of Nanoparticle Research* 15(5), 1–14.

Chen, G., Hu, J., Dong, L., Zeng, D., Guo, L. *et al.* (2019a) The tolerance of salinity in rice requires the presence of a functional copy of FLN2. *Biomolecules* 10(1), 17.

Chen, K., Wang, Y., Zhang, R., Zhang, H. and Gao, C. (2019b) CRISPR/Cas genome editing and precision plant breeding in agriculture. *Annual Review of Plant Biology* 70, 667–697.

Cheng, X., Chen, G. and Rodriguez, W.R. (2009) Micro- and nanotechnology for viral detection. *Analytical and Bioanalytical Chemistry* 393(2), 487–501.

Chowdhury, A.R., Kumar, R., Mahanty, A., Mukherjee, K., Kumar, S. *et al.* (2024) Inhibitory role of copper and silver nanocomposite on important bacterial and fungal pathogens in rice (*Oryza sativa*). *Scientific Reports* 14(1), 1779.

Clarance, P., Luvankar, B., Sales, J., Khusro, A., Agastian, P. *et al.* (2020) Green synthesis and characterization of gold nanoparticles using endophytic fungi *Fusarium solani* and its *in-vitro* anticancer and biomedical applications. *Saudi Journal of Biological Sciences* 27(2), 706–712.

Cunningham, F.J., Goh, N.S., Demirer, G.S., Matos, J.L. and Landry, M.P. (2018) Nanoparticle-mediated delivery towards advancing plant genetic engineering. *Trends in Biotechnology* 36(9), 882–897.

Das, A. and Dutta, P. (2021) Antifungal activity of biogenically synthesized silver and gold nanoparticles against sheath blight of rice. *Journal of Nanoscience and Nanotechnology* 21(6), 3547–3555.

Divya, K., Thampi, M., Vijayan, S., Varghese, S. and Jisha, M.S. (2020) Induction of defence response in *Oryza sativa* L. against *Rhizoctonia solani (Kuhn)* by chitosan nanoparticles. *Microbial Pathogenesis* 149, 104525.

Dong, B.R., Jiang, R., Chen, J.F., Xiao, Y., Lv, Z.Y. *et al.* (2023) Strategic nanoparticle-mediated plant disease resistance. *Critical Reviews in Biotechnology* 43(1), 22–37.

Dong, O.X., Yu, S., Jain, R., Zhang, N., Duong, P.Q. *et al.* (2020) Marker-free carotenoid-enriched rice generated through targeted gene insertion using CRISPR-Cas9. *Nature Communications* 11(1), 1178.

Dougan, J.A., Karlsson, C., Smith, W.E. and Graham, D. (2007) Enhanced oligonucleotide-nanoparticle conjugate stability using thioctic acid modified oligonucleotides. *Nucleic Acids Research* 35(11), 3668–3675.

Dutta, P., Kumari, A., Mahanta, M., Biswas, K.K., Dudkiewicz, A. *et al.* (2022) Advances in nanotechnology as a potential alternative for plant viral disease management. *Frontiers in Microbiology* 13, 935193.

Ebrahiminezhad, A., Zare-Hoseinabadi, A., Berenjian, A. and Ghasemi, Y. (2017) Green synthesis and characterization of zero-valent iron nanoparticles using stinging nettle (*Urtica dioica*) leaf extract. *Green Processing and Synthesis* 6(5), 469–475.

Edison, T.N.J.I. and Sethuraman, M.G. (2013) Electrocatalytic reduction of benzyl chloride by green synthesized silver nanoparticles using pod extract of *Acacia nilotica*. *ACS Sustainable Chemistry & Engineering* 1(10), 1326–1332.

El Gamal, A.Y., Tohamy, M.R., Abou-Zaid, M.I., Atia, M.M., El Sayed, T. *et al.* (2022) Silver nanoparticles as a viricidal agent to inhibit plant-infecting viruses and disrupt their acquisition and transmission by their aphid vector. *Archives of Virology* 167(1), 85–97.

El Sayed, M.T. and El-Sayed, A.S. (2020) Biocidal activity of metal nanoparticles synthesized by *Fusarium solani* against multidrug-resistant bacteria and mycotoxigenic fungi. *Journal of Microbiology and Biotechnology* 30(2), 226–236.

Elmer, W. and White, J.C. (2018) The future of nanotechnology in plant pathology. *Annual Review of Phytopathology* 56(1), 111–133.

El-Sheekh, M.M., El-Kassas, H.Y., Shams El-Din, N.G., Eissa, D.I., and El-Sherbiny, B.A. (2021) Green synthesis, characterization applications of iron oxide nanoparticles for antialgal and wastewater bioremediation using three brown algae. *International Journal of Phytoremediation* 23(14), 1538–1552.

Erikson, H., Sarapuu, A., and Tammeveski, K. (2019) Oxygen reduction reaction on silver catalysts in alkaline media: A minireview. *ChemElectroChem* 6(1), 73–86.

Eun, A.J.C., Huang, L., Chew, F.T., Li, S.F.Y. and Wong, S.M. (2002) Detection of two orchid viruses using quartz crystal microbalance (QCM) immunosensors. *Journal of Virological Methods* 99(1–2), 71–79.

Färkkilä, S.M.A., Mortimer, M., Jaaniso, R., Kahru, A., Kiisk, V. *et al.* (2023) Comparison of toxicity and cellular uptake of CdSe/ZnS and carbon quantum dots for molecular tracking using *Saccharomyces cerevisiae* as a fungal model. *Nanomaterials* 14(1), 10.

Fatima, R., Priya, M., Indurthi, L., Radhakrishnan, V. and Sudhakaran, R. (2020) Biosynthesis of silver nanoparticles using red algae *Portieria hornemannii* and its antibacterial activity against fish pathogens. *Microbial Pathogenesis* 138, 103780.

Feng, L., Xie, N. and Zhong, J. (2014) Carbon nanofibers and their composites: A review of synthesizing, properties and applications. *Materials* 7(5), 3919–3945.

Fesharaki, P.J., Nazari, P., Shakibaie, M., Rezaie, S., Banoee, M. *et al.* (2010) Biosynthesis of selenium nanoparticles using *Klebsiella pneumoniae* and their recovery by a simple sterilization process. *Brazilian Journal of Microbiology* 41(2), 461–466.

Fiorito, S., Serafino, A., Andreola, F., Togna, A. and Togna, G. (2006) Toxicity and biocompatibility of carbon nanoparticles. *Journal of Nanoscience and Nanotechnology* 6(3), 591–599.

Fister, A.S., Landherr, L., Maximova, S.N. and Guiltinan, M.J. (2018) Transient expression of CRISPR/Cas9 machinery targeting tcnpr3 enhances defense response in *Theobroma cacao*. *Frontiers in Plant Science* 9, 329023.

Folliero, V., Zannella, C., Chianese, A., Stelitano, D., Ambrosino, A. *et al.* (2021) Application of dendrimers for treating parasitic diseases. *Pharmaceutics* 13(3), 343.

Francis, D.V., Abdalla, A.K., Mahakham, W., Sarmah, A.K. and Ahmed, Z.F. (2024) Interaction of plants and metal nanoparticles: Exploring its molecular mechanisms for sustainable agriculture and crop improvement. *Environment International* 108859.

Frye, C.A., Tang, D. and Innes, R.W. (2001) Negative regulation of defense responses in plants by a conserved MAPKK kinase. *Proceedings of the National Academy of Sciences* 98(1), 373–378.

Gaba, S., Varma, A. and Goel, A. (2022) Protective and curative activity of biogenic copper oxide nanoparticles against Alternaria blight disease in oilseed crops: A review. *Journal of Plant Diseases and Protection* 129(2), 215–229.

Gaber, S.E., Hashem, A.H., El-Sayyad, G.S. and Attia, M.S. (2023) Antifungal activity of myco-synthesized bimetallic ZnO-CuO nanoparticles against fungal plant pathogen *Fusarium oxysporum*. *Biomass Conversion and Biorefinery* 14, 25395–25409.

Ghorbani, A., Astaraki, S., Rostami, M. and Pakdel, A. (2023) Unleashing the power of colloidal gold immunochromatographic assays for plant virus diagnostics. *MethodsX* 12, 102498.

Gohil, K., Dhakhda, S., Patel, V., Rathod, A. and Singh, P.K. (2024) Green synthesis of silver nanoparticles using various plant extracts and evaluation of their antimicrobial activities. *Biosciences Biotechnology Research Asia* 21(2).

González-Ballesteros, N. and Rodríguez-Argüelles, M.C. (2020) Seaweeds: A promising bionanofactory for ecofriendly synthesis of gold and silver nanoparticles. In: Torres, M.D., Kraan, S. and Dominguez, H. (eds) *Sustainable Seaweed Technologies*. Elsevier, pp. 507–541.

Gopinath, K., Gowri, S. and Arumugam, A. (2013) Phytosynthesis of silver nanoparticles using *Pterocarpus santalinus* leaf extract and their antibacterial properties. *Journal of Nanostructure in Chemistry* 3(1), 1–7.

Gul, M., Khan, R.S., Islam, Z.U., Khan, S., Shumaila, A. *et al.* (2024) Nanoparticles in plant resistance against bacterial pathogens: Current status and future prospects. *Molecular Biology Reports* 51(1), 92.

Guo, Y., Khan, R.A.A., Xiong, Y. and Fan, Z. (2022) Enhanced suppression of soil-borne phytopathogenic bacteria *Ralstonia solanacearum* in soil and promotion of tomato plant growth by synergetic effect of green synthesized nanoparticles and plant extract. *Journal of Applied Microbiology* 132(5), 3694–3704.

Hamida, R.S., Ali, M.A., Almohawes, Z.N., Alahdal, H., Momenah, M.A. *et al.* (2022) Green synthesis of hexagonal silver nanoparticles using a novel microalgae *Coelastrella aeroterrestrica* strain BA_Chlo4 and resulting anticancer, antibacterial, and antioxidant activities. *Pharmaceutics* 14(10), 2002.

Hebbalalu, D., Lalley, J., Nadagouda, M.N. and Varma, R.S. (2013) Greener techniques for the synthesis of silver nanoparticles using plant extracts, enzymes, bacteria, biodegradable polymers, and microwaves. *ACS Sustainable Chemistry & Engineering* 1(7), 703–712.

Hoang, A.S., Cong, H.H., Shukanov, V.P., Karytsko, L.A., Poljanskaja, S.N. *et al.* (2022) Evaluation of metal nano-particles as growth promoters and fungi inhibitors for cereal crops. *Chemical and Biological Technologies in Agriculture* 9(1), 12.

Hong, J., Wang, C., Wagner, D.C., Gardea-Torresdey, J.L., He, F. *et al.* (2021) Foliar application of nanoparticles: Mechanisms of absorption, transfer, and multiple impacts. *Environmental: Nano* 8(5), 1196–1210.

Hossain, A., Hong, X., Ibrahim, E., Li, B., Sun, G. *et al.* (2019) Green synthesis of silver nanoparticles with culture supernatant of a bacterium *Pseudomonas rhodesiae* and their antibacterial activity against soft rot pathogen *Dickeya dadantii*. *Molecules* 24(12), 2303.

Hossain, M.M., Hamza, A., Polash, S.A., Tushar, M.H., Takikawa, M. *et al.* (2024) Green synthesis of silver nanoparticles using *Phyllanthus emblica* extract: Investigation of antibacterial activity and biocompatibility *in vivo*. *RSC Pharmaceutics* 2024(1), 245–258.

Huang, J., Li, Q., Sun, D., Lu, Y., Su, Y. *et al.* (2007) Biosynthesis of silver and gold nanoparticles by novel sundried cinnamomum camphora leaf. *Nanotechnology* 18(10), 105104.

Hu, P., An, J., Faulkner, M.M., Wu, H., Li, Z. *et al.* (2020) Nanoparticle charge and size control foliar delivery efficiency to plant cells and organelles. *ACS Nano* 14(7), 7970–7986.

Huq, M.A., Khan, A.A., Alshehri, J.M., Rahman, M.S., Balusamy, S.R. *et al.* (2023) Bacterial mediated green synthesis of silver nanoparticles and their antibacterial and antifungal activities against drug-resistant pathogens. *Royal Society Open Science* 10(10), 230796.

Ibrahim, E., Zhang, M., Zhang, Y., Hossain, A., Qiu, W. *et al.* (2020) Green-synthesization of silver nanoparticles using endophytic bacteria isolated from garlic and its antifungal activity against wheat *fusarium* head blight pathogen *Fusarium graminearum*. *Nanomaterials* 10(2), 219.

Ibrahim, S., Saleem, B., Rehman, N., Zafar, S.A., Naeem, M.K. *et al.* (2022) CRISPR/Cas9 mediated disruption of inositol Pentakisphosphate 2-kinase 1 (*TaIPK1*) reduces phytic acid and improves iron and zinc accumulation in wheat grains. *Journal of Advanced Research* 37, 33–41.

Imada, K., Sakai, S., Kajihara, H., Tanaka, S. and Ito, S. (2016) Magnesium oxide nanoparticles induce systemic resistance in tomato against bacterial wilt disease. *Plant Pathology* 65(4), 551–560.

Imam, J., Singh, P.K. and Shukla, P. (2016) Plant microbe interactions in post genomic era: Perspectives and applications. *Frontiers in Microbiology* 7, 209205.

Iqtedar, M., Riaz, H., Kaleem, A., Abdullah, R., Aihetasham, A. *et al.* (2020) Biosynthesis, optimization and characterization of ZnO nanoparticles using *Bacillus cereus* MN181367 and their antimicrobial activity against multidrug resistant bacteria. *Revista Mexicana de Ingeniería Química* 19, 253–266.

Issaabadi, Z., Nasrollahzadeh, M. and Sajadi, S.M. (2017) Green synthesis of the copper nanoparticles supported on bentonite and investigation of its catalytic activity. *Journal of Cleaner Production* 142, 3584–3591.

Ivanova, N., Gugleva, V., Dobreva, M., Pehlivanov, I., Stefanov, S. *et al.* (2018) Silver nanoparticles as multi-functional drug delivery systems. In: Farrukh, M.A. (ed.) *Nanomedicines*. IntechOpen, pp. 71–92.

Jafarizad, A., Safaee, K., Gharibian, S., Omidi, Y. and Ekinci, D. (2015) Biosynthesis and *in-vitro* study of gold nanoparticles using *Mentha* and *Pelargonium* extracts. *Procedia Materials Science* 11, 224–230.

Jaganathan, D., Ramasamy, K., Sellamuthu, G., Jayabalan, S. and Venkataraman, G. (2018) CRISPR for crop improvement: An update review. *Frontiers in Plant Science* 9, 985.

Jain, D., Shivani, A.A., Singh, H., Daima, H.K., Singh, M. *et al.* (2020) Microbial fabrication of zinc oxide nanoparticles and evaluation of their antimicrobial and photocatalytic properties. *Frontiers in Chemistry* 8, 778.

Jalali, T., Ghanavati, F. and Osfouri, S. (2024) Green synthesis of ZnO nanoparticles using marine brown algae (*Cystoseira*) extract comprising sol-gel, and combustion techniques based on dye-sensitized solar cells application. *International Journal of Modern Physics B* 38(14), 2450178.

Jameel, M.S., Aziz, A.A., Dheyab, M.A., Khaniabadi, P.M., Kareem, A.A. *et al.* (2022) Mycosynthesis of ultrasonically-assisted uniform cubic silver nanoparticles by isolated phenols from *Agaricus bisporus* and its antibacterial activity. *Surfaces and Interfaces* 29, 101774.

Jamil, Y.M.S., Al-Hakimi, A.N., Al-Maydama, H.M.A., Almahwiti, G.Y., Qasem, A. *et al.* (2024) Optimum green synthesis, characterization, and antibacterial activity of silver nanoparticles prepared from an extract of *Aloe fleurentinorum*. *International Journal of Chemical Engineering* 2024(1), 2804165.

Ji, X., Zhang, H., Zhang, Y., Wang, Y. and Gao, C. (2015) Establishing a CRISPR-Cas-like immune system conferring DNA virus resistance in plants. *Nature Plants* 1(10), 1–4.

Jia, H., Orbovic, V., Jones, J.B. and Wang, N. (2016) Modification of the PthA4 effector binding elements in type I *CsLOB1* promoter using Cas9/sg RNA to produce transgenic Duncan grapefruit alleviating XccδpthA4: dcs LOB 1.3 infection. *Plant Biotechnology Journal* 14(5), 1291–1301.

Jiang, W., Zhou, H., Bi, H., Fromm, M., Yang, B. *et al.* (2013) Demonstration of CRISPR/Cas9/sgRNA-mediated targeted gene modification in *Arabidopsis*, tobacco, sorghum and rice. *Nucleic Acids Research* 41(20), e188–e188.

Jinek, M., Chylinski, K., Fonfara, I., Hauer, M., Doudna, J.A. *et al.* (2012) A programmable dual-RNA-guided DNA endonuclease in adaptive bacterial immunity. *Science* 337(6096), 816–821.

Kalishwaralal, K., Deepak, V., Pandian, S.R.K., Kottaisamy, M., BarathmaniKanth, S. *et al.* (2010) Biosynthesis of silver and gold nanoparticles using *Brevibacterium casei*. *Colloids and Surfaces B* 77(2), 257–262.

Kamel, S.M., Elgobashy, S.F., Omara, R.I., Derbalah, A.S., Abdelfatah, M. *et al.* (2022) Antifungal activity of copper oxide nanoparticles against root rot disease in cucumber. *Journal of Fungi* 8(9), 911.

Kang, M.G., Khan, F., Tabassum, N., Cho, K.J., Jo, D.M. *et al.* (2023) Inhibition of biofilm and virulence properties of pathogenic bacteria by silver and gold nanoparticles synthesized from *Lactiplantibacillus* sp. Strain C1. *ACS Omega* 8(11), 9873–9888.

Karmakar, A., Taufiqa, S., Baig, M.J. and Molla, K.A. (2022a) Increasing disease resistance in host plants through genome editing. *Proceedings of the Indian National Science Academy* 88(3), 417–429.

Karmakar, S., Das, P., Panda, D., Xie, K., Baig, M.J. *et al.* (2022b) A detailed landscape of CRISPR-Cas-mediated plant disease and pest management. *Plant Science* 323, 111376.

Karmakar, S., Panda, D., Panda, S., Dash, M., Saha, R. *et al.* (2024) A miniature alternative to Cas9 and Cas12: Transposon-associated tnpb mediates targeted genome editing in plants. *Plant Biotechnology Journal* 22(10), 2950–2953.

Kashyap, P., Shirkot, P., Das, R., Pandey, H. and Singh, D. (2023) Biosynthesis and characterization of copper nanoparticles from *Stenotrophomonas maltophilia* and its effect on plant pathogens and pesticide degradation. *Journal of Agriculture and Food Research* 13, 100654.

Kasithevar, M., Saravanan, M., Prakash, P., Kumar, H., Ovais, M. *et al.* (2017) Green synthesis of silver nanoparticles using *Alysicarpus monilifer* leaf extract and its antibacterial activity against MRSA and CoNS isolates in HIV patients. *Journal of Interdisciplinary Nanomedicine* 2(2), 131–141.

Kavitha, N.S., Venkatesh, K.S., Palani, N.S. and Ilangovan, R. (2020) Synthesis and characterization of zirconium oxide nanoparticles using fusarium solani extract. In: *AIP Conference Proceedings* 2265 (1).

Kaviya, S., Santhanalakshmi, J., Viswanathan, B., Muthumary, J. and Srinivasan, K. (2011) Biosynthesis of silver nanoparticles using citrus sinensis peel extract and its antibacterial activity. *Spectrochimica Acta. Part A, Molecular and Biomolecular Spectroscopy* 79(3), 594–598.

Kerdtoob, S., Chanthasena, P., Rosyidah, A., Limphirat, W., Penkhrue, W. *et al.* (2024) *Streptomyces monashensis* MSK03-mediated synthesis of gold nanoparticles: Characterization and antibacterial activity. *RSC Advances* 14(7), 4778–4787.

Khan, M., Khan, A.A., Parveen, A., Min, K., Yadav, V.K. *et al.* (2023) Mitigating the growth of plant pathogenic bacterium, fungi, and nematode by using plant-mediated synthesis of copper oxide nanoparticles (CuO NPs). *Green Chemistry Letters and Reviews* 16(1).

Kiew, P.L., Fauzi, N.A.M., Firdiani, S.A., Lam, L.K., Tan, L.S. *et al.* (2023) Iron oxide nanoparticles derived from *Chlorella vulgaris* extract: Characterization and crystal violet photodegradation studies. *Progress in Energy and Environment* 1–10.

Kim, Y.A., Moon, H., and Park, C.J. (2019) CRISPR/Cas9-targeted mutagenesis of Os8N3 in rice to confer resistance to *Xanthomonas oryzae* pv. *oryzae*. *Rice* 12(1), 13.

Klaine, S.J., Alvarez, P.J., Batley, G.E., Fernandes, T.F., Handy, R.D. *et al.* (2008) Nanomaterials in the environment: Behavior, fate, bioavailability, and effects. *Environmental Toxicology and Chemistry: An International Journal* 27(9), 1825–1851.

Klaus, T., Joerger, R., Olsson, E., and Granqvist, C.G. (1999) Silver-based crystalline nanoparticles, microbially fabricated. *Proceedings of the National Academy of Sciences* 96(24), 13611–13614.

Kora, A.J. and Rastogi, L. (2016) Catalytic degradation of anthropogenic dye pollutants using palladium nanoparticles synthesized by gum olibanum, a glucuronoarabinogalactan biopolymer. *Industrial Crops and Products* 81, 1–10.

Kumar, D.A., Palanichamy, V. and Roopan, S.M. (2014) Green synthesis of silver nanoparticles using alternanthera dentata leaf extract at room temperature and their antimicrobial activity. *Spectrochimica Acta. Part A, Molecular and Biomolecular Spectroscopy* 127, 168–171.

Kumar, B., Smita, K., Cumbal, L., Camacho, J., Hernández-Gallegos, E. *et al.* (2016a) One pot phytosynthesis of gold nanoparticles using *Genipa americana* fruit extract and its biological applications. *Materials Science and Engineering* 62, 725–731.

Kumar, V., Bano, D., Mohan, S., Singh, D.K. and Hasan, S.H. (2016b) Sunlight-induced green synthesis of silver nanoparticles using aqueous leaf extract of *Polyalthia longifolia* and its antioxidant activity. *Materials Letters* 181, 371–377.

Kuppusamy, P., Yusoff, M.M., Maniam, G.P. and Govindan, N. (2016) Biosynthesis of metallic nanoparticles using plant derivatives and their new avenues in pharmacological applications. An Updated Report. *Saudi Pharmaceutical Journal* 24(4), 473–484.

Kurczyńska, E., Godel-Jędrychowska, K., Sala, K. and Milewska-Hendel, A. (2021) Nanoparticles–plant interaction: What we know, where we are. *Applied Sciences* 11(12), 5473.

Landa, P., Dytrych, P., Prerostova, S., Petrova, S., Vankova, R. *et al.* (2017) Transcriptomic response of *Arabidopsis thaliana* exposed to CuO nanoparticles, bulk material, and ionic copper. *Environmental Science & Technology* 51(18), 10814–10824.

Larson, M.H., Gilbert, L.A., Wang, X., Lim, W.A., Weissman, J.S. *et al.* (2013) CRISPR interference (CRISPRi) for sequence-specific control of gene expression. *Nature Protocols* 8(11), 2180–2196.

Lassoued, R., Macall, D.M., Hesseln, H., Phillips, P.W. and Smyth, S.J. (2019) Benefits of genome-edited crops: Expert opinion. *Transgenic Research* 28(2), 247–256.

Lavanya, R. and Arun, V. (2021) Detection of Begomovirus in chilli and tomato plants using functionalized gold nanoparticles. *Scientific Reports* 11(1), 14203.

Lebaschi, S., Hekmati, M. and Veisi, H. (2017) Green synthesis of palladium nanoparticles mediated by black tea leaves (*Camellia sinensis*) extract: Catalytic activity in the reduction of 4-nitrophenol and Suzuki-Miyaura coupling reaction under ligand-free conditions. *Journal of Colloid and Interface Science* 485, 223–231.

Lee, H.J., Lee, G., Jang, N.R., Yun, J.H., Song, J.Y. *et al.* (2011) Biological synthesis of copper nanoparticles using plant extract. *Nanotechnology* 1(1), 371–374.

Lemire, J.A., Harrison, J.J. and Turner, R.J. (2013) Antimicrobial activity of metals: Mechanisms, molecular targets and applications. *Nature Reviews. Microbiology* 11(6), 371–384.

Li, S., Shen, L., Hu, P., Liu, Q., Zhu, X. *et al.* (2019) Developing disease-resistant thermosensitive male sterile rice by multiplex gene editing. *Journal of Integrative Plant Biology* 61(12), 1201–1205.

Li, G., Liu, X., Wang, H., Liang, S., Xia, B. *et al.* (2023) Detection, distribution and environmental risk of metal-based nanoparticles in a coastal bay. *Water Research* 242, 120242.

Li, Q., Feng, T., Li, H., Wang, Z., Wei, X. *et al.* (2024) Green synthesis of silver nanoparticles using endophytic bacterium *Bacillus zanthoxyli* GBE11 and their antimicrobial activity. *Biomass Conversion and Biorefinery* 14(12), 13173–13185.

Lin, Z., Ali, M.M., Yi, X., Zhang, L., Wang, S. *et al.* (2023) Unlocking the potential of carbon quantum dots for cell imaging, intracellular localization, and gene expression control in *Arabidopsis thaliana* (L.) *Heynh. International Journal of Molecular Sciences* 24(21), 15700.

Liu, S., Wei, L., Hao, L., Fang, N., Chang, M.W. *et al.* (2009) Sharper and faster 'nano darts' kill more bacteria: A study of antibacterial activity of individually dispersed pristine single-walled carbon nanotube. *ACS Nano* 3(12), 3891–3902.

Liu, Z., Li, Y., Sepúlveda, M.S., Jiang, Q., Jiao, Y. *et al.* (2021) Development of an adverse outcome pathway for nanoplastic toxicity in *Daphnia pulex* using proteomics. *Science of the Total Environment* 766, 144249.

Liu, H., Chen, W., Li, Y., Sun, L., Chai, Y. *et al.* (2022) CRISPR/Cas9 technology and its utility for crop improvement. *International Journal of Molecular Sciences* 23(18), 10442.

Loo, Y.Y., Chieng, B.W., Nishibuchi, M., and Radu, S. (2012) Synthesis of silver nanoparticles by using tea leaf extract from *Camellia sinensis*. *International Journal of Nanomedicine* 7, 4263–4267.

Luo, D., Wang, X., Burda, C. and Basilion, J.P. (2021a) Recent development of gold nanoparticles as contrast agents for cancer diagnosis. *Cancers* 13(8), 1825.

Luo, J., Li, S., Xu, J., Yan, L., Ma, Y. *et al.* (2021b) Pyramiding favorable alleles in an elite wheat variety in one generation by CRISPR-Cas9-mediated multiplex gene editing. *Molecular Plant* 14(6), 847–850.

Luu, D.D., Joe, A., Chen, Y., Parys, K., Bahar, O. *et al.* (2019) Biosynthesis and secretion of the microbial sulfated peptide RaxX and binding to the rice XA21 immune receptor. *Proceedings of the National Academy of Sciences* 116(17), 8525–8534.

Lv, Z., Jiang, R., Chen, J. and Chen, W. (2020) Nanoparticle-mediated gene transformation strategies for plant genetic engineering. *The Plant Journal: For Cell and Molecular Biology* 104(4), 880–891.

Ma, J., Chen, J., Wang, M., Ren, Y., Wang, S. *et al.* (2018) Disruption of OsSEFC3A increases the content of salicylic acid and induces plant defense responses in rice. *Journal of Experimental Botany* 69(5), 1051–1064.

Macovei, A., Sevilla, N.R., Cantos, C., Jonson, G.B., Slamet-Loedin, I. *et al.* (2018) Novel alleles of rice eIF4G generated by CRISPR/Cas9-targeted mutagenesis confer resistance to Rice tungro spherical virus. *Plant Biotechnology Journal* 16(11), 1918–1927.

Mageswari, A., Srinivasan, R., Subramanian, P., Ramesh, N. and Gothandam, K.M. (2016) Nanomaterials: Classification, biological synthesis and characterization. *Nanoscience in Food and Agriculture* 3, 31–71.

Mahfouze, H.A., El-Dougdoug, N.K. and Mahfouze, S.A. (2020) Virucidal activity of silver nanoparticles against *Banana bunchy top virus* (BBTV) in banana plants. *Bulletin of the National Research Centre* 44(1), 1–11.

Majumdar, R., Bag, B.G. and Ghosh, P. (2016) *Mimusops elengi* bark extract mediated green synthesis of gold nanoparticles and study of its catalytic activity. *Applied Nanoscience* 6(4), 521–528.

Malik, M.A.M., Haider, M.S., Zhai, Y., Khan, M.A.U. and Pappu, H.R. (2023) Towards developing resistance to chickpea chlorotic dwarf virus through CRISPR/Cas9-mediated gene editing using multiplexed gRNAs. *Journal of Plant Diseases and Protection* 130(1), 23–33.

Malik, M.A., Wani, A.H., Bhat, M.Y., Siddiqui, S., Alamri, S.A.M. *et al.* (2024) Fungal-mediated synthesis of silver nanoparticles: A novel strategy for plant disease management. *Frontiers in Microbiology* 15, 1399331.

Mallikarjuna, K., Narasimha, G., Dillip, G.R., Praveen, B., Shreedhar, B. *et al.* (2011) Green synthesis of silver nanoparticles using *Ocimum* leaf extract and their characterization. *Digest Journal of Nanomaterials and Biostructures* 6(1), 181–186.

Malnoy, M., Viola, R., Jung, M.-H., Koo, O.-J., Kim, S. *et al.* (2016) DNA-free genetically edited grapevine and apple protoplast using CRISPR/Cas9 ribonucleoproteins. *Frontiers in Plant Science* 7, 1904.

Miao, C., Xiao, L., Hua, K., Zou, C., Zhao, Y. *et al.* (2018) Mutations in a subfamily of abscisic acid receptor genes promote rice growth and productivity. *Proceedings of the National Academy of Sciences* 115(23), 6058–6063.

Molla, K.A. and Yang, Y. (2019) CRISPR/Cas-mediated base editing: Technical considerations and practical applications. *Trends in Biotechnology* 37(10), 1121–1142.

Molla, K.A. and Yang, Y. (2020) Predicting CRISPR/Cas9-induced mutations for precise genome editing. *Trends in Biotechnology* 38(2), 136–141.

Molla, K.A., Shih, J. and Yang, Y. (2020) Single-nucleotide editing for zebra3 and Wsl5 phenotypes in rice using CRISPR/Cas9-mediated adenine base editors. *Abiotech* 1(2), 106–118.

Molla, K.A., Sretenovic, S., Bansal, K.C. and Qi, Y. (2021) Precise plant genome editing using base editors and prime editors. *Nature Plants* 7(9), 1166–1187.

Mondal, S., Pramanik, K., Pal, P., Mitra, S., Ghosh, S.K. *et al.* (2023) Multifaceted roles of root exudates in light of plant-microbe interaction. In: Chandra, D. and Bhatt, P. (eds) *Unravelling Plant-Microbe Synergy*. Academic Press, pp. 49–76.

Mortezagholi, B., Movahed, E., Fathi, A., Soleimani, M., Forutan Mirhosseini, A. *et al.* (2022) Plant-mediated synthesis of silver-doped zinc oxide nanoparticles and evaluation of their antimicrobial activity against bacteria cause tooth decay. *Microscopy Research and Technique* 85(11), 3553–3564.

Mourdikoudis, S., Pallares, R.M. and Thanh, N.T. (2018) Characterization techniques for nanoparticles: Comparison and complementarity upon studying nanoparticle properties. *Nanoscale* 10(27), 12871–12934.

Muniyappan, N. and Nagarajan, N.S. (2014) Green synthesis of silver nanoparticles with *Dalbergia spinosa* leaves and their applications in biological and catalytic activities. *Process Biochemistry* 49(6), 1054–1061.

Mushtaq, M. and Molla, K.A. (2021) CRISPR technologies for plant biotechnology innovation. In: Ricroch, A., Chopra, S. and Kuntz, M. (eds) *Plant Biotechnology: Experience and Future Prospects*. Springer, Cham, Switzerland, pp. 51–67.

Nagajyothi, P.C., Muthuraman, P., Sreekanth, T.V.M., Kim, D.H. and Shim, J. (2017) Green synthesis: *In-vitro* anticancer activity of copper oxide nanoparticles against human cervical carcinoma cells. *Arabian Journal of Chemistry* 10(2), 215–225.

Nagy, E.D., Stevens, J.L., Yu, N., Hubmeier, C.S. and LaFaver, N. (eds) (2021) Novel disease resistance gene paralogs created by CRISPR/Cas9 in soy. *Plant Cell Reports* 40(6), 1047–1058.

Naika, H.R., Lingaraju, K., Manjunath, K., Kumar, D., Nagaraju, G. *et al.* (2015) Green synthesis of CuO nanoparticles using *Gloriosa superba* L. extract and their antibacterial activity. *Journal of Taibah University for Science* 9(1), 7–12.

Nair, B. and Pradeep, T. (2002) Coalescence of nanoclusters and formation of submicron crystallites assisted by *Lactobacillus* strains. *Crystal Growth & Design* 2(4), 293–298.

Namburi, K.R., Kora, A.J., Chetukuri, A., and Kota, V.S.M.K. (2021) Biogenic silver nanoparticles as an antibacterial agent against bacterial leaf blight causing rice phytopathogen *Xanthomonas oryzae* pv. *oryzae*. *Bioprocess and Biosystems Engineering* 44(9), 1975–1988.

Narasimhamurthy, K., Udayashankar, A.C., De Britto, S., Lavanya, S.N., Abdelrahman, M. *et al.* (2022) Chitosan and chitosan-derived nanoparticles modulate enhanced immune response in tomato against bacterial wilt disease. *International Journal of Biological Macromolecules* 220, 223–237.

Nekrasov, V., Staskawicz, B., Weigel, D., Jones, J.D., and Kamoun, S. (2013) Targeted mutagenesis in the model plant *Nicotiana benthamiana* using Cas9 RNA-guided endonuclease. *Nature Biotechnology* 31(8), 691–693.

Nekrasov, V., Wang, C., Win, J., Lanz, C., Weigel, D. *et al.* (2017) Rapid generation of a transgene-free powdery mildew resistant tomato by genome deletion. *Scientific Reports* 7(1), 482.

Ning, L., Zhu, B. and Gao, T. (2017) Gold nanoparticles: Promising agent to improve the diagnosis and therapy of cancer. *Current Drug Metabolism* 18(11), 1055–1067.

Ning, W., Luo, X., Zhang, Y., Tian, P., Xiao, Y. *et al.* (2024) Broad-spectrum nano-bactericide utilizing antimicrobial peptides and bimetallic Cu-Ag nanoparticles anchored onto multiwalled carbon nanotubes for sustained protection against persistent bacterial pathogens in crops. *International Journal of Biological Macromolecules* 265, 131042.

Noha, K., Bondok, A.M. and El-Dougdoug, K.A. (2018) Evaluation of silver nanoparticles as antiviral agent against ToMV and PVY in tomato plants. *Sciences* 8(1), 100–111.

Noor, S., Shah, Z., Javed, A., Ali, A., Hussain, S.B. *et al.* (2020) A fungal based synthesis method for copper nanoparticles with the determination of anticancer, antidiabetic and antibacterial activities. *Journal of Microbiological Methods* 174, 105966.

Oliva, R., Ji, C., Atienza-Grande, G., Huguet-Tapia, J.C., Perez-Quintero, A. *et al.* (2019) Broad-spectrum resistance to bacterial blight in rice using genome editing. *Nature Biotechnology* 37(11), 1344–1350.

Ortigosa, A., Gimenez-Ibanez, S., Leonhardt, N. and Solano, R. (2019) Design of a bacterial speck resistant tomato by CRISPR/Cas9-mediated editing of Sl JAZ2. *Plant Biotechnology Journal* 17(3), 665–673.

Pallath, N., Francis, B., Devanesan, S., Asemi, N., AlSalhi, M.S. *et al.* (2024) Synthesis of silver nanoparticles from marine bacteria and evaluation of antimicrobial, antifungal and cytotoxic effects. *Journal of King Saud University – Science* 36(2), 103073.

Parada, J., Rubilar, O., Fernández-Baldo, M.A., Bertolino, F.A., Durán, N, *et al.* (2019) The nanotechnology among us: Are metal and metal oxides nanoparticles a nano or mega risk for soil microbial communities? *Critical Reviews in Biotechnology* 39(2), 157–172.

Paraguay-Delgado, F., Hermida-Montero, L.A., Morales-Mendoza, J.E., Durán-Barradas, Z., Mtz-Enriquez, A.I. *et al.* (2022) Photocatalytic properties of cu-containing ZnO nanoparticles and their antifungal activity against agriculture-pathogenic fungus. *RSC Advances* 12(16), 9898–9908.

Paszkowski, J., Baur, M., Bogucki, A. and Potrykus, I. (1988) Gene targeting in plants. *The EMBO Journal* 7(13), 4021–4026.

Paul, B., Bhuyan, B., Purkayastha, D.D. and Dhar, S.S. (2016) Photocatalytic and antibacterial activities of gold and silver nanoparticles synthesized using biomass of *Parkia roxburghii* leaf. *Journal of Photochemistry and Photobiology B: Biology* 154, 1–7.

Pedone, D., Moglianetti, M., De Luca, E., Bardi, G. and Pompa, P.P. (2017) Platinum nanoparticles in nanobiomedicine. *Chemical Society Reviews* 46(16), 4951–4975.

Peng, Z. and Yang, H. (2009) Designer platinum nanoparticles: Control of shape, composition in alloy, nanostructure and electrocatalytic property. *Nano Today* 4(2), 143–164.

Peng, A., Chen, S., Lei, T., Xu, L., He, Y. *et al.* (2017) Engineering canker-resistant plants through CRISPR/Cas9-targeted editing of the susceptibility gene *CsLOB*1 promoter in citrus. *Plant Biotechnology Journal* 15(12), 1509–1519.

Philip, D. (2010) Green synthesis of gold and silver nanoparticles using *Hibiscus rosa sinensis*. *Physica E: Low-Dimensional Systems and Nanostructures* 42(5), 1417–1424.

Prabhu, D., Arulvasu, C., Babu, G., Manikandan, R. and Srinivasan, P. (2013) Biologically synthesized green silver nanoparticles from leaf extract of *Vitex negundo* L. induce growth-inhibitory effect on human colon cancer cell line HCT15. *Process Biochemistry* 48(2), 317–324.

Prihatna, C., Barbetti, M.J. and Barker, S.J. (2018) A novel tomato fusarium wilt tolerance gene. *Frontiers in Microbiology* 9, 1226.

Pyott, D.E., Sheehan, E. and Molnar, A. (2016) Engineering of CRISPR/Cas9-mediated potyvirus resistance in transgene-free *Arabidopsis* plants. *Molecular Plant Pathology* 17(8), 1276–1288.

Rad, F., Mohsenifar, A., Tabatabaei, M., Safarnejad, M.R., Shahryari, F. *et al.* (2012) Detection of *Candidatus Phytoplasma Aurantifolia* with a quantum dots fret-based biosensor. *Journal of Plant Pathology* 525–534.

Rajeshkumar, S., Malarkodi, C., Paulkumar, K., Vanaja, M., Gnanajobitha, G. *et al.* (2013) Intracellular and extracellular biosynthesis of silver nanoparticles by using marine bacteria vibrio alginolyticus. *Nanosci Nanotechnol* 3(1), 21–25.

Rastogi, A., Zivcak, M., Sytar, O., Kalaji, H.M., He, X. *et al.* (2017) Impact of metal and metal oxide nanoparticles on plant: A critical review. *Frontiers in Chemistry* 5, 78.

Ray, P., Sahu, D., Aminedi, R. and Chandran, D. (2022) Concepts and considerations for enhancing RNAi efficiency in phytopathogenic fungi for RNAi-based crop protection using nanocarrier-mediated dsRNA delivery systems. *Frontiers in Fungal Biology* 3, 977502.

Reddy, N.J., Nagoor Vali, D., Rani, M. and Rani, S.S. (2014) Evaluation of antioxidant, antibacterial and cytotoxic effects of green synthesized silver nanoparticles by *Piper longum* fruit. *Materials Science and Engineering: C* 34, 115–122.

Rodríguez-Serrano, C., Guzmán-Moreno, J., Ángeles-Chávez, C., Rodríguez-González, V., Ortega-Sigala, J.J. *et al.* (2020) Biosynthesis of silver nanoparticles by *Fusarium scirpi* and its potential as antimicrobial agent against uropathogenic escherichia coli biofilms. *PLoS One* 15(3), e0230275.

Roy, T., Basak, N., Mainak, S., Das, S., Ali, S.I. *et al.* (2024) *Burkholderia* sp. EIKU21 mediated synthesis of biogenic ZnO nanoparticle-based pigment for development of antibacterial cotton fabric through nanocoating. *Biomass Conversion and Biorefinery* 1–16.

Ruikar, S.S., Masurkar, S.A. and Ghatage, A.A. (2024) CRISPR-Cas9 in gene therapy: Ethical, legal, and social implications. *Naturalista Campano* 28(1), 183–188.

Saeed, S., Iqbal, A. and Ashraf, M.A. (2020) Bacterial-mediated synthesis of silver nanoparticles and their significant effect against pathogens. *Environmental Science and Pollution Research* 27(30), 37347–37356.

Safarpour, H., Safarnejad, M.R., Tabatabaei, M., Mohsenifar, A., Rad, F. *et al.* (2012) Development of a quantum dots FRET-based biosensor for efficient detection of *Polymyxa betae*. *Canadian Journal of Plant Pathology* 34(4), 507–515.

Salem, D.M., Ismail, M.M. and Tadros, H.R. (2020) Evaluation of the antibiofilm activity of three seaweed species and their biosynthesized iron oxide nanoparticles (Fe_3O_4-NPs). *Egyptian Journal of Aquatic Research* 46(4), 333–339.

Sander, J.D., Dahlborg, E.J., Goodwin, M.J., Cade, L., Zhang, F. *et al.* (2011) Selection-free zinc-finger-nuclease engineering by context-dependent assembly (CoDA). *Nature Methods* 8(1), 67–69.

Sankar, R., Karthik, A., Prabu, A., Karthik, S., Shivashangari, K.S. *et al.* (2013) *Origanum vulgare* mediated biosynthesis of silver nanoparticles for its antibacterial and anticancer activity. *Colloids and Surfaces B: Biointerfaces* 108(1), 80–84.

Santhoshkumar, T., Rahuman, A.A., Rajakumar, G., Marimuthu, S., Bagavan, A. *et al.* (2011) Synthesis of silver nanoparticles using *Nelumbo nucifera* leaf extract and its larvicidal activity against malaria and filariasis vectors. *Parasitology Research* 108(3), 693–702.

Santillán Martínez, M.I., Bracuto, V., Koseoglou, E., Appiano, M., Jacobsen, E. *et al.* (2020) CRISPR/Cas9-targeted mutagenesis of the tomato susceptibility gene *PMR4* for resistance against powdery mildew. *BMC Plant Biology* 20(1), 284.

Sathyavathi, R., Krishna, M.B., Rao, S.V., Saritha, R. and Rao, D.N. (2010) Biosynthesis of silver nanoparticles using *Coriandrum sativum* leaf extract and their application in nonlinear optics. *Advanced Science Letters* 3(2), 138–143.

Savitha, R. and Saraswathi, U. (2016) A study on the preventive effect of silver nano particles synthesized from *Millingtonia hortensis* in isoproterenol induced cardio toxicity in male wistar rats. *World Journal of Pharmacological Science* 5(8), 1442–1450.

Saxena, S., Sushmita, L.C., Babu, Y.M., Chavan, D.D., Venu, E. *et al.* (2024) Commercialization of CRISPR-edited crops: Opportunities and challenges. *CRISPRized Horticulture Crops* 471–486.

Seetharaman, P.K., Chandrasekaran, R., Periakaruppan, R., Gnanasekar, S., Sivaperumal, S. *et al.* (2021) Functional attributes of myco-synthesized silver nanoparticles from endophytic fungi: A new implication in biomedical applications. *Biology* 10(6), 473.

Shafie, R.M., Salama, A.M. and Farroh, K.Y. (2018) Silver nanoparticles activity against tomato spotted wilt virus. *Middle East Journal of Agricultural Research* 7, 1251–1267.

Shanmugam, R., Munusamy, T., Jayakodi, S., Al-Ghanim, K.A., Nicoletti, M. *et al.* (2023) Probiotic-bacteria (*Lactobacillus fermentum*)-wrapped zinc oxide nanoparticles: Biosynthesis, characterization, and antibacterial activity. *Fermentation* 9(5), 413.

Shao, M., Xiong, G. and He, G. (2023) Bacterial synthesis of metallic nanoparticles for biomedical applications. *Materials Chemistry Horizons* 2(4), 303–313.

Sharma, A., Sagar, A., Rana, J. and Rani, R. (2022) Green synthesis of silver nanoparticles and its antibacterial activity using fungus *Talaromyces purpureogenus* isolated from *Taxus baccata* linn. *Micro and Nano Systems Letters* 10(1), 2.

Sharmila, G., Farzana Fathima, M., Haries, S., Geetha, S., Manoj Kumar, N. *et al.* (2017) Green synthesis, characterization and antibacterial efficacy of palladium nanoparticles synthesized using *Filicium decipiens* leaf extract. *Journal of Molecular Structure* 1138, 35–40.

Silva, G., Poirot, L., Galetto, R., Smith, J., Montoya, G. *et al.* (2011) Meganucleases and other tools for targeted genome engineering: Perspectives and challenges for gene therapy. *Current Gene Therapy* 11(1), 11–27.

Singh, A., Singh, N.B., Hussain, I., Singh, H. and Singh, S.C. (2015) Plant-nanoparticle interaction: An approach to improve agricultural practices and plant productivity. *International Journal of Pharmacological Science Invention* 4(8), 25–40.

Singh, J., Dutta, T., Kim, K.H., Rawat, M., Samddar, P. *et al.* (2018) 'Green' synthesis of metals and their oxide nanoparticles: Applications for environmental remediation. *Journal of Nanobiotechnology* 16(1), 84.

Singh, A., Rajput, V., Varshney, A., Sharma, R., Ghazaryan, K. *et al.* (2024) Revolutionizing crop production: Nanoscale wonders – current applications, advances, and future frontiers. *Egyptian Journal of Soil Science* 64(1).

Slaymaker, I.M., Gao, L., Zetsche, B., Scott, D.A., Yan, W.X. *et al.* (2016) Rationally engineered Cas9 nucleases with improved specificity. *Science* 351(6268), 84–88.

Soares, C., Pereira, R. and Fidalgo, F. (2018) Metal-based nanomaterials and oxidative stress in plants: Current aspects and overview. In: Faisal, M., Saquib, Q., Alatar, A.A. and Al-Khedhairy, A.A. (eds) *Phytotoxicity of Nanoparticles*. Springer, pp. 197–227.

Srivastava, S.K. and Constanti, M. (2012) Room temperature biogenic synthesis of multiple nanoparticles (Ag, Pd, Fe, Rh, Ni, Ru, Pt, Co, and Li) by *Pseudomonas aeruginosa* SM1. *Journal of Nanoparticle Research* 14(4), 1–10.

Srivastava, P., Bragança, J., Ramanan, S.R. and Kowshik, M. (2013) Synthesis of silver nanoparticles using haloarchaeal isolate *Halococcus salifodinae* BK 3. *Extremophiles* 17(5), 821–831.

Strungaru, S.-A., Plavan, G., Ciobica, A., Nicoara, M., Robea, M.A. *et al.* (2018) Acute exposure to gold induces fast changes in social behavior and oxidative stress of zebrafish (*Danio rerio*). *Journal of Trace Elements in Medicine and Biology* 50, 249–256.

Suba, S., Vijayakumar, S., Vidhya, E., Punitha, V.N. and Nilavukkarasi, M. (2021) Microbial mediated synthesis of ZnO nanoparticles derived from *Lactobacillus* spp: Characterizations, antimicrobial and biocompatibility efficiencies. *Sensors International* 2, 100104.

Sweet, M.J., Chesser, A. and Singleton, I. (2012) Review: Metal-based nanoparticles: Size, function, and areas for advancement in applied microbiology. *Advances in Applied Microbiology* 80, 113.

Talebian, S., Shahnavaz, B., Nejabat, M., Abolhassani, Y. and Rassouli, F.B. (2023) Bacterial-mediated synthesis and characterization of copper oxide nanoparticles with antibacterial, antioxidant, and anticancer potentials. *Frontiers in Bioengineering and Biotechnology* 11, 1140010.

Tao, Z., Zhou, Q., Zheng, T., Mo, F. and Ouyang, S. (2023) Iron oxide nanoparticles in the soil environment: Adsorption, transformation, and environmental risk. *Journal of Hazardous Materials* 459, 132107.

Taylor, T.M., Weiss, J., Davidson, P.M. and Bruce, B.D. (2005) Liposomal nanocapsules in food science and agriculture. *Critical Reviews in Food Science and Nutrition* 45(7–8), 587–605.

Thomazella, P., Brail, Q., Dahlbeck, D. and Staskawicz, B. (2016) CRISPR-Cas9 mediated mutagenesis of a DMR6 ortholog in tomato confers broad-spectrum disease resistance. *BioRxiv*.

Tian, Y., Luo, J., Wang, H., Zaki, H.E.M., Yu, S. *et al.* (2022) Bioinspired green synthesis of silver nanoparticles using three plant extracts and their antibacterial activity against rice bacterial leaf blight pathogen *Xanthomonas oryzae* pv.. *Plants* 11(21), 2892.

Tiki, Y.L., Tolesa, L.D., Tiwikrama, A.H. and Chala, T.F. (2024) Ginger (*Zingiber officinale*)-mediated green synthesis of silver-doped tin oxide nanoparticles and evaluation of its antimicrobial activity. *ACS Omega* 9(10), 11443–11452.

Tortella, G., Rubilar, O., Pieretti, J.C., Fincheira, P., de Melo Santana, B. *et al.* (2023) Nanoparticles as a promising strategy to mitigate biotic stress in agriculture. *Antibiotics* 12(2), 338.

Tripathi, J.N., Ntui, V.O., Ron, M., Muiruri, S.K., Britt, A. *et al.* (2019) CRISPR/Cas9 editing of endogenous banana streak virus in the B genome of *Musa* spp. overcomes a major challenge in banana breeding. *Communications Biology* 2(1).

Turunc, E., Binzet, R., Gumus, I., Binzet, G. and Arslan, H. (2017) Green synthesis of silver and palladium nanoparticles using *Lithodora hispidula* (Sm.) Griseb. (*Boraginaceae*) and application to the electrocatalytic reduction of hydrogen peroxide. *Materials Chemistry and Physics* 202, 310–319.

Tyagi, S., Kumar, R., Kumar, V., Won, S.Y. and Shukla, P. (2021) Engineering disease resistant plants through CRISPR-Cas9 technology. *GM Crops & Food* 12(1), 125–144.

Usman, B., Nawaz, G., Zhao, N., Liao, S., Qin, B. *et al.* (2020) Programmed editing of rice (*Oryza sativa* L.) *OsSLP16* gene using CRISPR/Cas9 improves grain yield by modulating the expression of pyruvate enzymes and cell cycle proteins. *International Journal of Molecular Sciences* 22(1), 249.

Valkonen, J.P.T. (2014) The challenge of controlling plant viruses. *Annals of Applied Biology* 164(3), 313–317.

Vargas-Hernandez, M., Macias-Bobadilla, I., Guevara-Gonzalez, R.G., Rico-Garcia, E., Ocampo-Velazquez, R.V. *et al.* (2020) Nanoparticles as potential antivirals in agriculture. *Agriculture* 10(10), 444.

Vasanth, K., Ilango, K., MohanKumar, R., Agrawal, A. and Dubey, G.P. (2014) Anticancer activity of *Moringa oleifera* mediated silver nanoparticles on human cervical carcinoma cells by apoptosis induction. *Colloids and Surfaces B: Biointerfaces* 117, 354–359.

Vasanth, V., Murugesh, K.A. and Susikaran, S. (2022) Synthesis of titanium dioxide nanoparticles using *Spirulina platensis* algae extract. *The Pharma Innovation* 266–269.

Vimala, R.T.V., Sathishkumar, G. and Sivaramakrishnan, S. (2015) Optimization of reaction conditions to fabricate nano-silver using *Couroupita guianensis Aubl*. (leaf & fruit) and its enhanced larvicidal effect. *Spectrochimica Acta. Part A, Molecular and Biomolecular Spectroscopy* 135, 110–115.

Waghmode, P., Pitha, C.C. and Banka, V. (2024) The role of genome editing technologies in sustainable agriculture. *Journal of Advances in Biology & Biotechnology* 27(8), 543–558.

Wang, X., Liu, L.H., Ramström, O. and Yan, M. (2009) Engineering nanomaterial surfaces for biomedical applications. *Experimental Biology and Medicine* 234(10), 1128–1139.

Wang, X., Liu, X. and Han, H. (2013) Evaluation of antibacterial effects of carbon nanomaterials against copper-resistant *Ralstonia solanacearum*. *Colloids and Surfaces B: Biointerfaces* 103, 136–142.

Wang, X., Liu, X., Chen, J., Han, H. and Yuan, Z. (2014a) Evaluation and mechanism of antifungal effects of carbon nanomaterials in controlling plant fungal pathogen. *Carbon* 68, 798–806.

Wang, Y., Cheng, X., Shan, Q., Zhang, Y., Liu, J. *et al.* (2014b) Simultaneous editing of three homoeoalleles in hexaploid bread wheat confers heritable resistance to powdery mildew. *Nature Biotechnology* 32(9), 947–951.

Wang, F., Wang, C., Liu, P., Lei, C., Hao, W. *et al.* (2016) Enhanced rice blast resistance by CRISPR/Cas9-targeted mutagenesis of the ERF transcription factor gene OsERF922. *PLoS One* 11(4), e0154027.

Wang, X., Tu, M., Wang, D., Liu, J., Li, Y. *et al.* (2018) CRISPR/Cas9-mediated efficient targeted mutagenesis in grape in the first generation. *Plant Biotechnology Journal* 16(4), 844–855.

Wang, J.W., Grandio, E.G., Newkirk, G.M., Demirer, G.S., Butrus, S. *et al.* (2019) Nanoparticle-mediated genetic engineering of plants. *Molecular Plant* 12(8), 1037–1040.

Wang, Q., Zhang, P., Zhao, W., Li, Y., Jiang, Y. *et al.* (2023) Interplay of metal-based nanoparticles with plant rhizosphere microenvironment: Implications for nanosafety and nano-enabled sustainable agriculture. *Environmental Science* 10(2), 372–392.

Wei, Y., Zhang, H., Fan, J., Cai, Q., Zhang, Z. *et al.* (2024) Multiplex-genome-editing based rapid directional improvement of complex traits in rice. *Plant Biotechnology Journal* 22(9), 2624–2628.

Wu, K., Xu, C., Li, T., Ma, H., Gong, J. *et al.* (2023) Application of nanotechnology in plant genetic engineering. *International Journal of Molecular Sciences* 24(19), 14836.

Xie, K. and Yang, Y. (2013) RNA-guided genome editing in plants using a CRISPR-Cas system. *Molecular Plant* 6(6), 1975–1983.

Xin Lee, K., Shameli, K., Miyake, M., Kuwano, N., Bt Ahmad Khairudin, N.B. *et al.* (2016) Green synthesis of gold nanoparticles using aqueous extract of *Garcinia mangostana* fruit peels. *Journal of Nanomaterials* 2016(1), 1–7.

Xu, Z., Xu, X., Gong, Q., Li, Z., Li, Y. *et al.* (2019) Engineering broad-spectrum bacterial blight resistance by simultaneously disrupting variable TALE-binding elements of multiple susceptibility genes in rice. *Molecular Plant* 12(11), 1434–1446.

Xu, X., Yuan, Y., Feng, B. and Deng, W. (2020) CRISPR/Cas9-mediated gene-editing technology in fruit quality improvement. *Food Quality and Safety* 4(4), 159–166.

Xu, F., Li, Y., Zhao, X., Liu, G., Pang, B. *et al.* (2024) Diversity of fungus-mediated synthesis of gold nanoparticles: Properties, mechanisms, challenges, and solving methods. *Critical Reviews in Biotechnology* 44(5), 924–940.

Yan, Y., Zhu, X., Yu, Y., Li, C., Zhang, Z. *et al.* (2022) Nanotechnology strategies for plant genetic engineering. *Advanced Materials* 34(7), 2106945.

Yazdi, M.H., Sepehrizadeh, Z., Mahdavi, M., Shahverdi, A.R. and Faramarzi, M.A. (2016) Metal, metalloid, and oxide nanoparticles for therapeutic and diagnostic oncology. *Nano Biomedicine and Engineering* 8(4).

Yu, Z., Li, Q., Wang, J., Yu, Y., Wang, Y. *et al.* (2020) Reactive oxygen species-related nanoparticle toxicity in the biomedical field. *Nanoscale Research Letters* 15(1), 115.

Yuan, H., Hu, S., Huang, P., Song, H., Wang, K. *et al.* (2011) Single walled carbon nanotubes exhibit dual-phase regulation to exposed *Arabidopsis* mesophyll cells. *Nanoscale Research Letters* 6(1), 1–9.

Zeilmaker, T., Ludwig, N.R., Elberse, J., Seidl, M.F., Berke, L. *et al.* (2015) DOWNY MILDEW RESISTANT 6 and dmr6-LIKE OXYGENASE 1 are partially redundant but distinct suppressors of immunity in *Arabidopsis*. *The Plant Journal* 81(2), 210–222.

Zeng, Z., Han, N., Liu, C., Buerte, B., Zhou, C. *et al.* (2020) Functional dissection of HGGT and HPT in barley vitamin E biosynthesis via CRISPR/Cas9-enabled genome editing. *Annals of Botany* 126(5), 929–942.

Zhang, Y., Bai, Y., Wu, G., Zou, S., Chen, Y. *et al.* (2017a) Simultaneous modification of three homoeologs of TaEDR1 by genome editing enhances powdery mildew resistance in wheat. *The Plant Journal* 91(4), 714–724.

Zhang, Y., Ma, X., Xie, X. and Liu, Y.G. (2017b) CRISPR/Cas9-based genome editing in plants. *Progress in Molecular Biology and Translational Science* 149, 133–150.

Zhang, L., Wu, L., Si, Y. and Shu, K. (2018a) Size-dependent cytotoxicity of silver nanoparticles to *Azotobacter vinelandii*: Growth inhibition, cell injury, oxidative stress and internalization. *PLoS One* 13(12), e0209020.

Zhang, S., Wang, L., Zhao, R., Yu, W., Li, R. *et al.* (2018b) Knockout of slmapk3 reduced disease resistance to botrytis cinerea in tomato plants. *Journal of Agricultural and Food Chemistry* 66(34), 8949–8956.

Zhang, Y.Z., Shi, M. and Holmes, E.C. (2018c) Using metagenomics to characterize an expanding virosphere. *Cell* 172(6), 1168–1172.

Zhang, R., Liu, J., Chai, Z., Chen, S., Bai, Y. *et al.* (2019) Generation of herbicide tolerance traits and a new selectable marker in wheat using base editing. *Nature Plants* 5(5), 480–485.

Zhang, Y., Ren, Q., Tang, X., Liu, S., Malzahn, A.A. *et al.* (2021) Expanding the scope of plant genome engineering with Cas12a orthologs and highly multiplexable editing systems. *Nature Communications* 12(1).

Zheng, M., Jagota, A., Semke, E.D., Diner, B.A., McLean, R.S. *et al.* (2003) DNA-assisted dispersion and separation of carbon nanotubes. *Nature Materials* 2(5), 338–342.

Zhou, H., Liu, B., Weeks, D.P., Spalding, M.H. and Yang, B. (2014) Large chromosomal deletions and heritable small genetic changes induced by CRISPR/Cas9 in rice. *Nucleic Acids Research* 42(17), 10903–10914.

Zhou, J., Peng, Z., Long, J., Sosso, D., Liu, B. *et al.* (2015) Gene targeting by the TAL effector pthxo2 reveals cryptic resistance gene for bacterial blight of rice. *The Plant Journal* 82(4), 632–643.

4 Salinity-resilient Crops Reprogrammed Using Modern Molecular-nanotechnology Tools

Farzana Mustafa Era[1], Daniel Heltberg[2], Sumi Sarkar[1], Marium Khatun[1], Damien J. Platen[3] and A.K.M. Aminul Islam[1*]

[1]Department of Genetics and Plant Breeding, Gazipur Agricultural University, Gazipur, Bangladesh; [2]Rice Genetics Laboratory, School of Biological Sciences, University of Aberdeen, Scotland; [3]International Rice Research Institute, Los Baños, Laguna, Philippines

Abstract

Salinity is a global environmental challenge that can severely impact crop production and food security. As climate change intensifies, the frequency and severity of salt stress are expected to increase, necessitating innovative strategies to combat it. Modern molecular nanotechnology, particularly the use of nanoparticles, has emerged as a promising approach to address this challenge. Plants encounter physiological challenges such as osmotic stress, ionic imbalance, and reactive oxygen species (ROS) toxicity. Nanoparticles (NPs) have unique physical, chemical, and biological properties distinct from their bulk counterparts that can be used to improve the salt-tolerance mechanisms of plants. Plants uptake nanoparticles either from the soil or the air. Several NPs, mostly gold, silver, silicon, cerium, zinc, or iron, have been reported for their utilization in the development of salt-resistant crops. NPs have been found to expedite the expression of genes associated with salt tolerance in crops. Nanoparticles have demonstrated efficacy as effective carriers for genetic engineering technology. This chapter may help to understand the way to use nanoparticles to develop novel salt-resilient crop varieties.

Keywords: nanomaterials, nanoparticles, salinity stress, stress resilience

4.1 Introduction

Soil is the key source for feeding the ever-increasing global population, projected to exceed 9.8 billion by 2050 (FAO, 2020). The confluence of climate change, seawater intrusion, groundwater infiltration, as well as irrigation and fertilizer application, has led to a significant increase in soil salinity (Qian *et al.*, 2021; Li *et al.*, 2022). Soil salinity is a severe global issue, affecting over one billion hectares of land across 100 countries, with one-fifth of this being irrigated land (Hammam and Mohamed, 2020; Tian *et al.*, 2020; Hopmans *et al.*, 2021). If left unaddressed, half of the world's irrigated area will be salty within the next 30 years, resulting in a yearly $27 billion annual agricultural loss (Wang *et al.*, 2020,

*Corresponding author: aminulgpb@bsmrau.edu.bd

© CAB International 2025. *Stress-Resilient Crops: Coordinated omics-CRISPR-Nanotechnology Strategies* (ed. Jen-Tsung Chen)
DOI: 10.1079/9781800627307.0004

2021). Seed germination, plant growth, quality, and production are severely affected by soil (Yang and Guo, 2018; Singh, 2022).

Plants initially experience osmotic stress due to a lack of sufficient water in saline conditions, which triggers stomatal closure and ultimately leads to a reduction in photosynthesis (Sarwar *et al.*, 2022). Excessive accumulation of sodium ions (Na^+) within plant tissues can lead to ion toxicity. This ionic imbalance hinders the plant's ability to absorb and utilize potassium ions (K^+), which are essential for various physiological processes (Zhu, 2002; Li *et al.*, 2022). Moreover, ion toxicity leads to the production of reactive oxygen species (ROS) that damages the DNA and proteins (Hu *et al.*, 2021; Liu *et al.*, 2021a). In addition, the limited availability of CO_2 increases the photorespiration rate that generates H_2O_2 as a by-product (Bose *et al.*, 2014). Agronomic practices like irrigation or conventional breeding of salt-resistant crop varieties need time and it may not be sufficient to cope with demands. In this case, genetic engineering-mediated nanotechnology emerges as a promising frontier to enhance salt tolerance (Liu *et al.*, 2021a, b). Nanoscience covers the field of nanomaterials and their properties, while nanotechnology refers to the principle of nanoscience to use practically (Madkour, 2019; Wu *et al.*, 2020). Nanomaterials such as NPs, nanocapsules, or nanocrystals are either naturally occurring or artificially created particles ranging from 1 nm to 100 nm in size (Kalita *et al.*, 2021). Nanoparticles offer a wide range of surface area that helps to interact with the cellular and molecular level of plants to facilitate water retention, nutrient uptake as well as stress resilience. Nanoparticles can protect photosynthesis, aid in ROS detoxification, and reduce osmotic and ionic stress, leading to improved salinity tolerance in plants (Singh *et al.*, 2023). This review explored the types of nanoparticles that could be used and the potential benefits and challenges associated with their use. Through understanding and harnessing the power of nanotechnology, the researchers can develop nano-enhanced salt stress, thereby addressing the impending challenges posed by climate change.

4.2 Nanomaterials in Plant Improvement

The Greek *nanos* gave the idea of the word 'nano,' which means dwarf (Al-Khayri *et al.*, 2023). A human eye can only see up to 10,000 nm. A set of properties having one dimension out of three that ranged between 1 nm and 100 nm is termed 'nanomaterials' or 'nanoparticles' (Wu *et al.*, 2020). Organic, inorganic, and hybrid NPs are observed depending on their nature. Organic particles are biodegradable like starch, lignin, cellulose derivatives, and polysorbate, most widely used in agriculture. Inorganic particles are bio-undegradable like gold, silver, iron, zinc, or silica, and hybrids are the combination of both organic and inorganic materials that can mitigate the limitations of both types (Al-Khayri *et al.*, 2023). These particles also showed a vast range of applications for crop improvement (Table 4.1).

4.2.1 Particles

4.2.1.1 Gold

Gold particles (AuNPs) are commonly used for the transfer of genetic elements into plants, as they quickly adsorb DNA without causing toxicity in cells (Moaveni, 2011). A by-product of the chemical synthesis pathway of AuNPs poses a great risk to human health. So, the green way, more precisely plant-based AuNPs, is more demanding nowadays. Biomolecules of plant extracts from roots, stems, leaves, or pulps are used to reduce Au^{3+} to Au^0 and prevent coagulation (Qiao and Qi, 2021). Arora *et al.* (2012) found an effective increase in plant parameters like seed germination, plant height, leaf numbers, and seed yield by using the foliar spray of AuNPs. They also reported an increasing rate of reducing and total soluble sugars at the application of 25 ppm AuNPs in *Brassica juncea*.

4.2.1.2 Silver

Small silver nanoparticles (AgNPs) enter through the semi-permeable cell wall into the plant and are then transported through the

Table 4.1. Application of nanoparticles to mitigate salt stress.

NPs	Chemical structure	Form of NPs	Concentration	Salinity dose	Crop	References
Si	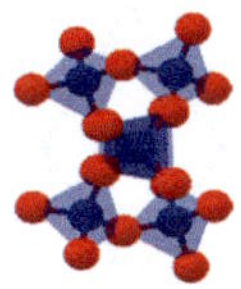 SiO_2	SiO_2	50 mg/L	100 mM	Potato	Gowayed *et al.*, 2017
		Na_2SiO_3	0.25 and 0.5 mM	100 mM	Wheat	Tuna *et al.*, 2008
		SiO_2	25 ppm	20% seawater equivalent to 100 mM NaCl	Rice	Matoh *et al.*, 1986
		SiO_2	50 mM	100 mM	Wheat	Mushtaq *et al.*, 2017
		Si $(OH)_4$	150 mg/L	Ec 6 ds/m	Okra	Abbas *et al.*, 2014
		Si	2 mmol/L	100 mM	Sugarcane	Ashraf *et al.*, 2010
		K_2SiO_3	2 and 4 mmol/L	20 mM, 60 mM and 100 mM	Wheat	Bybordi, 2004
Fe	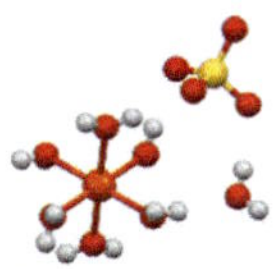 Fe_2SO_4	Fe_2SO_4	2g/L	0 and 100 mM NaCl	Sunflower	Torabian *et al.*, 2018
Zn	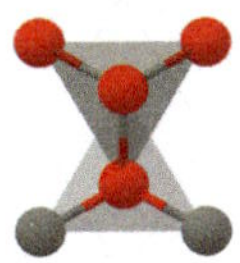 ZnO	Nano-Zn	100 and 200 ppm	10% and 20% of seawater	Cotton	Hussein and Abou-Baker, 2018
		ZnO mixed with Fe_2O_3	mixture 2 g/L	75 mM and 150 mM sodium chloride (NaCl)	Maize	Fathi *et al.*, 2017
		ZnO	10, 50, 100 mg/L	150 mM NaCl	Tomato	Faizan *et al.*, 2021
Cu		Cu	10 mg of Cu NP	100 mM NaCl	Tomato	Hernández-Hernández *et al.*, 2018
		Cu	250 mg L^{-1} Cu NP	50 mM NaCl	Tomato	Pérez-Labrada *et al.*, 2019
		Green Cu NPs	25, 50 and 100 mg kg^{-1} of soil	4 ds/m	Maize	Noman *et al.*, 2021

Ag	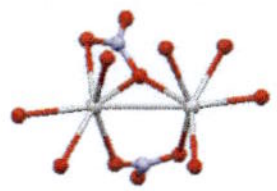 $AgNO_3$	Colloidal AgNPs	100 $mgdm^{-3}$	2.32 ds/m	Lilium	Byczyńska *et al.*, 2021
		$AgNO_3$	An aqueous solution of $AgNO_3$ (1 mM) was added dropwise into 50 mL of rosemary leaf extract	25, 50, 100 mM	Linseed	Khalofah *et al.*, 2021
Ti	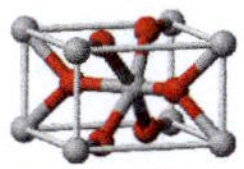 TiO_2	$nTiO_2$	(0·01%, 0·02% and 0·03%	80 mM of NaCl	Broad bean	Abdel Latef *et al.*, 2017
Ce	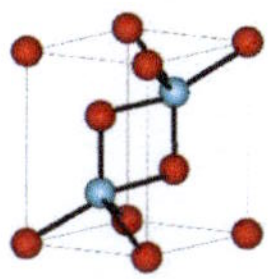 CeO_2	$Ce(NO_3)_3$	50 mg L^{-1} poly (acrylic acid) coated nanoceria	100 mM NaCl	*Arabidopsis thaliana*	Wu *et al.*, 2018a
		CeO_2	200 and 1000 mg kg^{-1}	100 mM NaCl,	Canola	Rossi *et al.*, 2016
		CeO_2	0.1 mM	200 mM NaCl	Rapeseed	Khan *et al.*, 2021
C		put-CQD	10 mg/L	100 mM NaCl)	Grapevine	Gohari *et al.*, 2021
Mn	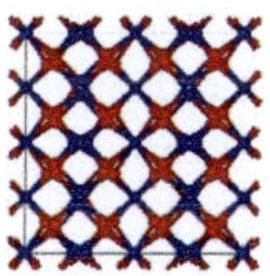 Mn_2O_3	Mn_2O_3	0.1, 0.5, 1 mg/L	100 mM NaCl	*Capsicum annuum* L.	Ye *et al.*, 2020
		Mn_3O_4	1 and 5 mg per plant	0.3% NaCl	Cucumber	Lu *et al.*, 2020
K		Nano-NPK	50 and 100 ppm	(60, 120 mM)	Pea	El-Hefnawy, 2020

apoplastic and symplastic pathways (Yan and Chen, 2019). AgNPs have been reported to reduce oxidative stress and Na^+ uptake and increase the K^+ uptake in pearl millet under salt stress (Khan *et al.*, 2020a). AgNPs can enhance the plant biomass in maize and nodulation in cowpeas (Sillen *et al.*, 2015; Pallavi, Mehta *et al.*, 2016).

4.2.1.3 Iron

Iron NPs are more stable than other NPs and can be released in soil under different pH levels. Shoot elongation, seed germination, and photosynthesis rate can be aided by iron NPs (Plaksenkova *et al.*, 2019; Rezayian *et al.*, 2023). Adrees *et al.* (2020) explained the effect of iron NPs on wheat grain yield under Cd-contaminated drought-stressed soil. They found that iron NPs increase the iron concentration in grain and reduce the Cd content. Iron NPs reduced the production of malondialdehyde (MDA) as well as hydrogen peroxide and induced the enzymatic and nonenzymatic reactions to protect the grape plants under drought conditions (Bidabadi *et al.*, 2023). Mazhar *et al.* (2022) also reported the activity of antioxidant enzymes under drought stress in flax plants. This phenomenon led the plant to withstand under saline conditions as well. Tawfik *et al.* (2021) found an increased rate of K^+ as well as reducing concentration of Na^+ under salt stress in *Moringaa oleifera.*

4.2.1.4 Zinc

ZnS, ZnSe, or ZnO are different types of Zn MPs whereas ZnO is the most widely used one (Sturikova *et al.*, 2018). ZnO NPs increase the production of proline along with the antioxidant activities under salt stress (Torabian *et al.*, 2016). ZnO NPs enhance the shoot growth and photosynthetic pigment formation as well as some phenolic compounds in *Lupinus termis* under salinity (Abdel Latef *et al.*, 2017). ZnO NPs-treated maize reported improved grain yield and quality under drought stress (Sun *et al.*, 2020).

4.2.1.5 Silica

Silica NPs or SiNPs can be used to improve the water-holding capacity of soil (Bapat *et al.*, 2020). The nontoxic nature of silica makes it more suitable for use in crop quality improvement as well as yield (Goswami *et al.*, 2022). Foliar spray of SiNPs helps to reduce the accumulation of heavy metals in plants as well as improve the K^+ and Na^+ ratio under heavy metal-contaminated saline conditions (El-Saadony *et al.*, 2021). Alsaeedi *et al.* (2019) also reported the perfect plant growth and production with the use of SiNPs under saline conditions. The application of SiO_2-NPs under salt-stress conditions boosted K^+ content in Banan by 10.54%, while also improving chlorophyll synthesis and lowering cell-wall damage (Mahmoud *et al.*, 2020).

4.2.1.6 Titanium

Titanium NPs have a positive effect against drought stress in wheat (Khan and Bano, 2019). The most used titanium NP is TiO_2 (Kalita *et al.*, 2021). The use of TiO_2 enhances the plant growth and photosynthetic rate under stressed conditions. The antioxidant enzyme activity, as well as the oil profile, has been improved with the use of 100mg/L TiO_2 in *Dracocephalum moldavica* under salinity (Gohari *et al.*, 2020). The amino acid concentration has been affected by TiO_2 application in barley (Mattiello and Marchiol, 2017).

4.2.1.7 Carbon

A carbon-based nanomaterial, polyhydroxy fullerene (PHF), has a profound effect on biomass improvement in bitter melon by activating antioxidants (Gharbi *et al.*, 2005; Kole *et al.*, 2013). PHF has influenced the root growth in barley under UV-B radiation and salt-stress conditions (Panova *et al.*, 2016). The shoot length, root number, root length, plantlet length, and different biochemicals like carotenoids, flavonoids, and nutrients have been reported to be increased by using carbon nanotubes (CNTs) in date palm (Taha *et al.*, 2016).

4.2.1.8 Magnetic NPs

Magnetic NPs are specified for a target-based remedy. Small Fe_3O_4 and Fe_2O_3 have large surface areas with high stability and low toxicity, which makes them more reliable magnetic NPs (Cardoso *et al.*, 2018). Magnetic NPs were integrated with an exogenous DNA and spotted into the pollen of cotton by creating a magnetic field. Through pollination, transgenic DNA enters the genome and is inherited in offspring (Zhao *et al.*, 2017).

4.2.1.9 Cerium

CeO_2 affects the tomato fruit yield (Wang *et al.*, 2012). The gluten and starch have also been reported to be decreased in rice by using CeO_2 (Rico *et al.*, 2013). CeO_2 also alters the nutritional value of soybean and cucumber as well as the dry biomass of lettuce (Trujillo-Reyes *et al.*, 2014; Peralta-Videa *et al.*, 2014; Hong *et al.*, 2016).

4.2.1.10 Biogenic materials

The environmentally safe way is to fabricate the NPs with plant-based materials that are termed green or biogenic materials. The synthesis of green NPs includes plant-part extracts like flowers, fruit leaves, or seeds, which are diluted with a polar solvent like water. Then these plant extracts are mixed with metal ion solvent to make an NP solution that is further purified and characterized. With the foliar spray, biogenic NPs enter the nucleus as well as the chloroplast and act as scavengers of ROS to mitigate the abiotic stress (Maqsood *et al.*, 2023). The lower-level application of biologically synthesized ZnO-NPs and TiO_2 NPs enhances the photosynthetic rate in maize and saline-treated wheat, respectively (Del Buono *et al.*, 2021; Mustafa *et al.*, 2021). Biogenic TiO_2-treated wheat was reported with high efficacy of POD, SOD, CAT, and APX (Satti *et al.*, 2022).

4.3 Uptake and Translocation Mechanism by Plants

4.3.1 From soil

Plants can absorb water and soluble substances from the soil by the root (Fig. 4.1) (Hong *et al.*,

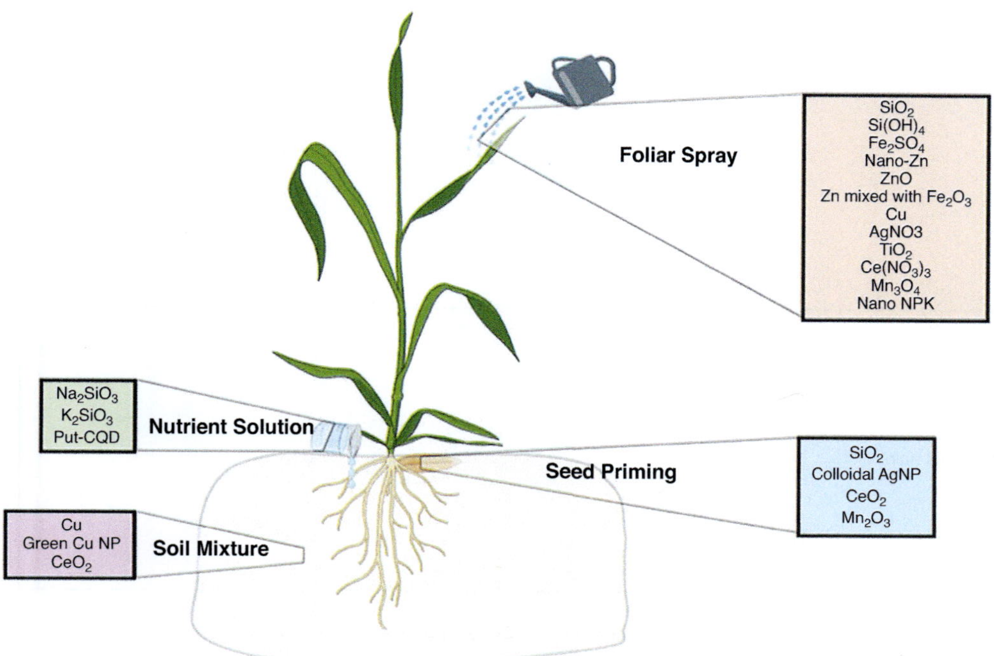

Fig. 4.1. Mode of application of various nanoparticles.

2021). So the NPs can also be entered through the root system towards the plant. The semi-permeable root cell wall has small pores that do not allow entry of large NPs into the cell (Pérez-de-Luque, 2017). Metal-based NPs like AgNPs, CeO_2, or TiO_2, of sizes from 10 nm to 140 nm can be translocated through the root system (Larue *et al.*, 2012; Zhao *et al.*, 2012; Hu *et al.*, 2022). Some studies showed that it is difficult for plants to uptake metal-based NPs that are more than 100 nm (Banerjee *et al.*, 2019). On the other hand, silicon-based or polymer-based NPs can exceed the limitation of 100 nm size (Slomberg and Schoenfisch, 2012; Prasad *et al.*, 2018). NPs move to the aerial parts of the plant both in apoplastic and symplastic pathways (Fig. 4.2). Through the apoplastic pathway, NPs run through the intercellular space and reach the xylem vessel (Sattelmacher, 2001). This pathway contains a Casperian strip that can be avoided in a symplastic way (Robards and Robb, 1972). In the symplastic pathway, NPs are absorbed by the plasmodesmata and run through the cytoplasm of the cell (Robards and Robb, 1972; Pérez-de-Luque, 2017). Sometimes NPs use a sieve tube of phloem to reach the nonphotosynthetic organs (Raliya *et al.*, 2016). Hydrophilic is another pathway that permits NPs into the cell, but due to its very tiny pores, the choice of particles is very limited (Li *et al.*, 2020). The protoplast of the cell can offer entry of particles sized >1 μm, which neutralize the suitability of NPs through endocytosis (Liu *et al.*, 2009; Ma and Yan, 2018; Lv *et al.*, 2019). Endocytosis is the process where particles are entered into the cell by forming a vesicle after binding with a specific receptor through the plasma membrane (Pastan and Willingham, 1985).

4.3.2 From air

Nanoparticles are mostly applied as a foliar spray on plants. Plant uptake NPs either by waxy cuticle or stomata of the leaf (Fig. 4.1; Wang *et al.*, 2023). The hydrophobic cuticle is composed of wax, cutin, and pectin that protects the leaf from losing access to water as well as entering the NPs (Rani *et al.*, 2023). The pore size of the cuticle is less than 5 nm allowing the NPs that are 0.6–4.8 nm in size

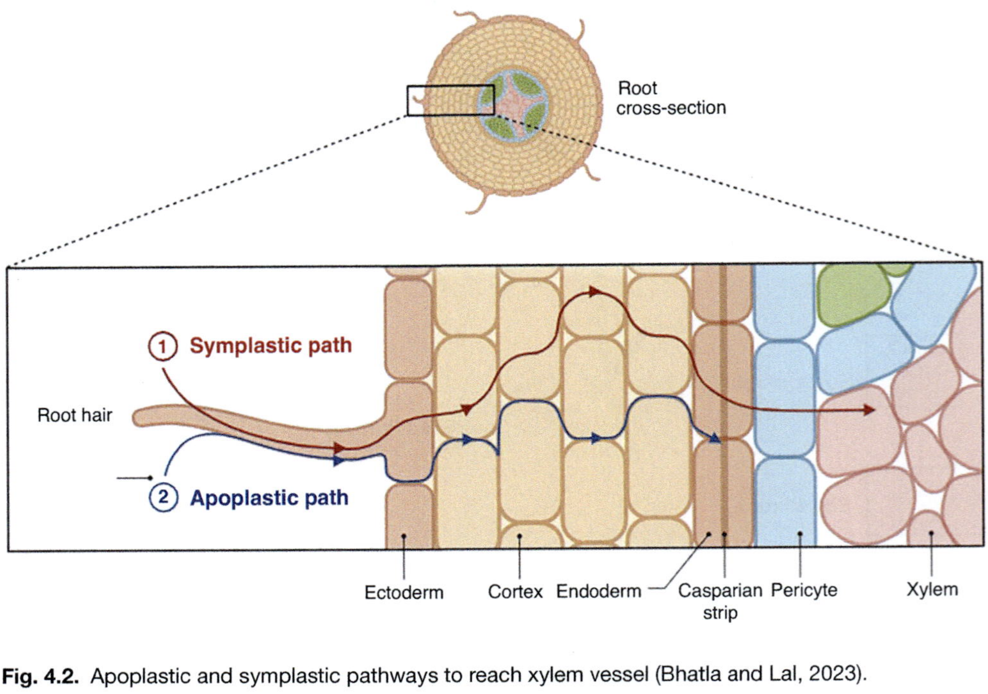

Fig. 4.2. Apoplastic and symplastic pathways to reach xylem vessel (Bhatla and Lal, 2023).

(Eichert *et al.*, 2008; Eichert and Goldbach, 2008; Banerjee *et al.*, 2019). There are two ways: (i) hydrophilic pathway; and (ii) lipophilic pathway through which NPs can beat the barriers of cuticle (Schönherr, 2002). In the hydrophilic pathway, NPs enter through the protein channels or pores that attract favorably with water molecules. On the other hand, NPs are diffused through the waxy cuticle in the lipophilic pathway (Pérez-de-Luque, 2017). The stomatal pathway has a wide range of size choices, i.e. >10 nm, which makes it most preferable to nanoparticle movement in leaves (Eichert *et al.*, 2008; Zhu *et al.*, 2020; Ha *et al.*, 2021). Stomata regulate the water and gas exchange of plants. After cellular entry, nanoparticles traverse from the mesophyll cell to the bundle sheath through both apoplastic and symplastic routes, ultimately reaching the sieve tubes of the phloem. Nanoparticles can spread throughout the entire plant by foliar spray, especially towards the sink organs, via the phloem tissue (Pérez-de-Luque, 2017; Rani *et al.*, 2023).

4.4 Salinity Responses in Plant

Salinity negatively impacts plants by altering root cell membrane permeability, leading to electrolyte leakage, reduced chlorophyll, and, ultimately, plant death (Fig. 4.3) (Zeng *et al.*, 2001; Khan *et al.*, 2020b; Sarwar *et al.*, 2022). High salt concentrations in the soil inhibit germination and seed emergence and accumulate toxic ions, impairing overall plant growth. Osmotic stress of the root forces the plant to reduce turgor pressure and close the stomata. The closed stomata cause excessive photorespiration that leads to the production of ROS (Hoshida *et al.*, 2000). ROS reduces the production of NADP, which hampers the transfer of electrons towards Photosystem II that hinder the photosynthesis system. When a plant takes an excessive amount

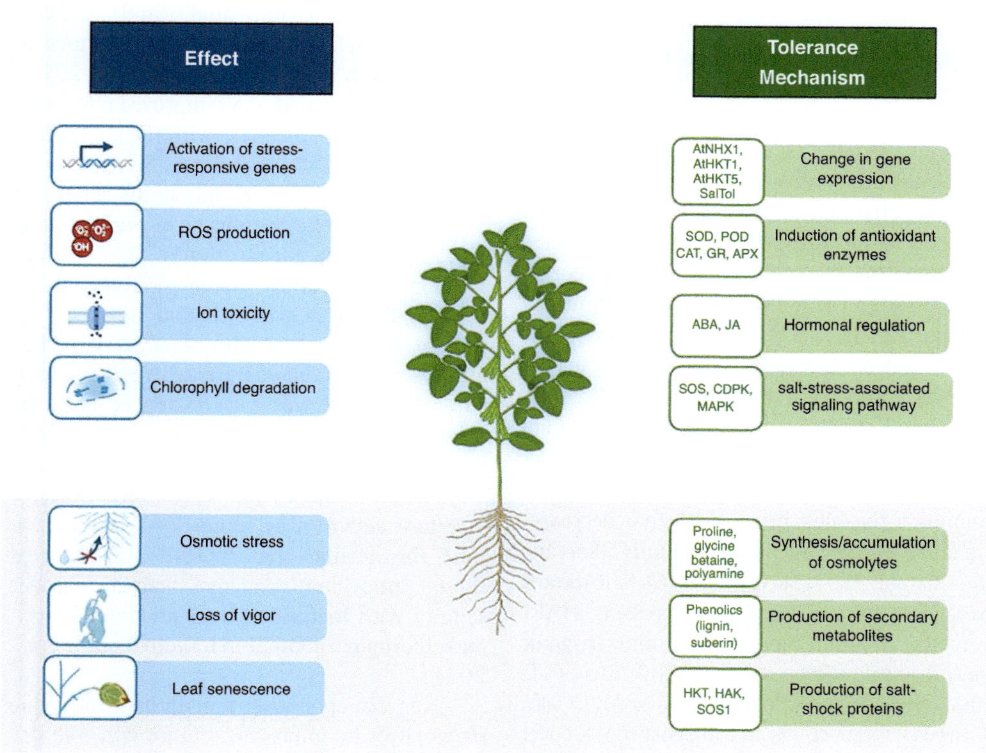

Fig. 4.3. Effects and mechanisms of salt stress tolerance in crops.

of Na^+, it tends to replace K^+, which disrupts the enzymatic activities. Plants accumulate various organic compounds called osmolytes, like proline, glycine betaine, sugars, and mannitol, under salt stress. These uncharged molecules do not interfere with the biomolecular process but help to maintain osmotic balance, scavenge ROS as well, as facilitate continuous water influx into the cell that might mitigate the detrimental effect of salinity (Sharma *et al.*, 2019). Maintaining ion homeostasis, particularly the regulation of sodium (Na^+) within the cell, is a critical process for plants to tolerate salt stress. The SOS pathway plays a vital role in knocking out the Na concentration from the cell. With the increase of salinity, the concentration of Ca^{2+}, a protein binder of SOS3, increases. This facilitates the interaction between SOS2 and SOS3. The SOS–SOS3 complex phosphorylates SOS1, a Na^+/H^+ antiporter located in the plasma membrane, which subsequently pumps Na out from the cell (Shi *et al.*, 2003; Chele *et al.*, 2021). Excess Na is transported across the Na^+/H^+ exchanger (NHX1) located in the tonoplast, the membrane surrounding the vacuole, and compartmentalized within the vacuoles. This process protects the plant from salinity stress as well (Blumwald and Poole, 1985; Zhu, 2003; Bernstein, 2019).

Genes can control the salt uptake and their transport channels to mitigate the saline effect. Na^+ can enter the cell through nonspecific cation channels, sometimes at high-salt conditions, through the nonselective high-affinity K^+ transporter (HKT). This transporter may allow the passage of sodium ions due to their similar chemical properties. Expression of HKT1 in wheat and AtHKT1 in *Arabidopsis* maintain the Na and K ratio in roots to survive under salinity (Munns, 2005). HAL1 gene from yeast maintains the high internal K^+ that decreases the intracellular Na^+ under salinity (Serrano and Gaxiola, 1994; Serrano, 1996). Salt-treated transgenic tomato lines overexpressing HAL1 were able to retain more K than nontransgenic control plants (Gisbert *et al.*, 2000; Rus *et al.*, 2001). *Arabidopsis* K^+ transporter or AKT1 also showed high selectivity to transport root K^+ over Na^+ under salt stress (Véry and Sentenac, 2003; Munns, 2005).

4.5 Nanoparticles in Salinity Tolerance

Nanoparticles could be a great choice to mitigate salt stress due to their preferential properties like size, high surface area, and physiological amenities. Researchers reported a number of NPs that can help the plant's mechanism to beat the salinity (Kundu *et al.*, 2023). The reduced concentration of SiO_2 exerted a discernible influence on the shoot and root length, root count, and fresh weight of callus in potatoes subjected to 50 mM and 100 mM NaCl concentrations. Increased antioxidant activities were also reported under the same concentration of SiO_2. This helps to improve nutrient absorption, reduces Na uptake, and decreases transpiration (Gowayed *et al.*, 2017). Tuna *et al.* (2008) also reported a notable increase in the use of Si, in the dry matter and chlorophyll contents in wheat under high-saline conditions that is supported by Abbas *et al.* (2014) in okra and Garg and Bhandari (2016) in *Cicer arietinum* L. The application of Si under salt stress has been reported to enhance juice quality and sugarcane yield, in correlation with the maintenance of an improved K^+/Na^+ ratio (Ashraf *et al.*, 2010). Ahmad (2014) reported an improved photosynthetic rate under salt stress with the application of 4 mmol/L K_2SO_3 in wheat. Si application improves the uptake of K^+ due to the activation of H-ATPase in the membrane of barley (Liang *et al.*, 1996). The introduction of Si in nutrient solution significantly increased shoot water consumption both in control and 100 mM NaCl salinity stress in wheat (Tahir *et al.*, 2012). They also observed decreased osmotic stress under salt stress. Torabian *et al.* (2018) applied both the normal and nano form of $FeSO_4$ in sunflowers under salt stress. They reported more enhanced antioxidant enzymes like CAT, PRO, POX, or invertase activities by nano $FeSO_4$ application than the normal one. ZnO-NP also reported those antioxidant enzyme activities under salinity with increased shoot and root length, and chlorophyll content in tomato (Faizan *et al.*, 2021).

Nano Zn application can mitigate the saline stress effect by improving the growth and dry weight in cotton (Hussein and Abou-Baker, 2018). A mixture of Fe_3O_4 and ZnO NPs spray, enhanced root dry weight by 18%, absorption rate by 8%, and sub-stomatal CO_2 concentration by 5% in maize grown in salinity (Fathi *et al.*, 2017). The application of Fe_3O_4 can counteract the negative effects of salinity stress during the tillering and flowering stages of wheat (El-Saber *et al.*, 2021). Cu NPs influenced the plant height by 6%, and stem diameter by 5% under NaCl-induced saline-prone tomato production. The experiment also revealed that Cu activates the antioxidant defense mechanism that initiates the expression of SOD, PR1, and JA gene (Hernández-Hernández *et al.*, 2018). Regarding induced systematic resistance against stress, JA signaling constitutes one of the pathways (Shoresh *et al.*, 2010). Pérez-Labrada *et al.* (2019) found that using Cu NPs lowered the Na^+/K^+ ratio while increasing 16% phenols, 104% phenylalanine ammonia lyase, ascorbate peroxidase, 26% glutathione peroxidase, 8% SOD, and 93% CAT in tomato production under salinity. The addition of 100 mg/kg green Cu NPs to saline soil boosted the maize plant growth as well as reduced oxidative damage by lowering reactive oxidative damage (Noman *et al.*, 2021). Byczyńska *et al.* (2021) reported the improved bulb yield while they were treated with AgNPs compared to control under salt stress. Foliar spray of $AgNO_3$ improved the levels of soluble sugars, proline, soluble proteins, H_2O_2, malondialdehyde, and both enzymatic and non-enzymatic antioxidants in NaCl-treated linseed plants (Khalofah *et al.*, 2021). The application of 0.01% $nTiO_2$ boosted enzymatic antioxidants and raised levels of soluble sugars, amino acids, and proline in salt-stressed plants compared to those without the supplementation (Abdel Latef *et al.*, 2017). Cerium oxide nanoparticles helped *Arabidopsis thaliana* leaves to retain more $K^{+,}$ which is the key function to combat salinity stress (Wu *et al.*, 2018a). The total chlorophyll of canola (*Brassica napus*) had increased by 27% when exposed to 200 mg/Kg CeO_2 NPs compared to the control.

It is noteworthy that exposure to CeO_2 appeared to reduce the plant's uptake of Na^+ (Rossi *et al.*, 2016). Khan *et al.* (2021) reported a 12% increase in the germination rate of rapeseed coupled with a 41% improvement in biomass production under salt-stress conditions when subjected to CeO_2 NPs treatment. They also found 29% K^+ retention as well as 18.5% accumulation of Na^+ under salinity. Putrescine-functionalized carbon quantum dot or put-CQD, improved potassium uptake in grapevines while boosting enzyme activities associated with salt tolerance (Gohari *et al.*, 2021). Priming seeds with MnNPs at 1 mg/L significantly increased root length by 55.7% under 100 mM of NaCl. Moreover, salt stress inhibits Mn uptake which could be facilitated by external application of Mn. MnSOD gene expression, which is crucial for defending cells against ROS, has also been increased by the application of MnNPs at 1 mg/L dose in *Capsicum annuum* L. (Gill and Tuteja, 2010; Ye *et al.*, 2020). MnNPs (Mn_3O_4) also showed significant improvement in plant health by enhancing leaf pigment content, photosynthesis rate, and plant biomass in cucumbers (Lu *et al.*, 2020). Foliar application of 100 ppm Nano-K along with nitrogen and phosphorus resulted in higher value of mitotic index (12.27%) in pea compared to control under 120 mM of NaCl. As K stimulates the assimilation of protein, it enhances plant growth as well (AbdElgawad *et al.*, 2016; El-Hefnawy, 2020). Selenium NPs can alter DNA methylation and induction of upregulation of WRKY1 transcription factor in bitter gourd. Additionally, Se-NPs also affected the expression of PAL and 4CL genes (Rajaee Behbahani *et al.*, 2020).

4.6 Molecular Acceleration of Plant Physiology under Salinity Using Nanomaterials

4.6.1 Na/K pathway

A high affinity K^+ transporter gene, HKT1, reported in a polyacrylic acid-coated nanoceria (PNC), mediated cotton under salinity (200 mM NaCl). HKT1 was significantly upregulated in stress-affected plants compared to the control at both its first and second leaf stage. On the other hand, the KOR gene that allows the movement of K^+ outside of the cell was significantly downward in the cotton plant under salt stress (Liu *et al.*, 2021b). Furthermore, the gene HAK5, responsible for maintaining potassium levels within mesophyll cells, exhibited a fourfold upregulation in *Arabidopsis thaliana* subjected to PNC application (Wu *et al.*, 2018b). Root and leaf potassium levels were found to be elevated following the application of PNC. Additionally, cucumber plants subjected to 75 mM NaCl for 7 days displayed a threefold increase in the expression of the CsAKT1 gene (Peng *et al.*, 2022). Ca^{2+} assisted NHX1 antiporter helps to move the Na^+ to the vacuole (Abou-Zeid *et al.*, 2021). Adding multi-walled carbon nanotubes or MWCNTs to rapeseed plants boosts the activity of antioxidant genes like APX and SOD. This leads to higher levels of their transcripts BnAPX, BnCu/ZnSOD, and BnMnSOD. As a result, the balance of ions and redox in the plant is restored because the level of harmful ROS decreases, as does the ratio of Na^+ to K^+ ions (Zhao *et al.*, 2019). SE NPs boost ATPase and SOS1, but significantly downregulate the expression of the *NHX1* gene in salt-stressed wheat leaves treated with 3000 ppm NaCl (Soliman *et al.*, 2023).

4.6.2 ABA synthesis pathway

Ag and Si NPs-treated *Salanum lycopersicum* led to changes in gene activity that affect water and salt balance by controlling stomatal closure under salt stress. Certain genes involved in ABA pathways, such as AREB, CRK1, and NCED3, became more active, while others, like TAS14, ZFHD1, DDF2, and specific MAPK variations, became less active (Al-Khayri *et al.*, 2023). TAS14 and NCED3 are the ABA-dependent stress-responsive proteins. Ag NPs downregulate the ABA-independent stress-responsive protein coding gene ZFHD1 (Zinc finger homeodomain 1) in *Arabidopsis* (Tran *et al.*, 2007). NPs decrease DDF2 (Integrase-type DNA-binding superfamily protein) that helps plants handle osmotic stress by lowering GA production, a major consequence of ABA signaling (Magome *et al.*, 2008).

4.6.3 SOS pathway

SOS2-SOS3 complex interacts with SOS1 to exclude Na^+ from the cell and maintain ion homeostasis. Iron oxide, consisting of a mixture of $FeCl_3.6H_2O$, and $FeCl_2.4H_2O$ in a 2:1 molecular ratio, can upregulate the salt-responsive genes like HKT1, SOS1, and NHX1 in *Eucalyptus tereticornis* subjected to salt stress induced by treatment with 300 nM NaCl (Singh *et al.*, 2021). The higher expression of SOS1 has also been reported in rice under 75 mM NaCl saline conditions following the application of PNC (polycoated cerium oxide nanoparticles) (Zhou *et al.*, 2021). Ag NPs-treated tomato showed 1.28-fold higher expression of SOS2 germinated under 200 mM NaCl compared to control (Almutairi, 2016).

4.6.4 Antioxidant enzyme activity

Plants use SOD, APX, CAT, POD/POX, DHAR, MDHAR, GPX, GR, and GST to rescue the oxidative damage caused by salinity. SOD (superoxide dismutase) is the primary defensive enzyme that plants use against ROS damage (Singh *et al.*, 2023). SOD effectively scavenges the excess ROS in tomato and chickpea roots under high salinity (Fu *et al.*, 2023). In *Trachyspermum ammi* L., foliar spray of nano Fe_2O_3 significantly boosts the levels of enzymes like SOD, CAT, PPO, and POX, especially when the plants are exposed to higher salt levels like eight ds/m and 12 ds/m (Abdoli *et al.*, 2020). Overproduction of MDA (malondialdehyde) may affect the plant growth. Under salt stress, MDA levels increase in wheat. However, with the spray of nano-Zn particles, especially up to 50 mg/L, the MDA levels drop significantly (Lalarukh *et al.*, 2022). Gaafar *et al.* (2020) also reported the lower level of MDA with Zn-NPs in soybean. When *Arabidopsis* leaves are stressed by 100 mM NaCl, those treated with PNC show less production of ROS (Wu *et al.*, 2018a). Ag-NPs-treated pearl millet grows better under salt stress due to the boosted activity of important antioxidant enzymes like SOD, CAT, and GPX. Wahid *et al.* (2020) also supported the findings that wheat cultivated under saline conditions of 100 mM NaCl exhibited the observed response with Ag-NPs.

4.7 Breeding New Nanoparticles-mediated Crops to Combat Salinity

The advent of biotechnology offers promising avenues to engineer salt tolerance in crops. Transgenic approaches, involving the introduction of specific salt-tolerant genes like

HKT, NHX, or SOS1 have emerged as a potent strategy to mitigate the detrimental effects of salinity stress in crops. A plant can combat salinity either by maintaining the Na^+/K^+ ratio or by excluding the Na from the cell. Previous study shows that high-affinity K^+ carrier gene like *AtHKT1* in *Arabidopsis* helps move Na and a bit of Potassium around. This AtHKT1 transports the Na^+ towards the shoot and leaves (Uozumi *et al.*, 2000; Berthomieu *et al.*, 2003). *Arabidopsis*-donated AtHKT1;one overexpressed in rice was reported with 125% higher shoot weight in 80 mM NaCl (Plett *et al.*, 2010). The HmHKT1;five and HvHKT1;five from the plant species *Hordeum mainum* (sea barley) and *Hordeum vulgare* (common barley), respectively, have been shown to improve salinity tolerance in rice grown in a 75 mM NaCl environment (Huang *et al.*, 2019) NHX gene from *Arabidopsis thaliana* introduced salt tolerance in wheat, tomato, and maize (Abideen *et al.*, 2022). The overexpression of SOS1 in *Arabidopsis* results in an enhancement of salt tolerance, thereby illustrating that the mitigation of sodium accumulation within plant cells can effectively confer improved resilience to salinity (Shi *et al.*, 2003). The SOD2 gene from yeast to rice plants resulted in a 77% lower concentration of Na^+ in the shoots of the modified rice plants (Zhao *et al.*, 2006).

Nanoparticles can be used as vehicles to deliver genetic materials such as DNA, RNA, or small interfering RNA into plant cells (Table 4.2). Nano-mediated gene delivery has applications in

Table 4.2. Nanoparticle-mediated gene transfer in crops.

Crop	Nanoparticles as carrier	Carry	Reference
Allium cepa 	Gold nanoparticle functionalized mesoporous silica nanoparticle (Au-MSN) plant	Bovine serum albumin (BSA) Enhanced green fluorescent protein (eGFP)	Martin-Ortigosa *et al.* (2012)
Zea mays 	Mesoporous silica nanoparticles (MSNs)	Cre protein	Martin-Ortigosa *et al.*, 2014
Nicotiana tabacum 	Mesoporous silica nanoparticles (MSNs) Multi-walled carbon nanotubes (MWCNTs)	Green fluorescent protein (GFP) gene Yellow fluorescent protein (YFP)	Torney *et al.*, 2007 Burlaka *et al.*, 2015
Gossypium hirsutum 	Silicon carbide whisker	T1AVP1 gene	Asad *et al.*, 2008

plant improvement including stress tolerance. The cell wall limits the uptake of NPs that are greater than 20 nm, but some evidence showed it can be up to 50 nm (Schwab *et al.*, 2016; Wang *et al.*, 2016; Tripathi *et al.*, 2017).

Mesoporous silica nanoparticles (MSNs) have earned interest owing to their porous architecture, exceptional thermal stability, and the capability to bind to a variety of functional groups on to their surface (Narayan *et al.*, 2018). The gold-capped MSN was reported as a carrier to deliver a plasmid coated with green fluorescent protein (GFP) gene to the DNA of *Nicotiana* (Torney *et al.*, 2007). This gold-capped MSN was also reported in *Allium cepa* and *Zea mays* as a genetic carrier of different genes (Martin-Ortigosa *et al.*, 2012; Martin-Ortigosa *et al.*, 2014). Burlaka *et al.* (2015) found multi-walled carbon nanotubes (MWCNTs) as genetic transporter of yellow fluorescent protein (YFP) to the *N.tabacum* magnetic gold NP. Other magnetic and metal NPs like AuNps, ZnSNPs, Fe_3O_4NPS, and Ca NPS composited carbon bullets can also be potential gene delivery tools in plants (Vijayakumar *et al.*, 2010; Yu-Qin *et al.*, 2012; Naqvi *et al.*, 2012; Hao *et al.*, 2013; Zhao *et al.*, 2017). Asad *et al.* (2008) used silicon carbide whiskers to transfer a gene T1AVP1 into cotton, resulting in enhanced salt tolerance compared to the control.

4.8 Backward to Use Nanoparticles

Nanoparticles (NPs) are beneficial for mitigating the toxic effects of ROS under salinity stress due to their ultra-small size, as noted by Rahman *et al.* (2020). However, the excessive use of NPs can lead to the production of ROS in plants which causes oxidative stress and cellular damage (McClements and Xiao, 2017; Xuan *et al.*, 2023). Although employed in agriculture to alleviate environmental stress, concerns have arisen regarding the accumulation of NPs into the food chain, posing risks to human and animal health (Priester *et al.*, 2012; Al-Khayri *et al.*, 2023). Inhaling NPs applied to plants can cause inflammatory responses, genotoxicity, as well as protein fibrillation, and may accumulate in organs like the lungs, kidneys, and liver (Singh *et al.*, 2023). Plant growth and root elongation can be retarded by using high doses of certain NPs like ZnO, TiO_2, and iron oxide NPs. For instance, the nitrogen-fixing process can be prevented in legumes with the use of nano-cerium oxide (Al-Khayri *et al.*, 2023). NPs can leach into the soil and water bodies that may contaminate the ecosystems and beneficial microorganisms (Ray *et al.*, 2009; Xuan *et al.*, 2023). Foliar sprays of NPs may cause harm to environmental enemies like bees, vital for pollination, that leads to the disruption of biodiversity (McClements and Xiao, 2017). Moreover, there are no proper guidelines currently in place for using and disposing of NPs (Xuan *et al.*, 2023). Insufficient documentation of these effects could raise the chances of unanticipated consequences. Therefore, to use NPs safely in agriculture, we need better regulations and more research on their long-term impacts.

4.9 Conclusion

Nanomaterials have demonstrated the ability to enhance plant tolerance to salt. Nanotechnology provides an innovative and highly efficient technique for improving agro-based product quality and modernizing farming processes that lead to meeting the world's growing population. Nanoparticles are tiny enough to enter plant tissues, thereby enhancing the plant's growth even under stress like salt. Their large surface area also helps plants to absorb and distribute nutrients effectively. Several NPs, mostly, Ag, Au, TiO_2, SiO_2, Zn, Fe, and Ce, have proved their ability to reduce the toxic effects of salt stress on plants, allowing them to maintain better growth even when exposed to high salinity levels in soil or water. Though nanotechnology shows promise in addressing salt stress in plants, it is still primarily confined to laboratory research. To use this in farming, there is a need for clear regulations and policies, particularly regarding the disposal of nanoparticles, which are currently insufficient. Therefore, it is imperative to continue the advancing research on molecule-based nanotechnology to ensure sustainable crop production under salt-stress conditions.

References

Abbas, T., Pervez, M.A., Ayyub, C.M., Shaheen, M.R., Tahseen, S. *et al.* (2014) Evaluation of different okra genotypes for salt tolerance. College of Agriculture, University of Sargodha, Pakistan. *International Journal of Plant, Animal and Environmental Sciences* 4(3), 23–30.

Abdel Latef, A.A.H., Abu Alhmad, M.F. and Abdelfattah, K.E. (2017) The possible roles of priming with ZnO nanoparticles in mitigation of salinity stress in lupine (*Lupinus termis*) plants. *Journal of Plant Growth Regulation* 36(1), 60–70.

AbdElgawad, H., Zinta, G., Hegab, M.M., Pandey, R., Asard, H. *et al.* (2016) High salinity induces different oxidative stress and antioxidant responses in maize seedlings organs. *Frontiers in Plant Science* 7, 276.

Abdoli, S., Ghassemi-Golezani, K. and Alizadeh-Salteh, S. (2020) Responses of ajowan (*Trachyspermum ammi* L.) to exogenous salicylic acid and iron oxide nanoparticles under salt stress. *Environmental Science and Pollution Research* 27(29), 36939–36953.

Abideen, Z., Hanif, M., Munir, N. and Nielsen, B.L. (2022) Impact of nanomaterials on the regulation of gene expression and metabolomics of plants under salt stress. *Plants* 11(5), 691.

Abou-Zeid, H.M., Ismail, G.S.M. and Abdel-Latif, S.A. (2021) Influence of seed priming with ZnO nanoparticles on the salt-induced damages in wheat (*Triticum aestivum* L.) plants. *Journal of Plant Nutrition* 44(5), 629–643.

Adrees, M., Khan, Z.S., Ali, S., Hafeez, M., Khalid, S. *et al.* (2020) Simultaneous mitigation of cadmium and drought stress in wheat by soil application of iron nanoparticles. *Chemosphere* 238, 124681.

Ahmad, B. (2014) Interactive effects of silicon and potassium nitrate in improving salt tolerance of wheat. *Journal of Integrative Agriculture* 13(9), 1889–1899.

Al-Khayri, J.M., Rashmi, R., Surya Ulhas, R., Sudheer, W.N., Banadka, A. *et al.* (2023) The role of nanoparticles in response of plants to abiotic stress at physiological, biochemical, and molecular levels. *Plants* 12(2), 292.

Almutairi, Z.M. (2016) Influence of silver nano-particles on the salt resistance of tomato (*Solanum lycopersicum*) during germination. *International Journal of Agriculture and Biology* 18(2), 449–457.

Alsaeedi, A., El-Ramady, H., Alshaal, T., El-Garawany, M., Elhawat, N. *et al.* (2019) Silica nanoparticles boost growth and productivity of cucumber under water deficit and salinity stresses by balancing nutrients uptake. *Plant Physiology and Biochemistry* 139, 1–10.

Arora, S., Sharma, P., Kumar, S., Nayan, R., Khanna, P.K. *et al.* (2012) Gold-nanoparticle induced enhancement in growth and seed yield of *Brassica juncea*. *Journal of Plant Growth Regulation* 66(3), 303–310.

Asad, S., Mukhtar, Z., Nazir, F., Hashmi, J.A., Mansoor, S. *et al.* (2008) Silicon carbide whisker-mediated embryogenic callus transformation of cotton (*Gossypium hirsutum* L.) and regeneration of salt tolerant plants. *Molecular Biotechnology* 40(2), 161–169.

Ashraf, M., Ahmad, R., Bhatti, A.S., Afzal, M., Sarwar, A. *et al.* (2010) Amelioration of salt stress in sugarcane (*Saccharum officinarum* L.) by supplying potassium and silicon in hydroponics. *Pedosphere* 20(2), 153–162.

Banerjee, K., Pramanik, P., Maity, A., Joshi, D.C., Wani, S.H. *et al.* (2019) Methods of using nanomaterials to plant systems and their delivery to plants (mode of entry, uptake, translocation, accumulation, biotransformation, and barriers). In: Ghorbanpour, M. and Wani, S.H. (eds) *Advances in Phytonanotechnology*. Academic Press, pp. 123–152.

Bapat, G., Zinjarde, S. and Tamhane, V. (2020) Evaluation of silica nanoparticle mediated delivery of protease inhibitor in tomato plants and its effect on insect pest *Helicoverpa armigera*. *Colloids and Surfaces B* 193, 111079.

Bernstein, N. (2019) Plants and salt: Plant response and adaptations to salinity. In: Seckbach, J. and Rampelotto, P. (eds) *Model Ecosystems in Extreme Environments*. Academic Press, pp. 101–112.

Berthomieu, P., Conéjéro, G., Nublat, A., Brackenbury, W.J., Lambert, C. *et al.* (2003) Functional analysis of AtHKT1 in *Arabidopsis* shows that Na+ recirculation by the phloem is crucial for salt tolerance. *The EMBO Journal* 22(9), 2004–2014.

Bhatla, S.C. and Lal, M.A. (2023) *Plant Physiology, Development and Metabolism*. Springer Nature.

Bidabadi, S.S., Sabbatini, P. and VanderWeide, J. (2023) Iron oxide (Fe_2o_3) nanoparticles alleviate PEG-simulated drought stress in grape (*Vitis vinifera* L.) plants by regulating leaf antioxidants. *Scientia Horticulturae* 312, 111847.

Blumwald, E. and Poole, R.J. (1985) Na^+/H^+ antiport in isolated tonoplast vesicles from storage tissue of beta vulgaris. *Plant Physiology* 78(1), 163–167.

Bose, J., Rodrigo-Moreno, A. and Shabala, S. (2014) ROS homeostasis in halophytes in the context of salinity stress tolerance. *Journal of Experimental Botany* 65(5), 1241–1257.

Burlaka, O.M., Pirko, Y.V., Yemets, A.I. and Blume, Y.B. (2015) Plant genetic transformation using carbon nanotubes for DNA delivery. *Cytology and Genetics* 49(6), 3–12.

Bybordi, A. (2004) Interactive effects of silicon and potassium nitrate in improving salt tolerance of wheat. *International Journal of Agriculture* 13, 1889–1899.

Byczyńska, A., Zawadzińska, A., Piechocki, R., Pietrak, A. and Salachna, P. (2021) Pre-planting application of colloidal silver nanoparticles enhances bulb yield of Lilium grown under NaC1 stress. *Proceedings of 1st International Electronic Conference on Agronomy 2021*.

Cardoso, V.F., Francesko, A., Ribeiro, C., Bañobre-López, M., Martins, P. *et al.* (2018) Advances in magnetic nanoparticles for biomedical applications. *Advanced Healthcare Materials* 7(5), 1700845.

Chele, K.H., Tinte, M.M., Piater, L.A., Dubery, I.A. and Tugizimana, F. (2021) Soil salinity, a serious environmental issue and plant responses: A metabolomics perspective. *Metabolites* 11(11), 724.

Del Buono, D., Michele, A., Costantino, F., Trevisan, M. and Lucini, L. (2021) Biogenic ZnO nanoparticles synthesized using a novel plant extract: Application to enhance physiological and biochemical traits in maize. *Nanomaterials* 11(5), 1270.

Eichert, T. and Goldbach, H.E. (2008) Equivalent pore radii of hydrophilic foliar uptake routes in stomatous and astomatous leaf surfaces: Further evidence for a stomatal pathway. *Physiologia Plantarum* 132(4), 491–502.

Eichert, T., Kurtz, A., Steiner, U. and Goldbach, H.E. (2008) Size exclusion limits and lateral heterogeneity of the stomatal foliar uptake pathway for aqueous solutes and water-suspended nanoparticles. *Physiologia Plantarum* 134(1), 151–160.

El-Saber, M.M., Mahdi, A.A., Hassan, A.H., Farroh, K.Y. and Osman, A. (2021) Effects of magnetite nanoparticles on physiological processes to alleviate salinity induced oxidative damage in wheat. *Journal of the Science of Food and Agriculture* 101(13), 5550–5562.

El-Hefnawy, S.F. (2020) Nano NPK and growth regulator promoting changes in growth and mitotic index of pea plants under salinity stress. *Journal of Agricultural Chemistry and Biotechnology* 11(9), 263–269.

El-Saadony, M.T., Desoky, E.S.M., Saad, A.M., Eid, R.S., Selem, E. *et al.* (2021) Biological silicon nanoparticles improve *Phaseolus vulgaris* L. yield and minimize its contaminant contents on a heavy metals-contaminated saline soil. *Journal of Environmental Sciences* 106, 1–14.

Faizan, M., Bhat, J.A., Chen, C., Alyemeni, M.N., Wijaya, L. *et al.* (2021) Zinc oxide nanoparticles (ZnO-nps) induce salt tolerance by improving the antioxidant system and photosynthetic machinery in tomato. *Plant Physiology and Biochemistry* 161, 122–130.

FAO (2020) *The State of Food Security and Nutrition in the World: Transforming food systems for affordable healthy diets*. FAO, Rome. DOI: 10.4060/ca9692en.

Fathi, A., Zahedi, M. and Torabian, S. (2017) Effect of interaction between salinity and nanoparticles (Fe_2O_3 and ZnO) on physiological parameters of Zea mays l. *Journal of Plant Nutrition* 40(19), 2745–2755.

Fu, C., Khan, M.N., Yan, J., Hong, X., Zhao, F. *et al.* (2023) Mechanisms of nanomaterials for improving plant salt tolerance. *Crop and Environment* 2(2), 92–99.

Gaafar, R., Diab, R., Halawa, M., Elshanshory, A., El-Shaer, A. *et al.* (2020) Role of zinc oxide nanoparticles in ameliorating salt tolerance in soybean. *Egyptian Journal of Botany* 60(3), 733–747.

Garg, N. and Bhandari, P. (2016) Silicon nutrition and mycorrhizal inoculations improve growth, nutrient status, K+/Na+ ratio and yield of *Cicer arietinum* L. genotypes under salinity stress. *Plant Growth Regulation* 78(3), 371–387.

Gharbi, N., Pressac, M., Hadchouel, M., Szwarc, H., Wilson, S.R. *et al.* (2005) Fullerene is a powerful antioxidant *in vivo* with no acute or subacute toxicity. *Nano Letters* 5(12), 2578–2585.

Gill, S.S. and Tuteja, N. (2010) Reactive oxygen species and antioxidant machinery in abiotic stress tolerance in crop plants. *Plant Physiology and Biochemistry* 48(12), 909–930.

Gisbert, C., Rus, A.M., Bolarín, M.C., López-Coronado, J.M., Arrillaga, I. *et al.* (2000) The yeast HAL1 gene improves salt tolerance of transgenic tomato. *Plant Physiology* 123(1), 393–402.

Gohari, G., Mohammadi, A., Akbari, A., Panahirad, S., Dadpour, M.R. *et al.* (2020) Titanium dioxide nanoparticles (TiO_2 NPs) promote growth and ameliorate salinity stress effects on essential oil profile and biochemical attributes of *Dracocephalum moldavica*. *Scientific Reports* 10(1), 912.

Gohari, G., Panahirad, S., Sadeghi, M., Akbari, A., Zareei, E. *et al.* (2021) Putrescine-functionalized carbon quantum dot (put-CQD) nanoparticles effectively prime grapevine (*Vitis vinifera cv.* 'Sultana') against salt stress. *BMC Plant Biology* 21(1), 120.

Goswami, P., Mathur, J. and Srivastava, N. (2022) Silica nanoparticles as novel sustainable approach for plant growth and crop protection. *Heliyon* 8(7), e09908.

Gowayed, M.H., Al-Zahrani, H.S.M. and Metwali, E.M.R. (2017) Improving the salinity tolerance in potato (*Solanum tuberosum*) by exogenous application of silicon dioxide nanoparticles. *International Journal of Agriculture and Biology* 19(1), 183–192.

Ha, N., Seo, E., Kim, S. and Lee, S.J. (2021) Adsorption of nanoparticles suspended in a drop on a leaf surface of *Perilla frutescens* and their infiltration through stomatal pathway. *Scientific Reports* 11(1), 11556.

Hammam, A.A. and Mohamed, E.S. (2020) Mapping soil salinity in the east nile delta using several methodological approaches of salinity assessment. *The Egyptian Journal of Remote Sensing and Space Science* 23(2), 125–131.

Hao, Y., Yang, X., Shi, Y., Song, S., Xing, J. *et al.* (2013) Magnetic gold nanoparticles as a vehicle for fluorescein isothiocyanate and DNA delivery into plant cells. *Botany* 91(7), 457–466.

Hernández-Hernández, H., Juárez-Maldonado, A., Benavides-Mendoza, A., Ortega-Ortiz, H., Cadenas-Pliego, G. *et al.* (2018) Chitosan-PVA and copper nanoparticles improve growth and overexpress the SOD and JA genes in tomato plants under salt stress. *Agronomy* 8(9), 175.

Hong, J., Wang, L., Sun, Y., Zhao, L., Niu, G. *et al.* (2016) Foliar applied nanoscale and microscale ceo_2 and cuo alter cucumber (*Cucumis sativus*) fruit quality. *Science of the Total Environment* 563, 904–911.

Hong, J., Wang, C., Wagner, D.C., Gardea-Torresdey, J.L., He, F. *et al.* (2021) Foliar application of nanoparticles: Mechanisms of absorption, transfer, and multiple impacts. *Environmental Science* 8(5), 1196–1210.

Hopmans, J.W., Qureshi, A.S., Kisekka, I., Munns, R., Grattan, S.R. *et al.* (2021) Critical knowledge gaps and research priorities in global soil salinity. *Advances in Agronomy* 169, 1–191.

Hoshida, H., Tanaka, Y., Hibino, T., Hayashi, Y., Tanaka, A. *et al.* (2000) Enhanced tolerance to salt stress in transgenic rice that overexpresses chloroplast glutamine synthetase. *Plant Molecular Biology* 43, 103–111.

Hu, D., Lv, G., Qie, Y., Wang, H., Yang, F. *et al.* (2021) Response of morphological characters and photosynthetic characteristics of *Haloxylon ammodendron* to water and salt stress. *Sustainability* 13(1), 388.

Hu, D., Bai, Y., Li, L., Ai, M., Jin, J. *et al.* (2022) Phytotoxicity assessment study of gold nanocluster on broad bean (*Vicia faba* L.) seedling. *Environmental Pollutants and Bioavailability* 34(1), 284–296.

Huang, L., Kuang, L., Wu, L., Wu, D., and Zhang, G (2019) Comparisons in functions of HKT1;5 transporters between *Hordeum marinum* and *Hordeum vulgare* in responses to salt stress. *Plant Growth Regulation* 89(3), 309–319.

Hussein, M.M. and Abou-Baker, N. (2018) The contribution of nano-zinc to alleviate salinity stress on cotton plants. *Royal Society Open Science* 5(8), 171809. DOI: 10.1098/rsos.171809.

Kalita, R., Saha, O., Rahman, N., Tiwari, S., and Phukon, M. (2021) Nanotechnology in agriculture. In: Al-Khayri, J., Ansari, M.I., and Singh, A.K. (eds) *Nanobiotechnology: Mitigation of Abiotic Stress in Plants*. Springer, pp. 101–116.

Khalofah, A., Kilany, M., and Migdadi, H. (2021) Phytostimulatory influence of *Comamonas testosteroni* and silver nanoparticles on *Linum usitatissimum* L. under salinity stress. *Plants* 10(4), 790.

Khan, N. and Bano, A. (2019) Exopolysaccharide producing rhizobacteria and their impact on growth and drought tolerance of wheat grown under rainfed conditions. *PLoS One* 14(9), e0222302.

Khan, I., Raza, M.A., Awan, S.A., Shah, G.A., Rizwan, M. *et al.* (2020a) Amelioration of salt induced toxicity in pearl millet by seed priming with silver nanoparticles (AgNPs): The oxidative damage, antioxidant enzymes and ions uptake are major determinants of salt tolerant capacity. *Plant Physiology and Biochemistry* 156, 221–232.

Khan, W.U.D., Tanveer, M., Shaukat, R., Ali, M. and Pirdad, F. (2020b) An overview of salinity tolerance mechanism in plants. In: Hassanuzzaman, M. and Tanveer, M. (eds) *Salt and Drought Stress Tolerance in Plants: Signaling Networks and Adaptive Mechanisms*. Springer, pp. 1–16.

Khan, M.N., Li, Y., Khan, Z., Chen, L., Liu, J. *et al.* (2021) Nanoceria seed priming enhanced salt tolerance in rapeseed through modulating ROS homeostasis and α-amylase activities. *Journal of Nanobiotechnology* 19(1), 276.

Kole, C., Kole, P., Randunu, K.M., Choudhary, P., Podila, R. *et al.* (2013) Nanobiotechnology can boost crop production and quality: First evidence from increased plant biomass, fruit yield and phytomedicine content in bitter melon (*Momordica charantia*). *BMC Biotechnology* 13, 1–10.

Kundu, P., Kalwan, G., Gill, R., Chauhan, N.S. and Gill, S.S. (2023) Effects of nanoparticles on the plant growth under salinity stress conditions. In: Chauhan, N.S. and Gill, S.S. (eds) *The Impact of Nanoparticles on Agriculture and Soil*. Academic Press, pp. 239–257.

Lalarukh, I., Zahra, N., Al Huqail, A.A., Amjad, S.F., Al-Dhumri, S.A. *et al.* (2022) Exogenously applied Zno nanoparticles induced salt tolerance in potentially high yielding modern wheat (*Triticum aestivum* L.) cultivars. *Environmental Technology & Innovation* 27, 102799.

Larue, C., Veronesi, G., Flank, A.M., Surble, S., Herlin-Boime, N. *et al.* (2012) Comparative uptake and impact of TiO_2 nanoparticles in wheat and rapeseed. *Journal of Toxicology and Environmental Health, Part A* 75(13–15), 722–734.

Li, L., Luo, Y., Li, R., Zhou, Q., Peijnenburg, W.J.G.M. *et al.* (2020) Effective uptake of submicrometre plastics by crop plants via a crack-entry mode. *Nature Sustainability* 3(11), 929–937.

Li, Z., Zhu, L., Zhao, F., Li, J., Zhang, X. *et al.* (2022) Plant salinity stress response and nano-enabled plant salt tolerance. *Frontiers in Plant Science* 13, 843994.

Liang, Y., Shen, Q., Shen, Z. and Ma, T. (1996) Effects of silicon on salinity tolerance of two barley cultivars. *Journal of Plant Nutrition* 19(1), 173–183.

Liu, Q., Chen, B., Wang, Q., Shi, X., Xiao, Z. *et al.* (2009) Carbon nanotubes as molecular transporters for walled plant cells. *Nano Letters* 9(3), 1007–1010.

Liu, J., Fu, C., Li, G., Khan, M.N. and Wu, H. (2021a) ROS homeostasis and plant salt tolerance: Plant nanobiotechnology updates. *Sustainability* 13(6), 3552.

Liu, J., Li, G., Chen, L., Gu, J., Wu, H. *et al.* (2021b) Cerium oxide nanoparticles improve cotton salt tolerance by enabling better ability to maintain cytosolic K^+/Na^+ ratio. *Journal of Nanobiotechnology* 19(1), 153.

Lu, L., Huang, M., Huang, Y., Corvini, P.F.X., Ji, R. *et al.* (2020) Mn_3O_4 nanozymes boost endogenous antioxidant metabolites in cucumber (*Cucumis sativus*) plant and enhance resistance to salinity stress. *Environmental Science: Nano* 7(6), 1692–1703.

Lv, J., Christie, P. and Zhang, S. (2019) Uptake, translocation, and transformation of metal-based nanoparticles in plants: Recent advances and methodological challenges. *Environmental Science* 6(1), 41–59.

Ma, X. and Yan, J. (2018) Plant uptake and accumulation of engineered metallic nanoparticles from lab to field conditions. *Current Opinion in Environmental Science & Health* 6, 16–20.

Madkour, L.H. (2019) *Nanoelectronic materials: Fundamentals and applications*, Vol. 116. Springer.

Magome, H., Yamaguchi, S., Hanada, A., Kamiya, Y. and Oda, K. (2008) The DDF1 transcriptional activator upregulates expression of a gibberellin-deactivating gene, GA2ox7, under high-salinity stress in *Arabidopsis*. *The Plant Journal* 56(4), 613–626.

Mahmoud, L.M., Dutt, M., Shalan, A.M., El-Kady, M.E., El-Boray, M.S. *et al.* (2020) Silicon nanoparticles mitigate oxidative stress of *in vitro*-derived banana (*Musa acuminata* 'grand nain') under simulated water deficit or salinity stress. *South African Journal of Botany* 132, 155–163.

Maqsood, M.F., Shahbaz, M., Khalid, F., Rasheed, Y., Asif, K. *et al.* (2023) Biogenic nanoparticles application in agriculture for ROS mitigation and abiotic stress tolerance: A review. *Plant Stress* 10, 100281.

Martin-Ortigosa, S., Valenstein, J.S., Lin, V.S.-Y., Trewyn, B.G. and Wang, K. (2012) Gold functionalized mesoporous silica nanoparticle mediated protein and DNA codelivery to plant cells via the biolistic method. *Advanced Functional Materials* 22(17), 3576–3582.

Martin-Ortigosa, S., Peterson, D.J., Valenstein, J.S., Lin, V.S.Y., Trewyn, B.G. *et al.* (2014) Mesoporous silica nanoparticle-mediated intracellular cre protein delivery for maize genome editing via *loxp* site excision. *Plant Physiology* 164(2), 537–547.

Matoh, T., Kairusmee, P. and Takahashi, E. (1986) Salt-induced damage to rice plants and alleviation effect of silicate. *Soil Science and Plant Nutrition* 32(2), 295–304.

Mattiello, A. and Marchiol, L. (2017) Application of nanotechnology in agriculture: Assessment of TiO_2 nanoparticle effects on barley. *Application of Titanium Dioxide* 23–39.

Mazhar, M.W., Ishtiaq, M., Maqbool, M., Akram, R., Shahid, A. *et al.* (2022) Seed priming with iron oxide nanoparticles raises biomass production and agronomic profile of water-stressed flax plants. *Agronomy* 12(5), 982.

McClements, D.J. and Xiao, H. (2017) Is nano safe in foods? Establishing the factors impacting the gastrointestinal fate and toxicity of organic and inorganic food-grade nanoparticles. *NPJ Science of Food* 1(1), 6.

Moaveni, P. (2011) The nanoparticles in plants. *Journal of Nanostructure in Chemistry* 2, 59–78.

Munns, R. (2005) Genes and salt tolerance: Bringing them together. *The New Phytologist* 167(3), 645–663.

Mushtaq, A., Jamil, N., Riaz, M., Hornyak, G.L., Ahmed, N, *et al.* (2017) Synthesis of silica nanoparticles and their effect on priming of wheat (*Triticum aestivum* L.) under salinity stress. *Biological Forum – An International Journal* 9, 150–157.

Mustafa, N., Raja, N.I., Ilyas, N., Ikram, M., Mashwani, Z. *et al.* (2021) Foliar applications of plant-based titanium dioxide nanoparticles to improve agronomic and physiological attributes of wheat (*Triticum aestivum* L.) plants under salinity stress. *Green Processing and Synthesis* 10(1), 246–257.

Naqvi, S., Maitra, A.N., Abdin, M.Z., Akmal, M.D., Arora, I. *et al.* (2012) Calcium phosphate nanoparticle mediated genetic transformation in plants. *Journal of Materials Chemistry* 22(8), 3500.

Narayan, R., Nayak, U.Y., Raichur, A.M. and Garg, S. (2018) Mesoporous silica nanoparticles: A comprehensive review on synthesis and recent advances. *Pharmaceutics* 10(3), 118. DOI: 10.3390/pharmaceutics10030118.

Noman, M., Ahmed, T., Shahid, M., Niazi, M.B.K., Qasim, M. *et al.* (2021) Biogenic copper nanoparticles produced by using the *Klebsiella pneumoniae* strain NST2 curtailed salt stress effects in maize by modulating the cellular oxidative repair mechanisms. *Ecotoxicology and Environmental Safety* 217, 112264.

Pallavi, Mehta, C.M., Srivastava, R., Arora, S. and Sharma, A.K. (2016) Impact assessment of silver nanoparticles on plant growth and soil bacterial diversity. *3 Biotech 6* 254. DOI: 10.1007/s13205-016-0567-7.

Panova, G.G., Ktitorova, I.N., Skobeleva, O.V., Sinjavina, N.G., Charykov, N.A. *et al.* (2016) Impact of polyhydroxy fullerene (fullerol or fullerenol) on growth and biophysical characteristics of barley seedlings in favourable and stressful conditions. *Plant Growth Regulation* 79(3), 309–317.

Pastan, I. and Willingham, M.C. (1985) The pathway of endocytosis. In: Pastan, I. and Willingham, M.C. (eds) *Endocytosis*. Plenum Press, New York and London, pp. 1–44.

Peng, Y., Chen, L., Zhu, L., Cui, L., Yang, L. *et al.* (2022) CsAKT1 is a key gene for the CeO_2 nanoparticle's improved cucumber salt tolerance: A validation from CRISPR-Cas9 lines. *Environmental Science: Nano* 9(12), 4367–4381.

Peralta-Videa, J.R., Hernandez-Viezcas, J.A., Zhao, L., Diaz, B.C., Ge, Y. *et al.* (2014) Cerium dioxide and zinc oxide nanoparticles alter the nutritional value of soil cultivated soybean plants. *Plant Physiology and Biochemistry* 80, 128–135.

Pérez-de-Luque, A. (2017) Interaction of nanomaterials with plants: What do we need for real applications in agriculture?. *Frontiers in Environmental Science* 5, 12.

Pérez-Labrada, F., López-Vargas, E.R., Ortega-Ortiz, H., Cadenas-Pliego, G., Benavides-Mendoza, A. *et al.* (2019) Responses of tomato plants under saline stress to foliar application of copper nanoparticles. *Plants* 8(6), 151.

Plaksenkova, I., Jermaļonoka, M., Bankovska, L., Gavarāne, I., Gerbreders, V. *et al.* (2019) Effects of Fe_3O_4 nanoparticle stress on the growth and development of rocket *eruca sativa*. *Journal of Nanomaterials* 2019(1), 2678247.

Plett, D., Safwat, G., Gilliham, M., Møller, I.S., Roy, S. *et al.* (2010) Improved salinity tolerance of rice through cell type-specific expression of AtHKT1;1. *PLoS One* 5(9), e12571. DOI: 10.1371/journal.pone.0012571.

Prasad, A., Astete, C.E., Bodoki, A.E., Windham, M., Bodoki, E. *et al.* (2018) Zein nanoparticles uptake and translocation in hydroponically grown sugar cane plants. *Journal of Agricultural and Food Chemistry* 66(26), 6544–6551.

Priester, J.H., Ge, Y., Mielke, R.E., Horst, A.M., Moritz, S.C. *et al.* (2012) Soybean susceptibility to manufactured nanomaterials with evidence for food quality and soil fertility interruption. *Proceedings of the National Academy of Sciences of the United States of America* 109(37), E2451–E2456.

Qian, D., Xiong, S., Li, M., Tian, L. and Qing Qu, L. (2021) OsFes1C, a potential nucleotide exchange factor for OsBiP1, is involved in the ER and salt stress responses. *Plant Physiology* 187(1), 396–408.

Qiao, J. and Qi, L. (2021) Recent progress in plant-gold nanoparticles fabrication methods and bio-applications. *Talanta* 223, 121396.

Rahman, M.S., Chakraborty, A., Mazumdar, S., Nandi, N.C., Bhuiyan, M.N.I. *et al.* (2020) Effects of poly(Vinylpyrrolidone) protected platinum nanoparticles on seed germination and growth performance of *pisum sativum*. *Nano-Structures & Nano-Objects* 21, 100408.

Rajaee Behbahani, S., Iranbakhsh, A., Ebadi, M., Majd, A. and Ardebili, Z.O. (2020) Red elemental selenium nanoparticles mediated substantial variations in growth, tissue differentiation, metabolism, gene transcription, epigenetic cytosine DNA methylation, and callogenesis in bittermelon (*Momordica charantia*): An in *vitro* experiment. *PLoS One* 15(7), e0235556.

Raliya, R., Franke, C., Chavalmane, S., Nair, R., Reed, N. *et al.* (2016) Quantitative understanding of nanoparticle uptake in watermelon plants. *Frontiers in Plant Science* 7, 1288.

Rani, S., Kumari, N. and Sharma, V. (2023) Uptake, translocation, transformation and physiological effects of nanoparticles in plants. *Archives of Agronomy and Soil Science* 69(9), 1579–1599.

Ray, P.C., Yu, H. and Fu, P.P. (2009) Toxicity and environmental risks of nanomaterials: Challenges and future needs. *Journal of Environmental Science and Health* 27(1), 1–35.

Rezayian, M., Niknam, V. and Arabloo, M. (2023) Iron nanoparticle regulate succinate dehydrogenase activity in canola plants under drought stress. *Scientific Reports* 13(1), 9628.

Rico, C.M., Morales, M.I., Barrios, A.C., McCreary, R., Hong, J. *et al.* (2013) Effect of cerium oxide nanoparticles on the quality of rice (*Oryza sativa* L.) grains. *Journal of Agricultural and Food Chemistry* 61(47), 11278–11285.

Robards, A.W. and Robb, M.E. (1972) Uptake and binding of uranyl ions by barley roots. *Science* 178(4064), 980–982.

Rossi, L., Zhang, W., Lombardini, L. and Ma, X. (2016) The impact of cerium oxide nanoparticles on the salt stress responses of *Brassica napus* L. *Environmental Pollution* 219, 28–36.

Rus, A.M., Estan, M.T., Gisbert, C., Garcia-Sogo, B., Serrano, R. *et al.* (2001) Expressing the yeast HAL1 gene in tomato increases fruit yield and enhances k^+/na^+ selectivity under salt stress. *Plant, Cell & Environment* 24(8), 875–880.

Sarwar, N., Rehman, A., Ahmad, S. and Hasanuzzaman, M. (2022) *Modern Techniques of Rice Crop Production*. Springer. DOI: 10.1007/978-981-16-4955-4.

Sattelmacher, B. (2001) The apoplast and its significance for plant mineral nutrition. *New Phytologist* 149(2), 167–192.

Satti, S.H., Raja, N.I., Ikram, M., Oraby, H.F., Mashwani, Z.U.R. *et al.* (2022) Plant-based titanium dioxide nanoparticles trigger biochemical and proteome modifications in *Triticum aestivum* L. under biotic stress of *puccinia striiformis*. *Molecules* 27(13), 4274.

Schönherr, J. (2002) A mechanistic analysis of penetration of glyphosate salts across Astomatous cuticular membranes. *Pest Management Science: Formerly Pesticide Science* 58(4), 343–351.

Schwab, F., Zhai, G., Kern, M., Turner, A., Schnoor, J.L. *et al.* (2016) Barriers, pathways and processes for uptake, translocation and accumulation of nanomaterials in plants: Critical review. *Nanotoxicology* 10(3), 257–278.

Serrano, R. (1996) Salt tolerance in plants and microorganisms: Toxicity targets and defense responses. *International Review of Cytology* 165, 1–52.

Serrano, R. and Gaxiola, R. (1994) Microbial models and salt stress tolerance in plants. *Critical Reviews in Plant Sciences* 13(2), 121–138.

Sharma, A., Shahzad, B., Kumar, V., Kohli, S.K., Sidhu, G.P.S. *et al.* (2019) Phytohormones regulate accumulation of Osmolytes under abiotic stress. *Biomolecules* 9(7), 285.

Shi, H., Lee, B., Wu, S.J. and Zhu, J.K. (2003) Overexpression of a plasma membrane Na^+/H^+ antiporter gene improves salt tolerance in *Arabidopsis thaliana*. *Nature Biotechnology* 21(1), 81–85.

Shoresh, M., Harman, G.E. and Mastouri, F. (2010) Induced systemic resistance and plant responses to fungal biocontrol agents. *Annual Review of Phytopathology* 48, 21–43.

Sillen, W.M., Thijs, S., Abbamondi, G.R., Janssen, J., Weyens, N. *et al.* (2015) Effects of silver nanoparticles on soil microorganisms and maize biomass are linked in the rhizosphere. *Soil Biology and Biochemistry* 91, 14–22.

Singh, A. (2022) Soil salinity: A global threat to sustainable development. *Soil Use and Management* 38(1), 39–67.

Singh, D., Sillu, D., Kumar, A. and Agnihotri, S. (2021) Dual nanozyme characteristics of iron oxide nanoparticles alleviate salinity stress and promote the growth of an agroforestry tree, *eucalyptus tereticornis sm*. *Environmental Science: Nano* 8(5), 1308–1325.

Singh, A., Rajput, V.D., Sharma, R., Ghazaryan, K. and Minkina, T. (2023) Salinity stress and nanoparticles: Insights into antioxidative enzymatic resistance, signaling, and defense mechanisms. *Environmental Research* 235, 116585.

Slomberg, D.L. and Schoenfisch, M.H. (2012) Silica nanoparticle phytotoxicity to *Arabidopsis thaliana*. *Environmental Science & Technology* 46(18), 10247–10254.

Soliman, M.H., Alnusairi, G.S.H., Khan, A.A., Alnusaire, T.S., Fakhr, M.A. *et al.* (2023) Biochar and selenium nanoparticles induce water transporter genes for sustaining carbon assimilation and grain production in salt-stressed wheat. *Journal of Plant Growth Regulation* 42(3), 1522–1543.

Sturikova, H., Krystofova, O., Huska, D. and Adam, V. (2018) Zinc, zinc nanoparticles and plants. *Journal of Hazardous Materials* 349, 101–110.

Sun, L., Song, F., Guo, J., Zhu, X., Liu, S. *et al.* (2020) Nano-ZnO-induced drought tolerance is associated with melatonin synthesis and metabolism in maize. *International Journal of Molecular Sciences* 21(3), 782.

Taha, R.A., Hassan, M.M., Ibrahim, E.A., Abou Baker, N.H. and Shaaban, E.A. (2016) Carbon nanotubes impact on date palm *in vitro* cultures. *Plant Cell, Tissue and Organ Culture (PCTOC)* 127(2), 525–534.

Tahir, M.A., Aziz, T., Farooq, M. and Sarwar, G. (2012) Silicon-induced changes in growth, ionic composition, water relations, chlorophyll contents and membrane permeability in two salt-stressed wheat genotypes. *Archives of Agronomy and Soil Science* 58(3), 247–256.

Tawfik, M.M., Mohamed, M.H., Sadak, M.S. and Thalooth, A.T. (2021) Iron oxide nanoparticles effect on growth, physiological traits and nutritional contents of *Moringa oleifera* grown in saline environment. *Bulletin of the National Research Centre* 45(1), 1–9.

Tian, F., Hou, M., Qiu, Y., Zhang, T. and Yuan, Y. (2020) Salinity stress effects on transpiration and plant growth under different salinity soil levels based on thermal infrared remote (TIR) technique. *Geoderma* 357, 113961.

Torabian, S., Zahedi, M. and Khoshgoftar, A.H. (2016) Effects of foliar spray of two kinds of zinc oxide on the growth and ion concentration of sunflower cultivars under salt stress. *Journal of Plant Nutrition* 39(2), 172–180.

Torabian, S., Farhangi-Abriz, S. and Zahedi, M. (2018) Efficacy of $FeSO_4$ nano formulations on osmolytes and antioxidative enzymes of sunflower under salt stress. *Indian Journal of Plant Physiology* 23(2), 305–315.

Torney, F., Trewyn, B.G., Lin, V.S.Y. and Wang, K. (2007) Mesoporous silica nanoparticles deliver DNA and chemicals into plants. *Nature Nanotechnology* 2, 295–300.

Tran, L.-S.P., Nakashima, K., Sakuma, Y., Osakabe, Y., Qin, F. *et al.* (2007) Co-expression of the stress-inducible zinc finger Homeodomain ZFHD1 and NAC transcription factors enhances expression of the ERD1 gene in *Arabidopsis*. *The Plant Journal* 49(1), 46–63.

Tripathi, D.K., Singh, S., Singh, S., Pandey, R., Singh, V.P. *et al.* (2017) An overview on manufactured nanoparticles in plants: Uptake, translocation, accumulation and phytotoxicity. *Plant Physiology and Biochemistry* 110, 2–12.

Trujillo-Reyes, J., Majumdar, S., Botez, C.E., Peralta-Videa, J.R. and Gardea-Torresdey, J.L. (2014) Exposure studies of core-shell Fe/Fe_3O_4 and Cu/CuO NPs to lettuce (*Lactuca sativa*) plants: Are they a potential physiological and nutritional hazard? *Journal of Hazardous Materials* 267, 255–263.

Tuna, A.L., Kaya, C., Higgs, D., Murillo-Amador, B., Aydemir, S. *et al.* (2008) Silicon improves salinity tolerance in wheat plants. *Environmental and Experimental Botany* 62(1), 10–16.

Uozumi, N., Kim, E.J., Rubio, F., Yamaguchi, T., Muto, S. *et al.* (2000) The *Arabidopsis* HKT1 gene homolog mediates inward na^+ currents in *Xenopus laevis oocytes* and Na^+ uptake in *Saccharomyces cerevisiae*. *Plant Physiology* 122(4), 1249–1260.

Véry, A.A. and Sentenac, H. (2003) Molecular mechanisms and regulation of K+ transport in higher plants. *Annual Review of Plant Biology* 54(1), 575–603.

Vijayakumar, P.S., Abhilash, O.U., Khan, B.M. and Prasad, B.L.V. (2010) Nanogold-Loaded sharp-edged carbon bullets as plant-gene carriers. *Advanced Functional Materials* 20(15), 2416–2423.

Wahid, I., Kumari, S., Ahmad, R., Hussain, S.J., Alamri, S. *et al.* (2020) Silver nanoparticle regulates salt tolerance in wheat through changes in ABA concentration, ion homeostasis, and defense systems. *Biomolecules* 10(11), 1506.

Wang, Q., Ma, X., Zhang, W., Pei, H. and Chen, Y. (2012) The impact of cerium oxide nanoparticles on tomato (*Solanum lycopersicum* L.) and its implications for food safety. *Metallomics* 4(10), 1105–1112.

Wang, P., Lombi, E., Zhao, F.J. and Kopittke, P.M. (2016) Nanotechnology: A new opportunity in plant sciences. *Trends in Plant Science* 21(8), 699–712.

Wang, F., Shi, Z., Biswas, A., Yang, S. and Ding, J. (2020) Multi-algorithm comparison for predicting soil salinity. *Geoderma* 365, 114211.

Wang, F., Yang, S., Wei, Y., Shi, Q. and Ding, J. (2021) Characterizing soil salinity at multiple depth using electromagnetic induction and remote sensing data with random forests: A case study in Tarim River Basin of Southern Xinjiang, China. *Science of the Total Environment* 754, 142030.

Wang, X., Xie, H., Wang, P. and Yin, H. (2023) Nanoparticles in plants: Uptake, transport and physiological activity in leaf and root. *Materials* 16(8), 3097.

Wu, H., Shabala, L., Shabala, S. and Giraldo, J.P. (2018a) Hydroxyl radical scavenging by cerium oxide nanoparticles improves *Arabidopsis* salinity tolerance by enhancing leaf mesophyll potassium retention. *Environmental Science: Nano* 5(7), 1567–1583.

Wu, H., Zhang, X., Giraldo, J.P. and Shabala, S. (2018b) It is not all about sodium: Revealing tissue specificity and signalling roles of potassium in plant responses to salt stress. *Plant and Soil* 431(1–2), 1–17.

Wu, Q., Miao, W., Zhang, Y., Gao, H. and Hui, D. (2020) Mechanical properties of nanomaterials: A review. *Nanotechnology Reviews* 9(1), 259–273.

Xuan, L., Ju, Z., Skonieczna, M., Zhou, P.K. and Huang, R. (2023) Nanoparticles-induced potential toxicity on human health: Applications, toxicity mechanisms, and evaluation models. *Medical Communications* 4(4), e327.

Yan, A. and Chen, Z. (2019) Impacts of silver nanoparticles on plants: A focus on the phytotoxicity and underlying mechanism. *International Journal of Molecular Science* 20, 1003.

Yang, Y. and Guo, Y. (2018) Elucidating the molecular mechanisms mediating plant salt-stress responses. *The New Phytologist* 217(2), 523–539.

Ye, Y., Cota-Ruiz, K., Hernández-Viezcas, J.A., Valdés, C., Medina-Velo, I.A. *et al.* (2020) Manganese nanoparticles control salinity-modulated molecular responses in *Capsicum annuum* L. through priming: A sustainable approach for agriculture. *ACS Sustainable Chemistry & Engineering* 8(3), 1427–1436.

Yu-Qin, F.U., Lu-Hua, L.I., Pi-Wu, W.A.N.G., Jing, Q.U., Yong-Ping, F.U. *et al.* (2012) Delivering DNA into plant cell by gene carriers of zns nanoparticles. *Chemical Research in Chinese Universities* 28(4), 672–676.

Zeng, L., Shannon, M.C. and Lesch, S.M. (2001) Timing of salinity stress affects rice growth and yield components. *Agricultural Water Management* 48(3), 191–206.

Zhao, F., Guo, S., Zhang, H. and Zhao, Y. (2006) Expression of yeast SOD2 in transgenic rice results in increased salt tolerance. *Plant Science* 170(2), 216–224.

Zhao, L., Peralta-Videa, J.R., Varela-Ramirez, A., Castillo-Michel, H., Li, C. *et al.* (2012) Effect of surface coating and organic matter on the uptake of CeO_2 NPs by corn plants grown in soil: Insight into the uptake mechanism. *Journal of Hazardous Materials* 225, 131–138.

Zhao, X., Meng, Z., Wang, Y., Chen, W., Sun, C. *et al.* (2017) Pollen magnetofection for genetic modification with magnetic nanoparticles as gene carriers. *Nature Plants* 3(12), 956–964.

Zhao, G., Zhao, Y., Lou, W., Su, J., Wei, S. *et al.* (2019) Nitrate reductase-dependent nitric oxide is crucial for multi-walled carbon nanotube-induced plant tolerance against salinity. *Nanoscale* 11(21), 10511–10523.

Zhou, H., Wu, H., Zhang, F., Su, Y., Guan, W. *et al.* (2021) Molecular basis of cerium oxide nanoparticle enhancement of rice salt tolerance and yield. *Environmental Science: Nano* 8(11), 3294–3311.

Zhu, J.K. (2002) Salt and drought stress signal transduction in plants. *Annual Review of Plant Biology* 53(1), 247–273.

Zhu, J.K. (2003) Regulation of ion homeostasis under salt stress. *Current Opinion in Plant Biology* 6(5), 441–445.

Zhu, J., Li, J., Shen, Y., Liu, S., Zeng, N. *et al.* (2020) Mechanism of zinc oxide nanoparticle entry into wheat seedling leaves. *Environmental Science: Nano* 7(12), 3901–3913.

5 Future Crop Designing: Anti-stress Capacities Gained by CRISPR-mediated Releasing the Potential of Functional Genomes

Rashi Bhavsar[1], Sanjay S.[2], Shreya Sen[3], Vandana Chendanda Devaiah[4], Akshatha Banadka[4,5]*, Praveen Nagella[5]* and Jameel M. Al-Khayri[6]*

[1]Department of Biotechnology, The Oxford College of Science, Bangalore, Karnataka, India; [2]Department of Biotechnology, School of Biological Sciences and Biotechnology, Goa University, India; [3]National Centre for Biological Sciences, Bengaluru, Karnataka, India; [4]Department of Biotechnology and Genetics, School of Sciences, Jain (deemed to be a university), Bengaluru, Karnataka, India; [5]Department of Life Sciences, School of Sciences, CHRIST (deemed to be a university), Bengaluru, Karnataka, India; [6]Department of Agricultural Biotechnology, College of Agriculture and Food Sciences, King Faisal University, Al-Ahsa, Saudi Arabia

Abstract

Abiotic stresses, including temperature fluctuations, salinity, and drought, as well as biotic stresses such as viral, bacterial, and fungal infections, exert detrimental effects on plant growth and development, thereby significantly impeding overall plant productivity and crop yield. Traditionally, the sustainable mitigation of abiotic stress has been achieved through the breeding of tolerant cultivars; however, this process is characterized by its time-consuming and labor-intensive nature, as well as its inherent lack of precision. Thus, there is a pressing need to adopt advanced genome technology to address these limitations and enhance the efficacy of stress-tolerance breeding efforts. This can be addressed by facilitating site-specific modifications of selected functional genomic elements, thus providing a potential avenue for introducing desired traits to combat adverse stress conditions. Among various genome engineering methodologies, CRISPR-Cas9 has emerged as the most promising genome-editing tool, attributed to its notable efficiency, precision, and rapidity. This study offers insights into the prospective trajectory of crop improvement through the advancement of crop enhancement strategies, employing CRISPR technology to enhance crop resilience against stress conditions by selectively modifying or activating specific functional genomes.

Keywords: crop stress, abiotic stress, biotic stress, CRISPR technology, CRISPR-Cas9, crop improvement, crop resilience

*Corresponding authors: b.akshatha@jainuniversity.ac.in, praveen.n@christuniversity.in and jkhayri@kfu.edu.sa

© CAB International 2025. *Stress-Resilient Crops: Coordinated omics-CRISPR-Nanotechnology Strategies* (ed. Jen-Tsung Chen)
DOI: 10.1079/9781800627307.0005

5.1 Introduction

Plants are commonly subjected to myriad environmental stresses spanning a broad spectrum of challenges. These stresses are mainly categorized into abiotic and biotic. Among the several abiotic stresses that affect crops each year are drought, salinity, heat, cold, nutrient concentration, high light intensity, ozone (O_3), anaerobic stresses, metals, and metalloids (Boyer, 1982; Wang *et al.*, 2003; Chaves and Oliveira, 2004; Nakashima and Yamaguchi-Shinozaki, 2006; Agarwal and Grover, 2006; Hirel *et al.*, 2007; Bailey-Serres and Voesenek, 2008; Pandey *et al.*, 2017).

Additionally, crops can also be vulnerable to biotic stresses such as infection by phytopathogens (including bacteria, fungi, viruses, and nematodes) and attack by herbivore pests (Atkinson and Urwin, 2012). In fact, studies show that several abiotic stress conditions deteriorate the plant defense system which leads to amplified sensitivity to biotic pathogen infections, hence illustrating their possible interconnectedness (Amtmann *et al.*, 2008; Goel *et al.*, 2008; Mittler and Blumwald, 2010; Atkinson and Urwin, 2012). For instance, elevated temperatures have been documented to promote the dissemination of pathogens (Bale *et al.*, 2002).

Often, a confluence of two or more of these stressors is also observed, further exacerbating the challenges to the plant's physiological, developmental, and growth processes, thereby impeding efficient functioning. In response to diverse stress conditions, plants have unique response mechanisms tailored to each specific stress condition encountered. This indicates a higher level of complexity in plant responses governed by disparate signaling pathways resulting in diminished efficiency and consequent implications for plant yield. Consequently, substantial losses ensue across agroecological contexts, agricultural productivity, and implications for climate change mitigation (Pandey *et al.*, 2017). These considerations are compounded by imperatives surrounding global food security and the imperative to augment agricultural yield by 85% in order to accommodate the continually expanding global population (Alexandratos and Bruinsma, 2012; Ray *et al.*, 2013).

While traditional breeding methodologies have achieved considerable success in enhancing agricultural productivity, they are plagued by challenges such as diminishing fitness and genetic diversity. Moreover, their reliance on natural allelic variations renders them inefficient in attaining desired traits aimed at promoting sustainability in production. Hence, this underscores the necessity for the implementation of advanced technological interventions aimed at mitigating stress conditions, hence improving the yield in plants (Wang *et al.*, 2019c).

Progress in technology has facilitated the utilization of genomic engineering methodologies to enact precise modifications within the plant genome, thereby enhancing adaptability, augmenting yield, and conferring resilience against diverse stressors. Several gene-engineering techniques like meganucleases (Puchta *et al.*, 1993), transcription activator-like nucleases (TALENs) (Zhang *et al.*, 2011, 2012), and zinc-finger nucleases (ZFNs) (Zhang *et al.*, 2010), have been developed and adopted over the past few years. However, Clustered Regularly Interspaced Short Palindromic Repeats-CRISPR-associated protein 9 (CRISPR-Cas9) has proved to be a widely used genome-editing tool due to its capacity to induce targeted mutations efficiently, cost-effectively, rapidly, and with precision (Jiang *et al.*, 2013). Furthermore, CRISPR-Cas9 has profoundly transformed biological research, offering extensive applicability to crop plants. This review highlights a prospective strategy for advancing crop enhancement and leveraging CRISPR technology to fortify crops against stressors by selectively modifying or activating specific functional genomes. This approach focusses on harnessing the capabilities of CRISPR-mediated genome editing to bolster crop resilience to adverse environmental conditions, thereby facilitating improved agricultural productivity, plant vitality, and sustainability (Chen *et al.*, 2019).

5.2 Understanding Diverse Stress Conditions in Crops and Their Implications

In the dynamic environment of agriculture, crops are exposed to myriad stress factors that can profoundly impact their growth, development, and overall productivity. These stress conditions, categorized as abiotic, biotic, environmental, and physiological stressors, pose significant challenges to farmers worldwide (Umar *et al.*, 2022). By comprehensively understanding these stressors and their implications, research-based agricultural practices can be adapted to mitigate their adverse effects and enhance crop resilience (Lamers *et al.*, 2020). (Refer to Fig. 5.1 for various stress factors influencing crops.)

Drought stress in crops results from inadequate water availability, adversely impacting plant growth, development, and productivity. Causes include reduced rainfall, high temperatures, and soil water deficits. Consequences encompass stomatal closure, reduced photosynthesis, osmotic stress, and cellular damage due to ROS (reactive oxygen species) accumulation. To mitigate drought stress, plants activate mechanisms like osmotic adjustments through compatible solutes, stomatal regulation for water conservation, and antioxidative defense systems to scavenge ROS (Gupta *et al.*, 2020). Additionally, plants deploy molecular mechanisms involving stress-responsive genes, hormones (e.g. abscisic acid), and signaling pathways to enhance drought tolerance by promoting water uptake, maintaining

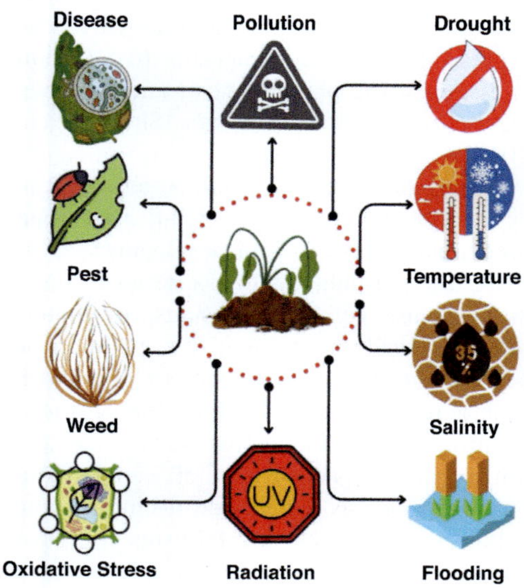

Fig. 5.1. Various stress factors influencing crops.

cellular turgor, and restoring cellular homeostasis (Kuromori *et al.*, 2021).

Temperature stress affects crops by altering physiological processes crucial for growth and development. Causes of temperature stress include extreme heat or cold spells, leading to disrupted enzymes, membrane integrity, and metabolic pathways. Consequences involve reduced photosynthesis, protein denaturation, impaired nutrient uptake, and yield losses (Hatfield and Prueger, 2015). To combat temperature stress, plants employ various mechanisms such as heat-shock proteins to protect proteins, osmolytes for osmotic regulation, and antioxidants to mitigate oxidative damage (Oroz *et al.*, 2017). Additionally, plants modulate gene expression for stress adaptation, adjust membrane fluidity, and activate signaling pathways like the CBF (C-repeat binding factor) pathway to enhance temperature tolerance and ensure optimal growth under varying temperature conditions (Ritonga and Chen, 2020).

Salinity stress in crops arises from soil or irrigation water with high salt concentrations, disrupting water and nutrient uptake mechanisms essential for plant growth. The causes include natural soil salinity, excessive fertilizer use, and poor irrigation practices. Consequences of salinity stress encompass osmotic stress, ion toxicity, nutrient imbalances, and reduced growth and yield. Plants counter salinity stress through mechanisms such as ion exclusion from roots, ion compartmentalization in vacuoles, osmotic adjustment via compatible solutes, and synthesis of antioxidants to combat oxidative damage. Additionally, plants activate signal transduction pathways like the SOS (salt overly sensitive) pathway to regulate ion homeostasis and enhance salt tolerance, ensuring sustained productivity in saline conditions (Van Zelm *et al.*, 2020).

Flooding stress in crops occurs when excess water hinders oxygen availability to roots, leading to impaired nutrient uptake and root damage. Causes include heavy rains, poor drainage, or waterlogged soil due to topography. Consequences of flooding stress involve decreased photosynthesis, nutrient deficiencies, root hypoxia, and susceptibility to diseases (Gibbs *et al.*, 2011). Plants combat flooding stress through mechanisms like aerenchyma formation for improved oxygen transport to submerged roots, ethylene signaling to regulate growth responses, and biochemical adaptations for anaerobic metabolism. Additionally, plants activate flooding-responsive genes, promote adventitious root formation, and modulate hormone levels to enhance tolerance to waterlogging, ensuring survival and productivity under flooded conditions (Sasidharan and Voesenek, 2015).

Disease stress in crops results from pathogenic microorganisms such as fungi, bacteria, viruses, nematodes, and other parasites, causing detrimental effects on plant health and productivity. Causes of disease stress include environmental conditions favoring pathogen proliferation, poor crop management practices, and lack of genetic resistance in plant varieties. Consequences encompass reduced growth, yield losses, quality deterioration, and even plant death (Singla and Krattinger, 2016). Plants develop tolerance to disease stress through mechanisms like inducing systemic acquired resistance (SAR), producing antimicrobial compounds, activating defense-related genes, and forming physical barriers like lignin and callose depositions to impede pathogen entry, thereby enhancing plant immunity and survival against various diseases (Wilson *et al.*, 2023).

Pest stress in crops arises from insect pests, mites, nematodes, rodents, and other organisms that feed on plants, leading to yield losses and quality deterioration. Causes include environmental conditions favoring pest outbreaks, monoculture farming practices, and a lack of natural enemies to control pest populations. Consequences of pest stress involve defoliation, stunted growth, transmission of diseases, and economic losses (Liu *et al.*, 2021b). Plants exhibit tolerance to pest stress through mechanisms like producing insecticidal compounds (e.g. protease inhibitors), emitting volatile compounds to attract natural enemies, and eliciting defense responses like phytohormone signaling pathways (e.g. jasmonic acid) to deter pests and enhance resistance, ensuring crop protection and sustainability (Kumari *et al.*, 2022).

Weed stress in crops originates from unwanted plants competing with cultivated crops for resources such as water, nutrients, and light, resulting in reduced crop growth and yield. Causes include weed seed dispersal, ineffective weed management practices, and the development of herbicide-resistant weed populations. Consequences of weed stress encompass yield reductions, crop damage, increased weed seed banks, and economic losses (Quan *et al.*, 2023). Plants develop tolerance to weed stress through mechanisms like allelopathy, where plants release biochemicals to suppress weed growth, competitive root exudation for resource utilization, and crop rotation to disrupt weed life cycles. Integrated weed management strategies involving cultural practices, herbicides, and crop diversity help mitigate weed stress and promote optimal crop growth (Scognamiglio and Schneider, 2020).

Pollution stress in crops results from contaminants such as heavy metals, air pollutants, pesticides, and industrial chemicals, adversely affecting plant physiology and productivity. Causes include industrial activities, vehicle emissions, improper waste disposal, and agricultural chemical use (Chen *et al.*, 2021). Consequences encompass reduced growth, chlorosis, leaf necrosis, and impaired photosynthesis. Plants respond to pollution stress through mechanisms like phytoremediation, where they absorb and detoxify pollutants, sequester heavy metals in vacuoles, and synthesize enzymes for detoxification. Additionally, plants deploy oxidative stress-response pathways, upregulate antioxidant enzymes, and activate gene expression to enhance tolerance to pollution stress, ensuring plant survival and minimizing pollution-related damage (Singh *et al.*, 2016).

Radiation stress on crops, largely induced by excessive exposure to UV-B rays, can hinder plant growth and development. This stress disrupts photosynthesis, damages DNA, and impairs cellular functions, ultimately reducing yields. To combat this, plants have evolved various mechanisms to cope with radiation stress. These include the synthesis of protective pigments like flavonoids and anthocyanins, activation of antioxidant enzymes such as superoxide dismutase and catalase, and repair mechanisms for DNA damage. Understanding these processes is crucial for developing resilient crop varieties that can tolerate radiation stress and thrive in challenging environments, ensuring food security for the future (Fu *et al.*, 2019).

Oxidative stress in crops arises from an imbalance between ROS production and antioxidant defense mechanisms due to various factors like drought, high temperatures, pollutants, and pathogens. This stress leads to cellular damage, lipid peroxidation, protein oxidation, DNA damage, and ultimately impaired growth and yield. However, plants have evolved complex mechanisms to tolerate oxidative stress, including enzymatic antioxidants (e.g. superoxide dismutase, catalase), nonenzymatic antioxidants (e.g. glutathione, ascorbate), and signaling molecules (e.g. salicylic acid, jasmonic acid). These mechanisms scavenge ROS, repair damaged biomolecules, and activate stress-responsive genes, hence enhancing the plant's resilience to oxidative stress (Kerchev and Van Breusegem, 2021). Table 5.1 enlists the plants under specific stress circumstances along with their resilience mechanisms.

5.3 CRISPR-based Precise Genome Editing

The CRISPR-Cas system represents a specialized two-component adaptive immune system identified within archaeal and bacterial genomes, employed as a defense mechanism against bacteriophages. To be specific, their genomes comprise distinctive CRISPR (Clustered Regularly Interspaced Short Palindromic Repeats) sequences and spacer sequences. Spacers, which range from 17 to 84 nucleotides, serve to complement the invading foreign genome (Marraffini and Sontheimer, 2010). When a bacterial genome encounters an invading viral genome, small segments of the foreign DNA are integrated into the CRISPR-spacer complex of the bacterial genome, resulting in the formation of new spacers at the leading strand of the array. This leading region encompasses a promoter, protein binding sites, and elements necessary for spacer integration. Spacers serve to accumulate the genetic memory of the cell by recording the infection, thereby preventing future infections from the same virus. This phase is commonly referred to as 'Adaptation'. Palindromes facilitate the integration process of spacers into the CRISPR system by providing both direction and position (Gupta *et al.*, 2019). Through successive transcription of the spacers, numerous short mature CRISPR RNAs (crRNAs) are produced, typically measuring 17–20 nucleotides, and these crRNAs complement the target DNA. Hence, during any subsequent viral infection, the protospacer (crRNA and viral DNA) instigates the breakdown of the viral DNA (Brouns *et al.*, 2008). The trans-activating crRNA (tracrRNA), a diminutive noncoding RNA molecule, associates with crRNA to create a binary hybrid entity recognized as single-guide RNA (sgRNA). TracrRNA fulfills two principal roles: assisting in the processing of pre-crRNA by RNase III and instigating the cleavage of DNA guided by crRNA via the cas9 protein (Jinek *et al.*, 2012). This cleavage is initiated by the identification of a complementary 20-nucleotide target sequence and adjacent PAM (protospacer adjacent motif), a 2–5 nucleotide sequence aiding in recognition and cleavage of the viral DNA which is upstream to the PAM motif (Gleditzsch *et al.*, 2019). Moreover, the crRNP (CRISPR-Cas9 ribonucleoprotein) complex is responsible for recognizing the target DNA during the cleavage process, while cleavage at specific sites is facilitated by nuclease domains similar to HNH and RuvC (Lone *et al.*, 2018). The cleavage of the target DNA within the genome results in the formation of a blunt-ended double-strand break (DSB). Subsequently, the host cellular machinery undertakes the repair of the DSB (Mei *et al.*, 2016).

5.4 Types of CRISPR Approaches

5.4.1 Gene knockout using CRISPR

In this technique, CRISPR is used to induce loss of function of a target gene. The technique utilizes non-homologous end-joining (NHEJ) DNA repair mechanism (Rai *et al.*, 2023) to induce small deletions or insertions (INDELs) to make the target gene nonfunctional. The knockout strategy helps in studying the downstream effect of inactivation of the gene

Table 5.1. Plants and their defense mechanisms under specific stress situations.

Plant	Stress condition	Mechanism of stress tolerance	References
Arabidopsis thaliana L.	Drought stress	Ethanol treatment-based regulation and enhancement of drought tolerance.	Bashir *et al.*, 2022
Arabidopsis thaliana L.	Heat stress	Phytochrome B activity inhibition by neighbour signals increases the quantity of Phytochrome Interacting Factors, Enhancing heat-stress tolerance and reducing damage. Plants developing in low-red/far-red environments have reduced Phytochrome B activity and thus enriched growth rates.	Arico *et al.*, 2019
Arabidopsis thaliana L.	Cold stress	TIR-NB-LRR gene RPP4 (Recognition of Peronospora Parasitica 4) exhibits effective resistance to cold stress; this gene's gain-of-function mutation is found in the chilling-sensitive mutant chs2-1 (rpp4-1d) in *Arabidopsis*.	Bao *et al.*, 2014
Arabidopsis thaliana L.	Disease stress	Pathogenesis-related (PR) genes are activated when tissues exposed to pathogens have higher amounts of salicylic acid. The broad collection of PR genes increases resistance to a variety of diseases by encoding several proteins with antibacterial action.	Verma *et al.*, 2016
Arabidopsis thaliana L.	High metal stress	Phytochelatins, which are found in plants, compete with heavy metals for binding sites, enhancing metal tolerance and detoxification. Phytochelatins have the ability to participate in both long-distance metal transport and detoxification.	Seregin and Kozhevnikova, 2023
Arabidopsis thaliana L.	Radiation stress	UV-B tolerance is facilitated by the COP1–HY5–RUPs light signaling module, regulating UVR8 activity through the CRY blue-light receptors and the Phy-red/far-red light receptors.	Chen and Liu, 2021

Continued

Table 5.1. Continued

Plant	Stress condition	Mechanism of stress tolerance	References
Citrullus lanatus (Thunb.) Matsum. & Nakai, *Cucumis melo* L.	Disease stress	The nitrogen-regulating gene ntrC contributes to nitrogen use and pathogenicity. The ntrC deletion mutant strain showed markedly reduced virulence, *in vitro* growth, and *in vivo* colonization ability and incapability to use nitrate. On the other hand, it showed noticeably increased biofilm development and stress tolerance to high salt, oxygen, and copper ions.	Liu *et al*., 2023
Fragaria ×ananassa	High Metal stress	In strawberries, exogenous melatonin increases the antioxidant defense system, so mitigating the growth and photosynthetic suppression caused by cadmium.	Saqib *et al*., 2023
Glycine max L.	Flooding stress	Soybean flooding tolerance was imparted by ABA via control of glycolysis and nuclear-localized zinc finger protein, cell division cycle 5, and transducin.	Komatsu *et al*., 2013
Glycine max L.	Weed stress	The inherent resistance of rice to the herbicide benzobicyclon, a selective triketone, was found to be attributed to the rice HIS1 gene. Mesotrione causes extreme sensitivity in conventional soybean. The rice HIS1 gene (TDO) can be expressed ectopically in soybean, where it imparts a commercial level of mesotrione resistance, making it distinct from weeds in the mesotrione environment.	Dai *et al*., 2022
Glycine soja Sieb. & Zucc.	Salinity stress	The genes GmSALT3 and GmSALT18, which are associated with salt tolerance, are present in the wild soybean line NY 36–87. This provides genetic material and a unique locus for breeding salt-tolerant soybeans.	Guo *et al*., 2021
Leguminous crops	Salinity stress	The salinity stress tolerance is enhanced by cell mutagenic autopolyploidy. By exhibiting a particular enrichment of DEG genes with different patterns of gene expression in both roots and shoots, polyploids demonstrated improved salt tolerance.	Mangena, 2023

Continued

Table 5.1. Continued

Plant	Stress condition	Mechanism of stress tolerance	References
Mangifera indica L.	Pest stress	An essential antinutritive function of polyphenol oxidase is seen in a plant's defense against interactions between pests and plant pathogens. PPOs help to reduce the availability of proteins to insect pests by catalysing the oxygen-dependent oxidation of phenols to quinones, which are reactive molecules that can covalently alter and cross-link a range of cellular nucleophiles, including side chains of amino acids.	Mani *et al.*, 2022
Oryza sativa L.	Drought stress	Drought mitigation is provided via the activation of acetic-acid-induced JA signaling in the root, which partially inhibits ABA signaling and modifies ABA responses in the shoot during drought conditions. The transpiration decrease that was seen after acetic acid treatment was lessened in shoots where JA signaling in the root tip is active in part.	Ogawa *et al.*, 2021
Oryza sativa L.	Cold stress	The members of the DREB1/CBF subfamily are expressed in response to cold stress, and ectopically induced expression of these members increases plant resistance to freezing and cold.	Abiri *et al.*, 2017
Oryza sativa L.	Flooding stress	The finding of genes like Sub1A, SNORKEL1/2, OsEIL1a, SD1, DEC1, and ACE1, which control the quiescence response or rapid internode elongation to aid in plant survival during submergence, is the basis for the phytohormone signaling that regulates responses to flooding.	Wang and Komatsu, 2022

Continued

Table 5.1. Continued

Plant	Stress condition	Mechanism of stress tolerance	References
Oryza sativa L., *Cenchrus americanus* (L.) Morrone, *Sorghum bicolor* L.	Flooding stress	A condition that occurs when two species – one drought-tolerant, one wetland – are planted together using the mixed-cropping approach, causing their root systems to become closely entangled, strengthening the upland crops' ability to withstand flooding. With this method, upland crops under flood stress (O_2-deficient nutrient culture) saw increases in their rates of photosynthetic and transpiration.	Iijima *et al.*, 2016
Pisum sativum L. *Hordeum vulgare* L.	Weed stress	Intercropping increases crop growth and lessens weed pressure when it comes to weed management. It operates on the basis of resource partitioning, which allows coexisting crop species with distinct resource acquisition methods to better utilize resources while depriving weeds.	Sharma *et al.*, 2021
Solanum habrochaites S. Knapp & D.M. Spooner	Pest stress	Tomato wild relatives, such as *S. habrochaites* and *S. pimpinellifolium*, have glandular trichomes and related allelochemicals, such as acyl sugars, sesquiterpenes, and methylketones. Numerous studies have confirmed the role of trichomes in tomato resistance against various insect pests.	Ferrero *et al.*, 2019
Solanum tuberosum L.	Disease stress	Five different concentrations of tricalcium phosphate, copper oxide, and silicon dioxide nanoparticles were tested for their ability to control the *in vivo* and *in vitro* infections of black scurf disease by impacting the activity of the enzymes polyphenol oxidase, peroxidase, catalase, and chitinase in the leaves and tubers.	El-Shewy, 2019
Sorghum bicolor L.	Pest stress	Many NAC genes were expressed in response to the greenbug attack, according to RNA-sequencing data. SbNAC genes may contribute to genetic resistance to greenbugs since they perform a variety of roles in response to greenbug infestations.	Baillo *et al.*, 2019

Continued

Table 5.1. Continued

Plant	Stress condition	Mechanism of stress tolerance	References
Triticum aestivum L.	Heavy metal stress	The AtSAP13 promoter interacted with numerous AP2/EREBP TFs, and those plants overexpressing AtSAP13 exhibited notable resistance to heavy metals (Zn, Cd, and AsIII).	Xin *et al.*, 2011
Vigna unguiculata (L.) Walp.	Salinity stress	Root-derived DEGs (GmsSOS1, SOS2, and OsNHX2) are the genes involved in redox processes, increase the activity of antioxidant enzymes and decrease the build-up of Na^+ under salt stress.	Pan *et al.*, 2019
Zea mays L.	Drought stress	Transgenic plants can be created using a variety of genetic engineering techniques, such as biolistic or *Agrobacterium*-mediated transformation, to add genes related to pathways that regulate water flow in plant cells and are implicated in drought-stress response, such as ABA signaling and aquaporin regulation.	Arshad *et al.*, 2022
Zea mays L.	Weed stress	Examining the transcriptome responses to weed pressures in both controlled and natural environments revealed a limited number of genes whose expression levels are strongly influenced by competition. These genes may act as competition markers and offer a mechanism for determining cis- and trans-signaling elements in maize that are responsive to competition. Monitoring competition response and cultivating weed tolerance may benefit from this information. Agronomically and ecologically beneficial species, especially winter brassica crops like canola and camelina, could benefit greatly from intercropping with maize if its response to weeds is altered.	Horvath *et al.*, 2019

of interest (Fig. 5.2). NHEJ represents a repair mechanism that is often characterized by its error-prone nature across various organisms due to its tendency to induce frameshift mutations, involving the deletion or insertion of a base. This process necessitates the involvement of nucleases for DNA sequence cleavage, polymerases for the introduction of

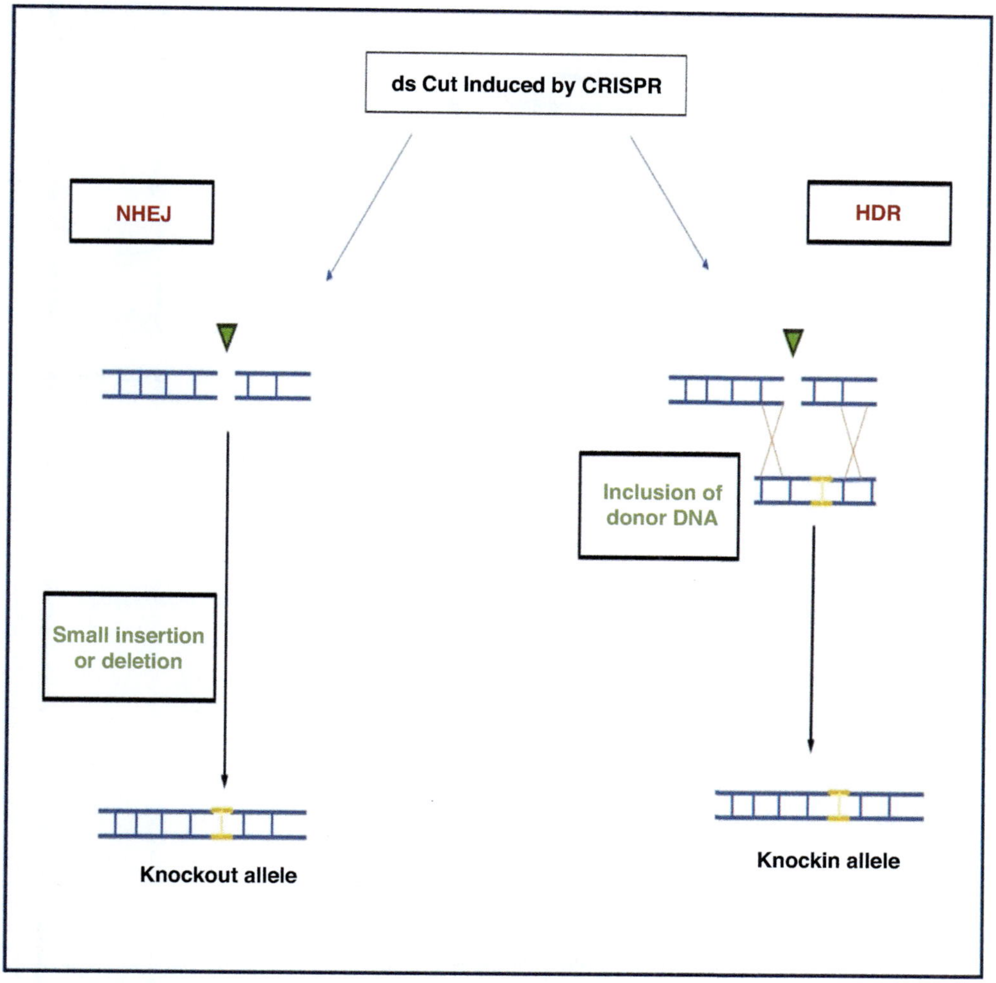

Fig. 5.2. Commonly employed CRISPR approaches in genome editing.

new DNA sequences, ligases for the joining of sequences, and numerous enzymes for repairing the double-stranded break (Lieber, 2010; Tang *et al.*, 2019; Ray and Raghavan, 2020).

5.4.2 Gene Knock-in Using CRISPR

In this technique, an allele present in a donor DNA is inserted into a target site cleaved by the CRISPR tool. Such an insertion is enabled via homologous recombination induced by the homology directed repair (HDR) mechanism (Rai *et al.*, 2023). Via such insertions, phenotypic variations can be studied and desired, and stress-tolerant traits can be imparted to the plant. These mechanisms exhibit a high degree of fidelity; however, they are unable to operate in the absence of a homologous sequence located near the site of the double-stranded break (Lieber, 2010).

The advent of the CRISPR/Cas editing system has paved the way for introducing novel possibilities in editing plant genomes, aiming to enhance desired traits and develop varieties resistant to various stresses (Chandrasekaran *et al.*, 2016; Baruah *et al.*, 2017; Chikkaputtaiah *et al.*, 2017; Hassan *et al.*, 2020). Research indicates that the CRISPR/Cas editing system can effectively

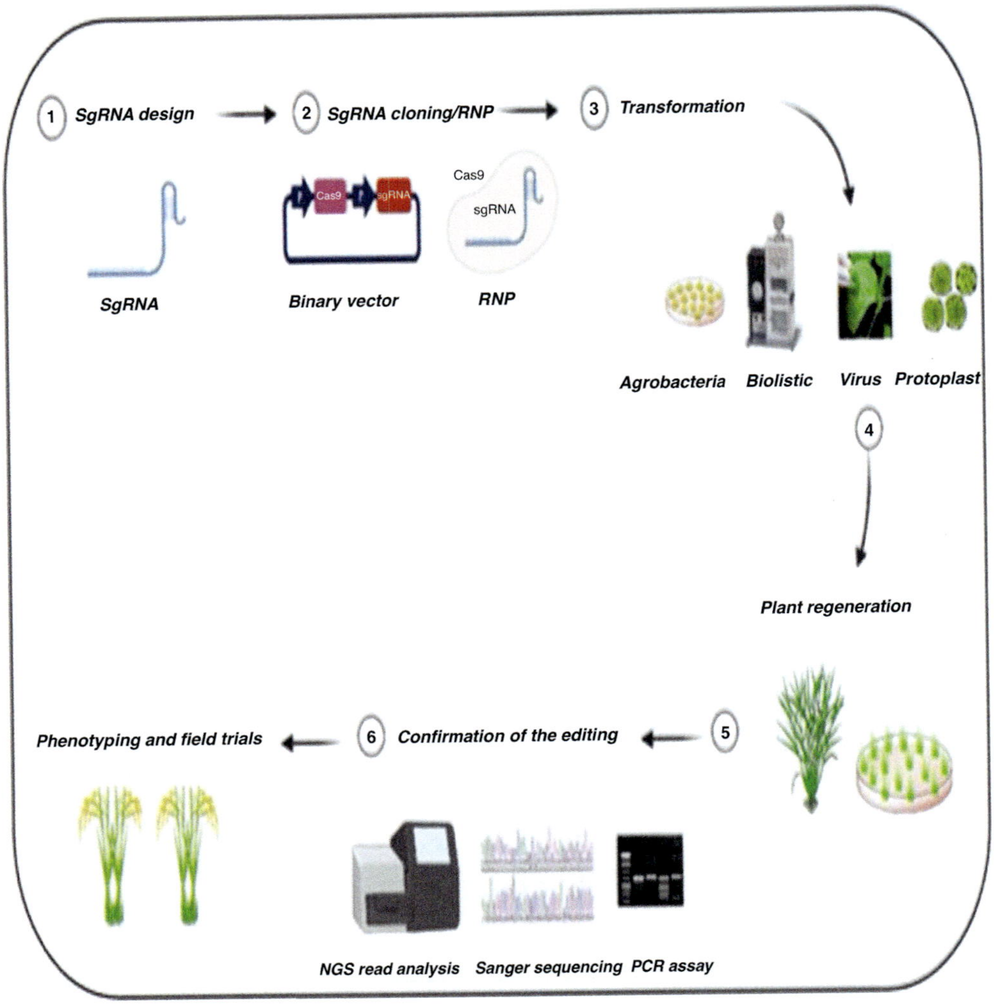

Fig. 5.3. Schematic representation of CRISPR-Cas9 genome editing in plants.

target complex quantitative genes associated with abiotic stress factors, either directly or indirectly (Mushtaq *et al.*, 2018). Fig. 5.3 shows a schematic display of CRISPR-Cas9 genome editing in plants.

5.4.3 CRISPR-mediated studies for crop-stress resilience

In the domain of contemporary biotechnology, particularly within the realm of genetic manipulation and crop enhancement, substantial progress has been achieved in augmenting stress resilience and resistance to pathogens in diverse plant species. Progress has been made through the innovative utilization of CRISPR/Cas9 technology, enabling researchers to pinpoint specific genes and transcription factors linked to stress responses, thereby laying the groundwork for the advancement of stress-tolerant crops.

For instance, genes like ART1 (Aluminum Resistant Transcription Factor) have been identified and altered to improve plants' capacity to thrive in environments rich in aluminium, due to their crucial role in aluminium tolerance. Similarly, the DRO1 (Deeper Rooting1) gene, responsible for regulating root-growth angle and aiding in drought resistance, has been a key focus in research efforts aimed at enhancing plants' ability to withstand water stress. Genes such as SUB1A associated with submergence tolerance, and PSTOL1 (Phosphorus Starvation Tolerance1) linked to phosphorus acquisition efficiency, have been subjects of genetic modifications with the help of CRISPR/Cas9 technology to boost plant survival in waterlogged conditions and low-phosphorus environments, respectively (Barrero *et al.*, 2022). The utilization of gene-editing techniques with CRISPR/Cas9 technology has opened avenues for scientists to investigate plant–pathogen interactions and devise innovative approaches to combat various diseases. In crops like watermelon, the Clpsk1 gene has been identified for imparting resistance to *Fusarium oxysporum* f.sp. *niveum*, a harmful fungal pathogen causing significant crop losses (Zhang *et al.*, 2020). Similarly, *Arabidopsis*, a model plant species, demonstrated enhanced resistance to turnip mosaic virus (TuMV) following manipulation of the eIF(iso)4E gene, showcasing the potential of targeted genetic alterations in boosting plant defenses against viral infections (Pyott *et al.*, 2016). Furthermore, research in rice has highlighted genes like Os8n3, offering resistance against bacterial pathogens such as *Xanthomonas oryzae* pv. oryzae, and OsERF922, known for its role in combating the fungus *Magnaporthe oryzae*, illustrating the diverse applications of genetic engineering in strengthening plant immunity against various microbial threats (Kim *et al.*, 2019: Wang *et al.*, 2016). Remarkably, the exploration of gene-editing technologies has ventured into enhancing resistance against bacterial pathogens in diverse plant species.

In tomatoes, genetic manipulation of the SlJAZ2 gene has displayed potential in conferring resistance to *Pseudomonas syringae* pv. tomato (Pto) DC3000, a common bacterial pathogen affecting tomato crops (Ortigosa *et al.*, 2019). Similarly, the eIF4E gene in cucumber has been associated with resistance against cucumber vein yellowing virus, zucchini yellow mosaic virus, and papaya ringspot mosaic virus-W (Chandrasekaran *et al.*, 2016). In cotton, the 14-3-3 gene has been linked to resistance against *Verticillium dahliae* fungus, while wheat's mlo gene provides resistance to powdery mildew (Wang *et al.*, 2014; Zhang *et al.*, 2018). Likewise, the SiMlo gene in tomatoes has exhibited effectiveness in providing resistance to powdery mildew (Nekrasov *et al.*, 2017).

Similarly, in citrus plants, targeting the CsLOB1 gene has been aimed at enhancing resistance to *Xanthomonas citri* subsp. *citri* (Xcc), a bacterium causing citrus canker disease and significant economic losses in citrus cultivation (Peng *et al.*, 2017). The upregulation of the CsWRKY22 gene in oranges has proved effective in combating Xcc, underscoring the potential of genetic engineering in reinforcing plant immunity against bacterial pathogens (Wang *et al.*, 2019b). Recent advancements in CRISPR/Cas9-based gene-editing technologies have not only transformed disease resistance but also broadened the horizons of stress tolerance in crops. By utilizing CRISPR/Cas9-mediated manipulations, scientists have enhanced drought tolerance in crops like soybean through the editing of genes such as GmHdz4, providing valuable insights into improving plant resilience to water scarcity (Zhong *et al.*, 2022). Similarly, the adjustment of the Squamosa Promoter-Binding Protein-Like eight gene in alfalfa has resulted in improved drought resistance, highlighting the importance of gene editing in creating stress-tolerant crop varieties (Singer *et al.*, 2021).

The genetic modification of NtAITRs in tobacco and OsDST in rice has exhibited promising outcomes in enhancing drought and salt tolerance in these crops, laying the foundation for the development of stress-resilient varieties capable of thriving in challenging environmental conditions (Li *et al.*, 2022a). The field of gene editing has not only concentrated on disease resistance and stress tolerance but has also delved into investigating crucial molecular mechanisms that govern stress responses in plants. The identification and subsequent manipulation of the BnaA10 ABF2-BnaA6.RGA protein complex has boosted the expression levels of drought-responsive genes like BnaC9 RAB18, offering significant insights into the essential roles of DELLA proteins in enhancing drought tolerance in crops like rapeseed (Wu *et al.*, 2020). Additionally, the elimination of the tomato Auxin Response Factor 4 (SlARF4) gene has shown increased resistance to water-deficit stress in tomatoes (Kumar *et al.*, 2023).

Likewise, the thorough examination of the functions of LHY1a and LHY1b genes in drought tolerance has presented alternative targets for CRISPR/Cas9-based genome editing to rapidly generate mutant alleles in elite soybean varieties, introducing new possibilities for enhancing drought resilience in these crops (Wang *et al.*, 2020). Moreover, the precise gene editing of the OsDST gene in the Indica mega rice cultivar MTU1010 has displayed the potential to improve drought and salt tolerance, demonstrating the efficacy of genetic engineering in producing stress-resilient rice varieties capable of enduring challenging environmental conditions (Kumar *et al.*, 2020). In rice, progress in gene editing has opened up new opportunities for enhancing stress responses and cultivating resilient crop varieties.

Exploring genetic targets such as OsmiR535 has provided fresh insights into enhancing drought and salinity stress tolerance in rice, offering a valuable framework for future gene-editing endeavors in stress-tolerant rice cultivars (Yue *et al.*, 2020). Significantly, the inhibition of DST-mediated activation of Nitrate Reductase 1.2 in rice under drought-stress conditions has resulted in reduced nitrogen assimilation, unveiling the intricate molecular regulatory mechanisms governing plant responses to water scarcity (Han *et al.*, 2022). Consequently, the CRISPR/Cas9-mediated manipulation of genes like ARE1 in wheat has revealed novel pathways for enhancing nitrogen-use efficiency and increasing grain yield potential, emphasizing the indispensable role of gene-editing technologies in crop enhancement for sustainable agriculture (Zhang *et al.*, 2021b).

Advances in gene-editing technologies have not only been limited to disease resistance and stress tolerance but have also expanded to metal-stress responses in plants, illuminating innovative strategies for fortifying plant resilience to heavy-metal toxicity. In wheat, targeted modifications of the TaARE1 protein have resulted in enhanced nitrogen utilization, delayed aging, and increased grain yield potential, providing valuable insights into improving yield and nutrient uptake in wheat varieties (Zhang *et al.*, 2021b). Additionally, the use of CRISPR/Cas9 to edit the Ospmei12 gene in rice has revealed its role in growth, cell-wall formation, and responses to hormones and heavy metals, representing a significant advance in understanding how rice tolerates heavy metals (Li *et al.*, 2022b). The knockout of the SpHMA1 gene in metallophyte *S. pumilum* through CRISPR/Cas9 has expanded our knowledge of heavy-metal-tolerance mechanisms in plants, offering important information for developing stress-resistant crop varieties (Zhao *et al.*, 2018).

Within *Oryza sativa*, genes such as OsHMA2, OsCCX2, CAL1, and OsHMA3 interact to regulate cadmium transport, while OsLCT1 and OsNramp5 play crucial roles in reducing cadmium levels through CRISPR/Cas9 manipulations. Furthermore, the essential role of OsNRAMP1 in regulating metal ions and reactive oxygen species has paved the way for creating heavy-metal-tolerant *O. sativa* strains. The function of the OsHAK1 gene in encoding a potassium transporter and the impact of CRISPR/Cas9-induced OsHAK1 knockout on cesium uptake elucidate complex mechanisms in ion transport regulation vital for plant growth and nutrition. Potassium (K^+), a vital macronutrient, plays a crucial part in plant growth, with K^+-deficient plants showing increased ROS levels, prompting gene-editing studies like those on OsPRX2 to shed light on pathways for enhancing K^+ deficiency tolerance by altering stomatal movement dynamics in *O. sativa* (Tang *et al.*, 2017; Joshi *et al.*, 2023).

Additionally, the role of OsARM1 as an arsenite-responsive MYB1 factor controlling arsenic-related transporter genes in *O. sativa* has been revealed through CRISPR/Cas9, demonstrating enhanced arsenic tolerance in knockout mutants (osarm1). In a significant application, scientists employed a CRISPR cytosine base editing system (CBE) to introduce a C-T point mutation in the OsNRT1.1B gene of *Japonica rice* cv. *Nipponbare*. This targeted edit resulted in the conversion of amino acid residue T327 to M327, representing a known allele variation between Nipponbare and IR24 rice varieties. The induced mutation in the OsNRT1.1B gene led to improved nitrogen utilization efficiency (NUE) compared to the wild type (Nieves-Cordones *et al.*, 2017). Going beyond species boundaries, the successful use of CRISPR/Cas9 to edit the ALS1 gene of *Glycine max* for chlorsulfuron resistance highlights the technology's adaptability in imparting herbicide resilience in various crops, while introducing precise mutations through adenine base editors in the OsTubA2 gene of *O. sativa* signifies a breakthrough in cultivating herbicide (dinitroaniline) tolerance (Li *et al.*, 2015; Liu *et al.*, 2020). The diverse applications of CRISPR/Cas9 also extend to gene editing for enhanced stress responses, such as manipulating the ACQOS/VICTR association with salt-stress resilience in *Arabidopsis thaliana*, enhancing salinity stress tolerance in soybean through GmAITR gene mutations, and uncovering distinct traits via mutations in an Inositol Trisphosphate 5/6 Kinase gene in barley for improved salt tolerance (Vlčko and Ohnoutková, 2020; Wang *et al.*, 2021; Kim *et al.*, 2021).

Genetic editing has not only been effective in enhancing drought-stress tolerance in *Arabidopsis* by merging CRISPR/dCas9 with a Histone Acetyltransferase beyond salt stress, but it has also illuminated new pathways for stress-response modulation in soybean (Paixão *et al.*, 2019). The role of GmpPLA genes in regulating flood-stress responses has been elucidated through the use of CRISPR/Cas9 technology, indicating progress in comprehending gene families and their functions under challenging conditions (Xiao *et al.*, 2021). The successful targeting of the Mildew Resistance Locus O (MLO) gene, an essential transmembrane protein involved in powdery mildew susceptibility, in cultivated tomato and wild species such as *S. peruvianum* has led to heightened resistance to *Oidium neolycopersici* without causing any undesired phenotypic alterations (Nekrasov *et al.*, 2017).

Moreover, the manipulation of the Downy Mildew Resistant 6 (DMR6) gene in *A.thaliana*, which encodes a 2-oxoglutarate Fe(II)-dependent dioxygenase linked to pathogen susceptibility, through CRISPR/Cas9 has resulted in enhanced resistance against various tomato pathogens while maintaining plant growth (Thomazella *et al.*, 2021). The identification of negative regulatory genes SlHyPRP1 and SlDEA1 responsive to abiotic and biotic stresses in *S. lycopersicum* prompted researchers to utilize CRISPR/Cas9 for genome editing, leading to improved tolerance to drought, salinity, and pathogenic stressors (Saikia *et al.*, 2024). Furthermore, the discovery of the drought-responsive 1R-MYB gene ZmRL6 in barley and *Aegilops tauschii* revealed that its overexpression enhances drought tolerance, whereas its knockout increases drought sensitivity (Zhang *et al.*, 2023).

In *L. esculentum*, the targeting of the heat-sensitive Slagamous-Like six gene using CRISPR/Cas9 has resulted in increased fruit setting under heat-stress conditions (Klap *et al.*, 2017). Three separate research studies have showcased the significant potential of CRISPR/Cas technology in augmenting herbicide resistance in rice plants. The initial study employed a synthetic complex of nuclease-deficient Cas9 fused with an activation-induced cytidine deaminase (AID) enzyme, known as target-AID, to induce a specific point mutation (C287T) in the ALS gene of rice, leading to resistance to the herbicide imazamox (Shimatani *et al.*, 2017). In the second study, mutant rice plants generated through CRISPR/Cas technology exhibited elevated expression of transcripts for CP12 and Ubiquitin two genes, along with a tenfold upregulation of HPPD and PPO1 genes, resulting in herbicide resistance without compromising yield or other crucial agronomic traits in *Oryza sativa* (Lu *et al.*, 2021). The final study concentrated on the herbicide resistance gene BEL, which provides resistance to bentazon and sulfonylurea herbicides in rice (Xu *et al.*, 2014). These findings collectively underscore the substantial potential of CRISPR/Cas technology in developing herbicide-resistant crops with enhanced agricultural sustainability and productivity. The improvement of plant tolerance to diverse stresses through genetic manipulation has been a key focus of recent research efforts. For example, the overexpression of ATP Sulfurylase and Selenocysteine Methyltransferase genes in *B.juncea* has been demonstrated to enhance selenium tolerance (LeDuc *et al.*, 2006). Additionally, the transfer of BcMT1 and BcMT2 metallothionein genes from *B. campestris* to *Arabidopsis* has resulted in increased resistance to copper and cadmium (Lv *et al.*, 2012).

These investigations collectively emphasize the crucial role of genetic modification in discovering new pathways to enhance plant responses to environmental pressures, diseases, and heavy-metal poisoning, laying the groundwork for creating stress-resistant and environmentally robust crop varieties with the goal of ensuring global food security and sustainability. By conducting a comprehensive examination of the various uses of genetic modification techniques in boosting stress tolerance, disease immunity, and nutrient uptake in different crop species, it becomes clear that the incorporation of CRISPR/Cas9 and other genetic-editing methods shows great potential in transforming agriculture and tackling the challenges presented by shifting environmental circumstances and evolving pathogen risks. The accuracy, adaptability, and effectiveness of genetic-editing technologies provide unparalleled prospects for crop enhancement, leading to the development of stress-tolerant, disease-resistant, and high-yielding crop varieties capable of flourishing in various agricultural settings. As scientists continue to elucidate the intricate molecular mechanisms governing plant responses to stress and perfect genetic-editing approaches for crop improvement, the agricultural sector is on the verge of groundbreaking advancements that could redefine the limits of crop productivity, sustainability, and resilience amidst global adversities. Fig. 5.4 elaborates on a simplified workflow for CRISPR-Cas9 plant genome editing.

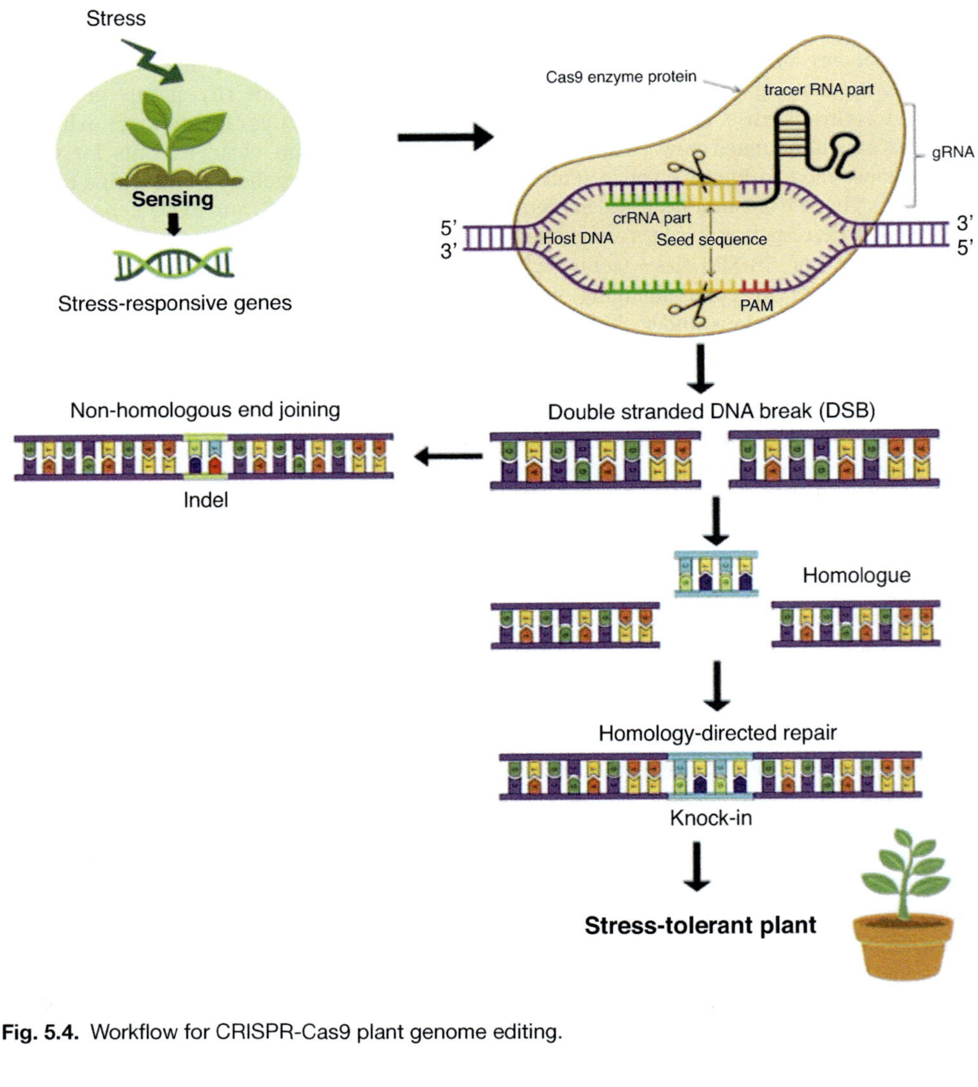

Fig. 5.4. Workflow for CRISPR-Cas9 plant genome editing.

5.5 Unleashing Crop-stress Tolerance through Precision Editing of Functional Genomes with CRISPR Technology

A functional gene is any gene that undergoes expression. The end product of a gene's expression is a protein. There are various methods to study the function of a gene. The most advanced of them is the genomic approach to studying gene functions (Bunnik and Le Roch, 2013). Some of the techniques in functional genomic studies involve real-time-PCR, NGS, mass spectrometry, and bioinformatics-based analysis. Model systems as well as CRISPR technology are extensively used to study variations in genomic sequences (Gasperskaja and Kučinskas, 2017). One can appreciate various different functions in an organism via the means of functional genomics.

5.5.1 Stress tolerance in plants

The first clue of functional genomes in plant-stress tolerance was generated after the complete genomic sequencing of *Arabidopsis thaliana*. The knowledge was further expanded following the genomic sequencing of several other plants, such as: rice, maize, tomato, lotus etc. Various interesting data have been generated through these processes. For instance, salt stress in rice induces transcriptional increase in genes associated with cell rescue, defense, transport, energy, and metabolism. Fifty percent of these genes were found to also be associated with detoxification, growth and development (Bohnert *et al.*, 2001).

Upon inducing cold stress to *Arabidopsis thaliana* pollens, a reduced expression was observed for genes that code for COR (cold-responsive), β-amylase, and lipid transfer protein. There was also a downregulation of genes related to photosynthesis (Lee and Lee, 2003). Genes coding for TFs like AP2/ERF, MYB, bHLH and WRKY were found to be expressed in high abundance upon induction of drought stress in the sunflower. Some of the pathways involved in battling drought stress in sunflowers, involve photosynthesis, carotenoid biosynthesis, starch and sucrose metabolism, protein phosphorylation, and phytohormone signaling (Shi *et al.*, 2024). In wheat, TaPR-1-1 (a member of the pathogenesis-related protein family) was seen to be expressed in order to confer tolerance to cold stress, saline stress, and osmotic stress (Wang *et al.*, 2019a). Similarly, various functional genes have been found to be associated in stress-tolerance response in different plants.

5.5.2 Precision targeting functional genomes using CRISPR

CRISPR-Cas9 is commonly used as a genome-editing tool. The CRISPR-Cas9 system is guided by an sgRNA that recognizes any target DNA which has a protospacer adjacent motif (PAM) adjacent to its 3' end. Such a system is highly specific, and it is easy to target specific regions of DNA by synthesizing the required sequence of the sgRNA (Jain, 2015). Thus, functional screening of genomes has become easier ever since the advent of the CRISPR technology. A functional gene of interest is identified and targeted using specifically designed sgRNA. This causes a perturbation in the normal phenotype of the gene in that organism. The perturbation can be studied for drawing further conclusions (Ford *et al.*, 2019). The CRISPR tool can be designed to target multiple genes at once. This is a massive advantage in the case of plants which have undergone whole genome duplications to give rise to multigene families with functional redundancy (Jain, 2015). This technique is widely applied in studying plant-stress response and also to modify plant genomes in order to make them resilient (Fig. 5.5).

5.5.3 Gene regulation using CRISPR

For the purpose of gene regulation, the composition of Cas9 is often altered to nullify the nuclease activity. It is then denoted as dCas9; dCas9 is fused to effector domains to either inhibit (CRISPRi) or activate (CRISPRa) transcription. CRISPRi inhibits the recruitment of RNA polymerase to ORFs whereas CRISPRa

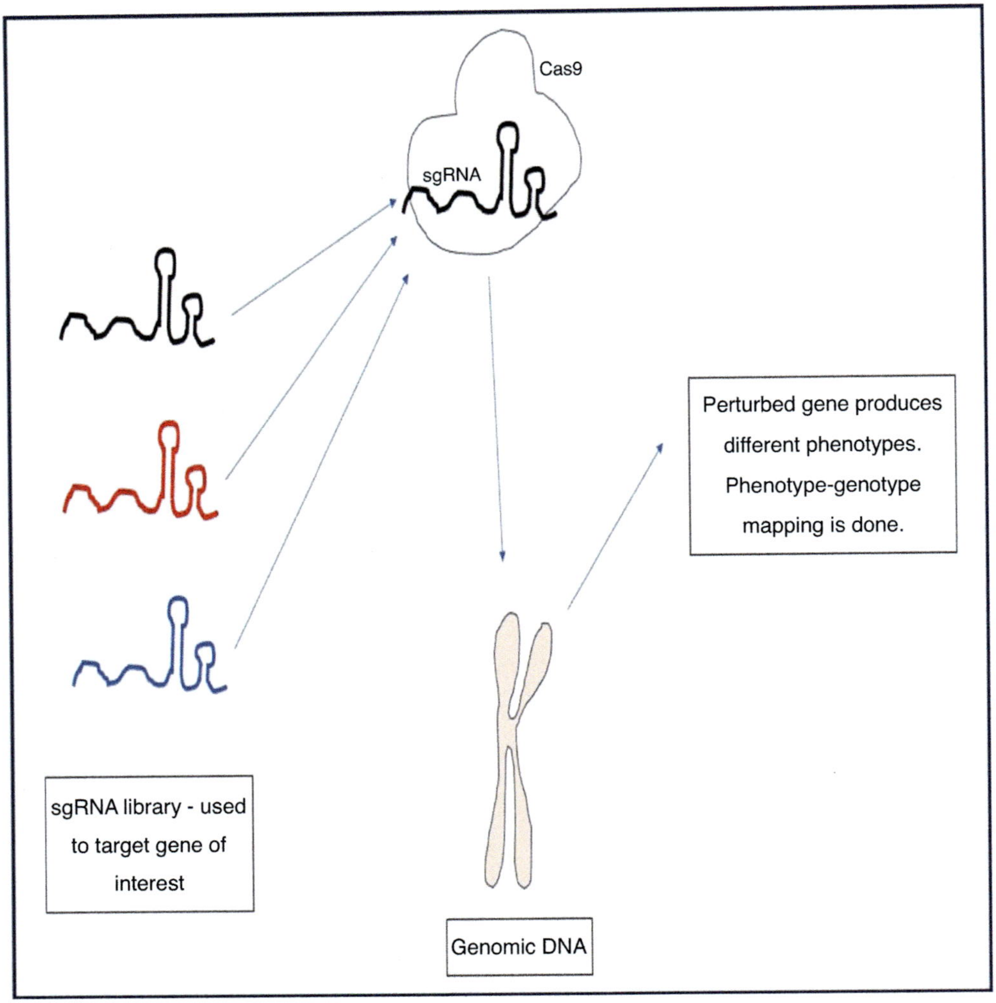

Fig. 5.5. An illustration of precision targeting of functional genomes using CRISPR-Cas9.

actively recruits RNA polymerase and other activators of transcription (McCarty *et al.*, 2020).

5.5.3.1 Targeting functional genomes associated with stress tolerance in plants using CRISPR

The transcription factor AREB1 was found to be important in drought-stress response in plants. CRISPR tool was designed to increase the responsiveness of the AREB1 promoter. This resulted in increased stomatal conductance and chlorophyll concentration, even in drought conditions. Thus, CRISPR recombinants showed better survival in drought conditions. This was established in *Arabidopsis* (Rai *et al.*, 2023). In another instance, a thermoresistant line of genic male sterile maize was produced using CRISPR-Cas9. The ZmTMS5 gene was knocked out using CRISPR to obtain the desired trait in the crop. The recombinants were found to be thermoresistant and exhibited genic male sterility, thus aiding in maintaining the hybrid purity of the line (Li *et al.*, 2017). A91 nucleotide base deletion was introduced in the bHLH transcription factor gene in rice using CRISPR-Cas9. These mutants exhibited tolerance to salt stress. The phenotypes observed were: improved oxidative stress control, shoot weight increase, increased chlorophyll content, and upregulated ion transport (Alam *et al.*, 2022).

5.5.4 Identification of stress-responsive genes

The development of various omics-based techniques has made it much easier and efficient to screen and identify genes associated with stress response in plants. Special emphasis must be laid on the techniques of transcriptomics, proteomics, and metabolomics in this regard.

5.5.4.1 Transcriptomics approach

This can be implemented via:

- First-generation sequencing: RNA is isolated to create cDNA library. These cDNAs are treated to hybridization or expression analysis sequencing can be done on these using the microarray technology (Kamali and Singh, 2023);
- next-generation sequencing: RNA is isolated to create a library. These molecules are sequenced and then expression analysis is performed (Kamali and Singh, 2023); and
- third-generation sequencing: RNA is isolated to prepare cDNA and then that is sequenced using techniques such as nanopore sequencing or PacBio SMRT. The sequencing is done for the full-length transcripts of the cDNA (Kamali and Singh, 2023).

Thus, the transcriptomics approach aids in selecting genes of interest based on specific expression patterns in the stress-tolerance pathways.

5.5.4.2 Proteomics approach

Proteome profiling is done for stress-responsive pathway proteins in plants. This can be implemented by techniques such as: 2D gel electrophoresis, MALDI-Tof analysis, mass spectrometry etc. The different proteome-related studies that are done involve: phosphoproteome, proteogenome, organellar, nuclear, and cell-wall proteome studies (Naik *et al.*, 2023).

5.5.4.3 Metabolomics approach

Targeted or untargeted quantification and documentation of metabolites are performed to understand metabolic pathways associated with stress response in plants. Such quantification is usually done using techniques such as LC-MS, GC-MS, NMR, or FITR etc. Clear understanding of the associated metabolites help identify the targets for further stress-response-related research (Naik *et al.*, 2023).

5.6 Exploring Novel Applications of CRISPR for Future Crop Designing

Exploring additional applications of CRISPR beyond stress tolerance is crucial for unlocking its full potential in agriculture. By targeting traits like yield, nutritional quality, and environmental adaptability, CRISPR can further enhance crop productivity and resilience (Achary *et al.*, 2021). CRISPR-Cas technology is assisting in the creation of climate-resilient crops while also enhancing various aspects of crop quality including their visual appeal, taste, nutritional content, and desired characteristics (Razzaq *et al.*, 2021). Synergizing CRISPR with other agricultural technologies, such as precision farming and biotechnology, can amplify its impact. This integration fosters innovation and enables holistic approaches to address the complex challenges facing modern agriculture, ensuring sustainable food production for a growing population (Badiyal *et al.*, 2024). Beyond stress tolerance, CRISPR technology offers a plethora of applications in agricultural development, revolutionizing crop improvement strategies. One promising avenue lies in enhancing nutritional content. CRISPR enables precise modifications to crop genomes, targeting genes responsible for nutrient biosynthesis and uptake. This holds immense potential for addressing malnutrition, as crops can be fortified with essential vitamins, minerals, and other beneficial compounds (Kumar *et al.*, 2022). CRISPR facilitates the development of crops with enhanced pest and disease resistance. By targeting genes involved in plant–pathogen interactions, CRISPR can confer robust immunity against devastating pathogens, reducing reliance on chemical pesticides and promoting sustainable agricultural practices (Khan *et al.*, 2022). Additionally, CRISPR-mediated modifications in yield and quality traits offer opportunities for improving crop productivity and market competitiveness. Targeted edits can optimize traits such as yield, flavor, shelf life, and postharvest characteristics, meeting consumer demands and ensuring food security (Kumar *et al.*, 2022).

5.6.1 Enhanced nutritional content

CRISPR technology offers an innovative approach to enhancing nutritional content in crops by precisely targeting specific genes. CRISPR-Cas systems have enabled the development of a range of genetically modified golden crops, enriched with carotenoids, aimed at addressing vitamin A deficiency (Kumar *et al.*, 2022). For instance, scientists have successfully used CRISPR to increase the iron and zinc content in rice, addressing micronutrient deficiencies prevalent in regions where rice is a dietary staple. The rice vacuolar iron transporter genes (OsVIT1 and OsVIT2) exemplify this category, playing a specific role in transporting zinc ions (Zn^{2+}) and ferrous ions (Fe^{2+}) into vacuoles through the tonoplast (Kim *et al.*, 2006). Reducing the activity of OsVIT genes in rice by knocking it down leads to elevated levels of iron and zinc in the grains while diminishing their presence in the flag leaves (Bashir *et al.*, 2013). Plants that have undergone genetic editing have been engineered to naturally reduce cytokinin levels, promoting the accumulation of phosphorus, calcium, sulfur, copper, manganese, iron, and zinc in the plant biomass (Ramireddy *et al.*, 2018). The Kitaake cultivar of Golden Rice was created using knock-in techniques, incorporating a 5.2-kb cassette for carotenogenesis containing CrtI and maize PSY genes. This variant boasts a β-carotene content of 7.9 μg/g dry weight (DW) in its endosperm (Dong *et al.*, 2020). This targeted gene editing holds immense potential in combating malnutrition globally, offering sustainable solutions to improve human health. By harnessing CRISPRs precision, researchers can tailor crops to contain essential nutrients, ensuring access to healthier diets, and addressing nutritional deficiencies in vulnerable populations, ultimately contributing to improved wellbeing and food security worldwide (Senguttuvel *et al.*, 2023).

5.6.2 Pest and disease resistance

CRISPR technology presents a game-changing strategy in agricultural pest and disease management by conferring resistance against pathogens and pests. In a study, researchers have used CRISPR to develop wheat varieties resistant to powdery mildew, a devastating fungal disease. Wheat plants with triple knockouts of TaMLO were resistant to powdery mildew and also exhibited leaf chlorosis (Wang *et al.*, 2014). An instance of two distinct resistances against the same disease, derived from CRISPR/Cas9 technology, can be seen in the knockouts of TaMLO (Wang *et al.*, 2014) and TaEDR1 (Zhang *et al.*, 2017), both providing wheat with resistance to powdery mildew. By precisely targeting genes responsible for susceptibility to pests and diseases, CRISPR enables the creation of resilient crops without chemical pesticides. This approach not only reduces environmental pollution but also promotes sustainable farming practices, ensuring crop health and yield stability. Harnessing CRISPR's precision in pest and disease resistance holds promise for enhancing agricultural sustainability and resilience in the face of evolving threats (Borrelli *et al.*, 2018).

5.6.3 Improved yield and quality

CRISPR technology precision in targeting genes associated with yield and quality traits revolutionizes crop improvement, offering substantial benefits to agriculture. Scientists have utilized CRISPR to enhance the yield potential of crops like tomatoes by modifying genes involved in fruit size and development (Li *et al.*, 2018). By precisely editing these genes, CRISPR enhances crop productivity and market value, ensuring higher yields of superior quality produce. This targeted approach not only boosts agricultural efficiency but also meets consumer demands for nutritious and visually appealing crops. CRISPR's role in improving yield and quality underscores its potential in driving agricultural innovation and sustainability (Liu *et al.*, 2021a).

5.6.4 Environmental adaptation

CRISPR technology holds promise in addressing the urgent need for environmental adaptation in agriculture by modifying genes to enable crops to thrive in diverse conditions. In a study, researchers have utilized CRISPR to enhance drought tolerance in crops like maize by targeting genes involved in water regulation. CRISPR-Cas enabled advanced breeding technology to generate novel variants of ARGOS8. Maize ARGOS8 functions as a suppressor of ethylene responses. In a field experiment, it was found that the ARGOS8 variants, when compared to the wild type, boosted grain yield by five bushels per acre in situations of flowering stress, while showing no reduction in yield under adequately watered conditions. These findings underscore the effectiveness of the CRISPR-Cas9 system in producing new genetic variations beneficial for cultivating drought-resistant crops (Shi *et al.*, 2017).

This targeted modification equips crops to withstand harsh environmental conditions, mitigating the impacts of climate change on agriculture. By bolstering resilience and adaptability, CRISPR offers a sustainable solution to ensure food security in the face of shifting climatic patterns, highlighting its pivotal role in safeguarding agricultural productivity and environmental sustainability (Sami *et al.*, 2021).

5.7 Synergistic Approaches in Crop Designing

5.7.1 Integration with traditional breeding methods

The integration of CRISPR with traditional breeding methods represents a synergistic approach to crop improvement, leveraging the strengths of each technique (Ahmar *et al.*, 2020). For example, researchers have combined CRISPR-mediated gene editing with conventional breeding in tomato plants to enhance disease resistance while preserving desirable traits like flavor and shelf life. This integration allows for precise modifications in specific genes, complementing the broader genetic diversity generated through traditional breeding (Tang *et al.*, 2023). By combining the efficiency of CRISPR with the genetic variability of traditional methods, this approach accelerates crop improvement programs, offering tailored solutions to agricultural challenges while maintaining genetic diversity and trait integrity.

5.7.2 Precision phenotyping technologies

Precision phenotyping technologies offer a transformative approach to guide CRISPR edits in crop improvement (Zhang *et al.*, 2021a). High-throughput phenotyping platforms enable rapid and accurate assessment of plant traits under various environmental conditions. By integrating these technologies with CRISPR, researchers can identify and edit target genes more effectively, enhancing the efficiency and accuracy of trait selection in breeding programs. This synergy between precision phenotyping and CRISPR accelerates crop improvement efforts, leading to the development of resilient and high-yielding varieties tailored to specific agricultural needs (Liu *et al.*, 2022).

5.7.3 Omics technologies

Omics technologies, including genomics, transcriptomics, and proteomics, significantly augment CRISPR-based genome editing by providing comprehensive insights into gene function and regulatory networks. As functional genomics and molecular biology advance, numerous genes responsible for crucial quality traits have been effectively identified and isolated, including Wx, BADH2, FAD2 etc. (Yang *et al.*, 2021). Integrating genomics data allows for the identification of target genes essential for desired traits, guiding CRISPR edits with precision. Transcriptomics and proteomics further enhance understanding by elucidating gene expression patterns and protein interactions, facilitating informed editing decisions. Combining transcriptomic analysis with CRISPR editing in rice led to the development of high-yielding varieties with improved grain quality (Yang *et al.*, 2005; Nawaz *et al.*, 2020). Leveraging omics technologies maximizes the efficiency and effectiveness of CRISPR-mediated crop improvement, driving advancements in agriculture.

5.8 Conclusion and Prospects

The utilization of CRISPR technology represents a promising and valuable tool for scientists in the development of stress-resistant crops, addressing critical challenges such as climate change, global food security, and agricultural sustainability. However, navigating the complexities it presents can prove daunting. For instance, stress tolerance, a trait governed by intricate functional genomes and pathways, may require more than single-gene editing to achieve optimal results. Moreover, the risk of off-target mutations remains a concern, despite advancements in enhancing efficiency and target specificity. Future strategies entail the exploration of nontransgenic approaches, as well as the establishment and refinement of precise, multiplex editing protocols to generate resilient crop varieties resistant to both biotic and abiotic stresses. Hence, while pursuing the enhancement of abiotic stress resistance through genetic modification, thorough examination of potential unintended consequences is imperative. Moving forward, advancements in genetic and protein engineering hold promise for elucidating plant resistance mechanisms, thereby bolstering the application of CRISPR/Cas genome editing in crop abiotic stress resistance. Furthermore, the integration of omics approaches, including genomics, transcriptomics, proteomics, phenomics, and metabolomics, presents an unparalleled opportunity to enhance the efficacy and applicability of currently employed CRISPR technology.

Acknowledgements

The authors express appreciation for the support provided by the Deanship of Scientific Research, Vice Presidency for Graduate Studies and Scientific Research, King Faisal University, Saudi Arabia [Grant No. KFU252368].

References

Abiri, R., Shaharuddin, N.A., Maziah, M., Yusof, Z.N.B., Atabaki, N. *et al.* (2017) Role of ethylene and the APETALA 2/ethylene response factor superfamily in rice under various abiotic and biotic stress conditions. *Environmental and Experimental Botany* 134, 33–44. DOI: 10.1016/j.envexpbot.2016.10.015.

Achary, V.M.M. and Reddy, M.K. (2021) CRISPR-Cas9 mediated mutation in grain width and weight2 (GW2) locus improves aleurone layer and grain nutritional quality in rice. *Scientific Reports* 11(1), 21941. DOI: 10.1038/s41598-021-00828-z.

Agarwal, S. and Grover, A. (2006) Molecular biology, biotechnology and genomics of flooding-associated low O_2 stress response in plants. *Critical Reviews in Plant Sciences* 25(1), 1–21. DOI: 10.1080/07352680500365232.

Ahmar, S., Gill, R.A., Jung, K.H., Faheem, A., Qasim, M.U. *et al.* (2020) Conventional and molecular techniques from simple breeding to speed breeding in crop plants: Recent advances and future outlook. *International Journal of Molecular Sciences* 21(7), 2590. DOI: 10.3390/ijms21072590.

Alam, M.S., Kong, J., Tao, R., Ahmed, T., Alamin, M. *et al.* (2022) CRISPR/Cas9 mediated knockout of the OsbHLH024 transcription factor improves salt stress resistance in rice (*Oryza sativa* L.). *Plants* 11(9), 1184. DOI: 10.3390/plants11091184.

Alexandratos, N. and Bruinsma, J. (2012) World agriculture towards 2030/2050: The 2012 revision [ESA Working Paper no.12-03]. FAO, Rome.

Amtmann, A., Troufflard, S. and Armengaud, P. (2008) The effect of potassium nutrition on pest and disease resistance in plants. *Physiologia Plantarum* 133(4), 682–691. DOI: 10.1111/j.1399-3054.2008.01075.x.

Arico, D., Legris, M., Castro, L., Garcia, C.F., Laino, A. *et al.* (2019) Neighbour signals perceived by phytochrome B increase thermotolerance in *Arabidopsis*. *Plant, Cell & Environment* 42(9), 2554–2566. DOI: 10.1111/pce.13575.

Arshad, A., Abbas, A. and Rehman, A. (2022) Mechanism of drought stress tolerance in maize. *Biological and Agricultural Sciences Research Journal* 2022(1), 3. DOI: 10.54112/basrj.v2022i1.3.

Atkinson, N.J. and Urwin, P.E. (2012) The interaction of plant biotic and abiotic stresses: From genes to the field. *Journal of Experimental Botany* 63(10), 3523–3543. DOI: 10.1093/jxb/ers100.

Badiyal, A., Mahajan, R., Rana, R.S., Sood, R., Walia, A. *et al.* (2024) Synergizing biotechnology and natural farming: Pioneering agricultural sustainability through innovative interventions. *Frontiers in Plant Science* 15, 1280846. DOI: 10.3389/fpls.2024.1280846.

Bailey-Serres, J. and Voesenek, L.A.C.J. (2008) Flooding stress: Acclimations and genetic diversity. *Annual Review of Plant Biology* 59(1), 313–339. DOI: 10.1146/annurev.arplant.59.032607.092752.

Baillo, E.H., Kimotho, R.N., Zhang, Z. and Xu, P. (2019) Transcription factors associated with abiotic and biotic stress tolerance and their potential for crops improvement. *Genes* 10(10), 771. DOI: 10.3390/genes10100771.

Bale, J.S., Masters, G.J., Hodkinson, I.D., Awmack, C., Bezemer, T.M. *et al.* (2002) Herbivory in global climate change research: Direct effects of rising temperature on insect herbivores. *Global Change Biology* 8(1), 1–16. DOI: 10.1046/j.1365-2486.2002.00451.x.

Bao, F., Huang, X., Zhu, C., Zhang, X., Li, X. *et al.* (2014) *Arabidopsis* HSP90 protein modulates RPP4-mediated temperature-dependent cell death and defense responses. *The New Phytologist* 202(4), 1320–1334. DOI: 10.1111/nph.12760.

Barrero, L.S., Willmann, M.R., Craft, E.J., Akther, K.M., Harrington, S.E. *et al.* (2022) Identifying genes associated with abiotic stress tolerance suitable for CRISPR/Cas9 editing in upland rice cultivars adapted to acid soils. *Plant Direct* 6(12), e469. DOI: 10.1002/pld3.469.

Baruah, I., Debbarma, J., Deka, H.P. and Keshavaiah, C. (2017) The DEAD-box RNA helicases and multiple abiotic stresses in plants: A systematic review of recent advances and challenges. *Plant Omics* 10(5), 252–262. DOI: 10.21475/poj.1005.17.pne855.

Bashir, K., Takahashi, R., Akhtar, S., Ishimaru, Y., Nakanishi, H. *et al.* (2013) The knockdown of OsVIT2 and MIT affects iron localization in rice seed. *Rice* 6(1), 31. DOI: 10.1186/1939-8433-6-31.

Bashir, K., Todaka, D., Rasheed, S., Matsui, A., Ahmad, Z. *et al.* (2022) Ethanol-mediated novel survival strategy against drought stress in plants. *Plant & Cell Physiology* 63(9), 1181–1192. DOI: 10.1093/pcp/pcac114.

Bohnert, H.J., Ayoubi, P., Borchert, C., Bressan, R.A., Burnap, R.L. *et al.* (2001) A genomics approach towards salt stress tolerance. *Plant Physiology and Biochemistry: PPB / Societe Francaise de Physiologie Vegetale* 39(3–4), 295–311. DOI: 10.1016/S0981-9428(00)01237-7.

Borrelli, V.M.G., Brambilla, V., Rogowsky, P., Marocco, A. and Lanubile, A. (2018) The enhancement of plant disease resistance using CRISPR/Cas9 technology. *Frontiers in Plant Science* 9, 1245. DOI: 10.3389/fpls.2018.01245.

Boyer, J.S. (1982) Plant productivity and environment. *Science* 218(4571), 443–448. DOI: 10.1126/science.218.4571.443.

Brouns, S.J., Jore, M.M., Lundgren, M., Westra, E.R., Slijkhuis, R.J. *et al.* (2008) Small CRISPR RNAs guide antiviral defense in prokaryotes. *Science* 321(5891), 960–964. DOI: 10.1126/science.1159689.

Bunnik, E.M. and Le Roch, K.G. (2013) An introduction to functional genomics and systems biology. *Advances in Wound Care: The Journal for Prevention and Healing* 2(9), 490–498. DOI: 10.1089/wound.2012.0379.

Chandrasekaran, J., Brumin, M., Wolf, D., Leibman, D., Klap, C. *et al.* (2016) Development of broad virus resistance in non-transgenic cucumber using CRISPR/Cas9 technology. *Molecular Plant Pathology* 17(7), 1140–1153. DOI: 10.1111/mpp.12375.

Chaves, M.M. and Oliveira, M.M. (2004) Mechanisms underlying plant resilience to water deficits: Prospects for water-saving agriculture. *Journal of Experimental Botany* 55(407), 2365–2384. DOI: 10.1093/jxb/erh269.

Chen, S. and Liu, H. (2021) How plants protect themselves from ultraviolet-B radiation stress. *Plant Physiology* 187(3), 1096–1103. DOI: 10.1093/plphys/kiab245.

Chen, K., Wang, Y., Zhang, R., Zhang, H. and Gao, C. (2019) CRISPR/Cas genome editing and precision plant breeding in agriculture. *Annual Review of Plant Biology* 70(1), 667–697. DOI: 10.1146/annurev-arplant-050718-100049.

Chen, Y.-G., He, X.-L.-S., Huang, J.-H., Luo, R., Ge, H.-Z. *et al.* (2021) Impacts of heavy metals and medicinal crops on ecological systems, environmental pollution, cultivation, and production processes in China. *Ecotoxicology and Environmental Safety* 219, 112336. DOI: 10.1016/j.ecoenv.2021.112336.

Chikkaputtaiah, C., Debbarma, J., Baruah, I., Havlickova, L., Deka Boruah, H.P. *et al.* (2017) Molecular genetics and functional genomics of abiotic stress-responsive genes in oilseed rape (*Brassica napus* L.): A review of recent advances and future. *Plant Biotechnology Reports* 11(6), 365–384. DOI: 10.1007/s11816-017-0458-3.

Dai, S., Georgelis, N., Bedair, M., Hong, Y.-J., Qi, Q. *et al.* (2022) Ectopic expression of a rice triketone dioxygenase gene confers mesotrione tolerance in soybean. *Pest Management Science* 78(7), 2816–2827. DOI: 10.1002/ps.6904.

Dong, O.X., Yu, S., Jain, R., Zhang, N., Duong, P.Q. *et al.* (2020) Marker-free carotenoid-enriched rice generated through targeted gene insertion using CRISPR-Cas9. *Nature Communications* 11(1), 1178. DOI: 10.1038/s41467-020-14981-y.

El-Shewy, E. (2019) The efficacy of copper oxide, tri-calcium phosphate and silicon dioxide nanoparticles in controlling black scurf disease of potato. *Ḥawliyyaāt Al-'ulūm Al-Zirā'iyyat' Bi-Muštuhur* 57(1), 129–138. DOI: 10.21608/assjm.2019.42223.

Ferrero, V., Baeten, L., Blanco-Sánchez, L., Planelló, R., Díaz-Pendón, J.A. *et al.* (2019) Complex patterns in tolerance and resistance to pests and diseases underpin the domestication of tomato. *The New Phytologist* 226(1), 254–266. DOI: 10.1111/nph.16353.

Ford, K., McDonald, D. and Mali, P. (2019) Functional genomics via CRISPR-Cas. *Journal of Molecular Biology* 431(1), 48–65. DOI: 10.1016/j.jmb.2018.06.034.

Fu, L., Ding, Z., Kumpeangkeaw, A., Sun, X. and Zhang, J. (2019) Gene coexpression analysis reveals dose-dependent and type-specific networks responding to ionizing radiation in the aquatic model plant *Lemna minor* using public data. Journal of Genetics 98(1). DOI: 10.1007/s12041-019-1063-8.

Gasperskaja, E. and Kučinskas, V. (2017) The most common technologies and tools for functional genome analysis. *Acta Medica Lituanica* 24(1), 1–11. DOI: 10.6001/actamedica.v24i1.3457.

Gibbs, D.J., Lee, S.C., Isa, N.M., Gramuglia, S., Fukao, T. *et al.* (2011) Homeostatic response to hypoxia is regulated by the N-end rule pathway in plants. *Nature* 479(7373), 415–418. DOI: 10.1038/nature10534.

Gleditzsch, D., Pausch, P., Müller-Esparza, H., Özcan, A., Guo, X. *et al.* (2019) PAM identification by CRISPR-Cas effector complexes: Diversified mechanisms and structures. RNA Biology 16(4), 504–517. DOI: 10.1080/15476286.2018.1504546.

Goel, A.K., Lundberg, D., Torres, M.A., Matthews, R., Akimoto-Tomiyama, C. *et al.* (2008) The *Pseudomonas syringae* type III effector HopAM1 enhances virulence on water-stressed plants. *Molecular Plant-Microbe Interactions* 21(3), 361–370. DOI: 10.1094/MPMI-21-3-0361.

Guo, X., Jiang, J., Liu, Y., Yu, L., Chang, R. *et al.* (2021) Identification of a novel salt tolerance-related locus in wild soybean (*Glycine soja* Sieb. & Zucc.). *Frontiers in Plant Science* 12, 791175. DOI: 10.3389/fpls.2021.791175.

Gupta, D., Bhattacharjee, O., Mandal, D., Sen, M.K., Dey, D. *et al.* (2019) CRISPR-Cas9 system: A new-fangled dawn in gene editing. *Life Sciences* 232(116636), 116636. DOI: 10.1016/j.lfs.2019.116636.

Gupta, A., Rico-Medina, A. and Caño-Delgado, A.I. (2020) The physiology of plant responses to drought. *Science* 368(6488), 266–269. DOI: 10.1126/science.aaz7614.

Han, M.-L., Lv, Q.-Y., Zhang, J., Wang, T., Zhang, C.-X. *et al.* (2022) Decreasing nitrogen assimilation under drought stress by suppressing DST-mediated activation of nitrate reductase 1.2 in rice. *Molecular Plant* 15(1), 167–178. DOI: 10.1016/j.molp.2021.09.005.

Hassan, M.M., Yuan, G., Chen, J.-G., Tuskan, G.A. and Yang, X. (2020) Prime editing technology and its prospects for future applications in plant biology research. *Biodesign Research* 9350905. DOI: 10.34133/2020/9350905.

Hatfield, J.L. and Prueger, J.H. (2015) Temperature extremes: Effect on plant growth and development. *Weather and Climate Extremes* 10, 4–10. DOI: 10.1016/j.wace.2015.08.001.

Hirel, B., Le Gouis, J., Ney, B. and Gallais, A. (2007) The challenge of improving nitrogen use efficiency in crop plants: Towards a more central role for genetic variability and quantitative genetics within integrated approaches. *Journal of Experimental Botany* 58(9), 2369–2387. DOI: 10.1093/jxb/erm097.

Horvath, D.P., Clay, S.A., Bruggeman, S.A., Anderson, J.V., Chao, W.S. *et al.* (2019) Varying weed densities alter the corn transcriptome, highlighting a core set of weed-induced genes and processes with potential for manipulating weed tolerance. *The Plant Genome* 12(3), 1–9. DOI: 10.3835/plantgenome2019.05.0035.

Iijima, M., Awala, S.K., Watanabe, Y., Kawato, Y., Fujioka, Y. *et al.* (2016) Mixed cropping has the potential to enhance flood tolerance of drought-adapted grain crops. *Journal of Plant Physiology* 192, 21–25. DOI: 10.1016/j.jplph.2016.01.004.

Jain, M. (2015) Function genomics of abiotic stress tolerance in plants: A CRISPR approach. *Frontiers in Plant Science* 6, 375. DOI: 10.3389/fpls.2015.00375.

Jiang, W., Zhou, H., Bi, H., Fromm, M., Yang, B. *et al.* (2013) Demonstration of CRISPR/Cas9/sgRNA-mediated targeted gene modification in *Arabidopsis*, tobacco, sorghum and rice. *Nucleic Acids Research* 41(20), e188–e188. DOI: 10.1093/nar/gkt780.

Jinek, M., Chylinski, K., Fonfara, I., Hauer, M., Doudna, J.A. *et al.* (2012) A programmable dual-RNA-guided DNA endonuclease in adaptive bacterial immunity. *Science* 337(6096), 816–821. DOI: 10.1126/science.1225829.

Joshi, A., Yang, S.Y., Song, H.G., Min, J. and Lee, J.H. (2023) Genetic databases and gene editing tools for enhancing crop resistance against abiotic stress. *Biology* 12(11), 1400. DOI: 10.3390/biology12111400.

Kamali, S. and Singh, A. (2023) Genomic and transcriptomic approaches to developing abiotic stress-resilient crops. *Agronomy* 13(12), 2903. DOI: 10.3390/agronomy13122903.

Kerchev, P.I. and Breusegem, F. (2021) Improving oxidative stress resilience in plants. *The Plant Journal* 109(2), 359–372. DOI: 10.1111/tpj.15493.

Khan, Z.A., Kumar, R. and Dasgupta, I. (2022) CRISPR/Cas-mediated resistance against viruses in plants. *International Journal of Molecular Sciences* 23(4), 2303. DOI: 10.3390/ijms23042303.

Kim, S.A., Punshon, T., Lanzirotti, A., Li, L., Alonso, J.M. *et al.* (2006) Localization of iron in *Arabidopsis* seed requires the vacuolar membrane transporter VIT1. *Science* 314(5803), 1295–1298. DOI: 10.1126/science.1132563.

Kim, Y.-A., Moon, H. and Park, C. (2019) CRISPR/Cas9-targeted mutagenesis of Os8N3 in rice to confer resistance to *Xanthomonas oryzae* pv. oryzae. *Rice* 12(1), 67. DOI: 10.1186/s12284-019-0325-7.

Kim, S.-T., Choi, M., Bae, S.-J. and Kim, J.-S. (2021) The functional association of ACQOS/VICTR with salt stress resistance in *Arabidopsis thaliana* was confirmed by CRISPR-mediated mutagenesis. *International Journal of Molecular Sciences* 22(21), 11389. DOI: 10.3390/ijms222111389.

Klap, C., Yeshayahou, E., Bolger, A.M., Arazi, T., Gupta, S.K. *et al.* (2017) Tomato facultative parthenocarpy results from SLAGAMOUS-LIKE 6 loss of function. *Plant Biotechnology Journal* 15(5), 634–647. DOI: 10.1111/pbi.12662.

Komatsu, S., Han, C., Nanjo, Y., Altaf-Un-Nahar, M., Wang, K. *et al.* (2013) Label-free quantitative proteomic analysis of abscisic acid effect in early-stage soybean under flooding. *Journal of Proteome Research* 12(11), 4769–4784. DOI: 10.1021/pr4001898.

Kumar, V.V.S., Verma, R.K., Yadav, S.K., Yadav, P., Watts, A. *et al.* (2020) CRISPR-Cas9 mediated genome editing of drought and salt tolerance (*OsDST*) gene in indica mega rice cultivar MTU1010. *Physiology and Molecular Biology of Plants* 26, 1099–1110. DOI: 10.1007/s12298-020-00819-w.

Kumar, D., Yadav, A., Ahmad, R., Dwivedi, U.N. and Yadav, K. (2022) CRISPR-based genome editing for nutrient enrichment in crops: A promising approach toward global food security. *Frontiers in Genetics* 13, 932859. DOI: 10.3389/fgene.2022.932859.

Kumar, M., Prusty, M.R., Pandey, M.K., Singh, P.K., Bohra, A. *et al.* (2023) Application of CRISPR/Cas9-mediated gene editing for abiotic stress management in crop plants. *Frontiers in Plant Science* 14, 1157678. DOI: 10.3389/fpls.2023.1157678.

Kumari, P., Jasrotia, P., Kumar, D., Kashyap, P.L., Kumar, S. *et al.* (2022) Biotechnological approaches for host plant resistance to insect pests. *Frontiers in Genetics* 13. DOI: 10.3389/fgene.2022.914029.

Kuromori, T., Fujita, M., Takahashi, F., Yamaguchi-Shinozaki, K. and Shinozaki, K. (2021) Inter-tissue and inter-organ signaling in drought stress response and phenotyping of drought tolerance. *The Plant Journal* 109(2), 342–358. DOI: 10.1111/tpj.15619.

Lamers, J., Meer, T. and Testerink, C. (2020) How plants sense and respond to stressful environments. *Plant Physiology* 182(4), 1624–1635. DOI: 10.1104/pp.19.01464.

LeDuc, D.L., AbdelSamie, M., Móntes-Bayon, M., Wu, C.P., Reisinger, S.J. *et al.* (2006) Overexpressing both ATP sulfurylase and selenocysteine methyltransferase enhances selenium phytoremediation traits in Indian mustard. *Environmental Pollution* 144(1), 70–76. DOI: 10.1016/j.envpol.2006.01.008.

Lee, J.-Y. and Lee, D.-H. (2003) Use of serial analysis of gene expression technology to reveal changes in gene expression in arabidopsis pollen undergoing cold stress. *Plant Physiology* 132(2), 517–529. DOI: 10.1104/pp.103.020511.

Li, Z., Liu, Z.-B., Xing, A., Moon, B.P., Koellhoffer, J.P. *et al.* (2015) Cas9-guide RNA directed genome editing in soybean. *Plant Physiology* 169(2), 960–970. DOI: 10.1104/pp.15.00783.

Li, J., Zhang, H., Si, X., Tian, Y., Chen, K. *et al.* (2017) Generation of thermosensitive male-sterile maize by targeted knockout of the ZmTMS5 gene. *Journal of Genetics and Genomics = Yi Chuan Xue Bao* 44(9), 465–468. DOI: 10.1016/j.jgg.2017.02.002.

Li, X., Wang, Y., Chen, S., Tian, H., Fu, D. *et al.* (2018) Lycopene is enriched in tomato fruit by CRISPR/Cas9-mediated multiplex genome editing. *Frontiers in Plant Science* 9, 559. DOI: 10.3389/fpls.2018.00559.

Li, G., Ma, Y., Wang, X., Cheng, N., Meng, D. *et al.* (2022a) CCRISPR/Cas9 gene editing of NTAITRs, a family of transcription repressor genes, leads to enhanced drought tolerance in tobacco. *International Journal of Molecular Sciences* 23(23), 15268. DOI: 10.3390/ijms232315268.

Li, Z., Rao, M.J., Li, J., Wang, Y., Chen, P. *et al.* (2022b) CRISPR/CAS9 mutant rice OSPMEI12 involved in growth, cell wall development, and response to phytohormone and heavy metal stress. *International Journal of Molecular Sciences* 23(24), 16082. DOI: 10.3390/ijms232416082.

Lieber, M.R. (2010) The mechanism of double-strand DNA break repair by the nonhomologous DNA end-joining pathway. *Annual Review of Biochemistry* 79(1), 181–211. DOI: 10.1146/annurev.biochem.052308.093131.

Liu, L., Kuang, Y., Yan, F., Li, S., Ren, B. *et al.* (2020) Developing a novel artificial rice germplasm for dinitroaniline herbicide resistance by base editing of OsTUBA2. *Plant Biotechnology Journal* 19(1), 5–7. DOI: 10.1111/pbi.13430.

Liu, Q., Yang, F., Zhang, J., Liu, H., Rahman, S. *et al.* (2021a) Application of CRISPR/Cas9 in crop quality improvement. *International Journal of Molecular Sciences* 22(8), 4206. DOI: 10.3390/ijms22084206.

Liu, Z., Wang, H., Xie, J., Lv, J., Zhang, G. *et al.* (2021b) The roles of cruciferae glucosinolates in disease and pest resistance. *Plants* 10(6), 1097. DOI: 10.3390/plants10061097.

Liu, H., Chen, W., Li, Y., Sun, L., Chai, Y. *et al.* (2022) CRISPR/Cas9 technology and its utility for crop improvement. *International Journal of Molecular Sciences* 23(18), 10442. DOI: 10.3390/ijms231810442.

Liu, D., Zhao, M., Qiao, P., Li, Z., Chen, G. *et al.* (2023) NtrC contributes to nitrogen utilization, stress tolerance, and virulence in *Acidovorax citrulli*. *Microorganisms* 11(3), 767. DOI: 10.3390/microorganisms11030767.

Lone, B.A., Karna, S.K.L., Ahmad, F., Shahi, N. and Pokharel, Y.R. (2018) CRISPR/Cas9 system: A bacterial tailor for genomic engineering. *Genetics Research International* 1–17. DOI: 10.1155/2018/3797214.

Lu, Y., Wang, J., Chen, B., Mo, S., Lian, L. *et al.* (2021) A donor-DNA-free CRISPR/Cas-based approach to gene knock-up in rice. *Nature Plants* 7(11), 1445–1452. DOI: 10.1038/s41477-021-01019-4.

Lv, Y., Deng, X., Quan, L., Xia, Y. and Shen, Z. (2012) Metallothioneins BcMT1 and BcMT2 from *Brassica campestris* enhance tolerance to Cadmium and Copper and decrease production of reactive oxygen species in *Arabidopsis thaliana*. *Plant and Soil* 367(1–2), 507–519. DOI: 10.1007/s11104-012-1486-y.

Mangena, P. (2023) Cell mutagenic autopolyploidy enhances salinity stress tolerance in leguminous crops. *Cells* 12, 16. DOI: 10.3390/cells12162082.

Mani, M., Natarajan, N., Hegde, R.D. and Tej, M.K. (2022) Host plant resistance to insect pests in horticultural crops. In: Mani, M. (ed.) *Trends in Horticultural Entomology*, Vol. 1. Springer, Singapore, pp. 335–386. DOI: 10.1007/978-981-19-0343-4_11.

Marraffini, L.A. and Sontheimer, E.J. (2010) Self versus non-self discrimination during CRISPR RNA-directed immunity. *Nature* 463(7280), 568–571. DOI: 10.1038/nature08703.

McCarty, N.S., Graham, A.E., Studená, L. and Ledesma-Amaro, R. (2020) Multiplexed CRISPR technologies for gene editing and transcriptional regulation. *Nature Communications* 11(1), 1–13. DOI: 10.1038/s41467-020-15053-x.

Mei, Y., Wang, Y., Chen, H., Sun, Z.S. and Ju, X.D. (2016) Recent progress in CRISPR/Cas9 technology. *Journal of Genetics and Genomics = Yi Chuan Xue Bao* 43(2), 63–75. DOI: 10.1016/j.jgg.2016.01.001.

Mittler, R. and Blumwald, E. (2010) Genetic engineering for modern agriculture: Challenges and perspectives. *Annual Review of Plant Biology* 61(1), 443–462. DOI: 10.1146/annurev-arplant-042809-112116.

Mushtaq, M., Bhat, J.A., Mir, Z.A., Sakina, A., Ali, S. *et al.* (2018) CRISPR/Cas approach: A new way of looking at plant-abiotic interactions. *Journal of Plant Physiology* 224–225, 156–162. DOI: 10.1016/j.jplph.2018.04.001.

Naik, B., Kumar, V., Rizwanuddin, S., Chauhan, M., Choudhary, M. *et al.* (2023) Genomics, proteomics, and metabolomics approaches to improve abiotic stress tolerance in tomato plant. *International Journal of Molecular Sciences* 24(3), 3025. DOI: 10.3390/ijms24033025.

Nakashima, K. and Yamaguchi-Shinozaki, K. (2006) Regulons involved in osmotic stress-responsive and cold stress-responsive gene expression in plants. *Physiologia Plantarum* 126(1), 62–71. DOI: 10.1111/j.1399-3054.2005.00592.x.

Nawaz, G., Usman, B., Zhao, N., Han, Y., Li, Z. *et al.* (2020) CRISPR/Cas9 directed mutagenesis of OsGA20ox2 in high yielding basmati rice (*Oryza sativa* L.) line and comparative proteome profiling of unveiled changes triggered by mutations. *International Journal of Molecular Sciences* 21(17), 6170. DOI: 10.3390/ijms21176170.

Nekrasov, V., Wang, C., Win, J., Lanz, C., Weigel, D. *et al.* (2017) Rapid generation of a transgene-free powdery mildew resistant tomato by genome deletion. *Scientific Reports* 7(1), 482. DOI: 10.1038/s41598-017-00578-x.

Nieves-Cordones, M., Mohamed, S., Tanoi, K., Kobayashi, N., Takagi, K. *et al.* (2017) Production of low-Cs + rice plants by inactivation of the K + transporter Os HAK 1 with the CRISPR-Cas System. *The Plant Journal* 92(1), 43–56. DOI: 10.1111/tpj.13632.

Ogawa, D., Suzuki, Y., Yokoo, T., Katoh, E., Teruya, M. *et al.* (2021) Acetic-acid-induced jasmonate signaling in root enhances drought avoidance in rice. *Scientific Reports* 11(1), 6280. DOI: 10.1038/s41598-021-85355-7.

Oroz, J., Kim, J.H., Chang, B.J. and Zweckstetter, M. (2017) Mechanistic basis for the recognition of a misfolded protein by the molecular chaperone Hsp90. *Nature Structural & Molecular Biology* 24(4), 407–413. DOI: 10.1038/nsmb.3380.

Ortigosa, A., Gimenez-Ibanez, S., Leonhardt, N. and Solano, R. (2019) Design of a bacterial speck resistant tomato by CRISPR/Cas9-mediated editing of SlJAZ2. *Plant Biotechnology Journal* 17(3), 665–673. DOI: 10.1111/pbi.13006.

Paixão, J.F.R., Gillet, F.-X., Ribeiro, T.P., Bournaud, C., Lourenço-Tessutti, I.T. *et al.* (2019) Improved drought stress tolerance in arabidopsis by CRISPR/dCas9 fusion with a Histone AcetylTransferase. *Scientific Reports* 9(1). DOI: 10.1038/s41598-019-44571-y.

Pan, L., Yu, X., Shao, J., Liu, Z., Gao, T. *et al.* (2019) Transcriptomic profiling and analysis of differentially expressed genes in asparagus bean (*Vigna unguiculata* ssp. *Sesquipedalis*) under salt stress. *PLoS One* 14(7), e0219799. DOI: 10.1371/journal.pone.0219799.

Pandey, P., Irulappan, V., Bagavathiannan, M.V. and Senthil-Kumar, M. (2017) Impact of combined abiotic and biotic stresses on plant growth and avenues for crop improvement by exploiting physio-morphological traits. *Frontiers in Plant Science* 8, 537. DOI: 10.3389/fpls.2017.00537.

Peng, A., Chen, S., Lei, T., Xu, L., He, Y. *et al.* (2017) Engineering canker-resistant plants through CRISPR/Cas9-targeted editing of the susceptibility gene CsLOB1 promoter in citrus. *Plant Biotechnology Journal* 15(12), 1509–1519. DOI: 10.1111/pbi.12733.

Puchta, H., Dujon, B. and Hohn, B. (1993) Homologous recombination in plant cells is enhanced by *in vivo* induction of double strand breaks into DNA by a site-specific endonuclease. *Nucleic Acids Research* 21(22), 5034–5040. DOI: 10.1093/nar/21.22.5034.

Pyott, D.E., Sheehan, E. and Molnar, A. (2016) Engineering of CRISPR/Cas9-mediated potyvirus resistance in transgene-free *Arabidopsis* plants. *Molecular Plant Pathology* 17(8), 1276–1288. DOI: 10.1111/mpp.12417.

Quan, L., Chen, K., Chen, T., Li, H., Li, W. *et al.* (2023) Monitoring weed mechanical and chemical damage stress based on chlorophyll fluorescence imaging. *Frontiers in Plant Science* 14, 1188981. DOI: 10.3389/fpls.2023.1188981.

Rai, G.K., Khanday, D.M., Kumar, P., Magotra, I., Choudhary, S.M. *et al.* (2023) Enhancing crop resilience to drought stress through CRISPR-Cas9 genome editing. *Plants* 12(12), 2306. DOI: 10.3390/plants12122306.

Ramireddy, E., Galuszka, P. and Schmülling, T. (2018) Zn-fortified cereal grains in field-grown barley by enhanced root cytokinin breakdown. *Plant Signaling & Behavior* 13(11), e1530023. DOI: 10.1080/15592324.2018.1530023.

Ray, U. and Raghavan, S.C. (2020) Modulation of DNA double-strand break repair as a strategy to improve precise genome editing. *Oncogene* 39(41), 6393–6405. DOI: 10.1038/s41388-020-01445-2.

Ray, D.K., Mueller, N.D., Paul, C.W. and Foley, J.A. (2013) *Yield trends are insufficient to double global rice production by 2050*. Public Library of Science (PLOS), San Francisco. DOI: 10.1371/journal.pone.0066428.

Razzaq, A., Kaur, P., Akhter, N., Wani, S.H. and Saleem, F. (2021) Next-generation breeding strategies for climate-ready crops. *Frontiers in Plant Science* 12, 620420. DOI: 10.3389/fpls.2021.620420.

Ritonga, F.N. and Chen, S. (2020) Physiological and molecular mechanism involved in cold stress tolerance in plants. *Plants* 9(5), 560. DOI: 10.3390/plants9050560.

Saikia, B., Debbarma, J., Maharana, J., Sastry, G.N., Chikkaputtaiah, C. *et al.* (2024) CRISPR/Cas9-based genome editing and functional analysis of SlHyPRP1 and SlDEA1genes of *Solanum lycopersicum* L. in imparting genetic tolerance to multiple stress factors. *Frontiers in Plant Science* 15, 1304381. DOI: 10.3389/fpls.2024.1304381.

Sami, A., Xue, Z., Tazein, S., Arshad, A., He Zhu, Z. *et al.* (2021) CRISPR-Cas9-based genetic engineering for crop improvement under drought stress. *Bioengineered* 12(1), 5814–5829. DOI: 10.1080/21655979.2021.1969831.

Saqib, M., Shahzad, U., Zulfiqar, F., Tiwari, R.K., Lal, M.K. *et al.* (2023) Exogenous melatonin alleviates cadmium-induced inhibition of growth and photosynthesis through upregulating antioxidant defense system in strawberry. *South African Journal of Botany* 157, 10–18. DOI: 10.1016/j.sajb.2023.03.039.

Sasidharan, R. and Voesenek, L. (2015) Ethylene-mediated acclimations to flooding stress. *Plant Physiology* 169(1), 3–12. DOI: 10.1104/pp.15.00387.

Scognamiglio, M. and Schneider, B. (2020) Identification of potential allelochemicals from donor plants and their synergistic effects on the metabolome of *Aegilops geniculate*. *Frontiers in Plant Science* 11, 1046. DOI: 10.3389/fpls.2020.01046.

Senguttuvel, P., Padmavathi, G., Jasmine, C., Sanjeeva Rao, D., Neeraja, C. *et al.* (2023) Rice biofortification: Breeding and genomic approaches for genetic enhancement of grain zinc and iron contents. *Frontiers in Plant Science* 14, 1138408. DOI: 10.3389/fpls.2023.1138408.

Seregin, I.V. and Kozhevnikova, A.D. (2023) Phytochelatins: Sulfur-containing metal(loid)-chelating ligands in plants. *International Journal of Molecular Sciences* 24(3), 2430. DOI: 10.3390/ijms24032430.

Sharma, G., Shrestha, S., Kunwar, S. and Tseng, T.-M. (2021) Crop diversification for improved weed management: A review. *Agriculture* 11(5), 461. DOI: 10.3390/agriculture11050461.

Shi, J., Gao, H., Wang, H., Lafitte, H.R., Archibald, R.L. *et al.* (2017) ARGOS8 variants generated by CRISPR-Cas9 improve maize grain yield under field drought stress conditions. *Plant Biotechnology Journal* 15(2), 207–216. DOI: 10.1111/pbi.12603.

Shi, H., Hou, J., Li, D., Hu, H., Wang, Y. *et al.* (2024) Comparative transcriptome and coexpression network analysis reveals key pathways and hub candidate genes associated with sunflower (*Helianthus annuus* L.) drought tolerance. *BMC Plant Biology* 24(1), 224. DOI: 10.1186/s12870-024-04932-w.

Shimatani, Z., Kashojiya, S., Takayama, M., Terada, R., Arazoe, T. *et al.* (2017) Targeted base editing in rice and tomato using a CRISPR-Cas9 cytidine deaminase fusion. *Nature Biotechnology* 35(5), 441–443. DOI: 10.1038/nbt.3833.

Singer, S.D., Burton Hughes, K., Subedi, U., Dhariwal, G.K., Kader, K. *et al.* (2021) The CRISPR/CAS9-mediated modulation of SQUAMOSA promoter-binding protein-like 8 in Alfalfa leads to distinct phenotypic outcomes. *Frontiers in Plant Science* 12, 774146. DOI: 10.3389/fpls.2021.774146.

Singh, S., Parihar, P., Singh, R., Singh, V.P. and Prasad, S.M. (2016) Heavy metal tolerance in plants: Role of transcriptomics, proteomics, metabolomics, and ionomics. *Frontiers in Plant Science* 6, 1143. DOI: 10.3389/fpls.2015.01143.

Singla, J. and Krattinger, S.G. (2016) Biotic stress resistance genes in wheat. In: Wrigley, C.W., Faubion, J., Corke, H. and Seetharaman, K. (eds) *Encyclopedia of Food Grains*, Vol. 1. Elsevier, pp. 1388–1392. DOI: 10.1016/B978-0-08-100596-5.00229-8.

Tang, L., Mao, B., Li, Y., Lv, Q., Zhang, L. *et al.* (2017) Knockout of OsNramp5 using the CRISPR/Cas9 system produces low Cd-accumulating Indica rice without compromising yield. *Scientific Reports* 7(1). DOI: 10.1038/s41598-017-14832-9.

Tang, X.-D., Gao, F., Liu, M.-J., Fan, Q.-L., Chen, D.-K. *et al.* (2019) Methods for enhancing clustered regularly interspaced short palindromic repeats/Cas9-mediated homology-directed repair efficiency. *Frontiers in Genetics* 10. DOI: 10.3389/fgene.2019.00551.

Tang, Q., Wang, X., Jin, X., Peng, J., Zhang, H. *et al.* (2023) CRISPR/Cas technology revolutionizes crop breeding. *Plants* 12, 3119. DOI: 10.3390/plants12173119.

Thomazella, D.P.T., Seong, K., Mackelprang, R., Dahlbeck, D., Geng, Y. *et al.* (2021) Loss of function of a DMR6 ortholog in tomato confers broad-spectrum disease resistance. *Proceedings of the National Academy of Sciences of the United States of America* 118(27), e2026152118. DOI: 10.1073/pnas.2026152118.

Umar, O.B., Ranti, L.A., Abdulbaki, A.S., Bola, A.L., Abdulhamid, A.K. *et al.* (2022) Stresses in plants: Biotic and abiotic. In: Ansari, M. (ed.) *Current Trends in Wheat Research.* IntechOpen. Pakistan.

Van Zelm, E., Zhang, Y. and Testerink, C. (2020) Salt tolerance mechanisms of plants. *Annual Review of Plant Biology* 71(1), 403–433. DOI: 10.1146/annurev-arplant-050718-100005.

Verma, V., Ravindran, P. and Kumar, P.P. (2016) Plant hormone-mediated regulation of stress responses. *BMC Plant Biology* 16(1), 86. DOI: 10.1186/s12870-016-0771-y.

Vlčko, T. and Ohnoutková, L. (2020) Allelic variants of CRISPR/Cas9 induced mutation in an inositol trisphosphate 5/6 kinase gene manifest different phenotypes in Barley. *Plants* 9(2), 195. 10.3390/plants9020195.

Wang, X. and Komatsu, S. (2022) The role of phytohormones in plant response to flooding. *International Journal of Molecular Sciences* 23(12), 6383. DOI: 10.3390/ijms23126383.

Wang, W., Vinocur, B. and Altman, A. (2003) Plant responses to drought, salinity and extreme temperatures: Towards genetic engineering for stress tolerance. *Planta* 218(1), 1–14. DOI: 10.1007/s00425-003-1105-5.

Wang, Y., Cheng, X., Shan, Q., Zhang, Y., Liu, J. *et al.* (2014) Simultaneous editing of three homoeoalleles in hexaploid bread wheat confers heritable resistance to powdery mildew. *Nature Biotechnology* 32(9), 947–951. DOI: 10.1038/nbt.2969.

Wang, F., Wang, C., Liu, P., Lei, C., Hao, W. *et al.* (2016) Enhanced rice blast resistance by CRISPR/Cas9-targeted mutagenesis of the ERF transcription factor gene OSERF922. *PLoS One* 11(4), e0154027. DOI: 10.1371/journal.pone.0154027.

Wang, J., Mao, X., Wang, R., Li, A., Zhao, G. *et al.* (2019a) Identification of wheat stress-responding genes and TaPR-1-1 function by screening a cDNA yeast library prepared following abiotic stress. *Scientific Reports* 9(1), 1–12. DOI: 10.1038/s41598-018-37859-y.

Wang, L., Chen, S., Peng, A., Xie, Z., He, Y. *et al.* (2019b) CRISPR/Cas9-mediated editing of CsWRKY22 reduces susceptibility to *Xanthomonas citri* subsp. citri in Wanjincheng orange (*Citrus sinensis* (L.) Osbeck). *Plant Biotechnology Reports* 13(5), 501–510. DOI: 10.1007/s11816-019-00556-x.

Wang, T., Zhang, H. and Zhu, H. (2019c) CRISPR technology is revolutionizing the improvement of tomato and other fruit crops. *Horticulture Research* 6(1), 1–13. DOI: 10.1038/s41438-019-0159-x.

Wang, K., Bu, T., Cheng, Q., Dong, L., Su, T. *et al.* (2020) Two homologous LHY pairs negatively control soybean drought tolerance by repressing the abscisic acid responses. *The New Phytologist* 229(5), 2660–2675. DOI: 10.1111/nph.17019.

Wang, T., Xun, H., Wang, W., Ding, X., Tian, H. *et al.* (2021) Mutation of GMAITR genes by CRISPR/Cas9 genome editing results in enhanced salinity stress tolerance in soybean. *Frontiers in Plant Science* 12. DOI: 10.3389/fpls.2021.779598.

Wilson, S.K., Pretorius, T. and Naidoo, S. (2023) Mechanisms of systemic resistance to pathogen infection in plants and their potential application in forestry. *BMC Plant Biology* 23(1), 404. DOI: 10.1186/s12870-023-04391-9.

Wu, J., Yan, G., Duan, Z., Wang, Z., Kang, C. *et al.* (2020) Roles of the *Brassica napus* DELLA protein BNaA6.RGA, in modulating drought tolerance by interacting with the ABA signaling component BNaA10.ABF2. *Frontiers in Plant Science* 11, 577. DOI: 10.3389/fpls.2020.00577.

Xiao, Y., Karikari, B., Wang, L., Chang, F. and Zhao, T. (2021) Structure characterization and potential role of soybean phospholipases a multigene family in response to multiple abiotic stress uncovered by CRISPR/Cas9 technology. *Environmental and Experimental Botany* 188, 104521. DOI: 10.1016/j.envexpbot.2021.104521.

Xin, M., Wang, Y., Yao, Y., Song, N., Hu, Z. *et al.* (2011) Identification and characterization of wheat long non-protein coding RNAs responsive to powdery mildew infection and heat stress by using microarray analysis and SBS sequencing. *BMC Plant Biology* 11(1), 61. DOI: 10.1186/1471-2229-11-61.

Xu, R., Li, H., Qin, R., Wang, L., Li, L. *et al.* (2014) Gene targeting using the *Agrobacterium tumefaciens*-mediated CRISPR-Cas system in rice. *Rice* 7(1). DOI: 10.1186/s12284-014-0005-6.

Yang, G., Inoue, A., Takasaki, H., Kaku, H., Akao, S. *et al.* (2005) Proteomic approach to analyze auxin- and zinc-responsive protein in rice. *Journal of Proteome Research* 4(2), 456–463. DOI: 10.1021/pr049801h.

Yang, Y., Xu, C., Shen, Z. and Yan, C. (2021) Crop quality improvement through genome editing strategy. *Frontiers in Genome Editing* 3, 819687. DOI: 10.3389/fgeed.2021.819687.

Yue, E., Cao, H. and Liu, B. (2020) Osmir535, a potential genetic editing target for drought and salinity stress tolerance in *Oryza sativa*. *Plants* 9(10), 1337. DOI: 10.3390/plants9101337.

Zhang, F., Maeder, M.L., Unger-Wallace, E., Hoshaw, J.P., Reyon, D. *et al.* (2010) High frequency targeted mutagenesis in *Arabidopsis thaliana* using zinc finger nucleases. *Proceedings of the National Academy of Sciences of the United States of America* 107(26), 12028–12033. DOI: 10.1073/pnas.0914991107.

Zhang, F., Cong, L., Lodato, S., Kosuri, S., Church, G.M. *et al.* (2011) Efficient construction of sequence-specific TAL effectors for modulating mammalian transcription. *Nature Biotechnology* 29, 149–153.

Zhang, Y., Zhang, F., Li, X., Baller, J.A., Qi, Y. *et al.* (2012) Transcription activator-like effector nucleases enable efficient plant genome engineering. *Plant Physiology* 161(1), 20–27. DOI: 10.1104/pp.112.205179.

Zhang, Y., Bai, Y., Wu, G., Zou, S., Chen, Y. *et al.* (2017) Simultaneous modification of three homoeologs of *TaEDR1* by genome editing enhances powdery mildew resistance in wheat. *The Plant Journal* 91(4), 714–724. DOI: 10.1111/tpj.13599.

Zhang, Z., Ge, X., Luo, X., Wang, P., Fan, Q. *et al.* (2018) Simultaneous editing of two copies of Gh14-3-3d confers enhanced transgene-clean plant defense against *Verticillium dahliae* in allotetraploid upland cotton. *Frontiers in Plant Science* 9, 842. DOI: 10.3389/fpls.2018.00842.

Zhang, M., Liu, Q., Yang, X., Xu, J., Liu, G. *et al.* (2020) CRISPR/Cas9-mediated mutagenesis of Clpsk1 in watermelon to confer resistance to *Fusarium oxysporum* f.sp. niveum. *Plant Cell Reports* 39(5), 589–595. DOI: 10.1007/s00299-020-02516-0.

Zhang, D., Zhang, Z., Unver, T. and Zhang, B. (2021a) CRISPR/Cas: A powerful tool for gene function study and crop improvement. *Journal of Advanced Research* 29, 207–221. DOI: 10.1016/j.jare.2020.10.003.

Zhang, J., Zhang, H., Li, S., Li, J., Yan, L. *et al.* (2021b) Increasing yield potential through manipulating of an ARE1 ortholog related to nitrogen use efficiency in wheat by CRISPR/Cas9. *Journal of Integrative Plant Biology* 63(9), 1649–1663. DOI: 10.1111/jipb.13151.

Zhang, P., Wang, T., Cao, L., Jiao, Z., Ku, L. *et al.* (2023) Molecular mechanism analysis of ZmRL6 positively regulating drought stress tolerance in maize. *Stress Biology* 3(1), 47. DOI: 10.1007/s44154-023-00125-x.

Zhao, H., Wang, L., Zhao, F., Wu, L., Liu, A. *et al.* (2018) SpHMA1 is a chloroplast cadmium exporter protecting photochemical reactions in the Cd hyperaccumulator *Sedum plumbizincicola*. *Plant, Cell & Environment* 42(4), 1112–1124. DOI: 10.1111/pce.13456.

Zhong, X., Hong, W., Shu, Y., Li, J., Liu, L, *et al.* (2022) CRISPR/cas9 mediated gene-editing of gmHdz4 transcription factor enhances drought tolerance in soybean (*Glycine max* [L.] Merr.). *Frontiers in Plant Science* 13, 988505. DOI: 10.3389/fpls.2022.988505.

6 Mitigating Abiotic Stress Using Nanoparticles and Involved Cellular Processes Studied by Multiple Omics

Varsha Toppo[1], Akash Shaikh[1], Jameel M. Al-Khayri[2] and Praveen Nagella[1]*
[1]Department of Life Sciences, School of Sciences, Christ University, Bengaluru, India; [2]Department of Agricultural Biotechnology, College of Agriculture and Food Sciences, King Faisal University, Al-Ahsa, Saudi Arabia

Abstract
Crop yield is greatly impacted by abiotic stress, necessitating creative methods to increase plant resilience. In order to reduce these stressors, this chapter focusses on the possibilities of integrating multi-omics technology with nanoparticle applications. Studies suggest that nanoparticles improve plant tolerance by improving antioxidant activity and minimizing oxidative damage. In addition, multi-omics techniques (genomics, transcriptomics, proteomics, and metabolomics) offer a comprehensive understanding of the molecular processes that stress initiates. The creation of nanoparticles designed to specifically target particular stress-response pathways is made possible by these findings. The chapter's case studies demonstrate how tailored nanoparticles made with multi-omics data might improve crop performance in difficult conditions. The chapter highlights a promising avenue for developing targeted, sustainable solutions to increase crop resilience under abiotic stress conditions by combining nanotechnology with omics techniques, providing long-term sustainability.

Keywords: abiotic stress, nanoparticles, omics approach, agriculture, climate change

6.1 Introduction

In the face of escalating global challenges such as climate change, population growth, and diminishing natural resources, the agricultural sector is under immense pressure to enhance crop productivity and sustainability. All of the above anthropologic factors are negatively impacting every aspect of biological life on Earth, and plants are no exception, coming under intense threat of abiotic stressors. Abiotic stress, defined as the negative impact of nonliving environmental factors on plants, is a major challenge to global agriculture as plants are sessile organisms and are constantly exposed to a variety of unnatural stresses, including drought, salinity, extreme temperatures, and heavy-metal toxicity (Zhang *et al.*, 2023; Bartas, 2024). These stressors adversely affect plant growth, development, and yield, posing a significant threat to global food security. Various strategies have been developed over the years to mitigate the detrimental effects of abiotic stress. Traditional approaches, such as selective breeding and genetic modification,

*Corresponding author: praveen.n@christuniversity.in; jkhayri@kfu.edu.sa

© CAB International 2025. *Stress-Resilient Crops: Coordinated omics-CRISPR-Nanotechnology Strategies* (ed. Jen-Tsung Chen)
DOI: 10.1079/9781800627307.0006

have been employed to produce stress-tolerant crop varieties (Rai *et al.*, 2021). These methods, while successful in some cases, are often time-consuming and limited by the complexity of plant stress responses, which are governed by intricate networks of genes, proteins, and metabolites (Hamadani *et al.*, 2021; Oyebamiji *et al.*, 2024). In addition, the reliance on chemical inputs such as fertilizers and pesticides, commonly used to improve plant growth under stress conditions, presents environmental challenges, including soil degradation, pollution, and loss of biodiversity (Savci, 2012; Bai *et al.*, 2020). These limitations highlight the need for innovative and sustainable solutions to address abiotic stress in agriculture.

Nanotechnology has emerged as a promising tool to address these challenges. Nanoparticles (NPs), due to their unique physicochemical properties such as high reactivity, large surface area, and ability to penetrate cellular barriers, offer innovative solutions to mitigate the adverse effects of abiotic stress on plants (Al-Khayri *et al.*, 2023; Mariyam *et al.*, 2024). By influencing various cellular processes and enhancing stress tolerance, nanoparticles can play a crucial role in improving plant resilience and productivity. However, the full potential of nanoparticles in agriculture remains to be realized, particularly in terms of understanding their mechanisms of action at the molecular level. This is where omics technologies, like genomics, transcriptomics, proteomics, and metabolomics, play a critical role (Vailati-Riboni *et al.*, 2017). Transcriptomic profiling is a useful tool for understanding global gene expression, but it does not give a full picture of cellular functions. Since transcriptomics is unable to identify these changes, proteins –the functional products of gene expression – go through a variety of co- and post-translational modifications that profoundly affect their activity, structure, and function. These alterations are essential for controlling how proteins behave, particularly when they are under stress from the environment.

It is crucial to research alterations in the protein composition of cells in order to completely comprehend how environmental influences, such as abiotic stress, impact protein behavior and stress-tolerance mechanisms at the molecular level. Proteomics provides a thorough understanding of protein abundance, changes, and how these factors affect an organism's adaptive responses. Proteomics is the study of the entire set of proteins (the proteome) expressed by an organism (Mondal *et al.*, 2024). Omics approaches provide comprehensive insights into how plants respond to nanoparticles at different biological levels, from gene expression and protein synthesis to metabolite production and cellular signaling pathways (Jha and Pudake, 2016; Joshi *et al.*, 2024; Mustafa *et al.*, 2024). By integrating nanoparticle applications with omics data, researchers can identify key molecular targets and pathways involved in stress tolerance, paving the way for the development of more effective and tailored nanoparticle-based interventions (Barkataki and Singh, 2019). In this context, the integration of nanoparticle technology with omics approaches holds great promise for advancing our understanding of plant responses to abiotic stress and improving crop resilience. This chapter explores the current state of research in nanoparticle applications for stress mitigation, highlights the role of omics technologies in elucidating the underlying mechanisms, and discusses future directions for optimizing the use of nanoparticles in agriculture.

6.2 The Role of Nanoparticles in Mitigating Abiotic Stress

The use of nanoparticles in agriculture is a possible approach for addressing abiotic stress, which has a considerable impact on crop production and food security. Nanoparticles have been proved in studies to improve plant resilience by increasing nutrient uptake, stress tolerance, and priming plants for better defense responses. However, despite these advantages, questions about their environmental impact and potential toxicity persist. To ensure the safe and sustainable use of nanoparticles in agriculture, continued study and appropriate application are required (Imran and Ortas I, 2024).

Nanoparticles such as zinc oxide (ZnO), silver, and copper oxide are gaining popularity for their capacity to improve crop tolerance to abiotic stressors such as salt, drought, and nutritional shortages. ZnO nanoparticles, for example, increase nutrient uptake and plant growth during stress by gradually releasing critical zinc, as seen in salt-stressed quinoa seedlings (Verma *et al.*, 2024). Silver nanoparticles, with their antibacterial capabilities, aid in pathogen reduction, allowing plants to concentrate more on stress response. Meanwhile, copper oxide nanoparticles improve photosynthetic efficiency and antioxidant enzyme activity, which are essential for countering oxidative stress. However, the possible environmental impact and toxicity of these nanoparticles necessitate careful study and additional research to ensure their safe use in agriculture (Rehman *et al.*, 2023a). Nanoparticles can help plants survive drought by causing physiological and biochemical changes. Silver nanoparticles (AgNPs) have been shown to boost antioxidant activity in wheat during drought circumstances, lowering oxidative stress and promoting healthier growth. When plants are stressed by drought, zinc oxide nanoparticles (ZnO NPs) can increase the activity of antioxidant enzymes such as catalase (CAT), superoxide dismutase (SOD), and peroxidase (POD). Reactive oxygen species (ROS) are frequently created during stress, and these enzymes aid in reducing their harmful effects to protect and increase the resilience of the plants (Rehman *et al.*, 2023b).

Plants that are stressed by salinity may suffer from nutritional imbalances and harmful ion accumulation. ZnO NPs, for example, are nanoparticles that can help plants withstand salt stress. For instance, ZnO NPs increase antioxidant enzyme activity, which helps reduce the negative effects of ROS, and hence, enhanced maize seedling growth under salt stress. By increasing antioxidant enzyme activity, reducing ROS levels, and enhancing nutrient uptake, AgNPs can mitigate the effects of salinity stress on plants and enhance their growth in saline environments (Abasi *et al.*, 2022). In stressed plants, certain nanoparticles can improve nutrient uptake. Because of their high surface area-to-volume ratio, which makes them more soluble and available to plants than traditional fertilizers, nanofertilizers are more effective. Carbon- or biochar-based nanoparticles can absorb heavy metals from the soil, lowering their toxicity and limit plant uptake. Furthermore, certain nanoparticles promote the synthesis of proteins called metallothioneins, which bind to heavy metals and reduce their detrimental effects on plants (Rai *et al.*, 2023).

Nanoparticles are proving to be a revolutionary tool in agriculture, particularly in reducing the effects of abiotic stress by enhancing nutrient delivery and antioxidant defense mechanisms. Their ability to improve the uptake and transport of vital nutrients, such as zinc, iron, and magnesium, ensures efficient fertilization and growth regulation while minimizing nutrient loss. Additionally, NPs help combat oxidative stress caused by drought and salinity by scavenging harmful ROS and boosting the production of beneficial secondary metabolites like phenolics and flavonoids. Despite these promising benefits, there remain concerns about the environmental impact and potential toxicity of NPs, highlighting the need for ongoing research to ensure their safe and sustainable application in agriculture (Fizzah Zulfiqar *et al.*, 2024). Plants have shown remarkable success with NPs in mitigating a range of abiotic stressors. Zinc oxide NPs (ZnO NPs), for example, increase the development of quinoa and maize under salt stress by decreasing oxidative damage and increasing antioxidant enzyme activity. Wheat's resistance to drought is enhanced by AgNPs, which lower ROS and boost antioxidant activity (Yadav *et al.*, 2024). Copper oxide NPs, or CuO NPs, increase the activity of antioxidant enzymes and photosynthetic efficiency, helping plants tolerate heat stress. Additionally, by reducing metal uptake and bolstering antioxidant defenses, silicon NPs help rice plants cope with heavy-metal stress, such as cadmium toxicity. These uses serve as examples of how plant resistance and production can be efficiently supported by NPs in harsh environmental circumstances (Singh *et al.*, 2024). Through their oxygen-vacant surface, cerium oxide NPs (CeO_2 NPs) scavenge radicals, oscillating between Ce^{3+} and Ce^{4+} states, lowering ROS and H_2O_2 levels in rice seedlings. Lower doses inhibited ROS in *M. arborea*, but larger doses imitated SOD activity. In *A.thaliana*, a similar effect was observed, but lipid peroxidation increased by a factor of 2.5 as well. TiO_2 NPs applied to spinach seeds and during UV-B-induced oxidative stress reduced O_{-2} radicals, H_2O_2 levels, and lipid peroxidation in chloroplasts. TiO_2 NPs work similarly to CeO_2 NPs in neutralizing ROS and H_2O_2 by alternately binding to the Ti^{+4} and Ti^{+3} states. They also improved the stress tolerance of chickpea chlorophyll by reducing electrolyte loss (Al-Khayri *et al.*, 2023).

6.3 Cellular Processes Influenced by Nanoparticles under Stress Conditions

Plants respond to abiotic stress in different ways. ROS have the ability to cause cellular damage while simultaneously signaling defense responses. On the other hand, tiny peptides trigger stress responses via receptor-like kinases (Datta *et al.*, 2024). Plants can adapt at the cellular level thanks to epigenetic modifications like DNA methylation and histone modifications, and resilience depends heavily on genes linked to stress. Woody plants physiologically alter characteristics such as stomatal density and root architecture to efficiently utilize water during times of stress. To fully comprehend the relationships and effects of coupled stresses on plant resilience, additional research is necessary due to their complexity (Rafique *et al.*, 2024). Through their effects on growth, metabolism, and stress-tolerance mechanisms, NPs can have a significant impact on how plants respond to stress. For example, metal-based NPs, such as copper and silver, can increase the activity of antioxidant enzymes, which helps plants better withstand oxidative stress and drought. In order to increase stress resilience, they also interact with biological functions such as glycolysis and photosynthesis. NPs have a dual nature, though, meaning that their type and concentration must be carefully considered as they have the capacity to both benefit and harm plant systems. When used appropriately, NPs have a bright future ahead of them in terms of improving plant stress management and sustainable agriculture (Venzhik *et al.*, 2024).

Plant tissues are permeable to NPs, which can alter cellular functions and cause noticeable physiological alterations. Via apoplastic and symplastic routes, they can penetrate plant cells and produce ROS, which can lead to oxidative stress. The way cells interact could be disrupted by this. High quantities of NPs, like AgNPs, can have harmful effects such as poor seed germination and DNA damage. NPs like AgNPs can restrict growth by altering photosynthesis, nutrient uptake, and hormone control. To balance their advantages and hazards in agriculture, careful management is required (Noori *et al.*, 2024). It has been discovered that certain NPs enhance the vitality, growth, and development of plants. The functioning of various NPs, such as magnetized iron (Fe NP), aluminum (AlNP), gold (AuNP), copper (CuNP), silver (AgNP), zinc (ZnNP), titanium dioxide (TiO_2 NP), zinc oxide (ZnO NP), and cerium oxide (CeO_2 NP), has been clarified by recent investigations. The potential of these materials to improve several aspects of plant health has been investigated (Basak *et al.*, 2023). Plant growth has been demonstrated to be enhanced by carbon-based nanomaterials such as MWCNTs and SWCNTs, fullerols, and C60/c70 fullerenes. By employing sophisticated mapping techniques, MWCNTs and SWCNTs were discovered to boost the biomass of the tomato plants. They were found in the roots, leaves, and fruits of the treated plants (Mathew and Victório, 2022). MWCNTs increased tobacco cell proliferation by 55–64% in different research. Furthermore, *Cicer arietinum* treated for ten days with CNTs coated with citric acid exhibited enhanced absorption of NPs and proliferation. The plants' vascular bundles developed an 'aligned network' as a result of the CNTs, improving the plants' capacity to absorb water (Rajput *et al.*, 2024).

NPs have the potential to impact mitochondrial function and detoxification by inducing oxidative stress. Plastic NPs, for instance, have the ability to injure cells by interfering with the stability of proteins. Nematic polymers (NPs) regulate ion transport and osmotic equilibrium to help plants better withstand salt stress. Moreover, they interact with proteins that bind metals, possibly changing how those proteins function. NPs have the ability to affect signaling pathways and gene expression linked to stress responses and growth. Notwithstanding their possible advantages, more research is necessary to determine their toxicity and long-term impacts on cellular health (Nakhaei *et al.*, 2024). Plant growth may be positively or negatively impacted by nanoparticles. Influenced by the NPs' characteristics and the structure of the plant, they enter plants through the roots and ascend to the shoots through the vascular system. NPs may have phytotoxic consequences such as slowed development, decreased biomass, and abnormalities in the roots. Nevertheless, rather than direct toxicity, growth suppression may be the consequence of physical interactions between NPs and plant cell transport channels, such as obstructing plasmodesmata or cell-wall pores (Abideen *et al.*, 2022).

Studies have demonstrated the ability of NPs to enhance plant nutrition. For example, plants like *Lolium multiflorum*, *Zea mays*, *Chrysanthemums*, *Triticum aestivum*, *Arachis hypogaea*, and *Vigna unguiculata* have shown improved growth, higher seed weight, increased yield, and improved quality when macronutrient NPs like nitrogen, phosphorus, potassium, calcium, and magnesium are applied (Venzhik *et al.*, 2024). Plant species, environmental factors, and NP characteristics all affect how plants and NPs interact. Studies point to a number of potential entry points, including air deposition, contaminated soil, direct application, and unintentional discharge, even though the precise route of entry is unknown. The pericycle, cortex, and xylem are examples of structures that aid in directing NP entrance into the roots. Important elements regulate particle access into plants, with the cell wall acting as the primary barrier. It functions as a filter, letting through only NPs with pore sizes that correspond to their own. NPs can pass through the cell wall and arrive at the cell membrane if their sizes are compatible (Aqeel *et al.*, 2022). Within plant cells, metal-based NPs interact with proteins to change into reactive ions that interrupt metabolic processes. They can absorb inorganic ions from the soil, thanks to their vast surface area, which can cause indirect toxicity symptoms including chlorosis and wilting. ROS, which include hydrogen peroxide (H_2O_2), superoxide radicals ($O2^{\bullet-}$), singlet oxygen ($^1O_2$), and hydroxyl radicals ($OH^{\bullet}$), are increased when nanoparticles cause toxicity. ROS are produced when electron transport is disrupted in mitochondria and chloroplasts, and the enzyme SOD converts $O2^{\bullet-}$ to H_2O_2. High ROS levels induce oxidative stress, which can lead to DNA damage, lipid peroxidation, protein oxidation, membrane damage, electrolyte leakage, and cell death (Shikari *et al.*, 2022). Conversely, low ROS levels can function as signaling molecules. Both nonenzymatic (such as ascorbate, glutathione, and proline) and enzymatic (like SOD, catalase, and peroxidase) components make up a plant's antioxidant system. Studies conducted on treated seeds of castor beans (*Ricinus communis* L.) showed elevated levels of ROS generation in conjunction with elevated activities of SOD and peroxidase (POD). As the main line of defense against oxidative stress, SOD plays a critical role in converting toxic superoxide radicals ($O2^{\bullet-}$) into less harmful hydrogen peroxide (H_2O_2) (Singh *et al.*, 2022).

NPs can help plants better withstand salt stress by activating genes that respond to stress through the modulation of gene expression. They also aid in controlling the water balance, which enables plants to stay hydrated and perform better. NPs also help plants produce more energy and have easier access to vital nutrients by improving photosynthesis and nutrient uptake. Overall, these processes promote plant development and productivity, making NPs an excellent tool for improving crop resilience to saline stress (Abideen *et al.*, 2022). Under flooding stress, soybeans were able to transition more quickly from anaerobic to aerobic metabolic processes thanks to the assistance of AgNPs, which also improved plant recovery. Likewise, ascorbate/glutathione pathway activation by aluminum oxide NPs improved ROS scavenging, an important process for reducing oxidative damage. NPs mostly affected proteins involved in stress response, cell-wall maintenance, protein synthesis, and signaling during the later stages of stress recovery. Through the enhancement of glycolysis and the tricarboxylic acid (TCA) cycle, two essential processes in energy metabolism, CuNPs in wheat increased the plant's resistance to salt stress. FeNPs, on the other hand, increased the growth of *Triticum aestivum* seedlings by controlling proteins involved in photosynthesis, demonstrating their function in enhancing the plant's capacity to store energy in times of stress (Mustafa and Komatsu, 2022).

6.4 Omics Approaches in Studying Plant Responses to Nanoparticles

6.4.1 An introduction to omics technologies and their relevance

Omics technologies – an umbrella term encompassing the fields of genomics, transcriptomics, proteomics, and metabolomics – represent a suite of high-throughput methodologies that facilitate a comprehensive understanding of biological systems at multiple levels. These technologies have revolutionized the field of biological research by allowing researchers to comprehensively research biological molecules and their interactions since each of the subfields, like genomics and metabolomics, provides unique insights into the molecular underpinnings of biological systems. Among the multiple omics fields, genomics may be regarded as one of the oldest and most well-explored. A genome refers to the entire genetic makeup of an organism, particularly the entire length of the DNA, which houses both coding and noncoding genes (Brosius, 2009). Genomics, thus, involves the study of an organism's entire genome, including the sequencing, mapping, and analysis of the genetic material (Robinson, 2002). This technology allows researchers to identify genetic variations and understand their implications in various biological processes. Genomics has been instrumental in identifying genes associated with stress responses and understanding the genetic basis of plant resilience to abiotic stressors. This was only possible due to the correlation between the molecules that are produced during stress and the genes that transcribe said proteins, the study of which comes under the field of transcriptomics. As suggested by the name, transcriptomics focusses on the complete set of RNA transcripts produced by the genome under specific conditions, which is called the transcriptome (Wang *et al.*, 2009). This technology provides insights into gene expression patterns and regulatory mechanisms. By analyzing transcriptomic data, researchers can identify stress-responsive genes and understand how plants modulate their gene expression in response to any stress, in this case, nanoparticle exposure (Wang *et al.*, 2020; Dai and Shen, 2022). However, simply studying the RNA sequences would yield only part of the information, because proteins regulate the stress response in plants. Hence, to get a bigger picture of the way in which plants respond to stress, the proteins need to be studied as well, which comes within the purview of the field of proteomics. This field involves the large-scale study of proteins, including their structures, functions, and interactions. Proteomic analyses help in understanding the functional aspects of the proteome, which is the term given to the entire array of proteins synthesized by an organism (Anderson and Anderson, 1998; Al-Amrani *et al.*, 2021). This field allows us to understand how protein expression and modifications change under stress conditions; thus, this technology is crucial for identifying stress-responsive proteins and elucidating their roles in stress mitigation. These proteins regulate the stress response pathway in conjunction with other molecules as well, primarily the secondary metabolites. Hence, the final piece of the puzzle to complete the whole picture of how plants respond to stress is the study of the metabolites of a plant, which is called metabolomics. Metabolites are small molecules involved in various biochemical pathways, and their levels can provide insights into the physiological state of an organism. Metabolomics is the newest emerging field under the umbrella of omics, and such studies help in understanding the metabolic adjustments plants make in response to NP-induced stress (Shulaev *et al.*, 2008; Clish, 2015). The data provided by all the different aspects of omics studies are helpful in their own regard but reach their full potential when combined and integrated with each other. This approach of integrating multiple omics studies is called systems biology, and it helps researchers gain a comprehensive understanding of the complex networks and pathways involved in stress responses. This integrated approach is essential for developing strategies to enhance plant resilience to abiotic stressors.

6.4.2 Insights from omics studies on plant responses to nanoparticle treatment

The application of omics technologies to study plant responses to NP treatment under abiotic stress conditions has provided significant insights into the underlying molecular mechanisms. Several studies have previously highlighted that the application of NPs to plants under stress conditions can improve the plants' tolerance towards said stress (Table 6.1). For example, AgNPs have been shown to alleviate salinity stress in *Satureja hortensis* L. (Nejatzadeh, 2021). Similarly, NPs have also been shown to have potential uses as a safeguard against drought stress (Kandhol *et al.*, 2022). While such studies have only shown the *in vivo* or *in vitro* effects of the application of NPs, genomic studies are going a step forward in also highlighting how these NPs can modulate specific gene expressions in plants under different stress conditions (Al-Khayri *et al.*, 2023; Chandrashekar *et al.*, 2023). This genomic modulation suggests that NPs can act as signaling molecules, triggering a cascade of gene expression changes that enable plants to cope better with adverse environmental conditions. Transcriptomics has also been utilized in parallel, to gain further insights into the specific pathways which are being regulated by NP application. In *Brassica chinensis* L., transcriptomic studies have helped reveal the exact pathways that were affected by foliar application of TiO_2 NPs, including simultaneous upregulation and downregulation of different sections of the genome (Wei *et al.*, 2024). While this particular study was not focussing on stress responses, the explored pathways led to the identification of the mechanisms through which the NPs were affecting the growth and yield of the plants, thus opening up a new avenue for research into NP-mediated stress responses. Similarly, both transcriptomic and proteomic studies have revealed how ZnO NPs affect the growth of tomato plants, including detailed insights into how NP application increased the expression of genes responsible for the production of antioxidative enzymes and regulators involved in the carbon/nitrogen metabolism (Sun *et al.*, 2020). Metabolomics has also played a critical role in this regard, including in important crop plants like maize (*Zea mays* L.), where metabolomic analysis has helped scientists determine the different levels of metabolites and minerals produced in response to different levels of iron NP application (Wang *et al.*, 2021). *In vitro* reaction of *Nardostachys jatamansi* to charged AuNPs was investigated in this work. Significant changes in gene expression were found by RNA sequencing of shoots treated with 60 μM citrate-AuNPs and 40 μM CTAB-AuNPs in contrast to controls. Specifically, 104, 1255, and 660 differentially expressed genes (DEGs) were found across different comparisons. CTAB-AuNPs had a stronger impact on biological and metabolic processes, according to gene ontology (GO) analysis. Growth, antioxidants, hormone control, metabolite production, and stress response have all been associated with key DEGs. Transcription factors such as Myb, WRKY, NAC, and ERF were found to be increased along pathways associated with photosynthesis, antioxidant activity, and defensive responses, as revealed by KEGG analysis (Joshi and Joshi, 2024). All these examples highlight the importance of each level of omics study in deciphering the ways in which NPs alter the genetic expression, protein transcription, and metabolite production in plants, which can be an important aspect of imparting stress tolerance in plants via NP application.

6.5 Challenges and Considerations in Using Nanoparticles

There are benefits and drawbacks to using NPs in agriculture with regard to their effects on the environment and human health. Although they can increase crop yields and decrease the need for chemicals, their possible consequences on ecosystems and human health should be carefully considered (Mohanty *et al.*, 2024). Although the discharge of NPs into the environment may cause bioaccumulation and upset ecosystems, they may also help clean up pollution and restore the health of the soil. Although they can increase agricultural production and nutrient uptake, there are health risks due to their potential toxicity. Therefore, in order to address the long-term impacts and guarantee safe use in agriculture, comprehensive risk evaluations and strong regulatory frameworks are crucial (Langangmeilu *et al.*, 2024).

There are difficulties in delivering and distributing NPs throughout plants when using them in agricultural applications. Although biological constraints and other obstacles prevent NPs from being absorbed through techniques like root or leaf treatments, they

Table 6.1. Effects of nanoparticle application on plants under abiotic stress as revealed by omics analysis.

No.	Nano-particle	Plant species	Type of stress	Effect of NP application	References
1.	TiO_2	*Linum usitatissimum* L.	Drought Stress	Increase in carotenoid and chlorophyll content; decrease in levels of H_2O_2 and malondialdehyde	Aghdam *et al*., 2016
2.	TiO_2	*Triticum aestivum* L.	Drought Stress	Increase in seedling dry weight, relative water content, carotenoid and chlorophyll levels, and activity of antioxidant enzymes	Faraji and Sepehri, 2018
3.	CuNPs	*Zea mays* L.	Drought Stress	Increase in anthocyanin, carotenoid, and chlorophyll content, along with higher leaf biomass and grain yield	Van Nguyen *et al*., 2022
4.	ZnO	*Glycine max* L.	Drought stress	Percentage of total germination and rate of germination increased	Sedghi *et al*., 2013
5.	ZnO	*Abelmoschuss esculentus* L.	Salinity stress	Increase in photosynthetic activity along with SOD and CAT enzymatic activity	Alabdallah and Alzahrani, 2020
6.	ZnO and Si	*Mangifera indica* L.	Salinity stress	Increase in plant growth and nutrient uptake	Elsheery *et al*., 2020
7.	SiO_2	*Solanum lycopersicum* L.	Salinity stress	Upregulation and downregulation of corresponding genes responsible for natural salt-stress response	Zainab, 2016
8.	AgNPs	*Triticum aestivum* L.	Heat stress	Increased morphological growth like leaf area, root, and shoot number	Iqbal *et al*., 2019
9.	Al_2O_3	*Glycine max* L.	Flooding stress	Increase in hypocotyl length and protein synthesis	Mustafa *et al*., 2015
10.	TiO_2	*Cicer arietinum* L.	Cold stress	Decrease in electrolyte leakage and malondialdehyde levels	Mohammadi *et al*., 2013

can increase crop resilience and yield. Using different techniques, such as coating stems with NPs, could increase their efficacy. Furthermore, excessive amounts of NPs can be poisonous, which can harm plants, and they frequently diffuse unevenly within plants. Resolving these problems is essential to optimize the advantages of NPs in farming (Sembada and Lenggoro, 2024).

There is no information on how nanomaterial-based agrochemicals infiltrate seeds and plants, or impact the food chain, nor is there any understanding of the mechanisms underlying their toxicity in plants. The hazardous effects of NPs on plants, including *Cucumis sativus*, *Zea mays*, *Allium cepa*, and *Lycopersicum esculentum*, have been the subject of certain studies. These investigations look at cytotoxic effects such as apoptosis, micronuclei production, and chromosomal abnormalities. By exposing seedlings to varying NP concentrations and tracking germination, growth, and NP accumulation, phytotoxicity can be evaluated both *in vitro* and *in vivo*. To completely comprehend the hazardous mechanisms and long-term impacts of NPs in agriculture, more study is necessary (Biswas and Pal, 2024). The soil microbiome and its essential processes may be significantly impacted by the presence of nanomaterials in the soil. These substances have the power to change soil microbial diversity, abundance, and general functioning, all of which are important for plant growth and nutrient cycling. Specifically, by interfering with enzymatic activities, NPs can potentially affect vital microbial functions like mineralization, nitrogen fixation, and plant growth promotion. Ammonification, denitrification, and phosphate solubilization are a few of the processes that may be impacted by this reduced capacity for nutrient recycling. It is important to comprehend the toxicity of nanomaterials and the impact of soil attributes in this context since these changes, especially at larger concentrations of NPs, may lead to decreased soil fertility and wider ecological effects (Khan *et al.*, 2021). Aquatic environments may become contaminated and experience bioaccumulation as a result of nanomaterials utilized in agrochemicals and other industrial items seeping into the food chain. Aquatic life may be at risk when industrial wastewater containing NPs leaks into bodies of water. Nanomaterials have the potential to enter aquatic species' systems and alter their physiology and eating habits. The ecology as a whole and aquatic ecosystems may be negatively impacted in a cascade manner. Tests for toxicology are useful instruments for assessing how NPs affect aquatic life. By assessing the possible harm posed by nanomaterials, these experiments can support the development of risk assessment frameworks. Regulatory bodies and industry can evaluate and control the dangers related to the use of nanomaterials in different ecosystems, especially water, by creating a database of toxicological data (Murali *et al.*, 2022).

The use of NP technologies in agriculture raises important ethical and regulatory issues related to transparency, safety, and environmental effects. These technologies present threats to ecosystems and human health even if they have the potential to increase agricultural productivity. For this reason, evaluating these hazards and guaranteeing safe usage requires the establishment of strong regulatory frameworks. Owing to the distinct characteristics of NPs, particular safety precautions are required to address any toxicity and environmental issues (Kaur *et al.*, 2024). In order to guarantee that the advantages of these technologies do not jeopardize public health or environmental wellbeing, several nations are creating rules specifically geared toward the use of nanotechnology in agriculture. These regulations center on safety evaluations and product approvals. Additionally important are ethical issues, which highlight the necessity of ethical behavior and open communication in order to build public confidence. Addressing societal issues and coordinating technology improvements with local values require active stakeholder engagement. Overall, while NP technologies have a lot to offer agriculture, their risks and moral repercussions require careful thought and tight laws to ensure their responsible and safe deployment (Kumari *et al.*, 2023). Developing localized, flexible nanomaterial systems and guaranteeing the high-quality, economical mass manufacturing of nano-based products are two of the main obstacles facing nanotechnology in sustainable crop production. Bridging the gap between basic research and practical applications is a key challenge, as is the safety of the environment and human health during their use and disposal. The widespread implementation of nanobiotechnology is further impeded by its high costs, hazards, and gaps in knowledge. Regulations, researchers, and industry must work together to develop guidelines for the sustainable, safe, and successful application of nanotechnology in agriculture (Fizzah Zulfiqar *et al.*, 2024).

6.6 Future Directions and Potential of Nanoparticle-Omics Integration

6.6.1 Emerging trends in nanoparticle technology and omics research for stress mitigation

The field of nanotechnology is rapidly evolving, offering novel tools to enhance plant resilience under abiotic stress conditions. A significant trend in recent years is the development of NPs specifically designed for agricultural applications, which are increasingly being integrated with omics technologies to uncover deeper insights into their mechanisms of action (Ul Haq *et al.*, 2020; Singh *et al.*, 2021). This convergence of NPs and multi-omics platforms can reshape our understanding of how plants respond to NPs at the molecular and cellular levels, particularly under adverse environmental conditions such as drought, salinity, and extreme temperatures. Some NPs can be engineered to possess unique physicochemical characteristics, including size, shape, surface charge, and functionalization, which significantly influence their interaction with plant systems. In this regard, one of the most promising approaches is the biosynthesis of NPs for agricultural applications, which involves using biotic sources to synthesize NPs (Nguyen *et al.*, 2022). The primary advantages of such an approach are that the NPs thus synthesized are environmentally friendly and sustainable and also that they are synthesized from existing microbes or plants, making them biologically compatible with agricultural crops (Chakraborty *et al.*, 2022). Moreover, biosynthesis also allows for precise control over the shape and size of the desired NPs, allowing scientists to tailor-make NPs to serve specific purposes for abiotic stress response, the nature of which can be determined by multi-omics studies. For example, metabolomics studies can reveal the exact molecular concentrations at which foreign nanoparticles can impact the ROS-scavenging pathway which gets activated in plants due to environmental stress (Kumari *et al.*, 2022). This knowledge can let the scientists determine the size and the nature of the NP to be synthesized, thus streamlining the process and leaving guesswork or trial-and-error out of the equation. This subclass of specifically designed NPs is sometimes referred to as engineered NPs or E-NPs, and there are some examples of it, like carbon nanotubes and fullerenes, that have already proved to be very effective in other biological fields like functional nanocomposites and nanomedicine (García-Ovando *et al.*, 2022). Building on the foundation laid by these fields, the field of agriculture can co-opt the processes and make use of biosynthesis techniques and engineered NPs to help combat specific abiotic stresses, as required by different geographical areas.

6.6.2 The potential for custom-designed NPs based on omics insights

Aside from focussing on the future prospects of NPs in combating abiotic stress in plants, the integration of multi-omics data with intelligent NP designing is also a big aspect of the projected success of such an approach. As explored above, the potential for custom-designed NPs based on omics insights is a rapidly evolving area of research with significant implications for agriculture. Omics technologies provide a wealth of data on the molecular and biochemical responses of plants to various stressors, enabling the identification of specific targets for NP intervention. By leveraging this information, researchers can design NPs with tailored properties to address specific stress-related challenges. One of the key advantages of custom-designed NPs is their ability to deliver targeted interventions at the molecular level. NPs can be functionalized with specific ligands or biomolecules that recognize and bind to target proteins or receptors involved in stress responses. This targeted delivery enhances the efficacy of the NPs and reduces off-target effects, thereby improving the overall stress tolerance of the plants. For example, for salt-stressed *Solanum lycopersicum* L., NP-mediated sucrose nonfermenting 1-related protein kinase (SnRK), an important part of the ABA response to osmotic stress, activated ABA Response Element Binding Protein (AREB), which helps regulate water and salt balance in the plant (Kadota *et al.*, 2014; Almutairi, 2016; Al-Khayri *et al.*, 2023). This response is brought about specifically by Ag and Si NPs, the selection of which was aided and narrowed down by the use of multi-omics data to first properly understand the ABA-response pathway in the plant when subjected to osmotic stress. Similarly, multi-omics data can inform the design of NPs with controlled release properties, particularly nano-encapsulated particles. By understanding the dynamics of stress responses, researchers can develop NPs that specifically target certain areas of the plant, or are activated only after receiving a specific molecular signal, ensuring targeted delivery (El-Saadony *et al.*, 2022; Sarraf *et al.*, 2022). Another promising application of custom-designed NPs is in the field of gene editing. Recent advances in CRISPR-Cas9 technology have enabled precise genome editing in plants, offering new opportunities for enhancing stress tolerance. NPs can be used as delivery vehicles for CRISPR-Cas9 components, facilitating efficient and targeted gene editing (Ran *et al.*, 2017; Azameti and Imoro, 2023). By integrating omics data with gene-editing techniques, researchers can identify and modify key genes involved in stress responses, creating plants with enhanced resilience to abiotic stress. Overall, the potential for custom-designed NPs based on omics insights represents a transformative approach to agriculture. By combining the precision of nanotechnology with the comprehensive data provided by omics technologies, researchers can develop innovative solutions to address the challenges posed by abiotic stress and improve the sustainability of agricultural systems.

6.7 Conclusion

Nanoparticles hold significant potential in mitigating abiotic stress in plants by enhancing growth and yield under challenging conditions such as drought and salinity. By modulating cellular processes, NPs improve plant stress tolerance, making them valuable tools for boosting agricultural productivity. Omics technologies, including metabolomics, transcriptomics, and proteomics, play a crucial role in understanding how NPs influence plant responses at the molecular level. These approaches help identify changes in gene expression, protein production, and metabolite levels, providing insights necessary for developing stress-resistant crops. The practical application of these two disciplines in the field of abiotic stress resistance in plants can be multiplied exponentially only by realizing the importance of interdisciplinary research that combines nanotechnology, plant biology, and omics studies to optimize the use of NPs in agriculture. This collaborative approach is essential for advancing sustainable farming practices. The most practical way of implementing such an interdisciplinary approach is to utilize the multi-omics data to identify key genes or metabolic pathways which are responsible for the natural plant response against abiotic stress, and then design engineered NPs to upregulate those genes or increase the metabolic activity of important metabolites. Based on the currently available data from various avenues of research pursued in the past decade, the future for NP applications in agriculture, particularly in enhancing crop resilience to environmental stresses, looks to be hopeful in supporting worldwide food security in the face of climate change. However, there is also a pressing need to address challenges such as understanding the long-term effects and safety of NPs on plant health and the environment to ensure their sustainable usage.

References

Abasi, F., Raja, N.I., Mashwani, Z.U.R., Amjad, M.S., Ehsan, M. *et al.* (2022) Biogenic silver nanoparticles as a stress alleviator in plants: A mechanistic overview. *Molecules* 27(11), 3378. DOI: 10.3390/molecules27113378.

Abideen, Z., Hanif, M., Munir, N. and Nielsen, B.L. (2022) Impact of nanomaterials on the regulation of gene expression and metabolomics of plants under salt stress. *Plants* 11(5), 691. DOI: 10.3390/plants11050691.

Aghdam, M.T.B., Mohammadi, H. and Ghorbanpour, M. (2016) Effects of nanoparticulate anatase titanium dioxide on physiological and biochemical performance of *Linum usitatissimum* (*Linaceae*) under well-watered and drought stress conditions. *Brazilian Journal of Botany* 39(1), 139–146. DOI: 10.1007/s40415-015-0227-x.

Alabdallah, N.M. and Alzahrani, H.S. (2020) The potential mitigation effect of ZnO nanoparticles on [Abelmoschus esculentus L. moench] metabolism under salt stress conditions. *Saudi Journal of Biological Sciences* 27(11), 3132–3137. DOI: 10.1016/j.sjbs.2020.08.005.

Al-Amrani, S., Al-Jabri, Z., Al-Zaabi, A., Alshekaili, J. and Al-Khabori, M. (2021) Proteomics: Concepts and applications in human medicine. *World Journal of Biological Chemistry* 12(5), 57–69. DOI: 10.4331/wjbc.v12.i5.57.

Al-Khayri, J.M., Rashmi, R., Surya Ulhas, R., Sudheer, W.N., Banadka, A. *et al.* (2023) The role of nanoparticles in response of plants to abiotic stress at physiological, biochemical, and molecular levels. *Plants* 12(2), 292. DOI: 10.3390/plants12020292.

Almutairi, Z.M. (2016) Influence of silver nano-particles on the salt resistance of tomato (*Solanum lycopersicum*) during germination. *International Journal of Agriculture and Biology* 18(2), 449–457. DOI: 10.17957/IJAB/15.0114.

Anderson, N.L. and Anderson, N.G. (1998) Proteome and proteomics: New technologies, new concepts, and new words. *Electrophoresis* 19(11), 1853–1861. DOI: 10.1002/elps.1150191103.

Aqeel, U., Aftab, T., Khan, M.M.A., Naeem, M. and Khan, M.N. (2022) A comprehensive review of impacts of diverse nanoparticles on growth, development and physiological adjustments in plants under changing environment. *Chemosphere* 291(Pt 1), 132672. DOI: 10.1016/j.chemosphere.2021.132672.

Azameti, M.K. and Imoro, A.-W.M. (2023) Nanotechnology: A promising field in enhancing abiotic stress tolerance in plants. *Crop Design* 2(2), 100037. DOI: 10.1016/j.cropd.2023.100037.

Bai, Y.-C., Chang, Y.-Y., Hussain, M., Lu, B., Zhang, J.-P. *et al.* (2020) Soil chemical and microbiological properties are changed by long-term chemical fertilizers that limit ecosystem functioning. *Microorganisms* 8(5), 694. DOI: 10.3390/microorganisms8050694.

Barkataki, M.P. and Singh, T. (2019) Plant-nanoparticle interactions: Mechanisms, effects, and approaches. *Engineered Nanomaterials and Phytonanotechnology: Challenges for Plant Sustainability* 55–83. DOI: 10.1016/bs.coac.2019.09.007.

Bartas, M. (2024) Abiotic stresses in plants: From molecules to environment. *International Journal of Molecular Sciences* 25(15), 8072. DOI: 10.3390/ijms25158072.

Basak, A., Amarshettiwar, S.B., Bordoloi, N., Shahni, Y.S., Mubeen, B.S. *et al.* (2023) Nanomaterial-induced changes in plant physiology and genetics: Implications for crop improvement strategies. *International Journal of Environment and Climate Change* 13(11), 2043–2048. DOI: 10.9734/ijecc/2023/v13i113364.

Biswas, A. and Pal, S. (2024) Plant-nano interactions: A new insight of nano-phytotoxicity. *Plant Physiology and Biochemistry: PPB / Societe Francaise de Physiologie Vegetale* 210(108646), 108646. DOI: 10.1016/j.plaphy.2024.108646.

Brosius, J. (2009) The fragmented gene. Annals of the New York academy of sciences 1, 186–193. DOI: 10.1111/j.1749-6632.2009.05004.x.

Chakraborty, S., Singh, A. and Roychoudhury, A. (2022) Biogenic nanoparticles and generation of abiotic stress-resilient plants: A new approach for sustainable agriculture. *Plant Stress* 6(100117), 100117. DOI: 10.1016/j.stress.2022.100117.

Chandrashekar, H.K., Singh, G., Kaniyassery, A., Thorat, S.A., Nayak, R. *et al.* (2023) Nanoparticle-mediated amelioration of drought stress in plants: A systematic review 13(10), 336. DOI: 10.1007/s13205-023-03751-4.

Clish, C.B. (2015) Metabolomics: An emerging but powerful tool for precision medicine. *Cold Spring Harbor Molecular Case Studies* 1(1), a000588. DOI: 10.1101/mcs.a000588.

Dai, X. and Shen, L. (2022) Advances and trends in omics technology development. *Frontiers in Medicine* 9, 911861. DOI: 10.3389/fmed.2022.911861.

Datta, T., Kumar, R.S., Sinha, H. and Trivedi, P.K. (2024) Small but mighty: Peptides regulating abiotic stress responses in plants. *Plant, Cell & Environment* 47(4), 1207–1223. DOI: 10.1111/pce.14792.

El-Saadony, M.T., Saad, A.M., Soliman, S.M., Salem, H.M., Desoky, E.-S.M. *et al.* (2022) Role of nanoparticles in enhancing crop tolerance to abiotic stress: A comprehensive review. *Frontiers in Plant Science* 13, 946717. DOI: 10.3389/fpls.2022.946717.

Elsheery, N.I., Helaly, M.N., El-Hoseiny, H.M. and Alam-Eldein, S.M. (2020) Zinc oxide and silicone nanoparticles to improve the resistance mechanism and annual productivity of salt-stressed mango trees. *Agronomy* 10(4), 558. DOI: 10.3390/agronomy10040558.

Faraji, J. and Sepehri, A. (2018) Titanium dioxide nanoparticles and sodium nitroprusside alleviate the adverse effects of cadmium stress on germination and seedling growth of wheat (*Triticum aestivum* L.). *Universitas Scientiarum* 23(1), 61. DOI: 10.11144/Javeriana.SC23-1.tdna.

Fizzah Zulfiqar, H., Afroze, B., Shakoor, S., Saad Bhutta, M., Ahmed, M. *et al.* (2024) Nanoparticles in agriculture: Enhancing crop resilience and productivity against abiotic stresses. In: Hasanuzzaman, M. and Nahar, K. (eds) *Abiotic Stress in Crop Plants: Ecophysiological Responses and Molecular Approaches.* IntechOpen. DOI: 10.5772/intechopen.114843.

García-Ovando, A.E., Ramírez Piña, J.E., Esquivel Naranjo, E.U., Cervantes Chávez, J.A. and Esquivel, K. (2022) Biosynthesized nanoparticles and implications by their use in crops: Effects over physiology, action mechanisms, plant stress responses and toxicity. *Plant Stress* 6(100109), 100109. DOI: 10.1016/j.stress.2022.100109.

Hamadani, H., Rashid, S.M., Parrah, J.D., Khan, A.A., Dar, K.A. *et al.* (2021) Traditional farming practices and its consequences. *Microbiota and Biofertilizers* 119–128. DOI: 10.1007/978-3-030-61010-4_6.

Imran and Ortas I (2024) Nano biotechnology: Exploring the potential of nano materials in enhancing crop health and combating abiotic stress. *Communications in Soil Science and Plant Analysis* 55(16), 2509–2528. DOI: 10.1080/00103624.2024.2358854.

Iqbal, M., Raja, N.I., Mashwani, Z.-U.-R., Wattoo, F.H., Hussain, M. *et al.* (2019) Assessment of agNPs exposure on physiological and biochemical changes and antioxidative defence system in wheat (*Triticum aestivum* L.) under heat stress. *IET Nanobiotechnology* 13(2), 230–236. DOI: 10.1049/iet-nbt.2018.5041.

Jha, S. and Pudake, R.N. (2016) Molecular mechanism of plant–nanoparticle interactions. In: Kole, C., Kumar, D.S. and Khodakovskaya, M.V. (eds) *Plant Nanotechnology: Principles and Practice*. Springer, Cham, Switzerland, pp. 155–181. DOI: 10.1007/978-3-319-42154-4_7.

Joshi, S. and Joshi, R. (2024) Comparative transcriptome profiling unveils the role of differentially charged gold nanoparticle in growth, metabolic regulation and defence response of micro-propagated *Nardostachys jatamansi* wall. ex DC. *South African Journal of Botany: Official Journal of the South African Association of Botanists* 167, 585–596. DOI: 10.1016/j.sajb.2024.02.052.

Joshi, S., Patil, S., Shaikh, A., Jamla, M. and Kumar, V. (2024) Modern omics toolbox for producing combined and multifactorial abiotic stress tolerant plants. *Plant Stress* 11(100301), 100301. DOI: 10.1016/j.stress.2023.100301.

Kadota, Y., Sklenar, J., Derbyshire, P., Stransfeld, L., Asai, S. *et al.* (2014) Direct regulation of the NADPH oxidase RBOHD by the PRR-associated kinase BIK1 during plant immunity. *Molecular Cell* 54(1), 43–55. DOI: 10.1016/j.molcel.2014.02.021.

Kandhol, N., Jain, M. and Tripathi, D.K. (2022) Nanoparticles as potential hallmarks of drought stress tolerance in plants. *Physiologia Plantarum* 174(2), e13665. DOI: 10.1111/ppl.13665.

Kaur, H., Thakur, A., Thakur, R.C. and Kumar, A. (2024) Ethical and safety considerations in nanomaterial use. In: Thakur, A. and Kumar, A. (eds) *Sustainability, Safety, and Applications of Nanomaterials-Based Corrosion Inhibitors.* IGI Global. pp. 326–356. DOI: 10.4018/979-8-3693-7640-9.ch013.

Khan, S.T., Adil, S.F., Shaik, M.R., Alkhathlan, H.Z., Khan, M. *et al.* (2021) Engineered nanomaterials in soil: Their impact on soil microbiome and plant health. *Plants* 11(1), 109. DOI: 10.3390/plants11010109.

Kumari, S., Khanna, R.R., Nazir, F., Albaqami, M., Chhillar, H. *et al.* (2022) Bio-synthesized nanoparticles in developing plant abiotic stress resilience: A new boon for sustainable approach. *International Journal of Molecular Sciences* 23(8), 4452. DOI: 10.3390/ijms23084452.

Kumari, R., Suman, K., Karmakar, S., Mishra, V., Lakra, S.G. *et al.* (2023) Regulation and safety measures for nanotechnology-based agri-products. *Frontiers in Genome Editing* 5, 1200987. DOI: 10.3389/fgeed.2023.1200987.

Langangmeilu, G.P., Moirangthem, A., Vani, N.U., Chavan, V.A. and Sagar, P.S. (2024) Nanotechnology in agriculture: Opportunity and challenges. In: *Futuristic Trends in Chemical Material Sciences & Nano Technology.* IIP. pp. 257–265. DOI: 10.58532/V3BECS24P5CH1.

Mariyam, S., Upadhyay, S.K., Chakraborty, K., Verma, K.K., Duhan, J.S. *et al.* (2024) Nanotechnology, a frontier in agricultural science, a novel approach in abiotic stress management and convergence with new age medicine. *The Science of the Total Environment* 912(169097), 169097. DOI: 10.1016/j.scitotenv.2023.169097.

Mathew, S. and Victório, C.P. (2022) Carbon nanotubes applications in agriculture. In: Abraham, J., Thomas, S. and Kalarikkal, N. (eds) *Handbook of Carbon Nanotubes*. Springer, Cham, Switzerland, pp. 1579–1593. DOI: 10.1007/978-3-030-91346-5_35.

Mohammadi, R., Maali-Amiri, R. and Abbasi, A. (2013) Effect of TiO_2 nanoparticles on chickpea response to cold stress. *Biological Trace Element Research* 152(3), 403–410. DOI: 10.1007/s12011-013-9631-x.

Mohanty, P., Singh, P.K., Lenka, B., Adhya, T.K., Verma, S.K. *et al.* (2024) Biofabricated nanomaterials in sustainable agriculture: Insights, challenges and prospects. *Biofabrication* 16(4), 042003. DOI: 10.1088/1758-5090/ad60f7.

Mondal, S., Seelam, R.T., Mondal, B. and Jespersen, D. (2024) Understanding abiotic stress tolerance mechanisms in non-food grass species through omics approaches. *Plant Stress* 13(100516), 100516. DOI: 10.1016/j.stress.2024.100516.

Murali, M., Gowtham, H.G., Singh, S.B., Shilpa, N., Aiyaz, M. *et al.* (2022) Fate, bioaccumulation and toxicity of engineered nanomaterials in plants: Current challenges and future prospects. *The Science of the Total Environment* 811(152249), 152249. DOI: 10.1016/j.scitotenv.2021.152249.

Mustafa, G. and Komatsu, S. (2022) Proteomics of plant-nanoparticle interaction mechanism. *Plant and Nanoparticles. Springer* 67–84. DOI: 10.1007/978-981-19-2503-0_3.

Mustafa, G., Sakata, K. and Komatsu, S. (2015) Proteomic analysis of flooded soybean root exposed to aluminum oxide nanoparticles. *Journal of Proteomics* 128, 280–297. DOI: 10.1016/j.jprot.2015.08.010.

Mustafa, G., Farooq, A., Roy, D., Adhikari, A. and Hossain, Z. (2024) Assessment of nanoparticle-plant interactions: An integrative proteomic approach. In: Biswas, J.K. and Rai, M. (eds) *Nanotechnology for Environmental Management*. CRC Press, Boca Raton, Florida, pp. 347–367. DOI: 10.1201/9781003350941-24.

Nakhaei, A., Raissi, H. and Farzad, F. (2024) Engineered nanoparticles as selinexor drug delivery systems across the cell membrane and related signaling pathways in cancer cells. *Journal of Molecular Graphics and Modelling* 131(108809), 108809. DOI: 10.1016/j.jmgm.2024.108809.

Nejatzadeh, F. (2021) Effect of silver nanoparticles on salt tolerance of *Satureja hortensis* i. during *in vitro* and *in vivo* germination tests. *Heliyon* 7(2), e05981. DOI: 10.1016/j.heliyon.2021.e05981.

Nguyen, N.T.T., Nguyen, L.M., Nguyen, T.T.T., Liew, R.K., Nguyen, D.T.C. *et al.* (2022) Recent advances on botanical biosynthesis of nanoparticles for catalytic, water treatment and agricultural applications: A review. *Science of the Total Environment* 827(154160), 154160. DOI: 10.1016/j.scitotenv.2022.154160.

Noori, A., Hasanuzzaman, M., Roychowdhury, R., Sarraf, M., Afzal, S. *et al.* (2024) Silver nanoparticles in plant health: Physiological response to phytotoxicity and oxidative stress. *Plant Physiology and Biochemistry* 209, 108538. DOI: 10.1016/j.plaphy.2024.108538.

Oyebamiji, Y.O., Adigun, B.A., Shamsudin, N.A.A., Ikmal, A.M., Salisu, M.A. *et al.* (2024) Recent advancements in mitigating abiotic stresses in crops. *Horticulturae* 10(2), 156. DOI: 10.3390/horticulturae10020156.

Rafique, S., Quadri, S.N. and Abdin, M.Z. (2024) Plant defense mechanism in combined stresses – cellular and molecular perspective. *Journal of Plant Stress Physiology* 33–42. DOI: 10.25081/jpsp.2024.v10.8790.

Rai, K.K., Kumar, A., Rai, A., Rai, V.P. and Rai, A.C. (2021) Conventional breeding approaches for abiotic stress management in horticultural crops. In: Rai, A., Rai, V.P., Chandra, A., Kumar, A. and Rai, K.K. (eds) *Stress Tolerance in Horticultural Crops. Elsevier*. pp. 21–32. DOI: 10.1016/B978-0-12-822849-4.00017-6.

Rai, P.K., Song, H. and Kim, K.-H. (2023) Nanoparticles modulate heavy-metal and arsenic stress in food crops: Hormesis for food security/safety and public health. *Science of the Total Environment* 902(166064), 166064. DOI: 10.1016/j.scitotenv.2023.166064.

Rajput, V.D., Singh, A., Tomar, B., Minkina, T., Movsesyan, H.S. *et al.* (2024) Nanoparticle-mediated approaches in agriculture addressing abiotic stress from soil to plant cells. *Nanotechnology Applications and Innovations for Improved Soil Health.* IGI Global 72–89. DOI: 10.4018/979-8-3693-1471-5.ch005.

Ran, Y., Liang, Z. and Gao, C. (2017) Current and future editing reagent delivery systems for plant genome editing. *Science China. Life Sciences* 60(5), 490–505. DOI: 10.1007/s11427-017-9022-1.

Rehman, A., Weng, J., Li, P., Shah, I.H. and Rahman, S. (2023a) Green synthesized zinc oxide nanoparticles confer drought tolerance in melon (*Cucumis melo* L). *Environmental and Experimental Botany* 212, 105384. DOI: 10.1016/j.envexpbot.2023.105384.

Rehman, A., Khan, S., Sun, F., Peng, Z., Feng, K. *et al.* (2023b) Exploring the nano-wonders: Unveiling the role of nanoparticles in enhancing salinity and drought tolerance in plants. *Frontiers in Plant Science* 14, 1324176. DOI: 10.3389/fpls.2023.1324176.

Robinson, R. (ed.) (2002) *Genetics*. Macmillan Library Reference, Farmington Hills, Michigan.

Sarraf, M., Vishwakarma, K., Kumar, V., Arif, N., Das, S. *et al.* (2022) Metal/metalloid-based nanomaterials for plant abiotic stress tolerance: An overview of the mechanisms. *Plants* 11(3), 316. DOI: 10.3390/plants11030316.

Savci, S. (2012) Investigation of effect of chemical fertilizers on environment. *APCBEE Procedia* 1, 287–292. DOI: 10.1016/j.apcbee.2012.03.047.

Sedghi, M., Hadi, M. and Toluie, S.G. (2013) Effect of nano zinc oxide on the germination of soybean seeds under drought stress. *Annals of West University of Timişoara, Ser. Biology* 2, 73–78.

Sembada, A.A. and Lenggoro, I.W. (2024) Transport of nanoparticles into plants and their detection methods. *Nanomaterials* 14(2), 2. DOI: 10.3390/nano14020131.

Shikari, A.B., Dikilitas, M., Simsek, E., Guldur, M.E., Simsek, U. *et al.* (2022) Nanotechnology and its role in cereal crops under abiotic stress. In: Latef, A.A.H.A. (ed.) *Sustainable Remedies for Abiotic Stress in Cereals.* Springer. pp. 675–687. DOI: 10.1007/978-981-19-5121-3_26.

Shulaev, V., Cortes, D., Miller, G. and Mittler, R. (2008) Metabolomics for plant stress response. *Physiologia Plantarum* 132(2), 199–208. DOI: 10.1111/j.1399-3054.2007.01025.x.

Singh, R.P., Handa, R. and Manchanda, G. (2021) Nanoparticles in sustainable agriculture: An emerging opportunity. *Journal of Controlled Release: Official Journal of the Controlled Release Society* 329, 1234–1248. DOI: 10.1016/j.jconrel.2020.10.051.

Singh, A., Mehta, S., Yadav, S., Nagar, G., Ghosh, R. *et al.* (2022) How to cope with the challenges of environmental stresses in the era of global climate change: An update on ROS stave off in plants. *International Journal of Molecular Sciences* 23(4), 4. DOI: 10.3390/ijms23041995.

Singh, S., Singh, O., Shahi, U.P., Singh, R., Singh, P.K. *et al.* (2024) Addressing abiotic stress in agriculture through nanotechnology solutions. In: Rajput, V.D. (ed.) *Harnessing NanoOmics and Nanozymes for Sustainable Agriculture.* IGI Global. pp. 320–347. DOI: 10.4018/979-8-3693-1890-4.ch016.

Sun, L., Wang, Y., Wang, R., Wang, R., Zhang, P. *et al.* (2020) Physiological, transcriptomic, and metabolomic analyses reveal zinc oxide nanoparticles modulate plant growth in tomato. *Environmental Science: Nano* 7(11), 3587–3604. DOI: 10.1039/D0EN00723D.

Ul Haq, I., Ijaz, S. and Khan, N.A. (2020) Application of nanotechnology for integrated plant disease management. In: Ul Haq, I. and Ijaz, S. (eds) *Plant Disease Management Strategies for Sustainable Agriculture through Traditional and Modern Approaches*. Springer, Cham, Switzerland, pp. 173–185. DOI: 10.1007/978-3-030-35955-3_8.

Vailati-Riboni, M., Palombo, V. and Loor, J.J. (2017) What are omics sciences? In: Ametaj, B.N. (ed.) *Periparturient Diseases of Dairy Cows*. Springer, Cham, Switzerland. DOI: 10.1007/978-3-319-43033-1_1.

Van Nguyen, D., Nguyen, H.M., Le, N.T., Nguyen, K.H., Nguyen, H.T. *et al.* (2022) Copper nanoparticle application enhances plant growth and grain yield in maize under drought stress conditions. *Journal of Plant Growth Regulation* 41(1), 364–375. DOI: 10.1007/s00344-021-10301-w.

Venzhik, Y., Deryabin, A. and Dykman, L. (2024) Nanomaterials in plant physiology: Main effects in normal and under temperature stress. *Plant Science: An International Journal of Experimental Plant Biology* 346(112148), 112148. DOI: 10.1016/j.plantsci.2024.112148.

Verma, N., Kaushal, P. and Sidhu, A.K. (2024) Harnessing biological synthesis: Zinc oxide nanoparticles for plant biotic stress management. *Frontiers in Chemistry* 12, 1432469. DOI: 10.3389/fchem.2024.1432469.

Wang, Z., Gerstein, M. and Snyder, M. (2009) RNA-seq: A revolutionary tool for transcriptomics. *Nature Reviews Genetics* 10(1), 57–63. DOI: 10.1038/nrg2484.

Wang, X., Li, N., Li, W., Gao, X., Cha, M. *et al.* (2020) Advances in transcriptomics in the response to stress in plants. *Global Medical Genetics* 7(2), 30–34. DOI: 10.1055/s-0040-1714414.

Wang, Y., Chen, S., Deng, C., Shi, X., Cota-Ruiz, K. *et al.* (2021) Metabolomic analysis reveals dose-dependent alteration of maize (*Zea mays* L.) metabolites and mineral nutrient profiles upon exposure to zerovalent iron nanoparticles. *NanoImpact* 23(100336), 100336. DOI: 10.1016/j.impact.2021.100336.

Wei, L., Ji, L., Rico, C., He, C., Shakoor, I. *et al.* (2024) Transcriptomics reveals the pathway for increasing *Brassica chinensis* L. yield under foliar application of titanium oxide nanoparticles. *Journal of Agricultural and Food Chemistry* 72(34), 18957–18970. DOI: 10.1021/acs.jafc.4c04075.

Yadav, N., Bora, S., Devi, B., Upadhyay, C. and Singh, P. (2024) Nanoparticle-mediated defense priming: A review of strategies for enhancing plant resilience against biotic and abiotic stresses. *Plant Physiology and Biochemistry* 213(108796), 108796. DOI: 10.1016/j.plaphy.2024.108796.

Zainab, A. (2016) Effect of nano-silicon application on the expression of salt tolerance genes in germinating tomato (Solanum lycopersicum L.) seedlings under salt stress. *Plant Omics Journal* 1, 106–114.

Zhang, Y., Xu, J., Li, R., Ge, Y., Li, Y. *et al.* (2023) Plants' response to abiotic stress: Mechanisms and strategies. *International Journal of Molecular Sciences* 24(13), 10915. DOI: 10.3390/ijms241310915.

7 Smart and Stress-resilient Agriculture Achieved by Engineered Nanomaterials and Nanobiotechnology

Zakir Ullah[1], Javed Iqbal[2*], Banzeer Ahsan Abbasi[3], Shumaila Ijaz[4], Shumaila Ahmad[1], Shoaib Khan[5], Ghulam Murtaza[6], Rashid Iqbal[7,8], Tayyaba Shahzadi[9], Shobana Sampath[10], Amin Fathi[11], Kashaf Khan[1], Sobia Kanwal[12] and Tariq Mahmood[1]

[1]Department of Plant Sciences, Faculty of Biological Sciences, Quaid-i-Azam University Islamabad, Pakistan; [2]Department of Botany, Bacha Khan University, Charsadda, Khyber Pakhtunkhwa, Pakistan; [3]Department of Botany, Rawalpindi Women's University, Rawalpindi, Pakistan; [4]Department of Burn and Plastic Surgery, Shenzhen Institute of Translational Medicine, First Affiliated Hospital of Shenzhen University, School of Biomedical Engineering, Shenzhen University Medical School, Shenzhen University, Shenzhen, China; [5]Department of Chemistry, Abbottabad University of Science and Technology (AUST), Abbottabad, Pakistan; [6]School of Agriculture, Yunnan University, Kunming, Yunnan, China; [7]Department of Agronomy, The Islamia University of Bahawalpur, Pakistan; [8]Department of Life Sciences, Western Caspian University, Baku, Azerbaijan; [9]Department of Chemistry, Government College for Women, University Sialkot, Pakistan; [10]Department of Biotechnology, Vel Tech Rangarajan Dr Sagunthala R&D Institute of Science and Technology, Chennai, India; [11]Department of Agronomy, Islamic Azad University, Amol, Iran; [12]Department of Biology and Environmental Sciences, Allama Iqbal Open University, Islamabad, Pakistan

Abstract

A growing subject known as 'smart and stress-resilient agriculture' combines innovative nanotechnology and nanobiotechnology to solve problems brought on by environmental stress, deteriorating soil fertility, and climate change. Crop resilience and productivity can be increased with the use of engineered nanomaterials (ENMs), such as metal oxide nanoparticles, carbon-based nanostructures, and nanocomposites. By modifying physiological and biochemical processes at the cellular level, these nanomaterials can enhance plant growth, nutrient absorption, and stress tolerance. Applications of nanobiotechnology, such as nanosensors and nanofertilizers, also optimize the usage of water, nutrients, and agrochemicals by offering accurate monitoring and targeted delivery systems. The potential of ENMs to improve plant stress resistance, lessen the effects of biotic and abiotic stressors, and support

*Corresponding author: javed89qau@gmail.com

© CAB International 2025. *Stress-Resilient Crops: Coordinated omics-CRISPR-Nanotechnology Strategies* (ed. Jen-Tsung Chen)
DOI: 10.1079/9781800627307.0007

sustainable agriculture is examined in this chapter. It also goes into the underlying molecular processes that nanoparticles use to improve stress tolerance, such as increased antioxidant defense systems, altered plant signaling pathways, and activation of genes that respond to stress. The study also discusses the safety and environmental issues related to using nanomaterials in agriculture, highlighting the necessity of eco-friendly and sustainable methods. A resilient and sustainable agricultural future appears to be within reach with the potential for current farming to be revolutionized using nanotechnology in agricultural methods.

Keywords: nanobiotechnology, stress-resilient agriculture, nanofertilizers, nanoparticles, sustainable agriculture, nanosensors

7.1 Introduction

Currently, the agricultural industry is facing unpredictable climate change, shrinking arable land, and soil pollution due to industrialization and urbanization, due to adverse effects of harmful chemicals from pesticides and synthetic fertilizers (Dimkpa and Bindraban, 2018). According to a recent United Nations analysis, the global population is projected to reach 8.5 billion by 2030 and approximately 9 billion by 2050 (Kah *et al.*, 2018). This increase implies a greater demand for food, requiring the same amount of agricultural land to meet supply needs. Although pesticides and fertilizers, particularly urea fertilizers, play a key role in enhancing food production (Kah *et al.*, 2018), their environmental impact remains a concern (Fraceto *et al.*, 2016). Chemicals known as fertilizers, which can be synthetic or organic, are added to the soil to supply essential nutrients for plant growth. The extensive use of these pesticides and fertilizers leads to many environmental problems, such as air, water, and soil contamination (Cunningham *et al.*, 2018). These pesticides are causing an increase in the poisoning of water supplies, and they represent a major hazard to the ecosystem even at low concentrations (Cunningham *et al.*, 2018). One of the primary causes of water pollution is nitrates, which are present in chemical fertilizers. The most common type of dissolved nitrogen in groundwater is nitrate (Kah *et al.*, 2019). In semi-arid and dry locations, irrigation has become increasingly common in recent years, leading to an increase in nitrate build-up in the soil (Research and Markets, 2021). Moreover, runoff and leakage of pesticides into nearby water bodies lead to a substantial accumulation of nitrogen and phosphorus compounds, which deteriorate water quality and contribute to pollution through the excessive growth of aquatic plants and algae, a phenomenon known as eutrophication (Kah *et al.*, 2019). Chemical fertilizers often contain nitrates along with other pollutants such as phosphates, chlorides, and heavy metals including mercury (Hg), arsenic (As), lead (Pb), cadmium (Cd), nickel (Ni), and copper (Cu), which further contaminate the water (Kahru and Ivask, 2013; Research and Markets, 2021).

Chemical pesticides and fertilizers damage the ecosystem by putting creatures that are not their intended targets in significant danger (Judy and Bertsch, 2014). The toxic substances included in fertilizers and pesticides frequently have a systemic effect and are absorbed by plants. Once in the food chain through vegetables, grains, and water, they can seriously impair human and higher animal health (Handford *et al.*, 2015). Research indicates that heavy metals found in fertilizers, such as lead, cadmium, mercury, and uranium, can negatively impact the kidneys, lungs, and liver, and may even lead to cancer (Ijaz *et al.*, 2023a). Prolonged use of chemical fertilizers can disrupt the soil's pH, electrical conductivity (EC), and salinity balance, while also reducing microbial activity. Additionally, some components of fertilizers pose risks to the respiratory system and skin (Prasad *et al.*, 2017). Chemical pesticides can biomagnify and pose a significant risk even at extremely low doses since they are often exceedingly refractory and hydrophobic. Most of them are also very poisonous to larger mammals (Judy and Bertsch, 2014). Thus, it is critical to lessen agriculture's reliance on artificial fertilizers and pesticides and instead explore affordable, sustainable, and environmentally friendly alternatives (Fig. 7.1).

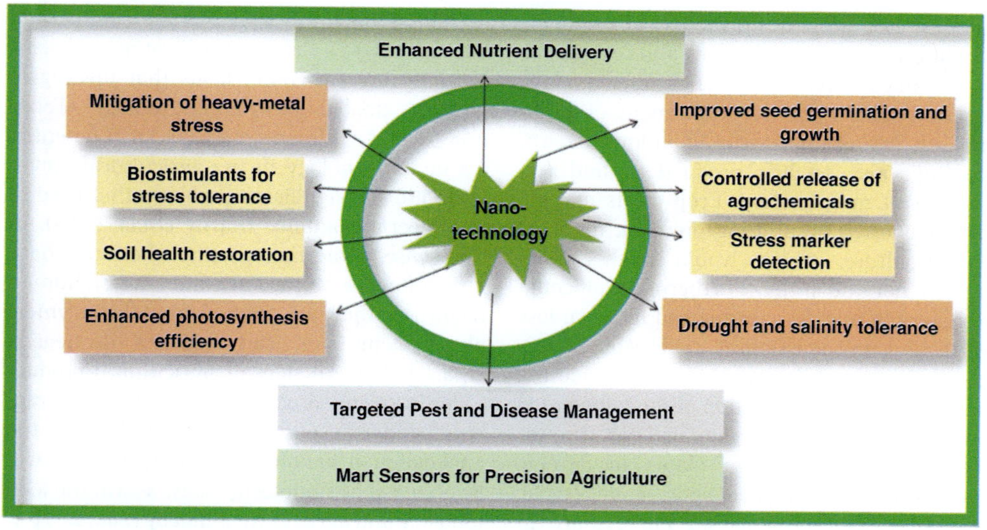

Fig. 7.1. Smart and stress-resilient agriculture achieved by nanobiotechnology.

7.2 Integrating Nanotechnology in Agriculture

In recent decades, there has been a significant rise in interest in nanotechnology due to its diverse applications in fields such as medicine, pharmaceuticals, energy, materials science, and catalysis (Giraldo *et al.*, 2014). Nanoparticles, typically ranging from 1 nm to 100 nm in size, hold potential for use in various sectors, including industry, healthcare, and agriculture. Researchers have explored multiple methods for nanoparticle synthesis, including chemical, biological, and physical approaches (Ullah *et al.*, 2023a). However, these techniques face notable challenges, such as scaling up the process, isolating and purifying nanoparticles from the micro-emulsion (comprising oil, surfactant, co-surfactant, and aqueous phases), and the extensive use of surfactants, which pose significant limitations (Kah *et al.* 2018, 2019).

The use of plant extracts in green techniques to synthesize nanoparticles is beneficial due to its simplicity, convenience, eco-friendliness, and reduced reaction time (Kahru and Ivask, 2013). The potential for enhancing fertilization, plant growth regulators, and insecticides in agriculture may be increased by using eco-friendly and green technologies to manufacture nanomaterials (Fraceto *et al.*, 2016). They also reduce the quantity of dangerous substances that contaminate the environment. As a result, this technology aids in the reduction of environmental contaminants (Cunningham *et al.*, 2018). Nanotechnology has also lately attracted attention because of its numerous uses in a variety of industries, including agriculture, the environment, and medicine (Kah *et al.*, 2019). The small nanoparticles' vast surface area makes them appealing for addressing problems that chemical and physical pesticides as well as biological control approaches cannot solve.

Even though many techniques fall under the purview of agriculture, nanotechnology in agriculture has advanced well over the past ten years thanks to a wealth of public financing (Fehér *et al.*, 2011). This might be explained by the special characteristics of agricultural production, which operates as an open system with a free flow of matter and energy. Nanotechnology promises to lessen the usage of pesticides by introducing novel agrochemical agents and innovative delivery mechanisms to boost agricultural yield (Kottegoda *et al.*, 2017). Utilization of nanotechnology includes:

Nanotechnology holds great potential to boost agricultural productivity through various applications, including the following: (i) nanoformulations of agricultural chemicals for improved fertilizers and pesticides to enhance crop growth; (ii) the use of nanosensors in crops to detect diseases and residues of agrochemicals; (iii) nanotechnology in plant genetic engineering; (iv) assessment of plant diseases; (v) advancements in animal health, breeding, and poultry production; and (vi) postharvest management. Crop yields may be further boosted by utilizing precision methods of agriculture, which would neither hurt the soil nor the water. Furthermore, it could decrease soil microorganisms and nitrogen loss by leaching and emissions (Giraldo *et al.*, 2014). The applications of nanomaterials include the preparation and preservation of food; increased longevity of goods; and gene or DNA transfer utilizing nanoparticles in plants to develop insect-resistant cultivars (Fehér *et al.*, 2011). Nanotechnology presents an opportunity to accelerate the growth of biomass-to-fuel production. According to experts, while the potential advantages of nanotechnology in food, fisheries, agriculture, and aquaculture are promising, they must be carefully weighed against concerns related to environmental impact, soil and water health, and occupational safety for workers (Kahru and Ivask, 2013). Food technology is now the focus of investigation, testing, and, in specific instances, application of nanotechnology (Judy and Bertsch, 2014). It is hypothesized that nanoparticles have specific mechanical, chemical, and physical features. Agricultural waste products have recently gained attention as a renewable source of raw materials that can be processed for various applications, including the production of nanomaterials (Kottegoda *et al.*, 2017). Insecticide resistance is a prime example of evolutionary change occurring on an ecological time scale (Ullah *et al.*, 2023a). Because of its economic significance and ability to shed light on real-time mechanisms at work, pesticide resistance research is essential. Insects are becoming a bigger issue for public health and agriculture. Numerous different types of selected regimes may be used in agricultural techniques (Kottegoda *et al.*, 2017). Applications of nanotechnology are being tried in agricultural and food technologies. The goals of using nanomaterials in agriculture are to boost plant yields and decrease the amount of plant protection agents sprayed (Ullah *et al.*, 2023a). Nanoparticles and nanocapsules are examples of devices used in nanotechnology for illness diagnosis and therapy (Raliya and Tarafdar, 2013). Genetic modification and plant breeding are two more fields that investigate gadgets created by nanotechnology (Fehér *et al.*, 2011). Although there is a lot of potential for nanotechnology in the agricultural sector, there are still significant problems that need to be properly examined (Karny *et al.*, 2018). In this sense, some nanoparticle chemicals have little influence on the environment and human health as they are formed from biopolymers like proteins and carbohydrates (Kottegoda *et al.*, 2017). All steps of agricultural product manufacturing, processing, preservation, wrapping, and transport can benefit through the application of nanotechnology. Agriculture will be changed by nanotechnology, as will the food industry and agriculture (Shalaby *et al.*, 2016).

The worldwide market for nanomaterials in agriculture is quickly rising, with forecasts estimating it will reach \$3.4 billion by 2027, growing at a CAGR of 16.3% from 2020 (Mozafari *et al.*, 2018; Research and Markets, 2021). This rise is driven by the rising need for greater crop protection, precision farming, and sustainable agriculture methods (Raliya and Tarafdar, 2013). Techniques enhance the ability of plants to take in vitamins and minerals, identify diseases, and control pests (Ullah *et al.*, 2023a). These materials can be utilized, for instance, in the development of smart nanosystems for nutrient immobilization and controlled release in the soil. Such strategies help minimize nutrient leaching, enhance crop nutrient uptake, and reduce eutrophication by limiting nitrogen absorption into groundwater (Raliya and Tarafdar, 2013; Karny *et al.*, 2018). Additionally, small particles can improve both the structure and function of pesticides by enhancing solubility, increasing resistance to hydrolysis and photodecomposition, and enabling more targeted and controlled application to specific organisms (Raliya and Tarafdar, 2013; Karny *et al.*, 2018).

7.2.1 Agricultural applications of nanomaterials

Nanoparticles are increasingly being used across various fields, including agriculture, where they aid farmers by improving soil conditions, increasing crop yields, and reducing water usage (Shalaby *et al.*, 2016). Major challenges to crop production include the reliance on pesticides and fertilizers, and abiotic stresses like drought and salinity, which significantly impact crop quality and output (Mozafari *et al.*, 2018). Drought stress, a major form of abiotic stress, leads to reduced plant growth and yield, primarily due to inadequate precipitation and fluctuations in light intensity and temperature, which create water deficit conditions (Wu *et al.*, 2017). Drought stress triggers a range of plant responses, including changes in growth, antioxidant levels, and the regulation of metabolites and proteins (Elmer and White, 2018; Ahmad *et al.*, 2022).

Phytogenic nanoparticles have been explored extensively for mitigating the detrimental effects of drought on plants. For example, citrate-coated iron nanoparticles synthesized by *Chaetomorpha antennina* have been shown to enhance drought resistance in *Setaria italica* by increasing levels of osmolytes and antioxidant enzymes like catalase (CAT), peroxidase (POX), and superoxide dismutase (SOD) (Nair *et al.*, 2011). Similarly, selenium nanoparticles derived from *Allium sativum* have improved drought-stressed wheat plant growth by enhancing development indices like leaf area, biomass, and height, and reducing damage parameters like ion leakage and lipid peroxidation (Ahmad *et al.*, 2022). Salinity stress disrupts the ionic and osmotic balance in plant cells, altering their primary metabolism due to the excessive accumulation of sodium and chloride ions. This imbalance disrupts cellular membranes, affects ionic equilibrium, and hampers plant growth and development (Elmer and White, 2018).

Phytogenic nanoparticles have been found to be effective in alleviating salt stress in plants. By improving plant growth traits like root and shoot length, fresh and dry weight, flavonoid and phenolic content, and photosynthetic pigment levels, and decreasing stress markers like hydrogen peroxide and malondialdehyde, magnesium nanoparticles made from *Hordeum vulgare* leaves help counteract salinity stress (Siddiqi and Husen, 2017). Similarly, in salt-stressed *Abelmoschus esculentus*, zinc nanoparticles derived from *Sorghum bicolor* leaves boost overall growth and photosynthetic indicators as well as the antioxidant defense system (Ghazi Hossaini *et al.*, 2021). Zinc oxide nanoparticles from *Phoenix dactylifera* extracts improve biomass and growth rate in *Vigna unguiculata* and *Abelmoschus esculentus* during salt stress by detoxifying reactive oxygen species (ROS) through overexpression of SOD and CAT enzymes (Guilherme *et al.*, 2015). Additionally, titanium oxide nanoparticles produced from *Buddleja asiatica* leaves have been found to decrease salinity-induced damage in wheat (Kah *et al.*, 2018).

Mentha piperita rhizome extracts have been used to synthesize gold-silver alloy nanoparticles that effectively mitigate salt stress-induced damage in *Mentha piperita* by improving biochemical and growth parameters (Ullah *et al.*, 2019). Calcium oxide nanoparticles synthesized from *Juglans regia* have been shown to improve germination rates and other key growth parameters in salt-stressed *Phaseolus vulgaris* (Rizwan *et al.*, 2017). *Punica granatum* fruit extract-derived calcium oxide nanoparticles also alleviate salt stress in *Triticale callus* tissues by initiating calcium-mediated stress response pathways (Prasad *et al.*, 2017). Nanoparticles protect cells from ionic toxicity during salinity stress, helping maintain nutritional homeostasis (Amenta *et al.*, 2015). For instance, photogenic gold nanoparticles applied to roots and shoots reduce ionic accumulation and enhance nitrogen metabolism and antioxidant levels, including SOD, ascorbate, glutathione peroxidase, and glutathione reductase, reducing ROS and lipid peroxidation at the cellular level (Demirer *et al.*, 2019). Sulfur nanoparticles derived from *Ocimum basilicum* have been found to improve nutrient uptake, sodium-to-potassium ion ratio, and enhance the absorption of cysteine, soluble proteins, and amino acids in wheat under salinity stress (Mitter *et al.*, 2017). Zinc oxide nanoparticles from *Carica papaya* influence ROS and antioxidant metabolism in *Carthamus tinctorius* under salt stress, increasing proline levels and antioxidant enzyme activity (Vinayaka and Thakur, 2010).

Compared to chemically synthesized nanoparticles, biogenic nanoparticles are more appealing due to their cost-effectiveness, non-toxicity, and eco-friendliness (Vural *et al.*, 2018). However, their large-scale application faces several barriers, including limited availability

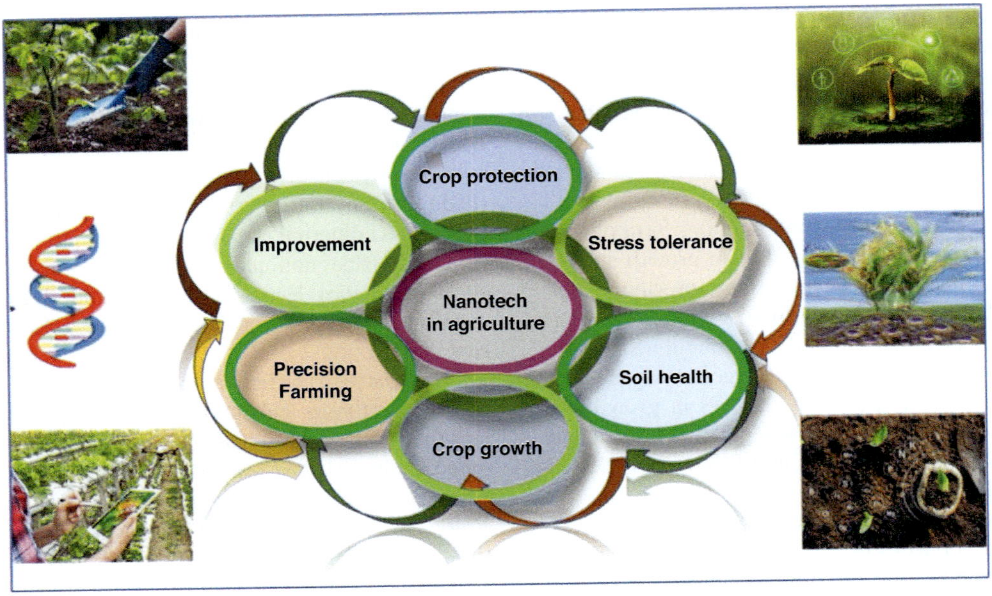

Fig. 7.2. Applications of nanotechnology in agriculture.

of raw materials, lengthy synthesis processes, and the need for chemical agents as oxidizing and reducing components (Raliya *et al.*, 2018). The necessity to characterize nanoparticles for size and shape to ensure quality also adds to production costs. Additionally, the generation and conversion rates of biogenic nanoparticles are lower than those of chemically synthesized ones, reducing their economic benefits (Hatami *et al.*, 2017). Addressing these limitations will be essential in enhancing the use of biogenic nanoparticles in agriculture. Overall, phytogenic nanoparticles offer promising solutions for enhancing plant resistance to abiotic stresses like drought and salinity, improving growth, and mitigating environmental stress markers. These advancements could bridge plant sciences and nanotechnology, contributing to sustainable agricultural practices (Fig. 7.2).

7.3 Nanoformulations for Agricultural Enhancement

The application of nanotechnology in agriculture has led to the creation of several ENMs developed to address specific agricultural concerns (de Oliveira *et al.*, 2014). These nanomaterials can be generically divided into nanoparticles, nanotubes, and nanofibers, each with distinct features and uses (Shalaby *et al.*, 2016). Nanoparticles are possibly the most utilized and researched nanomaterials in agriculture. They are generally spherical or near-spherical particles with dimensions ranging from 1 nm to 100 nm (Syu *et al.*, 2014).

Metal and metal oxide nanoparticles (MONP) are the most developing formulations of nanoparticles currently (Kah *et al.*, 2018). High reactivity, antibacterial capabilities, and photocatalytic activity are appealing aspects of these nanoparticles, which encouraged scientists to pay attention to potential implications in agriculture (Prasad *et al.*, 2017). Such MONPs include silver (Ag), zinc oxide (ZnO), titanium dioxide (TiO2), copper oxide (CuO), etc. These are powerful antibacterial agents for nutrition delivery and photocatalytic destruction of insecticides (Amenta *et al.*, 2015). Research has revealed that ZnO nanoparticles can promote seed germination and seedling development in peanuts (Zhu *et al.*, 2019). Silver nanoparticles have shown excellent antibacterial effects against plant pathogens (Zhu, 2016). Carbon-based nanoparticles are another formulation of manufactured nanoformulations which includes fullerenes and carbon dots and are essential owing to their large surface area and outstanding adsorption capabilities (Amenta *et al.*, 2015). These can be employed in agriculture for controlled release of agrochemicals and soil remediation. Carbon dots have demonstrated potential as nanocarriers for the regulated release of insecticides, minimizing environmental effects. Polymer nanoparticles include chitosan nanoparticles, and PLA (polylactic acid) nanoparticles due to their different qualities of biodegradability and biocompatibility, and are appropriate for encapsulating and transporting insecticides, fertilizers, and genetic material. Chitosan nanoparticles have been employed to carry genetic material for plant transformation with higher efficiency compared to traditional approaches.

Nanotubes are cylindrical structures with diameters of less than 100 nm and lengths that can reach several micrometers. Carbon nanotubes (CNTs) have many types, i.e. single-walled CNTs (SWCNTs) and multi-walled CNTs (MWCNTs), which are commonly used due to their high strength-to-weight ratio, excellent thermal and electrical conductivity, and applicable in the delivery of genetic material, water purification, and soil sensors. Studies have revealed that CNTs may penetrate plant cell walls and function as molecular transporters for DNA delivery. Halloysite nanotubes are natural clay nanotubes, biocompatible, have great loading capacity, and are employed for the controlled release of fertilizers, pesticides, and seed coatings (Cutler *et al.*, 2010). Halloysite nanotubes have been employed to generate slow-release fertilizer formulations, enhancing nutrient utilization efficiency (Cutler *et al.*, 2010). Nanofibers are fiber-like structures with dimensions smaller than 100 nm and high aspect ratios, which include electrospun polymer nanofibers and cellulose nanofibers. Electrospun polymer nanofibers feature a few popular examples like PVA (polyvinyl alcohol), PLA, and PCL (polycaprolactone) nanofibers and are essential owing to their high porosity and huge surface area-to-volume ratio (Wasternack, 2007). They are employed for seed coatings, controlled release of agrochemicals, and tissue culture scaffolds. Electrospun nanofibers have been employed to generate smart seed coverings that respond to environmental stimuli for the regulated release of agrochemicals (Cutler *et al.*, 2010). Cellulose nanofibers are derived from natural sources and are biodegradable, high-strength structures, which can be utilized for reinforcement of biodegradable plastics for agriculture. Cellulose nanofibers have shown potential as soil conditioners, improving water retention and nutrient availability (Ashraf and Foolad, 2007). Mesoporous silica nanoparticles (MSNs) have been employed to transport DNA into plant cells (Mittler, 2017); clay nanosheets to carry RNA interference (RNAi) molecules, enabling viral protection in plants for up to 20 days with a single application. Early diagnosis of stress is key for effective treatment (Wang *et al.*, 2016).

Nanobiosensors allow quick, sensitive, and specific detection of stress markers. Quantum dots (QDs) have been utilized to produce fluorescent biosensors for detecting plant diseases. (Kah *et al.*, 2013) described a QD-based biosensor for detecting *Pseudomonas syringae* in tomatoes. Graphene oxide (GO) has been utilized to produce sensors for sensing plant hormones related to stress reactions. Bortesi and Fischer (2015) created a GO-based electrochemical sensor for detecting abscisic acid, a crucial hormone in drought-stress response. Nanoemulsions have been utilized to generate effective delivery methods for plant growth regulators and stress-mitigating chemicals. Shinozaki and Yamaguchi-Shinozaki (2007) created a nanoemulsion-based delivery method for salicylic acid, boosting its effectiveness in improving plant stress tolerance. Shinozaki and Yamaguchi-Shinozaki (2007) created chitosan nanoparticles for encapsulating and administering abscisic acid, enhancing drought tolerance in maize. Nanoparticles can boost the efficiency of helpful microorganisms in improving plant stress tolerance. Fernández *et al.* (2013) found that silver nanoparticles boosted the capacity of *Trichoderma harzianum* to protect tomato plants against salt stress. Nanocarriers can be utilized to transmit microbial metabolites that boost plant stress tolerance. Agegnehu *et al.* created silica nanoparticles for delivering harpin proteins, which boost plant defensive responses against pathogens (Fernández *et al.*, 2013). CRISPR/Cas9 technology can be used to knock out or change genes implicated in negative stress responses, hence improving tolerance.

For example, altering genes associated with ABA sensitivity can increase drought resistance (Vessey, 2003). These techniques involve overexpressing or silencing certain TFs or regulatory genes to increase stress resistance. Transgenic tomato plants expressing DREB genes have shown increased drought and salinity tolerance (Ijaz *et al.*, 2023c).

7.3.1 Nanofertilizers

In recent decades, crop yields have risen dramatically, playing a crucial role in meeting global nutritional needs (Ijaz *et al.*, 2023c). The use of artificial fertilizers has significantly enhanced crop production. However, the performance of chemical fertilizers has been suboptimal, leading to a reduction in their usage due to environmental pollution and increased production costs. Studies indicate that 50–70% of nitrogen from chemical fertilizers is lost to the environment (Shaikh *et al.*, 2024), prompting a shift in research focus towards developing more efficient solutions (Ijaz *et al.*, 2023c). Nanotechnology is being explored to create slow-release fertilizers, minimize nutrient loss, and enhance the availability of less accessible nutrients. The effectiveness of these advancements is heavily influenced by the surface area of the nanoparticles (Wasternack, 2007).

7.3.1.1 Application of nanofertilizer in plants

Nanofertilizers are pivotal to agricultural sustainability due to their ability to enhance crop yields, improve nutrient quality, and reduce production costs (Kah *et al.*, 2013). Singh highlighted new methods that improve agricultural efficiency, noting that nanofertilizers boost crop production while lowering costs and reducing population risks. However, excessive concentrations or doses of nanofertilizers can diminish crop yields (Bortesi and Fischer, 2015). Kah *et al.* (2013) found that nanofertilizers increase median efficacy by 18–29% compared to standard fertilizers (Shinozaki and Yamaguchi-Shinozaki (2007) and Liu and Lal reported a 32% improvement in growth rates with phosphatic nanofertilizers (Fernández *et al.*, 2013). Nanofertilizers aid nutrient uptake through nano-cuticle pores and enhance plant metabolism. Traditional nitrogenous fertilizers have only 30–60% efficiency, and phosphatic fertilizers suffer from 80–90% loss due to chemical reactions in the soil. In contrast, nanofertilizers, such as those made from hydroxyapatite and urea composites, control nitrogen release, reduce ammonia volatilization, and maintain phosphorus availability over time. This minimizes nutrient loss to the environment and ensures nutrients are available where and when needed (Vessey, 2003).

Advanced nanofertilizers, designed to minimize unwanted volatilization, leverage slow-release chemicals to optimize nutrient delivery (Ijaz *et al.*, 2023b). These smart fertilizers are informed by the signaling between soil microorganisms and plant roots, with nanoparticles altering root signals and ethylene production in *Arabidopsis* (Chakraborty *et al.*, 2022; Shaikh *et al.*, 2024). Controlled release of nutrients based on internal root cues can be a breakthrough in addressing phosphorus and nitrogen deficiencies. Nanofertilizers are made from materials like zeolite, hydroxyapatite, polyacrylic acid, clay minerals, and chitosan. They can be applied via foliar treatments or soil incorporation (Chaffai *et al.*, 2024). Compared to conventional fertilizers, nanofertilizers like hydroxyapatite-urea composites offer extended nitrogen release due to their large surface area and strong interaction (Pudake *et al.*, 2024; Ullah *et al.*, 2024). Polyacrylamide polymers in potassic fertilizers provide slow potassium release, reducing leaching in sandy soils (Gill *et al.*, 2024). Nanotechnology's potential in compost generation and organic waste breakdown is promising, though significant results are still emerging (Ijaz *et al.*, 2024). Nanofertilizers demonstrate high efficacy and reduced nutrient loss, essential for sustainable agriculture. Evaluating the economic feasibility of these technologies remains crucial for their widespread adoption (Rajpal *et al.*, 2024).

7.3.1.2 Nanopesticides

In agriculture, pesticides are crucial for enhancing crop yields, with new and varied insecticides developed each year. However, only about 0.1% of these pesticides effectively target pests, while the remaining 99.99% can contaminate the environment, impacting human health and contributing to resistance in pests, weeds, and diseases (Bibi *et al.*, 2023). Biopesticides offer an alternative, but their effectiveness is often limited by environmental factors and slower development. Nanopesticides provide a promising solution by offering controlled, long-lasting pest control with reduced environmental impact (Fen *et al.*, 2022). Unlike traditional pesticides, nanopesticides are nanosized, enhancing the solubility of active components and thus minimizing environmental harm. They are delivered in smaller amounts, conserving water and energy (Fen *et al.*, 2022). Nanopesticides also increase crop production while lowering costs related to labor and water usage. By addressing the drawbacks of conventional pesticides and reducing their environmental hazards, nanopesticides represent a sustainable and efficient option for pest management (Ijaz *et al.*, 2023b).

7.3.1.3 Nanobiosensors

Nanobiosensors are advanced instruments designed to detect biological or chemical components with nanoscale precision, providing a digital electrical signal (Singla, 2020). These sensors use immobilized bioreceptor probes to identify target analytes such as diseases, pesticides, environmental contaminants, glucose, urea, and metabolites (Singla, 2020). Unlike traditional methods like PCR and bisulphite treatments for DNA methylation detection, the first nanowire field biosensor, based on a transistor, can detect DNA methylation as an electrical signal. Nanobiosensors can also identify specific proteins, crucial for detecting biomarkers, plant microbes, mineral deficiencies, and distinguishing between plant species (Ullah *et al.*, 2023b).

In agriculture, nanosensors are valuable for diagnosing soil diseases caused by fungi, viruses, bacteria, and other microorganisms. By measuring differences in oxygen consumption between harmful and beneficial microbes, nanosensors can identify suitable soil organisms (Bouri *et al.*, 2024). They are also used for detecting heavy metals in water through a noncontact sensor method involving electrode arrays on silicon chips. Heavy-metal ions create deposits on these chips, allowing detection through quantum jump conductance (Zenda *et al.*, 2023). Zeolites, when combined with nanosensors, enhance sustainable agriculture by improving fertilizer efficiency, water retention, and plant development. The integration of nanosensors with zeolites allows for monitoring water and nutrient discharge in soil and plants (Ullah *et al.*, 2023c).

The next generation of biosensors is expected to be nanobiosensors which combine sensitive components with physiochemical transducers for ultra-low concentration detection (Ullah *et al.*, 2023c). These sensors offer high sensitivity, rapid electron transfer, long-term stability, and a high surface-to-volume ratio compared to previous generations (Naaz *et al.*, 2023; Surendran *et al.*, 2024). Nanobiosensors can detect harmful contaminants in food, providing safer alternatives (Meena *et al.*, 2023). Fluorescence-based nanobiosensors, utilizing quantum dots and fluorescence resonance energy transfer (FRET), offer easy detection methods. However, the high cost of fluorescence-based sensors can be mitigated with metal-enhanced fluorescence (MEF) techniques (Bouri *et al.*, 2024).

Zinc oxide (ZnO) is another key material used in nanobiosensor development, particularly for detecting free radicals in food-packaging materials due to its strong catalytic power and high adsorption capability (Zenda *et al.*, 2023). Magnetoelastic materials, combining magnets with gold nanoparticles, offer benefits such as simple separation and reduced nonspecific binding, which improves signal-to-noise ratios. These materials include oxides, metals, and alloys (Ullah *et al.*, 2023c). Additionally, microfluidics technology, merging nanotechnology, biosensing, and microsystems, has been applied to detect foodborne allergens, pathogens, toxins, and other contaminants (Table 7.1) (Surendran *et al.*, 2024).

Most lipid membrane-based nanobiosensors are designed for detecting microorganisms, metals, and toxins. However, they face technological challenges such as reduced stability and potential damage outside of electrolytic solutions. To enhance stability, materials like glass, ZnO, and graphene can be employed (Wani, 2023). Nanobiosensors provide efficient and accurate detection, enabling continuous monitoring while reducing time and cost (Ahmed *et al.*, 2023). The excessive use of pesticides and chemicals has led to unsustainable agricultural practices and

Table 7.1. Nanoformulations for agricultural enhancement.

Nanobiosensors	Nanomaterial used	Sensor type	Applications	References
Quantum dot nanosensor	Gold particles	Immunobased sensor	Detection of mycotoxins ZEA, DON, FB1/FB2 in corn oats and barley	Janbandhu *et al.* (2024)
QD nanosensor	Quantum dots	Fluorescence	Finding of pathogens	Choudhury and Moulick (2022)
QD nanosensor	Quantum dots	Fluorescence	*Phytoplasma aurantifolia*	Meena *et al.* (2023)
DNA nanobiosensor	Carbon nanotubes	Immunosensor	*Ganoderma boninse* for infection of palm oil tree	Janbandhu *et al.* (2024)
Surface plasmon resonance	Multi-walled carbon nanotubes	SPR	Recognition of Cymbidium mosaic virus	Choudhury and Moulick (2022)
Artificial nose	Carbon nanomaterials	Volatile organic compound profile	Finding pathogens depending on the organic compounds released	Wani (2023)
Plant hormone nanobiosensor	Receptor DAD2 from Petunia hybrid and HTLT from *Striga hermohthica* with green fluorescent protein	Fluorescent based	Revealing strigolactones as signaling molecules for plant growth and parasitism	Ullah *et al.* (2023c)
Smart nanobiosensor	Zinc oxide and copper	Electrochemical	Increasing the growth of tomato chili and cucurbits in Mexico	Meena *et al.* (2023)
Acetylcholinesterase E on titanium oxide	Cholinergic enzyme	Differential pulse voltammetry	Exposure of dichlorvos	Surendran *et al.* (2024)
Acetylcholinesterase	Cholinergic enzyme	Amperometric	Revealing of chlorpyrifos	Meena *et al.* (2023)
Acetylcholinesterase onto polymeric surface and silver	Enzyme	Amperometric	Recognition of malathion	Choudhury and Moulick (2022)
Acetylcholinesterase on white paper using indophenol acetate	Enzyme	Colored reaction	Finding of chlorpyrifos	Wani (2023)

food production. In contrast, nanotechnology-based agricultural technologies have achieved higher yields (Kumar *et al.*, 2021). Significant research has been conducted on the application of biosensors (Kumar *et al.*, 2021). Nanoparticles incorporated into nanobiosensors act as bioreceptors within a transducer system. These sensors deliver recognition elements with the signal to detect single or multiple fragments. The integration of immobilization, miniaturization, and functionalization of biocomponents into a transduction system enhances the overall performance of nanomaterials (Hakim *et al.*, 2021).

7.4 Nanotechnology for the Management of the Food Supply Chain

Nanotechnology offers significant benefits in the development of analytical tools for quality control, biosecurity, and safety in agriculture and the food supply chain (Ullah *et al.*, 2023c). With the global population growing rapidly, the agricultural sector faces increased demand for food. Historically, chemical fertilizers were widely used to boost production, but studies show that less than half of these fertilizers are effectively utilized by plants, with the remainder contributing to soil pollution and runoff into water bodies. For instance, phosphorus, nitrogen, and potassium losses after application can reach 80–90%, 40–70%, and 50–90%, respectively (Naaz *et al.*, 2023; Meena *et al.*, 2023; Janbandhu *et al.*, 2024). This inefficiency has led to a need for technologies that control the release of fertilizers to improve crop yield and environmental sustainability. Nanoparticle-based fertilizers, or nanofertilizers, present a promising solution due to their high surface-to-volume ratio, controlled-release capabilities, and enhanced nutrient delivery (Choudhury and Moulick, 2022; Wani, 2023).

Nanobiosensors, which are analytical devices with dimensions of 100 nm or less, offer improved performance over traditional sensors. They use nanomaterials like nanotubes, nanowires, nanoparticles, or nanocrystals to enhance sensitivity, reduce response times, and improve detection limits (Kumar *et al.*, 2021; Hakim *et al.*, 2021). Despite significant research, practical applications for field analysis remain limited, suggesting a substantial opportunity for advancement. Nanotechnologies could enhance biosensor performance, enabling practical applications in agriculture by integrating with techniques like surface plasmon resonance, mass spectrometry, and microfluidics (Ullah *et al.*, 2023b; Zenda *et al.*, 2023; Bouri *et al.*, 2024). This integration promises automated systems that combine biosensing with data management, micro-/nanofluidics, and wireless control, critical for future agricultural applications (Bouri *et al.*, 2024; Ullah *et al.*, 2023c). Nanosensors could revolutionize agriculture by providing precise control and real-time monitoring of various parameters. They can measure soil pH, nutrient levels, residual pesticides, and humidity, and detect pathogens, contributing to better crop management and reducing the need for excessive fertilizers and pesticides (Naaz *et al.*, 2023; Meena *et al.*, 2023; Surendran *et al.*, 2024). The ability of nanosensors to monitor pests, pathogens, and chemical residues allows for more targeted use of inputs, potentially increasing productivity and reducing waste (Pudake *et al.*, 2024; Ullah *et al.*, 2023b). Automated irrigation systems, coupled with nanosensors that assess soil water tension, could improve water-usage efficiency, especially in regions with water scarcity (Naaz *et al.*, 2023; Surendran *et al.*, 2024).

In the food supply chain, nanosensors enhance food quality and safety by providing rapid, sensitive, and cost-effective detection of contaminants and monitoring of freshness (Ullah *et al.*, 2023b). They can track various food components, such as sugars, vitamins, and minerals, as well as detect pesticides, heavy metals, and toxins (Meena *et al.*, 2023). Intelligent packaging systems equipped with nanosensors can monitor factors like humidity, temperature, and gas composition, ensuring food safety and quality during storage and transport (Choudhury and Moulick, 2022; Wani, 2023). These advancements highlight the potential of nanosensors to transform both agricultural practices and food safety management (Fig. 7.3).

7.4.1 Systems for sustainable intensification in agriculture

Sustainable intensification refers to a production approach that aims to increase output without causing negative environmental impacts, all within the same agricultural region (Ahmed *et al.*, 2023). Sustainable intensification (SI) refers to a strategy or system that increases agricultural yields without causing negative impacts on the environment, while also allowing for the cultivation of larger areas of land (Ullah *et al.*, 2023c). The phrase is therefore quite extensive, as it does not explicitly express or favor any one perspective on agricultural production (Naaz *et al.*, 2023). The focus is on results rather than procedures, and it does not dictate the use of certain technology,

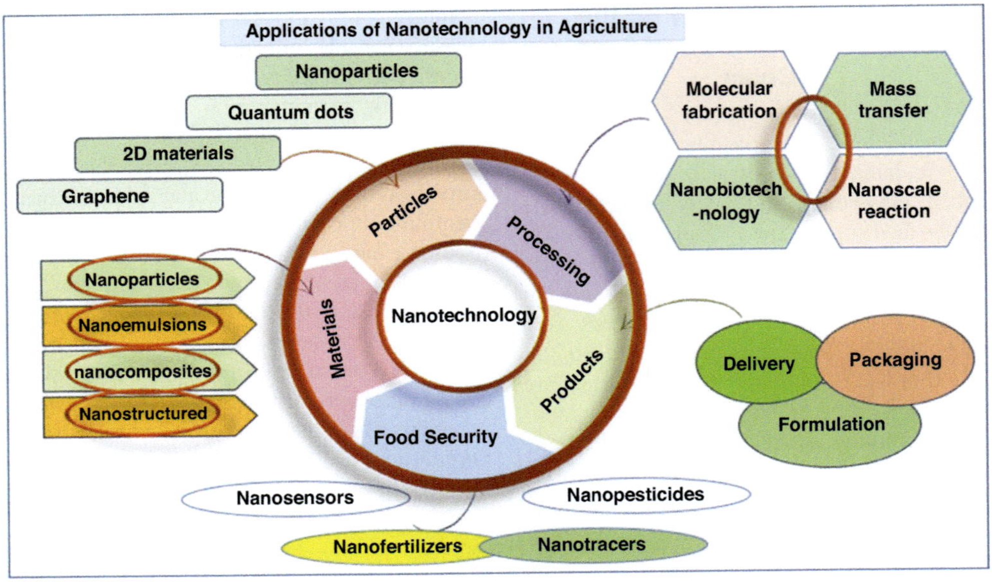

Fig. 7.3. Applications of nanotechnology in the agricultural field.

species combinations, or design elements. This paradigm offers a structure for examining the choice of the most effective combination of techniques for agricultural production, considering the impact of the current biophysical, social, cultural, and economic conditions (Zenda *et al.*, 2023). Within this framework, novel nanomaterials have been developed to enhance productivity by utilizing inorganic, polymeric, and lipid nanoparticles. These nanoparticles are formed through several methods such as emulsification, ionic gelation, polymerization, and oxydoreduction (Ullah *et al.*, 2023c). As societies accumulated surplus resources and diversified their economies, agriculture eventually became recognized as a separate economic sector, distinct from nature or the environment (Wang *et al.*, 2016). The Cartesian perspective, which views nature as a machine, has been bolstered by historical religious divides between humans and the natural world (Ullah *et al.*, 2023c). This has led to a gradual detachment from nature, resulting in the emergence of two separate realms: human-created systems of production and the wild natural environment. Recently, with increasing concern for sustainability, various frameworks have been developed to classify different levels of environmental consciousness, from superficial to deeply committed (Pudake *et al.*, 2024). For certain individuals, there is a fundamental division regarding the existence of nature: whether it is independent of human beings or whether it is a component of a postmodern context (Ullah *et al.*, 2024). Some have called for agriculture to become low-carbon (Kumar *et al.*, 2021) and nutrition-sensitive (Hakim *et al.*, 2021), given the necessity for the industry to solve global social and ecological concerns. Sustainable production systems must display three critical qualities at the production end of food systems (Sukhpal *et al.*, 2018). These systems often differ dramatically from previous methodologies. They are generally multifunctional, playing multiple roles within landscapes and economies (Gondalia *et al.*, 2022). They generate not just food and other commodities for farmers and markets but also provide significant public goods such as clean water, wildlife habitats, carbon sequestration, flood protection, groundwater recharge, picturesque landscapes, and possibilities for recreation and tourism (Mahanty *et al.*, 2023). Structurally, these systems leverage the synergies and efficiencies deriving from their complex ecological, social, and economic linkages (Kumar *et al.*, 2021).

Sustainable agricultural systems are diversified, synergistic, and adaptable to their specific social and ecological contexts (Ullah *et al.*, 2023c). There is no one-size-fits-all approach to achieving agricultural sustainability; it requires adjusting practices to fit the unique circumstances of each agricultural system (Hakim *et al.*, 2021). For instance, in the UK, cattle and dairy companies moving towards sustainability, faced difficulty in lowering pollutants while boosting yields (Ullah *et al.*, 2023b). These systems frequently entail complicated combinations of domesticated species and management approaches, necessitating greater skill levels from farmers (Kumar *et al.*, 2021). Effective and sustainable production requires farmers to understand how inputs like seeds, fertilizers, and pesticides interact with biological processes and ecosystem services (Sukhpal *et al.*, 2018).

Farmers must recognize that increased complexity and information can lead to significant improvements in production. Systems with robust social and human assets are better equipped to innovate amidst uncertainty (Gondalia *et al.*, 2022), and peer learning plays a crucial role in adopting context-specific, knowledge-intensive sustainable practices (Mahanty *et al.*, 2023). Traditionally, agricultural sustainability has been associated with reduced input consumption, which often meant larger areas were needed to produce the same amount of food. Organic systems, for example, may have lower yields per unit of land but aim to decrease input use and enhance natural capital. Despite this, biological systems can remain efficient if managed well, with knowledge and information compensating for reduced external inputs (Table 7.2) (Yoon *et al.*, 2021).

Recent research shows that effective agricultural sustainability programs often involve changes in production practices, such as replacing fertilizers with nitrogen-fixing legumes, using natural predators instead of pesticides, and adopting zero-tillage instead of plowing (Yoon *et al.*, 2021). A promising approach emphasizes intensifying resource use and optimizing existing resources (e.g. land, water, biodiversity) and technology (Irfan and Kumar, 2024). Some propose extending agriculture into nonagricultural areas to fulfill the food demands of rising populations and shifting consumption habits. However, if preserving biodiversity and ecosystem services – like carbon storage in forests – is given priority, competition for land from other users would make this more and more expensive (Kole, 2022).

Others propose enhancing yields by revisiting Green Revolution practices, focussing solely on biotechnology or adopting organic methods. Regardless of the approach, maximizing current

Table 7.2. Differences between sustainable intensification and historically conventional forms of agricultural intensification.

Properties	Conventional forms of agricultural intensification	Sustainable intensification
Primary goals of farmers	Increased crop and livestock yields	Improve productivity and incomes, develop natural capital in on- and off-farm environments, and build knowledge and social capital
Knowledge development	Tends to be solely 'expert'- driven	Agroecological design emerged largely via collaborations between 'experts' and other stakeholders; new technologies and practices are the result of participatory research and development.
Knowledge dissemination	Conventional extension chain from public or private research to farmers	Traditional extension methods, combined with participatory dissemination through peer-to-peer learning
Stewardship of ecosystem services	Focus on the provisioning services provided by agricultural landscapes, while relying on external inputs to compensate for regulating and supporting services; interactions with adjacent nonagricultural landscapes are often treated as externalities.	A deeper understanding of the diverse ecosystem services offered by agricultural landscapes and the recognition of the reciprocal relationship between agricultural and nonagricultural elements within these landscapes

agricultural land is crucial (Chandra *et al.*, 2021). Traditionally, there have been three methods to characterize agricultural intensification: (i) raising yields per hectare; (ii) intensifying cropping or livestock; and (iii) repurposing land for crops that are more valuable or nutrient-dense. While intensification aims to improve productivity, it has sometimes led to environmental degradation and social disruption (Bamal *et al.*, 2021). Nevertheless, sustainable intensification has the potential to promote greener economies and benefit from broader sectoral development (Rajpal *et al.*, 2024).

In the latter half of the 20th century, intensification rather than expansion was the main driver of increased global food production. Although agricultural land expanded by 11% and arable land by 10%, global food production surged by 145%, with significant increases in Asia and Latin America (Chandra *et al.*, 2021). China saw a fivefold growth, primarily in the 1980s and 1990s, while industrialized nations also experienced substantial productivity gains (Kole, 2022). The world population grew from 3 billion to over 7 billion, yet per capita agricultural productivity has surpassed population growth, providing 25% more food per person compared to 1960 (Kah *et al.*, 2013). Raising agricultural output is essential for reducing poverty, which is associated with hunger and environmental damage. Agricultural expansion is more successful in alleviating poverty than general economic growth, particularly in largely agricultural nations with high poverty levels (Kumar *et al.*, 2021). However, agricultural systems alone cannot alleviate global challenges, as conflicts have diminished output. Sub-Saharan African countries afflicted by conflicts produced 12% less food during wartime than during peacetime, with conflict-related losses estimated to have totaled $121 billion between 1970 and 1997 (Hakim *et al.*, 2021).

Evaluations of intensification practices often focus on specific techniques, such as conservation agriculture or integrated pest management, but may overlook broader outcomes and benefits (Gondalia *et al.*, 2022). The lack of comprehensive data on various intensification strategies limits understanding of their regional and global impacts (Sukhpal *et al.*, 2018). Additionally, nanobiosensors for detecting pesticides have limitations due to their cost and effectiveness, though advancements in multisampling and pretreatment may improve their efficiency (Irfan and Kumar, 2024). Despite progress and the rise of 'sustainable intensification,' ensuring that agricultural systems worldwide boost productivity while positively impacting natural and social capital remains a significant challenge (Kole, 2022).

7.4.2 systems to improve the availability of water

In the future decades, the share of global freshwater available for agriculture is predicted to fall because of increased demands from industry, energy generation, and household usage. Additionally, competition among diverse agricultural demands, evolving eating trends (such as increasing consumption of meat and sugar), and changes in the global energy balance (Chandra *et al.*, 2021) would affect water availability for food production (Bamal *et al.*, 2021). From 1990 to 2004, the yearly efficiency gain in agricultural water consumption was roughly 1% for both rainfed and irrigated regions (Liu *et al.*, 2023). At this rate, the industry could handle only roughly 20% of the expected demand-supply imbalance by 2030, with advances in supply covering an additional 20% (Kumar *et al.*, 2021). The most obvious and damaging inequalities are projected in nations experiencing rapid economic expansion alongside high poverty levels, such as India, China, Brazil, and South Africa (Ullah *et al.*, 2023b). Agricultural runoff from both agricultural and animal systems also adds to pollution, significantly lowering the supply of clean water (Ullah *et al.*, 2024). Given the increased likelihood of localized water shocks, such as droughts, floods, and dry spells, it is necessary to manage water resources at numerous scales, from small farms to whole watersheds and river basins (Ullah *et al.*, 2023b). Spatial planning is vital to maintaining multifunctional landscapes, which sustain varied ecosystems capable of buffering storm floods and encouraging groundwater recharge instead of surface runoff (Kumar *et al.*, 2021). Wetlands, meandering rivers, woodlands, and landscape mosaics are significant natural features that increase resilience. Effective watershed and river basin management is crucial for collecting rainfall (Bamal *et al.*, 2021). In many places, a considerable percentage of rainfall (sometimes over 50%) originates from local to mesoscale vapor fluxes, with moisture from upwind forests contributing to downstream rainfall (Liu *et al.*, 2023). This moisture feedback, found in locations like the Sahel where moisture from West African rainforests supports rainfall in the semi-arid savannah, highlights the necessity of managing water resources at the watershed and regional levels to assure future local food production (Liu *et al.*, 2023).

7.4.3 Systems to improve the quality of the soil

It has been demonstrated that hydrogels, nanoclays, and nanozeolites increase the soil's ability to retain water (Kumar *et al.*, 2021), acting as slow-release water sources to lessen dry spells during the crop season (Rizi *et al.*, 2022). These materials are excellent for both agricultural applications and the regeneration of degraded regions. Organic nanomaterials, such as polymer and carbon nanotubes, and inorganic ones, such nanometals and metal oxides, have also been employed to absorb environmental contaminants (Kumar *et al.*, 2021), boosting soil remediation effectiveness and lowering treatment times and costs. Natural and organic minerals in the soil can interact with nanomaterials, changing the distribution between aqueous and solid phases. While much research has focussed on the behavior and fate of nanomaterials in aquatic systems (Custodio *et al.*, 2023), there are limited data on their behavior and fate in soil systems (Rizi *et al.*, 2022).

7.4.4 identification of gaps and obstacles

Despite significant progress in exploring nanotechnology's potential in agriculture, several challenges must be addressed before this technology can make a substantial impact in the field (Custodio *et al.*, 2023). To effectively deliver active agents like fertilizers, pesticides, and nutrients while upholding environmental sustainability and green chemistry principles, some key areas need to be addressed. These include: (i) developing hybrid carriers that are specifically designed for this purpose; (ii) creating processes that are easily scaled up to an industrial level; (iii) comparing the effects of nanoformulations with current commercial products to show concrete benefits; (iv) expanding our understanding of risk and life-cycle assessments of nanomaterials, including their effects on nontarget organisms like other plants, soil microbiota, and bees (Ali *et al.*, 2023a); and (v) updating regulations related to the use of nanomaterials (Muttatettu and Anitha, 20232). The effective deployment of nanopesticides, such as atrazine, provides a great case study to highlight crucial characteristics for forecasting nanoparticle behavior in the environment (Cabrera-Barjas *et al.*, 2022). Research on the atrazine-nano pesticide system focussed on interactions with both target plants, like mustard, and non-target plants, like maize, and included risk assessments (Custodio *et al.*, 2023).

Implementing nanotechnology in agriculture also requires developing methods to assess nanoparticles present in various environmental compartments (Karumamkandathil *et al.*, 2022). Current methodologies often fall short in understanding nanomaterial dynamics, their interactions with target and nontarget species, and potential synergistic effects (Khan *et al.*, 2024). Advancements in these methods are crucial for conducting life cycle evaluations of newly developed nanomaterials (Mogazy, 2024). Additionally, research should address potential resistance mechanisms to nanomaterials by target organisms. New analytical approaches would aid in predicting, locating, and quantifying engineered nanomaterials in the environment (Meszaros *et al.*, 2023). Collaboration among scientists from diverse fields is essential in this regard (Naveen and Sontakke, 2024). However, the use of nanomaterials in agriculture also presents significant challenges (Ali *et al.*, 2023b). The long-term impacts of nanomaterials on human health and the environment are not fully understood, requiring extensive safety studies and regulatory frameworks (Mogazy, 2024). Further investigation is needed into nanomaterials' behavior in complex environmental systems, particularly their potential for bioaccumulation (Meszaros *et al.*, 2023). Consumer acceptance of nanotechnology in food production remains a concern, highlighting the need for transparent communication and engagement (Naveen and Sontakke, 2024). Scaling up nanomaterial manufacturing for agricultural use while maintaining cost-effectiveness continues to be a challenge (Ali *et al.*, 2023b).

Despite these obstacles, nanomaterials hold tremendous potential for agriculture. They offer the possibility of more efficient use of agricultural inputs, increased crop yields, and improved food quality (Bora *et al.*, 2023). As research progresses, nanomaterials are expected to play a crucial role in creating sustainable and resilient agricultural systems capable of meeting the needs of a growing global population amid climate change and resource limitations (Ahmad *et al.*, 2022). Integrating nanotechnology with emerging technologies like artificial intelligence and biotechnology is anticipated to enhance the impact of nanomaterials in agriculture. This multidisciplinary approach could lead to smart agricultural systems that optimize resource use, minimize environmental impact, and adapt to changing climate conditions with unprecedented precision (Cutolo *et al.*, 2023). As we approach this nano-agricultural revolution, ongoing research, responsible development, and collaboration among scientists, policymakers, and industry stakeholders will be key to realizing the full potential of nanomaterials for sustainable and productive agriculture (Govaichelvan *et al.*, 2024).

7.4.5 Concerns about health that the Environmental Production Agency declared

Researchers have discovered that nanopesticides can minimize the detrimental effects of bacteria on soil and plants (Zenda *et al.*, 2023; Naaz *et al.*, 2023]. Similarly, (110) revealed that polyethylene glycol-based nanoformulations of carbofuran and acephate decreased toxicity to nontarget species (Ali *et al.*, 2023b). This improvement in effectiveness is related to the gradual release of active components rather than enhanced absorption of nanoformulated substances. (Bora *et al.*, 2023) reported that nanoparticle-based silica-associated chlorfenapyr exhibited twice the insecticidal activity against cotton bollworms compared to microparticle-based formulations (Surendran *et al.*, 2024). These findings suggest that the controlled release of active components could be an effective future strategy for pest management.

Nanoparticles, such as those made from silver (Ag), copper (Cu), and aluminum (Al), are recognized for their antibacterial properties. For example, silver-silica nanoparticles have been shown to eliminate infections from damaged leaves within three days of application (Cutolo *et al.*, 2023). Copper nanoparticles also demonstrate significant antibacterial and antifungal activity (Choudhury and Moulick, 2022]. Biogenic nanoparticles, which are derived from natural sources, are gaining attention due to their low cost, nontoxic nature, and environmental friendliness compared to chemical nanoparticles (Govaichelvan *et al.*, 2024). Despite their promising potential, there are challenges to their widespread use, such as the limited availability of raw materials, longer reaction times, and the complexity of large-scale manufacturing (Zhang *et al.*, 2023). Additionally, characterizing these nanoparticles for quality assurance increases production costs, and their yield and conversion rates are currently lower than chemically synthesized nanoparticles (Kumar *et al.*, 2021; Ahmed *et al.*, 2023).

Weed management is a significant challenge in modern agriculture, and the effectiveness of nanoherbicides depends largely on biodegradable polymers (Choudhury and Moulick, 2022). For instance, atrazine encapsulated in a biopolymer has improved biocompatibility, bioavailability, and physicochemical properties, leading to enhanced stability, reduced soil mobility, and better herbicidal action compared to free atrazine (Ahmed *et al.*, 2023). Nanoformulated atrazine has proved more effective against specific weeds, such as hairy beggarticks and slender amaranth, with the ability to deliver herbicidal activity directly through the leaf vascular tissues (Kumar *et al.*, 2021).

However, the ecological impact of nanopesticides is not well understood. Inorganic nanoparticles, including Ni, Cu, SiO_2, Ag, ZrO_2, and TiO_2, can be toxic to earthworms, affecting their reproduction and behavior (Kumar *et al.*, 2021). Studies have shown that silver nanoparticles and silver nitrate ($AgNO_3$) can significantly reduce earthworm reproduction, and copper nanoparticles also have harmful effects on these organisms (Yoon *et al.*, 2021; Gondalia *et al.*, 2022; Mahanty *et al.*, 2023). Further study is needed to solve these ecological challenges and guarantee the proper use of nanotechnology in agriculture.

7.5 Nanotechnology and Its Use in the Food Sector

Nanotechnology, which focusses on microscopic materials, has a profound impact on the food industry (Cutolo *et al.*, 2023). It offers significant potential for enhancing nutritional values, flavor, and color, and extending the shelf life of food products (Govaichelvan *et al.*, 2024). This emerging technology can revolutionize the food sector through its application in food additives (incorporating nanomaterials) and food packaging (containing nanomaterials) (Zhang *et al.*, 2023). Notable advancements have been made in food processing through nanotechnology, addressing the rising demand for high-quality, low-calorie, and low-fat food products (Khan *et al.*, 2021). Nanofiltration is a prominent nanotechnology-based method with considerable promise in food processing. This technique employs nanoporous membranes for processes such as softening and water purification, effectively removing divalent ions (da Cunha *et al.*, 2023). In the dairy industry, nanotechnology is utilized for milk standardization, protein fractionation, and improving microbiological quality (Ramesh *et al.*, 2023a). Advantages of this approach include reduced processing time, lower energy consumption, improved product quality, and enhanced separation efficiency compared to traditional filtration methods.

Nanotechnology has also made significant strides in food safety. Rapid, robust, and sensitive detection of food contaminants is critical, and biosensing approaches are highly effective for this purpose (Cisneros *et al.*, 2023). These approaches incorporate a transducing mechanism linked with a specialized bioreceptor that binds tightly to target pollutants, creating biological or chemical signals (Chaudhry *et al.*, 2008; Sozer and Kokini, 2009). Transducers then convert these signals into measurable forms. For example, Agromicron's nanobioluminescence technology uses luminous proteins to detect harmful bacteria such as *Escherichia coli* or *Salmonella* by emitting visible light (Abbas *et al.*, 2009). In food management, ensuring quality and safety are primary concerns. Advanced instrumentation and analytical techniques are essential for detecting food quality and safety risks, even at very low concentrations (Thiruvengadam *et al.*, 2018). However, these tools are expensive and require skilled operators, which can delay results. Therefore, developing economical, field-applicable sensors with reliable analytical performance for food nanomaterial evaluation is necessary (Ranjan *et al.*, 2014).

7.5.1 Importance of stress resilience in modern agricultural practices

Stress resilience in agriculture has become increasingly crucial due to the rising challenges posed by climate change, environmental degradation, and food security issues (Cisneros *et al.*, 2023). Extreme weather events like droughts, floods, and heatwaves are becoming more frequent and severe, placing unprecedented stress on crops (Sozer and Kokini, 2009). According to the Food and Agriculture Organization (FAO), between 2008 and 2018, approximately 23% of the total damage and losses from climate-related disasters in developing countries occurred in the agriculture sector, amounting to $108 billion in lost crop and livestock production. At the same time, biotic stressors such as pests and diseases are evolving and spreading to new regions (Abbas *et al.*, 2009), with annual crop losses due to pests estimated at 20–40% of global agricultural production (Thiruvengadam *et al.*, 2018).

In this context, developing stress-resilient crop varieties is essential for maintaining and enhancing agricultural productivity (Ranjan *et al.*, 2014). Such crops can withstand extreme conditions and maintain production stability in variable environments, thereby supporting food security and improving farmer livelihoods (Kumar *et al.*, 2020). The World Bank projects that climate-smart agricultural practices, including the use of stress-resilient crops, could increase yields by an average of 17% and reduce food costs by 15% by 2050. Additionally, stress-resilient crops could contribute to sustainable agriculture by reducing the need for costly inputs like irrigation and pesticides, thus minimizing environmental impacts (Singh *et al.*, 2017).

The integration of advanced breeding techniques, such as genomic selection and gene editing, with traditional methods presents promising opportunities for developing crops with enhanced stress tolerance (Ummi and Siddiquee, 2019). In 2016, global investment in agricultural research and development, crucial for creating stress-resilient crops, reached $79 billion, with over 75% of this funding coming from high-income countries (Ummi and Siddiquee, 2019). As the global population grows and arable land becomes more limited, the importance of stress resilience in agriculture becomes even more critical, making it a fundamental component of future food production strategies (Berekaa, 2015). The UN projects that global food production will need to increase by 60% by 2050 to meet the demands of the growing population, highlighting the urgent need for more resilient and efficient agricultural systems (UN, 2017) (Ummi and Siddiquee, 2019).

7.6 Understanding Abiotic and Biotic Stresses in Agriculture

In agriculture, understanding abiotic and biotic stressors is vital for establishing robust crop systems and guaranteeing global food security. Abiotic stressors, such as drought, salt, severe temperatures, and nutrient deficits, are nonliving environmental conditions that greatly affect crop yield (Berekaa, 2015). According to the Food and Agriculture Organization (FAO), abiotic stressors can lower average yields for main crops by over 50%. Drought alone affects nearly 40% of the world's agricultural land, incurring yearly losses estimated at $80 billion (Mura *et al.*, 2014). Salinity, another important abiotic stress, affects around 20% of irrigated land globally, with an estimated economic effect of $27.3 billion per year (Sonkaria *et al.*, 2012). Climate warming is increasing these issues, with the Intergovernmental Panel on Climate Warming (IPCC) forecasting that every 1°C increase in global mean temperature will cut world yields of wheat by 6%, rice by 3.2%, maize by 7.4%, and soybean by 3.1% (Nile *et al.*, 2020). Biotic stressors, on the other hand, are induced by living creatures such as pests, diseases, and weeds (Berekaa, 2015). The Centre for Agriculture and Bioscience International (CABI) predicts that invasive species alone cost world agriculture more than $540 billion yearly (CABI, 2022) (Ummi and Siddiquee, 2019). Furthermore, plant diseases cause losses of roughly 10–16% in worldwide agricultural output, equal to over $220 billion yearly (King *et al.*, 2018). Insect pests are responsible for an extra 5–20% of agricultural losses, with recent research indicating that climate change might raise insect-driven losses of wheat, rice, and maize by 10–25% each degree Celsius of warming (King *et al.*, 2018).

The intricacy of these stressors is further increased by their interactions; for instance, drought-stressed plants typically become more sensitive to insect assaults (Masters, 2005). This phenomenon, known as cross-stress tolerance, is an emerging topic of research. Studies have indicated that exposure to one form of stress can occasionally confer tolerance to other stressors, giving possible routes for producing multi-stress-tolerant crops (Asma *et al.*, 2023). Recent developments in genetics, phenomics, and artificial intelligence are giving new methods for understanding and alleviating these pressures. For example, the use of CRISPR gene editing has shown promise in creating crops with greater tolerance to numerous stressors (Serraj *et al.*, 2019). The application of machine learning in agriculture, notably in stress prediction and management, is increasing quickly, with the AI in the agriculture industry predicted to reach $4 billion by 2026, rising at a CAGR of 25.5% (Tacon, 2023). Moreover, the global market for stress-tolerant seeds is predicted to reach $53.95 billion by 2026, expanding at a CAGR of 8.2% from 2021, showing increasing investment in this crucial field (Phour and Sindhu, 2023). This expansion is partially driven by public-private partnerships and international efforts. For instance, the Crops to End Hunger program, backed by CGIAR and national agricultural research institutions, intends to speed the development of climate-resilient, nutritious crop varieties for smallholder farmers in low- and middle-income countries (Kumar *et al.*, 2022).

As the world population is predicted to exceed 9.7 billion by 2050 (UN, 2019), recognizing and treating both abiotic and biotic stressors will be important in establishing sustainable agricultural systems capable of supplying future food demands while responding to changing environmental circumstances (Savary *et al.*, 2012). The UN Sustainable Development Goal 2 (Zero Hunger) highlights the need for resilient agricultural practices, intending to secure viable systems for food production while implementing resilient agricultural practices by 2030 (UN, 2015). Achieving this objective will need continuous investment in research, technology transfer, and capacity building, particularly in places most vulnerable to climate change and food insecurity (Deutsch *et al.*, 2018).

7.6.1 Abiotic stresses affecting crop productivity

Abiotic stressors are nonliving environmental conditions that negatively affect crop growth, development, and yield. While drought, salt, and temperature extremes are frequently regarded as the most serious risks, additional abiotic factors such as heavy-metal toxicity also play major roles in restricting agricultural output (Savary *et al.*, 2012). Drought is defined as a limitation of water availability, resulting in diminished soil moisture and hindered plant development. It is one of the most prevalent and severe abiotic stressors impacting agriculture globally. Drought can affect agricultural production by 50–80% depending on timing, length, and intensity (Deutsch *et al.*, 2018). Global economic losses due to drought are estimated at $80 billion yearly (Ryan *et al.*, 2023). Approximately 40% of the world's agricultural area is impacted by drought. The 2012 drought in the U.S. resulted in agricultural losses surpassing $30 billion. In Africa, drought produces crop decreases of 10–20% on average but can surpass 50% in severe circumstances. Salinity stress develops when there is an excessive concentration of soluble salts in the soil. This impedes water intake by plants and produces ionic toxicity. Salinity impacts around 20% of irrigated land globally, amounting to 45 million hectares (Kumar *et al.*, 2022). Annual economic losses owing to salinity are estimated at $27.3 billion (Kumar *et al.*, 2022). Salinity can impair agricultural production by 20–50% in impacted areas (Savary *et al.*, 2012). In Australia, 5.7 million hectares of agricultural land are damaged by salinity, with anticipated expenses of AUD 1.5 billion annually by 2050. In India, roughly 6.73 million hectares are salt-affected, incurring yearly losses of around $6.87 billion.

Salinity stress impacts plants through osmotic stress, ionic toxicity, and nutritional imbalance. It impairs membrane integrity, enzyme function, and photosynthesis (Deutsch *et al.*, 2018). Heat stress occurs when plants are subjected to temperatures over their optimum range for a protracted period. Every 1°C increase in global mean temperature is anticipated to lower world yields of wheat by 6%, rice by 3.2%, maize by 7.4%, and soybean by 3.1%. Heatwaves can induce yield decreases of up to 40% in key crops (Deutsch *et al.*, 2018). The 2003 European heatwave caused estimated agricultural losses of €13 billion (Ryan *et al.*, 2023). In India, heat stress is projected to cause wheat output losses of 4–6 million tons yearly (World Health Organisation, 2023). Heat stress impacts plant metabolism, lowers photosynthetic efficiency, and can lead to pollen sterility and fruit abortion (Lal, 2016). Cold stress, particularly chilling (0–15°C) and freezing (< 0°C) conditions, can severely harm crops, especially in temperate and high-altitude settings. Frost damage costs the U.S. agriculture industry an estimated $1–2 billion yearly. In Europe, late spring frosts produce average output losses of 34% in wheat and 21% in barley (Lal, 2016). The 2007 Easter freeze in the USA caused agricultural losses surpassing $2 billion, damaging crops from Kansas to North Carolina (Kumar *et al.*, 2022). In China, low-temperature stress damages more than 20 million hectares of winter wheat yearly (Lal, 2016). Cold stress can induce cellular dehydration, membrane damage, and disruption of metabolic processes. It also lowers development and delays crop maturity (Alexandratos and Bruinsma, 2012).

Heavy-metal stress occurs when harmful metals (e.g. cadmium, lead, mercury) accumulate in soils and plants, frequently owing to industrial activity, excessive use of fertilizers, or natural geological processes (Molotoks *et al.*, 2021). Approximately 80 million hectares of arable land globally are expected to be polluted by heavy metals (da Cunha Dias *et al.*, 2021).

Heavy-metal pollution can affect agricultural yields by 10–50%, depending on the metal and its concentration (da Cunha Dias *et al.*, 2021). In China, about 20 million hectares of cropland (about 20% of total arable land) are polluted with heavy metals, hurting food production worth $3 billion yearly (Kumar *et al.*, 2022). A study in India indicated that heavy-metal pollution in agricultural soils resulted in economic losses of nearly $3.7 billion per year (Berekaa, 2015). Heavy metals interfere with plant metabolism, alter enzyme activity, and can induce oxidative stress. They also compete with important nutrients for absorption, resulting in nutritional deficits (Ummi and Siddiquee, 2019). Heavy-metal pollution in agricultural soils is anticipated to rise, particularly in emerging nations, due to increased industrialization and urbanization. By 2050, it is anticipated that heavy-metal contamination might impact up to 25% of worldwide arable land (Mura *et al.*, 2014). These data underline the critical need for developing and implementing solutions to reduce the consequences of diverse abiotic stressors on crop yield (Sonkaria *et al.*, 2012). This involves breeding stress-tolerant crop types, upgrading soil and water management methods, adopting climate-smart agricultural techniques, and creating efficient remediation measures for polluted soils (Nile *et al.*, 2020). Additionally, the market for soil remediation is predicted to reach $44.5 billion by 2026, expanding at a CAGR of 9.7% from 2021 (King *et al.*, 2018), indicating the increased awareness of soil health in sustainable agriculture (Masters, 2005).

7.6.2 Biotic stresses (pests, diseases) affecting crop productivity

Stresses, induced by pests, diseases, and weeds constitute a serious threat to worldwide agricultural yield under biotic stresses (Cisneros *et al.*, 2023). These stressors not only affect yields but also impair the quality of agricultural goods, leading to considerable economic losses and food security issues globally. Insect pests are one of the major biotic stressors affecting crops globally (Demirer *et al.*, 2019). They inflict harm by direct feeding, spread of diseases, and loss in crop quality (Vural *et al.*, 2018). Insect pests are responsible for ruining roughly 18–20% of worldwide agricultural yield annually (Vinayaka and Thakur, 2010). The economic impact of agricultural losses due to insect pests is estimated at $470 billion each year (Vural *et al.*, 2018). In Africa, autumn armyworm infestations have damaged about 44 million hectares of maize, potentially incurring production losses of 8.3–20.6 million tons per year (Raliya *et al.*, 2018). In India, the cotton bollworm causes yearly losses of around $300 million. With rising global temperatures, insect pest damage is anticipated to rise by 10–25% per degree Celsius of warming for wheat, rice, and maize (de Oliveira *et al.*, 2014). Plant pathogens, including fungi, bacteria, viruses, and nematodes, cause different illnesses that adversely impair crop output. Plant diseases are responsible for losses of 10–16% in worldwide agricultural output, equal to roughly $220 billion yearly (de Oliveira *et al.*, 2014). The economic cost of plant pathogens in the USA alone is estimated at $33 billion each year (Kah *et al.*, 2018). Wheat rust infections afflict around 29 million hectares globally, producing yearly losses of about 15 million tons (Amenta *et al.*, 2015). Rice blast disease impacts 85 nations and produces losses sufficient to feed 60 million people yearly (Rizi *et al.*, 2022), and global banana output, possibly influencing 17 million tons of yearly production (Kumar *et al.*, 2021). Weeds compete with crops for resources such as water, nutrients, and light, greatly reducing agricultural output. Weeds cause an estimated 34% loss in crop output globally (Ullah *et al.*, 2023b). The economic impact of weeds on world agriculture is estimated at $95 billion yearly (Pudake *et al.*, 2024). In the USA, agricultural losses due to weeds are estimated at $33 billion yearly (Wang *et al.*, 2016). In Australia, the cost of weeds to Australian agriculture is estimated at AUD 4.8 billion yearly. Globally, 263 weed species have gained resistance to 167 different herbicides, creating considerable problems for weed control (Mitter *et al.*, 2017).

7.6.3 Combined and interacting biotic stresses

Biotic problems sometimes occur concurrently or in combination with abiotic forces, exacerbating their effect on crop productivity. The combination of drought stress and insect herbivory can damage crop production by up to 70% more than either stress alone (Ullah *et al.*, 2023b). Plant viruses carried by insect vectors can cause output losses of up to 80% in sensitive crop types (Pudake *et al.*, 2024). Total crop losses attributable to biotic stresses (pests, diseases, and weeds) are estimated at 20–40% of worldwide agricultural production (Wang *et al.*, 2016). The worldwide economic impact of invading insects alone is estimated at a minimum of $70 billion per year (Amenta *et al.*, 2015). Climate change is anticipated to impact the distribution and severity of biotic stresses, potentially resulting in yield losses of 6–10% per degree Celsius increase in world mean temperature (Pudake *et al.*, 2024). The worldwide agricultural biologicals market, which includes biopesticides and biocontrol agents, is anticipated to reach $18.9 billion by 2025, expanding at a CAGR of 13.6% from 2020 (Naaz *et al.*, 2023). The market for genetically modified seeds, many of which are engineered to withstand pests and diseases, is anticipated to reach $30.24 billion by 2026, rising at a CAGR of 6.5% from 2021 (Ullah *et al.*, 2023c).

7.7 Role of Nanomaterials in Enhancing Plant Growth, Nutrient Uptake, and Stress Tolerance

The introduction of nanomaterials in agriculture has opened new options for optimizing plant growth, improving nutrient absorption efficiency, and increasing stress tolerance (82). This chapter discusses the various functions of nanomaterials in these key elements of plant physiology and agriculture (Ahmed *et al.*, 2023). Nanomaterials have demonstrated extraordinary potential in boosting plant development through multiple processes. Nanoparticles, particularly metal and metal oxide nanoparticles, have been demonstrated to boost seed germination rates and stimulate early seedling development (Surendran *et al.*, 2024). For instance, zinc oxide (ZnO) nanoparticles have been observed to boost germination percentage and seedling vigor in peanuts (Choudhury and Moulick, 2022). The method involves increased water intake and activation of enzymes involved in seed germination. Some nanomaterials, such as carbon nanotubes (CNTs), have been shown to boost photosynthetic activity. Naaz *et al.* (2023) revealed that single-walled CNTs may enter chloroplast membranes and stimulate electron transport, possibly enhancing photosynthetic efficiency by up to three times. Nanoparticles can increase root growth and development (Sukhpal *et al.*, 2018). For example, titanium dioxide (TiO2) nanoparticles have been demonstrated to stimulate root growth in *Arabidopsis thaliana* by regulating root hair production and lateral root formation (Custodio *et al.*, 20230). Nanomaterials have a crucial role in increasing nutrient absorption and utilization in plants: nanoparticle-based fertilizers have shown greater nutrient-usage efficiency compared to traditional fertilizers. For instance, urea-coated hydroxyapatite nanoparticles displayed slower release and greater nitrogen-usage efficiency in wheat (Amenta *et al.*, 2015). Nanomaterials can effectively provide micronutrients to plants. Zinc oxide nanoparticles, for example, have exhibited better zinc bioavailability compared to standard zinc sulfate fertilizers in mung bean plants (de Oliveira *et al.*, 2014).

Some nanoparticles can function as transporters for nutrients, boosting their absorption and transfer inside plants. Amenta *et al.* (2015) demonstrated that mesoporous silica nanoparticles may efficiently transfer micronutrients including copper and zinc to wheat plants, increasing their nutritional status. Nanomaterials have shown potential in enhancing plant tolerance to several biotic and abiotic stresses: certain nanoparticles can boost plant drought resistance. For instance, silica nanoparticles have been demonstrated to increase water-usage efficiency and drought tolerance in tomato plants by modifying the antioxidant defense system (Amenta *et al.*, 2015). Nanomaterials can alleviate the effects of salt stress on plants. Hakim *et al.* (2021) found that silicon nanoparticles might reduce salt stress in strawberry plants by boosting antioxidant enzyme activities and lowering sodium intake. Some nanomaterials can improve plant tolerance to temperature extremes. For example, cerium oxide nanoparticles have been demonstrated to increase heat-stress tolerance in *Arabidopsis thaliana* by lowering oxidative stress (Singla, 2020). Nanomaterials, particularly metal nanoparticles, have shown potential in improving plant resistance against diseases. Silver nanoparticles, for instance, have proven antifungal effects against numerous plant diseases (Rizi *et al.*, 2022).

7.7.1 Mechanisms of action

The favorable benefits of nanomaterials on plant development, nutrient absorption, and stress tolerance can be linked to many mechanisms:

- enhanced nutrient solubility and bioavailability: nanoparticles can improve the solubility of nutrients, making them more accessible for plant absorption (Kumar *et al.*, 2021);
- modulation of gene expression: certain nanomaterials can alter the expression of genes related to plant development, nutrient absorption, and stress responses (Ullah *et al.*, 2023b);
- antioxidant activity: many nanomaterials contain antioxidant capabilities, helping plants resist oxidative damage under varied environmental stressors (Pudake *et al.*, 2024);
- hormonal regulation: some nanoparticles can alter plant hormone levels, consequently influencing growth and development (Rizwan *et al.*, 2017); and
- improved water retention: certain nanomaterials, including hydrogel nanocomposites, can promote soil-water retention, thereby benefitting plant development and stress tolerance (de Oliveira *et al.*, 2014).

While nanomaterials offer considerable potential in boosting plant growth, nutrient absorption, and stress tolerance, numerous difficulties need to be solved. The long-term implications of nanomaterials on plant health, soil ecosystems, and food safety deserve additional exploration (Amenta *et al.*, 2015). Determining optimum doses for various nanomaterials and crop species is critical since high concentrations can contribute to phytotoxicity (Wang *et al.*, 2016). Scaling up the manufacturing of nanomaterials for agricultural uses while preserving cost-effectiveness remains a problem. Developing proper laws for the use of nanomaterials in agriculture is vital to assure safety and public acceptability (Cisneros *et al.*, 2023). Nanomaterials provide enormous promise for promoting plant growth, improving nutrient absorption efficiency, and increasing stress tolerance in crops. As research in this sector continues, we may expect to see increasingly sophisticated and focussed uses of nanomaterials in agriculture (Sozer and Kokini, 2009). The combination of nanotechnology with other developing technologies, such as genetic engineering and precision agriculture, might lead to the production of extremely robust and productive crop types capable of solving the problems of global food security in a changing environment (Abbas *et al.*, 2009).

7.7.2 Stress perception and signal transduction

Plants detect abiotic stress through multiple sensors and receptors. These include membrane-bound receptors such as histidine kinases and G-protein coupled receptors (Abbas *et al.*, 2009). Once a stress signal is recognized, it begins a cascade of signaling processes including supplementary messengers such as calcium ions (Ca^{2+}), ROS, and phytohormones like abscisic acid (ABA). These chemical messengers activate pathways of signaling, especially the mitogen-activated protein kinase (MAPK pathways, which regulate the expression of genes that respond to stress) (Kumar *et al.*, 2020).

7.7.3 Transcriptional regulation

Transcription factors (TFs) are essential controllers of the expression of genes in response to stress:

- DREB (dehydration-responsive element binding) proteins bind to dehydration-responsive elements in the promoters of stress-responsive genes, boosting their expression (Kumar *et al.*, 2020).
- NAC (NAM, ATAF1/2, and CUC2): these TFs are control genes involved in osmotic balance, antioxidant defense, and growth regulation (Singh *et al.*, 2017).
- WRKY Proteins: these TFs control the expression of genes related to pathogen defense, secondary metabolism, and stress responses (Ummi and Siddiquee, 2019).
- Phytohormones such as ABA, ethylene, salicylic acid (SA), and jasmonic acid (JA) play essential roles in modulating stress responses. ABA is crucial to drought and

salinity-stress responses, prompting stomatal closure and the activation of stress-protective genes (Ummi and Siddiquee, 2019).

- Ethylene controls growth and senescence, while SA and JA are implicated in biotic and abiotic stress signaling (Berekaa, 2015).
- Osmolytes such as proline, glycine betaine, and trehalose build in response to osmotic stress, helping to maintain cellular osmotic equilibrium, stabilize proteins and membranes, and scavenge ROS (Sonkaria *et al.*, 2012).

Abiotic stressors typically lead to the overproduction of ROS, which can harm biological components. Plants boost their antioxidant defense mechanisms, including enzymes like superoxide dismutase (SOD), catalase (CAT), and peroxidases, to detoxify ROS and minimize oxidative damage (Nile *et al.*, 2020).

7.7.4 Targeted delivery systems for stress mitigation

7.7.4.1 Nanotechnology-based delivery systems

Nanotechnology enables accuracy and efficiency in delivering stress-mitigating chemicals directly to specific areas within plants (Kumar *et al.*, 2022):

- Nanoparticles: engineered nanoparticles can contain and distribute bioactive chemicals such as TFs, hormones, and antioxidants. For instance, silica nanoparticles can transport ABA to stomatal cells, improving drought resistance (Kumar *et al.*, 2022).
- Nanocarriers: lipid-based nanocarriers and polymeric nanoparticles promote the regulated release of stress-mitigating chemicals, enhancing their stability and bioavailability (Savary *et al.*, 2012).
- CRISPR/Cas9 and other gene-editing technologies allow precise modulation of stress-responsive genes.
- Traditional delivery techniques, like as foliar sprays and soil additives, are being refined for higher efficiency.
- Foliar sprays: these are the direct application of hormones, osmo-protectants, and nutritional solutions onto leaves. Techniques like electrostatic spraying enable uniform coverage and greater absorption (Deutsch *et al.*, 2018).
- Soil amendments: incorporating biochar, organic matter, and microbial inoculants into soil enhances water retention, nutritional availability, and stress tolerance (Ryan *et al.*, 2023).
- Microbial inoculants: beneficial microorganisms, such as mycorrhizal fungi and plant growth-promoting rhizobacteria (PGPR), boost plant stress tolerance by enhancing nutrient absorption, generating stress-related hormones, and inducing systemic resistance (Kumar *et al.*, 2022).

7.7.5 Role of engineered nanomaterials in stress resilience

Engineered nanomaterials (ENMs) are pivotal in the agri-tech revolution, addressing significant economic losses in agriculture caused by environmental conditions such as heat, drought, frost, and salinity. Climate change exacerbates these challenges by intensifying the frequency and severity of stressors that threaten sustainable crop production (Chandra *et al.*, 2021). To mitigate these effects, various nanomaterials have been employed to enhance plant growth and photosynthetic efficiency (Bamal *et al.*, 2021; Liu *et al.*, 2023). For instance, it has been demonstrated that cerium oxide nanoparticles increase photosynthetic efficiency by lowering the build-up of reactive oxygen species during stressful situations (Rizi *et al.*, 2022; Custodio *et al.*, 2023]. The nanoscale size of these compounds (< 100 nm) allows them to overcome biological barriers, enabling efficient delivery of insecticides and nutrients via foliar or root treatments (Handford *et al.*, 2015; Wu *et al.*, 2017; Rizwan *et al.*, 2017). Several ENMs function as nonviral nanocarriers with outstanding capacities for gene transport, alongside polymers, liposomes, and inorganic nanoparticles (Kahru and Ivask, 2013; Mozafari *et al.*, 2018; Kah *et al.*, 2019). ENMs provide a wide range of uses, including operating as nanobiosensors, coating materials for seeds, nanofertilizers to promote soil microbiota and health, nanoinsecticides for

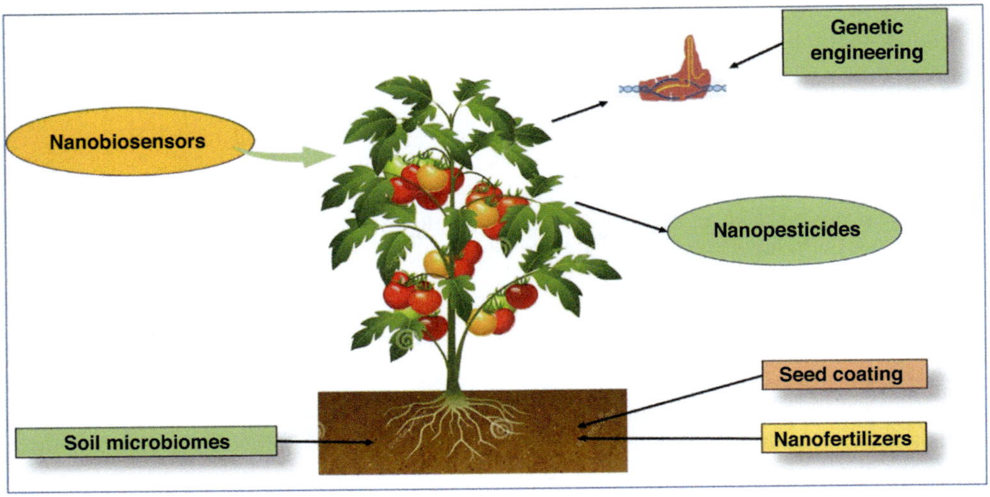

Fig. 7.4. Potential uses of ENMs in sustainable agriculture for boosting stress resistance of agricultural plants.

effective pathogen control, and nanocarriers to assist genetic engineering in crops (Fig. 7.4).

Optimizing dosage is a crucial challenge in mitigating the harmful effects of engineered nanoparticles in agriculture. Concentration-based studies conducted in natural environments are essential to determine the precise and nontoxic levels of these nanomaterials. One promising agro-product gaining attention is a polymeric ENM coated with a plant-based insecticide, noted for its nontoxic properties (Sukhpal *et al.*, 2018).

7.7.6 Enhancement of crop stress tolerance by nanomaterial-mediated techniques

The automation of agricultural operations has proved advantageous by guaranteeing excellent crop management and environmental sustainability with minimum human involvement (Wani, 2023). Plant nanosensors provide farmers with a tool to engage directly with their crops, boosting plant protection and sustainability (Vessey, 2003; Rajpal *et al.*, 2024). These integrated nanoparticles can detect signaling molecules associated with stress and nutritional deficits, transforming them into electrical or optical signals for analysis (Ijaz *et al.*, 2023b; Kumar *et al.*, 2022; Shaikh *et al.*, 2024).

7.8 Integration of Nanosensors and Nanodevices for Real-time Monitoring of Environmental Variables

Examples of nano-based technologies that are used for the automated evaluation of plant nutritional balance, stress levels, and general health include nanobiosensors and global positioning systems (GPS). These technologies assist the effective administration of numerous agricultural operations with the minimum requirement for manual inspection (Ijaz *et al.*, 2023b; Chakraborty *et al.*, 2022). Real-time monitoring of plant health enables efficient resource management, averting large losses and boosting plant survival and development (Savary *et al.*, 2012). This breakthrough promotes sustainable agriculture by enhancing control over resource management, soil health, crop development through genetic engineering, and the precise use of agrochemicals such as fertilizers (Deutsch *et al.*, 2018). Small sensors are particularly useful for detecting and monitoring pathogen infestations, as well as measuring a wide variety of parameters, including diseases, pesticides, herbicides, soil pH, moisture, and more (Ryan *et al.*, 2023). They also measure rhizosphere soil health by detecting gases, trace heavy metals, and micronutrients. The data produced by these sensors help prevent crops from illnesses and enhance output by giving exact nutrients to the soil (Chaffai *et al.*, 2024; Pudake *et al.*, 2024). The successful application of nanosensors can result in improved agricultural yields, helping farmers and increasing precision farming by developing old processes into smarter, more cost-efficient, and sustainable ways (Ullah *et al.*, 2024).

7.9 Environmental and Regulatory Considerations

Although there is considerable variability in the evidence regarding the eco-toxicological effects of nanoparticles, most scientific studies suggest mild to moderate toxicity in terrestrial plants (Savary *et al.*, 2012; Kumar *et al.*, 2022). Exposure to metallic nanoparticles can lead to oxidative damage in plants by generating ROS and activating antioxidant defense mechanisms [142]. Different nanomaterials activate various antioxidant enzymes: Fe_2O_4, CeO_2, and Co_3O_4 stimulate catalase production; Fe_3O_4, CeO_2, MnO_2, CuO, and Au boost guaiacol peroxidase levels; and CeO_2, Pt, and fullerene promote superoxide dismutase production (Lal, 2016). As ENMs are increasingly used and concentrated in various fields, they present significant environmental challenges and have become among the most difficult pollutants to manage (Ryan *et al.*, 2023). The toxicity of a nanoparticle is influenced by its solubility and its ability to specifically bind to biological targets (World Health Organisation, 2023).

7.10 Potential Environmental Implications of Nanomaterials in Agriculture

It is crucial to examine the bioaccumulation of ENMs in food chains and their interactions with other environmental pollutants, as they could significantly affect vital plant processes and threaten agricultural sustainability (World Health Organisation, 2023). The antibacterial properties of metallic ENMs could trigger a cascade of unknown changes in microbial populations in the environment (Lal, 2016), potentially harming both plants and humans. These nanoparticles may enter the food chain in various scenarios, leading to unforeseen consequences. For instance, the introduction of metal nanoparticles such as ZnO, CuO, and TiO into agricultural soil can alter the composition of microbial communities (Lal, 2016).

7.10.1 Limitations and limitations in applying nanotechnology in agriculture

Proper monitoring of the introduction of both chemical and green ENMs into agricultural fields is crucial. ENMs should be utilized in agriculture only after confirming their safety to ensure they improve yields and effectively address key challenges (Alexandratos and Bruinsma, 2012). One example is a polymeric ENM with a plant-based insecticide coating, which has become popular for its nontoxic nature (Molotoks *et al.*, 2021). Furthermore, it is crucial to understand the effects of ENMs on soil microorganisms to avoid adverse impacts on soil health and ecosystems (da Cunha Dias *et al.*, 2021). ENMs can have a considerable effect on ecosystem functioning, as their presence in treated soil may alter soil microbiota and affect related trophic species. Despite their potential, several challenges hinder the widespread adoption of ENM-based agricultural technologies (Sadigov, 2022). Moreover, the adoption of new technologies depends heavily on their practicality and cost-effectiveness, which is challenging in the low-margin agriculture industry (Hunter *et al.*, 2017).

7.11 Conclusion

ENMs and nanobiotechnology offer a promising pathway to enhance agricultural resilience against biotic and abiotic stresses, addressing key challenges in modern farming. ENMs such as nanofertilizers, nanopesticides, and nanoscale nutrient delivery systems have shown significant potential to improve plant growth, nutrient uptake efficiency, and stress tolerance by modulating physiological and biochemical pathways at the cellular level. Nanofertilizers enhance nutrient availability and reduce leaching losses, while nanopesticides provide targeted pest control, minimizing environmental contamination and nontarget effects. Additionally, nanosensors integrated into agricultural systems enable real-time monitoring of soil and plant health, facilitating precision agriculture by allowing timely detection of stress signals and nutrient deficiencies. Nanobiotechnology further enhances agricultural productivity through the development of gene delivery systems and nanosystems that improve plant genetic resilience against environmental extremes such as drought, salinity, and extreme temperatures. However, despite the potential benefits, the environmental and health implications of widespread ENM use remain under-researched. Concerns regarding nanoparticle accumulation, ecotoxicity, and potential impacts on human health necessitate the development of comprehensive risk assessment protocols and regulatory guidelines to ensure the safe deployment of these technologies. Future research should focus on elucidating the long-term interactions of nanomaterials with agricultural ecosystems, improving the biocompatibility and efficiency of ENMs, and scaling up nanotechnology applications for practical field use. By integrating ENMs and nanobiotechnology with conventional agricultural practices, a sustainable, smart, and stress-resilient agricultural system can be realized, contributing significantly to global food security and environmental sustainabilityUllah *et al.*, 2023a.

References

Abbas, K.A., Saleh, A.M., Mohamed, A. and MohdAzhan, N. (2009) The recent advances in the nanotechnology and its applications in food processing: A review. *Journal of Food Agriculture and Environment* 7(3–4), 14–17.

Ahmad, Y., Haider, S., Iqbal, J., Abbasi, B.A., Yaseen, T. *et al.* (2022) The mechanisms of genome editing technologies in crop plants. In: Prakash, C.S., Fiaz, S. and Fahad, S. (eds) *Principles and Practices of OMICS and Genome Editing for Crop Improvement*. Springer, Cham, Switzerland, pp. 295–313.

Ahmed, T., Luo, J., Noman, M., Ijaz, M., Wang, X. *et al.* (2023) Microbe-mediated nanoparticle intervention for the management of plant diseases. *Crop Health* 1(1), 3.

Alexandratos, N. and Bruinsma, J. (2012) *World agriculture towards 2030/2050: The 2012 revision*. ESA Working Paper no.12-03. FAO, Rome.

Ali, S., Mir, R.A., Tyagi, A., Manzar, N., Kashyap, A.S. *et al.* (2023a) Chromium toxicity in plants: Signaling mitigation, and future perspectives. *Plants* 12, 1502.

Ali, S., Tyagi, A., Park, S. and Bae, H. (2023b) Understanding the mechanobiology of phytoacoustics through molecular lens: Mechanisms and future perspectives. *Journal of Advanced Research* 65, 47–72.

Amenta, V., Aschberger, K., Arena, M., Bouwmeester, H., Botelho Moniz, F. *et al.* (2015) Regulatory aspects of nanotechnology in the agri/feed/food sector in EU and non-EU countries. *Regulatory Toxicology and Pharmacology* 73(1), 463–476.

Ashraf, M. and Foolad, M.R. (2007) Roles of glycine betaine and proline in improving plant abiotic stress resistance. *Environmental and Experimental Botany* 59(2), 206–216.

Asma, J., Subrahmanyam, D. and Krishnaveni, D. (2023) The global lifeline: A staple crop sustaining two thirds of the world's population. *Agriculture Archives*.

Bamal, D., Singh, A., Chaudhary, G., Kumar, M., Singh, M. *et al.* (2021) Silver nanoparticles biosynthesis, characterization, antimicrobial activities, applications, cytotoxicity and safety issues: An updated review. *Nanomaterials* 11(8), 2086.

Berekaa, M.M. (2015) Nanotechnology in food industry: Advances in food processing, packaging and food safety. *International Journal of Current Microbiology and Applied Sciences* 4(5), 345–357.

Bibi, S., Ullah, R., Burni, T., Ullah, Z. and Kazi, M. (2023) Impact of resorcinol and biochar application on the growth attributes, metabolite contents, and antioxidant systems of tomato (*Lycopersicon esculentum* Mill.). *ACS Omega* 8(48), 45750–45762.

Bora, S.S., Anshu, Deka, P. and Barooah, M. (2023) Drought-tolerant plant growth-promoting rhizobacteria associated with millets. In: Sharma, A.K., Pudake, R.N., Kumari, M. and Sapkal, D.R. (eds) *Millet Rhizosphere*. Springer, Singapore, pp. 83–96.

Bortesi, L. and Fischer, R. (2015) The CRISPR/Cas9 system for plant genome editing and beyond. *Biotechnology Advances* 33(1), 41–52.

Bouri, M., Taieb, K.H., Bolouri, P., Rhouma, A. and Şahin, F. (2024) Microbial biostimulants for improving crop nutrition, health, and productivity in climate-smart agriculture. In: Sayyed, R.Z. and Ilyas, N. (eds) *Plant Holobiome Engineering for Climate-Smart Agriculture*. Springer, pp. 235–255.

Cabrera-Barjas, G., Banerjee, A., Valdes, O., Moncada, M., Sirajunnisa, A.R. *et al.* (2022) Food biotechnology: Innovations and challenges. In: Bhat, R. (ed.) *Future Foods*. Academic Press, pp. 697–719.

Chaffai, R., Ganesan, M. and Cherif, A. (2024) Nanoparticles for sustainable agriculture and abiotic stress resistance. In: Ganesan, M. and Chaffai, R. (eds) *Plant Adaptation to Abiotic Stress: From Signaling Pathways and Microbiomes to Molecular Mechanisms*. Springer, pp. 195–208.

Chakraborty, S., Singh, A. and Roychoudhury, A. (2022) Biogenic nanoparticles and generation of abiotic stress-resilient plants: A new approach for sustainable agriculture. *Plant Stress* 6, 100117.

Chandra, A.K., Pandey, D., Tiwari, A., Gururani, K., Agarwal, A. *et al.* (2021) Metal based nanoparticles trigger the differential expression of key regulatory genes which regulate iron and zinc homeostasis mechanism in finger millet. *Journal of Cereal Science* 100, 103235.

Chaudhry, Q., Scotter, M., Blackburn, J., Ross, B., Boxall, A. *et al.* (2008) Applications and implications of nanotechnologies for the food sector. *Food Additives and Contaminants* 25(3), 241–258.

Choudhury, S. and Moulick, D. (2022) *Response of Field Crops to Abiotic Stress: Current Status and Future Prospects*. CRC Press, Boca Raton, Florida.

Cisneros, A., Martín-García, T., Primc, A., Orlando, F., Gomez, J.J.L. *et al.* (2023) Transgene-free, virus-based gene silencing in plants by artificial small RNAs derived from minimal precursors. *Nucleic Acids Research* 51, 19.

Cunningham, F.J., Goh, N.S., Demirer, G.S., Matos, J.L. and Landry, M.P. (2018) Nanoparticle-mediated delivery towards advancing plant genetic engineering. *Trends in Biotechnology* 36(9), 882–897.

Custodio, J.M., Concepcion, R., Roche, J. and Bandala, A. (2023) *In silico* root trait phenotyping for maize in c-k deficit and toxic conditions using convolutional neural networks. In: *2023 IEEE 15th International Conference on Humanoid, Nanotechnology, Information Technology, Communication and Control, Environment, and Management (HNICEM)*, IEEE, Coron, Palawan, Philippines, pp. 1–6.

Cutler, S.R., Rodriguez, P.L., Finkelstein, R.R. and Abrams, S.R. (2010) Abscisic acid: The emergence of a core signaling network. *Annual Review of Plant Biology* 61, 651–679.

Cutolo, E.A., Caferri, R., Campitiello, R. and Cutolo, M. (2023) The clinical promise of microalgae in rheumatoid arthritis: From natural compounds to recombinant therapeutics. *Marine Drugs* 21(12), 630.

da Cunha Dias, T.A., da, C., Lora, E.E.S., Maya, D.M.Y. and del Olmo, O.A. (2021) Global potential assessment of available land for bioenergy projects in 2050 within food security limits. *Land Use Policy* 105, 105346.

da Cunha, M.S., Rocha, J.A., da, S., Machado, A.L.S., Martins, G.B. *et al.* (2023) Applications and goals of bioeconomy: A synthesis of the scientific literature. *Revista de Gestão e Secretariado (Management and Administrative Professional Review)* 14(10), 16983–17009.

de Oliveira, J.L., Campos, E.V.R., Bakshi, M., Abhilash, P.C. and Fraceto, L.F. (2014) Application of nanotechnology for the encapsulation of botanical insecticides for sustainable agriculture: Prospects and promises. *Biotechnology Advances* 32(8), 1550–1561.

Demirer, G.S., Zhang, H., Matos, J.L., Goh, N.S., Cunningham, F.J. *et al.* (2019) High aspect ratio nanomaterials enable delivery of functional genetic material without DNA integration in mature plants. *Nature Nanotechnology* 14(5), 456–464.

Deutsch, C.A., Tewksbury, J.J., Tigchelaar, M., Battisti, D.S., Merrill, S.C. *et al.* (2018) Increase in crop losses to insect pests in a warming climate. *Science* 361(6405), 916–919.

Dimkpa, C.O. and Bindraban, P.S. (2018) Nanofertilizers: New products for the industry? *Journal of Agricultural and Food Chemistry* 66(26), 6462–6473.

Elmer, W.H. and White, J.C. (2018) The future of nanotechnology in plant pathology. *Annual Review of Phytopathology* 56(1), 111–133.

Fehér, B., Kozma-Bognár, L., Kevei, E., Hajdu, A., Binkert, M. *et al.* (2011) Functional interaction of the circadian clock and UV RESISTANCE LOCUS 8-controlled UV-B signaling pathways in *Arabidopsis thaliana*. *The Plant Journal* 67(1), 37–48.

Fen, L.B., Rashid, A.H.A., Nordin, N.I., Hossain, M.M., Uddin, S.M.K. *et al.* (2022) Applications of nanomaterials in agriculture and their safety aspect. In: Thangadurai, D., Islam, S., Sangeetha, J. and Cruz-Martins, N. (eds) *Biogenic Nanomaterials*. Apple Academic Press, pp. 243–299.

Fernández, V., Eichert, T., Del Río, V., López-Casado, G., Heredia-Guerrero, J.A. *et al.* (2013) Leaf structural changes associated with iron deficiency chlorosis in field-grown pear and peach: Physiological implications. *Plant and Soil* 359(1–2), 149–164.

Fraceto, L.F., Grillo, R., de Medeiros, G.A., Scognamiglio, V., Rea, G, *et al.* (2016) Nanotechnology in agriculture: Which innovation potential does it have? *Frontiers in Environmental Science* 4, 20.

Ghazi Hossaini, S.M., Sanders, J.P. and Huizen, R. (2021) Nanomaterials: An emerging tool for enhancing crop production. *Frontiers in Plant Science* 12, 758621.

Gill, R., Kalwan, G., Nehra, A., Shah, A.H., Tuteja, N. *et al.* (2024) Conclusion and future perspective on the role of nanotechnology in abiotic stress tolerance in global agriculture. In: Pudake, R.N., Tripathi, R.M. and Gill, S.S. (eds) *Nanotechnology for Abiotic Stress Tolerance and Management in Crop Plants*. Academic Press, pp. 335–342.

Giraldo, J.P., Landry, M.P., Faltermeier, S.M., McNicholas, T.P., Iverson, N.M. *et al.* (2014) Plant nanobionics approach to augment photosynthesis and biochemical sensing. *Nature Materials* 13(4), 400–408.

Gondalia, N., Vashi, R., Barot, V., Sharma, F., Anishkumar, P.K. *et al.* (2022) Genomic designing for abiotic stress tolerance in pea (*Pisum sativum* L.). In: Kole, C. (ed.) *Genomic Designing for Abiotic Stress Resistant Pulse Crops*. Springer, Cham, Switzerland, pp. 45–113.

Govaichelvan, K.N., Pathmanathan, D., Zainal-Abidin, R. and Abu, A. (2024) Machine learning for major food crops breeding: Applications, challenges, and ways forward. *Agronomy Journal* 116(3), 1112–1125.

Guilherme, M.R., Aouada, F.A., Fajardo, A.R., Martins, A.F., Paulino, A.T. *et al.* (2015) Superabsorbent hydrogels based on polysaccharides for application in agriculture as soil conditioner and nutrient carrier: A review. *European Polymer Journal* 72, 365–385.

Hakim, S., Naqqash, T., Nawaz, M.S., Laraib, I., Siddique, M.J. *et al.* (2021) Rhizosphere engineering with plant growth-promoting microorganisms for agriculture and ecological sustainability. *Frontiers in Sustainable Food Systems* 5, 617157.

Handford, C.E., Dean, M., Spence, M., Henchion, M., Elliott, C.T. *et al.* (2015) Awareness and attitudes towards the emerging use of nanotechnology in the agri-food sector. *Food Control* 57, 24–34.

Hatami, M., Hadian, J. and Ghorbanpour, M. (2017) Mechanisms underlying toxicity and stimulatory role of single-walled carbon nanotubes in *Hyoscyamus niger* during drought stress simulated by polyethylene glycol. *Journal of Hazardous Materials* 324, 306–320.

Hunter, M.C., Smith, R.G., Schipanski, M.E., Atwood, L.W. and Mortensen, D.A. (2017) Agriculture in 2050: Recalibrating targets for sustainable intensification. *BioScience* 67(4), 386–391.

Ijaz, S., Iqbal, J., Abbasi, B.A., Ullah, Z., Yaseen, T. *et al.* (2023a) Rosmarinic acid and its derivatives: Current insights on anticancer potential and other biomedical applications. *Biomedicine & Pharmacotherapy* 162, 114687.

Ijaz, M., Khan, F., Ahmed, T., Noman, M., Zulfiqar, F. *et al.* (2023b) Nanobiotechnology to advance stress resilience in plants: Current opportunities and challenges. *Materials Today. Bio* 22, 100759.

Ijaz, S., Iqbal, J., Abbasi, B.A., Tufail, A., Ullah, Z. *et al.* (2023c) Biofortification: Lessons from the past and strategies for future food security. In: Nadeem, M.A., Baloch, F.S., Fiaz, S., Aasim, M., Habyarimana, E. *et al.* (eds) *Legumes Biofortification*. Springer International Publishing, Cham, pp. 521–545.

Ijaz, S., Iqbal, J., Abbasi, B.A., Tufail, A., Ullah, Z. *et al.* (2024) Heavy metal-polluted arable land and its consequences: A global scenario. In: Nawaz, T., Zhou, R., Saud, S., Adnan, M., and Fahad, S. (eds) *Biochar-Assisted Remediation of Contaminated Soils Under Changing Climate*. Elsevier, pp. 71–99.

Irfan, M. and Kumar, P. (eds) (2024) *Bioengineered Fruit and Vegetables: Trends in Postharvest Quality Improvement*. CRC Press, Boca Raton, Florida.

Janbandhu, M.S., Mehta, A., Beese, S., Pandey, S.K., Singh, B. *et al.* (2024) Advances and emerging trends in horticultural production and management. *Journal of Experimental Agriculture International* 46(3), 47–69.

Judy, J.D. and Bertsch, P.M. (2014) Bioavailability, toxicity, and fate of manufactured nanomaterials in terrestrial ecosystems. *Advances in Agronomy* 123, 1–64.

Kah, M., Beulke, S., Tiede, K. and Hofmann, T. (2013) Nanopesticides: State of knowledge, environmental fate, and exposure modeling. *Critical Reviews in Environmental Science and Technology* 43(16), 1823–1867.

Kah, M., Kookana, R.S., Gogos, A. and Bucheli, T.D. (2018) A critical evaluation of nanopesticides and nanofertilizers against their conventional analogues. *Nature Nanotechnology* 13(8), 677–684.

Kah, M., Tufenkji, N. and White, J.C. (2019) Nano-enabled strategies to enhance crop nutrition and protection. *Nature Nanotechnology* 14(6), 532–540.

Kahru, A. and Ivask, A. (2013) Mapping the dawn of nanoecotoxicological research. *Accounts of Chemical Research* 46(3), 823–833.

Karny, A., Zinger, A., Kajal, A., Shainsky-Roitman, J. and Schroeder, A. (2018) Therapeutic nanoparticles penetrate leaves and deliver nutrients to agricultural crops. *Scientific Reports* 8(1), 7589.

Karumamkandathil, R., Jayasree, P.K., Radha, J., Uthup, T.K., Mathew, S.A. *et al.* (2022) Genetics and genomics of abiotic stress in rubber tree (*Hevea brasiliensis*). In: Kole, C. (ed.) *Genomic Designing for Abiotic Stress Resistant Technical Crops*. Springer, Cham, Switzerland, pp. 245–298.

Khan, S.A.R., Yu, Z., Panait, M., Janjua, L.R. and Shah, A. (eds) (2021) *Global Corporate Social Responsibility Initiatives for Reluctant Businesses*. IGI Global.

Khan, N., Choi, S.H., Lee, C.H., Qu, M. and Jeon, J.S. (2024) Photosynthesis: Genetic strategies adopted to gain higher efficiency. *International Journal of Molecular Sciences* 25(16), 8933.

King, T., Osmond-McLeod, M.J. and Duffy, L.L. (2018) Nanotechnology in the food sector and potential applications for the poultry industry. *Trends in Food Science & Technology* 72, 62–73.

Kole, C. (ed.) (2022) *Genomic Designing for Biotic Stress Resistant Pulse Crops*. Springer.

Kottegoda, N., Sandaruwan, C., Priyadarshana, G., Siriwardhana, A., Rathnayake, U.A. *et al.* (2017) Urea-hydroxyapatite nanohybrids for slow release of nitrogen. *ACS Nano* 11(2), 1214–1221.

Kumar, P., Mahajan, P., Kaur, R. and Gautam, S. (2020) Nanotechnology and its challenges in the food sector: A review. *Materials Today Chemistry* 17, 100332.

Kumar, J., Mir, R.R., Shafi, S., Sen Gupta, D., Djalovic, I. *et al.* (2021) Genomics associated interventions for heat stress tolerance in cool season adapted grain legumes. *International Journal of Molecular Sciences* 23(1), 399.

Kumar, L., Chhogyel, N., Gopalakrishnan, T., Hasan, M.K., Jayasinghe, S.L. *et al.* (2022) Climate change and future of agri-food production. In: Bhat, R. (ed.) *Future Foods*. Academic Press, pp. 49–79.

Lal, R. (2016) Feeding 11 billion on 0.5 billion hectare of area under cereal crops. *Food and Energy Security* 5(4), 239–251.

Liu, S., Zenda, T., Tian, Z. and Huang, Z. (2023) Metabolic pathways engineering for drought or/and heat tolerance in cereals. *Frontiers in Plant Science* 14, 1111875.

Mahanty, A., Mitra, T., Mohanty, I., Behera, L., Pati, S. *et al.* (2023) Nanotechnology: An answer for mitigating future challenges in aquaculture. In: Sarkar, B. and Sonawane, A. (eds) *Biological Applications of Nanoparticles*. Springer, pp. 149–161.

Masters, W.A. (2005) Paying for prosperity: How and why to invest in agricultural research and development in Africa. *Journal of International Affairs* 58(2), 35–64.

Meena, N.K., Koli, H.K. and Mondal, K. (2023) Genetic innovations in crop improvement. In: Choudhary, L., Reddy, K.V.N., Mondal, K., Huded, S. and S, M. (eds) *Modern Horizons in Agriculture*, Vol. 1. Stella International, pp. 184–208.

Meszaros, R.Z., Zakanyi, B. and Baracza, M.K. (2023) Applicability of mineral raw materials in the food industry and agriculture as nanomaterials: A review. *Bányászati És Kohászati Lapok-Kohászat* 156(1).

Mittler, R. (2017) ROS are good. *Trends in Plant Science* 22(1), 11–19.

Mitter, N., Worrall, E.A., Robinson, K.E., Li, P., Jain, R.G. *et al.* (2017) Clay nanosheets for topical delivery of RNAi for sustained protection against plant viruses. *Nature Plants* 3(2), 1–10.

Mogazy, A.M. (2024) The regulatory function of polyamines in plant abiotic stress. In: Abd-Elsalam, K.A. and Mohamed, H.I. (eds) *Plant Growth Regulators to Manage Biotic and Abiotic Stress in Agroecosystems*. CRC Press, Boca Raton, Florida, pp. 257–277.

Molotoks, A., Smith, P. and Dawson, T.P. (2021) Impacts of land use, population, and climate change on global food security. *Food and Energy Security* 10(1), e261.

Mozafari, A., Havas, F. and Ghaderi, N. (2018) Application of iron nanoparticles and salicylic acid in *in vitro* culture of strawberries (Fragaria × ananassa Duch.) to cope with drought stress. *Plant Cell, Tissue and Organ Culture (PCTOC)* 132(3), 511–523.

Mura, S., Carta, D., Roggero, P.P., Cheli, F. and Greppi, G.F. (2014) Nanotechnology and its applications in food and animal science. *Italian Journal of Food Science* 26(1), 91–102.

Naaz, S., Sachdev, S., Husain, R., Pandey, V. and Ansari, M.I. (2023) Nanomaterials and nanocomposites: Significant uses in plant performance, production, and toxicity response. In: Husen, A. (ed.)

Nanomaterials and Nanocomposites Exposures to Plants: Response, Interaction, Phytotoxicity and Defense Mechanisms. Springer, pp. 1–18.

Nair, R., Poulose, A.C., Nagaoka, Y., Yoshida, Y., Maekawa, T. *et al.* (2011) Uptake of FITC labeled silica nanoparticles and quantum dots by rice seedlings: Effects on seed germination and their potential as biolabels for plants. *Journal of Fluorescence* 21(6), 2057–2068.

Naveen, A.K. and Sontakke, M. (2024) A review on regulatory aspects, challenges and public perception in acceptance of genetically modified foods. *Food Science and Biotechnology* 33(4), 791–804.

Nile, S.H., Baskar, V., Selvaraj, D., Nile, A., Xiao, J. *et al.* (2020) Nanotechnologies in food science: Applications, recent trends, and future perspectives. *Nano-Micro Letters* 12(1), 45.

Phour, M. and Sindhu, S.S. (2023) Soil salinity and climate change: Microbiome-based strategies for mitigation of salt stress to sustainable agriculture. In: Parray, J.A. (ed.) *Climate Change and Microbiome Dynamics: Carbon Cycle Feedbacks*. Springer, Cham, Switzerland, pp. 191–243.

Prasad, R., Bhattacharyya, A. and Nguyen, Q.D. (2017) Nanotechnology in sustainable agriculture: Recent developments, challenges, and perspectives. *Frontiers in Microbiology* 8, 1014.

Pudake, R.N., Tripathi, R.M. and Gill, S.S. (eds) (2024) *Nanotechnology for Abiotic Stress Tolerance and Management in Crop Plants*. Elsevier.

Rajpal, V.R., Dhingra, Y., Khungar, L., Mehta, S., Minkina, T. *et al.* (2024) Current research in biotechnology. *Current Research in Biotechnology* 7, 100219.

Raliya, R. and Tarafdar, J.C. (2013) ZnO nanoparticle biosynthesis and its effect on phosphorous-mobilizing enzyme secretion and gum contents in Clusterbean (*Cyamopsis tetragonoloba* L.). *Agricultural Research* 2(1), 48–57.

Raliya, R., Saharan, V., Dimkpa, C. and Biswas, P. (2018) Nanofertilizer for precision and sustainable agriculture: Current state and future perspectives. *Journal of Agricultural and Food Chemistry* 66(26), 6487–6503.

Ramesh, A.S., Meghana, N.V.P., Tripathy, A., Karthik, L., Kiran, G. *et al.* (2023a) Future prospects of nanotechnology in aquaculture. In: Kirthi, A.V., Loganathan, K. and Karunasagar, I. (eds) *Nanotechnological Approaches to the Advancement of Innovations in Aquaculture*. Springer International Publishing, Cham, Switzerland, pp. 89–97.

Ranjan, S., Dasgupta, N., Chakraborty, A.R., Melvin Samuel, S., Ramalingam, C. *et al.* (2014) Nanoscience and nanotechnologies in food industries: Opportunities and research trends. *Journal of Nanoparticle Research* 16, 1–23.

Research and Markets (2021) Global nanotechnology in agriculture market (2020 to 2027) by type and geography. Available at: https://www.researchandmarkets.com/reports/5240109/ (accessed 1 April 2025).

Rizi, K.S., Rezayi, M., Aryan, E., Meshkat, Z. and Mobarhan, M.G. (2022) Nanotechnology applications in food and nutrition science. In: Visakh, P.M. (ed.) *Nanomaterials and Nanotechnology in Medicine*. Wiley, pp. 185–223.

Rizwan, M., Ali, S., Qayyum, M.F., Ok, Y.S., Adrees, M. *et al.* (2017) Effect of metal and metal oxide nanoparticles on growth and physiology of globally important food crops: A critical review. *Journal of Hazardous Materials* 322, 2–16.

Ryan, M., Isakhanyan, G. and Tekinerdogan, B. (2023) An interdisciplinary approach to artificial intelligence in agriculture. *NJAS: Impact in Agricultural and Life Sciences* 95(1), 2168568.

Sadigov, R. (2022) Rapid growth of the world population and its socioeconomic results. *The Scientific World Journal* 2022(1), 8110229.

Savary, S., Ficke, A., Aubertot, J.N. and Hollier, C. (2012) Crop losses due to diseases and their implications for global food production losses and food security. *Food Security* 4(4), 519–537.

Serraj, R., Krishnan, L. and Pingali, P. (2019) Agriculture and food systems to 2050: A synthesis. In: Serraj, R. and Pingali, P. (eds) *Agriculture & Food Systems to 2050*. CGIAR and World Scientific, pp. 3–45.

Shaikh, A., Jamla, M., Joshi, S., Patil, S., Oak, U. *et al.* (2024) Microbial nanotechnology for producing stress smart crops. *Plant Nano Biology* 7, 100063.

Shalaby, T.A., Bayoumi, Y., Abdalla, N., Taha, H., Alshaal, T. *et al.* (2016) Nanoparticles, soils, plants, and sustainable agriculture. In: Ranjan, S., Dasgupta, N. and Lichtfouse, E. (eds) *Nanoscience in Food and Agriculture 1*. Springer, Cham, pp. 283–312.

Shinozaki, K. and Yamaguchi-Shinozaki, K. (2007) Gene networks involved in drought stress response and tolerance. *Journal of Experimental Botany* 58(2), 221–227.

Siddiqi, K.S. and Husen, A. (2017) Plant response to engineered metal oxide nanoparticles. *Nanoscale Research Letters* 12(1), 1–18.

Singh, T., Shukla, S., Kumar, P., Wahla, V., Bajpai, V.K. *et al.* (2017) Application of nanotechnology in food science: Perception and overview. *Frontiers in Microbiology* 8, 1501.

Singla, P. (2020) Plant synthetic biology: A paradigm shift targeting stress mitigation, reduction of ecological footprints and sustainable transformation in agriculture. In: Biri, G. and Sharma, M.P. (eds) *Plant Stress Biology: Strategies and Trends*. Springer, pp. 435–489.

Sonkaria, S., Ahn, S.H. and Khare, V. (2012) Nanotechnology and its impact on food and nutrition: A review. *Recent Patents on Food, Nutrition & Agriculture* 4(1), 8–18.

Sozer, N. and Kokini, J.L. (2009) Nanotechnology and its applications in the food sector. *Trends in Biotechnology* 27(2), 82–89.

Sukhpal, H., Awasthy, S. and Sarethy, I.P. (2018) Nanotechnology in the food industry: Perspectives and prospects. In: Rani, V. and Yadav, U.C.S. (eds) *Functional Food and Human Health*. Springer, pp. 425–445.

Surendran, S., Mayekar, A. and Sharma, R. (2024) Nanotechnology for sustainable aquaculture. In: Sarma, D., Chandra, S. and Mallik, S.K. (eds) *Aquaculture and Conservation of Inland Coldwater Fishes*. Springer, pp. 127–138.

Syu, Y., Hung, J.-H., Chen, J.-C. and Chuang, H. (2014) Impacts of size and shape of silver nanoparticles on *Arabidopsis* plant growth and gene expression. *Plant Physiology and Biochemistry* 83, 57–64.

Tacon, A.G. (2023) Contribution of fish and seafood to global food and feed supply: An analysis of the FAO food balance sheet for 2019. *Reviews in Fisheries Science & Aquaculture* 31(2), 274–283.

Thiruvengadam, M., Rajakumar, G. and Chung, I.M. (2018) Nanotechnology: Current uses and future applications in the food industry. *3 Biotech* 8, 1–13.

Ullah, S., Ullah, Z., Sohil, M., Ullah, R., Ihsan, M. *et al.* (2019) Traditional herbal medicinal plants of Khyber Pakhtoonkhwa pakistan using for analgesic, anti-inflammatory, anti-cancer, anti-ulcer and antidiabetic activities. *International Journal of Horticulture and Food Science* 1(1), 90–98.

Ullah, Z., Gul, F., Iqbal, J., Abbasi, B.A., Kanwal, S. *et al.* (2023a) Biogenic synthesis of multifunctional silver oxide nanoparticles (Ag2ONPs) using *parieteria alsinaefolia* delile aqueous extract and assessment of their diverse biological applications. *Microorganisms* 11(4), 1069.

Ullah, R., Ullah, Z., Iqbal, J., Chalgham, W. and Ahmad, A. (2023b) Aspartic acid-based nano-copper induces resilience in *Zea mays* to applied lead stress via conserving photosynthetic pigments and triggering the antioxidant biosystem. *Sustainability* 15(16), 12186.

Ullah, Z., Iqbal, J., Abbasi, B.A., Akhtar, W., Kanwal, S. *et al.* (2023c) Assessment of gus expression induced by anti-sense OsPPO gene promoter and antioxidant enzymatic assays in response to drought and heavy metal stress in transgenic *Arabidopsis thaliana*. *Sustainability* 15(17), 12783.

Ullah, Z., Iqbal, J., Gul, F., Abbasi, B.A., Kanwal, S. *et al.* (2024) Biogenic synthesis, characterization, and *in vitro* biological investigation of silver oxide nanoparticles (AgONPs) using *Rhynchosia capitata*. *Scientific Reports* 14(1), 10484.

Ummi, A.S. and Siddiquee, S. (2019) Nanotechnology applications in food: Opportunities and challenges in food industry. In: Siddiquee, S., Melvin, G.J.H. and Rahman, Md.M. (eds) *Nanotechnology: Applications in Energy, Drug and Food*. Springer, pp. 295–308.

Vessey, J.K. (2003) Plant growth promotes rhizobacteria as biofertilizers. *Plant and Soil* 255(2), 571–586.

Vinayaka, A.C. and Thakur, M.S. (2010) Focus on quantum dots as potential fluorescent probes for monitoring food toxicants and foodborne pathogens. *Analytical and Bioanalytical Chemistry* 397(4), 1445–1455.

Vural, S., Açıkbaş, Y., Koçyiğit, Ü.M., Soylu, H.M., Gümüş, S. *et al.* (2018) Quercetin loaded graphene oxide/chitosan composite hydrogels: Preparation, characterization, and antibacterial activity. *Journal of Applied Polymer Science* 135(5), 45868.

Wang, P., Lombi, E., Zhao, F.J. and Kopittke, P.M. (2016) Nanotechnology: A new opportunity in plant sciences. *Trends in Plant Science* 21(8), 699–712.

Wani, S.H. (2023) Editorial: Mechanisms of abiotic stress responses and tolerance in plants: Physiological, biochemical and molecular interventions, volume II. *Frontiers in Plant Science* 14, 1272255.

Wasternack, C. (2007) Jasmonates: An update on biosynthesis, signal transduction and action in plant stress response, growth and development. *Annals of Botany* 100(4), 681–697.

World Health Organisation (2023) *The Growth of Ultra-processed Foods in India: An Analysis of Trends, Issues and Policy Recommendations*. WHO, Geneva, Switzerland.

Wu, H., Tito, N. and Giraldo, J.P. (2017) Anionic cerium oxide nanoparticles protect plant photosynthesis from abiotic stress by scavenging reactive oxygen species. *ACS Nano* 11(11), 11283–11297.

Yoon, B.K., Tae, H., Jackman, J.A., Guha, S., Kagan, C.R. *et al.* (2021) Entrepreneurial talent building for 21st century agricultural innovation. *ACS Nano* 15(7), 10748–10758.

Zenda, T., Wang, N., Yan, X., Dong, A., Yang, Q. *et al.* (2023) Opportunities and avenues for achieving crop climate resilience. *Environmental and Experimental Botany* 213, 105414.

Zhang, X., Liu, M., Zheng, F., Dong, Y., Hua, Y. *et al.* (2023) Optimizing sowing patterns in winter wheat can reduce N2O emissions and improve grain yield and NUE by enhancing N uptake. *Frontiers in Plant Science* 14.

Zhu, J.K. (2016) Abiotic stress signaling and responses in plants. *Cell* 167(2), 313–324.

Zhu, Y., Xu, F., Liu, Q., Chen, M., Liu, X. *et al.* (2019) Nanomaterials and plants: Positive effects, toxicity and the remediation of metal and metalloid pollution in soil. *Science of the Total Environment* 662, 414–421.

8 Abiotic Stress Resilience of CRISPR-edited Crops and Underlying Mechanisms Dissected by High-throughput Omics Technologies

Zunnorain Ali[1], Zakir Ullah[1], Javed Iqbal[2*], Banzeer Ahsan Abbasi[3], Shumaila Ijaz[4], Muhammad Awais Ibrahim[1], Shoaib Khan[5], Ghulam Murtaza[6], Rashid Iqbal[7,8], Shobana Sampath[9], Amin Fathi[10] and Tariq Mahmood[1]

[1]Department of Plant Sciences, Faculty of Biological Sciences, Quaid-i-Azam University, Islamabad, Pakistan; [2]Department of Botany, Bacha Khan University, Charsadda, Khyber Pakhtunkhwa, Pakistan; [3]Department of Botany, Rawalpindi Women University, Pakistan; [4]Department of Burn and Plastic Surgery, Shenzhen Institute of Translational Medicine, The First Affiliated Hospital of Shenzhen University, School of Biomedical Engineering, Shenzhen University Medical School, Shenzhen University, China; [5]Department of Chemistry, Abbottabad University of Science and Technology (AUST), Abbottabad, Pakistan; [6]School of Agriculture, Yunnan University, Kunming, Yunnan, China; [7]Department of Agronomy, Faculty of Agriculture and Environment, The Islamia University of Bahawalpur, Pakistan; [8]Department of Life Sciences, Western Caspian University, Baku, Azerbaijan; [9]Department of Biotechnology, Vel Tech Rangarajan Dr. Sagunthala R&D Institute of Science and Technology, Chennai, Tamil Nadu, India; [10]epartment of Agronomy, Ayatollah Amoli Branch, Islamic Azad University, Amol, Iran

---

Abstract

Abiotic stressors such as drought, salinity, temperature extremes, and heavy metals significantly impair agricultural productivity worldwide. Recent advances in CRISPR-Cas9 technology have revolutionized the field of crop improvement by enabling precise genome editing to enhance abiotic stress tolerance in crops. This chapter explores the application of CRISPR-Cas9 in developing stress-resilient crops, focussing on the underlying mechanisms revealed through high-throughput omics technologies such as genomics, transcriptomics, proteomics, and metabolomics. These omics approaches provide a comprehensive understanding of the complex molecular networks and gene regulatory mechanisms that govern plant responses to abiotic stress. The integration of CRISPR-Cas9 with omics technologies enables the identification of key stress-responsive genes and the precise editing of these targets to enhance crop resilience. The chapter discusses the potential of CRISPR-Cas9 in creating transgene-free, climate-resilient crops, addressing current challenges, and highlighting future directions for sustainable agriculture.

Keywords: CRISPR-Cas9, abiotic stress, crops, genome editing, omics, metabolomics, stress tolerance

---

*Corresponding author: javed89qau@gmail.com

© CAB International 2025. *Stress-Resilient Crops: Coordinated omics-CRISPR-Nanotechnology Strategies* (ed. Jen-Tsung Chen)

DOI: 10.1079/9781800627307.0008

8.1 Introduction

Abiotic stress is a term used to describe adverse environmental factors that have a detrimental effect on the growth and productivity of plants. It is considered a major obstacle nowadays in agriculture (Sahoo *et al.*, 2020). The presence of stressors such as drought, salt, extreme temperatures, and heavy metals can significantly diminish agricultural yields, hence posing a grave risk to world food security (Razzaq *et al.*, 2021). The swift progress in genetic-editing technologies, namely CRISPR-Cas systems, has created fresh opportunities for improving crop resilience to severe challenges. CRISPR-edited crops, when combined with high-throughput omics technologies that allow for thorough investigation of molecular responses at different levels, are becoming a hopeful solution to the difficulties presented by abiotic stress. This chapter summarizes the current knowledge regarding abiotic stress in crops, the resistance mechanisms provided by CRISPR editing, and the involvement of omics technology in analyzing these intricate responses (Rasheed *et al.*, 2021).

8.1.1 Overview of abiotic stress in crops

Abiotic stress refers to a range of environmental conditions that can affect the development and growth of plants. These include:

- **Drought stress.** Drought is a severe abiotic stress that greatly diminishes crop yield globally. Plants adapt to drought by modifying their physiological and molecular processes, such as closing their stomata, adjusting their osmotic balance, and increasing the production of protective proteins and metabolites (Razzaq *et al.*, 2021). Nevertheless, the intricate genetic make-up of drought tolerance has posed difficulties in creating drought-resistant crops using conventional breeding techniques.
- **Salinity stress.** Salinity impacts plants by inducing ionic and osmotic stress, resulting in ion toxicity and water shortage. These factors hinder plant growth and development. The concentration of sodium ions (Na^+) in the soil affects the ionic balance within plant cells, resulting in diminished water intake and photosynthesis. Plants have evolved numerous strategies to reduce the consequences of salt, including the exclusion of Na^+ from the root system and the segregation of harmful ions within vacuoles (Farooqi *et al.*, 2022).
- **Temperature stress.** Both fluctuating temperatures can significantly impact the metabolism of plants. Heat stress alters protein structure and function, causing damage to cells, while cold stress hinders the fluidity of the membrane and enzyme activity, resulting in diminished photosynthesis and growth. Plants respond to temperature extremes through the development of heat-shock proteins (HSPs), antioxidant substances, and other defensive molecules (Sahoo *et al.*, 2020).
- **Heavy-metal stress.** Heavy metals such as lead, cadmium, and mercury can be particularly hazardous to plants, halting cellular processes by creating reactive oxygen species (ROS) and affecting food intake. Plants evolved to develop detoxification techniques, such as the creation of metal-chelating polypeptides and the induction of antioxidant systems, to deal with stress from heavy metals (Bhat *et al.*, 2021).

8.1.2 Mechanisms of abiotic stress resilience in crops

Plants have developed extensive and diversified methods to cope with abiotic stress. These mechanisms work at the cellular, molecular, and physiological levels, and include osmotic adjustment in cases of water deficiency; plants store suitable solutes, such as proline and glycine betaine, to maintain the integrity of cells and safeguard cellular structures. These osmolytes serve as ROS scavengers, lowering oxidative damage (Razzaq *et al.*, 2021).

8.1.2.1 Ion homeostasis and transport

In response to saline stress, plants control the absorption and division of ions to maintain the internal ionic balance of the cell. It includes the activation of ion carriers, such as the high-affinity potassium transporter (HKT) and the sodium/hydrogen antiporter (NHX), which promote the exclusion of sodium from the cytoplasm or its retention into vacuoles (Farooqi *et al.*, 2022).

8.1.2.2 Antioxidant defense

Abiotic stress frequently leads to the excessive production of ROS, which can cause oxidative harm to biological components. Plants fight this by strengthening their antioxidant defense systems, which include enzymes such as superoxide dismutase (SOD), catalase (CAT), and peroxidases (POD). These enzymes operate in concert to neutralize ROS while shielding cells from damage caused by oxidative stress (Jain *et al.*, 2019).

8.1.2.3 Gene expression and signal transduction

Abiotic stress induces an induction of complicated signaling networks that influence the expression of genes. Transcription factors, such as dehydration-responsive element-binding (DREB) proteins and heat-shock factors (HSFs), exert vital functions in influencing the expression of stress-responsive genes. These genes code for peptides that shield cellular structures, filter chemicals that are harmful, and maintain cell equilibrium (Sahoo *et al.*, 2020).

8.1.3 The role of CRISPR-Cas systems in enhancing abiotic stress resilience

The system known as CRISPR-Cas emerged as a strong tool for accurate editing of genomes, enabling the targeted change of genes linked with abiotic stress response. By eliminating negative regulators or introducing positive genes, CRISPR can boost the tolerance of crops to environmental shocks. Recent research shows the successful application of CRISPR-Cas9 in modifying genes associated with drought, salinity, and heat tolerance in several crop species (Rasheed *et al.*, 2021):

- **Targeted mutagenesis**. CRISPR-Cas9 is a tool that can create specific mutations in genes that adversely impact stress tolerance. For example, the deletion of the ABA-hypersensitive germination1 (ahg1) gene in rice has been demonstrated to promote drought tolerance via modifying abscisic acid (ABA) signaling pathways (Bhat *et al.*, 2021).
- **Gene activation and repression**. CRISPR-Cas systems can potentially be programmed to induce or suppress genes implicated in stress responses. This is performed by CRISPR activation (CRISPRa) or CRISPR interference (CRISPRi) approaches, which use modified Cas9 proteins to control the expression of genes without generating breaks in double-strands (Ahmad *et al.*, 2022).
- **Multiplexed editing**. One of the benefits of CRISPR technology is the capacity to simultaneously focus on numerous genes, which is particularly valuable for increasing complicated features like tolerance to stress. Multiplied editing has been applied to change numerous genes that respond to stress in crops, resulting in greater tolerance to diverse abiotic challenges (Fig. 8.1) (Rasheed *et al.*, 2021).

8.1.4 The role of omics technologies in dissecting abiotic stress responses

High-throughput omics tools, including genomes, transcriptomics, proteomics, and metabolomics, have changed our view of plant reactions to abiotic stress. These developments enable the thorough investigation of molecular alterations at various stages, providing insights into the intricate networks that modulate stress tolerance:

- **Genomics**. Genome-wide association studies (GWAS) and quantitative trait loci (QTL) analysis have proved important in discovering genes related to abiotic stress endurance. These genetic techniques provide a framework for CRISPR-based editing by selecting candidate genes for focussed alteration (Al-Tamimi *et al.*, 2022).
- **Transcriptomics**. Transcriptomic investigations employing RNA sequencing (RNA-seq) have discovered critical systems

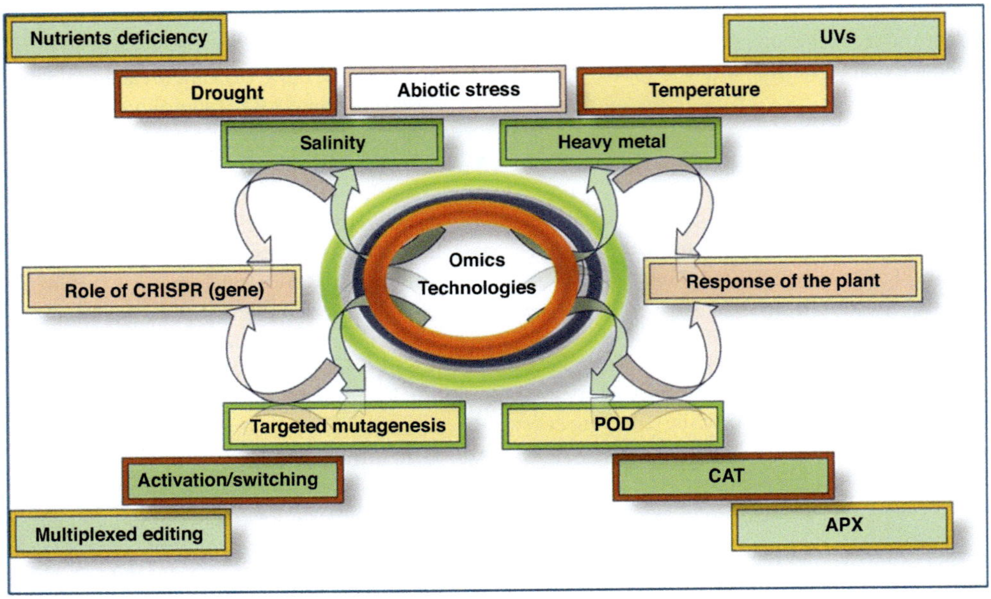

Fig. 8.1. Abiotic stress and role of CRISPR-Cas systems in enhancing abiotic stress resilience.

of regulation involved in stress responses. By comparing the transcripts of CRISPR-edited and wild-type plants under stressful circumstances, scientists can discover variously expressed genes and clarify the cellular processes driving increased stress tolerance (Farooqi *et al.*, 2022).

- **Proteomics.** Proteomic investigations provide explanations for the variations in how proteins express and modify after translation in reacting to abiotic stress. These investigations help to comprehend the physiological function of proteins in stress adaption and in identifying prospective possibilities for gene editing (Sahoo *et al.*, 2020).
- **Metabolomics.** Metabolomics encompasses an in-depth profile of metabolites, which are the end products of biological processes. Metabolomic investigations have discovered important metabolites that contribute to stress tolerance, including osmolytes, antioxidants, and secondary metabolites. These insights are critical for understanding the metabolic processes that promote stress resistance (Fig. 8.2) (Tinte *et al.*, 2021).

8.1.5 The role of CRISPR technology in enhancing stress resilience

The worldwide agriculture business is under unprecedented strain to develop crops that can tolerate diverse abiotic factors such as drought, salinity, severe temperatures, and heavy metals. Traditional breeding approaches have typically fallen insufficient due to the complex, multifaceted nature of such stress responses. However, the development of CRISPR-Cas gene-editing technology has changed the sector by enabling a precise, efficient, and cost-effective approach to boost stress resilience in crops. This section investigates how CRISPR technology is being employed to create crops that are more tolerant of environmental challenges, ultimately contributing to global food security.

8.1.5.1 CRISPR-Cas technology: a breakthrough in crop genome editing

CRISPR-Cas technique has emerged as a breakthrough tool for genome editing, permitting precise adjustments in the DNA of living creatures. In plants, CRISPR-Cas has been

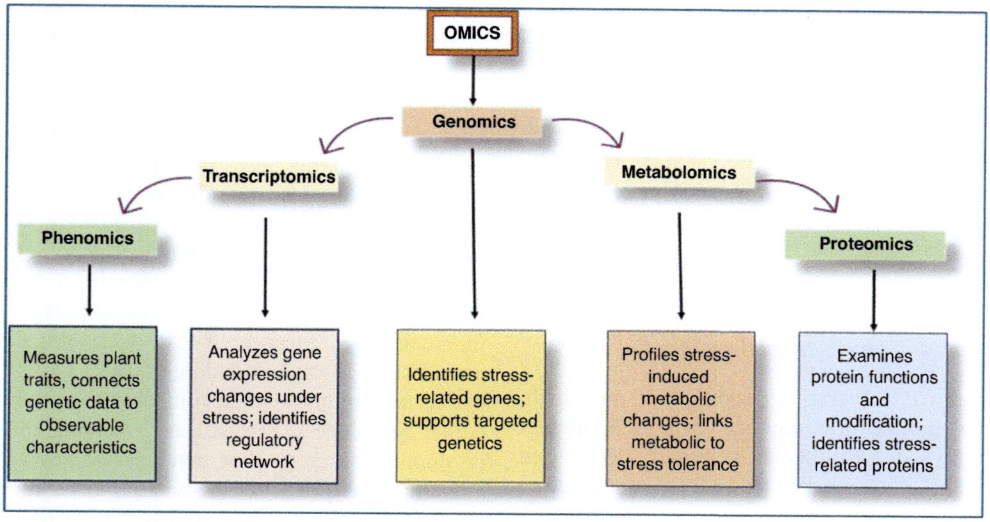

Fig. 8.2. Flow chart showing omics including possible technologies.

successfully utilized to target genes linked with abiotic stress tolerance, giving new paths for crop development (Rasheed *et al.*, 2021). The technology's simplicity, efficiency, and versatility make it preferable to prior genome-editing technologies like zinc finger nucleases (ZFNs) and transcription activator-like effector nucleases (TALENs). One of the key advantages of CRISPR-Cas is its capacity to specifically target specific genes important for stress responses. For example, CRISPR-Cas9 has been employed to knock out inhibitors of stress tolerance, improving the resistance of crops to unfavorable environments (Bhat *et al.*, 2021). This focussed method enables the generation of crop varieties with better tolerance to many stresses, including drought, salinity, and severe temperatures.

8.1.5.2 Enhancing drought tolerance through CRISPR-Cas

Drought constitutes one of the major abiotic factors affecting agricultural productivity worldwide. CRISPR technology has been essential in generating drought-resistant crops by changing genes involved in water-usage efficiency, stomatal control, and root architecture. For instance, CRISPR-Cas9 is being used to alter the genes involved in abscisic acid (ABA) signaling, a crucial mechanism in plant drought responses. By deleting ABA sensitivity-related genes, investigators have generated rice varieties with greater drought tolerance without losing yield. Likewise, CRISPR-Cas has permitted the multiplex editing of genes related to root development, leading to the generation of crops having larger and more efficient root systems. These alterations boost the plant's ability to acquire water from the deeper levels of soil, therefore improving drought resilience.

8.1.5.3 Salinity tolerance: CRISPR-Cas modifications for ionic homeostasis

Salt stress is another key trouble for the agricultural sector, particularly in semi-arid and arid locations where salinization of the soil is frequent. High levels of salt in the soil impair plant ionic balance, resulting in lower growth and productivity. CRISPR technology has been applied to modify genes involved in the transport of ions and homeostasis, thereby boosting tolerance to salinity in crops. One significant use is the use of CRISPR-Cas9 to knock off the gene that encodes the sodium transporter HKT1, which is responsible for sodium intake in plants. By lowering the expression of HKT1, researchers have generated rice and wheat cultivars that are more tolerant to saline environments, as these plants are more capable of excluding sodium from the roots (Rahman *et al.*, 2022). Moreover, CRISPR has been utilized to boost the expression of genes that regulate the vacuolar sequestration of sodium ions, significantly increasing the capacity of plants to manage salt stress (Ahmad *et al.*, 2022).

8.1.5.4 *Heat and cold tolerance: targeting stress-responsive genes*

Roaring temperatures, both low and high, can significantly damage crop development and yield. CRISPR-Cas technology has been employed to develop resistant-to-temperature crops by targeting genes that control heat-shock proteins (HSPs) and cold-responsive transcription factors (C-repeat binding factors, CBFs). For example, CRISPR has been exploited to boost the expression of HSPs, which protect protein molecules from denaturation at extreme conditions. Similarly, the overexpression of CBFs using CRISPR-mediated gene editing has enhanced cold tolerance in various varieties of crops, including tomatoes and maize (Haque *et al.*, 2018).

8.1.5.5 *Multiplex genome editing: engineering crops for multi-stress tolerance*

One of the most notable benefits of CRISPR technology is its ability to undertake multiplex editing of genomes, where many genes can be targeted concurrently. This is particularly important in producing crops that can survive a mix of abiotic stressors. Researchers have used CRISPR to simultaneously alter genes associated with drought, salinity, and heat tolerance in rice, leading to a variety that can grow under different stress conditions (Jaganathan *et al.*, 2018).

Multiplex editing also enables the stacking of advantageous features, such as greater nutrient usage effectiveness and enhanced ability to resist diseases and pests, alongside abiotic stress resistance. This comprehensive approach to agricultural modification holds considerable promise for producing robust crop types that can handle the threats of climate change (Zaidi *et al.*, 2020).

8.1.6 Importance of high-throughput omics technologies in studying crop resilience

The multifaceted nature of abiotic stress responses in plants needs modern approaches to unravel the underlying mechanisms and find targets for crop development. High-throughput omics technologies, encompassing genomes, transcriptomics, proteomics, metabolomics, and phenomics, have emerged as important tools in this respect. These technologies provide a deep understanding of the cellular, biological, and physiological mechanisms that confer resilience to stress, thereby aiding the production of crops that can tolerate severe climatic circumstances. This section dives into the significance of high-throughput omics technology in investigating and developing crop resistance.

8.1.6.1 *Genomics: foundation for understanding genetic responses to stress*

Genomics constitutes the core of crop development initiatives, offering important insights regarding the genetic basis of stress tolerance. Through large-scale sequencing initiatives, researchers have uncovered multiple quantitative trait loci (QTLs), genes, and alleles linked to stress resistance. For example, developments in next-generation sequencing (NGS) technologies have allowed the quick discovery of genetic variants that contribute to stress resistance in crops such as rice, wheat, and maize (Farooqi *et al.*, 2022). The combining of genomic data with additional omics layers has proved helpful in creating a thorough picture of how these variations in genetics emerge as phenotypic features under stressful situations.

8.1.6.2 *Transcriptomics: deciphering gene expression under stress*

Transcriptomics, the scientific investigation of RNA transcripts, offers an evolving view of gene expression variations in response to stress. RNA sequencing (RNA-seq) has become a critical method in finding genes responsive to stress and regulatory networks. For instance, transcriptome investigations have revealed the overexpression of HSPs and other defensive molecules in plants subjected to heat stress, offering insights into the molecular processes that enable heat tolerance (Zhou *et al.*, 2022). These investigations are useful for finding potential genes that can be targeted by CRISPR-Cas genome modification to boost stress resistance.

8.1.6.3 Proteomics: understanding protein-level responses

Proteomics refers to the large-scale study of protein molecules, which are the molecules that operate within cells. Proteomics provides knowledge of the post-translational changes and interactions between proteins that occur in reaction to stress. High-resolution proteome investigations have enabled the discovery of important stress-responsive proteins, such as those involved in antioxidant defense and signal transduction pathways, that are essential for plant survival under extreme circumstances (Dikobe *et al.*, 2023). The proteins in question serve as potential markers and sites for genetic engineering targeted at enhancing resistance to stress in crops.

8.1.6.4 Metabolomics: profiling stress-induced metabolic changes

Metabolomics emphasizes the complete investigation of metabolites, the small chemicals that play key functions in plant metabolism. Stress circumstances typically lead to considerable changes in the plant metabolites, with the build-up of osmoprotectants, antioxidants, and additional metabolites that assist ameliorate stress damage. Metabolomic studies have provided useful insights into the metabolic pathways triggered in responses to drought and salinity, hence suggesting prospective targets for crop development (Muthamilarasan *et al.*, 2019). These results help design crops with greater metabolic tolerance to abiotic stresses.

8.1.6.5 Phenomics: linking genotype to phenotype

Phenomics, the high-throughput assessment of phenotypes, is vital for correlating molecular and genetic data to observable features. Advanced phenotyping technologies, including image-based methods and automated sensors, enable the exact assessment of plant attributes like rate of growth, accumulation of biomass, and stress responses underregulated and field circumstances (Salas Fernandez *et al.*, 2017). The production of more resilient crop varieties is made possible by the identification of important features and the genetic factors that underlie them thanks to the integration of phenomics data with other omics layers.

8.1.7 Multi-omics integration: a holistic approach to crop improvement

The real power of omics technologies lies within their integration. By merging data from genomes, transcriptomics, proteomics, metabolomics, and phenomics, researchers may develop a holistic perspective of the complex networks that drive plant stress responses. This multi-omics method allows for the detection of critical regulatory sites and pathways that can be focussed for genetic enhancement. For example, integrating transcriptome and proteomic data has revealed that transcription factors and their subsequent targets are crucial for heat-stress tolerance in crops (Kaur *et al.*, 2021). This integrated research is crucial for producing crops with greater resilience to numerous stressors.

8.2 Abiotic Stress in Crops: Challenges and Impacts

Abiotic stresses, which encompass numerous nonliving components of the environment, have become a significant worry for worldwide agriculture, especially in the face of climate change. These pressures significantly affect crop development, growth, and yield, resulting in significant financial losses and compromising food security worldwide. Understanding the numerous forms of abiotic stresses and their distinct impacts on crops is vital for establishing resilient agricultural techniques and boosting crop output.

8.2.1 Types of abiotic stresses affecting crop productivity

8.2.1.1 Drought stress

Drought represents one of the most severe abiotic stresses, effecting practically every aspect of plant growth and development. It results in a reduction of water availability, which directly affects photosynthesis, stomatal conductance, and the absorption of nutrients. Under extreme drought, crops exhibit slow growth, decreased area of leaves, and impaired reproductive growth, ultimately leading to severe output losses (Joshi *et al.*, 2023). The stage of reproduction is particularly sensitive, as water deficiencies during this period can substantially diminish the number of grains and size, resulting in poor yields.

8.2.1.2 Salinity stress

Soil salinity is another major factor in abiotic stress, particularly in semi-arid and arid environments where irrigation methods often contribute to an increase of salts in the soil. Salinity stress affects ionic balance inside cells in plants, causing ion toxicity, osmotic stress, and nutrient shortages. These consequences lead to lower germination of seeds, impaired shoot and root growth, and reduced photosynthetic efficiency. Over time, continuous exposure to salinity might lead to the demise of the plant (Kumawat *et al.*, 2022). Rice and wheat, two of the world's primary cereal crops, are particularly susceptible to salinity, which can lead to large output decreases.

8.2.1.3 Temperature stress (heat and cold)

Temperature extremes, both low and high, are progressively damaging agricultural output as worldwide temperatures continue to change. Heat stress, characterized by temperatures over the ideal range for crop growth, leads to protein denaturation, membrane instability, and disruption of metabolic activities. This stress is particularly detrimental during the reproductive phase, where it may contribute to pollination sterility and decreased grain filling, significantly reducing output (Sehgal *et al.*, 2018). Conversely, cold stress, which occurs at temperatures below the optimum range, inhibits membrane fluidity, enzyme activity, and photosynthesis, leading to diminished growth and delayed development. Crops such as corn and rice, which are vulnerable to low temperatures, face large production losses when exposed to winter stress at critical phases of development.

8.2.1.4 Heavy-metal stress

Heavy metals, such as cadmium (Cd), lead (Pb), and mercury (Hg), impose a serious threat to crop output, particularly in areas plagued by pollution from factories or excessive consumption of chemical fertilization and insecticides. Heavy-metal stress leads to the formation of ROS, causing oxidative harm to biological components, including lipids, proteins, and DNA. This stress disturbs several physiological systems, including respiration, photosynthesis, and intake of nutrients, leading to lower production and development (Berni *et al.*, 2019). In crops like rice and wheat, contamination with heavy minerals not only affects yield but also poses significant risks to health due to the accumulation of hazardous metals in edible sections of the plants.

8.2.1.5 Flooding and waterlogging

Flooding and waterlogging are getting more frequent due to variations in rainfall and insufficient systems of drainage, particularly in shallow agricultural districts. Excess water surrounding the plant's root zone depletes oxygen levels, leading to hypoxic circumstances that affect root respiration and nutrient uptake. Flooding stress also alters hormonal homeostasis, leading to lower growth and delayed development (Akram *et al.*, 2019). For example, in rice production, waterlogging at the initial phases can lead to poor growth of roots, which severely affects total plant vigor and productivity.

8.2.1.6 Oxidative stress

Oxidative stress, typically an additional result of other abiotic stresses such as dehydration, salt, and heavy metal exposure, comprises an excessive production of ROS. While ROS plays a role in

Table 8.1. The physiological impacts of abiotic stress on plants, and mitigation strategies.

Abiotic stress	Physiological effects	Crops affected	Impact on crop yield	Common mitigation strategies
Drought	Reduced photosynthesis, stomatal closure, wilting	Maize, wheat, rice, soybean	Significant yield reduction due to water scarcity	Irrigation, drought-resistant varieties, mulching
Salinity	Ion toxicity, osmotic stress, reduced nutrient uptake	Rice, wheat, tomato, cotton	Decreased yield due to impaired growth and development	Salt-tolerant varieties, leaching, proper irrigation
Extreme heat	Protein denaturation, impaired pollination, accelerated senescence	Wheat, maize, rice, soybean	Reduced yield due to poor grain filling and sterility	Heat-tolerant varieties, shading, timely planting
Cold	Membrane damage, reduced enzymatic activity, poor germination	Wheat, rice, barley, canola	Lower yield due to frost damage and delayed growth	Cold-tolerant varieties, protective covers, adjusted planting times
Heavy metals	Disruption of cellular functions, oxidative stress	Rice, wheat, soybean, spinach	Yield loss due to toxicity and growth inhibition	Phytoremediation, chelation therapy, soil amendments
Flooding	Hypoxia, reduced root respiration, nutrient leaching	Rice, wheat, maize, soybean	Yield decline due to root damage and poor nutrient uptake	Drainage management, flood-tolerant varieties, raised beds

signaling between cells under normal settings, their excessive build-up under stress situations leads to oxidative damage, compromising cell structures and processes (Table 8.1). This stress increases senescence, lowers the rate of photosynthesis, and hinders the intake of nutrients, all of which leads to reduced agricultural output (Saikia *et al.*, 2020).

8.3 Current Strategies for Mitigating Abiotic Stress in Agriculture

Abiotic stresses, like drought, salt, severe temperatures, and toxicity from heavy metals, present considerable obstacles to agricultural output, especially amid climate change. The development of solutions to reduce the impacts of these pressures is crucial for guaranteeing food security and environmentally friendly agriculture. Recent breakthroughs in biotechnology, agricultural management, and the administration of beneficial microbes offer promising possibilities for boosting crop tolerance to abiotic challenges.

8.3.1 Genetic engineering and CRISPR-Cas9 technology

Among the most exciting possibilities for alleviating abiotic stress in crops is through the application of genetic engineering, notably the CRISPR-Cas9 system. This approach enables the accurate modification of genes involved with tolerance to stress, enabling the production of crop types that are better adapted to endure environmental challenges. For instance, CRISPR-Cas9 has been utilized to knock out or change genes involved in drought and salinity tolerance, resulting to enhanced efficiency in water use and ion homeostasis in crops like wheat and rice (Minhas *et al.*, 2017). These genetically altered crops have shown greater tolerance to many abiotic challenges, making them a significant component of current agricultural techniques.

8.3.2 Chemical priming for enhanced stress tolerance

Chemical priming is an alternate helpful strategy for minimizing abiotic stresses in crops. This strategy involves an external application of either synthetic or natural chemicals that prepare plants to better endure stress. These substances may stimulate physiological and molecular defense mechanisms, such as the build-up of osmoprotectants, antioxidants, and stress-responsive proteins. For example, treatment with compounds like salicylic acid, jasmonic acid, and glycine betaine has been proved to enhance salinity and drought tolerance in many crops (Sako *et al.*, 2021). The efficiency of chemical priming resides in its capacity to trigger a quick and strong response to stress, presenting crops with a further layer of defense against environmental difficulties.

8.3.3 Use of biostimulants and microbial inoculants

Biostimulants and microbe inoculants have attracted substantial interest as environmentally conscious and environmentally friendly techniques for relieving abiotic stress in crops. Biostimulants, produced from natural sources such as seaweed extracts, humic compounds, and microbial fermentation products, improve plant growth and stress tolerance by enhancing nutrient absorption, water efficiency, and the activation of shock-responsive pathways. Studies have demonstrated that the application of biostimulants can greatly improve the performance of plants under drought, salt, and heat-stress conditions (Ma *et al.*, 2022). Microbial inoculants, comprising plant growth-promoting rhizobacteria (PGPR) and mycorrhizal fungi, are integral to augmenting plant resilience against abiotic stresses. These advantageous microorganisms establish symbiotic associations with plant roots, facilitating nutrient uptake, hormone modulation, and the synthesis of compounds that mitigate stress. For example, PGPRs have been demonstrated to promote systemic tolerance in plants by influencing antioxidant mechanisms, enhancing water absorption, and optimizing root structure in conditions of drought and salinity stress (Munir *et al.*, 2022).

8.3.4 Agronomic management practices

Agronomic management strategies are pivotal for alleviating the effects of abiotic stressors and enhancing the resilience of crops. These strategies encompass optimizing plant population density, managing irrigation practices, applying soil amendments, and implementing crop rotation. For instance, the adoption of low-density planting practices can diminish inter-plant competition for water and nutrients during drought periods, while incorporating cover crops and organic mulches can bolster soil moisture retention and attenuate the adverse impacts of thermal stress (Mariani and Ferrante, 2017). The application of precision agriculture technologies, including remote sensing and soil moisture monitoring devices, is also fundamental in the management of abiotic stressors. Such technologies facilitate real-time monitoring of environmental variables, enabling farmers to make data-driven decisions regarding irrigation, fertilization, and other management interventions. By optimizing resource allocation and reducing stress exposure, precision agriculture has the potential to substantially improve crop resilience and overall productivity.

8.3.5 Nanotechnology in abiotic stress management

Nanotechnology has emerged as an innovative approach for addressing abiotic stress in agricultural contexts. Nanoparticles (NPs) can be utilized for the targeted delivery of nutrients, agrochemicals, and agents that alleviate stress directly to the plants, thereby enhancing their efficiency and efficacy (Ullah *et al.*, 2023a). For instance, nanofertilizers have demonstrated increased nutrient utilization efficiency and a reduction in the detrimental effects of salinity and heavy-metal toxicity on crop development (Manzoor *et al.*, 2022). Furthermore, nanoparticles can facilitate seed priming, which enhances germination rates and stress tolerance by optimizing the physiological mechanisms related to stress adaptation (Ullah *et al.*, 2023a; Ullah *et al.*, 2024).

8.3.6 Climate-smart agriculture and integrated approaches

Climate-smart agriculture (CSA) encompasses a range of strategies aimed at improving the resilience of agricultural systems in response to climate change and abiotic stressors. CSA incorporates the utilization of stress-resistant crop varieties, efficient water management techniques, and the conservation of natural resources. It also highlights the necessity for integrated methodologies that amalgamate genetic, agronomic, and technological solutions to confront the multifaceted challenges presented by abiotic stressors. For instance, the integration of CRISPR-Cas9 modified crops, biostimulants, and precision irrigation could foster a resilient system adept at simultaneously managing various stressors. By embracing a comprehensive approach CSA seeks to guarantee sustainable food production amidst escalating environmental challenges (Rajanna *et al.*, 2023).

8.4 CRISPR-Cas Systems: A Tool for Crop Genome Editing

The CRISPR-Cas system has fundamentally transformed the landscape of genetic engineering in numerous biological disciplines, with a notable impact on agriculture. This advanced technology has facilitated precise, efficient, and adaptable alterations of plant genomes, thereby enhancing the development of crops that exhibit improved resistance to abiotic stressors. In this section, we aim to elucidate the fundamental principles of CRISPR-Cas technology and its extensive applications within the realm of plant science.

8.4.1 Basics of CRISPR-Cas technology

Initially identified as an adaptive immune mechanism in prokaryotic organisms, the CRISPR-Cas system has been repurposed into a formidable genome-editing instrument. The system comprises primarily two essential components: the guide RNA (gRNA) and the CRISPR-associated protein (Cas). The gRNA is engineered to be complementary to the designated DNA sequence, directing the Cas protein to the specific genomic site where it catalyzes a double-strand break (DSB) (Knott and Doudna, 2018). The most prevalent Cas protein utilized in plant genome editing is Cas9, which is sourced from *Streptococcus pyogenes*. The primary function of Cas9 is to create a DSB at the target locus, which the cellular machinery subsequently repairs via one of two pathways: non-homologous end-joining (NHEJ) or homology-directed repair (HDR). NHEJ frequently results in insertions or deletions (indels) that can disrupt gene functionality, thus serving as a valuable mechanism for gene knockout. Conversely, HDR can be employed for precise gene editing through the provision of a repair template (Schindele *et al.*, 2020). Since its inception, CRISPR-Cas9 has undergone optimization and expansion, resulting in a plethora of applications in plant science. These applications encompass base editing, prime editing, and the creation of CRISPR interference (CRISPRi) and CRISPR activation (CRISPRa) systems, which enable the regulation of gene expression without necessitating alterations to the DNA sequence (Ansori *et al.*, 2023).

8.4.2 Applications of CRISPR-Cas in plant science

8.4.2.1 Gene knockout for stress resilience

One of the principal utilizations of CRISPR-Cas in botanical studies involves the knockout of genes that adversely affect stress resilience. By impairing these genes, scientists can augment the plant's capacity to endure abiotic stressors such as drought, salinity, and extreme temperature variations. For instance, CRISPR-Cas9 has been employed to disrupt the SlDFR gene in *Solanum lycopersicum*, resulting in cultivars exhibiting enhanced drought resilience due to improved water-use efficiency (Rogowsky, 2017).

8.4.2.2 Base editing for precision mutagenesis

Base editing is a highly refined variant of the CRISPR technique that enables the immediate transformation of one DNA base pair into another without producing a DSB. This method is particularly beneficial for inducing point mutations that can provide tolerance to abiotic stressors. For instance, base editing has been applied to create mutations in the OsHKT1;5 genes in rice, which boosts the plant's resistance to salinity by enhancing sodium ion exclusion from the roots (Huang and Puchta, 2021).

8.4.2.3 Prime editing for precision gene correction

Prime editing, an innovation in CRISPR technology, increases the capabilities of genome editing by permitting precise deletions, insertions, and base conversions. Unlike regular CRISPR, prime editing is not dependent on HDR, making it useful in nondividing cells where HDR is less efficient. This method has significant potential in rectifying genetic alterations related to abiotic stress sensitivity. For example, primary editing can be used to repair harmful genes in crop genes, thereby restoring or boosting stress tolerance (Li *et al.*, 2021).

8.4.2.4 CRISPRi and CRISPRa for gene regulation

CRISPR interference (CRISPRi) and CRISPR activation (CRISPRa) are methods that modify gene expression without affecting the DNA sequence. CRISPRi uses a catalytically inactive Cas9 (dCas9) to inhibit transcription by attaching to a gene's promoter or coding area, hence lowering gene expression. Conversely, CRISPR recruits transcriptional activators to boost the expression of target genes. These techniques are invaluable for investigating gene function and can be used to increase or decrease genes implicated in responses to stress, providing information about their roles in abiotic stress tolerance (Pandey *et al.*, 2019).

8.4.2.5 Multiplexed genome editing

One of the primary benefits of CRISPR technology is its capacity to undertake multiple genomes editing, where many genes can be targeted concurrently. This feature is particularly advantageous in crops when numerous genes contribute to stress tolerance. For example, multiplex editing has been employed in rice to concurrently knock out many genes involved with salinity and drought sensitivity, resulting in plants with improved tolerance to both conditions (Puchta, 2017).

8.4.2.6 Chromosome engineering

One of the primary benefits of CRISPR technology is its capacity to undertake multiple genomes editing, where many genes can be targeted concurrently. This feature is particularly advantageous in crops when numerous genes contribute to stress tolerance. For example, multiplex editing has been employed in rice to concurrently knock out many genes involved with salinity and drought sensitivity, resulting in plants with improved tolerance to both conditions (McCarty *et al.*, 2020).

8.4.2.7 Recent advancements in CRISPR-mediated genome editing for crops

The CRISPR-Cas9 method has revolutionized genome editing, permitting precise adjustments in the DNA of plants, which is crucial for increasing crop attributes such as stress resilience, yield, and nutritional content. Over the past several years, considerable improvements have been made in the application of CRISPR technology in plant science, resulting to the creation of more efficient, diverse, and precise editing approaches. This section discusses some of the most recent developments in CRISPR-mediated genome modification for crop enhancement.

ADVANCEMENTS IN CRISPR-CAS VARIANTS AND EFFICIENCY. Among the most important developments in CRISPR technology is the discovery of multiple Cas9 variants that give more specificity and fewer off-target effects. Researchers have designed Cas9 proteins, such as the high-fidelity versions SpCas9-HF1 and eSpCas9, which limit unintentional DNA cleavage, hence boosting the exactness of editing the genome (Jaganathan *et al.*, 2018). Additionally, other Cas proteins, such as Cpf1 (now known as Cas12a) and Cas12b, have been created, enabling other methods for editing the genome with unique PAM prerequisites and diverse cutting patterns, increasing the toolkit available to make plant genome changes (Zhang *et al.*, 2023). These findings are especially crucial for polyploid crops, such as wheat, where editing many homologous genes simultaneously is problematic due to the presence of numerous copies of each gene. The availability of Cas variants with varied particularities allows for targeted and effective editing, which is vital to enhancing complicated characteristics in these crops (Vats *et al.*, 2019).

BASE EDITING AND PRIME EDITING: PRECISION MUTAGENESIS. Another notable innovation is the introduction of base modifying and prime editing technologies that allow for more precise genomic alterations without the requirement for double-strand breaks (DSBs). Base editing enables the conversion of individual DNA bases, such as C to T, or A to G, with high precision. This approach is particularly useful for inserting specific mutations that provide desired features, like as herbicide resistance or better stress tolerance (Rao and Wang, 2021). Prime editing, a more recent breakthrough, expands these capabilities by permitting the direct insertion, deletion, or substitution of DNA sequences without relying on the cell's repair processes. Prime editing is less susceptible to off-target effects and can be utilized to produce sophisticated genetic modifications which were previously difficult or impossible to do with traditional CRISPR-Cas9 systems (Chen *et al.*, 2019). These developments considerably increase the potential applications of CRISPR technology in the breeding of plants, allowing for the exact engineering of crops with improved features.

MULTIPLEX GENOME EDITING: TARGETING MULTIPLE TRAITS SIMULTANEOUSLY. Multiplex editing of genomes, which allows for the simultaneous targeting of many genes, has become increasingly viable with the development of CRISPR-Cas technology. This technique is particularly advantageous for increasing complex traits that are driven by several genes, such as abiotic stress tolerance, where multiple pathways need to be adjusted for optimal performance (Rasheed *et al.*, 2021). For example, in rice, researchers have successfully employed multiplex editing to concurrently knock out many genes related in drought and salinity tolerance, producing in plants that demonstrate greater resistance to both stresses (Arora and Narula, 2017). This method is also being examined for enhancing nutritious content and resistance to diseases, where the ability to target several genes in a single editing session can considerably accelerate the production of new crop types.

CRISPR-ENABLED GENE REGULATION: BEYOND GENOME EDITING. CRISPR technology is not restricted to specific genome editing; it has also been used for the regulation of genes using CRISPR interference (CRISPRi) and CRISPR activation (CRISPRa) systems. CRISPRi uses a catalytically ineffective Cas9 (dCas9) to inhibit transcription, thereby silencing the expression of genes. Conversely, CRISPRa recruits transcriptional activators to promote gene expression (Nadakuduti and Enciso-Rodríguez, 2020). These methods are invaluable for understanding how genes work and for fine-tuning gene expression levels in plants. For instance, CRISPRa has been used to boost the expression of genes that contribute to drought resistance, resulting in enhanced water-use efficiency in plants (Khurshid *et al.*, 2018). CRISPRi, on the other hand, is being examined for its ability to downregulate unwanted features, like susceptibility to pests and diseases, without permanently modifying the genome (Fig. 8.3).

8.5 Delivery Systems and DNA-Free Editing

Getting CRISPR reagents into plant cells efficiently has been one of the main problems with genome editing. Improvements in delivery techniques have been the focus of recent developments, especially for crops that are challenging to change. While methods like particle bombardment, protoplast transfection, and agrobacterium-mediated transformation are still frequently used, novel strategies like viral vectors and delivery systems based on nanotechnology are being developed to improve efficacy and increase the number of editable

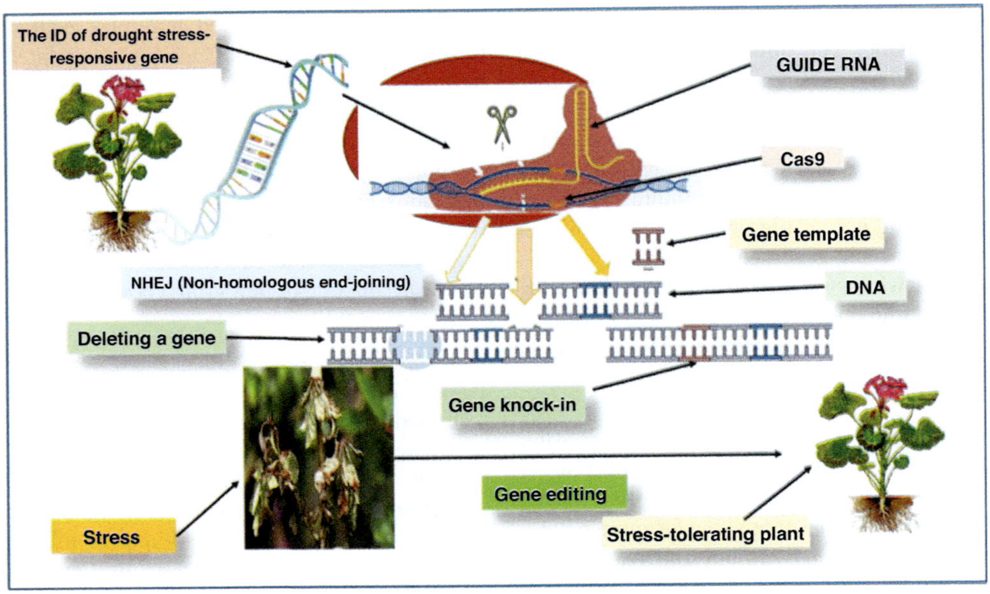

Fig. 8.3. CRISPR-enabled gene editing and regulation of plants.

species (Laforest and Nadakuduti, 2022; Ullah *et al.*, 2023). In addition, there has been a surge in the development of DNA-free editing techniques that use ribonucleoprotein (RNP) complexes to deliver CRISPR reagents. By lowering the possibility of introducing foreign DNA into the plant genome, these techniques increase the acceptance of the modified plants under regulatory frameworks that are less restrictive for nontransgenic crops (Li *et al.*, 2021). For crops meant for commercial use, DNA-free modification is especially attractive since it complies with consumer and regulatory expectations for non-GMO goods.

8.6 Physiological and Molecular Impacts of Abiotic Stress on Crops

Abiotic stress, covering elements such as drought, salt, severe temperatures, and toxicity from heavy metals, offers considerable difficulties to worldwide agricultural output. These challenges lead to a cascade of physiological and molecular reactions in plants, eventually impacting their growth, development, and production. Understanding these implications is crucial for creating ways to boost the resilience of crops through genetic engineering and sophisticated biotechnologies like CRISPR-Cas.

8.6.1 Physiological impacts of abiotic stress

8.6.1.1 Photosynthesis inhibition and chlorophyll degradation

Abiotic stressors, notably drought and salt, substantially degrade photosynthesis by diminishing the effectiveness of photosystems and decreasing chlorophyll concentration. Drought stress causes stomatal closure, lowering CO_2 availability, which in turn reduces the photosynthetic rate. This decline in photosynthesis is generally followed by the deterioration of chlorophyll, further lowering the plant's capacity to gather light energy (Francini and Sebastiani, 2019).

8.6.1.2 Alterations in water and nutrient uptake

Water stress, produced by drought or salt, directly inhibits the plant's capacity to absorb nutrients and water from the soil. Salinity stress, for example, leads to osmotic stress by lowering the ground's water potential, which makes it more difficult for plants to take up water. This state not only lowers water intake but also leads to the build-up of harmful ions like sodium (Na^+) and chloride (Cl^-), which affect nutritional balance and uptake (Zafar *et al.*, 2019). The ensuing nutritional deficits further hamper plant development and may result in symptoms such as chlorosis and stunted growth.

8.6.1.3 Reduction in growth and biomass accumulation

The combined impact of decreased photosynthesis and restricted water and nutrient absorption contribute to a considerable decrease in plant development and accumulation of biomass under abiotic stress (Ijaz *et al.*, 2023). Drought and salinity are particularly damaging at important growth phases, such as starting seeds and reproduction, where they can cause lasting damage. Heat stress also impacts development by interrupting enzymatic activity needed for cellular metabolism, resulting in diminished cell division and expansion (Debbarma *et al.*, 2019).

8.6.1.4 Impaired reproductive development

Abiotic stressors have severe impacts on the development of reproduction, leading to decreased fertility, poor seed set, and poorer grain production. High temperatures, for instance, can produce pollen sterility, whereas dryness throughout the flowering season might lead to floral abortion and poor pollination. Salinity stress similarly impacts reproductive organs, leading to lower fruit set and seed viability (Hyun, 2020). These impacts on reproduction are particularly problematic since they immediately translate into diminished agricultural yield and quality.

MOLECULAR IMPACTS OF ABIOTIC STRESS

Reactive oxygen species production and oxidative stress. ROS are molecules such as superoxide anions (O2-), hydrogen peroxide (H2O2), and hydroxyl radicals (OH•) that are produced in excess because of abiotic environments. While ROS plays a function in signaling under normal settings, their excessive build-up during stress leads to oxidative damage to proteins, lipids, and DNA. This oxidative stress is a main contributor to the breakdown of cellular components and the suppression of critical metabolic activities (Bhat *et al.*, 2021).

Altered gene expression and signaling pathways. Abiotic stressors generate considerable changes in gene expression, notably in genes implicated in stress detection, transmission of signals, and protective responses. For example, salinity and drought stress activate genes encoding for osmoprotectants, such as proline and glycine betaine, which assist to preserve cellular osmotic equilibrium. Heat stress stimulates the production of HSPs that operate as molecular chaperones to protect proteins from denaturation. These stress-induced changes in gene expression are generally regulated by signaling pathways including hormones like ABA and ethylene, which play essential roles in coordinating the plant's response to environmental stresses.

Epigenetic Modifications. Epigenetic alterations, including DNA methylation, histone modifications, and chromatin remodeling, play a vital role in regulating gene expression under abiotic stress conditions. These alterations can lead to the activation or suppression of stress-responsive genes, allowing plants to adapt to harsh environments. Epigenetic regulation is particularly crucial for the establishment of stress memory, where plants exposed to stress may 'remember' the event and respond more effectively to subsequent shocks (Ahmad *et al.*, 2022). This adaptive mechanism is being studied in crop breeding initiatives to boost stress tolerance.

Impact on Hormonal Balance. Abiotic stressors disturb the balance of plant hormones, which are key regulators of development and growth. For instance, drought and salt stress cause a rise in ABA levels, which increases stomatal closure and minimizes water loss but also limits development. Conversely, gibberellins (GAs), which stimulate growth, are commonly downregulated under stressful circumstances, leading to lower plant height and delayed development. The interaction between various hormones under stress is complicated and involves extensive signaling networks that influence the plant's overall stress response.

8.7 Ethical and Regulatory Considerations in the Application of CRISPR in Agriculture

CRISPR technology is a formidable tool in the field of genome editing, presenting great possibilities for agricultural innovation. However, its fast progress has prompted serious ethical and regulatory issues. As agricultural biotechnology increasingly relies on this gene-editing technique, recognizing the ethical implications and implementing suitable regulatory frameworks is vital to guarantee responsible use. This section discusses the ethical and regulatory concerns surrounding the implementation of CRISPR in agriculture, as emphasized by current studies.

8.7.1 Ethical considerations

The emergence of CRISPR technology has provoked serious ethical arguments, particularly in the context of food security and sustainability. The possibility of CRISPR to develop new types of crops with enhanced features has garnered attention to various ethical problems, including the modification of organisms without transgenic participation and the societal ramifications of such modifications.

8.7.2 Precision and nontransgenic editing

One of the primary ethical concerns focusses on CRISPR's potential to execute nontransgenic alterations, which fundamentally differs from previous genetic changes. Unlike prior gene-editing methods, CRISPR enables precise alterations that do not always necessitate the transfer of genes between species. This trait has raised disputes over whether such alterations should still be included within the umbrella of genetically modified organisms (GMOs). As Schultz-Bergin (2018) points out, the nontransgenic character of CRISPR might minimize criticism based on the ethical problem of 'crossing species boundaries'.

8.7.3 Environmental and animal welfare impacts

Furthermore, CRISPR's usage in enhancing animal welfare, notably in food production, poses ethical considerations regarding the amount to which these technologies should be employed. By reducing unwanted repercussions through precise gene modifications, CRISPR might increase animal health and wellbeing. However, the growing possibility of genetically altering more animals for agricultural reasons raises worries about animal exploitation (Schultz-Bergin, 2018).

8.7.4 Public perception and food ethics

The cultural acceptance of CRISPR-modified crops is another ethical challenge. Bartkowski *et al.* (2018) highlight the potential for CRISPR to alter agriculture but caution that the speed at which the technology evolves might exceed public discussions and governmental responses. This scenario needs a reevaluation of food ethics, particularly as consumers cope with the blurred boundaries between natural and technologically modified species (Bartkowski *et al.*, 2018).

8.7.5 Regulatory challenges

The fast development of CRISPR technology poses substantial difficulties to existing regulatory regimes. Unlike typical GMOs, CRISPR-edited crops generally do not transfer foreign genes, complicating their categorization within present legal systems. Regulatory agencies worldwide are wrestling with the job of changing rules to accommodate the unique traits of CRISPR-edited species.

8.7.6 Global regulatory divergence

Different nations have implemented varied regulatory approaches to CRISPR-edited crops. For example, although some governments regulate CRISPR crops similarly to GMOs, others are exploring more permissive frameworks due to the technology's nontransgenic capabilities. Ahmad *et al.* (2021b) underlines the need for a clear worldwide regulatory framework to guarantee that CRISPR technology is utilized safely and efficiently across borders. The authors add that while some CRISPR-edited crops have reached the market, variations in regulatory status between areas offer obstacles to international commerce and acceptability (Ahmad *et al.*, 2021b).

8.7.7 Risk assessment and safety concerns

An important component of regulatory frameworks is the risk assessment of CRISPR-edited crops. Regulatory agencies must analyze possible off-target impacts, environmental dangers, and unforeseen consequences linked with genome editing. Discuss the need to limit environmental and health concerns through greater specificity in CRISPR editing. Their analysis implies that increasing precision in gene modifications, coupled with comprehensive risk assessments, can limit the potential risks of CRISPR technology.

8.7.8 Ethical oversight and public involvement

As regulatory frameworks change, there is a growing awareness that ethical issues must be integral to decision-making processes. Shinwari *et al.* (2018) underline the need for public discourse and openness in regulating CRISPR uses. They call for the participation of different societal voices, including ethicists, law-makers, and the public, in deciding the future of CRISPR legislation. This strategy ensures that the technology progresses in a socially acceptable manner (Shinwari *et al.*, 2018).

8.8 CRISPR-edited Crops: Mechanisms of Abiotic Stress Resilience

8.8.1 Molecular mechanisms targeted by CRISPR for enhancing stress tolerance

Agricultural output is seriously threatened globally by abiotic factors like drought, salt, excessive temperatures, and heavy-metal toxicity. Traditional breeding techniques have struggled to generate stress-resilient crops at a scale needed to preserve food security as global climate change aggravates these circumstances. By permitting the exact modification of genes linked with stress tolerance, the development of CRISPR-Cas genome-editing technology has transformed plant biotechnology. Using current developments in the field, this chapter investigates the molecular pathways targeted by CRISPR in building abiotic stress resistance in crops.

8.8.2 Targeting regulatory and structural genes

In plants, abiotic stress tolerance sometimes requires a sophisticated network of structural and regulating genes. Targeted mutation of these genes made possible by CRISPR helps plants respond better to environmental pressures. To increase resilience, for instance, one can modify regulatory genes controlling the expression of related stress processes. Targeting ABA-induced transcription repressors (AITRs) in soybean, CRISPR has been utilized to help control ABA signaling pathways. Salinity tolerance is much improved by knockout mutations in the GmAITR gene family without sacrificing plant fitness (Wang *et al.*, 2021). Likewise, CRISPR has made the exact editing of hybrid proline-rich proteins (HyPRPs), negative stress-response regulators in crops like tomato possible. Eliminating certain SlHyPRP1 gene domains increased resistance to salt stress at both the germination and vegetative phases, therefore underlining the potential of CRISPR-based domain editing to produce stress-resistant crops.

8.8.3 Modulating stress-responsive pathways

Dissecting stress-responsive systems in plants has been made possible using CRISPR technology, which also helps researchers modify important components for increased tolerance. One important adaptation mechanism a plant utilizes to handle endoplasmic reticulum (ER) stress brought on by unfavorable environmental circumstances is the unfolded protein response (UPR) pathway. Recent studies show how well CRISPR targets UPR-related genes, hence enhancing plant tolerance to abiotic conditions such as heat and drought. Editing these genes will help scientists maximize the plant's capacity to control protein-folding stress, hence improving its development and growth under stressful circumstances (Vu *et al.*, 2023).

8.8.4 Engineering structural defense mechanisms

Beyond conventional routes of action, CRISPR has been used directly to strengthen plant structural defenses. Improved resilience in crops has resulted from editing genes that encode proteins resisting abiotic stressors. To increase salt tolerance in rice, for instance, CRISPR-mediated editing of genes encoding for ion transporters has been used. Using CRISPR has helped the NHX and SOS1 transporters – which are essential for preserving ion homeostasis under high salinity conditions – manageable to improve the plant's survival in salty soils (Farhat *et al.*, 2019). Apart from aiming at ion transporters, CRISPR has been applied to improve osmoprotectant synthesis and other suitable solute production. Under water-limited situations, these molecules preserve cellular architecture and aid in maintaining cell turgor. Editing genes linked to osmoprotectant production has shown promise for increasing drought tolerance in many different crops (Zafar *et al.*, 2019).

8.8.5 Multiplex genome editing for multi-stress tolerance

Multiplex genome editing, which lets several genes be simultaneously modified is among the most exciting developments in CRISPR technology. For growing crops that can endure several abiotic stresses – such as drought, salt, and heat – this method is very helpful. Multiplex editing presents a complete method for engineering resilience in crops by targeting numerous stress-responsive genes at once. Recent research, for example, has looked at using multiplex CRISPR/Cas9 to edit proline-rich proteins and others related to stress genes, therefore producing multi-stress-resistant crop variants. This method not only increases the plant's capacity to resist different environmental pressures but also raises its general production in unfavorable conditions (Saikia *et al.*, 2020).

8.8.6 Targeting Cis-regulatory elements for stress resilience

Targeting cis-regulatory elements – which affect the expression of stress-responsive genes – CRISPR technology has also been used. Editing these regulatory regions can cause important genes involved in abiotic stress tolerance to either up- or downregulate. Through exact modification of the expression levels of many genes, CRISPR enables a more exact regulation of plant stress reactions, hence improving resistance to severe environmental circumstances (Bhat *et al.*, 2021).

8.9 Prospects and Challenges

Although CRISPR technology has great possibilities to raise abiotic stress tolerance in crops, obstacles still must be solved. Though developments in CRISPR technology have lowered these dangers, off-target effects – where unintentional gene editing occurs – remain a worry. Crucially, public acceptability and governmental clearance of CRISPR-edited crops will determine how widely these developments are adopted (Pandita, 2021). Combining omics technology with CRISPR editing presents fresh opportunities for the identification of fresh genes and regulatory networks linked with abiotic stress resistance. High-throughput genomes, transcriptomics, proteomics, and metabolomics can offer a complete understanding of the biochemical reactions of plants to stress, therefore enabling the creation of more focussed and potent CRISPR techniques for crop improvement.

8.9.1 Case studies of CRISPR-edited crops with improved abiotic stress resilience

Globally, agricultural output is greatly reduced by abiotic stress including drought, salt, severe temperatures, and heavy-metal toxicity. Emerging as a potent technique for creating crops with increased tolerance to environmental pressures is the latest use of CRISPR-Cas genome editing. Focussing on certain stressors like drought, salt, heat, cold, oxidative stress, and heavy metals, this section features case studies of CRISPR-edited crops showing enhanced abiotic stress tolerance.

8.9.1.1 Drought tolerance

Among the most severe abiotic pressures influencing agricultural output globally is drought. Using CRISPR-Cas, the creation of drought-tolerant agricultural cultivars has demonstrated encouraging outcomes:

- **Rice**. Using rice, CRISPR-Cas9 was utilized to alter the OsDREB2A gene, which is vital for the response to drought stress. By stimulating downstream stress-responsive genes, the modified plants displayed higher drought tolerance, hence enhancing plant life and output under low water availability (Erdoğan *et al.*, 2023).
- **Tomato**. CRISPR technology is also effectively used in tomatoes to increase drought resistance. Researchers increased tomato plant drought tolerance by deleting the SlHYPRP1 gene, which serves as a negative regulator of stress responses. Under drought, this gene editing preserved growth and enhanced water retention.
- **Salinity tolerance**. Thousands of hectares of agricultural land worldwide are affected by high soil salinity; hence the creation of salt-tolerant crops is very necessary for food security.
- **Rice**. Researchers targeted the *OsRR22* gene in rice using CRISPR-Cas9, therefore controlling the salt-stress response. By improving the plant's capacity to preserve ionic equilibrium and lower sodium build-up in tissues, knockout mutations of the corresponding gene produced enhanced salt tolerance (Joshi *et al.*, 2020).
- **Soybean**. By aiming at the *GmAITR* gene family, soybean was also edited utilizing CRISPR-Cas9 to enhance salt tolerance. Through the regulation of ABA, a fundamental mechanism for stress adaptation, the mutations generated in this gene family have been demonstrated to improve salt tolerance (Wang *et al.*, 2021).

8.9.1.2 Heat- and cold-stress tolerance

Severe temperatures, both high and low, have significant consequences on plant development and output. Improved heat- and cold-stress tolerance has been engineered in crops using CRISPR technology:

- **Tomato (heat tolerance)**. Using CRISPR-Cas9, the HsfA2 gene – which expresses a heat-shock protein necessary for thermotolerance – was modified in tomato. Upregulating heat genes that respond to stress helped the modified plants show better heat tolerance, which improved survival and fruit set at high temperatures (Li *et al.*, 2022).
- ***Arabidopsis* (cold tolerance).** *Arabidopsis'* cold tolerance was raised by a CRISPR-mediated deletion of CBF2, a negative regulator of the cold-response pathway, which activated cold-responsive genes boosting freezing tolerance. In colder climes, this method has promise for crops like barley and wheat (Yousefi *et al.*, 2022).

8.9.1.3 Other abiotic stresses (e.g. oxidative stress, heavy metals)

In addition to drought, salinity, and temperature extremes, CRISPR has been employed to combat other abiotic stresses like oxidative stress and heavy-metal toxicity, which are prevalent in areas with industrial pollution or degraded soils.

- **Oxidative stress in rice**. Modifying the OsCYP96B gene, which codes a cytochrome P450 enzyme, CRISPR-Cas9 was utilized to improve oxidative stress endurance in rice. Under severe oxidative stress, the modified plants demonstrated lower oxidative damage and better survival, therefore proving the value of CRISPR in controlling ROS levels (Kumar *et al.*, 2023).
- **Heavy-metal tolerance in *Arabidopsis*.** *Arabidopsis* was also used with CRISPR technology to target the MTF1 gene, implicated in metal transport and sequestration, therefore conferring heavy-metal resistance. Reduced tolerance to cadmium and arsenic resulting from the knockout of this gene opened the path for related uses in food crops including rice and maize (Bhat *et al.*, 2021).

8.10 High-Throughput Omics Technologies: Unveiling the Underlying Mechanisms

High-throughput omics technologies – genomics, transcriptomics, proteomics, and metabolomics – have revolutionized our capacity to grasp the fundamental biological pathways of abiotic stress tolerance in crops modified by CRISPR. When combined, these technologies may present a whole picture of how plants react to environmental challenges as each one of them offers a distinct layer of knowledge about the biological processes under operation.

8.10.1 Genomics: examining the genetic manual

Genomics is centered on the thorough investigation of an organism's whole DNA sequence including all its genes. Within the framework of plant stress resistance, genomics offers an understanding of the genetic variants that support a plant's tolerance of unfavorable environmental circumstances. Identification of QTLs and genes linked with abiotic stress tolerance in crops has been much aided by high-throughput sequencing methods including next-generation sequencing (NGS). For maize, for instance, new genomic discoveries have revealed multiple QTLs connected to drought and heat tolerance, which are currently being used for breeding stress-resilient variants (Farooqi *et al.*, 2022). Doing genome-wide association studies (GWAS) helps us to better grasp how certain genetic variants relate to stress reactions. Moreover, pan-genomics, which investigates the collective genomes of a species, has proved essential in spotting stress-resilience genes found in wild crop cousins but usually lacking in domesticated types. By use of gene-editing technologies such as CRISPR-Cas (Zenda *et al.*, 2021), these genetic resources from wild relatives are quite beneficial in enhancing stress resistance.

8.10.2 Transcriptomics: stress affects gene expression

Transcriptomics explores under conditions the RNA transcripts generated by the genome. Analyzing gene expression patterns helps scientists to pinpoint which genes, in response to stress, are either active or inhibited. Understanding the dynamic reactions of plants to abiotic stimuli such as drought, salt, and severe temperatures depends especially on this. For example, complicated transcriptional networks controlling stress reactions have been exposed using RNA sequencing (RNA-seq). Under drought and salt stress, transcriptomics has found important stress-responsive TFs in wheat that control downstream genes engaged in osmotic balance, ion homeostasis, and antioxidative defenses (Shah *et al.*, 2018). These results have been vital for the creation of CRISPR-edited crops able to either activate or repress these networks, thereby improving tolerance to certain stressors. Furthermore, becoming increasingly useful for analyzing cell-specific stress responses is single-cell transcriptomics. By allowing researchers to examine the gene expression patterns of individual cells inside a plant, this method provides a hitherto unheard-of understanding of how various cell types within tissues react to stress (Zenda *et al.*, 2021).

8.10.3 Proteomics: stress response: protein dynamics

Although transcriptomics and genomes reveal information on the possibility of gene expression, proteomics presents a more direct assessment of cellular activity employing protein identification and quantification in a stressed-out plant. Proteomics studies the synthesis, modification, and interactions among proteins in response to environmental signals. High-throughput mass spectrometry-based proteomics has progressed our knowledge of how plants control protein-level stress. Among the important actors seen in crops under abiotic stress are proteins linked to antioxidant defenses, ion transport, and stress-signaling pathways. Proteomic studies under heat stress, for instance, have underlined HSPs and proteins involved in ROS detoxification in rice, which are important in reducing cellular damage (Kausar *et al.*, 2022). Investigating post-translational modifications (PTMs), which generally control protein function, localization, and stability, is one of proteomics' strongest features. Particularly in regulating protein–protein interactions engaged in signaling pathways, PTMs including phosphorylation and ubiquitination have been demonstrated to be essential in fine-tuning the plant's response to abiotic stressors (Razzaq *et al.*, 2021).

8.10.4 Metabolomics: linking genotype to phenotype

The thorough study of metabolites – small molecules essential for intermediates and products of biological processes – is known as metabolomics. Linking the genotype to the observable phenotype, metabolomics provides an understanding of how stress alters plant biochemical processes.

Metabolomics has found important osmoprotectants, antioxidants, and secondary metabolites in stress-tolerant crops that enable plants to negotiate demanding environmental circumstances. Drought-tolerant cultivars, for example, commonly accumulate metabolites such as polyamines, glycine betaine, and proline, which help in osmotic adjustment and shield cell structures from dehydration (Abdelrahman *et al.*, 2018). Furthermore, exposing how metabolic pathways are rearranged in stress reactions through metabolomic investigations. For maize, for instance, metabolomics has been utilized to find stress-related biochemical fingerprints that match rising photosynthetic efficiency and water-use efficiency during drought circumstances (Farooqi *et al.*, 2022). These results suggest fresh directions for metabolic engineering based on CRISPR to improve abiotic stress tolerance.

8.10.5 Combining multi-omics for all-inclusive views

Combining genomes, transcriptomics, proteomics, and metabolomics – collectively known as multi-omics – allows a comprehensive knowledge of the intricate relationships controlling stress reactions in plants. Using network models that capture the interactions between genes, proteins, and metabolites, multi-omics techniques help researchers discover important regulating nodes that might be targets for genetic modification. Crops like wheat and lentils have one of the most effective uses for multi-omics integration as the combination of omics data has helped to identify biomarkers for stress resilience. These biomarkers are currently being used for genome editing and marker-assisted selection (MAS), therefore hastening the creation of multi-stress-tolerant agricultural varieties (Tiwari *et al.*, 2022). Furthermore, being applied to forecast the results of genetic changes and to devise CRISPR-based therapies for plant resistance to environmental stress are systems biology techniques, which combine omics data with computer modeling (Yang *et al.*, 2021).

8.10.6 Integration of multi-omics data to decode stress responses in crops modified by CRISPR

The rapid development of CRISPR-Cas technology allows scientists to precisely modify genes linked to abiotic stress responses in agricultural plants. However, knowing how these changes result in better stress resistance calls for a comprehensive picture of the molecular processes involved. Here is where multi-omics integration – the integrated study of genomes, transcriptomics, proteomics, metabolomics, and other omics –fits. Integration of multi-omics data allows researchers to break apart the intricate networks and processes controlling plant responses to abiotic stressors, hence improving the effectiveness of crops modified using CRISPR-edited genes.

8.10.7 Genomics and transcriptomics: mapping stress-responsive genes

Multi-omics integration is fundamentally driven by the interaction between genomics and transcriptomics, which offers an understanding of the genetic and transcriptional environments of CRISPR-edited crops under stress. While transcriptomics shows how these genes are produced in response to different abiotic stimuli, genomics helps to identify stress-responsive genes and QTLs. For instance, combining genomes and transcriptomics in wheat has produced differently expressed transcripts (DETs) that are essential for stress acclimatization. This method helps separate important TFs including TaWRKY33, a critical hub gene implicated in response to heat, drought, and salt stressors (Da Ros *et al.*, 2023). Precise genetic treatments that increase stress tolerance are made possible by the integration of genomes and transcriptomics data, therefore enabling CRISPR applications to target these important regulators more effectively.

8.10.8 Proteomics: understanding protein dynamics under stress

Proteomics provides data on the quantity, alteration, and interaction of proteins in CRISPR-edited crops, therefore adding even another level of complexity to the multi-omics method. Understanding how genetic changes affect the protein machinery that controls stress responses requires this knowledge. Important functions for HSPs, antioxidant enzymes, and ion transporters in increasing stress resistance have been uncovered by proteomic research. Under drought stress, for example, proteome profiling of rice revealed numerous increased HSPs shielding cellular structures against heat-induced damage (Razzaq *et al.*, 2021). These results are vital for applications of CRISPR aiming to raise the quantity or activity of such proteins, thereby strengthening the plant's capacity to resist high temperatures and water shortage.

8.10.9 Metabolomics: linking genetic edits to phenotypic outcomes

Metabolomics offers a clear path from genetic changes to the physiological effects they generate. Metabolomics clarifies the metabolic changes in response to abiotic stimuli utilizing analysis of the tiny molecules acting as intermediates in biochemical processes. For instance, for research on drought-tolerant maize, metabolomics demonstrated an increase in osmoprotectants including proline and glycine betaine, which support cellular integrity under water-limited circumstances (Farooqi *et al.*, 2022). Integrating metabolomic data with genomic and transcriptomic results helps researchers to pinpoint important metabolic pathways sensitive to CRISPR-induced genetic changes, therefore enabling more focussed treatments to improve stress resistance. Integration of layers of multi-omics for holistic understanding combining multi-omics layers – genomics, transcriptomics, proteomics, and metabolomics – allows for a whole new knowledge of how CRISPR-edited crops respond to environmental stresses. Combining these many datasets, systems biology techniques offer insights into the regulatory networks controlling stress reactions. Using transcriptomics, proteomics, and metabolomics data, systems biology showed in wheat a co-expression network of over 150 hub genes, many of which were associated with ethylene-responsive transcription factors and other stress regulators (Zhou *et al.*, 2022). This network indicated possible targets for CRISPR editing that may be utilized to concurrently improve the resistance of the plant to many abiotic stressors. Lentils have also been used in such multi-omics integration; integrating transcriptomics, proteomics, and metabolomics data allowed the discovery of important stress-responsive genes and metabolites helping to provide drought and salinity resistance. Currently under use in breeding operations to create multi-stress-tolerant cultivars (Tiwari *et al.*, 2022) these multi-omics-characterized plants

8.10.10 Bioinformatics and data-integration challenges

Though multi-omics integration has great promise, data processing, normalizing, and integration still present difficult problems. Multi-omics studies produce an enormous amount of data, which can be daunting. Integrating these many datasets calls for sophisticated bioinformatics tools and knowledge. Multi-omics data have been combined and physiologically significant insights extracted using methods like network-based integration and machine learning. By combining omics data from several sources, machine learning models have been used to forecast, for example, how particular gene changes would affect the general phenotype of the plant (Bodein *et al.*, 2022). Furthermore, the integration of omics data has been demonstrated to increase the precision of prediction algorithms applied to choose the most likely genomic targets for CRISPR editing. These models can forecast how CRISPR changes may impact the response of the plant to abiotic stress by using data from gene expression, protein interactions, and metabolite levels (Yang *et al.*, 2021).

8.11 Case Studies: Omics-based Insights into Stress Resilience in CRISPR-Edited Crops

8.11.1 CRISPR-Cas

Combining technology with omics techniques like genomics, transcriptomics, proteomics, and metabolomics has opened the path for a thorough investigation of the molecular processes underneath abiotic stress tolerance in crops. Through genome editing, researchers may precisely target important genes, proteins, and metabolites engaged in stress responses by using these high-throughput methods. Several example studies illustrating the value of combining omics data with CRISPR-mediated gene editing for improving stress tolerance in crops are presented in this part.

8.11.1.1 Wheat: integration of genomics and transcriptomics to enhance drought and heat tolerance

Staple crop wheat is very vulnerable to abiotic conditions like heat and drought. Using CRISPR technology, a new study combining genomes and transcriptomics found and modified stress-responsive genes in wheat. Researchers found important transcription factors and regulating elements linked with increased drought tolerance employing GWAS and RNA sequencing. More specifically, the TaWRKY33 gene was identified as a significant regulator of drought and heat reactions, hence influencing downstream stress-protective pathways (He *et al.*, 2016). Negative regulatory domains within the TaWRKY33 gene were deleted using CRISPR-Cas9, therefore enhancing drought and heat tolerance in modified wheat lines. This study shows how integrating genomes and transcriptomics helps to identify stress-tolerance genes and promotes the creation of climate-resilient wheat variants.

8.11.1.2 Rice: proteomics and metabolomics to improve salt tolerance

Although rice is a staple crop worldwide, salt stress seriously reduces its yield. Combining metabolomics and proteomics with CRISPR-Cas genome editing has produced a significant new understanding of the molecular processes improving salt tolerance in rice. In one work, mass spectrometry-based proteomics was used to discover proteins involved in ion homeostasis and ROS detoxification, both important for salt-stress tolerance. Using a regulator of the salt-stress response, the OsRR22 gene was edited to increase the plant's capacity to control ion imbalances, hence improving salt tolerance (Joshi *et al.*, 2023). Further revealing an abundance of osmoprotectants such as proline and glycine betaine, which help to preserve cellular integrity under salty circumstances, were metabolomics. This work shows how, paired with CRISPR-mediated gene editing, proteomic and metabolic profiling may maximize stress tolerance in rice.

8.11.1.3 *Tomato: omics-assisted CRISPR editing for multi-stress tolerance*

Multiple abiotic stresses – including drought, salt, and heat – have been addressed in tomato through omics technologies in concert with CRISPR-Cas. Using transcriptome profiling, researchers found numerous stress-responsive genes involved in controlling ABA signaling pathways: SlDREB1 and SlAREB1. These genes were concurrently changed using CRISPR-Cas9, therefore enhancing tomato tolerance to salt and drought. In this work, multi-omics data – especially from transcriptomics and proteomics – were important for mapping out the intricate signaling networks engaged in stress responses (Ahmad *et al.*, 2021a). HSPs and antioxidant enzymes, among other stress-protective proteins, were found by proteomic analysis of the modified lines to be upregulated, therefore improving the plant's survival under demanding environmental circumstances.

8.11.1.4 *Rapeseed: enhancing cold and drought tolerance through multi-omics integration*

A major oilseed crop, rapeseed (*Brassica napus*) suffers greatly in output from abiotic stressors like cold and drought. Using a multi-omics approach – combining genomes, transcriptomics, proteomics, and metabolomics – a recent case study broke out the molecular pathways behind stress responses in rapeseed. Key genes involved in cold acclimation and drought response were found by QTL mapping and RNA sequencing: BnCBF1 controls cold tolerance and BnDREB2A modifies drought tolerance (Raza *et al.*, 2021). Improved stress tolerance came from eliminating negative regulators of these pathways using CRISPR-Cas9. The work also made use of metabolomics to pinpoint important metabolites such as antioxidants and soluble carbohydrates that build up in response to drought and cold pressures. Apart from offering an understanding of stress-response systems, this all-encompassing omics-based approach helped to create rapeseed cultivars with improved abiotic stress tolerance.

8.11.1.5 *Maize: omics and CRISPR-Cas for heat and drought resilience*

Combining genomic and transcriptomic data with CRISPR-Cas has allowed major progress in the creation of heat- and drought-tolerant maize cultivars. Using RNA sequencing, researchers found important genes controlling heat and drought responses, including ZmHsf1 and ZmDREB2A, both of which are transcription factors controlling paths of heat shock and dehydration response. Negative regulators of these pathways were eliminated using CRISPR-Cas9, therefore increasing the expression of stress-protective genes. While metabolomic research found an increase in osmoprotectants that support drought tolerance, proteomic profiling revealed the overexpression of HSPs and chaperones that minimize protein misfolding under heat stress (Kumar *et al.*, 2023). This multi-omics-driven CRISPR method offers a strong basis for creating maize crops resistant to climate change.

8.11.2 Case studies: omics-based insights into stress resilience in CRISPR-edited crops

CRISPR-Cas technology, combined with omics approaches such as genomics, transcriptomics, proteomics, and metabolomics, has paved the way for a detailed exploration of the molecular mechanisms underpinning abiotic stress resilience in crops. By leveraging these high-throughput technologies, researchers can identify key genes, proteins, and metabolites involved in stress responses and target them precisely through genome editing. This section highlights several case studies that demonstrate the utility of integrating omics data with CRISPR-mediated gene editing for enhancing stress tolerance in crops.

8.11.2.1 *Wheat: integration of genomics and transcriptomics to enhance drought and heat tolerance*

Wheat, a staple crop, is particularly susceptible to abiotic stresses such as drought and heat. A recent study integrated genomics and transcriptomics to identify and edit stress-responsive genes in wheat using CRISPR technology. Through GWAS and RNA sequencing, researchers identified key TFs and regulatory elements associated with enhanced drought tolerance. Specifically, the TaWRKY33 gene was found to be a major regulator of drought and heat responses, modulating downstream stress-protective pathways (Da Ros *et al.*, 2023). CRISPR-Cas9 was used to knock out negative regulatory domains within the TaWRKY33 gene, leading to improved drought and heat tolerance in edited wheat lines. This case illustrates how combining genomics, and transcriptomics enables the identification of stress-tolerance genes and facilitates the development of climate-resilient wheat varieties.

8.11.2.2 Rice: proteomics and metabolomics to improve salt tolerance

Rice is a vital global crop, but its productivity is severely affected by salinity stress. Integrating proteomics and metabolomics with CRISPR-Cas genome editing has yielded important insights into the molecular pathways that enhance salt tolerance in rice. In one study, researchers applied mass spectrometry-based proteomics to identify proteins involved in ion homeostasis and ROS detoxification, both critical for salt stress resilience. By editing the OsRR22 gene, a regulator of salt-stress response, CRISPR technology was able to enhance the plant's ability to manage ion imbalances, resulting in improved salt tolerance (Joshi *et al.*, 2023). Metabolomics further revealed an accumulation of osmoprotectants such as proline and glycine betaine, which contribute to maintaining cellular integrity under saline conditions. This study exemplifies how proteomic and metabolic profiling, combined with CRISPR-mediated gene editing, can optimize stress resilience in rice.

8.11.2.3 Tomato: omics-assisted CRISPR editing for multi-stress tolerance

In tomatoes, the integration of omics technologies with CRISPR-Cas has been used to address multiple abiotic stresses, including drought, salinity, and heat. Through transcriptomic profiling, researchers identified several stress-responsive genes, including SlDREB1 and SlAREB1, which are involved in regulating ABA signaling pathways. CRISPR-Cas9 was employed to simultaneously edit these genes, improving tomato resilience to both drought and salinity. In this study, multi-omics data, particularly from transcriptomics and proteomics, were critical for mapping out the complex signaling networks involved in stress responses (Ahmad *et al.*, 2021b). Proteomic analysis of the edited lines revealed the upregulation of stress-protective proteins, such as HSPs and antioxidant enzymes, which enhanced the plant's ability to survive under harsh environmental conditions.

8.11.2.4 Rapeseed: enhancing cold and drought tolerance through multi-omics integration

Rapeseed (*Brassica napus*), an important oilseed crop, faces significant productivity challenges due to abiotic stresses like cold and drought. A recent case study utilized a multi-omics approach – combining genomics, transcriptomics, proteomics, and metabolomics – to dissect the molecular mechanisms underlying stress responses in rapeseed. Through QTL mapping and RNA sequencing, key genes involved in cold acclimation and drought response were identified, including BnCBF1, which regulates cold tolerance, and BnDREB2A, which modulates drought tolerance (Raza *et al.*, 2021). CRISPR-Cas9 was employed to knock out negative regulators of these pathways, resulting in improved stress resilience. The study also utilized metabolomics to identify key metabolites, such as soluble sugars and antioxidants, which accumulate in response to cold and drought stresses. This comprehensive omics-based approach not only provided insights into stress-response mechanisms but also facilitated the development of rapeseed cultivars with enhanced abiotic stress tolerance.

8.11.2.5 Maize: omics and CRISPR-Cas for heat and drought resilience

In maize, integrating genomic and transcriptomic data with CRISPR-Cas has enabled significant advancements in the development of heat- and drought-tolerant varieties. By employing RNA sequencing, researchers identified key genes involved in heat- and drought-stress responses, including ZmHsf1 and

ZmDREB2A, both of which are transcription factors regulating heat shock and dehydration response pathways. CRISPR-Cas9 was utilized to knock out negative regulators of these pathways, leading to enhanced expression of stress-protective genes. Proteomic profiling revealed the upregulation of HSPs and chaperones that mitigate protein misfolding during heat stress, while metabolomic analysis identified an increase in osmoprotectants that aid in drought tolerance (Kumar *et al.*, 2023). This multi-omics-driven CRISPR approach provides a robust framework for developing climate-resilient maize cultivars.

8.12 Genomic Approaches: Identifying Stress-related Genes and Pathways

8.12.1 Genome-wide association studies (GWAS) and CRISPR applications

The swift progress of GWAS has greatly enhanced our comprehension of intricate characteristics of plants, specifically their ability to withstand abiotic stress. GWAS enables the detection of genetic loci associated with specific traits by examining the entire genome for connections between genetic variations and observable phenotypes. CRISPR-Cas9 gene editing can be used in conjunction with GWAS data to make specific genetic alterations that improve stress tolerance in crops. This section explores the amalgamation of GWAS and CRISPR technology to pinpoint and manipulate stress-related genes and pathways.

8.12.2 Role of GWAS in identifying stress-related genes

GWAS has become a crucial instrument in agricultural research for analyzing the genetic foundation of resistance to abiotic stress. GWAS offers valuable insights into the genetic make-up of stress responses by identifying the connections between single nucleotide polymorphisms (SNPs) and complex features like drought tolerance or salinity resistance. An example of this is in wheat, where a GWAS study discovered 153 QTLs that are linked to cold tolerance and resistance to diseases such as leaf rust and yellow rust. Following that, these genetic locations were further refined to identify specific candidate genes, which were then confirmed using CRISPR-Cas9 technology. This showcases the possibility of combining GWAS with gene-editing techniques to enhance crop development (Pang *et al.*, 2021).

8.12.3 Rice is a prime crop that is often used as a model in GWAS applications

Over ten years, extensive research using GWAS techniques in rice has resulted in the identification of genetic loci that are linked to several characteristics associated with abiotic stress, such as the ability to withstand drought and tolerate high levels of salt. These regions have been further examined through functional validation tests, generally utilizing CRISPR-mediated knockouts or insertions to confirm the functions of candidate genes in stress responses (Wang *et al.*, 2019). GWAS has not only helped locate essential genes but has also increased our understanding of the genetic diversity underlying stress resilience, which is critical for breeding programs aiming at producing stress-tolerant rice varieties.

8.12.4 GWAS in CRISPR-edited crops: from discovery to application

The translation of GWAS findings into practical applications requires rigorous methodologies for gene functional validation, a gap that CRISPR-Cas9 technology solves quickly. CRISPR provides precise editing of the genetic loci found through GWAS, allowing researchers to investigate the functional significance of specific alleles or regulatory areas. For example, in maize, GWAS found numerous loci linked with heat and drought tolerance, including transcription factors such as ZmHsf1 and ZmDREB2A. CRISPR was applied to knock out negative regulators within these regions, leading to enhanced stress tolerance and demonstrating the efficacy of genome editing in confirming GWAS results (Kumar *et al.*, 2023). In another study, tomato was the topic of a GWAS aimed at discovering genetic variants linked with salinity tolerance. After identifying many potential genes, including those involved in ion transport and osmoprotectant production, CRISPR-Cas9 was utilized to alter these loci. The altered plants demonstrated higher salinity tolerance, confirming the roles of the targeted genes and providing an avenue for further agricultural enhancement

8.12.5 Fine-mapping and functional validation of GWAS loci

While GWAS can uncover potential loci related to abiotic stress, fine mapping is sometimes required to define the genetic variations responsible for the characteristic. This approach entails narrowing down the region surrounding a big GWAS finding to locate the causal gene or variation. In the post-GWAS era, the integration of fine-mapping approaches with CRISPR has changed our ability to validate and describe the function of discovered loci. For instance, in a work involving *Arabidopsis*, fine-mapping paired with CRISPR-based genome editing was utilized to evaluate the function of noncoding variations connected to drought tolerance. The researchers employed CRISPR to induce mutations into the anticipated enhancer regions of these variants and observed alterations in gene expression and phenotypic consequences, thereby tying noncoding regulatory elements to stress-related features (Morris *et al.*, 2023). This research highlights how GWAS, when combined with genome editing, can identify not only coding variants but also regulatory regions that contribute to complicated stress-response systems.

8.12.6 CRISPR as a tool for enhancing GWAS utility in plant breeding

The combination of GWAS and CRISPR-Cas technologies is becoming increasingly essential in plant breeding projects. By integrating high-throughput genotyping and phenotyping data, breeders can swiftly identify and alter genes responsible for desirable traits such as drought tolerance and cold resilience. One significant application of this strategy is in the creation of wheat varieties with a higher cold tolerance. By exploiting GWAS data to identify QTLs connected to cold resistance and using CRISPR to alter potential genes within these loci, breeders have been able to generate wheat lines that display enhanced survival and production under cold-stress conditions (Pang *et al.*, 2021). Another example was rapeseed, where GWAS identified numerous QTLs connected with heat and drought tolerance. CRISPR-Cas9 was subsequently applied to target critical genes within these loci, leading to the production of novel rapeseed cultivars with greater tolerance to environmental stresses (Raza *et al.*, 2021). The integration of GWAS and CRISPR provides a powerful framework for accelerating the creation of stress-resilient crops, which is critical in the face of climate change and increasing food security challenges.

8.12.7 Identification of key genes and regulatory networks involved in stress responses

The capacity to identify important genes and regulatory networks involved in abiotic stress responses is a vital part of enhancing CRISPR-edited crops for stress tolerance. Through the integration of multi-omics technologies, researchers are increasingly able to determine the individual genes and pathways that enable plants to endure adverse climatic conditions such as drought, salinity, heat, and cold. This section analyzes the role of gene regulatory networks (GRNs), transcription factors (TFs), and other genetic components in abiotic stress resistance, with specific attention on the application of CRISPR-Cas9 technology for gene functional validation.

8.12.8 Transcription factors and regulatory networks

Transcription factors play key functions in influencing the expression of stress-responsive genes. Gene regulatory networks (GRNs) are complex, consisting of interactions between

TFs and their target genes. Recent research in *Arabidopsis thaliana* demonstrated that TFs involved in ROS signaling are crucial for plant responses to environmental stress. In a study combining diverse regulatory input networks, researchers discovered 1491 transcription factors and their 31,393 target genes, producing a comprehensive GRN that might be related to certain stress activities such as ROS control (De Clercq *et al.*, 2021). By identifying important regulators of ROS responses, the study suggests prospective targets for CRISPR-mediated gene editing to boost abiotic stress tolerance. Another study in poplar (*Populus simonii*) identified promoter upstream transcripts (PROMPTs) as carriers of MYB transcription factors. These PROMPTs were demonstrated to promote the expression of target genes involved in osmotic stress responses. The researchers employed a combination of strand-specific RNA sequencing and nanosheet-mediated nucleic acid delivery to modify these TFs, revealing their involvement in transcriptional control under osmotic stress (Song *et al.*, 2019). This research underlines the necessity of integrating transcriptomics data to find novel regulatory components and proposes possible CRISPR targets for enhancing stress resilience.

8.12.9 Regulatory networks and stress pathways

Gene networks governing stress responses frequently involve a range of TFs, metabolic pathways, and signaling cascades. A comprehensive investigation in *Arabidopsis thaliana* highlighted the role of transcriptional and post-transcriptional control in ROS management. By applying machine learning and supervised learning methodologies, researchers constructed an integrated gene regulatory network (iGRN) spanning hundreds of genes and TFs. This system-level approach enabled the identification of many TFs previously unlinked to ROS responses, which were validated using phenotypic tests (De Clercq *et al.*, 2021). The integration of CRISPR interference (CRISPRi) and CRISPR activation (CRISPRa) with transcriptional profiling has provided new insights into gene regulation under stress. For instance, a study employing CiBER-seq (CRISPR interference barcode expression reporter sequencing) in yeast demonstrates the use of deep sequencing to map genomic networks implicated in nutritional stress responses. This technique allows the identification of metabolic and stress-response pathways, verifying the activity of TFs and key regulators in these pathways (Muller *et al.*, 2020).

8.13 Functional Validation of Candidate Genes Using CRISPR

The functional validation of candidate genes found using multi-omics techniques is a vital step in turning genomic discoveries into practical agricultural applications. CRISPR-Cas9 technology provides a potent tool for gene deletion, knock-in, and transcriptional regulation, allowing researchers to explore the involvement of individual genes in stress responses.

8.13.1 CRISPR applications in stress gene validation

A prime example of CRISPR's utility in gene validation is demonstrated in research on noncoding genomic regions. One study used CRISPR-Cas9 to alter cis-regulatory sequences in the sea urchin *Strongylocentrotus purpurates*, targeting the promoter regions of the nodal gene. The research indicated that modifying cis-activator and cis-repressor sites might modulate transcriptional output, so proving the regulatory significance of noncoding areas in stress responses (Pieplow *et al.*, 2021). CRISPR-Cas9 has also been useful in verifying the roles of RNA-based regulators. A study in poplar uses CRISPR to alter long noncoding RNAs (lncRNAs) involved in osmotic stress responses. By knocking out certain lncRNA loci, researchers were able to establish their involvement in regulating important MYB TFs, leading to increased plant tolerance to water stress (Song *et al.*, 2019). Another CRISPR usage in stress resilience involves the sea urchin germline gene Nanos2. Researchers employed CRISPR to dissect regulatory networks driving germ cell identity and stress adaptation. The study demonstrated how CRISPR-mediated mutagenesis could uncover upstream regulatory processes, shedding information on genomic components crucial for stress resilience (Pieplow *et al.*, 2021).

8.13.2 Transcriptome profiling in CRISPR-edited crops under stress conditions

The field of transcriptomics gives vital insights into the dynamic gene expression changes that occur in plants under abiotic stress. This approach permits the identification of differentially expressed genes (DEGs) in response to stress and offers a better understanding of the complex regulatory networks involved. When paired with CRISPR-Cas9 gene editing, transcriptomics can help explain the molecular pathways that contribute to better stress tolerance in crops. In this section, we study the application of transcriptome profiling in CRISPR-edited crops subjected to diverse stress conditions, concentrating on how transcriptomics contributes to identifying the genetic basis of stress responses and confirming the impacts of CRISPR interventions.

8.13.3 Case study: salinity stress in CRISPR-edited tomato

A critical work employing tomato (*Solanum lycopersicum*) addressed the relevance of transcriptome profiling in understanding gene expression under salinity stress. Researchers employed CRISPR-Cas9 to knock out regions of the SlHyPRP1 gene, which is a negative regulator of salt tolerance. By utilizing RNA sequencing (RNA-seq), they detected significant changes in the expression of stress-responsive genes in CRISPR-edited tomato lines under salt-stress conditions. The altered plants displayed increased salinity tolerance due to the precise deletion of the negative regulatory regions, which led to the overexpression of osmoprotective genes and ion transporters (Tran *et al.*, 2021). The transcriptome data showed that genes involved in osmotic regulation, ROS detoxification, and ABA signaling were differentially expressed in the altered plants compared to the wild-type controls. This study illustrates how transcriptome profiling can disclose the molecular processes by which CRISPR-edited crops gain greater stress tolerance, demonstrating a direct relationship between gene editing and phenotypic consequences.

8.13.4 Transcriptomic insights into drought tolerance in rice

Another noteworthy application of transcriptomics in CRISPR-edited crops was demonstrated in a study on rice (*Oryza sativa*) under drought stress. Researchers employed RNA-seq to assess the global transcriptome profiles of CRISPR-edited rice lines, in which important drought-responsive genes were knocked out. The study focussed on the OsDREB1 and OsNAC5 TFs, which are known to regulate stress-inducible genes. Transcriptome analysis revealed that the altered plants exhibited higher expression of dehydration-responsive genes, stress-related TFs, and heat-shock proteins, which collectively improved their drought tolerance (Abdelrahman *et al.*, 2018). The overexpression of these stress-responsive genes in CRISPR-edited rice lines indicated that the modification of key TFs might successfully boost drought tolerance. Transcriptome analysis thus provides essential insights into the genetic changes that occur in response to stress, verifying the functional significance of altered genes and contributing to the development of more resilient crop varieties.

8.13.5 Multi-stress tolerance in maize: transcriptome dynamics

In maize (Zea mays), transcriptome profiling has been applied to explore the impact of CRISPR-mediated gene editing on multi-stress tolerance. A recent investigation focussed on heat and drought tolerance by altering genes associated with ROS homeostasis and heat-shock response pathways. Through RNA-seq, the researchers discovered significant differentially expressed transcripts (DETs) in CRISPR-edited maize plants subjected to combined heat and drought stress. The altered plants demonstrated higher expression of genes involved in antioxidant defense, protein folding, and water transport, which contributed to improved survival and growth under stress conditions (Li *et al.*, 2022). The integration of transcriptome data with CRISPR editing allowed the researchers to define precise regulatory pathways triggered by the altered genes, offering a thorough understanding of how these genetic alterations provide multi-stress tolerance. This case study illustrates the significance of transcriptome profiling in identifying important molecular players engaged in stress responses and demonstrating the effectiveness of CRISPR-based gene editing in boosting crop resilience.

8.13.6 Advanced transcriptomic techniques in CRISPR-edited crops

Recent developments in single-cell transcriptomics have greatly increased our ability to study stress-responsive gene expression in CRISPR-edited crops. Single-cell CRISPR interference (CRISPRi) screens, paired with RNA-seq, allow for the exact investigation of gene expression changes at the cellular level. This strategy successfully discovered stable transcriptome signatures that were not found by conventional methodologies, showing the power of single-cell transcriptomics for fine-tuned gene expression investigation (Truchi *et al.*, 2023). Although this technology has been predominantly employed in cancer research, it holds significant potential for plant biology as well. By applying single-cell transcriptomics in CRISPR-edited crops, researchers could acquire a more complete understanding of how various cell types respond to abiotic stimuli, yielding deeper insights into the cell-specific regulation of stress-responsive genes.

8.13.7 Integration of transcriptomics with other omics approaches

To fully untangle the intricacy of stress responses in CRISPR-edited crops, transcriptomics is commonly coupled with other omics technologies, such as proteomics and metabolomics. For example, a study in rapeseed (*Brassica napus*) paired RNA-seq with mass spectrometry-based proteomics to investigate the multi-omics profile of CRISPR-edited plants under cold stress. The combination of transcriptome and proteomic data revealed coordinated changes in gene expression and protein abundance, particularly in genes implicated in antioxidant defense and cold acclimation, offering a holistic perspective of the stress response network (Raza *et al.*, 2021). Such multi-omics techniques enable researchers to validate the functional roles of CRISPR-edited genes at several levels of biological control, from gene expression to protein function and metabolic output. This complete strategy is critical for producing crops with greater tolerance to abiotic challenges, guaranteeing that the altered genes not only modulate gene expression but also lead to real physiological and phenotypic changes.

8.13.8 Key stress-responsive genes and transcription factors identified through RNA-Seq

The discovery of important stress-responsive genes and TFs has been considerably advanced by RNA-Seq technologies, notably in CRISPR-edited crops. Recent research has revealed numerous TFs that play key roles in increasing abiotic stress tolerance, such as drought, salt, and temperature extremes. For instance, GmAITR genes in soybean have been discovered as essential regulators of abscisic acid (ABA) signaling, a hormone crucial for stress responses. The modification of these genes using CRISPR/Cas9 has shown to boost salinity tolerance without fitness costs, making them attractive candidates for improving agricultural resilience (Wang *et al.*, 2021). In solanaceous crops, TF families, including WRKY, MYB, NAC, and ERF, have been widely connected with abiotic stress response. RNA-Seq investigations have indicated that these TFs influence the expression of genes involved in osmotic stress, cold tolerance, and oxidative stress. The NAC and bZIP TFs have been demonstrated to control stress-responsive genes effectively, contributing to increased stress tolerance in these crops (Tolosa and Zhang, 2020). In rice, ZFP TFs were discovered as important regulators of stress responses through epigenetic changes. Genome-wide research demonstrated that these TFs undergo cytosine DNA methylation in response to salt and osmotic stressors, leading to the activation of stress-related genes. This shows that epigenetic control plays a substantial role in the stress adaptation of crops (Ahmad *et al.*, 2019). Furthermore, SlNAP1, a NAC TF discovered in tomato, was demonstrated to promote drought tolerance and resistance to bacterial infections by RNA-Seq research. This TF affects hormonal pathways, including gibberellins (GAs), salicylic acid (SA), and ABA, which are critical for plant development and stress responses (Wang *et al.*, 2020). Another notable result is the significance of bHLH TFs in rice, where the OsbHLH024 gene was found to be a negative regulator of salt stress. CRISPR/Cas9-mediated deletion of this gene led to increased salt tolerance, indicating its potential for agricultural enhancement (Alam *et al.*, 2022).

8.13.9 Differential gene expression analysis in CRISPR vs. non-CRISPR crops

Differential gene expression (DGE) comparison between CRISPR-edited and nonedited crops provides insights into how gene editing alters stress responses. For example, in *Arabidopsis*, CRISPR/Cas9-mediated deletion of CBF1/2/3 genes, which are crucial for cold sensitivity, led to a dramatic change in the expression levels of over 3000 cold-responsive (COR) genes. These alterations show that CRISPR editing can fine-tune stress responses by modifying the expression of critical regulatory genes (Shi *et al.*, 2017). Similarly, the OsbHLH024 gene in rice was knocked out using CRISPR/Cas9, resulting in a different expression profile compared to nonedited plants under salt stress. The altered plants demonstrated elevated expression of ion transporter genes such as OsHKT1;3 and OsSOS1, which are critical for maintaining ion homeostasis under stress conditions (Alam *et al.*, 2022). In rice, RNA-Seq study of CRISPR-edited SNAC1-overexpressing lines revealed the differential expression of almost 1000 genes relative to wild-type plants during drought stress. These genes were engaged in different stress-responsive pathways, including ABA signaling and water-loss response, demonstrating that CRISPR/Cas9 can drastically affect the stress response at the transcriptome level (Li *et al.*, 2019).

Moreover, in maize, the overexpression of ZmBES1/BZR1-5 via CRISPR/Cas9 led to a distinct gene expression profile under osmotic stress, with elevated expression of genes implicated in stress response and metabolic pathways. This underlines the potential of CRISPR/Cas9 in altering stress-related pathways to promote crop resilience (Sun *et al.*, 2020). The CRISPR/Cas9-mediated editing of ERF TFs in several crops has also been shown to change the expression of stress-responsive genes. For instance, in *Arabidopsis*, the overexpression of JcERFVII2 resulted in the activation of genes involved in fermentative metabolism and oxidative stress response, revealing the broad influence of CRISPR editing on gene networks (Juntawong *et al.*, 2020).

8.13.10 High-throughput proteomics in studying stress resilience

High-throughput proteomics has become a crucial tool in studying the intricate protein-level changes that occur in CRISPR-edited crops under abiotic stress conditions. This approach allows for the thorough investigation of the proteome, revealing important proteins that contribute to stress resilience. Recent breakthroughs in mass spectrometry-based proteomics have permitted the identification and quantification of thousands of proteins in a single experiment, revealing insights into the molecular pathways underlying stress tolerance. For instance, the use of Zeno SWATH MS, a data-independent acquisition approach, has considerably expanded the depth and dynamic range of proteome investigations, allowing for the precise quantification of stress-responsive proteins in minute samples. This method has been particularly effective in identifying proteins implicated in oxidative stress-response pathways, which are crucial for strengthening the resilience of crops to environmental stresses (Wang *et al.*, 2022a).

In the context of CRISPR-edited crops, proteomic studies have revealed insights into the post-translational modifications (PTMs) that alter protein function under stress circumstances. PTMs like phosphorylation and ubiquitination play crucial roles in regulating stress-responsive proteins, consequently altering the plant's ability to cope with abiotic stresses such as drought, salt, and severe temperatures. For example, research on *Arabidopsis thaliana* has shown that CRISPR-mediated deletion of certain genes leads to modifications in the phosphorylation state of essential proteins involved in ABA signaling, a vital pathway for drought tolerance (Kausar *et al.*, 2022).

Moreover, high-throughput proteomics has aided the identification of novel stress-responsive proteins that potentially serve as targets for CRISPR/Cas9-mediated genome editing to boost agricultural resilience. Proteomic research of CRISPR-edited rice, for instance, revealed the overexpression of HSPs and other chaperones that protect cellular proteins from degradation under heat stress. These findings underline the potential of proteomics in directing the identification of candidate genes for editing, hence enabling the production of crops with higher stress tolerance (Mustafa and Komatsu, 2021).

In addition to discovering stress-responsive proteins, high-throughput proteomics also contributes to understanding the crosstalk between distinct stress-response pathways. For example, investigations have demonstrated that proteins involved in ROS scavenging and osmoprotection commonly interact with those regulating hormonal pathways, such as ABA and ethylene. This connection emphasizes the extensive network of protein–protein interactions that underlie the plant's response to abiotic stress, which can be analyzed by proteomic techniques.

8.13.11 Key stress-responsive proteins and post-translational modifications in CRISPR-edited crops

The research of major stress-responsive proteins in CRISPR-edited crops has improved substantially with the development of high-throughput proteomic techniques. These proteins, which play key roles in the plant's response to abiotic stress, are commonly exposed to numerous post-translational modifications (PTMs) that control their activity, stability, and localization. A recent study has highlighted the role of PTMs such as phosphorylation, ubiquitination, and sumoylation in influencing the activity of stress-related proteins. For example, in tomato, the CRISPR/Cas9-mediated editing of SlHyPRP1 led to the precise deletion of its functional domains, resulting in increased salt-stress tolerance. The study indicated that the altered protein underwent considerable phosphorylation modifications, which were crucial for its role in stress response (Tran *et al.*, 2021).

Another essential protein alteration involved in stress reactions is ubiquitination. A complete investigation of the ubiquitin-proteasome system in potato found that during drought stress, the ubiquitination of essential proteins was considerably elevated, boosting the plant's ability to cope with water scarcity. The study found 314 lysine modification sites across 200 proteins, revealing the widespread significance of ubiquitination in stress tolerance (Tang *et al.*, 2021). Similarly, CRISPR/Cas9 technology has been utilized to modify the activity of selenoproteins, which are known to undergo multiple PTMs including phosphorylation and acetylation. The insertion of noncanonical amino acids in these proteins has allowed for the precise investigation of their function and regulation under stressful situations. These alterations have been demonstrated to have key roles in oxidative stress responses and immunological modulation in plants (Morosky *et al.*, 2023).

8.13.12 Integration of proteomic data with genomic and transcriptomic insights

The combination of proteomic data with genomic and transcriptome analysis has offered a complete knowledge of the molecular pathways behind abiotic stress resistance in CRISPR-edited crops. By combining these omics techniques, researchers may trace the impact of gene editing from the DNA sequence level to RNA transcription and eventually to the protein's functional role within the cell. In a study employing CRISPR-edited soybeans, researchers showed that the mutation of GmAITR genes resulted in increased salinity tolerance, as indicated by both transcriptome and proteomic analyses. The combination of these datasets revealed that the changed expression of stress-related proteins was strongly linked to changes in the transcriptional control of ABA signaling pathways (Wang *et al.*, 2021).

Moreover, the integration of multi-omics data has been crucial in finding major regulatory networks that control stress responses in plants. For instance, a study on the hybrid proline-rich proteins (HyPRPs) in tomato indicated that CRISPR-mediated alterations led to considerable modifications in the proteome, particularly in proteins related to cell-wall structure and stress-signaling pathways. These proteome modifications were aligned with transcriptome findings revealing the downregulation of negative stress- response regulators, hence boosting the plant's overall stress resilience (Saikia *et al.*, 2020). In another example, the use of CRISPR/Cas9 to regulate upstream open reading frames (uORFs) in rice revealed insights into the regulation of protein translation under salt stress. The combination of proteomic data with genomic changes revealed how certain uORFs regulate the expression of downstream stress-responsive proteins, giving new paths for increasing crop tolerance by targeted genome editing (Um *et al.*, 2021).

8.13.12.1 Metabolomics: deciphering metabolic pathways under stress

METABOLIC PROFILING OF CRISPR-EDITED CROPS UNDER ABIOTIC STRESS. The use of metabolomics to examine CRISPR-edited crops under abiotic stress has revealed remarkable insights into the metabolic pathways that are altered in response to environmental stresses. Metabolic profiling allows for the identification and quantification of metabolites, the tiny molecules that are the end products of cellular processes, delivering a picture of the physiological condition of the plant. High-throughput metabolomics, along with CRISPR editing, has uncovered important metabolic pathways that are altered to promote stress resilience. For instance, in rice, CRISPR/Cas9 has been exploited to modify genes involved in glutamine metabolism, a system crucial for abiotic stress tolerance. The combination of metabolomic and transcriptome data found specific glutamine metabolism genes that were increased in response to coupled abiotic stressors, such as drought and salinity. These genes have a critical role in maintaining nitrogen balance and osmotic control, which are necessary for plant life under stress conditions (Muthuramalingam *et al.*, 2020).

In tomato, CRISPR/Cas9-mediated mutation of the SlMAPK3 gene, which encodes a mitogen-activated protein kinase, led to dramatic alterations in primary and secondary metabolites under drought stress. Metabolic profiling demonstrated an increase in the amounts of osmoprotectants such as proline and soluble carbohydrates, which are critical for maintaining cellular homeostasis under water-deficit situations. This study emphasizes the utility of metabolomics in understanding how gene changes translate into metabolic modifications that promote stress tolerance (Saikia *et al.*, 2020). Moreover, metabolomic investigations of CRISPR-edited maize (*Zea mays*) targeting the ZmNF-YB2 gene, a TF involved in drought response, demonstrated a shift in the balance of primary metabolites, particularly in pathways connected to energy production and antioxidant defenses. The altered plants demonstrated a higher accumulation of metabolites such as ascorbate and glutathione, which are crucial to detoxifying ROS created under stress. This metabolic reprogramming was critical for enhancing drought tolerance in maize (Wang *et al.*, 2022b). In addition to primary metabolites, secondary metabolites such as flavonoids and alkaloids have also been demonstrated to have a substantial impact on the stress tolerance of CRISPR-edited crops. For example, in *Arabidopsis*, the editing of genes involved in flavonoid biosynthesis resulted in the accumulation of flavonoid molecules that contributed to increased tolerance against oxidative stress generated by high salt. This study emphasizes the relevance of secondary metabolites in the plant's defensive mechanisms and their potential as targets for CRISPR-based metabolic engineering (Li *et al.*, 2022).

Furthermore, the metabolic profile of CRISPR-edited wheat (*Triticum aestivum*) under cold stress revealed significant modifications in the levels of polyamines, which are known to be involved in cold acclimation. The altered lines demonstrated an enhanced build-up of spermidine and spermine, polyamines that are crucial for maintaining membranes and scavenging free radicals during cold exposure. These metabolic alterations were directly connected to the increased cold tolerance found in the CRISPR-edited wheat lines.

KEY METABOLITES AND PATHWAYS CONTRIBUTING TO STRESS RESILIENCE. The metabolic landscape of CRISPR-edited crops under abiotic stress reveals a plethora of important metabolites and pathways that are instrumental in boosting stress resilience. These metabolites, which include sugars, amino acids, organic acids, and secondary metabolites, perform vital functions in maintaining cellular homeostasis, defending against oxidative stress, and ensuring continuous development and productivity under adverse conditions. For instance, in CRISPR-edited rice lines, the disruption of the SST/OsSPL10 gene, which is implicated in salt-stress tolerance, led to dramatic modifications in the amounts of primary metabolites such as sugars and organic acids. These modifications were related to enhanced osmotic balance and reduced salt build-up, which are crucial for preserving cellular integrity under high-salinity settings (Lian *et al.*, 2020). In maize, CRISPR/Cas9 technology has been applied to target genes involved in the manufacture of zealexins, a group of sesquiterpenoid phytoalexins that contribute to disease resistance. The quadruple mutants, who cannot manufacture zealexins, demonstrated a broad-spectrum loss of disease resistance, underlining the crucial function of these secondary metabolites in stress resilience. The genetic redundancy within the zealexin biosynthesis pathway emphasizes the strength of maize's defense systems, which rely on a complex network of metabolic pathways to resist environmental stressors (Ding *et al.*, 2020). Another example is discovered in wheat, where CRISPR editing of genes related to polyamine biosynthesis resulted in increased amounts of spermidine and spermine, polyamines that are known to maintain cellular membranes and scavenge free radicals under cold stress. These metabolic alterations were directly connected to greater cold tolerance, revealing the role of polyamine metabolism in the plant's adaptive response to low-temperature conditions (Bhat *et al.*, 2021).

Additionally, the impact of amino acids like proline and glutamine in stress tolerance has been widely explored in CRISPR-edited crops. These amino acids function as osmoprotectants, shielding cells from dehydration and oxidative damage. For example, in tomato, CRISPR/Cas9-mediated deletion of the SlMAPK3 gene resulted in an accumulation of proline, which contributed to better drought tolerance by preserving osmotic balance and protecting cellular structures from stress-induced damage.

CORRELATION OF METABOLIC CHANGES WITH PHENOTYPIC OUTCOMES. The association between metabolic alterations and phenotypic results in CRISPR-edited crops provides a clear link between molecular mechanisms and observable traits. The ability to connect metabolic alterations to greater stress resilience allows for a fuller knowledge of how plants adapt to abiotic stress at the cellular level. In rice, the editing of genes involved in glutamine metabolism led to significant alterations in nitrogen metabolism, which were related to better growth and production under drought and salinity stress. The altered lines revealed elevated amounts of glutamine, which plays a vital role in nitrogen absorption and stress-induced signaling pathways. These metabolic alterations were linked with enhanced root architecture and water-use efficiency, revealing the direct impact of metabolic adjustments on phenotypic features (Muthuramalingam *et al.*, 2020). In another study, CRISPR/Cas9-edited maize targeting the ZmNF-YB2 gene, a TF involved in drought response, exhibited significant alterations in the accumulation of antioxidants such as ascorbate and glutathione. These alterations were directly connected to better drought tolerance, as indicated by enhanced leaf water retention and reduced oxidative damage. The association between enhanced antioxidant levels and improved stress tolerance underscores the role of redox homeostasis in the phenotypic expression of stress resilience (Wang *et al.*, 2022a). In wheat, the association between polyamine metabolism and cold tolerance was further explored by associating the increased accumulation of spermidine and spermine with greater membrane integrity and reduced cellular leakage during cold stress. The modified wheat lines displayed higher survival rates and decreased damage to cellular membranes, suggesting a strong relationship between metabolic alterations and phenotypic tolerance to cold stress (Bhat *et al.*, 2021)

8.14 Systems Biology Approaches: Integrating Multi-omics Data

Systems biology offers a comprehensive method for understanding the complicated biological processes of CRISPR-edited crops, particularly when coping with abiotic stress. The integration of multi-omics data – including genomes, transcriptomics, proteomics, and metabolomics – provides a holistic understanding of the molecular pathways driving stress responses. This integration is critical for identifying how different molecular levels interact and contribute to overall plant resilience. The advent of comprehensive systems biology frameworks has enabled researchers to systematically integrate multi-omics data, revealing complicated networks and pathways that drive stress responses. For instance, research by Yang *et al.* (2021) highlighted the relevance of integrating genomic, transcriptomic, and proteomic data to find stress-responsive genes and their regulatory networks in crops like wheat and rice. The work indicated that multi-omics integration could identify critical regulatory genes and pathways that were previously undetected using single-omics techniques.

Another notable innovation in systems biology is the development of genome-scale models (GEMs), which allow for the integration of omics data into a single, cohesive framework. GEMs have been used to mimic the complex connections between metabolic pathways and environmental stresses, providing insights into how CRISPR-edited crops might be tailored for enhanced stress resilience. Dahal *et al.* (2020) explored the application of GEMs in combining multi-omics data to model the metabolic responses of crops under stress circumstances, highlighting the potential of these models to anticipate the results of genetic changes and guide crop development efforts. The integration of multi-omics data also benefits from the use of modern computational techniques and platforms. Tripathi *et al.* (2019) stressed the relevance of bioinformatics in systems biology, notably in the integration of large-scale omics datasets. Their work described different computational methodologies that assist the integration of genomes, transcriptomics, and proteomics data, enabling the creation of comprehensive models that can predict plant responses to abiotic stress.

8.14.1 Network analysis and modelling of stress responses in CRISPR-edited crops

Network analysis is a strong approach within systems biology that allows researchers to understand the complicated relationships between diverse molecular entities in CRISPR-edited crops. By creating interaction networks, researchers can uncover critical regulators and pathways that contribute to stress resilience. One approach to network analysis involves the use of weighted gene co-expression network analysis (WGCNA), which identifies clusters of highly linked genes and their correlations with various attributes. A work by Da Ros *et al.* (2023) uses WGCNA to examine the transcriptome data of wheat under diverse abiotic stresses. The research showed critical hub genes involved in stress tolerance, including those controlled by CRISPR/Cas9, giving prospective targets for increasing crop resilience through genetic engineering (Da Ros *et al.*, 2023).

In addition to WGCNA, integrative network analysis has been applied to incorporate data from several omics platforms, yielding a more comprehensive perspective of the molecular relationships that drive stress responses. For example, created an integrative network analysis pipeline that incorporated transcriptome, proteomic, and metabolomic data to untangle the molecular signaling events of brassinosteroid response in *Arabidopsis*. This technique allowed for the discovery of important regulatory nodes and signaling cascades that were affected by CRISPR alterations, illustrating the relevance of multi-omics integration in comprehending complex biological processes. Moreover, Bayesian network modeling has been employed to combine omics data and anticipate the regulatory processes involved in stress reactions. Maity *et al.* (2020) employed Bayesian structural equation modeling to integrate genomic and transcriptome data, successfully predicting the regulatory networks involved with abiotic stress tolerance in crops. This technique not only discovered crucial regulatory genes but also provided insights into the dynamic interactions between distinct molecular layers, which are essential for producing stress resilient.

8.14.2 Case studies: multi-omics integration for a holistic understanding of abiotic stress resilience

8.14.2.1 Case study 1: integration of multi-omics techniques in wheat for abiotic stress resilience

A comprehensive case study in wheat highlights the power of multi-omics integration in understanding and strengthening abiotic stress resilience. Researchers adopted a systems biology method to integrate transcriptomics, metabolomics, and proteomics data to explore the molecular responses of wheat under combinatorial abiotic stimuli, including heat, drought, and salinity. The integration of these multi-omics datasets allowed for the identification of 152 hub genes, among which the TaWRKY33 gene emerged as a crucial regulator for stress tolerance. This integrative method not only revealed critical regulatory networks but also proposed prospective gene combinations for future crop development efforts focussed on boosting resilience to different environmental challenges (Da Ros *et al.*, 2023).

8.14.2.2 Case study 2: multi-omics integration in oil palm for drought-stress response

In oil palm (*Elaeis guineensis*), a large-scale multi-omics integration study focussed on the plant's response to drought stress, including transcriptomics, proteomics, and metabolomics data. This study identified critical enzymes involved in cysteine and methionine metabolism as crucial to the drought-response mechanism (Ijaz *et al.*, 2024). The integration of multi-omics data indicated that a considerable fraction of these enzymes exhibited identical expression profiles under both drought and salinity stress, emphasizing the overlapping molecular pathways implicated in osmotic stress responses. This case study underlines the potential of multi-omics integration in unraveling complicated stress responses and identifying candidate genes for developing stress-resilient crops (Leão *et al.*, 2022).

8.14.2.3 Case study 3: omics-driven crop improvement in Brassica napus

In *Brassica napus* (rapeseed), multi-omics methods have been extensively employed to improve abiotic stress tolerance. This includes integrating genomes, transcriptomics, proteomics, and metabolomics data to better understand the molecular mechanisms underpinning stress reactions. One study highlighted the use of CRISPR/Cas9 to alter important stress-related genes, with subsequent multi-omics analysis revealing significant changes in the expression of genes involved in antioxidant defense and osmoprotectant production. This integrated method provides insights into the precise processes that impart tolerance to stressors such as drought and salinity, suggesting prospective targets for genetic enhancement of rapeseed (Raza *et al.*, 2021).

8.14.2.4 Case study 4: multi-omics integration in lentils for climate resilience

A recent study on lentils used multi-omics technologies, including genomes, transcriptomics, proteomics, and metabolomics, to better understand the crop's responses to abiotic stresses including drought and salinity. The study discovered 329 differently expressed genes common to both heat and biotic challenges, and 19 metabolites similar across distinct legumes, which were critical for stress tolerance. The integration of these omics datasets permitted the identification of important molecular players in stress resilience and offered a framework for breeding lentils with increased tolerance to harsh environmental conditions (Tiwari *et al.*, 2022).

8.14.2.5 Case study 5: strategic omics integration for stress tolerance in underutilized crops

Underutilized agricultural species, generally disregarded in traditional breeding programs, have demonstrated potential for resilience to multiple abiotic stresses through multi-omics integration. A study concentrating on these species used genomes, transcriptomics, proteomics, and bionomics to investigate the molecular pathways underpinning stress tolerance. By combining omics data, researchers were able to discover and define stress-responsive genes and pathways that could be targeted for genetic enhancement. This case study highlights the usefulness of multi-omics in researching the genetic diversity of neglected crops to boost their resilience in the face of climate change (Muthamilarasan *et al.*, 2019).

8.15 Future Directions and Challenges

8.15.1 Emerging trends in CRISPR and omics technologies for crop improvement

Recent breakthroughs in CRISPR technology have transformed crop enhancement tactics by permitting precise genetic alterations, hence boosting crop resilience against abiotic challenges. CRISPR/Cas systems, notably CRISPR/Cas9, have become the most versatile and commonly used tools in plant genome editing due to their efficacy, cost-effectiveness, and ease of use. These systems have been effectively used for numerous crops to develop features such as drought tolerance, increased yield, and disease resistance (Rao and Wang, 2021).

One of the rising trends is the integration of CRISPR with multi-omics technologies, which provides a full understanding of how genetic changes effect not just individual genes but entire biochemical networks. The integration of genomes, transcriptomics, proteomics, and metabolomics through high-throughput approaches is now offering deeper insights into stress reactions at numerous biological levels. This integration has been particularly influential in crops such as wheat and rice, where multi-omics data has been utilized to identify and confirm target genes for CRISPR editing, leading to better abiotic stress resilience (Yang *et al.*, 2021). Moreover, improvements in next-generation CRISPR technologies, such as base editing and prime editing, are pushing the frontiers of precision genome editing. These techniques allow for single-nucleotide changes without producing double-strand breaks, decreasing off-target effects and boosting the accuracy of genetic modifications. This has major implications for producing crops that are not just stress-resistant but also meet the demanding safety and regulatory criteria for genetically modified organisms (Tang *et al.*, 2023).

8.15.2 Challenges in translating omics findings to field applications

Despite the great potential of CRISPR and omics technology, considerable obstacles remain in translating laboratory findings into field applications. One of the key challenges is the complexity of plant–environment interactions, which are difficult to recreate in controlled situations. Laboratory circumstances sometimes fail to capture the complete spectrum of environmental variables that crops encounter in the field, resulting in inconsistencies between observed stress tolerance in the lab and real performance in agricultural settings (Chen *et al.*, 2019). Another problem arises in the integration and interpretation of multi-omics data, which generates large volumes of complicated and multidimensional data (Bibi *et al.*, 2023). The ability to glean useful insights from this data requires advanced computational tools and models that can effectively forecast how genetic changes will impact phenotypic features in varied contexts. The lack of defined techniques for data integration and analysis further complicates this process, often leading to conflicting outcomes across different investigations (Pramanik *et al.*, 2021). Furthermore, legal and public acceptance difficulties also represent significant impediments to the introduction of CRISPR-edited crops. The regulatory landscape for genetically altered organisms varies greatly between countries, with some places adopting more rigorous guidelines that treat CRISPR-edited crops similarly to GMOs. Public perception, informed by worries over the safety and ethical consequences of genetic editing, can further limit the widespread adoption of these technologies.

8.15.3 Potential of precision agriculture and gene editing in developing climate-resilient crops

Precision agriculture, paired with gene editing, presents significant potential for generating climate-resilient crops. Precision agriculture involves the use of advanced technologies such as drones, satellite imaging, and sensors to monitor and regulate crop output at a granular level (Da Ros *et al.*, 2023). When coupled with CRISPR-based gene editing, precision agriculture can assist the targeted deployment of genetic alterations, guaranteeing that crops are tailored for specific environmental circumstances (Sampath *et al.*, 2023). The combination of these technologies permits the development of crops that are not only adaptable to climate change but also capable of surviving in varied agroecological zones (Kausar *et al.*, 2022). For example, CRISPR has been used to generate drought-tolerant rice and heat-resistant wheat, while precision agriculture methods have been applied to monitor these crops' performance in real time, enabling prompt interventions that boost yield (Zaidi *et al.*, 2020). Moreover, precision agriculture can help overcome some of the hurdles involved with converting omics results into field applications by providing precise environmental data that can be utilized to fine-tune genetic alterations. This strategy guarantees that crops are customized to the unique conditions of the fields in which they are cultivated, maximizing their resilience and yield (Biswas *et al.*, 2021).

8.16 Conclusion

The combination of CRISPR-based genome editing with high-throughput omics technologies has considerably increased our understanding of abiotic stress tolerance in crops. CRISPR/Cas systems, notably CRISPR/Cas9, have permitted precise and targeted alterations of genes linked with stress tolerance, paving the path to producing crops that are more adaptable to changing environmental conditions. Omics approaches – such as genomics, transcriptomics, proteomics, and metabolomics – have played a major role in finding essential stress-responsive genes, proteins, and metabolic pathways. For example, integrative multi-omics studies in crops like wheat and rice have discovered hub genes and regulatory networks that are critical for controlling stress responses, suggesting the potential for generating climate-resilient crop types. The merging of various omics datasets has enabled a more thorough knowledge of the molecular pathways underlying stress tolerance. For instance, transcriptome and proteomic investigations have identified essential TFs and protein changes that increase stress resilience, while metabolomics has revealed critical metabolites involved in osmotic adjustment and oxidative stress management. These findings underline the necessity of merging CRISPR technology with multi-omics techniques to develop crops that can tolerate abiotic stresses more efficiently.

8.16.1 The future impact of CRISPR-edited crops on global food security

CRISPR-edited crops show tremendous potential for increasing global food security, particularly in the face of climate change and environmental stressors. By enabling the precise editing of genes related to abiotic stress tolerance, CRISPR technology offers a valuable tool for designing crop types that can retain high yields under harsh conditions. This is particularly crucial as the world population continues to grow and the need for food increases. The development of crops with higher drought tolerance, salinity resistance, and temperature resilience might dramatically minimize crop losses, assuring a consistent food supply even in adverse circumstances (Rahman *et al.*, 2022). Furthermore, CRISPR technology could play a vital role in lowering the dependency on chemical inputs such as fertilizers and pesticides, which are not only costly but also have significant environmental implications. By producing crops that are naturally resistant to pests and diseases, CRISPR could help encourage more sustainable agricultural methods, contributing to long-term food security (Rehman *et al.*, 2022).

8.16.2 Final thoughts on the integration of advanced technologies in crop science

The integration of CRISPR technology with advanced omics techniques constitutes a paradigm change in crop science, presenting unprecedented prospects for crop enhancement. As the field continues to grow, the focus will likely shift towards more complex uses of CRISPR, such as base editing, prime editing, and epigenome editing, which give even greater accuracy and versatility. These developments, along with high-throughput phenotyping and machine-learning methods, will enable the rapid production of crop varieties that are customized to specific environmental circumstances and agricultural needs (Zenda *et al.*, 2021). However, the successful translation of these technologies into mainstream agricultural practice will require continuous work to overcome the problems related to data integration, regulatory approval, and public acceptance. As researchers and policy makers work together to overcome these hurdles, the potential for CRISPR-edited crops to contribute to global food security and sustainable agriculture will become increasingly realized, ultimately transforming the way we approach crop improvement and food production in the 21st century.

References

Abdelrahman, M., Burritt, D.J. and Tran, L.-S.P. (2018) The use of metabolomic quantitative trait locus mapping and osmotic adjustment traits for the improvement of crop yields under environmental stresses. *Seminars in Cell & Developmental Biology* 83, 86–94. DOI: 10.1016/j.semcdb.2017.06.020.

Ahmad, A., Ashraf, S., Munawar, N., Jamil, A., Ghaffar, A. *et al.* (2021a) CRISPR/Cas-mediated abiotic stress tolerance in crops. In: *CRISPR Crops: The Future of Food Security*. Springer, pp. 177–211. DOI: 10.1007/978-981-15-7142-8_6.

Ahmad, A., Munawar, N., Khan, Z., Qusmani, A.T., Khan, S.H. *et al.* (2021b) An outlook on global regulatory landscape for genome-edited crops. *International Journal of Molecular Sciences* 22(21), 11753. DOI: 10.3390/ijms222111753.

Ahmad, F., Farman, K., Waseem, M., Rana, R.M., Nawaz, M.A. *et al.* (2019) Genome-wide identification, classification, expression profiling and DNA methylation (5mc) analysis of stress-responsive ZFP transcription factors in rice (*Oryza sativa* L.). *Gene* 718, 144018. DOI: 10.1016/j.gene.2019.144018.

Ahmad, Y., Haider, S., Iqbal, J., Abbasi, B.A., Yaseen, T. *et al.* (2022) The mechanisms of genome editing technologies in crop plants. In: Prakash, C.S., Fiaz, S. and Fahad, S. (eds) *Principles and Practices of OMICS and Genome Editing for Crop Improvement*. Springer International Publishing, Cham, Switzerland, pp. 295–313.

Akram, R., Fahad, S., Masood, N., Rasool, A., Ijaz, M. *et al.* (2019) Plant growth and morphological changes in rice under abiotic stress. In: Hassanuzzaman, M., Fujita, M., Nahar, K. and Biswas, J.K.

(eds) *Advances in Rice Research for Abiotic Stress Tolerance*. Elsevier, pp. 69–85. DOI: 10.1016/b978-0-12-814332-2.00004-6.

Alam, M.S., Kong, J., Tao, R., Ahmed, T., Alamin, M. *et al.* (2022) CRISPR/Cas9 mediated knockout of the *OsbHIH024* transcription factor improves salt stress resistance in rice (*Oryza sativa* L.). *Plants* 11(9), 1184. DOI: 10.3390/plants11091184.

Al-Tamimi, N., Langan, P., Bernád, V., Walsh, J., Mangina, E. *et al.* (2022) Capturing crop adaptation to abiotic stress using image-based technologies. *Open Biology* 12(6), 210353. DOI: 10.1098/rsob.210353.

Ansori, A.N., Antonius, Y., Susilo, R.J., Hayaza, S., Kharisma, V.D. *et al.* (2023) Application of CRISPR-cas9 genome editing technology in various fields: A review. *Narra Journal* 3(2), e184. DOI: 10.52225/narra.v3i2.184.

Arora, L. and Narula, A. (2017) Gene editing and crop improvement using CRISPR-CAS9 system. *Frontiers in Plant Science* 8. DOI: 10.3389/fpls.2017.01932.

Bartkowski, B., Theesfeld, I., Pirscher, F. and Timaeus, J. (2018) Snipping around for food: Economic, ethical and policy implications of CRISPR/Cas genome editing. *Geoforum* 96, 172–180. DOI: 10.1016/j.geoforum.2018.07.017.

Berni, R., Luyckx, M., Xu, X., Legay, S., Sergeant, K. *et al.* (2019) Reactive oxygen species and heavy metal stress in plants: Impact on the cell wall and secondary metabolism. *Environmental and Experimental Botany* 161, 98–106. DOI: 10.1016/j.envexpbot.2018.10.017.

Bhat, M.A., Mir, R.A., Kumar, V., Shah, A.A., Zargar, S.M. *et al.* (2021) Mechanistic insights of CRISPR/Cas-mediated genome editing towards enhancing abiotic stress tolerance in plants. *Physiologia Plantarum* 172(2), 1255–1268. DOI: 10.1111/ppl.13359.

Bibi, S., Ullah, R., Burni, T., Ullah, Z. and Kazi, M. (2023) Impact of resorcinol and biochar application on the growth attributes, metabolite contents, and antioxidant systems of tomato (*Lycopersicon esculentum* Mill.). *ACS Omega* 8(48), 45750–45762. DOI: 10.1021/acsomega.3c06233.

Biswas, S., Zhang, D. and Shi, J. (2021) CRISPR/Cas systems: Opportunities and challenges for crop breeding. *Plant Cell Reports* 40(6), 979–998. DOI: 10.1007/s00299-021-02708-2.

Bodein, A., Scott-Boyer, M.-P., Perin, O., Lê Cao, K.-A. and Droit, A. (2022) Interpretation of network-based integration from multi-omics longitudinal data. *Nucleic Acids Research* 50(5), e27. DOI: 10.1093/nar/gkab1200.

Chen, K., Wang, Y., Zhang, R., Zhang, H. and Gao, C. (2019) CRISPR/CAS genome editing and precision plant breeding in agriculture. *Annual Review of Plant Biology* 70(1), 667–697. DOI: 10.1146/annurev-arplant-050718-100049.

Dahal, S., Yurkovich, J.T., Xu, H., Palsson, B.O. and Yang, L. (2020) Synthesizing systems biology knowledge from omics using genome-scale models. *Proteomics* 20(17–18), 17–18. DOI: 10.1002/pmic.201900282.

Da Ros, L., Bollina, V., Soolanayakanahally, R., Pahari, S., Elferjani, R. *et al.* (2023) Multi-omics atlas of combinatorial abiotic stress responses in wheat. *The Plant Journal* 116(4), 1118–1135. DOI: 10.1111/tpj.16332.

Debbarma, J., Sarki, Y.N., Saikia, B., Boruah, H.P.D., Singha, D.L. *et al.* (2019) Ethylene response factor (ERF) family proteins in abiotic stresses and CRISPR-CAS9 genome editing of ERFs for multiple abiotic stress tolerance in crop plants: A review. *Molecular Biotechnology* 61(2), 153–172. DOI: 10.1007/s12033-018-0144-x.

De Clercq, I., Velde, J., Luo, X., Liu, L., Storme, V. *et al.* (2021) Integrative inference of transcriptional networks in *Arabidopsis* yields novel ROS signalling regulators. *Nature Plants* 7(4), 500–513. DOI: 10.1038/s41477-021-00894-1.

Dikobe, T., Masenya, K. and Manganyi, M.C. (2023) Molecular technologies ending with 'Omics': The driving force toward sustainable plant production and protection 480. DOI: 10.12688/f1000research.131413.1.

Ding, Y., Weckwerth, P.R., Poretsky, E., Murphy, K.M., Sims, J. *et al.* (2020) Genetic elucidation of interconnected antibiotic pathways mediating maize innate immunity. *Nature Plants* 6(11), 1375–1388. DOI: 10.1038/s41477-020-00787-9.

Erdoğan, İ., Cevher-Keskin, B., Bilir, Ö., Hong, Y. and Tör, M. (2023) Recent developments in CRISPR/CAS9 genome-editing technology related to plant disease resistance and abiotic stress tolerance. *Biology* 12(7), 1037. DOI: 10.3390/biology12071037.

Farhat, S., Jain, N., Singh, N., Sreevathsa, R., Dash, P.K. *et al.* (2019) CRISPR-Cas9 directed genome engineering for enhancing salt stress tolerance in rice. *Seminars in Cell & Developmental Biology* 96, 91–99. DOI: 10.1016/j.semcdb.2019.05.003.

Farooqi, M.Q.U., Nawaz, G., Wani, S.H., Choudhary, J.R., Rana, M. *et al.* (2022) Recent developments in multi-omics and breeding strategies for abiotic stress tolerance in maize (*Zea mays* L.). *Frontiers in Plant Science* 13. DOI: 10.3389/fpls.2022.965878.

Francini, A. and Sebastiani, L. (2019) Abiotic stress effects on performance of horticultural crops. *Horticulturae* 5(4), 67. DOI: 10.3390/horticulturae5040067.

Haque, E., Taniguchi, H., Hassan, M.M., Bhowmik, P., Karim, M.R. *et al.* (2018) Application of CRISPR/Cas9 genome editing technology for the improvement of crops cultivated in tropical climates: Recent progress, prospects, and challenges. *Frontiers in Plant Science* 9, 617. DOI: 10.3389/fpls.2018.00617.

He, G.-H., Xu, J.-Y., Wang, Y.-X., Liu, J.-M., Li, P.-S. *et al.* (2016) Drought-responsive WRKY transcription factor genes TaWRKY1 and TaWRKY33 from wheat confer drought and/or heat resistance in *Arabidopsis*. *BMC Plant Biology* 16(1), 116. DOI: 10.1186/s12870-016-0806-4.

Huang, T.-K. and Puchta, H. (2021) Novel CRISPR/Cas applications in plants: From prime editing to chromosome engineering. *Transgenic Research* 30(4), 529–549. DOI: 10.1007/s11248-021-00238-x.

Hyun, T. (2020) CRISPR/Cas-based genome editing to improve abiotic stress tolerance in plants. *Botanica Serbica* 44(2), 121–127. DOI: 10.2298/BOTSERB2002121H.

Ijaz, S., Iqbal, J., Abbasi, B.A., Tufail, A., Ullah, Z. *et al.* (2023) Biofortification: Lessons from the past and strategies for future food security. In: Nadeem, M.A., Baloch, F.S., Aasim, M., Habyarimana, E., Sonmez, O. *et al.* (eds) *Legumes Biofortification*. Springer International Publishing, Cham, Switzerland, pp. 521–545.

Ijaz, S., Iqbal, J., Abbas, B.A., Tufail, A., Ullah, Z. *et al.* (2024) Heavy metal-polluted arable land and its consequences: A global scenario. In: Hameed, R., Abbas, A., Khan, I., Balooch, S. and Iqbal, A. (eds) *Biochar Assisted Remediation of Contaminated Soils Under Changing Climate*. Elsevier, pp. 71–99.

Jaganathan, D., Ramasamy, K., Sellamuthu, G., Jayabalan, S. and Venkataraman, G. (2018) CRISPR for crop improvement: An update review. *Frontiers in Plant Science* 9, 985. DOI: 10.3389/fpls.2018.00985.

Jain, D., Ashraf, N., Khurana, J.P. and Kameshwari, M.N.S. (2019) The 'OMICS' approach for crop improvement against drought stress. In: Rajpal, V.R., Sehgal, D., Kumar, A. and Raina, S.N. (eds) *Genetic Enhancement of Crops for Tolerance to Abiotic Stress*. Springer, pp. 183–204. DOI: 10.1007/978-3-319-91956-0_8.

Joshi, A., Yang, S., Song, H., Min, J. and Lee, J. (2023) Genetic databases and gene editing tools for enhancing crop resistance against abiotic stress. *Biology* 12(11), 1400. DOI: 10.3390/biology12111400.

Joshi, R.K., Bharat, S.S. and Mishra, R. (2020) Engineering drought tolerance in plants through CRISPR/cas genome editing. *Biotech* 10(9). DOI: 10.1007/s13205-020-02390-3.

Juntawong, P., Butsayawarapat, P., Songserm, P., Pimjan, R. and Vuttipongchaikij, S. (2020) Overexpression of jatropha curcas ERFVII2 transcription factor confers low oxygen tolerance in transgenic Arabidopsis by modulating expression of metabolic enzymes and multiple stress-responsive genes. *Plants* 9(9), 1068. DOI: 10.3390/plants9091068.

Kaur, B., Sandhu, K.S., Kamal, R., Kaur, K., Singh, J. *et al.* (2021) Omics for the improvement of abiotic, biotic, and agronomic traits in major cereal crops: Applications, challenges, and prospects. *Plants* 10, 10. DOI: 10.3390/plants10101989.

Kausar, R., Wang, X. and Komatsu, S. (2022) Crop proteomics under abiotic stress: From data to insights. *Plants* 11(21), 2877. DOI: 10.3390/plants11212877.

Khurshid, H., Jan, S.A., Shinwari, Z.K., Jamal, M. and Shah, S.H. (2018) An era of CRISPR/Cas9 mediated plant genome editing. *Current Issues in Molecular Biology* 26, 47–54. DOI: 10.21775/cimb.026.047.

Knott, G.J. and Doudna, J.A. (2018) CRISPR-Cas guides the future of genetic engineering. *Science* 361(6405), 866–869. DOI: 10.1126/science.aat5011.

Kumar, M., Prusty, M.R., Pandey, M.K., Singh, P.K., Bohra, A. *et al.* (2023) Application of CRISPR/Cas9-mediated gene editing for abiotic stress management in crop plants. *Frontiers in Plant Science* 14, 1157678. DOI: 10.3389/fpls.2023.1157678.

Kumawat, K.C., Sharma, B., Nagpal, S., Kumar, A., Tiwari, S. *et al.* (2022) Plant growth-promoting rhizobacteria: Salt stress alleviators to improve crop productivity for sustainable agriculture development. *Frontiers in Plant Science* 13, 1101862. DOI: 10.3389/fpls.2022.1101862.

Laforest, L.C. and Nadakuduti, S.S. (2022) Advances in delivery mechanisms of CRISPR gene-editing reagents in plants. *Frontiers in Genome Editing* 4, 830178. DOI: 10.3389/fgeed.2022.830178.

Leão, A.P., Bittencourt, C.B., Silva, T.L., Rodrigues Neto, J.C., Braga, Í. de O. *et al.* (2022) Insights from a multi-omics integration (MOI) study in oil palm (*Elaeis guineensis* Jacq.) response to abiotic stresses. Part 2: Drought. *Plants* 11(20), 2786. DOI: 10.3390/plants11202786.

Li, C., Brant, E., Budak, H. and Zhang, B. (2021) CRISPR/Cas: A nobel prize award-winning precise genome editing technology for gene therapy and crop improvement. *Journal of Zhejiang University SCIENCE B* 22(4), 253–284. DOI: 10.1631/jzus.B2100009.

Li, X., Chang, Y., Ma, S., Shen, J., Hu, H. *et al.* (2019) Genome-wide identification of SNAC1-targeted genes involved in drought response in rice. *Frontiers in Plant Science* 10. DOI: 10.3389/fpls.2019.00982.

Li, Y., Wu, X., Zhang, Y. and Zhang, Q. (2022) CRISPR/Cas genome editing improves abiotic and biotic stress tolerance of crops. *Frontiers in Genome Editing* 4, 987817. DOI: 10.3389/fgeed.2022.987817.

Lian, T., Huang, Y., Xie, X., Huo, X., Shahid, M.Q. *et al.* (2020) Rice *SST* variation shapes the rhizosphere bacterial community, conferring tolerance to salt stress through regulating soil metabolites. *mSystems* 5(6), e00721-20. DOI: 10.1128/mSystems.00721-20.

Ma, Y., Freitas, H. and Dias, M.C. (2022) Strategies and prospects for biostimulants to alleviate abiotic stress in plants. *Frontiers in Plant Science* 13. DOI: 10.3389/fpls.2022.1024243.

Maity, A.K., Lee, S.C., Mallick, B.K. and Sarkar, T.R. (2020) Bayesian structural equation modeling in multiple omics data with application to circadian genes. *Bioinformatics* 36(13), 3951–3958. DOI: 10.1093/bioinformatics/btaa286.

Manzoor, N., Ali, L., Ahmed, T., Noman, M., Adrees, M. *et al.* (2022) Recent advancements and development in nano-enabled agriculture for improving abiotic stress tolerance in plants. *Frontiers in Plant Science* 13, 951752. DOI: 10.3389/fpls.2022.951752.

Mariani, L. and Ferrante, A. (2017) Agronomic management for enhancing plant tolerance to abiotic stresses – drought, salinity, hypoxia, and lodging. *Horticulturae* 3(4), 52. DOI: 10.3390/horticulturae3040052.

McCarty, N.S., Graham, A.E., Studená, L. and Ledesma-Amaro, R. (2020) Multiplexed CRISPR technologies for gene editing and transcriptional regulation. *Nature Communications* 11(1), 1281. DOI: 10.1038/s41467-020-15053-x.

Minhas, P.S., Rane, J. and Pasala, R.K. (2017) *Abiotic Stress Management for Resilient Agriculture.* Springer eBooks. DOI: 10.1007/978-981-10-5744-1.

Morosky, P., Comyns, C., Nunes, L.G.A., Chung, C.Z., Hoffmann, P.R. *et al.* (2023) Dual incorporation of non-canonical amino acids enables production of post-translationally modified selenoproteins. *Frontiers in Molecular Biosciences* 10, 1096261. DOI: 10.3389/fmolb.2023.1096261.

Morris, J.A., Caragine, C., Daniloski, Z., Domingo, J., Barry, T. *et al.* (2023) Discovery of target genes and pathways at GWAS loci by pooled single-cell CRISPR screens. *Science* 380(6646), eadh7699. DOI: 10.1126/science.adh7699.

Muller, R., Meacham, Z.A., Ferguson, L. and Ingolia, N.T. (2020) CiBER-Seq dissects genetic networks by quantitative crispri profiling of expression phenotypes. *Science* 370(6522), eabb9662. DOI: 10.1126/science.abb9662.

Munir, N., Hanif, M., Abideen, Z., Sohail, M., El-Keblawy, A. *et al.* (2022) Mechanisms and strategies of plant microbiome interactions to mitigate abiotic stresses. *Agronomy* 12(9), 2069. DOI: 10.3390/agronomy12092069.

Mustafa, G. and Komatsu, S. (2021) Plant proteomic research for improvement of food crops under stresses: A review. *Molecular Omics* 17(6), 860–880. DOI: 10.1039/d1mo00151e.

Muthamilarasan, M., Singh, N.K. and Prasad, M. (2019) Multi-omics approaches for strategic improvement of stress tolerance in underutilized crop species: A climate change perspective. *Advances in Genetics* 103, 1–38. DOI: 10.1016/bs.adgen.2019.01.001.

Muthuramalingam, P., Jeyasri, R., Selvaraj, A., Pandian, S.K. and Ramesh, M. (2020) Integrated transcriptomic and metabolomic analyses of glutamine metabolism genes unveil key players in *Oryza sativa* (L.) to ameliorate the unique and combined abiotic stress tolerance. *International Journal of Biological Macromolecules* 164, 222–231. DOI: 10.1016/j.ijbiomac.2020.07.143.

Nadakuduti, S.S. and Enciso-Rodríguez, F. (2020) Advances in genome editing with CRISPR systems and transformation technologies for plant DNA manipulation. *Frontiers in Plant Science* 11, 637159. DOI: 10.3389/fpls.2020.637159.

Pandey, P.K., Quilichini, T.D., Vaid, N., Gao, P., Xiang, D. *et al.* (2019) Versatile and multifaceted CRISPR/Cas gene editing tool for plant research. *Seminars in Cell & Developmental Biology* 96, 107–114. DOI: 10.1016/j.semcdb.2019.04.012.

Pandita, D. (2021) CRISPR/Cas-mediated genome editing for improved stress tolerance in plants. In: Aftab, T. and Hakeem, K.R. (eds) *Frontiers in Plant-Soil Interaction*. Elsevier eBooks, pp. 259–291. DOI: 10.1016/b978-0-323-90943-3.00001-8.

Pang, Y., Wu, Y., Liu, C., Li, W., Amand, P. *et al*. (2021) High-resolution genome-wide association study and genomic prediction for disease resistance and cold tolerance in wheat. *Theoretical and Applied Genetics* 134(9), 2857–2873. DOI: 10.1007/s00122-021-03863-6.

Pieplow, A., Dastaw, M., Sakuma, T., Sakamoto, N., Yamamoto, T. *et al*. (2021) CRISPR-Cas9 editing of non-coding genomic loci as a means of controlling gene expression in the sea urchin. *Developmental Biology* 472, 85–97. DOI: 10.1016/j.ydbio.2021.01.003.

Pramanik, D., Shelake, R.M., Kim, M.J. and Kim, J.-Y. (2021) CRISPR-mediated engineering across the central dogma in plant biology for basic research and crop improvement. *Molecular Plant* 14(1), 127–150. DOI: 10.1016/j.molp.2020.11.002.

Puchta, H. (2017) Applying CRISPR/Cas for genome engineering in plants: The best is yet to come. *Current Opinion in Plant Biology* 36, 1–8. DOI: 10.1016/j.pbi.2016.11.011.

Rahman, M.-U., Zulfiqar, S., Raza, M.A., Ahmad, N. and Zhang, B. (2022) Engineering abiotic stress tolerance in crop plants through CRISPR genome editing. *Cells* 11(22), 3590. DOI: 10.3390/cells11223590.

Rajanna, G.A., Suman, A. and Venkatesh, P. (2023) Mitigating drought stress effects in arid and semi-arid agro-ecosystems through bioirrigation strategies. *Sustainability* 15(4), 3542. DOI: 10.3390/su15043542.

Rao, M.J. and Wang, L. (2021) CRISPR/Cas9 technology for improving agronomic traits and future prospective in agriculture. *Planta* 254(4), 68. DOI: 10.1007/s00425-021-03716-y.

Rasheed, A., Gill, R.A., Hassan, M.U., Mahmood, A., Qari, S. *et al*. (2021) A critical review: Recent advancements in the use of CRISPR/Cas9 technology to enhance crops and alleviate global food crises. *Current Issues in Molecular Biology* 43(3), 1950–1976. DOI: 10.3390/cimb43030135.

Raza, A., Razzaq, A., Mehmood, S.S., Hussain, M.A., Wei, S. *et al*. (2021) Omics: The way forward to enhance abiotic stress tolerance in *Brassica napus* L. *GM Crops & Food* 12(1), 251–281. DOI: 10.1080/21645698.2020.1859898.

Razzaq, M.K., Aleem, M., Mansoor, S., Khan, M.A., Rauf, S. *et al*. (2021) Omics and CRISPR-Cas9 approaches for molecular insight, functional gene analysis, and stress tolerance development in crops. *International Journal of Molecular Sciences* 22(3), 1292. DOI: 10.3390/ijms22031292.

Rehman, R.S., Zafar, S.A., Ali, M., Pasha, A.N., Naveed, M.S. *et al*. (2022) CRISPR-Cas mediated genome editing: A paradigm shift towards sustainable agriculture and biotechnology. *Asian Plant Research Journal* 27–49. DOI: 10.9734/aprj/2022/v9i130197.

Rogowsky, P.M. (2017) CRISPR-Cas technology in plant science. *Potato Research* 60(3–4), 353–360. DOI: 10.1007/s11540-018-9387-y.

Sahoo, J.P., Behera, L., Sharma, S.S., Praveena, J., Nayak, S.K. *et al*. (2020) Omics studies and systems biology perspective towards abiotic stress response in plants. *American Journal of Plant Sciences* 11(12), 2172–2194. DOI: 10.4236/ajps.2020.1112152.

Saikia, B., Singh, S., Debbarma, J., Velmurugan, N., Dekaboruah, H. *et al*. (2020) Multigene CRISPR/Cas9 genome editing of hybrid proline rich proteins (HyPRPs) for sustainable multi-stress tolerance in crops: The review of a promising approach. *Physiology and Molecular Biology of Plants* 26(5), 857–869. DOI: 10.1007/s12298-020-00782-6.

Sako, K., Nguyen, H.M., and Seki, M. (2021) Advances in chemical priming to enhance abiotic stress tolerance in plants. *Plant and Cell Physiology* 61(12), 1995–2003. DOI: 10.1093/pcp/pcaa119.

Salas Fernandez, M.G., Bao, Y., Tang, L. and Schnable, P.S. (2017) A high-throughput, field-based phenotyping technology for tall biomass crops. *Plant Physiology* 174(4), 2008–2022. DOI: 10.1104/pp.17.00707.

Sampath, V., Rangarajan, N., Sharanappa, C.H., Deori, M., Veeraragavan, M. *et al*. (2023) Advancing crop improvement through CRISPR technology in precision agriculture trends. *International Journal of Environment and Climate Change* 13(11), 4683–4694. DOI: 10.9734/ijecc/2023/v13i113647.

Schindele, A., Dorn, A., and Puchta, H. (2020) CRISPR/Cas brings plant biology and breeding into the fast lane. *Current Opinion in Biotechnology* 61, 7–14. DOI: 10.1016/j.copbio.2019.08.006.

Schultz-Bergin, M. (2018) Is CRISPR an ethical game changer? *Journal of Agricultural and Environmental Ethics* 31(2), 219–238. DOI: 10.1007/s10806-018-9721-z.

Sehgal, A., Sita, K., Siddique, K.H.M., Kumar, R., Bhogireddy, S. *et al.* (2018) Drought or/and heat-stress effects on seed filling in food crops: Impacts on functional biochemistry, seed yields, and nutritional quality. *Frontiers in Plant Science* 9, 1705. DOI: 10.3389/fpls.2018.01705.

Shah, T., Xu, J., Zou, X., Cheng, Y., Nasir, M. *et al.* (2018) Omics approaches for engineering wheat production under abiotic stresses. *International Journal of Molecular Sciences* 19(8), 2390. DOI: 10.3390/ijms19082390.

Shi, Y., Huang, J., Sun, T., Wang, X., Zhu, C. *et al.* (2017) The precise regulation of different COR genes by individual CBF transcription factors in *Arabidopsis thaliana*. *Journal of Integrative Plant Biology* 59(2), 118–133. DOI: 10.1111/jipb.12515.

Shinwari, Z.K., Tanveer, F. and Khalil, A.T. (2018) Ethical issues regarding CRISPR mediated genome editing. *Current Issues in Molecular Biology* 103–110. DOI: 10.21775/cimb.026.103.

Song, Y., Xuan, A., Bu, C., Ci, D., Tian, M. *et al.* (2019) Osmotic stress-responsive promoter upstream transcripts (prompts) act as carriers of MYB transcription factors to induce the expression of target genes in *Populus simonii*. *Plant Biotechnology Journal* 17(1), 164–177. DOI: 10.1111/pbi.12955.

Sun, F., Yu, H., Qu, J., Cao, Y., Ding, L. *et al.* (2020) Maize ZMBES1/BZR1-5 decreases ABA sensitivity and confers tolerance to osmotic stress in transgenic *Arabidopsis*. *International Journal of Molecular Sciences* 21(3), 996. DOI: 10.3390/ijms21030996.

Tang, Q., Wang, X., Jin, X., Peng, J., Zhang, H. *et al.* (2023) CRISPR/Cas technology revolutionizes crop breeding. *Plants* 12(17), 3119. DOI: 10.3390/plants12173119.

Tang, X., Ghimire, S., Liu, W., Fu, X., Zhang, H. *et al.* (2021) Genome-wide identification of u-box genes and protein ubiquitination under PEG-induced drought stress in potato. *Physiologia Plantarum* 174(1), e13475. DOI: 10.1111/ppl.13475.

Tinte, M.M., Chele, K.H., Hooft, J.J.J. and Tugizimana, F. (2021) Metabolomics-guided elucidation of plant abiotic stress responses in the 4IR era. *Metabolites* 11(7), 445. DOI: 10.3390/metabo11070445.

Tiwari, M., Singh, B., Min, D. and Jagadish, S.V.K. (2022) Omics path to increasing productivity in less-studied crops under changing climate: Lentil a case study. *Frontiers in Plant Science* 13. DOI: 10.3389/fpls.2022.813985.

Tolosa, L.N. and Zhang, Z. (2020) The role of major transcription factors in solanaceous food crops under different stress conditions: Current and future perspectives. *Plants* 9(1), 56. DOI: 10.3390/plants9010056.

Tran, M.T., Doan, D.T.H., Kim, J., Song, Y.J., Sung, Y.W. *et al.* (2021) CRISPR/Cas9-based precise excision of SlHyPRP1 domain(s) to obtain salt stress-tolerant tomato. *Plant Cell Reports* 40(6), 999–1011. DOI: 10.1007/s00299-020-02622-z.

Tripathi, L.P., Esaki, T., Itoh, M.N., Chen, Y. and Mizuguchi, K. (2019) Integrative analysis of multi-omics data. In: *Encyclopedia of Bioinformatics and Computational Biology*. pp. 194–199. DOI: 10.1016/b978-0-12-809633-8.20096-4.

Truchi, M., Lacoux, C., Gille, C., Fassy, J., Magnone, V. *et al.* (2023) Detecting subtle transcriptomic perturbations induced by lncRNAs knock-down in single-cell CRIPSRi screening using a new sparse supervised autoencoder neural dnetwork. *BioRxiv (Cold Spring Harbor Laboratory)*. DOI: 10.1101/2023.07.11.548494.

Ullah, Z., Gul, F., Iqbal, J., Abbasi, B.A., Kanwal, S. *et al.* (2023) Biogenic synthesis of multifunctional silver oxide nanoparticles (Ag2ONPs) using *Parieteria alsinaefolia* delile aqueous extract and assessment of their diverse biological applications. *Microorganisms* 11(4), 1069. DOI: 10.3390/microorganisms11041069.

Ullah, Z., Iqbal, J., Gul, F., Abbasi, B.A., Kanwal, S. *et al.* (2024) Biogenic synthesis, characterization, and *in vitro* biological investigation of silver oxide nanoparticles (AgoNPS) using *Rhynchosia capitata*. *Scientific Reports* 14(1), 10484. DOI: 10.1038/s41598-024-60694-3.

Um, T., Park, T., Shim, J.S., Kim, Y.S., Lee, G.-S. *et al.* (2021) Application of upstream open reading frames (UORFS) editing for the development of stress-tolerant crops. *International Journal of Molecular Sciences* 22(7), 3743. DOI: 10.3390/ijms22073743.

Vats, S., Kumawat, S., Kumar, V., Patil, G.B., Joshi, T. *et al.* (2019) Genome editing in plants: Exploration of technological advancements and challenges. *Cells* 8(11), 1386. DOI: 10.3390/cells8111386.

Vu, B.N., Vu, T.V., Yoo, J.Y., Nguyen, N.T., Ko, K.S. *et al.* (2023) CRISPR-Cas-mediated unfolded protein response control for enhancing plant stress resistance. *Frontiers in Plant Science* 14, 1271368. DOI: 10.3389/fpls.2023.1271368.

Wang, J., Zheng, C., Shao, X., Hu, Z., Li, J. *et al.* (2020) Transcriptomic and genetic approaches reveal an essential role of the NAC transcription factor SlNAP1 in the growth and defense response of tomato. *Horticulture Research* 7(1), 209. DOI: 10.1038/s41438-020-00442-6.

Wang, Q., Tang, J., Han, B. and Huang, X. (2019) Advances in genome-wide association studies of complex traits in rice. *Theoretical and Applied Genetics* 133(5), 1415–1425. DOI: 10.1007/s00122-019-03473-3.

Wang, T., Xun, H., Wang, W., Ding, X., Tian, H. *et al.* (2021) Mutation of GMAITR genes by CRISPR/Cas9 genome editing results in enhanced salinity stress tolerance in soybean. *Frontiers in Plant Science* 12. DOI: 10.3389/fpls.2021.779598.

Wang, Y., Zafar, N., Ali, Q., Manghwar, H., Wang, G. *et al.* (2022a) CRISPR/Cas genome editing technologies for plant improvement against biotic and abiotic stresses: Advances, limitations, and future perspectives. *Cells* 11(23), 3928. DOI: 10.3390/cells11233928.

Wang, Z., Mülleder, M., Batruch, I., Chelur, A., Textoris-Taube, K. *et al.* (2022b) *High-throughput proteomics of nanogram-scale samples with Zeno SWATH DIA*. BioRxiv (Cold Spring Harbor Laboratory). DOI: 10.1101/2022.04.14.488299.

Yang, Y., Saand, M.A., Huang, L., Abdelaal, W.B., Zhang, J. *et al.* (2021) Applications of multi-omics technologies for crop improvement. *Frontiers in Plant Science* 12. DOI: 10.3389/fpls.2021.563953.

Yousefi, S., Marchese, A., Salami, S.A., Benny, J., Giovino, A. *et al.* (2022) Identifying conserved genes involved in crop tolerance to cold stress. *Functional Plant Biology* 49(10), 861–873. DOI: 10.1071/FP21290.

Zafar, S.A., Zaidi, S.S., Gaba, Y., Singla-Pareek, S.L., Dhankher, O.P. *et al.* (2019) Engineering abiotic stress tolerance via CRISPR/Cas-mediated genome editing. *Journal of Experimental Botany* 71(2), 470–479. DOI: 10.1093/jxb/erz476.

Zaidi, S.S.-E.-A., Mahas, A., Vanderschuren, H. and Mahfouz, M.M. (2020) Engineering crops of the future: CRISPR approaches to develop climate-resilient and disease-resistant plants. *Genome Biology* 21(1), 289. DOI: 10.1186/s13059-020-02204-y.

Zenda, T., Liu, S., Dong, A., Li, J., Wang, Y. *et al.* (2021) Omics-facilitated crop improvement for climate resilience and superior nutritive value. *Frontiers in Plant Science* 12, 774994. DOI: 10.3389/fpls.2021.774994.

Zhang, L., He, W., Fu, R., Wang, S., Chen, Y. *et al.* (2023) Guide-specific loss of efficiency and off-target reduction with Cas9 variants. *Nucleic Acids Research* 51(18), 9880–9893. DOI: 10.1093/nar/gkad702.

Zhou, R., Jiang, F., Niu, L., Song, X., Yu, L. *et al.* (2022) Increase crop resilience to heat stress using OMIC strategies. *Frontiers in Plant Science* 13, 891861. DOI: 10.3389/fpls.2022.891861.

9 Food and Nutritional Security Achieved by Coordinated Nanotechnology-omics-CRISPR Strategies

Shumaila Ijaz[1], Banzeer Ahsan Abbasi[2*], Javed Iqbal[3], Zakir Ullah[4], Naila Ijaz[5], Navid Hussain Shah[6], Tabassum Yaseen[3], Faiza Munir[7], Akhtar Munir[8], Shoaib Khan[9], Zaman Ashraf[10], Saira Qadir[2], Ali Haider[8], Mohsin Kazi[11] and Tariq Mahmood[4]

[1]Department of Burn and Plastic Surgery, Shenzhen Institute of Translational Medicine, The First Affiliated Hospital of Shenzhen University, School of Biomedical Engineering, Shenzhen University Medical School, Shenzhen University, China; [2]Department of Botany, Rawalpindi Women University, Rawalpindi, Pakistan; [3]Department of Botany, Bacha Khan University, Charsadda, Khyber Pakhtunkhwa, Pakistan; [4]Department of Plant Sciences, Faculty of Biological Sciences, Quaid-i-Azam University Islamabad, Pakistan; [5]Nano Composite Laboratory, Department of Physics, University of Agriculture Faisalabad, Pakistan; [6]Optoelectronic Research Center, School of Science, Minzu University of China, Beijing; [7]Department of Agricultural Sciences and Technology, Atta-ur-Rahman School of Applied Biosciences, National University of Sciences and Technology, Islamabad, Pakistan; [8]Department of Chemistry, Quaid-I-Azam University, Islamabad, Pakistan; [9]Department of Chemistry, Abbottabad University of Science and Technology (AUST), Abbottabad, Pakistan; [10]Department of Chemistry, Rawalpindi Women University, Pakistan; [11]Department of Pharmaceutics, College of Pharmacy, King Saud University, Riyadh, Saudi Arabia

Abstract

Achieving food and nutritional security is a major worldwide issue, and novel solutions are emerging thanks to recent developments in CRISPR, omics, and nanotechnology. The ultimate goal of these integrated efforts is to contribute to global food security by improving crop quality, yields, and nutritional content. Plant genetics may now be precisely modified thanks to the advent of CRISPR technology, which has become an effective tool for genome editing. By using this technique to increase the synthesis of vital vitamins and minerals, crops can have a higher nutritional profile. For example, tocopherols and tocotrienols, which are essential for human health, have significantly increased in barley as a result of controlled overexpression of particular genes. Furthermore, CRISPR is being used to biofortify key crops like wheat and rice, which efficiently combats hunger. A comprehensive understanding of the genetic and biochemical composition of plants is possible because of omics technologies such as proteomics, metabolomics, and genomics. Through the integration of these technologies with CRISPR,

*Corresponding author: benazirahsanabbasi786@gmail.com

© CAB International 2025. *Stress-Resilient Crops: Coordinated omics-CRISPR-Nanotechnology Strategies* (ed. Jen-Tsung Chen)
DOI: 10.1079/9781800627307.0009

scientists may pinpoint and modify the genes accountable for advantageous characteristics including resilience to stress, increased productivity, and improved nutritional value. Crop development projects are more effective and the breeding process moves more quickly because of this multi-omics approach. CRISPR and omics are enhanced by nanotechnology, which makes it possible to create cutting-edge delivery methods for nutrition and genetic materials. By guaranteeing that gene editing takes place at specific locations within the plant genome, nanoparticles can improve the accuracy of CRISPR applications. Furthermore, by increasing the bioavailability of nutrients in food items, nanotechnology might increase consumer accessibility to these nutrients. In order to achieve food and nutritional security, the intersection of nanotechnology omics and CRISPR methods constitutes a revolutionary approach. By utilizing these cutting-edge technologies, scientists may create crops that are not only more durable and productive but also enhanced with vital nutrients, thereby tackling the problems of world hunger and malnutrition. In order to achieve the 2030 targets of ending hunger and guaranteeing sustainable food systems throughout the world, an integrated strategy is essential. This chapter provides a thorough analysis of the complex link between food and nutritional security. It starts with a summary of the world's food difficulties and emphasizes how vital nutritional security is. It explores the fundamentals and uses of nanotechnology in agriculture; the development of omics technologies including proteomics, metabolomics, and genomics; and the ground-breaking effects of CRISPR technology on crop genetics. Through the integration of these state-of-the-art methods, the chapter addresses the accompanying obstacles, ethical issues, and regulatory frameworks, as well as their synergistic potential to increase crop resilience and food production sustainability. It highlights the critical role that innovation and international cooperation play in attaining food security as it closes with future orientations and a call to action for stakeholders.

Keywords: food security, nutritional security, CRISPR-Cas, omics, nanotechnology, biofortification, crop genetics, sustainable food systems

9.1 Introduction

With a rising global population comes a growing and significant problem of malnutrition. By 2030, there will be 8.3 billion people on Earth. Ninety-eight percent of the world's hungry population, estimated at 800 million, reside in developing countries (Sinha *et al.*, 2019). Global governments, scientists, and agricultural stakeholders now view food and nutritional security as their top priority due to the world's fast-expanding population and the intensifying effects of climate change. Innovative strategies that make use of cutting-edge technologies are becoming crucial as traditional farming methods fail to fulfill the growing demand for wholesome, sustainable food. The combination of omics technologies, nanotechnology, and CRISPR (Clustered Regularly Interspaced Short Palindromic Repeats) editing stands out as a game-changing tactic to improve food quality and agricultural output. Through precision agriculture and effective nutrient delivery, nanotechnology presents a special opportunity to enhance food production systems by maximizing plant growth and resistance. Comprehensive insights into the molecular processes and interactions governing plant and animal metabolism are made possible by omics technologies such as proteomics, metabolomics, and genomes. These insights open the door to customized therapies aimed at improving nutritional profiles. Combining these with CRISPR-based gene-editing techniques makes it possible to precisely alter the genetic make-up of crops and livestock, leading to the creation of variants that are more productive while simultaneously being resistant to environmental challenges, illnesses, and pests (Ijaz *et al.*, 2023, 2024c; Ullah *et al.*, 2023).

Malnutrition includes undernutrition (stunting, and being underweight), inadequate vitamin and mineral intake, obesity, and the ensuing noncommunicable diseases linked to food. Additionally, deficiencies in one or more micronutrients, such as iodine, iron, vitamin A, zinc, and iodine, affect around 340 million individuals (Rao and Fisher, 2021). The two food enrichment technologies, fortification and biofortification, use different approaches (Fig. 9.1). Since developing a biofortified crop only requires a one-time investment and has low ongoing expenditures, biofortification is more cost-effective than fortification (Shoab

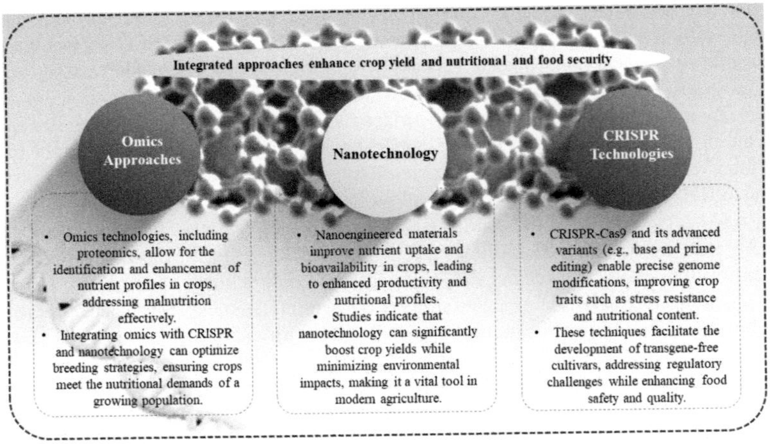

Fig. 9.1. Integrated approaches enhance crop yield and nutritional and food security.

and Hefferon, 2022). According to Ziarati *et al.* (2021), biofortified crops have a promising future in addressing nutritional issues. People can obtain micronutrient-rich crops through biofortification, which is thought to be a long-term and sustainable option (Van Der Straeten *et al.*, 2020). Through nutritional treatments and breeding, the crops are biofortified with the needed nutrients. The purposeful use of mineral fertilizers to raise a target mineral's concentration in crop parts that are edible is known as agronomic biofortification. Plant growth-promoting rhizobacteria (PGPR) and manufactured nanoparticles connected with fertilizers (nanofertilizers) are examples of advanced agronomic biofortification (Nayana *et al.*, 2020; Ijaz *et al.*, 2024d).

Furthermore, in order to enhance crop nutrient profiles, important quantitative trait loci (QTL) are also used in crop breeding projects. However, increasing agricultural production through plant breeding, particularly polyploid crop breeding, takes a lot of work and time (Nagamine and Ezura, 2022). Additionally, certain crops have had their genetic make-up altered using transgenic technology to biofortify them with desired nutrients. Genetically modified (GM) crops have negative effects on human nutrition and allergy responses. Additionally, they harm the ecosystem by dispersing pollutants into the ground, fostering pest resistance, and reducing crop diversification. Given the drawbacks associated with GM agriculture, genome-editing (GE) technology has clear benefits (Nagamine and Ezura, 2022). Therefore, with no integration of foreign gene sequences and with no off-target effects, genome editing results in predictable and inheritable mutations in specific genomic areas (Bhattacharya *et al.*, 2021). Before the discovery of CRISPR (Cas)-associated protein, systems including zinc-finger nucleases (ZFNs), transcription activator-like effector nucleases (TALENs), and homing endonucleases or meganucleases (HEs) were used as genome-editing tools (Wang *et al.*, 2016).

There are several engineered meganuclease variants that can cleave specific DNA sites to alter the genome and produce significant traits in crop species. The production of homing endonucleases and the creation of vectors for their entry into cells and potential off-target effects have been plagued by technical issues (Rey-Rico and Cucchiarini, 2016). ZFNs are a kind of nuclease that is made up of zinc-finger DNA-binding domains that have been designed and linked with a nuclease, usually the FokI nuclease. When cellular DNA repair systems are subjected to double-strand breaks generated by ZFN, it leads to amazingly focussed mutagenesis and targeted gene substitution. However, the ZFN technique has drawbacks including poor engineering feasibility, low specificity, normal effectiveness, complex design (requiring bespoke protein for each DNA sequence), and incapacity for RNA editing and gene knockdown (Davies *et al.*, 2017; Sun *et al.*, 2018).

The nonspecific DNA-cleaving nuclease known as TALENs is connected to a DNA-binding domain and may be engineered to target certain regions. TALENs, which are utilized to modify the genome, are composed of a modified array of TALE repeats joined to the FokI nuclease domain. However, there are drawbacks to TALENs as well, including their intricate design, lack of engineering practicality, limited specificity, low efficiency, and incapacity for RNA editing and gene knockdown (Sun *et al.*, 2018). With CRISPR-Cas technology, many of the aforementioned drawbacks of ZFNs, TALENs, and HEs have been overcome, making it a potential genome-editing tool. Because of its straightforward design, low cost, high efficiency, strong reproducibility, high engineering feasibility, capacity to produce gene knockout, RNA editing, and speedy cycle, it is presently the most widely used genome-editing approach in the world (Asmamaw and Zawdie, 2021). When combined, these well-thought-out tactics offer a comprehensive strategy to address the urgent problems of malnutrition and food insecurity. Through the use of nanotechnology, omics, and CRISPR, scientists and industry professionals have the opportunity to achieve previously unattainable breakthroughs in sustainable agriculture. These innovations will facilitate increased productivity, better nutritional value, and environmental sustainability. This introduction lays the groundwork for a thorough investigation of how the combination of these cutting-edge technologies might produce robust and fruitful food systems that can satisfy the dietary needs of coming generations.

9.2 Nutrient Deficiency in Crops

Crop nutrient deficits are a complicated, multifaceted issue that has a significant influence on food security, agricultural productivity, and human health (White and Brown, 2010). The fundamental issue here is the vital function that minerals like potassium, phosphorus, and nitrogen play in sustaining plant health and raising yield (White and Brown, 2010). Nutrient shortages have emerged as a major concern due to land-use changes, soil degradation, and climatic unpredictability; this reflects the complexity of our evolving agricultural system (White and Brown, 2010). Reduced crop yields and changed nutrient profiles have a domino impact on the whole agricultural supply chain as well as the larger public and human health system (White and Brown, 2010). The need for creative solutions that prioritize maximizing plant nutrition becomes more pressing as the world's population and climate continues to change, underscoring the significance of sustainable agricultural development and the general wellbeing of the world's population (White and Brown, 2010).

9.3 Nanotechnology to Enhance Nutrition and Food Security

By promoting food quality, safety, and availability, nanotechnology plays a critical role in improving nutrition and food security. The pressing demand for food security as the world's population rises can be met by genetically modifying crops through nanobreeding to maximize yield and nutritional value while reducing environmental impact (Sirajo *et al.*, 2024). Additionally, by using nanosensors and nanomaterials to identify hazardous compounds in the food supply chain, nanotechnology improves food safety and makes contaminant identification easier (Kiran *et al.*, 2024). Additionally, by increasing the bioavailability of vital nutrients, such as nanocapsules and nanoemulsions, nanofortification methods help prevent nutritional shortages, especially in populations that are more susceptible. As biodegradable materials and intelligent packaging are developed, food waste is decreased and shelf life is increased, all of which support sustainable resource usage (Shahid *et al.*, 2024) (Fig. 9.2). Collectively, these developments highlight how revolutionary nanotechnology may be in guaranteeing a safe and wholesome food supply in the future.

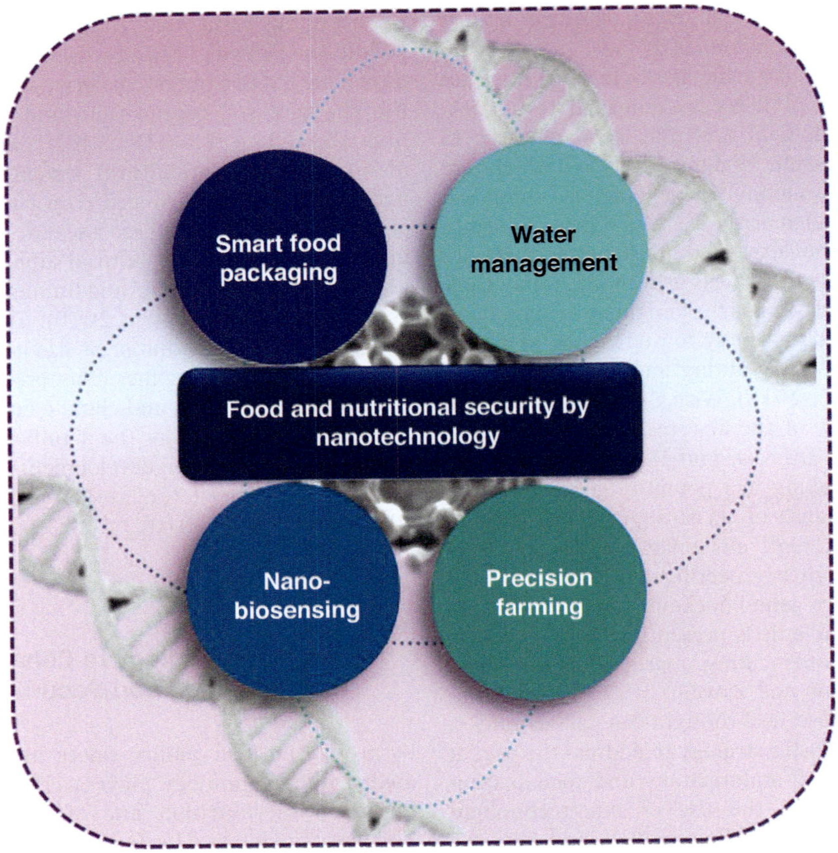

Fig. 9.2. Nanotechnology's role in ensuring food and nutritional security.

9.3.1 Nanopesticides for precision agriculture

The use of pesticides in agriculture presents a number of difficulties, such as environmental contamination, bioaccumulation, and the rise of insect resistance. Pest and fungal diseases pose a serious danger to crops worldwide, and there has been an increase in human poisonings from agricultural chemicals. Pesticides have been proved to be efficient against pests, but because of their harmful effects on people and the environment, it is important to use them sparingly in order to prevent environmental contamination and safeguard crops and stored goods. Therefore, with regard to the application of pesticides, there is a need for new technical developments that may offer a number of benefits, including increased durability, efficacy, and a decrease in the number of active ingredients needed (Rajak *et al.*, 2023).

In this regard, engineered nanostructures hold the potential for creating nanopesticides that include active chemicals inside of nanoscale systems. Numerous nanoformulations have been suggested, such as emulsions, nanoclays, inorganic metals and metal oxide nanoparticles, and polymeric nanocapsules (Fig. 9.3) (Shurfa *et al.*, 2023). These products, which are in different phases of research, have the potential to improve the environmental safety profiles or the efficacy of current pesticide-active components. These formulations play an important role in seed treatment, disease detection, pathogen monitoring with sensor technology, and obtaining predictive insights into environmental conditions, in addition to their ability to provide

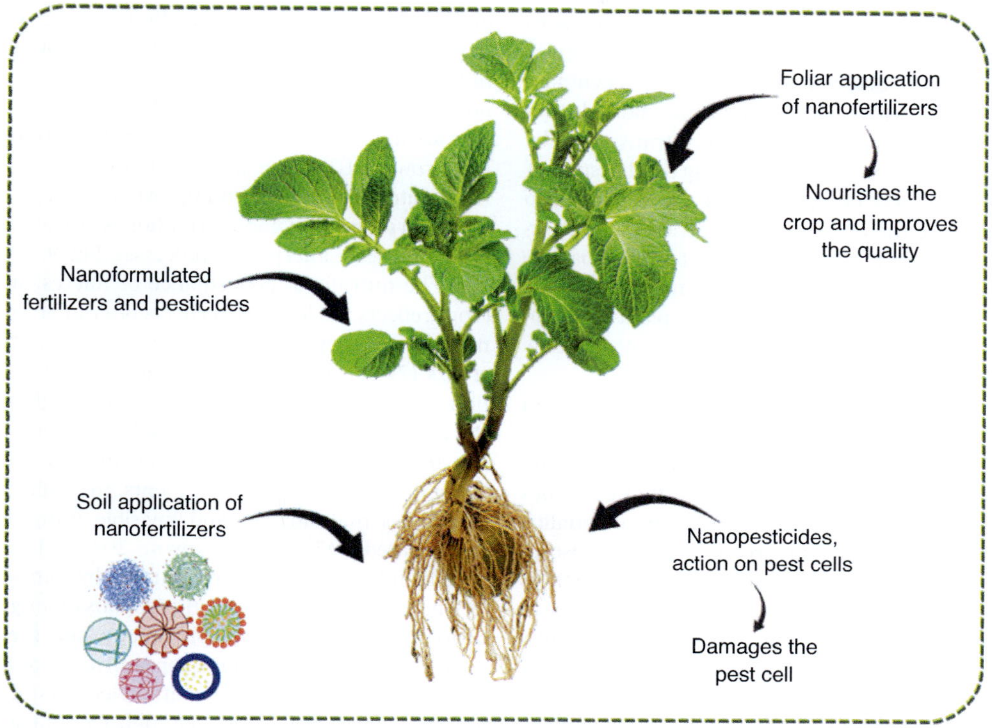

Fig. 9.3. Various applications of nanoformulated fertilizers and insecticides, along with their mechanistic models.

regulated or prolonged release patterns over an extended duration (Pathak *et al.*, 2022). Right now, the most promising method for sustainable pest management is nanoencapsulation or nanoformulation of insecticides.

9.4 Enhancing Nutrient Availability

The application of nanotechnology in agriculture has the potential to completely transform agricultural systems' nutrient availability. The capacity to develop materials that distribute nutrients in a controlled and continuous manner is at the forefront thanks to the application of the concepts of nanoscale interactions (Feynman, 1960; Vermeulen *et al.*, 2012). Furthermore, nutrients may be delivered and encapsulated using nanomaterials, shielding them from the environment and enabling plants to receive specific nutrients (Roco, 2003). With the help of these innovative techniques, a new era of targeted nutrient delivery will begin, ensuring that crops receive the nutrients they need to flourish to their full potential. Crop nutrition is changing dramatically with the application of nanotechnology to enhance plant nutrient uptake. Because of their distinct surface patterns, materials at the nanoscale can be tailored to improve nutrition and root-plant absorption (Feynman, 1960). By promoting nutrient transfer, their capacity to enhance root-to-microbe contact facilitates nutrient absorption even more (Vermeulen *et al.*, 2012). Through the utilization of these advancements, nanotechnology offers a means to boost crop development that is sustainable, lowers losses, and increases nutrient efficiency.

Through the utilization of these developments, nanotechnology offers a means to boost crop growth that is sustainable, decreases losses, and improve the efficiency of nutrient uptake. Plant nutrition uptake undergoes a fundamental change as a result of nanotechnology. By delivering nutrients to specific plant tissues with precision, nanoengineered materials maximize nutrient uptake and minimize waste (Feynman, 1960). Furthermore, bioactive compounds that initiate plant metabolism can be transported by nanoparticles, hence enhancing the digestion of nutrients (Whitesides, 2003). Furthermore, to enhance the absorption of nutrients, nanoparticles can function as transporters of bioactive substances that initiate plant metabolic pathways. Nanotechnology enhances plant nutrition and agronomic efficiency by opening up new channels for nutrient absorption.

A new chapter in agricultural innovation has begun with the discovery of nanomaterials, which have shown promise in treating crop nutritional deficits. This section examines three main areas: effects on nutritional quality, processes promoting greater nutrient absorption, and interactions between nanoparticles and plants. The main use of nanotechnology in agriculture is in the interactions between plants and nanoparticles. The notion that materials might be altered at the nanoscale level served as the foundation for Paul Feynman's concept of 'plenty of space at the bottom' (Feynman, 1960). Because nanoparticles function differently at this size, current evidence supports this theory (White and Brown, 2010). Whitesides' research on nanomachines emphasizes even more the potential and intricacy of these interactions (Roco, 2003). Complex surface contacts and particle characteristics that regulate the absorption pathway are some of the interactions that occur between nanoparticles and plant systems (White and Brown, 2010). When these findings are combined, nanomaterials are intelligently formulated to give nutrients to plants in the most efficient way possible.

Combining these understandings results in a sensible design for effective nanomaterials. A thorough grasp of the technology's revolutionary potential is provided by the molecular underpinnings of enhanced nutrient absorption by nanomaterials. The necessity to use these advancements for sustainable development rather than plant nutrient delivery is highlighted by Roco's description of the wider societal implications of nanotechnology (White and Brown, 2010). The need for innovative solutions is heightened by White and Brown's examination of the connection between plant nutrition and world health (White and Brown, 2010). Improved root surface contacts, improved nutrient transport into plant cells, and basic processes of interaction and solubility of nutrients are made possible by nanomaterials (White and Brown, 2010). The foundation for creating nanomaterials intended to improve and supply nutrition absorption and distribution is provided by this intricate interplay of processes (Fig. 9.4).

The nutritional performance of harvested crops reflects the long-term effects of enhanced nutrient delivery using nanomaterials. Smith *et al.*'s (2017) analysis of agricultural production and greenhouse gas emissions highlights the need for sustainable methods. By introducing improved nutritional profiles into crops using nanotechnological treatments, the ability to address malnutrition and invisible hunger is increased (White and Brown, 2010). The paradigm for enhancing crop quality by nanotechnology is shaped by the interactions among nutrient availability, plant physiology, and the resultant nutrient content. Crop nutrition and agronomic efficiency are essential elements of a secure global food supply. Enough food must be produced through sustainable agriculture to satisfy the demands of a world population that is growing at an accelerating rate. The study conducted by Tilman and Clark (2014) underscores the connection between environmental sustainability and agricultural efficiency, stressing the necessity for a well-rounded strategy that reduces negative environmental effects. Agronomic efficiency may be increased while minimizing resource depletion and environmental harm by implementing precision agriculture and agroecological practices (Ahmad *et al.*, 2022, 2024). The presence of micronutrients (vitamins, minerals, and bioactive substances) in harvested crops is one aspect of the multifaceted process that is crop quality.

In order to combat hidden hunger and malnutrition, Welch and Graham's research emphasizes the significance of micronutrient breeding for staple food crops (Welch and Graham, 2004). In addition, studies by Vaz Patto *et al.* (2015) have emphasized the significance of genetic diversity in enhancing nutritional quality. To increase the nutritional value of crops, particularly in regions with severe nutrient deficits, new technologies are being developed. Examples of these technologies include

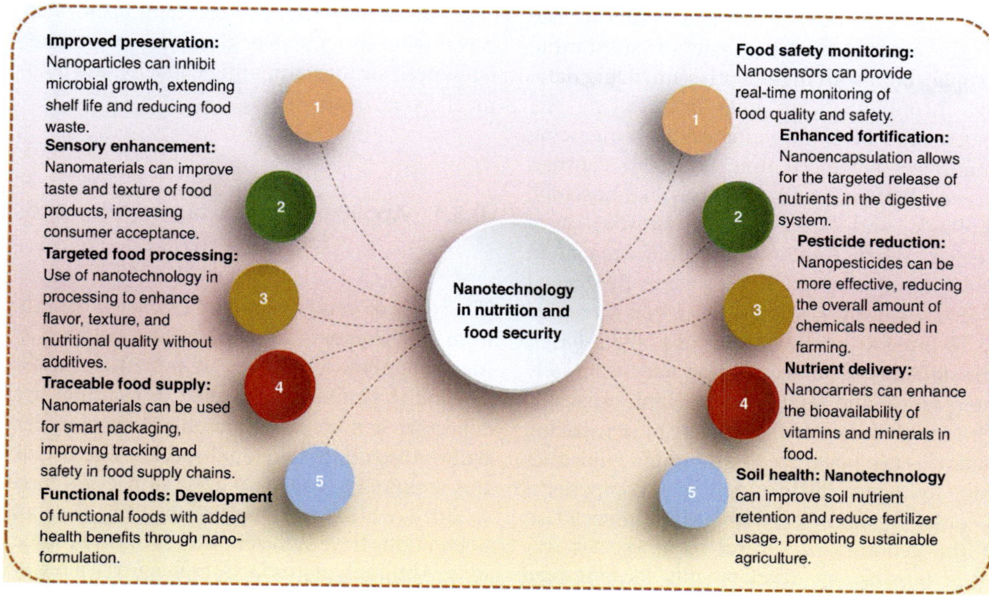

Fig. 9.4. Nanotechnology impacts on nutritional quality. This figure illustrates the transformative role of nanotechnology in enhancing food nutritional quality through several applications. It highlights nutrient delivery systems that utilize nanoparticles to protect and improve the bioavailability of vitamins and minerals, as well as fortification techniques involving nanoencapsulated nutrients that enrich food products.

biofortification through traditional breeding and biotechnology (Bouis and Saltzman, 2017). Finding a way to balance nutritional quality and agronomic efficiency is one of the major concerns. Sustainable intensification is a proposal put out by Pretty *et al.*, 2018 with the goal of raising yields while lowering adverse effects on the environment and society. The pursuit of high-yield, high-nutrient crop types remains a priority for crop breeding programs (Bouis and Saltzman, 2017). New agricultural technology and scientific advancements can help bridge the gap between the requirement for food security and the need for improved nutrition for people worldwide (Swaminathan and Bhavani, 2013).

9.5 Applications of Genomics in Food and Nutritional Security

Crop varieties with higher yields, higher nutritional contents, and greater resistance to environmental challenges may be developed thanks in large part to genomics. This increases food and nutritional security. The development of crops that are both high-yielding and enriched with vital micronutrients is made possible by the use of advanced genomic techniques, such as CRISPR-Cas9 and next-generation sequencing, which enable precise genetic modifications and help address the global malnutrition problem (Ceasar and Kavas, 2024). In addition, a thorough grasp of the genetic foundation of agricultural characteristics is made possible by the integration of omics techniques like transcriptomics, proteomics, and genomics, which is essential for successful breeding activities. Furthermore, the investigation of neglected crops, also known as orphan crops, increases genetic variety and provides substitute food sources, especially in areas with difficult agricultural environments. All of these genetic developments are critical to creating sustainable farming methods that guarantee food security for the world's expanding population (Majhi *et al.*, 2024).

As evidenced by the infamously complicated bread wheat genome (Zimin *et al.*, 2017), extremely long reads from TGS platforms like PacBio or the combined use of substantial mapping data with high-depth sequencing data, high-throughput, can help in resolving the complexities of plant genomes. As sequencing technology advanced over the 20th century, genome browsers and databases like Ensembl, GenBank, and UCSS Genome Browser were created. When GenBank was first created in 1982, it had around 500,000 bases. By the end of 2020, however, this quantity had increased by 18 times annually to over 723 billion bases. The difficulty is in sequencing and accurately interpreting and analyzing the data, especially with the abundance and diversity of sequencing platforms available. Specific tools and guidelines exist to detect distinct genomic components (Cagirici *et al.*, 2021). The whole association of the genomes of several plant species was made possible by developments in long-read sequencing and next-generation sequencing technology (Coletta *et al.*, 2021). The accessibility of genome sequences from several members of the same species highlighted questions about the adequacy of a single-reference genome and demonstrated the great degree of diversity among plant genomes (Danilevicz *et al.*, 2020).

First and foremost, the separation and characterization of the genome of interest regions is made easier by whole-genome sequencing, which also makes it possible to find genetic elements, such as noncoding regulatory elements and protein-coding genes. Genome sequencing offers more benefits than only defining the genetic components controlling significant agronomic features, while this is also valuable in breeding. A multitude of sequence-based molecular markers, including Insertion-Site-Based Polymorphisms (ISBPs) and Single Nucleotide Polymorphisms (SNPs) can now be predicted thanks to the increasing accessibility of NGS platforms. These markers can be validated and used to aid in marker-assisted selection, which is the process of selecting molecular markers associated with desired traits during the breeding process to produce improved varieties (Pont *et al.*, 2019). Lastly, the availability of genome sequences from several important species can provide light on the evolutionary histories behind agricultural plant diversity, which in turn can help develop breeding plans in response to changing needs. While genome sequencing can considerably improve breeding operations, it is crucial to remember that such procedures often need strong computing capacity and technical competence.

9.5.1 Application of metagenomics and food security

The study of genetic material extracted directly from environmental samples is known as metagenomics, and it is a potent scientific method that makes it possible to examine the collective genomes of microbial communities without isolating individual species. The use of this technology will have a big impact on food security, which is crucial to ensuring that all populations have sufficient access to wholesome food. Although plants face challenges as a result of their variety, there are also major advantages. Even though the genomes of many higher plants are somewhat vast, their metagenomes can have extremely rich information overall. Furthermore, the genomes of plants are quite conservative, and the potential for a wide range of micropartners from the soil organisms gives the systems the adaptability they need. These communities provide a vast array of possible functional capacities and may be thought of as the host's expanded genome (Schulz *et al.*, 2018).

Plant potential is greatly increased when microbes cooperate with them in a variety of ways, including feeding, growth, defense against stress, and tolerance to other environmental stresses. It was shown that plants have methods of proliferative and selection support of innocuous microbes, establishing particular communities, after characterizing the diverse microbiomes of the phyllosphere, endophytes, and rhizosphere of several plants (Berendsen *et al.*, 2012; Haider *et al.*, 2021, 2022). Compared to rhizoplane communities, endophytic compartment communities are substantially less complicated. But even though the endophytic microbiome can be changed by environmental factors and soil characteristics (Lundberg *et al.*, 2012), the immune system and biochemical characteristics of the plant are major determinants of the microbiome's structure. As a result, there is a general overlap in the endophytic community composition of plants that live in diverse environments (Santoyo *et al.*, 2016).

As of now, Oxford Nanopore Technologies (ONT) nanopore sequencing is the most appealing sequencing technique for metagenomics-based pathogen identification (Leggett and Clark, 2017). This method has the advantages of being quick, not requiring an amplification stage (direct sequencing), and adaptable to any pathogen aside from RNA viruses since it can identify and detect pathogens immediately. The lengthy read guaranteed by nanopore sequencing offers further benefits for predicting the species composition of microbial communities and reconstructing their functional pathways. The significant mistake rate of this technique is its primary disadvantage (Rang *et al.*, 2018). On the other hand, read accuracy (99.3%) has been steadily increasing. Furthermore, in metagenomic research, the integration of Illumina and ONT technologies can enhance the caliber of sequence assembly (Sevim *et al.*, 2019; Ciuffreda *et al.*, 2021).

More ideas for capitalizing on metagenomic data are becoming practical as it gets more accessible. It was recommended that next-generation agriculture should strive to customize the techniques and equipment adapted to the specific soil environment in accordance with the notion of customized medicine. The soil microbiome report and other preliminary molecular diagnostics may serve as the foundation for a well-informed decision about the best farming practices (Schlaeppi and Bulgarelli, 2015; Ijaz *et al.*, 2024b, 2024a). Metagenomics should enable target designation in this stage of breeding and provide details on the relative qualities of different varieties for this attribute. A combination of transcriptomic and genomic methods can be crucial for maximizing plant–microbial interactions and crop development; nevertheless, genomic analysis is unable to reveal information on the functional status of biological objects (Fig. 9.5).

9.5.2 Application of transcriptomic and food security

By offering insights into the genetic and molecular mechanisms underpinning crop development, stress responses, and yield potential, transcriptomics plays a critical role in improving food security. Researchers can find important genes linked to desired qualities like pest resistance and drought tolerance by combining transcriptomic data with other omics technologies. These genes are crucial for enhancing crop resilience in the face of climate

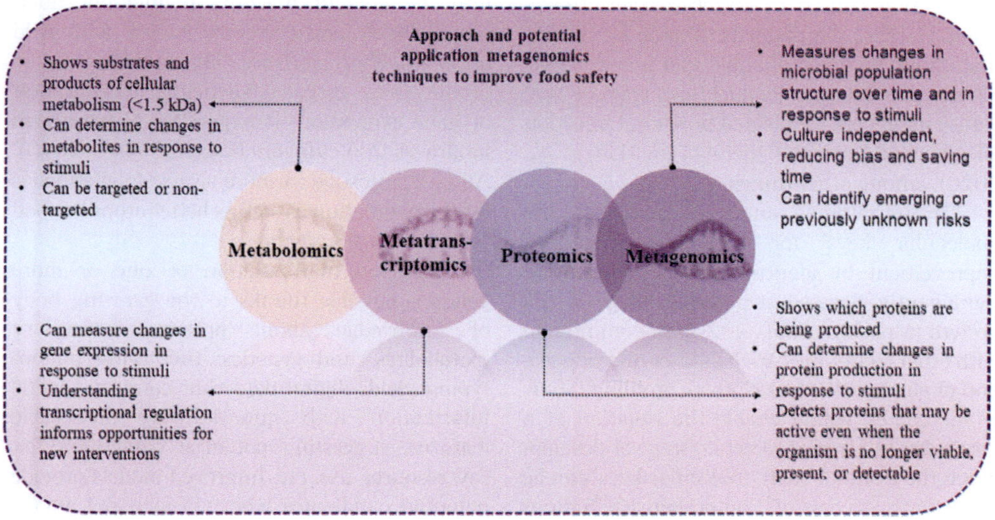

Fig. 9.5. Approach and potential application of metagenomics techniques to improve food safety.

change and rising food demand. Furthermore, precise genome editing to improve crop output and nutritional quality is made possible by the use of genome-editing techniques like CRISPR/Cas9 in combination with transcriptome profiling. This integrated strategy promotes the growth of conventional food plants as well as the production of transgenic plants with better features, which can provide vulnerable communities with more nutrient-dense food sources. In the end, utilizing transcriptomics in crop development techniques is essential to attaining worldwide sustainable food security.

Numerous plant science disciplines make extensive use of NGS-based transcriptomics. Plant cell growth and specialization processes are studied (Kozlova *et al.*, 2020), the molecular processes leading to the formation of mutualistic interactions between its symbionts and a plant are revealed (Xiong *et al.*, 2019), plants' circadian control is studied (Romanowski *et al.*, 2020), and so on. Stress and plant pathology research frequently make use of NGS-based transcriptomics. This method is crucial for understanding the signaling processes that occur during pathosystem formation and for describing the physiological response of plants to infections. For instance, NGS uncovered a potential gene network that underlies barley's resistance to *Blumeria graminis* (Li *et al.*, 2019), uncovered a wide range of possible molecular events linked to a susceptible plant response to an infection caused by *Pectobacterium atrosepticum*, clarified the patterns of gene expression that characterize compatible and incompatible interactions between *Phytophthora infestans* and potato plants, and indicated how viral infection affects phytohormone metabolism (Liu *et al.*, 2020), among other things.

Transcriptome analysis provided the 'molecular keys' for targeted agricultural improvement by identifying genes that were significantly elevated at a certain stage of cell growth (Mokshina *et al.*, 2020). In conjunction with other techniques, such as laser microdissection or single-cell transcriptome profiling (Shaw *et al.*, 2021), which enable the isolation of a particular cell type at a certain stage of development, the hunt for such 'molecular keys' can be particularly successful. Comprehensive findings are obtained when combining transcriptomics with metabolomic/biochemical tests to explore plant adaption mechanisms or metabolic pathways (Wang *et al.*, 2018). There are advantages and disadvantages to the enormous diversity of data that NGS transcriptomics offers. Although it is too big to be 'translated' into physiological language, it still accurately depicts the general picture of the target physiological activity. Because of this, interpreting NGS transcriptomic data is hard for higher plants and other species with complex genomes (Das *et al.*, 2020).

9.5.3 Techniques for gene editing to improve food and nutrition security

Traditional breeding procedures are insufficient to swiftly and properly generate new kinds to feed an expanding population under more complicated and unpredictable environmental and climatic circumstances. The use of molecular scissors in biotechnology to modify specific genes has created new opportunities and enhanced the toolkit available to breeders. Forage and food crops can be more effectively adapted to climate change with the help of technology. Regional variations can be shielded against novel infections that were previously irrelevant in the developing region more rapidly. With the 'linkage drag' that frequently happens in traditional mutant breeding taken into account, unique features may be incorporated into the current breeding population much more quickly and correctly. A crop with wild ancestral characteristics, like tomato, can be produced by targeted silencing or alteration of gene expression. It is possible to transform landraces into cultivars (Lacchini *et al.*, 2020). All of this results in more secure food supplies for the population as well as better crops for food and fodder.

Targeted manipulation of one or more genes is possible thanks to the growing body of knowledge about proteins, secondary metabolites, and genetics. Increasing phenotyping yields digital data that connect genetic information with quantitatively observable features, suggesting potential candidates for, say, resource use, etc. Improved models provide potential candidates who, via focussed alteration, can enhance the look, smell, or shelf life of meals. Plants have the ability to enhance specific advantageous components, such as the amount of oleic acid or anthocyanins, to improve animal or human health and nutritional value (Okuzaki *et al.*, 2018).

In the past ten years, a number of systems for targeted gene alteration have been built. Many arise from bacterial systems, although it has been noted that using individual components makes broad application conceivable. The FokI domain is coupled to DNA-binding domains to achieve the site-specificity. These are polypeptides found in ZFN, and each one attaches to a DNA triplet via a loop structure. Zinc serves as a bridge element to generate the loop structure, which is how these proteins received their name. The manufacturing of these zinc finger peptides is difficult, which restricts the technology's use despite the abundance of them. Following their discovery of *Xanthomonas* bacteria, TAL effectors were identified (Boch *et al.*, 2009). However, because one peptide needs to be connected to another for every target nucleotide, the TALEN modules' manufacturing is more complicated. All technologies aim to induce a double-strand break by first targeting a user-selected location in the target genome. Since double-strand breaks can also result from a variety of environmental conditions, eukaryotic cells have evolved distinct methods for error-free and effective double-strand break repair. This situation is used by molecular biologists to either better understand a gene's function through targeted knockouts or to change agricultural plants in certain ways.

Specificity may be affected by the target region choice. Sequence sections within gene families might differ from one another, but they also frequently contain conserved regions. There are two ways to apply this fact. If one wants to remove or modify a single gene from a gene family, one selects a distinct area. Conversely, a region that is conserved in all or most gene copies is chosen if an entire gene family like the wheat α-gliadins needs to be turned off (Sánchez-León *et al.*, 2018). This is another area where the platforms diverge. Consequently, the target zone is increased since two modules are required. As a region of 40–50 nucleotides is believed to be uncommon or perhaps unique within a particular genome, specificity rises and the likelihood of off-target events decreases. Since it is modular in nature, it can target any genomic sequence, unlike CRISPR/Cas technology which is restricted to those with a protospacer-associated motif next to the target sequence (Cong *et al.*, 2013). Unlike TALENs, CRISPR/Cas technology is significantly simpler to use in practically any laboratory for cloning; it typically combines two short oligos (Nishimasu *et al.*, 2018). Plants can rapidly correct partially enhanced off-target activity by choosing transgene-free offspring that carry the necessary changes. Such a plant can be readily removed if unwanted phenotypic alterations arise. Transferring a synthetic repair template simultaneously may offer an additional method for precise gene editing. This synthetic segment is used and integrated at the location of the double-strand break via homology-dependent repair. This enables the targeted substitution of desired gene variations at specific loci. The drawback of this strategy is that only a very small percentage of the mutant cells, roughly 5%, receive the intended alteration. A fluorescent reporter gene (GFP-YFP) was modified to change its emission spectrum in barley, serving as an example of this. The technique was established in corn by giving herbicide resistance (Svitashev *et al.*, 2015).

It is possible to manipulate many attributes at once by focussing on distinct genetic areas. In this sense, there have been several advancements and improvements. Sugarcane (*Saccharum officinarum*) has an astounding 107/109 genes producing caffeic acid O-methyltransferases that are mutable with a single TALEN pair. Several of the above-listed methods have been effectively utilized for a number of notable objectives. By using TALEN and CRISPR/Cas to specifically delete the MLO gene, for instance, wheat has been rendered agronomically significantly resistant to powdery mildew. Additional instances include lowering corn's phytic acid concentration and increasing tomato's anthocyanin content. Here is an overview of additional product attributes that have been enhanced by genome editing techniques.

The range of targeted gene-editing techniques has grown steadily in the last several years. Target sequence editing is made more accurate and consistent using base editors and prime editing. Chromosome engineering is one of the most recent advancements in the field of targeted genome manipulation. Here, certain chromosomal regions may invert as a result of deliberate induction of double-strand breaks.

For instance, it was demonstrated that the targeted application of SaCas9 could reverse a 1.1 Mb region on chromosome 4 in the model plant *Arabidopsis* (Schmidt *et al.*, 2020a). This opens up areas to the crossover process, which creates new variety through crossbreeding. In conclusion, breeders now have access to a new genetic variety pool thanks to gene editing. Furthermore, this method has two main benefits: accuracy and quickness. The three crops where technology has advanced the most are rice, wheat, and maize. The product quality and resistance to biotic and abiotic stressors have greatly improved in this instance. The paucity of genomic sequence data and effective transformation technologies in fruits and vegetables means that there is a need to catch up. Positive cases have also been documented, including modified mushrooms (Waltz, 2016), apples (Osakabe *et al.*, 2018), and tomato and virus-resistant cucumbers (Chandrasekaran *et al.*, 2016).

All things considered, there is still a lot of room for improvement. The disparities in opinions about crops that have undergone gene editing, however, pose a far greater issue. Plants that are identical to those obtained by traditional mutation breeding and do not carry extra DNA elements are not subject to excessive regulation in certain regions of the world. Almost entirely prohibiting their production, Europe has placed strict restrictions on them due to their general designation as genetically modified organisms (GMOs) (Schmidt *et al.*, 2020b). This adds needless ambiguity in an era of worldwide feed and food movements since it is also hard to confirm or track the origins of such plants. The use of these technologies in third-world nations is indirectly influenced by this European categorization, which increases unpredictability in food production.

9.5.4 Overview of the CRISPR-Cas systems

Crop improvement has undergone a revolution because of CRISPR-Cas technologies, which allow for precise genome editing to increase qualities like yield, quality, and stress tolerance. Although it was first known to cause random mutations, developments have produced advanced methods such as base editing, prime editing, and gene targeting that enable targeted changes to plant genomes. Breeders now choose to use CRISPR/Cas9 technology due to its adaptability, which allows for alterations in desirable features and nutritional components. Cas proteins that have been designed to reduce off-target effects and allow for heritable chromosomal rearrangements, a technique that efficiently manipulates genetic linkages, are among the recent discoveries. These advancements not only tackle obstacles in conventional breeding but also show potential for sustainable agriculture in light of worldwide concerns like food security and climate change (Jiang and Doudna, 2017).

The CRISPR-Cas system enables quick, site-specific modifications to the genome in a single cell or an entire organism. It alters epigenomes, regulates transcription, scans the whole genome, and pictures chromosomes. The numerous genome-editing applications of the CRISPR-Cas system have resulted in an upsurge in the production of modified crops in recent years. Using CRISPR-Cas, researchers have altered several agricultural systems for salinity, disease resistance, drought resistance, and thermotolerance. CRISPR-Cas is making it feasible to produce crops that are climate-ready and to increase agricultural quality metrics such as palatability, nutritional components, appearance, and other desirable traits (Liu *et al.*, 2021). The CRISPR-Cas system is a method of genome editing that is effective, practical, and affordable and can be used to biofortify key crops that lack certain vitamins or minerals. Furthermore, people in society readily accept the results of this technology since it does not involve the introduction of alien DNA and because there are fewer regulatory constraints. With the usage of this genome-editing technology, crop nutritional deficiencies might be completely eliminated, ensuring food security for the world's growing population. In addition, a proclamation supporting the use of gene editing for agricultural innovation was released by 13 nations that are members of the World Trade Organization. This is a first step toward creating a global regulatory framework.

9.5.5 Vitamin A-enriched crops

Isoprenoid pigments called carotenoids are extensively dispersed and vital to photosynthetic species. Although they cannot be produced from scratch, humans need carotenoids in their diet, mainly as precursors to β-carotene and vitamin A (Maoka, 2020). Vitamin A is necessary for biological functions such as light transduction in vision, embryonic development, maintaining the immune system, and function overall human health. Vitamin A deficiency (VAD) is one of the most important health issues facing the world today. It can lead to xerophthalmia, juvenile blindness, night blindness, and an increased risk of mortality and morbidity, particularly in young children (Reddy *et al.*, 2022). Consequently, attempts to combat VAD have concentrated on creating 'golden crops,' or staple foods enhanced with carotenoid content, either through traditional breeding or genetic engineering. The most widely used and successful genetic engineering technique at the moment seems to be the genome editing by CRISPR-based technique. It is a feasible alternative for producing 'golden' staple crops as it permits rapid, focussed, multiplex genetic modification of organisms without the requirement for DNA or transgenes (Zheng *et al.*, 2020, 2021). Numerous genome-edited golden crops, or carotenoid biofortified crops, have been produced using CRISPR-Cas-based technology to prevent vector-associated disease (VAD). For example, the 5.2-kb carotenogenesis cassette containing the maize PSY and CrtI genes was used to develop the golden rice cultivar 'Kitaake'. Dong *et al.* (2020) report that the endosperm of the cultivar has a dry weight of β-carotene of 7.9 μg/g (Fig. 9.6).

9.5.6 Vitamin E-enhanced foods

Tocopherol, often known as vitamin E, is a powerful antioxidant that is soluble in lipids and is crucial for human nutrition. Low vitamin E consumption has been linked to a number of human ailments, including some malignancies and cardiovascular conditions. Humans need between 15 mg and 30 mg of vitamin E per day (Ungurianu *et al.*, 2021). The targeted overexpression of *Hordeum vulgare homogentisate geranylgeranyltransferase* (*HvHGGT*) and *Hordeum vulgare homogentisate phytyltransferase* (*HvHPT*) with CRISPR-Cas9 technology resulted in a considerable increase in tocopherols and tocotrienol content (Zheng *et al.*, 2020). Other crops' vitamin E content can be increased by using these genes.

9.5.7 Crops rich in iron

Iron has a role in a number of metabolic processes, including the electron and oxygen

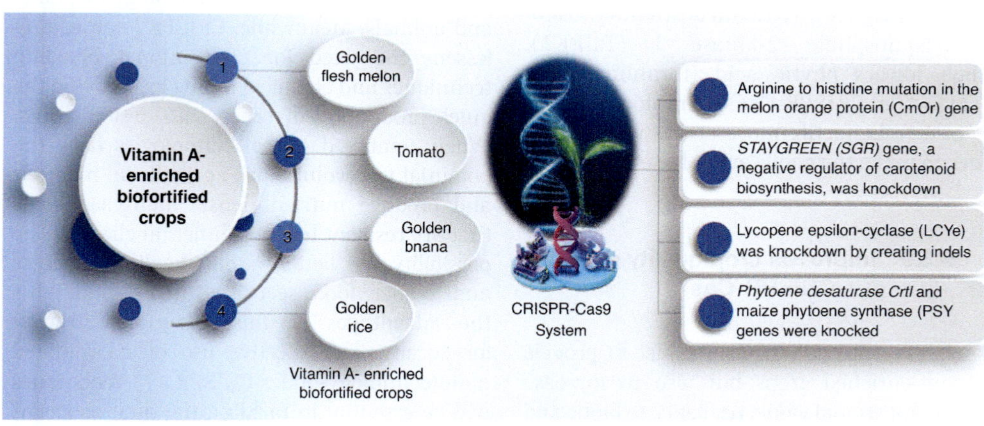

Fig. 9.6. CRISPR-Cas systems for biofortifying crops with enhanced vitamin A concentration.

transport chains. The most common illnesses in humans are caused by problems of iron metabolism, which include a wide range of ailments with different clinical manifestations, from anemia to neurodegenerative diseases. Humans can meet their iron needs by biofortifying their diets and crops. Lastly, there is CRISPR-Cas9-mediated biofortification. According to reports, this mechanism causes iron biofortification in wheat by interfering with the Inositol pentakisphosphate 2-kinase 1 (IPK1) gene (Ibrahim *et al.*, 2021).

9.5.8 Zinc-enhanced vegetables

About 100 enzymes' catalytic activity and a host of other cellular metabolic activities are influenced by the mineral zinc. It is involved in cell division, wound healing, immunological responses, protein synthesis, and DNA synthesis. Zinc supports the health of our immune system. The correct growth and activation of cells for both innate and adaptive immune responses are maintained by zinc. It helps maintain the epithelial barrier integrity, which is crucial for defending living things against harm and stopping the spread of infections. Furthermore, zinc has the ability to influence T cell growth and function, making it a promising option for immunomodulation. To the best of our knowledge, there is only one report on zinc augmentation in wheat. The CRISPR-Cas system in this crop improves zinc accumulation in wheat grains by interfering with Triticum aestivum Inositol Pentakisphosphate 2-kinase 1 (TaIPK1), which lowers phytic acid (Ibrahim *et al.*, 2021). The CRISPR-Cas technology has to be used to biofortify more crops with zinc in order to make the goods available everywhere.

9.5.9 Improved crop quality with CRISPR-Cas

Many crops are enhanced not just to provide nutrient-enriched crops but also to increase quality, nutritional value, resilience to biotic and abiotic stress, and productivity. Innovative agricultural technology has significantly increased crop yield over the course of several decades. Because crops include nutrients including vitamins, proteins, minerals, fiber, and bioactive compounds that are associated with human health, consumers are more interested in crop quality. CRISPR-based methods have significantly improved the precision and efficiency of cereal, oilseed, and horticultural crop quality as compared to traditional breeding. To the best of the reviewed literature, some crops have been reported to have improved in a variety of quality categories by genome editing mediated by CRISPR-Cas (Slavin and Lloyd, 2012; Ku and Ha, 2020).

9.6 Conclusion

In the search for improved food and nutritional security, the convergence of nanotechnology, omics technologies, and CRISPR methods offers a potential prospect. A comprehensive solution to guarantee sustained agricultural output and enhanced nutritional results is provided by the coordinated implementation of these cutting-edge techniques, while global issues like climate change, soil degradation, and population expansion continue to put pressure on our food systems. By enabling precise manipulations in nutrient supply and crop management, nanotechnology improves crop resilience and resource efficiency. Omics technologies enable the discovery of important features that may be targeted for improvement by providing a greater understanding of the biology of plants and animals. Meanwhile, CRISPR gene editing lessens the need for conventional breeding techniques and chemical inputs by enabling the quick production of crops with desired traits. When combined, these technologies have the potential to revolutionize agricultural methods and produce nutrient-dense, high-yield foods that are resilient to environmental change. The possibility for better public health outcomes and higher food security further highlights the advantages of this integrated strategy for society. The effective use of coordinated nanotechnology-omics-CRISPR techniques will be essential in tackling the dual concerns of food security and nutritional adequacy as research continues to progress and regulatory frameworks change. The full potential of these technologies may be realized by encouraging cooperation between scientists, decision makers, and agricultural stakeholders. This will guarantee a robust and sustainable food supply for future generations.

References

Ahmad, Y., Haider, S., Iqbal, J., Abbasi, B.A., Yaseen, T. *et al.* (2022) The mechanisms of genome editing technologies in crop plants. In: Prakash, C.S., Fiaz, S. and Fahad, S. (eds) *Principles and Practices of OMICS and Genome Editing for Crop Improvement*. Springer, Cham, Switzerland, pp. 295–313. DOI: 10.2147/BTT.S326422.

Ahmad, Y., Haider, S., Iqbal, J., Naseer, S., Attia, K.A. *et al.* (2024) *In-silico* analysis and transformation of OsMBY48 transcription factor driven by CaMV35s promoter in model plant – *Nicotiana tabacum* L. conferring abiotic stress tolerance. *GM Crops & Food* 15(1), 130–149.

Asmamaw, M. and Zawdie, B. (2021) Mechanism and applications of CRISPR/Cas-9-mediated genome editing. *Biological Targets and Therapy* 15, 353–361. DOI: 10.2147/BTT.S326422.

Berendsen, R.L., Pieterse, C.M. and Bakker, P.A. (2012) The rhizosphere microbiome and plant health. *Trends in Plant Science* 17(8), 478–486.

Bhattacharya, A., Parkhi, V. and Char, B. (2021) Genome editing for crop improvement: A perspective from India. *In Vitro Cellular & Developmental Biology – Plant* 57(4), 565–573. DOI: 10.1007/s11627-021-10184-2.

Boch, J., Scholze, H., Schornack, S., Landgraf, A., Hahn, S. *et al.* (2009) Breaking the code of DNA binding specificity of TAL-type III effectors. *Science* 326(5959), 1509–1512.

Bouis, H.E. and Saltzman, A. (2017) Improving nutrition through biofortification: A review of evidence from HarvestPlus, 2003 through 2016. *Global Food Security* 12, 49–58.

Cagirici, H.B., Sen, T.Z., and Budak, H. (2021) MirMachine: A one-stop shop for plant miRNA annotation. *Journal of Visualized Experiments* 171.

Ceasar, S.A. and Kavas, M. (2024) Plant genome editing to achieve food and nutrient security. *BMC Methods* 1(1), 3.

Chandrasekaran, J., Brumin, M., Wolf, D., Leibman, D., Klap, C. *et al.* (2016) Development of broad virus resistance in non-transgenic cucumber using CRISPR/Cas9 technology. *Molecular Plant Pathology* 17(7), 1140–1153.

Ciuffreda, L., Rodríguez-Pérez, H. and Flores, C. (2021) Nanopore sequencing and its application to the study of microbial communities. *Computational and Structural Biotechnology Journal* 19, 1497–1511.

Coletta, R.D., Qiu, Y., Ou, S., Hufford, M.B. and Hirsch, C.N. (2021) How the pan-genome is changing crop genomics and improvement. *Genome Biology* 22(1), 3.

Cong, L., Ran, F.A., Cox, D., Lin, S., Barretto, R. *et al.* (2013) Multiplex genome engineering using CRISPR/Cas systems. *Science* 339(6121), 819–823.

Danilevicz, M.F., Fernandez, C.G.T., Marsh, J.I., Bayer, P.E. and Edwards, D. (2020) Plant pangenomics: Approaches, applications and advancements. *Current Opinion in Plant Biology* 54, 18–25.

Das, S., McClain, C.J. and Rai, S.N. (2020) Fifteen years of gene set analysis for high-throughput genomic data: A review of statistical approaches and future challenges. *Entropy* 22(4), 427.

Davies, J.P., Kumar, S. and Sastry-Dent, L. (2017) Use of zinc-finger nucleases for crop improvement. *Progress in Molecular Biology and Translational Science* 149, 47–63. DOI: 10.1016/bs.pmbts.2017.03.006.

Dong, O.X., Yu, S., Jain, R., Zhang, N., Duong, P.Q. *et al.* (2020) Marker-free carotenoid-enriched rice generated through targeted gene insertion using CRISPR-cas9. *Nature Communications* 11(1), 1178. DOI: 10.1038/s41467-020-14981-y.

Feynman, R.P. (1960) There's plenty of room at the bottom. *Engineering and Science* 23(5), 22–36.

Haider, S., Iqbal, J., Naseer, S., Yaseen, T., Shaukat, M. *et al.* (2021) Molecular mechanisms of plant tolerance to heat stress: Current landscape and future perspectives. *Plant Cell Reports* 1–25.

Haider, S., Raza, A., Iqbal, J., Shaukat, M. and Mahmood, T. (2022) Analyzing the regulatory role of heat shock transcription factors in plant heat stress tolerance: A brief appraisal. *Molecular Biology Reports* 49(6), 5771–5785.

Ibrahim, S., Saleem, B., Rehman, N., Zafar, S.A., Naeem, M.K. *et al.* (2021) CRISPR/Cas9 mediated disruption of inositol pentakisphosphate 2-kinase 1 (TaIPK1) reduces phytic acid and improves iron and zinc accumulation in wheat grains. *Journal of Advanced Research* 37, 33–41. DOI: 10.3390/cells8080782.

Ijaz, S., Iqbal, J., Abbasi, B.A., Tufail, A., Ullah, Z. *et al.* (2023) Biofortification: lessons from the past and strategies for future food security. In: Nadeem, M.A., Baloch, F.S., Fiaz, S., Aasim, M., Habyarimana, E., Sonmez, O., and Zencirci, N. (eds) *Legumes Biofortification*. Springer, Cham, Switzerland, pp. 521–545.

Ijaz, S., Iqbal, J., Abbasi, B.A., Kanwal, S., Tavafoghi, M. *et al.* (2024a) Investigation of bioactive constituents and evaluation of different *in vitro* antimicrobial, antioxidant, and cytotoxicity potentials of different Portulacaria afra extracts. *Journal of King Saud University – Science* 36(2), 103033.

Ijaz, S., Iqbal, J., Abbasi, B.A., Kanwal, S., Tavafoghi, M. *et al.* (2024b) Investigation of bioactive constituents and evaluation of different *in vitro* antimicrobial, antioxidant, and cytotoxicity potentials of different *Portulacaria afra* extracts. *Journal of King Saud University – Science* 36(2), 103033.

Ijaz, S., Iqbal, J., Abbasi, B.A., Tufail, A., Ullah, Z. *et al.* (2024c) Heavy metal-polluted arable land and its consequences: A global scenario. In: Fahad, S., Adnan, M., Zhou, R., Nawaz, T. and Suad, S. (eds) *Biochar-Assisted Remediation of Contaminated Soils Under Changing Climate*. Elsevier, pp. 71–99.

Ijaz, S., Iqbal, J., Abbasi, B.A., Yaseen, T., Rehman, S. *et al.* (2024d) Role of OMICS-based technologies in plant sciences. In: Fiaz, S. and Prakash, C.S. (eds) *OMICs-based Techniques for Global Food Security*. Wiley, pp. 45–66.

Jiang, F. and Doudna, J.A. (2017) CRISPR-Cas9 structures and mechanisms. *Annual Review of Biophysics* 46, 505–529. DOI: 10.1146/annurev-biophys-062215-010822.

Kiran, V., Harini, K., Thirumalai, A., Girigoswami, K. and Girigoswami, A. (2024) Nanotechnology's role in ensuring food safety and security. *Biocatalysis and Agricultural Biotechnology* 58, 103220.

Kozlova, L.V., Nazipova, A.R., Gorshkov, O.V., Petrova, A.A. and Gorshkova, T.A. (2020) Elongating maize root: Zone-specific combinations of polysaccharides from type I and type II primary cell walls. *Scientific Reports* 10(1), 10956.

Ku, H.-K. and Ha, S.-H. (2020) Improving nutritional and functional quality by genome editing of crops: Status and perspectives. *Frontiers in Plant Science* 11, 577313. DOI: 10.3389/fpls.2020.577313.

Lacchini, E., Kiegle, E., Castellani, M., Adam, H., Jouannic, S. *et al.* (2020) CRISPR-mediated accelerated domestication of African rice landraces. *PLoS One* 15(3), e0229782.

Leggett, R.M. and Clark, M.D. (2017) A world of opportunities with nanopore sequencing. *Journal of Experimental Botany* 68(20), 5419–5429.

Li, Y., Guo, G., Zhou, L., Chen, Y., Zong, Y. *et al.* (2019) Transcriptome analysis identifies candidate genes and functional pathways controlling the response of two contrasting barley varieties to powdery mildew infection. *International Journal of Molecular Sciences* 21(1), 151.

Liu, Y., Liu, Y., Spetz, C., Li, L., and Wang, X. (2020) Comparative transcriptome analysis in *Triticum aestivum* infecting wheat dwarf virus reveals the effects of viral infection on phytohormone and photosynthesis metabolism pathways. *Phytopathology Research* 2(1), 1–13.

Liu, L., Gallagher, J., Arevalo, E.D., Chen, R., Skopelitis, T. *et al.* (2021) Enhancing grain-yield-related traits by CRISPR-Cas9 promoter editing of maize CLE genes. *Nature Plants* 7(3), 287–294.

Lundberg, D.S., Lebeis, S.L., Paredes, S.H., Yourstone, S., Gehring, J. *et al.* (2012) Defining the core *Arabidopsis thaliana* root microbiome. *Nature* 488(7409), 86–90.

Majhi, P.K., Bhoi, T.K., Sharma, P., Joshi, A., Samal, I. *et al.* (2024) Genetic diversity of food legumes and OMICS prospective to ensure nutritional security. In: Al-Khayri, J.M., Jain, S.M. and Penna, S. (eds) *Sustainable Utilization and Conservation of Plant Genetic Diversity*. Springer, pp. 107–125.

Maoka, T. (2020) Carotenoids as natural functional pigments. *Journal of Natural Medicines* 74(1), 1–16. DOI: 10.1007/s11418-019-01364-x.

Mokshina, N., Gorshkov, O., Galinousky, D. and Gorshkova, T. (2020) Genes with bast fiber-specific expression in flax plants: Molecular keys for targeted fiber crop improvement. *Industrial Crops and Products* 152, 112549.

Nagamine, A. and Ezura, H. (2022) Genome editing for improving crop nutrition. *Frontiers in Genome Editing* 4, 850104. DOI: 10.3389/fgeed.2022.850104.

Nayana, A.R., Joseph, B.J., Jose, A., and Radhakrishnan, E.K. (2020) Nanotechnological advances with PGPR applications. *Sustainable Agriculture Reviews* 41, 163–180. DOI: 10.1007/978-3-030-33996-8_9.

Nishimasu, H., Shi, X., Ishiguro, S., Gao, L., Hirano, S. *et al.* (2018) Engineered CRISPR-Cas9 nuclease with expanded targeting space. *Science* 361, 1259–1262.

Okuzaki, A., Ogawa, T., Koizuka, C., Kaneko, K., Inaba, M. *et al.* (2018) CRISPR/Cas9-mediated genome editing of the fatty acid desaturase 2 gene in *Brassica napus*. *Plant Physiology and Biochemistry* 131, 63–69.

Osakabe, Y., Liang, Z., Ren, C., Nishitani, C., Osakabe, K. *et al.* (2018) CRISPR-Cas9-mediated genome editing in apple and grapevine. *Nature Protocols* 13(12), 2844–2863.

Pathak, V.M., Verma, V.K., Rawat, B.S., Kaur, B., Babu, N. *et al.* (2022) Current status of pesticide effects on environment, human health and its eco-friendly management as bioremediation: A comprehensive review. *Frontiers in Microbiology* 13, 962619.

Pont, C., Wagner, S., Kremer, A., Orlando, L., Plomion, C. *et al.* (2019) Paleogenomics: reconstruction of plant evolutionary trajectories from modern and ancient DNA. *Genome Biology* 20(1), 29.

Pretty, J., Benton, T.G., Bharucha, Z.P., Dicks, L.V., Flora, C.B. *et al.* (2018) Global assessment of agricultural system redesign for sustainable intensification. *Nature Sustainability* 1(8), 441–446.

Rajak, P., Roy, S., Ganguly, A., Mandi, M., Dutta, A. *et al.* (2023) Agricultural pesticides – friends or foes to biosphere? *Journal of Hazardous Materials Advances* 10, 100264.

Rang, F.J., Kloosterman, W.P. and de Ridder, J. (2018) From squiggle to basepair: Computational approaches for improving nanopore sequencing read accuracy. *Genome Biology* 19(1), 90.

Rao, N. and Fisher, P.A. (2021) The impact of the COVID-19 pandemic on child and adolescent development around the world. *Child Development* 92(5), e738.

Reddy, G.B., Shalini, T., Ghosh, S., Pullakhandam, R., Kumar, B.N. *et al.* (2022) Prevalence of vitamin a deficiency and dietary inadequacy in Indian school-age children and adolescents. *European Journal of Nutrition* 61(1), 197–209. DOI: 10.1007/s00394-021-02636-7.

Rey-Rico, A. and Cucchiarini, M. (2016) Controlled release strategies for raav-mediated gene delivery. *Acta Biomaterialia* 29, 1–10. DOI: 10.1016/j.actbio.2015.10.015.

Roco, M.C. (2003) Broader societal issues of nanotechnology. *Journal of Nanoparticle Research* 5(34), 181–189.

Romanowski, A., Schlaen, R.G., Perez-Santangelo, S., Mancini, E. and Yanovsky, M.J. (2020) Global transcriptome analysis reveals circadian control of splicing events in *Arabidopsis thaliana*. *The Plant Journal* 103(2), 889–902.

Sánchez-León, S., Gil-Humanes, J., Ozuna, C.V., Giménez, M.J., Sousa, C. *et al.* (2018) Low-gluten, nontransgenic wheat engineered with CRISPR/Cas9. *Plant Biotechnology Journal* 16(4), 902–910.

Santoyo, G., Moreno-Hagelsieb, G., Orozco-Mosqueda, M.C. and Glick, B.R. (2016) Plant growth-promoting bacterial endophytes. *Microbiological Research* 183, 92–99.

Schlaeppi, K. and Bulgarelli, D. (2015) The plant microbiome at work. *Molecular Plant-Microbe Interactions* 28(3), 212–217.

Schmidt, C., Fransz, P., Rönspies, M., Dreissig, S., Fuchs, J. *et al.* (2020a) Changing local recombination patterns in *Arabidopsis* by CRISPR/Cas mediated chromosome engineering. *Nature Communications* 11(1), 4418.

Schmidt, S.M., Belisle, M. and Frommer, W.B. (2020b) The evolving landscape around genome editing in agriculture: Many countries have exempted or move to exempt forms of genome editing from GMO regulation of crop plants. *EMBO Reports* 21(6), e50680.

Schulz, T., Stoye, J. and Doerr, D. (2018) GraphTeams: A method for discovering spatial gene clusters in Hi-C sequencing data. *BMC Genomics* 19, 308.

Sevim, V., Lee, J., Egan, R., Clum, A., Hundley, H. *et al.* (2019) Shotgun metagenome data of a defined mock community using Oxford Nanopore, PacBio and illumina technologies. *Scientific Data* 6(1), 285.

Shahid, S.A., Ijaz, S., Iqbal, J., Khalil, A.T., and Ovais, M. (2024) Safety considerations of organic nanomaterials for phototheranostics. In: Abbas, M., Atiq, A., Ovais, M., and Hamblin, M.R. (eds) *Organic Nanomaterials for Cancer Phototheranostics*. Woodhead Publishing, Elsevier, pp. 233–252.

Shaw, R., Tian, X., and Xu, J. (2021) Single-cell transcriptome analysis in plants: Advances and challenges. *Molecular Plant* 14(1), 115–126.

Shoab, E. and Hefferon, K. (2022) Crop biofortification and food security. In: Kumar, V., Srivastava, A.K., Suprasanna, P. (eds) *Plant Nutrition and Food Security in the Era of Climate Change*. Academic Press, Cambridge, Massachusetts, pp. 423–436.

Shurfa, M.K., Girigoswami, A., Devi, R.S., Harini, K., Agraharam, G. *et al.* (2023) Combinatorial effect of doxorubicin entrapped in alginate-chitosan hybrid polymer and cerium oxide nanocomposites on skin cancer management in mice. *Journal of Pharmaceutical Sciences* 112(11), 2891–2900.

Sinha, P., Davis, J., Saag, L., Wanke, C., Salgame, P. *et al.* (2019) Undernutrition and tuberculosis: Public health implications. *The Journal of Infectious Diseases* 219(9), 1356–1363.

Sirajo, S.A., Kwon-Ndung, E.H. and Kana, H.A. (2024) Applications of nanobreeding for the enhancement of food security. *Fulafia Journal of Science and Technology* 8(1), 52–58.

Slavin, J.L. and Lloyd, B. (2012) Health benefits of fruits and vegetables. *Advances in Nutrition* 3(4), 506–516. DOI: 10.3945/an.112.002154.

Smith, M.R., Myers, S.S. and Huybers, P. (2017) Agricultural productivity and greenhouse gas emissions: Trade-offs or synergies between mitigation and food security. *Environmental Research Letters* 12(12), 12–24.

Sun, B., Yang, J., Yang, S., Ye, R.D., Chen, D. *et al.* (2018) A CRISPR-Cpf1-assisted non-homologous end joining genome editing system of mycobacterium smegmatis. *Biotechnology Journal* 13(9), e1700588. DOI: 10.1002/biot.201700588.

Svitashev, S., Young, J.K., Schwartz, C., Gao, H., Falco, S.C. *et al.* (2015) Targeted mutagenesis, precise gene editing, and site-specific gene insertion in maize using Cas9 and guide RNA. *Plant Physiology* 169(2), 931–945.

Swaminathan, M.S. and Bhavani, R.V. (2013) Food production and availability: Essential prerequisites for sustainable food security. *Indian Journal of Medical Research* 138(3), 383–391.

Tilman, D. and Clark, M. (2014) Global diets link environmental sustainability and human health. *Nature* 515(7528), 518–522.

Ullah, Z., Iqbal, J., Abbasi, B.A., Ijaz, S., Sharifi-Rad, J. *et al.* (2023) How genome editing can be helpful in the biofortification of legumes. In: Nadeem, M.A., M.A., F.S., Fiaz, S., Aasim, M., Habyarimana, E. *et al.* (eds) *Legumes Biofortification*. Springer, Cham, Switzerland, pp. 207–232.

Ungurianu, A., Zanfirescu, A., Nițulescu, G., and Margină, D. (2021) Vitamin E beyond its antioxidant label. *Antioxidants* 10(5), 634. DOI: 10.3390/antiox10050634.

Van Der Straeten, D., Bhullar, N.K., Steur, H., Gruissem, W., MacKenzie, D. *et al.* (2020) Multiplying the efficiency and impact of biofortification through metabolic engineering. *Nature Communications* 11(1), 5203. DOI: 10.1038/s41467-020-19020-4.

Vaz Patto, M.C., Amarowicz, R., Aryee, A.N.A., Boye, J.I., Chung, H.-J. *et al.* (2015) Achievements and challenges in improving the nutritional quality of food legumes. *Critical Reviews in Plant Sciences* 4(13), 105–143.

Vermeulen, S.J., Campbell, B.M. and Ingram, J.S.I. (2012) Climate change and food systems. *Annual Review of Environment and Resources* 37(1), 195–222.

Waltz, E. (2016) Gene-edited CRISPR mushroom escapes US regulation. *Nature* 532(7599), 293.

Wang, H., Russa, M. and Qi, L.S. (2016) CRISPR/Cas9 in genome editing and beyond. *Annual Review of Biochemistry* 85, 227–264. DOI: 10.1146/annurev-biochem-060815-014607.

Wang, R., Liu, P., Fan, J., and Li, L. (2018) Comparative transcriptome analysis two genotypes of *Acer truncatum* Bunge seeds reveals candidate genes that influences seed VLCFAs accumulation. *Scientific Reports* 8(1), 15504.

Welch, R.M. and Graham, R.D. (2004) Breeding for micronutrients in staple food crops from a human nutrition perspective. *Journal of Experimental Botany* 55(6), 353–364.

White, P.J. and Brown, P.H. (2010) Plant nutrition for sustainable development and global health. *Annals of Botany* 105(7), 1073–1080.

Whitesides, G.M. (2003) The once and future nanomachine. *Scientific American* 5(3), 78–83.

Xiong, B., Qiu, X., Huang, S., Wang, X., Zhang, X. *et al.* (2019) Physiological and transcriptome analyses of photosynthesis and chlorophyll metabolism in variegated citrus (*Shiranuhi* and *Huangguogan*) seedlings. *Scientific Reports* 9(1), 15670.

Zheng, X., Giuliano, G., and Al-Babili, S. (2020) Carotenoid biofortification in crop plants: Citius, altius, fortius. biochimica biophysica acta (BBA). *Molecular and Cell Biology of Lipids* 1865(11), 158664. DOI: 10.1016/j.bbalip.2020.158664.

Zheng, X., Kuijer, H.N.J., and Al-Babili, S. (2021) Carotenoid biofortification of crops in the CRISPR era. *Trends in Biotechnology* 39(9), 857–860. DOI: 10.1016/j.tibtech.2020.12.003.

Ziarati, P., Moradi, D., Rodriguez, L.C., Hochwimmer, B., Vambol, V. *et al.* (2021) Biofortification of *Oryza sativa* L. with agri-food waste to improve the dietary value of crops. *Ecological Questions* 33(1), 47–54. DOI: 10.12775/EQ.2022.004.

Zimin, A.V., Puiu, D., Hall, R., Kingan, S., Clavijo, B.J. *et al.* (2017) The first near-complete assembly of the hexaploid bread wheat genome *Triticum aestivum*. *Gigascience* 6, 1–7.

10 Exploring the Interaction of Nanomaterials with Crops Based on Multiple OMICS Tools

Shumaila Ijaz[1], Banzeer Ahsan Abbasi[2*], Javed Iqbal[3], Zakir Ullah[4], Naila Ijaz[5], Navid Hussain Shah[6], Shoaib Khan[7], Zaman Ashraf[8], Ghulam Murtaza[9], Rashid Iqbal[10,11], Tayyaba Shahzadi[12], Seweta Srivastava[13] and Tariq Mahmood[1]

[1]Department of Burn and Plastic Surgery, Shenzhen Institute of Translational Medicine, The First Affiliated Hospital of Shenzhen University, School of Biomedical Engineering, Shenzhen University Medical School, Shenzhen University, China; [2]Department of Botany, Rawalpindi Women University, Pakistan; [3]Department of Botany, Bacha Khan University, Charsadda, Khyber Pakhtunkhwa, Pakistan; [4]Department of Plant Sciences, Faculty of Biological Sciences, Quaid-i-Azam University Islamabad, Pakistan; [5]Nano Composite Laboratory, Department of Physics, University of Agriculture Faisalabad, Pakistan; [6]Optoelectronic Research Center, School of Science, Minzu University of China, Beijing, China; [7]Department of Chemistry, Abbottabad University of Science and Technology (AUST), Abbottabad, Pakistan; [8]Department of Chemistry, Rawalpindi Women University, Rawalpindi, Pakistan; [9]School of Agriculture, Yunnan University, Kunming, Yunnan, China; [10]Department of Agronomy, Faculty of Agriculture and Environment, The Islamia University of Bahawalpur, Pakistan; [11]Department of Life Sciences, Western Caspian University, Baku, Azerbaijan; [12]Department of Chemistry, Government College for Women, University Sialkot, Sialkot, Pakistan; [13]Department of Plant Pathology, School of Agriculture, Lovely Professional University, Phagwara, India

Abstract

The rapidly developing field of nanomaterials and crop interaction uses a range of omics approaches to enhance agricultural sustainability. Recent studies show the complex mechanisms via which nanoparticles impact crop development overall, stress responses in plants, and plant growth. Nanoparticles have the ability to change the gene expression and protein profiles of plants, revealing pathways that increase resistance to biotic and abiotic stresses. For instance, although nanoparticles improve metal homeostasis, carbon nanomaterials trigger immunological reactions linked to hormones. RNA-sequencing has discovered over 2200 differentially expressed genes relevant to developmental processes, and research on pepper plants has shown that iron oxide magnetic nanoparticles dramatically boost root and vegetative growth. For nanoparticles to effectively infiltrate plants, they must do this. Various methods, including direct application to the roots or leaves, have been studied, emphasizing the need for innovative delivery strategies to optimize absorption. Multi-omics methods need to be incorporated

*Corresponding author: benazirahsanabbasi786@gmail.com

© CAB International 2025. *Stress-Resilient Crops: Coordinated omics-CRISPR-Nanotechnology Strategies* (ed. Jen-Tsung Chen)
DOI: 10.1079/9781800627307.0010

in order to understand the intricate links that exist between crops and nanomaterials and to produce safer and more productive agricultural strategies. However, concerns over potential harm to plant health necessitate more investigation into the long-term impacts of nanoparticles. Scientists may investigate changes in gene expression, protein patterns, and metabolic pathways in crops exposed to these drugs by using omics methodologies such as transcriptomics, proteomics, metabolomics, and genomics. Nanotechnology finds application in agriculture in a number of ways, including sensors, nanoparticle-based fertilizers, pesticides, and herbicides. Agricultural production and sustainability can be enhanced by these applications by decreasing losses and improving input efficiency. Genetic engineering and agricultural improvement are made simpler by the use of nanomaterials to transfer genes to plants. This includes the delivery of nucleic acids via carbon nanotubes, magnetic nanoparticles, and mesoporous silica nanoparticles. Despite any potential benefits, there are grave worries about how nanoparticles may affect safety and the environment. Understanding how nanoparticles absorb, disperse, and internalize is necessary to assess their safety in plants. Other issues that must be resolved for nanoparticles to be used in practical fields include ensuring consumer acceptability and raising the cost of producing these materials. By acting as soil additives, seed coatings, and foliar sprays, nanomaterials can assist sustainable agriculture in meeting the need for traditional agrochemicals. The use of several omics technologies has greatly increased our understanding of how nanoparticles interact with crops. These tools have yielded significant new data about the molecular, physiological, and genetic responses of plants to designed nanomaterial exposure. Although there is much promise for nanotechnology to increase agricultural sustainability and production, addressing practical problems and safety concerns is necessary before it can be effectively incorporated into agriculture.

Keywords: nanomaterials, omics, agricultural sustainability, gene expression, protein profiles, nanotechnology, crop yields

10.1 Introduction

The science of nanotechnology is a fast-developing field with many facets and enormous potential applications. The scientific community's growing worry over the technology's effects on biota coincides with its fast rise (Patel *et al.*, 2021). The remarkable physicochemical properties of manufactured nanomaterials (NMs), also known as engineered nanoparticles or manufactured nanomaterials (MNMs), are being studied in precision agriculture because they are important in this rapidly developing field. These purposes include: (i) increasing crop productivity and nutritional quality; (ii) applying controlled release and targeted delivery of agrochemicals; (iii) protecting crops against pathogens; (iv) enhancing resilience to environmental stress; and (v) monitoring soil quality and detecting stressors using sensitive nanosensors (Giraldo *et al.*, 2019). Apart from their advantageous effects on the intended crop, they can also effectively minimize unwanted experiences to nontargeted species and leaching into surrounding waterways.

The scientific community is becoming concerned about the technology's effects on biota because of its widespread use in consumer items such as cosmetics and medicines, which inevitably results in enormous air releases. Because MNMs have a limited distribution, plant communities are especially vulnerable to them (Tripathi *et al.*, 2016). The positive and phytotoxic effects of nanoparticles (NPs) on plant communities have been extensively studied in the literature (Ijaz *et al.*, 2024a, b, c; Mathur *et al.*, 2023). Nevertheless, a lack of knowledge about how plants absorb and react physiologically to NPs has hindered the most effective use of nanotechnology in agricultural applications. The main factors that determine these NPs' impacts on plants are their kind, size, composition, and specific surface area (An *et al.*, 2022). The way that plants respond to NPs varies greatly and depends on a number of parameters, including the natural environment, the kind and age of the plants, the length and intensity of the exposure, and the characteristics of the NPs, such as size, shape, stability, or disintegration (Ma *et al.*, 2018).

The literature has carefully examined the combined consequences for certain NPs, including both phytotoxic and beneficial impacts on plants (Younis *et al.*, 2021). For instance, titanium oxide (TiO_2) NPs reduce oxidative stress caused by UV-B rays (Wang *et al.*, 2021); they also promote the conversion of inorganic nitrogen into organic form, facilitating the uptake and absorption of nitrate and boosting the rate of photosynthesis (Ijaz *et al.*, 2023a, b). On the other hand, TiO_2 NPs have a detrimental effect on the rate of transpiration and water acquisition in maize plants (Verma *et al.*, 2019). Comparably, tomato plant seedlings require water to germinate, and carbon nanotubes (CNTs) are among the NPs in this class that may penetrate the seed coat and help them acquire it (Verma *et al.*, 2019).

Numerous methods have been tried to synthesize NPs, which have a range of characteristics and intriguing consequences on biological systems (Prasad *et al.*, 2018). Similar to how several papers show the possible effects of NPs on plant morphology and physiology, very few studies have looked into the effects of MNMs on plant genetics. For instance, it has been demonstrated that cerium oxide (CeO_2) NPs are genotoxic to soybean seedlings, and that silver (Ag) NPs are genotoxic to onion seedlings; the impact of these NPs on seedling growth is determined at the gene level. In addition, adding carbon (C) NPs to natural organic matter prolonged the blooming period of *Oryza sativa* seedlings. Another example is zinc oxide (ZnO) NPs, which damage the cortical and epidermal cells of *Lolium perenne* (Amin *et al.*, 2023). Establishing the interaction process is essential to understanding how plants respond to NPs, particularly in connection to nutrient flow, photosynthetic efficiency, and defenses against oxidative stress. Transcriptomics studies have been conducted to evaluate the possible toxicity of NMs in plants, with a particular emphasis on the effects these materials may have on genes that react to stress (An *et al.*, 2023; Ijaz *et al.*, 2024d). The transcriptomic and proteomic investigations expand on our knowledge of plant–NP interactions by demonstrating how NPs impact gene expression. In research using tomato seedlings and carbon nanotubes (CNTs), overexpression of many genes implicated in stress-related processes was discovered by means of microarray analysis.

However, a comparative analysis of alterations in overall expression indicates a negligible impact on the transcriptome. Higher transcriptional responses than NPs may be caused by biotic and abiotic stressors such as infections, salt, or drought (Zhou *et al.*, 2022). It is clear that plants respond to new NMs by using regulatory circuits shared with other stress responses, rather than having the chance to develop distinct gene regulation. Abiotic stress often results in the silencing of genes involved in pathogen response and root growth. Zheng *et al.* (2023) have provided assessments of more papers that bolster these conclusions. Similarly, several studies have demonstrated that CNTs impacted the peroxide-scavenging enzyme gene expression in *Arabidopsis thaliana* seedlings. Manganese superoxide dismutase (Mn SOD1) and Ascorbate peroxidase (APX1) transcription levels were raised in leaves upon injection of single-walled carbon nanotubes (SWCNTs). ZnO and TiO_2 NPs were used in a study on *Arabidopsis thaliana*; the latter demonstrated a very significant positive impact on the plant, whereas the former demonstrated phytotoxicity in the studied species (Lee *et al.*, 2009). To determine NP absorption in plant tissues following root or foliar treatment, researchers have employed cutting-edge methodologies (Avellan *et al.*, 2019). These analyses have really shown that, contrary to their traditional counterparts, some NPs may enter plants through the roots or pass natural barriers to proceed within. The molecular mechanisms underlying the absorption and transport of NMs in plants, however, are yet unclear (Sanzari *et al.*, 2019). Yin *et al.* (2018) state that in order to fully utilize NPs for sustainable agricultural production, a comprehensive understanding of biological networks at multiple levels is required. In recent years, the study of NP–plant interaction has moved from a single end point, traditional methods to high-throughput, discovery-oriented system biology techniques, or 'omics.' Analytical methods and bioinformatics tools have become more sensitive and accurate (Quanbeck *et al.*, 2012).

The undifferentiated examination of an organism's biomolecules – specifically, its genes, mRNA, and proteins – is referred to as 'omics' (genomics, transcriptomics, and proteomics). Because mechanistic knowledge of complex agronomic aspects and crop response to NPs exposure is becoming more and more important, omics methods have gained popularity in agriculture and nanotoxicity research. This perspective looks at how biomolecules react when NPs disturb biological cycles, which helps in the creation of hypotheses. Then, a combination of global datasets onto pathways is used, utilizing advanced bioinformatics (Majumdar *et al.*, 2019). In order to determine the plant–NP interaction, researchers are anticipated to furnish more thorough knowledge of this method. The complex interactions between NMs and agricultural techniques are explored in this chapter, with a focus on the dual roles that NPs play in both improving plant systems and posing possible hazards. In a variety of environmental circumstances, the use of different NPs in agriculture has shown great potential for enhancing soil health and encouraging plant development. To maximize their advantages, however, a thorough knowledge of how these NPs are absorbed and distributed inside plant systems is necessary. By means of advanced transcriptome and proteomic studies, this chapter reveals the molecular aspects of plants exposed to NPs, providing insight into the gene regulatory mechanisms linked to NP-induced genotoxicity (Wahab *et al.*, 2024). Furthermore, the intricacies of the links between NPs and plants are revealed by genomic investigations, which provide insightful information about the genetic underpinnings of these interactions. It is becoming more crucial to balance the possible advantages of NPs in agriculture against the threats they may pose to ecosystem stability and plant health as research into their complex effects in agriculture advances.

10.2 Plant Interaction with Nanomaterials

Plants and NMs interact through intricate absorption mechanisms that are impacted by the size, surface charge, and coating of the NPs. The degree to which NPs can enter plant tissues and travel to different organs is determined by these characteristics. Plants exposed to NPs may experience major physiological changes, such as adjustments to transpiration rates, growth patterns, and photosynthesis. Depending on the kind of NM and its concentration, these impacts might differ significantly. It has been demonstrated that plants' resistance to abiotic stressors like drought and salt is increased by NMs. They have the power to modify stress-signaling pathways and enhance ion homeostasis, both of which are essential for preserving plant vitality in challenging circumstances. When applied in tiny volumes, certain NPs function as biostimulants, encouraging the growth and development of plants. They can promote metabolic processes and boost nutrient intake, which will increase crop yields. Although a lot of NPs are useful, there are worries about their possible genotoxicity. Certain NMs may cause harm to plant DNA, according to certain research, which emphasizes the importance of carefully evaluating their effects on the environment. Plant behavior and bioavailability of NMs can be influenced by their interactions with the surrounding environment, including soil and naturally occurring organic matter (Ahmad *et al.*, 2022; Sembada and Lenggoro, 2024) (Fig. 10.1). Interactions between plants and advantageous biomaterials, such as mycorrhizal fungus and bacteria that promote plant development, can greatly increase nutrient intake, according to multi-omics techniques. Improved plant growth and yield result from this, particularly in stressful environments like drought. Through the modulation of physiological responses, biomaterials can assist plants in overcoming abiotic stressors. As an example, certain NMs can improve the hydraulic conductivity of roots, which helps plants survive under harsh conditions and maintain osmotic equilibrium.

Understanding plant defense systems against pathogens is made easier with the inclusion of omics technology. By using this information, biomaterials that improve disease resistance can be created, decreasing the need for chemical pesticides. Because biomaterials replace chemical pesticides and fertilizers, they encourage sustainable agricultural practices. This reduces the negative environmental effects of conventional agriculture operations while simultaneously improving crop health. Researchers may address the issues of food security in a growing population by developing biomaterials that improve plant growth and production by utilizing insights from genomes, transcriptomics, and metabolomics (Kumar *et al.*, 2023). When present in high amounts, some biomaterials specifically, some NPs may be phytotoxic. This suggests that their use should be carefully considered as it may result in decreased plant development and production. The introduction of artificial biomaterials

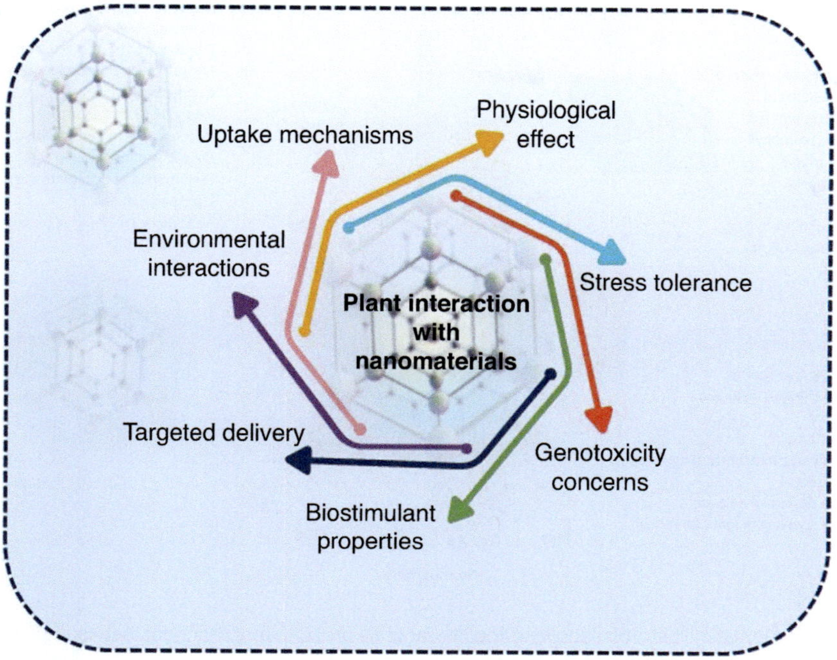

Fig. 10.1. Summary of plant interaction with nanomaterials.

has the potential to upset the microbial populations already present in the soil, which might have detrimental effects on plant development and soil health. It is essential to comprehend these interconnections using multi-omics in order to reduce such hazards. Farmers run the risk of relying too much on biomaterials to produce crops, ignoring biodiversity and conventional agricultural methods. Long-term sustainability problems might result from this. Plants and biomaterials can interact in a variety of intricate and context-specific ways. It is difficult to forecast the effects of biomaterials in various agricultural contexts since variables like plant species, soil type, and climate might affect the results (Egbuna *et al.*, 2021; Ali *et al.*, 2022).

The overall efficacy and safety of NPs in agricultural applications may be impacted by this dynamic. The controlled delivery of nutrients and genetic materials to certain plant organelles, including chloroplasts, is made possible by advancements in nanotechnology. This accuracy can improve plant functioning and increase the effectiveness of agricultural techniques. NMs sometimes display peculiar chemical and physical properties at the nanoscale, and they respond quite differently than they do at greater scales. This makes the use of nanotechnology to change matter a potential field for the development of novel products and processes. However, due to their unique characteristics, NMs might interact in unexpected ways with biological systems (Wang *et al.*, 2022; Ullah *et al.*, 2023). Different plants engage in chemical or physical interactions with NPs (Fig. 10.2).

Reactive oxygen species (ROS), which may cause major harm to plants, are produced by these interactions through physical and chemical communication (Elbasiouny *et al.*, 2022). These ROS change ion transport across the cell membrane, induce oxidative stress and damage, and promote fatty acid peroxidation. The properties of the NP, the surrounding conditions, and the kind of plant are the main determinants of plant–NP interaction (Haider *et al.*, 2021, 2022; Aqeel *et al.*, 2022). How the initial NP interaction site in plants selects the traffic path is yet unknown? However, some studies have offered some avenues for improving our understanding of

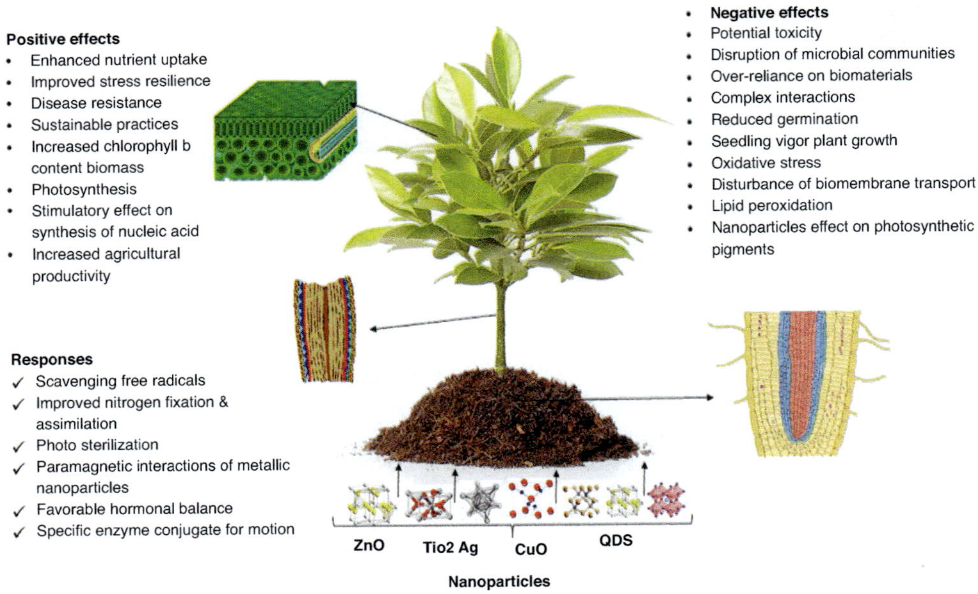

Fig. 10.2. A schematic illustration showing the interactions and different reactions between plants and NPs. (The negative effects highlight the detrimental effects of NP interactions in plants, whereas the positive effects and responses highlight the beneficial interactions of NPs in plant cells.)

the trafficking of NPs. The well-known pathways that influence the entry point include air fallout, unintentional release, contaminated soil or sediments, and direct application. According to Dev *et al.* (2017), the xylem, cortex, and pericycle are essential components that help NPs enter plants through the root. A small number of essential elements regulate particle entrance at different critical stages. The first barrier is the cell wall, which acts as a structural filter to let only NPs with suitable pore sizes pass through. The only NPs that could effectively pass through the cell wall and exit the cell membrane were those whose pore diameters were similar to those of the wall. Smaller NPs may reach plant cells rapidly through the root system, but bigger NPs can only enter through larger openings like stigma, hydathodes, and stomata. Larger NPs are seldom received by the cell; hence, cell metabolism remains unchanged. When metal-based NPs penetrate plant cells, they react with different functional protein groups to transform into reactive metal ions, which then impair the biochemical activity. Because of their enormous surface area, these NPs readily adsorb a variety of inorganic ions and compounds from the soil or nutritional media, resulting in indirect toxicity symptoms such as chlorosis and wilting (Dev *et al.*, 2018). The way that NPs interact with plants dictates how they accumulate and spread throughout different trophic levels.

10.3 Different NPs and Plant Interaction

Since animals consume plants and might potentially absorb NPs from them, plants play a crucial role in the food chain. They also influence how NPs are transported and behave within the environment. Plant absorption of NPs is influenced by their shape, concentration, coating, and functional group type. Enlarged and produced for agricultural applications are engineered nanomaterials (ENMs) because of their distinct chemical and physical characteristics and nanosize. NPs have a dynamic role in crop growth and span a diverse array of components (Agri *et al.*, 2021). NPs have advantages but also have the potential to be harmful. In contrast to a comparable category of NPs with different sizes and concentrations, different plant species behaved differently. Plants are altered genetically, morphologically, and physiologically by NPs. However, understanding the effects of various nanoparticles on plant health and production as well as the variety of soil microbes is essential to understanding the utilization of NPs in agriculture. It was discovered that nanochitosan helped the wheat crop, significantly enhancing seed germination (Li *et al.*, 2018). According to Rastogi *et al.* (2019), AgNPs dramatically enhanced the length of shoot/root, number of leaves, and dysgenic content in fenugreek plants. Increased stress-enzyme activity, sugar and proline accumulation, and plant growth development under salt stress have all been shown to benefit from TiO_2 NPs (Gohari *et al.*, 2020).

Allium cepa seeds germinate when exposed to low concentrations of ZnO NPs. Fe_2O_3 NPs were shown to have a beneficial influence on *Helianthus annuus* seed germination. Gold NPs were shown to have a positive effect on *Mimusops laurifolia*'s leaf size and quantity (Alshehddi and Bokhari, 2020). According to Chaudhary *et al.*, the use of nanochitosan improved maize production, agronomical/biochemical indices, and seed germination. Zinc oxide NPs improved plant height and seed germination in barley, according to Plaksenkova *et al.* (2020). Under salt stress, TiO_2 NPs enhanced the biochemical characteristics of *Moldavian balm* plants (Gohari *et al.*, 2020). By boosting photosynthesis and photoprotection mechanisms, SiO_2 NPs (300 ug/L) enhanced sugarcane development under chilling stress. Tauseef *et al.* (2021) discovered that cowpea growth and biochemical characteristics were enhanced by MgO NPs (0, 25 and 50 ppm). In rapeseed, the use of nanochitosan (10 mg/L) stimulated plantlet growth and microspore embryogenesis (Elsheery *et al.*, 2020).

Under phosphorus deficit, Zhang *et al.* (2019a) discovered that SiO_2 NPs increased antioxidant enzymes, stomatal conductance, and efficacy of water usage. The agronomic and biochemical characteristics of maize plants were enhanced by chitosan NPs (Choudhary *et al.*, 2017). ZnO NPs boost plant nutrient absorption efficiency and raise zinc levels, which in turn promote plant development. SiNPs decreased cadmium build-up and increased wheat grain production (Hussain *et al.*, 2019). The breakdown of organic molecules and the cycling of nutrients are largely dependent on soil microbial activity. According to various disciplines, it is crucial to demonstrate that the soils support a healthy microbial community (Sergaki *et al.*, 2018).

Therefore, it is essential to understand and quantify how various NPs affect microbes in order to maintain soil health. Under maize and fenugreek production, the application of agriusable nanocompounds like nanozeolite and nanochitosan enhanced the soil physicochemical parameters and soil enzyme activity. Kukreti *et al.* observed that nanosilicon dioxide has a positive impact on maize and soil health. Different beneficial bacterial isolates showed increased growth and protein content when exposed to nanozeolite and nanogypsum (Khati *et al.*, 2019). In bean plants, arbuscular mycorrhizal fungus containing TiO_2 improved the expression of chitin synthase genes in the roots when the plants were exposed to salt (Gohari *et al.*, 2020). By using nanozeolite, the bacterial and fungal populations in alfalfa straw were enhanced (Aminiyan *et al.*, 2018). Applying SiO_2 NPs to several crops under salt stress can promote antioxidant activity and seed germination (Siddiqui *et al.*, 2014). According to Sillen *et al.* (2015), there was a decrease in FDA activity in treated soil when they examined the impact of nanosilver on the FDA's monitoring of the maize rhizosphere in the USA. The dehydrogenase enzyme activity in the treated soil was decreased by AgNPs (McGee *et al.*, 2017). According to Chai *et al.*, oxide NPs enhanced the bacteria that solubilized phosphate and nitrogen. According to Metch *et al.* (2018), exposure to AuNPs caused certain changes in the functional microbial population. According to Kwak *et al.* (2017), the soil's dehydrogenase activity was enhanced by zinc oxide NPs. According to Karunakaran *et al.* (2013), SiO_2 NPs improved the amount of nutrients in the soil and encouraged the rhizobacterial population's growth, which in turn boosted maize seed germination.

10.4 Plant Uptake and NP Translocation

NPs penetrate the plant by stomatal openings, root pathways, or epidermal absorption in leaves because of their very small size. When it comes to roots, NPs enter the cell and travel along the symplastic or apoplastic pathway to reach different plant systems. Through the root's epidermis's cell wall, NPs pass through the cell membrane and are symplastically transferred to the vascular system. The size of the openings affects movement across the cell wall (Ali *et al.*, 2021). To get to the vascular system, NPs go not only via the symplastic routes but also the apoplastic pathway. Once inside, NPs are transported to the leaves of a plant for interaction with chloroplasts, causing the physiology of the plant to be disrupted (Kang *et al.*, 2023). Research on the absorption and translocation of NPs in plants is essential, especially in light of the growing application of NMs in agriculture. Numerous investigations have revealed that NPs, such as metal-based and polymeric NPs, may enter plants' tissues and move about, affecting their development and metabolism (Fig. 10.3).

10.4.1 Leaf

Substances can enter leaves through the stomata or by being absorbed through the epidermis of the leaf. In a study, it was demonstrated by confocal microscopy that NPs enter *Vicia faba* through the stomata at a certain size; NPs measuring 43 nm were able to pass through the cell system whereas NPs measuring 1.1 µm did not. Through the use of electronic and confocal microscopy, mass spectrometry, and *Zea mays* leaves, it was possible to detect that Cerium (37 nm) enters the plant system but does not go to the stem. However, this investigation was unable to determine whether the stomata or the leaf epidermis served as the entrance portals (Birbaum *et al.*, 2010). In a related study, NPs were introduced into the epidermis through the stomata of *Arabidopsis* seedlings using X-ray and electron fluorescence microscopy. After that, endocytic capillaries took them in order

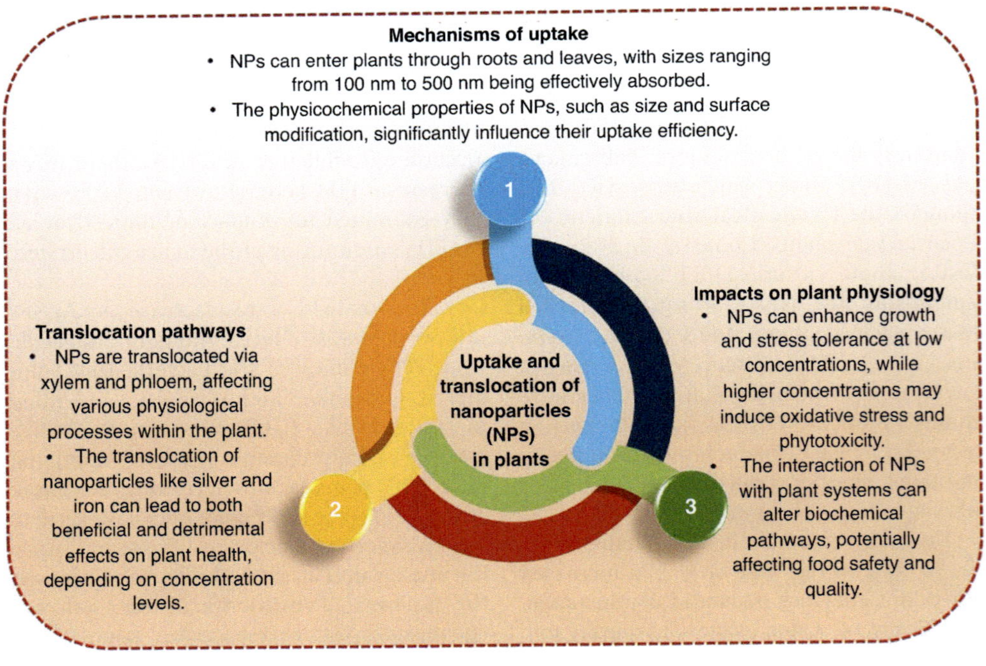

Fig. 10.3. Mechanism of uptake, translocation and impacts of nanoparticles in plants.

to transport them to the palisade layer (Wang *et al.*, 2013). In addition to studies on stomatal entrance, other studies have examined entry through the leaf's epidermis. Researchers studied the translocation of iron, zinc, and manganese after they were absorbed by the leaf epidermis using nonionic colloidal solutions (Taran *et al.*, 2014). A separate study found that bulk salt and AgNPs both had a hydrodynamic diameter of 47.9 nm and that they were absorbed by Lactuga through the cuticle and stomata before proceeding to the circulatory system. Additionally, the study showed that Ag^+ attachesto thiol groups and interconverts into AgNPs (Larue *et al.*, 2014).

10.4.2 Roots

The complex mechanism behind the NPs' initial contact, adsorption, and ultimate incorporation in plant roots is a complex web of routes and processes (Manna *et al.*, 2021). When NPs come into contact with plant roots, they first cling to the surface of the root by means of adsorption. The root surface is negatively charged due to the existence of root hairs, which release chemicals like mucus and organic acids. Positively charged NPs have a greater inclination to concentrate and then be absorbed by the surface of the root due to this electrostatic phenomenon. The appearance of lateral roots increases the number of possible surfaces that NPs may interact with and provides pathways for them to enter the architecture of the roots (Roy *et al.*, 2023). Two essential pathways the apoplastic and symplastic pathways are home to a dynamic symphony of NP mobility inside the structure of plant roots (Avellan *et al.*, 2021). Using the complex maze of cell walls as well as intercellular gaps to help in NP mobility, the apoplastic route meanders across the extracellular spaces. By passing through the permeable cell walls, NPs can use this path and arrive at the xylem tissue of the root. If the root does not have an exodermis layer, then NPs might enter the xylem or central column, which would increase the range of possible directions for their transport (Kurczyńska *et al.*, 2021; Murali *et al.*, 2022).

Studies that explore this process, including Su *et al.*'s (2020) work, clarify how AgNPs travel along the apoplastic channel in citrus roots. On the other hand, the symplastic route involves the passage of NPs via the channels of the plasmodesmata via the live components of plant cells. This route starts at the root epidermis, where small channels allow the NPs to pass through and enter the cytoplasm of the root cells (De Angelis *et al.*, 2023). NPs can travel through different cellular compartments thanks to symplastic mobility, which may have an impact on their absorption and physiological consequences on plants. After effectively penetrating plant tissue, NPs explore a wide range of absorption paths by plant cells. These channels include physical disruption, binding interactions with cell membrane proteins, endocytosis, and ion transport. The hydrophilic route offers one approach for NPs to enter plant cells; however, it is restricted by the size of the pores. One well-known process is endocytosis, which involves folding the plasma membrane inward to allow NPs to enter plant cells. This route demonstrates a lack of selectivity with respect to NP size, accepting particles smaller than 1 μm.

Additionally, plant roots' outer layer contains transport proteins, which NPs might interact with to enhance and expedite cellular absorption (Mitra *et al.*, 2023). The intricate process that NPs go through from their initial contact with the root surface to their internal absorption in the plant cell is shown by the coordinated activity of several channels and systems. An intricate schematic depiction of the intricate procedures involved in the absorption and transportation of NPs through root channels. A number of important parameters, such as surface charge, chemical makeup, and NP size, affect how well plant roots absorb NPs. Of them, the size of the NPs stands out as the main influence (Torrent *et al.*, 2020). Research has shown that maize roots and *Vicia faba* L. can absorb NPs smaller than 10 nm, such as CeO_2 NPs (< 25 nm) and AuNPs (3.5 nm) (Gui *et al.*, 2023). On the other hand, lower absorption is shown with bigger particle sizes, and root uptake is more challenging for particles larger than 100 nm (Prasad *et al.*, 2018).

The alterations and surface coatings of NPs have been identified as critical variables affecting the complex interactions between plants and NPs. These changes have the capacity to modify the course of interactions between NPs and roots, and therefore the absorption routes. Extensive studies have shed light on the complex impacts of surface changes on the interactions between plants and NPs. Cvjetko *et al.* (2017) found that plants of the *Allium cepa* genus absorbed AgNPs coated with cetyltrimethylammonium bromide (CTAB) more readily than those that were not, highlighting the impact of surface chemistry on the interaction between plants and NPs. The surface-coated NPs changed their transport behavior and might have an impact on the interactions between plant roots and bacteria by forming aggregates with them. Notably, research by Tan *et al.* showed that changes in surface charge might modify electrostatic interactions, which in turn can affect NP–root interactions (Spielman-Sun *et al.*, 2019). It has been demonstrated by researchers that uncharged and negatively charged NPs (22 nm) may enter the root xylem of *Arabidopsis thaliana* plants after being absorbed by their root cells. On the other hand, because positively charged NPs cannot enter the *Arabidopsis* root tissue, they only build up on the root epidermis. Root exudates, which comprise a wide range of chemical substances, have a substantial impact on the dynamics of NPs in the rhizosphere. These substances are released from plant roots and use chemotactic attraction or repulsion to precisely control the behavior of NPs (Wang and Si, 2023).

These chemotaxis-driven interactions highlight how NPs navigate in a sophisticated way inside the complex rhizosphere matrix. In addition to providing guidance, root exudates control the fate and dispersion of NPs throughout the rhizosphere. Root exudates have been shown by Khanna *et al.* (2021) to influence the aggregation behavior of NPs, which may result in the creation of aggregates of microbes and NPs. The mobility and availability of NPs for plants may be affected by these aggregates. The complex molecular interactions governing the fate of NPs at the soil–root interface are highlighted by the interaction between root exudates and NPs. Furthermore, the effect of root exudates on the interactions of NPs with soil constituents is increased (McManus *et al.*, 2018). The complex interactions between root exudates and NPs include chemotactic reactions, aggregation dynamics, and changes in soil properties. De La Rosa *et al.* state that these interactions offer a conducive environment for elucidating the mechanistic complexities of NP activity in the rhizosphere and its subsequent impacts on plant absorption.

Nanoparticles can enter plants through root hair routes, which make use of the root epidermal cell extensions as entry points. These fragile structures provide special areas of contact for the NPs in the soil solution, in addition to increasing the root surface area (Singh and Kumar, 2020). Researchers have explored the intriguing interaction between NPs and root hairs, providing a crucial understanding of NP absorption. It is possible that AgNPs will gather near the ends of maize plants' root hairs, suggesting a preferred entrance point (Yan and Chen, 2019; Ali *et al.*, 2021)

All of these investigations highlight the complex relationship that exists between root hairs and NPs, providing insight into a crucial process that aids in the absorption and transportation of NPs in plants (Ali *et al.*, 2021). The way in which NPs and microorganisms interact has a big impact on how plants will eventually absorb the particles. These connections begin when NPs come into contact with microbial surfaces in the rhizosphere. They are based on intricate physicochemical processes. NPs can group together with microbial products or stick to microbial cells. These relationships between NPs and microbes can affect the bioavailability and mobility of NPs in the soil matrix as a whole, which can affect how accessible they are to plant roots (Khan *et al.*, 2020; Rai and Golińska, 2023). Furthermore, these interactions could change the chemical signals that attract NPs by influencing root exudation patterns. These changes in the rhizosphere environment may help or hinder plant roots' ability to absorb NPs. Rajput *et al.* (2020) provide important background for comprehending the broader ecosystem dynamics by shedding light on how NP–microbe relationships affect the bioavailability of NPs to plants. Understanding the complex interactions that exist between microbes and NPs in the rhizosphere helps us better understand the mechanisms regulating plant uptake of NPs and provides opportunities for utilizing these interactions to support environmentally friendly practices and sustainable agriculture (Rajput *et al.*, 2020). It is crucial to stress that a complex interaction between the qualities of the NPs, the traits of the plants, the environment, and the length of exposure determines the precise paths via which NPs penetrate plant systems. These complex interactions are still being explored by ongoing research, which is expanding our knowledge of the possible impacts of NPs on ecosystems as well as plants.

10.4.3 Foliar uptake

Plant leaves can accumulate NPs, particularly those floating in air or water. From there, they may enter the leaf cuticle or be taken up by stomata, which are microscopic holes on the leaf surface.

10.4.3.1 Stomatal uptake

Stomata, which are tiny holes that control gas exchange, are the entry points for NPs that are deposited on leaf surfaces. Smaller NPs can pass through stomata that are open.

10.4.3.2 Cuticular penetration

A small number of NPs may be able to pass through the leaf surface's waxy cuticle. This is more typical for NPs that have surface changes to improve uptake (Avellan *et al.*, 2021). When NPs enter mesophyll cells through the stomata and cuticle, they begin a complex series of transport channels that allow them to travel across long distances throughout plants. The extracellular and plastid routes are among them; each has a unique function in the transport of NPs (Wu and Li, 2022). The primary purpose of the extracellular pathway is to traverse the extracellular space. NPs go via the area composed of the cell wall itself, the longitudinal channels that connect cell walls, the intermediate lamella, and the xylem. Particle size and surface charge are two elements that impact this trajectory and influence how successful the extracellular transport of the NPs is (Lew *et al.*, 2018). The plastid pathway, on the other hand, is dependent on intracellular channels known as plasmodesmata, which range in width from 3 nm to 20 nm. These channels allow NPs to collect in the cytoplasm and go to the plant's endothelial layer and Casparian strip, signifying significant locations along the plastid pathway (Lv *et al.*, 2019). Research has demonstrated that in plants, NPs smaller than 50 nm usually go through the plastid route (Newkirk *et al.*, 2021). Meanwhile, it has been found that the majority of 50–200 nm NPs follow the apoplast pathway. The vascular system's phloem transport pathways subsequently facilitate the movement of NPs from leaves to roots. Because of the enormous diameter of the phloem sieve tube, Nps as tiny as 0.405 µm may be efficiently transported, allowing the particles to disseminate from the leaves to other parts of the plant (Wang and Si, 2023; Rani *et al.*, 2023). The complicated nature of NP mobility inside plants is essentially highlighted by the interaction of these many transport channels, which also provides insight into the mechanisms of dispersion and their consequences for plant health and growth.

Plants transport NPs via a complicated process that is regulated by a number of critical parameters, including the properties of the NPs themselves, the features of the plant species, and the surrounding environment (Avellan *et al.*, 2021). Notably, the absorption of NPs is highly dependent on their size. Numerous investigations have demonstrated that metal-based NPs may penetrate plant leaves via the stomatal channel when their diameter is less than 50 nm (Wang and Si, 2023). The adsorption of positively charged NPs was controlled by the electrostatic interaction between them and the negatively charged plant cell walls. Significant differences in the absorption of NPs are also attributed to plant species. Variations in absorption efficiency are caused by different stomatal layouts, densities, and pore diameters in leaves. Temperature, humidity, and light are examples of environmental conditions that affect stomatal activity, which in turn affects the absorption of NPs (Hong *et al.*, 2021). The dynamic interaction of plant species, environmental factors, and NP characteristics shapes the complex process of NP absorption through leaves. In the end, this affects how

NPs are distributed and their potential effects on plants.

10.5 Nanoparticle Uptake in Hydroponic Farming

Plants are grown in nutrient solutions without soil in hydroponic systems, which offer a controlled setting for studying the dynamics of plant nanoparticle interactions. By avoiding the potentially confusing effects of soil characteristics, this technique provides clear benefits and enables targeted investigation of the interactions between plants and NPs. Plants grown hydroponically are exposed to fertilizer solutions that include NPs. This critical exposure phase allows NPs to directly engage with root surfaces, starting a sequence of events similar to those described in relation to root absorption mechanisms (Francis *et al.*, 2020). Simultaneously, ongoing studies have elaborated on the consequences of hydroponic absorption, demonstrating its suitability for deciphering NP–plant interactions. Research utilizing hydroponic techniques, for example, has shown that NPs, such as zinc oxide (ZnO) and copper oxide (CuO), may be directly added to nutritional solutions that support plant roots (Francis *et al.*, 2022). After conducting thorough examinations, it was shown that these NPs were able to migrate via root tissues and then to other regions of the plant, indicating the complex routes that allow NPs to move throughout plants. The hydroponic system's controlled environment makes it easier to analyze the interactions between NPs and roots, giving researchers insight into the variables influencing the absorption and intraplant mobility of these particles (Ristroph *et al.*, 2017). Moreover, the hydroponic method highlights the critical function of root exposure as a point of entry for NPs, providing insight into the complex interactions between plant and NP systems under well-regulated experimental conditions (Fig. 10.4).

10.6 Nanoparticles' Intracellular Journey

The introduction of NPs into the plant's cellular environment represents a critical turning point in the complex interplay between the plant and NP interactions. A molecular ballet unfolds in

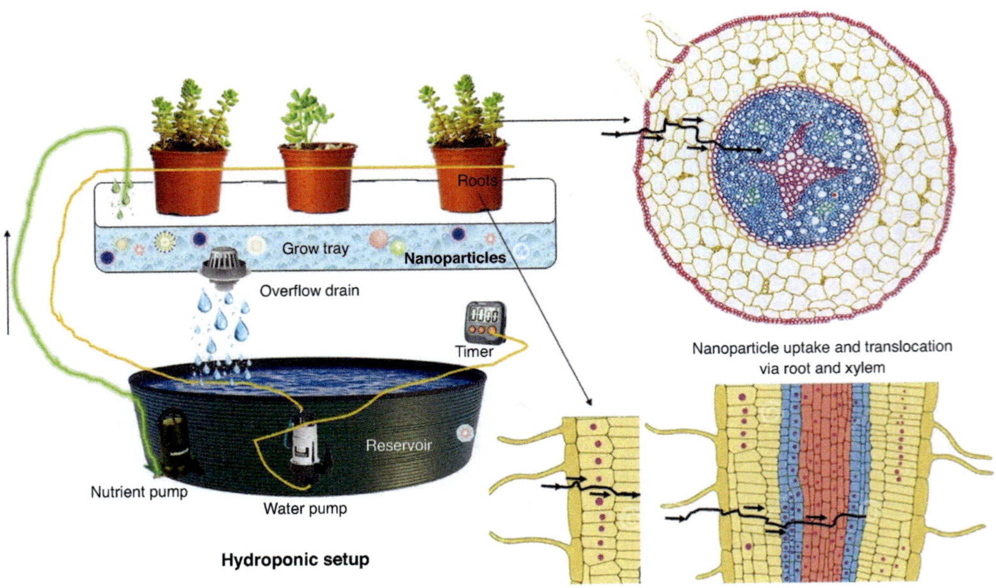

Fig. 10.4. Diagrammatic depiction of the absorption of NPs in a hydroponic setup.

the context of root cells, as NPs participate in a complex series of assimilation, intracellular travel, and interactions with cellular components. Significant research by Milewska-Hendel *et al.* (2021) examined the assimilation dynamics of AuNPs inside *Hordeum vulgare*, offering insights into the particles' internalization into root cells and subsequent navigation through the Golgi apparatus and endoplasmic reticulum. This voyage is not only an assimilation process; it resonates with far-reaching consequences throughout molecular landscapes. Once within root cells, NPs start a series of complex interactions. The work by Milewska-Hendel *et al.* (2021) highlights the confluence of AuNPs and cellular components while revealing the assimilation dynamics inside *Hordeum vulgare* roots. This close meeting is the beginning of a multifaceted journey that NPs take within the cellular havens. Surprisingly, the AuNPs go via the Golgi apparatus and endoplasmic reticulum, a voyage that has ramifications that go beyond simple absorption (Milewska-Hendel *et al.*, 2021).

The effects of ingesting NPs extend beyond the boundaries of root cells and ripple across the molecular landscape. The process of absorption turns into a conductor of molecular conversations outside of the cell. Gene expression patterns, metabolic landscapes, and molecular pathways are all impacted by the complex dance that occurs inside the Golgi apparatus and endoplasmic reticulum. The study of Manna *et al.* (2021) turns root cells into the focus of complex molecular conversations in the quest to comprehend NP assimilation. This trip illustrates the mutually reinforcing relationship between plant cells and NPs, spanning from assimilation to molecular symphonies (Manna *et al.*, 2021). Plant biology is affected by the molecular tangos that NPs create when they integrate and move through cells.

evaluations. Ag, SnO_2 and ZnO, NPs can change growth rate, cell viability, and chlorophyll at the cellular level in comparison to their bulk equivalents (Manzo *et al.*, 2013). To characterize protein profiles, post-translational modifications, and interactions, proteomics systems including gel, label, gel-free/label-free, and targeted have developed during the last 20 years. The many features of a proteomic approach, such as isolation, separation, visualization, and accurate protein identification, ultimately determine its success. Notwithstanding recent advancements, protein extraction techniques still require additional care, particularly for hydrophobic and big molecular mass proteins. Subcellular NP aggregation along the plasma membrane and cell wall, together with modifications to organelles such as chloroplasts and vacuoles, have been reported (Geisler-Lee *et al.*, 2013; Simon *et al.*, 2013). It was shown that 40 nm-sized NPs are absorbed by the root through the vascular system via the symplastic pathway; bigger NPs, on the other hand, accumulate via the apoplastic pathway in higher plants.

Maize roots' cell walls exhibited smaller porosity, decreased transpiration, and inhibited hydraulic conductivity following their exposure to clay NPs and TiO_2. C-based NM uptake, translocation, and generational transmission have all been shown in rice. Novel pathways of plant–NP interactions have been revealed by recent research on the gene expression of the *Arabidopsis thaliana* plant (Kaveh *et al.*, 2013). This research found alterations in gene expression when participants were exposed to TiO_2 NPs, AgNPs, ZnO NPs, and fullerene of varied sizes over an extended period of time. Furthermore, it was shown that the processes behind ZnO NP toxicity might be revealed by antioxidant enzymes and oxidative stress (Ghosh *et al.*, 2016).

10.7 Plant–nanoparticle Interaction with Respect to Proteome and Transcriptome

Since plants are seen to be the primary connection between the biosphere and the environment, it is essential to comprehend how NPs impact them in order to perform relevant environmental

10.7.1 Nanoparticle–plant interaction: genomic insight

Nanotechnology has applications in nearly every branch of research today since the scientific world has recently acknowledged it as the most advanced, labor-saving, and quickly evolving technology. Particularly with regard to plant performance, the toxicity brought on by the ongoing accumulation of NPs in the environment cannot be disregarded. The majority of the research shows how NPs are phytotoxic on a morphological and physiological level, but not much of it shows how NPs are toxic at the gene level. CuO NPs were shown to be hazardous to a number of agricultural crops, including *Lolium rigidum*, *Raphanus sativus*, *Lolium rigidum*, and *Lolium perenne*. Additionally, DNA was damaged after extended exposure to NPs. Plant development was rapidly disrupted due to the creation of mutagenic DNA lesions induced by the deposition of chemicals that had undergone oxidation. Similarly, the toxicity of ZnO NPs in the seedlings of wheat was investigated and it was shown that nitric oxide (NO) successfully reduces this toxicity in wheat. Furthermore, because NPs have been connected to the genotoxic effects of metals on plants, determining the NPs' genotoxic end points is a concern (Tripathi *et al.*, 2017).

10.7.2 Regulation of genes concerning NP-induced genotoxicity

The underlying mechanisms of NP-induced phytotoxicity are not properly known at the molecular level in higher plants. On the other hand, transcriptome analysis has revealed information on the up- and downregulation of several genes in response to NP exposure, which has given some insight into NP toxicity in higher plants at the genetic level. For instance, an investigation revealed that single-walled carbon nanotubes' (SWCNTs) exposure increased the expression of the genes for RTCS and SLR1 and decreased the expression of the genes for RTH1 and RTH3. Another important factor in the uptake of NPs is aquaporin. Numerous stress-inducible genes in plants are regulated or controlled by a range of stress conditions (Yan *et al.*, 2013). In a different study, it was shown that tomato seedlings' increased synthesis of aquaporins elevated the NtLRX1 and water channel genes. AgNPs were applied to *Eruca sativa* by certain researchers, who then noted the proteomic profile of the plants as well as altered gene and protein expressions. AgNPs changed the number of proteins in the endoplasmic reticulum (ER) and vacuole of plant cells, suggesting that these organelles are AgNPs' main targets (Vannini *et al.*, 2013). Similar expression findings for microRNAs (miRs) were seen when different quantities of AuNPs were applied to *Arabidopsis* plants.

The results of this study, however, significantly differ from those of Burklew and Ashlock (Kumar *et al.*, 2013) in relation to tomato plant seedlings. This demonstrates even further the phytotoxicity of NPs is species-specific. Furthermore, NPs may change gene expression even at extremely low doses, indicating that gene expression profiling may be a useful tool for assessing the cellular effects of NP-caused toxicity (Thomas *et al.*, 2011). Therefore, understanding the mechanism of NP toxicity and connecting any morphological or physiological alterations in plants to their genetic origins is made possible by the application of genomics analysis.

10.7.3 Nanoparticle–plant interaction: transcriptomic insight

Plant gene expression has been shown to be altered by NPs, according to several studies. Allegations have been made that ZnO and TiO_2 exposure changed the expression of some genes in *Arabidopsis thaliana* roots. They came to the conclusion that, of the three NPs listed above, ZnO NPs were the most harmful because they affected 660 and 826 genes related to stress regulation, respectively, whereas TiO_2 NPs, which were the least toxic, only affected 74 and 80 genes, respectively. Although the researchers have clearly shown that the phytotoxicity of NPs causes both up- and downregulation of genes, detailed descriptions have not yet been provided. In recent work, they observed the early alterations in gene expression by subjecting *A. thaliana* to abiotic and biotic challenges with NPs of different sizes and kinds. They detected downregulation of transcriptional repression and gene expression as a general response to stress generated by NPs (García-Sánchez *et al.*, 2015). In a different research, the impact of AuNPs on growth and development in *Arabidopsis* has been investigated by proteomic, ionomic, and transcriptome changes. A transcriptome study of *Arabidopsis* roots revealed variations in the expression of the BHLH, MYB, and WRKY gene families, which are implicated in the homeostasis of several critical metals and Fen (Tiwari *et al.*, 2016). It was concluded that Au has an effect on the regular homeostasis of important elements such as Mn, Fe, and Zn (Taylor *et al.*, 2014). Numerous studies have looked at the different kinetics that NMs use to swallow, translocate, and internalize NMs as well as the cytotoxicity seen at the gene expression level. These investigations also discovered that NMs had a variety of additional morphological and physiological impacts on plants, including growth and development. It has been noted that these effects are species-specific, despite the fact that the genetic mechanisms behind them are yet unknown.

10.7.4 Nanoparticle–plant interaction: proteomic insight

Numerous interdisciplinary researches conducted over the past few years on the effects of AgNPs on plants have demonstrated their toxicity on a variety of levels, including cellular, morphological, physiological, and others. On the other hand, very little research has looked at the harmful proteome impacts of AgNP stress on plants. Recently, an investigation using gel-based proteomics on *Oryza sativa* seedlings was carried out with the aim of eliminating the toxicity caused by AgNPs (Mirzajani *et al.*, 2014). According to reports, AgNP-receptive proteins are mostly linked to transcription, protein degradation, apoptosis, cell-wall formation, oxidative stress response pathway, Ca^{2+} signaling and regulation, and transcription.

According to earlier studies, released ions or AgNPs may attach to the secondary messenger receptors, Ca^{2+}/Na^{+} ATPases and Ca^{2+} ion channels, and disrupt the cell's metabolism. The structure of proteins linked to redox regulation can be changed by bulk metal and NPs, which can eventually upset cellular homeostasis, according to a proteomic study of *Eruca sativa* roots subjected to $AgNO_3$ and AgNP stress. Nonetheless, the fact that AgNP exposure is sufficient to change the endoplasmic reticulum and vacuole membrane proteins indicates that these NPs predominantly target these two particular organelles. These results suggest that the distinct chemical and physical characteristics of AgNPs, rather than their ongoing atmospheric emission, are what cause their phytotoxicity (Vannini *et al.*, 2013). Regarding the use of AuNPs, the agriculture industry is significant. A proteome study showed that glutathione S-transferases were induced in the shoot of *Arabidopsis thaliana* after it was exposed to AuNPs. Furthermore, studies have indicated that the plant hormone auxin is a major factor in determining the architecture of the Au-induced root system (Mustafa *et al.*, 2015; Tiwari *et al.*, 2016).

They showed that in contrast to bigger and smaller AgNPs, AgNPs with a particle size of 15 nm enhanced the germination and growth rate capacity of soybean seedlings under flood stress. The majority of the proteins impacted by AgNP exposure were connected with cellular metabolism, the detoxification pathway, and stress signaling. Under flood stress, the concentration of these proteins rose, but in response to AgNPs, the proteins quickly dropped. Furthermore, the application of AgNPs caused a change in metabolism from fermentative to cellular mechanisms.

The results show that when soybean seedlings are treated with AgNPs which have a 15 nm particle size and two ppm during water stress, there is a reduction in oxygen loss. Furthermore, it was discovered that at high concentrations of AgNP with particle sizes of 15 nm, soybean seedlings suffered damage (Mustafa *et al.*, 2015). Recently, Mustafa and Komatsu (2016) conducted a comparative investigation on two-day-old soybean seedlings subjected to water stress, using ZnO, Al_2O_3, and AgNPs. Remarkably, soybeans treated with 50 ppm Al_2O_3 NPs showed improved vigor and growth. These Al_2O_3NP-responsive proteins were mostly involved in protein synthesis, glycolysis, lipid metabolism, and protein breakdown.

A protein family with fivefold upregulation of the negative NmrA-like transcriptional regulatory was also noted upon treatment with Al_2O_3 NPs. According to the proteomic data, decreased root cell proliferation and controlled energy metabolism may be responsible for the development of soybean seedlings during flood stress. Different cell systems have been utilized to study the toxicities of different NPs, similar to AgNPs and AuNPs. For example, it has been shown that CeO_2 NPs suppress the production of stress-responsive proteins such as heat-shock protein, CAT, and APX, hence inhibiting the development of maize (Zhao *et al.*, 2012). In summary, proteome examination of the interactions between plants and NPs sheds light on the intricate processes that underlie plant reactions to NPs. The results show that proteins play a crucial role in mediating these interactions and have both positive and negative effects on the growth and health of plants. To maximize the use of NPs in agricultural and environmental applications and ensure that their potential benefits are maximized while limiting any negative effects, it is imperative to comprehend these proteome modifications. To help create sustainable practices in nanotechnology and agriculture, future research should clarify the precise pathways involved and the long-term effects of NP exposure on plant physiology.

10.8 Exploring the Potential Benefits along with Risks of Nanomaterials in Crops

Agricultural NMs have demonstrated encouraging promise for a range of uses, such as crop enhancement, disease control, and environmental clean-up. To guarantee their safe and long-lasting use, it is crucial to weigh the advantages and disadvantages of their usage. In order to provide important insights into the effects of metal NPs on crop production, soil health, and environmental sustainability, research must examine the possible advantages and hazards of these particles in agriculture. The possible advantages and disadvantages of using NPs in agriculture are depicted in Fig. 10.5.

10.8.1 Potential benefits

It has been documented that metal NPs increase agricultural production by encouraging nutrient absorption, root development, and seed

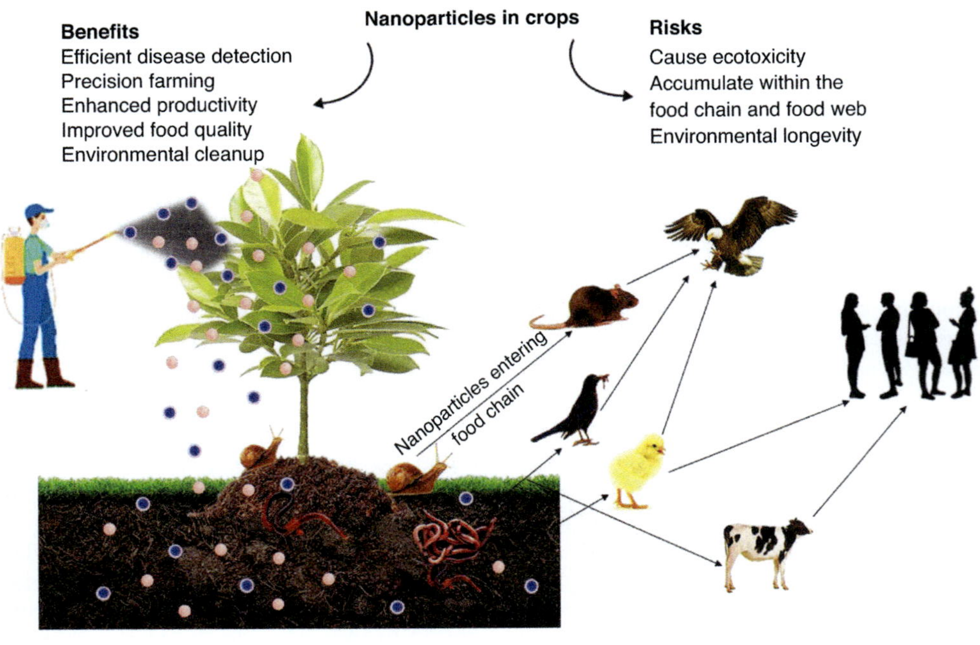

Fig. 10.5. Possible advantages and disadvantages of using NPs in agriculture along with the nanopesticides entering the food chain.

germination. According to Adil *et al.* (2022), ZnO NPs improved nutrient availability and absorption efficiency, which led to an improvement in wheat growth and output. Similar to this, it has been demonstrated that CuNPs increase nutrient absorption and photosynthetic efficiency, which in turn promotes tomato plant growth and yield (Wang *et al.*, 2021). According to Li *et al.* (2023), metal NPs have antibacterial qualities that make them viable options for managing plant diseases. Studies have indicated that they are efficacious against an extensive variety of plant diseases. For example, studies using AgNPs have demonstrated inhibitory effects against fungal infections, including *Alternaria solani*, *Fusarium* spp., and *Botrytis cinerea*. These NPs may be used to lessen the need for chemical treatments by substituting for traditional fungicides (Vera-Reyes *et al.*, 2022). Notably, broad-spectrum effective metal NPs, such as Cu, Ag, ZnO etc. suppress a variety of plant diseases and can help with flexible disease control (Kashyap *et al.*, 2023). Long-term disease management techniques may benefit from the many ways that NPs suppress plant diseases, since they may also slow the emergence of resistance in plant pathogens. Agricultural environmental remediation can make use of NPs, especially in soil and water systems. They have been used to get rid of pesticides, heavy metals, and organic contaminants. For example, arsenic-contaminated soil has been remedied using iron NPs, which successfully lowered the quantity of arsenic and improved the quality of the soil. These kinds of applications help lessen the negative effects that contaminants have on agricultural environments (Ji *et al.*, 2022).

10.8.2 Potential risks

NPs may be advantageous, but they may also be harmful to ecosystems and creatures that are not the intended targets (Côa *et al.*, 2020). According to studies, beneficial soil bacteria, earthworms, and aquatic species may be toxically affected by several NPs, including Ag and Cu NPs. According to Parada *et al.* (2019), these NPs may upset natural equilibriums and have a detrimental effect on biodiversity in agricultural environments. Concerns exist over the possible build-up of NPs in the food chain as well as the potential consequences for human health. Studies have indicated that plants may absorb and relocate NPs, which has led to concerns over the possibility of these particles building up in plant components that humans can consume. For instance, it has been demonstrated that CuNPs may build up in lettuce leaves and may have an effect on food safety. Consequently, it is critical to assess the dangers that might arise from consuming food tainted with NPs. There are still questions about the persistence and destiny of NPs in the environment. Long-term environmental effects might result from their persistence in sediments, water bodies, and soils (Tran *et al.*, 2024). Studies have demonstrated that NMs may interact and change with soil constituents, changing how they behave and their ability to move across the environment.

NMs in soils have the potential to change nutrient cycle processes and soil structure by interacting with organic matter and soil particles. For example, research has shown that at concentrations as low as 0.1 mg/kg, AgNPs may dramatically change the microbial communities in soil, affecting the fertility and health of the soil. Similarly, it has been demonstrated that CuNPs at 50 mg/kg impede plant root elongation and decrease soil enzyme activity, suggesting possible hazards to plant development and soil function (Xiao *et al.*, 2022). Likewise, NPs in sediments may affect species that live there and the stability of the substrate. AgNPs, for example, have been shown to lower the rates of reproduction and survival of the benthic organism *Lumbriculus variegatus*, suggesting possible ecological hazards (Rajala *et al.*, 2016). Determining their long-term effects on ecosystems requires an understanding of their environmental destiny. These quantitative investigations demonstrate the necessity of carefully evaluating NP concentrations and persistence in different environmental compartments in order to reduce possible dangers (Fig. 10.6).

10.9 Conclusion

In agriculture, NMs are starting to play a revolutionary role, improving crop yield and resilience while tackling environmental

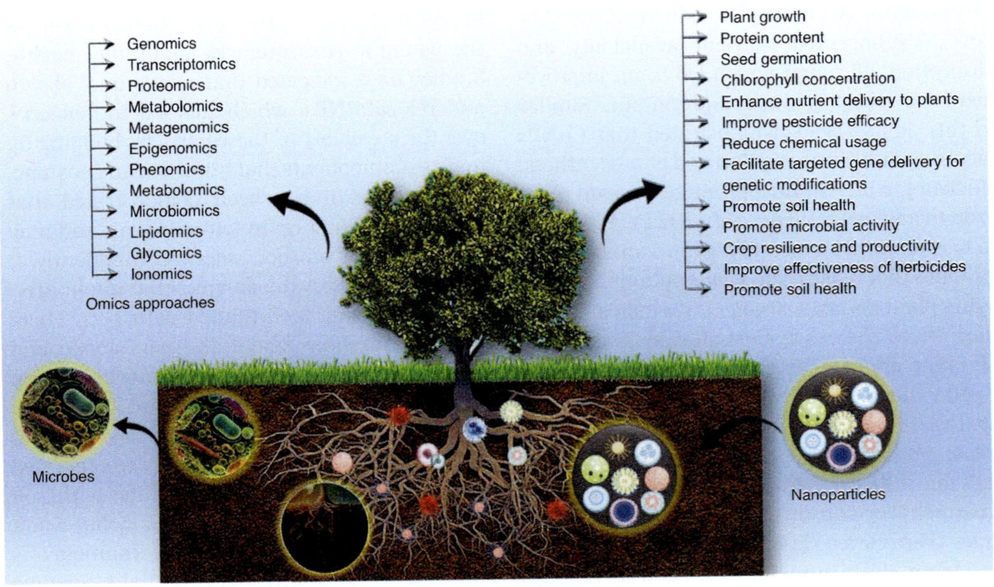

Fig. 10.6. The role of omics methods and NPs in agrobiotechnology.

issues. Their interactions with plants have the potential to greatly increase agricultural output overall, stress tolerance, and nutrient absorption. The implications for farming techniques and the application of nanotechnology in farming are highlighted in this summary of findings. Silver, zinc, and titanium dioxide NPs improve nutrient absorption and activate genes that respond to stress, which in turn promotes plant development. Research suggests that the use of NPs can result in an average yield gain of 20% and a reduction in illness incidence of up to 50%. They also improve soil structure and lessen soil toxicity, which supports sustainable land management. Minimizing the negative effects of conventional methods on the environment, such as fertilizer leaching. With the potential to significantly improve sustainable farming methods and food security, NMs have a bright future in agriculture. To guarantee safe application, more research and regulatory frameworks are necessary due to concerns over toxicity and environmental effects. NMs have a big impact on crop interactions; they can improve farming techniques but they can also cause problems. Although their distinct qualities raise questions regarding toxicity and environmental effects, they also promote nutrition intake, stress tolerance, and disease resistance. When utilized as antibacterial agents, NPs like zinc and silver can lower disease incidence by up to 50% and increase agricultural output by around 20%. Targeted nutrient delivery is made easier by NPs, which also increase soil availability and absorption efficiency. This is important for managing crop nutrient deficits. Research suggests that using nanofertilizers can improve the absorption of nutrients, resulting in higher crop yields and improved nutritional value. By triggering plant defense systems against abiotic stressors like salt and drought, NMs increase yield and resilience in plants. Additionally, they strengthen plants' antioxidant defenses, which lessens oxidative stress and boosts development in difficult environments. By increasing water retention and decreasing compaction, NPs enhance soil structure and fertility and promote crop growth and sustainability. Plant development can be positively impacted by the strengthening of symbiotic connections between soil microorganisms and NPs. Although there is great promise for using NMs in agriculture, issues with regulatory frameworks and their effects on the environment need to be resolved to ensure long-term use. NMs improve nutrient uptake and 30% reduce leaching by facilitating the effective delivery of fertilizers. They also improve plants' nutritional content without the need for conventional fertilizers. When taken at moderate concentrations, NPs accumulate within cells through the roots and leaves and do no harm. Additionally, they can strengthen plant defense mechanisms by turning on genes that respond to stress. NMs may increase crop resilience and stress tolerance by causing changes in gene expression and epigenetic variables. Although NMs have the potential to stimulate development, they may also cause oxidative stress, which can damage cellular biomolecules and be harmful. The influence of NPs on plant health is mostly determined by their physicochemical characteristics. Even though NPs have the potential to improve crop relationships and agricultural sustainability, their possible toxicity and environmental effects call for caution and more study.

References

Adil, M., Bashir, S., Bashir, S., Aslam, Z., Ahmad, N. *et al.* (2022) Zinc oxide nanoparticles improved chlorophyll contents, physical parameters, and wheat yield under salt stress. *Frontiers in Plant Science* 13, 932861.

Agri, U., Chaudhary, P. and Sharma, A. (2021) In vitro compatibility evaluation of agriusable nanochitosan on beneficial plant growth-promoting rhizobacteria and maize plant. *National Academy Science Letters* 44(6), 555–559.

Ahmad, Y., Haider, S., Iqbal, J., Abbasi, B.A., Yaseen, T. *et al.* (2022) The mechanisms of genome editing technologies in crop plants. In: Prakash, C.S., Fiaz, S. and Fahad, S. (eds) *Principles and Practices of OMICS and Genome Editing for Crop Improvement*. Springer, Cham, Switzerland, pp. 295–313.

Ali, S., Mehmood, A. and Khan, N. (2021) Uptake, translocation, and consequences of nanomaterials on plant growth and stress adaptation. *Journal of Nanomaterials* 2021(1), 1–17.

Ali, I., Salah, K.B.H., Sher, H., Ali, H., Ullah, Z. *et al.* (2022) Drought stress enhances the efficiency of floral dip method of agrobacterium-mediated transformation in *Arabidopsis thaliana*. *Brazilian Journal of Biology* 84, e259326.

Alshehddi, L.A.A. and Bokhari, N. (2020) Influence of gold and silver nanoparticles on the germination and growth of *Mimusops laurifolia* seeds in the South-Western regions in Saudi Arabia. *Saudi Journal of Biological Sciences* 27(1), 574–580.

Amin, H.H., Elsayed, A.B., Maswada, H.F. and Ibrahim Elsheery, N. (2023) Enhancing sugar beet plant health with zinc nanoparticles: A sustainable solution for disease management. *Journal of Soil, Plant and Environment* 2(1), 1–20.

Aminiyan, M.M., Hosseini, H. and Heydariyan, A. (2018) Microbial communities and their characteristics in a soil amended by nanozeolite and some plant residues: Short time *in-situ* incubation. *Eurasian Journal of Soil Science (EJSS)* 7(1), 9–19.

An, C., Sun, C., Li, N., Huang, B., Jiang, J. *et al.* (2022) Nanomaterials and nanotechnology for the delivery of agrochemicals: Strategies towards sustainable agriculture. *Journal of Nanobiotechnology* 20(1), 11.

An, X., Li, N., Zhang, S., Han, Y. and Zhang, Q. (2023) Integration of proteome and metabolome profiling to reveal heat stress response and tolerance mechanisms of *Serratia sp.* AXJ-M for the bioremediation of papermaking black liquor. *Journal of Hazardous Materials* 450, 131092.

Aqeel, U., Aftab, T., Khan, M.M.A., Naeem, M. and Khan, M.N. (2022) A comprehensive review of impacts of diverse nanoparticles on growth, development and physiological adjustments in plants under changing environment. *Chemosphere* 291(Pt 1), 132672.

Avellan, A., Yun, J., Zhang, Y., Spielman-Sun, E., Unrine, J.M. *et al.* (2019) Nanoparticle size and coating chemistry control foliar uptake pathways, translocation, and leaf-to-rhizosphere transport in wheat. *ACS Nano* 13(5), 5291–5305.

Avellan, A., Yun, J., Morais, B.P., Clement, E.T., Rodrigues, S.M. *et al.* (2021) Critical review: Role of inorganic nanoparticle properties on their foliar uptake and in planta translocation. *Environmental Science & Technology* 55(20), 13417–13431.

Birbaum, K., Brogioli, R., Schellenberg, M., Martinoia, E., Stark, W.J. *et al.* (2010) No evidence for cerium dioxide nanoparticle translocation in maize plants. *Environmental Science & Technology* 44(22), 8718–8723.

Choudhary, R.C., Kumaraswamy, R.V., Kumari, S., Sharma, S.S., Pal, A. *et al.* (2017) Cu-chitosan nanoparticle boost defense responses and plant growth in maize (*Zea mays* L). *Scientific Reports* 7(1), 9754.

Côa, F., Bortolozzo, L.S., Petry, R., Silva, G.H., Martins, C.H. *et al.* (2020) Environmental toxicity of nanopesticides against non-target organisms: The state of the art. In: Fraceto, L.F., Castro, V.L.S.S., Grillo, R., Avila, D. and Oliveira, H.C. (eds) *Nanopesticides: From Research and Development to Mechanisms of Action and Sustainable Use in Agriculture*. Springer, pp. 227–279.

Cvjetko, P., Milošić, A., Domijan, A.M., Vrček, I.V., Tolić, S. *et al.* (2017) Toxicity of silver ions and differently coated silver nanoparticles in *Allium cepa* roots. *Ecotoxicology and Environmental Safety* 137, 18–28.

De Angelis, G., Badiali, C., Chronopoulou, L., Palocci, C. and Pasqua, G. (2023) Confocal microscopy investigations of biopolymeric PLGA nanoparticle uptake in *Arabidopsis thaliana* L. *Cultured Cells and Plantlet Roots – Plants* 12(13), 2397.

Dev, A., Srivastava, A.K. and Karmakar, S. (2017) Uptake and toxicity of nanomaterials in plants. In: Ranjan, S., Dasgupta, N. and Lichtfouse, E. (eds) *Nanoscience in Food and Agriculture 5*. Springer, pp. 169–204.

Dev, A., Srivastava, A.K. and Karmakar, S. (2018) Nanomaterial toxicity for plants. *Environmental Chemistry Letters* 16(1), 85–100.

Egbuna, C., Parmar, V.K., Jeevanandam, J., Ezzat, S.M., Patrick-Iwuanyanwu, K.C. *et al.* (2021) Toxicity of nanoparticles in biomedical application: Nanotoxicology. *Journal of Toxicology* 2021(1), 9954443.

Elbasiouny, H., Elbehiry, F. and El-Ramady, H. (2022) Toxic effects of nanoparticles under combined stress on plants. In: Rajput, V.D., Minkina, T., Sushkova, S., Mandzhieva, S.S. and Rensing, C. (eds) *Toxicity of Nanoparticles in Plants. Elsevier*. pp. 109–129.

Elsheery, N.I., Sunoj, V.S.J., Wen, Y., Zhu, J.J., Muralidharan, G. *et al.* (2020) Foliar application of nanoparticles mitigates the chilling effect on photosynthesis and photoprotection in sugarcane. *Plant Physiology and Biochemistry* 149, 50–60.

Francis, D.V., Sood, N. and Gokhale, T. (2020) Applications of metal nanoparticles in agriculture. In: Pawar, S. (ed.) *Progress and Prospects in Nanoscience Today*. Nova Science.

Francis, D.V., Sood, N. and Gokhale, T. (2022) Biogenic CO and ZnO nanoparticles as nanofertilizers for sustainable growth of Amaranthus hybridus. *Plants* 11, 2776.

García-Sánchez, S., Bernales, I. and Cristobal, S. (2015) Early response to nanoparticles in the arabidopsis transcriptome compromises plant defence and root-hair development through salicylic acid signalling. *BMC Genomics* 16(1), 341.

Geisler-Lee, J., Wang, Q., Yao, Y., Zhang, W., Geisler, M. *et al.* (2013) Phytotoxicity, accumulation and transport of silver nanoparticles by *Arabidopsis thaliana*. *Nanotoxicology* 7(3), 323–337.

Ghosh, M., Jana, A., Sinha, S., Jothiramajayam, M., Nag, A. *et al.* (2016) Effects of zno nanoparticles in plants: Cytotoxicity, genotoxicity, deregulation of antioxidant defenses, and cell-cycle arrest. *Mutation Research/Genetic Toxicology and Environmental Mutagenesis* 807, 25–32.

Giraldo, J.P., Wu, H., Newkirk, G.M. and Kruss, S. (2019) Nanobiotechnology approaches for engineering smart plant sensors. *Nature Nanotechnology* 14(6), 541–553.

Gohari, G., Mohammadi, A., Akbari, A., Panahirad, S., Dadpour, M.R. *et al.* (2020) Titanium dioxide nanoparticles (TiO_2 NPs) promote growth and ameliorate salinity stress effects on essential oil profile and biochemical attributes of *Dracocephalum moldavica*. *Scientific Reports* 10(1), 912.

Gui, X., Dong, C., Fan, S., Jiao, C., Song, Z. *et al.* (2023) Effects of ceo_2 nanoparticles on nutritional quality of two crop plants, corn (*Zea mays* L.) and soybean (*Glycine max* L). *Molecules* 28(4), 1798.

Haider, S., Iqbal, J., Naseer, S., Yaseen, T., Shaukat, M. *et al.* (2021) Molecular mechanisms of plant tolerance to heat stress: Current landscape and future perspectives. *Plant Cell Reports* 40(12), 2247–2271.

Haider, S., Raza, A., Iqbal, J., Shaukat, M. and Mahmood, T. (2022) Analyzing the regulatory role of heat shock transcription factors in plant heat stress tolerance: A brief appraisal. *Molecular Biology Reports* 49(6), 5771–5785.

Hong, J., Wang, C., Wagner, D.C., Gardea-Torresdey, J.L., He, F. *et al.* (2021) Foliar application of nanoparticles: Mechanisms of absorption, transfer, and multiple impacts. *Environmental Science* 8(5), 1196–1210.

Hussain, A., Rizwan, M., Ali, Q. and Ali, S. (2019) Seed priming with silicon nanoparticles improved the biomass and yield while reduced the oxidative stress and cadmium concentration in wheat grains. *Environmental Science and Pollution Research* 26(8), 7579–7588.

Ijaz, S., Iqbal, J., Abbasi, B.A., Tufail, A., Ullah, Z. *et al.* (2023a) Biofortification: Lessons from the past and strategies for future food security. In: Nadeem, M.A., Baloch, F.S., Fiaz, S., Aasim, M. and Habyarimana, E. (eds) *Legumes Biofortification*. Springer, Cham, Switzerland, pp. 521–545.

Ijaz, S., Iqbal, J., Abbasi, B.A., Ullah, Z., Yaseen, T. *et al.* (2023b) Rosmarinic acid and its derivatives: Current insights on anticancer potential and other biomedical applications. *Biomedicine & Pharmacotherapy* 162, 114687.

Ijaz, S., Iqbal, J., Abbasi, B.A., Kanwal, S., Mahmoodi, M. *et al.* (2024a) Investigation of bioactive constituents and evaluation of *in vitro* bioactivities of different *Setaria glauca* extracts. *Journal of King Saud University – Science* 36(8), 103321.

Ijaz, S., Iqbal, J., Abbasi, B.A., Kanwal, S., Tavafoghi, M. *et al.* (2024b) Investigation of bioactive constituents and evaluation of different *in vitro* antimicrobial, antioxidant, and cytotoxicity potentials of different *Portulacaria afra* extracts. *Journal of King Saud University – Science* 36(2), 103033.

Ijaz, S., Iqbal, J., Abbasi, B.A., Tufail, A., Yaseen, T. *et al.* (2024c) Current stage of preclinical and clinical development of guggulsterone in cancers: Challenges and promises. *Cell Biology International* 48(2), 128–142.

Ijaz, S., Iqbal, J., Abbasi, B.A., Yaseen, T., Rehman, S. *et al.* (2024d) Role of OMICS-based technologies in plant sciences. In: Fiaz, S. and Prakash, C.S. (eds) *OMICs-based Techniques for Global Food Security. Wiley*. pp. 45–66.

Ji, M., Al-Huqail, A.A., Majdi, A., Liu, H., Ali, I. *et al.* (2022) Remediation of mine polluted soil with nano-enhanced materials: Development of extreme learning machine approaches. *Chemosphere* 307, 135772.

Kang, M.E., Weng, Y., Liu, Y., Wang, H., Ye, L. *et al.* (2023) A review on the toxicity mechanisms and potential risks of engineered nanoparticles to plants. *Reviews of Environmental Contamination and Toxicology* 261(1), 5.

Karunakaran, G., Manivasakan, P., Yuvakkumar, R., Prabu, P., Suriyaprabha, R. *et al.* (2013) Effect of nanosilica and silicon sources on plant growth promoting rhizobacteria, soil nutrients and maize seed germination. *IET Nanobiotechnology* 7(3), 70–77.

Kashyap, P., Shirkot, P., Das, R., Pandey, H. and Singh, D. (2023) Biosynthesis and characterization of copper nanoparticles from *Stenotrophomonas maltophilia* and its effect on plant pathogens and pesticide degradation. *Journal of Agriculture and Food Research* 13, 100654.

Kaveh, R., Li, Y.-S., Ranjbar, S., Tehrani, R., Brueck, C.L. *et al.* (2013) Changes in *Arabidopsis thaliana* gene expression in response to silver nanoparticles and silver ions. *Environmental Science & Technology* 47(18), 10637–10644.

Khan, M.R., Fromm, K.M., Rizvi, T.F., Giese, B., Ahamad, F. *et al.* (2020) Metal nanoparticle–microbe interactions: Synthesis and antimicrobial effects. *Particle & Particle Systems Characterization* 37(5), 1900419.

Khanna, K., Kohli, S.K., Handa, N., Kaur, H., Ohri, P. *et al.* (2021) Enthralling the impact of engineered nanoparticles on soil microbiome: A concentric approach towards environmental risks and cogitation. *Ecotoxicology and Environmental Safety* 222, 112459.

Khati, P., Sharma, A., Chaudhary, P., Singh, A.K., Gangola, S. *et al.* (2019) High-throughput sequencing approach to access the impact of nanozeolite treatment on species richness and evenness of soil metagenome. *Biocatalysis and Agricultural Biotechnology* 20, 101249.

Kumar, V., Guleria, P., Kumar, V. and Yadav, S.K. (2013) Gold nanoparticle exposure induces growth and yield enhancement in *Arabidopsis thaliana. The Science of the Total Environment* 461–468.

Kumar, U., Raj, S., Sreenikethanam, A., Maddheshiya, R., Kumari, S. *et al.* (2023) Multi-omics approaches in plant–microbe interactions hold enormous promise for sustainable agriculture. *Agronomy* 13(7), 1804.

Kurczyńska, E., Godel-Jędrychowska, K., Sala, K. and Milewska-Hendel, A. (2021) Nanoparticles–plant interaction: What we know, where we are. *Applied Sciences* 11(12), 5473.

Kwak, J., Yoon, S. and An, Y. (2017) Long-term effects of zno nanoparticles on exoenzyme activities in planted soils. *Environmental Engineering Research* 22(2).

Larue, C., Castillo-Michel, H., Sobanska, S., Cécillon, L., Bureau, S. *et al.* (2014) Foliar exposure of the crop *Lactuca sativa* to silver nanoparticles: Evidence for internalization and changes in Ag speciation. *Journal of Hazardous Materials* 264, 98–106.

Lee, C.W., Mahendra, S., Zodrow, K., Li, D., Tsai, Y.-C. *et al.* (2009) Developmental phytotoxicity of metal oxide nanoparticles to *Arabidopsis thaliana*. *Environmental Toxicology and Chemistry* 29(3), 669–675.

Lew, T.T.S., Wong, M.H., Kwak, S.Y., Sinclair, R., Koman, V.B. *et al.* (2018) Rational design principles for the transport and subcellular distribution of nanomaterials into plant protoplasts. *Small* 14(44), 1802086.

Li, Y., Jin, Q., Yang, D. and Cui, J. (2018) Molybdenum sulfide induces growth enhancement effect of rice (*Oryza sativa* L.) through regulating the synthesis of chlorophyll and the expression of aquaporin gene. *Journal of Agricultural and Food Chemistry* 66(16), 4013–4021.

Lv, J., Christie, P. and Zhang, S. (2019) Uptake, translocation, and transformation of metal-based nanoparticles in plants: Recent advances and methodological challenges. *Environmental Science: Nano* 6(1), 41–59.

Li, Y., Zhang, P., Li, M., Shakoor, N., Adeel, M. *et al.* (2023) Application and mechanisms of metal-based nanoparticles in the control of bacterial and fungal crop diseases. *Pest Management Science* 79(1), 21–36.

Ma, C., White, J.C., Zhao, J., Zhao, Q. and Xing, B. (2018) Uptake of engineered nanoparticles by food crops: Characterization, mechanisms, and implications. *Annual Review of Food Science and Technology* 9(1), 129–153.

Majumdar, S., Pagano, L., Wohlschlegel, J.A., Villani, M., Zappettini, A. *et al.* (2019) Proteomic, gene and metabolite characterization reveal the uptake and toxicity mechanisms of cadmium sulfide quantum dots in soybean plants. *Environmental Science: Nano* 6(10), 3010–3026.

Manna, I., Sahoo, S. and Bandyopadhyay, M. (2021) Effect of engineered nickel oxide nanoparticle on reactive oxygen species-nitric oxide interplay in the roots of *Allium cepa* L. *Frontiers in Plant Science* 12, 586509.

Manzo, S., Miglietta, M.L., Rametta, G., Buono, S. and Francia, G. (2013) Toxic effects of ZnO nanoparticles towards marine algae Dunaliella tertiolecta. *The Science of the Total Environment* 445, 371–376.

Mathur, P., Chakraborty, R., Aftab, T. and Roy, S. (2023) Engineered nanoparticles in plant growth: Phytotoxicity concerns and the strategies for their attenuation. *Plant Physiology and Biochemistry* 199, 107721.

McGee, C.F., Storey, S., Clipson, N. and Doyle, E. (2017) Soil microbial community responses to contamination with silver, aluminium oxide and silicon dioxide nanoparticles. *Ecotoxicology* 26(3), 449–458.

McManus, P., Hortin, J., Anderson, A.J., Jacobson, A.R., Britt, D.W. *et al.* (2018) Rhizosphere interactions between copper oxide nanoparticles and wheat root exudates in a sand matrix: Influences on copper bioavailability and uptake. *Environmental Toxicology and Chemistry* 37(10), 2619–2632.

Metch, J.W., Burrows, N.D., Murphy, C.J., Pruden, A. and Vikesland, P.J. (2018) Metagenomic analysis of microbial communities yields insight into impacts of nanoparticle design. *Nature Nanotechnology* 13(3), 253.

Milewska-Hendel, A., Sala, K., Gepfert, W. and Kurczyńska, E. (2021) Gold nanoparticles-induced modifications in cell wall composition in barley roots. *Cells* 10(8).

Mirzajani, F., Askari, H., Hamzelou, S., Schober, Y., Römpp, A. *et al.* (2014) Proteomics study of silver nanoparticles toxicity on *Oryza sativa* L. *Ecotoxicology and Environmental Safety* 108, 335–339.

Mitra, D., Adhikari, P., Djebaili, R., Thathola, P., Joshi, K. *et al.* (2023) Biosynthesis and characterization of nanoparticles, its advantages, various aspects and risk assessment to maintain the sustainable agriculture: Emerging technology in modern era science. *Plant Physiology and Biochemistry* 196, 103–120.

Murali, M., Gowtham, H.G., Singh, S.B., Shilpa, N., Aiyaz, M. *et al.* (2022) Fate, bioaccumulation and toxicity of engineered nanomaterials in plants: Current challenges and future prospects. *Science of the Total Environment* 811, 152249.

Mustafa, G. and Komatsu, S. (2016) Insights into the response of soybean mitochondrial proteins to various sizes of aluminum oxide nanoparticles under flooding stress. *Journal of Proteome Research* 15, 4464–4475.

Mustafa, G., Sakata, K., Hossain, Z. and Komatsu, S. (2015) Proteomic study on the effects of silver nanoparticles on soybean under flooding stress. *Journal of Proteomics* 122, 100–118.

Newkirk, G.M., De Allende, P., Jinkerson, R.E. and Giraldo, J.P. (2021) Nanotechnology approaches for chloroplast biotechnology advancements. *Frontiers in Plant Science* 12, 691295.

Parada, J., Rubilar, O., Fernández-Baldo, M.A., Bertolino, F.A., Durán, N, *et al.* (2019) The nanotechnology among us: Are metal and metal oxides nanoparticles a nano or mega risk for soil microbial communities? *Critical Reviews in Biotechnology* 39(2), 157–172.

Patel, M., Patel, R. and Rai, M. (2021) Nanomedicines: Applications and toxicological concerns. In: Rai, M., Patel, M. and Patel, R. (eds) *Nanotechnology in Medicine: Toxicity and Safety*, Vol. 1. John Wiley & Sons Ltd, Chichester, UK, pp. 1–28.

Plaksenkova, I., Kokina, I., Petrova, A., Jermaļonoka, M., Gerbreders, V. *et al.* (2020) The impact of zinc oxide nanoparticles on cytotoxicity, genotoxicity, and miRNA expression in barley (Hordeum vulgare L.) seedlings. *The Scientific World Journal* 1–13.

Prasad, A., Astete, C.E., Bodoki, A.E., Windham, M., Bodoki, E. *et al.* (2018) Zein nanoparticles uptake and translocation in hydroponically grown sugar cane plants. *Journal of Agricultural and Food Chemistry* 66(26), 6544–6551.

Quanbeck, S.M., Brachova, L., Campbell, A.A., Guan, X., Perera, A. *et al.* (2012) Metabolomics as a hypothesis-generating functional genomics tool for the annotation of Arabidopsis thaliana genes of 'unknown function'. *Frontiers in Plant Science* 3, 15.

Rai, M. and Golińska, P. (2023) *Myconanotechnology: Emerging Trends and Applications*. CRC Press, Boca Raton, Florida.

Rajala, J.E., Mäenpää, K., Vehniäinen, E.R., Väisänen, A., Scott-Fordsmand, J.J. *et al.* (2016) Toxicity testing of silver nanoparticles in artificial and natural sediments using the benthic organism *Lumbriculus variegatus*. *Archives of Environmental Contamination and Toxicology* 71(3), 405–414.

Rajput, V., Minkina, T., Ahmed, B., Sushkova, S., Singh, R. *et al.* (2020) Interaction of copper-based nanoparticles to soil, terrestrial, and aquatic systems: critical review of the state of the science and future perspectives. *Reviews of Environmental Contamination and Toxicology* 252, 51–96.

Rani, S., Kumari, N. and Sharma, V. (2023) Uptake, translocation, transformation and physiological effects of nanoparticles in plants. *Archives of Agronomy and Soil Science* 69(9), 1579–1599.

Rastogi, A., Tripathi, D.K., Yadav, S., Chauhan, D.K., Živčák, M. *et al.* (2019) Application of silicon nanoparticles in agriculture. *3 Biotech* 9(3), 90.

Ristroph, K.D., Astete, C.E., Bodoki, E. and Sabliov, C.M. (2017) Zein nanoparticles uptake by hydroponically grown soybean plants. *Environmental Science & Technology* 51(24), 14065–14071.

Roy, T., Dey, T.K. and Jamal, M. (2023) Microplastic/nanoplastic toxicity in plants: An imminent concern. *Environmental Monitoring and Assessment* 195(1), 27.

Sanzari, I., Leone, A. and Ambrosone, A. (2019) Nanotechnology in plant science: To make a long story short. *Frontiers in Bioengineering and Biotechnology* 7, 120.

Sembada, A.A. and Lenggoro, I.W. (2024) Transport of nanoparticles into plants and their detection methods. *Nanomaterials* 14(2), 131.

Sergaki, C., Lagunas, B., Lidbury, I., Gifford, M.L. and Schäfer, P. (2018) Challenges and approaches in microbiome research: From fundamental to applied. *Frontiers in Plant Science* 9, 1205.

Siddiqui, M.H., Al-Whaibi, M.H., Faisal, M. and Al Sahli, A.A. (2014) Nano-silicon dioxide mitigates the adverse effects of salt stress on *Cucurbita pepo* L. *Environmental Toxicology and Chemistry* 33(11), 2429–2437.

Sillen, W.M.A., Thijs, S., Abbamondi, G.R., Janssen, J., Weyens, N. *et al.* (2015) Effects of silver nanoparticles on soil microorganisms and maize biomass are linked in the rhizosphere. *Soil Biology and Biochemistry* 91, 14–22.

Simon, D.F., Domingos, R.F., Hauser, C., Hutchins, C.M., Zerges, W. *et al.* (2013) Transcriptome sequencing (RNA-seq) analysis of the effects of metal nanoparticle exposure on the transcriptome of *Chlamydomonas reinhardtii*. *Applied and Environmental Microbiology* 79(16), 4774–4785.

Singh, D. and Kumar, A. (2020) Understanding the effect of the interaction of nanoparticles with roots on the uptake in plants. *Environmental Nanotechnology* 277–304.

Spielman-Sun, E., Avellan, A., Bland, G.D., Tappero, R.V., Acerbo, A.S. *et al.* (2019) Nanoparticle surface charge influences translocation and leaf distribution in vascular plants with contrasting anatomy. *Environmental Science: Nano* 6(8), 2508–2519.

Taran, N., Batsmanova, L., Konotop, Y. and Okanenko, A. (2014) Redistribution of elements of metals in plant tissues under treatment by non-ionic colloidal solution of biogenic metal nanoparticles. *Nanoscale Research Letters* 9(1), 354.

Tauseef, A., Hisamuddin, K.A. and Uddin, I. (2021) Role of mgo nanoparticles in the suppression of *Meloidogyne incognita*, infecting cowpea and improvement in plant growth and physiology. *Experimental Parasitology* 220, 108045.

Taylor, A.F., Rylott, E.L., Anderson, C.W.N. and Bruce, N.C. (2014) Investigating the toxicity, uptake, nanoparticle formation and genetic response of plants to gold. *PLoS One* 9(4), e93793.

Thomas, C.R., George, S., Horst, A.M., Ji, Z., Miller, R.J. *et al.* (2011) Nanomaterials in the environment: From materials to high-throughput screening to organisms. *ACS Nano* 5, 13–20.

Tiwari, M., Krishnamurthy, S., Shukla, D., Kiiskila, J., Jain, A. *et al.* (2016) Comparative transcriptome and proteome analysis to reveal the biosynthesis of gold nanoparticles in *Arabidopsis*. *Scientific Reports* 6, 21733.

Torrent, L., Iglesias, M., Marguí, E., Hidalgo, M., Verdaguer, D. *et al.* (2020) Uptake, translocation and ligand of silver in *Lactuca sativa* exposed to silver nanoparticles of different size, coatings and concentration. *Journal of Hazardous Materials* 384, 121201.

Tran, T.K., Nguyen, M.K., Lin, C., Hoang, T.D., Nguyen, T.C. *et al.* (2024) Review on fate, transport, toxicity and health risk of nanoparticles in natural ecosystems: Emerging challenges in the modern age and solutions toward a sustainable environment. *The Science of the Total Environment* 912, 169331.

Tripathi, D.K., Singh, S., Singh, V.P., Prasad, S.M. and Chauhan, D.K. (2016) Silicon nanoparticles more efficiently alleviate arsenate toxicity than silicon in maize cultiver and hybrid differing in arsenate tolerance. *Frontiers in Environmental Science* 4, 46.

Tripathi, D.K., Mishra, R.K., Singh, S., Singh, S., Vishwakarma, K. *et al.* (2017) Nitric oxide ameliorates zinc oxide nanoparticles phytotoxicity in wheat seedlings: Implication of the ascorbate-glutathione cycle. *Frontiers in Plant Science* 8, 1.

Ullah, R., Ullah, Z., Iqbal, J., Chalgham, W. and Ahmad, A. (2023) Aspartic acid-based nano-copper induces resilience in *Zea mays* to applied lead stress via conserving photosynthetic pigments and triggering the antioxidant biosystem. *Sustainability* 15(16), 12186.

Vannini, C., Domingo, G., Onelli, E., Prinsi, B., Marsoni, M. *et al.* (2013) Morphological and proteomic responses of *Eruca sativa* exposed to silver nanoparticles or silver nitrate. *PLoS One* 8(7), e68752.

Vera-Reyes, I., Altamirano-Hernández, J., Reyes-De la, H., Granados-Echegoyen, C.A. and Loera-Alvarado, G. (2022) Inhibition of phytopathogenic and beneficial fungi applying silver nanoparticles *in vitro*. *Molecules* 27(23), 8147.

Verma, S.K., Das, A.K., Gantait, S., Kumar, V. and Gurel, E. (2019) Applications of carbon nanomaterials in the plant system: A perspective view on the pros and cons. *The Science of the Total Environment* 667, 485–499.

Wahab, A., Muhammad, M., Ullah, S., Abdi, G., Shah, G.M. *et al.* (2024) Agriculture and environmental management through nanotechnology: Eco-friendly nanomaterial synthesis for soil-plant systems, food safety, and sustainability. *The Science of the Total Environment* 926, 171862.

Wang, C. and Si, L. (2023) A bibliometric analysis of digital literacy research from 1990 to 2022 and research on emerging themes during the covid-19 pandemic. *Sustainability* 15(7), 5769.

Wang, W.-N., Tarafdar, J.C. and Biswas, P. (2013) Nanoparticle synthesis and delivery by an aerosol route for watermelon plant foliar uptake. *Journal of Nanoparticle Research* 15(1), 1417.

Wang, J., Li, M., Feng, J., Yan, X., Chen, H. *et al.* (2021) Effects of TiO2-NPs pretreatment on UV-b stress tolerance in Arabidopsis thaliana. *Chemosphere* 281, 130809.

Wang, W., Liu, H., Huang, Z., Fu, F., Wang, W. *et al.* (2022) The effect of organic ligand modification on protein corona formation of nanoscale metal organic frameworks. *Chinese Chemical Letters* 33(9), 4185–4190.

Wu, H. and Li, Z. (2022) Nano-enabled agriculture: How do nanoparticles cross barriers in plants. *Plant Communications* 3(6), 100346.

Xiao, Y., Ma, J., Xian, J., Peijnenburg, W.J.G.M., Du, Y. *et al.* (2022) Copper accumulation and physiological markers of soybean (*Glycine max*) grown in agricultural soil amended with copper nanoparticles. *Ecotoxicology and Environmental Safety* 229, 113088.

Yan, A. and Chen, Z. (2019) Impacts of silver nanoparticles on plants: A focus on the phytotoxicity and underlying mechanism. *International Journal of Molecular Sciences* 20(5), 1003.

Yan, W., Lien, H.-L., Koel, B.E. and Zhang, W. (2013) Iron nanoparticles for environmental clean-up: Recent developments and future outlook. *Environmental Science Processes & Impacts* 15, 63–77.

Yin, J., Wang, Y. and Gilbertson, L.M. (2018) Opportunities to advance sustainable design of nano-enabled agriculture identified through a literature review. *Environmental Science: Nano* 5(1), 11–26.

Younis, S.A., Kim, K.-H., Shaheen, S.M., Antoniadis, V., Tsang, Y.F. *et al.* (2021) Advancements of nanotechnologies in crop promotion and soil fertility: Benefits, life cycle assessment, and legislation policies. *Renewable and Sustainable Energy Reviews* 152, 111686.

Zhang, H., Lu, L., Zhao, X., Zhao, S., Gu, X. *et al.* (2019a) Metabolomics reveals the 'invisible' responses of spinach plants exposed to CeO_2 nanoparticles. *Environmental Science & Technology* 53(10), 6007–6017.

Zhao, L., Peng, B., Hernandez-Viezcas, J.A., Rico, C., Sun, Y. *et al*. (2012) Stress response and tolerance of Zea mays to CeO_2 nanoparticles: Cross talk among H_2O_2, heat shock protein, and lipid peroxidation. *ACS Nano* 6(11), 9615–9622.

Zheng, L., Cao, W., Dou, B., Zeng, X., Cao, M. *et al*. (2023) Ac36deoGlcNAz could selectively label O-GlcNAc modified proteins with minimal S-glyco-modification. *Chinese Chemical Letters* 34(2), 107598.

Zhou, L., Yan, X., Pei, X., Du, J., Ma, R. *et al*. (2022) The role of $NiFe_2O_4$ nanoparticle in the anaerobic digestion (AD) of waste activated sludge (WAS). *Chinese Chemical Letters* 33(1), 428–433.

11 Stress-resilient Crops: Advanced Breeding Using Coordinated Strategies against Changing Climate

Shoumik Saha[1], Nimitha K.[2], Shampa Purkayastha[3], Maksud Hasan Shaha[4*], Sk Md Ajahar Uddin[5], Sayed Abdul Akher[6], Sarfaraz Ahmad[7], Santanu Kundu[8], Shaukat Ali[9] and Akbar Hossain[10*]

[1]Assistant Professor (Genetics and Plant Breeding), Faculty of Agriculture, Rural and Tribal Development (ARTD), Ramakrishna Mission Vivekananda Educational and Research Institute (RKMVERI), Morabadi, Jharkhand, India; [2]Department of Genetics and Plant Breeding, Bidhan Chandra Krishi Viswavidyalaya, Mohanpur, Nadia, India; [3]Assistant Professor (Genetics and Plant Breeding), Centurion University of Technology and Management, Paralekhamundi, India; [4]Department of Agronomy, Bidhan Chandra Krishi Viswavidyalaya, Mohanpur, Nadia, West Bengal, India; [5]Department of Entomology, Bidhan Chandra Krishi Viswavidyalaya, Mohanpur, Nadia, West Bengal, India; [6]Department of Genetics and Plant Breeding, Tobacco Research Institute, Graduate School of Chinese Academy of Agricultural Science, China; [7]Guest Faculty (Genetics and Plant Breeding), College of Agriculture, Mandawa, Swami Keshewanand Rajasthan Agriculture University, Bikaner, Rajasthan, India; [8]Division of Agronomy, ICAR-Indian Agricultural Research Institute, New Delhi, India; [9]Department of Agronomy, SMS, KVK, Chandgothi, SKRAU, Bikaner, Rajasthan, India; [10]Soil Science Division, Bangladesh Wheat and Maize Research Institute, Dinajpur, Bangladesh

Abstract

Climate change has emerged as one of the biggest threats to global food security in recent times. To meet the demand of the ever-increasing world population, it is high time to develop climate-resilient crop varieties. Utilization of advanced plant breeding tools and techniques in a coordinated manner can help scientists breed crops adaptable to unpredictable weather patterns. With rising temperatures, erratic rainfall, prolonged drought, and extreme weather events becoming more frequent, agricultural productivity faces unprecedented challenges. This chapter scrutinizes the integration of advanced breeding techniques, such as genomic selection, marker-assisted selection, and CRISPR-Cas gene editing, in generating stress-resilient crop varieties. It highlights the importance of a coordinated approach that associates traditional breeding methods with modern breeding and biotechnological tools, leveraging genetic diversity and novel traits to enhance the resistance to biotic and abiotic stresses. Additionally, the chapter discusses the role of phenotyping, big data analytics, and simulation models in predicting crop performance under varying environmental conditions, ultimately aiming to ensure food security in a rapidly changing world.

*Corresponding authors: maksudhasanshah@gmail.com and akarhossainwrc@gmail.com

© CAB International 2025. *Stress-Resilient Crops: Coordinated omics-CRISPR-Nanotechnology Strategies* (ed. Jen-Tsung Chen)
DOI: 10.1079/9781800627307.0011

Keywords: climate, food security, drought, biotic stress

11.1 Introduction

The 'next generation green revolution' is required to meet the predicted global food demand of 2050 (Dhankher and Foyer, 2018). According to the United Nations, the global population will be 9.8 billion by 2050. To maintain a consistent supply of food for human survival, improvement in crop production as well as sustainable development of new agricultural concepts and approaches are necessary. In the present era, climate change poses a substantial threat to agriculture, with its effects already manifested in increased frequency of drought and extreme precipitation events, high temperatures, as well as increased salt and heavy-metal contamination of soils, which will often be accompanied by increased infestation by pests, and pathogens (Long *et al.*, 2015). Historically, farmers have adapted to environmental changes by adopting new crop varieties and modifying their agricultural practices. However, with global temperatures on the rise, the pace of environmental change is expected to accelerate beyond what has been previously experienced. Additionally, the expansion of crop cultivation into suboptimal environments and nonarable lands underscores the growing importance of developing climate-resilient crops to ensure food security (Kathuria *et al.*, 2007). Climate-resilient agriculture (CRA) is a strategy that aims to increase farm incomes and productivity over the long run while utilizing the natural resources that are already available through systems of crop production. Despite the obvious benefits, smallholder farmers' adoption rates of climate-resilient crops and crop types vary greatly. Climate-resilient crops and crop varieties have been proposed as a strategy for farmers to cope with or adapt to climate change. Abiotic stressors including heat, drought, floods, salt, and shorter growing seasons (early-maturing crops) as well as pests linked to alterations in weather or climate patterns (disease and pest resistance) were addressed by the adoption of climate-resilient crops and crop varieties. The adoption of climate-resilient crops and crop types aimed to tackle broader issues related to crop system sustainability and climate change, including enhancing soil moisture retention, improving soil quality, and mitigating erosion through the use of cover crops.

More and more, plants need to adjust to the effects of salt, drought, flooding, and other climatic conditions (Zhang *et al.*, 2022). Technology can assist in boosting plant resistance to such stresses, from traditional breeding methods to CRISPR technology for genetically engineered crops. In the meantime, agroforestry and other integrated farming systems are becoming more well-known as a way to boost crop resilience and address risks to food security (Ahmad *et al.*, 2021).

The International Rice Research Institute (IRRI) has worked with partners in Asia and Africa to produce a series of high-yielding rice cultivars known as GSR. GSR combines features including disease resistance, drought tolerance, and high yield potential to solve the problems of climate change and food security. The team found so-called 'green genes' in 2018 after analyzing and cloning hundreds of rice genetic sequences. These genes are segments of the genetic code that are associated with high stress tolerance and effective nutrient usage. The initiative has been successful in producing a number of green super rice types since its start, and they have been made available in nations including Vietnam, the Philippines, and India (Beale, 2005). In this chapter, all the advanced breeding methods with synchronized plans against climate change have been explained.

11.2 Traditional Breeding Approaches

Plant breeding has produced improved cultivars to sustain an expanding population, which has gradually enhanced crop yield. The function of plant breeding is receiving a lot of attention because of recent changes in the global environment, including global warming, resource depletion, pest and disease outbreaks, and diversity of consumer needs. The field of traditional breeding has experienced significant growth, leading to improvements in agricultural productivity, output, and nutrition. Since the 20th century, traditional breeding has been quite successful in producing contemporary cultivars with high levels of yield. It is insufficient, therefore, to fulfill the need for agricultural production brought on by the rapid increase in population. Plant breeding evolved as a result of attempts to cross cultivars with desirable features in order to create better-performing varieties (Acquaah, 2009). Phenotypes with desired characteristics, such as semi-dwarf and nutrient-efficient plants, are chosen by breeders. Plant breeding targets variety via novel variations in order to select, discover, and expand varieties with important qualities that breeders and consumers want to see in the following generations. Therefore, if appropriate individuals cannot be chosen for the following generation, breeding will fail. In order to ensure successful breeding, breeders researched several breeding techniques while taking into account a number of factors, including age at breeding, scale at breeding, combination ability, and manner of fertilization. Furthermore, the degree of gene-to-gene interaction and the genetic makeup of the cultivars' attributes dictated the breeding strategy (Kearsey, 1997). Systems for breeding should be adaptable and simple, to adjust to shifting objectives in breeding. An initiative to create superior oat cultivars combines a variety of breeding techniques. For many years, pedigree, bulk, recurrent selection, backcross, and single-seed descent have been the standard breeding techniques. Three categories may be used to categorize the enormous range of breeding techniques: (i) plant breeding based on controlled mating by selecting plants presenting recombination of desired genes from different parents; (ii) plant breeding based on observed variation by selecting plants based on natural variants occurring in nature or within traditional varieties; and (iii) plant breeding based on monitored recombination by selecting particular genes or marker profiles, using molecular tools for tracking within-genome variation. Within this framework, three primary plant breeding eras can be distinguished: (i) plant breeding through observational variant selection, regardless of origin; (ii) controlled mating generation and selection of expanded variation; and (iii) tracking the inheritance of within-genome variation and selection of particular recombinants. The introduction of new variations into genomes by genetic engineering can be regarded as the fourth step of plant breeding, which is not covered in this study (Breseghello and Coelho, 2013).

The breeder's visual judgment can be the sole basis for carrying out these procedures; the desired trait can be chosen directly, or indirectly, by choosing qualities that are connected to the wanted trait (Gjedrem and Thodesen, 2005). Furthermore, it is better if the needed characteristic is with high heritability or regulated by genes that are additive in nature since selection is done on the basis of the phenotype of the plant (Brown and Caligari, 2011).

For less heritability and quantitative characteristics (quality, yield, biotic and abiotic stress), breeding strategies based on classical phenotypic selection are inadequate since these traits are heavily impacted by genotype and environment interaction. Traditional breeding techniques are also expensive and land use inefficient, time-consuming, and labor-intensive. Additionally, these breeding techniques are less effective due to their poor precision and dependability. Therefore, to swiftly and precisely produce crops with high yield, superior quality, and tolerance to climate change, a new breeding technique is needed. To get beyond the drawbacks of conventional breeding technology, researchers successfully developed molecular breeding technology based on biotechnology. Predictive breeding, another name for genomic breeding technology, is currently being used. In conclusion, conventional to molecular breeding has been the evolution of plant breeding. Soon enough, predictive breeding will be accessible. The persistent use of conventional breeding techniques may result in a reduction of the gene pool, rendering crops more vulnerable to stressors and impeding development (Araus *et al.*, 2008). Continued use of conventional breeding techniques within a species may result in a reduction of the gene pool from which cultivars are selected, making crops more susceptible to biotic and abiotic challenges and impeding further advancement. There are a number of ways to introduce exotic diversity into top germplasm without causing unwanted side effects. Since there would be no gamete segregation in traditional breeding, identical parents would not create diversity (Acquaah, 2009). Conventional breeding is an applied science that relies heavily on the experiences, observations, and decision-making abilities of breeders (Allard, 1961). The strategy, though, is that if a characteristic is not advantageous to natural selection, its genotype may be lost in the first few generations. Conversely, through natural selection, the genotype associated with a competitive yet unfavorable trait may persist across future generations (Acquaah, 2009).

Finding, creating, choosing, and utilizing genetic diversity in agronomic qualities is the foundation of plant breeding. This may be accomplished in a variety of methods and at widely disparate rates. Our writers discuss a variety of methods and tools, each with advantages and disadvantages, from traditional plant breeding based on pollination to the CRISPR/Cas genetic scissors. Mass selection, pure line selection, pedigree breeding, and backcrossing are examples of traditional breeding strategies that are frequently employed in traditional plant breeding (Acquaah, 2015; Begna, 2021). By modifying the plant genome within naturally occurring genetic bounds, these techniques seek to enhance cultivars by emphasizing characteristics regulated by one or a small number of genes for simpler breeding than quantitative qualities regulated by a large number of genes. The law of natural inheritance is not violated by conventional breeding. In conventional breeding, high performance on preselected traits is the basis for the selection of pants, a selective breeding process. Through hybridization, the necessary traits are obtained from closely related individuals and combined into new cultivars (Al-Khayri *et al.*, 2018). Hybridization is the process by which conventional breeding achieves variability. The most significant genotype is found by phenotypic selection. Following evaluation, the chosen cultivars are issued as varieties. Since there would be no gamete segregation in traditional breeding, identical parents would not create diversity (Acquaah, 2009). A new cultivar is often released after more than ten years of conventional breeding, which is a more time-consuming procedure. The phenotypic manifestations of crops are too relied upon in conventional breeding to find superior ones. Because genotype by environment interactions strongly influence phenotypes, chosen cultivars are therefore not error-free (Lema, 2018). Conventional breeding is an applied science that depends on the judgment of breeders and their experiences, observations, and talents (Allard, 1961). The primary cause of conventional breeding's uneven results is an overreliance on subjective analysis. Modern plant breeding, on the other hand, is more scientific and less subjective. Consequently, contemporary breeding techniques are more productive and efficient (Jiang, 2013). Crop improvement has increased due to the extensive use of genetic and genomic approaches in conventional breeding. These days, staple crops including rice, wheat, sorghum, maize, etc. are frequently farmed using these contemporary methods (Rashid *et al.*, 2017)

The USDA also typically recognizes conventional breeding techniques as non-GMO. However, these methods could involve the following: the use of tissue and cell culture; methods for bridging genetic bridges; wide crosses and embryo rescue; translocation breeding; sophisticated pollination techniques and *in vitro* fertilization; methods for polyploidization; creation of seedless varieties; anther and microspore culture for haplodization; interspecific grafting; and the generation of inbred and hybrid lines.

Furthermore, two particular sophisticated conventional breeding techniques are (i) the use of chemical mutagens like sodium azide or alkylating agents (like EMS; ethyl methyl sulfonate) or x-ray or gamma ray mutagenesis; and (ii) cell-fusion procedures for somatic hybridization, such as protoplast fusions and cell sorting (Bahadur *et al.*, 2015).

These methods include whole chromosome transfers, cisgenics, and the use of zinc finger nucleases (ZNF), TALENS, and CRISPRs as mutagenesis agents. PBL expertise is highly suited to these applications as these approaches are often needed for the utilization of embryogenic cell cultures and their intermediates, as well as advanced tissue culture.

Including cutting-edge digital technologies will be beneficial in improving breeding results, especially for quantitative features that are challenging to breed. To meet the breeding technologies, a vast amount of data must be combined and managed using digital technology.

The loss of other advantageous features and the gradual reduction of genetic variety are the results of cultivating different crop varieties generation after generation for particular traits. Some crops, particularly those that are completely sterile without seeds, reproduce vegetatively. Examples of these crops are banana, apple, cassava, and sugarcane. It was necessary to create alternate methods for this significant population, notably somatic tissue modification techniques such as mutation breeding and biotechnology.

Plant mutation breeding, also known as variant breeding, is a technique for creating novel crop varieties that involve causing spontaneous genetic diversity in plants using chemical or physical techniques. The majority of genetic diversity and the force behind evolution may be attributed to 'mutation.'

A key strategy for resolving these issues and quickly producing unique germplasm is plant mutant breeding. One kind of breeding that is done to improve agricultural plants for human consumption is called mutation breeding. Crop plants have been improved via the application of several mutant breeding techniques. Through the use of genetic variants created early in the plant breeding process, mutation breeding can produce vast populations of better crops more quickly. These crops can then be further bred until the desired outcome is obtained. Plant breeders are left with little choice but to try mutation induction when they cannot find the gene or genes necessary for stress tolerance or resistance to a specific disease in the gene pool. When mutagens are used, chromosomes are broken or genes are changed. Natural mistakes in the replication of deoxyribonucleic acid (DNA) result in gene mutations. The majority of these mistakes are fixed, but some could survive the following cell division and manifest as spontaneous mutations in the plant's progeny. In order to produce more desirable and commercially valuable mutations, research concentrated on transforming 'random' mutation induction into a more controlled mutagenesis process.

Plants can be mutated by introducing active transposable elements, like mPing, using physical and chemical techniques (Saraswat *et al.*, 2023). Until now, physical mutagenesis has been carried out using radiation from particles such as ß and α particles, as well as radiation from electromagnetic sources including gamma rays, x-rays, and UV light. The most often employed mutagen to improve the crop out of all of this physical mutagenesis was radiation. Since radiation does not transfer to subsequent generations of the created variety, plant varieties generated using radiation are just as safe as those developed using traditional or marker-aided breeding. Plants can therefore be bred for many generations to get the greatest outcome without sustaining harm or turning radioactive. The mutant breeding approach benefits from several comparative advantages as a result, including being rapid, inexpensive, reliable, and transferable, as well as non-hazardous and ecologically benign. These plants underwent spontaneous mutation, which eliminated pod or head breaking and shortened seed dormancy periods. Gamma rays have been the most often employed mutagen; according to the IAEA, over 80% of mutant types originated from gamma-induced mutations. Of all the registered mutant variations, 71 are exclusive to the *Arachis hypogaea* species, according to the FAO/IAEA Mutant Variety Database). Through physical mutagenesis, 65 different kinds were produced from these: 10 x-ray varieties, 41 gamma ray variants, 12 β ray varieties, and 2 laser-influenced types. Radiations such as α and β rays, fast neutrons, thermal neutrons, x and y rays, and UV radiation are examples of physical mutagens. Almond, lima bean, watermelon, potato, eggplant, cabbage, and several types of nut are among the other plants that have spontaneous mutations (Mba, 2013). By incorporating spontaneously altered alleles of the genes that led to the green revolution, wheat types with high-yielding and lodging resistance were created, ensuring sustenance for millions of people worldwide. Other instances include the introduction of the rice variety IR8, which was created in the Philippines by the IRRI from a dwarf line, and the use of dwarf germplasm Dee-geo-woo-gen from China (Mba *et al.*, 2012). Plant breeders have been introducing genetic variety since the 1920s by using chemical and physical mutagenesis (Stadler, 1928). The green revolution was made possible by semi-dwarf rice varieties that were created independently of spontaneous and artificial mutations in the gibberellin 20-oxidase gene (Ashikari *et al.*, 2005). Breeding has used induced and spontaneous mutations in waxy homologs in hexaploid wheat to alter the starch quality (Slade *et al.*, 2005; Dong *et al.*, 2009).

Since genetic diversity is essential for both vegetative and sexually propagated plant breeding programs to be successful, several researchers (Mashenkov, 1986) have documented the importance of chemical mutagens in boosting genetic variability in higher plants (Kleinhofs *et al.*, 1978). To improve agricultural plants, chemical mutagens such as diethyl sulphate (DES), sodium azide (NaN3) treatment of seed, and ethyl methane sulphonate (EMS) were also employed. Since genetic variety is a necessary component of both vegetatively and sexually propagated plants' effective breeding programs, chemical mutagens are frequently employed to increase genetic variability in higher plants.

Numerous authors claimed that Huayu 16's flowers were injected with 0.3% EMS, and that selection thereafter allowed for the development of Huayu 40, a high-yielding peanut cultivar (Wang *et al.*, 2010; Wang *et al.*, 2011). Huayu 40, the enhanced cultivar, has a sequential branching pattern and an upright growth habit. Huayu 40 grows more quickly and has deeper green leaves than its wild counterpart, Huayu 16 (Wang *et al.*, 2011). Furthermore, Huayu 40's leaf-water content and its levels of chlorophyll a and b were found to be substantially greater than those of Huayu 16 (Wang *et al.*, 2010). A relatively high frequency of mutants for intraspecific differentiation in groundnut was produced by mutagenesis using EMS (Gowda *et al.*, 1996). There have also been reports of improved seed output and its constituent parts in mung bean and chickpea due to induced mutation (Kharkwal and Nagar, 2005). Improvements with 'golden' included larger seeds and more pods per plant. Large-seeded mutant breeding lines in the peanut cultivar known as 'Georgia Browne' were detected. Through groundnut-induced mutations, numerous seedcoat colors have previously been identified (Suvendu *et al.*, 2007).

In order to create new mutant cultivars that are high-yielding, resistant to biotic and abiotic stresses, have deeper rooting systems, enhanced uptake of specific metals, and modified oil, starch, and protein content that can boost industrial processing, it is necessary to establish mutant resources for various crop plants.

Still, it did not result in the necessary changes to the mutant spectrum. The low frequency of commercially valuable mutations and the simultaneous rise in plant damage with rising radiation doses were limitations. This prompted researchers to look for possibly more effective mutagens. This led to the discovery of novel radiation therapy techniques and mutagenic chemical agents. Plant breeding has benefited greatly from the combination of tissue culture, isotope methods, and radiation mutant breeding. They have sped up breeding, improved selection technologies, and created new methods for causing genetic variety. Next-generation sequencing (NGS) or microsatellite markers can be used to map the mutations linked to the advantageous phenotypes. A technique known as 'space mutagenesis' uses cosmic rays to create spontaneous mutations in space (Novak and Brunner, 1992).

11.3 Advanced Breeding Techniques

With revolutionary breakthroughs throughout the course of its almost 100-year existence, traditional breeding has freed humanity from famine. Notwithstanding these remarkable accomplishments, it came to a standstill because of the lengthy breeding period and the small number of crops that could be crossed. Amidst these challenges, a new idea emerged, replacing the existing phenotype-based breeding method with a molecular-based selection strategy. Breeders made breeding selections visually up until that point. However, with the development of molecular breeding, gene analysis technology made it feasible to ascertain if the parent generation's genes were carried over to the subsequent generation (Savadi, 2018). Plant genome sequencing using NGS makes it possible to identify molecular markers. Accurate selection is made feasible by molecular breeding. Using genomic markers that identify each individual's features, molecular breeding allows for the precise selection of desirable qualities in usable people without being significantly influenced by the environment (Metzker, 2010). To address the demands of the growing global population and large-scale plant breeding needs, current molecular breeding serves as a powerful precision tool, reducing the effort and time required to develop new varieties while enabling detailed and accurate research. Modern breeding techniques such as CRISPR-Cas-9, marker-assisted selection, and genomic selection offer more efficiency and precision in selecting superior plants based on genomics, enabling the identification of desirable traits and recessive alleles at an early developmental stage.

Because DNA markers (like RFLP, RAPD, AFLP etc.) are unaffected by the growing environment or heredity and do not change the phenotypic expression of plants, they are widely used (Ashraf and Foolad, 2013). Traits that are expressed only in the late reproductive stage, such as grain yields, flower colors, seed characteristics etc., can be chosen with the use of MAS. Even at an early stage of plant growth, these features may be determined in a genotype by the application of DNA markers (Madhusudhana, 2019). When choosing features that are less expensive, time-consuming, and heritable, MAS is a more successful method. Nevertheless, the approach works best when a small number of quantitative features are chosen, as they regulate a significant portion of the population's variance (Shu *et al.*, 2016). SSRs are widely used because they are inexpensive and simple to operate (Collard *et al.*, 2008). Due to their ability to modify plant phenotypic expression without changing it, DNA markers are widely used. They are co-dominant markers that are the most desirable in genetics and breeding due to their simplicity. However, because of high operating costs and the need for high-quality DNA, application in many industries has been limited.

The 20th century saw the commercialization of a backcrossing breeding technique. Using marker-assisted backcrossing (MABC), important genes that indicate better qualities may be quickly inserted into the best varieties or lines, producing a variety that has both the desired alleles and the transgene (Ragot *et al.*, 1995). MAS helps to genetically improve difficult-to-phenotype crops with beneficial features. Furthermore, transgenic fixation in elite varieties can happen quickly (Moose and Mumm, 2008). However, MAS has a notable impact on a smaller number of QTLs.

In plant breeding, the quantity of molecular markers for desirable features has been steadily rising recently. The foundation for creating the molecular markers utilized in MA is QTL mapping, which identifies genetic loci statistically related to attributes of interest (Ibrahim *et al.*, 2020). DNA markers are used to manipulate a genomic area that is important in quickly expressing a desired phenotype. This process is known as MAS. These initiatives ushered in a new era for the study of molecular breeding (Sharma *et al.*, 2021a, b). It is also used in plant breeding to increase crop production and quality as well as resistance to biotic or abiotic stress. For MAS to be possible, genes responsible for the required character and associated markers to QTLs must be identified (Khush, 2000). Four categories make up the MAS framework in plant breeding: (i) marker-based recurrent selection; (ii) marker-assisted backcrossing; (iii) marker-assisted pyramiding; and (iv) marker-assisted selection in early generation. These mechanisms substantially bind the breeding cycle and distinguish genetic material in initial generations of segregation (Nadeem *et al.*, 2018).

Molecular markers have facilitated the genetic recovery and rapid selection of recurring parents (Muranty *et al.*, 2020). They are routinely employed in popular varieties to remove undesired features (e.g. disease susceptibility) (Sharma *et al.*, 2021a, b). A method for introducing one or more significant genes into an elite variety's genetic background from a donor is known as marker-assisted backcrossing, and it was originally published in 1992.

Multiplier-assisted pyramiding is the technique that unites many genes into a single genotype. Combining many abiotic and biotic stress-resistance genes is the most popular pyramiding technique, which is used to stop resistance to certain illnesses or pressures from eroding. This technique pyramids the gene combinations to transfer genes into enhanced kinds effectively. The benefit of early-generation MAS is that it selects markers early, which allows for the examination of less significant lines in the following generation than the previous one (Collard *et al.*, 2008).

Enhancing desirable QTLs and alleles in the population prior to the inbred line extraction is the aim of marker-based recurrent selection (Bankole *et al.*, 2017). Because it may leverage small genes/QTLs that have little effect on the phenotype, this technique can efficiently breed complex phenotypes (Abdulmalik *et al.*, 2017). The inability to effectively select features made up of several modest impact alleles is the drawback of MAS in plant breeding. Furthermore, linkage drags impede the improvement of desired characteristics in every breeding effort.

Thus, to overcome the limitations of MAS, predictive breeding techniques utilizing agro-bigdata and sophisticated statistical models integrated into certain fundamental machine-learning algorithms are being employed.

11.4 Genomic Selection

Compared to traditional breeding methods, genomic selection (GS) yields better results faster by enhancing grain yield. Because it employs genome-wide markers to calculate projected breeding value by speculating on the impact of quantitative gene loci, it is an improved form of marker-assisted selection (Wang *et al.*, 2018a). In the early 21st century, large-scale genomic data became available because of the introduction of novel sequencing technologies, and predictive breeding techniques like GS were born. By employing several kinds of whole-genome scanning markers, GS has a high predictive capacity for genotype selection with interesting features, including quantitative traits, overcoming the limits of marker-assisted selection (MAS,) different statistical models used with GS, and techniques to boost GS effectiveness. According to a study by the International Maize and Wheat Improvement Center (CIMMYT), plants like maize can have breeding intervals as short as half those of conventional breeding (Crossa *et al.*, 2017). Unlike standard MAS, which uses a small number of markers, GS uses markers spread across the genome to determine the value of breeding (Meuwissen, 2009). MAS relies entirely on molecular markers linked to the desired qualities, whereas GS is more like traditional breeding techniques, which rely on phenotypes and the breeder's capacity for selection. Plant individuals are chosen by breeders in traditional breeding according to their experience and personal preferences. MAS, on the other hand, uses molecular markers to select plant individuals in an objective manner. The breeding or genetic population is trained mechanically using statistical models as the basis for the GS operations (training population). Subsequently, the breeding population (also known as the validation population) is subjected to selection using the model.

Based on anticipated breeding values, the finest individuals are chosen for breeding in this kind of contemporary breeding scheme. The worth of a chosen person based on the average performance of its progeny is known as breeding values. In contrast to linkage analysis and genome-wide association studies, the primary goal of GS is not to identify QTL but rather to infer future progeny performance from current DNA samples. Breeding values can be forecast using techniques like the Bayesian approach and penalized regression method. Nonetheless, the correlation between estimated and actual breeding values may be used to assess the accuracy of the forecast (Rabier *et al.*, 2016). The training population and validation population are the two types of datasets utilized in genomic selections that are used to determine these estimated and actual breeding values. Breeding values in the validation population are predicted by the training population, which has previously undergone genotyping and phenotyping (Dwivedi *et al.*, 2015). Small-effect genes have been successfully mapped via genomic selection. Small-effect genes with numerous QTLs have been effectively mapped using genomic selection. Since they control the majority of commercially and agronomically relevant characteristics, these modest-impact genes are significant in plant breeding (Robertsen *et al.*, 2019). Given the rising cost of crop phenotyping, genomic selection can be a useful strategy for improving quantitative features that are costly or difficult to identify through phenotypic expressions (Sweeney *et al.*, 2019).

GS may be done when MAS is done at a very high density. It primarily evaluates the existence of a large number of valuable markers and forecasts the degree to which the combination of markers and genes in a genotype might favorably impact one or more characteristics. The method is highly computationally costly, needs a reference population, and uses genome-wide analysis to forecast phenotypes in populations with two or more parents. The GS models at IRRI will benefit greatly from low-coverage sequencing of the complete gene-bank collections of rice accessions. This will make the models much more accurate, especially when paired with automated and mechanical phenotyping systems that process experiments and data at a high speed and accuracy (high-throughput phenotyping platforms). These issues are handled via a novel technique known as GS. Enhancement of intricate features necessitates phenotypic assessment across several sites and several years to validate the association between genotype and environment. That was challenging, though, because there was little money and labor. The advancement of NGS technology has resulted in decreased sequencing prices and easier access to high-resolution genomic data. GS is now feasible thanks to improvements in sequencing techniques (Gorjanc *et al.*, 2015). To choose GS crops, one can utilize any high-performance marker in the genome that is currently accessible (Jannink, 2010). GS models take into account all markers affecting a character, irrespective of a certain threshold, given a marker collection that spans the whole genome. While GS generates predictions by including every marker found in the genome in the model, MAS (traditional) concentrates on limited QTLs or key genes. The genetic variation loss is avoided by accounting for all genetic influences. For qualities governed by several markers, GS is therefore more successful than MAS (Meuwissen, 2009). The main benefit of GS is its ability to forecast mature individual phenotypic information from genetic data derived from seeds or seedlings. This procedure speeds up crop varietal creation and does away with the necessity for a thorough phenotypic assessment by year or environment (Bhat *et al.*, 2016). Past research found the genomic prediction (GP) accuracy for wheat is two times higher and three times greater for maize when compared to MAS (Heffner *et al.*, 2009). Consequently, there is a chance that the genome-based predictive selection may eventually supplant marker-assisted breeding or phenotypic selection.

To build a model, GS integrates genotype and phenotypic data from the training population. Next, the genomic estimated breeding values (GEBVs) of individuals in the breeding population. Consequently, GS uses genotype data to choose individuals from the breeding population according to their GEBVs. Phenotypic data of the population is not necessary in this instance (Meuwissen, 2009). Two components mainly constitute breeding values: parents' average breeding values and progeny's deviation from the parents' mean breeding values. Large-scale progeny phenotypes are not necessary as GS employs a higher density of markers to measure Mendelian sampling. Compared to thorough phenotyping, this method is more efficient and shortens the breeding cycle. Simple qualities with high heritability and complex ones with low heritability can both benefit from GS. Furthermore, it is feasible to lower breeding lines or a hybrid cost of development (Crossa *et al.*, 2017). Training models work similarly to how breeders get breeding resources. Consequently, GS is more akin to a traditional selection process. Conversely, selection may be conducted early in the development stage using trained models, such as MAS, which combines the benefits of traditional breeding with MAS.

A highly dense marker availability is beneficial in GS since it uses the LD between the markers and quantitative loci to evaluate the marker effects (Meuwissen *et al.*, 2001). If the marker intervals are evenly distributed, some efficiency may be ensured in GS for low-density markers also (Habier *et al.*, 2009; Spindel *et al.*, 2015). Furthermore, additive effect sizes are used to choose markers, and compared to uniformly spaced markers, oligogenic traits have higher predictability (Li *et al.*, 2018). High-density SNPs are a commonly utilized strategy in GS to improve accuracy. The training population's design also affects how accurate genetic forecasts are. In GS, training populations with a track record are crucial. Breeding lines containing genotype and phenotypic information for the desired qualities make up the training population. A predictive model is trained utilizing markers, and then the trained model is used to predict GEBVs in the test population (Akdemir and Isidro-Sánchez, 2019).

Prediction accuracy generally tends to increase with training population size. Breeders ought to select a training population that is intended to optimize prediction accuracy while minimizing phenotypic expenses (Lado *et al.*, 2013). Utilizing the test population's genetic data to create the training population is a further strategy for improving it. Applying a genetic algorithm to the test population's genetic information results in a notable improvement in prediction accuracy (Akdemir *et al.*, 2015; Lorenz and Smith, 2015). Because of less recombination between the QTLs and marker and a shared ancestor when individuals have a tight genetic distance, GP accuracy is increased when there is a genetic link present between the test and training populations (Wientjes *et al.*, 2020). Moreover, polymorphic loci resulting in genetic differences are shared by the two populations (Habier *et al.*, 2010). Furthermore, the variance of interaction between the genetic background and the QTLs is shared if the groups share the same genetic background (Lorenz and Cohen, 2012). So, if the breeders wish to improve GP accuracy, they should think about training the structure of the population of unrelated ones.

11.5 Quantitative Trait Loci (QTL) Mapping

In an effort to understand the genetic basis of variation in complex traits, phenotypic data (trait measurements) and genotypic data (typically molecular markers) are linked statistically in a process known as quantitative trait locus (QTL) analysis (Falconer and Mackay, 1996; Kearsey and Farquhar, 1998; Lynch and Walsh, 1998). A genome-wide inference of the association between genotype at different genomic regions and phenotype for a collection of quantitative characteristics in terms of the quantity, genomic positions, effects, and interaction of QTL is known as QTL mapping. Finding chromosomal areas that have a substantial impact on the variation of quantitative characteristics within a population is the main goal of QTL mapping. This localization is crucial for both our comprehension of the genetic mechanisms behind the variance and the final identification of the genes that are ultimately responsible (McCough and Doerge, 1995).

Researchers in agriculture may connect certain complex characteristics to particular chromosomal areas using QTL analysis. Determining whether phenotypic variations are mostly caused by several loci, each with minute effects or by a small number of loci with quite high effects has been a main objective of QTL analysis. Few loci of big impact appear to account for a significant percentage of the phenotypic variance in many quantitative variables; many loci of minor effect account for the remaining fraction (Remington and Purugganan, 2003; Roff, 2007).

For QTL mapping, a variety of genetic populations can be employed, including specific crosses between inbred and outbred strains: (i) parental strains that differ genetically for the trait such as lines produced by divergent artificial selection are necessary for quantitative trait locus (QTL) mapping; (ii) the parental lines (inbred strains) are crossed with different phenotypes can produce an F1 generation (all members of the F1 generation share the same genetic makeup, despite being a mixture of two homozygous genotypes); variants linked to the phenotypic characteristic will be segregated into an F2 generation through crosses between F1s or backcross offspring when they are crossed to one of the parent lines. Different portions of the genome from each parental line are present in the individuals or strains that are produced by these two crossings. Both the genotype of markers that differ between the parental strains and the phenotype of each of these recombinant individuals or lines are evaluated; (iii) while accounting for the impact of other markers on the trait, statistical methods like composite interval mapping assess the likelihood that a marker or the distance between two markers is connected to a QTL influencing the trait. Plotting the test statistic against the chromosomal map location in recombination units (cM) is how the findings of this analysis are displayed. The chromosomal location corresponds to the greatest significant likelihood ratio, which indicates the best estimate of QTL placements. Likelihood ratios above this line are officially significant (Mackay, 2004).

Historically, the barrier to QTL analysis has been the availability of sufficiently dense markers, or genotypes. However, high-throughput technology and genomics have begun to overcome this hurdle. As a result, the fundamental phenotyping-related constraints that still exist in QTL analysis are somewhat mitigated by the incorporation of genomic and proteomic data as phenotypes. Identifying markers associated with QTLs that control any trait expression in one or more parental mapping populations is one method that makes use of classical molecular markers (Semagn *et al.*, 2015). Different molecular markers were produced during the course of the 1980s and

1990s, allowing suitable genome coverage and marker density in QTL mapping (Cooper *et al.*, 2007).

Large sample numbers are necessary for QTL research, and they are limited to mapping differences between the original parental strains. Some loci will go undiscovered since it is improbable that these strains have segregating alleles of significant influence at every locus contributing to diversity in wild populations. Furthermore, natural populations could not benefit from the precise alleles that segregate, especially in inbred lines. However, it is conceivable that other alleles at these same locations will be of interest. Therefore, rather than focussing on specific alleles, the purpose of many investigations is to locate loci. Applied research in agriculture, which is frequently interested in particular segregating alleles, is one noteworthy exception.

Individual genes have still only sometimes been found as a result of QTL mapping studies. In fact, Roff (2007) provides a list of quantitative characteristics for which single genes have significant impacts and whose molecular basis has been investigated; nonetheless, he points out that this number pales in comparison to the work that goes into QTL research. The fact that many QTLs map to genomic areas that are under 20 centimorgans (cM) long and sometimes contain numerous loci that impact the same trait is one reason for this disparity (Price, 2006). Furthermore, establishing causality by methods such as positional cloning (Clee *et al.*, 2006) and targeted gene substitution (Young, 1996) is necessary to pinpoint the precise loci influencing a quantitative characteristic.

The discovery of *a priori* candidate genes by traditional reverse genetics or bioinformatics often aids in the search for specific genes within a QTL. The candidate gene and the QTL must then be shown to have a functional relationship. One way to do this is by using functional complementation, which is the process of rescuing a loss-of-function mutation or producing an alternative phenotype by introducing wild-type complementary DNA from the gene in question into the nucleus (Broman and Sen, 2009).

In genetic research, genome-wide association studies (GWAS) are becoming more and more common and are a great addition to QTL mapping. While GWAS yields numerous unlinked individual genes or even nucleotides, these studies are rife with a high predicted number of false positives. QTLs contain many linked genes, which are consequently difficult to separate (McCough and Doerge, 1995).

Understanding the number of QTLs that significantly contribute to trait variation in a population may also be gained through QTL mapping. The genetic architecture of quantitative characteristics in the population can be determined by the QTL's additive effects, its dominant and epistatic effects, the nature of close linkage, pleiotropy, or genetic association between distinct qualities within a genomic region and its interactions between environment or QTL (Gupta *et al.*, 2014).

According to Bernardo (2008), QTL mapping in plants contributes to our understanding of genetics in connected species by offering helpful markers for deciphering the genetic architecture of complicated characteristics. Thus, it is essential to create linkage maps to identify relationships between phenotype and genetic markers (Wang *et al.*, 2016). Numerous techniques are available to link genotypes with a wide range of quantitative traits, including complex interval mapping and relatively straightforward single-marker analysis (Haley *et al.*, 1994), multiple regression (Whittaker *et al.*, 1996), joint mapping (Kearsey and Hyne, 1994), and composite interval mapping (Zeng, 1994). Numerous QTLs controlling difficult phenotypic features in rice (Yano *et al.*, 2000), *Arabidopsis* (El-Assal *et al.*, 2004), maize (Yano *et al.*, 2000), and other crops have been found in recent decades through QTL mapping research.

According to Tiwari *et al.*, 2024, there are different genes which are responsible for salt stress in plants like OsSOS3/OsCBL4, OsSOS1, OsSOS2, and OsCIPK24. The absorption and transport of ions are regulated by the gene complex OsCBL1-OsCIPK23, and OsAKT1 aids in the absorption of K^+ to shield cells from the damaging effects of Na^+. Because salinity-stress tolerance is highly influenced by ecogeographic regions due to its complex phenotypic, physiological, and polygenic character, it is challenging to identify genes, QTLs, or genetic markers for MAS. Breeders can identify genic zones causing variations in the target trait by using QTL mapping. By using the information provided here, breeding programs can create crosses that will increase the lines' resistance to salt stress.

A GWAS was conducted on the characteristics that make up the drought-resistance index. On chromosomes 2, 6, 7, 8, 9, and 10, nine QTLs linked to characteristics associated with drought were found. These QTLs include qPH10.1 (plant height), qPL9.1 (panicle length), qLW7.1 (panicle width), qLA7.1 (leaf area), qLL7.1 (leaf length), and qTGW2.1, qTGW8.1, qTGW6.1 (thousand-grain weight). The percentage of phenotypic variance accounted for by a single QTL ranged from 10.6 to 13.9% (Yi *et al.*, 2023).

11.6 Genetic Engineering

Genetic engineering opens myriad opportunities to alter the crop's DNA to make it perform better under stress conditions. Targeted manipulation of plant genes helps plants to be more resilient to climate change. The recombinant DNA (rDNA) technique enables the transfer of genes between unrelated species or genera. Genetically modified organisms (GMOs) are organisms with genes received through rDNA technology (Key *et al.*, 2008). Crops like soybean, maize, cotton, and canola are the prime GM crops commercially grown around the world.

rDNA technology involves the union of segments of different DNA fragments from different organisms. The basic step in this method involves the isolation of transgene of interest from the source organism by cutting enzymes 'restriction endonucleases' at 'recognition sites,' its insertion into a 'cloning vector,' transfer of cloning vector inserted with transgene into the host cell (bacterium), and identification of transformed cells and their maintenance. The required gene is identified from the genomic or cDNA library and then transferred to the host organism. The transgene is transferred either directly through methods like particle bombardment and electroporation or indirectly through *Agrobacterium*-mediated transformation. Microprojectile bombardment or the biolistic method uses a particle gun to shoot the micron-sized tungsten or gold particles coated with specific DNA fragments at high energy to penetrate into the plant cells directly. Electroporation uses the electrical impulses by an electroporator to widen the plant protoplast pores and the intake of DNA fragments through these pores happens. Some other methods like micro-injection are also available for direct transformation. *Agrobacterium* tumefaciens is a tumor-inducing bacterium that has Ti plasmid which can insert certain DNA fragments in the T-region into the host chromosome upon infection. The region causing tumor induction is removed while designing the bacterium. Such *Agrobacterium* cells with required DNA are co-cultivated with plant explants (cell suspension, protoplasts, shoot meristem, immature embryos, or inflorescence) in a regeneration medium for a specific time period and then the transformed cells are selected using particular antibiotics or scorable markers like luciferase, green fluorescent protein, or β-glucuronidase (Acquaah, 2009).

Numerous crops are modified genetically with the primary goal of imparting insect resistance. *Bacillus thuringiensis* is a bacterium found in soil, some strains of which can make proteins that kill specific insects having alkaline digestive tracts. The heterologous expression of proteins from Bt in plants kills the insects feeding on them by disrupting their digestive system. The specificity of different strains of Bt killing different insects makes it precise in controlling lepidopterans like cotton bollworms, European corn borers, and some dipteran pests. Bt is nontoxic to other organisms including humans and was discovered in 1911 and registered as a biopesticide in 1961 in the USA (Webber, 1995). The common crops with Bt insect resistance are cotton resistant to bollworm, corn resistant to European corn borer, and potato against Colorado potato beetle.

Due to various barriers of political, social, and biological factors, GM crops commercialized for disease resistance are much less compared to for insect resistance (Collinge *et al.*, 2008). Virus resistance is the most successful disease resistance through transgenics. Disease resistance involves several types of biological principles (St. Clair, 2010). The largely used strategy is based on the use of 'virus coat protein' genes. Examples are papaya ringspot virus-resistant papaya and multiple viruses-resistant Cucurbita (Fuchs and Gonsalves, 2007). Pathogen-derived resistance or post-transcriptional gene silencing is the underlying mechanism in these cases. The examples of approved cultivars for virus resistance are Rainbow and SunUp varieties of papaya; Spunta cultivar of PVY-resistant potato (Bravo-Almonacid and Segretin, 2016); and varieties of summer squash for watermelon mosaic virus and zucchini yellow mosaic virus (de Oliveira *et al.*, 2003). Through *Agrobacterium*-mediated gene transfer, resistance to bacterial blight was reported in a rice cultivar (Wang *et al.*, 2005). Tobacco plants altered through genetic engineering with two bacterial genes for the improved production of salicylic acid also showed resistance to viral and fungal diseases (Verberne *et al.*, 2000).

Abiotic stress response involves complex response mechanisms involving several genes in plants. In rice, dehydration response element binding factors (DREB) have a key role in the regulation of gene expression in response to drought stress. AtDREB1A gene along with 35s CaMV promoters was transferred to two rice local cultivars, namely JP-5 and KSK-282, and found to exhibit higher drought tolerance (Hussain *et al.*, 2014). Late embryogenesis of abundant protein-producing gene OsLEA3-2 in transformed rice and *Arabidopsis* showed high salinity and osmotic tolerance and was capable of recovery after drought for 20 days (Duan and Cai, 2012). Arabidopsis Vacuolar Pyrophosphatase (AVP1) gene under 35SCaMV promoter transferred through *Agrobacterium* in sugarcane confers tolerance to drought and salinity (Kumar *et al.*, 2014). Similarly in another study conducted by Muppala *et al.* (2021), transgenic maize exhibited improved drought tolerance by the introduction of receptor-like protein kinase gene and 9-cis-epoxycarotenoid dioxygenase gene involved in the abscisic acid pathway. Other examples of genetically engineered crops for various stresses are given in Table 11.1.

Even though genetic engineering is a precise and efficient breeding strategy for stress tolerance, it is accompanied by several environmental and ethical concerns, safety aspects, and regulatory considerations which makes the process more responsible with transparent practices. Any GM crop has to be evaluated through multiple tests to verify its varietal efficacy as well as food and environmental safety by regulatory authorities before its commercial release.

11.6.1 Genome editing

Manipulation of specific genes involved in specific pathways accountable for the response to various stresses has been a breeding strategy in the recent past. This helps in transgene-free gene targeting and hence faces fewer regulatory considerations. Techniques like RNAi, CRISPR-Cas9, and TALENs are being utilized immensely for the precise manipulation of such stress-responsive genes. For better understanding, such new genomic techniques using advanced genetic engineering methods are explained in this section.

11.6.2 RNAi (RNA interference)

RNAi is one of the precise and effective gene regulatory tools in the mitigation of stress-resilience breeding. This technique is being advocated against a wide variety of stresses like drought, salinity, insects, viruses, and diseases. It involves the use of small RNA (sRNA) molecules to downregulate the expression of specific genes before or after transcription. Its basic principle relies on the eukaryotic cell self-protection mechanism upon the detection of a double-stranded RNA (dsRNA) caused by various pathogenic infections. dsRNAs are cleaved by ribonuclease III enzyme also known as 'dicer' or 'dicer-like enzymes' resulting in the formation of 20–30 nucleotide long siRNA (small interfering RNA). Upon association with Argonaute family proteins (AGOs), siRNA then forms a siRNA-AGO complex which is recruited for making RNA-induced silencing complex (RISC), which is the real participant in gene silencing with the help of several other proteins. Duplex RNA strands then unwind releasing passenger strands and the retained 'guide RNA' helps in targeting complementary RNA locations. When the target is located, depending on the context, DNA methylation occurs resulting in 'transcriptional silencing' or at the stage of translation due to RNA destabilization causing 'post-transcriptional silencing'. Another widely used small RNA molecule is 'microRNA' (miRNA) involved in gene regulation in responses to various environmental cues (Pradhan *et al.*, 2015). miRNA can downregulate various genes on the reception of stress messages causing post-transcriptional silencing, so are used in stress-resilience breeding. Various attempts using RNAi to create abiotic stress tolerance are summarized in Table 11.2.

Table 11.1. Genetically modified crops for stress tolerance.

Stress	Crop	Gene introduced	Gene source	Trait	Method	Reference
Insect	Sugarcane	*Cry1Ac*	*Bacillus thuringiensis* subsp. Kurstaki strain HD73	Resistance to sugarcane borer (*Diatraea saccharalis*)	*Agrobacterium*-mediated transformation	Gianotto *et al.*, 2019
	Maize	*Cry1B*; *Cry1Da*	*Bacillus thuringiensis*	Resistance to sugarcane borer (*Diatraea saccharalis*)	NA	Horikoshi *et al.*, 2022
Disease	Tobacco	*DrsB1-CBD* recombinant peptide	DrsB1 mature peptide (GenBank accession P80282) *C. fulvum* Avr4 effector protein	Fungal disease resistance (*Alternaria alternata, A. solani, Fusarium oxysporum,* and *F. solani*)	*Agrobacterium*-mediated transformation	Khademi *et al.*, 2020
Drought	Wheat	*Hahb-4*	*Helianthus annuus*	Drought-stress tolerance	Microparticle bombardment	Ayala *et al.*, 2019
Salinity	Rice	*SaSRP3-1, SaVHAc1*	*Spartina alterniflora*	Increased K^+/Na^+, chlorophyll content, RWC, shoot-root growth	*Agrobacterium*-mediated transformation (pCAMBIA1305 vector under the 35 S CaMV promoter)	Biradar *et al.*, 2018
Osmotic and salinity stress	Rice	*OsGS1, OsGS2*	Gene pyramiding of GS isoforms of rice into plant transformation vector *in vitro* and development of transgenic rice	Enhanced nitrogen metabolism, low malondialdehyde content, high proline levels and low electrolyte leakage	*Agrobacterium*-mediated transformation (pMDC99 vector under the actin promoter)	James *et al.*, 2018
Oxidative stress	Tomato	*GME, GMP, GGP, GPP*	NA	Enhanced ascorbic acid content	*Agrobacterium*-mediated transformation (pMV vector under the 35 S CaMV promoter)	Li *et al.*, 2019

NA = not available

Table 11.2. RNAi in stress-resilience breeding.

Crops	Targeted gene	Gene function	Stress	Reference
Rice	Dehydrin gene WZT2	Osmotic stress tolerance	Salinity	Yu *et al.*, 2019
Rice	ABA catabolic gene, *OsABA8ox1*	Increase of the levels of endogenous abscisic acid, decreased ROS accumulation, plasmalemma injury and Na^+/K^+ ratio	Salinity	Liu *et al.*, 2022
Rice	Squalene synthase or farnesyltransferase (SQS)	Declined stomatal conductance and maintenance of higher relative leaf-water content	Drought	Manavalan *et al.*, 2012
Rice	*OsABA8ox3, an ABA catabolic gene*	Elevated catalase and dismutase activities and reduced malondialdehyde content	Drought	Cai *et al.*, 2015
Wheat	miRNA399 altering gene TaUBC24 expression	CBF signaling pathway	Cold tolerance	Peng *et al.*, 2021

Pathogenic infections or pest infestation in plants are also controlled by RNAi strategies, i.e., 'host-induced gene silencing- HIGS' (Nowara *et al.*, 2010) and 'spray-induced gene silencing- SIGS' (Koch *et al.*, 2019). Werner *et al.* (2020) established the success of SIGS in protecting barley from *Fusarium graminearum* infection. In several perennial plants, RNAi has been used commercially against viruses (Scorza *et al.*, 2013; Khalid *et al.*, 2017). Exogenous application of dsRNAs as a biopesticide is considered as the most efficient strategy in addition to its improved biosafety than other categories.

11.7 CRISPR/Cas (Clustered Regularly Interspaced Short Palindromic Repeat/CRISPR Associated Protein)

CRISPR/Cas, which is considered a 'clean gene' technique, is one of the gene-editing tools that is precise, cost-effective, reproducible, and fast with its simple design (Bhat *et al.*, 2021). A designed 'guide RNA' specific to the gene of interest along with a specific protein, Cas9 can identify and cleave DNA (Tripathi *et al.*, 2022). These site-directed nucleases create incisions at specific genomic positions and the host cell DNA repair mechanism will repair the double-strand cut either by 'non-homologous end-joining' (NHEJ) or 'homology-directed repair' (HDR). The gene loses its function due to insertions or deletions in NHEJ since it is an error-prone procedure (Wang *et al.*, 2022).

Climate change compounds the risks of pests and pathogens by altering their interaction with plants. A rise in temperatures uplifts not only the pest reproduction rate but also the host vulnerability and vector range (Singh *et al.*, 2023). The availability of whole-genome sequence data along with functional information on several genes involved in plant innate immunity has made 'plant resistance' a major target trait in genome editing. In rice, a major disease bacterial blight resistance was acquired by altering the effector-binding elements in sugar transporter genes (SWEET family) which are hijacked by the pathogen during infection to grow bacterial cells (Zhou *et al.*, 2015; Xu *et al.*, 2019; Oliva *et al.*, 2019). Other successful CRISPR-Cas interventions for enhanced disease resistance against viral, bacterial, and fungal pathogens in various crops were reported in many studies, i.e. phytophthora blight in tomato (de Toledo *et al.*, 2016), citrus canker in citrus (Jia *et al.*, 2017), powdery mildew in wheat (Wang *et al.*, 2014), and cotton leaf curl disease in tobacco (Yin *et al.*, 2019). The CRISPR technology is also implemented in developing herbicide-tolerant crops (Wong *et al.*, 2022) as climate change also causes accelerated weed growth and development. Climate change also worsens the effects of abiotic stresses like salinity and drought. Salinity tolerance in rice was acquired by editing various transcription factors involved in cytokinin signal transduction (Zhang *et al.*, 2019; Liu *et al.*, 2020), ABA transcription repressors in soybean (Wang *et al.*, 2021), and transcriptional regulators of auxin in tomato (Bouzroud *et al.*, 2020). Drought tolerance and cold tolerance (Li *et al.*, 2018) were also successfully enhanced using CRISPR technology in various crops. Crop varieties developed through targeted mutagenesis or genetic alteration without the introduction of 'foreign DNA' like CRISPR are exempted from restrictive regulations like that of GMO, giving positive impacts to commercial release after varietal trials in the coming years.

11.8 Omics Technologies in Climate-resilient Crop Breeding

Understanding the stress-responsive mechanism, the adaptive traits, and the genes or genetic loci involved in the adaptation process is a prerequisite for climate-resilience breeding for which genomics can be considered a promising tool. The incorporation of genomics tools for enhancing breeding efficiency is known as 'genomics-assisted breeding' (Varshney *et al.*, 2021). Molecular breeding approaches assisted by high-throughput sequencing have revolutionized genomics-assisted breeding. Since it is basically the germplasm improvement through novel allele introduction, this method is easily acceptable, unlike transgenics. These breeding methods comprise marker-assisted selection, marker-assisted backcrossing, genomic selection, marker-assisted recurrent selection, etc. Marker-assisted selection not only speeds up the process of crop improvement but is also helpful in crossbreeding. Genomic selection is the use of phenotypic data and genome-wide markers for the accelerated identification of superior lines for significant enhancement in genetic gain (Meuwissen *et al.*, 2001). This has been used in several crops to improve abiotic and biotic stress-tolerance traits.

Mapping of genetic loci desirable in stress mitigation from accessions of stress-prone environments helps in their further use in varietal release or population improvement. Next-generation sequencing advancement helped association mapping and genomic selection by the provision of high marker density coverage (Kole *et al.*, 2015). For identification of superior alleles or lines resilient to climate change and for trait mapping, genotyping is a critical step. Many platforms for genotyping are currently available to find out single nucleotide polymorphisms based on sequencing like genotyping-by-sequencing, exome sequencing, double digest restriction-associated DNA, restriction-site-associated DNA sequencing, and whole-genome resequencing. Certain SNP arrays also provide higher throughput with less per-sample cost and minimal data analysis (Rasheed *et al.*, 2017). Phenotyping facilities are also developing in a rapid manner with the advent of remote sensing with spectroradiometry, fluorescence, thermal and 3D laser scanning, ground robots, and aerial vehicles with sensors to capture real-time pictures throughout the cropping period to collect massive data (Tardieu *et al.*, 2017). Such high-throughput technologies help to identify alleles that contribute to tolerance under harsh conditions. Several investigations are being done for germplasm characterization and in different specialized populations to identify stress-tolerant donors for heat, drought, and salinity in wheat (Neumann *et al.*, 2011), soybean (Do *et al.*, 2019), pigeon pea (Varshney *et al.*, 2017), and chickpea (Li *et al.*, 2018). Moreover, efforts to sequence whole genomes were also done with 3180 accessions of rice (Wang *et al.*, 2018b), which can be used in further breeding programs. Likewise, genome-wide association studies and linkage analysis have been widely advocated to detect genetic loci for disease and pest resistance. Integrated traditional bulk segregant analysis and biparental segregating populations with sequencing technique are equipped for mapping major resistance loci in rice for blast resistance (Takagi *et al.*, 2013), bacterial wilt in ground nut (Luo *et al.*, 2019), fusarium wilt in pigeon pea (Singh *et al.*, 2016), and yellow rust in wheat (Ramirez-Gonzalez *et al.*, 2015).

High-throughput omics technologies comprising genomics, transcriptomics, proteomics, and metabolomics have helped plant breeders to develop stress-resilient varieties by giving insights into the molecular mechanism of stress adaptation by exploring genetic structure, expression patterns, gene regulation, protein functions, and metabolic pathways in crop plants. The amalgamation of divergent datasets from various omics approaches aids in reduced number of breeding cycles and exploitation of untapped diversity from crop wild relatives and ultimately to the generation of high-yielding superior climate-resilient crops.

11.9 Coordinated Breeding Strategies

Unprecedented climatic repercussions like disastrous flooding and erosion, rising temperatures, melting glaciers, and uncertain precipitation patterns pose serious damage to agricultural production by accelerating biotic and abiotic stress factors on crop growth and development. Rather than dealing with individual stresses separately, there is a need to tackle them in combination, as most of the stresses occur simultaneously. Holistic breeding programs combining traditional as well as modern techniques should be coordinated in dealing with the challenges (Fig. 11.1). All the aforementioned strategies, when integrated together, can help to tackle the constraints of each individual approach. Tapping the potential of genetic diversity from wild germplasm with the help of high-throughput genomics, phenomics, and sequencing to speed up the breeding techniques incorporating genome-editing tools with the help of artificial intelligence and machine learning opens up a plethora of opportunities in smart breeding for climate resilience.

11.10 Field Trials and Phenotyping

High-throughput phenotyping is a screening technique that generates data on particular traits and attributes of vast populations using high-throughput technologies. Plant breeding has been more effective with the use of high-throughput phenotyping, resulting in crops that are perfect for both production and

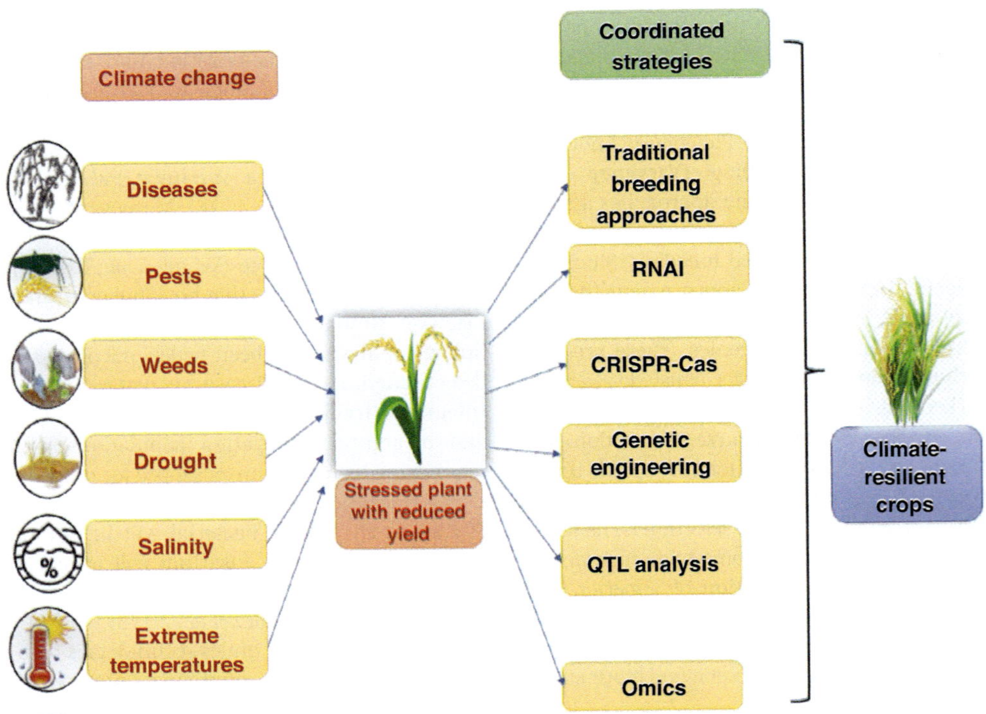

Fig. 11.1. Holistic breeding approach for climate resilience.

consumption. It has also contributed to the discovery of genes that may be targeted in an effort to alter or modify particular features. The advantages of high-throughput technologies are remarkably effective in terms of saving time and resources; quick data analysis enables quick discoveries to learn more about particular characteristics; automation and repeatability lead to more accurate outcomes. However, these studies can be costly to start, they can only examine one phenotype at a time, and they require equipment with extensive data-analysis skills.

With the help of genes and environmental variables, these phenotypes may be genetically tailored to provide the ideal growth factors for a variety of applications in plant, and cell biology. As more scientists employ high-throughput technology to conduct experiments more efficiently, more helpful and practical techniques are being created and applied to guarantee the highest possible quality in the phenotyping data. Errors and a lack of methodological repeatability might arise from manual phenotyping; the application of high-throughput phenotyping technologies can result in reduced variability across data and unbiased conclusions. One of the primary objectives of high-throughput phenotyping is genetic gain, which allows scientists to concentrate on learning more about various genomes and genetic diversity. High-throughput phenotyping can result in exponential advances in our understanding of population genetics (Araus *et al.*, 2018).

This is one of the main uses of the technology that researchers want to achieve. When combined with software and a database that can perform statistical tests and data analysis on the datasets, high-throughput phenotyping can be useful (Hossain *et al.*, 2004; Singh *et al.*, 2019). Furthermore, the setup and maintenance costs of doing high-throughput phenotyping are prohibitive for a large number of non-commercial researchers (Sozzani and Benfey, 2011; Andrade-Sanchez *et al.*, 2014).

In addition, the advantages of the new technology could not outweigh the setup fees and high-throughput phenotyping costs for a large number of non-commercial researchers. Every high-throughput application has a significant upfront cost. Finally, a lot of researchers believe that until procedures and practices are defined and accepted in every discipline, this technology will not be useful (Sturino *et al.*, 2010). This is a significant problem because there are several facilities using non-standard terminology and data-processing techniques; yet it is necessary to do this in order to create a worldwide database that can be used for further study and instruction. This will guarantee that methods are consistent, repeatable, and accessible to the public, while also greatly reducing the knowledge gap (Yu *et al.*, 2015).

11.11 Experimental Design and Data Analysis

Early in the 1930s, Fisher (1926) emphasized the value of experimental design in order to minimize bias and more accurately quantify the effects of interest. An essential factor in establishing selection circumstances that result in the distinct expression of genotype-to-genotype trait differences is appropriate agronomic management of breeding experiments. Effective trial management minimizes experimental error, which makes it easier to identify the optimal genotypes. The breeding program's target environment, including its soil and climate, photoperiod, and pest and disease prevalence, should be reflected at the field site. For all plants to grow under identical conditions, a field site must be uniform. Crop performance is subject to variability due to a combination of management and natural factors influencing field variability. Accurate agronomic management of breeding experiments can help to lessen crop performance variability as well as inherent field variability (Mathews and Crossa, 2022).

In the framework of classical designs, this chapter explains the fundamentals of experimental design. It also provides an overview of the emerging discipline of model-based design in the context of plant improvement research. Effective trial management makes it easier to identify the optimal genotypes by minimizing random variation, or experimental error, between plots, the smallest experimental unit. Data from field experiments still hold value in plant breeding, even if genetic and genomic data have been growing in importance. The plants must function in a variety of outdoor circumstances, and in the end, such data can only be collected through actual tests. Accurate field experimentation is necessary for techniques like genomic selection because field trial data are utilized to forecast the performance of untested populations (Honsdorf *et al.*, 2022).

Research in various domains has benefited from the use of concepts like replication, randomization, and blocking, which originated in agricultural science (Fisher, 1926; Preece, 1990). In situations when standard experimental designs are impractical or insufficiently address the breeding environment and objectives, new experimental designs have been devised with the aim of better fulfilling the objectives of plant breeding. Depending on whether a plant breeding experiment's objective is to identify potential lines in early generations or to provide final recommendations for commercialization with almost-finished cultivars, optimal approaches may vary.

There can be significant differences in the amount of accuracy required, the quantity of entries to be tested, and the resources available for these trials. When breeding choices require less accuracy, more lines can be evaluated with less replication to improve genetic gain. This is different from when farmers are recommending varieties when a very high degree of certainty is needed. In plant breeding projects, variations to the randomized complete block design (RCBD) have become commonplace because of the advantages of analyzing more lines with a given budget allocation. Understanding the advantages and disadvantages of each choice as well as the degrees of variance observed in experimental designs is necessary when choosing an experimental design (Shieh and Jan, 2004).

Less-sophisticated designs may be less-resilient to missing data, and simpler designs are typically easier to execute and evaluate. On the other hand, alternative designs that provide the evaluation of a bigger number of lines while maintaining adequate control over field error may yield higher genetic gains in resource-constrained systems. To find the best designs for their programs, breeders need to be ready to try new things (Kumar *et al.*, 2020).

Increasing selection frequently one of a plant breeding program's main objectives. Sites and the selection of entries are the factors that affect the experimental design. By lowering the number of chosen entrants or raising the number of tested entries, breeders can intensify selection. More entries are frequently used to boost intensity since most breeders want to maintain a minimum number of carefully chosen entries at each step of the breeding process (Masood *et al.*, 2008).

Selection intensity will be lowered by selecting top genotypes incorrectly. Two interactions exist between heritability and experimental design. First, compared to variables with lower heritability, highly heritable traits are less affected by environmental heterogeneity within the field. Reducing replication for highly heritable characteristics could result in a less-severe accuracy tradeoff than for traits with lower heritability. Second, an effective experimental design within sites will lower the unexplained variance contributing to phenotypic variation and enhance trial repeatability, enabling more precise identification of genetically better entries at any degree of heritability. The choice of parenting resources will contribute to a rise in additive variation. A careful selection of parents for crosses will guarantee the availability of meaningful genetic diversity for selection (Piepho *et al.*, 2003).

In terms of the number of entries to be reviewed, the experimental design and the decision of how many crosses to create, and how many progenies to examine each cross, interact. Plant breeding experimental design aims to minimize error while increasing the number of submissions. Numerous designs have been developed to assist in meeting the requirements of plant breeding experiments. Plots for non-experimental inputs, degrees of freedom utilized for randomization limitations, and tradeoffs between design and analytic complexity must all be considered when selecting a design. Numerous writers have proposed that larger levels of selection advantages result from increased selection intensity with enhanced or partially reproduced designs than from increased accuracy (Pearce, 1971).

Although randomized complete block design (RCBD) has been widely used in agricultural research, other experimental designs have been identified that might more effectively address the objectives of plant breeding. These other designs mostly deal with the problems that occur with the RCBD when blocks do not limit field variation enough. This happens when there are a lot of entries, which results in big block sizes and heterogeneity within blocks; it also happens when there are several directions of variation in the field; or when there are not enough seeds or other resources to duplicate every entry. In small-sized plots, it is more challenging to compare various field layouts of experimental designs and resource allocation. Field experiments are frequently used in plant breeding studies to evaluate genotypes (such as wheat varieties or hybrids of maize) and assess their potential for a variety of factors, including yield, throughout the whole breeding process. To make sure that only the best genotypes are retained for further selection, it is essential to estimate those factors accurately. Every genotype in a field experiment is assigned to one or more small plots within a field. The field (e.g. nutrition availability), measurement error, and genotype all have an impact on the measured performance on that plot.

11.12 Molecular Mechanisms Governing Climate-resilience in Plants

Nitric oxide (NOx), reactive oxygen species (ROS), and calcium (Ca^{2+}) are among the signaling molecules that are engaged in the perception and reaction to stress. These molecules are interrelated and regulate the growth, development, and defense responses of plants (Fichman and Mittler, 2021; Mittler *et al.*, 2022). Stress exposure causes quick local hikes in cytoplasmic levels of Ca^{2+}, which are detected and transmitted within the cell. This activation of the RBOHD (respiratory burst oxidase homolog D) proteins results in an apoplastic burst of ROS. Through a positive feedback loop mechanism, Ca^{2+} and ROS waves generate a continued cell-to-cell signaling chain of incidents that develop from the first tissue (the local tissue) that makes the entire plant feel stress (the systemic tissues). Rapid signaling in a stimulus-dependent manner is made possible by the systemically generating ROS and Ca^{2+} waves, which coordinate local responses (Fichman and Mittler, 2021).

Abiotic stressors, such as heat and dryness, target tissues that have not yet been exposed to the stress (Lindberg and Premkumar, 2023). Electrical waves associated with fast plasma membrane depolarization, the actions of plasmodesmata and glutamate receptor-like proteins are also connected to the systemic transfer of ROS and Ca^{2+} signals (Szechyńska-Hebda *et al.*, 2022; Lindberg and Premkumar, 2023). The diffusion and release of amino acids in the apoplast with stress stimulates the GLUTAMATE RECEPTOR-LIKE 3.3 calcium-permeable channel, which in turn triggers the transfer of Ca^{2+} waves and additional subsequent signaling events, according to recent research. Since the ROS/Ca^{2+} signaling system is fundamental to every abiotic stress response, distinctiveness in the signal must be established in order to modify these pathways (Choi *et al.*, 2012, 2017).

In spring wheat, the critical temperature barrier during anthesis is 25°C. Grain yields decline by 10% for each degree Celsius beyond this threshold, and they approach zero at 35°C. Investigations are still ongoing to identify the temperature-dependent response regulators during meiosis and in anthers (Lewandowska *et al.*, 2022).

Important developmental regulators that may be altered to improve gas exchange and increase water-usage efficiency include barley HvEPF1 (*Hordeum vulgare* epidermal patterning factor 1), maize ZmNAC4977 and wheat TaEPF1. Furthermore, it is believed that smaller stomata like those in barley help to promote a quick conductance response (Shen *et al.*, 2015). In contrast, as diversification in barley has favored stomatal density rise, features linked to the size of stomata are a suitable way to enhance rice water-usage efficacy (Chatterjee *et al.*, 2020). The resilience of crops produced on marginal terrain with few water supplies may also be increased by manipulating the leaf epidermal patterning (Dunn *et al.*, 2019). To sustain leaf-water content during a drought, for instance, barley HvCBP20 (cap-binding 20) gene mutation enhances epicuticular wax and alters the stomata patterning (Daszkowska-Golec *et al.*, 2017, 2020).

At the whole-plant level, less information is available about stress sensing despite the candidate genes characterization for drought and heat responses. Deeper root penetration and the collection of water and nutrients are made possible by the regulation of root system architecture (Comas *et al.*, 2013; Amelework *et al.*, 2015); however, the mechanisms behind these responses are not well understood and are still underutilized in agriculture (Comas *et al.*, 2013; Lynch, 2022). One potential method of customizing root system design to enhance drought avoidance is to manipulate the QTL DEEPER ROOTING 1 (DRO1) in rice, which governs the root development angle (Uga *et al.*, 2013; Roychoudhry and Kepinski, 2015).

Modified DRO1-containing maize roots may optimize water uptake because they possess steeper root development angles, less lateral branching, and axial roots. Plant breeders are still hampered by the technical difficulty of evaluating root features belowground, and it is still unclear how root architecture influences molecular reactions to several abiotic stressors (Maqbool *et al.*, 2022; Lynch, 2022).

11.13 Socio-economic and Policy Aspects

11.13.1 National programs for climate change adaptation

The Government of India (GOI) has started to integrate several policy programs and sectoral plans in order to guarantee coordination and efficient use of already-existing resources. One of the eight missions within the National Action Plan on Climate Change (NAPCC), the National Mission of Sustainable Agriculture, was established in 2010 to encourage the prudent usage of existing resources (Deshpande *et al.*, 2025). In order to address the problems with water resources and offer a long-term solution that envisions Per Drop More Crop, the Pradhan Mantri Krishi Sinchayee Yojana (PMKSY) was introduced in 2015. It encourages micro-/drip irrigation for the greatest possible water conservation (Tripathy *et al.*, 2023). In collaboration with the Indian government, the Paramparagat Krishi Vikas Yojana mission was carried out to significantly boost the adaptation of climate-smart practices and technology. Under the auspices of NAPCC, the GOI established the Green India Mission in 2014 with the primary goal of mitigating the consequences of climate change by conserving, restoring, and upgrading India's declining forest coverage.

In an effort to preserve soil health, the GOI introduced the Soil Health Card program (Patel and Chauhan, 2012), whose primary goal is to evaluate cluster soil samples and advise farmers on the fertility of their land. Neem-coated urea was also used in order to reduce the overabundance of urea fertilizers, safeguarding the health of the soil and providing nitrogen for plants. Initiatives like the National Project on Organic Farming and the National Agroforestry Policy were implemented in 2004 and 2014, respectively, to provide farmers with greater financial benefits and environmental preservation.

These regulations are intended to protect soil from erosion loss, increase the amount of carbon stored in the soil, and provide plant nutrients in the form of organic amendments. Several initiatives have already been launched by states like Andhra Pradesh, Himachal Pradesh, Sikkim etc. to support and encourage organic agricultural methods on a larger scale. Sikkim was recently declared an organic state by the GOI. For the past 7 years, the Indian Council of Agricultural Research (ICAR) has been preparing agriculture for climate change in around 650 districts of India by executing agricultural contingency plans through its network research institutes, state agriculture institutions, and all line departments (Meek and Anderson, 2020).

The SAARC nations use these models to prepare for the effects of climate change, including heatwaves, cyclones, floods, and saltwater intrusion. In 151 districts around India, ICAR has created climate-resilient villages, which state governments are now replicating in an effort to create carbon-positive villages as a whole. Initiated in 2005, the Mahatma Gandhi National Rural Job Guarantee Act (MGNREGA) aims to conserve the environment and provide job possibilities while also offering economic stability (Reddy *et al.*, 2014).

11.14 Conclusion

This chapter has explored the essential role of stress-resilient crops in safeguarding agricultural productivity amid the escalating challenges posed by climate change. It has highlighted the integration of advanced breeding techniques, including genomic selection, marker-assisted selection, and CRISPR-Cas9 gene editing, as pivotal in developing crops capable of withstanding various abiotic stresses such as drought, salinity, and heat. The importance of a coordinated approach that combines traditional breeding methods with modern biotechnological tools was emphasized, along with the critical role of phenotyping, big data analytics, and predictive modeling in enhancing the efficiency and effectiveness of breeding programs. The development of stress-resilient crops carries profound implications for global agriculture and food security. As climate change continues to threaten the stability of food systems, the creation and deployment of crops that can endure extreme environmental conditions are crucial for maintaining and increasing crop yields. This, in turn, supports the livelihoods of farmers, reduces the risk of food shortages, and helps stabilize food prices, particularly in regions most vulnerable to climate impacts. By adopting these advanced breeding strategies, the agricultural sector can better adapt to changing climate, thereby contributing to the long-term sustainability of global food production. As we look to the future, the need for continuous innovation in crop breeding becomes ever more apparent. The strategies discussed in this chapter provide a foundation for developing crops that are not only resilient to current climate challenges but also adaptable to future environmental changes. Continued research, investment, and international collaboration will be key in refining these approaches and ensuring their widespread implementation. The future of agriculture depends on our ability to effectively harness these advanced breeding techniques to create crops that can thrive in an unpredictable and changing world, ultimately ensuring food security for generations to come.

References

Abdulmalik, R.O., Menkir, A., Meseka, S.K., Unachukwu, N., Ado, S.G. *et al.* (2017) Genetic gains in grain yield of a maize population improved through marker assisted recurrent selection under stress and non-stress conditions in West Africa. *Frontiers in Plant Science* 8, 841.

Acquaah, G. (2009) *Principles of Plant Genetics and Breeding*. John Wiley & Sons.

Acquaah, G. (2015) Conventional plant breeding principles and techniques. In: Al-Khayri, J.M., Jain, S.M. and Johnson, D.V. (eds) *Advances in Plant Breeding Strategies: Breeding, Biotechnology and Molecular Tools*. Springer, pp. 115–158.

Ahmad, S., Sheng, Z., Jalal, R.S., Tabassum, J., Ahmed, F.K. *et al.* (2021) CRISPR–Cas technology: Towards improvement of abiotic stress tolerance in plants. In: Abd-Elsalam, K.A. and Lim, K.-T. (eds) *CRISPR and RNAi Systems. Elsevier*. pp. 755–772.

Akdemir, D. and Isidro-Sánchez, J. (2019) Design of training populations for selective phenotyping in genomic prediction. *Scientific Reports* 9(1), 1446.

Akdemir, D., Sanchez, J.I. and Jannink, J.L. (2015) Optimization of genomic selection training populations with a genetic algorithm. *Genetics, Selection, Evolution* 47(1), 38.

Al-Khayri, J.M., Jain, S.M. and Johnson, D. (eds) (2018) *Advances in Plant Breeding Strategies: Fruits*. Springer, Cham, Switzerland, pp. 727–771.

Allard, J.L. (1961) The electric potential in the neighbourhood of a thin slit. Doctoral dissertation, University of British Columbia.

Amelework, B., Shimelis, H., Tongoona, P. and Laing, M. (2015) Physiological mechanisms of drought tolerance in sorghum, genetic basis and breeding methods: A review. *African Journal of Agricultural Research* 10(31), 3029–3040.

Andrade-Sanchez, P., Gore, M.A., Heun, J.T., Thorp, K.R., Carmo-Silva, A.E. *et al.* (2014) Development and evaluation of a field-based high-throughput phenotyping platform. *Functional Plant Biology* 41(1), 68.

Araus, J.L., Slafer, G.A., Royo, C. and Serret, M.D. (2008) Breeding for yield potential and stress adaptation in cereals. *Critical Reviews in Plant Sciences* 27(6), 377–412.

Araus, J.L., Kefauver, S.C., Zaman-Allah, M., Olsen, M.S. and Cairns, J.E. (2018) Translating high-throughput phenotyping into genetic gain. *Trends in Plant Science* 23(5), 451–466.

Ashikari, M., Sakakibara, H., Lin, S., Yamamoto, T., Takashi, T. *et al.* (2005) Cytokinin oxidase regulates rice grain production. *Science* 309(5735), 741–745.
Ashraf, M. and Foolad, M.R. (2013) Crop breeding for salt tolerance in the era of molecular markers and marker-assisted selection. *Plant Breeding* 132(1), 10–20.
Ayala, F., Fedrigo, G.V., Burachik, M. and Miranda, P.V. (2019) Compositional equivalence of event IND-øø412-7 to non-transgenic wheat. *Transgenic Research* 28(2), 165–176.
Bahadur, B., Rajam, M.V., Sahijram, L. and Krishnamurthy, K. (2015) *Plant Biology and Biotechnology*. Springer, New Delhi, pp. 361–383.
Bankole, F., Menkir, A., Olaoye, G., Crossa, J., Hearne, S. *et al.* (2017) Genetic gains in yield and yield related traits under drought stress and favorable environments in a maize population improved using marker assisted recurrent selection. *Frontiers in Plant Science* 8, 808.
Beale, S.I. (2005) Green genes gleaned. *Trends in Plant Science* 10(7), 309–312.
Begna, T. (2021) Conventional breeding methods widely used to improve self-pollinated crops. *International Journal of Research Studies in Agricultural Sciences* 7(1), 1–16.
Bernardo, R. (2008) Molecular markers and selection for complex traits in plants: Learning from the last 20 years. *Crop Science* 48(5), 1649–1664.
Bhat, J.A., Ali, S., Salgotra, R.K., Mir, Z.A., Dutta, S. *et al.* (2016) Genomic selection in the *era* of next generation sequencing for complex traits in plant breeding. *Frontiers in Genetics* 7, 221.
Bhat, M.A., Mir, R.A., Kumar, V., Shah, A.A., Zargar, S.M. *et al.* (2021) Mechanistic insights of CRISPR/Cas-mediated genome editing towards enhancing abiotic stress tolerance in plants. *Physiologia Plantarum* 172(2), 1255–1268.
Biradar, H., Karan, R., and Subudhi, P.K. (2018) Transgene pyramiding of salt responsive protein 3-1 (SaSRP3-1) and SaVHaC1 from *Spartina alterniflora* L. enhances salt tolerance in rice. *Frontiers in Plant Science* 9, 389319.
Bouzroud, S., Gasparini, K., Hu, G., Barbosa, M.A.M., Rosa, B.L. *et al.* (2020) Down regulation and loss of *auxin response factor 4* function using CRISPR/Cas9 alters plant growth, stomatal function and improves tomato tolerance to salinity and osmotic stress. *Genes* 11(3), 272.
Bravo-Almonacid, F.F. and Segretin, M.E. (2016) Status of transgenic crops in Argentina. *Plant Pathogen Resistance Biotechnology* 273–283.
Breseghello, F. and Coelho, A.S.G. (2013) Traditional and modern plant breeding methods with examples in rice (*Oryza sativa* L). *Journal of Agricultural and Food Chemistry* 61(35), 8277–8286.
Broman, K.W. and Sen, S. (2009) *A Guide to QTL Mapping with R/QTL*, Vol. 46. Springer, New York.
Brown, J. and Caligari, P. (2011) *An introduction to Plant Breeding*. John Wiley & Sons.
Cai, S., Jiang, G., Ye, N., Chu, Z., Xu, X. *et al.* (2015) A key ABA catabolic gene, OsABA8Ox3, is involved in drought stress resistance in rice. *PLoS One* 10(2), e0116646.
Chatterjee, J., Thakur, V., Nepomuceno, R., Coe, R.A., Dionora, J. *et al.* (2020) Natural diversity in stomatal features of cultivated and wild *Oryza* species. *Rice* 13(1), 58.
Choi, W.G., Swanson, S.J. and Gilroy, S. (2012) High-resolution imaging of ca^{2+}, redox status, ROS and ph using GFP biosensors. *The Plant Journal* 70(1), 118–128.
Choi, W.G., Miller, G., Wallace, I., Harper, J., Mittler, R. *et al.* (2017) Orchestrating rapid long-distance signaling in plants with ca^{2+}, ROS and electrical signals. *The Plant Journal* 90(4), 698–707.
Clee, S.M., Yandell, B.S., Schueler, K.M., Rabaglia, M.E., Richards, O.C. *et al.* (2006) Positional cloning of sorcs1, a type 2 diabetes quantitative trait locus. *Nature Genetics* 38(6), 688–693.
Collard, B.C.Y., Vera Cruz, C.M., McNally, K.L., Virk, P.S. and Mackill, D.J. (2008) Rice molecular breeding laboratories in the genomics era: Current status and future considerations. *International Journal of Plant Genomics* 2008(1), 524847.
Collinge, D.B., Lund, O.S. and Thordal-Christensen, H. (2008) What are the prospects for genetically engineered, disease resistant plants. In: Collinge, D.B., Munk, L. and Cooke, B.M. (eds) *Sustainable Disease Management in a European Context*. Springer, pp. 217–231.
Comas, L.H., Becker, S.R., Cruz, V.M.V., Byrne, P.F. and Dierig, D.A. (2013) Root traits contributing to plant productivity under drought. *Frontiers in Plant Science* 4, 442.
Cooper, M., Podlich, D.W. and Luo, L. (2007) Modeling QTL effects and MAS in plant breeding. In: Varshney, R.K. and Tuberosa, R. (eds) *Genomics-Assisted Crop Improvement*. Springer, Dordrecht, The Netherlands, pp. 57–95.
Crossa, J., Pérez-Rodríguez, P., Cuevas, J., Montesinos-López, O., Jarquín, D. *et al.* (2017) Genomic selection in plant breeding: Methods, models, and perspectives. *Trends in Plant Science* 22(11), 961–975.

Daszkowska-Golec, A., Skubacz, A., Marzec, M., Slota, M., Kurowska, M. *et al.* (2017) Mutation in *HvCBP20* (cap binding protein 20) adapts barley to drought stress at phenotypic and transcriptomic levels. *Frontiers in Plant Science* 8, 942.

Daszkowska-Golec, A., Karcz, J., Plociniczak, T., Sitko, K. and Szarejko, I. (2020) Cuticular waxes: A shield of barley mutant in CBP20 (Cap-binding protein 20) gene when struggling with drought stress. *Plant Science* 300, 110593.

de Oliveira, A.C.B., Maluf, W.R., Pinto, J.E.B.P. and Azevedo, S.M. (2003) Resistance to papaya ringspot virus in summer squash Cucurbita pepo L. introgressed from an interspecific *C. pepo* × *C. moschata*. *Euphytica* 132(2), 211–215.

Deshpande, T., Mukherji, R. and Sastry, M. (2025) Policy styles and india's national action plan on climate change (NAPCC). *Policy Studies* 46(1), 46–63.

de Toledo, P., Thomazella, D., Brail, Q., Dahlbeck, D. and Staskawicz, B. (2016) CRISPR-Cas9 mediated mutagenesis of a *DMR6* ortholog in tomato confers broad-spectrum disease resistance. *BioRxiv*.

Dhankher, O.P. and Foyer, C.H. (2018) Climate resilient crops for improving global food security and safety. *Plant, Cell & Environment* 41(5), 877–884.

Do, T.D., Vuong, T.D., Dunn, D., Clubb, M., Valliyodan, B. *et al.* (2019) Identification of new loci for salt tolerance in soybean by high-resolution genome-wide association mapping. *BMC Genomics* 20(1), 1–16.

Dong, C., Dalton-Morgan, J., Vincent, K. and Sharp, P. (2009) A modified tilling method for wheat breeding. *The Plant Genome* 2(1), 39–47.

Duan, J. and Cai, W. (2012) OsLEA3-2, an abiotic stress induced gene of rice plays a key role in salt and drought tolerance. *PLoS One* 7(9), e45117.

Dunn, J., Hunt, L., Afsharinafar, M., Meselmani, M.A., Mitchell, A. *et al.* (2019) Reduced stomatal density in bread wheat leads to increased water-use efficiency. *Journal of Experimental Botany* 70(18), 4737–4748.

Dwivedi, S.L., Sahrawat, K.L., Upadhyaya, H.D., Mengoni, A., Galardini, M. *et al.* (2015) Advances in host plant and rhizobium genomics to enhance symbiotic nitrogen fixation in grain legumes. *Advances in Agronomy* 129, 1–116.

El-Assal, S.E.D., Alonso-Blanco, C., Hanhart, C.J. and Koornneef, M. (2004) Pleiotropic effects of the *Arabidopsis* cryptochrome 2 allelic variation underlie fruit trait-related QTL. *Plant Biology* 6(4), 370–374.

Falconer, D.S. and Mackay, T.F.C. (1996) *Introduction to Quantitative Genetics*, 4th edn. Longman, London.

Fichman, Y. and Mittler, R. (2021) Integration of electric, calcium, reactive oxygen species and hydraulic signals during rapid systemic signaling in plants. *The Plant Journal* 107(1), 7–20.

Fisher, R.A. (1926) On the capillary forces in an ideal soil: Correction of formulae given by W.B Haines. *The Journal of Agricultural Science* 16(3), 492–505.

Fuchs, M. and Gonsalves, D. (2007) Safety of virus-resistant transgenic plants two decades after their introduction: Lessons from realistic field risk assessment studies. *Annual Review of Phytopathology* 45, 173–202.

Gianotto, A.C., Rocha, M.S., Cutri, L., Lopes, F.C., Dal'Acqua, W. *et al.* (2019) The insect-protected CTC91087-6 sugarcane event expresses cry1AC protein preferentially in leaves and presents compositional equivalence to conventional sugarcane. *GM Crops & Food* 10(4), 208–219.

Gjedrem, T. and Thodesen, J. (2005) Selection. In: Gjedrem, T. (ed.) *Selection and Breeding Programs in Aquaculture*. Springer, Dordrecht, The Netherlands, pp. 89–111.

Gorjanc, G., Cleveland, M.A., Houston, R.D. and Hickey, J.M. (2015) Potential of genotyping-by-sequencing for genomic selection in livestock populations. *Genetics, Selection, Evolution* 47(1), 12.

Gowda, M.V.C., Nadaf, H.L. and Sheshagiri, R. (1996) The role of mutations in intraspecific differentiation of groundnut (*Arachis hypogaea* L). *Euphytica* 90(1), 105–113.

Gupta, P.K., Kulwal, P.L. and Jaiswal, V. (2014) Association mapping in crop plants: Opportunities and challenges. *Advances in Genetics* 85, 109–147.

Habier, D., Fernando, R.L. and Dekkers, J.C. (2009) Genomic selection using low-density marker panels. *Genetics* 182(1), 343–353.

Habier, D., Tetens, J., Seefried, F.R., Lichtner, P. and Thaller, G. (2010) The impact of genetic relationship information on genomic breeding values in German Holstein cattle. *Genetics Selection Evolution* 42, 1–12.

Haley, C.S., Knott, S.A. and Elsen, J.M. (1994) Mapping quantitative trait loci in crosses between outbred lines using least squares. *Genetics* 136(3), 1195–1207.

Heffner, E.L., Sorrells, M.E. and Jannink, J.L. (2009) Genomic selection for crop improvement. *Crop Science* 49(1), 1–2.

Honsdorf, N., Loon, J., Govaerts, B. and Verhulst, N. (2022) Crop management for breeding trials. In: Reynolds, M.P. and Braun, H.-J. (eds) *Wheat Improvement: Food Security in a Changing Climate*. Springer, Cham, Switzerland, pp. 257–274.

Horikoshi, R.J., Ferrari, G., Dourado, P.M., Climaco, J.I., Vertuan, H.V. *et al.* (2022) MON 95379 bt maize as a new tool to manage sugarcane borer (*Diatraea saccharalis*) in South America. *Pest Management Science* 78(8), 3456–3466.

Hossain, S.M., Wong, B.K.Y. and Simpson, E.M. (2004) The dark phase improves genetic discrimination for some high throughput mouse behavioral phenotyping. *Genes, Brain, and Behavior* 3(3), 167–177.

Hussain, Z., Ali, S., Hayat, Z., Zia, M.A., Iqbal, A. *et al.* (2014) Agrobacterium-mediated transformation of 'DREB1A' gene for improved drought tolerance in rice cultivars (Oryza sativa L). *Australian Journal of Crop Science* 8(7), 1114–1123.

Ibrahim, A.K., Zhang, L., Niyitanga, S., Afzal, M.Z., Xu, Y. *et al.* (2020) Principles and approaches of association mapping in plant breeding. *Tropical Plant Biology* 13(3), 212–224.

James, D., Borphukan, B., Fartyal, D., Ram, B., Singh, J. *et al.* (2018) Concurrent overexpression of osgs1; 1 and osgs2 genes in transgenic rice (*Oryza sativa* L.): Impact on tolerance to abiotic stresses. *Frontiers in Plant Science* 9, 307492.

Jannink, J.L. (2010) Dynamics of long-term genomic selection. *Genetics, Selection, Evolution* 42(1), 35.

Jia, H., Zhang, Y., Orbović, V., Xu, J., White, F.F. *et al.* (2017) Genome editing of the disease susceptibility gene Cs Lob1 in citrus confers resistance to citrus canker. *Plant Biotechnology Journal* 15(7), 817–823.

Jiang, G.L. (2013) Molecular markers and marker-assisted breeding in plants. In: Andersen, S.B. (ed.) *Plant Breeding from Laboratories to Fields.* InTech Open. pp. 45–83.

Kathuria, H., Giri, J., Tyagi, H. and Tyagi, A.K. (2007) Advances in transgenic rice biotechnology. *Critical Reviews in Plant Sciences* 26(2), 65–103.

Kearsey, M.J. (1997) Genetic resources and plant breeding. *Biotechnology and Plant Genetic Resources* 175.

Kearsey, M.J. and Farquhar, A.G.L. (1998) QTL analysis in plants: Where are we now. *Heredity* 80(2), 137–142.

Kearsey, M.J. and Hyne, V. (1994) QTL analysis: A simple 'marker-regression' approach. *Theoretical and Applied Genetics. Theoretische Und Angewandte Genetik* 89(6), 698–702.

Key, S., Ma, J.K.-C. and Drake, P.M. (2008) Genetically modified plants and human health. *Journal of the Royal Society of Medicine* 101(6), 290–298.

Khademi, M., Varasteh-Shams, M., Nazarian-Firouzabadi, F. and Ismaili, A. (2020) New recombinant antimicrobial peptides confer resistance to fungal pathogens in tobacco plants. *Frontiers in Plant Science* 11, 1236.

Khalid, A., Zhang, Q., Yasir, M. and Li, F. (2017) Small RNA based genetic engineering for plant viral resistance: Application in crop protection. *Frontiers in Microbiology* 8, 43.

Kharkwal, M.C. and Nagar, J.P. (2005) A high yielding chickpea (*Cicer arietinum* L.) mutant variety for late sown condition in North Western plain zone of India. *The Indian Journal of Genetics and Plant Breeding* 65(3), 229–230.

Khush, G.S. (2000) Strategies for increasing the yield potential of rice. In: Sheehy, J.E., Mitchell, P.L. and Hardy, B. (eds) *Redesigning Rice Photosynthesis to Increase Yield*, Vol. 7. Studies in Plant Science, Elsevier, pp. 207–212.

Kleinhofs, A., Warner, R.L., Murhlbauer, F.J. and Nilan, R.A. (1978) Induction and selection of specific gene mutations in hordeum and pisum. *Mutation Research/Fundamental and Molecular Mechanisms of Mutagenesis* 51(1), 29–35.

Koch, A., Höfle, L., Werner, B.T., Imani, J., Schmidt, A. *et al.* (2019) SIGS vs HIGS: A study on the efficacy of two dsRNA delivery strategies to silence fusarium FgCYP51 genes in infected host and non-host plants. *Molecular Plant Pathology* 20(12), 1636–1644.

Kole, C., Muthamilarasan, M., Henry, R., Edwards, D., Sharma, R. *et al.* (2015) Application of genomics-assisted breeding for generation of climate resilient crops: Progress and prospects. *Frontiers in Plant Science* 6, 563.

Kumar, T., Uzma, Khan, M.R., Abbas, Z. and Ali, G.M. (2014) Genetic improvement of sugarcane for drought and salinity stress tolerance using Arabidopsis vacuolar pyrophosphatase (AVP1) gene. *Molecular Biotechnology* 56(3), 199–209.

Kumar, A., Bharti, B., Kumar, J., Bhatia, D., Singh, G.P. *et al.* (2020) Improving the efficiency of wheat breeding experiments using alpha lattice design over randomised complete block design. *Cereal Research Communications* 48(1), 95–101.

Lado, B., Matus, I., Rodríguez, A., Inostroza, L., Poland, J. *et al.* (2013) Increased genomic prediction accuracy in wheat breeding through spatial adjustment of field trial data. *G3: Genes, Genomes, Genetics* 3(12), 2105–2114.

Lema, M. (2018) Marker assisted selection in comparison to conventional plant breeding: Review article. *Agricultural Research & Technology* 14(2), 555914.

Lewandowska, D., Orr, J., Schreiber, M., Colas, I., Ramsay, L. *et al.* (2022) The proteome of developing barley anthers during meiotic prophase. *Journal of Experimental Botany* 73(5), 1464–1482.

Li, Y., Ruperao, P., Batley, J., Edwards, D., Khan, T. *et al.* (2018) Investigating drought tolerance in chickpea using genome-wide association mapping and genomic selection based on whole-genome resequencing data. *Frontiers in Plant Science* 9, 190.

Li, X., Ye, J., Munir, S., Yang, T., Chen, W. *et al.* (2019) Biosynthetic gene pyramiding leads to ascorbate accumulation with enhanced oxidative stress tolerance in tomato. *International Journal of Molecular Sciences* 20(7), 1558.

Lindberg, S. and Premkumar, A. (2023) Ion changes and signaling under salt stress in wheat and other important crops. *Plants* 13(1), 46.

Liu, X., Wu, D., Shan, T., Xu, S., Qin, R. *et al.* (2020) The trihelix transcription factor osgtγ-2 is involved adaption to salt stress in rice. *Plant Molecular Biology* 103(4–5), 545–560.

Liu, X., Xie, X., Zheng, C., Wei, L., Li, X. *et al.* (2022) RNAi-mediated suppression of the abscisic acid catabolism gene OsABA8ox1 increases abscisic acid content and tolerance to saline–alkaline stress in rice (*Oryza sativa* L). *The Crop Journal* 10(2), 354–367.

Long, S.P., Marshall-Colon, A. and Zhu, X.G. (2015) Meeting the global food demand of the future by engineering crop photosynthesis and yield potential. *Cell* 161(1), 56–66.

Lorenz, A.J. and Smith, K.P. (2015) Adding genetically distant individuals to training populations reduces genomic prediction accuracy in barley. *Crop Science* 55(6), 2657–2667.

Lorenz, K. and Cohen, B.A. (2012) Small- and large-effect quantitative trait locus interactions underlie variation in yeast sporulation efficiency. *Genetics* 192(3), 1123–1132.

Luo, H., Pandey, M.K., Khan, A.W., Wu, B., Guo, J. *et al.* (2019) Next-generation sequencing identified genomic region and diagnostic markers for resistance to bacterial wilt on chromosome B02 in peanut (*Arachis hypogaea* L). *Plant Biotechnology Journal* 17(12), 2356–2369.

Lynch, J.P. (2022) Harnessing root architecture to address global challenges. *The Plant Journal* 109(2), 415–431.

Lynch, M. and Walsh, B. (1998) *Genetics and analysis of quantitative traits*, Vol. 1. Sinauer, Sunderland, Massachusetts, pp. 535–557.

Mackay, T.F. (2004) The genetic architecture of quantitative traits: Lessons from drosophila. *Current Opinion in Genetics & Development* 14(3), 253–257.

Madhusudhana, R. (2019) Marker-assisted breeding in sorghum. In: Aruna, C., Visarada, K.B.R.S., Bhat, B.V. and Tonapi, V.A. (eds) *Breeding Sorghum for Diverse End Uses*. Woodhead Publishing, pp. 93–114.

Manavalan, L.P., Chen, X., Clarke, J., Salmeron, J. and Nguyen, H.T. (2012) RNAi-mediated disruption of squalene synthase improves drought tolerance and yield in rice. *Journal of Experimental Botany* 63(1), 163–175.

Maqbool, S., Hassan, M.A., Xia, X., York, L.M., Rasheed, A. *et al.* (2022) Root system architecture in cereals: Progress, challenges and perspective. *The Plant Journal* 110(1), 23–42.

Mashenkov, A. (1986) Induced mutation process as a source of new mutants. *Maize Genetics Cooperation Newsletter* 60, 70–71.

Masood, M.A., Farooq, K., Mujahid, Y. and Anwar, M.Z. (2008) Improvement in precision of agricultural field experiments through design and analysis. *Pakistan Journal of Life and Social Science* 6(2), 89–91.

Mathews, K.L. and Crossa, J. (2022) Experimental design for plant improvement. In: Reynolds, M.P. and Braun, H.-J. (eds) *Wheat Improvement: Food Security in a Changing Climate*. Springer, Cham, Switzerland, pp. 215–235.

Mba, C. (2013) Induced mutations unleash the potentials of plant genetic resources for food and agriculture. *Agronomy* 3(1), 200–231.

Mba, C., Guimaraes, E.P. and Ghosh, K. (2012) Re-orienting crop improvement for the changing climatic conditions of the 21st century. *Agriculture & Food Security* 1(1), 1–17.
McCough, S.R. and Doerge, R.W. (1995) QTL mapping in rice. *Trends in Genetics* 11(12), 482–487.
Meek, D. and Anderson, C.R. (2020) Scale and the politics of the organic transition in Sikkim, India. *Agroecology and Sustainable Food Systems* 44(5), 653–672.
Metzker, M.L. (2010) Sequencing technologies – the next generation. *Nature Reviews. Genetics* 11(1), 31–46.
Meuwissen, T.H.E. (2009) Accuracy of breeding values of 'unrelated' individuals predicted by dense SNP genotyping. *Genetics, Selection, Evolution* 41(1), 35.
Meuwissen, T.H.E., Hayes, B.J. and Goddard, M.E. (2001) Prediction of total genetic value using genome-wide dense marker maps. *Genetics* 157(4), 1819–1829.
Mittler, R., Zandalinas, S.I., Fichman, Y. and Van Breusegem, F. (2022) Reactive oxygen species signalling in plant stress responses. *Nature Reviews. Molecular Cell Biology* 23(10), 663–679.
Moose, S.P. and Mumm, R.H. (2008) Molecular plant breeding as the foundation for 21st century crop improvement. Plant Physiology 147(3), 969–977.
Muppala, S., Gudlavalleti, P.K., Malireddy, K.R., Puligundla, S.K., and Dasari, P. (2021) Development of stable transgenic maize plants tolerant for drought by manipulating ABA signaling through *Agrobacterium*-mediated transformation. *Journal, Genetic Engineering & Biotechnology* 19(1), 96.
Muranty, H., Denancé, C., Feugey, L., Crépin, J.-L., Barbier, Y. *et al.* (2020) Using whole-genome SNP data to reconstruct a large multi-generation pedigree in apple germplasm. *BMC Plant Biology* 20(1), 2.
Nadeem, M.A., Nawaz, M.A., Shahid, M.Q., Doğan, Y., Comertpay, G. *et al.* (2018) DNA molecular markers in plant breeding: Current status and recent advancements in genomic selection and genome editing. *Biotechnology & Biotechnological Equipment* 32(2), 261–285.
Neumann, K., Kobiljski, B., Denčić, S., Varshney, R.K. and Börner, A. (2011) Genome-wide association mapping: A case study in bread wheat (*Triticum aestivum* L.). *Molecular Breeding* 27(1), 37–58.
Novak, F.J. and Brunner, H. (1992) Plant breeding: Induced mutation technology for crop improvement. *IAEA Bull* 4, 25–32.
Nowara, D., Gay, A., Lacomme, C., Shaw, J., Ridout, C. *et al.* (2010) HIGS: Host-induced gene silencing in the obligate biotrophic fungal pathogen *Blumeria graminis*. *The Plant Cell* 22(9), 3130–3141.
Oliva, R., Ji, C., Atienza-Grande, G., Huguet-Tapia, J.C., Perez-Quintero, A. *et al.* (2019) Broad-spectrum resistance to bacterial blight in rice using genome editing. *Nature Biotechnology* 37(11), 1344–1350.
Patel, J.K. and Chauhan, N.B. (2012) Attitude of farmers towards soil health card (SHC) programme. *Asian Journal of Soil Science* 7(1), 114–116.
Pearce, S.C. (1971) Precision in block experiments. *Biometrika* 58(1), 161–167.
Peng, K., Tian, Y., Sun, X., Song, C., Ren, Z. *et al.* (2021) Tae-miR399-*UBC24* module enhances freezing tolerance in winter wheat via a CBF signaling pathway. *Journal of Agricultural and Food Chemistry* 69(45), 13398–13415.
Piepho, H.P., Büchse, A. and Emrich, K. (2003) A hitchhiker's guide to mixed models for randomized experiments. *Journal of Agronomy and Crop Science* 189(5), 310–322.
Pradhan, A., Naik, N. and Kumar Sahoo, K. (2015) RNAi mediated drought and salinity stress tolerance in plants. *American Journal of Plant Sciences* 6(12), 1990–2008.
Preece, A.D. (1990) DISPLAN: Designing a usable medical expert system. In: Berry, D. and Hart, A. (eds) *Expert Systems: Human Issues*. Chapman & Hall / MIT Press, Massachusetts.
Price, A.H. (2006) Believe it or not, QTLs are accurate. *Trends in Plant Science* 11(5), 213–216.
Rabier, C.E., Barre, P., Asp, T., Charmet, G. and Mangin, B. (2016) On the accuracy of genomic selection. *PLoS One* 11(6), e0156086.
Ragot, M., Sisco, P.H., Hoisington, D.A. and Stuber, C.W. (1995) Molecular-marker-mediated characterization of favorable exotic alleles at quantitative trait loci in maize. *Crop Science* 35(5), 1306–1315.
Ramirez-Gonzalez, R.H., Segovia, V., Bird, N., Fenwick, P., Holdgate, S. *et al.* (2015) RNA-seq bulked segregant analysis enables the identification of high-resolution genetic markers for breeding in hexaploid wheat. *Plant Biotechnology Journal* 13(5), 613–624.
Rasheed, A., Hao, Y., Xia, X., Khan, A., Xu, Y. *et al.* (2017) Crop breeding chips and genotyping platforms: Progress, challenges, and perspectives. *Molecular Plant* 10(8), 1047–1064.
Rashid, B., Tariq, M., Khalid, A., Shams, F., Ali, Q. *et al.* (2017) Crop improvement: New approaches and modern technique. *Plant Gene and Trait* 8.

Reddy, D.N., Reddy, A.A. and Bantilan, M.C.S. (2014) The impact of Mahatma Gandhi National Rural Employment Guarantee Act (MGNREGA) on rural labor markets and agriculture. *India Review* 13(3), 251–273.

Remington, D.L. and Purugganan, M.D. (2003) Candidate genes, quantitative trait loci, and functional trait evolution in plants. *International Journal of Plant Sciences* 164(S3), S7–S20.

Robertsen, C.D., Hjortshøj, R.L. and Janss, L.L. (2019) Genomic selection in cereal breeding. *Agronomy* 9(2), 95.

Roff, D.A. (2007) A centennial celebration for quantitative genetics. *Evolution* 61(5), 1017–1032.

Roychoudhry, S. and Kepinski, S. (2015) Shoot and root branch growth angle control: The wonderfulness of lateralness. *Current Opinion in Plant Biology* 23, 124–131.

Saraswat, P., Chaturvedi, A., Singh, A., Tyagi, P., and Ranajan, R. (2023) Genome engineering in plants via transposable elements. In: Pandita, D. and Pandita, A. (eds) *Plant Transposable Elements*. Apple Academic Press, pp. 313–332.

Savadi, S. (2018) Molecular regulation of seed development and strategies for engineering seed size in crop plants. *Plant Growth Regulation* 84(3), 401–422.

Scorza, R., Callahan, A., Dardick, C., Ravelonandro, M., Polak, J. *et al.* (2013) Genetic engineering of plum pox virus resistance: 'Honeysweet' plum: From concept to product. *Plant Cell, Tissue and Organ Culture (PCTOC)* 115(1), 1–12.

Semagn, K., Beyene, Y., Babu, R., Nair, S., Gowda, M. *et al.* (2015) Quantitative trait loci mapping and molecular breeding for developing stress resilient maize for sub-Saharan Africa. *Crop Science* 55(4), 1449–1459.

Sharma, D., Chhabra, R., Muthusamy, V., Zunjare, R.U. and Hossain, F. (2021a) Molecular characterization of elite maize (*Zea mays* L.) inbreds using markers associated with iron and zinc transporter genes. *Genetic Resources and Crop Evolution* 68(4), 1545–1556.

Sharma, D., Prajapat, R.K., Mathur, M., Upadhyay, T.K., Maloo, S.R. *et al.* (2021b) Molecular marker-assisted breeding for crop improvement. In: Kumar, P. and Thakur, A.K. (eds) *Crop Improvement: Biotechnological Advances*. CRC Press, pp. 125–138.

Shen, L., Sun, P., Bonnell, V.C., Edwards, K.J., Hetherington, A.M. *et al.* (2015) Measuring stress signaling responses of stomata in isolated epidermis of graminaceous species. *Frontiers in Plant Science* 6, 533.

Shieh, G. and Jan, S.L. (2004) The effectiveness of randomized complete block design. *Statistica Neerlandica* 58(1), 111–124.

Shu, J., Liu, Y., Li, Z., Zhang, L., Fang, Z. *et al.* (2016) A generic SSR marker closely linked to a dominant genic male sterility gene (DGMS79-399-3) in broccoli (*Brassica oleracea* var. italica). *Molecular Breeding* 36(7), 1–8.

Singh, V.K., Khan, A.W., Saxena, R.K., Kumar, V., Kale, S.M. *et al.* (2016) Next-generation sequencing for identification of candidate genes for fusarium wilt and sterility mosaic disease in pigeonpea (*Cajanus cajan*). *Plant Biotechnology Journal* 14(5), 1183–1194.

Singh, D., Wang, X., Kumar, U., Gao, L., Noor, M. *et al.* (2019) High-throughput phenotyping enabled genetic dissection of crop lodging in wheat. *Frontiers in Plant Science* 10, 394.

Singh, B.K., Delgado-Baquerizo, M., Egidi, E., Guirado, E., Leach, J.E. *et al.* (2023) Climate change impacts on plant pathogens, food security and paths forward. *Nature Reviews Microbiology* 21(10), 640–656.

Slade, A.J., Fuerstenberg, S.I., Loeffler, D., Steine, M.N. and Facciotti, D. (2005) A reverse genetic, non-transgenic approach to wheat crop improvement by tilling. *Nature Biotechnology* 23(1), 75–81.

Sozzani, R. and Benfey, P.N. (2011) High-throughput phenotyping of multicellular organisms: Finding the link between genotype and phenotype. *Genome Biology* 12(3), 219.

Spindel, J., Begum, H., Akdemir, D., Virk, P., Collard, B. *et al.* (2015) Genomic selection and association mapping in rice (*Oryza sativa*): Effect of trait genetic architecture, training population composition, marker number and statistical model on accuracy of rice genomic selection in elite, tropical rice breeding lines. *PLoS Genetics* 11(2), e1004982.

Stadler, L.J. (1928) Genetic effects of x-rays in maize. *Proceedings of the National Academy of Sciences* 14(1), 69–75.

St. Clair, D.A. (2010) Quantitative disease resistance and quantitative resistance loci in breeding. *Annual Review of Phytopathology* 48(1), 247–268.

Sturino, J., Zorych, I., Mallick, B., Pokusaeva, K., Chang, Y.-Y. *et al.* (2010) Statistical methods for comparative phenomics using high-throughput phenotype microarrays. *The International Journal of Biostatistics* 6(1).

Suvendu, M., Badigannavar, A.M., Kale, D.M. and Murty, G.S. (2007) Induction of genetic variability in a disease resistant groundnut breeding line. *BARC Newsletter* 237–247.

Sweeney, D.W., Sun, J., Taagen, E. and Sorrells, M.E. (2019) Genomic selection in wheat. In: Miedaner, T. and Korzun, V. (eds) *Applications of Genetic and Genomic Research in Cereals*. Woodhead Publishing, pp. 273–302.

Szechyńska-Hebda, M., Lewandowska, M., Witoń, D., Fichman, Y., Mittler, R. *et al.* (2022) Aboveground plant-to-plant electrical signaling mediates network acquired acclimation. *The Plant Cell* 34(8), 3047–3065.

Takagi, H., Abe, A., Yoshida, K., Kosugi, S., Natsume, S. *et al.* (2013) QTL-seq: Rapid mapping of quantitative trait loci in rice by whole genome resequencing of DNA from two bulked populations. *The Plant Journal* 74(1), 174–183.

Tardieu, F., Cabrera-Bosquet, L., Pridmore, T. and Bennett, M. (2017) Plant phenomics, from sensors to knowledge. *Current Biology* 27(15), R770–R783.

Tiwari, K., Tiwari, S., Kumar, N., Sinha, S., Krishnamurthy, S.L. *et al.* (2024) QTLs and genes for salt stress tolerance: A journey from seed to seed continued. *Plants* 13(8), 1099.

Tripathi, L., Dhugga, K.S., Ntui, V.O., Runo, S., Syombua, E.D. *et al.* (2022) Genome editing for sustainable agriculture in Africa. *Frontiers in Genome Editing* 4.

Tripathy, M., Mishra, B., Satapathy, G.P. and Das, S. (2023) Analysis of constraints affecting beneficiaries under pradhan mantri krishi sinchayee yojana in Odisha. *Indian Journal of Extension Education* 59(3), 153–155.

Uga, Y., Sugimoto, K., Ogawa, S., Rane, J., Ishitani, M. *et al.* (2013) Control of root system architecture by DEEPER ROOTING1 increases rice yield under drought conditions. *Nature Genetics* 45(9), 1097–1102.

Varshney, R.K., Saxena, R.K., Upadhyaya, H.D., Khan, A.W., Yu, Y. *et al.* (2017) Whole-genome resequencing of 292 pigeonpea accessions identifies genomic regions associated with domestication and agronomic traits. *Nature Genetics* 49(7), 1082–1088.

Varshney, R.K., Bohra, A., Yu, J., Graner, A., Zhang, Q. *et al.* (2021) Designing future crops: Genomics-assisted breeding comes of age. *Trends in Plant Science* 26(6), 631–649.

Verberne, M.C., Verpoorte, R., Bol, J.F., Mercado-Blanco, J. and Linthorst, H.J. (2000) Overproduction of salicylic acid in plants by bacterial transgenes enhances pathogen resistance. *Nature Biotechnology* 18(7), 779–783.

Wang, Y., Xue, Y. and Li, J. (2005) Towards molecular breeding and improvement of rice in China. *Trends in Plant Science* 10(12), 610–614.

Wang, C.T., Wang, X.Z., Tang, Y.Y., Chen, D.X., Zhang, J.C. *et al.* (2010) High yielding mutants achieved by injecting EMS into peanut flower organs. *Journal of Nuclear Agricultural Sciences* 24(2), 239–242.

Wang, C.T., Wang, X.Z., Tang, Y.Y., Zhang, J.C., Chen, D.X. *et al.* (2011) Huayu 40, a groundnut cultivar developed through EMS mutagenesis. *Journal of SAT Agricultural Research* 9.

Wang, Y., Cheng, X., Shan, Q., Zhang, Y., Liu, J. *et al.* (2014) Simultaneous editing of three homoeoalleles in hexaploid bread wheat confers heritable resistance to powdery mildew. *Nature Biotechnology* 32(9), 947–951.

Wang, J., Sun, G., Ren, X., Li, C., Liu, L. *et al.* (2016) QTL underlying some agronomic traits in barley detected by SNP markers. *BMC Genetics* 17(1), 103.

Wang, L., Zhu, G., Johnson, W. and Kher, M. (2018a) Three new approaches to genomic selection. *Plant Breeding* 137(5), 673–681.

Wang, W., Mauleon, R., Hu, Z., Chebotarov, D., Tai, S. *et al.* (2018b) Genomic variation in 3,010 diverse accessions of Asian cultivated rice. *Nature* 557(7703), 43–49.

Wang, T., Xun, H., Wang, W., Ding, X., Tian, H. *et al.* (2021) Mutation of gmAITR Genes by CRISPR/Cas9 genome editing results in enhanced salinity stress tolerance in soybean. *Frontiers in Plant Science* 12, 779598.

Wang, Y., Zafar, N., Ali, Q., Manghwar, H., Wang, G. *et al.* (2022) CRISPR/Cas genome editing technologies for plant improvement against biotic and abiotic stresses: Advances, limitations, and future perspectives. *Cells* 11(23), 3928.

Webber, G.D. (1995) *Insect-Resistant Crops through Genetic Engineering*. University of Missouri Extension.

Werner, B.T., Gaffar, F.Y., Schuemann, J., Biedenkopf, D., and Koch, A.M. (2020) RNA-spray-mediated silencing of *Fusarium graminearum* AGO and DCL genes improve barley disease resistance. *Frontiers in Plant Science* 11, 476.

Whittaker, J.C., Thompson, R., and Visscher, P.M. (1996) On the mapping of QTL by regression of phenotype on marker-type. *Heredity* 77(1), 23–32.

Wientjes, Y.C.J., Bijma, P., Calus, M.P.L., Zwaan, B.J. and Van den, J. (2020) The evolution of the genetic architecture of traits under genomic selection. In: *ICQG 6 Abstracts 2020 6th International Conference of Quantitative Genetics*, Brisbane, pp. 2–13.

Wong, A.C.S., Massel, K., Lam, Y., Hintzsche, J. and Chauhan, B.S. (2022) Biotechnological road map for innovative weed management. *Frontiers in Plant Science* 13, 887723.

Xu, Z., Xu, X., Gong, Q., Li, Z., Li, Y. *et al.* (2019) Engineering broad-spectrum bacterial blight resistance by simultaneously disrupting variable TALE-binding elements of multiple susceptibility genes in rice. *Molecular Plant* 12(11), 1434–1446.

Yano, M., Katayose, Y., Ashikari, M., Yamanouchi, U., Monna, L. *et al.* (2000) Hd1, a major photoperiod sensitivity quantitative trait locus in rice, is closely related to the *Arabidopsis* flowering time gene CONSTANS. *The Plant Cell* 12(12), 2473–2484.

Yi, Y., Hassan, M.A., Cheng, X., Li, Y., Liu, H. *et al.* (2023) QTL mapping and analysis for drought tolerance in rice by genome-wide association study. *Frontiers in Plant Science* 14, 1223782.

Yin, K., Han, T., Xie, K., Zhao, J., Song, J. *et al.* (2019) Engineer complete resistance to cotton leaf curl multan virus by the CRISPR/Cas9 system in *Nicotiana benthamiana*. *Phytopathology Research* 1(1), 1–9.

Young, N.D. (1996) QTL mapping and quantitative disease resistance in plants. *Annual Review of Phytopathology* 34(1), 479–501.

Yu, S., Liao, K.P., Shaw, S.Y., Gainer, V.S., Churchill, S.E. *et al.* (2015) Toward high-throughput phenotyping: Unbiased automated feature extraction and selection from knowledge sources. *Journal of the American Medical Informatics Association* 22(5), 993–1000.

Yu, Z., Wang, X., Mu, X. and Zhang, L. (2019) RNAi mediated silencing of dehydrin gene WZY2 confers osmotic stress intolerance in transgenic wheat. *Functional Plant Biology* 46(10), 877–884.

Zeng, Z.B. (1994) Precision mapping of quantitative trait loci. *Genetics* 136(4), 1457–1468.

Zhang, A., Liu, Y., Wang, F., Li, T., Chen, Z. *et al.* (2019) Enhanced rice salinity tolerance via CRISPR/Cas9-targeted mutagenesis of the OsRR22 gene. *Molecular Breeding* 39, 1–10.

Zhang, H., Zhu, J., Gong, Z. and Zhu, J.-K. (2022) Abiotic stress responses in plants. *Nature Reviews. Genetics* 23(2), 104–119.

Zhou, J., Peng, Z., Long, J., Sosso, D., Liu, B. *et al.* (2015) Gene targeting by the TAL effector pthXo2 reveals cryptic resistance gene for bacterial blight of rice. *The Plant Journal* 82(4), 632–643.

12 Integrated Omics and CRISPR Tools for Speed and Precision Genetic Engineering in Crops

Gunna Sureshbabu Suruthi[1], Muthukrishnan Arun[2], Tajudheen Nargis Begum[3], Arockiya Anita Margret[4], Arokiaraj Sherlin Rosita[4], Asirvatham Alwin Robert[5], Arumugam Vijaya Anand and Palanisamy Sampathkumar[6*]

[1]Department of Human Genetics and Molecular Biology, Bharathiar University, Tamil Nadu, India; [2]Department of Biotechnology, Bharathiar University, Tamil Nadu, India; [3]Department of Biotechnology, Jamal Mohamed College, Tamil Nadu, India; [4]Department of Biotechnology and Bioinformatics, Bishop Heber College, Tamil Nadu, India; [5]Department of Endocrinology and Diabetes, Prince Sultan Military Medical City, Ministry of Defense, Riyadh, Saudi Arabia; [6]Department of Chemistry and Biosciences, SASTRA Deemed to be University, Srinivasa Ramanujan Centre, Tamil Nadu, India

Abstract

The integration of omics technologies and CRISPR/Cas9 genetic engineering presents a powerful approach to accelerate and enhance precision in crop improvement. This chapter focusses on the convergence of genomics, transcriptomics, proteomics, and metabolomics with CRISPR tools, highlighting their combined potential to revolutionize agricultural practices. By leveraging high-throughput sequencing, advanced bioinformatics, and molecular biology techniques, researchers can gain comprehensive insights into plant pathways and networks, facilitating targeted genetic modifications. The fusion of omics datasets enables the identification of key biomarkers and intricate molecular interactions, while CRISPR/Cas technology allows for precise genome editing. This synergy enhances the speed and accuracy of developing crops with improved traits such as disease resistance, yield, and nutritional value. The chapter also discusses technological innovations, successful case studies, and the role of bioinformatics in integrating these approaches. Despite challenges and limitations, the prospects of combining omics and CRISPR tools are promising, offering significant advancements in crop genetic engineering.

Keywords: integrated omics, CRISPR/Cas9, crop improvement, genetic engineering, genomics, transcriptomics, proteomics, metabolomics

*Corresponding author: pskumar16@yahoo.com

© CAB International 2025. *Stress-Resilient Crops: Coordinated omics-CRISPR-Nanotechnology Strategies* (ed. Jen-Tsung Chen)
DOI: 10.1079/9781800627307.0012

12.1 Introduction to Integrated Omics

Integrated omics, which involves the convergence of different omics datasets like metabolomics, genomics, transcriptomics, and proteomics, has become an important method in modern biological investigation (Misra *et al.*, 2019). By amalgamating data from different omics levels, scientists can achieve a holistic insight into complex biological systems, recognize key biomarkers, and unravel intricate molecular interactions and pathways that underpin diseases (El Bouhaddani *et al.*, 2024). This holistic approach is particularly valuable in fields, where omics resources are pivotal in understanding molecular, biochemical, and environmental interactions (Doolan *et al.*, 2022).

Genomics encompasses the investigation of an organism's entire DNA sequence, including all its genes. This field has advanced rapidly with the outcome of high-throughput sequencing techniques like next-generation sequencing (NGS), enabling detailed investigation of genetic variations and their impact on health. Genomics provides the foundational data required for other omics disciplines, offering insights into gene function, regulation, and interaction within the biological system (Varshney *et al.*, 2017). Transcriptomics focusses on the study of RNA transcripts made by the genome, providing a snapshot of gene expression at a given time. Techniques such as RNA sequencing (RNA-seq) enable researchers to quantify and analyze the entire transcriptome, revealing how various circumstances affect the regulation and expression of genes. By integrating transcriptomic data with genomic information, scientists can better understand gene expression patterns and their roles in disease mechanisms (Wang *et al.*, 2009).

Proteomics involves the comprehensive study of proteins, the functional molecules within cells that execute various biological processes. Techniques like mass spectrometry (MS) and protein microarrays are commonly used to identify and quantify proteins, as well as to study their interactions and modifications. Proteomic data help in understanding the functional outcomes of gene expression and the dynamic changes in protein activity under different physiological states (Aebersold and Mann, 2016). Metabolomics encompasses the thorough analysis of metabolites, the small molecules that participate in metabolism. A direct functional readout of physiological state and cellular activity is provided by this field. Utilizing techniques like liquid chromatography-mass spectrometry (LC-MS) and nuclear magnetic resonance (NMR) spectroscopy, metabolic profiling aids in the understanding of metabolic pathways and how they alter in response to genetic and environmental adaptations (Nicholson and Lindon, 2008).

Combining data from multiple omics has been shown to significantly improve predictive accuracy in various fields (Fig. 12.1). By integrating biomarkers across different omics systems, researchers can improve their capacity to predict outcomes and understand disease progression (Gholami *et al.*, 2022). Furthermore, the integration of multi-omics data is essential for studying complex biological processes comprehensively, highlighting the interrelationships between biomolecules and their functions (Subramanian *et al.*, 2020). Machine learning as well as deep learning approaches have been increasingly employed in the integrated research of multi-omics data. These computational methods play an important part in deciphering the vast amounts of data generated from multi-omics studies and extracting meaningful insights (Chen *et al.*, 2023). Integrative research that evaluates both omics profiles simultaneously is key, as it allows for a more complete knowledge of biological systems and disease mechanisms (Weiskittel *et al.*, 2021).

12.2 CRISPR/Cas9 Techniques

Clustered Regularly Interspaced Short Palindromic Repeats associated protein/CRISPR-associated protein 9 (CRISPR/Cas9) technology has transformed genome editing by delivering a clear and effective tool for targeted modifications in various organisms (El Bouhaddani *et al.*, 2024). The clarity and efficiency of CRISPR/Cas9 have enabled its widespread adoption in different fields, including marine diatoms, plant biology, precision genome editing in plants and yeast, as well as and genome engineering in

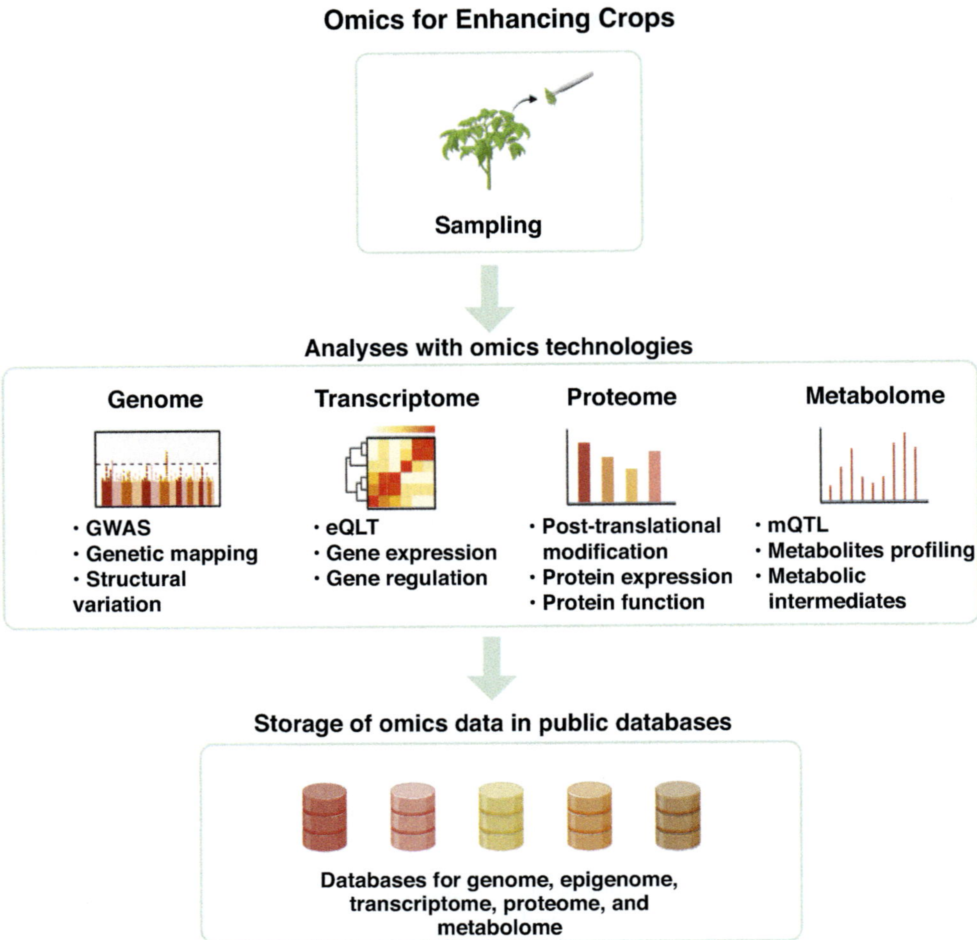

Fig. 12.1. Omics for enhancing crop breeding.

crops (Liu, 2023). This technology has been efficiently involved in diverse organisms, showcasing its versatility and effectiveness in genome editing (Razzaq *et al.*, 2021a).

CRISPR is a component of the immune system of bacteria, which protects against invading viruses (phages). When a bacterium is contaminated by a virus, it incorporates a segment of the viral DNA into its genome at the CRISPR locus. This segment, known as a spacer, serves as a genetic memory of the infection. If the bacterium encounters the same virus again, it transcribes the spacer into RNA, which leads the Cas9 nuclease to the corresponding viral DNA, enabling Cas9 to cleave and disable the virus (Barrangou *et al.*, 2007). Two essential elements of the CRISPR/Cas9 technique are a guide RNA (gRNA) and the Cas9 nuclease. The targeted gene's DNA sequence is matched by the gRNA. Upon binding to its corresponding DNA sequence, the gRNA guides the Cas9 enzyme to that exact region of the DNA. Next, Cas9 induces a double-strand break in the DNA (Fig. 12.2). This break is rectified by the cell's repair mechanisms, which either use homology-directed repair, which can be utilized to introduce precise genetic changes if a repair template is provided, or non-homologous endjoining, which can introduce small deletions or insertions that disrupt genes (Jinek *et al.*, 2012). This targeted approach enables precise modifications in the genetic code, allowing researchers to investigate

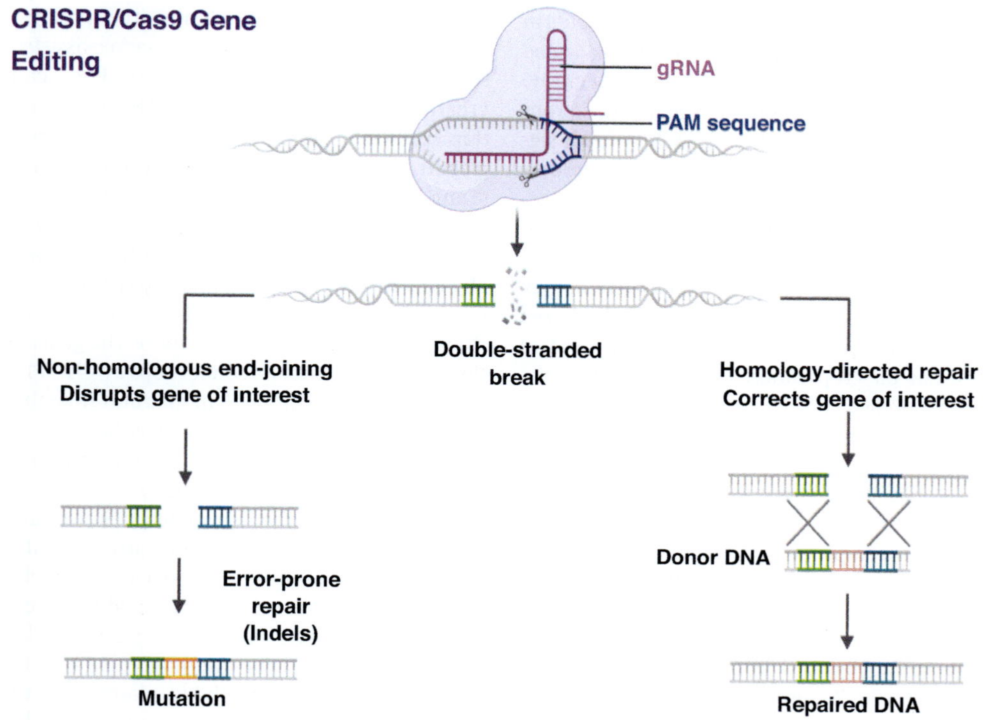

Fig. 12.2. CRISPR-Cas9 gene editing.

the functions of genes, create disease models, and potentially treat genetic disorders. The CRISPR/Cas9 approach's efficiency and versatility have established it as a valuable tool for genome manipulation across a diverse array of organisms, from bacteria to plants and animals. The technology's broad applications underscore its significance in advancing biological research and biotechnological innovations (Ansori *et al.*, 2023).

The capability to target specific DNA sequences with increased accuracy has made CRISPR/Cas9a valuable tool for generating genetically modified plants with desired traits (Peng *et al.*, 2020). Furthermore, the technology's flexibility permits multiplex genome editing, enabling the reconstruction of complex metabolic pathways in organisms like yeast (Hamdoun and Ehsan, 2017). The clarity and flexibility of CRISPR/Cas9 have facilitated the simultaneous editing of multiple genomic loci, expanding its applications in various research areas (Ma *et al.*, 2014). Despite its remarkable potential, CRISPR/Cas9 technology also presents obstacles like off-target effects and the need for continuous improvement in delivery methods and specificity. Evaluating nonspecific activity and enhancing the catalytic efficiency of Cas9 through protein evolution are areas of enduring research to address these challenges and improve the technology's precision. Additionally, the development of CRISPR/Cas9-based assays for nucleic acid detection and RNA strand displacement responsiveness demonstrates the continuous innovation and refinement of this genome-editing tool. These advancements aim to enhance the accuracy and controllability of CRISPR/Cas9 techniques for a comprehensive range of applications.

12.3 Combining Omics with CRISPR for Crop Improvement

The integration of omics approaches with CRISPR-associated genome-editing technology has become an effective strategy in crop development strategies. Recent advancements in CRISPR/Cas genome editing have transformed the field of agriculture by allowing accurate and efficient targeted changes in various crops (Ricroch *et al.*, 2017). This technology holds significant promise in accelerating crop improvement processes by enhancing traits like yield, quality, disease resistance, appearance, nutritional value, and stress tolerance (Gao and Chen, 2021). By leveraging CRISPR/Cas9 technology in combination with nanotechnology-based protocols, it is possible to develop transgenic-free solutions for crop enhancement, including the production of fertilizers, pesticides, and boosters to address global food scarcity issues (Naik *et al.*, 2022).

Genomics involves sequencing and analyzing the whole genome of a crop, with advances in high-throughput sequencing enabling the identification of genetic variations associated with having favorable characteristics including increased nutritional content, resilience to pests, and tolerance from drought (Varshney *et al.*, 2017). Transcriptomics examines gene expression levels under various conditions using RNA-seq to study profiles in response to environmental stresses or developmental stages, aiding in understanding regulatory mechanisms and identifying candidate genes for modification (Wang *et al.*, 2009). Proteomics concentrates on the investigation of proteins described by a crop, utilizing MS and protein microarrays to recognize and quantify proteins, study their interactions, and post-translational modifications, providing insights into gene expression's functional consequences and trait development mechanisms (Aebersold and Mann, 2016). Metabolomics comprehensively profiles metabolites produced during metabolism using techniques like NMR spectroscopy and LC-MS to analyze metabolic pathways and alterations, aiding in understanding the traits and identifying biomarkers for selection (Nicholson and Lindon, 2008).

The precise modification of crops using CRISPR technology, in conjunction with perfection agriculture approaches, offers a shift towards additionally sustainable and efficient farming practices, highlighting the enhancement of yield and environmental stress toleration in crops (Sampath *et al.*, 2023). Moreover, integrating CRISPR-Cas9 technology with traditional breeding methods and other progressive techniques, like molecular marker-assisted selection and double haploid breeding, can greatly advance crop breeding. This combination facilitates the development of new germplasms and enhances various desirable traits (Tang *et al.*, 2023). Using CRISPR systems in plant genome editing has revealed encouraging results in enhancing yield, bio-fortification, and abiotic and biotic stress tolerance in crops. Rice is one of the topmost researched crops in this area (Ricroch *et al.*, 2017). Furthermore, the generation of herbicide-invulnerable crops by CRISPR/Cas9-based gene editing demonstrates how this technique might speed up crop enhancement procedures by focussing on several features simultaneously (Dong *et al.*, 2021).

Additionally, growing cereal crops resistant to salt stress appears to be a promising use of the CRISPR/Cas genome-editing devices, thereby increasing agricultural resilience and positively impacting the environment and public health (Nazir *et al.*, 2022). The originality of CRISPR/Cas9 technology in modifying plant genomes offers new prospects for understanding the relationship between plants and pathogens, leading to improved crop protection strategies (Barakate and Stephens, 2016). The approach of CRISPR/Cas techniques has made new opportunities in plant improvement, with the possibility to revolutionize the editing of genomes in different crop species like rice, wheat, tomato, and maize (Maximiano and Franco, 2022). By harnessing CRISPR/Cas and similar genome-editing technologies, future agricultural needs can be met through precise and efficient genetic changes in crops and livestock (Bandyopadhyay *et al.*, 2020). The benefits of the CRISPR approach in enhancing crop productivity underscore its significance in achieving higher yields and ensuring food security (Sharma *et al.*, 2021).

12.4 Plant Pathways and Networks

The study of plant pathways and networks involves mapping and understanding the complex biological processes that control the growth and development of the plant, and responses to environmental triggers. This intricate web of interactions is crucial for optimizing crop traits including yield, stress tolerance, and nutritional value. The integration of advanced techniques like CRISPR-based editing and integrated omics approaches plays a key part in improving our understanding and manipulation of these pathways. Understanding the intricate biological pathways in plants involves the study of various molecular interactions and regulatory networks that govern physiological processes. These pathways include signal transduction cascades, metabolic networks, and gene regulatory networks, all of which play significant roles in the growth, development, and response of plants to their environmental stimuli.

Signal transduction pathways enable plants to perceive and respond to external signals such as light, hormones, and stress factors. These pathways involve a series of molecular events where a signal is transduced from the surface of the cell to the nucleus, leading to changes in gene expression. For example, the abscisic acid signaling pathway is crucial for drought response, regulating genes concerned with water conservation, and stress tolerance (Cutler *et al.*, 2010). Metabolic pathways encompass the biochemical reactions within cells that convert substrates into energy and biomolecules essential for growth and development. Central metabolic pathways like photosynthesis, glycolysis, and the tricarboxylic acid cycle are fundamental to plant energy metabolism. Secondary metabolic pathways, including the synthesis of flavonoids and alkaloids, are involved in plant defense and adaptation (Fernie and Stitt, 2012).

Gene regulatory networks consist of interactions between transcription factors and target genes, controlling the spatial and temporal expression of genes. These networks are pivotal in orchestrating developmental strategies and responses to environmental changes. For instance, the regulatory network involving WRKY transcription factors plays a noteworthy part in plant immune responses (Rushton *et al.*, 2010). By employing multi-omics technologies, researchers can achieve a thorough knowledge of the molecular mechanisms driving crop traits and the interactions within biological pathways (Zhang *et al.*, 2022). This integration allows for the prediction of complex traits in crops and the elucidation of regulatory networks essential for genetic modification (Yang *et al.*, 2021). Through the analysis of multi-omics datasets simultaneously, networks related to phenotypic traits can be constructed, incorporating features across various datasets to enhance crop traits (Acharjee *et al.*, 2016). By pinpointing candidate genes and pathways through multi-omics approaches, precision breeding can be achieved to enhance crop performance, especially at the beginning of changing climates (Anusha and Amir, 2024). Studying the functional networks of transcriptomes, proteomes, and metabolomes in response to stress conditions such as salinity can aid in identifying candidate genes concerned with salt-tolerance mechanisms in crops like grapevines (Daldoul *et al.*, 2014).

The integration of omics strategies suggests a revolutionary means to select crop varieties with desired traits, paving the way for more efficient and targeted breeding efforts (Mahmood *et al.*, 2022). Moreover, the benefit of CRISPR-Cas9-based multiplexed gene editing allows for the simultaneous modification of multiple genomic regions, enabling the regulation of various agronomic traits in crops such as wheat (Wang *et al.*, 2018). Moreover, by optimizing crop traits by omics integration, CRISPR-based genome editing plays a pivotal role in reshaping global agriculture by precisely targeting endogenous pathways or introducing new traits in plants (Begemann *et al.*, 2017). By using CRISPR-induced indels and base editing, researchers can achieve targeted modifications in plant genomes, facilitating both basic investigation and molecular crop breeding (Veillet *et al.*, 2020a). The CRISPR/Cas system has been instrumental in enhancing genetic diversity through directed evolution approaches, promising significant advancements in plant breeding and improvement (Ali *et al.*, 2023). Furthermore, the engineering of quantitative trait variation using genome-editing technologies presents opportunities to improve crop yields and address the increasing demand for food production (Rodríguez-Leal *et al.*, 2017). The capability to perform DNA-free editing of specific genes in crops like chickpea utilizing CRISPR/Cas9 in protoplasts showcases the potential for precise gene modifications in plant breeding (Badhan *et al.*, 2021).

12.5 Speed and Precision in Genetic Engineering

Genetic engineering has significantly impacted the field of biotechnology, delivering new opportunities for precise and rapid modifications to the genetic material of various organisms. CRISPR/Cas9 has evolved into an effective tool for genome engineering owing to its simplicity, design flexibility, and high efficiency (Pavese *et al.*, 2021). As an easier, more effective, and less expensive genetic modification technique than zinc finger nucleases and transcription activator-like effector nucleases, which rely on protein–DNA interactions for target explicitness, CRISPR systems use RNA–DNA interactions to direct DNA targeting and cleavage (Hassan *et al.*, 2021). The integration of CRISPR technology with precision agriculture has brought about a transformative shift in modern farming practices, aligning genetic engineering with data-driven agricultural approaches (Sampath *et al.*, 2023). This synergy between CRISPR and precision agriculture not only enhances crop improvement but also streamlines agricultural processes by leveraging omics data to identify target genes for precise editing (Sampath *et al.*, 2023). CRISPR/Cas9 has played an essential part in plant genetic engineering, providing a robust platform for the rapid and specific functional examination of candidate genes and mutations of interest (Niazian *et al.*, 2022).

When comparing the speed and precision of traditional genetic engineering methods with CRISPR/omics approaches, it is evident that CRISPR technology offers a significant advantage. Traditional methods often involve time-consuming and labor-intensive processes that may lack the precision and effectiveness of CRISPR-based techniques. CRISPR/Cas9 enables rapid and accurate genetic modifications, allowing for targeted edits with high explicitness and minimal off-target effects (Paudel *et al.*, 2021). This precision is crucial in applications such as agriculture, where the introduction of specific traits or modifications is essential for crop improvement (Sampath *et al.*, 2023). Historically, genetic modification depended on techniques such as chemical mutagenesis or transgenic approaches involving random insertion of foreign DNA. These methods lack precision, often resulting in unintended genomic alterations and lengthy screening processes to identify desired traits. CRISPR/Cas9 enables targeted modifications by guiding the Cas9 nuclease to specific DNA sequences complementary to a gRNA. This specificity minimizes off-target effects, enhancing precision in gene editing (Doudna and Charpentier, 2014). Moreover, the speed of CRISPR-mediated approaches is highlighted by their ability to enable the generation of animal models, functional genomic screening, and fast and accurate cell-line engineering (Cross *et al.*, 2016). The efficiency and versatility of CRISPR/Cas9 have been emphasized in various studies, underscoring its transformative potential for basic science, cellular engineering, and therapeutic applications (Mali *et al.*, 2013). By using the power of CRISPR technology, researchers can achieve precise and efficient genetic modifications that were previously challenging with traditional methods (Wei *et al.*, 2015).

12.6 Applications in Crop Traits Improvement

The integration of omics data with CRISPR/Cas9 technology has significantly impacted crop trait advancement by facilitating precise and targeted transformations in crop genomes to enhance various agronomic traits (Table 12.1). This synergy between multi-omics approaches and CRISPR/Cas9 has paved the way for precision breeding approaches that can address the challenges posed by changing climates and evolving agricultural needs. By pinpointing candidate genes and pathways through omics data, researchers can leverage CRISPR/Cas9 to edit these targets efficiently, leading to enhanced crop performance and resilience at the beginning of environmental stressors (Anusha and Amir, 2024). The versatility of CRISPR/Cas9 in enabling targeted edits in plant genomes has developed new opportunities for improving traits related to yield, nutritional quality, and consumer preference. This technology has been instrumental in revolutionizing horticultural food crops by facilitating rapid and accurate genetic modifications that cater to the specific needs of both producers and consumers (Wang

Table 12.1. Applications of CRISPR/Cas9 techniques in agriculture.

CRISPR/Cas9 application	Description	Examples in agriculture	Reference
Gene knockout	Inducing targeted mutations to deactivate genes	Knocking out genes reliable for vulnerability to diseases or pests	Zhang *et al.*, 2014
Gene knock-in	Introducing specific genetic sequences into target loci	Inserting genes for enhanced nutritional content or resistance traits	Li *et al.*, 2019
Gene editing for trait modification	Precise editing of DNA to alter traits like yield, quality, and stress tolerance	Modifying genes involved in photosynthesis or nutrient-uptake pathways	Liu *et al.*, 2021
Epigenome editing	Modifying gene expression without changing the sequence of DNA	Enhancing stress tolerance by altering epigenetic marks	Shin *et al.*, 2022
Base editing	Directly converting one base pair into another without inducing double-strand breaks	Correcting point mutations associated with disease resistance	Komor *et al.*, 2016
CRISPR interference (CRISPRi)	Temporarily repressing gene expression without changing the sequence of DNA	Suppressing genes involved in pathogen susceptibility during specific growth stages	Qi *et al.*, 2013

et al., 2021). This approach not only accelerates the breeding cycle but also allows for the development of novel variations that can contribute to the overall improvement of crop traits (Liu *et al.*, 2022).

In the context of leguminous crops, CRISPR/Cas9-mediated genome modification has appeared as an advanced breeding process that facilitates rapid gene manipulation and insertion, thereby revolutionizing crop improvement strategies. The efficiency and precision offered by CRISPR/Cas9 technology have enabled researchers to make targeted modifications in leguminous crop genomes, resulting in the development of enhanced types with enhanced traits. This approach highlights CRISPR/Cas9's revolutionary potential in promoting genetic improvement in crops (Roy and Sandhu, 2024). In addition, the benefit of CRISPR/Cas9 in crop quality improvement underscores its versatility in addressing various aspects of crop traits. By using CRISPR/Cas9 for targeted modifications, investigators can enhance the quality attributes of crops, including nutritional content, flavor profiles, and postharvest characteristics. This targeted approach to crop quality improvement showcases the precision and efficacy of CRISPR/Cas9 in driving advancements in agricultural biotechnology (Liu *et al.*, 2021).

The integration of omics data with CRISPR/Cas9 methods holds immense promise for understanding plant–pathogen interactions and enhancing crop defense mechanisms. By utilizing CRISPR-based tools to decipher omics data, investigators can gain an understanding of the molecular mechanisms underlying plant–pathogen interactions, resulting in the generation of enhanced crop protection strategies. This integrated approach highlights CRISPR/Cas9's potential to improve the knowledge of plant biology and enhance crop resilience against various abiotic and biotic stresses (Barakate and Stephens, 2016). In the realm of crop resilience, the viral dispersal of CRISPR/Cas9 elements has developed as a hopeful approach to rapidly improve crop traits and enhance resilience against abiotic and biotic stresses. Using viral vectors for targeted delivery, researchers can overcome traditional limitations associated with plant transformation and achieve precise changes in a wide spectrum of crop species. This innovative approach has demonstrated remarkable success in enhancing disease resistance, pest tolerance, and multiple trait improvements in diverse crops (Patil *et al.*, 2024). The efficiency of CRISPR/Cas9 in genome modification has been exemplified in the context of bread wheat, where DNA-free genome modification with CRISPR/Cas9 ribonucleoprotein complexes has paved the way for precise modifications in crop genomes. This approach not only streamlines the process of genome editing but also accelerates the development of new germplasm resources for improving important agronomic traits in crops. The DNA-free genome-editing strategy underscores the possibility of CRISPR/Cas9 in driving precision crop trait advancement (Liang *et al.*, 2017).

12.7 Technological Innovations and Tools

Genome editing has been transformed by breakthroughs in genetic engineering, especially with CRISPR/Cas systems. As researchers delve deeper into sequence databases, they are uncovering a variety of CRISPR variants, enhancing our ability to manipulate genetic material. This ongoing discovery process is steadily expanding the tools available for genome engineering (Koonin *et al.*, 2023). By identifying distinct protospacer adjacent motifs (PAMs), these novel CRISPR/Cas variants have improved editing efficiency and enabled more accurate targeting of particular genomic locations (Manghwar *et al.*, 2019). The creation of cytosine and adenine base editors has advanced genetic manipulation techniques. These tools allow researchers to make precise single-base changes in DNA within living cells and organisms. This capability provides a refined method for altering genetic information without causing larger DNA disruptions (Tang *et al.*, 2019).

To further demonstrate the ongoing development of CRISPR technology, scientists have created novel CRISPR base editor variations that maximize on-target effects while enabling accurate genomic editing (Collins and Beisel, 2020). The category and evolution of CRISPR-Cas systems have uncovered novel variants, such as Type VI systems, which specifically target RNA. This discovery has widened the range of genome-editing possibilities (Koonin *et al.*, 2017). Furthermore, the creation of Cas9 variants with flexible PAM distinction is essential for plant breeding and gene function study, underscoring the significance of customized CRISPR tools for specific applications (Veillet *et al.*, 2020b).

The evolution of CRISPR-Cas systems has resulted in the discovery of emanated variants linked to mobile genetic elements, offering new opportunities for developing versatile genome-editing devices (Makarova *et al.*, 2020). The advent of highly precise Cas9 variants, such as HypaCas9, highlights the importance of improving targeting accuracy while preserving elevated on-target activity, ensuring the accuracy of genetic alterations (Chen *et al.*, 2017). Furthermore, the comparison of newly created PAM-less CRISPR-SpRY systems with prominent CRISPR variants demonstrates the way genome-editing technologies are constantly evolving and improving editing capabilities (Riaz *et al.*, 2022). The incorporation of multi-omics platforms with CRISPR technologies has broadened the possibilities in genetic engineering. By characterizing unexplored structural variants from pangenomes, researchers can pinpoint unique target areas in variant-specific genes, thereby improving the efficiency of genome-editing techniques (Fernandez *et al.*, 2022).

12.8 Role of Bioinformatics in Integrating Omics and CRISPR

Bioinformatics plays a crucial role in integrating omics and CRISPR technologies to enhance crop improvement strategies. Omics studies produce extensive data that necessitate advanced data analysis and interpretation techniques (Liu *et al.*, 2023). Tools like Visual Omics offer comprehensive platforms for exploring and imagining various omics data, such as differential expression and protein–protein interaction analysis (Li *et al.*, 2023). These tools are crucial for understanding the complex biological processes underlying crop traits and their responses to genetic modifications. In CRISPR experiments, bioinformatics tools are essential for designing precise and efficient genetic-editing strategies while minimizing off-target impacts. These tools highlight the crucial role of bioinformatics in mitigating risks associated with CRISPR/Cas9-mediated genetic editing (Yuan, 2024). By leveraging bioinformatics technologies, researchers can improve the specificity and accuracy of CRISPR experiments, ensuring targeted genetic modifications in crops to enhance their traits.

Integrative databases and platforms are crucial for combining multi-omics and CRISPR data to achieve a comprehensive understanding of crop genetics and traits. Platforms like SoyOmics integrate diverse omics data, including genomes, transcriptomes, and phenotypic information, enabling thorough analyses (Liu *et al.*, 2023). Similarly, MOVICShiny provides interactive websites for multi-omics integration and visualization. These integrative tools are invaluable for researchers looking to leverage the integrated power of multi-omics and CRISPR technologies for crop improvement. Moreover, bioinformatics tools such as AIME and MSPL provide advanced data integration and feature selection methods for multi-omics analyses (Yu, 2022). These tools utilize deep learning algorithms and self-paced learning approaches to identify biological signatures and effectively integrate diverse omics data types (Yang *et al.*, 2019). By employing these cutting-edge technologies, researchers can streamline the integration of omics and CRISPR data, resulting in more precise and efficient crop improvement strategies.

12.9 Case Studies of Successful CRISPR and Omics Applications in Crops

The integration of CRISPR and omics technologies has markedly advanced crop improvement strategies, driving significant progress in genetic modifications across various crops. Landmark studies have showcased the versatility and efficacy of CRISPR-enabled genome editing in regulating transcription, modifying epigenomes, performing genome-wide screens, and imaging chromosomes (Barrangou and Doudna, 2016). This broad application of CRISPR technologies has facilitated rapid and precise genetic modifications in crops, improving their resilience to abiotic and biotic stresses (Patil *et al.*, 2024). Omics-established approaches have been crucial in advancing genetic improvements in crops, with successful applications across various species. Comprehensive omics analyses have provided beneficial understandings of the genetic makeup and physiological responses of crops, facilitating targeted genetic modifications to enhance traits and resilience. A notable success story in crop development is found in rice breeding, where omics-based approaches have played a necessary part in driving genetic improvements. The use of omics technologies in rice breeding has led to significant advancements in developing crops that are more resilient to environmental instability, representing a shift from the green revolution to the genome revolution (Chawla *et al.*, 2023). By leveraging omics data, investigators have achieved deeper learning of the genetic mechanisms underlying desirable traits in rice, allowing for precise genetic modifications that enrich crop resilience and productivity. Furthermore, the successful application of CRISPR/Cas9 technology in cucurbit crops has showcased its adaptability and effectiveness in enhancing desirable traits. The use of CRISPR/Cas9 in cucurbit crops marks a significant advancement in agricultural biotechnology, providing a powerful tool for precise genomic manipulation and targeted trait improvements (Xu *et al.*, 2024). This success highlights the transformative possibility of CRISPR technologies in advancing crop improvement efforts across a variety of plant species.

In vegetable crops, breeding for nutritional and organoleptic quality has been a key focus, with case studies in tomato and cauliflower illustrating the evolution of breeding methods for quality enhancements (Natalini *et al.*, 2021). By incorporating omics data and CRISPR technologies, researchers have targeted specific genes linked to nutritional and sensory attributes, resulting in the effect of improved varieties with enhanced quality traits (Natalini *et al.*, 2021). Additionally, the use of CRISPR/Cas9-refereed genome editing in leguminous crops like soybean, chickpea, lentil, and *Medicago truncatula* has demonstrated the technology's potential to revolutionize legume crop improvement. The successful genetic enhancements in these crops emphasize the versatility and efficacy of CRISPR/Cas9 methods in driving targeted trait improvements and enhancing crop resilience (Roy and Sandhu, 2024). The successful integration of CRISPR and omics technologies in crop development has paved the way for transformative advances in agriculture. By leveraging these cutting-edge tools, researchers have driven genetic improvements, enhanced crop resilience, and developed novel varieties with improved traits across a diverse array of crops.

12.10 Challenges and Limitations

Genetic engineering in crops has seen significant advancements with the emergence of tools like CRISPR/Cas and integrated omics technologies, offering precise and efficient methods for modifying crop genomes (Chen *et al.*, 2019). These technologies have the ability to revolutionize crop improvement by allowing targeted transformations that were previously challenging to achieve (Nerkar *et al.*, 2022). The ability to precisely edit crop genomes at specific sites has long been a goal of plant breeders, and genome-editing technologies now provide a means to achieve this with unprecedented accuracy (Avci and Sarahi, 2023). However, despite the promise these tools hold, some challenges and limitations need to be handled to completely harness their possibility for speed and precision genetic engineering in crops. The primary challenge facing the benefits of integrated omics and CRISPR tools in crop genetic engineering is the existing regulatory frameworks for genetically modified crops (Kuzma and Kokotovich, 2011). The current process-based regulations may not fully accommodate the intricacies of targeted genetic modifications, necessitating a reevaluation of regulatory policies to align with the capabilities of these advanced technologies. Additionally, the legal status of edited crops remains unclear, posing a barrier to the widespread adoption of genetically engineered plants (Scheben *et al.*, 2017).

Another limitation lies in the incomplete reference genomes and light knowledge of possible transformation targets in many crop species (Scheben *et al.*, 2017). Without a comprehensive understanding of the genetic modification of crops and suitable targets for modification, the full potential of integrated omics and CRISPR tools cannot be realized. Furthermore, the lack of specific understanding of functioning motifs in cis-regulatory elements inhibits the widespread application of editing these elements for crop refinement (Wolter and Puchta, 2018). One of the primary challenges lies in the technical complexity associated with integrating omics data with CRISPR/Cas9 technology. Each omics layer generates vast amounts of data, requiring sophisticated bioinformatics tools and expertise for meaningful interpretation and integration. The sheer volume and complexity of data make it challenging to identify precise targets for genetic editing. Moreover, the computational resources and algorithms needed to analyze multi-omics data are not always readily available or standardized, posing a significant barrier to researchers. While genome-editing technologies offer precise modifications, the efficiency of targeted modification in certain crops may still be a challenge (Chen *et al.*, 2019). The development of CRISPR/Cas9 technology has demonstrated promise in accelerating genetic advancement in perennial crops like sweet oranges (Tang *et al.*, 2021), but further optimization is needed to enhance gene-editing efficiency in a broader spectrum of crops. Additionally, the limited scope of genetic engineering, which currently focusses on modifying only a few genes, restricts the full potential of these technologies (Kan *et al.*, 2022).

In the realm of crop biotechnology, the protection of valuable crops from pests, diseases, and environmental stressors remains a critical objective (Abebe and Tafa, 2022). Genetic engineering plays an important position in developing biotic stress-resistant crops by expressing various protective proteins and utilizing genome-editing methods including CRISPR/Cas9 (Haroldsen *et al.*, 2012). The ability to pyramid multiple resistant genes utilizing marker-assisted preference and genetic engineering approaches has led to the development of crops with durable resistance to pathogens, insects, and herbicides, contributing to an increase in the production of the crops and global food safety (Mekonnen *et al.*, 2017). Moreover, the molecular characterization of genetic substances in plants is essential for molecular genetic engineering and controlled plant transformation (Heneen, 1985). Understanding the shape, expression, and regulation of genes in crops provides the foundation for leveraging tools like CRISPR/Cas and integrated omics approaches for precise genetic modifications. By unraveling the molecular mechanisms governing crop traits, researchers can tailor genetic engineering strategies to achieve desired outcomes in crop advancement. Despite the probable advantages of genetic engineering in crops, concerns about unintended health effects from genetically modified food crops persist (Lehrer and Bannon, 2005). The altered food supply resulting from genetic modifications raises apprehension about allergic reactions to biotech proteins in foods. Addressing these safety concerns and ensuring the regulatory management of genetically engineered crops are crucial steps in advancing the acceptance and adoption of genetically altered crops.

12.11 Future Prospects

The prospects of integrated omics and CRISPR tools for speed and precision genetic engineering in crops hold immense promise for revolutionizing agriculture and managing global food safety challenges. As advancements in omics technologies continue to unfold, the integration of multi-omics techniques is poised to play a pivotal role in identifying prospective genes and pathways for accurate breeding to improve crop performance, especially in the face of altering climates (Anusha and Amir, 2024). This integration will enable researchers to leverage a thorough knowledge of the molecular mechanisms controlling crop traits to tailor genetic engineering strategies for optimal outcomes in crop improvement (Williams *et al.*, 2024). CRISPR technology, with its precision and efficiency in genome editing, is arranged to transform the landscape of crop breeding by allowing immediate and prompt genetic modification of different crops in the field. The ability to carry out detailed and effective target modifications using CRISPR/Cas9 opens up unknown possibilities for developing unique plant variations through insertions, deletions, and substitutions, thereby accelerating the enhancement of crops like rice (Zegeye *et al.*, 2022). Furthermore, the incorporation of CRISPR-Cas strategy into plant biotechnology carries the promise of making crops with improved resistance to environmental stresses, increased productivity, improved nutritional quality, and improved quality traits, thereby revolutionizing agriculture (Vu *et al.*, 2023).

The future of CRISPR in exactitude agriculture not only promises improved crop types but also signifies a paradigm change towards more data-driven, customized, and environmentally cognizant farming procedures (Sampath *et al.*, 2023). By integrating pivotal molecular approaches like genomic selection, CRISPR-Cas9, and epigenetic engineering with the wealth of data from omics investigation, a network of systemic breeding methods can be developed to accelerate crop improvement programs (Williams *et al.*, 2024). This holistic approach to crop breeding, encompassing various omics technologies and CRISPR tools, will pave the way for the evolution of disease-resistant crops and sustainable agriculture practices (Shaw *et al.*, 2021). The potential of genome-editing technologies, such as CRISPR/Cas9, to make designer crops by presenting accurate and targeted changes in the genome holds significant promise for achieving global food security amidst climate adaptation and a growing population (Ahmad *et al.*, 2023). The integration of speed breeding with phenomic and genomic implements offers the opportunity for rapid gene identification and the acceleration of crop improvement programs (Razzaq *et al.*, 2021b). Moreover, the utilization of CRISPR-Cas9 to facilitate prompt genome-wide studies of gene function through the generation of large gRNA libraries presents a possibility for large-scale deployment of precision omics methods for the genetic engineering of seed vigour (Daniel, 2017).

In the kingdom of plant breeding, the integration of omics methods with NGS and the most delinquent results in genome-editing mechanisms has paved the way for targeted mutagenesis, meaning further prospects for specific genome engineering and the outcome of valuable crop traits (Razzaq *et al.*, 2019). The development of possible means for organelle genome editing in plants represents another crucial future direction in improving crop characteristics and resilience (Tian *et al.*, 2021). Additionally, the advancement of CRISPR-mediated tools offers new possibilities for comprehending plant–pathogen interactions and improving crop security through holistic approaches that transcend individual genes or pathogens (Barakate and Stephens, 2016).

12.12 Conclusion

Combining omics technologies with CRISPR/Cas9 genome modification represents a powerful synergy that is revolutionizing crop improvement. This combination offers unprecedented speed, precision, and efficiency in genetic engineering, addressing critical challenges in agriculture such as yield enhancement, stress tolerance, and nutritional improvement. The multi-omics approach provides a thorough comprehension of intricate biological systems in crops, from genomics to metabolomics. This wealth of data, when coupled with the precise genome-editing capabilities of CRISPR/Cas9, enables researchers to aim at specific genes and pathways with remarkable accuracy. The result is a more focussed and efficient crop improvement process, significantly reducing the amount of time and resources required for developing new varieties with desired traits.

While challenges remain, including off-target effects, delivery methods, technical complexities in data integration, and regulatory hurdles, the prospects of this integrated approach are highly promising. Developments in CRISPR technology, like base editing and prime editing, along with improved bioinformatics tools for analyzing omics data, are continually expanding the possibilities for crop enhancement. As we face global challenges like climate adaptation and food safety, the combination of integrated omics and CRISPR/Cas9 offers a powerful toolkit for developing resilient, high-yielding, and nutritionally enhanced crops. This approach accelerates the breeding cycle and opens up new avenues for creating non-transgenic crops with novel traits. As these technologies grow and mature, they indicate the possibility to transform agriculture, contributing significantly to sustainable food production and global food security in the light of population growth and climatic change.

Acknowledgement

The authors thank their respective universities and Institutions for their valuable support.

References

Abebe, A. and Tafa, Z. (2022) Review on: Impact of genetic engineering in biotic stresses resistance crop breeding. *International Journal for Research in Applied Sciences and Biotechnology* 9(2), 243–250.

Acharjee, A., Kloosterman, B., Visser, R.G. and Maliepaard, C. (2016) Integration of multi-omics data for prediction of phenotypic traits using random forest. *BMC Bioinformatics* 17(363), 373.

Aebersold, R. and Mann, M. (2016) Mass-spectrometric exploration of proteome structure and function. *Nature* 537(7620), 347–355.

Ahmad, A., Jamil, A. and Munawar, N. (2023) GMOs or non-GMOs? The CRISPR conundrum. *Frontiers in Plant Science* 14, 1232938.

Ali, A., Zafar, M.M., Farooq, Z., Ahmed, S.R., Ijaz, A. *et al.* (2023) Breakthrough in CRISPR/Cas system: Current and future directions and challenges. *Biotechnology Journal* 18(8), e2200642.

Ansori, A.N., Antonius, Y., Susilo, R.J., Hayaza, S., Kharisma, V.D. *et al.* (2023) Application of CRISPR-Cas9 genome editing technology in various fields: A review. *Narra Journal* 3(2), e184.

Anusha, M.R. and Amir, M. (2024) Integrating omics approaches for climate-resilient crops: A comprehensive review. *Journal of Advances in Biology & Biotechnology* 27(6), 351–363.

Avci, U. and Sarahi, H. (2023) Genome editing of gene families for crop improvement. In: Ricroch, A., Eriksson, D., Miladinovich, D., Sweet, J., Laere, K. *et al.* (eds) *A Roadmap for Plant Genome Editing*. Springer, Cham, Switzerland, pp. 3–15.

Badhan, S., Ball, A.S. and Mantri, N. (2021) First report of CRISPR/Cas9 mediated DNA-free editing of 4CL and RVE7 genes in chickpea protoplasts. *International Journal of Molecular Sciences* 22(1), 396.

Bandyopadhyay, A., Kancharla, N., Javalkote, V.S., Dasgupta, S. and Brutnell, T.P. (2020) CRISPR-Cas12a (Cpf1): A versatile tool in the plant genome editing tool box for agricultural advancement. *Frontiers in Plant Science* 11, 584151.

Barakate, A. and Stephens, J. (2016) An overview of CRISPR-based tools and their improvements: New opportunities in understanding plant–pathogen interactions for better crop protection. *Frontiers in Plant Science* 7, 765.

Barrangou, R. and Doudna, J.A. (2016) Applications of CRISPR technologies in research and beyond. *Nature Biotechnology* 34(9), 933–941.

Barrangou, R., Fremaux, C., Deveau, H., Richards, M., Boyaval, P. *et al.* (2007) CRISPR provides acquired resistance against viruses in prokaryotes. *Science* 315(5819), 1709–1712.

Begemann, M.B., Gray, B.N., January, E., Gordon, G.C., He, Y. *et al.* (2017) Precise insertion and guided editing of higher plant genomes using Cpf1 CRISPR nucleases. *Scientific Reports* 7(1), 11606.

Chawla, R., Poonia, A., Samantara, K., Mohapatra, S.R., Naik, S.B. *et al.* (2023) Green revolution to genome revolution: Driving better resilient crops against environmental instability. *Frontiers in Genetics* 14, 1204585.

Chen, J.S., Dagdas, Y.S., Kleinstiver, B.P., Welch, M.M., Sousa, A.A. *et al.* (2017) Enhanced proofreading governs CRISPR-Cas9 targeting accuracy. *Nature* 550(7676), 407–410.

Chen, K., Wang, Y., Zhang, R., Zhang, H. and Gao, C. (2019) CRISPR/Cas genome editing and precision plant breeding in agriculture. *Annual Review of Plant Biology* 70(1), 667–697.

Chen, C., Wang, J., Pan, D., Wang, X., Xu, Y. *et al.* (2023) Applications of multi-omics analysis in human diseases. *MedComm* 4(4), 315.

Collins, S.P. and Beisel, C.L. (2020) Your base editor might be flirting with single (Stranded) DNA: Faithful on-target CRISPR base editing without promiscuous deamination. *Molecular Cell* 79(5), 703–704.

Cross, B.C., Lawo, S., Archer, C.R., Hunt, J.R., Yarker, J.L. *et al.* (2016) Increasing the performance of pooled CRISPR-Cas9 drop-out screening. *Scientific Reports* 6(1), 31782.

Cutler, S.R., Rodriguez, P.L., Finkelstein, R.R. and Abrams, S.R. (2010) Abscisic acid: Emergence of a core signaling network. *Annual Review of Plant Biology* 61(1), 651–679.

Daldoul, S., Amar, A.B., Guillaumie, S. and Mliki, A. (2014) Integration of omics and system biology approaches to study grapevine (*Vitis vinifera* L.) response to salt stress: A perspective for functional genomics: A review. *Oeno One* 48(3), 189.

Daniel, I.O. (2017) Biology of seed vigor in the light of omics tools. *Advances in Seed Biology* 263–278.

Dong, H., Huang, Y. and Wang, K. (2021) The development of herbicide resistance crop plants using CRISPR/Cas9-mediated gene editing. *Genes* 12(6), 912.

Doolan, B.J., McGrath, J.A. and Onoufriadis, A. (2022) A clinician's guide to omics resources in dermatology. *Clinical and Experimental Dermatology* 47(5), 858–866.

Doudna, J.A. and Charpentier, E. (2014) The new frontier of genome engineering with CRISPR-Cas9. *Science* 346(6213), 1258096.

El Bouhaddani, S., Höllerhage, M., Uh, H.W., Moebius, C., Bickle, M. *et al.* (2024) Statistical integration of multi-omics and drug screening data from cell lines. *PLoS Computational Biology* 20(1), e1011809.

Fernandez, C.G.T., Nestor, B.J., Danilevicz, M.F., Marsh, J.I., Petereit, J. *et al.* (2022) Expanding gene-editing potential in crop improvement with pangenomes. *International Journal of Molecular Sciences* 23(4), 2276.

Fernie, A.R. and Stitt, M. (2012) On the discordance of metabolomics with proteomics and transcriptomics: Coping with increasing complexity in logic, chemistry, and network interactions scientific correspondence. *Plant Physiology* 158(3), 1139–1145.

Gao, C. and Chen, J. (2021) CRISPR adventures in China. *The CRISPR Journal* 4(3), 304–306.

Gholami, N., Haghparast, A., Alipourfard, I. and Nazari, M. (2022) Prostate cancer in omics era. *Cancer Cell International* 22(1), 274.

Hamdoun, Z. and Ehsan, H. (2017) Aftermath of the human genome project: An era of struggle and discovery. *Turkish Journal of Biology* 41(3), 403–418.

Haroldsen, V.M., Paulino, G., Chi-Ham, C.L. and Bennett, A.B. (2012) Research and adoption of biotechnology strategies could improve California fruit and nut crops. *California Agriculture* 66(2), 62–69.

Hassan, M.M., Zhang, Y., Yuan, G., De, K., Chen, J.G. *et al.* (2021) Construct design for CRISPR/Cas-based genome editing in plants. *Trends in Plant Science* 26(11), 1133–1152.

Heneen, W.K. (1985) Molecular biology and plant breeding. *Hereditas* 103, 109–128.

Jinek, M., Chylinski, K., Fonfara, I., Hauer, M., Doudna, J.A. *et al.* (2012) A programmable dual-RNA-guided DNA endonuclease in adaptive bacterial immunity. *Science* 337(6096), 816–821.

Kan, M., Huang, T. and Zhao, P. (2022) Artificial chromosome technology and its potential application in plants. *Frontiers in Plant Science* 13, 970943.

Komor, A.C., Kim, Y.B., Packer, M.S., Zuris, J.A. and Liu, D.R. (2016) Programmable editing of a target base in genomic DNA without double-stranded DNA cleavage. *Nature* 533(7603), 420–424.

Koonin, E.V., Makarova, K.S. and Zhang, F. (2017) Diversity, classification and evolution of CRISPR-Cas systems. *Current Opinion in Microbiology* 37, 67–78.

Koonin, E.V., Gootenberg, J.S. and Abudayyeh, O.O. (2023) Discovery of diverse CRISPR-Cas systems and expansion of the genome engineering toolbox. *Biochemistry* 62(24), 3465–3487.

Kuzma, J. and Kokotovich, A. (2011) Renegotiating GM crop regulation: Targeted gene-modification technology raises new issues for the oversight of genetically modified crops. *EMBO Reports* 12(9), 883–888.

Lehrer, S.B. and Bannon, G.A. (2005) Risks of allergic reactions to biotech proteins in foods: Perception and reality. *Allergy* 60(5), 559–564.

Li, S., Li, J., He, Y., Xu, M., Zhang, J. *et al.* (2019) Precise gene replacement in rice by RNA transcript-templated homologous recombination. *Nature Biotechnology* 37(4), 445–450.

Li, H., Shi, M., Ren, K., Zhang, L., Ye, W. *et al.* (2023) Visual omics: A web-based platform for omics data analysis and visualization with rich graph-tuning capabilities. *Bioinformatics* 39(1), btac777.

Liang, Z., Chen, K., Li, T., Zhang, Y., Wang, Y. *et al.* (2017) Efficient DNA-free genome editing of bread wheat using CRISPR/Cas9 ribonucleoprotein complexes. *Nature Communications* 8(1), 14261.

Liu, Y. (2023) Application and prospect of CRISPR/Cas9 technology in marine diatoms. *Theoretical and Natural Science* 4(1), 13–17.

Liu, Q., Yang, F., Zhang, J., Liu, H., Rahman, S. *et al.* (2021) Application of CRISPR/Cas9 in crop quality improvement. *International Journal of Molecular Sciences* 22(8), 4206.

Liu, H., Chen, W., Li, Y., Sun, L., Chai, Y. *et al.* (2022) CRISPR/Cas9 technology and its utility for crop improvement. *International Journal of Molecular Sciences* 23(18), 10442.

Liu, Y., Zhang, Y., Liu, X., Shen, Y., Tian, D. *et al.* (2023) SoyOmics: A deeply integrated database on soybean multi-omics. *Molecular Plant* 16(5), 794–797.

Ma, Y., Zhang, L. and Huang, X. (2014) Genome modification by CRISPR/Cas9. *The FEBS Journal* 281(23), 5186–5193.

Mahmood, U., Li, X., Fan, Y., Chang, W., Niu, Y. *et al.* (2022) Multi-omics revolution to promote plant breeding efficiency. *Frontiers in Plant Science* 13, 1062952.

Makarova, K.S., Wolf, Y.I., Iranzo, J., Shmakov, S.A., Alkhnbashi, O.S. *et al.* (2020) Evolutionary classification of CRISPR-Cas systems: A burst of class 2 and derived variants. *Nature Reviews Microbiology* 18(2), 67–83.

Mali, P., Esvelt, K.M. and Church, G.M. (2013) Cas9 as a versatile tool for engineering biology. *Nature Methods* 10(10), 957–963.

Manghwar, H., Lindsey, K., Zhang, X. and Jin, S. (2019) CRISPR/Cas system: Recent advances and future prospects for genome editing. *Trends in Plant Science* 24(12), 1102–1125.

Maximiano, M.R. and Franco, O.L. (2022) CRISPR/Cas: The new frontier in plant improvement. *ACS Agricultural Science & Technology* 2(2), 202–214.

Mekonnen, T., Haileselassie, T. and Tesfaye, K. (2017) Identification, mapping and pyramiding of genes/quantitative trait loci (qtls) for durable resistance of crops to biotic stresses. *Journal of Plant Pathology & Microbiology* 8(6).

Misra, B.B., Langefeld, C., Olivier, M. and Cox, L.A. (2019) Integrated omics: Tools, advances and future approaches. *Journal of Molecular Endocrinology* 62(1), R21–R45.

Naik, B.J., Shimoga, G., Kim, S.C., Manjulatha, M., Subramanyam Reddy, C. *et al.* (2022) CRISPR/Cas9 and nanotechnology pertinence in agricultural crop refinement. *Frontiers in Plant Science* 13, 843575.

Natalini, A., Acciarri, N. and Cardi, T. (2021) Breeding for nutritional and organoleptic quality in vegetable crops: The case of tomato and cauliflower. *Agriculture* 11(7), 606.

Nazir, R., Mandal, S., Mitra, S., Ghorai, M., Das, N. *et al.* (2022) Clustered regularly interspaced short palindromic repeats (CRISPR)/CRISPR-associated genome-editing toolkit to enhance salt stress tolerance in rice and wheat. *Physiologia Plantarum* 174(2).

Nerkar, G., Devarumath, S., Purankar, M., Kumar, A., Valarmathi, R. *et al.* (2022) Advances in crop breeding through precision genome editing. *Frontiers in Genetics* 13, 880195.

Niazian, M., Belzile, F. and Torkamaneh, D. (2022) CRISPR/Cas9 in planta hairy root transformation: A powerful platform for functional analysis of root traits in soybean. *Plants* 11(8), 1044.

Nicholson, J.K. and Lindon, J.C. (2008) Metabonomics. *Nature* 455(7216), 1054–1056.

Patil, A.M., Wagh, S.G., Janvale, G.B., Pawar, B.D. and Daspute, A.A. (2024) Viral delivery of CRISPR/Cas9 genome editing for rapid crop improvement: A promising approach to enhance crop resilience against biotic and abiotic stresses. *International Journal of Advanced Biochemistry Research* 8(3S), 782–796.

Paudel, N., Park, S., Jang, Y. and Rajametov, S.N. (2021) CRISPR/Cas9 trends and future scheme for genome engineering in plants.

Pavese, V., Moglia, A., Corredoira, E., Martínez, M.T., Torello Marinoni, D. *et al.* (2021) First report of CRISPR/Cas9 gene editing in castanea sativa mill. *Frontiers in Plant Science* 12, 728516.

Peng, H., Le, C., Wu, J., Li, X.F., Zhang, H. *et al.* (2020) A genome-editing nanomachine constructed with a clustered regularly interspaced short palindromic repeats system and activated by near-infrared illumination. *ACS Nano* 14(3), 2817–2826.

Qi, L.S., Larson, M.H., Gilbert, L.A., Doudna, J.A., Weissman, J.S. *et al.* (2013) Repurposing CRISPR as an RNA-guided platform for sequence-specific control of gene expression. *Cell* 152(5), 1173–1183.

Razzaq, A., Saleem, F., Kanwal, M., Mustafa, G., Yousaf, S. *et al.* (2019) Modern trends in plant genome editing: An inclusive review of the CRISPR/Cas9 toolbox. *International Journal of Molecular Sciences* 20(16), 4045.

Razzaq, A., Kaur, P., Akhter, N., Wani, S.H. and Saleem, F. (2021a) Next-generation breeding strategies for climate-ready crops. *Frontiers in Plant Science* 12, 620420.

Razzaq, M.K., Aleem, M., Mansoor, S., Khan, M.A., Rauf, S. *et al.* (2021b) Omics and CRISPR-Cas9 approaches for molecular insight, functional gene analysis, and stress tolerance development in crops. *International Journal of Molecular Sciences* 22(3), 1292.

Riaz, A., Kanwal, F., Ahmad, I., Ahmad, S., Farooq, A. *et al.* (2022) New hope for genome editing in cultivated grasses: CRISPR variants and application. *Frontiers in Genetics* 13, 866121.

Ricroch, A., Clairand, P. and Harwood, W. (2017) Use of CRISPR systems in plant genome editing: Toward new opportunities in agriculture. *Emerging Topics in Life Sciences* 1(2), 169–182.

Rodríguez-Leal, D., Lemmon, Z.H., Man, J., Bartlett, M.E. and Lippman, Z.B. (2017) Engineering quantitative trait variation for crop improvement by genome editing. *Cell* 171(2), 470–480.

Roy, A. and Sandhu, R. (2024) Advancements in genetic enhancement: CRISPR/Cas-mediated genome editing in leguminous crops. *Journal of Advances in Biology & Biotechnology* 27(6), 670–681.

Rushton, P.J., Somssich, I.E., Ringler, P. and Shen, Q.J. (2010) WRKY transcription factors. *Trends in Plant Science* 15(5), 247–258.

Sampath, V., Rangarajan, N., Sharanappa, C.H., Deori, M., Veeraragavan, M. *et al.* (2023) Advancing crop improvement through CRISPR technology in precision agriculture trends: A review. *International Journal of Environment and Climate Change* 13(11), 4683–4694.

Scheben, A., Wolter, F., Batley, J., Puchta, H. and Edwards, D. (2017) Towards CRISPR/Cas crops – bringing together genomics and genome editing. *New Phytologist* 216(3), 682–698.

Sharma, S.K., Gupta, O.P., Pathaw, N., Sharma, D., Maibam, A. *et al.* (2021) CRISPR-Cas-led revolution in diagnosis and management of emerging plant viruses: New avenues toward food and nutritional security. *Frontiers in Nutrition* 8, 751512.

Shaw, R.K., Shen, Y., Wang, J., Sheng, X., Zhao, Z. *et al.* (2021) Advances in multi-omics approaches for molecular breeding of black rot resistance in *Brassica oleracea* L. *Frontiers in Plant Science* 12, 742553.

Shin, H., Choi, W.L., Lim, J.Y. and Huh, J.H. (2022) Epigenome editing: Targeted manipulation of epigenetic modifications in plants. *Genes & Genomics* 1–9.

Subramanian, I., Verma, S., Kumar, S., Jere, A. and Anamika, K. (2020) Multi-omics data integration, interpretation, and its application. *Bioinformatics and Biology Insights* 14, 1177932219899051.

Tang, J., Lee, T. and Sun, T. (2019) Single-nucleotide editing: From principle, optimization to application. *Human Mutation* 40(12), 2171–2183.

Tang, X., Chen, S., Yu, H., Zheng, X., Zhang, F. *et al.* (2021) Development of a gRNA–tRNA array of CRISPR/Cas9 in combination with grafting technique to improve gene-editing efficiency of sweet orange. *Plant Cell Reports* 40(12), 2453–2456.

Tang, Q., Wang, X., Jin, X., Peng, J., Zhang, H. *et al.* (2023) CRISPR/Cas technology revolutionizes crop breeding. *Plants* 12(17), 3119.

Tian, Z., Wang, J.W., Li, J. and Han, B. (2021) Designing future crops: Challenges and strategies for sustainable agriculture. *The Plant Journal* 105(5), 1165–1178.

Varshney, R.K., Shi, C., Thudi, M., Mariac, C., Wallace, J. *et al.* (2017) Pearl millet genome sequence provides a resource to improve agronomic traits in arid environments. *Nature Biotechnology* 35(10), 969–976.

Veillet, F., Kermarrec, M.P., Chauvin, L., Chauvin, J.E. and Nogué, F. (2020a) CRISPR-induced indels and base editing using the staphylococcus aureus Cas9 in potato. *PLoS One* 15(8), e0235942.

Veillet, F., Perrot, L., Guyon-Debast, A., Kermarrec, M.P., Chauvin, L. *et al.* (2020b) Expanding the CRISPR toolbox in *P. Patens* using spcas9-NG variant and application for gene and base editing in *Solanaceae* crops. *International Journal of Molecular Sciences* 21(3), 1024.

Vu, B.N., Vu, T.V., Yoo, J.Y., Nguyen, N.T., Ko, K.S. *et al.* (2023) CRISPR-Cas-mediated unfolded protein response control for enhancing plant stress resistance. *Frontiers in Plant Science* 14, 1271368.

Wang, T., Zhang, C., Zhang, H. and Zhu, H. (2021) CRISPR/Cas9-mediated gene editing revolutionizes the improvement of horticulture food crops. *Journal of Agricultural and Food Chemistry* 69(45), 13260–13269.

Wang, Z., Gerstein, M. and Snyder, M. (2009) RNA-seq: A revolutionary tool for transcriptomics. *Nature Reviews. Genetics* 10(1), 57–63.

Wang, W., Pan, Q., He, F., Akhunova, A., Chao, S. *et al.* (2018) Transgenerational CRISPR-Cas9 activity facilitates multiplex gene editing in allopolyploid wheat. *The CRISPR Journal* 1(1), 65–74.

Wei, J., Wagner, S., Lu, D., Maclean, P., Carlson, D.F. *et al.* (2015) Efficient introgression of allelic variants by embryo-mediated editing of the bovine genome. *Scientific Reports* 5(1), 11735.

Weiskittel, T.M., Correia, C., Yu, G.T., Ung, C.Y., Kaufmann, S.H. *et al.* (2021) The trifecta of single-cell, systems-biology, and machine-learning approaches. *Genes* 12(7), 1098.

Williams, K., Subramani, M., Lofton, L.W., Penney, M., Todd, A. *et al.* (2024) Tools and techniques to accelerate crop breeding. *Plants* 13(11), 1520.

Wolter, F. and Puchta, H. (2018) Application of CRISPR/Cas to understand cis-and trans-regulatory elements in plants. *Plant Transcription Factors: Methods and Protocols* 23–40.

Xu, X., Yang, X. and Chen, X. (2024) Genomic advances in Cucurbitaceae: Implications for crop improvement and breeding. *International Journal of Horticulture* 14(3), 111–120.

Yang, Z.Y., Xia, L.Y., Zhang, H. and Liang, Y. (2019) MSPL: Multimodal self-paced learning for multi-omics feature selection and data integration. *IEEE Access* 7, 170513–170524.

Yang, Y., Saand, M.A., Huang, L., Abdelaal, W.B., Zhang, J. *et al.* (2021) Applications of multi-omics technologies for crop improvement. *Frontiers in Plant Science* 12, 563953.

Yu, T. (2022) AIME: Autoencoder-based integrative multi-omics data embedding that allows for confounder adjustments. *PLoS Computational Biology* 18(1), e1009826.

Yuan, S. (2024) Mitigating the off-target effects in CRISPR/Cas9-mediated genetic editing with bioinformatic technologies. *Transactions on Materials, Biotechnology and Life Sciences* 3, 318–326.

Zegeye, W.A., Tsegaw, M., Zhang, Y. and Cao, L. (2022) CRISPR-based genome editing: Advancements and opportunities for rice improvement. *International Journal of Molecular Sciences* 23(8), 4454.

Zhang, H., Zhang, J., Wei, P., Zhang, B., Gou, F. *et al.* (2014) The CRISPR/Cas9 system produces specific and homozygous targeted gene editing in rice in one generation. *Plant Biotechnology Journal* 12(6), 797–807.

Zhang, R., Zhang, C., Yu, C., Dong, J. and Hu, J. (2022) Integration of multi-omics technologies for crop improvement: Status and prospects. *Frontiers in Bioinformatics* 2, 1027457.

13 Reprogramming Epigenomes by CRISPR-based Technology against Plant Diseases

Gunna Sureshbabu Suruthi[1], Ahamed Kabeer Umera Begam[2*], Muthukrishnan Arun[3], Asirvatham Alwin Robert[4], Palanisamy Sampathkumar[5], Pemula Gowtham[1], Boro Arthi[1] and Arumugam Vijaya Anand[1]

[1]Department of Human Genetics and Molecular Biology, Bharathiar University, Coimbatore, Tamil Nadu, India; [2]Department of Biochemistry, Holy Cross College Affiliated to Bharathidasan University, Tiruchirappalli, Tamil Nadu, India; [3]Department of Biotechnology, Bharathiar University, Coimbatore, Tamil Nadu, India; [4]Department of Endocrinology and Diabetes, Prince Sultan Military Medical City, Ministry of Defense, Riyadh, Saudi Arabia; [5]Department of Chemistry and Biosciences, SASTRA Deemed to be University, Srinivasa Ramanujan Centre, Kumbakonam, Tamil Nadu, India

Abstract

Epigenomics, the study of epigenetic modifications across the genome, has become crucial in understanding plant responses to diseases. This chapter explores the innovative use of CRISPR-based technology for reprogramming epigenomes to combat plant diseases. It explores the basic principles of epigenomics, emphasizing how DNA methylation, histone modifications, and small noncoding RNAs are essential components of plant defense mechanisms. The integration of CRISPR/Cas systems for targeted epigenome editing is discussed, emphasizing its potential to enhance plant resilience against abiotic and biotic pressures. The article reviews various mechanisms of epigenetic regulation, including the pivotal role of chromatin remodeling in plant immunity. It also presents case studies demonstrating successful CRISPR-based epigenome reprogramming for disease resistance. Additionally, it addresses the challenges and limitations of this technology, providing insights into prospects for advancing crop improvement and sustainable agriculture through precise epigenetic modifications.

Keywords: CRISPR-Cas, epigenome reprogramming, chromatin remodeling, disease resistance, crop improvement, DNA methylation

*Corresponding author: umera150176@gmail.com

© CAB International 2025. *Stress-Resilient Crops: Coordinated omics-CRISPR-Nanotechnology Strategies* (ed. Jen-Tsung Chen)

DOI: 10.1079/9781800627307.0013

13.1 Introduction to Epigenomics and Plant Diseases

Epigenomics, the investigation of epigenetic changes on a genome-wide scale, has appeared as a potent tool for understanding plant responses to diseases, encompassing both abiotic and biotic strains (Jain *et al.*, 2024). The integration of epigenetics with genomics, forming epigenomics, has delivered an understanding of the molecular basis underlying plant disease resistance (Shaw *et al.*, 2022). Epigenetic processes like DNA methylation, histone modifications, and chromatin remodeling are crucial for controlling plant defense responses and can offer new avenues for enhancing crop disease resistance (Zhi and Chang, 2021). DNA methylation concerns adding a methyl group to the DNA, generally at cytosine in the context of CpG dinucleotides. In general, DNA methylation suppresses the transcription of genes. In plants, DNA methylation is necessary for regulating genes involved in pathogen defense (Dowen *et al.*, 2012). Proteins called histones are encircled by DNA to create chromatin. Post-translational changes of histones, like acetylation, methylation, phosphorylation, and ubiquitination, can either promote or inhibit the transcription of genes. For instance, histone acetylation is often related to gene activation, while methylation can trigger or suppress transcription depending on the context (Boyko and Kovalchuk, 2008). Small noncoding RNAs, including small interfering RNAs and microRNAs, can direct the modification of chromatin structure or directly degrade mRNA, thereby regulating gene expression. In plant defense, RNAi can target viral RNAs and prevent their replication (Luna *et al.*, 2012).

Epigenetic regulation also extends to fungal plant pathogens, where mechanisms like DNA methylation and modification of histones control the growth and pathogenicity of fungi like *Magnaporthe oryzae* (Dubey and Jeon, 2017). Histone acetylation, a critical epigenetic modification, has been recognized as a key regulator of plant immunity, with studies demonstrating how fungal effectors can suppress rice immunity by generating host histone hypoacetylation (Chen *et al.*, 2022). While the role of epigenetic principles in plant disease opposition against typical pathogens has been explored, its impact on the arrangement of the root microbiota remains an area requiring further investigation (Masenya *et al.*, 2024). DNA methylation mechanisms in plants have been extensively studied, with discussions on their underlying mechanisms and potential applications in breeding for improved plant traits (Shaikh *et al.*, 2022). Plant pathogens deploy effector proteins to modulate host immunity, showcasing the intricate interplay between pathogen effectors and host epigenetic responses (Soyer *et al.*, 2014). Epigenetic factors like DNA methylation and histone modifications have been recognized for their roles in modulating plant biotic interactions, memory, and priming (Alonso *et al.*, 2019). Epigenomic features have been instrumental in understanding molecular functions and differentiating germplasm in plants, offering valuable insights for crop improvement and breeding programs (Orantes-Bonilla *et al.*, 2023). Multi-omics technologies have been pivotal in elucidating the associations between crop genomes and phenotypes in different environmental circumstances, aiding in crop advancement efforts (Yang *et al.*, 2021). Plant defense opposed to fungal pathogens involves the modulation of chromatin accessibility for defense gene expression through epigenetic mechanisms, highlighting the plant's use of epigenetic weapons in combating pathogens (Mierziak and Wojtasik, 2024). Retrotransposons have been repurposed to regulate plant immune system genes, showcasing a notable mode of action through which transposable elements can influence plant immunity (McDowell and Meyers, 2013).

13.2 CRISPR-Cas: Overview

CRISPR-Cas methods, which consist of Clustered Regularly Interspaced Short Palindromic Repeats (CRISPR) and CRISPR-associated (Cas) proteins, have appeared as useful tools in genetic engineering and genome editing (Xu and Li, 2020). These prokaryotic adaptive immune systems offer protection against invasive nucleic acids, including viruses and plasmids, by identifying, binding, and cleaving them with high specificity (Dong *et al.*, 2023). The CRISPR-Cas method has found widespread applications in different fields, such as agriculture, human disease research, and biotechnology, due to its clarity, cost-effectiveness, and versatility (Xu and Li, 2020). In agriculture, the CRISPR-Cas approach has significantly influenced crop improvement by enhancing yield, resistance to diseases, and toleration of stress in crops like rice, maize, soybean etc. Researchers have utilized CRISPR-Cas to present complex genetic transformations, like gene replacements, insertions, and deletions, in crops, directing the development of novel germplasms with desirable traits (Ntsomboh-Ntsefong *et al.*, 2023). The benefit of CRISPR-Cas in plants has seen a significant increase in recent years, with successful genome editing achieved in various dicot and monocot species for both fundamental research and crop enhancement (Hassan *et al.*, 2021).

The versatility of CRISPR-Cas approaches extends beyond agriculture and human health, with applications in microbial viability, antimicrobial development, and microbiome composition alteration. These systems have been utilized as potent and programmable antimicrobials, facilitating the selective removal of bacterial pathogens and precise manipulation of microbiomes (Bikard and Barrangou, 2017). The adaptability of CRISPR-Cas strategies in diverse fields underscores its importance as a transformative tool with broad implications for genetic engineering and biological research. Moreover, the evolution and ecology of CRISPR-Cas methods have been extensively studied, highlighting their role in providing microbes with adaptive immunity against foreign DNA through the use of concise sequences comprehended as spacers to guide Cas proteins in cleaving invasive genetic material (Westra *et al.*, 2016). The ongoing exploration of CRISPR-Cas methods from uncultivated microbes has revealed new CRISPR-Cas variants, expanding the range of tools available for genome engineering and biotechnological applications (Burstein *et al.*, 2017).

13.3 CRISPR-based Epigenome Editing

CRISPR-associated epigenome editing in plants is a cutting-edge approach with the potential to manipulate plant traits through precise modifications at the epigenetic level (Ghoshal *et al.*, 2020). This technology enables stable and highly site-specific histone modifications, allowing for the targeted modulation of gene expression. By utilizing tools like CRISPR/Cas9, researchers can achieve locus-specific exploitation of DNA methylation, paving the way for epigenetic-associated molecular breeding in significant crop plants (Gahlaut *et al.*, 2020). The benefits of CRISPR-based epigenome alteration in plants propose a hopeful avenue for improving plant resilience to various stresses, including salt stress (Chaudhry *et al.*, 2022). This technology can help protect plants from the harmful consequences of environmental stressors, ultimately improving their overall health and productivity (Fig. 13.1). Pathogens often manipulate host gene expression to facilitate infection. By using CRISPR-based epigenome editing, researchers can modify the expression of genes implicated in plant immune responses. For instance, targeting genes related to systemic acquired resistance can improve the plant's ability to fend off different pathogens (Luna *et al.*, 2012). Manipulating the expression of genes implicated in development can lead to improved yield and quality of nutrition in crops. By targeting regulatory elements of key metabolic pathways, CRISPR-based epigenome editing can increase the production of essential nutrients or enhance biomass production, contributing to food safety and agricultural sustainability (Zhang *et al.*, 2018).

The versatility of CRISPR-associated genome-altering tools extends beyond traditional gene editing, encompassing epigenetic changes that can impact gene expression and plant phenotype (Thakore *et al.*, 2015). By combining CRISPR/deadCas9 with epigenetic-state-altering enzymes, researchers can target genes or cis-regulatory factors to regulate plant gene expression, providing a novel approach to fine-tuning plant traits (Giudice *et al.*, 2021). This approach shows significant promise for advancing plant biotechnology and crop improvement efforts by offering a means to precisely regulate gene expression at the epigenetic level (Fig. 13.2). The integration of CRISPR-associated epigenome editing with other omics techniques has further enhanced its utility in plant research and breeding programs (Wang *et al.*, 2023). By merging CRISPR/Cas9-mediated epigenetic modifications with multi-omics techniques, researchers can gain a

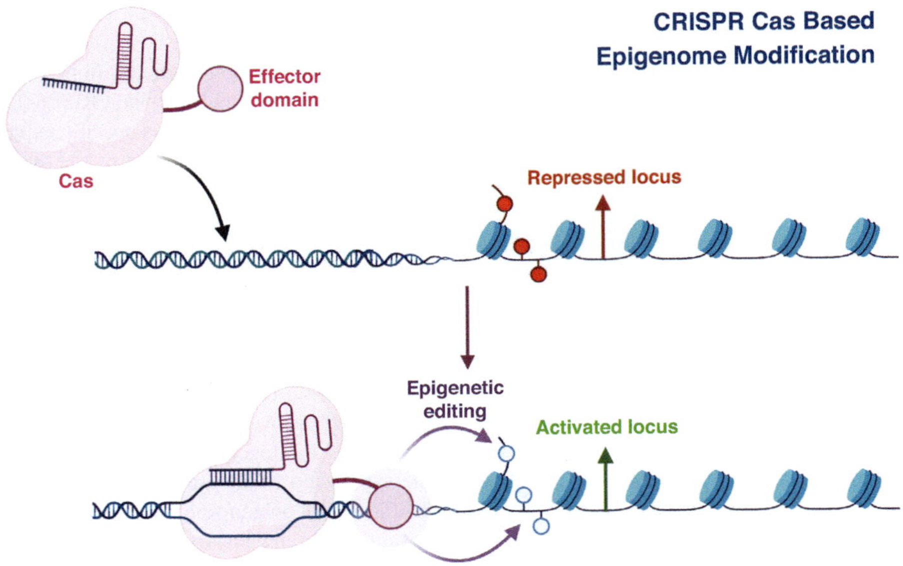

Fig. 13.1. Applications of CRISPR-Cas in crop improvement.

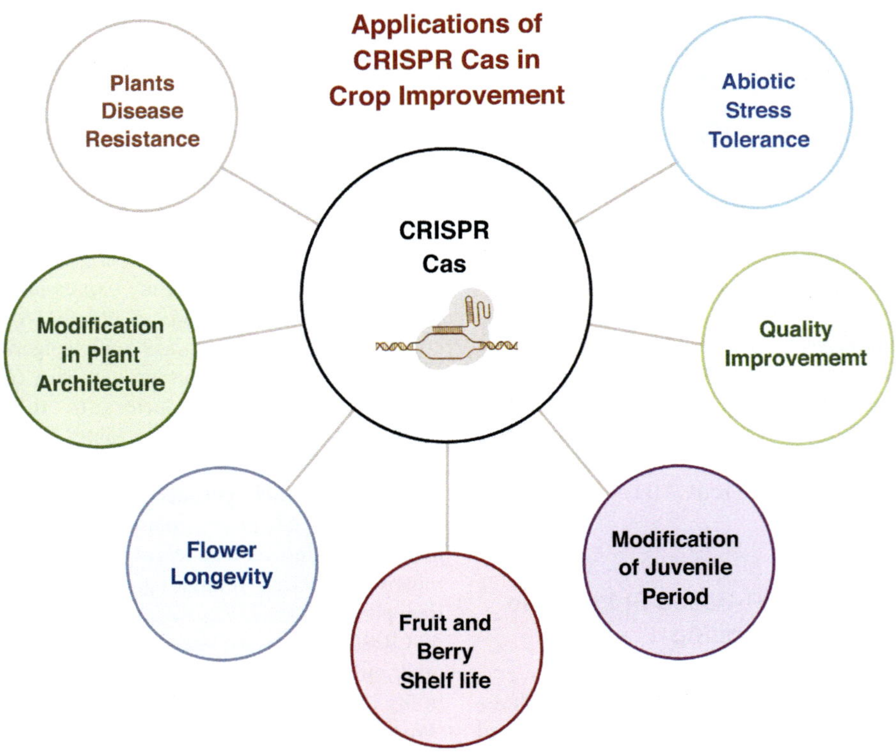

Fig. 13.2. CRISPR-Cas-based epigenome modification.

comprehensive understanding of plant genomes and phenotypes under varying environmental conditions, facilitating more informed crop improvement strategies (Yang *et al.*, 2021). This comprehensive approach to plant research highlights the importance of leveraging cutting-edge technologies to address complex agricultural challenges.

CRISPR technique has appeared as an adaptable tool for genome editing in different organisms, like plants, offering a broad scope of applications including gene knockout, epigenetic modification, transcriptional regulation, base editing, nucleic acid probing, and RNA editing (Prado *et al.*, 2024). The continuous advancements in CRISPR-associated tools, like base editors, CRISPR interference and activation, as well as plant epigenetic editors, have significantly expanded the capabilities of gene editing in plants, enabling researchers to elucidate gene function and regulatory mechanisms with unprecedented precision (Zhang *et al.*, 2022). In the realm of plant molecular biology, CRISPR-based epigenome editing holds immense potential for revolutionizing crop improvement strategies by harnessing epigenetic variability to enhance plant traits. Leveraging the robustness of CRISPR/Cas systems in crop plants, researchers can envision a future where site-specific epigenetic modifications become more feasible in crop models, offering new avenues for enhancing crop productivity and resilience (Kim, 2021). The ability to monitor and establish epigenetic modifications in plants through CRISPR/Cas9-based technologies denotes a noteworthy advancement in plant genome-editing methodologies (Khan *et al.*, 2021).

13.4 Mechanisms of Epigenetic Regulation in Plants

Epigenetic regulation in plants involves a complex interplay of various mechanisms that control gene expression in response to developmental signals and environmental triggers. DNA methylation, histone post-translational modifications, and noncoding RNAs are key players in this regulatory network (Corbin and Lopez, 2020). These mechanisms are required for plant development, stress responses, and adaptation to changing environmental conditions (Chang *et al.*, 2020). The intricate interactions between histone transformations, DNA methylation, and noncoding RNAs create a sophisticated circuitry that enables plants to control gene expression in a robust and precise way (Kim *et al.*, 2021).

Recent investigation has emphasized the significance of epigenetic regulation in plants under stress conditions, revealing the intricate and critical functions of epigenetic mechanisms in mediating responses to environmental cues (Abdulraheem *et al.*, 2024). Epigenetic modifications, particularly DNA methylation, have been extensively studied and are understood to control gene expression in response to developmental and environmental signals, eventually affecting the plant's phenotype (Corbin and Lopez, 2020). These modifications allow plants to respond to abiotic stresses without altering their DNA sequences, highlighting the flexibility and adaptability of epigenetic regulation (Singroha and Sharma, 2019). The regulation of intragenic transposable elements by epigenetic mechanisms influences gene transcription in plants, further emphasizing the broad-reaching effects of epigenetic regulation on plant biology (Kumar, 2017).

Epigenetic regulation plays a necessary part in plant responses to different stresses, including excessive temperatures, drought, salinity, and nutrient limitations. The epigenetic mechanisms involved in stress memory enable crops to remember and react more effectively to recurring stress conditions (Chang *et al.*, 2020). Additionally, epigenetic modifications have been implicated in plant–virus interactions, where epigenetically regulated genes pose as virus-triggered-induced resistance (Corrêa *et al.*, 2024). The part of epigenetic modifications in regulating the appearance of quality-related metabolites in plants has been increasingly recognized. These modifications are involved in secondary metabolism regulation and can influence the production of metabolites important for plant defense and adaptation to stress (Bhattarai *et al.*, 2021). Furthermore, epigenetic regulation has been linked to the control of apomixis and abiotic stress toleration in plants, highlighting its significance in plant reproduction and stress responses (Saravana Kumar *et al.*, 2020).

The coordination between phytohormones and epigenetic processes integrates multiple signals to regulate various plant traits, highlighting the complexity and interconnectedness of plant developmental processes (Kumar and Singh, 2016). Epigenetic modifications, like histone methylation, are involved in fundamental processes in plant development, providing an understanding of the molecular processes underlying the growth and differentiation of the plant (Le Vève *et al.*, 2024). The identification of key epigenetic regulators, such as PHF6, underscores the importance of epigenetic mechanisms in activating specific genes essential for plant development (Yong-Villalobos *et al.*, 2015).

13.5 Targeting Epigenetic Modifications to Combat Plant Diseases

Epigenetic changes are pivotal in plant immunity and responses to various strains, encompassing biotic and abiotic factors. These modifications involve methylation of DNA, post-translational modifications of histone, and remodeling of chromatin, which is crucial for regulating gene expression related to defense responses (Zhi and Chang, 2021). The dynamic interplay between epigenetic mechanisms and regulatory networks in plants governs their development and transformation to varying environmental conditions, underscoring the significance of targeting epigenetic modifications to enhance plant traits (Ahmed *et al.*, 2023). Research indicates that epigenetic modifications, particularly DNA methylation and small RNA-mediated modifications, are fundamental in plants' adaptive responses to stresses like nitrogen limitation, contributing to enhanced use efficiency and adaptation in crops (Hua *et al.*, 2020). Additionally, epigenetic regulation through histone modifications and remodeling of chromatin significantly impacts plant immunity against disease-causing agents, stressing the importance of understanding these molecular processes for disease resistance (Ding and Wang, 2015).

In the context of combating plant diseases induced by fungi, exploring epigenetic modifications presents a promising avenue. Epigenetic regulation in fungal plant pathogens remains an area with untapped potential, offering opportunities for novel insights into combating these detrimental organisms (Dubey and Jeon, 2017). By delving into the unexplored aspects of epigenetic modifications, researchers can uncover new strategies to understand and counteract fungal plant diseases effectively. Epigenetic modifications extend beyond plants' responses to biotic stresses to their adaptation to environmental changes. Modulating plant growth and physiological reactions through epigenetic variety can optimize crop traits, especially in facing global environmental transformation challenges (Kakoulidou *et al.*, 2021). This underscores the importance of leveraging epigenetic mechanisms to enhance crop resilience and productivity under changing climatic conditions.

Moreover, epigenetic modifications have been implicated in the rapid adaptation and stress reactions of non-model plants, demonstrating how these mechanisms contribute to generating stable phenotypes in a few generations. This phenomenon emphasizes the function of epigenetics in promoting plant stability and survival in diverse habitats (Thiebaut *et al.*, 2019). The heritable changes in gene expression mediated by epigenetic modifications offer plants a mechanism to swiftly adapt to fluctuating growth conditions and environmental stresses, ensuring their continued survival and evolution (Faltus, 2023). The complicated interplay between epigenetic modifications and plant responses to stresses, diseases, and environmental changes underscores the importance of targeting these mechanisms to enhance crop resilience and combat plant diseases. By delving into the diverse facets of epigenetics, researchers can unlock new opportunities to understand and manipulate gene expression patterns, paving the way for innovative solutions in agriculture.

13.6 CRISPR/dCas9 in Reprogramming Plant Epigenomes

Recently, the expansion and utilization of CRISPR-based devices have greatly impacted the field of genetic engineering, providing precise and efficient methods for targeted transcriptional and epigenetic regulation in crops (Lee *et al.*, 2019). CRISPR/dCas9 systems have enabled researchers to manipulate gene expression and epigenetic status with high specificity, offering a powerful tool for reprogramming plant epigenomes (Qu *et al.*, 2018). Through the claim of the CRISPR/Cas9 technique, researchers can modulate the expression of genes, direct cell phenotype, and examine the causal epigenetic processes of the regulation of genes in plants (Thakore *et al.*, 2015). One of the important benefits of CRISPR/dCas9 technology is its capability to induce targeted DNA methylation, thereby re-establishing epigenetic control of specific genes in plants (Qu *et al.*, 2018). This targeted epigenome editing approach shows promise in reversing aberrant gene expression patterns induced by environmental factors or stresses, providing a potential strategy for enhancing plant traits and combating diseases. By leveraging CRISPR/dCas9-mediated epigenetic modifications, researchers can activate or repress specific genes implicated in plant growth, stress reactions, and disease resistance, reshaping the plant epigenome for desired outcomes (Alejandra *et al.*, 2020).

The reprogramming is achieved through targeted repression of key regulatory genes, highlighting the prospect of CRISPR-mediated epigenome editing for inducing long-lasting modifications in the expression of gene and chromatin structure in plant cells (Gemberling *et al.*, 2021). Furthermore, the fusion of chromatin-altering domains to dCas9 has enabled precise epigenome editing in both plant cells and animals, offering a versatile platform for studying gene regulation and chromatin dynamics (Lau and Suh, 2018). The benefit of the CRISPR/dCas9 approach in reprogramming plant epigenomes extends beyond basic research to potential agricultural applications. By targeting specific regulatory elements and modulating gene expression, CRISPR-based epigenetic editing can be harnessed to enhance crop traits, improve stress tolerance, and increase yield under changing environmental conditions. Moreover, the ability to activate or repress endogenous genes through CRISPR/dCas9-mediated epigenetic modifications offers a promising avenue for developing genetically modified crops with preferred characteristics, like disease resistance or improved nutritional content (Anton and Bultmann, 2017).

In the context of crop breeding and trait manipulation, CRISPR/dCas9 systems provide a valuable tool for precise epigenetic editing, enabling researchers to engineer targeted changes in gene expression and chromatin states (Wu *et al.*, 2021). By interrogating enhancer function and manipulating the epigenome at specific loci, CRISPR-based epigenetic editing systems offer a means to unravel the regulatory networks governing plant development and stress responses. This targeted approach to epigenome engineering holds great promise for advancing precision agriculture and crop improvement strategies (Gemberling *et al.*, 2021).

13.7 CRISPR Delivery System in Plants

CRISPR-based gene editing has greatly impacted plant biotechnology by supplying exact and efficient genetic modification methods (Ebrahimi and Hashemi, 2024). Among the various delivery systems used for CRISPR/Cas9 in crops, *Agrobacterium*-mediated transformation is a popular choice due to its effectiveness in delivering CRISPR/Cas9 DNA components to plants (Zhang *et al.*, 2019). Still, the greatest possibility of CRISPR/Cas9 in plant genome editing is limited by the lack of efficient delivery systems (Gan and Ling, 2022). Studies have revealed the adaptability of the CRISPR/Cas9 method in plants, showcasing its ability to deliver molecular immunity against intruding DNA viruses (Ali *et al.*, 2015). The versatility of the CRISPR/Cas9 approach extends to conferring immunity to specific plant pathogens like geminiviruses (Mehta *et al.*, 2019). Furthermore, this system has shown promise in producing drought-resistant plants, improving stress toleration, engineering disease resistance, and enhancing plant architecture (Şimşek *et al.*, 2024). By engineering resistance against specific plant viruses like the cotton-leaf curl Kokhran virus, the CRISPR/Cas9 technique has effectively protected plants against viral infections (Rana *et al.*, 2021). Additionally, the CRISPR/Cas9 method has been operated for targeted genome mutagenesis in various plants, including petunia, demonstrating its potential for creating gene knockouts (Zhang *et al.*, 2016).

The widespread adoption of the *Streptococcus pyogenes* Cas9-based method in plant biotechnology underscores the importance of CRISPR/Cas9 in genome editing (Murovec *et al.*, 2017). Investigators have explored utilizing CRISPR/Cas approaches to improve plant tolerance to stress, highlighting its utility in agricultural biotechnology (Yuan *et al.*, 2024). The CRISPR/Cas9 technique can be involved in developing enhanced fruits, vegetables, and decorative crops, showcasing its versatility across different plants (Erpen-Dalla Corte *et al.*, 2019). Moreover, the system has been customized for specific applications, such as a pollen-specific CRISPR/Cas9 method designed to improve heritable gene transformations in maize (Jing *et al.*, 2021). To facilitate the widespread benefit of CRISPR/Cas9 in plant investigation, a complete molecular toolbox has been developed for multiplexed plant genome editing and transcriptional regulation (Lowder *et al.*, 2015). Innovative methods like temperature-regulated atmospheric pressure plasma have been explored for introducing the Cas9/single-guide RNA (gRNA) complex into plant cells, enabling genome editing without the need for DNA introduction (Yanagawa *et al.*, 2023). However, challenges such as rare off-target mutations have been identified, emphasizing the importance of refining CRISPR/Cas9 techniques for precise genome alternation in crops (Wang *et al.*, 2021).

The possibility of the CRISPR/Cas9 approach in enhancing plant disease resistance, alleviating stress burdens, improving nutritional content, and enhancing crop quality highlights its significance in plant biotechnology (Paudel *et al.*, 2021). By engineering resistance against pathogens like the tomato yellow leaf curl virus, the CRISPR/Cas9 technique offers a promising approach to enhancing plant immunity (Tashkandi *et al.*, 2018). Furthermore, studies have linked CRISPR/Cas9 intrusion in cassava to the expansion of editing-opposed geminiviruses, illustrating the dynamic interplay between genome-editing technologies and viral evolution (Mehta *et al.*, 2019). In the realm of crop improvement, genome-editing techniques like CRISPR/Cas9 have been instrumental in improving crop traits in cucurbitaceous crops and allopolyploid wheat, demonstrating the overall relevance of CRISPR/Cas9 across different crop species (Ram *et al.*, 2019). Detailed methodologies and troubleshooting guides have been developed to streamline the benefit of CRISPR/Cas9 in rice genome editing, emphasizing the importance of methodological precision in plant genome editing (Kumari *et al.*, 2022). Efforts to optimize DNA-free genome editing employing the CRISPR/Cas9 ribonucleoprotein method in bread wheat highlight ongoing advancements in precision crop breeding (Liang *et al.*, 2017).

The utility of CRISPR/Cas9 extends beyond crop improvement to fundamental research, where it has been employed to investigate gene function in model plants like *Arabidopsis thaliana* (Ruppel *et al.*, 2019). Additionally, CRISPR/Cas-associated genome altering has been instrumental in unraveling the functional features of plant root systems, showcasing its role in advancing plant-functioning genomics. The continuous development and refinement of CRISPR/Cas9 techniques for functional genomics and gene editing in plants highlight the transformative possibility of this technology in shaping agriculture and plant biotechnology in future (Kiryushkin *et al.*, 2023).

13.8 Applications of Epigenome Editing for Disease Resistance

Epigenome editing has appeared as a powerful device in plant biotechnology, showing a favorable avenue for enhancing disease resistance in plants. By harnessing technologies like CRISPR/Cas9, investigators have been able to exactly manipulate the genetic makeup of plants to grant opposition against various pathogens. The application of epigenome editing in plants has shown significant potential in combating viral diseases, such as the pepper mottle virus in tomatoes. This selected genome-editing approach permits the introduction of specific genetic changes now into elite plant strains, expediting the evolution of disease-resistant crops (Yoon *et al.*, 2020). Moreover, the benefits of CRISPR/Cas9-associated gene editing have been demonstrated to give resistance to viral diseases like turnip mosaic virus in cabbage. Through accurate editing host factors, researchers have been able to develop recessive genetic antagonism in plants towards viruses, showcasing the prospect of genome editing in enhancing crop immunity (Lee *et al.*, 2023). Additionally, genome-editing techniques have been instrumental in enhancing plant resistance against both abiotic and biotic strains, with examples already available to demonstrate the efficacy of this approach (Giudice *et al.*, 2021).

Progress in genome-editing tools, like base editing and prime editing, has further expanded the scope of generating disease-resistant plant varieties (Huang and Hong, 2024). By targeting disease-susceptibility genes through precise editing, researchers have been able to create plants with enhanced resistance to pathogens, offering a long-lasting choice in plant breeding for disease opposition (Peng *et al.*, 2017). The outcome of general virus opposition in nontransgenic cucumber exploiting CRISPR/Cas9 methods exemplifies how genome editing can introduce resistance alleles directly into crop plants, streamlining the breeding process (Chandrasekaran *et al.*, 2016). Epigenome editing, in particular, holds great promise for manipulating plant traits through CRISPR-associated transcriptional activation and targeted demethylation of DNA (Ghoshal *et al.*, 2020). This approach offers a novel way to modify plant genomes globally or at specific loci, creating new genetic variableness and potentially improving agronomic traits (Dhakate *et al.*, 2022). By leveraging epigenome editing tools like CRISPR/Cas9, researchers can manipulate DNA methylation and other epigenetic marks to facilitate molecular breeding of crop plants (Gahlaut *et al.*, 2020).

The benefit of dCas9 epigenome editing with the *Arabidopsis* histone acetyltransferase 1 has been reported as an encouraging approach for enhancing stress toleration in crops by activating stress-response genes. This approach demonstrates the prospect of CRISPR-mediated epigenome editing for manipulating plant traits through targeted modifications. Additionally, the delivery of viral gRNA systems for CRISPR-associated transcriptional activation and selected DNA demethylation in plants has shown great promise for precise and heritable epigenome editing (Ghoshal *et al.*, 2020). The benefit of epigenome editing in improving cassava bacterial blight resistance highlights its utility in crop improvement, demonstrating the potential for enhancing disease resistance through targeted epigenetic modifications (Veley *et al.*, 2023). Additionally, the development of an efficient *Agrobacterium*-associated genetic transformation technique for gene editing in strawberries underscores the creativity of CRISPR/Cas9 in enabling targeted genome editing across different plants (Akter *et al.*, 2024).

13.9 Case Studies of Successful Epigenome Reprograming by CRISPR against Plant Diseases

To explore successful cases of epigenome reprogramming by CRISPR against plant diseases, several studies deliver helpful understandings of the application of CRISPR/Cas technology in enhancing the resistance against plant diseases through selected genome editing. Parmar *et al.* (2017) highlight the efficient benefit of CRISPR/Cas9 in various horticultural crops for gene mutation, activation, repression, and epigenome editing, showcasing its versatility in enhancing stress tolerance and quality traits. This shows the possibility of CRISPR/Cas9 in precisely modifying plant genomes to confer resistance against abiotic and biotic strains, including plant diseases. Yıldırım *et al.* (2023) present a study on developing antagonism to beet curly top Iran virus in sugar beet using the CRISPR/Cas9 method, demonstrating the technique's possibility to create BCTD-resistant crops through straight cleavage of the viral genome or transformation of the plant's genome for viral immunity. This case study underscores the significance of CRISPR/Cas in conferring antagonism to specific plant viruses, showcasing its utility in combating viral diseases in crops.

Furthermore, Paul *et al.* (2021) provide an outline of analyses utilizing CRISPR/Cas9-associated genome editing to enhance crop disease opposition against oomycete and fungal pathogens, emphasizing the prospect of the CRISPR approach in enhancing disease antagonism in host crops. By targeting both host plants and pathogens, CRISPR/Cas9 offers a comprehensive approach to enhancing plant immunity and combating plant diseases effectively. Moreover, Pramanik *et al.* (2021) discuss the generation of pathogen-resistant tomatoes hostile to tomato yellow leaf curl virus and powdery mildew using CRISPR/Cas9, highlighting the valuable role of CRISPR/Cas as a genome-editing mechanism for developing disease-resistant plants. This case study exemplifies the prosperous application of the CRISPR/Cas approach in creating resistance to diseases in crops through selected genetic modifications. Additionally, Borrelli *et al.* (2018) delve into the improvement of plant disease resistance utilizing the CRISPR/Cas9 procedure, focussing on strategies utilized in model species to target viral genomes and induce targeted mutations in host plant genomes for viral disease resistance. This investigation underscores the precision and efficacy of CRISPR/Cas in giving antagonism to plant viruses, showcasing its potential to revolutionize plant disease management.

13.10 Challenges and Limitations

Challenges and limitations exist in reprogramming epigenomes exploiting CRISPR-associated techniques to improve plant disease resistance. CRISPR/Cas methods suggest a favorable strategy for improving disease resistance in plants, but several obstacles ought to be addressed to completely identify its ability (Huang and Hong, 2024). One significant challenge lies in the conveyance of CRISPR/Cas chemicals into plant cells, which is hindered by tissue culture limitations (Zhang *et al.*, 2022). Additionally, the benefit of CRISPR-based epigenome editing in crops is not without its pitfalls and constraints, usually associated with locus- and context-dependent characteristics such as DNA accessibility, chromatin state, and cellular dynamics (Dubois and Roudier, 2021).

Moreover, the complexity of data analysis, concerns regarding off-target effects, and ethical considerations pose additional challenges when utilizing CRISPR/Cas technology for enhancing plant disease resistance (Malook *et al.*, 2023). While newer technologies embedded in the CRISPR/Cas tool aim to confound some of these constraints and offer increased precision in genome restructuring in plants, there are still hurdles to be addressed (Dhakate *et al.*, 2022). The integration of CRISPR/Cas9 in the breeding of plants has indeed facilitated the development of cultivars with heritable disease resistance; however, challenges persist in achieving optimal outcomes (Ghimire, 2017).

Furthermore, the multiplex genome-altering approach in CRISPR/Cas9, while advantageous in simultaneously mutating multiple target genes to confer resistance against different plant viruses, also presents technical challenges that need to be carefully navigated (Hinge *et al.*, 2021). The management of plant viruses exploiting the CRISPR/Cas9 method in food crops faces obstacles such as resistance mechanisms, research approaches, and challenges specific to economically essential crop species (Shahriar *et al.*, 2021). Additionally, the execution of genome editing in polyploid crops poses unique challenges that require innovative solutions (Vats *et al.*, 2019).

The precision and efficacy of CRISPR-associated technologies in enhancing plant disease antagonism are also influenced by factors like the conveyance of CRISPR/Cas chemicals, the prospect for off-target effects, and the need for precise targeting of disease-related genes (Maximiano and Franco, 2022). The threats related to CRISPR-associated treatments, like off-target and on-target changes, underscore the importance of addressing safety concerns when utilizing these technologies for plant disease resistance (Binnie *et al.*, 2021). Therefore, while CRISPR/Cas systems hold immense promise for revolutionizing plant biotechnology, addressing the challenges and limitations associated with their application is crucial for realizing their full potential in enhancing plant disease resistance (Table 13.1).

13.11 Future Prospects

The future potential of reprogramming epigenomes using CRISPR-based technology against plant diseases holds significant promise in revolutionizing plant biology and agriculture. CRISPR-associated epigenome editing and transcriptional initiation in plants have shown significant possibilities for exploiting plant traits (Ghoshal *et al.*, 2020). While traditional gene-editing methods have been successful in creating new traits like disease resistance and increased yield in plants, the advancement of CRISPR/Cas technologies has introduced more precise tools for altering plant genomes through homologous recombination (Schindele *et al.*, 2018). This

Table 13.1. Key challenges and limitations associated with the use of integrated omics and CRISPR/Cas9 for genetic engineering in crops.

Challenge/limitation	Description	Reference
Technical complexity	Integrating various omics data with CRISPR/Cas9 requires advanced computational tools and expertise.	Xu and Li, 2020
Off-target effects	CRISPR/Cas9 can introduce unintended mutations, affecting plant genomes in unpredictable ways.	Zhang *et al.*, 2022
Regulatory hurdles	Stringent regulations on GMOs and genome-edited crops can delay or hinder the deployment of these technologies in agriculture.	Ahmed *et al.*, 2023
Ethical and public acceptance issues	Public concerns about GM crops and ethical issues around genome editing pose significant challenges.	Shaw *et al.*, 2022
Resource-intensive	High costs and resource demands for integrating omics data with CRISPR/Cas9 make it less accessible to developing countries.	Gahlaut *et al.*, 2020
Data management	Managing and analyzing large-scale omics datasets require robust bioinformatics infrastructure.	Chaudhry *et al.*, 2022
Limited understanding of epigenetics	Incomplete knowledge about plant epigenomes and their regulation can limit the effective application of these technologies.	Bhattarai *et al.*, 2021
Scalability issues	Scaling up CRISPR/Cas9 applications from model plants to crops can be challenging due to differences in genome complexity.	Faltus, 2023
Ecological impacts	Potential unintended ecological consequences of releasing genetically edited crops into the environment	Zhi and Chang, 2021

precision allows for epigenome editing, programmable genome editing, and transcriptome regulation in crops, offering a broad collection of possibilities for enhancing crop characteristics (Rehman *et al.*, 2022).

The benefit of CRISPR/Cas9 in plant biology opens up avenues for DNA labeling, epigenome editing, and the development of optimized gene-editing procedures (Liu *et al.*, 2017). Moreover, the use of dCas9-based gene regulation enables multiplex gene initiation, repression, and epigenome editing, paving the way for the construction of complex gene circuits in plants (Chen *et al.*, 2019). The capability to perform multiplex editing of epigenome through CRISPR-based technologies allows for precise and stable modifications in gene expression, offering a powerful tool for manipulating plant traits (Stover *et al.*, 2023).

CRISPR-associated epigenome editing has been shown to facilitate stable and highly site-specific histone modifications, enabling the modulation of gene expression in plants (Stover *et al.*, 2017). By leveraging CRISPR/Cas-associated genome-editing technologies, precise manipulation of plant genomes has become achievable, leading to the creation of plants with desirable traits (Zhang *et al.*, 2022). The unprecedented control offered by CRISPR-Cas9-mediated epigenome editing strategies presents new opportunities for precise gene activation and epigenetic-based therapeutics in plants (Batra *et al.*, 2023). The integration of CRISPR-mediated epigenome- and genome-editing tools into research can advance our knowledge of genetic and epigenetic mechanisms underlying plant diseases and aging-associated issues. These tools offer a programmable medium for editing the eukaryotic epigenome and genome, providing researchers with versatile methods for genetic and epigenetic management of the expression gene (Clement *et al.*, 2020).

The versatility of the CRISPR/Cas approaches in the pathogenesis of crops and the management of disease is evident in its applications for nucleic acid manipulation and detection, accelerating plant breeding and reshaping crop production (Wheatley and Yang, 2021). The utilization of CRISPR/Cas9 ribonucleoproteins has been shown to efficiently edit genes in various plant species, indicating the broad applicability of this technology in achieving precise genome restructuring (Malnoy *et al.*, 2016). Furthermore, the benefit of CRISPR/Cas for fungal genome editing presents a unique device for managing diseases associated with plants, highlighting the prospect of a CRISPR-mediated approach in agriculture (Muñoz *et al.*, 2019).

As the field of CRISPR genome and epigenome editing continues to evolve rapidly, advancements in CRISPR specificity and the effect of new gene-editing devices are anticipated, further enhancing the precision and efficiency of genome-editing methods (Luo, 2019). Using CRISPR/Cas9 approaches in the management of disease through gene activation provides unexplored genetic regulators for controlling plant diseases, showcasing the possible impact of CRISPR-based technologies in agriculture (Prabhukarthikeyan *et al.*, 2020). The effective DNA-free genome editing of plants utilizing CRISPR/Cas9 ribonucleoprotein complexes demonstrates the feasibility of achieving clear and pure genome editing in different kinds of crops (Liang *et al.*, 2017).

13.12 Conclusion

Epigenomics and CRISPR-based epigenome editing represent groundbreaking approaches to understanding and enhancing plant disease resistance. The integration of epigenetics with genomics has provided essential knowledge of the molecular events underlying plant defense reactions. CRISPR/Cas methods, particularly CRISPR/dCas9, have appeared as powerful devices for the precise manipulation of plant epigenomes, offering unprecedented control over gene expression and chromatin structure. The application of these technologies has shown significant potential in combating various plant diseases, from viral infections to fungal pathogens. By targeting specific epigenetic modifications, researchers can reprogram plant genomes to enhance immunity, stress tolerance, and overall crop resilience. Case studies have revealed the successful implementation of CRISPR-based strategies in producing disease-resistant varieties across multiple crop species.

Nevertheless, challenges remain, like delivery techniques, off-target impacts, and ethical concerns. Despite these limitations, the prospects of epigenome reprogramming using CRISPR technology are immensely promising. As the field continues to advance, we can anticipate more precise, efficient, and widespread applications of these techniques in agriculture. The convergence of epigenomics and CRISPR approach is poised to revolutionize plant enhancement strategies, potentially leading to more sustainable agricultural techniques and enhanced food safety. As research progresses, these tools may play a significant part in handling the challenges due to climate change and evolving plant pathogens, ultimately contributing to the establishment of resilient and high-yielding crops.

Acknowledgement

The authors thank their respective universities and institutions for their valuable support.

References

Abdulraheem, M.I., Xiong, Y., Moshood, A.Y., Cadenas-Pliego, G., Zhang, H. *et al.* (2024) Mechanisms of plant epigenetic regulation in response to plant stress: Recent discoveries and implications. *Plants* 13(2), 163.

Ahmed, M., Tariq, H., Haroon, F., Malik, S., Dar, M.S. *et al.* (2023) Harnessing epigenetic modifications for targeted trait manipulation in Pakistani wheat. *International Journal of Precision Farming* 1(1), 50–55.

Akter, F., Wu, S., Islam, M.S., Kyaw, H., Yang, J. *et al.* (2024) An efficient agrobacterium-mediated genetic transformation system for gene editing in strawberry (*Fragaria × ananassa*). *Plants* 13(5), 563.

Alejandra, G.C., Lucia, C., Ho, H.S., Luis, G., Juan, R.P. *et al.* (2020) Activation of pancreatic β-cell genes by multiplex epigenetic CRISPR-editing. *BioRxiv*.

Ali, Z., Abulfaraj, A., Idris, A., Ali, S., Tashkandi, M. *et al.* (2015) CRISPR/Cas9-mediated viral interference in plants. *Genome Biology* 16, 1–11.

Alonso, C., Ramos-Cruz, D. and Becker, C. (2019) The role of plant epigenetics in biotic interactions. *New Phytologist* 221(2), 731–737.

Anton, T. and Bultmann, S. (2017) Site-specific recruitment of epigenetic factors with a modular CRISPR/Cas system. *Nucleus* 8(3), 279–286.

Batra, S.S., Cabrera, A., Spence, J.P., Hilton, I.B., Song, Y.S. *et al.* (2023) Predicting the effect of CRISPR-Cas9-based epigenome editing. *BioRxiv*.

Bhattarai, K., Maharjan, B., Acharya, S., Bigyan, K.C., Pandit, R. *et al.* (2021) Epigenetic modifications and its basic mechanism. *Journal of Innovative Agriculture* 8(1), 19–25.

Bikard, D. and Barrangou, R. (2017) Using CRISPR-Cas systems as antimicrobials. *Current Opinion in Microbiology* 37, 155–160.

Binnie, A., Fernandes, E., Almeida-Lousada, H., Mello, R.A. and Castelo-Branco, P. (2021) CRISPR-based strategies in infectious disease diagnosis and therapy. *Infection* 49(3), 377–385.

Borrelli, V.M., Brambilla, V., Rogowsky, P., Marocco, A. and Lanubile, A. (2018) The enhancement of plant disease resistance using CRISPR/Cas9 technology. *Frontiers in Plant Science* 9, 1245.

Boyko, A. and Kovalchuk, I. (2008) Epigenetic control of plant stress response. *Environmental and Molecular Mutagenesis* 49(1), 61–72.

Burstein, D., Harrington, L.B., Strutt, S.C., Probst, A.J., Anantharaman, K. *et al.* (2017) New CRISPR-Cas systems from uncultivated microbes. *Nature* 542(7640), 237–241.

Chandrasekaran, J., Brumin, M., Wolf, D., Leibman, D., Klap, C. *et al.* (2016) Development of broad virus resistance in non-transgenic cucumber using CRISPR/Cas9 technology. *Molecular Plant Pathology* 17(7), 1140–1153.

Chang, Y.N., Zhu, C., Jiang, J., Zhang, H., Zhu, J.K. *et al.* (2020) Epigenetic regulation in plant abiotic stress responses. *Journal of Integrative Plant Biology* 62(5), 563–580.

Chaudhry, U.K., Gökçe, Z.N. and Gökçe, A.F. (2022) Salt stress and plant molecular responses. In: Kimatu, J.N. (ed.) *Plant Defense Mechanisms*. InTech Open, Ch.7.

Chen, K., Wang, Y., Zhang, R., Zhang, H. and Gao, C. (2019) CRISPR/Cas genome editing and precision plant breeding in agriculture. *Annual Review of Plant Biology* 70(1), 667–697.

Chen, X., Duan, Y., Qiao, F., Liu, H., Huang, J. *et al.* (2022) A secreted fungal effector suppresses rice immunity through host histone hypoacetylation. *The New Phytologist* 235(5), 1977–1994.

Clement, K., Hsu, J.Y., Canver, M.C., Joung, J.K. and Pinello, L. (2020) Technologies and computational analysis strategies for CRISPR applications. *Molecular Cell* 79(1), 11–29.

Corbin, K.R. and Lopez, C.M.R. (2020) Library preparation for whole genome bisulfite sequencing of plant genomes. In: Budak, M. and Yildiz, M. (eds) *DNA Methylation Mechanism*. IntechOpen, ch.8.

Corrêa, R.L., Kutnjak, D., Ambrós, S., Bustos, M. and Elena, S.F. (2024) Identification of epigenetically regulated genes involved in plant-virus interaction and their role in virus-triggered induced resistance. *BMC Plant Biology* 24(1), 172.

Dhakate, P., Sehgal, D., Vaishnavi, S., Chandra, A., Singh, A. *et al.* (2022) Comprehending the evolution of gene editing platforms for crop trait improvement. *Frontiers in Genetics* 13, 876987.

Ding, B. and Wang, G.L. (2015) Chromatin versus pathogens: The function of epigenetics in plant immunity. *Frontiers in Plant Science* 6, 675.

Dong, J., Du, Y. and Zhou, L. (2023) Research progress of CRISPR/Cas systems in nucleic acid detection of infectious diseases. *ILABMED* 1(1), 58–74.

Dowen, R.H., Pelizzola, M., Schmitz, R.J., Lister, R., Dowen, J.M. *et al.* (2012) Widespread dynamic DNA methylation in response to biotic stress. *Proceedings of the National Academy of Sciences* 109(32), E2183–E2191.

Dubey, A. and Jeon, J. (2017) Epigenetic regulation of development and pathogenesis in fungal plant pathogens. *Molecular Plant Pathology* 18(6), 887–898.

Dubois, A. and Roudier, F. (2021) Deciphering plant chromatin regulation via CRISPR/dCas9-based epigenome engineering. *Epigenomes* 5(3), 17.

Ebrahimi, V. and Hashemi, A. (2024) CRISPR-based gene editing in plants: Focus on reagents and their delivery tools. *BioImpacts*.

Erpen-Dalla Corte, L., Mahmoud, L.M., Moraes, T.S., Mou, Z., Grosser, J.W. *et al.* (2019) Development of improved fruit, vegetable, and ornamental crops using the CRISPR/Cas9 genome editing technique. *Plants* 8(12), 601.

Faltus, T. (2023) The applicability of the european GMO legislation to epigenetically modified organisms. *Frontiers in Bioengineering and Biotechnology* 11, 1124131.

Gahlaut, V., Zinta, G., Jaiswal, V. and Kumar, S. (2020) Quantitative epigenetics: A new avenue for crop improvement. *Epigenomes* 4(4), 25.

Gan, W.C. and Ling, A.P.K. (2022) CRISPR/Cas9 in plant biotechnology: Applications and challenges. *BioTechnologia* 103(1), 81–93.

Gemberling, M.P., Siklenka, K., Rodriguez, E., Tonn-Eisinger, K.R., Barrera, A. *et al.* (2021) Transgenic mice for *in vivo* epigenome editing with CRISPR-based systems. *Nature Methods* 18(8), 965–974.

Ghimire, B. (2017) Use of *Crispr*/Cas9 for development of disease resistant cultivars in plant breeding. *International Journal of Applied Sciences and Biotechnology* 5(4), 403–409.

Ghoshal, B., Vong, B., Picard, C.L., Feng, S., Tam, J.M. *et al.* (2020) A viral guide RNA delivery system for CRISPR-based transcriptional activation and heritable targeted DNA demethylation in *Arabidopsis thaliana*. *PLoS Genetics* 16(12), e1008983.

Giudice, G., Moffa, L., Varotto, S., Cardone, M.F., Bergamini, C. *et al.* (2021) Novel and emerging biotechnological crop protection approaches. *Plant Biotechnology Journal* 19(8), 1495–1510.

Hassan, M.M., Zhang, Y., Yuan, G., De, K., Chen, J.G. *et al.* (2021) Construct design for CRISPR/Cas-based genome editing in plants. *Trends in Plant Science* 26(11), 1133–1152.

Hinge, V.R., Chavhan, R.L., Kale, S.P., Suprasanna, P. and Kadam, U.S. (2021) Engineering resistance against viruses in field crops using CRISPR-Cas9. *Current Genomics* 22(3), 214–231.

Hua, Y.P., Zhou, T., Huang, J.Y., Yue, C.P., Song, H.X. *et al.* (2020) Genome-wide differential DNA methylation and miRNA expression profiling reveals epigenetic regulatory mechanisms underlying nitrogen-limitation-triggered adaptation and use efficiency enhancement in allotetraploid rapeseed. *International Journal of Molecular Sciences* 21(22), 8453.

Huang, W. and Hong, Z. (2024) Genetic mechanism of cassava disease resistance: From traditional breeding to CRISPR/Cas application. *Genomics and Applied Biology* 15.

Jain, A., Sarsaiya, S., Singh, R., Gong, Q., Wu, Q. *et al.* (2024) Omics approaches in understanding the benefits of plant-microbe interactions. *Frontiers in Microbiology* 15, 1391059.

Jing, C., Wei, M., Fang, P., Song, R. and Qi, W. (2021) Pollen-specific CRISPR/Cas9 system to increase heritable gene mutations in maize. *Agriculture* 11(8), 751.

Kakoulidou, I., Avramidou, E.V., Baránek, M., Brunel-Muguet, S., Farrona, S. *et al.* (2021) Epigenetics for crop improvement in times of global change. *Biology* 10(8), 766.

Khan, S.U., Khan, M.H.U., Ahmar, S. and Fan, C. (2021) Comprehensive study and multipurpose role of the CLV3/ESR-related (CLE) genes family in plant growth and development. *Journal of Cellular Physiology* 236(4), 2298–2317.

Kim, J.H. (2021) Multifaceted chromatin structure and transcription changes in plant stress response. *International Journal of Molecular Sciences* 22(4).

Kim, E.Y., Kim, K.D. and Cho, J. (2021) Harnessing epigenetic variability for crop improvement: Current status and future prospects. *Genes & Genomics* 44(3), 259–266.

Kiryushkin, A.S., Ilina, E.L. and Demchenko, K.N. (2023) Study of functional features of plant root systems using CRISPR/Cas-mediated genome editing. *Ecological Genetics* 21, 28–29.

Kumar, S. (2017) Epigenetic control of apomixis: A new perspective of an old enigma. *Advances in Plants & Agriculture Research* 7, 243.

Kumar, S. and Singh, A. (2016) Epigenetic regulation of abiotic stress tolerance in plants. *Advances in Plants & Agriculture Research* 5(5), 00179.

Kumari, D., Prasad, B. and Dwivedi, P. (2022) Genome editing platforms in rice (*Oryza sativa* L.): Basic methodology and troubleshooting. DOI: 10.21203/rs.3.rs-1961833/v1.

Lau, C.H. and Suh, Y. (2018) *In vivo* epigenome editing and transcriptional modulation using CRISPR technology. *Transgenic Research* 27(6), 489–509.

Le Vève, A., Sammarco, I., Latzel, V. and Lafon Placette, C. (2024) DNA methylation shows footprints of altitude selection in the clonal plant species *Fragaria vesca*. *BioRxiv*.

Lee, J.E., Neumann, M., Duro, D.I. and Schmid, M. (2019) CRISPR-based tools for targeted transcriptional and epigenetic regulation in plants. *PLoS One* 14(9), e0222778.

Lee, Y.R., Siddique, M.I., Kim, D.S., Lee, E.S., Han, K. *et al.* (2023) CRISPR/Cas9-mediated gene editing to confer turnip mosaic virus (TuMV) resistance in Chinese cabbage (*Brassica rapa*). *Horticulture Research* 10(6), uhad078.

Liang, Z., Chen, K., Li, T., Zhang, Y., Wang, Y. *et al.* (2017) Efficient DNA-free genome editing of bread wheat using CRISPR/Cas9 ribonucleoprotein complexes. *Nature Communications* 8(1), 14261.

Liu, X., Wu, S., Xu, J., Sui, C. and Wei, J. (2017) Application of CRISPR/Cas9 in plant biology. *Acta Pharmaceutica Sinica B* 7(3), 292–302.

Lowder, L.G., Zhang, D., Baltes, N.J., Paul, J.W.I. and Tang, X. (2015) A CRISPR/Cas9 toolbox for multiplexed plant genome editing and transcriptional regulation. *Plant Physiology* 169(2), 971–985.

Luna, E., Bruce, T.J., Roberts, M.R., Flors, V. and Ton, J. (2012) Next-generation systemic acquired resistance. *Plant Physiology* 158(2), 844–853.

Luo, Y. (2019) Refining CRISPR-based genome and epigenome editing off-targets. *Cell Biology and Toxicology* 35(4), 281–283.

Malnoy, M., Viola, R., Jung, M.H., Koo, O.J., Kim, S. *et al.* (2016) DNA-free genetically edited grapevine and apple protoplast using CRISPR/Cas9 ribonucleoproteins. *Frontiers in Plant Science* 7.

Malook, M.B., Aslam, S. and Ammar, A. (2023) Plant pathology in genome era new insight into disease resistance. *International Journal of Research and Advances in Agricultural Sciences* 2(3), 27–38.

Masenya, K., Manganyi, M.C. and Dikobe, T.B. (2024) Exploring cereal metagenomics: Unravelling microbial communities for improved food security. *Microorganisms* 12(3), 510.

Maximiano, M.R. and Franco, O.L. (2022) CRISPR/Cas: The new frontier in plant improvement. *ACS Agricultural Science & Technology* 2(2), 202–214.

McDowell, J.M. and Meyers, B.C. (2013) A transposable element is domesticated for service in the plant immune system. *Proceedings of the National Academy of Sciences* 110(37), 14821–14822.

Mehta, D., Stürchler, A., Anjanappa, R.B., Zaidi, S.S.E.A., Hirsch-Hoffmann, M. *et al.* (2019) Linking CRISPR-Cas9 interference in cassava to the evolution of editing-resistant geminiviruses. *Genome Biology* 20(1), 80.

Mierziak, J. and Wojtasik, W. (2024) Epigenetic weapons of plants against fungal pathogens. *BMC Plant Biology* 24(1), 175.

Murovec, J., Pirc, Ž. and Yang, B. (2017) New variants of CRISPR RNA-guided genome editing enzymes. *Plant Biotechnology Journal* 15(8), 917–926.

Muñoz, I.V., Sarrocco, S., Malfatti, L., Baroncelli, R. and Vannacci, G. (2019) CRISPR-Cas for fungal genome editing: A new tool for the management of plant diseases. *Frontiers in Plant Science* 10, 135.

Ntsomboh-Ntsefong, G., Seyum, E., Kingsley, T., Kassie, F., Gabriel, M. *et al.* (2023) CRISPR-cas-based genome editing for crop improvement: Progress, challenges and future prospects. *Global Journal Of Botanical Science* 11, 28–33.

Orantes-Bonilla, M., Wang, H., Lee, H.T., Golicz, A.A., Hu, D. *et al.* (2023) Transgressive and parental dominant gene expression and cytosine methylation during seed development in *Brassica napus* hybrids. *Theoretical and Applied Genetics* 136(5), 113.

Parmar, N., Singh, K.H., Sharma, D., Singh, L., Kumar, P. *et al.* (2017) Genetic engineering strategies for biotic and abiotic stress tolerance and quality enhancement in horticultural crops: A comprehensive review 1–35.

Paudel, N., Park, S., Jang, Y. and Rajametov, S.N. (2021) CRISPR/Cas9 trends and future scheme for genome engineering in plants. DOI: 10.20944/preprints202103.0277.v1.

Paul, N.C., Park, S.W., Liu, H., Choi, S., Ma, J. *et al.* (2021) Plant and fungal genome editing to enhance plant disease resistance using the CRISPR/Cas9 system. *Frontiers in Plant Science* 12, 700925.

Peng, A., Chen, S., Lei, T., Xu, L., He, Y. *et al.* (2017) Engineering canker-resistant plants through CRISPR/Cas9-targeted editing of the susceptibility gene Cs LOB 1 promoter in citrus. *Plant Biotechnology Journal* 15(12), 1509–1519.

Prabhukarthikeyan, S.R., Parameswaran, C., Keerthana, U., Teli, B., Jagannadham, P.T. *et al.* (2020) Understanding the plant-microbe interactions in CRISPR/Cas9 era: Indeed, a sprinting start in marathon. *Current Genomics* 21(6), 429–443.

Prado, G., Rocha, D.C., Santos, L.N.D., Contiliani, D.F., Nobile, P.M. *et al.* (2024) CRISPR technology towards genome editing of the perennial and semi-perennial crops citrus, coffee and sugarcane. *Frontiers in Plant Science* 14, 1331258.

Pramanik, D., Shelake, R.M., Park, J., Kim, M.J., Hwang, I. *et al.* (2021) CRISPR/Cas9-mediated generation of pathogen-resistant tomato against tomato yellow leaf curl virus and powdery mildew. *International Journal of Molecular Sciences* 22(4), 1878.

Qu, J., Zhu, L., Zhou, Z., Chen, P., Liu, S. *et al.* (2018) Reversing mechanoinductive DSP expression by CRISPR/dCas9-mediated epigenome editing. *American Journal of Respiratory and Critical Care Medicine* 198(5), 599–609.

Ram, C., Berwal, M.K., Gora, J.S., Kumar, R., Haldhar, S.M. *et al.* (2019) Genomic and biotechnological interventions for crop improvement in cucurbitaceous crops: A review. *Journal of Agriculture and Ecology* 7(1), 1–15.

Rana, M.H., Khan, M.Z., Mustafa, R., Kamal, H., Hussain, A. *et al.* (2021) Engineering resistance against cotton leaf curl kokhran virus-burewala strain using CRISPR-Cas9 system in *Nicotiana benthamiana*. DOI: 10.21203/rs.3.rs-604666/v1.

Rehman, R.S., Ali, M., Zafar, S.A., Ahmad, M., Pasha, A.N. *et al.* (2022) Tapping into the unsung potential of CRISPR/Cas technology in agriculture. *Asian Journal of Biochemistry, Genetics and Molecular Biology* 10(4), 1–26.

Ruppel, N.J., Estell, L.E., Jackson, R.I. and Wolyniak, M.J. (2019) An undergraduate research project utilizing CRISPR-Cas9 gene editing technology to study gene function in *Arabidopsis thaliana*. *Journal of Microbiology & Biology Education* 20(2), 10–1128.

Saravana Kumar, R.M., Wang, Y., Zhang, X., Cheng, H., Sun, L. *et al.* (2020) Redox components: Key regulators of epigenetic modifications in plants. *International Journal of Molecular Sciences* 21(4), 1419.

Schindele, P., Wolter, F. and Puchta, H. (2018) Transforming plant biology and breeding with CRISPR/Cas9, Cas12 and Cas13. *FEBS Letters* 592(12), 1954–1967.

Shahriar, S.A., Islam, M.N., Chun, C.N.W., Rahim, M.A., Paul, N.C. *et al.* (2021) Control of plant viral diseases by CRISPR/Cas9: Resistance mechanisms, strategies and challenges in food crops. *Plants* 10(7), 1264.

Shaikh, A.A., Chachar, S., Chachar, M., Ahmed, N., Guan, C. *et al.* (2022) Recent advances in DNA methylation and their potential breeding applications in plants. *Horticulturae* 8(7), 562.

Shaw, R.K., Shen, Y., Yu, H., Sheng, X., Wang, J. *et al.* (2022) Multi-omics approaches to improve clubroot resistance in brassica with a special focus on *Brassica oleracea* L. *International Journal of Molecular Sciences* 23(16), 9280.

Şimşek, Ö., Isak, M.A., Dönmez, D., Dalda Şekerci, A., İzgü, T. *et al.* (2024) Advanced biotechnological interventions in mitigating drought stress in plants. *Plants* 13(5), 717.

Singroha, G. and Sharma, P. (2019) Epigenetic modifications in plants under abiotic stress. *Epigenetics* 10.

Soyer, J.L., El Ghalid, M., Glaser, N., Ollivier, B., Linglin, J. *et al.* (2014) Epigenetic control of effector gene expression in the plant pathogenic fungus *Leptosphaeria maculans*. *PLoS Genetics* 10(3), e1004227.

Stover, J.D., Farhang, N., Berrett, K.C., Gertz, J., Lawrence, B. *et al.* (2017) CRISPR epigenome editing of AKAP150 in DRG neurons abolishes degenerative IVD-induced neuronal activation. *Molecular Therapy* 25(9), 2014–2027.

Stover, J.D., Trone, M.A., Lawrence, B. and Bowles, R.D. (2023) Multiplex epigenome editing of ion channel expression in nociceptive neurons abolished degenerative IVD-conditioned media-induced mechanical sensitivity. *JOR Spine* 6(2), e1253.

Tashkandi, M., Ali, Z., Aljedaani, F., Shami, A. and Mahfouz, M.M. (2018) Engineering resistance against Tomato yellow leaf curl virus via the CRISPR/Cas9 system in tomato. *Plant Signaling & Behavior* 13(10), e1525996.

Thakore, P.I., D'Ippolito, A.M., Song, L., Safi, A., Shivakumar, N.K. *et al.* (2015) Highly specific epigenome editing by CRISPR-Cas9 repressors for silencing of distal regulatory elements. *Nature Methods* 12(12), 1143–1149.

Thiebaut, F., Hemerly, A.S. and Ferreira, P.C.G. (2019) A role for epigenetic regulation in the adaptation and stress responses of non-model plants. *Frontiers in Plant Science* 10, 246.

Vats, S., Kumawat, S., Kumar, V., Patil, G.B., Joshi, T. *et al.* (2019) Genome editing in plants: Exploration of technological advancements and challenges. *Cells* 8(11), 1386.

Veley, K.M., Elliott, K., Jensen, G., Zhong, Z., Feng, S. *et al.* (2023) Improving cassava bacterial blight resistance by editing the epigenome. *Nature Communications* 14(1), 85.

Wang, M., Li, R. and Zhao, Q. (2023) Multi-omics techniques for forest plant breeding. *Forests* 14(6).

Wang, X., Tu, M., Wang, Y., Yin, W., Zhang, Y. *et al.* (2021) Whole-genome sequencing reveals rare off-target mutations in CRISPR/Cas9-edited grapevine. *Horticulture Research* 8(1).

Westra, E.R., Dowling, A.J., Broniewski, J.M. and Houte, S. (2016) Evolution and ecology of CRISPR. *Annual Review of Ecology, Evolution, and Systematics* 47(1), 307–331.

Wheatley, M.S. and Yang, Y. (2021) Versatile applications of the CRISPR/Cas toolkit in plant pathology and disease management. *Phytopathology®* 111(7), 1080–1090.

Wu, H., Wang, F. and Jiang, J.H. (2021) Inducible CRISPR-dCas9 transcriptional systems for sensing and genome regulation. *ChemBioChem* 22(11), 1894–1900.

Xu, Y. and Li, Z. (2020) CRISPR-Cas systems: Overview, innovations and applications in human disease research and gene therapy. *Computational and Structural Biotechnology Journal* 18, 2401–2415.

Yanagawa, Y., Suenaga, Y., Iijima, Y., Endo, M., Sanada, N. *et al.* (2023) Genome editing by introduction of Cas9/sgRNA into plant cells using temperature-controlled atmospheric pressure plasma. *PLoS One* 18(2), e0281767.

Yang, Y., Saand, M.A., Huang, L., Abdelaal, W.B., Zhang, J. *et al.* (2021) Applications of multi-omics technologies for crop improvement. *Frontiers in Plant Science* 12, 563953.

Yong-Villalobos, L., González-Morales, S.I., Wrobel, K., Gutiérrez-Alanis, D., Cervantes-Peréz, S.A. *et al.* (2015) Methylome analysis reveals an important role for epigenetic changes in the regulation of the *Arabidopsis* response to phosphate starvation. *Proceedings of the National Academy of Sciences* 112(52), E7293–E7302.

Yoon, Y.J., Venkatesh, J., Lee, J.H., Kim, J., Lee, H.E. *et al.* (2020) Genome editing of *eIF4E1* in tomato confers resistance to pepper mottle virus. *Frontiers in Plant Science* 11, 1098.

Yuan, C., Zhang, T. and Huang, C. (2024) Editorial: Applications of novel gene editing tools and approaches in plants. *Frontiers in Genome Editing* 6, 1404959.

Yıldırım, K., Kavas, M., Küçük, İ.S., Seçgin, Z. and Saraç, Ç.G. (2023) Development of highly efficient resistance to *beet curly top iran virus* (*Becurtovirus*) in sugar beet (*B. vulgaris*) via CRISPR/Cas9 system. *International Journal of Molecular Sciences* 24(7), 6515.

Zhang, B., Yang, X., Yang, C., Li, M. and Guo, Y. (2016) Exploiting the CRISPR/Cas9 system for targeted genome mutagenesis in petunia. *Scientific Reports* 6(1), 20315.

Zhang, C., Liu, S., Li, X., Zhang, R. and Li, J. (2022) Virus-induced gene editing and its applications in plants. *International Journal of Molecular Sciences* 23(18), 10202.

Zhang, H., Lang, Z. and Zhu, J.K. (2018) Dynamics and function of DNA methylation in plants. *Nature Reviews. Molecular Cell Biology* 19(8), 489–506.

Zhang, R.X., Li, B.B., Yang, Z.G., Huang, J.Q., Sun, W.H. *et al.* (2022) Dissecting plant gene functions using CRISPR toolsets for crop improvement. *Journal of Agricultural and Food Chemistry* 70(24), 7343–7359.

Zhang, Z., Hua, L., Gupta, A., Tricoli, D., Edwards, K.J. *et al.* (2019) Development of an agrobacterium-delivered CRISPR/Cas9 system for wheat genome editing. *Plant Biotechnology Journal* 17(8), 1623–1635.

Zhi, P. and Chang, C. (2021) Exploiting epigenetic variations for crop disease resistance improvement. *Frontiers in Plant Science* 12, 692328.

14 Profiling Plant–pathogen Interactions by Integrative Omics Tools Leads to Speed Breeding of CRISPR-edited Disease-tolerant Crops

Khadija Benamar[1,2], Rachid Ezzouggari[1,3], Mohammed Taoussi[1,4] and Rachid Lahlali[1]*

[1]Department of Plant Protection, Phytopathology Unit, Ecole Nationale d'Agriculture de Meknès, Meknès, Morocco; [2]Microbial Biotechnology and Bioactive Molecules Laboratory, Sciences and Technologies Faculty, Sidi Mohamed Ben Abdellah University, Fez, Morocco; [3]Laboratory of Biotechnology, Conservation, and Valorization of Natural Resources (LBCVNR), Faculty of Sciences Dhar El Mehraz, Sidi Mohamed Ben Abdallah University, Fez, Morocco; [4]Environment and Valorization of Microbial and Plant Resources Unit, Faculty of Sciences, Moulay Ismail University, Meknès, Morocco

Abstract

Profiling plant–pathogen interactions through integrative omics tools accelerates the breeding of CRISPR-edited disease-tolerant crops by providing a comprehensive analysis of resistance molecular mechanisms. By combining metabolomics, proteomics, transcriptomics, and genomics, researchers can identify key genetic factors and pathways associated with plant defense responses. This holistic approach facilitates the precise editing of target genes using CRISPR technology, enhancing crop resilience to specific pathogens. The integration of omics data, CRISPR technology and speed breeding allows for rapid identification of beneficial traits and their incorporation into breeding programs, significantly reducing the time needed to develop disease-resistant varieties. This strategy not only improves crop productivity and sustainability but also addresses challenges posed by emerging plant diseases in a timely manner.

Keywords: crop breeding, defense mechanism, genome editing, stress response

14.1 Introduction

Plant–pathogen interactions are an important research area for ensuring sustainable agriculture. With the development of integrative omics tools, such as metabolomics, proteomics, transcriptomics, and genomics, scientists are now equipped to dissect complex interactions at unprecedented levels of detail (Kumar *et al.*, 2021). These multi-omics approaches allow the comprehensive profiling of plant–pathogen interactions, revealing the intricate molecular

*Corresponding author: rlahlali@enameknes.ac.ma

© CAB International 2025. *Stress-Resilient Crops: Coordinated omics-CRISPR-Nanotechnology Strategies* (ed. Jen-Tsung Chen)
DOI: 10.1079/9781800627307.0014

dialogs that underpin plant defense mechanisms and pathogens strategies (Astapati and Nath, 2023).

Integrating these omics tools enables a more holistic understanding of plant responses to different stresses, and the determination of main genes and pathways included in resistance (Swarupa *et al.*, 2016). For example, recent searches have revealed that the integration of host and microbial omics can elucidate the dynamics of the plant microbiome, transforming our understanding of plant health and resilience against diseases (Agrahari *et al.*, 2020). This is especially relevant in facing emerging pathogens and the challenges related to climate change, which can exacerbate disease outbreaks and threaten crop yields.

The application of CRISPR/Cas9 in plant breeding has led to the development of disease-tolerant crops. Indeed, CRISPR facilitates the introduction of traits that enhance resistance to specific pathogens. The combination of integrative omics and CRISPR technology accelerates the breeding process and improves the efficiency of developing crops that can withstand biotic stresses (Bhimani *et al.*, 2024). This synergy between advanced molecular techniques and traditional breeding practices is paving the way for a new era of speed breeding, where the time needed to develop resilient crop varieties is significantly reduced (Kumar *et al.*, 2021).

The integration of omics tools in profiling plant–pathogen interactions, coupled with CRISPR-based editing strategies, represents a transformative approach leading to the enhancement of the resistance of crops to pathogens (Agrahari *et al.*, 2020). The future of agricultural research lies in the ability to integrate these diverse datasets and apply them to real-world challenges in food production and environmental sustainability (Astapati and Nath, 2023). As omics technologies are still evolving, they hold a high potential for the optimization of agricultural strategies and pave the way for more resistant and sustainable agricultural systems (Jain *et al.*, 2024).

This chapter will explore how these innovative methodologies allow the rapid development of disease-tolerant crops, ultimately enhancing agricultural productivity and sustainability in an increasingly challenging environmental landscape.

14.2 Integrative Omics Profiling of Plant–pathogen Interactions

Integrative omics profiling in plant–pathogen interactions represents a global approach combining various 'omics' technologies enabling understanding of the interaction of plants and pathogens at multiple biological levels. This method uses data from omic tools, including metabolomics, proteomics, transcriptomics, and genomics, to provide an insight into the interaction's dynamics (Swarupa *et al.*, 2016) (Fig. 14.1).

14.2.1 Genomics

The genomics field relies on sequencing and analysis of pathogenic organisms' genomes. It

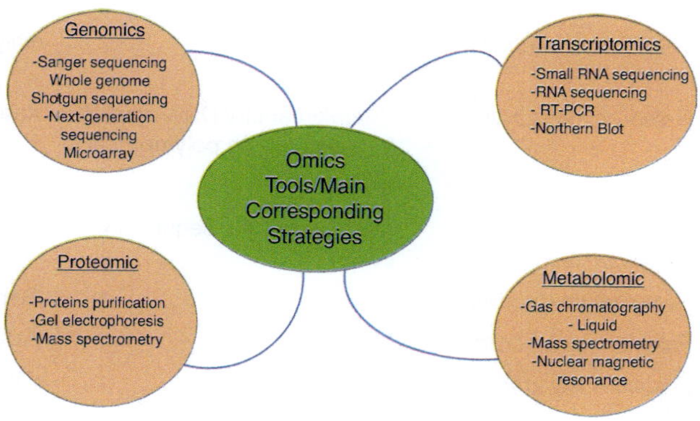

Fig. 14.1. Omics tools used in profiling plant–pathogen interactions and their main strategies.

is widely used for the study of organism's evolution and biodiversity (Weissenbach, 2016). It plays an important role in phytopathology, since it allows genomic determination of plant resistance genes as well as plant invaders' pathogenic genes (Bartoli and Roux, 2017). Genome sequencing can be performed through the use of sanger sequencing technology. However, nowadays, the use of next-generation sequencing has highly emerged (Sai Reddy *et al.*, 2022).

14.2.2 Transcriptomics

Transcriptomics is the set of RNA transcripts contained in a cell. Transcriptomics is based on the study of transcriptomes and measures their expression and variation under different stress conditions (Swarupa *et al.*, 2016). This technology has exhibited a high importance in the study of plant and pathogen interactions.

Traditionally, the study of transcriptomes was performed using complementary deoxyribonucleic acid in order to obtain the expression of sequence tags which are then analyzed through a serial study of gene expression, or by using sanger sequencing techniques. However, currently, direct RNA sequencing and DNA microarray techniques are more used in transcriptomics tools (Sai Reddy *et al.*, 2022).

14.2.3 Application of genomics and transcriptomics for understanding plant resistance to pathogens

Several studies have concerned plant genomics and transcriptomics in response to different pathogens including fungi, bacteria, and nematodes. Table 14.1 shows examples of recent studies using genomics and transcriptomics for profiling plant fungal pathogens interactions.

Indeed, Xue *et al.* (2015) have determined defense genes in *Phaseolus vulgaris* after its interaction with the fungal pathogen *Fusarium oxysporum* (*F. oxysporum*). For this purpose, they have adopted the technique of complementary DNA amplified fragment polymorphism. Their results revealed the distribution of 122 defense gene fragments across the genome in response to *F.oxysporum* infection (Xue *et al.*, 2015). For the same plant species, Padder *et al.* (2016) investigated the expression of defense genes in susceptible and resistant lines to the pathogenic fungi *Colletotrichum lindemuthianum*. They found that the resistant interaction was associated with an upregulation of genes encoding for peroxidases, lipoxygenases, and pathogenesis-related proteins, whereas in the susceptible lines, an upregulation of genes encoding sugar transporters was observed (Padder *et al.*, 2016).

Ishiga *et al.* (2015) studied the molecular mechanism of interaction between *Medicago*

Table 14.1. Examples of recent studies using genomics and transcriptomics for profiling plant fungal pathogens interactions.

Plant species	Fungi	Methods	Reference
Glycine max	*Phytophthora sojae*	Microarray	Lin *et al.*, 2014
Medicago truncatula	*Phakopsora pachyrhizi*	Microarray	Ishiga *et al.*, 2015
Phaseolus vulgaris L.	*Fusarium oxysporum*	Complementary DNA-amplified fragment length polymorphism	Xue *et al.*, 2015
Glycine max	*Fusarium oxysporum*	NGS	Lanubile *et al.*, 2015
Phaseolus vulgaris L.	*Colletotrichum Lindemuthianum*	Next-generation sequencing (NGS)	Padder *et al.*, 2016
Chickpea	*Fusarium oxysporum*	NGS	Gupta *et al.*, 2017
Chickpea	*Fusarium oxysporum*	Microarray	Ashraf *et al.*, 2018
Glycine max	*Fusarium oxysporum*	NGS	Chang *et al.*, 2019
Lentil	*Colletotrichum lentis*	NGS	Bhadauria *et al.*, 2017

truncatula and the fungal foliar pathogen *Phakopsora pachyrhizi*. Their findings showed the induction of several genes related to metabolic pathways, such as the genes associated with phytoalexin medicarpin production and chlorophyll catabolism (Ishiga *et al.*, 2015). In another study, Lin *et al.* (2014) used a comparative analysis of molecular reactions in ten soybean (*Glycine max*) near-isogenic lines, inoculated with the Oomycete *Phytophtora soje* (Lin *et al.*, 2014). They detected a variation in defense genes related to reactive oxygen species (ROS), jasmonic acid, and ethylene production, between the tested lines. Lanubile *et al.* (2015) showed higher activation of defense genes in soybean inoculated with pathogenic *Fusarium oxysporum* isolates, in comparison with its inoculation with non-pathogenic strains (Lanubile *et al.*, 2015). However, Gupta *et al.* (2017) demonstrated the activation of resistance genes in chickpea after *Fusarium oxysporum* attack. This activation was associated with the production of several defense molecules, like MAP kinase and serine-threonine kinase (Gupta *et al.*, 2017). In another study, 76 genes related to immunity and disease response to *Fusarium oxysporum* were identified in Chikpea (Ashraf *et al.*, 2018). Copley *et al.* (2017) noticed the alteration of a high number of genes responsible for cell-wall synthesis in wild soybean compared with cultivated soybean, after the infection by *Fusarium oxysporum* (Copley *et al.*, 2017).

Regarding plant genomics and transcriptomics in response to bacteria (Table 14.2), several studies have exhibited plant defense genes and their expression after bacterial infections.

Thus, for example, Nemchinov *et al.* (2017) investigated the responses of *Medicago sativa* (alfalfa) to the bacteria *Pseudomonas syringae* pv. *Syringae* in susceptible and resistant genotypes. They found a high representation of resistance genes in the resistant genotype, including two genes of CCR domains, and the RPM1 gene (Nemchinov *et al.*, 2017). The resistance in these genotypes was achieved by the WRKY transcription factors family, and genes included in isoflavonoids synthesis, and protein phosphorylation. In another study, Jiang *et al.* (2023) reported the revelation of more than 40 resistance genes in rice against bacterial blight caused by several *Xanthomonas oryzae* species (Jiang *et al.*, 2023). Xue *et al.* (2019) identified 54 *WRKY* genes in *Ziziphus jujuba*, responsible for its resistance to phytoplasma invasion. They showed that the majority of these genes were intensively expressed in different *Ziziphus jujuba* tissues, after phytoplasma infection (Xue *et al.*, 2019).

More recently, Qian *et al.*, 2021 identified 23 quantitative trait loci associated with sugarcane disease caused by *Leifsonia xyli* subsp. *Xyli* bacteria. Thus, their study is of great interest for developing breeding programs against this bacterial disease (You *et al.*, 2021).

As for plant genomics related to nematode infection, many studies have focussed on plant genomics in response to the nematodes *Heterodera glycines*, *Meloidogyne arenaria*, and *Meloidogyne incognita*. Indeed, for instance, Guimaraes *et al.* (2015) identified the main genes related to the resistance of wild peanuts to *Meloidogyne arenaria*. These genes were involved in salicylic and jasmonic acids pathways and were engaged in hormonal balance (Guimaraes *et al.*, 2015). Postnikova *et al.* (2015) analyzed genes in *Medicago* sativa-resistant and -susceptible cultivars, after *Meloidogyne incognita* infection (Postnikova *et al.*, 2015). Their results revealed a differentiated expression of several genes in the studied cultivars. Similarly, a difference in resistance genes and their expression

Table 14.2. Examples of recent studies using genomics and transcriptomics for profiling plant bacterial pathogens interactions.

Plant species	Bacteria	Methods	Reference
Peanut	*Ralstonia solanacearum*	NGS	Chen *et al.*, 2014
Medicago sativa L.	*Pseudomonas syringae*	NGS	Nemchinov *et al.*, 2017
Jujube	Phytoplasma	Genome-wide analysis	Xue *et al.*, 2019
Saccharum spp.	*Leifsonia xyli* subsp. *xyli*	Single nucleotide polymorphism array	You *et al.*, 2021

was observed between beans susceptible and resistant genotypes infected by the nematode *Heterodera glycines* (Jain *et al.*, 2016).

The use of proteomics is important for the study of plant pathogen interactions. Some examples are presented below.

14.2.4 Proteomics

Proteomics relies on the study of an organism's proteins, their abundance, and their modification (Swarupa *et al.*, 2016). These tools are based on the collection of samples, the isolation of proteins, then their extraction and fractionation, followed by their identification (Sharma *et al.*, 2020). Proteomic studies involve the use of mass spectrometry for protein quantification and identification as well as for the determination of post-translational modification. The strategies employed for this purpose are the bottom-up and top-down strategies. The first strategy is based on the analysis of protein mixtures which could present limitations related to the impossible investigation of protein mutation, whereas the second strategy concerns the analysis of only individual proteins (Sai Reddy *et al.*, 2022).

Other methods could be used such as western blot analysis (Yang *et al.*, 2015). For protein separation, chromatographical techniques can be used, as well as electrophoresis and ultrafiltration methods (Liu *et al.*, 2020).

14.2.5 Use of proteomics for the investigation of plant–pathogen interactions

Many studies have revealed the role of proteomics in understanding plant–pathogen interactions (Table 14.3). For instance, Yang *et al.* (2015) have shown, through proteomic study, the rapid response of wheat to the pathogenic fungi *Zymoseptoria tritici* (*Z.tritici*) during its biotrophic growth (Yang *et al.*, 2015). Indeed, wheat resistance was found to be associated with the enhancement of carbohydrate, stress, and apoplastic proteins, in addition to the reinforcement of cell walls. Proteomic analysis has allowed the identification of *Z. tritici* proteins associated with the incompatible interaction with wheat. These proteins are linked with signal transduction, as well as remodeling. Zhang *et al.* (2015) have used proteomics for the analysis of *Malus domestica* leaves proteins resulting from the interaction with *Alternaria alternate*. Their results identified some pathogenesis-related proteins in susceptible and resistant *Malus domestica* cultivars. These proteins included 1,3-glucanase, glutathione peroxidase, and

Table 14.3. Examples of recent studies using proteomics for profiling plant–pathogen interactions.

Plants	Pathogens	Methods	Reference
Wheat	*Zymoseptoria tritici*	Western blot analysis	Yang *et al.*, 2015
Apple (*Malus domestica*)	*Alternaria alternata*	Two-dimensional electrophoresis/mass spectrometry	Zhang *et al.*, 2015
Populus tomentosa and *Populus beijingensis*	*Botryosphaeria dothidea*	Nanoflow liquid chromatography/tandem mass spectrometry	Li *et al.*, 2019
Tomato	*Verticillium dahliae*	Multidimensional chromatography coupled with tandem mass spectrometry	Hu *et al.*, 2019
Rice	*Magnaporthe oryzae*	Mass spectrometry	Meng *et al.*, 2019
Cacao	*Moniliophthora perniciosa*	Electrophoresis/mass spectroscopy	Dos Santos *et al.*, 2020

ascorbate peroxidase, in addition to the protein Mal d1. The presence of different levels of this latter (major allergin protein Mal d1) between the susceptible and resistant cultivars, may be responsible of their differential response to *Alternaria alternate* (Zhang *et al.*, 2015). Similarly, Li *et al.* (2019) performed a proteomic study for two poplar ecotypes; the susceptible and the resistant to *Botryosphaeria dothidea*. Indeed, they determined 588 proteins included in many processes, such as oxidoreductases, hydrolytic and metabolic enzymes. Besides, unlike the susceptible genotype, an upregulation of 22 proteins was observed in the resistant genotype resulting in its fight against *Botryosphaeria dothidea* (Li *et al.*, 2019). The defense activities of these enzymes were related to phenyl propanoid biosynthesis, phenylalanine metabolism, photosynthesis, carbon fixation, oxidoreductase activity, and glyco-(lysis and genesis). In the same context, a proteomic study of proteins included in the compatible and incompatible interaction of tomato with two races of *Verticillium dahliae* was performed by Hu *et al.* (2019), their findings showed the presence of 814 and 584 proteins, and 456 and 637 proteins, in the incompatible and the compatible interactions with the races 1 and 2, respectively (Hu *et al.*, 2019). Additionally, the incompatible interaction was associated with the presence of certain proteins absent in the compatible interaction, like phenylalanine ammonia lyase, NADP-dependent dehydratase, and epimerase, as well as ammonia lyase, which were introduced in the resistance mechanism of tomato, by interfering with secondary metabolism system and strengthening cellular walls. Dos Santos *et al.* (2020) performed a comparative proteomic study of cacao-resistant and susceptible cultivars to *Moniliophthora perniciosa*. They revealed the presence of 246 and 308 proteins in the susceptible and the resistant genotypes, respectively. The resistant genotype exhibited more proteins related to stress and defense, like stand-out pathogenesis proteins, trypsin inhibitors, and oxidative stress control-associated proteins (Dos Santos *et al.*, 2020). Recently, a compilation of proteomic studies related to the resistance of rice to *Magnaporthe oryzae* was performed by Meng *et al.* (2019), this review has reported the main proteins involved in rice–*Magnaporthe* interaction, namely receptor-like kinases, proteins in relation to photosynthesis, hormone signaling, nitrogen activated protein kinase, and receptor-like kinase (Meng *et al.*, 2019). Indeed, these proteins are responsible for pathogen detection, signal transduction, and metabolism regulation. Post-translational changes like ubiquitin and phosphoproteomics also contributed to rice's defense against this pathogen.

14.2.6 Metabolomics

Metabolomics is the investigation of small molecules produced during cellular metabolism. Plants release particular metabolites having an important ecological role and essential function in coordinating the interaction between their life and the surrounding environment (Kimotho and Maina, 2024). Indeed, metabolites carry out different tasks interfering in plant–pathogen interactions, including the detection of pathogens attacks, signal transmission, enzyme modulation, and microorganism fight. Thus, the metabolomic approach allows the recognition of metabolites produced by plant hosts as well as by pathogens during their interaction (Castro-Moretti *et al.*, 2020). In metabolomics, three aims can be achieved, either the qualification of known and unknown metabolites, the quantification of metabolites, or the metabolic profiling of all metabolites of the concerned sample (Sai Reddy *et al.*, 2022). Different techniques are used in metabolomic studies, among them high-performance liquid chromatography, mass spectroscopy, gas chromatography, nuclear magnetic resonance, and a combination of these techniques (Swarupa *et al.*, 2016).

14.2.7 Application of metabolomic for profiling plant–pathogen interactions

Several studies were performed to determine the metabolomic profile of plants in response to several pathogens such as fungi, insects, and bacteria. Indeed, Qian *et al.* (2015) have performed a comparative metabolomic study in *Arabidopsis* before and after its infection with the pathogenic bacteria *Pseudomonas syringae* pv. *tomato*. They found that the level of soluble sugars including tagatofuranose, melibose, galactose, and fructose, was increased after infection with the studied bacteria (Qian *et al.*, 2015). In another study, Macel *et al.* (2019) investigated metabolites in 11 pepper species after their infection with the insect *Frankliniella occidentalis*. The results of LC-MS analysis revealed that metabolites were different according to the developmental stages and the tested species. Dimer and monomer diterpene glycosides; capsianosides were highlighted as the major metabolites responsible for pepper species resistance to *Frankliniella occidentalis*; whereas malonylated glycosides were associated with their susceptibility (Macel *et al.*, 2019). Additionally, Sanchez-Arcos *et al.* (2019) explored the metabolic profile of *Pisum sativum*, *Trifolium pratense*, and *Medicago sativa* after their infestation with various races of *Acyrthosiphon pisum* (native and non-native). Mass spectrometry analysis revealed that metabolites were different in the studied plant species, and dependent on the pathogen race. Besides, the identified metabolites were determined as saponins, amino acids, flavonoids, and peptides (Sanchez-Arcos *et al.*, 2019). de Falco *et al.* (2019) compared metabolites in different tomato genotypes (susceptible and tolerant), after their interaction with the insect *Tuta absoluta*. Results of nuclear magnetic resonance (NMR) spectroscopy reflected the high levels of aminobutyric acid in infected samples as well as an important abundance of organic acids, acyl sugars, fatty acids, and feruloyl quinic acid in the tolerant genotype (de Falco *et al.*, 2019). Interestingly, more studies have been dedicated to the investigation of plant metabolites in response to various pathogenic fungi. For instance, Galeano Garcia *et al.* (2018) used LC-MS and matrix-assisted laser desorption ionization mass spectroscopy for the early detection of tomato infection by *Phytophtora infestans*. Their findings revealed that tomatidine is a major marker of infection (Galeano Garcia *et al.*, 2018). For the same plant species, LC-MS analysis has allowed the detection of phenylpropanoids as a main compound related to tomato defense after its infection by *Ralstonia solanacearum* (Zeiss *et al.*, 2019). Moreover, a combined proteomic and metabolomic analysis of tomato infected by *Verticillium dahliae* has demonstrated the involvement of phenyl alanine ammonia lyase and enzymes related to secondary metabolism compounds in the resistance of tomato against this pathogenic fungus (Hu *et al.*, 2019). Zhu *et al.* (2022) showed through metabolomic analysis that the resistance of potato to *Phytophtora infestans* was highlighted by high levels of salicylic acid and other metabolites included in thickening the cell wall and having antimicrobial activities, such as ferulic, caffeoylnolic, and lucidenic acids (Zhu *et al.*, 2022). Besides, Lekota *et al.* (2020) elucidated metabolites interfering in potato resistance to *Spongospora subterranea* infection. Indeed, the abundant metabolites were found to be sugars, phenols, fatty, organic, and amino acids (Lekota *et al.*, 2020).

Another study combining metabolomics and transcriptomics showed the upregulation of genes associated with pathogenesis and carbon metabolism after the interaction of strawberries with powdery mildew. The mainly identified metabolites were flavonoids, amino acids, phenols, and carbohydrates (Duan *et al.*, 2022).

14.3 CRISPR Technology for Precision Editing

14.3.1 Principles of CRISPR-Cas9

CRISPR constitute a family of DNA fragments present in archea and bacteria genomes. Palindromic repeats are short sequences originating from the virus's DNA which has caused infection in predecessor cells. Cas-9 represents a DNA endonuclease. Indeed, in bacteria and archaea, the role of the CRISPR-Cas 9 system lies in providing adaptative antiviral properties toward DNA viruses (El-Mounadi *et al.*, 2020). This system involves three major components: CRISPR-related protein (Cas9) and non-coding RNAs, namely trans-activating CRISPR RNA (tracrRNA), and a precursor CRISPR RNA (pre-crRNA). Cas9 protein is a DNA nuclease containing HNH- and RuvC-like nuclease domains. It is included in CRISPR RNA maturation and CRISPR RNA-guided DNA cleavage. The tracrRNA is a small RNA with a sequence presenting a complementarity to the pre-crRNA repeats, permitting good RNA duplex formation, and the induction of CRISPR RNA-guided DNA cleavage (Zhang *et al.*, 2017). Thus, the Cas9 system can cut DNA specifically and induce the activation of double-stranded break repair. This repair takes place thanks to the mechanism of cellular non-homologous end-joining. As a result, deletions or insertions occur. In another case, if there is a similarity between the target locus and donor template, double-stranded breaks may be repaired by a homology-directed repair pathway (Noman *et al.*, 2016).

14.3.2 Use of CRISPR system for enhancing plant disease tolerance

CRISPR-Cas9 technology has been successfully applied against plant diseases, either by affecting the viral genome, or the host plant susceptibility factors; indeed, among examples of CRISPR/Cas9 system application against plant viral diseases. Ji *et al.* (2015) performed transient tests in *Nicatiana benthamiana* (*N. benthamiana*) and revealed that the single-guide RNA (Sg RNA)-Cas9 system induced inhibition of virus accumulation and resulted in mutations at the target sites (Ji *et al.*, 2015). They showed that transgenic *Arabidopsis* exhibited a high resistance to viral infection due to an overexpression of Sg RNA. Similarly, Baltes *et al.* (2015) have used CRISPR-Cas9 as a tool for the destruction of the bean yellow dwarf virus (BYDV). In fact, the use of BYDV replicas-based transient tests has exhibited that the reagents of CRISPR-Cas caused mutations in the viral genome and decreased the number of virus copies. Besides, the expression of CRISPR-Cas reagents in transgenic plants exposed to BYDV has diminished the viral infection and symptoms (Baltes *et al.*, 2015). In another study, Ali *et al.* (2015) proved that the CRISPR-Cas9 system has allowed the degradation of tomato yellow leaf curl virus, and induced mutations at the target sites. In *N. benthamiana* plants characterized by CRISPR/Cas 9 expression, a deletion or a reduction of viral DNA accumulation was observed with a significant decrease of infection symptoms (Ali *et al.*, 2015). Pyott *et al.* (2016) have successfully used CRISPR Cas-9 technology in order to induce specific mutations in *Arabidopsis thaliana*. Thus, the obtained plants were completely resistant to turnip mosaic virus, the main pathogen of vegetable crops (Pyott *et al.*, 2016). The application of CRISPR/Cas 9 technology has enabled the production of nontransgenic cucumber plants resistant to different viruses (Chandrasekaran *et al.*, 2016). More recently, Zhang *et al.* (2018) demonstrated the importance of CRISPR-Cas9 use in *Arabidopsis* and *N. benthamiana*. Indeed, this technology resulted in high resistance to tobacco and cucumber mosaic viruses through a reduction of viral RNA accumulation and a mitigation of infection symptoms. Additionally, the resistance was found to be inheritable to the progenies (Zhang *et al.*, 2018).

The efficiency of the CRISPR-Cas9 system was highlighted also towards other pathogens like bacteria and fungi. For instance, Malnoy *et al.* (2016) explored the efficiency of CRISPR-Cas9 ribonucleoproteins delivery to the apple cultivar Golden Delicious for desired mutagenesis. They targeted susceptible genes in order to enhance resistance to fire blight disease caused by *Erwinia amylovora*. Their results were positive demonstrating the targeted gene editing and the occurrence of fire blight disease resistance (Malnoy *et al.*, 2016). Besides, de oledo Thomazella *et al.* (2016) edited disease resistance in tomatoes against *Phytophthora capsici*, *Pseudomonas syringae*, and *Xanthomonas* spp. using the CRISPR-Cas9 system. Indeed, they induced small deletions of the tomato gene SlDMR6-1, resulting in disease resistance against the targeted bacteria without affecting tomato development and growth under greenhouse conditions (de oledo Thomazella *et al.*, 2016). Kim *et al.* (2019) employed CRISPR-Cas9 technology to confer resistance to rice against *Xanthomonas oryzae* pv. *oryzae*. In their study, this technology allowed to knockout Os8N3 gene in rice, and the obtained resistance in the transgenic plants was transmitted to the next generations (Kim *et al.*, 2019).

As regards fungal diseases, several studies proved the efficient use of the CRISPR-Cas9 system for their management. Thus, for example, Nekrasov *et al.* (2017) have obtained a nontransgenic tomato resistant to powdery mildew caused by *Oidium neolycopersici* through the application of CRISPR technology. Indeed, this tool has enabled genome deletion and induced, therefore, resistance against the pathogenic fungus (Nekrasov *et al.*, 2017). Whole-genome sequencing analysis showed that the studied tomato plants did not contain any foreign DNA fragments. In another study, Fister *et al.* (2018)

used *Agrobacterium*-based transformation in order to introduce the CRISPR-Cas9 system into cacao leaves for defense enhancement against *Phytophthora tropicalis*. Their work revealed that 27% of the gene TcNPR3 copies responsible for pathogenesis defense response suppression, were deleted in the treated tissues. Besides, high-throughput sequencing demonstrated the absence of mutations above the error rates of background sequencing (Fister *et al.*, 2018). Additionally, Wang *et al.* (2018) showed that the CRISPR-Cas9 system has enabled precise genome editing in grapes, resulting in its resistance to *Botrytis cinereal* (Wang *et al.*, 2018). Lu *et al.* (2020) engineered ethylene responsive 19 genes mutants (ERF019) in *Arabidopsis thaliana* using CRISPR-Cas 9 technology. They found that ERF019 mutants had a higher resistance to the pathogen *Phytophthora parasitica* in comparison with *Arabidopsis* lines overexpressing ERF019. RT-qPCR analyses allowed us to realize that the expression of marker genes responsible for multiple resistance mechanisms was upregulated in a higher manner in the mutant than in the wild lines during the fungal infection (Lu *et al.*, 2020).

14.4 Speed Breeding

14.4.1 Principles of speed breeding

Speed breeding is a tool of plant breeding aiming at accelerating crop development through the modification of growth conditions. Indeed, it allows the optimization of plant growth environment, permitting breeders and researchers to speed up the breeding cycle, by obtaining many generations even in one year (Kumar and Walia, 2024).

14.4.2 Techniques of speed breeding

Speed breeding techniques rely on the modification of variables related to plant growth such as photoperiod, soil moisture management, temperature regulation, and plant hormonal therapies (Fig. 14.2).

14.4.2.1 Photoperiod modification

This technique is based on the modification of the daily duration of plants' exposure to dark and light regimes in order to promote their rapid and fast growth, flowering, and development. The quality of the light system could also have an influence on photosynthetic activity, transpiration rate, and stomatal conductance (Wanga *et al.*, 2021). However, excessive photoperiod application could display some disadvantages, like the increase of hormonal stress and slowing down plant development (Potts *et al.*, 2023).

14.4.2.2 Temperature regulation

The temperature needs to be maintained at the optimum values for plant species in order to provide a good condition for plant development. It was claimed that warmer temperatures usually result in rapid plant growth. Vegetative growth occurs at a higher optimal temperature in comparison with reproductive growth. Indeed, plants are sensitive to a great temperature during the reproductive stage, since the elevated

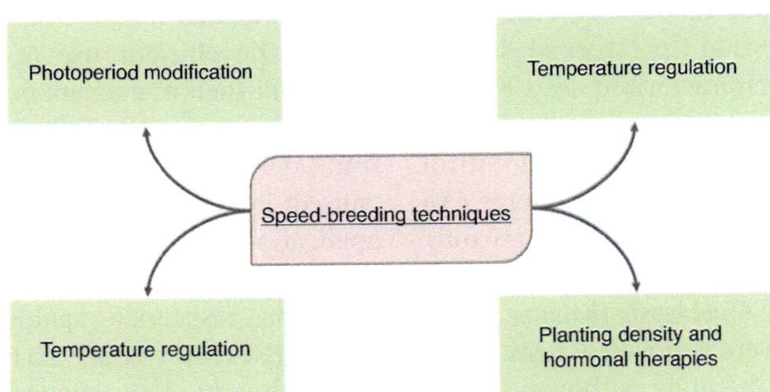

Fig. 14.2. Speed breeding techniques.

temperatures may impact meiosis as well as pollen viability (Pasala *et al.*, 2024). Thus, high temperatures could be employed to accelerate vegetative growth and low temperatures could be employed in reproductive development so as to reach an optimization of speed-breeding procedures.

14.4.2.3 Soil moisture management

Soil moisture can have an effect on plant development by affecting plant seed maturity, height and flowering. Thus, flooding or drought stresses could be used in speed breeding in order to induce early flowering and maturation. In several crops, the reduction of irrigation frequency was found to hasten the breeding of several crops (Kumar, 2022).

14.4.2.4 Planting density

High planting density is based on light competition of plants, causing a fast conversion from the vegetative state to the reproductive one. This technique is important since it allows early flowering, an increase in generation cycle numbers in one year, and permits early plant maturity (Wanga *et al.*, 2021).

14.4.2.5 Plant hormonal therapies

Plant hormonal therapies, in addition to other methods such as plant organ culture, have shown a successful result in accelerating plant development, through flowering speeding and positive response to the regulation of plant growth (Potts *et al.*, 2023).

14.4.3 Advantages of speed breeding in accelerating crop breeding cycles

Several studies have shown the successful application of speed breeding techniques for accelerating their breeding cycles. For instance, Mobini *et al.* (2015) have employed rapid generation technology (RGT) for quick generation development in faba bean and lentil. Indeed, they have used flurpirimidol concentrations, in addition to plant hormones combinations of auxines and cytokinin. Their results revealed that flurpirimidol, indol-3-acetic acid and zeatin at the concentration of 0.3 μM, 5.7 μM, and 2.3 μM, respectively, induced 90% setting seed, and 100% of faba plants flowering (Mobini *et al.*, 2015). Under rapid generation technology, one generation cycle was reached after 45 days for lentil and after 54 days for faba. Similarly, Mobini and Warkentin (2016) have used RGT for *Pisum sativum* L. two cultivars. In fact, using specific growth conditions, namely 266 plants in one square meter, furprimidol at the concentration of 0.6 μM, a photoperiod of 20 hour duration, and temperatures of 21 and 16°C in light and dark, respectively, high intensity of 500 μM m^{-2} s^{-1}, vermiculite substrate, and use of scheduled fertilizer, allowed 100% of plants flowering after 33.4 days on average (Mobini and Warkentin, 2016). Moreover, the seed setting took place in 97.8% of plants after 68.4 days per generation. The efficiency of rapid generation technology was also revealed by Collard *et al.* (2017) on rice, with a good effect on flowering time and plant height (Collard *et al.*, 2017). In the same context, they have used optimized growth conditions for soybean accelerated breeding. Thus, they obtained a reduced soybean generation time by employing a fluorescent lamp, a temperature of 30 and 25°C in the light and dark, respectively, and a dioxide carbon supplementation at a concentration of 400 ppm. Thus, five soybean generations were obtained in one year instead of only one or two generations. In another study, Croser *et al.* (2016) assessed the effect of light on lupin flowering. They found that a positive flowering response was obtained at red to far red (R/FR) declined ratios (Croser *et al.*, 2016). In the R/FR ratios below 3.5, the highest induction was induced. Jähne *et al.* (2020) have shown that a 10 hour photoperiod employing blue light enriched and deprived of the far-red light spectrum has resulted in soybean growth facilitation by allowing flowering after 23 days and maturity after 77 days of sowing (Jähne *et al.*, 2020). Rice and amaranth flowering were obtained after 35 and 60 days of sowing, respectively. Interestingly, Pazos-Navarro *et al.* (2017) used an assisted single-seed descent technique in order to accelerate *Trifolium subterraneum* L. (*T. subterraneum* L.) breeding cycle. Their findings revealed the development of multiple generations (2.7–6.1) in different *T. subterraneum* cultivars (Pazos-Navarro *et al.*, 2017). Recently, Samineni *et al.* (2020) have demonstrated that the increase of photoperiod allows early flowering and seed germination in chikpea. Overall, all these studies show the importance of speed breeding techniques in accelerating breeding cycles (Samineni *et al.*, 2020).

14.4.4 Application for disease phenotyping and tolerance in crops

Speed breeding plays an interesting role in plant disease phenotyping and resistance. Indeed, it was found that speed breeding contributed to cassava resistance against brown streak disease, through speeding cassava breeding strategies, by stimulating flowering and increasing viable flower numbers (Bizimana *et al.*, 2024). Speed breeding has been also employed to enhance wheat resistance to leaf rust (Jamil *et al.*, 2020). The use of speed breeding has allowed a rapid transfer of genes encoding for multiple disease resistance in barley (Hickey *et al.*, 2017). Similarly, speed breeding application has resulted in the rapid transfer of genes related to multiple quantitative traits in *Triticum aestivum*, including resistance to leaf rust and crown rot (Alahmad *et al.*, 2018). Speed breeding has permitted quick phenotyping of adult wheat plant resistance to tan spots. Indeed, using constant and controlled light and temperatures, wheat cultivars with moderate and high resistance to tan spots were observed (Dinglasan *et al.*, 2016). Rapid phenotyping of wheat leaf rust resistance was also achieved under speed breeding conditions including controlled light and temperature (Riaz *et al.*, 2016).

14.4.5 Integration of speed breeding with omics-guided approaches

14.4.5.1 Integration with gene mapping and marker-assisted selection

Gene mapping allows the determination of main genes and quantitative trait loci related to a phenotypic variation, permitting more overview of the genetic control of important agronomical traits, and providing essential data for marker-assisted selection studies (Dimitrijevic and Horn, 2017; Trkulja *et al.*, 2019). The accessibility to new genotyping and sequencing technologies has facilitated the assessment of genetic material in different seasons and environments for a specific agronomical trait. However, the development of different mapping populations needs more time for parent selection and crossing. These tasks can be carried out using speed breeding, which allows the production of wide and genetically different populations at less time and cost (Watson *et al.*, 2018). The integration of gene mapping tools and marker-assisted selection with speed breeding techniques could shorten plant breeding durations and provide insurance for food security and the growth of climate-resistant varieties (Watson *et al.*, 2018; Raza *et al.*, 2023). The combination of speed breeding with gene mapping may help the discovery and insertion of resistant genes to biotic and abiotic stresses in different crops (Devi *et al.*, 2023). Besides, based on genomics and speed breeding, Riaz *et al.* (2017) have succeeded in the phenotyping and determination of wheat-resistant genes to *Puccinia triticina* (Riaz *et al.*, 2017). Croser *et al.* (2021) have also revealed the beneficial use of speed breeding and marker-assisted selection for the development of herbicide-tolerant chickpea (Croser *et al.*, 2021).

14.4.5.2 Integration with genomic selection

Unlike marker-assisted selection which relies on only some main effect loci to carry out selection, genomic selection includes the use of genome-wide markers in order to evaluate complex traits controlled by several loci (Meuwissen *et al.*, 2001).

Genomic selection permits a quick rate of genetic trait selection. However, it becomes more efficient when combined with techniques that reduce generation duration (Hickey *et al.*, 2017). This approach has been used in many crops. For instance, it has allowed diverse genetic complex traits in spring wheat (Voss-Fels *et al.*, 2019). Speed breeding in addition to artificial intelligence is able to facilitate genomic understanding through machine learning for genomic selection (Niazian and Niedbała, 2020).

14.4.5.3 Integration with genome editing

Genetic engineering involves the insertion or deletion of gene fragments in a crop through the use of biotechnology. These techniques reflect many benefits in comparison with traditional breeding approaches (Rai, 2022). The combination of speed breeding with genome editing has led to fascinating results in crops by reducing the time of plant development, resistance to pathogens, and enhanced quality. Indeed, CRISPR enables the rapid introduction of specific traits or modifications (Schindele *et al.*, 2020). Speed breeding allows these modified plants to go through several generations rapidly; thus, researchers can evaluate the effects of the changes in a short time. Besides, using CRISPR to introduce desirable traits and speed breeding helps researchers to quickly determine and select the best-performing plants with the desired trait (Samantara *et al.*, 2022). The combination of CRISPR technology with speed breeding leads to an optimization of breeding programs. In fact, CRISPR can make targeted genetic changes in crops, and speed breeding can, in a short period, validate these changes, allowing for more efficient and effective breeding programs (Begna, 2022).

14.5 Challenges and Future Directions

14.5.1 Challenges and future directions in omics tools

Despite the important role of omics tools in profiling plant–pathogen interactions, their application faces several challenges, leading to future directions for their optimal use. Indeed, the major challenge in genomic studies is the difficulty in defining species and microbial communities due to genomics complexity and horizontal gene transfer (Crandall *et al.*, 2020). As for transcriptomics, the interpretation of transcriptomic data is the most challenging task, needing specific and robust analysis platforms for the best data interpretation (Gautam *et al.*, 2022). Concerning proteomics, since it relies on complex analytical workflow, the results obtained by these studies can be limited by each step; namely, sample preparation, which is challenged by the high amounts of lipids, carbohydrates, and organic acids, which could impact proteins extraction, separation of peptides through liquid chromatography, mass spectroscopy analysis, and the interpretation of informatics data (Liu *et al.*, 2019). Besides, proteomics is challenged by the obstacles related to the differentiation between plants and microbial proteins, as well as by the problems associated with post-translational modifications which can affect protein structure stability and activity (Jayaraman *et al.*, 2012). A set of future directions is promising to overcome challenges in omics tools. Thus, for instance, single-cell and spatial profiling allow a better understanding of plant–pathogen interactions, through RNA sequencing in single cells, and examining interactions across different spatial scales within plant tissues. Long-read sequencing development will allow for enhancement of the ability to resolve complex genomic structures, like repetitive regions, as well as structural variations in plants and pathogens (Hamim *et al.*, 2022). Besides, machine learning and data integration use are interesting for the analysis of wide range of transcriptomics and genomics datasets, as they facilitate understanding the molecular mechanism involved in plants pathogens interactions (Mishra *et al.*, 2019). For sample preparation in proteomic studies, different methods can be applied for efficient proteomic analysis, including the use of protein precipitation by acetone and trichloroacetic (Liu *et al.*, 2019). As for metabolomic studies, the pilar future directions for better and deep metabolomics profiling are the use of mass spectrometry imaging, matrix-assisted laser desorption, and ionization, as well as desorption electrospray ionization (Walker, 2021; Sivanesan *et al.*, 2023; Parker *et al.*, 2023). Indeed, these methods will enable a comprehensive determination of metabolites included in plant–pathogen interactions. Therefore, these future directions are very important for better use of omics tools in the study of plant–pathogen interactions.

14.5.2 Challenges and future directions in speed breeding

Although speed breeding exhibits several advantages, it presents some limitations. Indeed, excessive conditions used for plant growth can lead to adverse effects by limiting their growth and causing a decrease in seed yield. For instance, the intensive photoperiod can induce slowing plant growth, and the elevation of stress hormones (Ćeran *et al.*, 2024). Speed breeding is also challenged by the need for automated platforms lasers, thermometers, hyperspectral sensors, and humidity meters for accurate measurement (Sharma *et al.*, 2023). Besides, maintaining controlled conditions in speed breeding poses the problem of high energy consumption, and leads to an increase in the cost of speed breeding applications. Speed breeding technology needs effective expertise in biotechnology and plant breeding for optimal use of speed breeding techniques (Wanga *et al.*, 2021). Thus, several future directions are intended according to these challenges, such as the enhancement of plant adaptability to changing environments, through molecular methods and by development of plants tolerant to climate change for example. The use of artificial intelligence will help handle large and complex datasets, facilitating the right decisions and strategies for successful speed breeding (Pasala *et al.*, 2024). Also, the use of sustainable energy, like solar panels, constitutes a potential solution regarding the high cost of energy consumption for maintaining the appropriate conditions. Finally, training personnel on the knowledge needed to ensure sustainable speed breeding conditions constitutes a good strategy for better use of speed breeding in crops (Ćeran *et al.*, 2024).

14.5.3 Challenges and future directions in the CRISPR system

Despite the benefits of the use of CRISPR-Cas9 for the development of disease-tolerant crops, this technology is limited by some challenges. Indeed, the application of CRISPR-Cas9 can lead sometimes to editing undesirable regions of the genome. These off-target effects may cause a potential unintended change affecting crop characteristics or safety (Guo *et al.*, 2023). CRISPR components delivery into plant cells is also challenging and can be technically demanding to ensure that CRISPR-Cas9 reaches the targeted cells and induces the intended edits (Wang *et al.*, 2022). Long-term stability is also an important issue that should be guaranteed since unexpected genetic changes could occur over time, and result in the alteration of plant disease resistance; thus, the edited traits must remain stable across generations (He *et al.*, 2018). Furthermore, regulations constitute a challenge for CRISPR-edited crops, since they can prevent their commercialization or their acceptance by consumers (Zhang *et al.*, 2020). Several future directions are established for the improvement of CRISPR system application in editing disease-tolerant crops. Thus, developing a better delivery method for CRISPR components and the enhancement of CRISPR editing specificity for minimizing off-target effects constitute good ways to improve CRISPR system precision and efficiency (Naeem *et al.*, 2020; Hasanzadeh *et al.*, 2022). Making more effort to identify genes resistant to different pathogens will allow the design of crops resistant to a wide range of pathogens (Tyagi *et al.*, 2021). Multiple genes edition is also an important technique that enables engineering different traits, including plant pathogen resistance (Abdelrahman *et al.*, 2021).

Finally, increasing people's acceptance of CRISPR-edited crops is also important; thus, efforts in communication and transparency are needed to move forward.

14.6 Global Implications for Sustainable Agriculture and Food Safety

14.6.1 Omics tools implications for sustainable agriculture

Metabolomics, proteomics, transcriptomics, and genomics approaches contribute to the development of agriculture, and to ensuring food safety, through several implications. Indeed, genomic tools, genetic mapping, and the acceleration of breeding allow the determination of genes related to desirable traits, like resistance to pathogens and abiotic stress tolerance (Dolatabadian *et al.*, 2020; Li *et al.*, 2020). Hence, marker-assisted selection and genomic selection enable rapid breeding of plants (Hasan *et al.*, 2021; Krishnappa *et al.*, 2021). Genomic studies are also important for the determination of genetic diversity, and they help ensure more resistant crops, as well as the preservation of genetic resources (Govindaraj *et al.*, 2015; Wani *et al.*, 2022).

As for transcriptomic studies, they enable the analysis of plant responses to biotic stresses, as well as their reactions and tolerance mechanisms to abiotic stresses, including heavy metals, salt, temperature, and water (Wang *et al.*, 2020). Transcriptomics can be applied also to understand gene expression patterns involved in different developmental stages and to have an insight into the control of different properties in plants during their productivity and growth (Rangan *et al.*, 2017).

Concerning proteomics, it provides a global view of plant proteins engaged in several processes, like growth, diseases, and stress responses (Ma *et al.*, 2016; Mustafa and Komatsu, 2021). It aids in the determination of protein biomarkers for the early detection of plant diseases or their nutrient deficiencies (Cheng *et al.*, 2019; Kgang *et al.*, 2023).

Proteomic studies help to assess food quality and security, through the detection of the presence of allergens, contaminants, and adulterants (Afzaal *et al.*, 2022). Likewise, metabolomics can also be applied for the detection and quantification of harmful substances, like mycotoxins, or pesticides in food products (López-Ruiz *et al.*, 2019). Metabolomics analysis is interesting for the identification of key metabolites contributing to crop nutritional quality (Singh *et al.*, 2022); hence, it guides breeding programs regarding the enhancement of food nutritional content. Additionally, metabolomics can be used for maintaining metabolic changes in plants in response to various stresses, helping the development of crops with the ability to withstand adverse conditions (Khan *et al.*, 2017). Thus, omics tools play an important role in the agricultural field and food security.

14.6.2 Implications of speed breeding

Speed breeding exhibits several implications in the agricultural field. Indeed, it contributes to the enhancement of crop yield by enabling more crop cycles per year; thus, the time required by plants to reach maturity and produce seeds is shortened, and crop yield from the same plot of land is increased (Begna, 2022). Speed breeding allows developing crop varieties exhibiting resistance to climate change, and overcomes challenges related to climate change, such as the increased occurrence of diseases and pests (Rai, 2022). It enables exploration of a wide range of genetic diversity in a short time, which is important for the development of novel traits able to improve crop quality and productivity (Potts *et al.*, 2023). Speed breeding leads to a decrease in pesticide use, through an enhancement of resistance to pests, and contributes to the optimization of the use of fertilizers and water, playing an important role in sustainable agriculture (Pasala *et al.*, 2024).

14.6.3 Implications of CRISPR for agriculture and food security

CRISPR technology represents a potential tool for agriculture and food security. It is involved in crop improvement by enhancing plants' resistance to biotic and abiotic stresses, as well as by increasing the nutritional content of crops, including vitamin and mineral levels (Kumar *et al.*, 2022; Wang *et al.*, 2022). It constitutes a very interesting technology for increasing plant photosynthesis, through editing genes associated with the photosynthesis process, boosting crop yield and productivity (Khumsupan *et al.*, 2019). The CRISPR system allows the reduction of agricultural waste by extending postharvest fruit shelf life and storage properties, and by altering ripening genes, such as those responsible for ethylene production, fruit softening, and tissue browning (Hemalatha *et al.*, 2023). CRISPR is involved in ensuring sustainable farming practices, such as reducing the use of chemical fertilizers and pesticides, thanks to its contribution to nutrient enrichment and plant tolerance to pathogens (Ahmad *et al.*, 2021). The ability of CRISPR technology to perform precise and targeted changes in plant genomes is highly important for precision breeding, which makes it an efficient tool for sustainable farming practices (Nerkar *et al.*, 2022).

14.7 Conclusion

Omics tools integration in profiling plant–pathogen interactions has enabled our understanding of the complex dialogs that underpin disease resistance in crops. By leveraging metabolomics, proteomics, genomics, and transcriptomics, researchers can now dissect these interactions at unprecedented levels of detail, revealing the main genes and pathways included in plant defense mechanisms. This comprehensive profiling of host–pathogen interactions is important for developing efficient strategies to combat emerging pathogens and reduce the bad effects of biotic stresses on global food production. CRISPR/Cas9 tool has further amplified the potential of integrative omics in crop improvement. By enabling precise genome editing, CRISPR allows the rapid introduction of disease-resistance traits into elite crop varieties. The synergy between omics-guided insights and CRISPR-based editing strategies has given rise to speed breeding, a transformative approach that significantly accelerates the development of resilient crops. This integration of advanced molecular techniques and traditional breeding practices represents a paradigm shift in agricultural research, enabling a sustainable and food-secure future.

Acknowledgments

This research was financially supported by the Phytopathology Unit, Department of Plant Protection and Environment, Ecole Nationale d'Agriculture de Meknes, Morocco.

References

Abdelrahman, M., Wei, Z., Rohila, J.S. and Zhao, K. (2021) Multiplex genome-editing technologies for revolutionizing plant biology and crop improvement. *Frontiers in Plant Science* 12, 721203.

Afzaal, M., Saeed, F., Hussain, M., Shahid, F., Siddeeg, A. *et al.* (2022) Proteomics as a promising biomarker in food authentication, quality and safety: A review. *Food Science & Nutrition* 10(7), 2333–2346.

Agrahari, R.K., Singh, P., Koyama, H. and Panda, S.K. (2020) Plant-microbe interactions for sustainable agriculture in the post-genomic era. *Current Genomics* 21(3), 168–178.

Ahmad, S., Tang, L., Shahzad, R., Mawia, A.M., Rao, G.S. *et al.* (2021) CRISPR-based crop improvements: A way forward to achieve zero hunger. *Journal of Agricultural and Food Chemistry* 69(30), 8307–8323.

Alahmad, S., Dinglasan, E., Leung, K.M., Riaz, A., Derbal, N. *et al.* (2018) Speed breeding for multiple quantitative traits in durum wheat. *Plant Methods* 14, 36.

Ali, Z., Abulfaraj, A., Idris, A., Ali, S., Tashkandi, M. *et al.* (2015) CRISPR/Cas9-mediated viral interference in plants. *Genome Biology* 16, 238.

Ashraf, N., Basu, S., Narula, K., Ghosh, S., Tayal, R. *et al.* (2018) Integrative network analyses of wilt transcriptome in chickpea reveal genotype dependent regulatory hubs in immunity and susceptibility. *Scientific Reports* 8(1), 6528.

Astapati, A.D. and Nath, S. (2023) The complex interplay between plant-microbe and virus interactions in sustainable agriculture: Harnessing phytomicrobiomes for enhanced soil health, designer plants, resource use efficiency, and food security. *Crop Design* 2(1), 100028.

Baltes, N.J., Hummel, A.W., Konecna, E., Cegan, R., Bruns, A.N. *et al.* (2015) Conferring resistance to geminiviruses with the CRISPR-Cas prokaryotic immune system. *Nature Plants* 1(10), 1–4.

Bartoli, C. and Roux, F. (2017) Genome-wide association studies in plant pathosystems: Toward an ecological genomics approach. *Frontiers in Plant Science* 8, 763.

Begna, T. (2022) Speed breeding to accelerate crop improvement. *International Journal of Agricultural Science and Food Technology* 8(2), 178–186.

Bhadauria, V., Vijayan, P., Wei, Y. and Banniza, S. (2017) Transcriptome analysis reveals a complex interplay between resistance and effector genes during the compatible lentil-*Colletotrichum lentis* interaction. *Scientific Reports* 7, 42338.

Bhimani, P., Mahavar, P., Rajguru, B., Bhatt, V.D., Nathani, N.M. *et al.* (2024) Unveiling the green dialogue: Advancements in omics technologies for deciphering plant–microbe interactions in soil. *Discover Plants* 1(1).

Bizimana, J.P., Ngapout, Y., Nyirakanani, C., Shakir, S., Kanju, E. *et al.* (2024) Breeding strategies for mitigating cassava brown streak disease in Africa. *Tropical Plants* 3(1). DOI: 10.48130/Tp-0024-0006.

Castro-Moretti, F.R., Gentzel, I.N., Mackey, D. and Alonso, A.P. (2020) Metabolomics as an emerging tool for the study of plant–pathogen interactions. *Metabolites* 10, 52.

Ćeran, M., Miladinović, D., Đorđević, V., Trkulja, D., Radanović, A. *et al.* (2024) Genomics-assisted speed breeding for crop improvement: Present and future. *Frontiers in Sustainable Food Systems* 8, 1383302.

Chandrasekaran, J., Brumin, M., Wolf, D., Leibman, D., Klap, C. *et al.* (2016) Development of broad virus resistance in non-transgenic cucumber using CRISPR/Cas9 technology. *Molecular Plant Pathology* 17(7), 1140–1153.

Chang, C., Tian, L., Ma, L., Li, W., Nasir, F. *et al.* (2019) Differential responses of molecular mechanisms and physiochemical characters in wild and cultivated soybeans against invasion by the pathogenic *Fusarium oxysporum schltdl*. *Physiologia Plantarum* 166(4), 1008–1025.

Cheng, L., Zhang, S., Yang, L., Wang, Y., Yu, B. *et al.* (2019) Comparative proteomics illustrates the complexity of Fe, Mn and Zn deficiency-responsive mechanisms of potato (*Solanum tuberosum* L.) plants *in vitro*. *Planta* 250(1), 199–217.

Chen, Y., Ren, X., Zhou, X., Huang, L., Yan, L. *et al.* (2014) Dynamics in the resistant and susceptible peanut (*Arachis hypogaea* L.) root transcriptome on infection with the *Ralstonia solanacearum*. *BMC Genomics* 15, 1–16.

Collard, B.C.Y., Beredo, J.C., Lenaerts, B., Mendoza, R., Santelices, R. *et al.* (2017) Revisiting rice breeding methods – evaluating the use of rapid generation advance (RGA) for routine rice breeding. *Plant Production Science* 20(4), 337–352.

Copley, T.R., Aliferis, K.A., Kliebenstein, D.J. and Jabaji, S.H. (2017) An integrated RNAseq-1 H NMR metabolomics approach to understand soybean primary metabolism regulation in response to rhizoctonia foliar blight disease. *BMC Plant Biology* 17, 1–84.

Crandall, S.G., Gold, K.M., Jiménez-Gasco, M.D.M., Filgueiras, C.C. and Willett, D.S. (2020) A multi-omics approach to solving problems in plant disease ecology. *PLoS One* 15(9), e0237975.

Croser, J.S., Pazos-Navarro, M., Bennett, R.G., Tschirren, S., Edwards, K. *et al.* (2016) Time to flowering of temperate pulses *in vivo* and generation turnover *in vivo–in vitro* of narrow-leaf lupin accelerated by low red to far-red ratio and high intensity in the far-red region. *Plant Cell, Tissue and Organ Culture* 127(3), 591–599.

Croser, J.S., Mao, D., Dron, N., Michelmore, S., McMurray, L. *et al.* (2021) Evidence for the application of emerging technologies to accelerate crop improvement – A collaborative pipeline to introgress herbicide tolerance into chickpea. *Frontiers in Plant Science* 12, 779122.

de Falco, B., Manzo, D., Incerti, G., Garonna, A.P., Ercolano, M. *et al.* (2019) Metabolomics approach based on NMR spectroscopy and multivariate data analysis to explore the interaction between the leafminer *Tuta absoluta* and tomato (*Solanum lycopersicum*). *Phytochemical Analysis* 30(5), 556–563.

de oledo Thomazella, D.P., Brail, Q., Dahlbeck, D. and Staskawicz, B. (2016) CRISPR-Cas9 mediated mutagenesis of a DMR6 ortholog in tomato confers broad-spectrum disease resistance. *BioRxiv*.

Devi, O.B., Haokip, S.W. and Singh, Y.S. (2023) Speed breeding: An effective strategy for breeding of crop new kinds and its future prospects. *Vigyan Varta* 4, 83–86.

Dimitrijevic, A. and Horn, R. (2017) Sunflower hybrid breeding: From markers to genomic selection. *Frontiers in Plant Science* 8, 2238.

Dinglasan, E., Godwin, I.D., Mortlock, M.Y. and Hickey, L.T. (2016) Resistance to yellow spot in wheat grown under accelerated growth conditions. *Euphytica* 209(3), 693–707.

Dolatabadian, A., Bayer, P.E., Tirnaz, S., Hurgobin, B., Edwards, D. *et al.* (2020) Characterization of disease resistance genes in the *Brassica napus* pangenome reveals significant structural variation. *Plant Biotechnology Journal* 18(4), 969–982.

Dos Santos, E.C., Pirovani, C.P., Correa, S.C., Micheli, F. and Gramacho, K.P. (2020) The pathogen *Moniliophthora perniciosa* promotes differential proteomic modulation of cacao genotypes with contrasting resistance to witches' broom disease. *BMC Plant Biology* 20, 1–21.

Duan, W., Peng, L., Jiang, J., Zhang, H. and Tong, G. (2022) Combined transcriptome and metabolome analysis of strawberry fruits in response to powdery mildew infection. *Agronomy Journal* 114(2), 1027–1039.

El-Mounadi, K., Morales-Floriano, M.L. and Garcia-Ruiz, H. (2020) Principles, applications, and biosafety of plant genome editing using CRISPR-Cas9. *Frontiers in Plant Science* 11, 56.

Fister, A.S., Landherr, L., Maximova, S.N. and Guiltinan, M.J. (2018) Transient expression of CRISPR/Cas9 machinery targeting TcNPR3 enhances defense response in *Theobroma cacao*. *Frontiers in Plant Science* 9, 268.

Galeano Garcia, P., Neves Dos Santos, F., Zanotta, S., Eberlin, M.N. and Carazzone, C. (2018) Metabolomics of *Solanum lycopersicum* infected with *Phytophthora infestans* leads to early detection of late blight in asymptomatic plants. *Molecules* 23, 3330.

Gautam, P., Kumar, R., Feroz, Z., Vijayaraghavalu, S. and Kumar, M. (2022) RNA interference technology in plants: Mechanisms and applications in crop improvement. In: Singh, R.L., Mondal, S., Parihar, A. and Singh, P.K. (eds) *Plant Genomics for Sustainable Agriculture*. Springer, pp. 265–290.

Govindaraj, M., Vetriventhan, M. and Srinivasan, M. (2015) Importance of genetic diversity assessment in crop plants and its recent advances: An overview of its analytical perspectives. *Genetics Research International* 431487.

Guimaraes, P.M., Guimaraes, L.A., Morgante, C.V., Silva, O.B., Jr, Araujo, A.C.G. *et al.* (2015) Root transcriptome analysis of wild peanut reveals candidate genes for nematode resistance. *PLoS One* 10, e0140937.

Guo, C., Ma, X., Gao, F. and Guo, Y. (2023) Off-target effects in CRISPR/Cas9 gene editing. *Frontiers in Bioengineering and Biotechnology* 11, 1143157.

Gupta, S., Bhar, A., Chatterjee, M., Ghosh, A. and Das, S. (2017) Transcriptomic dissection reveals wide spread differential expression in chickpea during early time points of *Fusarium oxysporum* f.sp. *ciceri* Race 1 attack. *PLoS One* 12, e0178164.

Hamim, I., Sekine, K.-T. and Komatsu, K. (2022) How do emerging long-read sequencing technologies function in transforming the plant pathology research landscape?. *Plant Molecular Biology* 110(6), 469–484.

Hasan, N., Choudhary, S., Naaz, N., Sharma, N. and Laskar, R.A. (2021) Recent advancements in molecular marker-assisted selection and applications in plant breeding programmes. *Journal Genetic Engineering & Biotechnology* 19(1), 128.

Hasanzadeh, A., Noori, H., Jahandideh, A., Haeri Moghaddam, N., Kamrani Mousavi, S.M. *et al.* (2022) Smart strategies for precise delivery of CRISPR/Cas9 in genome editing. *ACS Applied Bio Materials* 5(2), 413–437.

He, Y., Zhu, M., Wang, L., Wu, J., Wang, Q. *et al.* (2018) Programmed self-elimination of the CRISPR/Cas9 construct greatly accelerates the isolation of edited and transgene-free rice plants. *Molecular Plant* 11(9), 1210–1213.

Hemalatha, P., Abda, E.M., Shah, S., Prabhu, S.V., Jayakumar, M. *et al.* (2023) Multi-faceted CRISPR-Cas9 strategy to reduce plant based food loss and waste for sustainable bio-economy: A review. *Journal of Environmental Management* 332, 117382.

Hickey, L.T., Germán, S.E., Pereyra, S.A., Diaz, J.E., Ziems, L.A. *et al.* (2017) Speed breeding for multiple disease resistance in barley. *Euphytica* 213, 1–14.

Hu, X., Puri, K.D., Gurung, S., Klosterman, S.J., Wallis, C.M. *et al.* (2019) Proteome and metabolome analyses reveal differential responses in tomato–*Verticillium dahlia* interactions. *Journal of Proteomics* 207, 103449.

Ishiga, Y., Uppalapati, S.R., Gill, U.S., Huhman, D., Tang, Y. *et al.* (2015) Transcriptomic and metabolomic analyses identify a role for chlorophyll catabolism and phytoalexin during Medicago nonhost resistance against Asian soybean rust. *Scientific Reports* 5, 1–17.

Jähne, F., Hahn, V., Würschum, T. and Leiser, W.L. (2020) Speed breeding short-day crops by LED-controlled light schemes. *Theoretical and Applied Genetics* 133(8), 2335–2342.

Jain, S., Chittem, K., Brueggeman, R., Osorno, J.M., Richards, J. *et al.* (2016) Comparative transcriptome analysis of resistant and susceptible common bean genotypes in response to soybean cyst nematode infection. *PLoS One* 11, e0159338.

Jain, A., Sarsaiya, S., Singh, R., Gong, Q., Wu, Q. *et al.* (2024) Omics approaches in understanding the benefits of plant-microbe interactions. *Frontiers in Microbiology* 15, 1391059.

Jamil, S., Shahzad, R., Ahmad, S., Fatima, R., Zahid, R. *et al.* (2020) Role of genetics, genomics, and breeding approaches to combat stripe rust of wheat. *Frontiers in Nutrition* 7, 580715.

Jayaraman, D., Forshey, K.L., Grimsrud, P.A. and Ané, J.-M. (2012) Leveraging proteomics to understand plant-microbe interactions. *Frontiers in Plant Science* 3, 44.

Jiang, N., Yan, J., Liang, Y., Shi, Y., He, Z. *et al.* (2023) Resistance genes and their interactions with bacterial blight/leaf streak pathogens (*Xanthomonas oryzae*) in rice (*Oryza sativa* L.): An updated review. *Oxford English Dictionary*.

Ji, X., Zhang, H., Zhang, Y., Wang, Y. and Gao, C. (2015) Establishing a CRISPR-Cas-like immune system conferring DNA virus resistance in plants. *Nature Plants* 1, 15144.

Kgang, I.E., Klein, A., Husselmann, L., Nkomo, A., Mathabe, P.M.K. *et al.* (2023) Bioassays and proteomics as early detection tools in postharvest management of table grapes (*Vitis vinifera* L.) diseases – a review. *Food Bioscience* 53, 102645.

Khan, N., Ali, S., Shahid, M.A. and Kharabian-Masouleh, A. (2017) Advances in detection of stress tolerance in plants through metabolomics approaches. *Plant Omics* 10(3), 153–163.

Khumsupan, P., Donovan, S. and McCormick, A.J. (2019) CRISPR/Cas in *Arabidopsis*: Overcoming challenges to accelerate improvements in crop photosynthetic efficiencies. *Physiologia Plantarum* 166(1), 428–437.

Kimotho, R.N. and Maina, S. (2024) Unraveling plant-microbe interactions: Can integrated omics approaches offer concrete answers? *Journal of Experimental Botany* 75(5), 1289–1313.

Kim, Y.-A., Moon, H. and Park, C.-J. (2019) CRISPR/Cas9-targeted mutagenesis of Os8N3 in rice to confer resistance to *Xanthomonas oryzae* pv. *oryzae*. *Rice* 12(1), 1–13.

Krishnappa, G., Savadi, S., Tyagi, B.S., Singh, S.K., Mamrutha, H.M. *et al.* (2021) Integrated genomic selection for rapid improvement of crops. *Genomics* 113(3), 1070–1086.

Kumar, V. (2022) Speed breeding: Accelerated plant breeding. *Journal of Agriculture Research and Technology* 36–39.

Kumar, K.P. and Walia, P. (2024) Speed breeding in cereal crops: Accelerating genetic improvement for rapid agricultural advancement. *Journal of Experimental Agriculture International* 46(6), 465–477.

Kumar, A., Kumar, R., Shukla, P. and Patel, H.K. (2021) *Omics Technologies for Sustainable Agriculture and Global Food Security (Vol. II)*. Springer.

Kumar, D., Yadav, A., Ahmad, R., Dwivedi, U.N. and Yadav, K. (2022) CRISPR-based genome editing for nutrient enrichment in crops: A promising approach toward global food security. *Frontiers in Genetics* 13, 932859.

Lanubile, A., Muppirala, U.K., Severin, A.J., Marocco, A. and Munkvold, G.P. (2015) Transcriptome profiling of soybean (*Glycine max*) roots challenged with pathogenic and non-pathogenic isolates of *Fusarium oxysporum*. *BMC Genomics* 16, 1089.

Li, Y., Feng, Y., Lü, Q., Yan, D., Liu, Z. *et al.* (2019) Comparative proteomic analysis of plant-pathogen interactions in resistant and susceptible poplar ecotypes infected with *Botryosphaeria dothidea*. *Phytopathology* 109(12), 2009–2021.

Lekota, M., Modisane, K.J., Apostolides, Z. and Waals, J.E. (2020) Metabolomic fingerprinting of potato cultivars differing in susceptibility to *Spongospora subterranea* f. sp *subterranea* root Infection. *SuInternational Journal of Molecular Sciences* 21(11), 3788.

Li, J., Liu, H., Yang, C., Wang, J., Yan, G. *et al.* (2020) Genome-wide identification of MYB genes and expression analysis under different biotic and abiotic stresses in *Helianthus annuus* L. *Industrial Crops and Products* 143, 111924.

Lin, F., Zhao, M., Baumann, D.D., Ping, J., Sun, L. *et al.* (2014) Molecular response to the pathogen *phytophthora sojae* among ten soybean near isogenic lines revealed by comparative transcriptomics. *BMC Genomics* 15, 1–13.

Liu, Y., Lu, S., Liu, K., Wang, S., Huang, L. *et al.* (2019) Proteomics: A powerful tool to study plant responses to biotic stress. *Plant Methods* 15, 135.

Liu, S., Li, Z., Yu, B., Wang, S., Shen, Y. *et al.* (2020) Recent advances on protein separation and purification methods. *Advances in Colloid and Interface Science* 284, 102254.

López-Ruiz, R., Romero-González, R. and Garrido Frenich, A. (2019) Metabolomics approaches for the determination of multiple contaminants in food. *Current Opinion in Food Science* 28, 49–57.

Lu, W., Deng, F., Jia, J., Chen, X., Li, J. *et al.* (2020) The *Arabidopsis thaliana* gene aterf019 negatively regulates plant resistance to *Phytophthora parasitica* by suppressing PAMP-triggered immunity. *Molecular Plant Pathology* 21(9), 1179–1193.

Ma, R., Sun, L., Chen, X., Mei, B., Chang, G. *et al.* (2016) Proteomic analyses provide novel insights into plant growth and ginsenoside biosynthesis in forest cultivated panax ginseng (*F. Ginseng*). *Frontiers in Plant Science* 7, 1.

Macel, M., Visschers, I.G.S., Peters, J.L., Kappers, I.F., Vos, R.C.H. *et al.* (2019) Metabolomics of thrips resistance in pepper (*Capsicum* spp.) reveals monomer and dimer acyclic diterpene glycosides as potential chemical defenses. *Journal of Chemical Ecology* 45(5–6), 490–501.

Malnoy, M., Viola, R., Jung, M.-H., Koo, O.-J., Kim, S. *et al.* (2016) DNA-free genetically edited grapevine and apple protoplast using CRISPR/Cas9 ribonucleoproteins. *Frontiers in Plant Science* 7, 1904.

Meng, Q., Gupta, R., Min, C.W., Kwon, S.W., Wang, Y, *et al.* (2019) Proteomics of rice-magnaporthe oryzae interaction: What have we learned so far? *Frontiers in Plant Science* 10, 1383.

Meuwissen, T.H.E., Hayes, B.J. and Goddard, M.E. (2001) Prediction of total genetic value using genome-wide dense marker maps. *Genetics* 157(4), 1819–1829.

Mishra, B., Kumar, N. and Mukhtar, M.S. (2019) Systems biology and machine learning in plant-pathogen interactions. *Molecular Plant-Microbe Interactions* 32(1), 45–55.

Mobini, S.H. and Warkentin, T.D. (2016) A simple and efficient method of *in vivo* rapid generation technology in pea (*Pisum sativum* L.). *Vitro Cellular & Developmental Biology – Plant* 52(5), 530–536.

Mobini, S.H., Lulsdorf, M., Warkentin, T.D. and Vandenberg, A. (2015) Plant growth regulators improve *in vitro* flowering and rapid generation advancement in lentil and faba bean. *Vitro Cellular & Developmental Biology – Plant* 51(1), 71–79.

Mustafa, G. and Komatsu, S. (2021) Plant proteomic research for improvement of food crops under stresses: A review. *Molecular Omics* 17(6), 860–880.

Naeem, M., Majeed, S., Hoque, M.Z. and Ahmad, I. (2020) Latest developed strategies to minimize the off-target effects in CRISPR-Cas-mediated genome editing. *Cells* 9(7), 1608.

Nekrasov, V., Wang, C., Win, J., Lanz, C., Weigel, D. *et al.* (2017) Rapid generation of a transgene-free powdery mildew resistant tomato by genome deletion. *Scientific Reports* 7(1), 482.

Nemchinov, L.G., Shao, J., Lee, M.N., Postnikova, O.A. and Samac, D.A. (2017) Resistant and susceptible responses in alfalfa (*Medicago sativa*) to bacterial stem blight caused by *Pseudomonas syringae* pv. *syringae*. *PLoS One* 12(12), e0189781.

Nerkar, G., Devarumath, S., Purankar, M., Kumar, A., Valarmathi, R. *et al.* (2022) Advances in crop breeding through precision genome editing. *Frontiers in Genetics* 13, 880195.

Niazian, M. and Niedbała, G. (2020) Machine learning for plant breeding and biotechnology. *Agriculture* 10, 436.

Noman, A., Aqeel, M. and He, S. (2016) CRISPR-Cas9: Tool for qualitative and quantitative plant genome editing. *Frontiers in Plant Science* 7, 1740.

Padder, B.A., Kamfwa, K., Awale, H.E. and Kelly, J.D. (2016) Transcriptome profiling of the *Phaseolus vulgaris-Colletotrichum lindemuthianum* pathosystem. *PLoS One* 11(11), e0165823.

Parker, G.D., Hanley, L. and Yu, X.-Y. (2023) Mass spectral imaging to map plant-microbe interactions. *Microorganisms* 11(8), 2045.

Pasala, R., Chennamsetti, M., Patil, B., Kadirvel, P., Geethanjali, S. *et al.* (2024) Revolutionizing crop production: The imperative of speed breeding technology in modern crop improvement. *Crop Breeding, Genetics and Genomics* 6(2), 1–23.

Pazos-Navarro, M., Castello, M., Bennett, R.G., Nichols, P. and Croser, J. (2017) *In vitro*-assisted single-seed descent for breeding-cycle compression in subterranean clover (*Trifolium subterraneum* L.). *Crop and Pasture Science* 68(11), 958.

Postnikova, O.A., Hult, M., Shao, J., Skantar, A. and Nemchinov, L.G. (2015) Transcriptome analysis of resistant and susceptible alfalfa cultivars infected with root-knot nematode *Meloidogyne incognita*. *PLoS One* 10(2), e0118269.

Potts, J., Jangra, S., Michael, V.N. and Wu, X. (2023) Speed breeding for crop improvement and food security. *Crops* 3(4), 276–291.

Pyott, D.E., Sheehan, E. and Molnar, A. (2016) Engineering of CRISPR/Cas9-mediated potyvirus resistance in transgene-free *Arabidopsis* plants. *Molecular Plant Pathology* 17(8), 1276–1288.

Qian, Y., Tan, D.-X., Reiter, R.J. and Shi, H. (2015) Comparative metabolomic analysis highlights the involvement of sugars and glycerol in melatonin-mediated innate immunity against bacterial pathogen in *Arabidopsis*. *Scientific Reports* 5, 15815.

Qian, Y., Sood, S., Luo, Z., Liu, H., Islam, M.S. *et al.* (2021) Identifying genomic regions controlling ratoon stunting disease resistance in sugarcane (*Saccharum* spp.) clonal F1 population. *The Crop Journal* 9(5), 1070–1078.

Rai, K.K. (2022) Integrating speed breeding with artificial intelligence for developing climate-smart crops. *Molecular Biology Reports* 49(12), 11385–11402.

Rangan, P., Furtado, A. and Henry, R.J. (2017) The transcriptome of the developing grain: A resource for understanding seed development and the molecular control of the functional and nutritional properties of wheat. *BMC Genomics* 18(1), 766.

Raza, A., Mubarik, M.S., Sharif, R., Habib, M., Jabeen, W. *et al.* (2023) Developing drought-smart, ready-to-grow future crops. *The Plant Genome* 16, e20279.

Riaz, A., Periyannan, S., Aitken, E. and Hickey, L. (2016) A rapid phenotyping method for adult plant resistance to leaf rust in wheat. *Plant Methods* 12, 1–10.

Riaz, A., Athiyannan, N., Periyannan, S., Afanasenko, O., Mitrofanova, O. *et al.* (2017) Mining Vavilov's treasure chest of wheat diversity for adult plant resistance to *Puccinia triticina*. *Plant Disease* 101(2), 317–323.

Sai Reddy, M.S., Karthik, S., Raju, B.J. and Yashaswini, G. (2022) Multi-omics approaches in insect-plant interactions BT – molecular advances. In: Tanda, A.S. (ed.) *Insect Resistance of Field Crops: Modern and Applied Approaches*. Springer, Cham, Switzerland, pp. 335–368.

Samantara, K., Bohra, A., Mohapatra, S.R., Prihatini, R., Asibe, F. *et al.* (2022) Breeding more crops in less time: A perspective on speed breeding. *Biology* 11(2), 275.

Samineni, S., Sen, M., Sajja, S.B. and Gaur, P.M. (2020) Rapid generation advance (RGA) in chickpea to produce up to seven generations per year and enable speed breeding. *The Crop Journal* 8(1), 164–169.

Sanchez-Arcos, C., Kai, M., Svatoš, A., Gershenzon, J. and Kunert, G. (2019) Untargeted metabolomics approach reveals differences in host plant chemistry before and after infestation with different pea aphid host races. *Frontiers in Plant Science* 10, 188.

Schindele, A., Dorn, A. and Puchta, H. (2020) CRISPR/Cas brings plant biology and breeding into the fast lane. *Current Opinion in Biotechnology* 61, 7–14.

Sharma, M., Sudheer, S., Usmani, Z., Rani, R. and Gupta, P. (2020) Deciphering the omics of plant-microbe interaction: Perspectives and new insights. *Current Genomics* 21(5), 343–362.

Sharma, S., Kumar, A., Dhakte, P., Raturi, G., Vishwakarma, G. *et al.* (2023) Speed breeding opportunities and challenges for crop improvement. *Journal of Plant Growth Regulation* 42(1), 46–59.

Singh, D.P., Bisen, M.S., Shukla, R., Prabha, R., Maurya, S. *et al.* (2022) Metabolomics-driven mining of metabolite resources: Applications and prospects for improving vegetable crops. *International Journal of Molecular Sciences* 23(20), 12062.

Sivanesan, I., Gopal, J., Hasan, N. and Muthu, M. (2023) A systematic assessment of matrix-assisted laser desorption/ionization time of flight mass spectrometry (MALDI-TOF MS) application for rapid identification of pathogenic microbes that affect food crops: Delivered and future deliverables. *RSC Advances* 13(25), 17297–17314.

Swarupa, V., Pavitra, K., Shivashankara, K.S. and Ravishankar, K.V. (2016) Omics-driven approaches in plant–microbe interaction. In: Singh, D.P., Singh, H.B. and Prabha, R. (eds) *Microbial Inoculants in Sustainable Agricultural Productivity: Vol. 1: Research Perspectives*. Springer, New Delhi, pp. 61–84.

Trkulja, D., Kondić-Špika, A., Brbaklić, L., Kobiljski, B., Mikić, S. *et al.* (2019) Genetic structure and allelic richness of the wheat core collection for association mapping of yield. *Zemdirbyste-Agriculture* 106(3), 257–264.

Tyagi, S., Kumar, R., Kumar, V., Won, S.Y. and Shukla, P. (2021) Engineering disease resistant plants through CRISPR-Cas9 technology. *GM Crops & Food* 12(1), 125–144.

Voss-Fels, K.P., Stahl, A., Wittkop, B., Lichthardt, C., Nagler, S. *et al.* (2019) Breeding improves wheat productivity under contrasting agrochemical input levels. *Nature Plants* 5(7), 706–714.

Walker, H.J. (2021) Metabolite imaging by mass spectrometry: A new discovery tool. In: *Advances in Botanical Research*. Elsevier, pp. 271–296.

Wang, X., Tu, M., Wang, D., Liu, J., Li, Y. *et al.* (2018) CRISPR/Cas9-mediated efficient targeted mutagenesis in grape in the first generation. *Plant Biotechnology Journal* 16(4), 844–855.

Wang, X., Li, N., Li, W., Gao, X., Cha, M. *et al.* (2020) Advances in transcriptomics in the response to stress in plants. *Global Medical Genetics* 7(2), 30–34.

Wang, Y., Zafar, N., Ali, Q., Manghwar, H., Wang, G. *et al.* (2022) CRISPR/Cas genome editing technologies for plant improvement against biotic and abiotic stresses: Advances, limitations, and future perspectives. *Cells* 11(23), 3928.

Wanga, M.A., Shimelis, H., Mashilo, J. and Laing, M.D. (2021) Opportunities and challenges of speed breeding: A review. *Plant Breeding* 140(2), 185–194.

Wani, S.H., Khan, H., Riaz, A., Joshi, D.C., Hussain, W. *et al.* (2022) Genetic diversity for developing climate-resilient wheats to achieve food security goals. *Advances in Agronomy* 171, 255–303.

Watson, A., Ghosh, S., Williams, M.J., Cuddy, W.S., Simmonds, J. *et al.* (2018) Speed breeding is a powerful tool to accelerate crop research and breeding. *Nature Plants* 4(1), 23–29.

Weissenbach, J. (2016) L'essor de la génomique. *Comptes Rendus – Biol* 339, 231–239.

Xue, R., Wu, J., Zhu, Z., Wang, L., Wang, X. *et al.* (2015) Differentially expressed genes in resistant and susceptible common bean (*Phaseolus vulgaris* L.) genotypes in response to *Fusarium oxysporum* f. sp. *phaseoli*. *PLoS One* 10(6), e0127698.

Xue, C., Li, H., Liu, Z., Wang, L., Zhao, Y. *et al.* (2019) Genome-wide analysis of the WRKY gene family and their positive responses to phytoplasma invasion in Chinese jujube. *BMC Genomics* 20(1), 464.

Yang, F., Li, W., Derbyshire, M., Larsen, M.R., Rudd, J.J. *et al.* (2015) Unraveling incompatibility between wheat and the fungal pathogen *Zymoseptoria tritici* through apoplastic proteomics. *BMC Genomics* 16(1), 362.

You, Q., Sood, S., Luo, Z., Liu, H., Sariful Islam, M. *et al.* (2021) Identifying genomic regions controlling ratoon stunting disease resistance in sugarcane (*Saccharum* spp.) clonal F1 population. *The Crop Journal* 9(5), 1070–1078.

Zeiss, D.R., Mhlongo, M.I., Tugizimana, F., Steenkamp, P.A., and Dubery, I.A. (2019) Metabolomic profiling of the host response of tomato (*Solanum lycopersicum*) following infection by *Ralstonia solanacearum*. *International Journal of Molecular Sciences* 20(16), 3945.

Zhang, C., Tian, Y., and Cong, P. (2015) Proteome analysis of pathogen-responsive proteins from apple leaves induced by the alternaria blotch *Alternaria alternata*. *PLoS One* 10(6), e0122233.

Zhang, H., Zhang, J., Lang, Z., Botella, J.R. and Zhu, J.-K. (2017) Genome editing: Principles and applications for functional genomics research and crop improvement. *Critical Reviews in Plant Sciences* 36(4), 291–309.

Zhang, T., Zheng, Q., Yi, X., An, H., Zhao, Y. *et al.* (2018) Establishing RNA virus resistance in plants by harnessing CRISPR immune system. *Plant Biotechnology Journal* 16(8), 1415–1423.

Zhang, D., Hussain, A., Manghwar, H., Xie, K., Xie, S. *et al.* (2020) Genome editing with the CRISPR-cas system: An art, ethics and global regulatory perspective. *Plant Biotechnology Journal* 18(8), 1651–1669.

Zhu, J., Tang, X., Sun, Y., Li, Y., Wang, Y. *et al.* (2022) Comparative metabolomic profiling of compatible and incompatible interactions between potato and *Phytophthora infestans*. *Frontiers in Microbiology* 13, 857160.

15 Revealing Complicated Plant–microbial–nanoparticle Interactions Using Bioinformatics and Multi-omics Tools

Rahul Gogoi[1], Priyadarshini Deka[2], Madhurjya Ranjan Sharma[1], Jitul Bora[1], Anshu[3] and Sudipta Sankar Bora[4]*

[1]Department of Agricultural Biotechnology, Assam Agricultural University, Jorhat, Assam, India; [2]Hereditary Biosciences, Guwahati, Assam, India; [3]Department of Medical Oncology, All India Institute of Medical Sciences, New Delhi, India; [4]Multidisciplinary Research Unit, Jorhat Medical College and Hospital, Jorhat, Assam, India

Abstract

The interaction of plants and microorganisms is a complex natural phenomenon that helps in growth and development. In the genomics era, the development of sequencing technologies and various omics tools has sped up biological science. By giving deeper insights, research into the intricate structure of plant–microbe interactions can offer numerous techniques to boost plant productivity in an environmentally benign way. Furthermore, it is thought to be economical and environmentally beneficial to synthesize plant-microbial nanoparticles utilizing nanotechnology. Because of their unique characteristics and small size (1–100 nm), nanoparticles (NPs) have potential applications in a variety of fields, including food, medicine, and agriculture. Understanding the intricate interactions between plants and microbes can be aided by the combination of bioinformatics and multi-omics technologies. The bioinformatics and omics technologies utilized to comprehend the molecular interactions between plants and microbes are highlighted in this chapter, which also emphasizes the benefits of employing this technology in a sustainable way for crop improvement. It also highlights the drawbacks and potential applications of this technology.

Keywords: complex, genomics, boost, economical, environmentally, nanotechnology, sustainable

15.1 Introduction

Microorganisms, with their vast diversity, colonize a wide range of environments including soil, plants, and animals. Plant–microbe interaction is indeed a fundamental process that has been evolving since plants first colonized Earth. These interactions are highly complex and dynamic, and involve continuous communication and quick adaptation between plants and several microbial communities. Plant–microbe interaction plays a crucial role in enhancing the overall capability of their plant hosts. These interactions, which include bacteria, fungi, archaea, and viruses, provide several benefits to plants like encouragement of growth, absorption of

*Corresponding author: sudip.asm@gmail.com

© CAB International 2025. *Stress-Resilient Crops: Coordinated omics-CRISPR-Nanotechnology Strategies* (ed. Jen-Tsung Chen)
DOI: 10.1079/9781800627307.0015

nutrients, tolerance to stress, and pathogen resistance, etc. These interactions are crucial for sustainable agriculture, as they overall decrease the need for chemical fertilizers as well as pesticides, contributing to healthier and stronger crop production systems (Berg, 2009). The recent and advanced technologies have contributed much to studying microbial communities in greater detail, revealing the complex interactions between microbes and their hosts, as well as their roles in various ecosystems. Plant–microbe symbioses are indeed ancient and pivotal associations and plants recognize the symbionts by sharing some molecular machinery with pathogenic responses, reflecting the complex and finely tuned interactions between plants and microorganisms. Plants use some sophisticated molecular interactions to distinguish between beneficial symbionts, like mycorrhizal fungi and nitrogen-fixing bacteria, and harmful pathogens. This recognition relies on a group of receptors called pattern recognition receptors (PRRs) with the ability to recognize microbial-associated molecular patterns (MAMPs). Also, the composition of plant-exuded chemicals is modified during plant–microbe interactions (Dolatabadian, 2020).

However, the specific molecular interactions between plants and the symbiotic microbes show much complexity in understanding the fundamental aspects of early symbioses. The interactions are highly specialized and involve complex signaling mechanisms, recognition pathways, and mutualistic adaptations that are unique to particular host-microbe pairs.

However, the study of plant–microbe interactions is important for understanding plant health and ecosystem dynamics, and for developing sustainable agricultural practices, especially in terms of increasing environmental stresses and the need for enhanced crop growth and development. So, despite the complexity, it is very crucial to understand the mechanism for any further modifications. The recent utilization of high-resolution omics technologies (Fig. 15.1) has allowed to dissect the complex interactions within plant-associated microbiomes by recreating simplified, controlled communities of microbes that interact with their host plants (Kumar *et al.*, 2023; Jain *et al.*, 2024). By combining advanced omics technologies in plant–microbe interaction studies, it is now precise and specific, and easy to explore how specific metabolic traits and interactions contribute to the assembly and stability of microbial communities in the rhizosphere (Bastas and Kumar, 2023). Moreover, omics along with bioinformatics have been suggested by several researchers to be the most efficient tools for studying and predicting the interactions between microbes and plants (Lima-Mendez *et al.*, 2015; Frantzeskakis *et al.*, 2020). This advancement could enlighten strategies for manipulating microbiomes to enhance plant health, productivity, and resilience to environmental stresses for sustainable crop improvement.

15.2 Omics Technologies and Bioinformatics

The advent of high-throughput sequencing technologies has revolutionized biological research, providing an unparalleled understanding of the complex interactions between plants and microorganisms. These technologies, collectively known as 'omics,' include various methods, including proteomics, metabolomics, transcriptomics, and genomics, each offering a different perspective on the molecular landscape.

15.2.1 High-throughput sequencing technologies

High-throughput sequencing (HTS) technologies, also referred to as next-generation sequencing (NGS) or massively parallel sequencing (MPS), have revolutionized the pace and scale of genetic data acquisition. These advanced sequencing methods, which enable the analysis of millions of nucleotides in a remarkably short period (Villamor *et al.*, 2019), offer unprecedented opportunities for research and diagnostics, particularly in the fields of plant pathology and microbial ecology. Since their initial application in 2009 for detecting plant viruses (Al Rwahnih *et al.*, 2009), HTS technologies have become integral tools in the diagnosis and study of plant pathogens (Al Rwahnih *et al.*, 2009; Bag *et al.*, 2015). These technologies allow researchers to sequence entire genomes or transcriptomes simultaneously, providing

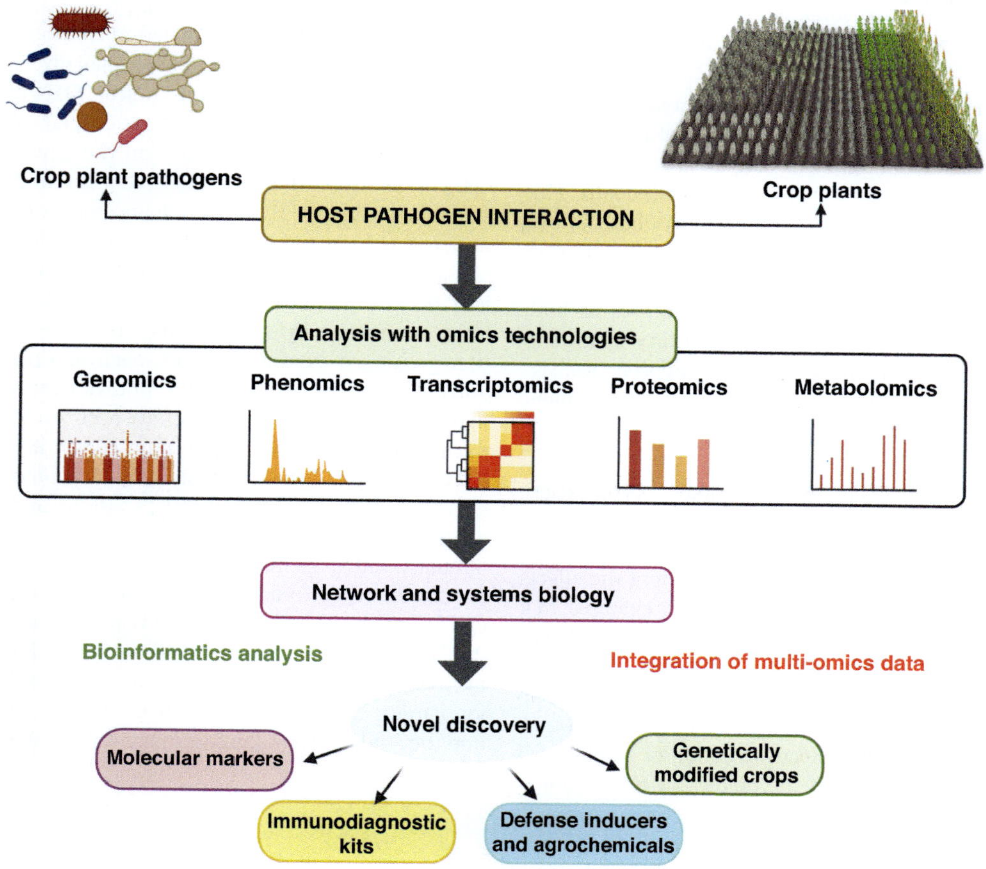

Fig. 15.1. Applications of omics technologies.

comprehensive datasets that are foundational for omics studies (Churko *et al.*, 2013). The ability to generate vast amounts of data quickly has opened new avenues for exploring the genetic and functional diversity within plant and microbial communities.

15.2.1.1 Applications of HTS in plant–microbial–nanoparticle interactions

Detection of plant viruses. HTS has emerged as a powerful tool for the detection of plant viruses, which pose significant threats to plant health. It serves as a robust alternative or complement to traditional diagnostic methods, particularly in identifying unexpected or unknown viruses that could threaten agricultural systems (Minicka *et al.*, 2020). By providing detailed insights into virus occurrence and infestation levels across various crops, weeds, trees, and ornamental plants, HTS facilitates the development of targeted regulations and novel strategies for plant protection.

Impact on soil microorganisms. The influence of genetically modified (GM) crops on soil microorganisms, which play crucial roles in soil ecosystems and agricultural productivity, can be effectively studied using HTS. This technology enables a comprehensive analysis of microbial communities, shedding light on the ecological impacts of GM crops on soil health (Jiang, 2023).

Analysis of bacterial diversity. HTS allows for the detailed examination of bacterial diversity and community composition in soil and plant roots. This application is vital for understanding the factors that shape bacterial communities and for promoting sustainable agricultural practices.

Identification of plant pathogens. HTS has revolutionized the identification of plant pathogens, including previously undetected ones, by leveraging its extensive sequencing depth. This capability allows for the detection of both known and novel pathogens, even those present in trace amounts. HTS, by sequencing genetic material directly from infected samples, serves as both an alternative and a supplementary method to traditional pathogen-detection approaches, significantly enhancing plant disease management strategies (Nizamani, 2023).

Microbial composition and nanoparticle impact. HTS is also employed to accurately and economically determine microbial composition, which is essential for understanding the effects of NPs on functional bacteria. For example, in a study by researchers (Cao *et al.*, 2022), HTS was utilized to generate high-throughput, precise microbial composition data. Additionally, the use of PICRUSt2 (Phylogenetic Investigation of Communities by Reconstruction of Unobserved States) allowed for the prediction of functional genes, thereby identifying key metabolic pathways involved in nitrogen and phosphorus removal across different constructed wetlands.

15.2.2 Multi-omics approaches

15.2.2.1 Genomics

Genomics focusses on the comprehensive analysis of an organism's DNA and plays a critical role in comprehending the intricate relationships that exist between microorganisms and plants, particularly in the presence of NPs. By utilizing advanced DNA sequencing and bioinformatics, researchers can identify essential genes and pathways involved in these interactions, such as PRRs, resistance genes (R-genes), and signaling pathways like PAMP-triggered immunity (PTI) and effector-triggered immunity (ETI). Metagenomics, a subset of genomics, further enhances our understanding by enabling the real-time analysis of microbial communities in diverse environments. This approach has been particularly useful in studying the effects of NPs on soil and root-associated microbiota, providing valuable insights into microbial diversity, function, and adaptation (Mitra, 2019). Understanding these interactions is crucial in agriculture, where they can inform the development of crops that are more resilient and sustainable. Despite the rapid advancements in nanotechnology and its widespread application across various scientific fields, the potential toxicity of NPs, particularly their impact on plant DNA, cannot be overlooked. Studies have shown that prolonged exposure to NPs, such as CuO and ZnO, can lead to significant genotoxic effects in plants. For example, CuO NPs have been found to cause DNA damage in agricultural crops like *Lolium rigidum*, *Lolium perenne*, and *Raphanus sativus*, leading to mutagenic DNA lesions and disruptions in plant growth (Atha *et al.*, 2012). Similarly, ZnO NPs have been shown to induce toxicity in wheat seedlings, though this effect can be mitigated by nitric oxide (Tripathi *et al.*, 2017). The genotoxic effects of NPs are complex and often species-specific, underscoring the need for careful consideration of their application and the study of their long-term environmental impact.

15.2.2.2 Transcriptomics

Transcriptomics focusses on the complete set of RNA transcripts produced by a genome under specific conditions, making it a powerful tool for exploring the dynamic nature of plant–microbe interactions, particularly in response to NP exposure. By analyzing gene expression patterns, researchers can uncover key regulatory networks and molecular mechanisms that govern these interactions. Studies have demonstrated that NPs can significantly alter gene expression in plants. For instance, exposure to ZnO and TiO_2 NPs in *Arabidopsis thaliana* roots resulted in substantial changes in gene expression, with ZnO NPs showing the highest toxicity. Specifically, ZnO NPs caused the upregulation of 660 genes and downregulation of 826 genes related to stress regulation, while TiO_2 NPs affected only 74 and 80 genes, respectively (Landa *et al.*, 2012). Further research has revealed that NP-induced stress generally leads to transcriptional repression in plants. In *A. thaliana*, exposure to various NPs under abiotic and biotic stress conditions resulted in early changes in gene expression, characterized by widespread downregulation. Additionally, transcriptomic studies of *Arabidopsis* exposed to Au NPs highlighted disruptions in the homeostasis of essential metals like Mn, Fe, and Zn. This was reflected in the altered expression of the MYB, BHLH, and WRKY gene families, as well as significant changes in the transcription of cation transporters and aquaporin genes (Tiwari *et al.*, 2016).

15.2.2.3 Proteomics

Large-scale protein research known as proteomics sheds light on the molecular processes underpinning the interactions between plants and microbes, especially in the presence of NPs. By examining the expression, structure, and function of proteins, proteomics links genetic information to actual biochemical processes, revealing how specific proteins mediate these complex interactions. Studies have shown that AgNPs can be toxic to plants at various levels, including cellular and physiological. However, fewer studies have focussed on the proteomic effects of AgNP stress. A recent gel-based proteomic study on *Oryza sativa* seedlings exposed to AgNPs identified proteins involved in calcium signaling, oxidative stress response, transcription, and cell division, among others, as being responsive to AgNP exposure. These findings suggest that AgNPs induce oxidative stress and disrupt cellular homeostasis by affecting Ca^{2+} signaling and redox regulation (Mirzajani *et al.*, 2014). Similarly, proteomic analyses of *Eruca sativa* roots exposed to AgNPs and $AgNO_3$ revealed that both NPs and bulk metals can alter proteins associated with redox regulation, further disrupting cellular homeostasis. Notably, AgNPs were found to target membrane proteins of vacuoles and the endoplasmic reticulum, indicating that these organelles are primary sites of NP-induced damage (Vannini *et al.*, 2013). In the agricultural sector, gold NPsc (AuNPs) have also been studied for their impact on plant proteomes. For instance, exposure of *Arabidopsis thaliana* to AuNPs resulted in the induction of glutathione S-transferases, with implications for root system architecture and stress responses. Further, experiments on soybean seedlings demonstrated that AgNPs of specific sizes (15 nm) and concentrations (two ppm) could enhance growth under flood stress, while higher concentrations caused damage, highlighting the size- and concentration-dependent effects of NPs on plant metabolism and stress responses (Mustafa and Komatsu, 2016). In addition to Ag^- and AuNPs, other NPs, such as Al_2O_3 and CeO_2, have been shown to impact plant growth and protein expression. For example, Al_2O_3 NPs were found to improve growth in flood-stressed soybean seedlings by regulating energy metabolism and reducing root cell proliferation. In contrast, CeO_2 NPs inhibited maize growth by affecting stress-responsive proteins like catalase and heat-shock proteins (Zhao *et al.*, 2012).

15.2.2.4 Metabolomics

Metabolomics, the study of metabolites (the small molecules produced during cellular processes) provides a comprehensive view of the metabolic state of a cell or organism. This approach offers critical insights into the functional outcomes of gene and protein activities, particularly in the context of plant–microbe interactions. By analyzing the metabolite profiles of plants and microbes, researchers can identify key compounds that mediate these interactions, such as signaling molecules and secondary metabolites that influence plant health and productivity. Metabolomics has been widely applied to study plant–microbe interactions, particularly in beneficial relationships like those involving mycorrhizae, plant growth-promoting bacteria (PGPB), and various fungi, including *Trichoderma* species. These studies have highlighted the role of metabolomics in distinguishing between symbiotic and pathogenic interactions by identifying specific metabolites involved in nutrient uptake, receptor recognition, and growth promotion. For instance, research on *Arabidopsis* interacting with PGPB and other microbes has demonstrated the integration of various omics data to illustrate these complex relationships. Despite significant advancements, the field of metabolomics continues to evolve with the development of new methodologies. Metabolomics has been employed to study the chemical exudation process from plant roots during development, revealing its impact on the composition of the rhizosphere community. For example, organic acids were identified as primary exuded metabolites consumed by bacteria in the substrate (Zhalnina *et al.*, 2018). Flavonoids in root exudates were also shown to induce bacterial nod genes, initiating nodule formation (Mhlongo *et al.*, 2018). Additionally, untargeted metabolomics studies have identified lipid markers in grapevines inoculated with *Plasmopara viticola*, illustrating the potential of metabolomics in disease studies (Negrel *et al.*, 2018). Metabolomics has also demonstrated the specificity of plant–PGPB interactions. For example, Chamam *et al.* (2013) found that two *Azospirillum* strains enhanced the growth of two *Oryza sativa* cultivars but differentially affected their metabolomes. In another study, Walker *et al.* (2011) used LC-MS and NMR to study the metabolism of two maize cultivars inoculated with three different *Azospirillum* strains, showing that these bacteria affected only secondary metabolism in a strain-dependent manner. Furthermore, Brusamarello-Santos *et al.* (2017) identified plant compounds influenced by bacterial nitrogen fixation in maize genotypes, demonstrating how metabolomics can uncover the metabolic effects of microbial interactions.

15.2.3 Bioinformatics tools for data analysis and integration

The massive datasets generated by omics technologies require sophisticated bioinformatics tools for analysis and integration. Bioinformatics plays a crucial role in processing and interpreting these data, allowing researchers to uncover patterns and relationships that would be otherwise obscured.

15.2.3.1 Data processing and management

Bioinformatics tools are critical for managing the large volumes of data produced by high-throughput sequencing. Essential tools for data preprocessing include fastp, Trimmomatic, Cutadapt, and BBDuk, which are widely used to assemble, annotate, and compare genomes, as well as to quantify and analyze transcriptomic, proteomic, and metabolomic data. Proper data management is crucial for ensuring the accuracy and reproducibility of research findings. Each sample that is sequenced using the FASTQ format is created by demultiplexing the binary base call sequence files (BCL) produced by NGS. Nucleotide sequences and their related quality ratings are intended to be stored in the text-based format known as FASTQ. The initial step in all bioinformatics pipelines involves assessing the quality of the data – a critical phase given the large volumes of NGS data and their susceptibility to problems like cross-contamination between samples, adapter contamination, and sequencing mistakes. Sequence quality control ensures that there are no technical artifacts that could taint downstream analysis and that the sequencing data are correct and dependable. It involves identifying low-quality bases, sequence bias, and the over-representation of certain sequences. Tools like fastp and FastQC are commonly used for quality assessment.

15.2.3.2 Data integration and visualization

Integrating data from multiple omics layers, such as genomics, transcriptomics, proteomics, and metabolomics, is essential for a thorough comprehension of the relationships between microbes and plants. Bioinformatics platforms play a crucial role in merging these diverse datasets, enabling researchers to construct detailed models of biological processes. Visualization tools further enhance the interpretation of complex data, helping to identify key trends and interactions. The integration of different omics profiles allows for greater understanding and more comprehensive images of molecular and biological processes for researchers. This integrative study was used to link omics elements to specific phenotypes, such as cardiovascular disease, cancer, or responses to treatments and interventions. The necessity for data integration arises from the complex biological processes that involve genetic variations, microbes, post-transcriptional changes, and metabolic activities – all of which interact to ascertain an organism's biological condition (Lindon and Nicholson, 2008). Since the early 2000s, multi-omics studies integrating genomic and metabolomic data have identified several alleles that account for significant variations in metabolic profiles (Gieger *et al.*, 2008). Sequence variants and alterations in lipid profiles have also been connected by extensive population investigations. Metabolites, being fundamental components of cellular processes, play critical roles in reactions that influence cell structure and function. Many computational and mathematical tools, such as deep learning models, Bayesian models, and least-squares-based models, have been created to accomplish these integrated assessments (Liang, 2015). Understanding the foundations of multi-omics data integration is crucial in the age of high-throughput data, providing early-career scientists who receive the required mentoring for developing effective integration methods.

15.2.3.3 Predictive modelling and network analysis

Bioinformatics plays a pivotal role in developing predictive models and conducting network analyses to simulate plant–microbe interactions under various conditions. These tools allow researchers to predict how alterations in one component, such as a gene or metabolite, might influence the overall system. Such predictions provide valuable insights for experimental design and hypothesis testing.

Predictive modeling in bioinformatics involves the use of algorithms to analyze data and build models that can forecast outcomes based on identified patterns and relationships. Techniques commonly employed in predictive modeling include:

- **linear regression:** a straightforward and interpretable method for predicting outcomes based on linear relationships in the data;
- **decision trees:** useful for handling complex data, decision trees provide intuitive insights by breaking down data into manageable segments;
- **random forest:** an advanced algorithm that combines multiple decision trees to enhance classification accuracy across large datasets;
- **Generalized Linear Model (GLM):** this algorithm focuses on narrowing down variables to find the best statistical fit; and
- **data integration and fusion:** a technique that combines multiple data sources to create a more accurate and comprehensive model of biological systems.

These predictive models have broad applications, including modeling viral replication and pathogenesis, predicting moonlighting DNA-binding proteins, and forecasting gene expression regulation (Regondi *et al.*, 2021).

Network analysis is a systems biology approach that utilizes graph theory and mathematical techniques to simplify and interpret complex biological data. By representing relationships between biological features as a graph, network analysis helps researchers understand the structure and behavior of biological systems. Applications of network analysis in bioinformatics include:

- **protein–protein interaction (PPI) networks:** these networks illustrate the interactions between proteins, which are crucial for nearly all cellular functions. PPI networks map out these interactions, ranging from the assembly of cellular components like the cytoskeleton to processes such as transcription, translation, and active transport. In PPI networks, nodes represent proteins, while edges define the interactions between them (Kurzbach, 2016);
- **human microarray datasets:** network analysis has been applied to human microarray datasets, particularly in studies related to the immune system and cancer;
- **oncological diseases:** a multiple network-based bioinformatics pipeline has been used to explore molecular mechanisms in oncological diseases, providing insights into disease-related pathways and biological processes with greater speed and accuracy (Dotolo *et al.*, 2021).

By combining predictive modeling with network analysis, bioinformatics provides a powerful framework for understanding and simulating the intricate interactions within biological systems, ultimately driving advances in experimental research and applied biology.

15.3 Deciphering Plant–microbe Interactions

The functioning of an ecosystem is largely dependent on interactions between plants and microbes, influencing plant health, soil fertility, and overall biodiversity. These interactions, often intricate and multifaceted, involve a dynamic exchange of signals and molecules between plants and their associated microbial communities. It is essential to comprehend the chemical underpinnings of these interactions for harnessing their potential in sustainable agriculture and environmental management.

15.3.1 Molecular interactions between plants and microbes

Plant–microbe interactions are governed by a series of intricate molecular events, beginning with recognition and communication between the organisms. These interactions are central to both defensive and symbiotic relationships, playing a crucial role in plant health, growth, and adaptation to environmental conditions. Plants detect microbial presence through specific receptors and signaling molecules, initiating a cascade of molecular events. For instance, plants recognize MAMPs using PRRs, which activate defense mechanisms like antimicrobial compound production and cell-wall reinforcement. Conversely, beneficial microbes, such as rhizobia and mycorrhizal fungi, engage in symbiosis with plants through signal exchange, leading to mutual benefits. In legumes, rhizobia bacteria initiate root nodule formation by secreting nod factors, which induce root hair curling and nodule development, while the plant provides carbohydrates necessary for nitrogen fixation.

15.3.1.1 Types of molecular interactions

1. **RNA silencing:** small RNA molecules inhibit gene expression at various levels, including transcription, mRNA stability, and translation, playing a critical role in regulating plant–microbe interactions;
2. **cross-kingdom transfers:** small RNAs (sRNAs) can be transferred from parasites to their plant hosts, leading to gene silencing that influences the interaction's outcome;
3. **symbiosis and mutualism:** in symbiotic and mutualistic interactions, plants and microbes exchange resources for mutual benefit; for example, nitrogen-fixing bacteria convert soil nitrogen into more accessible forms, enhancing plant growth, while mycorrhizal fungi improve phosphorus uptake and protect plants against pathogens;
4. **rhizosphere effect:** in the rhizosphere, the soil region near plant roots, plants release nutrients and signaling molecules that foster a diverse microbial community. This community, in turn, improves nutrient availability, suppresses diseases, and enhances plant resistance to various stresses. The rhizosphere effect also contributes to the biodegradation of organic contaminants, aiding in phytoremediation.

15.3.1.2 Interaction types

Plant–microbe interactions can be categorized into various types, including:

- **competition:** both species compete for the same resources, often resulting in detrimental effects for one or both;
- **commensalism:** one species benefits without harming or helping the other;
- **mutualism:** both species benefit, as seen in symbiotic relationships where plants and microbes exchange nutrients and other resources; and
- **parasitism:** one organism benefits at the expense of the other.

Commensalism and mutualism are the most common interactions, where one or both species gain from the relationship. Understanding these interactions is crucial for advancing our knowledge of plant health and productivity.

15.3.1.3 Molecular mechanisms in plant–microbe interactions

Specific molecular mechanisms underpin these interactions:

- **phytohormones:** microbes like *Rhizobium leguminosarum* produce phytohormones (e.g. auxins, cytokinins) that promote the development and uptake of nutrients by plants (Meena *et al.*, 2023);

- **exopolysaccharides:** produced by microorganisms like *Pseudomonas fluorescens*, that assist in biofilm development, adhesion to plant surfaces, and defense against environmental stress (Niazi *et al.*, 2023);
- **nod factors:** *Bradyrhizobium japonicum* secretes rhizobia, such as nod elements in legumes that trigger nodulation signaling cascades and create symbiotic connections (Grundy *et al.*, 2023);
- **pathogen effectors:** pathogens like *Pseudomonas syringae* produce effectors that alter the biological processes of the host, modifying plant immunity and encouraging infection (Zhang, 2022);
- **quorum-sensing molecules:** *Agrobacterium tumefaciens* is one type of microbe that produces quorum-sensing molecules that cause systemic plant reactions, and plant defense mechanisms are triggered by lipopolysaccharides from bacteria such as *Pseudomonas aeruginosa* (Elfaky, 2024);
- **volatile organic compounds (VOCs):** microbes like *Trichoderma harzianum* produce VOCs and can cause systemic reactions in plants (Chiquito-Contreras *et al.*, 2024);
- **extracellular enzymes:** *Fusarium oxysporum* produces extracellular enzymes that facilitate nutrient acquisition and microbial colonization (Niazi *et al.*, 2023).

15.3.2 Plant–microbe interactions: an understanding through omics technologies

Omics technologies have transformed our understanding of plant–microbe interactions at the molecular level. These high-throughput approaches (spanning genomics, transcriptomics, proteomics, and metabolomics) allow for a comprehensive exploration of the biological processes underlying these interactions, offering a holistic view of the molecular mechanisms involved.

15.3.2.1 Genomics and transcriptomics

Genomics and transcriptomics are essential tools for identifying the genes and regulatory networks that govern plant–microbe interactions. Genomic studies, which involve sequencing the complete genomes of both plants and microbes, help uncover genetic determinants of compatibility and resistance. NGS technology advancements have made it possible to conduct whole-genome sequencing, resequencing of many strains within a species, and *de novo* assemblies, revealing the genetic processes in charge of vital processes including nitrogen fixing and the creation of substances that promote growth (Crandall *et al.*, 2020; Carper *et al.*, 2022). Transcriptomics, which detects and measures RNA molecules using NGS, provides insights into how gene expression is modulated in response to microbial presence. This approach helps elucidate the temporal and spatial dynamics of plant–microbe interactions. For example, transcriptomic studies have shown that certain plant genes are upregulated in response to beneficial microbes, enhancing nutrient uptake and growth. Pathogens, on the other hand, can manipulate host gene expression to suppress plant immune responses, a process that can be detailed through transcriptomic analysis (Jeon, 2021).

15.3.2.2 Proteomics and metabolomics

Together, proteomics and metabolomics provide a direct link between gene expression and the functional outcomes of plant–microbe interactions. Proteomic studies identify specific proteins produced during these interactions, such as enzymes, structural proteins, and signaling molecules, which play critical roles in nutrient exchange, defense, and symbiosis. For instance, proteomic analysis has revealed the part host proteins and pathogen effectors play in the development of disease, such as those caused by *Phytophthora infestans* in potato and *Erwinia amylovora* in appl (Jain, 2021). Metabolomics, which provides both qualitative and quantitative insights into the chemical dialogue between plants and microbes, focusses on metabolites such as phytohormones, secondary metabolites, and signaling molecules. These metabolites mediate interactions and determine outcomes, whether leading to symbiosis, mutualism, or antagonism. For example, plants produce secondary metabolites like flavonoids to attract beneficial microbes, while pathogens may produce toxins to suppress plant defenses (Chouchani, 2022a). Although proteomics and metabolomics are underutilized compared to genomics and transcriptomics, especially in root–pathogen interaction studies, they provide invaluable insights on interactions between microbes and plants in a variety of tissues. This knowledge is crucial for developing improved methods for protecting crops and plants (Demiwal, 2024).

15.4 Applications for Sustainable Crop Improvement

15.4.1 Enhancing plant productivity and resilience

Plants are in constant contact with an uncountable number of microorganisms that live around, on, and within them. The intricate web of relationships that exists between bacteria, plants, and their surroundings has been acknowledged in recent years as a key element influencing plant health and productivity (Astapati, 2023). Microbes are vital to plants because they provide them with physiologically accessible forms of critical nutrients. They also help plants respond to weather extremes like heat and drought and protect against viruses and pests (Koza *et al.*, 2022). Plant rhizospheres are significant niches that are home to a wide variety of microorganisms. They have the ability to eliminate harmful substances, reduce disease, promote plant development, and help plants absorb nutrients. Utilizing such beneficial microorganisms for crop productivity provides an efficient technique to modulate crop production as well as output by maintaining plant health and quality through bioformulations (Shah, 2021). However, the ability to colonize plant environments is essential for all successful plant–microbe interactions. Colonization involves multiple steps, including recognition, adhesion, colonization and growth, invasion (limited to endophytes and diseases), and multiple methods of establishing relationships. Plant growth processes are influenced by phytohormones, such as the formation of ethylene, gibberellins, cytokinins, and indole-3-acetic acid (IAA) (Sosnowski, 2023). These hormones can be produced by both the plants and the microbes that live with them. Interestingly, bacteria can reduce ethylene levels in the following method. In plants, the molecule 1-aminocyclopropane-1-carboyclic acid (ACC) functions as a precursor to ethylene (Van de Poel and Van Der Straeten, 2014). Because ACC deaminase-producing bacteria can breakdown this molecule, it is taken up and stored in the root less effectively. Plant-associated microbes can offer a balance of macro- and micronutrients. The most well-known instance is nitrogen fixation by microorganisms. One of the best examples of rhizobacteria that promote plant growth is the symbiosis that exists between them and the legume host plants they inhabit (PGPR). This group of bacteria breaks down root exudates (carbohydrates), which provides the plant with nitrogen for amino acid synthesis (Fahde *et al.*, 2023). Bacteria may indirectly stimulate plant growth by releasing phosphorous from organic molecules, such as phytates, thereby contributing to plant nutrition (Alori *et al.*, 2017). In addition to causing microbial antagonists, plant-associated bacteria may also decrease the growth of pathogenic microorganisms by inducing plant defense mechanisms, a process known as 'induced systemic resistance'. Pathogens respond differentially to the resistance that these signaling pathways have triggered. Every component – the pathogen, the PGPR strain, and the host plant – interacts differently in these ways. They recognize one another through chemical signaling, which can be carried out by both root exudates and microbial metabolites. ISR mechanisms include: (i) developmental escape, which is connected to the stimulation of growth; (ii) physiological tolerance, which is connected to the modulation of symptom expression; (iii) environmental, which is connected to microbial antagonists in the rhizosphere; and (iv) biochemical resistance, which is connected to the induction of phytoalexins, pathogenesis-related proteins, cell-wall reinforcement, and 'priming' of defense responses (Chouchani, 2022b). PGPM inoculants help improve plant growth and tolerance to abiotic challenges by raising antioxidant levels and decreasing reactive oxygen species (ROS) and oxidative stress. In warm weather, a variety of microorganisms, including *Klebsiella*, *Pseudomonas*, *Aeromonas*, *Azospirillum*, *Achromobacter*, *Variovorax*, *Enterobacter*, *Bacillus*, and *Azotobacter*, promote plant growth.

15.4.2 Developing environmentally friendly agricultural practices

Agriculture has a crucial role in achieving global priorities such poverty eradication, rapid industrialization, economic diversification, sustainable resource exploitation, investments, and environmental management. Plant microbiomes have the potential to enhance plant resilience and yields in farming systems. Chemical-free agriculture is becoming increasingly important due to their negative impact on the environment and human health. Intensive agricultural practices relying on inorganic fertilizers, pesticides, and chemicals have led to increased yield but also soil degradation, biodiversity loss, crop susceptibility to pests/pathogens, and negative environmental impacts (Tudi *et al.*, 2021). These issues have significant implications for human health and food security. Microbiology has various potential benefits, including crop pest resistance, increased output, and reduced use of chemical pesticides (Sundh *et al.*, 2021). Agricultural advances that led to a 'green revolution' in the mid-20th century have resulted in severe environmental costs, including pollution, climate change, and biodiversity loss. Therefore, microbial biotechnology is receiving increased interest for its potential uses in sustainable agriculture and environmental health.

Sustainable agriculture has the capacity to satisfy future agricultural needs, unlike conventional farming methods. To ensure sustainable crop production, it is necessary to integrate plant growth-promoting microorganisms without compromising soil health. Microorganisms play a crucial role in sustainable agriculture by increasing plant nutrient availability and efficiency (Suman *et al.*, 2022). Biofertilizers, biopesticides, and bioherbicides are becoming increasingly popular in agriculture due to food safety and environmental concerns. Plant growth-promoting rhizobacteria (PGPR) and their antibacterial substances are environmentally safe and biodegradable (Chandran *et al.*, 2021). Microorganisms utilized chemical pollutants as a source of energy during the microbiological process. PGPR enriches soil with micro- and macronutrients through nitrogen fixation, phosphate/potassium/zinc solubilization, release of plant growth regulators, and organic matter degradation. It can also control plant pathogens, particularly fungi, by producing antifungal metabolites or inducing systemic resistance (Hasan *et al.*, 2024). The potential of beneficial rhizobacteria to reduce the use of synthetic compounds and fertilizer and to provide a sustainable agriculture is a component of an overall management system (Vejan *et al.*, 2016). Men employ naturally occurring organisms to create biofertilizers and insecticides that promote plant growth and control weeds, pests, and illnesses.

Soil microorganisms enhance plant nutrient absorption. Plants and beneficial microorganisms work together to recycle nutrients (Jacoby *et al.*, 2017a). Microbes aid the plant in absorbing essential energy sources. Plants provide the microorganisms with their waste byproducts in exchange for food. Plants require phosphorus and nitrogen for proper growth. Plants have little ability to extract these naturally occurring chemicals from their environment. Phosphate is essential for crop quality, maturity, stress tolerance, and nitrogen fixation, either directly or indirectly. *Penicillium bilaii* is a fungus that helps release phosphorus from soil. It produces an organic acid that dissolves phosphate in the soil, allowing roots to absorb it (Jacoby *et al.*, 2017b). In comparison to nonmicorrhizal controls, mycorrhizal plants exhibit increased growth and are generally more resilient to adverse conditions, including drought, soil pathogens, transplantation, low soil nutrient status, and soil pollution (Huey *et al.*, 2020). The large hyphal network enhances water and nutrient uptake, leading to improved plant development and tolerance to harsh conditions by increasing root absorptive area. Plants that have alternative ways of getting P from the soil typically lose out on the advantages of mycorrhizas (Etesami *et al.*, 2021). The activity of mycorrhizal fungi is ignored by the use of fertilizers in conventional cultivation. This could have significant long-term impact on crop productivity.

15.4.3 Potential for nanoparticles synthesis using plant–microbial interactions

Nanotechnology is an emerging cutting-edge technology in a variety of diverse sectors of research, including biology, chemistry, and material science. Nanotechnology has gained popularity for its agricultural uses, in especially the application of NPs (size <100 nm) for disease control and soil conditioning (Saritha *et al.*, 2022). Several studies have demonstrated that the application of NPs considerably promotes plant growth and development by modifying plant defense mechanisms, nutritional availability, and by interacting directly with the environment challenges in both field and greenhouse settings, whether biotic or abiotic (Rasheed *et al.*, 2022). In contrast, NPs produced via microbe-mediated biosynthesis are much more capable of being nontoxic, easily scaled up, stable over time, and environmentally friendly than NPs produced by chemical or physical means (Gomaa, 2022). Various microorganisms (for example, yeast, microalgae, bacteria, and fungus) have been found to be promising nanofactories for the manufacture of metallic nanoparticles (Ghosh *et al.*, 2021). Microbe-mediated NPs are synthesized employing microorganisms and a variety of processes, including extracellular synthesis, bioreduction, and green synthesis to produce NPs with desired characteristics. Many infections, such as *Xanthomonas oryzae* pv. *oryzae*, *Alternaria solani*, and *Fusarium oxysporum*, have caused a variety of plant diseases that these NPs have shown encouraging results in treating (Elmer *et al.*, 2018). A viable and environmentally beneficial method of producing NPs with a range of uses in the agricultural industry is the use of bacteria to manufacture NPs (Singh *et al.*, 2023). Due to their capacity to generate extracellular enzymes that can convert metal ions to their corresponding NPs, bacteria have been used to synthesize NPs. Bacteria use intracellular biosynthesis to produce NPs within their cells. Metal ions are taken up by the bacterial cells and then reduced to NPs by internal enzymes.

Fungi are a large category of organisms which have been widely utilized for NP generation because of their ability to create multiple extracellular enzymes along with metabolites that can convert metal ions to their equivalent NPs (Li *et al.*, 2022). Fungal NPs can be synthesized through a variety of methods, including fungal biomass and biosynthesis both within and outside of cells (Dorcheh and Vahabi, 2016). Much research have demonstrated that the synthesis of NPs using fungi has a number of benefits, including low cost, high production rate, and ease of growth (Karunakaran *et al.*, 2023). Yeasts, like other microorganisms, have been widely used for large-scale NP manufacturing (Boroumand Moghaddam *et al.*, 2015). Nanopesticides are any plant protectants that use nanomaterials as active components to enhance the effectiveness of disease control and biocidal properties (Zobir *et al.*, 2021). The remarkable microcidal activity of NPs has demonstrated their effectiveness as antibacterial treatments against plant diseases. Using microbe-oriented NPs is a unique way to prevent plant pathogenic illnesses (Wang *et al.*, 2017). The microbes utilized in these NPs have a variety of modes of action, including antibiosis, rivalry for nutrients and resources, and the development of plant systemic resistance. Additionally, biogenic NPs can be made to target particular phytopathogens by modifying their size, shape, and surface characteristics, which increases their effectiveness and selectivity. Silver, gold, manganese, zinc, copper, titanium, and other metallic NPs have been extensively studied (Mittal *et al.*, 2020). Microbe-mediated NPs have become a unique and successful solution owing to their capacity to administer targeted doses with greater efficacy, to plant disease management. Microorganisms have the opportunity to resolve a few of the shortcomings of conventional chemical-based pesticides, including their nonspecificity and detrimental environmental effects, through the production of NPs. Microbial-based NPs are able to target particular illnesses while using less pesticide, reducing environmental contamination.

15.5 Nanoparticles From Plant–microbe Interaction

15.5.1 Synthesis using nanoparticles

Natural combinations of microorganisms offer a critical ability to enhance NP synthesis without utilizing hazardous and expensive chemicals. The conventional techniques for creating NPs include electrospinning, inert condensation, microwave, radiolysis, spray pyrolysis, sol-gel approach, and chemical reduction (Lahiri *et al.*, 2021). However, the urgent demand for a faster, cheaper, more productive, non-toxic, and ecologically friendly approach has shifted the focus to greener methods of production. Microbes are known as small 'nanofactories,' and the production of microbial NPs has combined nanotechnology, microbiology, and biotechnology together to form the interesting topic of 'nanobiotechnology.' Making NPs from biological sources, such as bacteria, actinobacteria, algae, fungi, yeast, etc., is significantly safer than utilizing chemical and physical techniques (Pandit *et al.*, 2022). While high-purity and target-size NPs have been effectively created using physical and chemical processes, these processes are typically costly and require toxic materials. The chemical synthesis process may result in specific hazardous chemical species adhering to the surface of NPs, which could have negative impacts in therapeutic settings; these NPs could potentially interact directly with the human body. Furthermore, microbial synthesis enables exact control over NPs' composition, size, and form, which could have a big impact on their qualities and their applications.

15.5.2 Synthesis of nanoparticles using bacteria

Bacterial NP generation is a potential and sustainable method of creating NPs for a range of agricultural applications (Borehalli Mayegowda *et al.*, 2023). Due to their capacity to generate extracellular enzymes that can convert metal ions into their corresponding NPs, bacteria have been used to synthesize NPs. Bacteria can synthesize NPs utilizing a variety of methods, including intracellular biosynthesis, extracellular biosynthesis, and bioaccumulation (Koul *et al.*, 2021). Bacteria use intracellular biosynthesis to produce NPs within their cells. Metal ions are taken up into bacterial cells and reduced to NPs by intracellular enzymes. Once cells have been lysed, NPs are released into the extracellular space (Behzadi *et al.*, 2017). Extracellular biosynthesis is the process by which bacteria create NPs outside of their cells. Extracellular enzymes are secreted throughout the process, which can convert metal ions into their equivalent NPs. Several bacterial strains, such as *B. amyloliquefaciens*, *Acinetobacter calcoaceticus*, *P. stutzeri*, *Escherichia coli*, and *Lactobacillus* sp. were previously employed for microbe-mediated NP production (Ahmed *et al.*, 2023).

15.5.3 Fungal-based nanoparticle synthesis

Throughout history, fungi have been a diverse range of species exploited for NP production because of their capacity to generate a wide range of extracellular metabolites and enzymes that can reduce metal ions to their equivalent NPs (Li *et al.*, 2022). Fungal NPs can be synthesized through a variety of methods, including fungal biomass and both extracellular and intracellular biosynthesis (Dorcheh and Vahabi, 2016). Many researchers have revealed that the synthesis of NPs using fungi has a number of benefits, including simplicity of production, cheap cost, and high rate of culture (El-Moslamy *et al.*, 2023). Taken together, fungal-mediated NP synthesis provides a sustainable, cost-effective, and adaptable route to green NP synthesis with distinct features and prospective agricultural applications.

15.5.4 Microalgae-based nanoparticle synthesis

The NPs' biogenesis from microalgae provides various advantages, including low cost, high output rate, and ease of cultivation (Moraes *et al.*, 2021). Furthermore, microalgae are easily harvested and processed into NPs. Additionally, the synthesis of NPs from microalgae is environmentally favorable because it avoids the usage of toxic chemicals and waste generation. Through the use of extracellular or intracellular enzymes, photosynthetic microalgae can change metal ions into their equivalent NPs (Nitnavare *et al.*, 2022). For example, da Silva Ferreira *et al.*, 2017 produced silver chloride NPs employing the microalgal species *Chlorella vulgaris*, and found that they had antibacterial properties against harmful bacteria.

15.5.5 Potential applications in agriculture

Agriculture is practiced by producing food by cultivating various crops and keeping livestock. For most developing nations, it is recognized as the foundation of the economy, contributing significantly to advancement and growth. By altering traditional methods for assessing environmental problems and applying them to increase productivity, nanotechnology can be used as a tool to restructure agricultural sectors and assist in the understanding of the biochemical processes of crops. It is suggested that the application of NPs in agriculture eliminates the need for chemical spraying by providing a consistent supply of energetic components. Enhancing water and ingredient management, it can boost harvests and decrease nutrient waste during fertilizer application. The utilization of nanotechnology can manage plant maladies, plant pests, and diseases, which cause 20–40% of agricultural losses worldwide each year. Pest management in modern farming is mainly reliant on the use of pesticides such as insecticides, fungicides, and herbicides (Karlsson Green *et al.*, 2020). It is imperative that low-cost, high-performing pesticides with less environmental risk be developed. Innovative ideas like nanotechnology can provide pesticides with benefits like decreased toxicity, longer shelf life, and increased solubility of pesticides that are not very soluble in water, all of which could have positive effects on the environment (Worrall *et al.*, 2018). Nanotechnology can increase soil quality and fertilizer distribution. Enhancing crop productivity requires the application of nanotechnology in crop management. In agricultural research, nanomaterials and nanostructures like carbon nanotubes, nanofibers, and quantum dots are currently employed as biosensors to evaluate the distribution of fertilizer and the condition of the soil (Nongbet *et al.*, 2022). By reducing nutrient loss during fertilization and increasing the quality and yield with the right nutrients, NPs aim to lessen the number of chemicals distributed. Enhancing crop productivity requires the usage of nanofertilizer. Nanofertilizer is a nanometer-sized material that increases the supply to plants and controls how slowly nutrients are released into the soil over time into plants and manages the slow release of nutrients into the soil gradually in a carefully regulated way, therefore eliminating eutrophication and contamination of water (Haydar *et al.*, 2024). Fertilizers at the nanoscale can reduce the toxicity of the soil and therefore the potential adverse effects associated with high dosage are reduced. Food is covered with a variety of edible nanocoatings to regulate the postharvest quality of newly produced items. Silver NPs have recently received increased interest due to their antibacterial qualities, which are vital for food preparation (Bruna *et al.*, 2021). The application of AgNPs based on PVP to asparagus significantly slowed the growth of microorganisms, reducing weight loss, and a decline in skin colour changes. TiO_2 NPs have shown a significant effect on maize crop growth; additionally, SiO_2 with TiO_2 NPs improved the absorption capacity of plants and the activity of nitrate through the efficient use of water and fertilizer (Farahi *et al.*, 2023). The use of biosensors in conjunction with nanomaterials has a considerable impact on agricultural productivity. Precision farming utilizes nanosensors to guarantee the most effective use of natural agricultural resources including water, fertilizers, and pesticides (Duhan *et al.*, 2017). Additionally, they can be used to track bacterial activity in the environment and forecast when soil illnesses will manifest (Hermans *et al.*, 2017).

15.6 Conclusion

This chapter offers cutting-edge knowledge on the interactions between plants and microorganisms, highlighting both the benefits and limitations of these relationships. It also explores the variations in factors associated with either plants or microbes that influence these interactions, as well as the mechanisms underlying plant defense. The combination of transcriptomics, proteomics, metabolomics, and genomics technologies indicates a revolutionary advancement in decoding the challenges of plant–microbe interactions. These technologies also provide comprehensive insights into the molecular mechanisms and dynamic networks that support these interactions. These omics approaches offer a multidimensional perspective of the complex and interdependent relationships between plants and their associated microbes. The integration of multi-omics and bioinformatics is another approach to understanding and modifying these interaction systems.

References

Ahmed, T., Luo, J., Noman, M., Ijaz, M., Wang, X. *et al.* (2023) Microbe-mediated nanoparticle intervention for the management of plant diseases. *Crop Health* 1(1), 3. DOI: 10.1007/s44297-023-00006-9.

Al Rwahnih, M., Daubert, S., Golino, D. and Rowhani, A. (2009) Deep sequencing analysis of RNSs from a grapevine showing syrah decline symptoms reveals a multiple virus infection that includes a novel virus. *Virology* 387(2), 395–401. DOI: 10.1016/j.virol.2009.02.028.

Alori, E.T., Glick, B.R. and Babalola, O.O. (2017) Microbial phosphorus solubilization and its potential for use in sustainable agriculture. *Frontiers in Microbiology* 8, 971. DOI: 10.3389/fmicb.2017.00971.

Astapati, A.D. (2023) The complex interplay between plant-microbe and virus interactions in sustainable agriculture: Harnessing phytomicrobiomes for enhanced soil health, designer plants, resource use efficiency, and food security. *Science Direct*. Available at: https://www.sciencedirect.com/science/article/pii/S277289942300006X?via%3Dihub (accessed 9 September 2024).

Atha, D.H., Wang, H., Petersen, E.J., Cleveland, D., Holbrook, R.D. *et al.* (2012) Copper oxide nanoparticle mediated DNA damage in terrestrial plant models. *Environmental Science & Technology* 46(3), 1819–1827. DOI: 10.1021/es202660k.

Bag, S., Al Rwahnih, M., Li, A., Gonzalez, A., Rowhani, A. *et al.* (2015) Detection of a new luteovirus in imported nectarine trees: A case study to propose adoption of metagenomics in post-entry quarantine. *Phytopathology* 105(6), 840–846. DOI: 10.1094/PHYTO-09-14-0262-R.

Bastas, K.K. and Kumar, A. (2023) Omics technologies in the plant–microbe interactions. In: Bastas, K.K., Kumar, A. and Sivakumar, U. (eds) *Microbial Biocontrol: Molecular Perspective in Plant Disease Management*, Vol. 49. Springer Nature, Singapore, pp. 257–282. DOI: 10.1007/978-981-99-3947-3_13.

Behzadi, S., Serpooshan, V., Tao, W., Hamaly, M.A., Alkawareek, M.Y. *et al.* (2017) Cellular uptake of nanoparticles: Journey inside the cell. *Chemical Society Reviews* 46(14), 4218–4244. DOI: 10.1039/C6CS00636A.

Berg, G. (2009) Plant–microbe interactions promoting plant growth and health: Perspectives for controlled use of microorganisms in agriculture. *Applied Microbiology and Biotechnology* 84(1), 11–18. DOI: 10.1007/s00253-009-2092-7.

Borehalli Mayegowda, S., Roy, A., N.G., M., Pandit, S., Alghamdi, S. *et al.* (2023) Eco-friendly synthesized nanoparticles as antimicrobial agents: An updated review. *Frontiers in Cellular and Infection Microbiology* 13. DOI: 10.3389/fcimb.2023.1224778.

Boroumand Moghaddam, A., Namvar, F., Moniri, M., Md. Tahir, P., Azizi, S. *et al.* (2015) Nanoparticles biosynthesized by fungi and yeast: A review of their preparation, properties, and medical applications. *Molecules* 20(9), 16540–16565. DOI: 10.3390/molecules200916540.

Bruna, T., Maldonado-Bravo, F., Jara, P. and Caro, N. (2021) Silver nanoparticles and their antibacterial applications. *International Journal of Molecular Sciences* 22(13), 7202. DOI: 10.3390/ijms22137202.

Brusamarello-Santos, L.C., Gilard, F., Brulé, L., Quillere, I., Gourion, B. *et al.* (2017) Metabolic profiling of two maize (*Zea mays* L.) inbred lines inoculated with the nitrogen fixing plant-interacting bacteria *Herbaspirillum seropedicae* and *Azospirillum brasilense*. *PLoS One* 12(3), e0174576. DOI: 10.1371/journal.pone.0174576.

Cao, C., Huang, J. and Yan, C.-N. (2022) Unveiling changes of microbial community involved in N and P removal in constructed wetlands with exposing to silver nanoparticles. *Journal of Hazardous Materials* 432, 128642. DOI: 10.1016/j.jhazmat.2022.128642.

Carper, D.L., Appidi, M.R., Mudbhari, S., Shrestha, H.K., Hettich, R.L. *et al.* (2022) The promises, challenges, and opportunities of omics for studying the plant holobiont. *Microorganisms* 10, 10. DOI: 10.3390/microorganisms10102013.

Chamam, A., Sanguin, H., Bellvert, F., Meiffren, G., Comte, G. *et al.* (2013) Plant secondary metabolite profiling evidences strain-dependent effect in the *Azospirillum-oryza sativa* association. *Phytochemistry* 87, 65–77. DOI: 10.1016/j.phytochem.2012.11.009.

Chandran, H., Meena, M. and Swapnil, P. (2021) Plant growth-promoting rhizobacteria as a green alternative for sustainable agriculture. *Sustainability* 13(19), 10986. DOI: 10.3390/su131910986.

Chiquito-Contreras, C.J., Meza-Menchaca, T., Guzmán-López, O., Vásquez, E.C. and Ricaño-Rodríguez, J. (2024) Molecular insights into plant–microbe interactions: A comprehensive review of key mechanisms. *Frontiers in Bioscience-Elite* 16(1), 9. DOI: 10.31083/j.fbe1601009.

Chouchani, E.T. (2022a) Logic and mechanisms of metabolite signalling. *Nature Reviews. Endocrinology* 18(2), 71–72. DOI: 10.1038/s41574-021-00618-7.

Chouchani, E.T. (2022b) Logic and mechanisms of metabolite signalling. *Nature Reviews. Endocrinology* 18(2), 71–72. DOI: 10.1038/s41574-021-00618-7.

Churko, J.M., Mantalas, G.L., Snyder, M.P. and Wu, J.C. (2013) Overview of high throughput sequencing technologies to elucidate molecular pathways in cardiovascular diseases. *Circulation Research* 112(12), 1613–1623. DOI: 10.1161/CIRCRESAHA.113.300939.

Crandall, S.G., Gold, K.M., Jiménez-Gasco, M.D.M., Filgueiras, C.C. and Willett, D.S. (2020) A multi-omics approach to solving problems in plant disease ecology. *PLoS One* 15(9), e0237975. DOI: 10.1371/journal.pone.0237975.

da Silva Ferreira, V., ConzFerreira, M.E., Lima, L.M.T.R., Frasés, S., de Souza, W. *et al.* (2017) Green production of microalgae-based silver chloride nanoparticles with antimicrobial activity against pathogenic bacteria. *Enzyme and Microbial Technology* 97, 114–121. DOI: 10.1016/j.enzmictec.2016.10.018.

Demiwal, P. (2024) A metabolomics perspective on root-derived plant immunity and phytohormone interaction. Wiley Online Library. Available at: https://onlinelibrary.wiley.com/doi/abs/10.1111/ppl.14150 (accessed 9 September 2024).

Dolatabadian, A. (2020) Plant-microbe interaction. *Biology* 10(1), 15. DOI: 10.3390/biology10010015.

Dorcheh, S.K. and Vahabi, K. (2016) Biosynthesis of nanoparticles by fungi: Large-scale production. In: Mérillon, J.-M. and Ramawat, K.G. (eds) *Fungal Metabolites*. Springer, Cham, Switzerland, pp. 1–20. DOI: 10.1007/978-3-319-19456-1_8-1.

Dotolo, S., Marabotti, A., Rachiglio, A.M., Esposito Abate, R., Benedetto, M. *et al.* (2021) A multiple network-based bioinformatics pipeline for the study of molecular mechanisms in oncological diseases for personalized medicine. *Briefings in Bioinformatics* 22(6). DOI: 10.1093/bib/bbab180.

Duhan, J.S., Kumar, R., Kumar, N., Kaur, P., Nehra, K. *et al.* (2017) Nanotechnology: The new perspective in precision agriculture. *Biotechnology Reports* 15, 11–23. DOI: 10.1016/j.btre.2017.03.002.

Elfaky, M.A. (2024) Unveiling the hidden language of bacteria: anti-quorum sensing strategies for gram-negative bacteria infection control. PubMed. Available at: https://pubmed.ncbi.nlm.nih.gov/38409503/ (accessed 8 September 2024).

Elmer, W., Ma, C. and White, J. (2018) Nanoparticles for plant disease management. *Current Opinion in Environmental Science & Health* 6, 66–70. DOI: 10.1016/j.coesh.2018.08.002.

El-Moslamy, S.H., Yahia, I.S., Zahran, H.Y. and Kamoun, E.A. (2023) Novel biosynthesis of MnO NPs using Mycoendophyte: Industrial bioprocessing strategies and scaling-up production with its evaluation as anti-phytopathogenic agents. *Scientific Reports* 13(1). DOI: 10.1038/s41598-023-28749-z.

Etesami, H., Jeong, B.R. and Glick, B.R. (2021) Contribution of arbuscular mycorrhizal fungi, phosphate-solubilizing bacteria, and silicon to P uptake by plant. *Frontiers in Plant Science* 12, 699618. DOI: 10.3389/fpls.2021.699618.

Fahde, S., Boughribil, S., Sijilmassi, B. and Amri, A. (2023) Rhizobia: A promising source of plant growth-promoting molecules and their non-legume interactions: Examining applications and mechanisms. *Agriculture* 13(7), 1279. DOI: 10.3390/agriculture13071279.

Farahi, S.M.M., Taghavizadeh Yazdi, M.E., Einafshar, E., Akhondi, M., Ebadi, M. *et al.* (2023) The effects of titanium dioxide (TiO_2) nanoparticles on physiological, biochemical, and antioxidant properties of Vitex plant (*Vitex agnus - Castus* L.). *Heliyon* 9(11), e22144. DOI: 10.1016/j.heliyon.2023.e22144.

Frantzeskakis, L., Pietro, A., Rep, M., Schirawski, J., Wu, C.-H. *et al.* (2020) Rapid evolution in plant-microbe interactions – a molecular genomics perspective. *The New Phytologist* 225(3), 1134–1142. DOI: 10.1111/nph.15966.

Ghosh, S., Ahmad, R., Zeyaullah, M. and Khare, S.K. (2021) Microbial nano-factories: Synthesis and biomedical applications. *Frontiers in Chemistry* 9, 626834. DOI: 10.3389/fchem.2021.626834.

Gieger, C., Geistlinger, L., Altmaier, E., Angelis, M., Kronenberg, F. *et al.* (2008) Genetics meets metabolomics: A genome-wide association study of metabolite profiles in human serum. *PLoS Genetics* 4(11), e1000282. DOI: 10.1371/journal.pgen.1000282.

Gomaa, E.Z. (2022) Microbial mediated synthesis of zinc oxide nanoparticles, characterization and multifaceted applications. *Journal of Inorganic and Organometallic Polymers and Materials* 32(11), 4114–4132. DOI: 10.1007/s10904-022-02406-w.

Grundy, E.B., Gresshoff, P.M., Su, H. and Ferguson, B.J. (2023) Legumes regulate symbiosis with Rhizobia via their innate immune system. *International Journal of Molecular Sciences* 24(3), 2800. DOI: 10.3390/ijms24032800.

Hasan, A., Tabassum, B., Hashim, M. and Khan, N. (2024) Role of Plant Growth Promoting Rhizobacteria (PGPR) as a plant growth enhancer for sustainable agriculture: A review. *Bacteria* 3(2), 59–75. DOI: 10.3390/bacteria3020005.

Haydar, M.S., Ghosh, D. and Roy, S. (2024) Slow and controlled release nanofertilizers as an efficient tool for sustainable agriculture: Recent understanding and concerns. *Plant Nano Biology* 7, 100058. DOI: 10.1016/j.plana.2024.100058.

Hermans, S.M., Buckley, H.L., Case, B.S., Curran-Cournane, F., Taylor, M. *et al.* (2017) Bacteria as emerging indicators of soil condition. *Applied and Environmental Microbiology* 83(1), e02826–16. DOI: 10.1128/AEM.02826-16.

Huey, C.J., Gopinath, S.C.B., Uda, M.N.A., Zulhaimi, H.I., Jaafar, M.N. *et al.* (2020) Mycorrhiza: A natural resource assists plant growth under varied soil conditions. *3 Biotech* 10(5), 204. DOI: 10.1007/s13205-020-02188-3.

Jacoby, R., Peukert, M., Succurro, A., Koprivova, A. and Kopriva, S. (2017a) The role of soil microorganisms in plant mineral nutrition – current knowledge and future directions. *Frontiers in Plant Science* 8, 1617. DOI: 10.3389/fpls.2017.01617.

Jacoby, R., Peukert, M., Succurro, A., Koprivova, A. and Kopriva, S. (2017b) The role of soil microorganisms in plant mineral nutrition – current knowledge and future directions. *Frontiers in Plant Science* 8, 1617. DOI: 10.3389/fpls.2017.01617.

Jain, A. (2021) Deciphering plant-microbe crosstalk through proteomics studies. PubMed. Available at: h ttps://pubmed.ncbi.nlm.nih.gov/33022544/ (accessed 9 September 2024).

Jain, A., Sarsaiya, S., Singh, R., Gong, Q., Wu, Q. *et al.* (2024) Omics approaches in understanding the benefits of plant-microbe interactions. *Frontiers in Microbiology* 15, 1391059. DOI: 10.3389/fmicb.2024.1391059.

Jeon, J. (2021) Alternative splicing diversifies the transcriptome and proteome of the rice blast fungus during host infection. Available at: https://www.tandfonline.com/doi/full/10.1080/15476286.2022.2043040 (accessed 9 September 2024).

Jiang, J. (2023) High throughput sequencing technology facility research of genomic modification crop cultivation influencing soil microbe – PMC. Available at: https://www.ncbi.nlm.nih.gov/pmc/articles/PMC10264764/ (accessed 8 September 2024).

Karlsson Green, K., Stenberg, J.A. and Lankinen, Å. (2020) Making sense of integrated pest management (IPM) in the light of evolution. *Evolutionary Applications* 13(8), 1791–1805. DOI: 10.1111/eva.13067.

Karunakaran, G., Sudha, K.G., Ali, S. and Cho, E.-B. (2023) Biosynthesis of nanoparticles from various biological sources and its biomedical applications. *Molecules* 28(11), 4527. DOI: 10.3390/molecules28114527.

Koul, B., Poonia, A.K., Yadav, D. and Jin, J.-O. (2021) Microbe-mediated biosynthesis of nanoparticles: Applications and future prospects. *Biomolecules* 11(6), 886. DOI: 10.3390/biom11060886.

Koza, N.A., Adedayo, A.A., Babalola, O.O. and Kappo, A.P. (2022) Microorganisms in plant growth and development: Roles in abiotic stress tolerance and secondary metabolites secretion. *Microorganisms* 10(8), 1528. DOI: 10.3390/microorganisms10081528.

Kumar, U., Raj, S., Sreenikethanam, A., Maddheshiya, R., Kumari, S. *et al.* (2023) Multi-omics approaches in plant–microbe interactions hold enormous promise for sustainable agriculture. *Agronomy* 13(7), 1804. DOI: 10.3390/agronomy13071804.

Kurzbach, D. (2016) Network representation of protein interactions: Theory of graph description and analysis. *Protein Science* 25(9), 1617–1627. DOI: 10.1002/pro.2963.

Lahiri, D., Nag, M., Sheikh, H.I., Sarkar, T., Edinur, H.A. *et al.* (2021) Microbiologically-synthesized nanoparticles and their role in silencing the biofilm signaling cascade. *Frontiers in Microbiology* 12, 636588. DOI: 10.3389/fmicb.2021.636588.

Landa, P., Vankova, R., Andrlova, J., Hodek, J., Marsik, P. *et al.* (2012) Nanoparticle-specific changes in *Arabidopsis thaliana* gene expression after exposure to ZnO, TiO_2, and fullerene soot. *Journal of Hazardous Materials* 241–242, 55–62. DOI: 10.1016/j.jhazmat.2012.08.059.

Liang, M. (2015) Integrative data analysis of multi-platform cancer data with a multimodal deep learning approach. Available at: https://www.researchgate.net/publication/281681183_Integrative_Data_Analysis_of_Multi-Platform_Cancer_Data_with_a_Multimodal_Deep_Learning_Approach (accessed 8 September 2024).

Lima-Mendez, G., Faust, K., Henry, N., Decelle, J., Colin, S. *et al.* (2015) Determinants of community structure in the global plankton interactome. *Science* 348(6237), 1262073. DOI: 10.1126/science.1262073.

Lindon, J.C. and Nicholson, J.K. (2008) Spectroscopic and statistical techniques for information recovery in metabonomics and metabolomics. *Annual Review of Analytical Chemistry* 1, 45–69. DOI: 10.1146/annurev.anchem.1.031207.113026.

Li, Q., Liu, F., Li, M., Chen, C. and Gadd, G.M. (2022) Nanoparticle and nanomineral production by fungi. *Fungal Biology Reviews* 41, 31–44. DOI: 10.1016/j.fbr.2021.07.003.

Meena, M., Mehta, T., Nagda, A., Yadav, G., Sonigra, P. *et al.* (2023) PGPR-mediated synthesis and alteration of different secondary metabolites during plant-microbe interactions. In: Swapnil, P. (ed.) *Plant-Microbe Interaction – Recent Advances in Molecular and Biochemical Approaches*. Academic Press, pp. 229–255. DOI: 10.1016/B978-0-323-91875-6.00002-5.

Mhlongo, M.I., Piater, L.A., Madala, N.E., Labuschagne, N., Dubery, I.A. *et al.* (2018) The chemistry of plant-microbe interactions in the rhizosphere and the potential of metabolomics to reveal signaling related to defense priming and induced systemic resistance. DOI: 10.3389/fpls.2018.00112.

Minicka, J., Zarzyńska-Nowak, A., Budzyńska, D., Borodynko-Filas, N. and Hasiów-Jaroszewska, B. (2020) High-throughput sequencing facilitates discovery of new plant viruses in Poland. *Plants* 9(7), 820. DOI: 10.3390/plants9070820.

Mirzajani, F., Askari, H., Hamzelou, S., Schober, Y., Römpp, A. *et al.* (2014) Proteomics study of silver nanoparticles toxicity on *Oryza sativa* L. *Ecotoxicology and Environmental Safety* 108, 335–339. DOI: 10.1016/j.ecoenv.2014.07.013.

Mitra, S. (2019) Multiple Data Analyses and Statistical Approaches for Analyzing Data from Metagenomicstudies and Clinical Trials. SpringerLink. Available at: https://link.springer.com/protocol/10.1007/978-1-4939-9074-0_20 (accessed 8 September 2024).

Mittal, D., Kaur, G., Singh, P., Yadav, K. and Ali, S.A. (2020) Nanoparticle-based sustainable agriculture and food science: Recent advances and future outlook. *Frontiers in Nanotechnology* 2. DOI: 10.3389/fnano.2020.579954.

Moraes, L.C., Figueiredo, R.C., Ribeiro-Andrade, R., Pontes-Silva, A.V., Arantes, M.L. *et al.* (2021) High diversity of microalgae as a tool for the synthesis of different silver nanoparticles: A species-specific green synthesis. *Colloid and Interface Science Communications* 42, 100420. DOI: 10.1016/j.colcom.2021.100420.

Mustafa, G. and Komatsu, S. (2016) Insights into the response of soybean mitochondrial proteins to various sizes of aluminum oxide nanoparticles under flooding stress. *Journal of Proteome Research* 15(12), 4464–4475. DOI: 10.1021/acs.jproteome.6b00572.

Negrel, L., Halter, D., Wiedemann-Merdinoglu, S., Rustenholz, C., Merdinoglu, D. *et al.* (2018) Identification of lipid markers of *Plasmopara viticola* infection in grapevine using a non-targeted metabolomic approach. *Frontiers in Plant Science* 9, 360. DOI: 10.3389/fpls.2018.00360.

Niazi, P., Monib, A.W., Ozturk, H., Mansoor, M., Azizi, A. *et al.* (2023) Review on surface elements and bacterial biofilms in plant-bacterial associations. *Journal for Research in Applied Sciences and Biotechnology* 2(1), 204–214. DOI: 10.55544/jrasb.2.1.30.

Nitnavare, R., Bhattacharya, J., Thongmee, S. and Ghosh, S. (2022) Photosynthetic microbes in nanobiotechnology: Applications and perspectives. *The Science of the Total Environment* 841, 156457. DOI: 10.1016/j.scitotenv.2022.156457.

Nizamani, M.M. (2023) High-throughput sequencing in plant disease management: a comprehensive review of benefits, challenges, and future perspectives. *Phytopathology Research*. Available at: https://phytopatholres.biomedcentral.com/articles/10.1186/s42483-023-00199-5 (accessed 8 September 2024).

Nongbet, A., Mishra, A.K., Mohanta, Y.K., Mahanta, S., Ray, M.K. *et al.* (2022) A smart and sustainable attribute to modern agriculture. *Plants* 11(9). DOI: 10.3390/plants11192587.

Pandit, C., Roy, A., Ghotekar, S., Khusro, A., Islam, M.N. *et al.* (2022) Biological agents for synthesis of nanoparticles and their applications. *Journal of King Saud University – Science* 34(3), 101869. DOI: 10.1016/j.jksus.2022.101869.

Rasheed, A., Li, H., Tahir, M.M., Mahmood, A., Nawaz, M. *et al.* (2022) The role of nanoparticles in plant biochemical, physiological, and molecular responses under drought stress: A review. *Frontiers in Plant Science* 13. DOI: 10.3389/fpls.2022.976179.

Regondi, C., Fratelli, M., Damia, G., Guffanti, F., Ganzinelli, M. *et al.* (2021) Predictive modeling of gene expression regulation. *BMC Bioinformatics* 22(1), 571. DOI: 10.1186/s12859-021-04481-1.

Saritha, G.N.G., Anju, T. and Kumar, A. (2022) Nanotechnology – big impact: How nanotechnology is changing the future of agriculture. *Journal of Agriculture and Food Research* 10, 100457. DOI: 10.1016/j.jafr.2022.100457.

Shah, A. (2021) PGPR in agriculture: A sustainable approach to increasing climate change resilience. Available at: https://www.frontiersin.org/journals/sustainable-food-systems/articles/10.3389/fsufs.2021.667546/full (accessed 9 September 2024).

Singh, N.A., Narang, J., Garg, D., Jain, V., Payasi, D. *et al.* (2023) Nanoparticles synthesis via microorganisms and their prospective applications in agriculture. *Plant Nano Biology* 5, 100047. DOI: 10.1016/j.plana.2023.100047.

Sosnowski, J. (2023) PGPR in agriculture: A sustainable approach to increasing climate change resilience. Available at: https://doaj.org/article/149e869fcc0945ffb02f0fb4dba9ecec (accessed 9 September 2024).

Suman, J., Rakshit, A., Ogireddy, S.D., Singh, S., Gupta, C. *et al.* (2022) Microbiome as a key player in sustainable agriculture and human health. *Frontiers in Soil Science* 2. DOI: 10.3389/fsoil.2022.821589.

Sundh, I., Del Giudice, T. and Cembalo, L. (2021) Reaping the benefits of microorganisms in cropping systems: Is the regulatory policy adequate? *Microorganisms* 9(7), 1437. DOI: 10.3390/microorganisms9071437.

Tiwari, M., Krishnamurthy, S., Shukla, D., Kiiskila, J., Jain, A. *et al.* (2016) Comparative transcriptome and proteome analysis to reveal the biosynthesis of gold nanoparticles in *Arabidopsis*. *Scientific Reports* 6, 21733. DOI: 10.1038/srep21733.

Tripathi, D.K., Mishra, R.K., Singh, S., Singh, S., Vishwakarma, K. *et al.* (2017) Nitric oxide ameliorates zinc oxide nanoparticles phytotoxicity in wheat seedlings: Implication of the ascorbate-glutathione cycle. *Frontiers in Plant Science* 8. DOI: 10.3389/fpls.2017.00001.

Tudi, M., Daniel Ruan, H., Wang, L., Lyu, J., Sadler, R. *et al.* (2021) Agriculture development, pesticide application and its impact on the environment. *International Journal of Environmental Research and Public Health* 18(3), 1112. DOI: 10.3390/ijerph18031112.

Van de Poel, B. and Van Der Straeten, D. (2014) 1-aminocyclopropane-1-carboxylic acid (ACC) in plants: More than just the precursor of ethylene! *Frontiers in Plant Science* 5. DOI: 10.3389/fpls.2014.00640.

Vannini, C., Domingo, G., Onelli, E., Prinsi, B., Marsoni, M. *et al.* (2013) Morphological and proteomic responses of *Eruca sativa* exposed to silver nanoparticles or silver nitrate. *PLoS One* 8(7), e68752. DOI: 10.1371/journal.pone.0068752.

Vejan, P., Abdullah, R., Khadiran, T., Ismail, S. and Nasrulhaq Boyce, A. (2016) Role of plant growth promoting rhizobacteria in agricultural sustainability: A review. *Molecules* 21(5), 573. DOI: 10.3390/molecules21050573.

Villamor, D.E.V., Ho, T., Al Rwahnih, M., Martin, R.R. and Tzanetakis, I.E. (2019) High throughput sequencing for plant virus detection and discovery. *Phytopathology* 109(5), 716–725. DOI: 10.1094/PHYTO-07-18-0257-RVW.

Walker, V., Bertrand, C., Bellvert, F., Moënne-Loccoz, Y., Bally, R. *et al.* (2011) Host plant secondary metabolite profiling shows a complex, strain-dependent response of maize to plant growth-promoting rhizobacteria of the genus *Azospirillum*. *New Phytologist* 189(2), 494–506. DOI: 10.1111/j.1469-8137.2010.03484.x.

Wang, L., Hu, C. and Shao, L. (2017) The antimicrobial activity of nanoparticles: Present situation and prospects for the future. *International Journal of Nanomedicine* 12, 1227–1249. DOI: 10.2147/IJN.S121956.

Worrall, E., Hamid, A., Mody, K., Mitter, N. and Pappu, H. (2018) Nanotechnology for plant disease management. *Agronomy* 8(12), 285. DOI: 10.3390/agronomy8120285.

Zhalnina, K., Louie, K.B., Hao, Z., Mansoori, N., Rocha, U.N. *et al.* (2018) Dynamic root exudate chemistry and microbial substrate preferences drive patterns in rhizosphere microbial community assembly. *Nature Microbiology* 3(4), 470–480. DOI: 10.1038/s41564-018-0129-3.

Zhang, S. (2022) Action mechanisms of effectors in plant-pathogen interaction – PMC. Available at: https://www.ncbi.nlm.nih.gov/pmc/articles/PMC9224169/ (accessed 8 September 2024).

Zhao, L., Peng, B., Hernandez-Viezcas, J.A., Rico, C., Sun, Y. *et al.* (2012) Stress response and tolerance of *Zea mays* to CeO_2 nanoparticles: Cross talk among H_2O_2, heat shock protein, and lipid peroxidation. *ACS Nano* 6(11), 9615–9622. DOI: 10.1021/nn302975u.

Zobir, S.A.M., Ali, A., Adzmi, F., Sulaiman, M.R. and Ahmad, K. (2021) A review on nanopesticides for plant protection synthesized using the supramolecular chemistry of layered hydroxide hosts. *Biology* 10(11), 1077. DOI: 10.3390/biology10111077.

16 Dissecting Stress-tolerant Traits of CRISPR-edited Crops by High-throughput Omics Technologies

Rinkee Kumari[1], Ekta Pandey[1], Shahla Faizan[1], Sayyada Bushra[1], Alisha Hussain[1], Parul Verma[1] and Mohd. Sayeed Akhtar[2*]

[1]Department of Botany, Aligarh Muslim University, Aligarh, Uttar Pradesh, India; [2]Department of Botany, Gandhi Faiz-e-Aam College, Shahjahanpur, Uttar Pradesh, India

Abstract

In response to resource limitations and climate change, addressing food security is crucial for a growing global population. High-throughput omics technologies, and CRISPR/Cas9 gene editing, offer innovative ways to bolster agricultural resilience. Scientists have made significant progress in understanding how plants react to various stressors, resulting in the development of resilient crop varieties. Case studies on soybean, tomato, rice, coffee, and chickpea demonstrate successful real-world applications. By combining omics data with CRISPR/Cas9, researchers are creating crops better equipped to withstand harsh conditions and threats, paving the way for revolutionary advancements in sustainable agriculture and food security. Therefore, the aim of this chapter is to investigate the amalgamation of CRISPR/Cas9 with diverse omics methodologies in order to elucidate and augment characteristics pertinent to the resilience of crops under stress conditions.

Keywords: CRISPR/Cas9, food security, high-throughput, omics, stress tolerance

16.1 Introduction

Currently, the prevailing circumstances indicate that meeting the nutritional requirements of a burgeoning human population constitutes one of the foremost challenges facing humanity. Projections suggest that by the year 2050, it will be essential for global food production to escalate by 60–100% in order to sustain a population anticipated to be near 10 billion individuals. Significant impediments to agricultural practices and food production encompass an ever-growing human populace, extreme climatic events, diminishing arable land, and escalating levels of both abiotic and biotic stressors (Jaganathan *et al.*, 2018). One possible approach to tackling these difficulties is the development of robust crops that can endure stress using a cutting-edge gene-editing technique called CRISPR. CRISPR was first discovered in 2012 and is presently at the cutting edge of genome-editing research (Kafle, 2023). Zinc finger nucleases (ZFNs) are engineered to create a specific DNA double-strand break by using a series of DNA-binding zinc motifs together with FokI endonucleases. Transcription activator-like

*Corresponding author: sayeedbot@gmail.com

© CAB International 2025. *Stress-Resilient Crops: Coordinated omics-CRISPR-Nanotechnology Strategies* (ed. Jen-Tsung Chen)
DOI: 10.1079/9781800627307.0016

effector nucleases (TALENs) are produced by combining unique DNA endonucleases with transcription activator-like effectors. Such a combination enables the precise binding and breaking of DNA at specified locations. The use of Type II CRISPR/CRISPR-associated protein 9 (Cas9) systems has enabled the development of a technique to accurately remove individual nucleotides in crops, therefore improving the quality of crops. These genome-editing techniques utilize programmable nucleases to improve the accuracy of targeting a specific region (Arora and Narula, 2017).

Such technical advancements can also recognize and clip complementary DNA sequences throughout the whole genome. It is now seen as an adaptive protective mechanism, together with CRISPR/Cas9 CAS enzymes. In an attempt to allay societal worries about the involvement of 'foreign DNAs,' this technique may make it easier and more acceptable to market genetically altered food and agriculture products. Although several countries, like Argentina, have already started cultivating crops with modified genomes, other countries are still weighing the advantages and disadvantages (Kumar *et al.*, 2023). The Japanese market has recently seen the commercialization of CRISPR/Cas9-modified tomatoes, which have greater concentrations of aminobutyric acid (GABA) than non-edited ones (Waltz, 2022). The CRISPR/Cas9 system has been implemented in various plant species, including model plants such as sugarcane (Oz *et al.*, 2021), foxtail millet (Lin *et al.*, 2018), *Nicotiana benthamiana* (Li *et al.*, 2013), peanut (Yuan *et al.*, 2019; Raza *et al.*, 2021), *N. tabacum* (Donovan *et al.*, 2020; Kumar *et al.*, 2022a), and *Arabidopsis* (Li *et al.*, 2013; Nekrasov *et al.*, 2013), as well as in important crop plants such as maize (Liang *et al.*, 2014), rice (Miao *et al.*, 2013), wheat (Shan *et al.*, 2013), barley (Zeng *et al.*, 2020; Garcia-Gimenez and Jobling, 2022), pepper (Park *et al.*, 2021), tomato (Brooks *et al.*, 2014), banana (Tripathi *et al.*, 2019), soybean (Jacobs *et al.*, 2015), and sweet orange (Jia and Wang, 2014). Furthermore, CRISPR/Cas9-based multiplexing has been shown to target multiple genes in a single organism in maize (Char *et al.*, 2017), cotton (Gao *et al.*, 2017), rice (Miao *et al.*, 2013), and wheat (Wang *et al.*, 2018). CRISPR/Cas9 technology can be used to target genes involved in stress-related gene regulatory networks, signal transmission, and metabolite synthesis to develop agricultural plants that are more resistant to stress (Jain, 2015). Researchers have shown that altering specific genes in rice using CRISPR/Cas9 technology may increase the crop's resilience to abiotic stress (Shan *et al.*, 2013; Lou *et al.*, 2017). The AP2/ERF superfamily transcriptional factor, ethylene-responsive factor (ERF), could be used to improve plants' ability to deal with abiotic stress (Debbarma *et al.*, 2019). Crop breeding projects may benefit from studying CRISPR/Cas to enhance responses to diverse abiotic factors (Nascimento *et al.*, 2023). To improve agricultural plants' resistance to environmental stresses, the CRISPR-Cas genome editing mechanism has been brought up to date (Hossain *et al.*, 2023). The advancements and constraints of CRISPR-Cas-mediated genome editing in the context of enhancing climate-resilient wheat breeding to ensure worldwide food security and nutritional sustainability have also been discussed (Afroz *et al.*, 2023).

In addition to CRISPR technology, omics-based approaches such as transcriptomics, proteomics, phenomics, and metabolomics have emerged as useful tools for exploring the molecular mechanisms behind crop growth and stress factor regulation. These technologies provide researchers with a comprehensive understanding of plant systems at the molecular level and enable the development of new crop varieties with improved traits as well as the identification of genes related to stress tolerance (Muthamilarasan *et al.*, 2019). Various omics methodologies, including phenomics, genomics, proteomics, transcriptomics, and many more, have been used to identify the relevant molecular biological component that is integrated with plant systems. Using next-generation sequencing (NGS) technology, data on genomes, transcriptomes, proteomes, metabolomes, and phenomes may be effectively and rapidly produced (Yang *et al.*, 2021). The functions and interplay of genes in response to physiological and environmental stresses may be uncovered by integrating omics methods. Researchers can change agricultural science and enhance crop stress tolerance by integrating CRISPR and omics techniques. Integrating cutting-edge technology might lead to innovative cultivars of agricultural plants with higher yields, nutrient

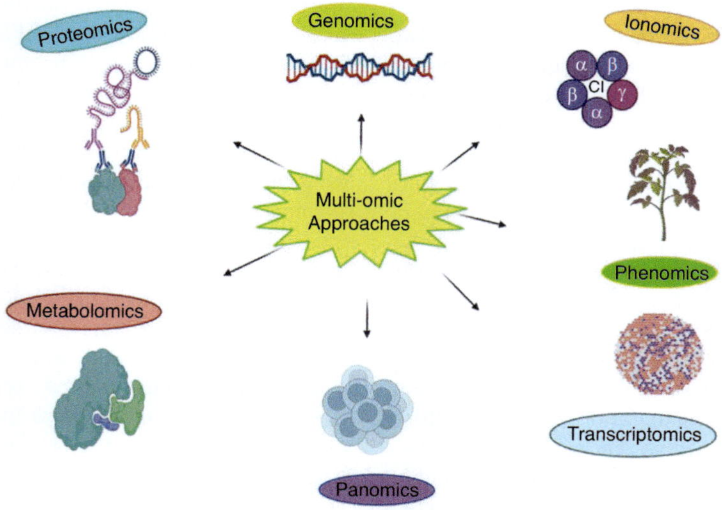

Fig. 16.1. Overview of multi-omics approaches: integrating genomics, proteomics, metabolomics, iononomics, phenomics, transcriptomics, and panomics for comprehensive biological insights.

value, and environmental stress tolerance (Großkinsky *et al.*, 2018). Therefore, the aim of this chapter is to investigate the amalgamation of CRISPR/Cas9 with diverse omics methodologies in order to elucidate and augment characteristics pertinent to the resilience of crops under stress conditions.

16.2 Omics Technologies for Stress-tolerance Analysis

Unraveling the complex responses of plants to stresses and pathogens requires a multifaceted approach. By leveraging various cutting-edge 'omics' technologies, scientists can better understand the intricate molecular mechanisms. Each 'omics' field offers a distinct perspective on plant stress research, providing valuable insights into the genomic, transcriptomic, proteomic, metabolomic, ionomics, phenomics, and phenomics dynamics that shape plant resilience and defense strategies (Fig. 16.1).

16.2.1 Genomics

An organism's genome is its whole genetic makeup, made up of all of its genes. The study of genomes, or genomics, includes their mapping, evolution, structure, function, and alteration. The current state of molecular biology has expedited the study of gene expression, profiling, and genome sequencing considerably. Next-generation sequencing technology facilitates the rapid decoding of genomes through processes like DNA extraction, PCR amplification, sequencing, and evaluating quality. This process allows for large-scale investigations into gene function and interactions, shedding light on cellular and organismal processes (Brown, 2023). The field of genomics is categorized into three sections, which are discussed below.

16.2.1.1 Structural genomics

Functional genomics explores gene function and interactions, while structural genomics determines genes' 3D structure and chromosomal location. Together, they reveal the intricate relationships between sequence and structure, providing a comprehensive understanding of the genome and its role in numerous physiological processes (Mohanta *et al.*, 2021). Genomic selection (GS) enables rapid identification of superior genotypes by leveraging dense genomic markers. This innovative approach improves quantitative characteristics by using marker and phenotypic data from reference populations to assess the impacts of all genetic loci (Alemu *et al.*, 2024). The GS methodology relies on two discrete datasets: the validation and training collections. Marker effects are estimated using the training set, which includes the reference population. In contrast, the validation set is composed of preselected candidates that have been genotyped, allowing for the evaluation of their potential (Tayeh *et al.*, 2015).

Compared to Sanger sequencing, NGS technologies such as Roche 454GS FLX Titanium and Illumina Solexa Genome Analyzer have lowered DNA sequencing cost and time (Płoski, 2016). *Arabidopsis*, soybean, maize, sorghum, rice, wheat, tomato, and other agricultural plants have had whole-genome sequencing. This has illuminated genomic properties including coding and non-coding genes, GC content, repetitive regions, and regulatory sequences (Usha *et al.*, 2022). For animals with smaller genomes, genome sequencing has revolutionized molecular breeding but is limited. Bacterial artificial chromosomal (BAC) libraries can map enormous chromosomes like the 1 Gb wheat chromosomal chromosome chromosome-by-chromosome (Varshney *et al.*, 2019). Furthermore, QTL-seq, a technique for identifying quantitative trait loci (QTLs), has been used to generate whole-genome sequences and pinpoint genomic areas linked to certain qualities, such as drought, salinity, heat, and cold tolerance. QTL mapping has granted valuable insights into the arrangement and development of chromosomes, revealing many QTLs associated with different tolerances to non-living environmental factors (Raj and Nadarajah, 2022). DNA markers, or genetic markers, are short DNA sequences that identify differences in the DNA of a population, including variants such as base deletions, insertions, and substitutions. The markers provide an association between genomic features; for instance, the ability to fight diseases and withstand environmental factors, and particular genetic loci (Ramesh *et al.*, 2020). The genetic enhancement of stress-related characteristics like as drought and salt tolerance has been greatly transformed by recent breakthroughs in molecular markers. Various molecular markers have been developed to identify plant polymorphisms, including RAPD, RFLP, AFLP, SNP, SSR, and STS (Chugh *et al.*, 2022). RFLP detects polymorphisms from mutations or deletions/insertions, while RAPD uses random primers to identify complementary sites. AFLP combines restriction digestion and PCR amplification to identify linkages (Alhasnawi *et al.*, 2024). SSR markers measure genetic diversity and differentiate homozygotic and heterozygotic alleles. SNPs characterize germplasm and map genes, enabling the understanding of complex traits using microarrays. Allelic markers are used in marker-assisted selection (MAS) to breed desired features. High-throughput DNA sequencing and genotyping have improved databases and datasets, simplifying SSR and SNP marker generation. Genomic-assisted breeding and SNP difference-based haplotype mapping have improved stress-tolerant crops (Hasan *et al.*, 2021).

16.2.1.2 Functional genomics

Structural genomics provides extensive resources and knowledge for functional genomics. The creation of thorough gene function experiments is included. There are several biotechnological technologies to find, isolate, clone, and describe genes, and implement functional transgenic studies by overexpressing or knocking out lines (Ma *et al.*, 2021). Pre-genome sequencing methods were arduous, including suppression subtractive hybridization (SSH), expressed sequence tag (EST), and cDNA-AFLP to identify candidate genes. However, NGS has greatly decreased the difficulties of these techniques (Chandrasekaran *et al.*, 2019). Crop genome sequencing has revealed genes related to disease resistance, stress tolerance, and yield (Chaudhary *et al.*, 2019; Jha *et al.*, 2020a). Precision genome editing tools like CRISPR/Cas9 and TALEN are also being used to improve agricultural properties. These genome-editing methods may enhance agricultural attributes without adding foreign DNA, perhaps increasing production by adding pest and disease resistance (Saeed *et al.*, 2022). TALEN and CRISPR/Cas9 have been utilized to develop mildew-resistant and other mutants (Dayani *et al.*, 2019; Ezin and Symonds, 2023).

Utilizing important crops, CRISPR/Cas9 technology has demonstrated its versatility and application to several crucial plant species (Montecillo *et al.*, 2020). The reverse genetic approach virus-induced gene silencing (VIGS) is used to explore gene functions related to biotic and abiotic stress (Bekele *et al.*, 2019). Comparative genomics has found mutations that affect crop growth, development, and stress tolerance (Shi *et al.*, 2023). Genome-wide association studies (GWAS) have identified DNA variations associated with specific traits like abiotic tolerance (Uffelmann *et al.*, 2021). GWAS aims to find genomic areas connected to agronomic traits, phenotypes, or QTL for gene identification, introgressive hybridization, and marker-assisted breeding. QTLomes, complete QTL characterizations, are the outcome of genomics and phenomics advances (Rahimi *et al.*, 2023). The QTLome concept is used to find trait-associated QTL alleles. Statistics, like meta-QTL analysis, have made it easier to gather and analyze QTL data from several researches to discover exact QTL areas.

16.2.1.3 Comparative genomics

Comparative genomics is a field of study that compares entire genomes or genome segments to identify conserved sequences, uncover evolutionary relationships, and understand biological similarities and differences between organisms. By aligning biological sequences, comparative genomics detects conserved regions, revealing significant synteny in related species. This method also detects minor variations within genetic material, enabling the examination of protein-coding segments and their influence on the structure and function of proteins. Furthermore, comparative genomics has led to the development of the 'genome zipper' concept, which predicts gene order and organization in partially sequenced genomes by linking annotated and fully sequenced genomes of related species, such as sorghum and rice, to less-studied species (González Guzmán *et al.*, 2022).

16.2.2 Transcriptomics

Transcriptomics analyzes a cell's or tissue's genome's entire RNA transcripts. For assessing transcriptional response to diverse responses to stimuli across specific times, this dynamic method is useful (Tyagi *et al.*, 2022). How genes are variably expressed *in vitro* may reveal their original functioning. As NGS, RNA seq, SAGE, microarrays, and digital gene expression profiling improve transcriptome analysis resolution, traditional methodologies have been replaced (Buccitelli and Selbach, 2020). Soybean and barley gene expression varies during development and reproduction under drought stress, according to a microarray study. Dehydration stress altered soybean gene expression (Xuan *et al.*, 2022). In *Arabidopsis*, soybean, and rice, abiotic stress affects gene expression and transcription factors. In rice, salt, and drought stress controlled stomatal aperture via novel transcription factors such as Cys2/C_2H_2-ZF TF and DST (Liu *et al.*, 2022). WRKY TFs also help wheat deal with abiotic stress (Gupta *et al.*, 2019). The regulation of gene expression in response to abiotic stress in crops like rice, tomato, and wheat involves a complex interplay of phytohormones, noncoding RNAs, and short peptides (Jha *et al.*, 2020b). In rice, the DREB2 and DRE/CRT transcription factors are crucial in mediating phytohormone-independent abiotic stress responses (Premkumar, 2023). Transcriptome investigations in sorghum and rice have revealed stress- and hormone-responsive genes, revealing crop growth and stress response. Transcriptomics is crucial to understanding the growth of plants and stress response (Punia *et al.*, 2021; Sharma *et al.*, 2021a). New transcriptomics methods like *in situ* RNA-seq and spatially resolved transcriptomics allow researchers to examine gene expression in residing cells and tissues and gain biological insights. RNA-seq investigations in several crops have shown tissue-specific stress and development responses, identifying trait-linked genes (Bashir *et al.*, 2019). RNA-seq has helped uncover drought-responsive genes in rapeseed oil, maize, and rice (Liu *et al.*, 2020; Yu *et al.*, 2020; and; Jiang *et al.*, 2024). In response to stress, comparative transcriptomics can identify genes across crop species. Davidson *et al.* (2012) revealed that common genes in cereals respond to heat stress, and cross-talk pathways in cotton and potatoes correlated with biotic and abiotic stress, which might improve crop resilience. Alternative splicing transcriptomics has produced distinct abiotic stress-responsive transcripts in rice, maize, and sorghum. This strategy illuminates splicing factor control of abiotic stress responses and allows gene expression modification to improve agricultural attributes.

16.2.3 Proteomics

A multifaceted strategy named proteomics studies an organism's total expressed protein. Sequence, structural, functional, and expression proteomics each help comprehend an organism's protein landscape (Monti *et al.*, 2019). Sequence proteomics utilizes high-performance liquid chromatography (HPLC) to identify amino acid sequences. Structural proteomics uses computer modeling, electron microscopy, NMR, crystallization, and protein crystal x-ray diffraction to study protein structure and function (Matthiesen and Bunkenborg, 2013). Functional proteomics combines yeast-two-hybrid experiments and protein microarray profiling to study protein functions. Recent advances in plant proteome research have improved expression proteomics, which measures protein expression under specific circumstances (Rasti and Vogiatzis, 2019). Two-dimensional gel electrophoresis (2-DE) and SDS-PAGE are essential for protein identification and quantification. Stable isotope labeling, ELISA, and western blotting have improved protein analysis accuracy and efficiency (Bergman and Bergquist, 2014). Proteomics has helped identify agricultural stress-responsive proteins, comprehend plant defense mechanisms, and characterize protein responses to abiotic stressors (Ahmad *et al.*, 2016). In particular, phosphoproteomics, which studies protein phosphorylation, has helped reveal pathways involved in numerous physiological activities and stress-related proteins in crops (Kosová *et al.*, 2018). Proteomic techniques have been used on chickpea, *Medicago* spp., wheat, grapevine, and legumes under different stress circumstances (Sui *et al.*, 2018; Pichereaux *et al.*, 2019; Azri *et al.*, 2020; Makonya *et al.*, 2021; Jan *et al.*, 2023). These studies uncovered stress-responsive proteins and shed light on the molecular processes involved in plants' stress response. Proteomics is thereby assisting in the understanding of plant stress responses, which may enhance agricultural output and stress tolerance.

16.2.4 Metabolomics

Metabolomics examines metabolites in biological systems containing cells. However, the metabolome is the whole collection of plant metabolic pathway metabolites (Sharma *et al.*, 2021b). NGS technologies are promising for analyzing gene expression regulation and cellular responses in crops subjected to environmental stresses. Using metabolomics and NGS, an organism's genome sequence may predict its early metabolic network (Ashraf *et al.*, 2022). Metabolism may measure protein expression metabolically, revealing key biochemical processes involved in gene function (Saito and Matsuda, 2010). Proteomics locates gene products. An organism's metabolic phenotype is reflected by its metabolites. Plant samples' metabolic profiles require liquid chromatography-mass spectrometry, gas chromatography-mass spectrometry, thin-layer chromatography, NMR, and other separation and analytical approaches due to their diverse chemical and physical properties. Metabolite identification and analysis need annotation and publication of metabolomics data (Madhu *et al.*, 2015). Quantifying LC/GC-MS-based metabolomics data and guidelines for annotating, including replication, quantification, sample preparation, recovery and recombination, peak misidentification, and ion suppression, could improve metabolite data acquisition and reporting (Chen *et al.*, 2020). The wide range of metabolites generated in plant systems aids environmental stress responses, making metabolomics crucial. Metagenomics can help understand plant stress responses, such as rice (Vo *et al.*, 2021), maize (Christensen *et al.*, 2021), barley (Hamany Djande *et al.*, 2021), and wheat (Razzaq *et al.*, 2021). Proteomics, transcriptomics, and metabolomics together offer an interconnected understanding of the relationships and phenotypes between the genome and the metabolome. The reactions of crops and plants to abiotic stress are highly correlated with these integrated omics. Metabolic engineering has increased the nutritional value and secondary metabolite production of genetically engineered crops.

16.2.5 Phenomics

Phenomics involves characterizing phenotypes by gathering comprehensive phenotypic data on a broad scale. The plant phenome is affected by the interactions among the management (M), environment (E), and genome (G), which are often called G-P-E interactions (Zhao *et al.*, 2019). Precise phenotyping is essential for gene and QTL mapping, facilitating genetic improvement in agricultural plants. Advanced imaging technologies enable non-invasive and rapid phenotyping using various methods like color imaging, lidar, and magnetic resonance imaging to assess traits such as biomass and root systems (Xiao *et al.*, 2022). Infrared thermography and chlorophyll fluorescence imaging are applied to evaluate crop stress responses. These methods are vital for understanding the effects of environmental stress on crops (Pérez-Bueno *et al.*, 2019). Field phenotyping techniques, including multispectral and hyperspectral technologies, are essential for analyzing agronomic traits. Computational tools like support vector machines and artificial neural networks help analyze phenotyping data. Challenges in plant phenomics include organizing information systems for easy reanalysis and generating new insights (Gill *et al.*, 2022). Recent advances like the high-throughput rice phenotyping facility (HRPF) and the rice automatic phenotyping (RAP) system have revolutionized plant breeding methods (Yang *et al.*, 2014). Ontology-driven systems and the Internet of Things (IoT) technologies have further enhanced crop phenotyping, enabling real-time monitoring of crop conditions (Neveu *et al.*, 2019). Future advancements in phenomics, involving new sensors and automation, aim to boost crop productivity, stress resilience, and overall management strategies. Integrating phenomics with proteomics, metabolomics, and genomics offers a comprehensive insight into crop interaction, aiding crop improvement strategies.

16.2.6 Ionomics

Ionomics includes all minerals and tracks the fundamental composition of the inorganic components of plant cells. Various physiological stimuli, genetic mutations, or developmental factors may induce changes in mineral composition, which can be identified by quantitatively assessing an organism's elemental composition (Singh *et al.*, 2022). To get rapid multi-element analysis of plant samples, several analytical techniques have been used, such as x-ray crystallography, neutron activation analysis (NAA), and inductively coupled plasma-mass/optical emission spectroscopy (ICP-MS/OES). As an example, inductively coupled plasma-mass spectrometry (ICP-MS) can run hundreds of samples per day with great sensitivity and is a very cost-effective method. In contrast, ICP-optical emission spectroscopy (ICP-OES) can detect elements with slightly less sensitivity but at a higher throughput. X-ray fluorescence is faster for localizing elements but less cost-effective than ICP-MS/OES (Yang *et al.*, 2021). On the other hand, NAA is slower and costly but can detect more than 30 elements simultaneously. Ionomics has been used to study the basic components of cereal grains, seeds, and leaves. Genetic mapping, the measurement of genetic variation in mineral elements, and the identification of ionomic regulatory networks in a range of crop species, such as rice, corn, barley, soybean, tomato, and several fruit variants, have all resulted from this (Huang and Salt, 2016). Studies such as this one have demonstrated the significance of ionomics in understanding crop response to biotic and abiotic stressors and in developing strategies to improve them. Someone suggested combining ionomics with other omics, such as metabolomics or genomics, as an efficient method of identifying genes and gene networks that could confer greater resistance to environmental and physiological stress on crops.

16.2.7 Panomics

Panomics is a new discipline that combines several different 'omics' domains, such as proteomics, metabolomics, genomes, PTM proteomics, and epigenomics phenomics. A comprehensive understanding of biological systems is made possible by this integration (Khan *et al.*, 2023). Researchers aim to elucidate the intricate molecular interplay between a plant's traits and its environment by integrating environmental factors with these omics. This integration also aids in reducing false positives, enhancing the prediction of complex traits, and identifying genes and markers to improve crop varieties (Kaur *et al.*, 2021). To analyze integrated data, specialized tools like PAINTOMICS, KaPPA-view, and COVAIN merge multi-omics datasets before interpretation. These tools provide visualizations of data from transcriptomics, metabolomics, and other datasets, displaying them on pathway maps and aiding in statistical analysis (Tyagi *et al.*, 2023). Novel genes and functional pathways promoting complex characteristics in crops have been identified by the combination of GWAS with panomics, which includes transcriptomics, proteomics, and metabolomics (James *et al.*, 2021). To further improve crop traits linked to nutritional value, plant fitness, and yield, a model for precision breeding is being considered that combines panomics with advanced genetically engineered technologies like TALENs and CRISPR/Cas9. This model would seek to identify phenotypic variances that are currently unexplained. The ultimate goal of combining multi-omics with the genotype-to-phenotype idea is to enhance crop breeding efforts via the development of quantitative and qualitative characteristics through the incorporation of epigenetic regulation caused by environmental variables (Daniel *et al.*, 2023; Fig. 16.2).

16.3 CRISPR Crops with Stress-tolerant Traits

The CRISPR/Cas system exploits base pairing between guide RNA and target sequence to identify particular locations, assisted by the protospacer-adjacent motif (PAM). This system, with components like sgRNA and Cas endonuclease, is known for its simplicity, flexibility, efficiency, and stability in genome editing (Saraswat and Ranjan, 2022). CRISPR/Cas, a bacterial defense mechanism, quickly replaced older techniques like ZFN and TALEN. Through sgRNA recruitment, Cas endonuclease induces DNA double-stranded breaks (DSBs). NHEJ, MMEJ, and HDR repair approaches enable gene deletion, knockout, fragment insertion, and replacement (Pandita and Pandita, 2023). Currently, CRISPR/Cas has been improved to target genomes. Cas-NG, xCas9, and, SpCas9 interact with plant PAM sites (Li *et al.*, 2022).

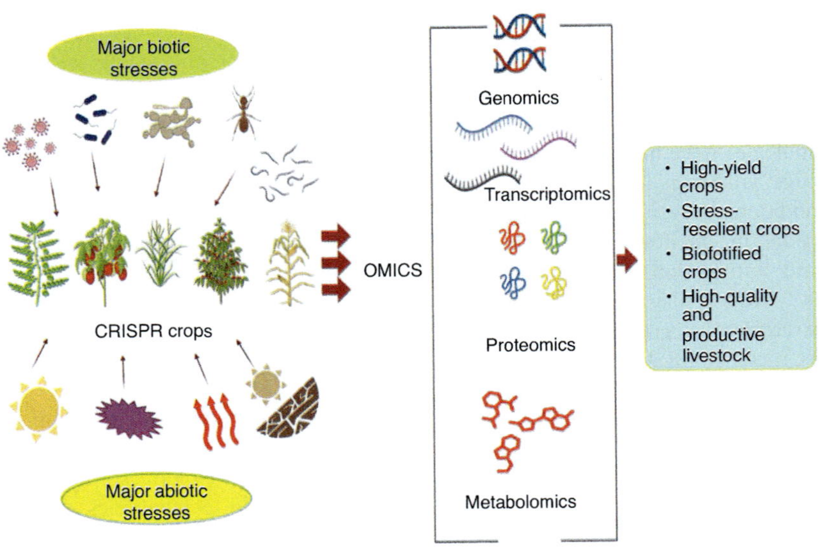

Fig. 16.2. Stress tolerance traits of CRISPR crops using multi-omics approaches. This depiction showcases how various omics disciplines such as genomics, proteomics, and metabolomics enhance the stress resilience of CRISPR-edited crops.

ScCas9, SaCas9, St1Cas9, and BlatCas9 modify plant genomic loci with particular PAM sequences (Zhong *et al.*, 2023). Cas12a and Cas12b from different bacteria exhibit AT-rich PAM selectivity and have been employed in targeted plant genome editing (Yang *et al.*, 2023). CRISPR/Cas has enabled alteration and regulation of genomes in many animals. CRISPR/Cas has great promise for crop genetics despite being the dominant genetic modification technique in plant sciences (Li *et al.*, 2022). Widely used CRISPR/Cas technology boosts agricultural productivity, quality, abiotic and biotic stress tolerance, pesticide resistance, and *de novo* crop domestication (Saeed *et al.*, 2024). Wang *et al.* (2020) found that editing 25 conserved amino acid sequences at the C-terminal of the rice cytokinin-activation enzyme-like gene lonely guy (oslogl5) boosted grain production under different geographical situations. Using CRISPR/Cas9 to impair BETAINE ALDEHYDE DEHYDROGENASE 2 (BADH2) activity created novel rice, maize, and sorghum genotypes with higher 2-acetyl-1-pyrroline (2AP) levels, a main aromatic chemical (Manchanda *et al.*, 2022). Singh *et al.* (2024a) generated three distinct paraquat tolerance 3 (pqt3) functional deficit mutants of rice using CRISPR/Cas9, which enhanced germination and development under oxidative and salt stress. Similarly, Khabbazi *et al.* (2021) employed CRISPR/Cas9-mediated HDR to replace endogenous 5-enolpyruvylsh ikimate-3-phosphate synthase (EPSPS) with new variations, resulting in stable inheritance and glyphosate-resistant rapeseed. Singh *et al.* (2024b) used CRISPR/Cas9 multi-gene editing to target genes affecting key agronomic characteristics to accelerate wild rice domestication.

A number of environmental conditions, such as heavy metals, drought, salt, and extreme temperatures, can cut agricultural output by 50%. Although conventional breeding has increased yields in the past, it also runs the danger of reducing genetic variety (Villalobos-López *et al.*, 2022). Furthermore, the development of desirable features for sustainable production may be hampered by its dependence on spontaneous allelic variations. To address these issues, the emphasis has switched to CRISPR/Cas-mediated crop editing. This cutting-edge technology presents a viable way to improve crops' resistance and adaptation (Table 16.1) so they may flourish in a range of environmental circumstances (Gajardo *et al.*, 2023). Biogenic stressors, including viral, fungal, and bacterial diseases, significantly impact worldwide agriculture (García-Montelongo *et al.*, 2023). For the food crisis and agricultural production, researchers have used CRISPR/Cas9 technology to confer resistance to pathogens including fungi, bacteria, viruses, and insects. This innovative approach has shown great potential in developing plants with heightened resistance to various agricultural threats, paving the way for more sustainable and robust crop production systems (Muha-ud-din *et al.*, 2023).

16.3.1 *Cicer arietinum*

Chickpea (*Cicer arietinum* L.) is a self-pollinating diploid legume with $2n=16$ chromosomes and a genome size of 738 Mb, containing 28,269 genes. It is the second most important legume crop and provides economic advantages. India is the largest producer, accounting for 65% of global chickpea production, with an annual yield of 9 million tons from 12 million hectares of cultivated area. Chickpea is a nutritious crop, rich in protein, carbohydrates, vitamins, minerals, fiber, folate, β-carotene, antioxidants, and health-enhancing fatty acids. However, growing climatic uncertainty, diseases, pests, and abiotic stresses make chickpea cultivation challenging globally, and conventional breeding techniques cannot meet the increasing demand (Koul *et al.*, 2022). The advent of genomic technologies, particularly CRISPR/Cas9, has accelerated agricultural molecular breeding projects, including genetic transformation and genome editing of chickpea cultivars (Jha *et al.*, 2019). Over the past 25 years, more than 55.1% of scientific studies have focussed on genetic transformation and *in vitro* regeneration of transgenic plants, 30.6% on chickpea insect pest resistance, and 12.2% on abiotic stress tolerance. However, only 2% of these studies have used genome editing to modify chickpea, while 97.5% have used stable genetic transformations and 8.1% have used composite plants (hairy root genetic transformation) (Hauptmann *et al.*, 1988; Amack *et al.*, 2022). The majority of these studies

Table 16.1. An overview of employed omics technologies in CRISPR-crops and their stress-resilience mechanisms against various stress factors.

CRISPR crops	Stress	Omics technologies	Target genes	Stress-tolerant traits	References
Alfalfa	Salt	Transcriptomic	MsSOS1	Salinity resistance	Acharya *et al.*, 2021
Banana	*Fusarium wilt*	Genomics	RGA2	Increased disease resistance	Soni *et al.*, 2024
Cassava	Sri Lankan cassava mosaic virus (SLCMV)	Metabolomics and proteomics	AT2G20050, D-serine, WRKY33; A0A140H8T1, A0A2C9WKV2, A0A251K1B9, CAT; A9YME8, A0A2C9UA53	Proteins and metabolites orchestrate and activate biochemical pathways in plants, enabling them to respond defensively against viral invaders.	Siriwan *et al.*, 2023
Chickpea	Drought and oxidative stress	Phosphoproteomic and secretomics	CaDREPP1, CaRRP1, CaFer1	Protein phosphorylation is triggered by water scarcity, stress-related secretomes, and extracellular matrix metabolic network functions, while CaFer one aids in iron buffering and oxidative stress coping.	Subba *et al.*, 2013; Parveen *et al.*, 2016
Cotton	Heat	Transcriptomics and genomics	GhHB12	GhHB12 has been identified and modified to enhance its heat tolerance.	Billah *et al.*, 2021
Grape	Cold	Transcriptomics and metabolomics	CBF (C-Repeat Binding factor) gene	Enhanced resistance to low temperatures, which enables grape cultivation in a cooler climate	Tillett, 2011
Cucumber	Powdery mildew	Genomics	CsWRKY20	CRISPR-mediated mutation of CsWRKY genes helps in developing disease-resistant varieties.	Najafi *et al.*, 2024
Maize	Drought	Genomics	ZmWRKY40 and ZmNF-YB2	Promotes photosynthetic activity, chaperone, and antioxidant synthesis	Farooqi *et al.*, 2022
Peanut	*Aspergillus flavus* infection	Transcriptomics and metabolomics	2 9S-LOX, 413S-LOX, and DEGs	Early infection led to the concentration of DEGs and metabolites in pectinase catabolism, hydrogen peroxide breakdown, and cell-wall tissue activities.	Wang *et al.*, 2023
Pigeon pea	Abiotic stress	Genomics and transcriptomic	TERMINAL FLOWER1b (TFL1b)	Promotes shoot apical meristem cell growth indeterminately	Meyer and Purugganan (2013)
Potato	*Phytophthora infestans*	Genomics	Rpi	RenSeq and genetic mapping detect the late blight-resistant gene.	Witek *et al.*, 2016
Quinoa	Salt	Proteomics	CqSOS1	Enhanced salt tolerance	Kiani-Pouya *et al.*, 2020
Rape-seed	Cold stress	Transcriptomics	DREB1/BNCBF	Improved cold tolerance and controlled chloroplast development	Raza *et al.*, 2021

Rice	Drought stress	Transcriptomics	OsDREB1A, OsDREB1B, OsDREB1C, OsDREB1D, and OsDREB2A	Regulation of stress-responsive genes and the accumulation of osmoprotectants stabilizes cells, enhances water absorption, and minimizes transcriptional water loss.	Rakkammal *et al.*, 2022
Sorghum	Drought	Metabolomics	SbDREB2	SbDREB2 is being targeted for improving drought tolerance, using metabolomics to understand its regulatory effects.	Prusty *et al.*, 2024
Soybean	Salt stress	Transcriptomics	GmLecRlk (Glyma. 07G005700, GSPRX9	Increased antioxidant gene expression and ROS resistance enhanced salt tolerance. Upregulation of GmDREB2, GmERF3, and GmbHLH30; and downregulation of GmLAMP1, GmGH3.6, and GmPUB8 improved soybean salt tolerance and various aspects of growth and enzyme activities.	Zhang *et al.*, 2019a
Sugarcane	Cold	Transcriptomic	MYB, E3 SUMO-Protein ligase, DEGs	Enhanced abiotic and biotic stress tolerance	Ali *et al.*, 2019
Sweet potato	*Fusarium oxysporum*	Transcriptomics	CERK1, MAPK, WRKY, NAC, MYB, ERF, PR	Enhanced the mechanism of disease resistance	Lin *et al.*, 2017
Tomato	Cadmium and drought stress	Metabolomics	SIMAPK3	Transgenic plants improve germination and seedling growth by increasing chlorophyll and root biomass under Cd^{2+} stress.	Muhammad *et al.*, 2019
Wheat	Salt stress	Genomics and transcriptomics	NRAMP-2 and OPAQUE1	Upon confirmation of functional grain quality characteristics, gene-specific markers could be developed to improve salinity tolerance.	Kumar *et al.*, 2022b

(85.7%) have used *Agrobacterium tumefaciens* to deliver T-DNA, minimal expression cassettes, or CRISPR/Cas9 elements, while other methods such as biolistics (12.2%), *A. rhizogenes* (8.1%), PEG-transformation of chickpea protoplasts (2%), and electroporation of embryogenic axes (2%) have also been employed (Hauptmann *et al.*, 1988; Amack *et al.*, 2022). The constitutive cauliflower mosaic virus (CaMV) 35S promoter was used in 65.7% of the chickpea transformation genetic constructs, while the nopaline synthase (NOS) promoter was used in 20.9% and other promoter sequences in 13.3% (Amack *et al.*, 2022). Transgenic methods have increased chickpea's abiotic stress tolerance and insect pest resistance by overexpressing genes like AtDREB1a, AtBAG4, TlBAG, CaHDZ12, CaWRKY70, P5CS, P5CSF129A, cryIIAa, cry1Aabc, cry1Ac, cryIAa3, cry2Aa, cryIAc, ASAL, cry1Ab, αAI1, and α-amylase inhibitor genes (Mehrotra *et al.*, 2011). The efficiency of CRISPR/Cas9 NHEJ in editing the chickpea genome and knocking out the 4CL and RVE7 genes in protoplasts suggests that strong plant regeneration methods are needed for genome editing (Badhan *et al.*, 2021).

16.3.2 *Solanum lycopersicum*

Tomato (*Solanum lycopersicum* L.) is an important fresh and processed vegetable and economic driver. Its minerals, vitamins A and C, and antioxidants like beta-carotene lycopene provide a nutritious meal (Kumar *et al.*, 2020). Diseases, insect pests, heat, drought, cold, and salt affect tomato. These issues have been addressed by modifying several tomatoes including using CRISPR/Cas9 tools (Erdoğan *et al.*, 2023). Cas9 modification of the Pelo and Mlo1 genes has resulted in significant TYLCV (Ty-5) and powdery mildew resistance (Pramanik *et al.*, 2021). Several CRISPR/Cas9-edited disease-resistant mutants have been created, including those that resist *Xanthomonas* spp., *P. syringae*, and *P. capsici* DMR6-1 upregulation after pathogen infection boosted disease resistance and salicylic acid (SA) in these mutants (Djennane *et al.*, 2024). Late blight (*Phytophthora infestans*)-resistant mutants were created via CRISPR/Cas9 editing of the MYBS2 gene. Using CRISPR/Cas9, tomato with targeted SlLBD40, a lateral organ boundaries domain transcription factor, were more drought-resistant than transgenic overexpression and WT tomato plants (Pandita, 2023). CRISPR/Cas9 eliminated SlLBD40 in tomato plants, increasing drought tolerance. Using CRISPR/Cas9 to edit tomato (*S. lycopersicum*) resulted in mutant plants with a knocked-out SlMAPK3 gene, showcasing an improved response to drought stress. In instances of drought, these modified plants displayed pronounced wilting symptoms, heightened H_2O_2 levels, reduced antioxidants, and increased membrane damage (Gajardo *et al.*, 2023). These findings suggest that SlMAPK3 protects the plant's cell membrane under drought stress (Majeed *et al.*, 2023). Water shortage resistance is increased by knocking down the tomato auxin response factor (SlARF4) gene. Modifying the pathogenesis-related 1 (NPR1) gene in tomato using CRISPR/Cas9 increased stomatal aperture, MDA levels, H_2O_2 content, and ion leakage in response to drought (Angon *et al.*, 2023). However, antioxidant activity decreased more than in wild-type plants. SlNPR1 regulates drought-stress responses in tomato and other crops (Krishna *et al.*, 2022). Genome editing may produce several SlNPR1 variations, improving drought tolerance over a range. Tomato may improve grain quality, which renders it an ideal model for CRISPR/Cas9-based genome editing (Li *et al.*, 2019a). Genome editing of the heat-sensitive gene Slagamous-LIKE 6 (SlAGL6) in tomato improved heat tolerance and fruit set under heat stress (Kumar *et al.*, 2023). Tomato's mitogen-activated protein kinase gene SlMAPK3 is essential for environmental stress response (Muhammad *et al.*, 2019). CRISPR/Cas9-edited slmapk3 mutants had better thermotolerance relative to wild-type plants, suggesting they negatively regulate heat tolerance (Marwein *et al.*, 2019). By favorably regulating ROS generation in the apoplastic area, BRZ1 helps tomato tolerate heat. By downregulating Respiratory Burst Oxidase Homolog 1 (RBOH1), CRISPR/Cas9-based editing of bzr1 mutants lowered apoplast H_2O_2 synthesis and heat tolerance (Yin *et al.*, 2018). Compared to wild-type tomato, CRISPR/Cas-mediated HSA1 (heat-stress sensitive albino 1) mutants enhanced heat stress sensitivity. Cold damage may also occur in tomato fruits due to chilling stress (Singh *et al.*, 2022). CRISPR/

Cas9-based cbf1 mutants showed that CBF1 protects tomato plants against cold damage and electrolyte leakage. These mutants have higher hydrogen peroxide and indole acetic acid levels, enabling cold-stress tolerance (Kumar *et al.*, 2023). Under low temperatures (4°C), tomato SlNPR1 expression and protein increased. CRISPR/Cas9 genome editing of SlNPR1 triggered chilling injury in tomato plants, which leads to H_2O_2, O_2^-, MDA, reduced proline, antioxidant enzymatic activities, and soluble protein accumulation (Shu *et al.*, 2023).

16.3.3 *Oryza sativa*

The rice harvest feeds more than half of the globe. CRISPR-Cas has been employed by scientists to enhance rice yield and quality. With regard to enhancing yield, resilience to biological and environmental stresses, and product quality/nutritional features in rice, this approach has been beneficial in creating novel variants for desirable qualities that were previously unavailable (Ahmad *et al.*, 2022). Utilizing CRISPR-Cas technology in rice improvement has led to notable successes. By targeting and mutating genes such as Pi21, Bsr-dl, and ERF922, researchers have produced single or triple mutants that exhibit enhanced resistance to blast disease (Dhadge and Walia, 2024). Precision targeting of the OsSEC3A gene using CRISPR-Cas yielded rice plants with increased immune responses against blast disease. Notably, CRISPR/Cas12a technology has enabled the creation of targeted mutations in the UPT box, enhancing resistance to bacterial blight (Mishra *et al.*, 2021). Furthermore, the OsPFT1 gene's regulatory function in rice sheath blight disease was confirmed through CRISPR-Cas9 editing. Mutants with knocked-out OsPFT1 showed resistance to sheath blight. Enhancing HDR repair efficiency, the fusion of Cas9 with VirD2 protein enabled precise modifications in genes like OsALS and OsCCD7 (Shah *et al.*, 2019). Herbicide tolerance in rice was achieved by the introduction of an in-frame epitope tag fusion at OsHDT. Reduced expression of genes that act as negative regulators of drought tolerance, including OsDIS1, OsDST, and OsSRFP1, was shown to enhance rice plants' drought tolerance (Sami *et al.*, 2021). Also, by rendering plants more responsive to ABA and closing their stomata more tightly during drought, altering the OsERA1 gene improved their ability to withstand drought stress. Plants engineered by CRISPR/Cas9 with loss-of-function mutations in OsSAPK2, which is involved in ABA-mediated stress tolerance, showed increased drought sensitivity compared to plants in their wild-type state (Ogata *et al.*, 2020). Research has also shown the significance of genes such as OsSRL1 and OsSRL2, which encode characteristics of the plant's leaf tissue; plants in genome-edited lines bearing homozygous mutants showed different physiological characteristics than plants in the wild-type condition (Tabassum *et al.*, 2021). Results from a study by Pandita (2023) showed that mutants produced by disrupting the activities of the SnRK2, SAPK-1, and SAPK-2 genes exhibited resistance to salt stress, suggesting that CRISPR-mediated gene expression modulation improved salt-stress tolerance. According to Nazir *et al.* (2022), CRISPR has helped rice plants become more resistant to salt stress by altering genes including OsDST, OsNAC45, AGO2, OsRR22, and OsBBS1. The growth and yield of rice are affected by cold stress. It has been shown that genes such as OsMYB30 have a detrimental effect on cold tolerance (Lv *et al.*, 2017). Conversely, OsAnn3 and OsAnn5, two plant annexin genes, positively control seedling rice's ability to withstand cold stress eliminating OsPRP1-induced cold sensitivity in rice to lower proline levels and antioxidant activity (Singh *et al.*, 2022). CRISPR/Cas9-mediated mutations in genes such as OsNramp1, OsNramp5, OsCd1, and OsLCT1 have changed cadmium levels in rice and provided prospective ways to lessen toxicity with regard to cadmium absorption in rice roots. These developments highlight the transformative potential of CRISPR-Cas technology in revolutionizing rice agriculture for improved quality, yield, and stress resilience (Li *et al.*, 2023).

16.3.4 *Coffea arabica*

Coffee, a globally popular beverage, was produced at a volume equivalent to 167 million 60 kg bags in 2020/21. *Coffea* plants, such as *Coffea arabica* and *C. canephora*, are the primary species responsible for producing the coffee beans consumed worldwide (Rendón Mera, 2023). Through breeding, these species have been cultivated to create commercial cultivars adaptable to various environments, highly productive, and resistant to stress while possessing high organoleptic quality and specific chemical attributes (Martinez *et al.*, 2023). Utilizing CRISPR/Cas9 technology to silence specific genes plays a crucial role in enhancing plants for better resistance against pathogens and increased tolerance to environmental stresses. Efforts to produce low-caffeine coffee have included RNA interference (RNAi) to suppress the XANTHOSINE METHYLTRANSFERASE gene (CaMXMT1), reducing caffeine and theobromine levels by up to 70% (Rajput *et al.*, 2021). However, this approach has limitations. Alternatively, CRISPR/Cas9 technology potentially targets genes like CaMXMT1 to achieve reduced caffeine and theobromine content, offering a new approach with promising results. To improve coffee quality, one must understand the chemical constituents and metabolic mechanisms that determine quality (Santoso, and Sisharmini, 2022). Research has concentrated on chlorogenic acid (CGA), sucrose, and trigonelline as coffee-quality components. The quality of coffee is improved by increasing sugar concentration in beans (Hall *et al.*, 2022). By knocking down the sucrose synthase (CcSUS2) gene from *C. canephora* using CRISPR/Cas9, coffee beans can have more sucrose. CRISPR/Cas9-targeted mutation of CGA production genes may improve coffee's disease resistance and antioxidant qualities (Li *et al.*, 2021). CRISPR/Cas9-mediated genome editing may introduce point mutations in the ACETOLACTATE SYNTHASE (ALS) gene to construct herbicide-resistant coffee plants (Prajapat *et al.*, 2021). The CRISPR/Cas9 technology can change transcriptional regulation and signal transduction genes to improve coffee drought tolerance. This may be done by targeting DREB, ERF, and MAPK (Santoso and Sisharmini, 2022). Regulating genes involved in the latter stages of ripening allows for consistent coffee cherry ripening. Using CRISPR/Cas9 to knock off the gene that triggers an early ethylene burst could regulate coffee fruit ripening (Maghuly *et al.*, 2020).

16.3.5 *Glycine max*

Recent developments in CRISPR/Cas9 technology have revolutionized targeted genome editing and gene function exploration, offering a new approach to reverse genetics research. This has significant implications for soybean, which holds substantial economic importance because of its high protein and oil content (Aslam *et al.*, 2022). In order to boost production and improve quality, it is critical to clarify gene functions, advance functional gene research, and strengthen breeding initiatives as the world's demand for soybean rises (Anderson *et al.*, 2019). It is imperative to tackle the issue of water scarcity in soybean, since it may seriously hinder its growth and result in yield losses of as much as 40%. Recent studies have demonstrated the potential for improving soybean's resistance to water shortage by genetic alterations utilizing CRISPR/Cas9. For instance, manipulating specific genes such as GmPLA and GmNF-YC14 using CRISPR/Cas9 has shown promise in conferring soybean with improved resilience to phosphorus deficiency and water scarcity (Yao *et al.*, 2023). Similarly, targeting genes such as GmNAC12 and GmHSP26 using CRISPR/Cas9 has been linked to improved drought tolerance and increased soybean yield. Salinity also significantly threatens soybean seed yield and quality (Sami *et al.*, 2021). Initiatives to tackle this obstacle have been concentrated on cultivating soybean strains resilient to saline conditions and investigating the genetic underpinnings of salt-stress tolerance. Research has indicated that modifying genes such as lncRNA77580 and GmAITR using CRISPR/Cas9 technology can enhance soybean's tolerance to saline conditions (Wang *et al.*, 2021). Furthermore, studies have underscored the importance of genes like GmSOS1 in maintaining sodium homeostasis for improved salt tolerance in soybean. In addition to environmental stressors, soybean plants are vulnerable to diverse stressors and pathogens, presenting a significant challenge to soybean production (Zhang *et al.*, 2019a). Gene-modification tools offer the ability to enhance soybean resilience to these pathogens. For instance, several studies have successfully modified genes such as GmLMM2, GmF3H1, GmF3H2, GmFNSII-1, Rsc4-1, Rsc4-2, Rsc4-3, Avr1b-1, GmDRR1, GmTCP19L, Avr45a, Glyma.07g110300, and GmTAP1 using CRISPR/Cas9 gene editing, resulting in improved disease resistance to various pathogens in soybean plants (Zhang *et al.*, 2020). Additionally, researchers have addressed the soybean cyst nematode (SCN) by targeting genes in the syntaxin family and cqSCN-006, demonstrating the impact of gene editing on soybean resistance to this devastating pest (Peng *et al.*, 2021).

16.4 Integration of CRISPR and Omics for Stress Tolerance

Integrating omics technologies and CRISPR-Cas9 gene editing holds immense potential for advancing our understanding and engineering of crop stress tolerance in biotic and abiotic stress. Modern techniques may identify important stress-responsive genes, respond with molecular insights into complicated stress-response systems, and allow precision genome editing to create stress-tolerant plants. Improved signaling molecules and metabolic by-products from drought-responsive genes and transcription factors increase plant drought tolerance. Hormonal instability, decreased antioxidant activity, and increased ROS production enhance drought vulnerability in S genes, drought-resistant genes, and transcription factors (Singh *et al.*, 2019). While AREB1 overexpression improved drought-stress tolerance, the knockout mutant was more vulnerable. SlLBD40 gene suppression using CRISPR/Cas9 increased tomato drought resilience compared to overexpressing transgenic and wild-type plants (Mohamed *et al.*, 2024). In CRISPR/Cas9-edited tomato plants with a SlMAPK3 gene deletion, drought stress produced wilting, increased H_2O_2, and reduced antioxidants and membrane damage. Knockout of the SlARF4 gene improves tomato drought resilience (Kumar *et al.*, 2023). Roca Paixão *et al.* (2019) found that fusing CRISPR/dCas9 with the Histone Acetyl Transferase (AtHAT) gene increased drought-stress tolerance in *Arabidopsis*. By using CRISPR/Cas9 to edit the tomato pathogenesis-related 1 (NPR1) gene, stomatal aperture, MDA, H_2O_2, and ion leakage were raised while antioxidant levels were lowered. Reduced H_2O_2, enhanced drought tolerance, and improved antioxidant enzyme levels were seen in rice with silenced OsDIS1, OsDST, and OsSRFP1 (Kumar *et al.*, 2023). *Arabidopsis* AtOST2 gene alterations improved stomatal response and closure by CRISPR/Cas9. Transpiration rate, chlorophyll content, stomata number, conductance, and vascular bundles are decreased in rice genome-modified lines with homozygous SRL1 and SRL2 mutants (Singh *et al.*, 2022). CRISPR/Cas9-based genome editing enabled drought-tolerant maize plants with higher grain yields. This increased drought tolerance by increasing ARGOS8 gene expression, which inhibits ethylene response (Shi *et al.*, 2017). Additionally, the WRKY3 and WRKY4 genes regulate plant drought-stress tolerance. Studying gene structure and function using omics helps enhance crops (Wani *et al.*, 2021).

Genetic resources such as mutant libraries, cDNAs, expression profiles, sequencing data, and QTLs are vital for plant biology structural and genomics investigations. Many QTLs have illuminated the complicated processes of drought-stress tolerance in crops. These QTLs provide increased grain production and other secondary features that help plants tolerate drought in rainfed uplands and lowlands (Tuberosa, 2012). High-throughput molecular breeding techniques such as MAS, SNP marker analysis, and GWAS are expected to improve our understanding of crop resilience to environmental stresses like drought stress (Sinha *et al.*, 2023). ABA-dependent drought response is enhanced by genes like OsCPK9 and AtCPK10. Zhu *et al.* (2024) and Bundó and Coca (2017) found that the OsCPK10 gene protects cellular membranes from ROS and drought stress. Bip (binding protein) molecules, molecular chaperones, help agricultural plants cope with drought. Biochemical and proteomic investigations indicated that drought-induced transgenic soybean plants overexpress bip (Vaseva *et al.*, 2012).

Transgenic rice overexpressing the OsSKIPa gene upregulates stress-related genes such as PP2C, CBF2, RD22, and SNAC1 and improves ROS scavenging. Rice can withstand drought due to transcription factors. ABA-dependent and ABA-independent TFs occur. The basic leucine zipper (bZIP) and NAM, ATAF, and CUC2 (NAC) families of ABA-dependent TFs mediate dehydration-induced signaling and physiological responses under drought stress (Manna *et al.*, 2021). Drought resistance is better conferred by bZIP TFs such as ABF-3 than by DREB1A/CBF3. The zinc finger and AP2/ERF families of ABA-independent TFs also help drought resistance. For instance, overexpressing the STZ gene increases spikelet fertility and grain output under drought-stressed circumstances (Sahoo *et al.*, 2013; Zargar *et al.*, 2022). Chilling and freezing temperatures below 20 and 0°C hinder plant growth and development, hindering their geographical spread and agricultural production (Zahra *et al.*, 2021). Low temperatures produce osmotic, oxidative, and metabolic stress in plants. Zeng *et al.* (2020) found that CRISPR/Cas9-engineered ospin5b, gs3, and osmyb30 mutants produced longer spikes, larger grains, and better cold tolerance. High temperatures affect the whole crop life cycle, especially heat-sensitive phases including early establishment, blooming, and gametophytogenesis (Hasanuzzaman *et al.*, 2013). CRISPR/Cas9-mediated slmapk3 mutants in tomatoes maintained ROS equilibrium by modulating antioxidant enzymes and HSPs/HSFs genes, improving high-temperature tolerance (Yu *et al.*, 2019). The CRISPR/Cas9-edited pyl1/4/6 triple knockout rice had higher yields, temperature tolerance, and lower preharvest germination than wild types (Li *et al.*, 2022).

Phenomics, genome sequencing, and molecular breeding promise crop progress but are expensive. SSR markers found stress-related QTLs in rice while SNP markers found climatic QTLs in maize (Raj and Nadarajah, 2022). While markers increase legume crops, chickpea cold stress research is restricted. In chickpea, SSR markers combined with DArT and CAPS markers identified beneficial QTLs for abiotic stress. The marker-assisted introgression of these advantageous QTLs has produced improved chickpea varieties (Gujaria, 2012). Marker-trait association (MTA) or GWAS studies also determine the connection between desired attributes and crop genome molecular markers (Mohammadi *et al.*, 2020). A recent GWAS investigation on white clover explored cold tolerance loci. The research found 53 cold tolerance-related loci in 192 distinct germplasm from Patagonia, South America (Kuo *et al.*, 2024). In addition, differentially expressed transcription factors, including AP2/ERF, WRKY, and bHLH, help plants tolerate cold temperatures. The AP2/ERF transcription factor family, especially CBFs/DREB1s, is essential for plant cold tolerance (Jiang *et al.*, 2020). Cold stress overexpresses CBFs in 15 min. Various legume species are more cold-tolerant because these CBFs directly influence cold-responsive (COR) gene expression (Li *et al.*, 2019b). To modulate legume CBF expression under cold stress, cis-elements in CBF promoters must be identified and combined with transcription factors. ICE1, a MYC-like bHLH activator, binds to CBF gene promoters during cold stress to stimulate gene expression (Jiang *et al.*, 2020). Proteins are essential for plant cold-stress resistance. As molecular chaperones, enzymes, and transcription factors, they regulate stress-signaling pathways and protect plants from cold stress (Chen *et al.*, 2021). Proteomics methods like 2DE, MS, LC-MS, MALDI, and iTRAQ have been used to analyze protein profiles in plant tissues or organelles under cold stress (Wu *et al.*, 2006).

Salt stress causes osmotic, ion, and secondary stress in plants, reducing crop output and quality. Salt tolerance in crop species has been improved by genetic engineering, notably the CRISPR/Cas9 system (Farhat *et al.*, 2019). Deleting the SlHyPRP1 gene's negative-response domain(s) increased tomato germination and vegetative salt tolerance (Tran *et al.*, 2021). CRISPR/Cas9 deletion of the OsbHLH024 gene in rice increased the expression of ion transporter genes such as OsHKT1:3, OsHAK7, and OsSOS1, improving salt tolerance (Alam *et al.*, 2022). The CRISPR/Cas9-induced OsRR22 gene mutation in rice increased salt tolerance without impacting other agronomic features. Rice with *OsRAV2* mutated gene by CRISPR/Cas9 survived better under salt-stress conditions (Zhang *et al.*, 2019b). Knockout of genes like OsDST, OsNAC041 (Wang *et al.*, 2016), OsmiR535 (Yue *et al.*, 2020), HvITPK1 (Vlčko and Ohnoutková, 2020), and SlARF4 (tomato) improves crop salt-stress tolerance. Omics techniques, such as AFLPs, RFLPs, SSRs, and SNP markers, have been employed to identify QTLs associated with salinity tolerance in various rice cultivars (Ghomi *et al.*, 2013). Rice salinity tolerance is connected to many QTLs. Several notable QTL mapping experiments employed genomics. Most salt-tolerance QTLs are found in rice seedlings and mature plants, with less germination (Singh *et al.*, 2007). The research found two SUMO proteases (OsSUMO1 and OsSUMO2) and a K^+ transporter gene (OsHAK6) that may be linked to rice salinity tolerance. These findings enable molecular breeding of rice for salt tolerance and high yield. A GWAS found that locus 167 candidate genes Os09g0278000, Os09g0279400,

Os09g0279100, Os09g0279500, and Os09g0280300 affect early seedling vigor, leading to rice salt tolerance (Haque *et al.*, 2021).

Heavy metals disrupt plant physiology and biochemistry, threatening sustainable agricultural growth. Metabolic homeostasis, cytotoxicity, osmotic stress, ion imbalance, oxidative stress, and membrane tissue disorder disturbance in plants may result from these metals (Goyal *et al.*, 2020). Researchers utilized CRISPR/Cas9 to fix this. Eliminating OsNramp5 and OsLCT1 genes decreased rice cadmium (Cd) accumulation. Similarly, knocking down the OsNRAMP1 gene reduced rice grain Cd and Pb (Jing *et al.*, 2023). The Cd-reducing activities of OsNRAMP5 and OsNRAMP1 are not redundant (Chang *et al.*, 2020). Researchers also employed CRISPR/Cas9 to eliminate the R2R3 MYB transcription factor OsARM1 to prevent rice from absorbing and transporting arsenic (Wang *et al.*, 2017). To generate low-Cs rice plants, the CRISPR/Cas9 technology inactivated the K^+ transporter OsHAK1. These CRISPR/Cas9-based methods may generate crop varieties with lower heavy metal build-up, promoting sustainable agriculture and food safety (Li *et al.*, 2022). The 'omics' technique has also been acclimated to study the plant's physiological interactions and molecular pathways in response to heavy metal and metalloid toxicity. Cadmium resilience in barley plants was explained by two QTLs, one minor and one large, which explained 47.24 and 38.59% of the phenotypic variation, respectively (Raza *et al.*, 2022). Additionally, 16 Cd tolerance candidate genes were found. A marker-trait association analysis of 704 SNPs found 9 QTLs explaining 8.6–12.6% of phenotypic variation for arsenic (As) attributes. Transcription regulation for toxicity-related includes was found in 25 candidate genes connected with these QTLs on chromosomes 2, 5, 6, 8, and 9. A composite interval mapping technique found four phytotoxicity tolerance QTLs on chromosomes 8 and 12. Forty-eight areas in rice that tolerate aluminum (Al) account for 57% of phenotypic variance (Rahman *et al.*, 2022). Four of these sites co-localized with previously known QTLs, while three novel regions (ART1, STAR2, Nrat1) were discovered on chromosomes 1, 9, and 12 by mapping using Indica/Japonica backgrounds (Famoso *et al.*, 2011). In addition, rice QTL mapping for toxic metal/metalloid stress tolerance revealed one QTL for copper (Cu) and mercury (Hg), three for arsenic (As), and two for iron (Fe) and zinc (Zn). Rice stress tolerance was expected to increase by candidate genes underpinning the QTLs for LOC_Os01g18584, LOC_Os01g20160), Zn (LOC_Os01g14440, and Fe (LOC_Os04g34600) (Hossain *et al.*, 2023).

Plant defense genes can be precisely edited using CRISPR-Cas9. CRISPR may target and change pattern recognition receptor (PRR) genes, which recognize pathogen-associated molecular patterns, and activate the frontline defense (Tyagi *et al.*, 2021). CRISPR may also alter the expression of TFs such as WRKY, MYB, and bZIP, which activate defense genes (Li *et al.*, 2024). Omics technologies reveal the molecular underpinnings of biotic stress tolerance (Mosa *et al.*, 2017). Genomic investigations may discover disease resistance or insect tolerance genes, whereas transcriptome and proteomic analysis show biotic stress-induced gene expression and protein patterns (Li *et al.*, 2019b). Metabolomics may explain changes in antimicrobial and defense-related metabolite synthesis (Serag *et al.*, 2023). Researchers may boost crop yield and sustainable agriculture by merging CRISPR-based genetic alterations with multi-omics methods to improve plant biotic and abiotic stress tolerance.

16.5 Conclusion and Future Perspective

Crop research is undergoing a revolution because of the combination of CRISPR/Cas9 technology and high-throughput omics techniques, which provide hitherto unattainable accuracy and insight into plant stress responses. These integrated techniques allow for the discovery and modification of important genes and regulatory networks that control stress tolerance, resulting in the development of improved plant breeding types. Chickpea, tomato, rice, coffee, and soybean are among the crops that show notable advancements in applying these technologies to overcome various obstacles such as salt, drought, metal toxicity, and high temperatures. Growing populations and environmental change are placing increasing strain on the world's food supply, but omics and CRISPR/Cas9 together offer a strong foundation for improving crop resilience, productivity, and quality. In future, the development of creative and sustainable agriculture techniques to guarantee global food security will depend heavily on the ongoing improvements and integration of these technologies.

References

Acharya, B.R., Sandhu, D. and Ferreira, J.F. (2021) Physiological, morphological, biochemical, and genetic responses of alfalfa to salinity. In: Yu, L.X. and Kole, C. (eds) *The Alfalfa Genome. Compendium of Plant Genomes*. Springer, Cham, Switzerland, pp. 145–157. DOI: 10.1007/978-3-030-74466-3_9.

Afroz, N., Ansary, M.W.R. and Islam, T. (2023) CRISPR-Cas genome editing for the development of abiotic stress-tolerant wheat. In: Khan, M.K., Pandey, A., Hamurcu, M., Gupta, O.P. and Gezgin, S. (eds) *Abiotic Stresses in Wheat*. Academic Press, Florida, pp. 195–207. DOI: 10.1016/B978-0-323-95368-9.00014-X.

Ahmad, P., Abdel Latef, A.A.H., Rasool, S., Akram, N.A., Ashraf, M. *et al*. (2016) Role of proteomics in crop stress tolerance. *Frontiers in Plant Science* 7, 1336. DOI: 10.3389/fpls.2016.01336.

Ahmad, S., Nisar, A., Jamil, S., Shahzad, R., Sheng, Z. *et al*. (2022) Harnessing CRISPR/Cas system for resistance against abiotic stresses in rice: Recent advances, applications, challenges, and prospects. In: Aftab, T. and Hakeem, K.R. (eds) *Plant Abiotic Stress Physiology*. Molecular Advancements. Apple Academic Press, Florida, pp. 241–261. DOI: 10.1201/9781003180579-10.

Alam, M.S., Kong, J., Tao, R., Ahmed, T., Alamin, M. *et al*. (2022) CRISPR/Cas9 mediated knockout of *OsBHLH024* transcription factor improves salt stress resistance in rice (*Oryza sativa* L.). *Plants* 11(9), 1184. DOI: 10.3390/plants11091184.

Alemu, A., Åstrand, J., Montesinos-Lopez, O.A., Sanchez, J.I., Fernandez-Gonzalez, J. *et al*. (2024) Genomic selection in plant breeding: Key factors shaping two decades of progress. *Molecular Plant* 17(4), 552–578. DOI: 10.1016/j.molp.2024.03.007.

Alhasnawi, A.N., Alasadiy, Y.D.K. and Doni, F. (2024) Assessment of the genetic diversity in plants using molecular markers: A review and perspective. *Tropical Agriculture* 101(1), 120–134. DOI: 10.37234.

Ali, A., Khan, M., Sharif, R., Mujtaba, M. and Gao, S.J. (2019) Sugarcane omics: an update on the current status of research and crop improvement. *Plants* 8(9), 344. DOI: 10.3390/plants8090344.

Amack, S.C., Ferreira, S.S. and Antunes, M.S. (2022) Tuning the transcriptional activity of the CaMV 35s promoter in plants by single-nucleotide changes in the TATA box. *ACS Synthetic Biology* 12(1), 178–185. DOI: 10.1021/acssynbio.2c00457.

Anderson, E.J., Ali, M.L., Beavis, W.D., Chen, P., Clemente, T.E. *et al*. (2019) Soybean [*Glycine max* L. Merr.] breeding: History, improvement, production and future opportunities. In: Al-Khayri, J., Jain, S. and Johnson, D. (eds) *Advances in Plant Breeding Strategies: Legumes*. Springer, Cham, Switzerland, pp. 431–516. DOI: 10.1007/978-3-030-23400-3_12.

Angon, P.B., Mondal, S., Akter, S., Sakil, M.A. and Jalil, M.A. (2023) Roles of CRISPR to mitigate drought and salinity stresses on plants. *Plant Stress* 8, 100169. DOI: 10.1016/j.stress.2023.100169.

Arora, L. and Narula, A. (2017) Gene editing and crop improvement using CRISPR-Cas9 system. *Frontiers in Plant Science* 8. DOI: 10.3389/fpls.2017.01932.

Ashraf, M.F., Hou, D., Hussain, Q., Imran, M., Pei, J. *et al*. (2022) Entailing the next-generation sequencing and metabolome for sustainable agriculture by improving plant tolerance. *International Journal of Molecular Sciences* 23(2), 651. DOI: 10.3390/ijms23020651.

Aslam, S., Munir, A., Aslam, H.M.U., Khan, S.H. and Ahmad, A. (2022) Genome editing advances in soybean improvement against biotic and abiotic stresses. In: Wani, S.H., Sofi, N.U.R., Bhat, M.A. and Lin, F. (eds) *Soybean Improvement: Physiological, Molecular and Genetic Perspectives*. Springer, Cham, Switzerland, pp. 241–274. DOI: 10.1007/978-3-031-12232-3_13.

Azri, W., Cosette, P., Guillou, C., Rabhi, M., Nasr, Z. *et al*. (2020) Physiological and proteomic responses to drought stress in leaves of two wild grapevines (*Vitis sylvestris*): A comparative study. *Plant Growth Regulation* 91(1), 37–52. DOI: 10.1007/s10725-020-00586-4.

Badhan, S., Ball, A.S. and Mantri, N. (2021) First report of CRISPR/Cas9 mediated DNA-free editing of 4CL and RVE7 genes in chickpea protoplasts. *International Journal of Molecular Sciences* 22(1), 396. DOI: 10.3390/ijms22010396.

Bashir, K., Matsui, A., Rasheed, S. and Seki, M. (2019) Recent advances in the characterization of plant transcriptomes in response to drought, salinity, heat, and cold stress. *F1000Research* 8. DOI: 10.12688/f1000research.184.

Bekele, D., Tesfaye, K. and Fikre, A. (2019) Applications of virus induced gene silencing (VIGS) in plant functional genomics studies. *Journal of Plant Biochemistry & Physiology* 7(1), 1000229. DOI: 10.4172/2329-9029.1000229.

Bergman, N. and Bergquist, J. (2014) Recent developments in proteomic methods and disease biomarkers. *The Analyst* 139(16), 3836–3851. DOI: 10.1039/c4an00627e.

Billah, M., Li, F. and Yang, Z. (2021) Regulatory network of cotton genes in response to salt, drought and wilt diseases (*Verticillium* and *Fusarium*): Progress and perspective. *Frontiers in Plant Science* 12, 759245. DOI: 10.3389/fpls.2021.759245.

Brooks, C., Nekrasov, V., Lippman, Z.B. and Eck, J. (2014) Efficient gene editing in tomato in the first generation using the clustered regularly interspaced short palindromic repeats/CRISPR-associated 9 system. *Plant Physiology* 166(3), 1292–1297. DOI: 10.1104/pp.114.247577.

Brown, T.A. (2023) *Genomes*, 5th edn. CRC Press, Boca Raton, Florida. DOI: 10.1201/9781003133162.

Buccitelli, C. and Selbach, M. (2020) MRNAs, proteins and the emerging principles of gene expression control. *Nature Reviews Genetics* 21(10), 630–644. DOI: 10.1038/s41576-020-0258-4.

Bundó, M. and Coca, M. (2017) Calcium-dependent protein kinase OsCPK10 mediates both drought tolerance and blast disease resistance in rice plants. *Journal of Experimental Botany* 68(11), 2963–2975. DOI: 10.1093/jxb/erx145.

Chandrasekaran, M., Raman, C., Karthikeyan, K. and Paramasivan, M. (2019) Functional annotation of hypothetical proteins derived from Suppressive Subtraction Hybridization (SSH) analysis shows NPR1 (non-pathogenesis related)-like activity. *Agronomy* 9(2), 57. DOI: 10.3390/agronomy9020057.

Chang, J.D., Huang, S., Yamaji, N., Zhang, W., Ma, J.F. *et al.* (2020) OsNRAMP1 transporter contributes to cadmium and manganese uptake in rice. *Plant, Cell & Environment* 43(10), 2476–2491. DOI: 10.1111/pce.13843.

Char, S.N., Neelakandan, A.K., Nahampun, H., Frame, B., Main, M. *et al.* (2017) An *Agrobacterium*-delivered CRISPR/Cas9 system for high-frequency targeted mutagenesis in maize. *Plant Biotechnology Journal* 15(2), 257–268. DOI: 10.1111/pbi.12611.

Chaudhary, J., Khatri, P., Singla, P., Kumawat, S., Kumari, A. *et al.* (2019) Advances in omics approaches for abiotic stress tolerance in tomato. *Biology* 8(4), 90. DOI: 10.3390/biology8040090.

Chen, L., Zhong, F. and Zhu, J. (2020) Bridging targeted and untargeted mass spectrometry-based metabolomics via hybrid approaches. *Metabolites* 10(9), 348. DOI: 10.3390/metabo10090348.

Chen, X., Ding, Y., Yang, Y., Song, C., Wang, B. *et al.* (2021) Protein kinases in plant responses to drought, salt, and cold stress. *Journal of Integrative Plant Biology* 63(1), 53–78. DOI: 10.1111/jipb.13061.

Christensen, S.A., Santana, E.A., Alborn, H.T., Block, A.K. and Chamberlain, C.A. (2021) Metabolomics by UHPLC-HRMS reveals the impact of heat stress on pathogen-elicited immunity in maize. *Metabolomics* 17(1), 6. DOI: 10.1007/s11306-020-01739-2.

Chugh, V., Kaur, D., Purwar, S., Kaushik, P., Sharma, V. *et al.* (2022) Applications of molecular markers for developing abiotic-stress-resilient oilseed crops. *Life* 13(1), 88. DOI: 10.3390/life13010088.

Daniel, M.A., Sebastin, R., Yu, J.K., Soosaimanickam, M.P. and Chung, J.W. (2023) Enhancing horticultural crops through genome editing: Applications, benefits, and considerations. *Horticulturae* 9(8), 884. DOI: 10.3390/horticulturae9080884.

Davidson, R.M., Gowda, M., Moghe, G., Lin, H., Vaillancourt, B. *et al.* (2012) Comparative transcriptomics of three *Poaceae* species reveals patterns of gene expression evolution. *The Plant Journal* 71(3), 492–502. DOI: 10.1111/j.1365-313X.2012.05005.x.

Dayani, S., Sabzalian, M.R. and Mazaheri-Tirani, M. (2019) CRISPR/Cas9 genome editing in bread wheat (*Triticum aestivum* L.) genetic improvement. In: Al-Khayri, J., Jain, S. and Johnson, D. (eds) *Advances in Plant Breeding Strategies: Cereals*. Springer, Cham, Switzerland, pp. 453–469. DOI: 10.1007/978-3-030-23108-8_12.

Debbarma, J., Sarki, Y.N., Saikia, B., Boruah, H.P.D., Singha, D.L. *et al.* (2019) Ethylene Response Factor (ERF) family proteins in abiotic stresses and CRISPR-Cas9 genome editing of ERFs for multiple abiotic stress tolerance in crop plants: A review. *Molecular Biotechnology* 61(2), 153–172. DOI: 10.1007/s12033-018-0144-x.

Dhadge, K. and Walia, P. (2024) Applications of CRISPR/Cas9 for biotic and abiotic stress resistance of rice (*Oryza sativa*). *Journal of Advances in Biology & Biotechnology* 27(7), 172–179. DOI: 10.9734/jabb/2024/v27i7977.

Djennane, S., Gersch, S., Le-Bohec, F., Piron, M.-C., Baltenweck, R. *et al.* (2024) CRISPR/Cas9 editing of downy mildew resistant 6 (DMR6-1) in grapevine leads to reduced susceptibility to *Plasmopara viticola*. *Journal of Experimental Botany* 75(7), 2100–2112. DOI: 10.1093/jxb/erad487.

Donovan, S., Mao, Y., Orr, D.J., Carmo-Silva, E. and McCormick, A.J. (2020) CRISPR-Cas9-mediated mutagenesis of the Rubisco small subunit family in *Nicotiana tabacum*. *Frontiers in Genome Editing* 2, 605614. DOI: 10.3389/fgeed.2020.605614.

Erdoğan, İ., Cevher-Keskin, B., Bilir, Ö., Hong, Y. and Tör, M. (2023) Recent developments in CRISPR/Cas9 genome-editing technology related to plant disease resistance and abiotic stress tolerance. *Biology* 12(7), 1037. DOI: 10.3390/biology12071037.

Ezin, V. and Symonds, R.C. (2023) Genome editing of tomato crop with CRISPR/Cas systems: Past, present, and future. In: Ezin, V. and Symonds, R.C. (eds) *CRISPR/Cas-Mediated Genome Editing in Plants*. Apple Academic Press, Florida, pp. 191–238.

Famoso, A.N., Zhao, K., Clark, R.T., Tung, C.-W., Wright, M.H. *et al.* (2011) Genetic architecture of aluminum tolerance in rice (*Oryza sativa*) determined through genome-wide association analysis and QTL mapping. *PLoS Genetics* 7(8), e1002221. DOI: 10.1371/journal.pgen.1002221.

Farhat, S., Jain, N., Singh, N., Sreevathsa, R., Dash, P.K. *et al.* (2019) CRISPR-Cas9 directed genome engineering for enhancing salt stress tolerance in rice. *Seminars in Cell and Developmental Biology* 96, 91–99. DOI: 10.1016/j.semcdb.2019.05.003.

Farooqi, M.Q.U., Nawaz, G., Wani, S.H., Choudhary, J.R., Rana, M. *et al.* (2022) Recent developments in multi-omics and breeding strategies for abiotic stress tolerance in maize (*Zea mays* L.). *Frontiers in Plant Science* 13, 965878. DOI: 10.3389/fpls.2022.965878.

Gajardo, H.A., Gómez-Espinoza, O., Boscariol Ferreira, P., Carrer, H., and Bravo, L.A. (2023) The potential of CRISPR/Cas technology to enhance crop performance on adverse soil conditions. *Plants* 12(9), 1892. DOI: 10.3390/plants12091892.

Gao, W., Long, L., Tian, X., Xu, F., Liu, J. *et al.* (2017) Genome editing in cotton with the CRISPR/Cas9 system. *Frontiers in Plant Science* 8, 1364. DOI: 10.3389/fpls.2017.01364.

Garcia-Gimenez, G. and Jobling, S.A. (2022) Gene editing for barley grain quality improvement. *Journal of Cereal Science* 103, 103394. DOI: 10.1016/j.jcs.2021.103394.

García-Montelongo, A.M., Montoya-Martínez, A.C., Morales-Sandoval, P.H., Parra-Cota, F.I. and de los Santos-Villalobos, S. (2023) Beneficial microorganisms as a sustainable alternative for mitigating biotic stresses in crops. *Stresses* 3(1), 210–228. DOI: 10.3390/stresses3010016.

Ghomi, K., Rabiei, B., Sabouri, H., and Sabouri, A. (2013) Mapping QTLs for traits related to salinity tolerance at seedling stage of rice (*Oryza sativa* L.): An agrigenomics study of an Iranian rice population. *Omics* 17(5), 242–251. DOI: 10.1089/omi.2012.0097.

Gill, T., Gill, S.K., Saini, D.K., Chopra, Y., de Koff, J.P. *et al.* (2022) A comprehensive review of high throughput phenotyping and machine learning for plant stress phenotyping. *Phenomics* 2(3), 156–183. DOI: 10.1007/s43657-022-00048-z.

González Guzmán, M., Cellini, F., Fotopoulos, V., Balestrini, R. and Arbona, V. (2022) New approaches to improve crop tolerance to biotic and abiotic stresses. *Physiologia Plantarum* 174(1), e13547. DOI: 10.1111/ppl.13547.

Goyal, D., Yadav, A., Prasad, M., Singh, T.B., Shrivastav, P. *et al.* (2020) Effect of heavy metals on plant growth: An overview. In: Naeem, M., Ansari, A. and Gill, S. (eds) *Contaminants in Agriculture: Sources, Impacts and Management*. Springer, Cham, Switzerland, pp. 79–101. DOI: 10.1007/978-3-030-41552-5_4.

Großkinsky, D.K., Syaifullah, S.J. and Roitsch, T. (2018) Integration of multi-omics techniques and physiological phenotyping within a holistic phenomics approach to study senescence in model and crop plants. *Journal of Experimental Botany* 69(4), 825–844. DOI: 10.1093/jxb/erx333.

Gujaria, N. (2012) Development of functional markers and transcript map of chickpea (*Cicer arietinum* L.). Doctoral thesis, Dr. Hari Singh Gour University Sagar, Madhya Pradesh, India.

Gupta, S., Mishra, V.K., Kumari, S. and Raavi, Chand, R. (2019) Deciphering genome-wide WRKY gene family of *Triticum aestivum* L. and their functional role in response to abiotic stress. *Genes & Genomics* 41, 79–94. DOI: 10.1007/s13258-018-0742-9.

Hall, R.D., Trevisan, F. and Vos, R.C. (2022) Coffee berry and green bean chemistry – opportunities for improving cup quality and crop circularity. *Food Research International* 151, 110825. DOI: 10.1016/j.foodres.2021.110825.

Hamany Djande, C.Y., Piater, L.A., Steenkamp, P.A., Tugizimana, F. and Dubery, I.A. (2021) A metabolomics approach and chemometric tools for differentiation of barley cultivars and biomarker discovery. *Metabolites* 11(9), 578. DOI: 10.3390/metabo11.

Haque, M.A., Rafii, M.Y., Yusoff, M.M., Ali, N.S., Yusuff, O. *et al.* (2021) Advanced breeding strategies and future perspectives of salinity tolerance in rice. *Agronomy* 11(8), 1631. DOI: 10.3390/agronomy11081631.

Hasan, N., Choudhary, S., Naaz, N., Sharma, N. and Laskar, R.A. (2021) Recent advancements in molecular marker-assisted selection and applications in plant breeding programmes. *Journal of Genetic Engineering and Biotechnology* 19(1), 128. DOI: 10.1186/s43141-021-00231-1.

Hasanuzzaman, M., Nahar, K., Alam, M.M., Roychowdhury, R. and Fujita, M. (2013) Physiological, biochemical, and molecular mechanisms of heat stress tolerance in plants. *International Journal of Molecular Sciences* 14(5), 9643–9684. DOI: 10.3390.

Hauptmann, R.M., Vasil, V., Ozias-Akins, P., Tabaeizadeh, Z., Rogers, S.G. *et al.* (1988) Evaluation of selectable markers for obtaining stable transformants in the *Gramineae*. *Plant Physiology* 86(2), 602–606. DOI: 10.1104/pp.86.2.602.

Hossain, M.A., Hossain, A.Z., Bourgerie, S., Fujita, M. and Dhankher, O.P. (eds) (2023) *Heavy Metal Toxicity and Tolerance in Plants: A Biological, Omics, and Genetic Engineering Approach*. John Wiley & Sons, Chichester, UK.

Huang, X.Y. and Salt, D.E. (2016) Plant ionomics: From elemental profiling to environmental adaptation. *Molecular Plant* 9(6), 787–797. DOI: 10.1016/j.molp.2016.05.003.

Jacobs, T.B., LaFayette, P.R., Schmitz, R.J. and Parrott, W.A. (2015) Targeted genome modifications in soybean with CRISPR/Cas9. *BMC Biotechnology* 15, 16. DOI: 10.1186/s12896-015-0131-2.

Jaganathan, D., Ramasamy, K., Sellamuthu, G., Jayabalan, S. and Venkataraman, G. (2018) CRISPR for crop improvement: An update review. *Frontiers in Plant Science* 9, 985. DOI: 10.3389/fpls.2018.00985.

Jain, M. (2015) Function genomics of abiotic stress tolerance in plants: A CRISPR approach. *Frontiers in Plant Science* 6, 375. DOI: 10.3389/fpls.2015.00375.

James, D., Rennya, P.R., Mallavarapu, M.D., Panigrahi, R.C. and Patel, H.K. (2021) Integrating pan-omics data in a systems approach for crop improvement: Opportunities and challenges. In: Kumar, A., Kumar, R., Shukla, P. and Patel, H.K. (eds) *Omics Technologies for Sustainable Agriculture and Global Food Security*, Vol. II. Springer, Singapore, pp. 215–246.

Jan, N., Rather, A.M., John, R., Chaturvedi, P., Ghatak, A. *et al.* (2023) Proteomics for abiotic stresses in legumes: Present status and future directions. *Critical Reviews in Biotechnology* 43(2), 171–190. DOI: 10.1080/07388551.2021.2025033.

Jha, U.C., Bohra, A., Nayyar, H., Rani, A., Devi, P. *et al.* (2019) Breeding and genomics approaches for improving productivity gains in chickpea under changing climate. In: Kole, C. (ed.) *Genomic Designing of Climate-Smart Pulse Crops*. Springer, Cham, Switzerland, pp. 135–164. DOI: 10.1007/978-3-319-96932-9_3.

Jha, U.C., Bohra, A., Pandey, S. and Parida, S.K. (2020a) Breeding, genetics, and genomics approaches for improving *Fusarium* wilt resistance in major grain legumes. *Frontiers in Genetics* 11, 1001. DOI: 10.3389/fgene.2020.01001.

Jha, U.C., Nayyar, H., Jha, R., Khurshid, M., Zhou, M. *et al.* (2020b) Long non-coding RNAs: Emerging players regulating plant abiotic stress response and adaptation. *BMC Plant Biology* 20(1), 466. DOI: 10.1186/s12870-020-02595-x.

Jia, H. and Wang, N. (2014) Targeted genome editing of sweet orange using Cas9/sgRNA. *PLoS One* 9(4), e93806. DOI: 10.1371/journal.pone.0093806.

Jiang, C., Zhang, H., Ren, J., Dong, J., Zhao, X. *et al.* (2020) Comparative transcriptome-based mining and expression profiling of transcription factors related to cold tolerance in peanut. *International Journal of Molecular Sciences* 21(6). DOI: 10.3390/ijms21061921.

Jiang, H., Zhang, Y., Li, J., Tang, R., Liang, F. *et al.* (2024) Genome-wide identification of SIMILAR to RCD ONE (SRO) gene family in rapeseed (*Brassica napus* L.) reveals their role in drought stress response. *Plant Signaling & Behavior* 19(1), 2379128. DOI: 10.1080/15592324.2024.2379128.

Jing, H., Yang, W., Chen, Y., Yang, L., Zhou, H. *et al.* (2023) Exploring the mechanism of Cd uptake and translocation in rice: Future perspectives of rice safety. *Science of the Total Environment* 897, 165369. DOI: 10.1016/j.scitotenv.2023.165369.

Kafle, S. (2023) CRISPR/CAS9: A new paradigm for crop improvement revolutionizing agriculture. *Journal of Agriculture and Food Research* 11, 100484. DOI: 10.1016/j.jafr.2022.100484.

Kaur, B., Sandhu, K.S., Kamal, R., Kaur, K., Singh, J. *et al.* (2021) Omics for the improvement of abiotic, biotic, and agronomic traits in major cereal crops: Applications, challenges, and prospects. *Plants* 10, 10. DOI: 10.3390/plants10101989.

Khabbazi, S.D., Khabbazi, A.D., Cevik, V. and Ergül, A. (2021) CRISPR/Cas9 system, an efficient approach to genome editing of plants for crop improvement. In: Tang, G., Teotia, S., Tang, X. and Singh, D. (eds) *RNA-Based Technologies for Functional Genomics in Plants*. Springer, Cham, Switzerland, pp. 369–391. DOI: 10.1007/978-3-030-64994-4_16.

Khan, M.K.R., Ditta, A., Wang, B., Fang, L., Anwar, Z. *et al.* (2023) The intervention of multi-omics approaches for developing abiotic stress resistance in cotton crop under climate change. In: Prakash, C.S., Fiaz, S., Nadeem, M.A., Baloch, F.S. and Qayyum, A. (eds) *Sustainable Agriculture in the Era of the OMICs Revolution*. Springer, Cham, Switzerland, pp. 37–82. DOI: 10.1007/978-3-031-15568-0_3.

Kiani-Pouya, A., Rasouli, F., Shabala, L., Tahir, A.T., Zhou, M. *et al.* (2020) Understanding the role of root-related traits in salinity tolerance of quinoa accessions with contrasting epidermal bladder cell patterning. *Planta* 251(5), 103. DOI: 10.1007/s00425-020-03395-1.

Kosová, K., Vítámvás, P., Urban, M.O., Prášil, I.T. and Renaut, J. (2018) Plant abiotic stress proteomics: The major factors determining alterations in cellular proteome. *Frontiers in Plant Science* 9, 122. DOI: 10.3389/fpls.2018.00122.

Koul, B., Sharma, K., Sehgal, V., Yadav, D., Mishra, M. *et al.* (2022) Chickpea (*Cicer arietinum* L.) biology and biotechnology: From domestication to biofortification and biopharming. *Plants* 11(21), 2926. DOI: 10.3390/plants11212926.

Krishna, R., Ansari, W.A., Soumia, P.S., Yadav, A., Jaiswal, D.K. *et al.* (2022) Biotechnological interventions in tomato (*Solanum lycopersicum*) for drought stress tolerance: Achievements and future prospects. *Biotech* 11(4), 48. DOI: 10.3390/biotech11040048.

Kumar, A., Kumar, V., Gull, A. and Nayik, G.A. (2020) Tomato (*Solanum lycopersicon*). In: Nayik, G.A. and Gull, A. (eds) *Antioxidants in Vegetables and Nuts-Properties and Health Benefits*. Springer, Singapore, pp. 191–207. DOI: 10.1007/978-981-15-7470-2_10.

Kumar, M., Ayzenshtat, D., Marko, A. and Bocobza, S. (2022a) Optimization of T-DNA configuration with UBIQUITIN10 promoters and tRNA–sgRNA complexes promotes highly efficient genome editing in allotetraploid tobacco. *Plant Cell Reports* 41, 175–194. DOI: 10.1007/s00299-021-02796-0.

Kumar, P., Choudhary, M., Halder, T., Prakash, N.R., Singh, V. *et al.* (2022b) Salinity stress tolerance and omics approaches: Revisiting the progress and achievements in major cereal crops. *Heredity* 128(6), 497–518. DOI: 10.1038/s41437-022-00516-2.

Kumar, M., Prusty, M.R., Pandey, M.K., Singh, P.K., Bohra, A. *et al.* (2023) Application of CRISPR/Cas9-mediated gene editing for abiotic stress management in crop plants. *Frontiers in Plant Science* 14, 1157678. DOI: 10.3389/fpls.2023.1157678.

Kuo, W.H., Zhong, L., Wright, S.J., Goad, D.M. and Olsen, K.M. (2024) Beyond cyanogenesis: Temperature gradients drive environmental adaptation in North American white clover (*Trifolium repens* L.). *Molecular Ecology* 33(17), e17484. DOI: 10.1111/mec.17484.

Li, J.-F., Norville, J.E., Aach, J., McCormack, M., Zhang, D. *et al.* (2013) Multiplex and homologous recombination-mediated genome editing in *Arabidopsis* and *Nicotiana benthamiana* using guide RNA and Cas9. *Nature Biotechnology* 31(8), 688–691. DOI: 10.1038/nbt.2654.

Li, R., Liu, C., Zhao, R., Wang, L., Chen, L. *et al.* (2019a) CRISPR/Cas9-mediated SlNPR1 mutagenesis reduces tomato plant drought tolerance. *BMC Plant Biology* 19(1), 38. DOI: 10.1186/s12870-018-1627-4.

Li, T., Wang, Y.H., Liu, J.X., Feng, K., Xu, Z.S. *et al.* (2019b) Advances in genomic, transcriptomic, proteomic, and metabolomic approaches to study biotic stress in fruit crops. *Critical Reviews in Biotechnology* 39(5), 680–692. DOI: 10.1080/07388551.2019.1608153.

Li, Z., Zhang, C., Zhang, Y., Zeng, W. and Cesarino, I. (2021) Coffee cell walls: Composition, influence on cup quality and opportunities for coffee improvements. *Food Quality and Safety* 5, fyab012. DOI: 10.1093/fqsafe/fyab012.

Li, Y., Wu, X., Zhang, Y. and Zhang, Q. (2022) CRISPR/Cas genome editing improves abiotic and biotic stress tolerance of crops. *Frontiers in Genome Editing* 4, 987817. DOI: 10.3389/fgeed.2022.987817.

Li, Y., Ding, L., Zhou, M., Chen, Z., Ding, Y. *et al.* (2023) Transcriptional regulatory network of plant cadmium stress response. *International Journal of Molecular Sciences* 24(5), 4378. DOI: 10.3390/ijms24054378.

Li, S., Khoso, M.A., Wu, J., Yu, B., Wagan, S. *et al.* (2024) Exploring the mechanisms of WRKY transcription factors and regulated pathways in response to abiotic stress. *Plant Stress* 12, 100429. DOI: 10.1016/j.stress.2024.100429.

Liang, Z., Zhang, K., Chen, K., and Gao, C. (2014) Targeted mutagenesis in *Zea mays* using TALENs and the CRISPR/Cas system. *Journal of Genetics and Genomics* 41(2), 63–68. DOI: 10.1016/j.jgg.2013.12.001.

Lin, Y., Zou, W., Lin, S., Onofua, D., Yang, Z. *et al.* (2017) Transcriptome profiling and digital gene expression analysis of sweet potato for the identification of putative genes involved in the defense response against *Fusarium oxysporum* f. sp. *batatas*. *PLoS One* 12(11), e0187838. DOI: 10.1371/journal.pone.0187838.

Lin, C.S., Hsu, C.T., Yang, L.H., Lee, L.Y., Fu, J.Y. *et al.* (2018) Application of protoplast technology to CRISPR/Cas9 mutagenesis: From single-cell mutation detection to mutant plant regeneration. *Plant Biotechnology Journal* 16(7), 1295–1310. DOI: 10.1111/pbi.12870.

Liu, S., Zenda, T., Li, J., Wang, Y., Liu, X. *et al.* (2020) Comparative transcriptomic analysis of contrasting hybrid cultivars reveal key drought-responsive genes and metabolic pathways regulating drought stress tolerance in maize at various stages. *PLoS One* 15(10), e0240468. DOI: 10.1371/journal.pone.0240468.

Liu, Y., Khan, A.R., and Gan, Y. (2022) C_2H_2 zinc finger proteins response to abiotic stress in plants. *International Journal of Molecular Sciences* 23(5), 2730. DOI: 10.3390/ijms23052730.

Lou, D., Wang, H., Liang, G. and Yu, D. (2017) OsSAPK2 confers abscisic acid sensitivity and tolerance to drought stress in rice. *Frontiers in Plant Science* 8, 993. DOI: 10.3389/fpls.2017.00993.

Lv, Y., Yang, M., Hu, D., Yang, Z., Ma, S. *et al.* (2017) The OsMYB30 transcription factor suppresses cold tolerance by interacting with a JAZ protein and suppressing β-amylase expression. *Plant Physiology* 173(2), 1475–1491. DOI: 10.1104/pp.16.01725.

Ma, S., Wang, M., Wu, J., Guo, W., Chen, Y. *et al.* (2021) WheatOmics: A platform combining multiple omics data to accelerate functional genomics studies in wheat. *Molecular Plant* 14(12), 1965–1968. DOI: 10.1016/j.molp.2021.10.006.

Madhu, B., Narita, M., Jauhiainen, A., Menon, S., Stubbs, M. *et al.* (2015) Metabolomic changes during cellular transformation monitored by metabolite-metabolite correlation analysis and correlated with gene expression. *Metabolomics* 11(6), 1848–1863. DOI: 10.1007/s11306-015-0838-z.

Maghuly, F., Jankowicz-Cieslak, J. and Bado, S. (2020) Improving coffee species for pathogen resistance. *CABI Reviews*. DOI: 10.1079/PAVSNNR202015009.

Majeed, Y., Zhu, X., Zhang, N., Ul-Ain, N., Raza, A. *et al.* (2023) Harnessing the role of mitogen-activated protein kinases against abiotic stresses in plants. *Frontiers in Plant Science* 14, 932923. DOI: 10.3389/fpls.2023.932923.

Makonya, G.M., Ogola, J.B.O., Gabier, H., Rafudeen, M.S., Muasya, A.M. *et al.* (2021) Proteome changes and associated physiological roles in chickpea (*Cicer arietinum*) tolerance to heat stress under field conditions. *Functional Plant Biology* 49(1), 13–24. DOI: 10.1071/FP21148.

Manchanda, P., Sharma, D., Kaur, A. and Singh, A. (2022) Improvement of genetic variation for nutrients and bioactive food components in cereal crops. In: Bangar, S.P. and Siroha, A.K. (eds) *Functional Cereals and Cereal Foods: Properties, Functionality and Applications*. Springer Cham, Switzerland, pp. 51–80. DOI: 10.1007/978-3-031-05611-6_3.

Manna, M., Thakur, T., Chirom, O., Mandlik, R., Deshmukh, R. *et al.* (2021) Transcription factors as key molecular target to strengthen the drought stress tolerance in plants. *Physiologia Plantarum* 172(2), 847–868. DOI: 10.1111/ppl.13268.

Martinez, H.E.P., Andrade, S.A.L., Santos, R.H.S., Baptistella, J.L.C. and Mazzafera, P. (2023) Agronomic practices toward coffee sustainability. A review. *Scientia Agricola* 81, e20220277. DOI: 10.1590/1678-992x-2022-0277.

Marwein, R., Debbarma, J., Sarki, Y.N., Baruah, I., Saikia, B. *et al.* (2019) Genetic engineering/genome editing approaches to modulate signaling processes in abiotic stress tolerance. In: Khan, M.I.R., Reddy, P.S., Ferrante, A. and Khan, N.A. (eds) *Plant Signaling Molecules*. Woodhead Publishing, pp. 63–82. DOI: 10.1016/B978-0-12-816451-8.00004-6.

Matthiesen, R. and Bunkenborg, J. (2013) Introduction to mass spectrometry-based proteomics. In: Matthiesen, R. (ed.) *Mass Spectrometry Data Analysis in Proteomics. Methods in Molecular Biology*, Vol. 1007. Humana Press, Totowa, New Jersey, pp. 1–45. DOI: 10.1007/978-1-62703-392-3_1.

Mehrotra, M., Singh, A.K., Sanyal, I., Altosaar, I. and Amla, D.V. (2011) Pyramiding of modified cry1Ab and cry1Ac genes of *Bacillus thuringiensis* in transgenic chickpea (*Cicer arietinum* L.) for improved resistance to pod borer insect *Helicoverpa armigera*. *Euphytica* 182(1), 87–102. DOI: 10.1007/s10681-011-0501-3.

Meyer, R.S. and Purugganan, M.D. (2013) Evolution of crop species: Genetics of domestication and diversification. *Nature Reviews Genetics* 14(12), 840–852. DOI: 10.1038/nrg3605.

Miao, J., Guo, D., Zhang, J., Huang, Q., Qin, G. *et al.* (2013) Targeted mutagenesis in rice using CRISPR-Cas system. *Cell Research* 23(10), 1233–1236. DOI: 10.1038/cr.2013.123.

Mishra, R., Zheng, W., Joshi, R.K., and Kaijun, Z. (2021) Genome editing strategies towards enhancement of rice disease resistance. *Rice Science* 28(2), 133–145. DOI: 10.1016/j.rsci.2021.01.003.

Mohamed, H.I., Khan, A., and Basit, A. (2024) CRISPR-Cas9 system mediated genome editing technology: An ultimate tool to enhance abiotic stress in crop plants. *Journal of Soil Science and Plant Nutrition* 24(2), 1799–1822. DOI: 10.1007/s42729-024-01778-x.

Mohammadi, M., Xavier, A., Beckett, T., Beyer, S., Chen, L. *et al.* (2020) Identification, deployment, and transferability of quantitative trait loci from genome-wide association studies in plants. *Current Plant Biology* 24, 100145. DOI: 10.1016/j.cpb.2020.100145.

Mohanta, T.K., Mishra, A.K., and Al-Harrasi, A. (2021) The 3D genome: From structure to function. *International Journal of Molecular Sciences* 22(21), 11585. DOI: 10.3390/ijms222111585.

Montecillo, J.A.V., Chu, L.L., and Bae, H. (2020) CRISPR-Cas9 system for plant genome editing: Current approaches and emerging developments. *Agronomy* 10(7), 1033. DOI: 10.3390/agronomy10071033.

Monti, C., Zilocchi, M., Colugnat, I., and Alberio, T. (2019) Proteomics turns functional. *Journal of Proteomics* 198, 36–44. DOI: 10.1016/j.jprot.2018.12.012.

Mosa, K.A., Ismail, A., and Helmy, M. (2017) *Plant Stress Tolerance: An Integrated Omics Approach*. Springer, Cham, Switzerland. DOI: 10.1007/978-3-319-59379-1.

Muhammad, T., Zhang, J., Ma, Y., Li, Y., Zhang, F. *et al.* (2019) Overexpression of a mitogen-activated protein kinase SlMAPK3 positively regulates tomato tolerance to cadmium and drought stress. *Molecules* 24(3), 556. DOI: 10.3390/molecules24030.

Muha-ud-din, G., Ali, F., Hameed, A., Naqvi, S.A.H., Nizamani, M.M. *et al.* (2023) CRISPR/Cas9-based genome editing: A revolutionary approach for crop improvement and global food security. *Physiological and Molecular Plant Pathology* 129, 102191. DOI: 10.1016/j.pmpp.2023.102191.

Muthamilarasan, M., Singh, N.K., and Prasad, M. (2019) Multi-omics approaches for strategic improvement of stress tolerance in underutilized crop species: A climate change perspective. *Advances in Genetics* 103, 1–38. DOI: 10.1016/bs.adgen.2019.

Najafi, M., Nasr-Esfahani, M., Vatandoost, J., Hassanzade-Khankahdani, H. and Moeini, M.J. (2024) Transcriptome-based analysis of candidate gene markers associated with resistance mechanism to *Pytophthora melonis* that causes root and crown rot in pumpkin. *Functional Plant Biology* 51(2), FP23038. DOI: 10.1071/FP23038_CO.

Nascimento, F.D.S., Rocha, A.D.J., Soares, J.M.D.S., Mascarenhas, M.S., Ferreira, M.D.S. *et al.* (2023) Gene editing for plant resistance to abiotic factors: A systematic review. *Plants* 12(2), 305. DOI: 10.3390/plants12020305.

Nazir, R., Mandal, S., Mitra, S., Ghorai, M., Das, N. *et al.* (2022) Clustered regularly interspaced short palindromic repeats (CRISPR)/CRISPR-associated genome-editing toolkit to enhance salt stress tolerance in rice and wheat. *Physiologia Plantarum* 174(2). DOI: 10.1111/ppl.13642.

Nekrasov, V., Staskawicz, B., Weigel, D., Jones, J.D.G. and Kamoun, S. (2013) Targeted mutagenesis in the model plant *Nicotiana benthamiana* using Cas9 RNA-guided endonuclease. *Nature Biotechnology* 31(8), 691–693. DOI: 10.1038/nbt.2655.

Neveu, P., Tireau, A., Hilgert, N., Nègre, V., Mineau-Cesari, J. *et al.* (2019) Dealing with multi-source and multi-scale information in plant phenomics: The ontology-driven phenotyping hybrid information system. *The New Phytologist* 221(1), 588–601. DOI: 10.1111/nph.15385.

Ogata, T., Ishizaki, T., Fujita, M. and Fujita, Y. (2020) CRISPR/Cas9-targeted mutagenesis of OsERA1 confers enhanced responses to abscisic acid and drought stress and increased primary root growth under nonstressed conditions in rice. *PLoS One* 15(12), e0243376. DOI: 10.1371/journal.pone.0243376.

Oz, M.T., Altpeter, A., Karan, R., Merotto, A. and Altpeter, F. (2021) CRISPR/Cas9-mediated multi-allelic gene targeting in sugarcane confers herbicide tolerance. *Frontiers in Genome Editing* 3, 673566. DOI: 10.3389/fgeed.2021.673566.

Pandita, D. (2023) CRISPR/Cas mediated genome editing for improving stress resilience in plants. In: Aftab, T. (ed.) *New Frontiers in Plant-Environment Interactions: Innovative Technologies and Developments*. Springer Nature, Cham, Switzerland, pp. 143–167. DOI: 10.1007/978-3-031-43729-8_7.

Pandita, D. and Pandita, A. (eds) (2023) *CRISPR/Cas-mediated Genome Editing in Plants*. Apple Academic Press, Florida. DOI: 10.1201/9781003331759.

Park, S.I., Kim, H.B., Jeon, H.J. and Kim, H. (2021) *Agrobacterium*-mediated *Capsicum annuum* gene editing in two cultivars, hot pepper CM334 and bell pepper dempsey. *International Journal of Molecular Sciences* 22(8), 3921. DOI: 10.3390/ijms22083921.

Parveen, S., Gupta, D.B., Dass, S., Kumar, A., Pandey, A. *et al*. (2016) Chickpea ferritin caFer1 participates in oxidative stress response, and promotes growth and development. *Scientific Reports* 6(1), 31218. DOI: 10.1038/srep31218.

Peng, D., Jiang, R., Peng, H. and Liu, S. (2021) Soybean cyst nematodes: A destructive threat to soybean production in China. *Phytopathology Research* 3(1), 19. DOI: 10.1186/s42483-021-00095-w.

Pérez-Bueno, M.L., Pineda, M. and Barón, M. (2019) Phenotyping plant responses to biotic stress by chlorophyll fluorescence imaging. *Frontiers in Plant Science* 10, 1135. DOI: 10.3389/fpls.2019.01135.

Pichereaux, C., Laurent, E.-A., Gargaros, A., Viudes, S., Durieu, C. *et al*. (2019) Analysis of durum wheat proteome changes under marine and fungal biostimulant treatments using large-scale quantitative proteomics: A useful dataset of durum wheat proteins. *Journal of Proteomics* 200, 28–39. DOI: 10.1016/j.jprot.2019.03.003.

Płoski, R. (2016) Next generation sequencing – general information about the technology, possibilities, and limitations. In: Demkov, U. and Polski, R. (eds) *Clinical Applications for Next-Generation Sequencing*. Academic Press, Florida, pp. 1–18. DOI: 10.1016/B978-0-12-801739-5.00001-5.

Prajapat, R.K., Mathur, M., Upadhyay, T.K., Lal, D., Maloo, S.R. *et al*. (2021) Genome editing for crop improvement. In: Kumar, P. and Thakur, A.K. (eds) *Crop Improvement: Biotechnological Advances*. CRC Press, Boca Raton, Florida, pp. 111–123.

Pramanik, D., Shelake, R.M., Park, J., Kim, M.J., Hwang, I. *et al*. (2021) CRISPR/Cas9-mediated generation of pathogen-resistant tomato against tomato yellow leaf curl virus and powdery mildew. *International Journal of Molecular Sciences* 22(4), 1878. DOI: 10.3390/ijms22041878.

Premkumar, A. (2023) Review on plant response to abiotic stress. In: Walunjkar Babasaheb Changdeo, W.B. (ed.) *Essentials of Plant Biotechnology*. AkiNik Publications, New Delhi, pp. 73–92.

Prusty, A., Panchal, A., Singh, R.K., and Prasad, M. (2024) Major transcription factor families at the nexus of regulating abiotic stress response in millets: A comprehensive review. *Planta* 259(5), 118. DOI: 10.1007/s00425-024-04394-2.

Punia, H., Tokas, J., Malik, A., Sangwan, S., Rani, A. *et al*. (2021) Genome-wide transcriptome profiling, characterization, and functional identification of *NAC* transcription factors in sorghum under salt stress. *Antioxidants* 10(10), 1605. DOI: 10.3390/antiox10101605.

Rahimi, Y., Khahani, B., Jamali, A., Alipour, H., Bihamta, M.R. *et al*. (2023) Genome-wide association study to identify genomic loci associated with early vigor in bread wheat under simulated water deficit complemented with quantitative trait loci meta-analysis. *G3* 13(2), jkac320. DOI: 10.1093/g3journal/jkac320.

Rahman, S.U., Nawaz, M.F., Gul, S., Yasin, G., Hussain, B. *et al*. (2022) State-of-the-art OMICS strategies against toxic effects of heavy metals in plants: A review. *Ecotoxicology and Environmental Safety* 242, 113952. DOI: 10.1016/j.ecoenv.2022.113952.

Rajput, M., Choudhary, K., Kumar, M., Vivekanand, V., Chawade, A. *et al*. (2021) RNA interference and CRISPR/Cas gene editing for crop improvement: Paradigm shift towards sustainable agriculture. *Plants* 10(9). DOI: 10.3390/plants10091914.

Raj, S.R.G. and Nadarajah, K. (2022) QTL and candidate genes: Techniques and advancement in abiotic stress resistance breeding of major cereals. *International Journal of Molecular Sciences* 24(1), 6. DOI: 10.3390/ijms24010006.

Rakkammal, K., Priya, A., Pandian, S., Maharajan, T., Rathinapriya, P. *et al*. (2022) Conventional and omics approaches for understanding the abiotic stress response in cereal crops – an updated overview. *Plants* 11(21), 2852. DOI: 10.3390/plants11212852.

Ramesh, P., Mallikarjuna, G., Sameena, S., Kumar, A., Gurulakshmi, K. *et al*. (2020) Advancements in molecular marker technologies and their applications in diversity studies. *Journal of Biosciences* 45(1), 123. DOI: 10.1007/s12038-020-00089-4.

Rasti, S. and Vogiatzis, C. (2019) A survey of computational methods in protein–protein interaction networks. *Annals of Operations Research* 276, 35–87. DOI: 10.1007_s10479-018-2956-2.

Raza, A., Razzaq, A., Mehmood, S.S., Hussain, M.A., Wei, S. *et al*. (2021) Omics: The way forward to enhance abiotic stress tolerance in *Brassica napus* L. *GM Crops & Food* 12(1), 251–281. DOI: 10.1080/21645698.2020.1859898.

Raza, A., Tabassum, J., Zahid, Z., Charagh, S., Bashir, S. *et al.* (2022) Advances in 'omics' approaches for improving toxic metals/metalloids tolerance in plants. *Frontiers in Plant Science* 12, 794373. DOI: 10.3389/fpls.2021.794373.

Razzaq, A., Guul, W., Khan, M.S. and Saleem, F. (2021) Metabolomics: A powerful tool to study the complexity of wheat metabolome. *Protein and Peptide Letters* 28(8), 878–895. DOI: 10.2174/0929866528666210127153532.

Rendón Mera, A.M. (2023) Quality of cauca coffee (*Coffea arabica* L.) under different agricultural management practices. Doctoral thesis, Universidad de Antioquia, Medellín, Colombia.

Roca Paixão, J.F., Gillet, F.-X., Ribeiro, T.P., Bournaud, C., Lourenço-Tessutti, I.T. *et al.* (2019) Improved drought stress tolerance in *Arabidopsis* by CRISPR/dCas9 fusion with a histone acetyltransferase. *Scientific Reports* 9(1), 8080. DOI: 10.1038/s41598-019-44571-y.

Saeed, F., Shah, T., Bukhat, S., Munsif, F., Ahmad, I. *et al.* (2022) Genome engineering as a tool for enhancing crop traits: Lessons from CRISPR/Cas9. In: Wani, S.H. and Hensel, G. (eds) *Genome Editing: Current Technology Advances and Applications for Crop Improvement*. Springer, Cham, Switzerland, pp. 3–25. DOI: 10.1007/978-3-031-08072-2_1.

Saeed, S., Khan, S.U., Afzal, R., Umar, F. and Ali, A. (2024) CRISPR-Cas technologies for food and nutritional security. In: Abd-Elsalam, K.A., Ahmad, A. and Zhang, B. (eds) *CRISPRized Horticulture Crops: Genome Modified Plants and Microbes in Food and Agriculture*. Academic Press, Florida, pp. 143–158. DOI: 10.1016/B978-0-443-13229-2.00019-3.

Sahoo, K.K., Tripathi, A.K., Pareek, A. and Singla-Pareek, S.L. (2013) Taming drought stress in rice through genetic engineering of transcription factors and protein kinases. *Plant Stress* 7(1), 60–72.

Saito, K. and Matsuda, F. (2010) Metabolomics for functional genomics, systems biology, and biotechnology. *Annual Review of Plant Biology* 61(1), 463–489. DOI: 10.1146/annurev.arplant.043008.092035.

Sami, A., Xue, Z., Tazein, S., Arshad, A., He Zhu, Z. *et al.* (2021) CRISPR–Cas9-based genetic engineering for crop improvement under drought stress. *Bioengineered* 12(1), 5814–5829. DOI: 10.1080/21655979.2021.1969831.

Santoso, T.J. and Sisharmini, A. (2022) Application of CRISPR/Cas9 technology to improve the important traits in coffee. In: *IOP Conference Series: Earth and Environmental Science. The 2nd International Conference on Sustainable Plantation*, Vol. 974. p. 012082. DOI: 10.1088/1755-1315/974/1/012082.

Saraswat, P. and Ranjan, R. (2022) CRISPR-Cas system: A precise tool for plant genome editing. *The Nucleus* 65(1), 81–98. DOI: 10.1007/s13237-021-00353-4.

Serag, A., Salem, M.A., Gong, S., Wu, J.L. and Farag, M.A. (2023) Decoding metabolic reprogramming in plants under pathogen attacks: A comprehensive review of emerging metabolomics technologies to maximize their applications. *Metabolites* 13(3), 424. DOI: 10.3390/metabo13030424.

Shah, P.R., Varanavasiappan, S., Kokiladevi, E., Ramanathan, A. and Kumar, K.K. (2019) Genome editing of rice PFT1 gene to study its role in rice sheath blight disease resistance. *International Journal of Current Microbiology and Applied Sciences* 8(6), 2356–2364. DOI: 10.20546/ijcmas.2019.806.281.

Shan, Q., Wang, Y., Li, J., Zhang, Y., Chen, K. *et al.* (2013) Targeted genome modification of crop plants using a CRISPR-Cas system. *Nature Biotechnology* 31(8), 686–688. DOI: 10.1038/nbt.2650.

Sharma, E., Borah, P., Kaur, A., Bhatnagar, A., Mohapatra, T. *et al.* (2021b) A comprehensive transcriptome analysis of contrasting rice cultivars highlights the role of auxin and ABA responsive genes in heat stress response. *Genomics* 113(3), 1247–1261. DOI: 10.1016/j.ygeno.2021.03.007.

Sharma, V., Gupta, P., Kagolla, P., Hangargi, B., Veershetty, A. *et al.* (2021a) Metabolomics intervention towards better understanding of plant traits. *Cells* 10(2), 346. DOI: 10.3390/cells10020346.

Shi, J., Gao, H., Wang, H., Lafitte, H.R., Archibald, R.L. *et al.* (2017) ARGOS8 variants generated by CRISPR-Cas9 improve maize grain yield under field drought stress conditions. *Plant Biotechnology Journal* 15(2), 207–216. DOI: 10.1111/pbi.12603.

Shi, J., Wang, X. and Wang, E. (2023) Mycorrhizal symbiosis in plant growth and stress adaptation: From genes to ecosystems. *Annual Review of Plant Biology* 74(1), 569–607. DOI: 10.1146/annurev-arplant-061722-090342.

Shu, P., Li, Y., Xiang, L., Sheng, J. and Shen, L. (2023) SlNPR1 modulates chilling stress resistance in tomato plant by alleviating oxidative damage and affecting the synthesis of ferulic acid. *Scientia Horticulturae* 307, 111486. DOI: 10.1016/j.scienta.2022.111486.

Singh, R.K., Gregorio, G.B., and Jain, R.K. (2007) QTL mapping for salinity tolerance in rice. *Physiology and Molecular Biology of Plants* 13(2), 87–99.

Singh, P.K., Srivastava, D., Tiwari, P., Tiwari, M., Verma, G. *et al.* (2019) Drought tolerance in plants: Molecular mechanism and regulation of signaling molecules. In: Khan, M.I.R., Reddy, P.S., Ferrante, A. and Khan, N.A. (eds) *Plant Signaling Molecules*. Woodhead Publishing, pp. 105–123.

Singh, A., Jaiswal, A., Singh, A., Tomar, R.S. and Kumar, A. (2022) Plant ionomics: Toward high-throughput nutrient profiling. In: Kumar, V., Srivastava, A.K. and Suprasanna, P. (eds) *Plant Nutrition and Food Security in the Era of Climate Change*. Academic Press, Florida, pp. 227–254. DOI: 10.1016/B978-0-12-822916-3.00015-9.

Singh, P.K., Devanna, B.N., Dubey, H., Singh, P., Joshi, G. *et al.* (2024a) The potential of genome editing to create novel alleles of resistance genes in rice. *Frontiers in Genome Editing* 6, 1415244. DOI: 10.3389/fgeed.2024.1415244.

Singh, S., Chaudhary, R., Chaturvedi, S., and Tiwari, S. (2024b) Deciphering the role of CRISPR/Cas9 in the amelioration of abiotic and biotic stress conditions. In: Kumar, A., Arora, S., Ogita, S., Yau, Y.Y., and Mukherjee, K. (eds) *Gene Editing in Plants: CRISPR-Cas and Its Applications*. Springer, Singapore, pp. 193–226. DOI: 10.1007/978-981-99-8529-6_8.

Sinha, D., Maurya, A.K., Abdi, G., Majeed, M., Agarwal, R. *et al.* (2023) Integrated genomic selection for accelerating breeding programs of climate-smart cereals. *Genes* 14(7), 1484. DOI: 10.3390/genes14071484.

Siriwan, W., Vannatim, N., Chaowongdee, S., Roytrakul, S., Charoenlappanit, S. *et al.* (2023) Integrated proteomic and metabolomic analysis of cassava cv. Kasetsart 50 infected with Sri Lankan cassava mosaic virus. *Agronomy* 13(3), 945. DOI: 10.3390/agronomy13030945.

Soni, K.B., Anuradha, T., Jadhav, P.R., and Alex, S. (2024) Genetic improvement of banana. In: Tiwari, S. and Koul, B. (eds) *Genetic Engineering of Crop Plants for Food and Health Security*, Vol. 1. Springer Nature, Singapore, pp. 305–329. DOI: 10.1007/978-981-99-5034-8_15.

Subba, P., Barua, P., Kumar, R., Datta, A., Soni, K.K. *et al.* (2013) Phosphoproteomic dynamics of chickpea (*Cicer arietinum* L.) reveals shared and distinct components of dehydration response. *Journal of Proteome Research* 12(11), 5025–5047. DOI: 10.1021/pr400628j.

Sui, X., Wu, Q., Chang, W., Fan, X. and Song, F. (2018) Proteomic analysis of the response of *Funnelifor mismosseae/Medicago sativa* to atrazine stress. *BMC Plant Biology* 18(1), 289. DOI: 10.1186/s12870-018-1492-1.

Tabassum, J., Ahmad, S., Hussain, B., Mawia, A.M., Zeb, A. *et al.* (2021) Applications and potential of genome-editing systems in rice improvement: Current and future perspectives. *Agronomy* 11(7), 1359. DOI: 10.3390/agronomy11071359.

Tayeh, N., Klein, A., Le Paslier, M.C., Jacquin, F., Houtin, H. *et al.* (2015) Genomic prediction in pea: Effect of marker density and training population size and composition on prediction accuracy. *Frontiers in Plant Science* 6, 941. DOI: 10.3389/fpls.2015.00941.

Tillett, R.L. (2011) Transcriptomic and transgenic approaches to understanding abiotic stress response in wine grape (*Vitis vinifera*). Doctoral thesis, University of Nevada.

Tran, M.T., Doan, D.T.H., Kim, J., Song, Y.J., Sung, Y.W. *et al.* (2021) CRISPR/Cas9-based precise excision of SlHyPRP1 domain(s) to obtain salt stress-tolerant tomato. *Plant Cell Reports* 40(6), 999–1011. DOI: 10.1007/s00299-020-02622-z.

Tripathi, J.N., Ntui, V.O., Ron, M., Muiruri, S.K., Britt, A. *et al.* (2019) CRISPR/Cas9 editing of endogenous *Banana streak virus* in the B genome of *Musa* spp. overcomes a major challenge in banana breeding. *Communications Biology* 2(1), 46. DOI: 10.1038/s42003-019-0288-7.

Tuberosa, R. (2012) Phenotyping for drought tolerance of crops in the genomics era. *Frontiers in Physiology* 3, 347. DOI: 10.3389/fphys.2012.00347.

Tyagi, S., Kumar, R., Kumar, V., Won, S.Y., and Shukla, P. (2021) Engineering disease resistant plants through CRISPR-Cas9 technology. *GM Crops & Food* 12(1), 125–144. DOI: 10.1080/21645698.2020.1831729.

Tyagi, P., Singh, D., Mathur, S., Singh, A. and Ranjan, R. (2022) Upcoming progress of transcriptomics studies on plants: An overview. *Frontiers in Plant Science* 13, 1030890. DOI: 10.3389/fpls.2022.1030890.

Tyagi, A., Ali, S., Park, S. and Bae, H. (2023) Exploring the potential of multiomics and other integrative approaches for improving waterlogging tolerance in plants. *Plants* 12(7), 1544. DOI: 10.3390/plants12071544.

Uffelmann, E., Huang, Q.Q., Munung, N.S., de Vries, J., Okada, Y. *et al.* (2021) Genome-wide association studies. *Nature Reviews Methods Primers* 1(1), 59. DOI: 10.1038/s43586-021-00056-9.

Usha, T., Panda, P., Goyal, A.K., Kukanur, A.A., Kamala, A. *et al.* (2022) Whole-genome sequencing of plants: past, present, and future. In: Singh, R.L., Mondal, S., Parihar, A., and Singh, P.K. (eds) *Plant Genomics for Sustainable Agriculture*. Springer, Singapore, pp. 191–239. DOI: 10.1007/978-981-16-6974-3_8.

Varshney, R.K., Pandey, M.K., Bohra, A., Singh, V.K., Thudi, M. *et al.* (2019) Toward the sequence-based breeding in legumes in the post-genome sequencing era. *Theoretical and Applied Genetics* 132(3), 797–816. DOI: 10.1007/s00122-018-3252-x.

Vaseva, I., Sabotič, J., Šuštar-Vozlič, J., Meglič, V., Kidrič, M. *et al.* (2012) The response of plants to drought stress: The role of dehydrins, chaperones, proteases and protease inhibitors in maintaining cellular protein function. In: Neves, D.F. and Sanz, J.D. (eds) *Droughts: New Research*, Vol. 1. Nova Science Publishers Inc., New York, pp. 1–45.

Villalobos-López, M.A., Arroyo-Becerra, A., Quintero-Jiménez, A. and Iturriaga, G. (2022) Biotechnological advances to improve abiotic stress tolerance in crops. *International Journal of Molecular Sciences* 23(19), 12053. DOI: 10.3390/ijms231912053.

Vlčko, T. and Ohnoutková, L. (2020) Allelic variants of CRISPR/Cas9 induced mutation in an inositol trisphosphate 5/6 kinase gene manifest different phenotypes in barley. *Plants* 9(2), 195. DOI: 10.3390/plants9020195.

Vo, K.T.X., Rahman, M.M., Rahman, M.M., Trinh, K.T.T., Kim, S.T. *et al.* (2021) Proteomics and metabolomics studies on the biotic stress responses of rice: An update. *Rice* 14(1), 30. DOI: 10.1186/s12284-021-00461-4.

Waltz, E. (2022) GABA-enriched tomato is first CRISPR-edited food to enter market. *Nature Biotechnology* 40(1), 9–11. DOI: 10.1038/d41587-021-00026-2.

Wang, H., La Russa, M. and Qi, L.S. (2016) CRISPR/Cas9 in genome editing and beyond. *Annual Review of Biochemistry* 85(1), 227–264. DOI: 10.1146/annurev-biochem-060815-014607.

Wang, F.Z., Chen, M.X., Yu, L.J., Xie, L.J., Yuan, L.B. *et al.* (2017) OsARM1, an R2R3 MYB transcription factor, is involved in regulation of the response to arsenic stress in rice. *Frontiers in Plant Science* 8, 1868. DOI: 10.3389/fpls.2017.01868.

Wang, W., Pan, Q., He, F., Akhunova, A., Chao, S. *et al.* (2018) Transgenerational CRISPR-Cas9 activity facilitates multiplex gene editing in allopolyploid wheat. *The CRISPR Journal* 1(1), 65–74. DOI: 10.1089/crispr.2017.0010.

Wang, C., Wang, G., Gao, Y., Lu, G., Habben, J.E. *et al.* (2020) A cytokinin-activation enzyme-like gene improves grain yield under various field conditions in rice. *Plant Molecular Biology* 102(4), 373–388. DOI: 10.1007/s11103-019-00952-5.

Wang, T., Xun, H., Wang, W., Ding, X., Tian, H. *et al.* (2021) Mutation of GmAITR genes by CRISPR/Cas9 genome editing results in enhanced salinity stress tolerance in soybean. *Frontiers in Plant Science* 12, 779598. DOI: 10.3389/fpls.2021.779598.

Wang, Y., Liu, D., Yin, H., Wang, H., Cao, C. *et al.* (2023) Transcriptomic and metabolomic analyses of the response of resistant peanut seeds to *Aspergillus flavus* infection. *Toxins* 15(7), 414. DOI: 10.3390/toxins15070414.

Wani, S.H., Anand, S., Singh, B., Bohra, A. and Joshi, R. (2021) WRKY transcription factors and plant defense responses: Latest discoveries and future prospects. *Plant Cell Reports* 40(7), 1071–1085. DOI: 10.1007/s00299-021-02691-8.

Witek, K., Jupe, F., Witek, A.I., Baker, D., Clark, M.D. *et al.* (2016) Accelerated cloning of a potato late blight-resistance gene using RenSeq and SMRT sequencing. *Nature Biotechnology* 34(6), 656–660. DOI: 10.1038/nbt.3540.

Wu, W.W., Wang, G., Baek, S.J. and Shen, R.F. (2006) Comparative study of three proteomic quantitative methods, DIGE, CiCAT, and iTRAQ, using 2D gel- or LC–MALDI TOF/TOF. *Journal of Proteome Research* 5(3), 651–658. DOI: 10.1021/pr050405o.

Xiao, Q., Bai, X., Zhang, C. and He, Y. (2022) Advanced high-throughput plant phenotyping techniques for genome-wide association studies: A review. *Journal of Advanced Research* 35, 215–230. DOI: 10.1016/j.jare.2021.05.002.

Xuan, H., Huang, Y., Zhou, L., Deng, S., Wang, C. *et al.* (2022) Key soybean seedlings drought-responsive genes and pathways revealed by comparative transcriptome analyses of two cultivars. *International Journal of Molecular Sciences* 23(5), 2893. DOI: 10.3390/ijms23052893.

Yang, W., Guo, Z., Huang, C., Duan, L., Chen, G. *et al.* (2014) Combining high-throughput phenotyping and genome-wide association studies to reveal natural genetic variation in rice. *Nature Communications* 5(1), 5087. DOI: 10.1038/ncomms6087.

Yang, Y., Saand, M.A., Huang, L., Abdelaal, W.B., Zhang, J. *et al.* (2021) Applications of multi-omics technologies for crop improvement. *Frontiers in Plant Science* 12, 563953. DOI: 10.3389/fpls.2021.563953.

Yang, Y., Wang, D., Lü, P., Ma, S., and Chen, K. (2023) Research progress on nucleic acid detection and genome editing of CRISPR/Cas12 system. *Molecular Biology Reports* 50(4), 3723–3738. DOI: 10.1007/s11033-023-08240-8.

Yao, D., Zhou, J., Zhang, A., Wang, J., Liu, Y. *et al.* (2023) Advances in CRISPR/Cas9-based research related to soybean [*Glycine max* Linn. Merr] molecular breeding. *Frontiers in Plant Science* 14, 1247707. DOI: 10.3389/fpls.2023.1247707.

Yin, Y., Qin, K., Song, X., Zhang, Q., Zhou, Y. *et al.* (2018) BZR1 transcription factor regulates heat stress tolerance through FERONIA receptor-like kinase-mediated reactive oxygen species signaling in tomato. *Plant and Cell Physiology* 59(11), 2239–2254. DOI: 10.1093/pcp/pcy146.

Yu, B., Liu, J., Wu, D., Liu, Y., Cen, W. *et al.* (2020) Weighted gene coexpression network analysis-based identification of key modules and hub genes associated with drought sensitivity in rice. *BMC Plant Biology* 20(1), 478. DOI: 10.1186/s12870-020-02705-9.

Yuan, M., Zhu, J., Gong, L., He, L., Lee, C. *et al.* (2019) Mutagenesis of FAD2 genes in peanut with CRISPR/Cas9 based gene editing. *BMC Biotechnology* 19, 24. DOI: 10.1186/s12896-019-0516-8.

Yue, E., Cao, H. and Liu, B. (2020) OsmiR535, a potential genetic editing target for drought and salinity stress tolerance in *Oryza sativa*. *Plants* 9(10), 1337. DOI: 10.3390/plants9101337.

Yu, W., Wang, L., Zhao, R., Sheng, J., Zhang, S. *et al.* (2019) Knockout of SlMAPK3 enhances tolerance to heat stress involving ROS homeostasis in tomato plants. *BMC Plant Biology* 19(1), 354. DOI: 10.1186/s12870-019-1939-z.

Zahra, N., Shaukat, K., Hafeez, M.B., Raza, A., Hussain, S. *et al.* (2021) Physiological and molecular responses to high, chilling, and freezing temperature in plant growth and production: Consequences and mitigation possibilities. In: Husen, A. (ed.) *Harsh Environment and Plant Resilience: Molecular and Functional Aspects*. Springer, Cham, Switzerland, pp. 235–290. DOI: 10.1007/978-3-030-65912-7_10.

Zargar, S.M., Mir, R.A., Ebinezer, L.B., Masi, A., Hami, A. *et al.* (2022) Physiological and multi-omics approaches for explaining drought stress tolerance and supporting sustainable production of rice. *Frontiers in Plant Science* 12, 803603. DOI: 10.3389/fpls.2021.803603.

Zeng, Y., Wen, J., Zhao, W., Wang, Q., and Huang, W. (2020) Rational improvement of rice yield and cold tolerance by editing the three genes OsPIN5b, GS3, and OsMYB30 with the CRISPR–Cas9 system. *Frontiers in Plant Science* 10, 1663. DOI: 10.3389/fpls.2019.01663.

Zhang, W., Liao, X., Cui, Y., Ma, W., Zhang, X. *et al.* (2019a) A cation diffusion facilitator, GmCDF1, negatively regulates salt tolerance in soybean. *PLoS Genetics* 15(1), e1007798. DOI: 10.1371/journal.pgen.1007798.

Zhang, A., Liu, Y., Wang, F., Li, T., Chen, Z. *et al.* (2019b) Enhanced rice salinity tolerance via CRISPR/Cas9-targeted mutagenesis of the OsRR22 gene. *Molecular Breeding* 39, 47. DOI: 10.1007/s11032-019-0954-y.

Zhang, P., Du, H., Wang, J., Pu, Y., Yang, C. *et al.* (2020) Multiplex CRISPR/Cas9-mediated metabolic engineering increases soya bean isoflavone content and resistance to soya bean mosaic virus. *Plant Biotechnology Journal* 18(6), 1384–1395. DOI: 10.1111/pbi.13302.

Zhao, C., Zhang, Y., Du, J., Guo, X., Wen, W. *et al.* (2019) Crop phenomics: Current status and perspectives. *Frontiers in Plant Science* 10, 714. DOI: 10.3389/fpls.2019.00714.

Zhong, Z., Liu, G., Tang, Z., Xiang, S., Yang, L. *et al.* (2023) Efficient plant genome engineering using a probiotic sourced CRISPR-Cas9 system. *Nature Communications* 14(1), 6102. DOI: 10.1038/s41467-023-41802-9.

Zhu, C., Jing, B., Lin, T., Li, X., Zhang, M. *et al.* (2024) Phosphorylation of sugar transporter TST2 by protein kinase CPK27 enhances drought tolerance in tomato. *Plant Physiology* 195(2), 1005–1024. DOI: 10.1093/plphys/kiae124.

17 Molecular Insights Into CRISPR-edited Pest-resistant Crops Studied by Multiple Omics Approaches

Ripu Daman Parihar[1], Urvashi Dhiman[2,3], Sonali Sharma[1], Sunali Bandral[1], Nimrita Singh[1], Vaishali Sharma[1], Shivam Jasrotia[4], Roohi Sharma[5], Prasoon Gupta[2,3] and Puja Ohri[5]*

[1]Nematology Laboratory, Department of Zoology, University of Jammu, India; [2]Natural Products and Medicinal Chemistry Division, CSIR-Indian Institute of Integrative Medicine, Jammu, India; [3]Academy of Scientific and Innovative Research (AcSIR), Ghaziabad, India; [4]Department of Biosciences, Chandigarh University, Mohali, India; [5]Nematology Laboratory, Department of Zoology, Guru Nanak Dev University, Amritsar, India

Abstract

Pest infestation is one of the dreadful biotic stresses to crops, leading to loss to worldwide agricultural societies and hampering food security. Various plant pests such as fungi, bacteria, and viruses result in different diseases, significantly reducing yield; furthermore, herbivores like insects also decrease yield by damaging the plant directly and acting as disease vectors. Over the past years, genome-edited technologies such as Clustered Regularly Interspaced Short Palindromic Repeats (CRISPR) have brought tremendous change in developing a pest-resistant variety of crops of great economic importance. Thus, CRISPR technology has significant potential for creating pest-resistant crops, offering a more precise and sustainable approach than traditional methods. This chapter discusses the commonly used CRISPR system CRISPR-Cas9 and its modification route in the crops involving alteration in the gene of interest. Seven food crops – rice, potato, cotton, mustard, tomato, soybean, and wheat – have been discussed, emphasizing their principal genes responsible for pest resistance and enhancement of crop yield. Therefore, these gene-editing omics tools play a pivotal role in pest management and minimizing hazardous chemical pesticide use.

Keywords: CRISPR-Cas9, crops genome, infestation, pest resistance, omics

17.1 Introduction

The global population is growing expeditiously, with an anticipated 10 billion people by 2050 (FAO, 2017), bringing unprecedented challenges to agriculture and food security. Therefore, food security has become a primary concern for the stakeholders at the global level. The consideration of sustaining the increasing global population while concurrently upholding environmental equilibrium has become critically imperative (Pandit *et al.*, 2022). Factors like

*Corresponding author: ohri_puja_11@rediffmail.com

© CAB International 2025. *Stress-Resilient Crops: Coordinated omics-CRISPR-Nanotechnology Strategies* (ed. Jen-Tsung Chen)
DOI: 10.1079/9781800627307.0017

groundwater depletion, global warming, and limited arable land have significantly impacted crop production. Furthermore, the frequency of diseases, arthropod pests, and weeds endangers plant growth, and the scarcity of long-lasting, broad-spectrum-resistant crop varieties exacerbates production problems.

Plant diseases are an unavoidable danger to agricultural output and food security globally. Microbial infections pose an ongoing danger to agricultural production, the economy, and food security (Manzoor *et al.*, 2024). Phytopathogens endanger global food sustainability and reduce agricultural productivity. Plant pests and diseases cause global output losses of up to 30.3% in rice and wheat, 22.6% in maize, and 17.2% in potatoes (Savary *et al.*, 2012). Infections in cereals due to fungus are responsible for yield reductions ranging from 15% to 2%, with potential losses reaching 60% in rigorous cases (Różewicz *et al.*, 2021). The financial burden of these losses is significant, averaging $220 billion per year (FAO, 2019). The agricultural sector relies heavily on chemical management to combat infections. Chemicals have spoiled the water, soil ecosystem, and air, causing bioaccumulation at greater trophic levels. In recent years, advances in genome editing have transformed plant biology, allowing for the exact modification of crops' genetic makeup to achieve aspired fruitful traits. Genome editing (or gene editing) refers to a process that allows genetic information to be inserted, deleted, or modified at specific locations inside the genome. New breeding techniques (NBT) are sophisticated, precise molecular approaches targeting one or more genes. Sequence-specific nucleases such as Clustered Regularly Interspaced Short Palindromic Repeats (CRISPR), zinc finger nucleases (ZFNs), meganucleases (MNs), transcription activator-like effector nucleases (TALENs) and base editing are extensively used in genome editing. CRISPR is a more precise and cost-effective method for modifying plant genomes (Belhaj *et al.*, 2015).

CRISPR technology was initially discovered in early 1987 and has been recognized in various different bacterial species. In 2013, CRISPR technology was successfully employed to modify vegetative genes, rapidly establishing itself as the preeminent gene-editing technique (Nekrasov *et al.*, 2017). Prokaryotes naturally use CRISPR-Cas systems to resist mobile genetic elements like plasmids and phages. The CRISPR genomic region consists of short sequences of foreign origin, known as spacers. Spacers play a relevant role in enabling the organism to identify and distinguish previously encountered mobile genetic elements (MGEs) (Vercoe *et al.*, 2013). The spacers are separated by repeats, which are regulatory sequences that appear repeatedly. The CRISPR array is a critical constituent of the CRISPR-Cas system, consisting of the leader sequence spacers, and repetitions (Fig. 17.1).

The structure and function of Cas proteins are key to CRISPR/Cas system automation (Khan *et al.*, 2022). CRISPR/Cas systems are categorized into Class I (types I, III, IV) and Class II (types II, V, VI) based on genomic analysis (Makarova *et al.*, 2022). Class II systems, including types II and V, are known for DNA cleavage, while Type VI specializes in RNA cleavage (Gupta *et al.*, 2019). Class I systems have multi-subunit Cas complexes, whereas Class II uses a single Cas protein. Type II CRISPR/Cas9, widely used in genetic engineering, involves Cas9 for DNA cleavage. Type V uses Cas12, which processes RNA independently of tracrRNA, making it ideal for multiplex editing. Type VI CRISPR systems, utilizing Cas13, target ssRNA with distinct catalytic sites for processing and cleavage (Meeske *et al.*, 2019; Yan *et al.*, 2021; Wei *et al.*, 2021).

The CRISPR mechanism involves several key stages: identifying the protospacer adjacent motif (PAM), synthesizing and cloning single-guide RNA (sgRNA), introducing it into the host cell, and verifying genomic modifications. The engineered sgRNA targets the gene of interest, while Cas9 induces double-stranded breaks, which are then repaired via non-homologous end-joining or homology-directed repair.

17.2 Omics Approaches in CRISPR/Cas Genome Editing

Plants are continuously exposed to abiotic and biotic stressors, which have a negative impact on agricultural production. Statistical approaches, computational skills, mass spectrometry, nucleic acid sequencing, and peptide-sequencing systems have all contributed to the growth of omics over the last two decades. Functional

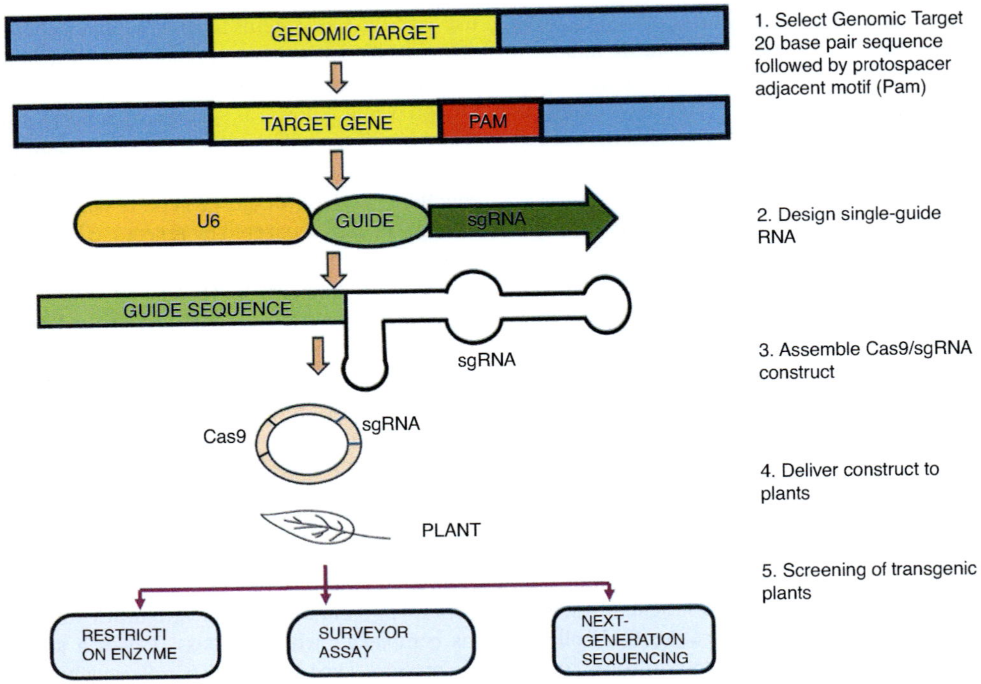

Fig. 17.1. Generalized diagram showing steps involved in CRISPR/Cas editing in plants.

genomics, specifically metabolomics, transcriptomics, and proteomics, have yielded significant advancements in understanding plant molecular stress responses.

17.2.1 Genomics

In the previous decade, the progress and innovation in next-generation sequencing (NGS) technologies have significantly revolutionized sequencing-dependent biological fields by identifying potential genes and enhancing yields. Complete sequencing of the *Arabidopsis thaliana* genome was documented and published (Kaul *et al.*, 2000). In the preceding decade, comprehensive genome sequences of vegetative crops, which include cucumber, potato (Potato Genome Sequencing Consortium, 2011), and pigeon pea, have been disseminated (Huang *et al.*, 2009; Varshney *et al.*, 2012). Functional genomics in plants has found numerous genes which control living and non-living stress reactions (Zhang *et al.*, 2017a; Wang *et al.*, 2018). In the *Solanum americanum*, combined genetic mapping and RenSeq were utilized for identifying genomic locus conferring resilience to late blight (Witek *et al.*, 2016). MutRenSeq, which is a novel version of RenSeq, has been utilized to isolate the R genes in wheat giving resilience against stem rust (Steuernagel *et al.*, 2016).

17.2.2 Transcriptomics

Transcriptomic technologies are utilized to analyze the transcriptome of an organism, which is a collection of entire RNA transcripts. Transcriptomics provides a comprehensive perspective of both active and dormant biological activities. The study of gene expression in the tissues of an organism reveals gene regulation and detailed biology. Two key contemporary approaches to the subject are RNA sequencing and microarrays (Lowe *et al.*, 2017). RNA sequencing captures and quantifies transcripts in RNA extracts by high-throughput sequencing and computational techniques (Ozsolak and Milos, 2011). Microarrays are made up of 'probes,' which are small nucleotide oligomers organized on a solid substrate (Romanov *et al.*, 2014). A comparative microarray technique revealed an interaction between functional genes and pathways in different stress transcriptome

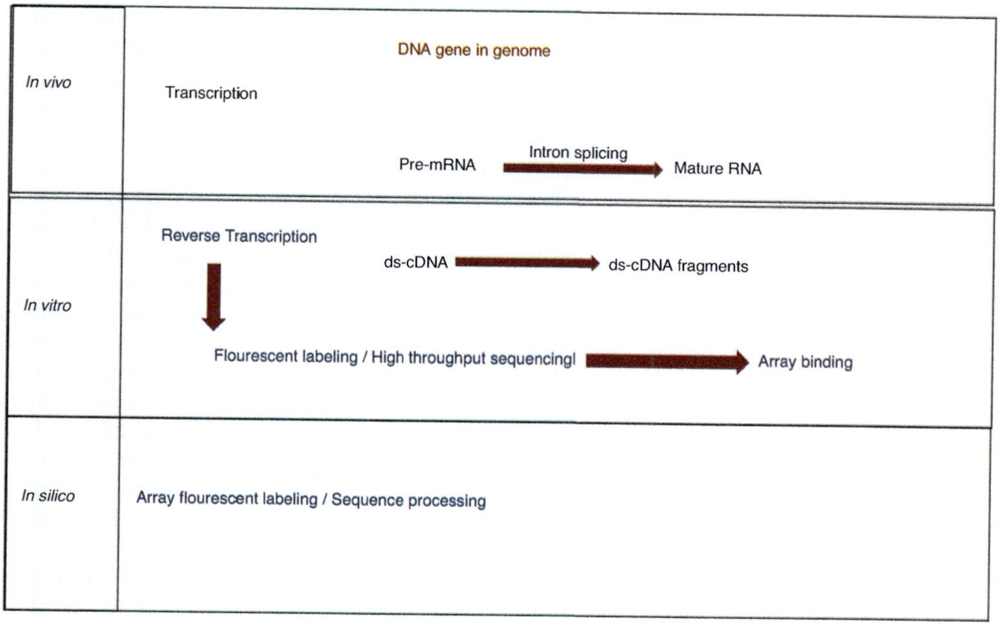

Fig. 17.2. A general procedure for transcriptomics (RNA sequencing and microarray DNA). The organism's mRNA is extracted and reverse transcriptase is employed to replicate it into stable double-stranded cDNA. In microarrays, ds-cDNA is fragmented and fluorescently tagged, whereas in RNA sequencing, the ds- cDNA is sequenced through high-throughput sequencing. In microarray, the tagged fragments bind to an ordered array of complementary oligonucleotides, and measuring the fluorescence intensity throughout the array reveals the abundance of a preset set of sequences.

investigations in cotton (Fig. 17.2) (Zhu *et al.*, 2013). Transcriptomics databases (genome-wide) for stress reactions in plants are available online (Zhang *et al.*, 2018a; Zhang *et al.*, 2019). BSR-seq, which is a merger of the RNA-seq with the bulked segregant examination, has the potential to improve tension tolerance in plants. For example, in *Brassica rapa*, *Bra019410* and *Bra019409* were potential candidate genes for the clubroot resistance (Hilton *et al.*, 2015; Kearns *et al.*, 2015).

17.2.3 Proteomics

Proteomics involves the quantifiable and qualitative examination of all proteins that are present in any cell, tissue, or organism. Protein levels and expression alter significantly during stress. Thus, studying proteome profiles during stress can give precise information on individual protein alteration associated with responses. Proteomic profiling is an effective method for determining connections between protein accumulation and characteristics in certain situations (such as drought tolerance). Mass spectrometry of several types was utilized recently for profiling the proteome in response to abiotic challenges (Shao *et al.*, 2015; Luan *et al.*, 2018). Proteome profiles can be analyzed to ascertain specific proteins' role in living and non-living-induced stress signaling, along with differently expressed stress-resistant proteins. Recent research on tomato identified 11 drought-related proteins utilizing 2D gel electrophoresis and MALDI-TOF-MS (Çelik *et al.*, 2021).

17.2.4 Metabolomics

Metabolomics deals with all the low-molecular-weight molecules and metabolites of an organism. Two rice genotypes (32R and 29S) infected with *Rhizoctonia solani* were examined by Suharti *et al.* (2016). According to their metabolomic research, the sensitive cultivar 29S increased inosine monophosphate whereas the resistant cultivar 32R increased levels of ADP, glyceric acid, mucic acid, and jasmonic acid (JA). This suggests that to resist *R. solani*, different cultivars employ varied metabolic methods. Enhancing rice's resistance to this fungus may be possible by breeding and selecting these compounds (Suharti *et al.*, 2016). The metabolomic analysis showed that the FHB-resistant recombinant inbred lines (RILs) exhibited elevated levels of several key compounds, including flavonoids, fatty acids, glycerophospholipids, phenylpropanoids, terpenoids, and lignins. These increased metabolite levels suggest a potential metabolic adaptation contributing to their resistance to FHB. The study identified many resistance-gene candidates within *QTL-Fhb2* allele, including 4CL (4-coumarate-CoA ligase), *ABC4* (ABC transporter-4) and *CAD* (cinnamyl alcohol dehydrogenase) (Dhokane *et al.*, 2016). NMR-based metabolic analysis of *Fusarium graminearum*-infected wheat spikelet and rachis showed that resistant plants exhibited greater amounts of putrescine, spermine, γ-aminobutyric acid (GABA), lactic acid, and lactic acid (Cuperlovic-Culf *et al.*, 2016). In breeding programs, these metabolites could be utilized to select for higher concentrations of these beneficial chemicals or as indicators for disease resistance.

17.2.5 Multi-omics approach

A combination of datasets from different omic groups during analysis is referred to as a multi-omic approach. Multi-omics gives a more complete understanding of cellular and organismal function. Kernal bunt (KB), caused by *Tilletia indica*, is a significant wheat disease that infects florets, leading to kernel damage and a foul odor due to fungal teliospores. In a study, multi-omics used mass spectrometry (MALDI-TOF/TOF) to compare highly and weakly virulent isolates of *T. indica*. This analysis identified 21 differentially expressed proteins in the more virulent isolates. Sequencing data revealed that some of these proteins are involved in oxalic acid production, highlighting how combining proteomics with metabolomics can effectively identify virulence-related compounds in plant pathogens (Pandey *et al.*, 2018). Numerous metabolomics investigations have examined stresses and pathogens in a range of crops, including potatoes and beans. Using *de novo* transcriptomics and metabolomics, the root metabolites of *Phaseolus vulgaris* infected by the *Fusarium oxysporum* f. sp. *Phaseoli* were examined. Expression of genes linked to pathogenicity differed between non-infected and infected plants, according to RNA sequencing data. Significant alterations in metabolites linked to the jasmonate, ethylene, flavonoid, and salicylic acid (SA), and pathways were shown by UPLC-MS analysis because of the disease (Chen *et al.*, 2019). Potato blight is a major worldwide potato disease that is caused by *Phytophthora infestans*. Using transcriptomics and metabolomics, responses in susceptible and tolerant potato varieties were investigated, and 4216 genes with differential expression were found in resistant vs susceptible lines by their transcriptome analysis. A total of 589 of the 4811 metabolites found by LC-HRMS-based metabolomics were identified, indicating notable alterations in terpenoids, phenylpropanoids, alkaloids, and flavonoids. These results offer fresh perspectives on the mechanisms underlying potato blight resistance and offer recommendations for creating potato variants resistant to the illness (Yogendra and Kushalappa, 2016). In 2016, Dhokane and his team studied RILs of wheat and analyzed them using both transcriptomics and metabolomics to find genes associated with *Fusarium* head blight resistance in the *QTL-Fhb2* allele. FHB (*Fusarium graminearum*) infection in wheat was investigated using metabolomics combined with transcriptome profiling. In addition to discovering over 100 genes linked to FHB resistance, they identified 789 metabolites, which included flavonoids, derivatives of tryptamine, phytohormones, and phenolamides. FHB infection was obtained by mutating or silencing the *TaTIR1* gene (Su *et al.*, 2021). Transcriptomics and metabolomics were used to examine the *Rp1-D* gene's reaction to *Puccinia sorghi*. They found that defense genes were activated in pathways related to secondary metabolism, including, flavonoids, phenylpropanoids, and terpenoids. Crucial transcription factors, i.e. BZIP84, MYB100, and WRKY53, have been correlated (Kim *et al.*, 2021).

17.3 Mechanism of CRISPR/CAS-9 Genome Editing

This mechanism comprises three basic steps: recognition, cleavage, and repair. The CRISPR/Cas9 system relies on two key components: guide RNA (gRNA) and the Cas9 protein. Guide RNA is composed of CRISPR RNA (crRNA) and transactivating CRISPR RNA (tracrRNA), where it is a long-loop structure and serves as a binding scaffold for the Cas9 nuclease, while crRNA, a 16–20 bp molecule, identifies and targets the DNA sequence. In prokaryotes, gRNA directs Cas9 to viral DNA, but in gene editing, tracrRNA and crRNA are combined to create single-guide RNA (sgRNA), which targets specific gene sequences for editing (Mei *et al.*, 2016). The sgRNA guides Cas9 to the target site, with Cas9 remaining inactive without it. Cas9 induces DSBs at a location of three base pairs upstream of the PAM, a short, conserved DNA sequence (Ceasar *et al.*, 2016). The PAM sequence varies in length and is species-dependent, typically between two and five base pairs. Cas9 protein recognizes the PAM sequence, initiates DNA melting, and forms an RNA-DNA hybrid, though the exact mechanism of melting is not fully understood. Cas9 then cleaves the DNA, with its HNH domain cutting the complementary strand and the RuvC domain cutting the noncomplementary strand, resulting in blunt-ended DSBs. These breaks are repaired by the host cell through homology-directed repair (HDR) or non-homologous end-joining (NHEJ) (Mei *et al.*, 2016; Jiang and Doudna, 2017; Liu *et al.*, 2019).

Non-homologous end-joining is the primary cellular repair mechanism for DSBs, effectively reconnecting DNA fragments without a homologous template and active throughout all cell cycle phases. In contrast, HDR is highly precise, utilizing a homologous DNA template, and mainly occurs during the late S and G2 phases. In CRISPR, HDR requires multiple donor DNA templates to accurately insert or replace genes at the DSB site (Fig. 17.3) (Liu *et al.*, 2019; Yang *et al.*, 2020).

17.4 CRISPR-Cas 9 Edited Crops

17.4.1 Rice

Over 50% of the global food population consumes rice, making it one of the most significant staple foods. However, several infections pose a serious threat to its quality and output. *Xanthomonas oryzae* pv. *oryzae* (Xoo) and *Magnaporthe oryzae* are two prominent pathogens that cause bacterial blight and rice blasts, respectively (Yin *et al.*, 2018). Gene-editing technology has increased pest resistance in rice plants by modifying various genes. Editing the RESISTANCE TO BLAST1 (*RBL1*) gene in *Oryza sativa* conferred broad-spectrum disease resistance but reduced yield by nearly 20-fold. *RBL1* encodes a key enzyme for phospholipid biosynthesis, and its mutation decreases phosphatidylinositol and phosphatidylinositol 4,5-bisphosphate levels (Sha *et al.*, 2023). Using gene-editing technology, two significant conventional rice varieties, Huanghuazhan (HHZ, Xian/indica) and Kongyu131 (KY131, Geng/japonica), have been successfully transformed (Li *et al.*, 2022). In Asia, Africa, and Australia, *X. oryzae* pv. oryzicola (Xoc) was a causative agent of the devastating rice disease known as bacterial leaf streak (BLS) (Ji *et al.*, 2014). Like other Xanthomonads, Xoc suppresses plant immunity by directly translocating effector proteins (EPs) into the host cells through T3SS (type III secretion system) (Yuan *et al.*, 2021). Super basmati rice was subjected to CRISPR/Cas9-mediated genome editing to alter four effector binding elements (EBEs) in the promoter region of *SWEET14* gene. This eventuated in the resistance to the strains of Xoo that carry their particular TALEs (AvrXa7, PthXo3, and TalF) (Zafar *et al.*, 2020). In plant cell nuclei, TALEs (transcription activator-like effectors) operate as eukaryotic transcription factors and act as virulence factors. They bind to EBEs of the targeted plant gene promoters through CRR (a TALE-encoded central repeat region).

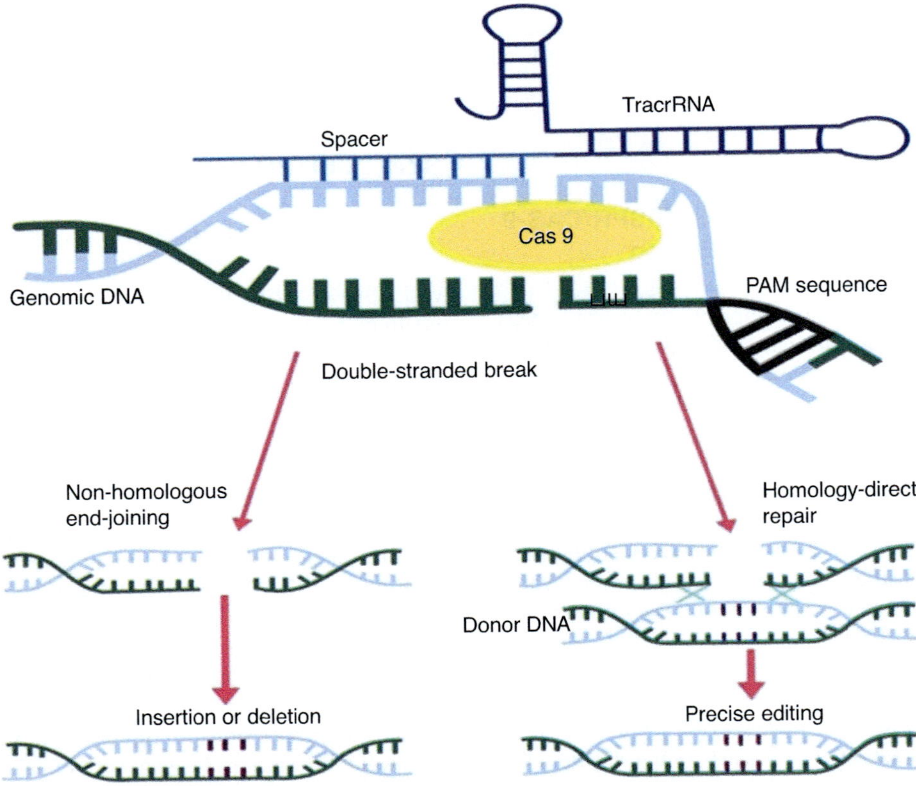

Fig. 17.3. Using CRISPR-Cas9 for genome engineering. The crRNA-tracrRNA, also known as sgRNA, directs the Cas9 enzyme to target certain DNA sequences and generate a double-strand break (DSB). DNA DSBs can be repaired using non-homologous end-joining (NHEJ) or homology-directed repair (HDR).

In Xoc strain *BLS256*, *Tal2g* upregulates the TALE-targeted gene *OsSULTR3;six* where it is regarded as a susceptibility (S) gene for BLS and encodes a putative sulfate transporter in rice (Cernadas *et al.*, 2014). Broad-spectrum resistance to Xoo is conferred by the loss of TALE-binding regions in three S genes in the Xoo rice pathosystem (Oliva *et al.*, 2019; Xu *et al.*, 2019). To increase the resistance level of rice to Xoo, the CRISPR/Cas9 system was utilized to delete Os8N3rice. Upon analyzing the genotypes and the edited Os8N3 in transgenic rice plants T0, T1, T2, and T3, it was noted that the mutations were inherited by the next generations, with homozygous mutants exhibiting notably increased resistance to the Xoo. T1 generation's segregation demonstrated the stable transmission of CRISPR/Cas9-mediated *Os8N3* gene editing without requiring the transfer of T-DNA. *Os8N3* gene editing in rice was produced with great efficiency by the CRISPR/Cas9 system (Kim *et al.*, 2019). Furthermore, five mutations were simultaneously incorporated in the promoter region of the *SWEET14* and *SWEET13* genes of the rice types IR64 and CiherangSub1, as well as the rice lines Kitaake and Japonica. To the majority of the Xoo strains, the homologous mutant plant demonstrated strong broad-spectrum resistance (Oliva *et al.*, 2019). Pinzhan (offspring of Pin13 and Huazhan) was successfully transformed into a new thermosensitive genic male sterility line (named PinzhanS) with elevated resilience to the rice blast and the bacterial blight by introducing specific mutations into *Xa13*, *TMS5*, and *Pi21* genes using the CRISPR/Cas9 multiplex genome-editing system (Li *et al.*, 2019a). Tropical Asia's rice output is severely hampered by rice tungro disease (RTD). The interplay between rice tungro bacilliform virus and the rice tungro spherical virus (RTSV) results in the development of RTD. The incidence of RTD in the field has decreased partly because of RTSV resistance in conventional cultivars. The translation initiation factor four gamma gene (*eIF4G*) is accountable for controlling the recessive trait of natural RTSV resistance. It is well known that eIF4G residues Y 1059V 1060V 1061 are connected to the reactions to RTSV. Using the CRISPR/Cas9 technology, mutations in *eIF4G* were developed (Macovei *et al.*, 2018). The expression of the *OsSEC3A* gene is found in multiple organs, and it may play a role in the exocyst complex through interactions with multiple other exocyst subunits. Dwarf stature and a lesion-mimic phenotype were brought about by *OsSEC3A* disruption via CRISPR/Cas9. Increased levels of SA upregulated transcript levels of genes related to pathogenesis and SA synthesis, and improved defiance to fungal pathogen *Magnaporthe oryzae* are all indications that the *Ossec3a* mutant displayed enhanced defense responses (Ma *et al.*, 2018). These findings indicate a feasible route toward developing rice cultivars resistant to various illnesses (Zhou *et al.*, 2018).

17.4.2 Potato

Globally, potatoes (*Solanum tuberosum* L.) are a staple crop, and one of the main obstacles to their production is the potato virus Y (PVY). PVY can result in up to 80% crop losses and impacts yield and tuber quality (Quenouille *et al.*, 2013). Technology CRISPR-Cas has revolutionized plant genome editing, enhancing gene understanding and crop traits (Zhu *et al.*, 2020; Zhang *et al.*, 2021a). Modifying the *eIF4E* gene family can potentially make crops immune to viruses like potato virus Y (PVY). A study was conducted on *eIF4E1*, a host protein necessary for PVY replication. The goal of this gene mutation study was to create potatoes that were resistant to PVY. Only two eIF4E1 alleles were mutated in the original transgenic potatoes, making them vulnerable to PVYNTN. Plants that underwent further rounds of genetic engineering developed partial resistance to PVYNTN because of complete knockouts of both eIF4E1 alleles (Lucioli *et al.*, 2022). Potato virus Y resistance is increased when the *eIF4E-VPg* connection is disrupted via CRISPR/Cas9-mediated mutagenesis. It stopped the virus from using the host translation machinery to replicate and propagate further. Because of their inability to connect to the viral VPg protein, the resulting mutant *eIF4E* proteins successfully stopped the viral infection cycle. The technique has the potential to be a useful method for cultivars of potatoes to acquire sustainable resistance against PVY (Noureen *et al.*, 2022). Late blight resistance of the Russet Burbank potato variety was significantly increased due to allelic replacement or single nucleotide polymorphism (SNP) correction in *StCCoAOMT*. Using CRISPR-Cas9 mediated homology-directed repair (HDR), a premature stop codon was removed, and full-length protein was reinstated in the caffeoyl CoA O methyltransferase (*StCCoAOMT*) gene of the RB potato through precision SNP mutation correction (Hegde *et al.*, 2021). The most dangerous disease that affects potato crops globally is late blight. *Phytophthora infestans*, an oomycete pathogen that can infect potato plants' stems, tubers, and leaves, is the cause. Potatoes with enhanced resistance against late blight were produced by functional knockouts of *DMG400000582* (*StDMR6-1*), *StDND1*, and *StCHL*. *StCHL1* and *StDMR6-1* are promising targets for gene editing to increase late blight resistance (Kieu *et al.*, 2021). In order to modify the potato coilin gene for PVY resistance, Cas9 has also been used. Coilin, present in cajal bodies of plants, are structurally and functionally important proteins and are involved in the metabolism of RNA, the nucleolus, and the connection between plants and viruses. In general, Cas9 might modify the coilin and *eIF4E*/*eIF*(iso)*4E* genes to give potatoes tolerance against different RNA viruses (Makhotenko *et al.*, 2019). With the ability to cause up to 80% damage, PVY is a destructive infection that severely reduces potato crop output and quality. Researchers have developed potato plants that are resistant to PVY by using the CRISPR/Cas system, notably the *Cas13a* version from *Leptotrichia shahii*. Transgenic potato lines showed significant resistance to several strains of PVY by focussing on conserved areas in the viral RNA genome. Remarkably, the Cas13a/sgRNA transgene expression was positively linked with the degree of resistance. This promising use of technology could develop broad-spectrum resistance against various viruses (Zhan *et al.*, 2019).

17.4.3 Cotton

A significant economic crop that yields natural fiber is cotton. Farmers frequently apply broad-spectrum insecticides that contain chemicals to nontarget pests, which has a negative effect on public wellbeing, the natural environment, and the economic system because pests reduce the amount and quality of fiber produced (Digilio *et al.*, 2012). As a result, the cotton farming sector now strongly emphasizes breeding for pest resistance (Zhang *et al.*, 2022a). As a consequence, scientists are now more focussed on developing breeding methods that are safer, more effective, and more stable than those of the past (Zhang *et al.*, 2022b). An endogenous noncoding RNA regulatory molecule with a fully developed structure and length longer than 200 nucleotides is known as lncRNA (long non-coding RNA) (Ulitsky and Bartel, 2013; Jandura and Krause, 2017). Previous research indicates that lncRNA has several functions depending on where it is located in the genome (Ma *et al.*, 2013). Because of its excellent efficiency, extreme stability, simplicity, and minimal inaccurate effect, the gene-editing technology has completely changed functional genomics research (Li *et al.*, 2019b; Manghwar *et al.*, 2020). A study spotlighted the function of long noncoding RNAs (lncRNAs) in cotton's defensive mechanism against sap-sucking insects, particularly whitefly infestation. *lncA07* and *lncD09* were identified as putative hub genes in cotton's defensive mechanisms by the co-expression network analysis. Reduced levels of JA were seen in *lncD09* and *lncA07* CRISPR/Cas9 knockout mutants, which increased their vulnerability to insect invasion. Genes that control development and stimulus resistance were found to be more abundant in wild-type plants compared to lncRNA knockout plants. *Ghir_A01G022270*, *Ghir_A01G022270*, and *Ghir_D04G014430*, and are among the notable candidate genes that have been linked to the JA-mediated signaling pathway. These discoveries highlight the crucial regulatory roles that lncRNAs play in cotton's defense mechanism against pests and present fresh opportunities for lncRNA genetic manipulation-based pest resistance enhancement (Zhang *et al.*, 2022c).

Genome editing has transformed plant biology with CRISPR/Cas9, especially in terms of building protection against viral infections. A study in *N. benthamiana* plants and cotton seedlings revealed multiplex CRISPR/Cas-mediated genome editing against CLCuD (cotton leaf curl disease) through temporary transformation. In *N. benthamiana* plants, the study discovered delayed symptoms, enhanced resistance to CLCuD, and low viral titer in infected plants co-infiltrated with Cas9-gRNA. For sustainable cotton, the method can be used to help develop cotton resistant to CLCuD. The research emphasizes how multiplex CRISPR/Cas might help evolve resistance against CLCuD (Binyameen *et al.*, 2021). The genome-editing system was also used in *G. hirsutum* to modify the *Gh14-3-3d* gene that was involved in the *Verticillium dahlia* infection mechanism (Zhang *et al.*, 2018b).

In a study by Quin-Hao, cotton *miR482* mutants were made by CRISPR/Cas9 genome editing by the transformation of pooled gRNAs. The biogenesis of *miR482* is disrupted by vast mutations seen in most isoforms of the *miR482*, including small substitutions and indels. In an experiment with *Verticillium dahliae*, all nine populations of mutants showed a lesser disease index in comparison to the population of control, with four populations exhibiting notably reduced symptoms of disease (Zhu *et al.*, 2020).

17.4.4 Mustard

Oilseed crops are one of the most essential food ingredients for human and livestock consumption. Mustard seeds are widely used in food manufacturing, and they belong to the family Brassicaceae or Cruciferaceae. They contain an abundant chemical content. Traditional oilseed crops include mustard, soybean, linseed, sunflower, safflower, and garden cress. Due to the rising need for both refined food oil and renewable commercial oils, sunflower (*Helianthus annus*), Indian mustard (*Brassica juncea*), soybean (*Glycine max*) and canola or rape (*Brassica napus*) are currently dominating the global oilseed market (Gupta *et al.*, 2021). Many abiotic factors, such as excessive or insufficient water availability, high soil and irrigation water salinity, and extremely high or low temperatures can impact mustard productivity (Mollika *et al.*, 2011). Moreover, plants usually deal with high chemical levels and mineral deficiency or toxicities in soil. These abiotic variables frequently coexist and have a negative impact on the crops. A key element to the rate of performance of mustard crops is the unavoidable presence of stressors. The flowering periods, silique development, germination of seeds, yield of seeds, oil production, and seed quality of mustard are directly impacted by abiotic stresses (Hasanuzzaman *et al.*, 2021). One of the major diseases affecting brown mustard, Alternaria blight disease, is caused by *Alternaria brassicicola* and *A. brassicae* has been documented from every continent across the globe. It can result in 10–80% loss in yield. Both yellow and field mustard are impacted by *Alternaria* sp. Fungicides such as iprodione or mancozeb are commonly used to prevent plant diseases caused by fungi; nevertheless, their widespread use is uneconomical and hazardous to the environment (Singh *et al.*, 2022). *Brassica juncea* has demonstrated resistance to *Sclerotinia sclerotiorum*, which often causes sclerotinia stem rot disease, using the CRISPR/Cas9 knockdown of the *BjuXLG1* as well as *BjuXLG2* genes. Knockdown of the above genes led to decreased glucosinolate accumulation, enhanced host defense mechanisms, and resistance to stem rot disease (Tiwari *et al.*, 2021).

Another set of the *BnWRKY11* and *BnWRKY70* genes in rapeseed plants were successfully edited using knockout, producing mutants with increased resistance to the pathogen *Sclerotinia*. This suggests that *BnWRKY70* may be involved in negatively regulating resistance to *Sclerotinia* and that the system is useful in the hope of creating rapeseed germplasm with high resilience to developing stem rot disease (Sun *et al.*, 2018).

17.4.5 Wheat

Common wheat, a grass that is extensively grown for seeds and cereal grain is a common food source worldwide. It is the primary supplier of processed cereal-based goods, including bread, biscuits, and noodles, and supplies more than 30% of the calories consumed globally (Petersen *et al.*, 2006). Due to various factors such as climate change, population growth, and water scarcity in arid and semi-arid regions, wheat faces unprecedented challenges even with notable increases in production following the Green Revolution in the 1960s (Peng *et al.*, 1999) and the introduction of marker-assisted molecular breeding (Hickey *et al.*, 2019). In addition, the extensive use of various agrochemicals aggravates ecological degradation and pollution (Tian *et al.*, 2020; Liu *et al.*, 2021). Furthermore, wheat's gene functional redundancy and hexaploid nature make breakthroughs in genetic approaches for choosing desirable phenotypes time-consuming, if not impossible, due to gene drag or gene linkage (Li *et al.*, 2020). To ensure global food security and ecosystem stability, advanced technologies must enhance wheat resilience and reduce environmental impact. Wheat, crucial for 2.5 billion people across 89 countries, faces numerous stresses and diseases each season (CGIAR, 2019; Simón *et al.*, 2021). The integration of gene-editing technologies, such as Cas12 and Cas9, into agricultural development and plant biology is highly significant. These tools have been utilized for precise genome editing in wheat, enabling the creation of novel alleles and the disruption of specific gene functions, thereby advancing crop improvement and functional genomics (Kefale and Getahun, 2022). Cas9 and Cas12a, with their distinct advantages and disadvantages, have greatly expanded their use. Cas12a has an advantage over Cas9 in that it can be utilized for multiplex genome editing and produces staggered DSBs, facilitating HDR rather than NHEJ. The International Wheat Genome Sequencing Consortium (IWGSC) services are also known to simplify gene discovery and character deconstruction, which provides a foundation for target selection and off-target analysis in *T. aestivum* genome editing (Hao *et al.*, 2020; Zhang *et al.*, 2021b).

The knockdown of 33S genes by gene-mediated genome editing showed improved resistance to diseases like stripe rust and powdery mildew (Taj *et al.*, 2022). The current technology has also been used in many laboratories to boost resistance against wheat disease. Editing *TaMLO* gene in wheat resulted in powdery mildew-resistant plants (Shan *et al.*, 2013; Wang *et al.*, 2014). It was also used to eliminate EDR1, a negative defense response to powdery mildew, resulting in wheat plants with increased resistance (Zhang *et al.*, 2017b). Fusarium head blight is a serious disease that significantly impacts wheat quality as well as output. In crops, disrupting S-genes is an appealing breeding approach for granting immunity to disease. Susceptibility genes, however, are connected to numerous crucial biological processes, therefore the removal of these genes usually has unintended pleiotropic effects. Mutations result in loss of function of one such S gene, mildew resistance. Powdery mildew resistance is long-lasting and broad-spectrum conferred by locus O (MLO) in several kinds of plants. Nevertheless, MLO-associated resistance is additionally correlated with yield losses and growth penalties, therefore reducing its extensive application in farming (Li *et al.*, 2021).

Utilizing CRISPR/Cas9 to knock down TaeIF4E in hexaploid wheat resulted in enhanced resistance to WYMV. Additionally, editing the TaPDIL5-1 homoeoalleles achieved complete resistance to WYMV. Research demonstrated that plants with triple mutations in either TaPDIL5-1 (Kan *et al.*, 2022) or TaeIF4E (Kan *et al.*, 2023) exhibited resistance to WYMV, whereas those with single or double mutations remained susceptible. This observation underscores the functional redundancy of these host factor genes and their role in counterbalancing each other, which facilitates the infection process of WYMV.

17.4.6 Tomato

Tomato (*Solanum lycopersicum*), a globally important crop, is frequently threatened by a variety of pests, leading to significant yield losses and increased reliance on chemical pesticides. Traditional breeding methods for pest resistance are often time-consuming and limited by the genetic diversity available within the species. In contrast, CRISPR-Cas9 offers a rapid and efficient approach to enhance pest resistance in tomato by directly targeting and modifying genes associated with defense mechanisms. Wild tomato species, i.e. *S. habrochaites* and *S. pennellii*, are naturally more resistant to diseases and pests compared to the cultivated varieties of tomato. These cultivated varieties, which are prevalent in global industrialized horticulture, were selectively bred to boost yields and produce larger fruits. However, during this breeding process, genes involved in natural defenses were weakened or lost. Consequently, the widely grown cultivated tomatoes today require more intensive pesticide use (Balcke *et al.*, 2017). A cutting-edge approach in tomato breeding is the use of editing technology. This method aims to reactivate or incorporate necessary genetic material into the high-yielding tomato cultivars, enhancing their resilience and resistance to pests and diseases. Ultimately, these tomato varieties (CRISPR GE) are expected to significantly reduce the need for crop protection products in tomato cultivation. Tomato yellow leaf curl virus (TYLCV), a highly impactful monopartite virus within the *Begomovirus* genus, poses a severe economic threat to tomato cultivation, potentially resulting in complete crop loss (Pietersen and Smith, 2002). Infected tomato plants display symptoms such as stunted growth, rubbery, cup-shaped leaves, and yellowing along the leaf margins, which contribute to substantial yield reductions (Moriones and Navas-Castillo, 2000). As a major global threat to tomato production, TYLCV significantly undermines crop yields and represents a critical challenge for growers worldwide. Traditional methods of control, such as chemical treatments and breeding for resistance, have limitations and are not always effective. Targeting the TYLCV genome, specifically the *Rep* gene, is crucial for viral replication, by designing CRISPR/Cas9 constructs to incorporate precise mutations in the *Rep* gene, that tend to disrupt the virus's ability to replicate within the tomato plants. Technology was highly effective in targeting and editing the TYLCV genome. The resulting mutations in the *Rep* gene significantly impaired the virus's replication capability. Transgenic tomato plants expressing the CRISPR/Cas9 constructs exhibited strong resistance to TYLCV infection. These plants showed lower viral loads and milder disease symptoms compared to nontransgenic control plants (Tashkandi *et al.*, 2018).

The fungal pathogen *Oidium neolycopersici* causes powdery mildew in tomato, significantly reducing fruit quality and yield (Jones *et al.*, 2001). Certain members of the Mildew Locus O (MLO) transmembrane protein family are linked to susceptibility to *O. neolycopersici*. Among the 16 MLO genes in tomato, *SlMLO1* has been identified as the primary susceptibility gene. Natural loss-of-function mutants of SlMLO1 have shown resistance to powdery mildew, highlighting the gene's role in susceptibility to this disease (Zheng *et al.*, 2016). The edited mutants with homozygous *Slmlo1* alleles, featuring a 48-bp deletion from the wild-type *SlMLO1*, showed complete resistance to the infection. Importantly, these *Slmlo1* plants lacked any foreign T-DNA sequences, making them indistinguishable from naturally occurring *Slmlo1* mutants (Nekrasov *et al.*, 2017).

Another approach to combat powdery mildew involves targeting the tomato *SlPMR4* gene, which encodes an enzyme for callose synthesis (Weis *et al.*, 2014). Knockout of *SlPMR4* triggers HR-like cell death at fungal penetration sites, conferring resistance to *O. neolycopersici* infection (Santillán Martínez *et al.*, 2020).

17.4.7 Soybean

Soybean (*Glycine max* L.) is a crucial crop for plant-based proteins, serving as both a vital oilseed and an important industrial raw material. However, soybean production is significantly impacted by plant diseases, which adversely affect both yield and quality. Key pathogens such as the soybean cyst nematode and charcoal rot are major contributors to economic losses, causing approximately 33% of the financial impact on soybean crops per hectare (Bandara *et al.*, 2020). These diseases compromise plant health and productivity, underscoring the need for effective management strategies to mitigate their detrimental effects and secure soybean yield and quality. The gene knockout technology is a reliable tool for developing disease-resistant soybean, as it has already shown effectiveness in many crops (Pompili *et al.*, 2020). It has been utilized to create resistance against certain pathogens by targeting either the pathogen's genome or the specific host gene involved in plant defense mechanisms with limited research using CRISPR/Cas9, but this technique holds significant potential for enhancing disease resistance.

Phytophthora soya, which causes root rot, lowers productivity by 10–40%, with some cases exceeding 100% (Chandra *et al.*, 2022). Soybean is vulnerable to a variety of stressors. The removal of *GmARM* by using CRISPR/Cas9 technology enhanced strain resilience in soybean. Transgenic soybean crops outperformed the wild type in terms of salt and alkali resistance. Simultaneously, transgenic soybean plants outperformed the wild type in terms of *Phytophthora soya* resistance. As a result, *GmARM* research provides a promising avenue for developing soybean varieties that are resistant to numerous stressors (Luo *et al.*, 2024).

Powdery mildew is a significant pathogen that reduces soybean output and seed quality. Disease-resistant cultivars, created by lowering the influence of S genes in the host plants through mutation, might be an effective method of protecting crops against this disease. Mildew Locus O (*MLO*) genes are well-known susceptibility genes in powdery mildew. Using gene-editing-based technology, specific mutations were induced in soybean *GmMLO* genes. A dual-sgRNA CRISPR/Cas9 construct was introduced into the Vietnamese soybean cultivar *DT26* via *Agrobacterium tumefaciens*-mediated transformation. At the T2 generation, *Gmmlo19/Gmmlo02/Gmmlo23* triple and *Gmmlo02/Gmmlo19/Gmmlo20/Gmmlo23* quadruple knockout mutants were isolated. When challenged with *Erysiphe difusa*, all mutants showed improved resistance, especially the quadruple mutant (Bui *et al.*, 2023).

Soybean root rot is a major disease, caused by *Phytophthora sojae*. It causes significant soybean crop losses that are difficult to treat with pesticides. *P. sojae* secretes a vast number of effectors that target host factors and encourage pathogenicity. The genetic engineering of such host targets is a viable way to increase soybean resistance. Although CRISPR/Cas9-mediated gene editing is utilized in crop disease-resistance breeding, there is a lack of reports on modifying soybean S genes to promote resistance to soybean root rot disease. Studies discovered that the major *P. sojae* effector *PsAvh52* lowers soybean immunity by targeting *GmTAP1*, increasing soybean sensitivity to *P. sojae*. CRISPR/Cas9 gene-editing technique knocks out GmTAP1 in soybean. *GmTAP1* deficiency increased defiance to three *P. sojae* strains (P233, P231, and P234). Their analysis of ROS generation, PTI-responsive gene expression, and MAPK activity revealed that *GmTAP1* loss of function had no impact on plant basal immunity. In the field, tap1 mutants did not show significant differences in pod number, plant height, hundred grain yield, or weight. In summary, this study has developed novel soybean lines that were resistant to various *P. sojae* strains while incurring no agronomic penalty in the field (Cai *et al.*, 2021).

Soybean mosaic virus (SMV) is a major pathogen that severely affects soybean production (Liu *et al.*, 2016). To address this issue, the CRISPR/Cas9 gene-editing system was employed to target three isoflavonoid-synthesizing genes, i.e. *GmF3H1*, *GmF3H2*, and *GmFNSII-1* in soybean root hairs and plants. This targeted approach led to an increase in isoflavone content in the leaves and seeds of the genetically modified soybean plants. The elevated levels of isoflavones in these plants conferred resistance to the SMV strain SC7. Furthermore, the coat protein content of SMV was significantly reduced in the triple mutants following infection with strain SC7 (Zhang *et al.*, 2022c). This study highlights the effectiveness of gene-editing technology in enhancing crop resistance to pathogens. For additional context, Table 17.1 provides a summary of various case studies involving prominent crops that have been modified through gene-knockout technology, showcasing the diverse applications and outcomes of these genetic interventions.

17.5 Future Prospects and Conclusion

Genome-editing technologies have steadily and consistently emerged as a tool for enhanced crop improvement, yield, and protection. However, the convoluted balance between modifying the defense system and response and preserving crop yield is challenging. These omics approaches have gained the attention of researchers because of their high specificity, efficiency, and cost and time effectiveness. CRISPR-based gene-editing systems have created crop varieties with enhanced agronomic traits and nutritional value. With significant potential for crop improvement and widespread agricultural use, the technology is expected to contribute to achieving sustainable development goals by 2030. Hence, such research strategies will help manage pest infestations, bolstering crop productivity and resilience to stress, and have underpinned a transformative shift in agricultural practices to achieve unadulterated and pesticide-free food.

Table 17.1. Cumulative list of CRISPR-Cas-modified pest-resistant crops.

Plant disease	Causative agent	Genes of interest	Mutation	Nuclease used	Resultant phenotype	Reference
			Oryza sativa (Common name: rice)			
Bacterial pathogen	*Magnoporthe oryzae* and *Xanthomonas oryzae*	*rbl1*	29-base-pair deletion	CRISPR/Cas9	Showed enhanced resistance to both *M. oryzae* and *Xanthomonas oryzae* pv. *oryzae*	Sha *et al.*, 2023
Bacterial blight	*X. oryzae* pv. *oryzae*	*Xa13*	149 nucleotides specifically deleted from Xa13 gene promoter region	CRISPR/Cas9	Enhanced disease resistance	Li *et al.*, 2022
Bacterial blight	*X. oryzae* pv. *oryza*	*Bsr-d1, Pi21* and *ERF922*	Frameshift mutations	CRISPR/Cas9	*ERF922* mutants exhibited the highest level of blast resistance	Zhou *et al.*, 2022
Bacterial leaf streak	*X. oryzae*	*OsSULTR3;6*	Frameshift mutations	CRISPR/Cas9	Enhanced disease resistance	Xu *et al.*, 2021
Bacterial blight	*X. oryzae* pv. *oryzae*	*SWEET14*/promoter	Deletions and Insertions	CRISPR/Cas9	Increased disease resistance	Zafar *et al.*, 2020
Bacterial blight	*X. oryzae* pv. *oryzae*	*Os8N3*	Frameshift mutations	CRISPR/Cas9	Enhanced disease resistance	Kim *et al.*, 2019
Bacterial blight	*X. oryzae* pv. *oryzae*	*SWEET11*, *SWEET13* and *SWEET14*/promoter	Frameshift mutations	CRISPR/Cas9	Enhanced disease resistance	Oliva *et al.*, 2019
Bacterial blight	*X. oryzae* pv. *oryzae*	*OsSWEET11* and *OsSWEET14*/promoter region	Frameshift mutations	CRISPR/Cas9	Enhanced disease resistance	Xu *et al.*, 2019
Rice blast	*M. oryzae*	*ms5/pi21/xa13* gene	Frameshift mutations	CRISPR/Cas9	Increased resistance against rice blast	Li *et al.*, 2019b

Continued

Table 17.1. Continued

Plant disease	Causative agent	Genes of interest	Mutation	Nuclease used	Resultant phenotype	Reference
Bacterial blight	*X. oryzae* pv. *oryzae*	*bsr-k1*	Frameshift mutations	CRISPR/Cas9	*bsr-k1* mutant confers broad-spectrum resistance to Xoo	Zhou *et al.*, 2018
Rice tungro disease	Rice tungro spherical virus	Initiation factor four gamma gene	Frameshift mutations	CRISPR/Cas9	Rice tungro spherical virus resistance	Macovei *et al.*, 2018
Rice blast	*M. oryzae*	*OsSEC3A*	Frameshift mutations Disruption of OsSEC3A	CRISPR/Cas9	Enhanced resistance to the fungal pathogen *Magnaporthe oryzae*	Ma *et al.*, 2018
Solanum tuberosum (Common name: potato)						
Viral infection	Potato virus Y	*eIF4E1*	Knockout	Cas9	Provided resistance to Potato virus Y	Lucioli *et al.*, 2022
Viral infection	Potato virus Y	*eIF4E*		Cas 9	Enhanced resistance to PVY^{O} strain.	Noureen *et al.*, 2022
Late blight	*Phytophthora infestans*	*CCOAOMT*	Knock-in	Cas9	Enhanced resistance against late blight	Hegde *et al.*, 2021
Late blight	*P. infestans*	*StDND1*, *StCHL1*, and *DMG400000582* (*StDMR6-1*)	Knockout	CRISPR/Cas9	Resistance to *Phytophthora infestans*	Kieu *et al.*, 2021
Viral infection	Potato virus Y	*P3*, *NIB*, *CI*, *CP*		Cas13a	Enhanced resistance to PVY	Zhan *et al.*, 2019
Viral infection	Potato virus Y	Coilin	Knockout	Cas9	Improved resistance	Makhotenko *et al.*, 2019
Gossypium hirsutum (Common name: cotton)						

Cotton *Verticillium wilt*	*Verticillium dahlia Gh14-3-3d*	*Gh14-3-3d*	Insertions and deletions (indels) in Gh14-3-3d	CRISPR/Cas9	Led to bronzing of veins, interveinal chlorosis, the yellowing of leaves, leaves become dry, giving scorched appearance	Man *et al*., 2022
Herbivory	Sap-sucking insects, particularly whitefly infestation	*IncD09* and *IncA07*	In IncD09 plants: indels and substitution (deletion mainly). IncA07: similar pattern of deletion as in IncD09	CRISPR/Cas9	Gene knockout of *IncA07* and *IncD09* genes led to decreased resilience against the herbivorous sucking insect	Zhang *et al*., 2022a
Cotton leaf curl disease	Cotton leaf curl viruses (CLCuVs)	Three regions in the viral genome	Viral interference while using sgRNA	Cas9-gRNA	Helped to mitigate vein-thickening, the curling of leaves downwards or upwards, hindered development and the growth of leaf-like enations	Binyameen *et al*., 2021
Brassica juncea (Common name: mustard)						
Sclerotinia stem rot	*Sclerotinia sclerotiorum*	*BnaA06. MPK3/BnaC03.MPK3* and *BnaA03.MKK5*		CRISPR/ Cas9	Improved the plants' defenses against pathogen invasion	Zhang *et al*., 2022b
Sclerotinia stem rot	*S. sclerotiorum*	*BjuXLG1* and *BjuXLG2*		CRISPR/Cas9	Decreased accumulation of glucosinolate, enhanced host defense mechanism and resistance to star rot disease	Tiwari *et al*., 2021

Continued

Table 17.1. Continued

Plant disease	Causative agent	Genes of interest	Mutation	Nuclease used	Resultant phenotype	Reference
Sclerotinia stem rot	*S. sclerotiorum*	*BnWRKY11*, and *BnWRKY70*		CRISPR/Cas9	*BnWRKY70* mutants: increased resistance to fungal infection. *BnWRK11* mutants: showed no significant difference in fungal infection resistance	Sun *et al.*, 2018
			Triticum aestivum (Common name: wheat)			
Stripe rust	*Puccinia striiformis* f. sp. *tritici* (Pst) is a biotrophic fungal pathogen	33 Susceptible genes	CRISPR/Cas knockdowns	Cas9 and Cas12a	Improved resistance	Taj *et al.*, 2022
Yellow mosaic disease	Wheat yellow mosaic virus (WYMV)	*TaPDIL5-1*	Knockout	CRISPR/Cas9-based genome editing	Improved resistance against virus	Kan *et al.*, 2022
Powdery mildew-resistant	Fungal pathogen *Blumeria graminis* f. sp. tritici	*TaMLO 1*gene	Knockout	CRISPR/Cas9	Improved resistance against virus	Li *et al.*, 2021
Fusarium headblight	*Fusarium graminearum*	*TaHRC*	Knockout	Cas-9	Improved resistance against virus	Su *et al.*, 2019
Aphid infestation	*Sitobian avenae*	*Pinellia pedatisecta* agglutinin (*PPA*) gene	Transgene particle bombardment		Resistance to aphids	Duan *et al.*, 2018
Powdery mildew resistance	*Agrobacterium*	*TaEDR1* gene	Knockout	Cas-9	Provided resistance	Zhang *et al.*, 2017b
			Solanum lycopersicum (Common name: tomato)			
Powdery mildew	Tomato yellow leaf curl virus	*Pelo*, *Mlo1*	Knockout	Cas9	Provided high resistance	Pramanik *et al.*, 2021
Powdery mildew	Tomato yellow leaf curl virus	*JAZ2*	Knockout	Cas9	Improved resistance	Pramanik *et al.*, 2021

Light blight	*P. infestans*	*MYBS2*	Knockout	Cas9	High resistance	Liu *et al*., 2021
Gray mold	*Botrytis cinerea*	*MYC2, MAPK3*	Knockout	Cas9	Decreased disease resistance	Zhang *et al*., 2018c
Powdery mildew	*Oidium* sp.	*Mlo1*	Knockout	Cas9	Resistance	Nekrasov *et al*., 2017
PVY, CMV		*eIF4E1* (host gene)	Knockout	Cas9	PVY and CMV resistant mutants	Nekrasov *et al*., 2017
Downy mildew		*DMR6-1*	Knockout	Cas9	High resistance	Thomazella *et al*., 2016
Downy mildew; Powdery mildew		*SlDMR6-1*	Knockout	CRISPR/SpCas9	Disease resistance to different pathogens, including *Xanthomonas* spp., *P. syringae* and *P. capsici*	Thomazella *et al*., 2016
			Glycine max (Common name: soybean)			
Phytophthora root rot	*P. soya*	*GmARM*	Knockout	Cas9	Resistance to *Phytophthora soya*	Luo *et al*., 2024
Soybean root rot	*P. sojae*	*GmTAP1*	Knockout	Cas9	Enhanced resistance to *P. sojae* strains	Liu *et al*., 2023
Powdery mildew	*Erysiphe difusa* (fungus)	*GmMLO*	Knockout	Cas9	Enhanced resistance to the pathogen	Bui *et al*., 2023
	Soybean mosaic virus (SMV)	*GmF3H2 GmF3H1 GmFNSII-1*	Knockout	Cas9	Resistance to soybean mosaic virus strain SC7	Zhang *et al*., 2022c

References

Balcke, G.U., Bennewitz, S., Bergau, N., Athmer, B., Henning, A. *et al.* (2017) Multi-omics of tomato glandular trichomes reveals distinct features of central carbon metabolism supporting high productivity of specialized metabolites. *The Plant Cell* 29(5), 960–983. DOI: 10.1105/tpc.17.00060.

Bandara, A.Y., Weerasooriya, D.K., Bradley, C.A., Allen, T.W. and Esker, P.D. (2020) Dissecting the economic impact of soybean diseases in the United States over two decades. *PLoS One* 15(4), e0231141. DOI: 10.1371/journal.pone.0231141.

Belhaj, K., Chaparro-Garcia, A., Kamoun, S., Patron, N.J. and Nekrasov, V. (2015) Editing plant genomes with CRISPR/Cas9. *Current Opinion in Biotechnology* 32, 76–84. DOI: 10.1016/j.copbio.2014.11.007.

Binyameen, B., Khan, Z., Khan, S.H., Ahmad, A., Munawar, N. *et al.* (2021) Using multiplexed CRISPR/Cas9 for suppression of cotton leaf curl virus. *International Journal of Molecular Sciences* 22(22), 12543. DOI: 10.3390/ijms222212543.

Bui, T.P., Le, H., Ta, D.T., Nguyen, C.X., Le, N.T. *et al.* (2023) Enhancing powdery mildew resistance in soybean by targeted mutation of MLO genes using the CRISPR/Cas9 system. *BMC Plant Biology* 23(1), 533. DOI: 10.1186/s12870-023-04549-5.

Cai, M., Zhang, C., Wang, W., Peng, Q., Song, X. *et al.* (2021) Stepwise accumulation of mutations in CesA3 in phytophthora sojae results in increasing resistance to CAA fungicides. *Evolutionary Applications* 14(4), 996–1008. DOI: 10.1111/eva.13176.

Ceasar, S.A., Rajan, V., Prykhozhij, S.V., Berman, J.N. and Ignacimuthu, S. (2016) Insert, remove or replace: A highly advanced genome editing system using CRISPR/Cas9. *Biochimica et Biophysica Acta (BBA)-Molecular Cell Research* 1863(9), 2333–2344. DOI: 10.1016/j.bbamcr.2016.06.009.

Çelik, Ö., Ayan, A., Meriç, S. and Atak, Ç. (2021) Comparison of tolerance related proteomic profiles of two drought tolerant tomato mutants improved by gamma radiation. *Journal of Biotechnology* 330, 35–44. DOI: 10.1016/j.jbiotec.2021.02.012.

Cernadas, R.A., Doyle, E.L., Niño-Liu, D.O., Wilkins, K.E., Bancroft, T. *et al.* (2014) Code-assisted discovery of TAL effector targets in bacterial leaf streak of rice reveals contrast with bacterial blight and a novel susceptibility gene. *PLoS Pathogens* 10(2), e1003972. DOI: 10.1371/journal.ppat.1003972.

CGIAR (Consultative Group for International Agricultural Research) (2019) *Wheat in the World*. CGIAR Research Program on Wheat.

Chandra, S., Maranna, S., Saini, M., Kumawat, G., Nataraj, V. *et al.* (2022) Achievements, challenges and prospects of hybrid soybean. In: Bohra, A., Parihar, A.K., Naik, S. and Chandra, A. (eds) *Plant Male Sterility Systems for Accelerating Crop Improvement*. Springer, pp. 167–193. DOI: 10.1007/978-981-19-3808-5_8#DOI.

Chen, L., Wu, Q., He, W., He, T., Wu, Q. *et al.* (2019) Combined de novo transcriptome and metabolome analysis of common bean response to *Fusarium oxysporum* f. sp. *Phaseoli Infection. International Journal of Molecular Sciences* 20(24), 6278. DOI: 10.3390/ijms20246278.

Cuperlovic-Culf, M., Wang, L., Forseille, L., Boyle, K., Merkley, N. *et al.* (2016) Metabolic biomarker panels of response to Fusarium head blight infection in different wheat varieties. *PLoS One* 11(4), e0153642. DOI: 10.1371/journal.pone.0153642.

Dhokane, D., Karre, S., Kushalappa, A.C. and McCartney, C. (2016) Integrated metabolo-transcriptomics reveals Fusarium head blight candidate resistance genes in wheat QTL-Fhb2. *PLoS One* 11(5), e0155851. DOI: 10.1371/journal.pone.0155851.

Digilio, M.C., Sasso, R., Di Leo, M.G., Iodice, L., Monti, M.M. *et al.* (2012) Interactions between Bt-expressing tomato and non-target insects: The aphid *Macrosiphum euphorbiae* and its natural enemies. *Journal of Plant Interactions* 7(1), 71–77. DOI: 10.1080/17429145.2011.632695.

Duan, X., Hou, Q., Liu, G., Pang, X., Niu, Z. *et al.* (2018) Expression of *Pinellia pedatisecta* lectin gene in transgenic wheat enhances resistance to wheat aphids. *Molecules* 23(4), 748. DOI: 10.3390/molecules23040748.

FAO (2017) *The Future of Food and Agriculture: Trends and Challenges*. Food and Agriculture Organization of the United Nations, Rome.

FAO (2019) *New Standards to Curb the Global Spread of Plant Pests and Diseases*. Food and Agriculture Organization of the United Nations, Rome.

Gupta, D., Bhattacharjee, O., Mandal, D., Sen, M.K., Dey, D. *et al.* (2019) CRISPR-Cas9 system: A new-fangled dawn in gene editing. *Life Sciences* 232, 116636. DOI: 10.1016/j.lfs.2019.116636.

Gupta, S., Kumar, A., Patel, R. and Kumar, V. (2021) Genetically modified crop regulations: Scope and opportunity using the CRISPR-Cas9 genome editing approach. *Molecular Biology Reports* 48(5), 4851–4863. DOI: 10.1007/s11033-021-06477-9.

Hao, C., Jiao, C., Hou, J., Li, T., Liu, H. *et al.* (2020) Resequencing of 145 landmark cultivars reveals asymmetric sub-genome selection and strong founder genotype effects on wheat breeding in China. *Molecular Plant* 13(12), 1733–1751. DOI: 10.1016/j.molp.2020.09.001.

Hasanuzzaman, M., Parvin, K., Bardhan, K., Nahar, K., Anee, T.I. *et al.* (2021) Biostimulants for the regulation of reactive oxygen species metabolism in plants under abiotic stress. *Cells* 10(10), 2537. DOI: 10.3390/cells10102537.

Hegde, N., Joshi, S., Soni, N. and Kushalappa, A.C. (2021) The caffeoyl-CoA O-methyltransferase gene SNP replacement in russet burbank potato variety enhances late blight resistance through cell wall reinforcement. *Plant Cell Reports* 40(1), 237–254. DOI: 10.1007/s00299-020-02629-6.

Hickey, L.T., Hafeez, A.N., Robinson, H., Jackson, S.A., Leal-Bertioli, S.C.M. *et al.* (2019) Breeding crops to feed 10 billion. *Nature Biotechnology* 37(7), 744–754. DOI: 10.1038/s41587-019-0152-9.

Hilton, I.B., D'Ippolito, A.M., Vockley, C.M., Thakore, P.I., Crawford, G.E. *et al.* (2015) Epigenome editing by a CRISPR-Cas9-based acetyltransferase activates genes from promoters and enhancers. *Nature Biotechnology* 33(5), 510–517. DOI: 10.1038/nbt.3199.

Huang, S., Li, R., Zhang, Z., Li, L., Gu, X. *et al.* (2009) The genome of the cucumber, *Cucumis sativus* L. *Nature Genetics* 41(12), 1275–1281. DOI: 10.1038/ng.475.

Jandura, A. and Krause, H.M. (2017) The new RNA world: Growing evidence for long noncoding RNA functionality. *Trends in Genetics* 33(10), 665–676. DOI: 10.1016/j.tig.2017.08.002.

Ji, Z.Y., Zakria, M., Zou, L.F., Xiong, L., Li, Z. *et al.* (2014) Genetic diversity of transcriptional activator-like effector genes in Chinese isolates of *Xanthomonas oryzae* pv. *oryzicola*. *Phytopathology* 104(7), 672–682. DOI: 10.1094/PHYTO-08-13-0232-R.

Jiang, F. and Doudna, J.A. (2017) CRISPR-Cas9 structures and mechanisms. *Annual Review of Biophysics* 46(1), 505–529. DOI: 10.1146/annurev-biophys-062215-010822.

Jones, H., Whipps, J.M. and Gurr, S.J. (2001) The tomato powdery mildew fungus *Oidium neolycopersici*. *Molecular Plant Pathology* 2(6), 303–309. DOI: 10.1046/j.1464-6722.2001.00084.x.

Kan, J., Cai, Y., Cheng, C., Jiang, C., Jin, Y. *et al.* (2022) Simultaneous editing of host factor geneTaPDIL5-1 homoeoalleles confers wheat yellow mosaic virus resistance in hexaploid wheat. *The New Phytologist* 234(2), 340–344. DOI: 10.1111/nph.18002.

Kan, J., Cai, Y., Cheng, C., Chen, S., Jiang, C. *et al.* (2023) CRISPR/Cas9-guided knockout of eIF4E improves wheat yellow mosaic virus resistance without yield penalty. *Plant Biotechnology Journal* 21(5), 893. DOI: 10.1111/pbi.14002.

Kaul, S., Koo, H.L., Jenkins, J., Rizzo, M., Rooney, T. *et al.* (2000) Analysis of the genome sequence of the flowering plant *Arabidopsis thaliana*. *Nature* 408(6814), 796–815. DOI: 10.1038/35048692.

Kearns, N.A., Pham, H., Tabak, B., Genga, R.M., Silverstein, N.J. *et al.* (2015) Functional annotation of native enhancers with a Cas9-histone demethylase fusion. *Nature Methods* 12(5), 401–403. DOI: 10.1038/nmeth.3325.

Kefale, H. and Getahun, S. (2022) Gene editing improves the agronomic important traits of wheat–CRISPR-Cas9 and Cas12/Cpf1. In: Ansari, M.-u.R. (ed.) *Wheat: Recent Advances*. Intech Open. DOI: 10.5772/intechopen.103867.

Khan, Z., Ali, Z., Khan, A.A., Sattar, T., Zeshan, A. *et al.* (2022) History and classification of CRISPR/Cas system. In: Ahmad, A., Khan, S.H. and Khan, Z. (eds) *The CRISPR/Cas Tool Kit for Genome Editing*. Springer, pp. 29–52. DOI: 10.1007/978-981-16-6305-5_2.

Kieu, N.P., Lenman, M., Wang, E.S., Petersen, B.L. and Andreasson, E. (2021) Mutations introduced in susceptibility genes through CRISPR/Cas9 genome editing confer increased late blight resistance in potatoes. *Scientific Reports* 11(1), 4487. DOI: 10.1038/s41598-021-83972-w.

Kim, Y.A., Moon, H. and Park, C.J. (2019) CRISPR/Cas9-targeted mutagenesis of Os8N3 in rice to confer resistance to *Xanthomonas oryzae* pv. *oryzae*. *Rice* 12(1), 67. DOI: 10.1186/s12284-019-0325-7.

Kim, S.B., Van den Broeck, L., Karre, S., Choi, H., Christensen, S.A. *et al.* (2021) Analysis of the transcriptomic, metabolomic, and gene regulatory responses to *Puccinia sorghi* in maize. *Molecular Plant Pathology* 22(4), 465–479. DOI: 10.1111/mpp.13040.

Li, J., Manghwar, H., Sun, L., Wang, P., Wang, G. *et al.* (2019a) Whole genome sequencing reveals rare off-target mutations and considerable inherent genetic or/and somaclonal variations in CRISPR/Cas9-edited cotton plants. *Plant Biotechnology Journal* 17(5), 858–868. DOI: 10.1111/pbi.13020.

Li, S., Shen, L., Hu, P., Liu, Q., Zhu, X. *et al.* (2019b) Developing disease-resistant thermosensitive male sterile rice by multiplex gene editing. *Journal of Integrative Plant Biology* 61(12), 1201–1205. DOI: 10.1111/jipb.12774.

Li, C., Zhang, R., Meng, X., Chen, S., Zong, Y. *et al.* (2020) Targeted, random mutagenesis of plant genes with dual cytosine and adenine base editors. *Nature Biotechnology* 38(7), 875–882. DOI: 10.1038/s41587-019-0393-7.

Li, S., Zhang, C., Li, J., Yan, L., Wang, N. *et al.* (2021) Present and future prospects for wheat improvement through genome editing and advanced technologies. *Plant Communications* 2(4), 100211. DOI: 10.1016/j.xplc.2021.100211.

Li, C., Zhou, L., Wu, B., Li, S., Zha, W. *et al.* (2022) Improvement of bacterial blight resistance in two conventionally cultivated rice varieties by editing the noncoding region. *Cells* 11(16), 2535. DOI: 10.3390/cells11162535.

Liu, J.Z., Fang, Y. and Pang, H. (2016) The current status of the soybean-soybean mosaic virus (SMV) pathosystem. *Frontiers in Microbiology* 7, 1906.

Liu, M., Rehman, S., Tang, X., Gu, K., Fan, Q. *et al.* (2019) Methodologies for improving HDR efficiency. *Frontiers in Genetics* 9, 691. DOI: 10.3389/fgene.2018.00691.

Liu, Y., Wang, H., Jiang, Z., Wang, W., Xu, R. *et al.* (2021) Genomic basis of geographical adaptation to soil nitrogen in rice. *Nature* 590(7847), 600–605. DOI: 10.1038/s41586-020-03091-w.

Liu, T., Ji, J., Cheng, Y., Zhang, S., Wang, Z. *et al.* (2023) CRISPR/Cas9-mediated editing of GmTAP1 confers enhanced resistance to *Phytophthora sojae* in soybean. *Journal of Integrative Plant Biology* 65(7), 1609–1612. DOI: 10.1111/jipb.13476.

Lowe, R., Shirley, N., Bleackley, M., Dolan, S. and Shafee, T. (2017) Transcriptomics technologies. *PLoS Computational Biology* 13(5), e1005457. DOI: 10.1371/journal.pcbi.1005457.

Luan, H., Shen, H., Pan, Y., Guo, B., Lv, C. *et al.* (2018) Elucidating the hypoxic stress response in barley (*Hordeum vulgare* L.) during waterlogging: A proteomics approach. *Scientific Reports* 8(1), 9655. DOI: 10.1038/s41598-018-27726-1.

Lucioli, A., Tavazza, R., Baima, S., Fatyol, K., Burgyan, J. *et al.* (2022) CRISPR-Cas9 targeting of the eIF4E1 gene extends the potato virus Y resistance spectrum of the *Solanum tuberosum* L. cv. Desirée. *Frontiers in Microbiology* 13, 873930. DOI: 10.3389/fmicb.2022.873930.

Luo, T., Ma, C., Fan, Y., Qiu, Z., Li, M. *et al.* (2024) CRISPR-Cas9-mediated editing of GmARM improves resistance to multiple stresses in soybean. *Plant Science* 346, 112147. DOI: 10.1016/j.plantsci.2024.112147.

Ma, J., Chen, J., Wang, M., Ren, Y., Wang, S. *et al.* (2018) Disruption of OsSEC3A increases the content of salicylic acid and induces plant defense responses in rice. *Journal of Experimental Botany* 69(5), 1051–1064. DOI: 10.1093/jxb/erx458.

Ma, L., Bajic, V.B. and Zhang, Z. (2013) On the classification of long non-coding RNAs. *RNA Biology* 10(6), 925–933. DOI: 10.4161/rna.24604.

Macovei, A., Sevilla, N.R., Cantos, C., Jonson, G.B., Slamet-Loedin, I. *et al.* (2018) Novel alleles of rice eIF4G generated by CRISPR/Cas9-targeted mutagenesis confer resistance to Rice tungro spherical virus. *Plant Biotechnology Journal* 16(11), 1918–1927. DOI: 10.1111/pbi.12927.

Makarova, K.S., Wolf, Y.I. and Koonin, E.V. (2022) Evolutionary classification of CRISPR-Cas systems. *CRISPR: Biology and Applications*. DOI: 10.1002/9781683673798.

Makhotenko, A.V., Khromov, A.V., Snigir, E.A., Makarova, S.S., Makarov, V.V. *et al.* (2019) Functional analysis of coilin in virus resistance and stress tolerance of potato *Solanum tuberosum* using CRISPR-Cas9 editing. *Doklady Biochemistry and Biophysics* 484(1), 88–91. DOI: 10.1134/S1607672919010241.

Man, M., Zhu, Y., Liu, L., Luo, L., Han, X. *et al.* (2022) Defense mechanisms of cotton *Fusarium* and *Verticillium* wilt and comparison of pathogenic response in cotton and humans. *International Journal of Molecular Sciences* 23(20), 12217. DOI: 10.3390/ijms232012217.

Manghwar, H., Li, B., Ding, X., Hussain, A., Lindsey, K. *et al.* (2020) CRISPR/Cas systems in genome editing: Methodologies and tools for sgRNA design, off-target evaluation, and strategies to mitigate off-target effects. *Advanced Science* 7(6), 1902312. DOI: 10.1002/advs.201902312.

Manzoor, S., Nabi, S.U., Rather, T.R., Gani, G., Mir, Z.A. *et al.* (2024) Advancing crop disease resistance through genome editing: A promising approach for enhancing agricultural production. *Frontiers in Genome Editing* 6, 1399051. DOI: 10.3389/fgeed.2024.1399051.

Meeske, A.J., Nakandakari-Higa, S. and Marraffini, L.A. (2019) Cas13-induced cellular dormancy prevents the rise of CRISPR-resistant bacteriophage. *Nature* 570(7760), 241–245. DOI: 10.1038/s41586-019-1257-5.

Mei, Y., Wang, Y., Chen, H., Sun, Z.S. and Ju, X.D. (2016) Recent progress in CRISPR/Cas9 technology. *Journal of Genetics and Genomics* 43(2), 63–75. DOI: 10.1016/j.jgg.2016.01.001.

Mollika, S.R., Sarker, R.H. and Hoque, M.I. (2011) *In vitro* plant regeneration in *Brassica* spp. *Plant Tissue Culture and Biotechnology* 21(2), 127–134. DOI: 10.3329/ptcb.v21i2.10235.

Moriones, E. and Navas-Castillo, J. (2000) Tomato yellow leaf curl virus, an emerging virus complex causing epidemics worldwide. *Virus Research* 71(1–2), 123–134. DOI: 10.1016/s0168-1702(00)00193-3.

Nekrasov, V., Wang, C., Win, J., Lanz, C., Weigel, D. *et al.* (2017) Rapid generation of a transgene-free powdery mildew resistant tomato by genome deletion. *Scientific Reports* 7(1), 482. DOI: 10.1038/s41598-017-00578-x.

Noureen, A., Khan, M.Z., Amin, I., Zainab, T. and Mansoor, S. (2022) CRISPR/Cas9-mediated targeting of susceptibility factor eIF4E-enhanced resistance against potato virus Y. *Frontiers in Genetics* 13, 1035804. DOI: 10.3389/fgene.2022.1035804.

Oliva, R., Ji, C., Atienza-Grande, G., Huguet-Tapia, J.C., Perez-Quintero, A. *et al.* (2019) Broad-spectrum resistance to bacterial blight in rice using genome editing. *Nature Biotechnology* 37(11), 1344–1350. DOI: 10.1038/s41587-019-0267-z.

Ozsolak, F. and Milos, P.M. (2011) RNA sequencing: Advances, challenges and opportunities. *Nature Reviews. Genetics* 12(2), 87–98. DOI: 10.1038/nrg2934.

Pandey, V., Singh, M., Pandey, D. and Kumar, A. (2018) Integrated proteomics, genomics, metabolomics approaches reveal oxalic acid as pathogenicity factor in *Tilletia indica* inciting karnal bunt disease of wheat. *Scientific Reports* 8(1), 7826. DOI: 10.1038/s41598-018-26257-z.

Pandit, M.A., Kumar, J., Gulati, S., Bhandari, N., Mehta, P. *et al.* (2022) Major biological control strategies for plant pathogens. *Pathogens* 11(2), 273. DOI: 10.3390/pathogens11020273.

Peng, J., Richards, D.E., Hartley, N.M., Murphy, G.P., Devos, K.M. *et al.* (1999) 'Green revolution' genes encode mutant gibberellin response modulators. *Nature* 400(6741), 256–261. DOI: 10.1038/22307.

Petersen, G., Seberg, O., Yde, M. and Berthelsen, K. (2006) Phylogenetic relationships of *Triticum* and *Aegilops* and evidence for the origin of the A, B, and D genomes of common wheat (*Triticum aestivum*). *Molecular Phylogenetics and Evolution* 39(1), 70–82. DOI: 10.1016/j.ympev.2006.01.023.

Pietersen, G. and Smith, M.F. (2002) Tomato yellow leaf curl virus resistant tomatoes show resistance to Tomato curly stunt virus. *Plant Disease* 86(5), 528–534. DOI: 10.1094/PDIS.2002.86.5.528.

Pompili, V., Dalla Costa, L., Piazza, S., Pindo, M. and Malnoy, M. (2020) Reduced fire blight susceptibility in apple cultivars using a high-efficiency CRISPR/Cas9-FLP/FRT-based gene editing system. *Plant Biotechnology Journal* 18(3), 845–858. DOI: 10.1111/pbi.13253.

Potato Genome Sequencing Consortium (2011) Genome sequence and analysis of the tuber crop potato. *Nature* 475, 189–195. DOI: 10.1038/nature10158.

Pramanik, D., Shelake, R.M., Park, J., Kim, M.J., Hwang, I. *et al.* (2021) CRISPR/Cas9-mediated generation of pathogen-resistant tomato against *Tomato yellow leaf curl virus* and powdery mildew. *International Journal of Molecular Sciences* 22(4), 1878. DOI: 10.3390/ijms22041878.

Quenouille, J., Vassilakos, N. and Moury, B. (2013) Potato virus Y: A major crop pathogen that has provided major insights into the evolution of viral pathogenicity. *Molecular Plant Pathology* 14(5), 439–452. DOI: 10.1111/mpp.12024.

Romanov, V., Davidoff, S.N., Miles, A.R., Grainger, D.W., Gale, B.K. *et al.* (2014) A critical comparison of protein microarray fabrication technologies. *The Analyst* 139(6), 1303–1326. DOI: 10.1039/c3an01577g.

Różewicz, M., Wyzińska, M. and Grabiński, J. (2021) The most important fungal diseases of cereals: Problems and possible solutions. *Agronomy* 11(4), 714. DOI: 10.3390/agronomy11040714.

Santillán Martínez, M.I., Bracuto, V., Koseoglou, E., Appiano, M., Jacobsen, E. *et al.* (2020) CRISPR/Cas9-targeted mutagenesis of the tomato susceptibility gene PMR4 for resistance against powdery mildew. *BMC Plant Biology* 20(1), 284. DOI: 10.1186/s12870-020-02497-y.

Savary, S., Ficke, A., Aubertot, J.N. and Hollier, C. (2012) Crop losses due to diseases and their implications for global food production losses and food security. *Food Security* 4(4), 519–537. DOI: 10.1007/s12571-012-0200-5.

Sha, G., Sun, P., Kong, X., Han, X., Sun, Q. *et al.* (2023) Genome editing of a rice CDP-DAG synthase confers multipathogen resistance. *Nature* 618(7967), 1017–1023. DOI: 10.1038/s41586-023-06205-2.

Shan, Q., Wang, Y., Li, J., Zhang, Y., Chen, K. *et al.* (2013) Targeted genome modification of crop plants using a CRISPR-Cas system. *Nature Biotechnology* 31(8), 686–688. DOI: 10.1038/nbt.2650.

Shao, S., Guo, T. and Aebersold, R. (2015) Mass spectrometry-based proteomic quest for diabetes biomarkers. *Biochimica et Biophysica Acta (BBA) – Proteins and Proteomics* 1854(6), 519–527. DOI: 10.1016/j.bbapap.2014.12.012.

Simón, M.R., Börner, A. and Struik, P.C. (2021) Fungal wheat diseases: Etiology, breeding, and integrated management. *Frontiers in Plant Science* 12, 671060. DOI: 10.3389/fpls.2021.671060.

Singh, R.K., Prasad, A., Maurya, J. and Prasad, M. (2022) Regulation of small RNA-mediated high temperature stress responses in crop plants. *Plant Cell Reports* 41(3), 765–773. DOI: 10.1007/s00299-021-02745-x.

Steuernagel, B., Periyannan, S.K., Hernández-Pinzón, I., Witek, K., Rouse, M.N. *et al.* (2016) Rapid cloning of disease-resistance genes in plants using mutagenesis and sequence capture. *Nature Biotechnology* 34(6), 652–655. DOI: 10.1038/nbt.3543.

Su, Z., Bernardo, A., Tian, B., Chen, H., Wang, S. *et al.* (2019) A deletion mutation in taHRC confers Fhb1 resistance to Fusarium head blight in wheat. *Nature Genetics* 51(7), 1099–1105. DOI: 10.1038/s41588-019-0425-8.

Su, P., Zhao, L., Li, W., Zhao, J., Yan, J. *et al.* (2021) Integrated metabolo-transcriptomics and functional characterization reveals that the wheat auxin receptor TIR1 negatively regulates defense against *Fusarium graminearum*. *Journal of Integrative Plant Biology* 63(2), 340–352. DOI: 10.1111/jipb.12992.

Suharti, W.S., Nose, A. and Zheng, S.H. (2016) Metabolomic study of two rice lines infected by *Rhizoctonia solani* in negative ion mode by CE/TOF-MS. *Journal of Plant Physiology* 206, 13–24. DOI: 10.1016/j.jplph.2016.09.004.

Sun, Q., Lin, L., Liu, D., Wu, D., Fang, Y. *et al.* (2018) CRISPR/Cas9-mediated multiplex genome editing of the BnWRKY11 and BnWRKY70 genes in *Brassica napus* L. *International Journal of Molecular Sciences* 19(9), 2716. DOI: 10.3390/ijms19092716.

Taj, M., Sajjad, M., Li, M., Yasmeen, A., Mubarik, M.S. *et al.* (2022) Potential targets for CRISPR/Cas knockdowns to enhance genetic resistance against some diseases in wheat (*Triticum aestivum* L.). *Frontiers in Genetics* 13, 926955. DOI: 10.3389/fgene.2022.926955.

Tashkandi, M., Ali, Z., Aljedaani, F., Shami, A. and Mahfouz, M.M. (2018) Engineering resistance against tomato yellow leaf curl virus via the CRISPR/Cas9 system in tomato. *Plant Signaling & Behavior* 13(10), e1525996. DOI: 10.1080/15592324.2018.1525996.

Thomazella, P. de T., Brail, Q., Dahlbeck, D. and Staskawicz, B. (2016) CRISPR-Cas9 mediated mutagenesis of a DMR6 ortholog in tomato confers broad-spectrum disease resistance. In: *Proceedings of the National Academy of Sciences*. DOI: 10.1073/pnas.2026152118.

Tian, D., Li, Z., O'Connor, D. and Shen, Z. (2020) The need to prioritize sustainable phosphate-based fertilizers. *Soil Use and Management* 36(3), 351–354. DOI: 10.1111/sum.12578.

Tiwari, R., Kaur, J. and Bisht, N.C. (2021) Extra-large G-proteins influence plant response to *Sclerotinia sclerotiorum* by regulating glucosinolate metabolism in *Brassica juncea*. *Molecular Plant Pathology* 22(10), 1180–1194. DOI: 10.1111/mpp.13096.

Ulitsky, I. and Bartel, D.P. (2013) LincRNAs: Genomics, evolution, and mechanisms. *Cell* 154(1), 26–46. DOI: 10.1016/j.cell.2013.06.020.

Varshney, R.K., Chen, W., Li, Y., Bharti, A.K., Saxena, R.K. *et al.* (2012) Draft genome sequence of pigeonpea (*Cajanus cajan*), an orphan legume crop of resource-poor farmers. *Nature Biotechnology* 30(1), 83–89. DOI: 10.1038/nbt.2022.

Vercoe, R.B., Chang, J.T., Dy, R.L., Taylor, C., Gristwood, T. *et al.* (2013) Cytotoxic chromosomal targeting by CRISPR/Cas systems can reshape bacterial genomes and expel or remodel pathogenicity islands. *PLoS Genetics* 9(4), e1003454. DOI: 10.1371/journal.pgen.1003454.

Wang, Y., Cheng, X., Shan, Q., Zhang, Y., Liu, J. *et al.* (2014) Simultaneous editing of three homoeoalleles in hexaploid bread wheat confers heritable resistance to powdery mildew. *Nature Biotechnology* 32(9), 947–951. DOI: 10.1038/nbt.2969.

Wang, P., Su, L., Gao, H., Jiang, X., Wu, X. *et al.* (2018) Genome-wide characterization of bHLH genes in grape and analysis of their potential relevance to abiotic stress tolerance and secondary metabolite biosynthesis. *Frontiers in Plant Science* 9, 64. DOI: 10.3389/fpls.2018.00064.

Weis, C., Hildebrandt, U., Hoffmann, T., Hemetsberger, C., Pfeilmeier, S. *et al.* (2014) CYP83A1 is required for metabolic compatibility of *Arabidopsis* with the adapted powdery mildew fungus *Erysiphe cruciferarum*. *The New Phytologist* 202(4), 1310–1319. DOI: 10.1111/nph.12759.

Wei, Y., Yang, Z., Zong, C., Wang, B., Ge, X. *et al.* (2021) Trans single-stranded DNA cleavage via CRISPR/Cas14a1 activated by target RNA without destruction. *Angewandte Chemie* 133(45), 24443–24449. DOI: 10.1002/ange.202110384.

Witek, K., Jupe, F., Witek, A.I., Baker, D., Clark, M.D. *et al.* (2016) Accelerated cloning of a potato late blight-resistance gene using RenSeq and SMRT sequencing. *Nature Biotechnology* 34(6), 656–660. DOI: 10.1038/nbt.3540.

Xu, Z., Xu, X., Gong, Q., Li, Z., Li, Y. *et al.* (2019) Engineering broad-spectrum bacterial blight resistance by simultaneously disrupting variable TALE-binding elements of multiple susceptibility genes in rice. *Molecular Plant* 12(11), 1434–1446. DOI: 10.1016/j.molp.2019.08.006.

Xu, X., Xu, Z., Li, Z., Zakria, M., Zou, L. *et al.* (2021) Increasing resistance to bacterial leaf streak in rice by editing the promoter of susceptibility gene OsSULRT3.6. *Plant Biotechnology Journal* 19(6), 1101–1103. DOI: 10.1111/pbi.13602.

Yan, D., Ren, B., Liu, L., Yan, F., Li, S. *et al.* (2021) High-efficiency and multiplex adenine base editing in plants using new TaDA variants. *Molecular Plant* 14(5), 722–731. DOI: 10.1016/j.molp.2021.02.007.

Yang, H., Ren, S., Yu, S., Pan, H., Li, T. *et al.* (2020) Methods favoring homology-directed repair choice in response to CRISPR/Cas9 induced-double strand breaks. *International Journal of Molecular Sciences* 21(18), 6461. DOI: 10.3390/ijms21186461.

Yin, X., Zou, B., Hong, X., Gao, M., Yang, W. *et al.* (2018) Rice copine genes Os BON 1 and Os BON 3 function as suppressors of broad-spectrum disease resistance. *Plant Biotechnology Journal* 16(8), 1476–1487. DOI: 10.1111/pbi.12890.

Yogendra, K.N. and Kushalappa, A.C. (2016) Integrated transcriptomics and metabolomics reveal induction of hierarchies of resistance genes in potato against late blight. *Functional Plant Biology* 43(8), 766. DOI: 10.1071/FP16028.

Yuan, M., Ngou, B.P.M., Ding, P. and Xin, X.F. (2021) PTI-ETI crosstalk: An integrative view of plant immunity. *Current Opinion in Plant Biology* 62, 102030. DOI: 10.1016/j.pbi.2021.102030.

Zafar, K., Khan, M.Z., Amin, I., Mukhtar, Z., Yasmin, S. *et al.* (2020) Precise CRISPR-Cas9 mediated genome editing in super basmati rice for resistance against bacterial blight by targeting the major susceptibility gene. *Frontiers in Plant Science* 11, 575. DOI: 10.3389/fpls.2020.00575.

Zhan, X., Zhang, F., Zhong, Z., Chen, R., Wang, Y. *et al.* (2019) Generation of virus-resistant potato plants by RNA genome targeting. *Plant Biotechnology Journal* 17(9), 1814–1822. DOI: 10.1111/pbi.13102.

Zhang, L., Li, X., Ma, B., Gao, Q., Du, H. *et al.* (2017a) The tartary buckwheat genome provides insights into rutin biosynthesis and abiotic stress tolerance. *Molecular Plant* 10(9), 1224–1237. DOI: 10.1016/j.molp.2017.08.013.

Zhang, Y., Bai, Y., Wu, G., Zou, S., Chen, Y. *et al.* (2017b) Simultaneous modification of three homoeologs of Ta EDR1 by genome editing enhances powdery mildew resistance in wheat. *The Plant Journal* 91(4), 714–724. DOI: 10.1111/tpj.13599.

Zhang, J., Li, Y., Guo, J., Du, B., He, G. *et al.* (2018a) Lipid profiles reveal different responses to brown planthopper infestation for pest susceptible and resistant rice plants. *Metabolomics* 14(9), 120. DOI: 10.1007/s11306-018-1422-0.

Zhang, X., Yao, C., Fu, S., Xuan, H., Wen, S. *et al.* (2018b) Stress2TF: A manually curated database of TF regulation in plant response to stress. *Gene* 638, 36–40. DOI: 10.1016/j.gene.2017.09.067.

Zhang, S., Wang, L., Zhao, R., Yu, W., Li, R. *et al.* (2018c) Knockout of SIMAPK3 reduced disease resistance to *Botrytis cinerea* in tomato plants. *Journal of Agricultural and Food Chemistry* 66(34), 8949–8956. DOI: 10.1021/acs.jafc.8b02191.

Zhang, X., Xu, Y. and Huang, B. (2019) Lipidomic reprogramming associated with drought stress priming-enhanced heat tolerance in tall fescue (*Festuca arundinacea*). *Plant, Cell & Environment* 42(3), 947–958. DOI: 10.1111/pce.13405.

Zhang, D., Zhang, Z., Unver, T. and Zhang, B. (2021a) CRISPR/Cas: A powerful tool for gene function study and crop improvement. *Journal of Advanced Research* 29, 207–221. DOI: 10.1016/j.jare.2020.10.003.

Zhang, L., Dong, C., Chen, Z., Gui, L., Chen, C. *et al.* (2021b) WheatGmap: A comprehensive platform for wheat gene mapping and genomic studies. *Molecular Plant* 14(2), 187–190. DOI: 10.1016/j.molp.2020.11.018.

Zhang, J., Li, J., Saeed, S., Batchelor, W.D., Alariqi, M. *et al.* (2022a) Identification and functional analysis of lncRNA by CRISPR/Cas9 during the cotton response to sap-sucking insect infestation. *Frontiers in Plant Science* 13, 784511. DOI: 10.3389/fpls.2022.784511.

Zhang, Y., Guo, W., Chen, L., Shen, X., Yang, H. *et al.* (2022b) CRISPR/Cas9-mediated targeted mutagenesis of gmugt enhanced soybean resistance against leaf-chewing insects through flavonoids biosynthesis. *Frontiers in Plant Science* 13, 802716. DOI: 10.3389/fpls.2022.802716.

Zhang, Y., Song, J., Wang, L., Yang, M., Hu, K. *et al.* (2022c) Identifying quantitative trait loci and candidate genes conferring resistance to soybean mosaic virus SC7 by quantitative trait loci-sequencing in soybean. *Frontiers in Plant Science* 13, 843633. DOI: 10.3389/fpls.2022.843633.

Zheng, Z., Appiano, M., Pavan, S., Bracuto, V., Ricciardi, L. *et al.* (2016) Genome-wide study of the tomato S1MLO gene family and its functional characterization in response to the powdery mildew fungus *Oidium neolycopersici*. *Frontiers in Plant Science* 7, 380. DOI: 10.3389/fpls.2016.00380.

Zhou, X., Liao, H., Chern, M., Yin, J., Chen, Y. *et al.* (2018) Loss of function of a rice TPR-domain RNA-binding protein confers broad-spectrum disease resistance. *Proceedings of the National Academy of Sciences of the United States of America* 115(12), 3174–3179. DOI: 10.1073/pnas.1705927115.

Zhou, Y., Xu, S., Jiang, N., Zhao, X., Bai, Z. *et al.* (2022) Engineering of rice varieties with enhanced resistances to both blast and bacterial blight diseases via CRISPR/Cas9. *Plant Biotechnology Journal* 20(5), 876–885. DOI: 10.1111/pbi.13766.

Zhu, H., Li, C. and Gao, C. (2020) Applications of CRISPR-Cas in agriculture and plant biotechnology. *Nature Reviews Molecular Cell Biology* 21(11), 661–677. DOI: 10.1038/s41580-020-00288-9.

Zhu, Y.N., Shi, D.Q., Ruan, M.B., Zhang, L.L., Meng, Z.H. *et al.* (2013) Transcriptome analysis reveals crosstalk of responsive genes to multiple abiotic stresses in cotton (*Gossypium hirsutum* L.). *PLoS One* 8(11), e80218. DOI: 10.1371/journal.pone.0080218.

18 Multiple Omics Evaluation on Stress-resilient CRISPR-edited Crops in Single-Cell and Spatial Resolution

Roohi Sharma[1], Sandeep Kour[1], Deepak Kumar[1], Parkirti[1], Vikram[1], Anchita Singh[1], Alisha Sharma[1], Nitika Kapoor[2] and Puja Ohri[1*]

[1]Department of Zoology, Guru Nanak Dev University, Amritsar, Punjab, India; [2]Department of Botanical and Environmental Sciences, Guru Nanak Dev University, Amritsar, Punjab, India

Abstract

The integration of CRISPR/Cas9 technology into plant genome engineering has ushered in a new era of agricultural innovation, particularly in enhancing stress resilience in crops. By enabling precise modifications at the genetic level, CRISPR/Cas9 empowers researchers to target specific genes associated with stress adaptation, thereby bolstering crop yield, quality, and resilience against environmental challenges. A significant advancement lies in the multidisciplinary approach that combines various omics techniques – such as genomics, transcriptomics, and proteomics – to comprehensively analyze stress-resilient CRISPR-edited crops at single-cell and spatial resolutions. This sophisticated methodology allows researchers to delve deep into the molecular and cellular changes occurring in crops under stress conditions, particularly those engineered using CRISPR gene editing. At the single-cell level, analyses uncover intricate cellular responses, shedding light on how individual cells react to stress and express edited traits. Moreover, spatial omics techniques offer insights into stress-related pathway dynamics within plant tissues, elucidating gene distribution patterns in CRISPR-edited crops. The present chapter provides a comprehensive exploration of stress-resilient CRISPR-edited crops, delving into the nuances of multiple omics analysis at both single-cell and spatial resolutions. By employing cutting-edge techniques, researchers can unravel the complexities of stress adaptation mechanisms in plants, paving the way for the development of resilient crops vital for global food security in light of evolving environmental circumstances.

Keywords: CRISPR/Cas9, Multiple omics, Plant stress, Single cell, Spatial resolution

18.1 Introduction

Scientists prioritize addressing abiotic stresses like drought, salinity, temperature extremes, and climate change, and biotic stress like weeds, insects, and pests in crop development that involves selecting high-yielding varieties with stability across different environments (Razzaq *et al.*, 2019). The most crucial step of developing resilient germplasm is crucial for sustainable agriculture (Langridge *et al.*, 2021). 'Omics' techniques encompass genomics, metabolomics, transcriptomics, and proteomics, providing a comprehensive understanding of biological

*Corresponding author: ohri_puja_11@gmail.com

© CAB International 2025. *Stress-Resilient Crops: Coordinated omics-CRISPR-Nanotechnology Strategies* (ed. Jen-Tsung Chen)
DOI: 10.1079/9781800627307.0018

molecules (Hasin *et al.*, 2017). Online databases play a vital role in plant genomics, aiding in the identification of stress-responsive genes globally. Functional genomics facilitates the discovery of stress-related genes in crops (Singh *et al.*, 2019). Access to whole genome sequences and genomic advancements promise efficient methods for pinpointing stress-responsive genes across genomes. Genome-wide association studies (GWAS) have pinpointed stress-responsive genes and beneficial alleles for complex traits. Using synteny and homology analysis, genetic databases and bioinformatics tools aid in the identification of stress-resistant gene families across plant species (Sharma and Pandey, 2017).

Currently, three methods are utilized for targeted genome editing: transcription-activator-like effector nucleases (TALEN), Clustered Regularly Interspaced Short Palindromic Repeats (CRISPR), and zinc finger nucleases (ZFN) (Khan *et al.*, 2019). Among these, CRISPR-Cas9 stands out for its affordability, simplicity, speed, and efficacy in gene knockout within cells (Shan *et al.*, 2013). It has been effectively used for genome engineering in bacteria, plants, and mammals, allowing for single-stranded RNA editing, live-cell labeling, high-throughput screening, gene knockdown, and endogenous gene-expression modulation.

CRISPR-Cas9 has revolutionized molecular biology by facilitating the creation of disease models and enhancing our understanding of disease mechanisms and management. Despite its remarkable advantages, challenges remain to be addressed (Chandrasegaran and Carroll, 2016). CRISPR-Cas9's superiority over TALEN and ZFN lies in its ease of target design, capability for mutation induction via guided RNA and Cas9 protein insertion, and the capacity to multiplex multiple genes simultaneously. Today, omics and CRISPR-Cas9 technologies are set to uncover genes that are resilience to stress, providing insights into molecular mechanisms and enabling genome engineering for crop-stress endurance (Mao *et al.*, 2013). Ongoing advancements in genome-modification technologies, coupled with the growing volume of molecular biological data, will propel the development of new strategies to enhance and develop new cultivars of significant crops and select commercially beneficial plant features (Razzaq *et al.*, 2021).

Understanding multicellular plants with CRISPR is challenging due to diverse organs, cell types, and physiological states. Recent progress in metabolomic technologies, combined with insights from functional genomics and bioinformatics, has empowered us with methodologies to extract information from individual cells or specific cell types to analyze metabolites within specific tissue subregions and to explore the spatiotemporal dynamics of metabolites during development (Sun *et al.*, 2019; Nakabayashi *et al.*, 2021). This capability enables researchers to investigate functional diversity and epigenetic variability within cell populations. While multi-omics single-cell studies have been gaining momentum recently, there is a scarcity of high-throughput, high-resolution methodologies capable of assessing molecular conditions with spatial precision. By profiling metabolites at the cellular level and employing quantitative analyses, new avenues are opening up for a deeper understanding of plant metabolic pathways and the engineering of plant metabolites. Recent studies have also integrated spatial transcriptomics with single-cell metabolomics techniques to address various biological inquiries (Rao *et al.*, 2021).

This chapter explores recent advances in the sampling of single-cell types in plant research, available analytical platforms, applications of single-cell metabolomics, associated challenges, and future prospects for leveraging single-cell metabolomics in enhancing crop plants. We emphasize the advantages of integrating these methodologies to study transcriptomic variations across various cell types during development and to identify key regulatory genes.

18.2 Multi-Omics Technology for Evaluating CRISPR-edited Crops to Combat Stress

18.2.1 Multi-omics technology

The availability of sequencing technology for the majority of species and the development of genomics techniques have made functional genomic analysis possible in many crops. Functional genomics has revealed a number of genes in plants that regulate responses to both biotic and abiotic stress (Thudi *et al.*, 2021). In agricultural plants, some of the relevant genes have been manipulated to increase tolerance to biotic and abiotic stresses. From wild crop-related genomics, several novel candidate genes for crop stress (abiotic and biotic) tolerance have been found. Fosmid DNA libraries, Hi-C internet-accessible sequencing data, and a high-density buckwheat whole-genome sequencing genomic map are a few examples. In addition to predicting approximately 33,500 genes, the authors identified whole-genome duplication and several candidate genes for resistance to chilling stress, drought, and heavy metals (Razzaq *et al.*, 2021). For instance, based on the *AGPv4* version of the maize *B73* genome, 214 genes encoding *ZmAP2/ERF* proteins with the whole (APETALA2/ethylene responsive factor) *AP2/ERF* domain were ultimately discovered in a study. Based on the quantity of *AP2/ERF* domains and amino acid (AA) sequence similarities between *AP2/ERF* proteins from rice, maize, and *Arabidopsis*, all 214 putative *ZmAP2/ERF* proteins were divided into three different families: the (related to *ABI3/VP1*) RAV family (4), the ERF family (166), and the AP2 family (44). Additionally, 27 chosen *ZmAP2/ERF* genes were further validated to respond to three distinct abiotic challenges by quantitative real-time polymerase chain-reaction analysis, indicating their potential roles in a variety of abiotic stress responses (Zhang *et al.*, 2022). In another experiment, a comprehensive molecular and functional characterization of the cotton plant under abiotic stress was conducted to elucidate the roles of the *LEA3* (late embryogenesis abundant) gene. Researchers identified 88 genes in cotton, categorizing them into six clades. Additionally, several ABA (abscisic acid)/stress-associated genes were downregulated, while a key gene, Gh_A08G0694, which was the most upregulated, was knocked out in tetraploid cotton. This greatly enhanced the vulnerability of cotton plants to salt and drought conditions. In a similar vein, overexpressing the crucial gene markedly improved the overexpressed plants' resistance to salt and drought stress. Main abiotic stress-tolerance genes strongly interact with the crucial gene homologous to *GhLEA3* protein (Shiraku *et al.*, 2022). Further, genomic sequence data on fruit crops has been available (Ahmad *et al.*, 2021), opening the door to molecular insights into plant characteristics and fruit quality based on transposon insertions, whole-genome or single-gene duplication, and gene function. Numerous fruit crops' red fruit color has been demonstrated to be caused by certain myeloblastosis viral oncogene homologs (MYBs) that regulate the levels of various anthocyanins, phlobaphenes, and betalains (Allan and Espley, 2018). Furthermore, flavor, texture, taste, scent, astringency, and piquancy are all influenced by MYBs in addition to carotenoid synthesis (Cna'ani *et al.*, 2015). The overexpression method can also be used to increase the levels of flavonoids. For instance, *MYB10* enhanced the anthocyanins in apples such as epicatechin, procyanidin B2, and quercetin glycosides. When mice were fed the apples that had been engineered with *MYB10*, the expression of inflammatory genes changed, and the mice showed less inflammation (Espley *et al.*, 2014). According to the findings, MYB transcription factors (TF) may be used to modify the nutritional quality of fruit in order to improve health. Although most genome-engineering research has focussed on biotic stresses, such as pathogens causing diseases, CRISPR-Cas9 genome editing has been successfully applied to a number of crops, including cotton, maize, rice, and wheat. For instance, in *O. sativa*, the method was successful in eliminating all three enhanced disease resistance 1 (EDR1) homologs to produce plants (*Taedr1*) that are more resilient to powdery mildew (Zhang *et al.*, 2017). Fig. 18.1 represents multi-omics and CRISPR techniques in crop-stress resilience.

Further, transcriptome research approaches have been able to progress from a basic DNA microarray platform to RNA-seq technology thanks to recent advancements in sequencing technology. RNA-seq is a relatively new method that measures the overall biological quantities of the transcriptome (Nejat *et al.*, 2018; Seyfferth *et al.*, 2021). Among its numerous advantages is the ability to evaluate an entire transcriptome without requiring a genomic reference sequence due to its high sensitivity, high throughput, and efficacy (Negi *et al.*, 2022). A common sequencing method in molecular biology, biotechnology, and bioinformatics is called RNA-seq technology. The transcriptome is the total amount of RNA molecules – messenger RNA and non-coding RNA – that are transcribed

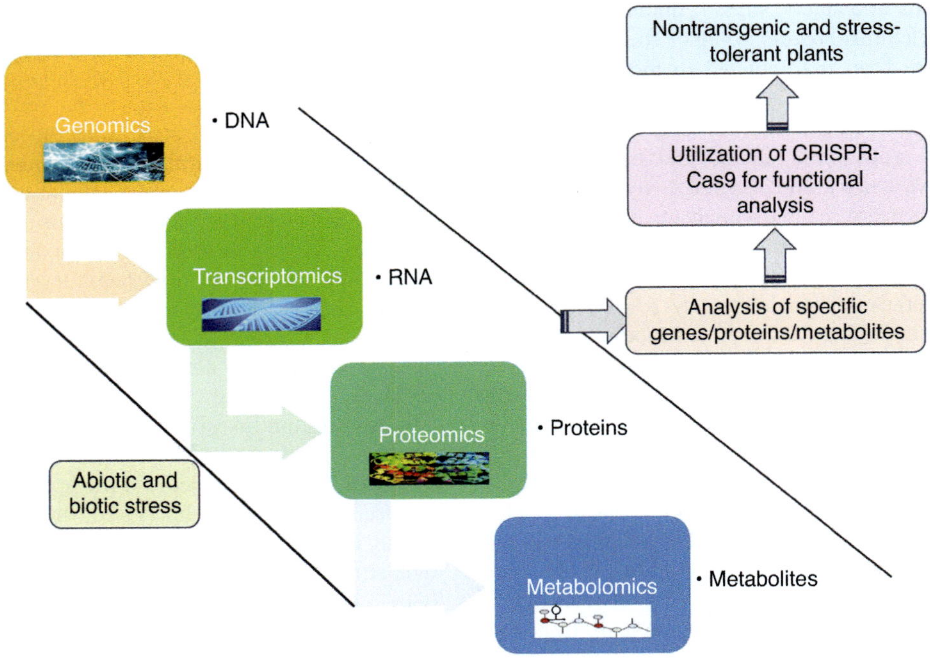

Fig. 18.1. Multi-omics and CRISPR technique in crop stress resilience.

from a given tissue or cell at a given functional or developmental stage (Tyagi *et al.*, 2022). They are referred to as 'bridges' because they carefully regulate the genetic information's transit from DNA to protein. On the other hand, noncoding RNA has a multifaceted impact on gene expression, protein synthesis, and several physiological processes. Transcriptomics knowledge thus advances research on the functions of tissues, cells, and organisms. According to one of the first comparative transcriptomic studies conducted on rice, the sensitive genotype is more susceptible to water stress-induced gene expression fluctuations linked to leaf senescence and degradation processes than the tolerant genotype (Degenkolbe *et al.*, 2009). Furthermore, Lenka *et al.* (2011) found that the *landrace Nagina22* significantly upregulated genes associated with the α-linolenic acid metabolic pathway, which is compatible with its high tolerance to dehydration. The transcriptome response of a drought-tolerant introgression line, designated *H471*, to water stress was compared in another research to that of the drought-resistant donor *PSBRC28* and the drought-susceptible recurrent parent (Huang *et al.*, 2014).

As cellular components, proteins play important functions in controlling plant stress. They also control plant gene series, chemical metabolites, and the functioning of nucleic acids or histones. Proteomics is the study of the whole expressed protein array of an organism across place and time (Mergner and Kuster, 2022; Naik *et al.*, 2023). Proteomics analyses yield a wealth of information on the expressed proteins in the form of proteome spectra that vary depending on the organs, age, environmental conditions, developmental stage, and dietary adequacy. Additionally, it aids in the recognition and evaluation of several stress-sensitive proteins (temperature, salinity, and water), helping to understand their role in abiotic stress-induced signaling (Rai *et al.*, 2021). The total proteome was frequently used to assess plant stress tolerance, following research on other aspects of the proteome, such as the organellar proteome, phosphorproteome, nuclear proteome, cell-wall proteome, and proteogenome (Mehta *et al.*, 2019a). A comparative physiological and proteome investigation was performed on two wheat cultivars that showed different responses to drought stress. These two cultivars' roots and leaves responded differently to water scarcity based on differences in the amount of drought-responsive proteins (DRPs). The majority of the 73 DRPs that were found in the leaves of cultivars that were both drought-tolerant and -sensitive were linked to the metabolism of carbohydrates and energy (34%), photosynthesis (23%), and cell-wall biosynthesis and breakdown (17%) (Faghani *et al.*, 2015). Another comparative proteome analysis revealed the presence of many antioxidant proteins in rice and chickpea under drought stress, which is linked to cell defense and rescue. For instance, betaine aldehyde dehydrogenase-1, 2-methylene-furan-3-one reductase, likely NAD(P)H dehydrogenase, lactoylglutathione-lyase, MLP-like protein-34, P23 protein, chitinase 2-like, Kunitz proteinase inhibitor-1, and heat-shock proteins (HSP) were the key proteins in chickpea under stress. Other proteins were found in high abundance under drought-treated leaves in various crops, while proteins such as T-complex protein, *HSP70*, and *HSP90* proteins were found in lower abundance. These proteins included ER-luminal binding protein, trigger factor-like protein, HSPs, and other molecular chaperones (i.e. calnexin, endoplasmin). These polypeptides play a variety of functions in the body, including maintaining protein folding and renaturing damaged proteins caused by stress (Agrawal *et al.*, 2016; Gupta *et al.*, 2020).

Metabolomics is regarded as an essential area of systems biology and an effective means for studying how organisms react at the metabolite level to outside stimuli. Understanding the chemical signals that plants receive as they develop and grow depends on metabolomics (Nakabayashi and Saito, 2015; Carrera *et al.*, 2021; Vo *et al.*, 2021). However, because the metabolome is not always linked to the plant genome like the transcriptome and proteomes, it is challenging to determine the entire significance of metabolomics for evaluating plant responses to stress. In the fields of plant physiology and biochemistry, metabolomic research is becoming more and more widespread (Ghatak *et al.*, 2018). Also, Kumar *et al.* (2023) found a general increase of flavonoids and phenols in tomato seedlings infected with *Meloidogyne incognita*. Further, in *Mentha spicata* plants, the co-occurrence of saline stress and copper exposure modified the major metabolic pathways to promote the formation of several secondary metabolites, including limonene (Chrysargyris *et al.*, 2019). Additionally, depending on the plant under study, the combination of heat- and drought-activated secondary metabolites resulted in either resistance or vulnerability to this stress combination. For instance, in the cotton genotype *Zhongmian 23*, tolerance to a combination of heat stress and drought was linked to higher levels of flavonoids and phenols (Ibrahim *et al.*, 2019). Furthermore, metabolomics signatures shared by Nipponbare inoculated with 10 PGPR strains included decreased alkylresorcinol (5-tridecyl resorcinol, 5-pentadecyl resorcinol, 5 [12-heptadecyl] resorcinol) levels and the varying activation of two antimicrobial compounds, N-p-coumaroylputrescine and N-feruloylputrescine, albeit in different ways (Valette *et al.*, 2020). Since *Pseudomonas* is antagonistic to other bacteria and fungi, it is employed as a biocontrol agent against rice disease. High-performance liquid chromatography (HPLC) examination of the roots and root exudates of rice afflicted with *P. putida* demonstrated the induction of salicylic acid (Kandaswamy *et al.*, 2019).

18.3 CRISPR Technology

CRISPR technology is a straightforward yet effective genome-editing method. It makes it simple for researchers to change the functions of genes and DNA sequences. The correction of genetic abnormalities, the treatment and prevention of disease, and the enhancement of crops are among its possible uses (Vidyasagar, 2018). Through the development of genome-editing techniques using CRISPR, along with improvements in computing and imaging power, a new era of disease diagnosis and personalized genetic susceptibility prediction has begun. We can quickly identify and change the genes governing plant properties, which will change the rate at which agricultural research and plant breeding are conducted (Wang and Doudna, 2023). CRISPR-Cas and its related protein caspase-9 (Cas9) provide potential answers to a range of genetic issues. It directs the DNA-cleaving enzyme Cas9 to its genetic target DNAs using short-guide RNAs (gRNAs). The nuclease and recognition lobes of Cas9 are connected by a highly conserved arginine-rich helix to establish interactions with the (single-guide RNA) sgRNA (Bravo *et al.*, 2022). The RuvC (crossover junction endodeoxyribonuclease RuvC) and HNH (an endonuclease domain named for characteristic histidine and asparagine residues) domains are located in the nuclease lobe and recognition lobe, respectively. The cutting site is three base pairs upstream of the (protospacer-adjacent motif) PAM, and 8–12 base pairs (bps) upstream of the PAM is crucial for identification by the Cas9-sgRNA complex (Zheng *et al.*, 2020). As a genome-editing technique, it has immense potential, but there are still a few issues that need to be resolved before use. One of the biggest problems with CRISPR gene editing is off-target DNA cleavage, which happens when the molecular scissors cut the wrong part of the host genome, changing the genome in a way that is non-specific, not desirable, unanticipated, or even harmful (Kotagama *et al.*, 2019). Rather than correcting these mistakes, the mechanism creates genetic mutations. Many techniques, including sgRNA optimization, off-target identification, prime editing, Cas9 nuclease modification, and others, have been developed to lessen the impacts of off-targeting (Asmamaw and Zawdie, 2021).

18.3.1 CRISPR/Cas9 system in genome-editing mechanism

Because of the Cas9 enzyme, CRISPR/Cas9 is incredibly effective, precise, accurate, and selective. Cas9 is essentially a bilobed structure with two grooves for nucleic acid binding that is recognition lobe (REC), which is large and has active sites and the nuclease lobe (NUC). A helix bridge connects them both (Anders *et al.*, 2014). Cas9's specificity is dependent on REC, however, NUC has an additional two domains. RuvC and HNH come first, followed by the PAM interacting domain. Caspase-9 is dormant by nature and only becomes active when its REC lobe attaches to sgRNA. Because the target sequence is complementary to the sgRNA sequence and the PAM sequence is present downstream of the target site, a complex consisting of Cas9 and sgRNA forms, which then scans the target DNA sequence and recognizes the target site with ease. Next, sgRNA binds to the target sequence. Following this, HNH cleaves the target sequence strand, RuvC cleaves the opposite strand, and DABs are formed (Ousterout *et al.*, 2015). A sophisticated kind of CRISPR/Cas9 has been created by researchers that can recognize 2–5 nucleotides of the PAM sequence and 20–24 nucleotides of the target site's complementary sgRNA sequence. CRISPR/Cas9 can therefore target a region that is between 22 and 29 nucleotides long (Hsu *et al.*, 2013; Mali *et al.*, 2013). There are now two different kinds of repair processes that are crucial to the mending of DSBs. The two methods are non-homologous end-joining (NHEJ) and homologous directed repair (HDR). Homologous-directed repair connects only homologous templates and is an error-free approach frequently employed in plants for gene knock-in, but more difficult than NHEJ (Schiml *et al.*, 2014). Despite its error-prone nature, NHEJ rejoins the ends by using DNA ligase. It causes Indels to mutate and insert or delete a certain gene sequence. Thus, it is frequently employed for gene knockout (Gratz *et al.*, 2014) and to facilitate the CRISPR/Cas9 system (Fig. 18.2).

18.3.2 Stress-resilient CRISPR-edited crops in agriculture

18.3.2.1 CRISPR/Cas9 for biotic stress

Despite two different types of defense mechanisms, namely effectors-triggered immunity (ETI) responses and PAMP (pathogen-associated molecular patterns), triggered immunity (PTI) plants are prone to attack from phytopathogens like bacteria, fungi, viruses, insects, weeds, nematodes, mites, etc. (Ghosh and Dey, 2022). It has been reported that approximately 20–40% of global crop loss is due to biotic factors (Erdoğan *et al.*, 2023). CRISPR/Cas9, a revolutionary genome-editing technology has emerged as a powerful, efficient, practical and time-saving approach to successfully produce desirable modifications in target genes. This technology has great application potential such as producing disease and pest-resistant crop varieties, improving plant immune systems to obtain long-duration resistance against

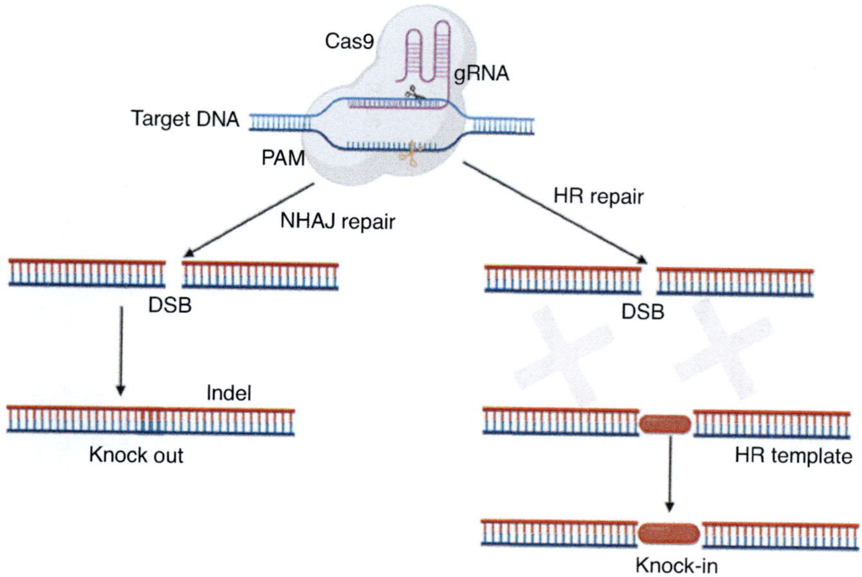

Fig. 18.2. Illustration of CRISPR-Cas9 editing tool.

phytopathogens, and improving crop yield and quality (Li *et al.*, 2022a) (Table 18.1).

Bacteria in plants have become increasingly difficult to manage due to continuous genetic evolution, delayed disease diagnosis, and the lack of effective chemical controls. CRISPR/Cas9 has been widely used to modulate genes to get resistance against different bacterial diseases. For instance, in rice, *SWEET11*, *SWEET13*, *SWEET 14* (sugar transporters) genes were manipulated through CRISPR/Cas9 to provide resistance against *Xanthomonas oryzae* (Oliva *et al.*, 2019). Similarly, to increase resistivity against *Phytophthora capsici* without harming the health of a plant, *SlDMR6-1* gene has been knocked out using this technology (Zeilmaker *et al.*, 2015). In tomato, *DMR6* gene, which acts as a negative regulator of host plant immunity, is inactivated using CRISPR/Cas9 to provide resistance against *Pseudomonas syringae* (Langner *et al.*, 2018).

Fungi possess high genetic flexibility, allowing them to readily adapt and infect host plants with ease, making them formidable pathogens in agricultural systems. In wheat, grapevine, and tomato, *MLO* (mildew resistance locus) gene was modified using the prominent gene technology to provide resistance against *Blumeria gaminis*, which causes powdery mildew disease (Nekrasov *et al.*, 2017). In *O. sativa*, CRISPR/Cas9 is employed to genetically manipulate *OsERF922* and *EDR1* genes to develop ethylene signaling and resistance against *Magnaporthe oryzae*, the rice blast fungus (Wang *et al.*, 2016). In wheat, a mutation of *Taedr1* gene to develop resistance against fungal phytopathogen like *Erysiphe cichoracearum* using gene knockout technology was carried out (Zhang *et al.*, 2017).

Viruses are invasive obligatory parasites, which have a high ability to cause infection in host plants leading to a huge loss of global crop yield. As viruses are dependent on host plant machinery for replication and multiplication, resistance against them can be developed by targeting genes related to pathogenicity or by modifying host susceptibility. CRISPR/Cas9 is widely used to provide direct resistance against both DNA and RNA viruses by rupturing viral genomes to build resistance against viruses like the bean yellow dwarf virus, beet severe curly top virus, and tomato leaf curl virus (Mushtaq *et al.*, 2020). CRISPR/Cas9 was used for the first time in *Arabidopsis thaliana* and *Nicotiana benthamiana* (Hirano *et al.*, 2016). Resistant transgenic cassava varieties were developed using gene

Table 18.1. CRISPR genome editing to boost crop resistance to biotic stress.

Host plant	Stress	Disease	Gene targeted	Effects	Reference
Hordeum vulgare	*Blumeria gaminis*	Powdery mildew	*Mlo* (mildew resistance locus)	Provided resistance to powdery mildew disease	Erdoğan *et al.*, 2023
Glycine max	*Agrotis ipsilon*	Cutting of seedlings common cutworm	*GmCDPK38* (calcium-dependent protein kinase 38)	Provided high resistance to common cutworm	Li *et al.*, 2022b
Arabidopsis thaliana	*Fusarium graminearum*	Fusarium head blight	*At2OGO* (2-oxoglutarate Fe (II)-dependent oxygenase)	Enhanced resistance to fg and upregulation of defense genes	Bashar and Hanif, 2021
Solanum lycopersicum	*Botrytis cinerea*	Gray mold	*SlPL* (pectate lyase)	Reduced incidence of gray mold	Silva *et al.*, 2021
Brassica napus	*Verticillium longisporum*	Verticillium stem striping	*BnCRT1a* (calreticulin)	Increased in biosynthesis and signaling genes of ethylene as defense mechanism	Bashar and Hanif, 2021
Beta vulgaris	*Meloidogyne incognita*	Root galls and infestation	*Mi-msp5*	Provided resistance to nematode infection	Joshi *et al.*, 2020
B. vulgaris	Beet necrotic yellow vein virus (BNYVV)	Rhizomania	*Rz1*	Enhanced tolerance to beet necrotic yellow vein virus (BNYVV)	Liebe *et al.*, 2019
Vitis vinifera	*Uncinula necator*	Powdery mildew	*VvMOL3* (mildew resistance locus O)	Provided resistance to powdery mildew	Wan *et al.*, 2020
Arabidopsis thaliana	*Phytophthora parasitica*	Root infection	*ERF019* (ethylene-responsive transcription factor 19)	Provided high resistance to *Phytophthora parasitica*	Lu *et al.*, 2020
O. sativa	Chilo suppressalis	Stem borer	*OsHPL2* (hydroperoxide lyase gene)	Enhanced resistance to stem borers	Li *et al.*, 2020
S. lycopersicum	*Xanthomonas* spp.	Bacterial leaf spot	*SlJAZ2* (jasmonate ZIM-domain-2 ortholog)	Enhanced resistance to bacterial leaf spot disease	Ortigosa *et al.*, 2019

S. lycopersicum	*Phelipanche aegyptiaca*	Cause damage to tomato plant	*CCD8* (carotenoid cleavage dioxygenase 8)	Developed weed-resistant crop	Bari *et al*., 2019
O. sativa	Rice tungro spherical virus (RTSV)	Rice tungro	*eIF4G* (eukaryotic translation initiation factor 4E)	Provides resistance to RTSV	Macovei *et al*., 2018
S. lycopersicum	Tomato mosaic virus (ToMV)	Stem necrosis, stunted growth	*SlDCL2b* (dicer-like 2b)	Enhanced resistance to tomato mosaic virus (ToMV)	Wang *et al*., 2018
Citrus sinensis	*Xanthomonas citri*	Citrus bacterial canker	*CsLOB1* (lateral organ boundary 1)	Improved resistance to citrus canker disease	Peng *et al*., 2017
S. lycopersicum	*Oidium neolycopersici*	Powdery mildew	*SlMlo1* (mildew-resistant locus O)	Providedresistance to powdery mildew	Nekrasov *et al*., 2017
Triticum aestivum	*Blumeria graminis*	Powdery mildew	*TaEDR1* (enhanced disease resistance 1)	Provided resis-tance to powdery mildew	Zhang *et al*., 2017
O. sativa	*Magnaporthe oryzae*	Rice blast	*OsERF922*(ethylene-responsive transcription factor 922) *OsSEC3A*	Provided resistance to rice blast disease	Wang *et al*., 2016
Cucumis sativus	*Bemisia tabaci*	Cucumber vein yellowing virus	*eIF4E* (eukaryotic translation initiation factor 4E)	Provided resistance to potyviruses	Chandrasekaran *et al*., 2016
O. sativa	*Xanthomonas oryzae*	Proteobacterium	*OsSWEET13* (sugars will eventually be exported transporter 13)	Provided resistance to bacterial blight	Zhou *et al*., 2015

knockout against African cassava mosaic virus (ACMV) (Mehta *et al.*, 2019b).

Insect resistance development using CRISPR/Cas9 is challenging due to the unavailability of an embryonic microinjection system and inadequate guidance on the genomic sequence (Ghosh and Dey, 2022). When Potato Protease Inhibitor II (*pinII*) gene is knocked in, resistance against Lepidopteran insect is achieved in *Arabidopsis*. In transgenic peanut expressing *Cry1Ab* transgene, *Bacillus thuringiensis* developed resistance against *Spodoptera litura*. *Cry1Ab* also provides resistance against the insect *Plutella xylostella* in broccoli (Zhu-Salzman *et al.*, 2004).

Maximum damage to plants is caused by weeds as they compete with plants for food, water shelter, sunlight, space, and fertilizers inhibiting proper growth and development of plants. Some pathogenic weeds infect both aerial parts and the root system of the host plant. Time and again, reusing the same herbicides leads to the development of resistivity in weeds (Shaner, 2000). The most appropriate approach is to use modern gene-editing tools. In tomato, CRISPR/Cas9-mediated mutagenesis of *CCD8* gene provided resistance against *Phelipanche aegyptiaca* weed (Bari *et al.*, 2019). CRISPR/Cas9-mediated missense point mutations of acetolactate synthetase (ALS) in *Arabidopsis thaliana* induced resistance against herbicide and hence promoted capacity against weeds (Yang *et al.*, 2020).

18.3.2.2 CRISPR/Cas9 for abiotic stress

Abiotic stress in the environment poses a threat to food security as it significantly reduces crop yields, often decreasing their annual production globally. Various researchers and plant breeders are adopting many advanced methodologies and approaches, which have been employed to boost stress resilience in multiple crop species (Gupta *et al.*, 2023). So, CRISPR/Cas9 technology offers a platform for generating genetically edited plants without transgene, akin to mutagenized crops that facilitate the development of abiotic stress-tolerant crop plants, ensuring sustained crop productivity in dynamic environmental conditions (Shendekar *et al.*, 2024). For example, the current research endeavors aim to assess the response of *S. lycopersicum* L. to drought and salinity stress by employing CRISPR/Cas9 genome editing on two potential negative regulatory genes, namely *SlHyPRP1* and *SlDEA1*. It has been noticed that the combining effect of loss-of-function in these two genes is imparting abiotic stress tolerance in tomato crops (Saikia *et al.*, 2024). In the case of moss *Bryum argenteum*, the CRISPR/Cas9 genetic-editing system targets the *BaABI3A* and *BaABI3B* genes, implying that these two genes likely have redundant roles in desiccation stress and potentially enhance the plant's response to desiccation stress (Leng *et al.*, 2024). Moreover, in *Saccharum* spp. hybrids, an upregulation in transcript levels of *ScNsLTP* gene was observed, which suggested the involvement of this gene in responding to chilling and drought stress. With the efficient utilization of CRISPR/Cas9 technology, it is possible to enhance the expression of the *ScNsLTP* gene, offering a potential avenue to confer resistance against both stresses in sugarcane (Mall *et al.*, 2024). In maize (*Zea mays* L.), the gene knockout was used to edit *ZmHDT103* gene, and it has been noticed that the mutant type that was achieved after editing significantly lowered water-loss rate in drought conditions as compared to the wild type, which indicated that the *ZmHDT103* gene served as a negative regulator of maize's resistance to water-deficit conditions (Wang *et al.*, 2024). Also, adverse conditions like temperature variations, limited water, saltiness, metal toxicity, and ultraviolet radiation threaten sugar beet yield and development globally. The gene knockout technology can enhance sugar beet resilience to these stressors, boosting productivity and sustainability in cultivation (Misra *et al.*, 2023). Through the utilization of CRISPR/Cas9 technology, researchers have the capability to pinpoint specific genes within stress-response pathways, allowing for targeted modifications that further enhance tolerance to a multitude of biotic and abiotic stresses (Kumar *et al.*, 2023) (Table 18.2).

18.4 Single-cell Resolution Techniques in Omics Analysis

Omics technologies refer to high-throughput experimentation methods that can characterize precise molecular information.

Table 18.2. CRISPR genome editing to boost crop resistance to abiotic stress.

Host plant	Stress	Gene target	Effects	Reference
Solanum lycopersicum	Drought and salinity	*SlHyPRP1* (*S. lycopersicum* hybrid proline-rich protein) *and SlDEA1* (*S. lycopersicum* arachidonic acid 1)	Created positive morphological and phenotypical responses	Saikia *et al*., 2024
Bryum argenteum	Desiccation stress	BaABI3A (*B. argenteum* ABA-insensitive 3A) and BaABI3B (*B. argenteum* ABA-insensitive 3B)	Enhanced the plant response to desiccation	Leng *et al*., 2024
Saccharum spp. hybrids	Chilling and drought stresses	ScNsLTP (*Saccharum* non-specific lipid transfer protein)	Provided resistance against chilling drought	Mall *et al*., 2024
Zea mays	Drought stress	ZmHDT103 (*Z. mays* histone deacetylase 103)	Increased drought tolerance in maize	Wang *et al*., 2024
O. sativa	Drought stress and salinity stress	*OsSAKP2* (*O. sativa* osmotic stress/ABA-activated protein kinase 3)	Increased drought tolerance and salinity tolerance	Lou *et al*., 2023
Cicer arietinum	Drought stress	*4CL* (4-coumarate ligase) and *RVE7* (Reveille 7)	Improved drought tolerance by knocking out these genes	Razzaq *et al*., 2022
Triticum aestivum	Salinity stress	*TaHAG1* (*T. aestivum* histone acetyltransferase of GNAT family 1)	Increased the salt tolerance of plants	Zheng *et al*., 2021
O. sativa	Drought stress	*OsERA1* (*O. sativa*-enhanced response to ABA 1)	Enhanced response to drought stress	Tripathy *et al*., 2021
Glycine max	ABA and salinity stress	*GmAITRs* (*G. max* ABA-induced transcription repressors)	Mutant genes show enhanced salt tolerance	Wang *et al*., 2021
O. sativa	Salinity tolerance	*OsRR22* (*O. sativa* response regulator 22)	Improvement in salinity tolerance	Srikanth *et al*., 2021
O. sativa	Drought and heat stress	*OsNAC006*	Maintained high levels of antioxidant enzymes associated with tolerance to stress	Wang *et al*., 2020
Brassica napus	Drought stress	*BnaA6.RGA* (*B. napus* A6 repressor of ga1-3)	Increased the expression level of the other water-deficit-responsive gene	Wu *et al*., 2020
O. sativa	Salinity stress	*OsPQT3* (*O. sativa* paraquat tolerance 3)	Improved tolerance to salt and oxidative stress	Alfatih *et al*., 2020

Continued

Table 18.2. Continued

Host plant	Stress	Gene target	Effects	Reference
O. sativa	Chilling stress	*OsMYB30* (*O. sativa* myeloblastosis30)	Increased panicle length, grain size, and excellent cold tolerance	Zeng *et al.*, 2020
O. sativa	Drought tolerance	*OsPYL9* (*O. sativa* pyrabactin resistance 1-like 9)	Mutants showed an increase in grain yield under both drought and well-watered field conditions	Usman *et al.*, 2020
S. lycopersicum	Drought stress	*SlARF4* (*S. lycopersicum* ADP-ribosylation factor 4)	Helped in increasing drought tolerance	Bouzroud *et al.*, 2020
S. lycopersicum	UV stress	*SlUVR8* (*S. lycopersicum* UV resistance locus 8)	Increased UV-B stress tolerance in plants	Liu *et al.*, 2020a
O. sativa	Chilling stress	*OsAnn3* (*O. sativa* annexins 3)	Provided cold tolerance	Romero and Gatica-Arias, 2019
O. sativa	Alkaline stress	*OsPPa6* (*O. sativa* pyrophosphorylase 6)	Maintained osmotic regulation	Wang *et al.*, 2019

The subjects of investigation in omics technologies vary from genes to metabolites to molecular interactions, signifying genomics, metabolomics, and interactomics, respectively (Depuydt *et al.*, 2023). The functioning of a cell is determined majorly by the interactions within the cell's transcriptome, genome, proteome, and epigenome. The separation of single cells and investigation of infinitesimal amounts of molecules contained within each cell has enabled the study of these omic layers at single-cell resolution (Chappell *et al.*, 2018). Single-cell omics techniques allow the analysis of variabilities within cells and among various populations that were earlier thought to be homogenous (Gay *et al.*, 2016; Wang *et al.*, 2017; Moore and Jessberger, 2017).

18.4.1 Introduction to single-cell omics technologies

Several single-cell techniques exist for all kinds of molecule that may chatacterize a specific omic layer; however, these techniques vary in terms of precision, specificity, efficiency, and accessibility. Some single-cell omics methods like single-cell RNA sequencing (scRNA-seq) are well developed whereas others are comparatively novel (Jean-Baptiste *et al.*, 2019; Graeff *et al.*, 2021; Shahan *et al.*, 2022). The advancement of omics techniques has enabled the development of multi-omics techniques for single cells that can acquire numerous omic layers from one cell (Chappell *et al.*, 2018). These single-cell multi-omic technologies include single-cell genome and transcriptome sequencing (scG&T-seq), single-cell methylome and transcriptome sequencing (scM&T-seq), single-cell triple omics sequencing (scTrio-seq), and single-cell chromatin overall omic-scale landscape sequencing (scCOOL-seq) (Hu *et al.*, 2016; Angermueller *et al.*, 2016; Hou *et al.*, 2016; Macaulay *et al.*, 2017). By profiling numerous omic layers, a more comprehensive set of data on individual cells may be obtained than by profiling a single omic layer only. This data more accurately reflect the intricate webs of connections that govern cellular processes. Thus, single-cell multi-omics methods might be more effective than mono-omics methods executed on a set of individual cells or distinct mono-omics methods executed on different sets of comparable single cells (Chappell *et al.*, 2018).

18.4.2 Applications and advantages of single-cell resolution in crop research

Profiling of single-cell genes during the last decade was possible, firstly in microbes and animals and later in plants (Mo and Jiao, 2022). However, recently, characterization of the single-cell transcriptome has gained much traction in the field of vegetative research. In contrast to conventional transcriptomics, single-cell sequencing of plants can resolve their heterogeneity (Chen *et al.*, 2023a). For instance, researchers detected the activation of stress-defense genes at the organ level and inhibition of growth-related, energy-intensive activities when using conventional transcriptomics to assess drought-induced transcriptome alterations. But, single-cell transcriptomics data revealed that in response to minor drought stress, while the majority of gene activities were elevated in the epidermis, the mesophyll displayed a considerable downregulation of genes. Understanding the intricate reaction of plants to environmental cues requires an understanding of single-cell networks. Single-cell transcriptomics is a useful tool for identifying tissue-specific stress-responsive pathways that can eventually be precisely modified in imperative tissues, without triggering unforeseen negative effects on non-target tissues or organs (Tenorio Berrío *et al.*, 2022).

By extricating the roles of novel genes, scRNA-seq has helped greatly in extending our knowledge of plant biology (Rich-Griffin *et al.*, 2020; Shaw *et al.*, 2021; Minne *et al.*, 2022). For instance, by using scRNA-seq, a study identified the cell-specific expressions of genes induced by *Arabidopsis TMO5/LHW* and revealed that *TMO5/LHW*-dependent biosynthetic pathway of cytokinesis in plant vascular cells was triggered by phosphate deficiency (Wendrich *et al.*, 2020). Grounded on correlations between single-cell expression patterns and differential expression (DE) study, a scRNA-seq investigation in rice found two genes, namely, *Os01g0934800* and *Os01g0949900*, that were controlled by the TF *OsNAC78*. These findings were then validated experimentally by utilizing the yeast one-hybrid system (Xie *et al.*, 2020). The recently discovered target genes of TF *OsNAC78* were earlier linked to detoxification and ROS scavenging, respectively, suggesting possible roles of *OsNAC78*. Further, the integration of data obtained by profiling the singe-cell transcriptome of maturing ears in maize with supplementary omics data showed that productivity characters in maize were regulated by cell-specific gene markers, and three possible productivity-trait contributing genes were characterized (Xu *et al.*, 2021).

Lately, the evaluation of single-cell transcriptomes has been keenly utilized to define the identities of cells, create gene expression lines, and analyze their regulatory networks (Denyer *et al.*, 2019; Liu *et al.*, 2020b; Zhang *et al.*, 2021; Dorrity *et al.*, 2021; Roszak *et al.*, 2021). When combined with transcriptome analysis, examination of genome-wide chromatin accessibility in particular cells has shown promise in elucidating the regulatory processes (Dorrity *et al.*, 2021). Moreover, numerous studies focussing on single cells have brought attention to the significance of epigenomic regulation in diverse processes, like cell differentiation in *Zea mays* (Satterlee *et al.*, 2020). These methods ought to offer novel avenues for research focussing on genome regulation in plants. There is still a lot of room for single-cell omics research in plants compared to that done in animals.

18.4.3 Limitations

Plant research has clearly benefited from the implementation of individual omics technologies, particularly in the past ten years (Hurgobin and Lewsey, 2022). However, every technology has unique data analysis and technological difficulties, which are intensified at the single-cell level. For example, in order to release cells for scRNA-seq, the polysaccharide cell wall of plant cells needs to be removed, which takes place via protoplast isolation (Bezrutczyk *et al.*, 2021; Conde *et al.*, 2021; Farmer *et al.*, 2021). However, variations in the thickness and composition of plant cell walls among various species, tissues, stages of development, and environmental stressors, make the isolation of protoplast a difficult task in every new experimental setting (Shulse *et al.*, 2019). Moreover, the isolation of protoplast leads to the induction of undesirable metabolic and transcriptional modifications in plant cells (Shaw *et al.*, 2021). This, along with the fact that plant cells are typically larger (100 mm) than those of mammals (10–30 mm) and that the sizes of cells vary across species of plants, their organs, and tissues might make it difficult for the most widely used single-cell methods to capture and profile some cell types, which might introduce biases into experimental processes (Zhang *et al.*, 2019; Ryu *et al.*, 2019; Shaw *et al.*, 2021; Hurgobin and Lewsey, 2022).

Various locations of cells within plants also result in variations in the area accessible to enzymes that are used to break down these cells; for instance, certain cells within the plant parts that are encircled by other cells are difficult to break down. As these cells are eliminated during the filtering process, they tend to go unnoticed during the analysis. Since it has been shown that obtaining core stele cells is more difficult than obtaining epidermal cells, the protoplast preparation is focussed on the cells that are external to plant materials and are sufficiently exposed to digestive enzymes (Denyer *et al.*, 2019; Farmer *et al.*, 2021). Aside from the location of cells, their growth stage also affects the breakdown. In comparison to mature cells, younger cells are easily broken down, leading to their increased participation in the examinations (Jean-Baptiste *et al.*, 2019; Ryu *et al.*, 2019). Hence, the development of improved techniques for the separation of plant tissues and cells that can further be employed for a broad variety of floral species is necessary to enhance the accessibility and usage of scRNA-seq analysis among researchers.

Additionally, to obtain more accurate sequencing data, single-cell sequencing techniques need to be improved. For instance, the RNA-seq library is generated by a droplet-based technique like the 10x chromium method which can be utilized for analyzing thousands of cells in a single experimental setup. However, occasionally, a droplet contains more than one cell or nucleus, or a poor quantity of beads, which can interfere with the results of the analysis (Lareau *et al.*, 2020). Hence, improvement in the relevant technique is critical to guarantee that only a single cell or nucleus is present in the droplet to attain greater scalability (Chen *et al.*, 2023a). In genomics studies, the use of multi-omics methods is becoming a standard practice. Earlier research frequently conducted separate analyses of 'omics datasets' before combining the findings to offer a deeper understanding of the living systems under analysis (Tarazona *et al.*, 2021). However, an integrative examination of these distinct modalities can more precisely represent these processes (Gomez-Cabrero *et al.*, 2014; Walley *et al.*, 2016; Clark *et al.*, 2021). This is quite challenging and necessitates giving thoughtful contemplation to a number of aspects, including the absence of heterogeneity among omics technologies and the calculation load related to the storage and scrutinization of different kinds of data (Duan *et al.*, 2021). Additional analysis obstacles, such as the prevalence of missing data and randomness among modalities, are posed by the budding curiosity in single-cell analysis and the enormous potential offered by single-cell multi-omics technologies (Tarazona *et al.*, 2021). Although there is constant development and improvement in specialized programs and workflow that tackle these problems, it is ambiguous how well they will operate with datasets of individual cells from plants as well as how they will deal with the evaluation of new modalities for plant single-cell analysis.

18.5 Spatially Resolved Omics Techniques in Crop Evaluation

Spatial multi-omics methodologies are promptly developing to empower the study of various molecular analytes to the extent of subcellular resolution within the context of their native tissue. The characterization of a sample by spatial multi-ome is achieved frequently by inducing combinations in spatial mono-omics methods (Vandereyken *et al.*, 2023). Multi-omics approaches have played a significant role with high-throughput techniques in numerous crops like wheat, millet, rice, cotton, etc., in association with spatially resolved omic techniques. The creation of multi-omics resources can facilitate the precise and effective identification of superior alleles where future research on multiple omics approaches holds promise for effectively enhancing crops (Cao *et al.*, 2022). Single cells taken from plant tissues can now be studied from a spatial perspective because of recent developments in spatial transcriptomics. Reconstructing cell-fate lineages, identifying cell types and revealing cell-to-cell interactions have all been accomplished with success using this technology. The difficulties in data analysis, sample processing, and multiple-omics technology integration will be resolved by future technological developments. For successive approaches, the cell-sorting technologies and the isolation of nuclei tagged in specific cell types (INTACT) have been coupled recently with bulk RNA sequencing to investigate the spatial expression of target genes in particular cell types (Moreno-Romero *et al.*, 2017; Yin *et al.*, 2023).

18.5.1 Overview of spatially resolved omics technologies

The advancement of spatial transcriptomics, particularly with techniques like Slide-seq and 10x Visium, has significantly propelled biological research forward. These technologies enable high-resolution mapping of gene expression within tissues, offering insights into the spatial organization and interactions of cells. The 10x Visium method involves meticulous sample preparation, starting with the freezing of plant tissue samples, followed by the construction of a library through mRNA tagging and cDNA synthesis. This systematic approach culminates in comprehensive data analysis (Fig. 18.3) (Yu *et al.*, 2023).

Spatial omics techniques are crucial for elucidating the complex molecular hierarchies from genome to phenome in individual cells. By integrating multi-omics methodologies with spatial resolution, researchers can now study intermolecular dynamics and assess how genetic variations influence cellular functions more accurately (Chen *et al.*, 2023b). This capability enhances our understanding of cellular interactions and organization within specific tissues, providing a more detailed view of biological processes. These assays not only identify interactions between different tissue regions with varying transcript or protein abundance but also complement other techniques for characterizing cellular phenotypes. As a result, spatial transcriptomics can define cellular behavior and tissue architecture with unprecedented precision (Yang *et al.*, 2021).

The landscape of spatial transcriptomics is diverse, with various methodologies differing in several key aspects. Coverage refers to the breadth of genomic regions that can be analyzed, while scale or throughput indicate the volume of data that can be processed. Spatial resolution pertains to the level of detail at which cellular components can be distinguished, and multiplexing capability describes the ability to analyze multiple targets simultaneously (Park *et al.*, 2022). Each technique's unique features make them suitable for different research needs. For example, higher spatial resolution is critical for studies requiring detailed cellular mapping, whereas higher throughput may be essential for large-scale tissue analyses. Understanding these

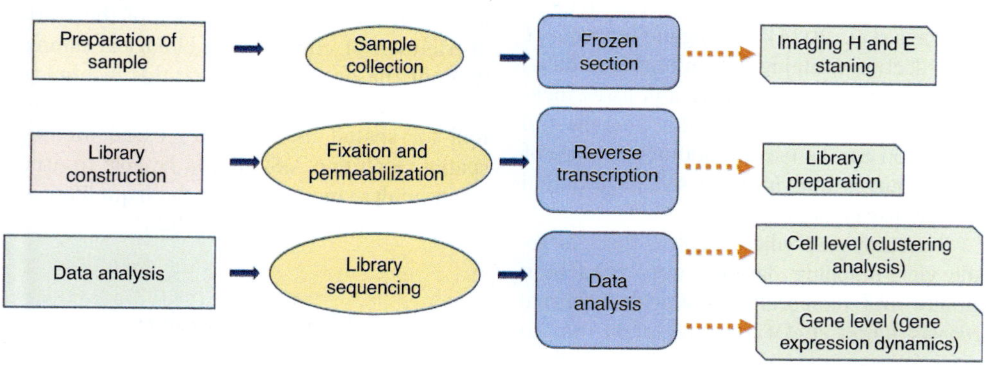

Fig. 18.3. Overview of 10x Visium spatial transcriptomics analysis of a sample.

differences helps researchers select the appropriate method for their specific scientific questions, thus driving more precise and comprehensive biological insights.

18.5.2 Importance of spatial resolution in understanding crop responses to stress

Understanding plant responses to external stimuli and their interaction with the environment is crucial for optimizing agricultural practices. This requires detailed knowledge of the spatial distribution of metabolites within plants. Dai *et al.* (2020) introduced a cutting-edge spatial-resolution targeted metabolomics technique that quantifies the distribution of 56 endogenous metabolites, which included amino acids, phenolic acids, alkaloids, theanine, theaflavins, flavonol/flavone glycosides, and catechins within a single tea leaf. This novel approach provides a deeper insight into the metabolic processes and responses of plants at a cellular level, which can inform better crop management and improvement strategies. For effective crop monitoring in the field, managing spatial variability in both time and space is essential. High-resolution spectral data from satellites such as Sentinel (via Copernicus), VenXIIs, PlanetScope, and Proba offer significant advantages for the agricultural sector. These platforms provide data with varying spatial resolutions (from 1 km to less than 10 m) and spectral configurations, covering bands in the visible (VIS), near-infrared (NIR), thermal infrared (TIR), and shortwave infrared (SWIR) regions (Kganyago *et al.*, 2024). The regular intervals of data collection (ranging from monthly to daily) enable continuous monitoring and assessment of crop conditions and soil health. This spectral information can be used to optimize the use of farm resources and mitigate soil degradation processes like erosion.

In a study on predicting cotton yield, large-scale yield-mapping datasets were used both spatially and temporally to provide accurate forecasts. Filippi *et al.* (2020) highlight that, to date, empirical and data-driven methods have largely dominated yield prediction studies. These methods have successfully produced forecasts with the spatial resolution and extent necessary for cotton growers to make informed management decisions. Integrating these predictions with high-resolution satellite data and metabolomic insights can enhance the precision of yield forecasts, leading to more efficient and sustainable agricultural practices.

18.5.3 Limitations

In situ hybridization (ISH), one of the major categories of spatially resolved transcriptomics (SRT) is limited by a strong background of autofluorescence. Spatial transcriptomics can only resolve spatial information up to a distance of 100 μm between centers in mapping the spatial profiles of different types of tissues (Ramnauth *et al.*, 2023). Due to widely available technology, there will be more difficulties with data analysis like the amount of SRT data being enormous. Future research must improve storage solutions and accelerate SRT data pre-processing to handle high-density datasets and mitigate batch effects. One of the biggest challenges still is figuring out how to appropriately adjust for batch differences so that different slices can be compared to one another to get more accurate and trustworthy analysis results. In general, there are a number of issues that must be resolved, including data normalization, batch effect elimination, data registration across slices, and information gap-filling between sections (Cheng *et al.*, 2023). One major challenge in spatial metabolomics, which relies on mass spectrometry imaging (MSI), is its increasing complexity despite its growing popularity for biological applications. While MSI allows for the direct visualization of the spatial distribution of chemicals in biological samples without the need for labeling, this technology faces significant limitations. Therefore, an ideal tool for spatial metabolomics based on high spatial resolution MSI should be able to accurately analyze most chemicals both qualitatively and quantitatively (Ma *et al.*, 2023).

18.6 Conclusion

The integration of multiple omics analyses, coupled with single-cell and spatial resolution techniques, has unveiled intricate molecular mechanisms underlying stress resilience in CRISPR-edited crops. Through this comprehensive approach, we have elucidated dynamic gene expression patterns, metabolic responses, and epigenetic modifications crucial for plant adaptation to environmental challenges. These findings not only deepen our understanding of crop resilience but also offer invaluable insights for precision breeding strategies aimed at developing robust agricultural systems resilient to diverse stressors. Harnessing the power of multi-omics technologies holds immense promise for accelerating the development of resilient crop varieties essential for sustainable food security in a changing climate.

References

Agrawal, L., Gupta, S., Mishra, S.K., Pandey, G., Kumar, S. *et al.* (2016) Elucidation of complex nature of PEG induced drought-stress response in rice root using comparative proteomics approach. *Frontiers in Plant Science* 7, 221500.

Ahmad, R., Anjum, M.A., Naz, S. and Balal, R.M. (2021) Applications of molecular markers in fruit crops for breeding programs: A review. *Phyton* 90(1), 17–34.

Alfatih, A., Wu, J., Jan, S.U., Zhang, Z.S., Xia, J.Q. *et al.* (2020) Loss of rice PARAQUAT TOLERANCE 3 confers enhanced resistance to abiotic stresses and increases grain yield in field. *Plant, Cell & Environment* 43(11), 2743–2754.

Allan, A.C. and Espley, R.V. (2018) MYBs drive novel consumer traits in fruits and vegetables. *Trends in Plant Science* 23(8), 693–705.

Anders, C., Niewoehner, O., Duerst, A. and Jinek, M. (2014) Structural basis of PAM-dependent target DNA recognition by the Cas9 endonuclease. *Nature* 513(7519), 569–573.

Angermueller, C., Clark, S.J., Lee, H.J., Macaulay, I.C., Teng, M.J. *et al.* (2016) Parallel single-cell sequencing links transcriptional and epigenetic heterogeneity. *Nature Methods* 13(3), 229–232.

Asmamaw, M. and Zawdie, B. (2021) Mechanism and applications of CRISPR/Cas-9-mediated genome editing. *Biologics: Targets and Therapy* 353–361.

Bari, V.K., Nassar, J.A., Kheredin, S.M., Gal-On, A., Ron, M. *et al.* (2019) CRISPR/Cas9-mediated mutagenesis of CAROTENOID CLEAVAGE DIOXYGENASE 8 in tomato provides resistance against the parasitic weed *Phelipanche aegyptiaca*. *Scientific Reports* 9(1), 11438.

Bashar, K.K. and Hanif, M.A. (2021) Crop gene editing against biotic stresses via CRISPR/Cas9 tools: A review. *Archives of Phytopathology and Plant Protection* 54(15–16), 1159–1181.

Bezrutczyk, M., Zöllner, N.R., Kruse, C.P.S., Hartwig, T., Lautwein, T. *et al.* (2021) Evidence for phloem loading via the abaxial bundle sheath cells in maize leaves. *The Plant Cell* 33(3), 531–547.

Bouzroud, S., Gasparini, K., Hu, G., Barbosa, M.A.M., Rosa, B.L. *et al.* (2020) Down regulation and loss of *auxin response factor 4* function using CRISPR/Cas9 alters plant growth, stomatal function and improves tomato tolerance to salinity and osmotic stress. *Genes* 11(3), 272.

Bravo, J.P.K., Liu, M.S., Hibshman, G.N., Dangerfield, T.L., Jung, K. *et al.* (2022) Structural basis for mismatch surveillance by CRISPR-Cas9. *Nature* 603(7900), 343–347.

Cao, P., Zhao, Y., Wu, F., Xin, D., Liu, C. *et al.* (2022) Multi-omics techniques for soybean molecular breeding. *International Journal of Molecular Sciences* 23(9), 4994.

Carrera, F.P., Noceda, C., Maridueña-Zavala, M.G. and Cevallos-Cevallos, J.M. (2021) Metabolomics, a powerful tool for understanding plant abiotic stress. *Agronomy* 11(5), 824.

Chandrasegaran, S. and Carroll, D. (2016) Origins of programmable nucleases for genome engineering. *Journal of Molecular Biology* 428(5), 963–989.

Chandrasekaran, J., Brumin, M., Wolf, D., Leibman, D., Klap, C. *et al.* (2016) Development of broad virus resistance in non-transgenic cucumber using CRISPR/Cas9 technology. *Molecular Plant Pathology* 17(7), 1140–1153.

Chappell, L., Russell, A.J.C. and Voet, T. (2018) Single-cell (multi)omics technologies. *Annual Review of Genomics and Human Genetics* 19, 15–41.

Chen, C., Ge, Y. and Lu, L. (2023a) Opportunities and challenges in the application of single-cell and spatial transcriptomics in plants. *Frontiers in Plant Science* 14, 1185377.

Chen, J., Wang, Y. and Ko, J. (2023b) Single-cell and spatially resolved omics: Advances and limitations. *Journal of Pharmaceutical Analysis* 13(8), 833–835.

Cheng, M., Jiang, Y., Xu, J., Mentis, A.F.A., Wang, S. *et al.* (2023) Spatially resolved transcriptomics: A comprehensive review of their technological advances, applications, and challenges. *Journal of Genetics and Genomics* 50(9), 625–640.

Chrysargyris, A., Papakyriakou, E., Petropoulos, S.A. and Tzortzakis, N. (2019) The combined and single effect of salinity and copper stress on growth and quality of *Mentha spicata* plants. *Journal of Hazardous Materials* 368, 584–593.

Clark, N.M., Nolan, T.M., Wang, P., Song, G., Montes, C. *et al.* (2021) Integrated omics networks reveal the temporal signaling events of brassinosteroid response in *Arabidopsis*. *Nature Communications* 12(1), 5858.

Cna'ani, A., Spitzer-Rimon, B., Ravid, J., Farhi, M., Masci, T. *et al.* (2015) Two showy traits, scent emission and pigmentation, are finely coregulated by the MYB transcription factor PH4 in petunia flowers. *The New Phytologist* 208(3), 708–714.

Conde, D., Triozzi, P.M., Balmant, K.M., Doty, A.L., Miranda, M. *et al.* (2021) A robust method of nuclei isolation for single-cell RNA sequencing of solid tissues from the plant genus populus. *PLoS One* 16(5), e0251149.

Dai, W., Hu, Z., Xie, D., Tan, J. and Lin, Z. (2020) A novel spatial-resolution targeted metabolomics method in a single leaf of the tea plant (*Camellia sinensis*). *Food Chemistry* 311, 126007.

Degenkolbe, T., Do, P.T., Zuther, E., Repsilber, D., Walther, D. *et al.* (2009) Expression profiling of rice cultivars differing in their tolerance to long-term drought stress. *Plant Molecular Biology* 69, 133–153.

Denyer, T., Ma, X., Klesen, S., Scacchi, E., Nieselt, K. *et al.* (2019) Spatiotemporal developmental trajectories in the *Arabidopsis* root revealed using high-throughput single-cell RNA sequencing. *Developmental Cell* 48(6), 840–852.

Depuydt, T., De Rybel, B. and Vandepoele, K. (2023) Charting plant gene functions in the multi-omics and single-cell era. *Trends in Plant Science* 28(3), 283–296.

Dorrity, M.W., Alexandre, C.M., Hamm, M.O., Vigil, A.L., Fields, S. *et al.* (2021) The regulatory landscape of *Arabidopsis thaliana* roots at single-cell resolution. *Nature Communications* 12(1), 3334.

Duan, R., Gao, L., Gao, Y., Hu, Y., Xu, H. *et al.* (2021) Evaluation and comparison of multi-omics data integration methods for cancer subtyping. *PLoS Computational Biology* 17(8), e1009224.

Erdoğan, İ., Cevher-Keskin, B., Bilir, Ö., Hong, Y. and Tör, M. (2023) Recent developments in CRISPR/Cas9 genome-editing technology related to plant disease resistance and abiotic stress tolerance. *Biology* 12(7), 1037.

Espley, R.V., Butts, C.A., Laing, W.A., Martell, S., Smith, H. *et al.* (2014) Dietary flavonoids from modified apple reduce inflammation markers and modulate gut microbiota in mice. *The Journal of Nutrition* 144(2), 146–154.

Faghani, E., Gharechahi, J., Komatsu, S., Mirzaei, M., Khavarinejad, R.A. *et al.* (2015) Comparative physiology and proteomic analysis of two wheat genotypes contrasting in drought tolerance. *Journal of Proteomics* 114, 1–15.

Farmer, A., Thibivilliers, S., Ryu, K.H., Schiefelbein, J. and Libault, M. (2021) Single-nucleus RNA and ATAC sequencing reveals the impact of chromatin accessibility on gene expression in *Arabidopsis* roots at the single-cell level. *Molecular Plant* 14(3), 372–383.

Filippi, P., Whelan, B.M., Vervoort, R.W. and Bishop, T.F.A. (2020) Mid-season empirical cotton yield forecasts at fine resolutions using large yield mapping datasets and diverse spatial covariates. *Agricultural Systems* 184, 102894.

Gay, L., Baker, A.M. and Graham, T.A. (2016) Tumour cell heterogeneity. *F1000Research* 5, 238.

Ghatak, A., Chaturvedi, P. and Weckwerth, W. (2018) Metabolomics in plant stress physiology. *Plant Genetics and Molecular Biology* 187–236.

Ghosh, S. and Dey, G. (2022) Biotic and abiotic stress tolerance through CRISPR-Cas mediated genome editing. *Journal of Plant Biochemistry and Biotechnology* 31(2), 227–238.

Gomez-Cabrero, D., Abugessaisa, I., Maier, D., Teschendorff, A., Merkenschlager, M. *et al.* (2014) Data integration in the era of omics: Current and future challenges. *BMC Systems Biology* 8, 1–10.

Graeff, M., Rana, S., Wendrich, J.R., Dorier, J., Eekhout, T. *et al.* (2021) A single-cell morpho-transcriptomic map of brassinosteroid action in the *Arabidopsis* root. *Molecular Plant* 14(12), 1985–1999.

Gratz, S.J., Ukken, F.P., Rubinstein, C.D., Thiede, G., Donohue, L.K. *et al.* (2014) Highly specific and efficient CRISPR/Cas9-catalyzed homology-directed repair in Drosophila. *Genetics* 196(4), 961–971.

Gupta, S., Mishra, S.K., Misra, S., Pandey, V., Agrawal, L. *et al.* (2020) Revealing the complexity of protein abundance in chickpea root under drought-stress using a comparative proteomics approach. *Plant Physiology and Biochemistry* 151, 88–102.

Gupta, N.K., Shavrukov, Y., Singhal, R.K. and Borisjuk, N. (eds) (2023) *Multiple Abiotic Stress Tolerances in Higher Plants: Addressing the Growing Challenges*. CRC Press, Boca Raton, Florida.

Hasin, Y., Seldin, M. and Lusis, A. (2017) Multi-omics approaches to disease. *Genome Biology* 18(1), 83.

Hirano, H., Gootenberg, J.S., Horii, T., Abudayyeh, O.O., Kimura, M. *et al.* (2016) Structure and engineering of *Francisella novicida* Cas9. *Cell* 164(5), 950–961.

Hou, Y., Guo, H., Cao, C., Li, X., Hu, B. *et al.* (2016) Single-cell triple omics sequencing reveals genetic, epigenetic, and transcriptomic heterogeneity in hepatocellular carcinomas. *Cell Research* 26(3), 304–319.

Hsu, P.D., Scott, D.A., Weinstein, J.A., Ran, F.A., Konermann, S. *et al.* (2013) DNA targeting specificity of RNA-guided Cas9 nucleases. *Nature Biotechnology* 31(9), 827–832.

Hu, Y., Huang, K., An, Q., Du, G., Hu, G. *et al.* (2016) Simultaneous profiling of transcriptome and DNA methylome from a single cell. *Genome Biology* 17, 88.

Huang, L., Zhang, F., Zhang, F., Wang, W., Zhou, Y. *et al.* (2014) Comparative transcriptome sequencing of tolerant rice introgression line and its parents in response to drought stress. *BMC Genomics* 15(1), 1026.

Hurgobin, B. and Lewsey, M.G. (2022) Applications of cell- and tissue-specific omics to improve plant productivity. *Emerging Topics in Life Sciences* 6(2), 163–173.

Ibrahim, W., Zhu, Y.M., Chen, Y., Qiu, C.W., Zhu, S. *et al.* (2019) Genotypic differences in leaf secondary metabolism, plant hormones and yield under alone and combined stress of drought and salinity in cotton genotypes. *Physiologia Plantarum* 165(2), 343–355.

Jean-Baptiste, K., McFaline-Figueroa, J.L., Alexandre, C.M., Dorrity, M.W., Saunders, L. *et al.* (2019) Dynamics of gene expression in single root cells of *Arabidopsis thaliana*. *The Plant Cell* 31(5), 993–1011.

Joshi, I., Kumar, A., Kohli, D., Singh, A.K., Sirohi, A. *et al.* (2020) Conferring root-knot nematode resistance via host-delivered RNAi-mediated silencing of four mi-msp genes in *Arabidopsis*. *Plant Science* 298, 110592.

Kandaswamy, R., Ramasamy, M.K., Palanivel, R. and Balasundaram, U. (2019) Impact of *Pseudomonas putida* RRF3 on the root transcriptome of rice plants: Insights into defense response, secondary metabolism and root exudation. *Journal of Biosciences* 44(4), 1–13.

Kganyago, M., Adjorlolo, C., Mhangara, P. and Tsoeleng, L. (2024) Optical remote sensing of crop biophysical and biochemical parameters: An overview of advances in sensor technologies and machine learning algorithms for precision agriculture. *Computers and Electronics in Agriculture* 218, 108730.

Khan, N., Bano, A., Rahman, M.A., Rathinasabapathi, B. and Babar, M.A. (2019) UPLC-HRMS-based untargeted metabolic profiling reveals changes in chickpea (*Cicer arietinum*) metabolome following long-term drought stress. *Plant, Cell & Environment* 42(1), 115–132.

Kotagama, O.W., Jayasinghe, C.D. and Abeysinghe, T. (2019) Era of genomic medicine: A narrative review on CRISPR technology as a potential therapeutic tool for human diseases. *BioMed Research International* 1369682.

Kumar, M., Prusty, M.R., Pandey, M.K., Singh, P.K., Bohra, A. *et al.* (2023) Application of CRISPR/Cas9-mediated gene editing for abiotic stress management in crop plants. *Frontiers in Plant Science* 14, 1157678.

Langner, T., Kamoun, S. and Belhaj, K. (2018) CRISPR crops: Plant genome editing toward disease resistance. *Annual Review of Phytopathology* 56, 479–512.

Langridge, P., Braun, H., Hulke, B., Ober, E. and Prasanna, B.M. (2021) Breeding crops for climate resilience. *Theoretical and Applied Genetics* 134(6), 1607–1611.

Lareau, C.A., Ma, S., Duarte, F.M. and Buenrostro, J.D. (2020) Inference and effects of barcode multiplets in droplet-based single-cell assays. *Nature Communications* 11(1), 866.

Leng, F., Zhou, G., Shi, R., Liu, C., Lin, Y. *et al.* (2024) Development of PEG-mediated genetic transformation and gene editing system of *Bryum argenteum* as an abiotic stress tolerance model plant. *Plant Cell Reports* 43(3), 63.

Lenka, S.K., Katiyar, A., Chinnusamy, V. and Bansal, K.C. (2011) Comparative analysis of drought-responsive transcriptome in indica rice genotypes with contrasting drought tolerance. *Plant Biotechnology Journal* 9(3), 315–327.

Li, H., Wang, Z., Han, K., Guo, M., Zou, Y. *et al.* (2020) Cloning and functional identification of a *Chilo suppressalis*-inducible promoter of rice gene, OsHPL2. *Pest Management Science* 76(9), 3177–3187.

Li, Y., Wu, X., Zhang, Y. and Zhang, Q. (2022a) CRISPR/Cas genome editing improves abiotic and biotic stress tolerance of crops. *Frontiers in Genome Editing* 4, 987817.

Li, X., Hu, D., Cai, L., Wang, H., Liu, X. *et al.* (2022b) CALCIUM-DEPENDENT PROTEIN KINASE38 regulates flowering time and common cutworm resistance in soybean. *Plant Physiology* 190(1), 480–499.

Liebe, S., Wibberg, D., Maiss, E. and Varrelmann, M. (2019) Application of a reverse genetic system for beet necrotic yellow vein virus to study Rz1 resistance response in sugar beet. *Frontiers in Plant Science* 10, 1703.

Liu, X., Zhang, Q., Yang, G., Zhang, C., Dong, H. *et al.* (2020a) Pivotal roles of tomato photoreceptor SlUVR8 in seedling development and UV-B stress tolerance. *Biochemical and Biophysical Research Communications* 522(1), 177–183.

Liu, Z., Zhou, Y., Guo, J., Li, J., Tian, Z. *et al.* (2020b) Global dynamic molecular profiling of stomatal lineage cell development by single-cell RNA sequencing. *Molecular Plant* 13(8), 1178–1193.

Lou, D., Lu, S., Chen, Z., Lin, Y., Yu, D. *et al.* (2023) Molecular characterization reveals that OsSAPK3 improves drought tolerance and grain yield in rice. *BMC Plant Biology* 23(1), 53.

Lu, W., Deng, F., Jia, J., Chen, X., Li, J. *et al.* (2020) The *Arabidopsis thaliana* gene AtERF019 negatively regulates plant resistance to *Phytophthora parasitica* by suppressing PAMP-triggered immunity. *Molecular Plant Pathology* 21(9), 1179–1193.

Ma, S., Leng, Y., Li, X., Meng, Y., Yin, Z. *et al.* (2023) High spatial resolution mass spectrometry imaging for spatial metabolomics: Advances, challenges, and future perspectives. *TrAC Trends in Analytical Chemistry* 159, 116902.

Macaulay, I.C., Ponting, C.P. and Voet, T. (2017) Single-cell multiomics: Multiple measurements from single cells. *Trends in Genetics* 33(2), 155–168.

Macovei, A., Sevilla, N.R., Cantos, C., Jonson, G.B., Slamet-Loedin, I. *et al.* (2018) Novel alleles of rice eIF4g generated by CRISPR/Cas9-targeted mutagenesis confer resistance to rice tungro spherical virus. *Plant Biotechnology Journal* 16(11), 1918–1927.

Mali, P., Aach, J., Stranges, P.B., Esvelt, K.M., Moosburner, M. *et al.* (2013) CAS9 transcriptional activators for target specificity screening and paired nickases for cooperative genome engineering. *Nature Biotechnology* 31(9), 833–838.

Mall, A.K., Manimekalai, R., Misra, V., Pandey, H., Srivastava, S. *et al.* (2024) CRISPR/Cas-mediated genome editing for sugarcane improvement. *Sugar Tech* 1–13.

Mao, Y., Zhang, H., Xu, N., Zhang, B., Gou, F. *et al.* (2013) Application of the CRISPR-Cas system for efficient genome engineering in plants. *Molecular Plant* 6(6), 2008–2011.

Mehta, D., Stürchler, A., Anjanappa, R.B., Zaidi, S.S.E.A., Hirsch-Hoffmann, M. *et al.* (2019a) Linking CRISPR-Cas9 interference in cassava to the evolution of editing-resistant geminiviruses. *Genome Biology* 20(1), 80.

Mehta, S., James, D. and Reddy, M.K. (2019b) Omics technologies for abiotic stress tolerance in plants: Current status and prospects. In: Wani, S.H. (ed.) *Recent Approaches in Omics for Plant Resilience to Climate Change*. Springer, pp. 1–34.

Mergner, J. and Kuster, B. (2022) Plant proteome dynamics. *Annual Review of Plant Biology* 73, 67–92.

Minne, M., Ke, Y., Saura-Sanchez, M. and De Rybel, B. (2022) Advancing root developmental research through single-cell technologies. *Current Opinion in Plant Biology* 65, 102113.

Misra, V., Mall, A.K., Pandey, H., Srivastava, S. and Sharma, A. (2023) Advancements and prospects of CRISPR/Cas9 technologies for abiotic and biotic stresses in sugar beet. *Frontiers in Genetics* 14, 1235855.

Mo, Y. and Jiao, Y. (2022) Advances and applications of single-cell omics technologies in plant research. *The Plant Journal* 110(6), 1551–1563.

Moore, D.L. and Jessberger, S. (2017) Creating age asymmetry: Consequences of inheriting damaged goods in mammalian cells. *Trends in Cell Biology* 27(1), 82–92.

Moreno-Romero, J., Santos-González, J., Hennig, L. and Köhler, C. (2017) Applying the INTACT method to purify endosperm nuclei and to generate parental-specific epigenome profiles. *Nature Protocols* 12(2), 238–254.

Mushtaq, M., Mukhtar, S., Sakina, A., Dar, A.A., Bhat, R. *et al.* (2020) Tweaking genome-editing approaches for virus interference in crop plants. *Plant Physiology and Biochemistry* 147, 242–250.

Naik, B., Kumar, V., Rizwanuddin, S., Chauhan, M., Choudhary, M. *et al.* (2023) Genomics, proteomics, and metabolomics approaches to improve abiotic stress tolerance in tomato plant. *International Journal of Molecular Sciences* 24(3), 3025.

Nakabayashi, R. and Saito, K. (2015) Integrated metabolomics for abiotic stress responses in plants. *Current Opinion in Plant Biology* 24, 10–16.

Nakabayashi, R., Hashimoto, K., Mori, T., Toyooka, K., Sudo, H. *et al.* (2021) Spatial metabolomics using imaging mass spectrometry to identify the localization of asparaptine A in *Asparagus officinalis*. *Plant Biotechnology* 38(3), 311–315.

Negi, A., Shukla, A., Jaiswar, A., Shrinet, J. and Jasrotia, R.S. (2022) Applications and challenges of microarray and RNA-sequencing. *Bioinformatics* 91–103.

Nejat, N., Ramalingam, A. and Mantri, N. (2018) Advances in transcriptomics of plants. *Plant Genetics and Molecular Biology* 161–185.

Nekrasov, V., Wang, C., Win, J., Lanz, C., Weigel, D. *et al.* (2017) Rapid generation of a transgene-free powdery mildew resistant tomato by genome deletion. *Scientific Reports* 7(1), 482.

Oliva, R., Ji, C., Atienza-Grande, G., Huguet-Tapia, J.C., Perez-Quintero, A. *et al.* (2019) Broad-spectrum resistance to bacterial blight in rice using genome editing. *Nature Biotechnology* 37(11), 1344–1350.

Ortigosa, A., Gimenez-Ibanez, S., Leonhardt, N. and Solano, R. (2019) Design of a bacterial speck resistant tomato by CRISPR/Cas9-mediated editing of SI JAZ 2. *Plant Biotechnology Journal* 17(3), 665–673.

Ousterout, D.G., Kabadi, A.M., Thakore, P.I., Majoros, W.H., Reddy, T.E. *et al.* (2015) Multiplex CRISPR/Cas9-based genome editing for correction of dystrophin mutations that cause Duchenne muscular dystrophy. *Nature Communications* 6(1), 6244.

Park, J., Kim, J., Lewy, T., Rice, C.M., Elemento, O. *et al.* (2022) Spatial omics technologies at multimodal and single cell/subcellular level. *Genome Biology* 23(1), 256.

Peng, A., Chen, S., Lei, T., Xu, L., He, Y. *et al.* (2017) Engineering cankerâ resistant plants through CRISPR/Cas9â targeted editing of the susceptibility gene CsLOB1 promoter in citrus. *Plant Biotechnology Journal* 15(12), 1509–1519.

Rai, G.K., Parveen, A., Jamwal, G., Basu, U., Kumar, R.R. *et al.* (2021) Leaf proteome response to drought stress and antioxidant potential in tomato (*Solanum lycopersicum* L.). *Atmosphere* 12(8), 1021.

Ramnauth, A.D., Tippani, M., Divecha, H.R., Papariello, A.R., Miller, R.A. *et al.* (2023) Spatially-resolved transcriptomics of human dentate gyrus across postnatal lifespan reveals heterogeneity in markers for proliferation, extracellular matrix, and neuroinflammation. *BioRxiv* 2023–11.

Rao, A., Barkley, D., França, G.S. and Yanai, I. (2021) Exploring tissue architecture using spatial transcriptomics. *Nature* 596(7871), 211–220.

Razzaq, M.K., Rauf, S., Khurshid, M., Iqbal, S., Bhat, J.A. *et al.* (2019) Pollen viability an index of abiotic stresses tolerance and methods for the improved pollen viability. *Pakistan Journal of Agricultural Research* 32(4).

Razzaq, M.K., Aleem, M., Mansoor, S., Khan, M.A., Rauf, S. *et al.* (2021) Omics and CRISPR-Cas9 approaches for molecular insight, functional gene analysis, and stress tolerance development in crops. *International Journal of Molecular Sciences* 22(3), 1292.

Razzaq, M.K., Akhter, M., Ahmad, R.M., Cheema, K.L., Hina, A. *et al.* (2022) CRISPR-Cas9 based stress tolerance: New hope for abiotic stress tolerance in chickpea (*Cicer arietinum*). *Molecular Biology Reports* 49(9), 8977–8985.

Rich-Griffin, C., Stechemesser, A., Finch, J., Lucas, E., Ott, S. *et al.* (2020) Single-cell transcriptomics: A high-resolution avenue for plant functional genomics. *Trends in Plant Science* 25(2), 186–197.

Romero, F.M. and Gatica-Arias, A. (2019) CRISPR/Cas9: Development and application in rice breeding. *Rice Science* 26(5), 265–281.

Roszak, P., Heo, J.O., Blob, B., Toyokura, K., Sugiyama, Y. *et al.* (2021) Cell-by-cell dissection of phloem development links a maturation gradient to cell specialization. *Science* 374(6575), eaba5531.

Ryu, K.H., Huang, L., Kang, H.M. and Schiefelbein, J. (2019) Single-cell RNA sequencing resolves molecular relationships among individual plant cells. *Plant Physiology* 179(4), 1444–1456.

Saikia, B., Debbarma, J., Maharana, J., Sastry, G.N. and Chikkaputtaiah, C. (2024) CRISPR/Cas9-based genome editing and functional analysis of SlHyPRP1 and SlDEA1 genes of *Solanum lycopersicum* L. in imparting genetic tolerance to multiple stress factors. *Frontiers in Plant Science* 15, 1304381.

Satterlee, J.W., Strable, J. and Scanlon, M.J. (2020) Plant stem-cell organization and differentiation at single-cell resolution. *Proceedings of the National Academy of Sciences* 117(52), 33689–33699.

Schiml, S., Fauser, F. and Puchta, H. (2014) The CRISPR/Cas a system can be used as nuclease for in planta gene targeting and as paired nickases for directed mutagenesis in *Arabidopsis* resulting in heritable progeny. *The Plant Journal* 80(6), 1139–1150.

Seyfferth, C., Renema, J., Wendrich, J.R., Eekhout, T., Seurinck, R. *et al.* (2021) Advances and opportunities in single-cell transcriptomics for plant research. *Annual Review of Plant Biology* 72, 847–866.

Shahan, R., Hsu, C.-W., Nolan, T.M., Cole, B.J., Taylor, I.W. *et al.* (2022) A single-cell *Arabidopsis* root atlas reveals developmental trajectories in wild-type and cell identity mutants. *Developmental Cell* 57(4), 543–560.

Shan, Q., Wang, Y., Li, J., Zhang, Y., Chen, K. *et al.* (2013) Targeted genome modification of crop plants using a CRISPR-Cas system. *Nature Biotechnology* 31(8), 686–688.

Shaner, D.L. (2000) The impact of glyphosate-tolerant crops on the use of other herbicides and on resistance management. *Pest Management Science: Formerly Pesticide Science* 56(4), 320–326.

Sharma, M. and Pandey, G.K. (2017) Editorial: Genomics and functional genomics of stress-mediated signaling in plants. vloume i. *Current Genomics* (6), 467–468.

Shaw, R., Tian, X. and Xu, J. (2021) Single-cell transcriptome analysis in plants: Advances and challenges. *Molecular Plant* 14(1), 115–126.

Shendekar, S., Mangla, S., Gore, V., Meshram, M. and Raut, D. (2024) The CRISPR revolution: Editing plant genomes for abiotic stress resilience and a better tomorrow. *Biotechnology and Biological Sciences* 93.

Shiraku, M.L., Magwanga, R.O., Zhang, Y., Hou, Y., Kirungu, J.N. *et al.* (2022) Late embryogenesis abundant gene LEA3 (GH_A08g0694) enhances drought and salt stress tolerance in cotton. *International Journal of Biological Macromolecules* 207, 700–714.

Shulse, C.N., Cole, B.J., Ciobanu, D., Lin, J., Yoshinaga, Y. *et al.* (2019) High-throughput single-cell transcriptome profiling of plant cell types. *Cell Reports* 27(7), 2241–2247.

Silva, C.J., Abeele, C., Ortega-Salazar, I., Papin, V., Adaskaveg, J.A. *et al.* (2021) Host susceptibility factors render ripe tomato fruit vulnerable to fungal disease despite active immune responses. *Journal of Experimental Botany* 72(7), 2696–2709.

Singh, B., Salaria, N., Thakur, K., Kukreja, S., Gautam, S. *et al.* (2019) Functional genomic approaches to improve crop plant heat stress tolerance. *F1000Research* 8, 1721.

Srikanth, M., Tata, S.S. and Godi, S. (2021) Abiotic stress tolerance enhancement in rice through CRISPR/Cas9 approach. *Research Journal of Agricultural Sciences* 12(5), 1543–1546.

Sun, C., Li, T., Song, X., Huang, L., Zang, Q. *et al.* (2019) Spatially resolved metabolomics to discover tumor-associated metabolic alterations. *Proceedings of the National Academy of Sciences* 116(1), 52–57.

Tarazona, S., Arzalluz-Luque, A. and Conesa, A. (2021) Undisclosed, unmet and neglected challenges in multi-omics studies. *Nature Computational Science* 1(6), 395–402.

Tenorio Berrío, R., Verstaen, K., Vandamme, N., Pevernagie, J., Achon, I. *et al.* (2022) Single-cell transcriptomics sheds light on the identity and metabolism of developing leaf cells. *Plant Physiology* 188(2), 898–918.

Thudi, M., Palakurthi, R., Schnable, J.C., Chitikineni, A., Dreisigacker, S. *et al.* (2021) Genomic resources in plant breeding for sustainable agriculture. *Journal of Plant Physiology* 257, 153351.

Tripathy, S.K., Majhi, P.K. and Ahamad, A. (2021) Achieving drought tolerance in rice by targeted genome editing. *Biotica Research Today* 3(11), 1065–1067.

Tyagi, P., Singh, D., Mathur, S., Singh, A. and Ranjan, R. (2022) Upcoming progress of transcriptomics studies on plants: An overview. *Frontiers in Plant Science* 13, 1030890.

Usman, B., Nawaz, G., Zhao, N., Liu, Y. and Li, R. (2020) Generation of high yielding and fragrant rice (*Oryza sativa* L.) lines by CRISPR/Cas9 targeted mutagenesis of three homoeologs of cytochrome P450 gene family and OsBADH2 and transcriptome and proteome profiling of revealed changes triggered by mutations. *Plants* 9(6), 788.

Valette, M., Rey, M., Gerin, F., Comte, G. and Wisniewski-Dyé, F. (2020) A common metabolomic signature is observed upon inoculation of rice roots with various rhizobacteria. *Journal of Integrative Plant Biology* 62(2), 228–246.

Vandereyken, K., Sifrim, A., Thienpont, B. and Voet, T. (2023) Methods and applications for single-cell and spatial multi-omics. *Nature Reviews Genetics* 24(8), 494–515.

Vidyasagar, A. (2018) *What is CRISPR*. Live Science.

Vo, K.T.X., Rahman, M.M., Rahman, M.M., Trinh, K.T.T., Kim, S.T. *et al.* (2021) Proteomics and metabolomics studies on the biotic stress responses of rice: An update. *Rice* 14, 1–16.

Walley, J.W., Sartor, R.C., Shen, Z., Schmitz, R.J., Wu, K.J. *et al.* (2016) Integration of omic networks in a developmental atlas of maize. *Science* 353(6301), 814–818.

Wan, D.Y., Guo, Y., Cheng, Y., Hu, Y., Xiao, S. *et al.* (2020) CRISPR/Cas9-mediated mutagenesis of VvMLO3 results in enhanced resistance to powdery mildew in grapevine (*Vitis vinifera*). *Horticulture Research* 7.

Wang, J.Y. and Doudna, J.A. (2023) CRISPR technology: A decade of genome editing is only the beginning. *Science* 379(6629), eadd8643.

Wang, F., Wang, C., Liu, P., Lei, C., Hao, W. *et al.* (2016) Enhanced rice blast resistance by CRISPR/Cas9-targeted mutagenesis of the ERF transcription factor gene OsERF922. *PLoS One* 11(4), e0154027.

Wang, R., Jin, C. and Hu, X. (2017) Evidence of drug-response heterogeneity rapidly generated from a single cancer cell. *Oncotarget* 8(25), 41113–41124.

Wang, T., Deng, Z., Zhang, X., Wang, H., Wang, Y. *et al.* (2018) Tomato DCL2b is required for the biosynthesis of 22-nt small RNAs, the resulting secondary siRNAs, and the host defense against ToMV. *Horticulture Research* 5, 62.

Wang, B., Xie, G., Liu, Z., He, R., Han, J. *et al.* (2019) Mutagenesis reveals that the OsPPA6 gene is required for enhancing the alkaline tolerance in rice. *Frontiers in Plant Science* 10, 759.

Wang, B., Zhong, Z., Wang, X., Han, X., Yu, D. *et al.* (2020) Knockout of the OsNAC006 transcription factor causes drought and heat sensitivity in rice. *International Journal of Molecular Sciences* 21(7), 2288.

Wang, T., Xun, H., Wang, W., Ding, X., Tian, H. *et al.* (2021) Mutation of gmaitr genes by CRISPR/Cas9 genome editing results in enhanced salinity stress tolerance in soybean. *Frontiers in Plant Science* 12, 779598.

Wang, X., Guo, Y., Wang, Y., Peng, Y., Zhang, H. *et al.* (2024) ZmHDT103 negatively regulates drought stress tolerance in maize seedlings. *Agronomy* 14(1), 134.

Wendrich, J.R., Yang, B., Vandamme, N., Verstaen, K., Smet, W. *et al.* (2020) Vascular transcription factors guide plant epidermal responses to limiting phosphate conditions. *Science* 370(6518), eaay4970.

Wu, J., Kang, C., Guo, L., Liu, K., Tu, J. *et al.* (2020) Roles of the *Brassica napus* DELLA protein BnaA6.RGA, in modulating drought tolerance by interacting with the ABA signaling component BnaA10.ABF2. *Frontiers in Plant Science* 11, 527615.

Xie, Y., Jiang, S., Li, L., Yu, X., Wang, Y. *et al.* (2020) Single-cell RNA sequencing efficiently predicts transcription factor targets in plants. *Frontiers in Plant Science* 11, 603302.

Xu, X., Crow, M., Rice, B.R., Li, F., Harris, B. *et al.* (2021) Single-cell RNA sequencing of developing maize ears facilitates functional analysis and trait candidate gene discovery. *Developmental Cell* 56(4), 557–568.

Yang, W., Qi, W., Li, Y., Wang, J. and Jiang, L. (2020) Programmed sequential cutting endows Cas9 versatile base substitution capability in plants. *Science China Life Sciences* 12(1), 388–397.

Yang, Y., Saand, M.A., Huang, L., Abdelaal, W.B., Zhang, J. *et al.* (2021) Applications of multi-omics technologies for crop improvement. *Frontiers in Plant Science* 12, 563953.

Yin, R., Xia, K. and Xu, X. (2023) Spatial transcriptomics drives a new era in plant research. *The Plant Journal* 116(6), 1571–1581.

Yu, X., Liu, Z. and Sun, X. (2023) Single-cell and spatial multi-omics in the plant sciences: Technical advances, applications, and perspectives. *Plant Communications* 4(3), 100508.

Zeilmaker, T., Ludwig, N.R., Elberse, J., Seidl, M.F., Berke, L. *et al.* (2015) DOWNY MILDEW RESISTANT 6 and DMR6-LIKE OXYGENASE 1 are partially redundant but distinct suppressors of immunity in *Arabidopsis*. *The Plant Journal* 81(2), 210–222.

Zeng, Y., Wen, J., Zhao, W., Wang, Q. and Huang, W. (2020) Rational improvement of rice yield and cold tolerance by editing the three genes OsPIN5b, GS3, and OsMYB30 with the CRISPR–Cas9 system. *Frontiers in Plant Science* 10, 478091.

Zhang, Y., Bai, Y., Wu, G., Zou, S., Chen, Y. *et al.* (2017) Simultaneous modification of three homoeologs of Ta EDR1 by genome editing enhances powdery mildew resistance in wheat. *The Plant Journal* 91(4), 714–724.

Zhang, T.Q., Xu, Z.G., Shang, G.D. and Wang, J.W. (2019) A single-cell RNA sequencing profiles the developmental landscape of *Arabidopsis* root. *Molecular Plant* 12(5), 648–660.

Zhang, T.Q., Chen, Y. and Wang, J.W. (2021) A single-cell analysis of the *Arabidopsis* vegetative shoot apex. *Developmental Cell* 56(7), 1056–1074.

Zhang, J., Liao, J., Ling, Q., Xi, Y. and Qian, Y. (2022) Genome-wide identification and expression profiling analysis of maize AP2/ERF superfamily genes reveal essential roles in abiotic stress tolerance. *BMC Genomics* 23(1), 125.

Zheng, N., Li, L. and Wang, X. (2020) Molecular mechanisms, off-target activities, and clinical potentials of genome editing systems. *Clinical and Translational Medicine* 10(1), 412–426.

Zheng, M., Lin, J., Liu, X., Chu, W., Li, J. *et al.* (2021) Histone acetyltransferase TaHAG1 acts as a crucial regulator to strengthen salt tolerance of hexaploid wheat. *Plant Physiology* 186(4), 1951–1969.

Zhou, J., Peng, Z., Long, J., Sosso, D., Liu, B. *et al.* (2015) Gene targeting by the TAL effector PthXo2 reveals cryptic resistance gene for bacterial blight of rice. *The Plant Journal* 82(4), 632–643.

Zhu-Salzman, K., Salzman, R.A., Ahn, J.E. and Koiwa, H. (2004) Transcriptional regulation of sorghum defense determinants against a phloem-feeding aphid. *Plant Physiology* 134(1), 420–431.

19 Transgene-free CRISPR-edited Crops for Stress-resilient Agriculture: Technological Advancements and Applications

Saniya[1], Sarfraz Ahmad[2*], Maksud Hasan Shah[3], Mohammad Ghani Raghib[4], Vikas Belwal[5] and Pankaj Kumar[6]

[1]Department of Horticulture, ICAR-National Research Centre for Grapes, Pune, Maharashtra, India; [2]Department of Genetics and Plant Breeding, College of Agriculture Mandawa, Swami Keshwanand Rajasthan Agriculture University, Bikaner, Rajasthan, India; [3]Department of Agronomy, College of Agriculture Chandgothi, Swami Keshwanand Rajasthan Agriculture University, Bikaner, Rajasthan, India; [4]Department of Horticulture, Faculty of Agriculture, Nangarhar University, Afghanistan; [5]Department of Tree Improvement, College of Forestry Ranichauri, VCSG Uttarakhand University of Horticulture and Forestry, Bharsar, Uttarakhand, India; [6]Department of Crop Improvement, College of Forestry Ranichauri, VCSG Uttarakhand University of Horticulture and Forestry, Bharsar, Uttarakhand, India

Abstract

Recent advancements in genome-editing technologies, particularly the clustered regularly interspaced short palindromic repeats (CRISPR) system, have revolutionized agricultural science. However, regulatory concerns persist due to the involvement of transgenes in some applications. The emergence of transgene-free CRISPR-edited crops has created new opportunities for developing stress-resilient agricultural systems. These crops are engineered to withstand various abiotic and biotic stresses, such as drought, salinity, pests, and diseases, without introducing foreign genetic material. This transgene-free approach addresses regulatory challenges and improves public acceptance, making it a promising strategy for sustainable agriculture. This chapter explores the paradigm shift in agricultural biotechnology through the adoption of transgene-free CRISPR-edited crops to enhance stress resilience in the face of growing environmental challenges. It provides a comprehensive review of the technological advancements that enable precise genome editing without the incorporation of foreign genes. The chapter highlights the evolution of CRISPR-Cas technology and its crucial role in developing stress-resistant crops, emphasizing the advantages of transgene-free methods. Case studies are included to demonstrate the successful application of transgene-free CRISPR editing in major crops, showcasing improved yields, enhanced nutritional profiles, and increased sustainability. By synthesizing recent technological innovations, practical applications, and regulatory considerations, this chapter contributes to the ongoing dialog on using cutting-edge biotechnological solutions to address global food security challenges.

*Corresponding author: sarfraz_firoz@rediffmail.com

© CAB International 2025. *Stress-Resilient Crops: Coordinated omics-CRISPR-Nanotechnology Strategies* (ed. Jen-Tsung Chen)
DOI: 10.1079/9781800627307.0019

Keywords: CRISPR-Cas9, Drought, Transgene-free, Transgene killer CRISPR, Ribonucleoprotein, Stress-resilient

19.1 Introduction

Agriculture, as the foundation of food production, is highly susceptible to environmental changes. Rising global temperatures, increased frequency of droughts, and soil salinization caused by erratic rainfall patterns are leading to declining crop yields in many regions (FAO, 2021). Researchers are developing drought-resistant crops by studying secondary metabolites (SMs) and phytohormones (PHs) involved in stress signaling, as crops unable to tolerate these stresses often exhibit stunted growth or fail to mature (Ahmad *et al.*, 2023). For example, maize and rice, two of the world's most important crops, show considerable yield reduction when subjected to water stress and high salinity (Muthu *et al.*, 2020; Liao *et al.*, 2022). The Food and Agriculture Organization (FAO) predicts that the negative impacts of climate change on food production could push millions of people into hunger by 2050 (FAO, 2021). Therefore, enhancing crop resilience to abiotic (e.g. drought, salinity, and temperature extremes) and biotic (e.g. pests and diseases) stresses has become an urgent priority in the effort to ensure global food security.

Historically, crop improvement relied on traditional breeding techniques such as selective breeding and hybridization. These methods have contributed immensely to increasing crop yields and pest resistance. However, conventional breeding is often slow, requiring multiple generations to develop desired traits. Moreover, its reliance on natural genetic diversity means that breeders may not always be able to identify and propagate specific traits such as drought tolerance or disease resistance. Modern breeding techniques, such as marker-assisted selection (MAS), have accelerated trait selection, but even MAS is limited by the availability of markers closely linked to desired traits. Moreover, many important agronomic traits are controlled by multiple genes, making their improvement more challenging using traditional approaches.

Genetic engineering (GE) has introduced new possibilities for crop improvement by enabling the direct introduction of genes into plants. GE crops, such as Bt cotton and Roundup Ready soybeans, have provided benefits such as pest resistance and herbicide tolerance (I.S.A.A.A, 2019). However, the introduction of foreign DNA (transgenes) into plants has been controversial, with concerns over environmental safety, potential health risks, and ethical issues regarding the creation of 'unnatural' organisms (Vega Rodríguez *et al.*, 2022). Regulatory barriers are also a significant obstacle for transgenic crops. Many countries, particularly in the European Union, have strict regulations on the cultivation and commercialization of genetically modified organisms (GMOs). Public opposition to transgenic crops remains strong, particularly in regions where agriculture plays a central role in the economy and food security.

The discovery of CRISPR-Cas9, a revolutionary genome-editing tool, has transformed plant biotechnology by offering a precise, efficient, and cost-effective method for modifying plant genomes (Ding *et al.*, 2016). Unlike transgenic methods, which introduce foreign DNA into the plant genome, CRISPR-Cas9 allows for the alteration of native genes, leading to site-specific mutations or gene knockouts without the integration of foreign DNA (Jacinto *et al.*, 2020). This transgene-free approach is particularly appealing for developing crops that meet regulatory and public acceptance criteria. Several countries, including the USA and Japan, have ruled that gene-edited crops without foreign DNA are not subject to GMO regulations (Turnbull *et al.*, 2021). This opens up new opportunities for the deployment of stress-resilient crops without the socio-political challenges associated with transgenic plants.

This chapter explores the technological advancements in transgene-free CRISPR editing and their applications in developing stress-resilient crops. It provides a comprehensive overview of the current state of CRISPR technology, highlights the latest innovations in the field, and discusses the potential implications for sustainable agriculture. As the global agricultural landscape continues to evolve, transgene-free CRISPR-edited crops offer a promising solution for addressing the pressing challenges of food security and environmental sustainability.

19.2 CRISPR-Cas9 Technology: An Overview

CRISPR-Cas9 was first identified as part of an adaptive immune system in bacteria and archaea, where it provides defense against invading viruses (Tyagi *et al.*, 2021). The system consists of two main components: a short RNA molecule called guide RNA (gRNA) and the Cas9 nuclease enzyme. The gRNA guides the Cas9 enzyme to a specific DNA sequence, where Cas9 introduces a double-strand break (DSB). The DSB is repaired by the plant's DNA repair machinery, resulting in either targeted mutations (via non-homologous end-joining) or precise corrections (via homology-directed repair) (Doudna and Charpentier, 2014).

In plant systems, CRISPR-Cas9 can be deployed in a variety of ways. The most common method involves delivering the gRNA and Cas9 components either through *Agrobacterium*-mediated transformation, particle bombardment, or protoplast transfection (Wu *et al.*, 2020). In the case of transgene-free editing, ribonucleoprotein (RNP) complexes consisting of Cas9 protein, and gRNA are delivered directly to plant cells. These components act transiently, editing the target DNA without integrating into the plant's genome, ensuring no foreign DNA is left behind (Gu *et al.*, 2021). CRISPR-induced mutations can lead to gene knockouts, which are particularly useful for conferring traits like pest resistance or improved stress tolerance. For example, knocking out susceptibility genes (S-genes) involved in pathogen infection can lead to enhanced resistance against bacterial and fungal diseases (Debbarma *et al.*, 2023).

19.3 Technological Advancements in CRISPR-Cas9 for Transgene-free Crop Editing

In the rapidly evolving field of genetic engineering, the development of transgene-free CRISPR-edited crops represents a significant milestone, particularly for stress-resilient agriculture. Transgene-free crops are genetically modified to possess beneficial traits, such as increased resistance to biotic and abiotic stresses, without introducing foreign DNA from other species. This makes them more acceptable in terms of regulatory frameworks and public perception, as they closely mimic natural mutations and conventional breeding outcomes. Fig. 19.1 presents an overview of the steps involved in the transgene-free CRISPR-edited technology.

19.3.1 The concept of transgene-free CRISPR editing

The concept of transgene-free CRISPR editing revolves around the ability to introduce precise mutations in a plant genome without inserting exogenous genes. Traditional genetic modification techniques involve the introduction of foreign DNA into a plant, often from unrelated species, which raises environmental and safety concerns. In contrast, CRISPR-based technologies enable highly targeted mutations by utilizing the cell's native DNA repair mechanisms.

In transgene-free CRISPR editing, the CRISPR-Cas9 system is often delivered transiently to the plant cells, and once the desired genetic modification is made, the CRISPR machinery is removed. The resulting plants do not carry any foreign genes, but the desired trait remains due to the precise editing event (Woo *et al.*, 2015). A brief comparison among transgene-free and transgenic CRISPR-edited crops is presented in Table 19.1.

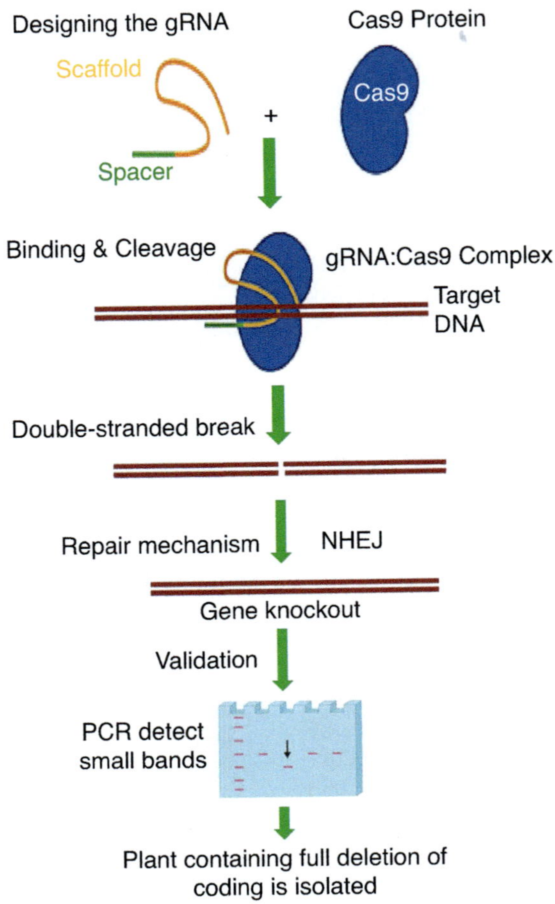

Fig. 19.1. An overview of the steps involved in the transgene-free CRISPR editing technology.

Table 19.1. The difference between transgene-free and transgenic CRISPR-edited crops.

Particulars	Transgene-free CRISPR	Transgenic CRISPR
Introduction of foreign DNA	No permanent integration of foreign DNA	Permanent integration of foreign DNA
Delivery method	Cas9 and gRNA delivered transiently (as RNA, RNPs)	Delivered as DNA via plasmids or vectors
Regulation	Less regulated (depending on the country)	More strictly regulated (considered GMO)
Selection markers	Often no need for selection markers	Typically includes selection markers
Public perception	Considered non-GMO, higher acceptance	Considered GMO, lower acceptance
Application	Gene knockout, gene correction, non-GMO crops	Creation of transgenic organisms, stable trait integration

19.3.2 Techniques for developing transgene-free CRISPR-edited crops

The primary techniques that have been developed to ensure this transgene-free outcome are detailed below.

19.3.2.1 Ribonucleoprotein (RNP) delivery

One of the most effective methods for achieving transgene-free crops involves the delivery of pre-assembled CRISPR-Cas9 ribonucleoproteins (RNPs) directly into plant cells. This approach bypasses the need to integrate any foreign DNA into the genome, as the CRISPR-Cas9 complex is directly introduced as a protein-RNA complex. Once inside the plant cell, the Cas9 enzyme makes a precise cut at the targeted location in the genome, and the plant's natural repair mechanisms, such as non-homologous end-joining (NHEJ) or homology-directed repair (HDR), fix the break, introducing the desired mutation (Najafi *et al.*, 2023).

RNP delivery is particularly advantageous for genome editing because the editing machinery degrades rapidly within the cell, reducing the likelihood of off-target effects. As no transgene integration occurs, plants regenerated from edited cells are free of foreign DNA, which minimizes regulatory hurdles. This method has already been successfully applied in crops like wheat (Liang *et al.*, 2017), demonstrating its potential for broader agricultural use. Additionally, the pre-assembled Cas9 and guide RNA in RNPs allow for immediate editing activity, enhancing efficiency. Since RNPs do not replicate within the cell, the risks of long-term genome alterations and immune responses are minimized. This versatile and precise method can be applied across a wide range of species, making it a valuable tool for safe and efficient genome editing (Park and Choe, 2019).

19.3.2.2 Virus-induced gene editing (VIGE)

VIGE is another innovative approach to developing transgene-free crops. In this technique, plant viruses are engineered to carry gRNAs specific to the target gene. Once the virus infects the plant, it induces CRISPR-mediated editing at the desired location without integrating any foreign DNA into the plant's genome. Importantly, the virus can be eliminated from the plant following the editing process, leaving behind only the desired genetic modification (Son and Park, 2022).

This method is particularly useful for crops that are difficult to transform using traditional methods, and it also allows for multiplexed gene editing, where multiple genes can be targeted simultaneously. VIGE has the potential to be a rapid and cost-effective method for developing stress-resilient crops, particularly in regions where conventional breeding is limited by environmental conditions.

19.3.2.3 Protoplast transformation and regeneration

Protoplasts, which are plant cells that have had their cell walls removed, can be used as a platform for introducing CRISPR-Cas9 components without the integration of foreign DNA. In this method, protoplasts are transiently transformed with CRISPR-Cas9 plasmids, RNA, or protein complexes, and the desired edits are made before the protoplasts are regenerated into whole plants (Najafi *et al.*, 2023). Because the editing components do not integrate into the plant genome, the regenerated plants are transgene-free.

This method has shown promise in crops such as rice and tobacco (Lin *et al.*, 2020), although the regeneration of protoplasts into viable plants can be challenging for some species. Nevertheless, advances in protoplast culture techniques continue to improve the efficiency of this approach, making it a viable option for developing stress-resilient crops.

19.3.2.4 Particle bombardment

Particle bombardment, also known as biolistic transformation, involves using high-velocity microparticles coated with CRISPR components (e.g. DNA plasmids or RNPs) to penetrate plant cells. This method can be applied to a wide range of plant species, especially those that are recalcitrant to *Agrobacterium*-mediated transformation (Sandhya *et al.*, 2020).

While biolistic transformation traditionally involved the delivery of DNA, advancements in RNP-based systems have allowed for transgene-free editing using this technique. For example, Dong *et al.* (2021) demonstrated the successful introduction of gene edits in maize using RNPs delivered via particle bombardment.

19.3.2.5 Agrobacterium-mediated delivery

Agrobacterium tumefaciens, a natural plant pathogen, has long been used as a vector for introducing foreign DNA into plant cells. Recent advancements, however, have focussed on using *Agrobacterium* for transient CRISPR component expression, where no foreign DNA is permanently integrated into the plant genome (Veillet *et al.*, 2019).

In this approach, CRISPR-Cas9 components are expressed temporarily, creating edits before the foreign DNA is naturally degraded by the plant's cellular mechanisms. This transient method reduces the risk of off-target effects and enables the production of transgene-free plants, although optimizing the transient expression system is still a challenge.

19.3.2.6 Transgene killer CRISPR (TKC) system

It is an innovative approach designed to eliminate transgenes from GMOs. This system takes advantage of the CRISPR-Cas technology to remove or inactivate transgenes introduced during genetic engineering processes, ensuring that the resulting organism no longer carries any foreign DNA. The TKC system relies on the CRISPR-Cas9 machinery, but it is specially engineered to target and cleave transgene sequences introduced into the organism's genome.

Yubing *et al.* developed a TKC system in rice that facilitates the active and automatic removal of transgenes in edited progeny. This TKC technology operates by incorporating two cassettes, which express the suicide genes *barnase* and *CMS2* into the CRISPR/Cas9 construct (Yubing *et al.*, 2019).

19.3.3 Base editing and prime editing: next-generation precision tools

Beyond conventional CRISPR-Cas9, which primarily induces double-strand breaks (DSBs) followed by NHEJ or HDR, newer genome-editing tools like base editing and prime editing offer greater precision without the need for DSBs. These innovations are essential for developing transgene-free crops with minimal off-target effects.

19.3.3.1 Base editing

Base editing, a recent innovation in genome editing, enables the conversion of a single nucleotide base into another without introducing DSBs (Lee *et al.*, 2021). Base editors consist of a catalytically inactive Cas9 protein (Cas9 nickase) fused to a deaminase enzyme, which can convert one nucleotide base into another. The most common types are cytosine base editors (CBEs) and adenine base editors (ABEs). For example, cytosine base editors convert cytosine (C) into thymine (T), while adenine base editors convert adenine (A) into guanine (G). These base changes can create targeted mutations that enhance stress tolerance in crops. Importantly, base editing avoids the creation of DSBs, reducing the risk of unintended large-scale genome rearrangements.

Base editing has been applied to confer abiotic stress tolerance in rice by editing genes involved in salinity and drought resistance (Lou *et al.*, 2017). In particular, editing OsSAPK2 in rice using base editors improved salt tolerance by enhancing the plant's ability to cope with ion imbalances under saline conditions.

19.3.3.2 Prime editing

Prime editing represents the latest advancement in precision genome editing. Unlike base editing, which is limited to specific nucleotide transitions, prime editing allows for more complex insertions, deletions, and substitutions without the need for DSBs (Kantor *et al.*, 2020). Prime editors use a Cas9 nickase fused to a reverse transcriptase enzyme and a prime editing guide RNA (pegRNA) that directs the precise edits.

This technique has immense potential for plant genome editing, as it enables the introduction of precise modifications that could enhance stress tolerance, such as inserting drought-resistance alleles or correcting mutations that reduce crop productivity under stress conditions. Prime editing has not yet been widely applied in plants due to challenges in delivering complex machinery efficiently. However, its potential for producing transgene-free crops with precisely targeted traits holds promise for future crop improvement efforts.

19.3.4 Minimizing off-target effects

One of the concerns associated with CRISPR-Cas9 editing is the possibility of off-target effects, where the Cas9 enzyme cuts DNA at unintended sites. These off-target effects can lead to unintended mutations, which may affect the plant's phenotype or introduce regulatory concerns. To minimize off-target effects, researchers have developed several strategies:

1. *Improved guide RNA design:* By optimizing the design of gRNAs to ensure high specificity for the target sequence, researchers can reduce the likelihood of off-target binding. Tools such as CRISPR-P and CRISPR RGEN tools allow for the computational prediction and minimization of off-target sites (Manghwar *et al.*, 2020).
2. *Modified Cas9 variants*: High-fidelity (HF) Cas9 variants have been engineered to enhance the specificity of DNA cleavage, significantly reducing off-target effects while maintaining editing efficiency (Kleinstiver *et al.*, 2016).
3. *Cas9 nuclease variants*: Cas9 nickase and dCas9-FokI fusion proteins are modified forms of the Cas9 enzyme that introduce nicks or single-strand breaks instead of DSBs. These variants are less likely to cause large-scale genome rearrangements or off-target mutations (Naeem *et al.*, 2020).

By incorporating these strategies, researchers can make genome edits that are both precise and safe, which is critical for the regulatory approval and public acceptance of CRISPR-edited crops.

19.4 Applications of Transgene-free CRISPR-edited Crops

The advent of transgene-free CRISPR-edited crops has opened up new possibilities for developing stress-resilient agricultural systems. These crops are designed to withstand various abiotic and biotic stresses, such as drought, salinity, pests, and diseases, without the introduction of foreign genetic material. Applications of transgene-free CRISPR-edited technique in the development of stress-resistant crops are presented in Table 19.2.

19.4.1 Drought tolerance

Drought is one of the most significant environmental stresses affecting global agricultural productivity. As climate change intensifies, water scarcity is becoming an increasingly critical issue for crop survival. Several transgene-free CRISPR-edited crops have been developed to improve drought tolerance by targeting key genes involved in water-use efficiency, root structure, and stress-response pathways.

19.4.1.1 CRISPR-mediated modifications in rice

Rice (*Oryza sativa*), a staple food for over half of the global population, is highly susceptible to drought stress, which can severely impact yields. Using CRISPR-Cas9, researchers have edited several genes that play essential roles in the plant's drought response. For instance, the gene *OsSAPK2* (SNF1-related protein kinase2) was edited to improve drought tolerance. The knockout of this gene increased the plant's ability to maintain water uptake and photosynthetic efficiency under drought conditions (Sami *et al.*, 2021).

Another example is the modification of the *OsDREB2A* gene, a transcription factor involved in the drought response. CRISPR-Cas9 was used to enhance the expression of this gene, leading to improved drought resistance through better regulation of stress-related proteins (Du *et al.*, 2023).

Gene editing of drought tolerance-related genes, such as Reveille 7 (RVE7) and 4-coumarate ligase (4CL), has shown promising results in chickpea. RVE7, an MYB transcription factor regulating the circadian rhythm, and 4CL, involved in the lignin biosynthesis pathway, were targeted in chickpea using DNA-free CRISPR/Cas9. The editing achieved greater precision in

Table 19.2. Applications of transgene-free CRISPR-edited crops for stress-resilient agriculture.

Type of stress	Crop	Target gene	CRISPR strategy	Observed effects	Reference
Drought	Arabidopsis	*miR169a*	Knock out the regulatory gene *miR169a*	Improved drought tolerance	Zhao *et al.*, 2016
Drought	Arabidopsis	*STL1*	Knock out the *STL1*	Improved drought tolerance	Wang *et al.*, 2022
Drought	Rice	*OsPYL1*	Knockout of ABA receptor gene	Enhanced drought tolerance through improved water-use efficiency and stomatal regulation	Chakraborty *et al.*, 2024
Drought	Rice	*SRL1, SRL2*	Knockout of *SRL1* & *SRL2* genes	Enhanced drought tolerance by regulating leaf morphology	Liao *et al.*, 2019
Drought	Rice	*ERA1*	Knockout of regulatory gene *ERA1*	Enhanced drought tolerance through regulation of ABA signaling and the dehydration response	Ogata *et al.*, 2020
Drought	Wheat	*DREB2A*	Knock-in of drought-responsive transcription factor	Enhanced drought tolerance through improved root architecture and water retention	Hrmova and Hussain, 2021
Salinity	Rice	*GI*	Knockout of regulatory gene	Enhanced salinity tolerance	Wang *et al.* 2021b
Salinity	Rice	*SLR1*	Knockout of gibberellin signaling gene	Enhanced salinity tolerance by regulating ion homeostasis and sodium exclusion	Liu *et al.*, 2021
Salinity	Rice	*RAV2*	Knockout of the promoter element	Enhanced salinity through the regulation of developmental processes	Duan *et al.*, 2016
Salinity	Rice	*RR9, RR10*	Knockout of regulatory gene	Enhanced salinity through cytokinin signaling modification	Wang *et al.*, 2019
Salinity	Wheat	*HKT1*	Knockout of sodium transporter gene	Improved salinity tolerance by restricting harmful sodium ion uptake	Singh *et al.*, 2021

Continued

Table 19.2. Continued

Type of stress	Crop	Target gene	CRISPR strategy	Observed effects	Reference
Salinity	Soybean	*GmAITR*	Knockout of transcription factor and regulatory gene	Improved salinity tolerance by plant development and seed size regulation	Wang *et al.* 2021a
Salinity	Tomato	*HyPRP1*	Knockout of regulatory gene	Enhanced salinity tolerance by modulating trans-membrane protein	Tran *et al.*, 2021
Salinity	Tomato	*ABIG1*	Knockout of transcription factor and regulatory gene	Improved salinity tolerance	Ding *et al.*, 2022
Drought and salinity	Cotton	*ABI5*	Knockout of *ABI5* transcription factor	Improved tolerance to drought and salinity through altered ABA signaling pathways	Luo *et al.*, 2020
Biotic stress (pathogen)	Soybean	*EDS1*	Knockout of *EDS1* gene	Enhanced resistance to multiple pathogens through modulation of immune signaling pathways	Liu *et al.*, 2015
Powdery mildew (disease)	Barley	*MLO*	Knockout of susceptibility gene	Durable resistance to powdery mildew without transgene insertion	Kusch and Panstruga, 2017
Fusarium head blight (disease)	Barley	*RGA*	Knockout of *RGA* gene	Enhanced resistance to Fusarium head blight by disrupting fungal pathogen interaction	Cheng *et al.*, 2022
Bacterial blight (disease)	Rice	*SWEET*	Knockout of susceptibility gene	Increased resistance to bacterial blight by inhibiting pathogen entry	Zafar *et al.*, 2020
Heat	Maize	*ZmTMM*	Deletion of stomatal regulatory gene	Improved heat tolerance through reduced stomatal density and better temperature regulation	Hasanuzzaman *et al.*, 2023

Continued

Table 19.2. Continued

Type of stress	Crop	Target gene	CRISPR strategy	Observed effects	Reference
Cold	Tomato	*CBF*	Knock-in of cold-responsive transcription factor	Increased cold tolerance through upregulated expression of cold-responsive pathways	Li *et al.*, 2018
Nitrogen deficiency	Rice	*NRT1.1B*	Knockout of nitrogen transporter gene	Improved nitrogen-use efficiency and better growth in nitrogen-poor soils	Aluko *et al.*, 2023
Multiple stresses	Rice	*TIR1*	Knockout of *TIR1* auxin receptor gene	Increased resilience to various abiotic stresses by modifying auxin signaling pathways	Wu *et al.*, 2024
Multiple stresses	Potato	*FT*	Multiplex CRISPR editing targeting flowering genes	Extended growing season and improved resilience to abiotic stresses	Tiwari *et al.*, 2020
General stress marker	Tomato	*PDS*	Knockout of *PDS* gene	Used as a marker for successful CRISPR editing; not directly linked to stress resilience	Saikia *et al.*, 2024

knocking out the RVE7 gene compared to the 4CL gene (Jain *et al.*, 2023). These modifications demonstrate the potential of CRISPR-edited crop varieties in mitigating the adverse effects of water scarcity.

19.4.1.2 Wheat for drought and heat resilience

Wheat (*Triticum aestivum*) is another crop that faces significant challenges under drought conditions. CRISPR-Cas9 has been used to edit genes that control root architecture and osmotic stress responses, improving water uptake and enhancing drought resilience. For example, the *TaDREB3* gene, involved in the abscisic acid (ABA)-mediated drought-response pathway, was targeted for mutation, resulting in wheat plants with deeper root systems and increased drought tolerance (Trono and Pecchioni, 2022).

By targeting genes like *OsNAC14* and *OsNAC29*, which regulate root length and stress-response mechanisms, transgene-free wheat plants have been generated with enhanced adaptability to both drought and heat stresses. Such genetic improvements in wheat can help sustain yields in water-scarce environments.

19.4.2 Salt tolerance

Saline soils affect crop productivity by disrupting water absorption and causing ionic imbalances within plant cells. Enhancing salinity tolerance is essential for improving crop resilience, particularly in coastal and arid regions where soil salinization is prevalent. CRISPR-Cas9 technology has been employed to edit genes that mediate ion transport and stress-signaling pathways, leading to enhanced salt tolerance.

19.4.2.1 Rice salt tolerance via CRISPR

Rice is particularly vulnerable to high salt concentrations in the soil, leading to decreased growth and grain production. CRISPR-Cas9 has been used to edit genes such as *OsHKT1* (high-affinity potassium transporter) that control sodium ion homeostasis, helping the plant maintain lower sodium levels in the shoot. This modification enhances the plant's ability to survive in saline conditions without accumulating toxic sodium levels (Alam *et al.*, 2022). Additionally, the knockout of the *SLR1* gene in rice led to enhanced salinity tolerance by regulating the expression of stress-responsive genes and improving sodium exclusion from the roots (Liu *et al.*, 2021).

Moreover, CRISPR has been applied to modify the *OsRR22* gene, which regulates the plant's response to salt stress. Knockout mutations in this gene have been shown to improve the ability of rice plants to tolerate high salinity by modulating the plant's salt-exclusion mechanisms (Zhang *et al.*, 2019).

19.4.2.2 Tomato salt tolerance

Tomato (*Solanum lycopersicum*), one of the world's most cultivated crops, suffers significant yield losses under salt stress. Using CRISPR-Cas9, researchers have edited the *SlHKT1;2* gene, responsible for controlling sodium transport in tomato plants. This mutation has been shown to improve salt tolerance by limiting sodium uptake in the roots and reducing its translocation to the shoots (Wang *et al.*, 2020).

By reducing the salt sensitivity of key crops like tomato and rice, CRISPR-Cas9 is playing a crucial role in developing varieties that can thrive in saline soils, offering a solution to the growing problem of soil degradation in arid regions.

19.4.3 Cold- and heat-stress tolerance

Temperature extremes, such as cold and heat, have a significant impact on crop productivity. Heat stress, especially during key growth stages like flowering and grain filling, can significantly reduce crop yields. CRISPR-Cas9 has been used to edit heat-responsive genes, enhancing a plant's ability to tolerate higher temperatures and maintain productivity.

19.4.3.1 Maize and heat stress

Maize (*Zea mays*), a major global food crop, is highly sensitive to heat stress during the reproductive phase. Using CRISPR-Cas9, researchers have targeted the *ZmSPX1* gene, which plays a critical role in phosphate metabolism and stress response under high temperatures. The edited plants showed improved tolerance to heat stress by maintaining higher grain yield and better pollen viability at elevated temperatures (Luo *et al.*, 2024).

19.4.3.2 Soybean heat resistance

Soybean (*Glycine max*) is another crop that suffers significant yield losses under heat stress. CRISPR-Cas9 has been used to edit the *GmHSFA1* gene, a heat-shock transcription factor involved in the plant's ability to respond to elevated temperatures. Knockout of this gene has resulted in soybeans that exhibit greater resilience to high temperatures, particularly during the flowering and pod development stages (Jianing *et al.*, 2022). The development of heat-tolerant maize and soybean varieties is vital for sustaining yields in regions experiencing rising temperatures due to global warming.

19.4.4 Nutrient deficiency resistance

Crops growing in nutrient-deficient soils often suffer from poor growth and reduced yields. CRISPR-edited crops have been engineered to overcome these limitations by enhancing nutrient uptake and utilization. For instance, CRISPR editing of the gene in rice has improved nitrogen-use efficiency, enabling plants to grow in nitrogen-deficient soils (Fiaz *et al.*, 2021). This innovation reduces the dependence on synthetic fertilizers, contributing to more environmentally friendly agricultural practices.

19.4.5 Biotic stress resistance: disease and pest tolerance

In addition to abiotic stresses, crops also face constant pressure from pests and diseases, which can devastate yields. CRISPR-Cas9 has been applied to develop disease-resistant crops by targeting susceptibility genes (S-genes) that are exploited by pathogens to facilitate infection.

19.4.5.1 Bacterial blight resistance in rice

Bacterial blight, caused by *Xanthomonas oryzae*, is a significant disease affecting rice production in many parts of Asia. CRISPR-Cas9 has been used to knock out the *OsSWEET13* gene, which is a susceptibility gene targeted by the pathogen. Edited rice plants lacking this gene have shown strong resistance to bacterial blight, offering a sustainable solution to controlling this devastating disease (Xu *et al.*, 2019).

19.4.5.2 Powdery mildew resistance in wheat

Powdery mildew is a common fungal disease affecting wheat, often leading to significant yield losses. Using CRISPR-Cas9, researchers have successfully knocked out the *MLO* gene, which is involved in the susceptibility of wheat to powdery mildew. The resulting plants exhibit strong resistance to the pathogen without any adverse effects on growth or development (Wang *et al.*, 2014).

19.4.5.3 Mildew-resistant grapevine

Grapevine (*Vitis vinifera*), which is susceptible to downy mildew, has been successfully edited using CRISPR-Cas9 to knock out the *MLO* susceptibility gene. This modification has conferred resistance to the fungal pathogen, reducing the need for fungicide applications and enhancing sustainability in viticulture (Wan *et al.*, 2020).

19.4.5.4 Insect-resistant crops

In addition to fungal and bacterial pathogens, CRISPR-Cas9 has also been applied to develop insect-resistant crops. For example, in cotton, the knockout of the *GhWRKY70* gene has resulted in improved resistance to aphids by enhancing the plant's defense response mechanisms (Long *et al.*, 2020). Such advances in pest-resistant crops are crucial for reducing pesticide use and promoting sustainable agricultural practices.

19.4.5.5 Herbicide-tolerant canola

A major success story in CRISPR-edited, transgene-free crops is the development of herbicide-tolerant canola (*Brassica napus*). By editing the *ACCase* gene, which is involved in lipid biosynthesis, researchers produced canola plants with enhanced tolerance to herbicides, allowing for more efficient weed management. The edited canola varieties have been commercialized in the USA without the regulatory burden faced by GMO crops (Subedi *et al.*, 2020).

19.5 Regulatory and Ethical Considerations

The regulatory framework surrounding gene-edited crops varies across regions. In some countries, transgene-free CRISPR-edited crops are not classified as GMOs, which significantly eases the approval process for commercialization. For instance, the U.S. Department of Agriculture (USDA) has exempted certain CRISPR-edited crops from GMO regulations, provided they do not contain foreign DNA (Gupta *et al.*, 2021). Similarly, Japan and Australia have adopted more lenient regulatory approaches toward gene-edited crops, distinguishing them from traditional GMOs. However, the European Union has taken a more stringent stance, categorizing all gene-edited crops, including transgene-free varieties, as GMOs and subjecting them to strict regulations (Zhang *et al.*, 2020).

While transgene-free genome editing holds promise for agriculture, public acceptance remains a critical challenge. Many consumers remain cautious about the safety of gene-edited crops, even in the absence of foreign DNA. Ethical concerns related to biodiversity, environmental safety, and corporate control of seed markets continue to shape the debate around the use of CRISPR technology in agriculture (Munawar *et al.*, 2024). Educational initiatives and transparent communication about the benefits and safety of transgene-free crops are essential for building public trust and fostering acceptance of this technology.

19.6 Conclusion

Transgene-free CRISPR-edited crops represent a groundbreaking advancement in agricultural biotechnology. By allowing for the precise, targeted modification of plant genomes without the introduction of foreign DNA, these crops offer a powerful tool for improving stress resilience in agriculture. Techniques such as ribonucleoprotein delivery, virus-induced gene editing, and protoplast transformation have proved to be successful in producing transgene-free crops that are more resilient to drought, salinity, disease, and nutrient deficiencies. As global food security becomes increasingly challenged by climate change and environmental degradation, transgene-free CRISPR-edited crops will play a pivotal role in the development of sustainable agricultural systems.

References

Ahmad, S., Belwal, V., Punia, S.S., Ram, M.D., Dailp *et al.* (2023) Role of plant secondary metabolites and phytohormones in drought tolerance: A review. *Gesunde Pflanzen* 75(4), 729–746. DOI: 1007/s10343-022-00795-z.

Alam, M.S., Kong, J., Tao, R., Ahmed, T., Alamin, M. *et al.* (2022) CRISPR/Cas9 mediated knockout of the *OsbHIH024* transcription factor improves salt stress resistance in rice (*Oryza sativa* L.). *Plants* 11(9), 1184. DOI: 3390/plants11091184.

Aluko, O.O., Kant, S., Adedire, O.M., Li, C., Yuan, G. *et al.* (2023) Unlocking the potentials of nitrate transporters at improving plant nitrogen use efficiency. *Frontiers in Plant Science* 14, 1074839. DOI: 3389/fpls.2023.1074839.

Chakraborty, A., Choudhury, S., Kar, S.R., Deb, P. and Wylie, S.J. (2024) Gene editing for tolerance to temperature stress in plants: A review. *Plant Gene* 37(18), 100439. DOI: 1016/j.plgene.2023.100439.

Cheng, J., Hill, C.B., Shabala, S., Li, C. and Zhou, M. (2022) Manipulating GA-related genes for cereal crop improvement. *International Journal of Molecular Sciences* 23(22), 14046. DOI: 3390/ijms232214046.

Debbarma, J., Saikia, B., Singha, D.L., Das, D., Keot, A.K. *et al.* (2023) CRISPR/Cas9-mediated mutation in XSP10 and SlSAMT genes impart genetic tolerance to fusarium wilt disease of tomato (*Solanum lycopersicum* L.). *Genes* 14(2), 488. DOI: 3390/genes14020488.

Ding, Y., Li, H., Chen, L.L. and Xie, K. (2016) Recent advances in genome editing using CRISPR/Cas9. *Frontiers in Plant Science* 7, 703. DOI: 3389/fpls.2016.00703.

Ding, F., Qiang, X., Jia, Z., Li, L., Hu, J. *et al.* (2022) Knockout of a novel salt responsive gene SlABIG1 enhance salinity tolerance in tomato. *Environmental and Experimental Botany* 200, 104903. DOI: 1016/j.envexpbot.2022.104903.

Dong, S., Qin, Y.L., Vakulskas, C.A., Collingwood, M.A., Marand, M. *et al.* (2021) Efficient targeted mutagenesis mediated by CRISPR-Cas12a ribonucleoprotein complexes in maize. *Frontiers in Genome Editing* 3, 670529. DOI: 3389/fgeed.2021.670529.

Doudna, J.A. and Charpentier, E. (2014) The new frontier of genome engineering with CRISPR-Cas9. *Science* 346(6213), 1258096. DOI: 1126/science.1258096.

Du, F., Wang, Y., Wang, J., Li, Y., Zhang, Y. *et al.* (2023) The basic helix-loop-helix transcription factor gene, OsbHLH38, plays a key role in controlling rice salt tolerance. *Journal of Integrative Plant Biology* 65(8), 1859–1873. DOI: 1111/jipb.13489.

Duan, Y.B., Li, J., Qin, R.Y., Xu, R.F., Li, H. *et al.* (2016) Identification of a regulatory element responsible for salt induction of rice OsRAV2 through *ex situ* and *in situ* promoter analysis. *Plant Molecular Biology* 90(1–2), 49–62. DOI: 1007/s11103-015-0393-z.

FAO (2021) *World food and agriculture-statistical yearbook 2021; FAO statistical yearbook—world food and agriculture*. Food and Agriculture Organization of the United Nations, Rome.

Fiaz, S., Wang, X., Khan, S.A., Ahmar, S., Noor, M.A. *et al.* (2021) Novel plant breeding techniques to advance nitrogen use efficiency in rice: A review. *GM Crops & Food* 12(2), 627–646. DOI: 1080/21645698.2021.1921545.

Gu, X., Liu, L. and Zhang, H. (2021) Transgene-free genome editing in plants. *Frontiers in Genome Editing* 3, 805317. DOI: 3389/fgeed.2021.805317.

Hasanuzzaman, M., Zhou, M. and Shabala, S. (2023) How does stomatal density and residual transpiration contribute to osmotic stress tolerance? *Plants* 12(3), 494. DOI: 3390/plants12030494.

Gupta, S., Kumar, A., Patel, R. and Kumar, V. (2021) Genetically modified crop regulations: Scope and opportunity using the CRISPR-Cas9 genome editing approach. *Molecular Biology Reports* 48(5), 4851–4863. DOI: 1007/s11033-021-06477-9.

Hrmova, M. and Hussain, S.S. (2021) Plant transcription factors involved in drought and associated stresses. *International Journal of Molecular Sciences* 22(11), 5662. DOI: 3390/ijms22115662.

I.S.A.A.A. (2019) *Global Status of Commercialized Biotech/GM Crops: 2016*, Vol. 54. ISAAA Br, pp. 3–13.

Jacinto, F.V., Link, W. and Ferreira, B.I. (2020) CRISPR/Cas9-mediated genome editing: From basic research to translational medicine. *Journal of Cellular and Molecular Medicine* 24(7), 3766–3778. DOI: 1111/jcmm.14916.

Jain, S.K., Wettberg, E.J. von, Punia, S.S., Parihar, A.K., Lamichaney, A. *et al.* (2023) Genomic-mediated breeding strategies for global warming in chickpeas (*Cicer arietinum* L.). *Agriculture* 13(9), 1721. DOI: 3390/agriculture13091721.

Jianing, G., Yuhong, G., Yijun, G., Rasheed, A., Qian, Z. *et al.* (2022) Improvement of heat stress tolerance in soybean (*Glycine max* L.), by using conventional and molecular tools. *Frontiers in Plant Science* 13, 993189. DOI: 3389/fpls.2022.993189.

Kantor, A., McClements, M.E. and MacLaren, R.E. (2020) CRISPR-Cas9 DNA base-editing and prime-editing. *International Journal of Molecular Sciences* 21(17), 6240. DOI: 3390/ijms21176240.

Kleinstiver, B.P., Pattanayak, V., Prew, M.S., Tsai, S.Q., Nguyen, N.T. *et al.* (2016) High-fidelity CRISPR-Cas9 nucleases with no detectable genome-wide off-target effects. *Nature* 529(7587), 490–495. DOI: 1038/nature16526.

Kusch, S. and Panstruga, R. (2017) Mlo-based resistance: An apparently universal 'weapon' to defeat powdery mildew disease. *Molecular Plant-Microbe Interactions* 30(3), 179–189. DOI: 1094/MPMI-12-16-0255-CR.

Lee, H.K., Oh, Y., Hong, J., Lee, S.H. and Hur, J.K. (2021) Development of CRISPR technology for precise single-base genome editing: A brief review. *BMB Reports* 54(2), 98–105. DOI: 5483/BMBRep.2021.54.2.217.

Li, R., Zhang, L., Wang, L., Chen, L., Zhao, R. *et al.* (2018) Reduction of tomato-plant chilling tolerance by CRISPR-Cas9-mediated slcbf1 mutagenesis. *Journal of Agricultural and Food Chemistry* 66(34), 9042–9051. DOI: 1021/acs.jafc.8b02177.

Liang, Z., Chen, K., Li, T., Zhang, Y., Wang, Y. *et al.* (2017) Efficient DNA-free genome editing of bread wheat using CRISPR/Cas9 ribonucleoprotein complexes. *Nature Communications* 8(1), 14261. DOI: 1038/ncomms14261.

Liao, S., Qin, X., Luo, L., Han, Y., Wang, X. *et al.* (2019) CRISPR/Cas9-induced mutagenesis of semi-rolled leaf1,2 confers curled leaf phenotype and drought tolerance by influencing protein expression patterns and ROS scavenging in rice (*Oryza sativa* L.). *Agronomy* 9(11), 728. DOI: 3390/agronomy9110728.

Liao, Q., Gu, S., Kang, S., Du, T., Tong, L. *et al.* (2022) Mild water and salt stress improve water use efficiency by decreasing stomatal conductance via osmotic adjustment in field maize. *The Science of the Total Environment* 805, 150364. DOI: 1016/j.scitotenv.2021.150364.

Lin, Q., Zong, Y., Xue, C., Wang, S., Jin, S. *et al.* (2020) Prime genome editing in rice and wheat. *Nature Biotechnology* 38(5), 582–585. DOI: 1038/s41587-020-0455-x.

Liu, J.Z., Graham, M.A., Pedley, K.F. and Whitham, S.A. (2015) Gaining insight into soybean defense responses using functional genomics approaches. *Briefings in Functional Genomics* 14(4), 283–290. DOI: 1093/bfgp/elv009.

Liu, Y., Chen, X., Xue, S., Quan, T., Cui, D. *et al.* (2021) SET DOMAIN GROUP 721 protein functions in saline-alkaline stress tolerance in the model rice variety kitaake. *Plant Biotechnology Journal* 19(12), 2576–2588. DOI: 1111/pbi.13683.

Long, L., Xu, F.C., Zhao, J.R., Li, B., Xu, L. *et al.* (2020) GbMPK3 overexpression increases cotton sensitivity to verticillium dahliae by regulating salicylic acid signaling. *Plant Science* 292, 110374. DOI: 1016/j.plantsci.2019.110374.

Lou, D., Wang, H., Liang, G. and Yu, D. (2017) OsSAPK2 confers abscisic acid sensitivity and tolerance to drought stress in rice. *Frontiers in Plant Science* 8, 993. DOI: 3389/fpls.2017.00993.

Luo, X., Li, C., He, X., Zhang, X. and Zhu, L. (2020) ABA signaling is negatively regulated by GbWRKY1 through JAZ1 and ABI1 to affect salt and drought tolerance. *Plant Cell Reports* 39(2), 181–194. DOI: 1007/s00299-019-02480-4.

Luo, B., Sahito, J.H., Zhang, H., Zhao, J., Yang, G. *et al.* (2024) SPX family response to low phosphorus stress and the involvement of *zmspx*1 in phosphorus homeostasis in maize. *Frontiers in Plant Science* 15, 1385977. DOI: 3389/fpls.2024.1385977.

Manghwar, H., Li, B., Ding, X., Hussain, A., Lindsey, K. *et al.* (2020) CRISPR/Cas systems in genome editing: Methodologies and tools for sgrna design, off-target evaluation, and strategies to mitigate off-target effects. *Advanced Science* 7(6), 1902312. DOI: 1002/advs.201902312.

Munawar, N., Ahsan, K. and Ahmad, A. (2024) CRISPR-edited plants' social, ethical, policy, and governance issues. In: Abd-Elsalam, K.A. and Ahmad, A. (eds) *Global Regulatory Outlook for CRISPRized Plants*. Academic Press, pp. 367–396.

Muthu, V., Abbai, R., Nallathambi, J., Rahman, H., Ramasamy, S. *et al.* (2020) Pyramiding QTLs controlling tolerance against drought, salinity, and submergence in rice through marker assisted breeding. *PLoS One* 15(1), e0227421. DOI: 1371/journal.pone.0227421.

Naeem, M., Majeed, S., Hoque, M.Z. and Ahmad, I. (2020) Latest developed strategies to minimize the off-target effects in CRISPR-Cas-mediated genome editing. *Cells* 9(7), 1608. DOI: 3390/cells9071608.

Najafi, S., Bertini, E., D'Incà, E., Fasoli, M. and Zenoni, S. (2023) DNA-free genome editing in grapevine using CRISPR/Cas9 ribonucleoprotein complexes followed by protoplast regeneration. *Horticulture Research* 10(1), uhac240. DOI: 1093/hr/uhac240.

Ogata, T., Ishizaki, T., Fujita, M. and Fujita, Y. (2020) CRISPR/Cas9-targeted mutagenesis of osera1 confers enhanced responses to abscisic acid and drought stress and increased primary root growth under nonstressed conditions in rice. *PLoS One* 15(12), e0243376. DOI: 1371/journal.pone.0243376.

Park, J. and Choe, S. (2019) DNA-free genome editing with preassembled CRISPR/Cas9 ribonucleoproteins in plants. *Transgenic Research* 28(Suppl 2), 61–64. DOI: 1007/s11248-019-00136-3.

Saikia, B., Remya, S., Debbarma, J., Maharana, J., Sastry, G.N. *et al.* (2024) CRISPR/Cas9-based genome editing and functional analysis of *slhyprp1* and *sldea1* genes of *Solanum lycopersicum* L. in imparting genetic tolerance to multiple stress factors. *Frontiers in Plant Science* 15, 1304381. DOI: 3389/fpls.2024.1304381.

Sami, A., Xue, Z., Tazein, S., Arshad, A., He Zhu, Z. *et al.* (2021) CRISPR-Cas9-based genetic engineering for crop improvement under drought stress. *Bioengineered* 12(1), 5814–5829. DOI: 1080/21655979.2021.1969831.

Sandhya, D., Jogam, P., Allini, V.R., Abbagani, S. and Alok, A. (2020) The present and potential future methods for delivering CRISPR/Cas9 components in plants. *Journal, Genetic Engineering & Biotechnology* 18(1), 25. DOI: 1186/s43141-020-00036-8.

Singh, M., Nara, U., Kumar, A., Choudhary, A., Singh, H. *et al.* (2021) Salinity tolerance mechanisms and their breeding implications. *Journal, Genetic Engineering & Biotechnology* 19(1), 173. DOI: 1186/s43141-021-00274-4.

Son, S. and Park, S.R. (2022) Challenges facing CRISPR/Cas9-based genome editing in plants. *Frontiers in Plant Science* 13, 902413. DOI: 3389/fpls.2022.902413.

Subedi, U., Ozga, J.A., Chen, G., Foroud, N.A. and Singer, S.D. (2020) CRISPR/Cas-mediated genome editing for the improvement of oilseed crop productivity. *Critical Reviews in Plant Sciences* 39(3), 195–221. DOI: 1080/07352689.2020.1782568.

Tiwari, J.K., Challam, C., Chakrabarti, S.K. and Feingold, S.E. (2020) Climate-smart potato: An integrated breeding, genomics, and phenomics approach. *Genomic Designing of Climate-Smart Vegetable Crops* 1–46. DOI: 1007/978-3-319-97415-6_1.

Tran, M.T., Doan, D.T.H., Kim, J., Song, Y.J., Sung, Y.W. *et al.* (2021) CRISPR/Cas9-based precise excision of SlHyPRP1 domain(s) to obtain salt stress-tolerant tomato. *Plant Cell Reports* 40(6), 999–1011. DOI: 1007/s00299-020-02622-z.

Trono, D. and Pecchioni, N. (2022) Candidate genes associated with abiotic stress response in plants as tools to engineer tolerance to drought, salinity and extreme temperatures in wheat: An overview. *Plants* 11(23), 3358. DOI: 3390/plants11233358.

Turnbull, C., Lillemo, M. and Hvoslef-Eide, T.A.K. (2021) Global regulation of genetically modified crops amid the gene edited crop boom – a review. *Frontiers in Plant Science* 12, 630396. DOI: 3389/fpls.2021.630396.

Tyagi, S., Kumar, R., Kumar, V., Won, S.Y. and Shukla, P. (2021) Engineering disease resistant plants through CRISPR-Cas9 technology. *GM Crops & Food* 12(1), 125–144. DOI: 1080/21645698.2020.1831729.

Vega Rodríguez, A., Rodríguez-Oramas, C., Sanjuán Velázquez, E., Torre, A., Rubio Armendáriz, C. *et al.* (2022) Myths and realities about genetically modified food: A risk-benefit analysis. *Applied Sciences* 12(6), 2861. DOI: 3390/app12062861.

Veillet, F., Perrot, L., Chauvin, L., Kermarrec, M.P., Guyon-Debast, A. *et al.* (2019) Transgene-free genome editing in tomato and potato plants using *Agrobacterium*-mediated delivery of a CRISPR/Cas9 cytidine base editor. *International Journal of Molecular Sciences* 20(2), 402. DOI: 3390/ijms20020402.

Wan, D.Y., Guo, Y., Cheng, Y., Hu, Y., Xiao, S. *et al.* (2020) CRISPR/Cas9-mediated mutagenesis of *VvMLO3* results in enhanced resistance to powdery mildew in grapevine (*Vitis vinifera*). *Horticulture Research* 7, 116. DOI: 1038/s41438-020-0339-8.

Wang, Y., Cheng, X., Shan, Q., Zhang, Y., Liu, J. *et al.* (2014) Simultaneous editing of three homoeoalleles in hexaploid bread wheat confers heritable resistance to powdery mildew. *Nature Biotechnology* 32(9), 947–951. DOI: 1038/nbt.2969.

Wang, W.C., Lin, T.C., Kieber, J. and Tsai, Y.C. (2019) Response regulators 9 and 10 negatively regulate salinity tolerance in rice. *Plant & Cell Physiology* 60(11), 2549–2563. DOI: 1093/pcp/pcz149.

Wang, Z., Hong, Y., Zhu, G., Li, Y., Niu, Q. *et al.* (2020) Loss of salt tolerance during tomato domestication conferred by variation in a NA^+/K^+ transporter. *The EMBO Journal* 39(10), e103256.

Wang, T., Xun, H., Wang, W., Ding, X., Tian, H. *et al.* (2021a) Mutation of gmaitr genes by CRISPR/Cas9 genome editing results in enhanced salinity stress tolerance in soybean. *Frontiers in Plant Science* 12, 779598. DOI: 3389/fpls.2021.779598.

Wang, X., He, Y., Wei, H. and Wang, L. (2021b) A clock regulatory module is required for salt tolerance and control of heading date in rice. *Plant, Cell & Environment* 44(10), 3283–3301. DOI: 1111/pce.14167.

Wang, Y., Cao, Y., Liang, X., Zhuang, J., Wang, X. *et al.* (2022) A dirigent family protein confers variation of casparian strip thickness and salt tolerance in maize. *Nature Communications* 13(1), 2222. DOI: 1038/s41467-022-29809-0.

Woo, J.W., Kim, J., Kwon, S.I., Corvalán, C., Cho, S.W. *et al.* (2015) DNA-free genome editing in plants with preassembled CRISPR-Cas9 ribonucleoproteins. *Nature Biotechnology* 33(11), 1162–1164. DOI: 1038/nbt.3389.

Wu, S., Zhu, H., Liu, J., Yang, Q., Shao, X. *et al.* (2020) Establishment of a PEG-mediated protoplast transformation system based on DNA and CRISPR/Cas9 ribonucleoprotein complexes for banana. *BMC Plant Biology* 20(1), 425. DOI: 1186/s12870-020-02609-8.

Wu, D., Cao, Y., Wang, D., Zong, G., Han, K. *et al.* (2024) Auxin receptor OstTIR1 mediates auxin signaling during seed filling in rice. *Plant Physiology* 194(4), 2434–2448. DOI: 1093/plphys/kiae013.

Xu, Z., Xu, X., Gong, Q., Li, Z., Li, Y. *et al.* (2019) Engineering broad-spectrum bacterial blight resistance by simultaneously disrupting variable TALE-binding elements of multiple susceptibility genes in rice. *Molecular Plant* 12(11), 1434–1446. DOI: 1016/j.molp.2019.08.006.

Yubing, H.E., Min, Z.H.U., Lihao, W., Junhua, W.U., Qiaoyan, W. *et al.* (2019) Improvements of TKC technology accelerate isolation of transgene-free CRISPR/Cas9-edited rice plants. *Rice Science* 26(2), 109–117. DOI: 1016/j.rsci.2018.11.001.

Zafar, K., Khan, M.Z., Amin, I., Mukhtar, Z., Yasmin, S. *et al.* (2020) Precise CRISPR-Cas9 mediated genome editing in super basmati rice for resistance against bacterial blight by targeting the major susceptibility gene. *Frontiers in Plant Science* 11, 575. DOI: 3389/fpls.2020.00575.

Zhang, A., Liu, Y., Wang, F., Li, T., Chen, Z. *et al.* (2019) Enhanced rice salinity tolerance via CRISPR/Cas9-targeted mutagenesis of the *OsRR22* gene. *Molecular Breeding* 39, 1–10. DOI: 1007/s11032-019-0954-y.

Zhang, D., Hussain, A., Manghwar, H., Xie, K., Xie, S. *et al.* (2020) Genome editing with the CRISPR-Cas system: An art, ethics and global regulatory perspective. *Plant Biotechnology Journal* 18(8), 1651–1669. DOI: 1111/pbi.13383.

Zhao, Y., Zhang, C., Liu, W., Gao, W., Liu, C. *et al.* (2016) An alternative strategy for targeted gene replacement in plants using a dual-sgRNA/Cas9 design. *Scientific Reports* 6(1), 23890. DOI: 1038/srep23890.

20 Resilient Capacity against Rising Temperature Gained by Coordinated Approaches of CRISPR/Cas with Functional Genomics

Parul Verma, Sana Choudhary*, Nidhi Sharma, Ekta Pandey and Rinkee Kumari
Department of Botany, Aligarh Muslim University, Aligarh, India

Abstract
Globally, climate change impacts agricultural productivity. Changes in the ideal temperature range have put many plant species in danger of becoming extinct, and this has resulted in an acceleration of the loss of biodiversity and a gradual modification of ecosystem frameworks. Climate change is currently a major danger to world agriculture since it is expected to lower productivity. Global food insecurity increases as a result of lower production, which also raises food prices (FAO, 2014). If this situation continues, food prices will rise even more, possibly causing famine and societal unrest in some cases. The food supply will be impacted by climate change unless steps are taken to make more resilient crops. CRISPR genome-editing technology provides new avenues for improving plant heat tolerance by enabling targeted changes to heat-stress-related genes. By combining CRISPR with functional genomics, researchers can determine the functions of different genes in heat-tolerance pathways. A focussed approach to improving crop types is made possible by the understanding of how particular genes contribute to physiological responses.

Keywords: climate change, biodiversity, resilient crops, CRISPR, functional genomics

20.1 Introduction

The phrase 'climate change' denotes the drastic alteration in our climate that is caused by global warming. Climate conditions have changed, resulting in changes to land use, soil microbiota, and water equilibrium. These changes include temperature variations, unpredictable rainfall patterns, and fluctuations in atmospheric gas composition (Zhao *et al.*, 2017). High temperature (HT) is a major abiotic stressor to plants, posing serious challenges to their growth and output due to rising global temperatures brought by climate change (Chen *et al.*, 2022a). The average temperature is predicted to rise by 2.5–4.5°C by the end of the century as a result of the steady increase in greenhouse gases (GHGs) (Bernstein *et al.*, 2008). The changing climate presents a substantial risk to worldwide food supply (Tripathi *et al.*, 2016). Nearly 815 million people around the world suffer from malnutrition, which is obstructing efforts to achieve the goal of

*Corresponding author: sana.12amu@gmail.com

© CAB International 2025. *Stress-Resilient Crops: Coordinated omics-CRISPR-Nanotechnology Strategies* (ed. Jen-Tsung Chen)
DOI: 10.1079/9781800627307.0020

ending hunger by 2030 (Richardson *et al.*, 2018). Another research study indicates that over the past three decades climate change has resulted in a 1–5% decrease in global agricultural productivity each decade (Mashizha, 2019). The three major cereal grains – maize (*Zea mays* L.), rice (*Oryza sativa*), and wheat (*Triticum aestivum* L.) – are the main crops consumed worldwide as essential foods by a majority of people (Wang *et al.*, 2018a). The impact of this climate change scenario on wheat production until 2050 is likely to have an adverse effect on wheat production. The spatial and temporal variability observed in different locations under study demonstrated that yield reductions occur with increasing temperatures and elevated carbon dioxide concentrations (Martre *et al.*, 2015; Asseng *et al.*, 2017). Rising temperatures can decrease the rate of carbon assimilation and lead to lower maize grain yields (Sage and Kubien, 2007). Numerous studies highlight the potential effects of raised temperatures on the growth and development, trait performance, yield components, and grain quality of rice (Krishnan *et al.*, 2011). One of the main causes of rice floret sterility under high temperatures is the pollen grains' reduced ability to absorb water, resulting in insufficient opening of the anthers (Matsui *et al.*, 2000). Another researcher found that a 2°C increase led to a 15–17% reduction in grain yields for both rice and wheat (Aggarwal and Mall, 2002). For example, the International Rice Research Institute (IRRI) estimates that one hectare of arable land is diminished every 7.7 s, and this rate of decline may accelerate as global temperatures rise (Stamm *et al.*, 2011). Therefore, enhancing crop resilience to climate change is essential for safeguarding food and nutritional security, attainable through molecular breeding strategies or genetic engineering techniques.

Plants have developed a variety of defense mechanisms, such as the production of antioxidants and enzymatic reactions, to withstand adverse light and temperature conditions, with their effectiveness influenced by stress duration and species-specific tolerance. For instance, elevated temperatures may trigger stomatal closure to avert water loss, serving as a crucial defense mechanism for tolerating high heat (Liu *et al.*, 2020). Stomata improve heat tolerance in climate-resilient crops by controlling transpiration for cooling, enabling gas exchange for photosynthesis, enhancing water-use efficiency, and promoting the production of protective heat-shock proteins.

The latest advancement in genome-editing technology is the CRISPR/Cas system. The field of genome engineering has seen a significant transformation by the discovery of CRISPR (Clustered Regularly Interspaced Short Palindromic Repeats) and Cas9 (CRISPR-associated nuclease 9), which function as defense mechanisms providing acquired resistance to viruses in prokaryotes. These discoveries have also opened the door to numerous applications across various industries (Doudna and Charpentier, 2014). Recent reviews provide detailed descriptions and illustrations of CRISPR/Cas9 technology (Musunuru, 2017). A genomics-led breeding strategy for developing new 'climate change-ready' cultivars begins by identifying the precise stressors that are expected to impact various climate change scenarios (Varshney *et al.*, 2005). CRISPR/Cas9-mediated modification of OsEPF1 in rice variety ASD 16 led to increased stomatal density and enhanced stomatal conductance, photosynthetic rate, and transpiration, showcasing its potential to improve climate resilience (Rathnasamy *et al.*, 2023). CRISPR/Cas technology and functional genomics contribute to the resilience of plants under heat-stress conditions (Fig. 20.1). Functional genomic methods, including genome-wide association studies (GWAS) and gene expression analysis via microarrays, can facilitate the identification of new genes related to heat stress. Additionally, suppression subtractive hybridization (SSH) serves as an effective technique for screening and cloning genes or expressed sequence tags (ESTs) that show differential expression under heat-stress conditions

20.2 CRISPR-Cas System

The three endonucleases utilized in gene editing are transcription activator-like effector nucleases (TALENs), zinc finger nucleases, and

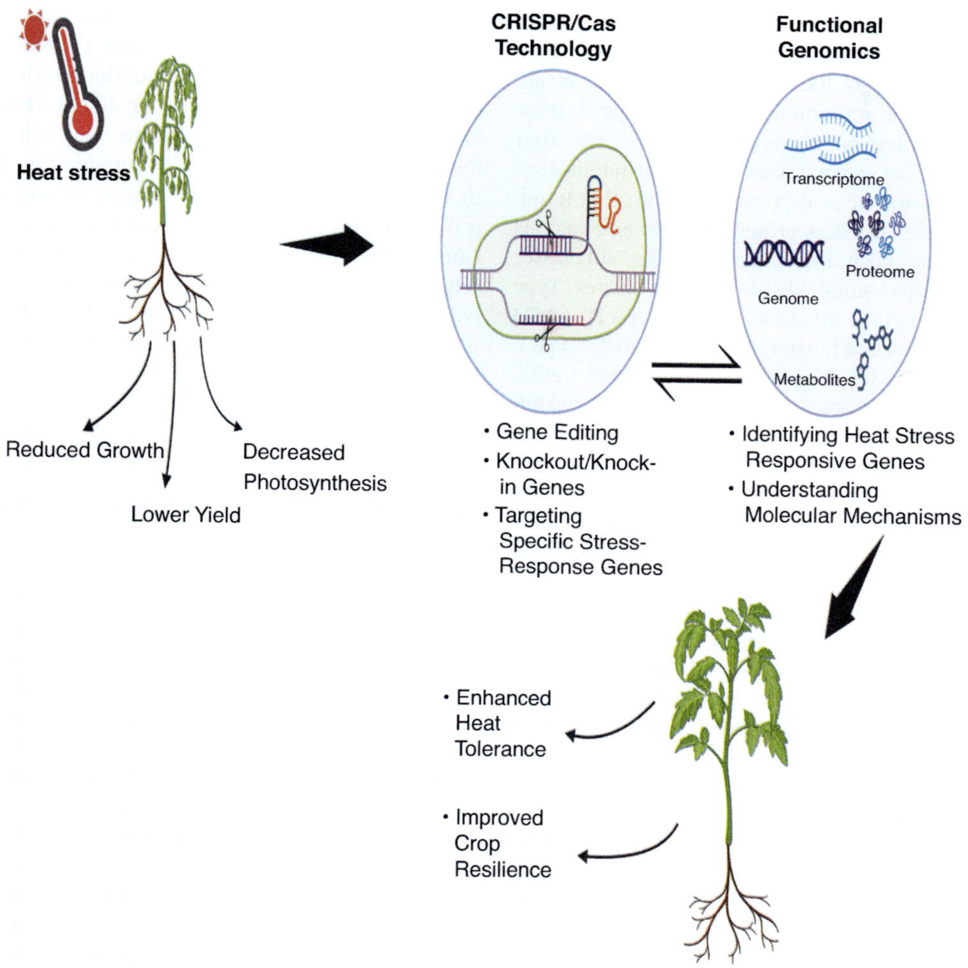

Fig. 20.1. CRISPR/Cas technology and functional genomics contribute to enhancing the resilience of plants against heat stress.

CRISPR-Cas9 (Malzahn *et al.*, 2017). CRISPR/Cas is a potent genetic engineer capable of improving existing crop varieties and genetic resources. In 1987, the CRISPR-Cas system was first identified by a team of scientists from Japan (Ishino *et al.*, 1987), although they were unclear what this entity represented and how it functioned. Its role in immunity was not understood until after 2005 when researchers revealed that significant sequences included in the genomes of bacteria were derived from viruses or plasmids (Doudna and Charpentier, 2014). The method was recommended for useful gene modification and editing in 2012 (Jinek *et al.*, 2012). Later, the CRISPR-Cas-engineered system proved to be valuable in numerous ways. It is already employed for editing plant genomes (Schaeffer and Nakata, 2015). The fundamental structure of CRISPR includes operons of Cas genes, along with spacers and repeats. For instance, each array contains repeats of 25–35 bp that are interspersed with spacers of 30–40 bp, complemented by multiple well-organized clusters of Cas genes typically accompanied by accessory genes.

There are two important classes of CRISPR systems, with Class II being the most typical, characterized specifically via the nuclease

Cas9 (Makarova *et al.*, 2015). There are no Type II CRISPR-Cas systems in archaea; they are exclusively found in bacteria (Haft *et al.*, 2005). For antiviral defense, Class I produces a CRISPR-associated complex, or CASCADE, that consists of several Cas proteins that destroy foreign nucleic acids. Class II, on the other hand, uses one large Cas protein with several regions to accomplish the same goal. Six subclasses are further subdivided from these classes: Type I (Cas3), Type III (Cas10), and Type IV (Csf1) comprise Class I, whereas Type II (Cas9), Type V (Cas12a, Cas12b, Cas12c, Cas12d/CasY, CasX), and Type VI (Cas13a, Cas13b, Cas13c, Cas13d) are included in Class II (Liu *et al.*, 2016). Furthermore, these kinds are classified into 33 subtypes according to the CRISPR array's shape and the distinct signature genes linked to each subtype (Liu *et al.*, 2016). In addition, these kinds are classified into 33 subtypes based on the CRISPR array's shape and the existence of particular signature genes that are particular to each subtype. The AT-rich leader sequence of the CRISPR locus consists of straight short repeats (28–37 bp) dotted with distinct spacers (32–38 bp) (Barrangou and Marraffini, 2014). In a CRISPR array, repeat-spacer sequences usually consist of less than 50 units. In addition, these CRISPR arrays are frequently surrounded by tiny clusters of Cas genes. Three stages make up the CRISPR-based defense mechanism: expression, interference, and adaptation. The first phase, called adaptation, involves adding new spacers to the CRISPR array one at a time. Cas genes and precursor CRISPR-RNA (pre-crRNA) are activated in the next expression stage. Cas proteins aid in the conversion of pre-crRNA into mature crRNA, together with other elements such as Type II RNase III. Ultimately, the target sequence may be identified and cut during the interference phase thanks to the cooperation of the crRNA and Cas proteins (Bortesi and Fischer, 2015).

20.2.1 Three phases of CRISPR/Cas

After an infection, little sections of the invader's genome identified as spacers are extracted and incorporated into the CRISPR locus to start the defense process. These spacers are translated into CRISPR RNAs (crRNAs) during the subsequent targeting phase, which direct CRISPR-associated (Cas) nucleases to cut out the invasive DNA or RNA (Hille and Charpentier, 2016). This discussion emphasizes the various molecular mechanisms involved in CRISPR targeting and their link to the immunization phase through a third phase called primed spacer acquisition. During this phase, Cas proteins assist in the crRNA-guided acquisition of extra spacers, bolstering the immune response and providing faster and more effective protection for the population.

20.2.2 How CRISPR/Cas acts as a gene-editing tool

The most adaptable and broadly utilized system for genome editing is the Type II CRISPR/Cas9 system derived from the bacterium *Streptococcus pyogenes*. The Type II systems represent the typical and well-understood Class II systems, employing the Cas9 protein as their effector (Sapranauskas *et al.*, 2011). The RuvC-like endonuclease domain, which is composed of RuvC-I, -II, and -III subdomains, and the HNH (His-Asn-His) endonuclease domain are both present in Cas9 orthologs. Together with crRNA and a transactivating crRNA (tracrRNA), Cas9 forms a ribonucleoprotein (RNP) complex. This complex operates by cleaving DNA in a manner guided by the crRNA (Gasiunas *et al.*, 2012; Jinek *et al.*, 2012). The tracrRNA and crRNA can be joined to form a single-guide RNA (sgRNA) in systems that have been changed, such as those used for genome editing (Mali *et al.*, 2013). Among the several Cas9 variations found in public databases, SpCas9 – which comes from the *Streptococcus pyogenes* strain SF370 – has received the most extensive investigation of any ortholog (Deltcheva *et al.*, 2011; Jinek *et al.*, 2013). A short protospacer adjacent motif (PAM) that is close to the target sequence is necessary for Cas9–gRNA complexes to identify and cut the target DNA. SpCas9 is a large 1368 amino acid DNA endonuclease that functions across multiple domains. Its principal role in the genome is to cleave target DNA to produce a blunt-ended double-strand break (DSB). The host's cellular apparatus then fixes this DSB (Mei *et al.*, 2016).

There are two main mechanisms to repair the DSB that Cas9 created. One of these is the prone-to-error non-homologous end-joining, which frequently causes sporadic, tiny insertions and deletions (indels) at the cleavage site (Lieber, 2010). Alternatively, by using a homologous repair template, high-fidelity homology-directed repair can be used to repair the DSB. This method permits exact alterations at the genomic location (San Filippo *et al.*, 2008). The Cas9 protein and sgRNA make up the CRISPR/Cas9 system. The two parts of the sgRNA are transactivating RNA (tracrRNA) and CRISPR RNA (crRNA). The 20-nucleotide sequence of the crRNA is complementary to the target sequence of interest. It forms a double-stranded RNA structure when it pairs with tracrRNA (Jinek *et al.*, 2012). Pre-crRNA is activated by RNase III, which then transforms it into mature crRNA (JIANG). Six domains make up the Cas9 nuclease: a protospacer adjacent motif (PAM) interaction (PI) domain, two nuclease domains (HNH and RuvC) that cleave complementary and non-complementary strands at the target site, two recognition domains (REC I and REC II) that are important for binding to sgRNA and DNA, and a bridge helix domain that helps nuclease activity commence (Jinek *et al.*, 2012). For the purpose of designing sgRNA(s) that maximize on-target effects while diminishing off-target effects, a number of web tools have been established (Hsu *et al.*, 2013; Montague *et al.*, 2014). Subsequently, CRISPR-array-generated RNA (gRNA) was modified by Jennifer Doudna and Emmanuelle Charpentier (2020 Nobel Laureates) to accurately target particular genomic regions in both prokaryotic and eukaryotic cells for genome editing. The result of this innovation was the creation of the CRISPR/Cas9 system, a precise RNA-guided gene-editing technique that has revolutionized the biological sciences (Jinek *et al.*, 2012) (Fig. 20.2). Bioinformatics methods were then used to find different forms of Cas nucleases (such as Cas12, Cas12a, and Cas10) that target RNA, DNA, and even proteins with high efficiency by analyzing matched spacer sequences across distinct genomes. Consequently, bioinformatics has been crucial in the development of CRISPR/Cas9 technology, both in terms of discovery and improvement.

20.2.3 Methods for delivering CRISPR/Cas9 constructs in plant cells

CRISPR/Cas delivery methods are categorized into indirect and direct approaches. Indirect methods, including the floral dip technique and *Agrobacterium tumefaciens*-mediated delivery, are predominantly employed for genetic transformation in plants. Direct methods encompass polyethylene glycol (PEG)-mediated delivery and bombardment-mediated delivery, primarily used for transient gene expression (Alghuthaymi *et al.*, 2021).

20.3 CRISPR/Cas-mediated Editing of Heat-Stress Response Genes

Heat stress majorly happens due to an increment in mean temperature, i.e. minimum and maximum temperature for about 15 days (Hatfield and Prueger, 2015; Buragohain *et al.*, 2017). This heat stress can have a negative effect on plant health and can lead to various morphological, biochemical, and physiological changes (Wahid and Shabbir, 2005). Heat stress negatively impacts various biological processes, including turgor pressure, chlorophyll content, carbon assimilation, photosynthetic rate, cellular metabolism, nutrient acquisition, and gas exchange at the leaf surface. Collectively, these negative effects result in reduced crop yields (Hussain *et al.*, 2018; Nadeem *et al.*, 2019). The reproductive stage is significantly impacted by heat stress, affecting ovule and pollen development, communication between male and female elements, fertilization, and embryo development (Chaturvedi *et al.*, 2021). Rice yields are expected to rise by 0.5/ha for every increase of 75 ppm in CO_2 levels, while a temperature rise of 1°C could result in a reduction of 0.6 tons/ha in yields. Heat stress significantly reduces growth indices and photosynthetic rate in maize and sugarcane (Teng *et al.*, 2022).

Over 60% of a crop's yield is diminished due to the negative effects of heat stress, with these effects being most severe during the early stages of reproduction and flowering (Begcy *et al.*, 2019). In reaction to differing intensities and extent of heat stress, plants utilize a variety of responses, including modifications in the expression of genes responsible for producing

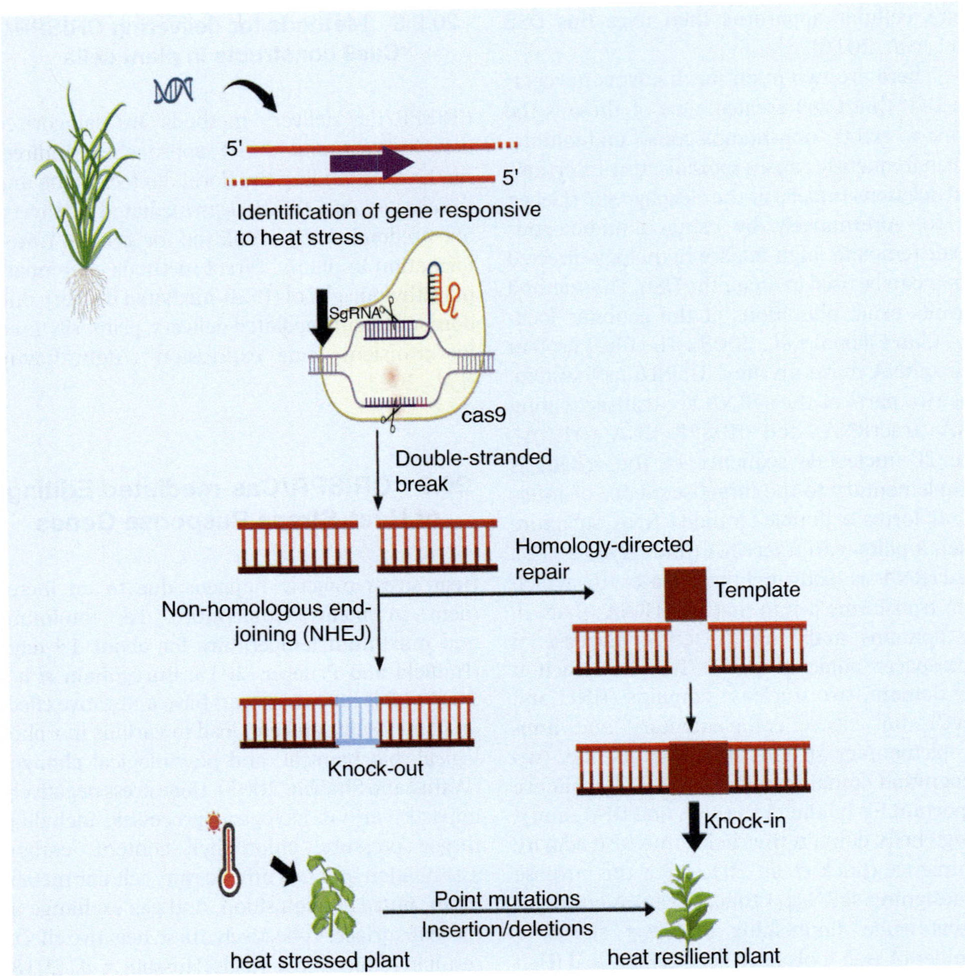

Fig. 20.2. Genome editing using CRISPR/Cas technology enhances heat-stress tolerance in plants.

scavenger proteins, heat-shock proteins, enzymes that create reactive oxygen species (ROS), and enzymes that help accumulate metabolites essential for normal cellular functions. The plant's defensive response is made up of all these processes (Fragkostefanakis *et al.*, 2015).

The advancement of genome editing has unlocked a world of possibilities for fine-tuning genetic characteristics that improve resistance to heat stress (Table 20.1). Tools for editing genomes selectively add, replace, or remove DNA at specific locations (Chen *et al.*, 2019). Heat stress has an initial effect on the plasma membrane, which serves as a trigger for subsequent signaling pathways (Kan and Lin, 2021). In maize, the drought-responsive WRKY gene ZmWRKY106 functions as a positive regulator in responses to abiotic stress by modifying genes linked to stress and reducing ROS. This enhances tolerance to heat and drought (Wang *et al.*, 2018b). Reduction of ROS and upregulation of genes linked to heat-shock proteins and heat-stress transcription factors were the results of the CRISPR/Cas9 deletion of SlMAPK3, which led to enhanced heat tolerance (Yu *et al.*, 2019). With the help of CRISPR-Cas9, the hybrid proline-rich protein 1 (SlHyPRP1) gene was precisely edited, producing variants that confer multi-stress tolerance and suggesting its applicability for stress breeding programs (Tran *et al.*, 2023). In tomato, thermotolerance was attained by modifying the expression of BZR1 (Brassinazole-Resistant 1) through CRISPR. The mutant BZR1 lines exhibited decreased hydrogen peroxide (H_2O_2) production in the apoplast and reduced activation of Respiratory Burst Oxidase Homolog 1 (RBOH1), highlighting the critical role of BZR1 in the heat-stress response (Yin *et al.*, 2018).

Heat-tolerant plants bearing parthenocarpic fruits were created by means of CRISPR/Cas-mediated mutation of the tomato Agamous-Like 6 (AGL6) gene (Klap *et al.*, 2017). Similar to this, CRISPR/Cas technology was employed to alter the Thermosensitive Genic Male Sterile 5 (TGMS5) gene in maize, resulting in the production of thermosensitive male sterile plants (Li *et al.*, 2017). The creation of CRISPR/Cas-mediated HSA1 (heat-stress sensitive albino 1) mutants in tomato revealed that these mutants were more sensitive to heat stress than wild-type plants (Qiu *et al.*, 2018). In lettuce, the knockout of NCED4, an essential regulatory enzyme in ABA biosynthesis, enabled seed germination at elevated temperatures. Consequently, LsNCED4 mutants may prove economically beneficial in regions with high temperatures (Bertier *et al.*, 2018). To enhance a plant's resistance to heat, the hsps gene, which boosts osmolyte levels and protects cellular proteins from damage, can be overexpressed (Debbarma *et al.*, 2019). The transcription factor OsbZIP46CA1 and the protein kinase SAPK6 in rice also enhance the plant's resistance to heat stress (Chang *et al.*, 2017). Stable mutant soybean lines generated via CRISPR that overexpress GmHsp90A2 demonstrated enhanced tolerance to heat stress, reflected in elevated chlorophyll content and reduced MDA levels when compared to control plants (Huang *et al.*, 2019).

Table 20.1. Some examples of potential heat-tolerant genes that have been successfully validated in major crops.

Plant species	Target gene	Gene function	Method of delivery	Reference
Rice (*Oryza sativa*)	OsOr-R115H	Transcription factor	*Agrobacterium*-based transformation	Jung *et al.*, 2021
Rice (*Oryza sativa*)	OsERF115/ AP2EREBP110	Transcription factor	*Agrobacterium*-based transformation	Park *et al.*, 2021
Turnip (*Brassica rapa* L.)	OsTOGR1	DEAD-box RNA helicase	*Agrobacterium*-mediated vacuum infiltration transformation	Yarra and Xue, 2020
Tomato (*Solanum lycopersicum*)	*SlBBX17*	B-Box (BBX) protein	*Agrobacterium*-mediated transformation	Xu *et al.*, 2022
Tomato (*Solanum lycopersicum*)	SlIAA9	Transcriptional repressor	*Agrobacterium*-mediated transformation	Rahmat *et al.*, 2023
Tomato (*Solanum lycopersicum*)	SlWRKY3	Group I WRKY transcription factor	*Agrobacterium*-mediated transformation	Wang *et al.*, 2022
Wheat (*Triticum aestivum*)	TaHsfA1	Transcription factor	*Agrobacterium tumefaciens*-mediated gene transformation	Wang *et al.*, 2023
Maize (*Zea mays* L.)	ZmNAC074	NAC transcription factor	*Agrobacterium tumefaciens* strain GV3101-mediated gene transformation	Xi *et al.*, 2022
Cucumber (*Cucumis sativus* L.)	*CsSPT*	Transcription factor	*Agrobacterium tumefaciens*	Liang *et al.*, 2024
Lettuce (*Lactuca sativa* L.)	LsMYB15	Transcription factor	*Agrobacterium* strain GV3101	Chen *et al.*, 2022b

20.3.1 Creation of mutant libraries using CRISPR

Recently, the programmable capabilities of the CRISPR-Cas system have enabled the creation of extensive knockout mutagenized libraries for forward genetic screening. One of the primary benefits of utilizing CRISPR for screening is that it is a simple method for determining causative genes with a reasonable degree of specificity (Meng *et al.*, 2017). Moreover, the capacity to multiplex permits the simultaneous targeting of multiple genes, which can aid in overcoming genetic overlap and make saturation mutagenesis easier with reduced population sizes. The adaptability of the CRISPR toolbox permits scientists to produce diverse genomic modifications, including specific single nucleotide variants (SNVs) and transcriptional alterations, in addition to indels and loss-of-function mutations. These modifications are not attainable with alternative mutagenic or knockdown approaches (Gaudelli *et al.*, 2017). A pooled CRISPR library was explored by two research groups in the field of rice. In order to identify 12,802 rice genes, some researchers synthesized 25,604 gRNAs, which produced 14,000 initial transformants (T0 lines) (Meng *et al.*, 2017). Meanwhile, Lu *et al.* (2017) achieved 84,384 T0 lines through collective modification of a library that included 88,541 gRNAs. Nevertheless, in both studies, only a small subset of T0 lines had gRNA (target gene) identities assigned. The transformation of tomatoes was accomplished by pooling CRISPR libraries to create a set of mutant lines more quickly and with fewer transformation tries. A single transformation aimed toward immunity-related leucine-rich repeat subfamily XII genes resulted in heritable mutations in 15 of the 54 targeted genes. A second library was developed using three sgRNAs per construct to address 18 putative transporter genes, increasing and producing persistent mutations in 15 of the 18 genes (Jacobs *et al.*, 2017). Multiplex mutagenesis was produced using a pooled CRISPR-Cas9 in soybean, targeting 102 candidate genes across 70 vectors. The approach yielded 407 T0 lines with a 59.2% average mutagenesis frequency, leveraging gRNAs to address gene redundancy. Double mutants gmric1/gmric2 increased nodule numbers, while triple mutants gmrdn1-1/1-2/1-3 showed reduced nodulation (Bai *et al.*, 2020). An incredibly successful CRISPR/Cas9 gene-editing system was implemented in sweet basil to target the susceptibility gene ObDMR1, resulting in 56 transgenic lines with a 92.6% mutation rate at the target site. This system produced complete knockout mutants and transgene-free homozygous plants by the second generation, facilitating advancements in basil functional genomics and breeding (Navet and Tian, 2020).

Broadly, mutant libraries represent a valuable resource for enhancing genome-editing techniques focussed on addressing abiotic stressors in crops. By examining these collections for stress resilience and pinpointing specific targets for genetic modification, researchers can advise more improved strategies to boost stress tolerance in plants. Utilizing the CRISPR/Cas9 system, we created specific mutations in the TMS5 gene, a key thermosensitive genic male sterility (TGMS) factor, and successfully developed 11 new 'transgene clean' TGMS lines for hybrid rice breeding within a year. This efficient approach not only speeds up the creation of sterile lines but also enhances the potential for exploiting heterosis in rice subspecies (Zhou *et al.*, 2016). A mutant tomato line exhibiting parthenocarpy was identified through screening of an EMS-mutagenized population, with a mutation in the SlAGAMOUS-LIKE 6 (SlAGL6) gene confirmed via CRISPR/Cas9. This mutant produces high-quality seedless fruits under heat stress without affecting pollen viability or sexual reproduction, positioning SlAGL6 as a key regulator of fruit set transition (Klap *et al.*, 2017). CRISPR-induced mutations in NCED4 in lettuce led to mutant lines that showed over 70% seed germination at 37°C, which was notably better than that of wild-type plants (Bertier *et al.*, 2018).

20.3.2 Functional genomics

Functional genomics includes characterizing gene functions and providing comprehensive sequence, mapping, and other data to reveal the functions of individual genes. Small RNA analysis and the related marker information serve as highly effective tools for functional genomics, assisting in the understanding of gene regulation at both genetic and expression levels, and facilitating their application for crop improvement. Functional genomics has been useful in identifying the genes linked to agricultural stress responses (Sharma and Pandey, 2017; Singh *et al.* 2019a, b). CRISPR-Cas9 may be modified for novel methods, such as RNA cleavage, chromatin imaging, and epigenomic regulation (O'Connell *et al.*, 2014; Kearns *et al.*, 2015; Hilton *et al.*, 2015). The advancement of functional genomics is expected to rely on CRISPR-Cas9 technology due to its adaptability, ease of use, effectiveness, and adaptability. Comparative analysis of model plants through functional genomics techniques is expected to advance the goal of developing crop species that can adapt to environmental stress. Identifying novel genes, understanding how these genes express themselves in response to various stressors, and gaining insights into their role in the stress-tolerance process will enable the effective development of strategies for producing stress-tolerant transgenic crops. The use of CRISPR-Cas systems has attracted a lot of attention and has significant functional genomics implications (Zhang *et al.*, 2017; Wang *et al.*, 2018c). In crop plants, certain genes have been manipulated to create tolerance to biotic and abiotic stresses. The implementation of CRISPR-Cas systems has generated a lot of interest and has numerous implications for functional genomics.

20.3.3 Functional genomics of heat-stress response

Due to earlier domestication, the Green Revolution, and traditional breeding prioritizing improving yield and quality qualities, modern crop types have poor heat tolerance. To increase thermotolerance, however, one must now be aware of the genes, markers, and QTL regions linked to heat tolerance. Previous research revealed that the germplasms of different crops still contain a significant amount of genetic variation. Genome-wide association studies (GWAS) are a robust method for determining genotype–phenotype relationships in wild populations and are utilized to investigate the genetic makeup of populations. Numerous rice features have been linked to loci recognized by GWAS, and few significant genes have been effectively validated through follow-up transgenic research (Wang *et al.*, 2019; Chen *et al.*, 2022c). Additionally, this method has been utilized for major food crops such as wheat, rice, maize, sorghum, and *Brassica napus* L. to recognize natural variations associated with heat stress and to gain insights into their genetic foundations.

A transcriptomic study is another effective method for identifying and elucidating the key molecular mechanisms that plants employ in a reaction to extreme heat (Gao *et al.*, 2022). Plants respond to heat stress by activating a range of heat-responsive genes, making transcriptomic analyses valuable for screening potential candidate genes associated with heat tolerance. Microarray studies facilitate the identification of genes based on their expression profiles during specific developmental stages of plants under stress conditions (Ghorbani *et al.*, 2019). Some researchers examined the heat-responsive genes involved in potato tuberization (Singh *et al.*, 2015). Furthermore, suppression subtractive hybridization (SSH) makes it easier to identify the genes linked to the heat-stress responses and permits the tracking of their activity (Huang Min *et al.*, 2009). This method has been effectively utilized in other plant species facing various abiotic stresses, such as *Sedum alfredii* (Liu *et al.*, 2019), *Triticum aestivum* (Wang *et al.*, 2019), and *Lolium perenne* (Chen *et al.*, 2023), producing noteworthy results. VIGS is a multifunctional genomics tool that permits the silencing of single genes or multiple genes simultaneously in a single plant. Additionally, it can be utilized in a high-throughput manner to target genes across multiple plants (Lu *et al.*, 2003). VIGS is a eukaryotic reverse genetics technique based on RNA that utilizes various classes of short RNAs with distinct functions in gene silencing (Zhang *et al.*, 2016). These RNAs have been used, either singly or in combination with extensive, large-scale techniques, to explore the function of genes in many plant species (Liu *et al.*, 2016). VIGS has been used in various plant species to examine genes related to a variety of processes, including plant development, disease resistance, tolerance to abiotic stress, and resistance to nematodes and insects (Senthil-Kumar and Mysore, 2011). This dsRNA is then processed by the enzyme Dicer, which cleaves it into small interfering RNAs (siRNAs) approximately 21 nucleotides in length. These siRNAs cleave into two single-stranded RNAs, one of which joins the RNA-induced silencing complex (RISC) and the other of which is destroyed. RISC then targets and degrades the corresponding endogenous gene, allowing researchers to observe the effects of gene knockdown through phenotypic analysis (Robertson, 2004). VIGS-mediated silencing of the fatty acid desaturase (FAD7) gene enhanced membrane stability and enhanced high-temperature stress tolerance in *Nicotiana benthamiana*, indicating that FAD7 plays a crucial role in maintaining thermotolerance (Hiremath *et al.*, 2017). Silencing of the GhMAP3K65 gene in cotton increased resilience to pathogen infection and heat stress, indicating that GhMAP3K65 acts as a negative regulator of plant defense responses (Zhai *et al.*, 2017). VIGS-mediated silencing of NbGAE6 and other GAE family genes significantly reduced GalA levels in *Nicotiana benthamiana*, indicating that these genes play a critical role in pectin synthesis and could be crucial for enhancing plant resilience in response to environmental stress (Ahmed *et al.*, 2020). VIGS has employed the role of small heat-shock proteins (CaHSP16.4 and CaHsp25.9) in conferring high-temperature stress (Feng *et al.*, 2019). A functional genomics study revealed that ZmCDPK7 enhances thermotolerance in maize by phosphorylating sHSP17.4 and RBOHB, resulting in improved photosynthetic rates and antioxidant activity, while its expression is regulated by abscisic acid under heat stress (Zhao *et al.*, 2021).

20.4 Omics Approach

Omics techniques are crucial for identifying differentially expressed genes, proteins, and metabolites related to thermal stress responses in diverse cultivars, such as resilient and susceptible varieties. Approaches like transcriptomics, metabolomics, and proteomics facilitate the understanding of adaptation and tolerance mechanisms by investigating the key genes, enzymes, and metabolites that regulate essential pathways involved in heat-stress responses in crops.

20.4.1 Transcriptomics

Transcriptomics studies the complete set of RNA molecules in living organisms, focussing on the overall number of transcripts, their levels in specific cells, and any modifications that occur after transcription (Hussain, 2018). In *Arabidopsis*, a transcriptomic study conducted under drought and heat stress revealed approximately 770 transcripts that remained unchanged, alongside the identification of 53 distinct proteins that exhibited differences. Key proteins linked to protein folding, photosynthesis, and ROS scavenging, as well as metabolites like galactinol and tyrosine, were identified, underscoring their importance in osmotic balance and guiding the creation of heat-tolerant jojoba cultivars (Zheng *et al.*, 2022). The transcriptome comparison of hybrid rice with its parents under 42°C heat stress revealed unique gene-expression patterns and distinct proteins, highlighting the HSF–Hsp regulatory network's role in enhancing photosynthesis and antioxidant defenses for improved heat tolerance (Wang *et al.*, 2020). In tomato, the analysis identified differentially expressed genes and proteins under high-temperature stress, indicating that the 'protein processing in the endoplasmic reticulum' pathway plays a crucial role in heat tolerance and emphasizing the potential of the SlBAG5 gene in thermotolerance mechanisms (Ding *et al.*, 2020). Some researchers investigated the impact of water and heat stress on soybean, identifying proteins that fluctuate under stress conditions and highlighting the significance of MED37C in enhancing stress resilience (Katam *et al.*, 2020). Transcriptomic analysis in GB (glycerol)-primed Chinese cabbage (*Brassica rapa* L. ssp. pekinensis) identified 582 differentially expressed genes, revealing enhanced thermotolerance through activation of ABA and SA signaling pathways under heat stress (Quan *et al.*, 2023). Transcriptome profiling of drought- and heat-stressed barley genotypes revealed over 900 differentially expressed genes, with the stress-sensitive line exhibiting a greater transcriptional response than the stress-tolerant line, highlighting potential targets for breeding climate-resilient varieties (Mahalingam *et al.*, 2022).

20.4.2 Proteomics

Proteomics examines the roles, structures, functions, locations, and interactions of proteins, as well as their responses to stress or normal conditions. A comprehensive knowledge of plant stress signaling, essential proteins, and their metabolic pathways contributes to the creation of biotechnological instruments targeted at improving stress resistance (Aslam *et al.*, 2016). Integrative transcriptomic and proteomic analyses of banana peels stored at 20 and 30°C identified key proteins and genes affected by high-temperature stress, revealing impacts on chlorophyll metabolism, fruit firmness, and hormone signaling, ultimately leading to increased senescence and loss of firmness in the banana peel (Li *et al.*, 2019). Another study utilized 2D-DIGE proteomics to analyze the effects of drought and heat stresses on soybean leaves, revealing differential expression of 25 photosynthesis-related proteins that regulate key processes, along with evidence that increased carbonic anhydrase accumulation contributes to ROS resistance despite elevated H_2O_2 levels (Das *et al.*, 2016). This study utilized iTRAQ-based quantitative proteomics to investigate heat-stress responses in pepper seedlings, identifying 1591 heat-responsive proteins and linking enhanced heat tolerance to improved photosynthesis, ROS scavenging, signal transduction, stress-defense mechanisms, and carbohydrate metabolism (Wang *et al.*, 2021)

20.4.3 Metabolomics

Recent trends in metabolomics have provided deeper insights into the survival mechanisms of plants in response to heat stress at the metabolite level (Bokszczanin and Fragkostefanakis, 2013). So far, metabolomics has been minimally utilized in exploring the heat-stress tolerance mechanisms in grain legumes. Given the sensitivity of seed germination and development to heat stress, metabolite profiling of soybean genotypes exposed that antioxidants like tocopherols, flavonoids, ascorbate, and phenylpropanoids help mitigate heat stress in tolerant varieties (Chebrolu *et al.*, 2016). Another study utilized non-targeted metabolomics to profile soybean leaf metabolites under heat and drought stress, revealing differential accumulation of key metabolites involved in carbohydrate and amino acid metabolism, and highlighting the importance of maintaining metabolic homeostasis for improving stress resilience in soybean (Das *et al.*, 2017). This study characterized high-temperature stress-induced modifications in morpho-physicochemical traits and gene expression in chickpea, revealing enhanced thermotolerance in a tolerant cultivar through increased water-use efficiency, osmolytes, and secondary metabolites, while identifying 49 HTS-responsive metabolites in the metabolome landscape that may confer thermotolerance and aid in developing heat-resistant crops (Pareek *et al.*, 2019). In transgenic wheat lines expressing the maize phosphoenolpyruvate carboxylase gene, 25 distinct metabolites were identified under high-temperature conditions, including three polyols, four sugars, six organic acids, and 12 amino acids, compared to the wild type (Qi *et al.*, 2017). Overall, metabolomics is essential for enhancing our understanding of other omics disciplines and developing comprehensive models of biological systems, providing insights into heat tolerance that may not be addressed by other omics approaches (Paupière *et al.*, 2017).

20.5 Conclusion

In current years, CRISPR/Cas has developed as a groundbreaking and highly effective tool for targeted genome editing, rapidly advancing through the development of new CRISPR-based editing techniques. These innovations aim to enhance various aspects of genome engineering, including, crop yield, enhanced nutrient efficiency, increased pathogen resistance and tolerance to abiotic stress in crop species. While the main focus of this tool has primarily been on creating gene knockouts, exploring its other applications will be crucial for advancing stress biology. The CRISPR-Cas9 system has yet to be utilized to investigate the pathways associated with responses and adaptations to abiotic stress. Producing crops that can withstand high temperatures will be made possible by combining genetic engineering and biotechnology advancements with a deep comprehension of plant omics. In summary, a multidisciplinary approach combining genetic, molecular, and agronomic approaches is crucial in order to protect food security against climate change. Gene-editing technologies, particularly CRISPR/Cas9, have profoundly transformed both fundamental plant research and crop enhancement. The ability to easily remove the transgenes used to induce genomic alterations from the genome by genetic segregation is a major advantage of these abilities, as it produces gene-edited varieties that are exactly equivalent to those created through conventional breeding methods.

References

Aggarwal, P.K. and Mall, R.K. (2002) Climate change and rice yields in diverse agro-environments of India. II. Effect of uncertainties in scenarios and crop models on impact assessment. *Climatic Change* 52(3), 331–343.

Ahmed, R.I., Ren, A., Yang, D., Ding, A. and Kong, Y. (2020) Identification and characterization of pectin related gene NbGAE6 through virus-induced gene silencing in *Nicotiana benthamiana*. *Gene* 741, 144522.

Alghuthaymi, M.A., Ahmad, A., Khan, Z., Khan, S.H., Ahmed, F.K. *et al.* (2021) Exosome/liposome-like nanoparticles: New carriers for CRISPR genome editing in plants. *International Journal of Molecular Sciences* 22(14), 7456.

Aslam, B., Basit, M., Nisar, M.A., Khurshid, M. and Rasool, M.H. (2016) Proteomics: Technologies and their applications. *Journal of Chromatographic Science* 55(2), 182–196.

Asseng, S., Cammarano, D., Basso, B., Chung, U., Alderman, P.D. *et al.* (2017) Hot spots of wheat yield decline with rising temperatures. *Global Change Biology* 23(6), 2464–2472.

Bai, M., Yuan, J., Kuang, H., Gong, P., Li, S. *et al.* (2020) Generation of a multiplex mutagenesis population via pooled CRISPR-Cas9 in soya bean. *Plant Biotechnology Journal* 18(3), 721–731.

Barrangou, R. and Marraffini, L.A. (2014) CRISPR-Cas systems: Prokaryotes upgrade to adaptive immunity. *Molecular Cell* 54(2), 234–244.

Begcy, K., Nosenko, T., Zhou, L.Z., Fragner, L., Weckwerth, W. *et al.* (2019) Male sterility in maize after transient heat stress during the tetrad stage of pollen development. *Plant Physiology* 181(2), 683–700.

Bernstein, L., Bosch, P., Canziani, O., Chen, Z., Christ, R. *et al.* (2008) *Climate change 2007: Synthesis Report: An Assessment of the Intergovernmental Panel on Climate Change*. IPCC.

Bertier, L.D., Ron, M., Huo, H., Bradford, K.J., Britt, A.B. *et al.* (2018) High-resolution analysis of the efficiency, heritability, and editing outcomes of CRISPR/Cas9-induced modifications of NCED4 in lettuce (*Lactuca sativa*). *G3: Genes, Genomes, Genetics* 8(5), 1513–1521.

Bokszczanin, K.L. and Fragkostefanakis, S. (2013) Solanaceae Pollen Thermotolerance Initial Training Network (SPOT-ITN) Consortium perspectives on deciphering mechanisms underlying plant heat stress response and thermotolerance. *Frontiers in Plant Science* 4, 315.

Bortesi, L. and Fischer, R. (2015) The CRISPR/Cas9 system for plant genome editing and beyond. *Biotechnology Advances* 33, 41–52.

Buragohain, S., Sarma, B., Nath, D.J., Gogoi, N., Meena, R.S. *et al.* (2017) Effect of 10 years of biofertiliser use on soil quality and rice yield on an Inceptisol in Assam, India. *Soil Research* 56(1), 49.

Chang, Y., Nguyen, B.H., Xie, Y., Xiao, B., Tang, N. *et al.* (2017) Co-overexpression of the constitutively active form of OsbZIP46 and ABA-activated protein kinase SAPK6 improves drought and temperature stress resistance in rice. *Frontiers in Plant Science* 8, 1102.

Chaturvedi, P., Wiese, A.J., Ghatak, A., Záveská Drábková, L., Weckwerth, W. *et al.* (2021) Heat stress response mechanisms in pollen development. *The New Phytologist* 231(2), 571–585.

Chebrolu, K.K., Fritschi, F.B., Ye, S., Krishnan, H.B., Smith, J.R. *et al.* (2016) Impact of heat stress during seed development on soybean seed metabolome. *Metabolomics* 12(2), 1–14.

Chen, K., Wang, Y., Zhang, R., Zhang, H. and Gao, C. (2019) CRISPR/Cas genome editing and precision plant breeding in agriculture. *Annual Review of Plant Biology* 70(1), 667–697.

Chen, Z., Galli, M. and Gallavotti, A. (2022a) Mechanisms of temperature-regulated growth and thermotolerance in crop species. *Current Opinion in Plant Biology* 65, 102134.

Chen, L., Xu, M., Liu, C., Hao, J., Fan, S. *et al*. (2022b) LsMYB15 regulates bolting in leaf lettuce (*Lactuca sativa* L.) under high-temperature stress. *Frontiers in Plant Science* 13, 921021.

Chen, R., Deng, Y., Ding, Y., Guo, J., Qiu, J. *et al*. (2022c) Rice functional genomics: Decades' efforts and roads ahead. *Science China. Life Sciences* 65(1), 33–92.

Chen, Y., Sun, Y., Wan, H., Chen, L., Cao, L. *et al*. (2023) Integrative analysis of transcriptome and yeast screening system identified heat stress-responding genes in ryegrass. *Environmental and Experimental Botany* 210, 105333.

Das, A., Eldakak, M., Paudel, B., Kim, D.W., Hemmati, H. *et al*. (2016) Leaf proteome analysis reveals prospective drought and heat stress response mechanisms in soybean. *BioMed Research International* 2016(1), 6021047.

Das, A., Rushton, P.J. and Rohila, J.S. (2017) Metabolomic profiling of soybeans (*Glycine max* L.) reveals the importance of sugar and nitrogen metabolism under drought and heat stress. *Plants* 6(2), 21.

Debbarma, J., Sarki, Y.N., Saikia, B., Boruah, H.P.D., Singha, D.L. *et al*. (2019) Ethylene Response Factor (ERF) family proteins in abiotic stresses and CRISPR-Cas9 genome editing of ERFs for multiple abiotic stress tolerance in crop plants: A review. *Molecular Biotechnology* 61(2), 153–172.

Deltcheva, E., Chylinski, K., Sharma, C.M., Gonzales, K., Chao, Y. *et al*. (2011) CRISPR RNA maturation by trans-encoded small RNA and host factor RNase III. *Nature* 471(7340), 602–607.

Ding, H., Mo, S., Qian, Y., Yuan, G., Wu, X. *et al*. (2020) Integrated proteome and transcriptome analyses revealed key factors involved in tomato (*Solanum lycopersicum*) under high temperature stress. *Food and Energy Security* 9(4), e239.

Doudna, J.A. and Charpentier, E. (2014) The new frontier of genome engineering with CRISPR-Cas9. *Science* 346(6213), 1258096.

Feng, X.H., Zhang, H.X., Ali, M., Gai, W.X., Cheng, G.X. *et al*. (2019) A small heat shock protein CaHsp25.9 positively regulates heat, salt, and drought stress tolerance in pepper (*Capsicum annuum* L.). *Plant Physiology and Biochemistry* 142, 151–162.

Fragkostefanakis, S., Simm, S., Paul, P., Bublak, D., Scharf, K. *et al*. (2015) Chaperone network composition in *Solanum lycopersicum* explored by transcriptome profiling and microarray meta-analysis. *Plant, Cell & Environment* 38(4), 693–709.

Gao, C., Mumtaz, M.A., Zhou, Y., Yang, Z., Shu, H. *et al*. (2022) Integrated transcriptomic and metabolomic analyses of cold-tolerant and cold-sensitive pepper species reveal key genes and essential metabolic pathways involved in response to cold stress. *International Journal of Molecular Sciences* 23(12), 6683.

Gasiunas, G., Barrangou, R., Horvath, P. and Siksnys, V. (2012) Cas9-crRNA ribonucleoprotein complex mediates specific DNA cleavage for adaptive immunity in bacteria. *Proceedings of the National Academy of Sciences of the United States of America* 109(39), E2579–E2586.

Gaudelli, N.M., Komor, A.C., Rees, H.A., Packer, M.S., Badran, A.H. *et al*. (2017) Programmable base editing of A•T to G•C in genomic DNA without DNA cleavage. *Nature* 551(7681), 464–471.

Ghorbani, R., Alemzadeh, A. and Razi, H. (2019) Microarray analysis of transcriptional responses to salt and drought stress in *Arabidopsis thaliana*. *Heliyon* 5(11), e02614.

Haft, D.H., Selengut, J., Mongodin, E.F. and Nelson, K.E. (2005) A guild of 45 CRISPR-associated (Cas) protein families and multiple CRISPR/Cas subtypes exist in prokaryotic genomes. *PLoS Computational Biology* 1(6), e60.

Hatfield, J.L. and Prueger, J.H. (2015) Temperature extremes: Effect on plant growth and development. *Weather and Climate Extremes* 10, 4–10.

Hille, F. and Charpentier, E. (2016) CRISPR-Cas: Biology, mechanisms and relevance. *Philosophical Transactions of the Royal Society B* 371(1707), 20150496.

Hilton, I.B., D'Ippolito, A.M., Vockley, C.M., Thakore, P.I., Crawford, G.E. *et al*. (2015) Epigenome editing by a CRISPR-Cas9-based acetyltransferase activates genes from promoters and enhancers. *Nature Biotechnology* 33(5), 510–517.

Hiremath, S.S., Sajeevan, R.S., Nataraja, K.N., Chaturvedi, A.K., Chinnusamy, V. *et al*. (2017) Silencing of Fatty Acid Desaturase (FAD7) gene enhances membrane stability and photosynthetic efficiency under heat stress in tobacco (*Nicotiana benthamiana*). *Indian Journal of Experimental Biology* 55(8), 532–541.

Hsu, P.D., Scott, D.A., Weinstein, J.A., Ran, F.A., Konermann, S. *et al*. (2013) DNA targeting specificity of RNA-guided Cas9 nucleases. *Nature Biotechnology* 31(9), 827–832.

Huang, Y., Xuan, H., Yang, C., Guo, N., Wang, H. *et al*. (2019) GmHsp90A2 is involved in soybean heat stress as a positive regulator. *Plant Science* 285, 26–33.

Huang Min, H.M., Wang Yan, W.Y., Pan Lin, P.L., Yang ChengMing, Y.C., Sun Qun, S.Q. *et al.* (2009) Screening and identification of heat tolerance genes of hybrid rice. *Southwest China Journal of Agricultural Sciences* 22(5), 1199–1203.

Hussain, S. (2018) Native RNA-sequencing throws its hat into the transcriptomics ring. *Trends in Biochemical Sciences* 43(4), 225–227.

Hussain, M., Farooq, S., Hasan, W., Ul-Allah, S., Tanveer, M. *et al.* (2018) Drought stress in sunflower: Physiological effects and its management through breeding and agronomic alternatives. *Agricultural Water Management* 201, 152–166.

Ishino, Y., Shinagawa, H., Makino, K., Amemura, M. and Nakata, A. (1987) Nucleotide sequence of theiap gene, responsible for alkaline phosphatase isozyme conversion in *escherichia coli*, and identification of the gene product. *Journal of Bacteriology* 169(12), 5429–5433.

Jacobs, T.B., Zhang, N., Patel, D. and Martin, G.B. (2017) Generation of a collection of mutant tomato lines using pooled CRISPR libraries. *Plant Physiology* 174(4), 2023–2037.

Jinek, M., Chylinski, K., Fonfara, I., Hauer, M., Doudna, J.A. *et al.* (2012) A programmable dual-RNA-guided DNA endonuclease in adaptive bacterial immunity. *Science* 337(6096), 816–821.

Jinek, M., East, A., Cheng, A., Lin, S., Ma, E. *et al.* (2013) RNA-programmed genome editing in human cells. *eLife* 2, e00471.

Jung, Y.J., Go, J.Y., Lee, H.J., Park, J.S., Kim, J.Y. *et al.* (2021) Overexpression of orange gene (OsOr-R115H) enhances heat tolerance and defense-related gene expression in rice (*Oryza sativa* L.). *Genes* 12(12), 1891.

Kan, Y. and Lin, H.X. (2021) Molecular regulation and genetic control of rice thermal response. *The Crop Journal* 9(3), 497–505.

Katam, R., Shokri, S., Murthy, N., Singh, S.K., Suravajhala, P. *et al.* (2020) Proteomics, physiological, and biochemical analysis of cross tolerance mechanisms in response to heat and water stresses in soybean. *PLoS One* 15(6), e0233905.

Kearns, N.A., Pham, H., Tabak, B., Genga, R.M., Silverstein, N.J. *et al.* (2015) Functional annotation of native enhancers with a Cas9-histone demethylase fusion. *Nature Methods* 12(5), 401–403.

Klap, C., Yeshayahou, E., Bolger, A.M., Arazi, T., Gupta, S.K. *et al.* (2017) Tomato facultative parthenocarpy results from SlAGAMOUS-LIKE 6 loss of function. *Plant Biotechnology Journal* 15(5), 634–647.

Krishnan, P., Ramakrishnan, B., Reddy, K.R. and Reddy, V.R. (2011) High-temperature effects on rice growth, yield, and grain quality. *Advances in Agronomy* 111, 87–206.

Li, J., Zhang, H., Si, X., Tian, Y., Chen, K. *et al.* (2017) Generation of thermosensitive male-sterile maize by targeted knockout of the ZmTMS5 gene. *Journal of Genetics and Genomics* 44(9), 465–468.

Li, T., Wu, Q., Duan, X., Yun, Z. and Jiang, Y. (2019) Proteomic and transcriptomic analysis to unravel the influence of high temperature on banana fruit during postharvest storage. *Functional & Integrative Genomics* 19(3), 467–486.

Liang, Y., Yang, C., Ming, F., Yu, B., Cheng, Z. *et al.* (2024) A bHLH transcription factor, CsSPT, regulates high-temperature resistance in cucumber. *Horticultural Plant Journal* 10(2), 503–514.

Lieber, M.R. (2010) The mechanism of double-strand DNA break repair by the nonhomologous DNA end-joining pathway. *Annual Review of Biochemistry* 79(1), 181–211.

Liu, N., Xie, K., Jia, Q., Zhao, J., Chen, T. *et al.* (2016) *Foxtail mosaic virus*-induced gene silencing in monocot plants. *Plant Physiology* 171(3), 1801–1807.

Liu, M., He, X., Feng, T., Zhuo, R., Qiu, W. *et al.* (2019) CDNA library for mining functional genes in *Sedum alfredii* hance related to cadmium tolerance and characterization of the roles of a novel SaCTP2 gene in enhancing cadmium hyperaccumulation. *Environmental Science & Technology* 53(18), 10926–10940.

Liu, J.J., Orlova, N., Oakes, B.L., Ma, E., Spinner, H.B. *et al.* (2020) CRISPR-Casx is an RNA-dominated enzyme active for human genome editing. *Nature* 566(7743), 218.

Lu, R., Malcuit, I., Moffett, P., Ruiz, M.T., Peart, J. *et al.* (2003) High throughput virus-induced gene silencing implicates heat shock protein 90 in plant disease resistance. *The EMBO Journal* 22(21), 5690–5699.

Lu, Y., Ye, X., Guo, R., Huang, J., Wang, W. *et al.* (2017) Genome-wide targeted mutagenesis in rice using the CRISPR/Cas9 system. *Molecular Plant* 10(9), 1242–1245.

Mahalingam, R., Duhan, N., Kaundal, R., Smertenko, A., Nazarov, T. *et al.* (2022) Heat and drought induced transcriptomic changes in barley varieties with contrasting stress response phenotypes. *Frontiers in Plant Science* 13, 1066421.

Makarova, K.S., Wolf, Y.I., Alkhnbashi, O.S., Costa, F., Shah, S.A. *et al.* (2015) An updated evolutionary classification of CRISPR-Cas systems. *Nature Reviews. Microbiology* 13(11), 722–736.

Mali, P., Yang, L., Esvelt, K.M., Aach, J., Guell, M. *et al.* (2013) RNA-guided human genome engineering via Cas9. *Science* 339(6121), 823–826.

Malzahn, A., Lowder, L. and Qi, Y. (2017) Plant genome editing with TALEN and CRISPR. *Cell & Bioscience* 7, 1–18.

Martre, P., Wallach, D., Asseng, S., Ewert, F., Jones, J.W. *et al.* (2015) Multimodel ensembles of wheat growth: Many models are better than one. *Global Change Biology* 21(2), 911–925.

Mashizha, T.M. (2019) Building adaptive capacity: Reducing the climate vulnerability of smallholder farmers in Zimbabwe. *Business Strategy & Development* 2(3), 166–172.

Matsui, T., Omasa, K. and Horie, T. (2000) High temperature at flowering inhibits swelling of pollen grains, a driving force for thecae dehiscence in rice (*Oryza sativa* L.). *Plant Production Science* 3(4), 430–434.

Mei, Y., Wang, Y., Chen, H., Sun, Z.S. and Ju, X.D. (2016) Recent progress in CRISPR/Cas9 technology. *Journal of Genetics and Genomics* 43(2), 63–75.

Meng, X., Yu, H., Zhang, Y., Zhuang, F., Song, X. *et al.* (2017) Construction of a genome-wide mutant library in rice using CRISPR/Cas9. *Molecular Plant* 10(9), 1238–1241.

Montague, T.G., Cruz, J.M., Gagnon, J.A., Church, G.M. and Valen, E. (2014) CHOPCHOP: A CRISPR/Cas9 and TALEN web tool for genome editing. *Nucleic Acids Research* 42(W1), W401–W407.

Musunuru, K. (2017) The hope and hype of CRISPR-cas9 genome editing: a review. *JAMA Cardiology* 2(8), 914–919.

Nadeem, M., Li, J., Yahya, M., Wang, M., Ali, A. *et al.* (2019) Grain legumes and fear of salt stress: Focus on mechanisms and management strategies. *International Journal of Molecular Sciences* 20(4), 799.

Navet, N. and Tian, M. (2020) Efficient targeted mutagenesis in allotetraploid sweet basil by CRISPR/Cas9. *Plant Direct* 4(6), e00233.

O'Connell, M.R., Oakes, B.L., Sternberg, S.H., East-Seletsky, A., Kaplan, M. *et al.* (2014) Programmable RNA recognition and cleavage by CRISPR/Cas9. *Nature* 516(7530), 263–266.

Pareek, A., Rathi, D., Mishra, D., Chakraborty, S. and Chakraborty, N. (2019) Physiological plasticity to high temperature stress in chickpea: Adaptive responses and variable tolerance. *Plant Science* 289, 110258.

Park, S.I., Kwon, H.J., Cho, M.H., Song, J.S., Kim, B.G. *et al.* (2021) The OsERF115/AP2EREBP110 transcription factor is involved in the multiple stress tolerance to heat and drought in rice plants. *International Journal of Molecular Sciences* 22(13), 7181.

Paupière, M.J., Müller, F., Li, H., Rieu, I., Tikunov, Y.M. *et al.* (2017) Untargeted metabolomic analysis of tomato pollen development and heat stress response. *Plant Reproduction* 30, 81–94.

Qiu, Z., Kang, S., He, L., Zhao, J., Zhang, S, *et al.* (2018) The newly identified heat-stress sensitive albino 1 gene affects chloroplastdevelopment in rice. *Plant Science* 267, 168–179.

Qi, X., Xu, W., Zhang, J., Guo, R., Zhao, M. *et al.* (2017) Physiological characteristics and metabolomics of transgenic wheat containing the maize C4 Phosphoenolpyruvate Carboxylase (PEPC) gene under high temperature stress. *Protoplasma* 254, 1017–1030.

Quan, J., Li, X., Li, Z., Wu, M., Zhu, B. *et al.* (2023) Transcriptomic analysis of heat stress response in *Brassica rapa* L. ssp. pekinensis with improved thermotolerance through exogenous *Glycine betaine*. *International Journal of Molecular Sciences* 24(7), 6429.

Rahmat, B.P.N., Octavianis, G., Budiarto, R., Jadid, N., Widiastuti, A. *et al.* (2023) SlIAA9 mutation maintains photosynthetic capabilities under heat-stress conditions. *Plants* 12(2), 378.

Rathnasamy, S.A., Kambale, R., Elangovan, A., Mohanavel, W., Shanmugavel, P. *et al.* (2023) Altering stomatal density for manipulating transpiration and photosynthetic traits in rice through CRISPR/Cas9 mutagenesis. *Current Issues in Molecular Biology* 45(5), 3801–3814.

Richardson, K.J., Lewis, K.H., Krishnamurthy, P.K., Kent, C., Wiltshire, A.J. *et al.* (2018) Food security outcomes under a changing climate: Impacts of mitigation and adaptation on vulnerability to food insecurity. *Climatic Change* 147(1–2), 327–341.

Robertson, D. (2004) VIGS vectors for gene silencing: Many targets, many tools. *Annual Review of Plant Biology* 55(1), 495–519.

Sage, R.F. and Kubien, D.S. (2007) The temperature response of C3 and C4 photosynthesis. *Plant, Cell & Environment* 30(9), 1086–1106.

San Filippo, J., Sung, P. and Klein, H. (2008) Mechanism of eukaryotic homologous recombination. *Annual Review of Biochemistry* 77(1), 229–257.

Sapranauskas, R., Gasiunas, G., Fremaux, C., Barrangou, R., Horvath, P. *et al.* (2011) The streptococcus thermophilus CRISPR/Cas system provides immunity in *Escherichia coli*. *Nucleic Acids Research* 39(21), 9275–9282.

Schaeffer, S.M. and Nakata, P.A. (2015) CRISPR/Cas9-mediated genome editing and gene replacement in plants: Transitioning from lab to field. *Plant Science* 240, 130–142.

Senthil-Kumar, M. and Mysore, K.S. (2011) New dimensions for VIGS in plant functional genomics. *Trends in Plant Science* 16(12), 656–665.

Sharma, M. and Pandey, G.K. (2017) Editorial: Genomics and functional genomics of stress-mediated signaling in plants: Volume I. *Current Genomics* 18(6), 467–468.

Singh, A., Siddappa, S., Bhardwaj, V., Singh, B., Kumar, D, *et al.* (2015) Expression profiling of potato cultivars with contrasting tuberization at elevated temperature using microarray analysis. *Plant Physiology and Biochemistry* 97, 108–116.

Singh, B., Kukreja, S., Salaria, N., Thakur, K., Gautam, S. *et al.* (2019a) VIGS: A flexible tool for the study of functional genomics of plants under abiotic stresses. *Journal of Crop Improvement* 33(5), 567–604.

Singh, B., Salaria, N., Thakur, K., Kukreja, S., Gautam, S. *et al.* (2019b) Functional genomic approaches to improve crop plant heat stress tolerance. *F1000Research* 8, 1721.

Stamm, P., Ramamoorthy, R. and Kumar, P.P. (2011) Feeding the extra billions: Strategies to improve crops and enhance future food security. *Plant Biotechnology Reports* 5(2), 107–120.

Teng, L.I., Zhang, X.P., Qing, L.I.U., Jin, L.I.U., Chen, Y.Q, *et al.* (2022) Yield penalty ofmaize (*Zea mays* L.) under heat stress in different growth stages: a review. *Journal of Integrative Agriculture* 21(9), 2465–2476.

Tran, M.T., Son, G.H., Song, Y.J., Nguyen, N.T., Park, S, *et al.* (2023) CRISPR-Cas9-based precise engineering of *slhyprp1* proteintowards multi-stress tolerance in tomato. *Frontiers in Plant Science* 14, 1186932.

Tripathi, A., Tripathi, D.K., Chauhan, D.K., Kumar, N. and Singh, G.S. (2016) Paradigms of climate change impacts on some major food sources of the world: A review on current knowledge and future prospects. *Agriculture, Ecosystems & Environment* 216, 356–373.

Varshney, R.K., Graner, A. and Sorrells, M.E. (2005) Genomics-assisted breeding for crop improvement. *Trends in Plant Science* 10(12), 621–630.

Wahid, A. and Shabbir, A. (2005) Induction of heat stress tolerance in barley seedlings by pre-sowing seed treatment with glycinebetaine. *Plant Growth Regulation* 46(2), 133–141.

Wang, J., Vanga, S.K., Saxena, R., Orsat, V. and Raghavan, V. (2018a) Effect of climate change on the yield of cereal crops: A review. *Climate* 6(2), 41.

Wang, C.T., Ru, J.N., Liu, Y.W., Li, M., Zhao, D. *et al.* (2018b) Maize WRKY transcription factor ZmwRKY106 confers drought and heat tolerance in transgenic plants. *International Journal of Molecular Sciences* 19(10), 3046.

Wang, P., Su, L., Gao, H., Jiang, X., Wu, X. *et al.* (2018c) Genome-wide characterization of bHLH genes in grape and analysis of their potential relevance to abiotic stress tolerance and secondary metabolite biosynthesis. *Frontiers in Plant Science* 9, 64.

Wang, J., Mao, X., Wang, R., Li, A., Zhao, G. *et al.* (2019) Identification of wheat stress-responding genes and TaPR-1-1 function by screening a cDNA yeast library prepared following abiotic stress. *Scientific Reports* 9(1), 141.

Wang, Y., Yu, Y., Huang, M., Gao, P., Chen, H. *et al.* (2020) Transcriptomic and proteomic profiles of II YOU 838 (*Oryza sativa*) provide insights into heat stress tolerance in hybrid rice. *PeerJ* 8, e8306.

Wang, J., Liang, C., Yang, S., Song, J., Li, X. *et al.* (2021) ITRAQ-based quantitative proteomic analysis of heat stress-induced mechanisms in pepper seedlings. *PeerJ* 9, e11509.

Wang, Y., Gai, W., Yuan, L., Shang, L., Li, F. *et al.* (2022) Heat-inducible SlWRKY3 confers thermotolerance by activating the SlGRXS1 gene cluster in tomato. *Horticultural Plant Journal* 10(2), 515–531.

Wang, H., Feng, M., Jiang, Y., Du, D., Dong, C. *et al.* (2023) Thermosensitive SUMOylation of TaHsfA1 defines a dynamic ON/OFF molecular switch for the heat stress response in wheat. *The Plant Cell* 35(10), 3889–3910.

Xi, Y., Ling, Q., Zhou, Y., Liu, X. and Qian, Y. (2022) ZmNAC074, a maize stress-responsive NAC transcription factor, confers heat stress tolerance in transgenic *Arabidopsis*. *Frontiers in Plant Science* 13, 986628.

Xu, X., Wang, Q., Li, W., Hu, T., Wang, Q. *et al.* (2022) Overexpression of SlBBX17 affects plant growth and enhances heat tolerance in tomato. *International Journal of Biological Macromolecules* 206, 799–811.

Yarra, R. and Xue, Y. (2020) Ectopic expression of nucleolar DEAD-box RNA helicase OstOGR1 confers improved heat stress tolerance in transgenic Chinese cabbage. *Plant Cell Reports* 39(12), 1803–1814.

Yin, Y., Qin, K., Song, X., Zhang, Q., Zhou, Y. *et al.* (2018) BZR1 transcription factor regulates heat stress tolerance through FERONIA receptor-like kinase-mediated reactive oxygen species signaling in tomato. *Plant and Cell Physiology* 59(11), 2239–2254.

Yu, W., Wang, L., Zhao, R., Sheng, J., Zhang, S. *et al.* (2019) Knockout of S1mAPK3 enhances tolerance to heat stress involving ROS homeostasis in tomato plants. *BMC Plant Biology* 19(1), 354.

Zhai, N., Jia, H., Liu, D., Liu, S., Ma, M. *et al.* (2017) GhMAP3K65, a cotton Raf-like MAP3K gene, enhances susceptibility to pathogen infection and heat stress by negatively modulating growth and development in transgenic *Nicotiana benthamiana*. *International Journal of Molecular Sciences* 18(11), 2462.

Zhang, Z., Liu, X., Guo, X., Wang, X.J. and Zhang, X. (2016) *Arabidopsis* AGO3 predominantly recruits 24-nt small RNAs to regulate epigenetic silencing. *Nature Plants* 2(5), 1–7.

Zhang, L., Li, X., Ma, B., Gao, Q., Du, H. *et al.* (2017) The tartary buckwheat genome provides insights into rutin biosynthesis and abiotic stress tolerance. *Molecular Plant* 10(9), 1224–1237.

Zhao, C., Liu, B., Piao, S., Wang, X., Lobell, D.B. *et al.* (2017) Temperature increase reduces global yields of major crops in four independent estimates. *Proceedings of the National Academy of Sciences of the United States of America* 114(35), 9326–9331.

Zhao, Y., Du, H., Wang, Y., Wang, H., Yang, S. *et al.* (2021) The calcium-dependent protein kinase ZmCDPK7 functions in heat-stress tolerance in maize. *Journal of Integrative Plant Biology* 63(3), 510–527.

Zheng, L., Wu, W., Chen, Q., Zhang, G., Gao, F. *et al.* (2022) Integrated transcriptomics, proteomics, and metabolomics identified biological processes and metabolic pathways involved in heat stress response in jojoba. *Industrial Crops and Products* 183, 114946.

Zhou, H., He, M., Li, J., Chen, L., Huang, Z. *et al.* (2016) Development of commercial thermo-sensitive genic male sterile rice accelerates hybrid rice breeding using the CRISPR/Cas9-mediated TMS5 editing system. *Scientific Reports* 6(1), 37395.

21 Multi-stress-tolerant Crops: Identification and Functional Analysis of Hub Genes and Regulating Responses of Complicated Stressors

Sirajudeen Benazir Begum[1], Habeebuheen Fathima Seyed Mafia Haniff[2], Tajudheen Nargis Begum[3]*, Anbazhagan Priya[1], Asirvatham Alwin Robert[4], Manoharan Rajesh[5], Gunna Sureshbabu Suruthi[6] and Arumugam Vijaya Anand[6]

[1]PG and Research Department of Zoology, V.O. Chidambaram College, Thoothukudi, Tamil Nadu, India; [2]Department of Zoology, Jamal Mohamed College, Tiruchirapalli, Tamil Nadu, India; [3]Department of Biotechnology, Jamal Mohamed College, Tiruchirapalli, Tamil Nadu, India; [4]Department of Endocrinology and Diabetes, Prince Sultan Military Medical City, Ministry of Defense, Riyadh, Kingdom of Saudi Arabia; [5]Department of Biochemistry, Bharathiar University, Coimbatore, Tamil Nadu, India; [6]Department of Human Genetics and Molecular Biology, Bharathiar University, Coimbatore, Tamil Nadu, India

Abstract

Plant stress responses are regulated by intricate signaling networks and entail a variety of modifications at the molecular and cellular levels. Thus, it is becoming more essential to study protein–protein interactions (PPIs) and the networks that result in comprehending these reactions. The hubs or hub proteins are called key proteins in scale-free PPI networks with the highest degree of connectivity. The significance of structural analysis in the study of hub proteins is that it provides insights into their evolutionary conservation, interaction dynamics, and functional versatility. We concentrate on the definitions of hub proteins as they currently exist, including the characteristics that are typically thought to define a hub as well as the difficulties and methods involved in identifying hub proteins. Proteomics and interactomics, two high-throughput screening methods, have made it conceivable to recognize the protein interaction networks that moderate signal processing from information to suitable outcomes. The topology of such networks can be instantly visualized and analyzed using nodes and edges in graphical representations. Plant networks can be created using the already-existing public. Case studies from various crop species exemplify the application of these approaches, highlighting conserved structural motifs and network rewiring strategies underlying multi-stress tolerance. Emerging approaches like predictive modeling of protein interactions and network-based breeding strategies can potentially speed up the application of hub protein knowledge in agricultural biotechnology. To advance sustainable agriculture practices in a changing environment, this chapter offers a thorough framework for academics and practitioners who wish to use structural biology to understand the intricacies of multi-stress-tolerant crops.

*Corresponding author: nargisalmaas@gmail.com

© CAB International 2025. *Stress-Resilient Crops: Coordinated omics-CRISPR-Nanotechnology Strategies* (ed. Jen-Tsung Chen)

DOI: 10.1079/9781800627307.0021

Keywords: plant stress response, protein–protein interactions, hub proteins, transcriptomics, hub protein identification

21.1 Introduction

The transition of agriculture has evolved to ensure food and nutritional safety in the new era due to the problems of the 21st century. To feed a growing population sustainably, farmers now confront enormous hurdles (Firbank *et al.*, 2018). Achieving a high plant yield in agricultural activities is not always possible since it depends on several variables, such as the quality of the cultivar, the availability of nutrients and water, the level of pathogen infection, natural disasters, and soil conditions that impact plant growth and development. More notably, during unexpected global climate changes, the emergence of more severe or numerous stressors, and land restrictions for agriculture, scientists are grappling with the challenge of tolerable plant productivity to provide for humans (Thao and Tran, 2016). Plants are constantly subjected to different environmental strains. These can have a biotic source, such as disease or herbivore assaults, or they can be abiotic, like dryness, salinity, and unfavorable light or temperature conditions. Because of this, the plant strain response is extremely complicated, and even while the molecular pathways that arise upon stress identification have been well-researched, there is still much to learn (Duque *et al.*, 2013; Ben Rejeb *et al.*, 2014).

It is crucial to analyze systems as a whole rather than just their parts in the modern era of systems biology-related investigation. As a result, researching the collection of physical protein interactions within a plant cell, or the plant–protein interactome, can yield important new understandings of biological processes (Vandereyken *et al.*, 2018). Plant protein–protein interaction (PPI) information has become more accurate during the past decade due to developments in high-throughput proteomics and interactomics technology (Braun *et al.*, 2013; Xing *et al.*, 2016). Research indicates that the majority of biological networks (Fig. 21.1), such as PPI networks, are scale-free, indicating that a smaller number of proteins cooperate with a considerable number of different proteins (Fig. 21.2). These proteins are known as hub proteins, or hubs, due to their strong interconnectedness. PPI networks and entire biological systems depend heavily on them (Zafar *et al.*, 2020).

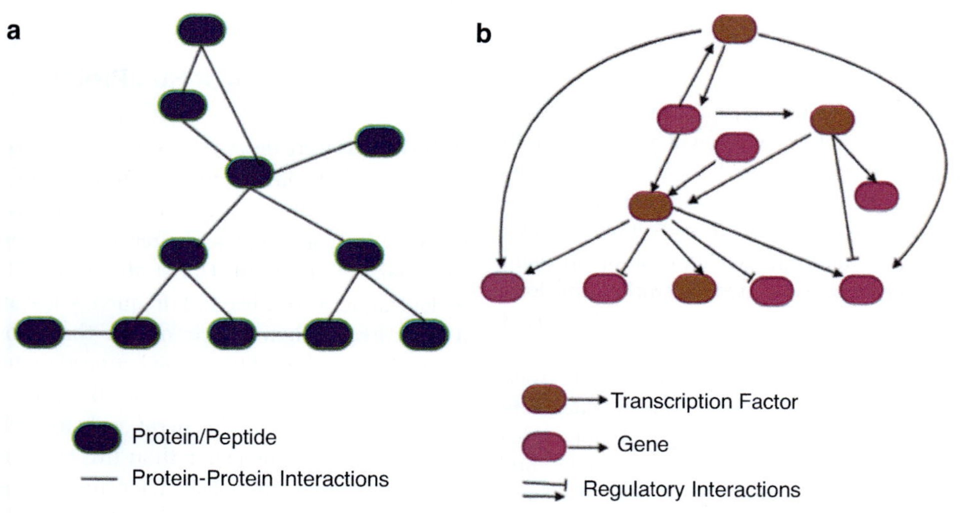

Fig. 21.1. Graphical model of gene regulatory network and protein–protein interaction network.

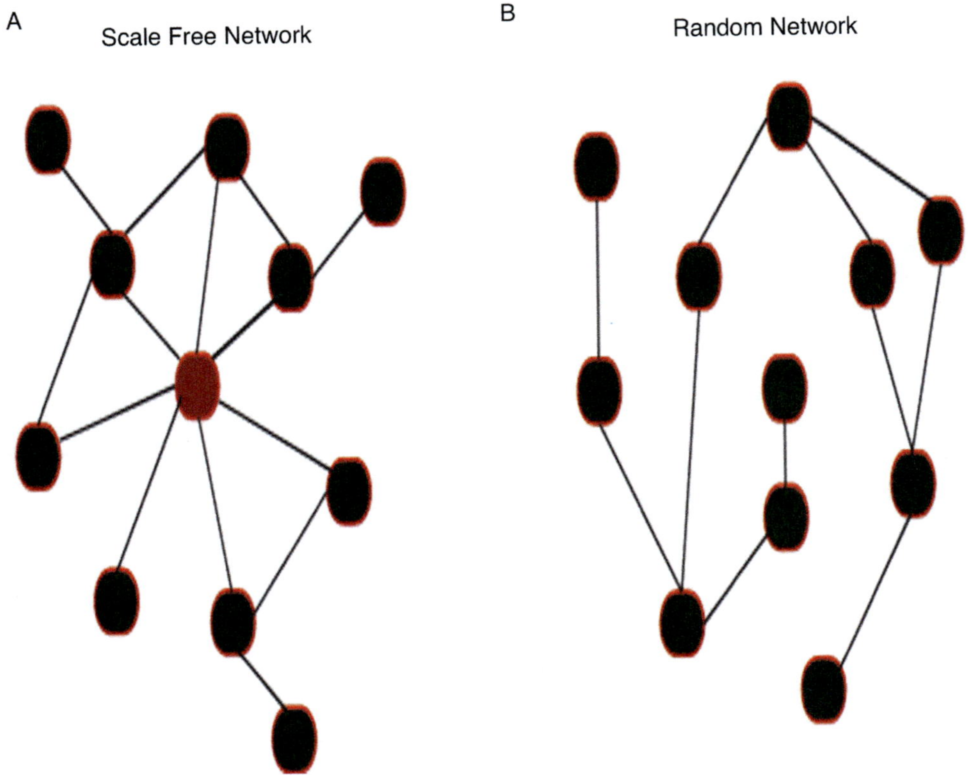

Fig. 21.2. (A) In a scale-free network, most nodes communicate with a small number of different nodes (green), but a small percentage of nodes communicate with a considerable number of nodes and operate as the hubs of the network (red). (B) In a random network, the nodes (red) are associated with a constant chance, thus the prevalence of nodes has an equal number of connections.

The removal or disruption of a hub will cause a significant alteration in the biological networks' structure, which will have a profound effect on the fitness of the organism (He and Zhang, 2006). Specifically, hubs are thought to be crucial in the signal transduction pathway that occurs after stress identification in plant stress response-associated networks, and loss-of-function mutants frequently exhibit aberrant stress reactions (Dietz *et al.*, 2010). Hub proteins are still not well understood, in spite of their significance and significant advancements in the interactome area. This is demonstrated by the disparities in definitions, traits, and categorization schemes. With a particular focus on hubs connected to plant strain response in considerable-scale plant PPI networks, this study critically evaluates the facts currently known in the area of plant interactome investigation.

21.2 Properties of Hub Proteins

Hub proteins were described as positively linked core nodes that connect many network modules and possess a substantial number of interaction partners in a frequent scale-free PPI network (Bertolazzi *et al.*, 2013; Ota *et al.*, 2016). The biological networks, like PPI, frequently feature a scale-free design, which means that most network nodes have little or no connection and a few numbers have a lot of connectivity; these are known as hubs. In general scale-free PPI networks, a point of better than five relations has already been established for the change between high-, medium-, or low-degree nodes (Vallabhajosyula *et al.*, 2009). This is sure to be quite lowly, though, for hubs, which are the proteins in extensive PPI networks with the highest degree of connectivity. Many proteins have more than five interactors in large-scale plant PPI networks, suggesting that this criterion is not warranted.

Differentiating between minor (6–10), moderate (11–50), large (51–100), and super hubs (> 100) is another recommendation made by certain research. However, as PPI networks get larger, more interchanges will be discovered for a single protein, which makes a limited threshold appear less appropriate. A floating cut-off may appear more appropriate and adaptable in the context of the idea that the hub part is a characteristic of the specific interactome network rather than being specified at the status of particular proteins (Dosztányi *et al.*, 2006). Hub protein designation established on interaction degree is further complicated by the quality and source of the PPI information (Bertolazzi *et al.*, 2013). Thus, it is difficult to determine a universal cut-off for plant hub proteins.

21.2.1 Network properties of hub genes

The network properties of hub genes refer to their characteristics within biological networks, like gene regulatory networks or PPI interaction networks (Table 21.1). Hub genes typically exhibit specific properties that make them central and influential within these networks. The network properties of hub genes provide insights into their critical roles in biological systems, particularly in multi-stress tolerance in crops. Understanding these properties helps researchers decipher the molecular mechanisms underpinning stress responses and identify potential targets for genetic improvement. The most widely used metrics for centrality in biological networks include (i) 'degree centrality,' which is merely the number of relations a protein is present in a network; (ii) 'betweenness centrality,' which is established on the briefest ways of connecting proteins in a network and describes the frequency of a protein in refereeing

Table 21.1. Network and structural characteristics of hub genes.

Network features	
Degree	Number of direct connections in the network
Between-ness Centrality	Hub genes often have high between-ness centrality, indicating they play a pivotal part in the communication and flow of information within the network.
Closeness centrality	The measure of proximity to all further genes in the network
Eigenvector centrality	Influence established on the associations of neighboring genes
Clustering coefficient	Hub genes can have low clustering coefficients, as they often connect too many genes that are not themselves connected, though this can vary.
Structural properties	
Protein domains	Hub genes often encode proteins with multiple or highly versatile domains, enabling diverse interactions.
PPI interfaces	Hub genes frequently have extensive PPI interfaces, facilitating multiple interactions.
Highly charged surfaces	Short-hub proteins frequently retain positively charged externals to promote binding with numerous associates.
Single/multiple interfaces	Hubs are separated into two types, based on the existence of a single or multiple interaction areas.
3D structure	Known 3D structures of hub proteins can reveal interaction hotspots and are crucial for understanding their multifunctionality.

associations to different proteins; and (iii) 'eigenvector centrality,' which describes a protein's influence in a network, not just by the number of connections but also via the significance of these associations (Liseron-Monfils and Ware, 2015).

PPI network structure, interactome functionality, and stability are also frequently attributed to hub proteins with a high capacity for connection and centrality (Dietz *et al.*, 2010).

According to several studies (Mukhtar *et al.*, 2011; Weßling *et al.*, 2014; Garbutt *et al.*, 2014; Ota *et al.*, 2016), hub proteins may also be weak points in PPI networks and can completely collapse a system if they are removed or are the target of infection. However, there has been much discussion over the association with one hub degree, centrality, and essentiality. The association between centrality and mortality is substituted by one with pleiotropy and centrality when datasets that are less systematically biased are employed. Thus, rather than being connected with essentiality or lethality, the number of physical relations that a protein mediates appears to be reasonably associated considering how many biological processes it participates in (Yu *et al.*, 2008). Indeed, many investigations have shown that hubs frequently take part in crucial complexes and functions and moderate crucial interactions. As a result, when they are knocked out, numerous phenotypic features are affected, though mortality is not always the outcome.

According to certain research, hub proteins also have lower interconnectedness than non-hub proteins (Vallabhajosyula *et al.*, 2009). Conversely, ironically, a growing number of investigations characterize PPI network hubs as positively interconnected proteins, suggesting that hubs frequently form bigger complexes (Piya *et al.*, 2014; Lumba *et al.*, 2014; McCormack *et al.*, 2016). Finally, a considerable bimodal distribution known as co-expression bimodality is frequently observed in the co-expression of hub proteins with their interaction members (Vallabhajosyula *et al.*, 2009). These suggest the presence of two types of hub proteins known as date hubs and party hubs, one with low and the other with increased median interactor co-expression significance (Agarwal *et al.*, 2010; Chang *et al.*, 2013)

21.2.2 Date vs. party hubs in networks

Hubs have generally been regarded as static, essentially nonbiochemical entities in the importance of investigations on PPI networks, with little or no consideration given to temporal patterns of expression of genes, concentrations of protein, binding constants, and post-translational transformations. By comparing filtered protein–protein interactome data from yeast with mRNA expression information for interacting associates, Han *et al.* (2004) endeavored to overcome at least one of these problems. The idea of the date and party hub was the result of this computational study. The party hub is characterized by constant, simultaneous interaction between the hub protein and its associates. Data demonstrate a population of hubs in which the hub and its interaction partners displayed comparable designs for intracellular localization and expression of mRNA. Date hubs, on the other hand, interact with partners at separate times and places, as evidenced by the majority of the time by the shortage of co-expression and co-localization. The investigators proposed that party hubs operate within modules, while date hubs moderate contact between modules within the intracellular network.

Jin *et al.* (2007) offered more evidence for the party and date hub concept recently; however, they recommended using the words 'motif party hub' and 'motif date hub' instead of concentrating on individual proteins because they represent significant components in network module organization. A party hub in plant biology is the transcriptionally active chromosome (TAC), a multiprotein molecular engine that powers transcription inside the chloroplast. According to Pfalz *et al.* (2006), the TAC proteome is made up of 35 proteins, the majority of which have unidentified roles. By co-localizing and co-expressing these TAC proteins, a complex party motif is showcased. An intriguing finding is that thioredoxin (Trx), a recently discovered Trx linked to the TACs, has been revealed to be important for the normal maturation of *Arabidopsis*, as evidenced by the severe albino phenotype displayed by plants that were phenotypically deficient in the Trx (Arsova *et al.*, 2010).

21.2.3 Structural properties of hub genes

The capacity of hubs to selectively attach to several associates simultaneously in an identical place and time has prompted some significant biochemical inquiries about the structural underpinnings of seeming promiscuity in binding combined with specificity. Certain characteristics relating to structure are considerably more frequently attributed to and hence proposed to be typical of hub proteins (Table 21.1). Hub-labeled proteins have a substantially higher percentage of multi-domain proteins – long proteins with repetitive domains that allow them to bind multiple partners – than non-hub proteins. According to Ekman *et al.* (2006), these protein-binding domains are typically connected to globular or organized portions of the proteins. On the other hand, several investigations indicate that eukaryotic hub proteins frequently have disordered regions (Haynes *et al.*, 2006; Vincent and Schnell, 2016). The proteins that don't fully fold and maintain their flexibility are understood as intrinsically disordered proteins or protein regions (IDPs or IDPRs). IDPRs function as flexible extended interaction surfaces, facilitating the binding and recognition of many proteins. IDPRs frequently have short linear motifs in their core. These brief segments of protein sequence, which are usually between three and 11 residues long, are particularly important in signaling networks with high connectivity proteins because they mediate interactions by causing the formation of secondary structures on binding along with a structured associate or globular protein part.

According to Dosztányi *et al.* (2006), these areas are distinguished by low amino acid complexity and typically develop via the development of internal repetition areas. However, some smaller hub proteins have very few or no disordered regions and are too short to have many binding domains. Instead, many of these tiny hubs feature positively charged surfaces that make binding with numerous partners easier. These surfaces include arginine, tyrosine, histidine, and methionine. Due to these structural hub properties, hub proteins are now categorized into two groups: single interface hubs (SIHs) as well as multiple interface hubs (MIHs), which are distinguished by the existence of one or more exchange spots for partner binding, respectively. The SIH/MIH and date/party categorization of hub proteins have a significant though incomplete overlap, even when connected, and are thought to fall short of general hub classification methods in terms of comprehensiveness (Vandereyken *et al.*, 2018).

The genetic origin of multi-stress toleration in crops involves the intricate interplay of different genes and regulatory networks that enable plants to withstand and adapt to diverse environmental challenges (Fig. 21.3). Higher species have evolved incredibly intricate signaling pathways and processes that enable them to react to a broad range of shifting environmental requirements, invasions of microorganisms, and developmental obstacles (Verma *et al.*, 2016). These systems are also closely regulated to respond appropriately to certain difficulties with the least amount of energy wasted (Tomé *et al.*, 2014). This innate complexity contributes to the fact that many genes and proteins are currently unknown, poorly understood, or lack proper characterization. Research that uses interactomics-based methods may be able to close these gaps and help to further understand how plants respond to stress. Plants are home to enormous and dynamic protein networks; in proteomes comprising 30,000–40,000 proteins, the number of protein exchange pairings is predicted to vary from 75,000 to 299,000 (Sheth and Thaker, 2014). These proteins also differ in their subcellular locations, expression levels, physicochemical characteristics, and reaction kinetics. During the last ten years, large-scale plant PPI networks have mainly been rebuilt; however, the information is still outlying from illustrating international PPI maps in a cell of the plant (Hao *et al.*, 2016). Thus, in future years, plant PPI maps will grow wider, more correct, and more dependable, making the designation of hub proteins simpler and more accurate. Identifying 'hub genes' that play pivotal roles in coordinating these responses is crucial for understanding and enhancing crop resilience.

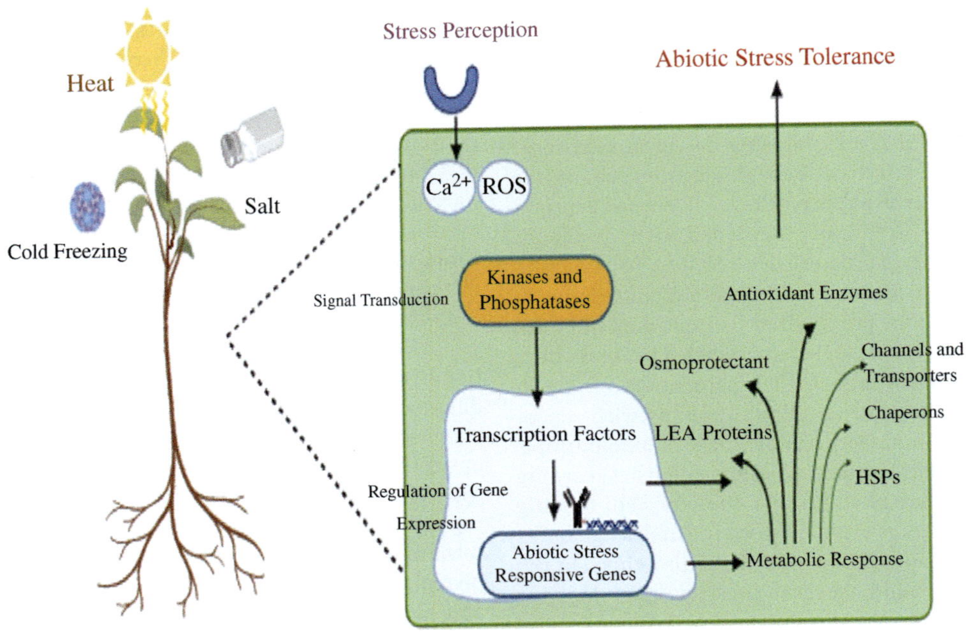

Fig. 21.3. Genetic basis of hub genes in multi-stress-tolerant crops.

21.3 Genetic Diversity and Screening

Genetic diversity plays a crucial part in the evolution and screening of hub genes in multi-stress-resistant crops. Multi-stress resistance guides the capability of plants to resist and adjust to different environmental strains, like drought, saltiness, heat, and pathogens (Fig. 21.4). Genetic diversity provides a reservoir of different genetic traits within a crop species. This diversity allows for the selection of individuals or populations that exhibit resilience to multiple stresses. By studying diverse genetic backgrounds, researchers can identify genes that confer broad-spectrum resistance or tolerance to different stressors.

21.3.1 Breeding strategies

Plant breeding programs often rely on genetic diversity to introduce desirable traits into crop varieties. Screening for hub genes involved in stress-response pathways can aid in selecting breeding lines with enhanced multi-stress resistance. These hub genes may regulate key mechanisms such as stress signaling, detoxification pathways, or physiological adaptations.

21.3.2 Genomic approaches

Advances in genomics, like next-generation sequencing and genome-wide association studies (GWAS), allow researchers to analyze genetic diversity comprehensively. Through GWAS, associations can be made between genetic variants (including those in hub genes) and phenotypic traits related to stress resistance.

21.3.3 Network analysis

Hub genes are identified based on their central role in biological networks. In multi-stress-resistant crops, these networks may involve pathways related to stress signaling (e.g. MAPK pathways), osmotic regulation, ROS scavenging,

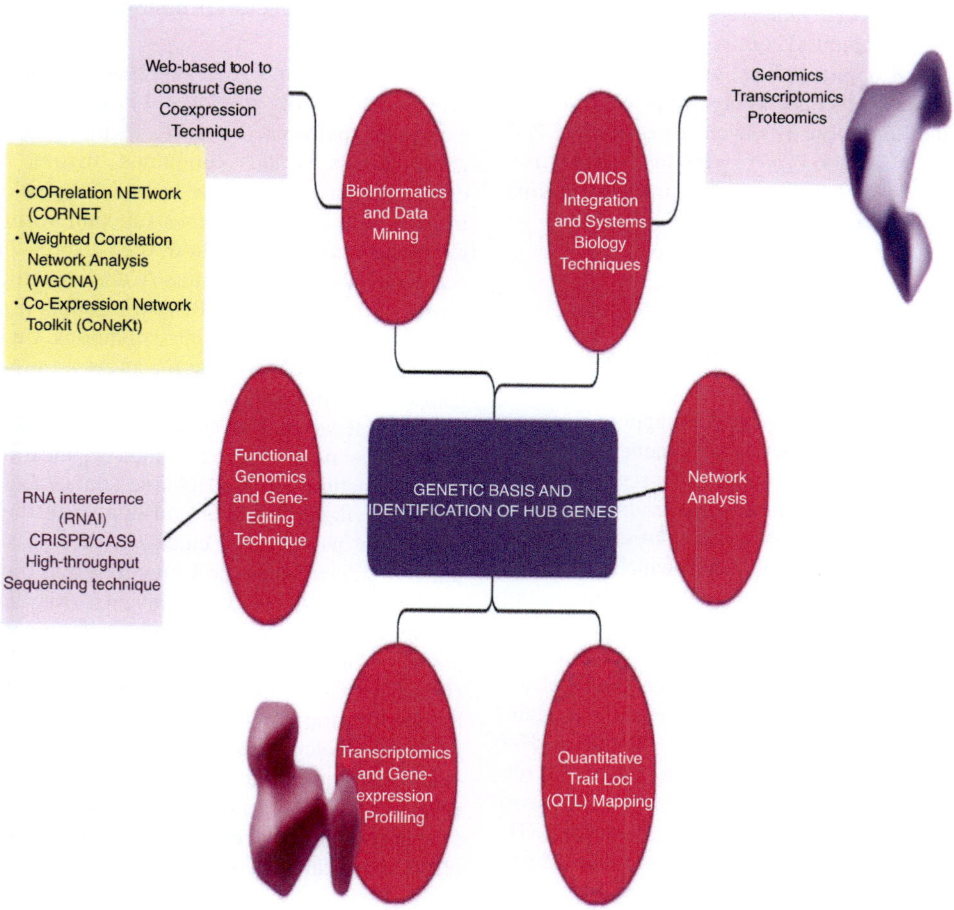

Fig. 21.4. Screening of hub genes in multi-stress-tolerant crops.

or hormonal responses. Integration of omics data (transcriptomics, proteomics) with network analysis helps pinpoint hub genes that are pivotal in coordinating responses to multiple stresses. Once potential hub genes are identified, functional validation is crucial to confirm their roles in stress resistance. Techniques such as gene editing (e.g. CRISPR/Cas9) or transgenic approaches can be used to manipulate the manifestation of hub genes in crop plants and assess their impact on stress tolerance. Studying natural genetic variation across diverse populations of crop species provides insights into the distribution and prevalence of beneficial alleles associated with stress tolerance. Hub genes showing high variability across populations may indicate their importance in adaptive responses to varying environmental conditions.

Consider a research project focussed on developing drought- and salt-tolerant rice varieties. Researchers collect genetic samples from a diverse set of rice accessions from different geographical regions. Through GWAS and transcriptomic analyses under controlled stress conditions, they identify several hub genes involved in water-use efficiency, ion homeostasis, and ROS detoxification pathways. These hub genes are further validated through gene-editing techniques to create transgenic rice lines with improved tolerance to drought and saltiness. The resulting varieties demonstrate improved yield stability under challenging environmental conditions, showcasing the practical application of genetic diversity and hub gene screening in crop improvement. The studies like those by Cavanagh *et al.* (2013) in wheat and Zhao *et al.* (2016) in rice have used GWAS to uncover genetic loci linked to multiple stress tolerances.

21.3.4 Quantitative trait loci mapping

Quantitative trait loci (QTL) mapping is a powerful strategy employed in genetics and plant breeding to determine genomic regions related to variation in quantitative traits, like yield, resistance to diseases, and stress toleration. When it comes to hub genes, which are central regulators in biological networks, integrating QTL mapping can provide insights into how genetic variation in these hub genes influences complex traits of interest. QTL mapping typically involves mapping regions of the genome related to phenotypic divergence in a population. Hub genes, identified through network analysis, represent key regulators within biological pathways that influence trait variation. Integrating QTL mapping with hub genes helps pinpoint genetic loci that directly influence the expression or function of these central regulators, shedding light on their roles in trait variation.

To comprehend the genetic cause of complex traits in plants, Li *et al.* (2016) explore the integration of QTL mapping with network biology techniques, such as the discovery of hub genes. They demonstrate how QTL mapping may be used to analyze the genetic regulation of hub genes and provides instances of how these genes have been linked to trait variation. Network biology and QTL mapping in crop systems genetics are reviewed by Serin *et al.* (2017). They investigate how our knowledge of the genetic architecture of complex traits might be improved by network-based methods, such as the discovery of hub genes. To find important regulators, especially hub genes underpinning trait variation, highlights the significance of integrating QTL mapping with gene network research.

21.3.5 Transcriptomics and gene-expression profiling

Transcriptomics and gene expression profiling are powerful tools employed to analyze the expression patterns of genes, including hub genes, across various conditions, tissues, or developmental stages. These techniques provide insights into how hub genes function within biological networks and how their expression correlates with complex traits. Transcriptomics involves the study of all RNA transcripts produced in a cell under specific conditions. Gene expression profiling assesses the levels of mRNA transcripts to understand which genes are active and to what extent. Hub genes are central nodes in biological networks, regulating key pathways and interactions. Transcriptomic studies focus on identifying hub genes and analyzing their expression patterns to elucidate their functions in different biological functions and their association with complex traits. An overview of RNA-seq technology, which is frequently employed in transcriptomic research, is discussed by Ozsolak and Milos (2011). It goes over the identification and description of hub genes in diverse biological situations, as well as the investigation of gene expression profiles using RNA-seq. Interpreting gene expression data related to hub genes requires an understanding of the technical aspects of RNA-sequence.

21.3.6 Functional genomics and gene editing

To understand the functions of hub genes, which are essential elements of biological networks, functional genomics and gene editing are essential. The study of functional genomics focusses on the role that genes play in cells and organisms. High-throughput approaches are frequently employed to investigate gene expression, protein interactions, and phenotypic effects. Researchers can directly modify the genome using gene-editing techniques like CRISPR/Cas9 to investigate gene function and confirm the functions of particular genes, particularly hub genes, in a range of biological processes. By utilizing methods like RNA interference, CRISPR/Cas9, and high-throughput sequencing, functional genomics seeks to comprehend gene function on a broad scale.

Understanding the roles of genes within biological networks entails analyzing patterns of gene expression, interactions between proteins, and other molecular activities. Targeting specific DNA sequences, gene-editing tools, like CRISPR/Cas9, allow for precise genome alterations. This makes it possible for scientists to alter or even delete specific genes, such as hub genes found through network analysis. The application of gene editing and functional genomics technologies in plants is explained by Liu *et al.* (2017). They explained how these tools are used to study gene function, including hub genes, and their prospect for crop advancement and genetic engineering. The article highlights examples where functional genomics and gene editing have been applied to confirm the roles of hub genes in stress reactions, development, and other biological processes.

21.3.7 Omics integration and systems biology

Understanding the functions of hub genes within biological networks requires an application of systems biology and omics integration techniques. These methods entail building comprehensive models of biological systems by combining data from multiple high-throughput techniques, including proteomics, metabolomics, transcriptomics, and genomes. Large-scale datasets that capture various facets of cellular and molecular processes are produced using omics technology. By combining several data formats, scientists can create comprehensive models of biological systems, such as hub gene-based regulatory networks. Instead of concentrating on individual parts, systems biology seeks to comprehend biological systems as a whole. The study employs computational modeling, network analysis, and statistical techniques to clarify how hub genes interact with other molecules and pathways to control various biological processes.

The integration of omics techniques (genomics, transcriptomics, and proteomics) to research oilseed crop seed development is covered by Wang *et al.* (2017). Although lipid and protein content is the main emphasis, this study shows how hub genes involved in metabolic pathways and regulatory networks controlling seed development may be studied through the application of omics integration. Not exclusively focussed on hub genes, it highlights integrative approaches in plant signaling pathways that are often regulated by hub genes. It discusses how omics technologies, coupled with computational modeling, have advanced our understanding of signaling networks implicated in plant growth and stress reactions (Kamburova *et al.*, 2017). Shah *et al.* (2018) provide insights into integrative omics analysis in wheat, focussing on genomics and transcriptomics. They discuss how these strategies have been used to decide hub genes and decipher their roles in wheat development, disease resistance, and environmental adaptation. The article demonstrates the utility of omics integration in studying complex traits and regulatory networks in crop plants, including the role of hub genes.

21.3.8 Bioinformatics and data mining

Advanced bioinformatics tools and databases are utilized for data mining, gene annotation, and comparative analyses across large datasets. These computational approaches facilitate the identification of candidate genes and regulatory factors implicated in multi-stress tolerance. Large PPI networks are often produced by *in silico* predictions, which are frequently based on sequence alignment, co-expression of the related genes, protein structural data, and/or evolutionary conservation (Vella *et al.*, 2017). Established and understood sequence resemblances between interacting proteins in different species, also known as conserved interacting orthologs or interologs,mean that many computationally expected interactome maps are generated.

For *Arabidopsis thaliana*, the first large-scale species interactome map was constructed utilizing interolog mapping (Geisler-Lee *et al.*, 2007). PPI network reconstruction also makes use of genetic algorithms, most of which are domain-based and include the Elucidating Network Topology utilizing the sequence method. This method, which has already indicated huge PPI networks for different organisms, like *A. thaliana* and *Populus trichocarpa*, combines a pairwise association of maintained domains and predictions of subcellular protein distribution as information characteristics (Rodgers-Melnick *et al.*, 2013; Hao *et al.*, 2016).

These expected PPIs offer very valuable information and are frequently a critical first step in mapping large-scale PPI networks, but the techniques are unable to identify relations between proteins that lack conserved domains or distinct interacting domains. Additionally, interolog-based networks and other computationally expected networks established on publications data are constrained by what is understood at a particular time and by the information that is incorporated in the studies. As a result, they are unable to accurately represent the molecular connectivity. Thus, anticipated PPIs can provide details on the prospect for protein interactions. Nevertheless, it is only through experimental methods, ideally in plants, that it will be possible to verify if these interactions occur in a cell or tissue, or at a certain period.

21.4 Web-based Tool to Construct Gene Co-expression Network

21.4.1 Cornet

A web-based application called CORrelation NETworks (CORNET) is used to build and visualize co-expressed genes derived from PPIs, regulatory relations, and microarray data (De Bodt *et al.*, 2012). The microarray datasets (such as abiotic stress and AtGenExpress) were obtained from the TAIR10 databases. Thus far, CORNET has been created for GCN in maize (CORNET Maize 1.0) and *A. thaliana* (CORNET 2.0). Numerous features have been created, including the 'TF tool,' 'PPI tool,' 'Co-expression tool,' and 'Browse Experiments.' Using PCC or Spearman, users can compute the correlation coefficients for the genes they are interested in by entering them into the 'Co-expression tool.' Gene network outcome will be produced in either Cytoscape or tab-delimited arrangement following the establishment of the gene co-expression network. Through Cytoscape, users may see the network. In addition, CORNET offers additional capabilities such as 'User-defined,' which enables users to connect and analyze their microarray datasets. Users can merge the PPI network and the co-expression network by using the function under the 'PPI tool' section. Investigating the relationship between genes and proteins and obtaining the functioning annotation of the uncharacterized genes are made possible by the integration of the two distinct data types.

21.4.2 Weighted correlation network analysis

The first R tool for building a correlation network is weighted correlation network analysis (WGCNA). WGCNA is established on the idea of a scale-free network, which finds gene-expression clusters in a dataset by a sequence of correlations. It can identify highly associated gene clusters, compute topological features, and summarize the clusters using an intramodular hub gene or the module eigengene (Langfelder and Horvath, 2008). Additional features of WGCNA include comparisons across modules of various species, data visualization, and data modeling. Even though utilizing the WGCNA needs command-line proficiency, a wealth of in-depth online tutorials is accessible. According to a prior study, WGCNA performs better than different clustering techniques in identifying non-overlapping clusters and is better sensitive to noise in datasets (Saelens *et al.*, 2018). Much literature associated with crop progress has been reported on the WGCNA (Ma *et al.*, 2019; Jia *et al.*, 2021).

21.4.3 Co-expression network toolkit

Co-Expression Network Toolkit (CoNekT) is an online platform designed to offer data on co-expression networks and gene expression in certain plant species (Proost and Mutwil, 2018), such as flowering plants, seed plants, and green algae. The CoNekT contains three crop species: *Oryza sativa*, *Solunam lycopersicum*, and *Zea mays*. With the use of this established database, users can look for co-expressed genes and the areas around them, as well as compare co-expression networks between other species. The RNA-seq

Table 21.2. List of co-expression tools for analyzing the gene co-expression networks in crops.

Types	Co-expression network tool	Description	Reference
Web-based tool	CORNET 2.0	An integrating device for networks of plant co-expression	De Bodt *et al.*, 2012
	PlaNet	A tool for analyzing comparative co-expression networks	Proost and Mutwil, 2017
	RECoN	A method for identifying co-expressed genes in the abiotic stress response using co-expression	Krishnan *et al.*, 2017
	webCeToolmi	An online resource for locating co-expression modules inside a certain co-expression network	Cardozo *et al.*, 2019
	CoExp	An online tool for co-expression network exploitation	García-Ruiz *et al.*, 2021
Command line-based tool	WGCNA	Weighted Correlation Network Analysis with the R package	Langfelder and Horvath, 2008
	Petal	Co-expression network modeling using the R package	Petereit *et al.*, 2016
	LSTrAP	A method for building co-expression networks using RNA-seq data	Bozhilova *et al.*, 2021
	COGENT	An R program that builds gene co-expression networks without requiring additional validation data or annotation	Lemoine *et al.*, 2021
	Juxtapose	A tool for comparing networks of gene co-expression (GCNs)	Ovens *et al.*, 2021

datasets from the SRA database were analyzed by CoNekT. In the GCN analysis, the heuristic clustering chiseling algorithm was utilized for clustering identification, and the highest reciprocal rank metric score was employed to quantify the correlation linkages.

The CoNekT has three functions: Species, Tools, and Search. After selecting 'Species,' users are presented with a species list along with transcript, profile, and network statistics (Table 21.2). The simplest method is to choose the species name, after which a list of gene IDs and descriptions will appear. By clicking on the gene ID, users can see the specifics of co-expressed genes and expression patterns. Then, the neighborhood gene and cluster ID associated with the co-expressed genes are provided by CoNekT. Fascinatingly, users have the option to present the co-expression results using a variety of visualization methods, including tables, charts, and networks. Furthermore, the user has the option to download and store CoNekT on a local server. By letting users choose crops or plants of interest, this tool provides the advantage of searching through extensive expression data.

21.5 Experimental Approaches

High-throughput technologies such as split-ubiquitin systems and *in vivo* yeast two-hybrid (Y2H) are commonly employed to produce experimental plant PPI data. The latter approach is especially beneficial for the identification of membrane protein relations (Brückner *et al.*, 2009; Zhang *et al.*, 2010). Using a systematic Y2H-based method, the first testing called 'proteome-scale' interactome map for *A. thaliana* was created. Further routine procedures possess the classic *in vitro* protein arrays (Feilner, 2005; Popescu *et al.*, 2009) or the nucleic acid programmable protein arrays (NAPPA), which produce proteins by employing cell-free expression techniques and plasmid DNA to replace immobilized pure proteins on the array surface (Miersch and LaBaer, 2011). Proteome scale screenings are now possible thanks to a recent breakthrough in protein array technology, the HaloTag-NAPPA approach.

Even though all of these experimental methods typically produce excellent results, *in planta* validation of the interactions is still necessary because they are established on yeast or *in vitro* approaches. Fluorescence resonance energy transfer analyses (Bhat *et al.*, 2006; Long *et al.*, 2017), bimolecular fluorescence complementation (Miller *et al.*, 2015), split-luciferase techniques (Fujikawa and Kato, 2007; Li *et al.*, 2011), and co-immunoprecipitation (Roux *et al.*, 2011) are among the various targeted PPI methods that can be employed for confirmation. Finally, tag-established affinity purification techniques in conjunction with mass spectrometry are two methods particularly utilized for the identification of protein complexes in plants (Dedecker *et al.*, 2015). Other approaches include traditional immunoprecipitation tests. Particularly because of its great specificity, tandem affinity purification pursued by mass spectrometry is an extremely effective and popular technique. Protein complexes in *O. sativa and A. thaliana*, in enduringly changed cell suspension cultures and seedlings, as well as in *Medicago truncatula* hairy roots, have already been found using this technique. However, to significantly increase enthusiasm in specified interaction partners without losing potentially significant interactors, quantitative AP-MS techniques are exploited more these days. These methods compare the abundance of proteins to a negative control that co-purify with the bait (Meyer and Selbach, 2015).

21.6 Conclusion

The exploration of hub genes concerned with the regulation of multi-stress toleration in crops underscores their pivotal role in mitigating the unfavorable consequences of diverse environmental stressors. Through sophisticated genomic and functional analyses, researchers have elucidated intricate molecular pathways and signaling networks that orchestrate plant responses to multiple stresses. This knowledge not only enhances our fundamental understanding of plant biology but also provides a basis for targeted breeding techniques desired to develop robust crop varieties capable of thriving under challenging growing conditions. Moving forward, continued interdisciplinary research efforts are essential to harnessing the full potential of these hub genes for sustainable agriculture and food safety in the face of climate modification and other environmental pressures.

Acknowledgement

All the authors are thankful to their respective universities and institutions for their valuable support.

References

Agarwal, S., Deane, C.M., Porter, M.A. and Jones, N.S. (2010) Revisiting date and party hubs: Novel approaches to role assignment in protein interaction networks. *PLoS Computational Biology* 6(6), e1000817. DOI: 10.1371/journal.pcbi.1000817.

Arsova, B., Hoja, U., Wimmelbacher, M., Greiner, E., Ustün, S. *et al.* (2010) Plastidial thioredoxin Z interacts with two fructokinase-like proteins in a thiol-dependent manner: Evidence for an essential role in chloroplast development in *Arabidopsis* and *Nicotiana benthamiana*. *The Plant Cell* 22(5), 1498–1515. DOI: 10.1105/tpc.109.071001.

Ben Rejeb, I., Pastor, V. and Mauch-Mani, B. (2014) Plant responses to simultaneous biotic and abiotic stress: Molecular mechanisms. *Plants* 3(4), 458–475. DOI: 10.3390/plants3040458.

Bertolazzi, P., Bock, M.E. and Guerra, C. (2013) On the functional and structural characterization of hubs in protein-protein interaction networks. *Biotechnology Advances* 31(2), 274–286. DOI: 10.1016/j.biotechadv.2012.12.002.

Bhat, R.A., Lahaye, T. and Panstruga, R. (2006) The visible touch: In planta visualization of protein-protein interactions by fluorophore-based methods. *Plant Methods* 2, 1–14. DOI: 10.1186/1746-4811-2-12.

Bozhilova, L.V., Pardo-Diaz, J., Reinert, G. and Deane, C.M. (2021) COGENT: Evaluating the consistency of gene co-expression networks. *Bioinformatics* 37(13), 1928–1929. DOI: 10.1093/bioinformatics/btaa787.

Braun, P., Aubourg, S., Leene, J., Jaeger, G. and Lurin, C. (2013) Plant protein interactomes. *Annual Review of Plant Biology* 64(1), 161–187. DOI: 10.1146/annurev-arplant-050312-120140.

Brückner, A., Polge, C., Lentze, N., Auerbach, D. and Schlattner, U. (2009) Yeast two-hybrid, a powerful tool for systems biology. *International Journal of Molecular Sciences* 10(6), 2763–2788. DOI: 10.3390/ijms10062763.

Cardozo, L.E., Russo, P.S.T., Gomes-Correia, B., Araujo-Pereira, M., Sepúlveda-Hermosilla, G. *et al.* (2019) WebCEMiTool: Co-expression modular analysis made easy. *Frontiers in Genetics* 10, 146. DOI: 10.3389/fgene.2019.00146.

Cavanagh, C.R., Chao, S., Wang, S., Huang, B.E., Stephen, S. *et al.* (2013) Genome-wide comparative diversity uncovers multiple targets of selection for improvement in hexaploid wheat landraces and cultivars. *Proceedings of the National Academy of Sciences* 110(20), 8057–8062. DOI: 10.1073/pnas.1217133110.

Chang, X., Xu, T., Li, Y. and Wang, K. (2013) Dynamic modular architecture of protein-protein interaction networks beyond the dichotomy of 'date' and 'party' hubs. *Scientific Reports* 3(1), 1691. DOI: 10.1038/srep01691.

De Bodt, S., Hollunder, J., Nelissen, H., Meulemeester, N. and Inzé, D. (2012) CORNET 2.0: Integrating plant coexpression, protein-protein interactions, regulatory interactions, gene associations and functional annotations. *The New Phytologist* 195(3), 707–720. DOI: 10.1111/j.1469-8137.2012.04184.x.

Dedecker, M., Leene, J. and Jaeger, G. (2015) Unravelling plant molecular machineries through affinity purification coupled to mass spectrometry. *Current Opinion in Plant Biology* 24, 1–9. DOI: 10.1016/j.pbi.2015.01.001.

Dietz, K.J., Jacquot, J.P. and Harris, G. (2010) Hubs and bottlenecks in plant molecular signalling networks. *The New Phytologist* 188(4), 919–938. DOI: 10.1111/j.1469-8137.2010.03502.x.

Dosztányi, Z., Chen, J., Dunker, A.K., Simon, I. and Tompa, P. (2006) Disorder and sequence repeats in hub proteins and their implications for network evolution. *Journal of Proteome Research* 5(11), 2985–2995. DOI: 10.1021/pr060171o.

Duque, A.S., Almeida, A.M., Silva, A.B., Silva, J.M., Farinha, A.P. *et al.* (2013) Abiotic stress responses in plants: Unraveling the complexity of genes and networks to survive. In: Vahdati, K. and Leslie, C. (eds) *Abiotic Stress-Plant Responses and Applications in Agriculture*. InTech Open. pp. 49–101.

Ekman, D., Light, S., Björklund, A.K. and Elofsson, A. (2006) What properties characterize the hub proteins of the protein-protein interaction network of *Saccharomyces cerevisiae*? *Genome Biology* 7(6), 1–13. DOI: 10.1186/gb-2006-7-6-r45.

Feilner, T. (2005) High throughput identification of potential *Arabidopsis* mitogen-activated protein kinases substrates. *Molecular & Cellular Proteomics* 4(10), 1558–1568. DOI: 10.1074/mcp.M500007-MCP200.

Firbank, L.G., Attwood, S., Eory, V., Gadanakis, Y., Lynch, J.M. *et al.* (2018) Grand challenges in sustainable intensification and ecosystem services. *Frontiers in Sustainable Food Systems* 2, 7. DOI: 10.3389/fsufs.2018.00007.

Fujikawa, Y. and Kato, N. (2007) TECHNICAL ADVANCE: Split luciferase complementation assay to study protein-protein interactions in *Arabidopsis* protoplasts. *The Plant Journal* 52(1), 185–195. DOI: 10.1111/j.1365-313X.2007.03214.x.

Garbutt, C.C., Bangalore, P.V., Kannar, P. and Mukhtar, M.S. (2014) Getting to the edge: Protein dynamical networks as a new frontier in plant-microbe interactions. *Frontiers in Plant Science* 5, 312. DOI: 10.3389/fpls.2014.00312.

García-Ruiz, S., Gil-Martínez, A.L., Cisterna, A., Jurado-Ruiz, F., Reynolds, R.H. *et al.* (2021) CoExp: A web tool for the exploitation of co-expression networks. *Frontiers in Genetics* 12, 630187. DOI: 10.3389/fgene.2021.630187.

Geisler-Lee, J., O'Toole, N., Ammar, R., Provart, N.J., Millar, A.H. *et al.* (2007) A predicted interactome for *Arabidopsis*. *Plant Physiology* 145(2), 317–329. DOI: 10.1104/107.103465.

Han, J.-D.J., Bertin, N., Hao, T., Goldberg, D.S., Berriz, G.F. *et al.* (2004) Evidence for dynamically organized modularity in the yeast protein-protein interaction network. *Nature* 430(6995), 88–93. DOI: 10.1038/nature02555.

Hao, T., Peng, W., Wang, Q., Wang, B. and Sun, J. (2016) Reconstruction and application of protein-protein interaction network. *International Journal of Molecular Sciences* 17(6), 907. DOI: 10.3390/ijms17060907.

Haynes, C., Oldfield, C.J., Ji, F., Klitgord, N., Cusick, M.E. *et al.* (2006) Intrinsic disorder is a common feature of hub proteins from four eukaryotic interactomes. *PLoS Computational Biology* 2(8), e100. DOI: 10.1371/journal.pcbi.0020100.

He, X. and Zhang, J. (2006) Why do hubs tend to be essential in protein networks? *PLoS Genetics* 2(6), e88. DOI: 10.1371/journal.pgen.0020088.

Jia, X., Feng, H., Bu, Y., Ji, N., Lyu, Y. *et al.* (2021) Comparative transcriptome and weighted gene co-expression network analysis identify key transcription factors of *Rosa chinensis* 'Old blush' after exposure to a gradual drought stress followed by recovery. *Frontiers in Genetics* 12, 690264. DOI: 10.3389/fgene.2021.690264.

Jin, G., Zhang, S., Zhang, X.S. and Chen, L. (2007) Hubs with network motifs organize modularity dynamically in the protein-protein interaction network of yeast. *PLoS One* 2(11), e1207. DOI: 10.1371/journal.pone.0001207.

Kamburova, V.S., Nikitina, E.V., Shermatov, S.E., Buriev, Z.T., Kumpatla, S.P. *et al.* (2017) Genome editing in plants: An overview of tools and applications. *International Journal of Agronomy* 2017(1), 1–15. DOI: 10.1155/2017/7315351.

Krishnan, A., Gupta, C., Ambavaram, M.M. and Pereira, A. (2017) RECoN: Rice environment coexpression network for systems level analysis of abiotic-stress response. *Frontiers in Plant Science* 8, 1640. DOI: 10.3389/fpls.2017.01640.

Langfelder, P. and Horvath, S. (2008) WGCNA: An R package for weighted correlation network analysis. *BMC Bioinformatics* 9, 1–13. DOI: 10.1186/1471-2105-9-559.

Lemoine, G.G., Scott-Boyer, M.P., Ambroise, B., Périn, O. and Droit, A. (2021) GWENA: Gene co-expression networks analysis and extended modules characterization in a single bioconductor package. *BMC Bioinformatics* 22(1), 267. DOI: 10.1186/s12859-021-04179-4.

Li, J.F., Bush, J., Xiong, Y., Li, L. and McCormack, M. (2011) Large-scale protein-protein interaction analysis in *Arabidopsis* mesophyll protoplasts by split firefly luciferase complementation. *PLoS One* 6(11), e27364. DOI: 10.1371/journal.pone.0027364.

Li, C., Dong, Y., Zhao, T., Li, L., Li, C. *et al.* (2016) Genome-wide SNP linkage mapping and QTL analysis for fiber quality and yield traits in the upland cotton recombinant inbred lines population. *Frontiers in Plant Science* 7, 1356. DOI: 10.3389/fpls.2016.01356.

Liseron-Monfils, C. and Ware, D. (2015) Revealing gene regulation and associations through biological networks. *Current Plant Biology* 3–4, 30–39. DOI: 10.1016/j.cpb.2015.11.001.

Liu, S., Liu, Y., Zhao, J., Cai, S., Qian, H. *et al.* (2017) A computational interactome for prioritizing genes associated with complex agronomic traits in rice (*Oryza sativa*). *The Plant Journal* 90(1), 177–188. DOI: 10.1111/tpj.13475.

Long, Y., Stahl, Y., Weidtkamp-Peters, S., Postma, M., Zhou, W. *et al.* (2017) *In vivo* FRET-FLIM reveals cell-type-specific protein interactions in *Arabidopsis* roots. *Nature* 548(7665), 97–102. DOI: 10.1038/nature23317.

Lumba, S., Toh, S., Handfield, L.-F., Swan, M., Liu, R. *et al.* (2014) A mesoscale abscisic acid hormone interactome reveals a dynamic signaling landscape in *Arabidopsis*. *Developmental Cell* 29(3), 360–372. DOI: 10.1016/j.devcel.2014.04.004.

Ma, S., Lv, L., Meng, C., Zhou, C., Fu, J. *et al.* (2019) Genome-wide analysis of abscisic acid biosynthesis, catabolism, and signaling in Sorghum bicolor under saline-alkali stress. *Biomolecules* 9(12), 823. DOI: 10.3390/biom9120823.

McCormack, M.E., Lopez, J.A., Crocker, T.H. and Mukhtar, M.S. (2016) Making the right connections: Network biology and plant immune system dynamics. *Current Plant Biology* 5, 2–12. DOI: 10.1016/j.cpb.2015.10.002.

Meyer, K. and Selbach, M. (2015) Quantitative affinity purification mass spectrometry: A versatile technology to study protein-protein interactions. *Frontiers in Genetics* 6, 237. DOI: 10.3389/fgene.2015.00237.

Miersch, S. and LaBaer, J. (2011) Nucleic acid programmable protein arrays: Versatile tools for array-based functional protein studies. *Current Protocols in Protein Science* 64(1), 27–2. DOI: 10.1002/0471140864.ps2702s64.

Miller, K.E., Kim, Y., Huh, W.K. and Park, H.O. (2015) Bimolecular Fluorescence Complementation (BiFC) analysis: Advances and recent applications for genome-wide interaction studies. *Journal of Molecular Biology* 427(11), 2039–2055. DOI: 10.1016/j.jmb.2015.03.005.

Mukhtar, M.S., Carvunis, A.-R., Dreze, M., Epple, P., Steinbrenner, J. *et al.* (2011) Independently evolved virulence effectors converge onto hubs in a plant immune system network. *Science* 333(6042), 596–601. DOI: 10.1126/science.1203659.

Ota, M., Gonja, H., Koike, R. and Fukuchi, S. (2016) Multiple-localization and hub proteins. *PLoS One* 11(6), e0156455. DOI: 10.1371/journal.pone.0156455.

Ovens, K., Maleki, F., Eames, B.F. and McQuillan, I. (2021) Juxtapose: A gene-embedding approach for comparing co-expression networks. *BMC Bioinformatics* 22(1), 125. DOI: 10.1186/s12859-021-04055-1.

Ozsolak, F. and Milos, P.M. (2011) RNA sequencing: Advances, challenges and opportunities. *Nature Reviews Genetics* 12(2), 87–98. DOI: 10.1038/nrg2934.

Petereit, J., Smith, S., Harris, F.C. and Schlauch, K.A. (2016) Petal: Co-expression network modelling in R. *BMC Systems Biology* 10, 181–195. DOI: 10.1186/s12918-016-0298-8.

Pfalz, J., Liere, K., Kandlbinder, A., Dietz, K.J. and Oelmüller, R. (2006) PTAC2, -6, and -12 are components of the transcriptionally active plastid chromosome that are required for plastid gene expression. *The Plant Cell* 18(1), 176–197. DOI: 10.1105/tpc.105.036392.

Piya, S., Shrestha, S.K., Binder, B., Stewart, C.N. and Hewezi, T. (2014) Protein-protein interaction and gene co-expression maps of ARFs and Aux/IAAs in *Arabidopsis*. *Frontiers in Plant Science* 5, 744. DOI: 10.3389/fpls.2014.00744.

Popescu, S.C., Popescu, G.V., Bachan, S., Zhang, Z., Gerstein, M. *et al.* (2009) MAPK target networks in *Arabidopsis thaliana* revealed using functional protein microarrays. *Genes & Development* 23(1), 80–92. DOI: 10.1101/gad.1740009.

Proost, S. and Mutwil, M. (2017) PlaNet: Comparative co-expression network analyses for plants. In: Van Dijk, A.D.J. (ed.) *Plant Genomics Databases: Methods and Protocols*. Springer, pp. 213–227. DOI: 10.1007/978-1-4939-6658-5_12.

Proost, S. and Mutwil, M. (2018) CoNekT: An open-source framework for comparative genomic and transcriptomic network analyses. *Nucleic Acids Research* 46(W1), W133–W140. DOI: 10.1093/nar/gky336.

Rodgers-Melnick, E., Culp, M. and DiFazio, S.P. (2013) Predicting whole genome protein interaction networks from primary sequence data in model and non-model organisms using ENTS. *BMC Genomics* 14, 1–17. DOI: 10.1186/1471-2164-14-608.

Roux, M., Schwessinger, B., Albrecht, C., Chinchilla, D., Jones, A. *et al.* (2011) The *Arabidopsis* leucine-rich repeat receptor-like kinases BAK1/SERK3 and BKK1/SERK4 are required for innate immunity to hemibiotrophic and biotrophic pathogens. *The Plant Cell* 23(6), 2440–2455. DOI: 10.1105/tpc.111.084301.

Saelens, W., Cannoodt, R. and Saeys, Y. (2018) A comprehensive evaluation of module detection methods for gene expression data. *Nature Communications* 9(1), 1090. DOI: 10.1038/s41467-018-03424-4.

Serin, E.A.R., Snoek, L.B., Nijveen, H., Willems, L.A.J., Jiménez-Gómez, J.M. *et al.* (2017) Construction of a high-density genetic map from RNA-seq data for an *Arabidopsis* bay-0 × Shahdara RIL population. *Frontiers in Genetics* 8, 201. DOI: 10.3389/fgene.2017.00201.

Shah, T., Xu, J., Zou, X., Cheng, Y., Nasir, M. *et al.* (2018) Omics approaches for engineering wheat production under abiotic stresses. *International Journal of Molecular Sciences* 19(8), 2390. DOI: 10.3390/ijms19082390.

Sheth, B.P. and Thaker, V.S. (2014) Plant systems biology: Insights, advances and challenges. *Planta* 240(1), 33–54. DOI: 10.1007/s00425-014-2059-5.

Thao, N.P. and Tran, V.L.S.P. (2016) Enhancement of plant productivity in the post-genomics era. *Current Genomics* 17(4), 295–296. DOI: 10.2174/138920291704160607182507.

Tomé, F., Nägele, T., Adamo, M., Garg, A., Marco-Llorca, C. *et al.* (2014) The low energy signaling network. *Frontiers in Plant Science* 5, 353. DOI: 10.3389/fpls.2014.00353.

Vallabhajosyula, R.R., Chakravarti, D., Lutfeali, S., Ray, A. and Raval, A. (2009) Identifying hubs in protein interaction networks. *PLoS One* 4(4), e5344. DOI: 10.1371/journal.pone.0005344.

Vandereyken, K., Leene, J., Coninck, B. and Cammue, B.P.A. (2018) Hub protein controversy: Taking a closer look at plant stress response hubs. *Frontiers in Plant Science* 9, 694. DOI: 10.3389/fpls.2018.00694.

Vella, D., Zoppis, I., Mauri, G., Mauri, P. and Silvestre, D. (2017) From protein-protein interactions to protein co-expression networks: A new perspective to evaluate large-scale proteomic data. *EURASIP Journal on Bioinformatics & Systems Biology* 1–6. DOI: 10.1186/s13637-017-0059-z.

Verma, V., Ravindran, P. and Kumar, P.P. (2016) Plant hormone-mediated regulation of stress responses. *BMC Plant Biology* 16, 1–10. DOI: 10.1186/s12870-016-0771-y.

Vincent, M. and Schnell, S. (2016) A collection of intrinsic disorder characterizations from eukaryotic proteomes. *Scientific Data* 3(1), 1–9. DOI: 10.1038/sdata.2016.45.

Wang, H., Seo, J.K., Gao, S., Cui, X. and Jin, H. (2017) Silencing of *AtRAP*, a target gene of a bacteria-induced small RNA, triggers antibacterial defense responses through activation of LUS2 and down-regulation of *GLK1*. *The New Phytologist* 215(3), 1144–1155. DOI: 10.1111/nph.14654.

Weßling, R., Epple, P., Altmann, S., He, Y., Yang, L. *et al.* (2014) Convergent targeting of a common host protein-network by pathogen effectors from three kingdoms of life. *Cell Host & Microbe* 16(3), 364–375. DOI: 10.1016/j.chom.2014.08.004.

Xing, S., Wallmeroth, N., Berendzen, K.W. and Grefen, C. (2016) Techniques for the analysis of protein-protein interactions *in vivo*. *Plant Physiology* 171(2), 727–758. DOI: 10.1104/16.00470.

Yu, H., Braun, P., Yildirim, M.A., Lemmens, I., Venkatesan, K. *et al.* (2008) High-quality binary protein interaction map of the yeast interactome network. *Science* 322(5898), 104–110. DOI: 10.1126/science.1158684.

Zafar, S.A., Zaidi, S.S.E.A., Gaba, Y., Singla-Pareek, S.L., Dhankher, O.P. *et al.* (2020) Engineering abiotic stress tolerance via CRISPR/Cas-mediated genome editing. *Journal of Experimental Botany* 71(2), 470–479. DOI: 10.1093/jxb/erz476.

Zhang, Y., Gao, P. and Yuan, J.S. (2010) Plant protein-protein interaction network and interactome. *Current Genomics* 11(1), 40–46. DOI: 10.2174/138920210790218016.

Zhao, D., Shen, F., Zeng, J., Huang, R., Yu, Z. *et al.* (2016) Network analysis reveals seasonal variation of co-occurrence correlations between Cyanobacteria and other bacterioplankton. *The Science of the Total Environment* 573, 817–825. DOI: 10.1016/j.scitotenv.2016.08.150.

22 Climate-smart Crops: Delivering Tolerant Genes Based on Engineered Nanocarriers and CRISPR/Cas Genome-editing System

Naglaa A. Abdallah[1,2*], Aladdin Hamwieh[3], Khaled Radwan[2,4], Nourhan Fouad[3] and Michael Baum[3]

[1]Department of Genetics Faculty of Agriculture, Cairo University, Giza, Egypt; [2]National Biotechnology Network of Expertise, ASRT, Cairo, Egypt; [3]Department of Biotechnology, International Centre for Agricultural Research in the Dry Areas (ICARDA), Giza, Egypt; [4]Department of Biotechnology, Agricultural Genetic Engineering Research Institute (AGERI), ARC, Giza, Egypt

Abstract

Crop productivity is seriously threatened by climate changes such as global warming, which causes biotic and abiotic stress in crops. It is important to develop environmentally tolerant crop varieties to mitigate environmental stresses and feed the growing population. This chapter focusses on the combination of CRISPR/Cas technology and nanovehicles in plant breeding to improve agricultural productivity. The chapter begins with a general introduction to gene delivery through various nanocarriers, including liposomal, polymeric, and other nanoparticles, including the mechanism of catalase. The use of this approach is beneficial for several aspects related to CRISPR/Cas-mediated application in crops, such as increased bioavailability, specificity of target genes and controlled release or reduction of toxicity, probably due to the role of nanocarriers. This is illustrated by examples of applications where this convergence can lead to a significant improvement in disease resistance, stress tolerance, and yield. In addition, the chapter deals with the regulation and ethics of genetically modified plants and nanotechnology. This includes the existing regulations, their description, as well as safety and environmental concerns and challenges (such as public perception or ethics). We emphasize the need for a rigorous safety assessment and open, transparent communication between all stakeholders to build trust and promote the integration of these technologies. Finally, the chapter proposes an agenda for further research, including nanocarriers that respond to a specific trigger and targeted components, new types of effective carriers, and the combination of CRISPR/Cas with other strategies such as conventional breeding. In this way, researchers can utilize the incorporation of CRISPR/Cas technology into nanocarriers to address current obstacles in agriculture that will ultimately bring great benefits to global food security and sustainable agricultural practices.

Keywords: climate-smart, nanocarriers, CRISPR/Cas, genome editing, ribonucleoprotein, regulatory framework

*Corresponding author: naglaa.abdallah@agr.cu.edu.eg

© CAB International 2025. *Stress-Resilient Crops: Coordinated omics-CRISPR-Nanotechnology Strategies* (ed. Jen-Tsung Chen)
DOI: 10.1079/9781800627307.0022

22.1 Introduction

Global warming is widely recognized as one of the main factors affecting agricultural productivity. Crop growth and yields are severely affected by rising temperatures and erratic rainfall patterns, impacting food security and livelihoods in rural areas (Ferrer *et al.*, 2023). In addition, climate change is causing a geographical shift in croplands, leading to an increased number of droughts and floods (Ojoko *et al.*, 2017; Prasad and Cole, 2023). Such climatic challenges require revolutionary agricultural techniques that can function under a variety of environmental conditions. Climate-smart agriculture (CSA), which emphasizes sustainable practices that can improve productivity and increase the resilience of agroecosystems to climate change, has been proposed as a possible solution (Kurgat *et al.*, 2020; Shahzad *et al.*, 2021). In response, the development of climate-adapted crops has been highlighted as a necessity to cope with these changing conditions and ensure food security for future generations (Mujeyi *et al.*, 2021; Khan *et al.*, 2023).

Climate-smart crops have evolved from the realization that conventional agricultural practices are not sufficient to adapt to climate change. Heat-resistant and drought-tolerant crop varieties are needed to ensure the sustainability of agricultural production under unfavorable conditions (Khan *et al.*, 2023; Kottegoda *et al.*, 2023). In recent years, the advent of CRISPR/Cas genome-editing technology has led to groundbreaking opportunities for precise and efficient editing of a wide range of plant traits (Razzaq *et al.*, 2021; Zhou *et al.*, 2021). This chapter explains how both tools can help deliver tolerant genes (structurally modified or edited in a predictable way to achieve the desired effect) and develop climate-friendly crops that become resistant to various environmental stresses.

The importance of climate-adapted crops lies not only in increasing agricultural productivity, but also has broader socio-economic impacts – from cities to remote villages. Climate-adapted agricultural practices have the potential to improve household livelihoods and resilience, especially among smallholder farmers who are particularly vulnerable to climate variability (Wekesa *et al.*, 2018; Mujeyi *et al.*, 2021). These practices can improve food security and alleviate poverty in rural communities by promoting crop resilience and productivity (Mujeyi *et al.*, 2021; Ferrer *et al.*, 2023). In addition, climate-smart strategies could reduce greenhouse gas emissions in agriculture, bringing agricultural practices in line with global sustainability goals (Fawole and Aderinoye-Abdulwahab, 2021; Shahzad *et al.*, 2021). This chapter also highlights how climate-smart crops contribute to sustainable development and emphasizes that agricultural innovation requires a holistic approach.

Overall, we highlight the importance of biotechnological interventions in tackling the problems caused by climate change in agriculture. By exploring the convergence of engineered nanocarriers and genome-editing technologies, we examine how such nano-biosynthetic approaches can contribute to the development of climate-smart crops. This initiative plays a critical role in ensuring food security and improving the livelihoods of millions of smallholder farmers around the world, which is consistent with both sustainable development goals and climate-resilient strategies.

22.2 Overview of Climate-smart Crops

22.2.1 Definition and characteristics of climate-smart crops

Climate-smart crops: one of the building blocks of climate-smart agriculture (CSA), an approach that sustainably increases agricultural productivity and improves resilience to climate change while reducing greenhouse gas emissions. The Food and Agriculture Organization of the United Nations (FAO) defines CSA as focussing on food security and development goals in the context of climate change (Kurgat *et al.*, 2020; Balo and Mahata, 2022; Phiri *et al.*, 2022). The nature of climate-smart crops is multidimensional and encompasses a broad spectrum that is aligned with the genetic and agronomic systems in the environment in which they are found to withstand fluctuations in climate variability.

Drought, heat, and salt resistance are some of the most important traits that climate-smart crops carry genetically. These crops are often developed using sophisticated breeding strategies that include both traditional selective breeding and modern biotechnological tools and technologies such as CRISPR/Cas genome editing (Chandra and Shmelev, 2017; Kaptymer *et al.*, 2019). In one such case, drought-tolerant maize hybrids were bred to stabilize grain yields under water-deficit conditions and ensure food security for farmers in arid areas (Manda *et al.*, 2016; Gairhe *et al.*, 2021). These genetic modifications help to improve the resilience of crops under suboptimal conditions, which reduces the vulnerability of agricultural systems to the effects of climate change (Muzorewa and Chitakira, 2021).

The climate-intelligent plants are also resistant and have a genetic resilience to climatic fluctuations. The most important feature, however, is that they go hand-in-hand with sustainable agricultural practices such as soil-health management and the promotion of agrobiodiversity. The use of conservation agriculture, characterized by minimal soil disturbance and permanent ground cover, and other practices such as crop rotation, are common companions of climate-smart crops (Thierfelder *et al.*, 2016; De Pinto and Ulimwengu, 2017). These practices not only improve soil fertility and structure, but also carbon sequestration, making them a successful method for reducing greenhouse gas emissions (Arenas-Calle *et al.*, 2019). The positive relationship between climate-friendly crops and sustainable agricultural practices shows that systemic solutions for agriculture are needed: productivity through environmental friendliness.

Furthermore, climate-smart crops are designed to respond and adapt to changing climatic conditions, which is necessary for sustainable crop production under variable environmental conditions. Increased resilience is reflected in the selection of new plant varieties that can withstand more extreme and frequent climatic-stress events such as floods or heatwaves (Aryal *et al.*, 2018; Teklewold *et al.*, 2019). The development of heat-tolerant rice varieties, for example, has led to higher cereal yields in locations with rising temperatures and supported the food spring (Gairhe *et al.*, 2021). Their ability to adapt to different climate scenarios contributes to their ability to be grown alternatively in different agroecological zones.

In fact, climate-friendly crops also have far-reaching socio-economic consequences. Grain legumes are important as they contribute to resilience to climate change and secure the livelihoods of smallholder farmers who are most vulnerable under future climate conditions (Ekpa *et al.*, 2018; Kurgat *et al.*, 2020). For example, climate-smart agricultural practices and the production of climate-smart crops appear to make a positive contribution to increasing household income and reducing poverty, both at the local farming system level (Phiri *et al.*, 2022) and at the farm level (Phiri *et al.*, 2022; Belay *et al.*, 2023). These crops can also help align agricultural practices with broader development goals by contributing to national food security goals (Ekpa *et al.*, 2023).

22.2.2 Examples of climate-smart crops

Climate-smart crops (CSC) are resilient to climate-related stresses (drought, heat, and salinity) and have been developed largely through the application of advanced plant breeding or genetics and best sustainable agricultural practices. The three pillars of CSC are: sustainable increase productivity and income; strengthening resilience to climate change and variability (adaptation and mitigation); and reducing agricultural contribution to climate change (reduce greenhouse gases). Breakthroughs in molecular biology, such as integrating high-throughput genome sequencing and characterization, were integrated into breeding programs and are expected to yield successful development of climate-smart crops (Fig. 22.1). Below are examples of how climate-smart crops are used to increase productivity.

Drought-tolerant maize

Water scarcity in arid regions has favored the development of drought-tolerant maize varieties. The mutants are designed for yield stability under water scarcity and are of great importance in times when food security is a major challenge in several arid and semi-arid regions. The availability of drought-tolerant maize in areas with unreliable rainfall patterns such as Zambia can potentially increase farmers' productivity and

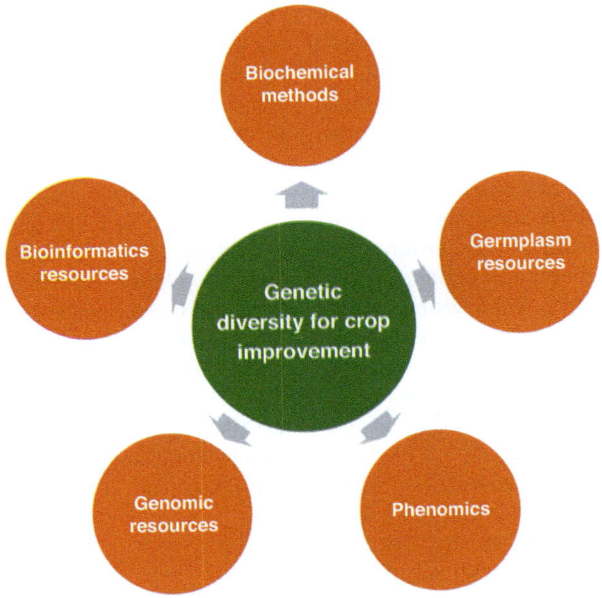

Fig. 22.1. Crop breeding using economic resources to improve productivity by incorporation of modern molecular technologies, including phenomics platforms, biotechnological tools, genomic resources, germplasm resources, and bioinformatics resources, in addition to the combination of genomic and bioinformatics resources.

yields, as studies have shown (Manda *et al.*, 2015). The Drought Tolerant Maize for Africa project was led by the International Maize and Wheat Improvement Center (CIMMYT). The varieties developed increased maize yields for more than 30 million farmers in 14 African countries, reduced dependence on water-intensive irrigation methods, increased yields and improved food security. These varieties fit into existing cropping systems, differentiate cropping patterns and reduce the vulnerability of agricultural systems to climatic changes.

Heat-resistant rice

Rice is one of the world's most important staple foods. Environmental changes significantly affect the different growth stages of rice and its yield due to heat stress. In Asia, the breeding of heat-resistant rice varieties is mainly focussed on heat stress, which limits rice cultivation in many regions of the world due to the rise in temperature as a result of global warming. These are bred to be productive in a future where high temperatures will be more common due to the warming of the planet. According to López-Ridaura *et al.* (2018), the results of some studies suggest that heat-tolerant rice varieties could play an important role in increasing yields and improving food security in countries such as India and Bangladesh, where rice is a staple food. Therefore, the adaptability of these crops to the changing climate is an important factor for agricultural productivity.

Conservation agriculture with legume crops

Conservation agriculture practices such as zero tillage and crop rotation are usually combined with the cultivation of legumes such as pigeon pea (*Cajanus cajan*) and cowpea (*Vigna unguiculata*). Nitrogen-fixing legumes such as chickpea are important components of climate-friendly cultivation systems that not only withstand drought but also promote soil fertility. In Mozambique, the cultivation of these legumes as part of conservation agriculture has been shown to increase resilience to climate-related losses and food security in the event of major crop failures (Thierfelder *et al.*, 2016). This is a perfect example of how crop choice and cultivation methods go hand-in-hand.

Sorghum and millet

Sorghum and millet are traditional, climate-tolerant crops that thrive better in dry conditions. As they are drought-resistant food crops and can grow in nutrient-poor soils, they are also ideal for areas with water scarcity. The promotion and cultivation of sorghum and millet have been promoted in several parts of Africa as a solution for improved food security and thus as a way to solve the productivity vs resilience problem (Mujeyi *et al.*, 2021). This makes them important players in climate-smart agriculture and their ability to withstand harsh conditions is outstanding.

Climate-smart jute

Since this fiber plant requires very little effort and can thrive in different climate zones, it is considered a climate-friendly plant. The study by Singh *et al.* (2019) reports that the adoption of integrated cropping practices (ICM) for jute cultivation, such as nutrient management and sustainable crop management, in rural eastern India led to an increase in productivity with the same resource input, i.e. higher productivity and resource efficiency were observed. The promotion of jute as a climate-friendly crop not only serves to enrich the local economy but also underlines the importance of more sustainable agricultural practices.

Small-scale irrigation crops

Small-scale irrigation technologies are important climate-smart practices that allow farmers to grow crops in places that would otherwise be unproductive due to low water availability. Tomatoes, peppers, and many other vegetables are also irrigated on a small scale to achieve higher productivity, especially in regions with unstable climates. Studies have shown that smallholder irrigation can help mitigate the effects of climate change on crop yields in southern Africa by increasing capacity and redistributing cropland (Mango *et al.*, 2018).

Agroforestry systems

Another example of climate-smart agriculture is agroforestry, in which trees and shrubs are integrated into the agricultural landscape. This leads to a reversal of biodiversity, improves soil health, and helps to create alternative income opportunities for farmers who produce wood and non-timber forest products. Incorporating trees into cropping systems can be a win-win for climate change as they store carbon and reduce soil erosion by providing shade. In countries such as East Africa, agroforestry systems are being considered as an option to improve agricultural resilience and sustainability (Kaptymer *et al.*, 2019).

The above examples of climate-smart crops illustrate the diverse strategies being used to improve agricultural resilience to climate change. By integrating advanced breeding techniques, sustainable agricultural practices and traditional crop varieties, these crops contribute to food security, improved livelihoods, and environmental sustainability. The continued development and adoption of climate-smart crops is critical to addressing the challenges of climate change and ensuring a sustainable agricultural future.

22.2.3 Benefits and impact of climate-smart crops on agriculture

Climate-smart crops will be necessary to increase the resilience and productivity of agriculture in the face of climate change. Their benefits and impacts relate to a range of dimensions, from better food security, higher farmer incomes, safer climate through improved environmental sustainability, to responding to changing climatic conditions. Here are the critical views on the various benefits and impacts on agriculture from climate-smart crops.

A key benefit of these climate-smart crops is their value for food security. Farmers could maintain or increase their yields under difficult conditions by developing crop varieties that are resistant to climate-related stresses such as drought, heat, and flooding. Drought-tolerant maize and heat-resistant rice varieties can thus improve the availability of food in regions with strong climate fluctuations (Khan *et al.*, 2023). The cultivation of these plants and their ability to survive harsh conditions make them an important source of food for rural and urban populations.

Farmers who adopt climate-smart agricultural practices, such as growing climate-smart crops, have been able to increase their income like never before. Research also shows that farmers who adopt these practices achieve higher yields and have easier access to markets – all of which lead to better economic performance (Fawole and Aderinoye-Abdulwahab, 2021). In Ghana, for example, the adoption of climate-smart practices has increased farmers' incomes, confirming that the adaptation approach is economically sustainable (Agbenyo *et al.*, 2022). It not only improves farmers' incomes but also plays a role in rural development and poverty eradication.

Climate-smart crops are designed to be integrated into sustainable agricultural practices that promote environmental health. These practices often include crop rotation, cover cropping, and conservation agriculture, which improve soil health and biodiversity while reducing the need for chemical inputs (Kaptymer *et al.*, 2019; Prasad and Cole, 2023). The adoption of climate-smart crops can lead to improved soil quality, reduced erosion and increased carbon sequestration, thus mitigating the effects of climate change on agricultural ecosystems (Tadesse *et al.*, 2021; Galani *et al.*, 2022). For example, conservation agriculture practices have been shown to improve soil nutrient content and water retention, which are crucial for sustainable crop production (Singh *et al.*, 2019; Galani *et al.*, 2022).

The development and breeding of climate-smart crops is an important way of adapting crops to changing climatic conditions. These crop niches are designed to withstand the vagaries of a planet to which they adapt unlike humanity or livestock, which they improve year after year over two full decades, making them an important aspect to consider as humanity pursues agricultural productivity under uncertainty (Shahzad *et al.*, 2021; Khan *et al.*, 2023). For example, the introduction of climate-adapted groundnut varieties in West Africa has enabled smallholder farmers to increase agricultural production and improve food security (Tabe-Ojong *et al.*, 2023). Climate-smart crops improve the adaptability of agricultural systems to climate variability.

Climate-smart research, including the use of climate-smart crops and practices, also contributes to limiting greenhouse gas emissions from the agricultural sector. They minimize the carbon footprint of agriculture by advocating reduced tillage, cover cropping, and efficient nutrient management (Shahzad *et al.*, 2021; Fawole and Aderinoye-Abdulwahab, 2021). For example, it has been observed that the introduction of climate-smart crops through agroforestry systems can be positively associated with carbon stocks and additional benefits such as increased biodiversity and soil health (Teklewold *et al.*, 2017; Manaye *et al.*, 2021). This could be a double-plus for achieving global climate goals, as people with more energy tend to support clean electricity.

In addition, the introduction of climate-smart crops helps to improve the overall resilience of agricultural systems. For example, if more than one crop species is grown in the fields and farmers adopt climate-smart practices, they are likely to be resilient to both climate change shocks and market fluctuations (Makate *et al.*, 2016; Kalinga *et al.*, 2021). Furthermore, it has been emphasized that crop diversification is the crucial element of resilient agriculture as it allows farmers to reduce risks and adapt to changing conditions. Therefore, this resilience is crucial for food security and sustainable livelihoods in times of climate change.

22.3 Gene-editing Technologies for Crop Improvement

The development of gene-editing technologies has changed the landscape of plant improvement techniques by largely simplifying (neat) and improving the modification of plant genomes. Of these technologies, the CRISPR/Cas9 system is presented as a revolutionary approach or tool to amplify specific genetic changes that enable researchers and breeders to modify plant-intensive traits such as disease resistance, abiotic stress tolerance, and nutrient bonus (Liu *et al.*, 2021; Wei *et al.*, 2023). This chapter provides an overview of the basic principles of gene editing and its applications, and discusses the current and future implications for agricultural technology to meet the urgent need to improve food security and help mitigate climate change.

Gene editing encompasses a variety of techniques designed to effect specific changes at the level of an organism's genome, in contrast to many traditional breeding strategies that rely on random mutations and extensive backcrossing (Liu *et al.*, 2022b; Yadav *et al.*, 2023). The CRISPR/Cas9 system has been widely used to generate targeted mutants due to its simplicity, flexibility, and specificity with low off-target effects (Naim *et al.*, 2018; Zhang *et al.*, 2014). By introducing specific double-strand breaks (cleavages) at predetermined sites in DNA, this technology facilitates the induction of targeted changes, which can range from insertions and deletions to substitutions in the host genome through cellular repair mechanisms (Xie and Yang, 2013; Ahmad *et al.*, 2021).

Gene-editing technologies are key components of modern plant improvement. In light of the growing world population and the challenges in agriculture caused by climate change, the demand for high-yielding plant varieties that are also resilient is increasing (Lassoued *et al.*, 2019; Avci *et al.*, 2023). In this sense, gene editing is seen as a means to accelerate the breeding and production of crops that are resistant to environmental stresses such as drought, salinity, and heat stress, which affect food security in underserved areas (Zenda *et al.*, 2021; Zhu, 2022). Recent work has shown that the use of CRISPR/Cas9 to improve abiotic stress tolerance in rice and wheat are good examples of how these technologies are consistent with sustainable agricultural approaches (Liu *et al.*, 2021; Kalaitzandonakes *et al.*, 2023).

Gene-editing technologies are not only ideal for improving stress tolerance, but they are also indispensable for achieving desirable traits for the nutritional quality of crops. By using a precise method of gene targeting in the context of nutrient biosynthesis, the preparation of biofortified varieties with essential vitamins and minerals can lead to the misjudgment of nutritional problems in individual staple-dependent populations (Odipio *et al.*, 2017; Nerkar *et al.*, 2022). This is a major advance in the development of nutrient-rich plants, as perfecting and adapting metabolic pathways through gene editing is now possible.

Despite the promising prospects, there are still some obstacles standing in the way of gene editing. The regulatory framework that has been created for genetically modified organisms (GMOs) ranges from very strict to relatively lax regulations, which affect both the commercialization and acceptance of gene-edited plants (Larriba *et al.*, 2024). In addition, public attitudes and ethical considerations about gene editing continue to influence conversations about its use in agriculture (Xie and Yang, 2013; Salava *et al.*, 2021). Overcoming these challenges will be key to realizing the full potential of gene editing for much-needed agricultural innovation.

Therefore, gene editing offers a new approach to crop improvement that could enable the development of higher-stress, higher-yielding crops and better nutrition. As agriculture faces challenges it has never had to overcome before, the role of these technologies in maintaining food security and sustainability through climate change is becoming increasingly important. This chapter thoroughly characterizes gene-editing technologies and explains their application, the general structure of the article, and the issues of crop improvement; a step towards the future where they are shown to be both promising and revolutionary for global agriculture.

22.3.1 Applications of CRISPR/Cas in crop improvement

Stress tolerance

The use of CRISPR/Cas9 technology in the construction of stress-resistant plants has proved to be an important innovation in agricultural biotechnology. It opens the door to precise interventions in the plant genome that allow researchers to improve traits such as drought resistance and salt tolerance, in addition to general resistance to abiotic stress. With the climate crisis threatening global food security, editing stress-tolerance traits through CRISPR/Cas9 can be an important addition to maintaining sustainability in agriculture.

Drought tolerance

CRISPR/Cas9 is a powerful tool for improving drought resistance in crops, one of the most important practical applications. For example, Kim *et al.* (2023) reported that knockout of the OsPUB7 gene in rice leads to improved drought tolerance, among other abiotic stress factors. This gene belongs to the class of genes involved in regulation and stress response, and its disruption leads to improved physiological performance under drought conditions. Similarly, Li *et al.*, 2022a, b used CRISPR/Cas9 to edit the NtAITR gene family in tobacco. They showed that these edited plants exhibited increased drought tolerance, suggesting a critical strategy for editing regulatory stress-responsive genes.

Salinity resistance

In addition to drought, salinity is another abiotic stress that has a major impact on plant productivity. Zhu *et al.* (2015) identified the SlDEAD31 gene in tomato to regulate salt and drought tolerance. Editing this gene with CRISPR/Cas9 enabled the researchers to develop plants that are more resistant to salt stress, thus increasing their overall robustness. In addition, Wang *et al.* (2021) used CRISPR/Cas9 to edit GmAITR genes in soybean and improved their tolerance to salt stress. These gene-clubbing plants showed better growth performance and physiological responses to salinity, highlighting the potential of genome editing in developing salt-resistant crops.

Heat-stress adaptation

Alongside drought and salinity, heat stress is one of the biggest challenges for crop production worldwide. For example, Shi *et al.* (2017) have used CRISPR/Cas9 to create variants of the ARGOS8 gene that increase corn grain yield under drought-stress conditions in the field, where elevated temperatures are also common. This gene is known to modulate plant responses to heat stress, and its manipulation has already been shown to increase both yield and resilience in maize. This suggests that it is a good candidate for adaptation to high-temperature stress.

Multi-stress tolerance

Another exciting application of CRISPR/Cas9 technology is the ability to confer multi-stress tolerance. Tran *et al.* (2023) generated the SlHyPRP1 protein in tomato, which conferred a degree of tolerance to various abiotic stresses such as drought and heat. In King's model, this would amount to creating plants with enhanced defense traits or stress tolerance by targeting one of these downstream nodes; a strategy that may prove increasingly useful as climate change exacerbates plant survival under various environmental stressors.

Mechanistic insights and future directions

The mechanism by which CRISPR/Cas9-induced modifications lead to improved stress tolerance is key for future applications. Nascimento *et al.* (2023) have produced another systematic review of gene editing for abiotic resistance in plants, focussing on the identification of target-related genes for stress tolerance. In the future, as this tool becomes more integrated with other high-throughput biotechnological tools such as transcription and proteomics, CRISPR/cas9-associated complex networks will have better resolution in stress responses, which would allow genetic engineers to create crops that are perfectly tailored to focal resilience traits.

The use of CRISPR/Cas9 has thus marked a major leap from normal plant stress biology to the interaction of development with adversity in agricultural and crop biotechnology. With the ability to make precise changes in the plant genome, this capability has the potential to improve drought resistance, tolerance to salinity, and other abiotic stress responses. Given the challenges of a rapidly changing global agricultural landscape in the face of these climate change-induced stressors, CRISPR/Cas9-mediated plant breeding will be critical to maintaining food security in a sustainable manner.

22.3.2 Yield enhancement

CRISPR/Cas technology enables an increase in the yield of most major crops, which has a major impact on agricultural biotechnology. Through precise genome modifications in plants, this novel gene-editing tool enables researchers to improve various agronomic traits associated with increased yield. Given the increasing global demand for food, modern, environmentally friendly, and efficient agriculture is essential. The development of high-yielding varieties appears to be possible through the application of the newly developed gene-editing technology CRISPR/Cas.

Enhancing yield potential through targeted gene editing

One of the most important applications of CRISPR/Cas technology is the targeted modification of genes associated with yield potential. For example, researchers have successfully used CRISPR/Cas9 to modify the OsSPL14 gene in rice, which is known to regulate the number of pollen, a key determinant of yield (Marshall, 2023). By knocking out this gene, the scientists were able to increase the number of pollen per plant, resulting in significantly higher grain yields. This example underlines the potential of CRISPR/Cas to improve yield-related traits through precise genetic intervention.

Improving nutritional quality and yield

Aside from the conflicting yield efforts, CRISPR/Cas technology can improve the nutritional quality of crops, which is an important pillar in alleviating global malnutrition. For example, CRISPR/Cas9-mediated multiplex gene delivery in tomato was shown to further favor the accumulation of lycopene, an antioxidant substance for humans (Li *et al.*, 2018). The increased lycopene content thus led to better nutritional status of the fruit and higher market value for farmers. The improvement in yield and nutritional quality is a dual benefit of CRISPR/Cas applications in crop improvement.

Multi-gene editing for yield improvement

CRISPR/Cas technology makes it possible to modify several genes simultaneously. Theoretically, multiplex genome editing can target numerous nuclear genes responsible for different agronomic traits of maize (Wang *et al.*, 2022). By activating multiple yield genes, scientists can generate new germplasm that performs better and is more resilient. This accelerates the breeding process, resulting in high-yielding varieties that can cope with changing environmental conditions.

Enhancing stress tolerance to boost yield

CRISPR/Cas technology also offers the possibility of developing stress-tolerant plants, which would be very important for maintaining yields under adverse conditions. Zhang *et al.* (2019) showed that knockout of the isomeric SUMO protease gene in rice contributed to salt tolerance and increased growth and productivity under saline conditions. CRISPR/Cas applications to improve crop resilience to abiotic stresses such as water scarcity and salinity will help ensure yield stability under unfavorable climatic conditions, which supports global food security.

Case studies of successful yield enhancement

Many successful case studies have demonstrated the efficiency of CRISPR/Cas technology to increase crop productivity. For example, targeted mutagenesis of the TMS5 gene with CRISPR has eliminated the complications in thermosensitive male sterile rice to accelerate the development of hybrid rice, ultimately leading to higher productivity (Barman *et al.*, 2019). This new system can shorten the breeding cycle and is important for the efficient production of commercial hybrid rice seed. It is also very effective in producing high-yielding inbred lines quickly, which are necessary to meet future demand with a rapid increase.

The application of CRISPR/Cas technology to increase crop yield has a very long history, with some applications still being put into practice today. The technology is capable of improving crop yield, nutritional quality, and stress tolerance through precise genetic modification. CRISPR/Cas has now become an important technological tool that can revolutionize plant breeding to produce food crops that can meet the challenges of climate change and global population growth, thus promoting sustainable agriculture. Ongoing research and development will be crucial to fully realize the potential of

CRISPR/Cas technology to address future food security challenges.

22.3.3 Disease and pest resistance

In recent years, biotechnology has developed a powerful and transformative use of CRISPR/Cas technology for genome editing of plants to make them more resistant to disease or pest attack. This novel gene-editing tool can introduce specific changes into a plant's genome, offering researchers the opportunity to improve traits that confer resistance to various biotic stressors such as viral pathogens, bacterial blight and gall diseases, or the globally prevalent fungal wilt and rust. As natural ecosystems are disrupted, global agricultural systems are under increasing pressure from pests and diseases, making the development of resistant plant varieties using CRISPR/Cas technology essential for maintaining food security and sustainable agricultural practices.

Enhancing viral resistance

The development of virus-resistant crops is definitely one of the most important applications of CRISPR/Cas technology. For example, the eIF4E gene in cucumber, which is required for the potyvirus life cycle, has been successfully studied using CRISPR/Cas9 technology (Chandrasekaran *et al.*, 2016). Using CRISPR/Cas9 to edit this gene allowed the researchers to generate universal resistance to multiple potyviruses. This is another example of how satisfactory virus-resistant flowers can now be created. The potential of CRISPR/Cas9 for creating banana varieties with increased disease resistance has also been demonstrated by targeting genes known to confer susceptibility to viral infections (Tripathi *et al.*, 2020). Taken together, these examples show how CRISPR/Cas technology can be used to increase virus resistance in crops with significant economic impact.

Bacterial resistance

There are also examples of the use of CRISPR/Cas in various crops to increase their bacterial resistance. Recently, targeted mutagenesis of the OsERF922 gene in rice was reported to increase resistance to bacterial blight caused by *Xanthomonas oryzae* (Wang *et al.*, 2019). These changes led to significantly improved plant health and a higher yield of the manipulated rice plants with fungal infection. Thus, this work underlines that CRISPR/Cas9 is an attractive genome-editing tool to permanently control bacterial pathogens in globally important crops.

Fungal resistance

Fungal diseases are a major problem for plant production around the world. The use of CRISPR/Cas technology to improve fungal resistance has yielded positive results. The claim that editing the eIF4E gene in cucumber leads to resistance to fungal diseases is unfounded; rather, this gene mainly contributes to virus resistance (Chandrasekaran *et al.*, 2016). In addition, the role of CRISPR/Cas in enhancing the plant immune system beyond loss of susceptibility has been emphasized in improving plant disease resistance (Schenke and Cai, 2020). Importantly, these advances demonstrate how CRISPR/Cas technology can be used flexibly against various threats to crops.

Multi-pathogen resistance

One of the main advantages of CRISPR/Cas technology is that it enables resistance to multiple pathogens simultaneously. The applications of genome editing in plant disease resistance were discussed and common goals were discussed to provide a basis for the development of precision components by developing crops with resistance to multiple pathotypes using the direct design approach of Remove Function Revolvent (Yin and Qiu, 2019). This in turn could increase the overall resistance of such plants and minimize their dependence on synthetic chemical pesticides as we can simultaneously target multiple susceptibility genes.

Case studies and future directions

Here we emphasize the effective use of CRISPR/Cas technology in crisis management for isolating gene segments from human-insured crops and include some case studies. For example, the CRISPR/Cas9 system was used to edit susceptibility genes in apple trees with the aim of altering fire blight resistance (Malnoy *et al.*, 2016). This novel method increased disease resistance and was not associated with the same regulatory issues as GMOs. As research continues to progress, the application of CRISPR/Cas technology, along with other biotechnological tools, such as RNA interference and traditional breeding methods, will be instrumental in creating crops that are resistant to disease.

Overall, CRISPR/Cas technology offers promising applications in various crops and can dramatically increase resistance to diseases and pests. All this enables genetic modification with very high precision and thus offers great opportunities for improving the resistance of crops to viral, bacterial, or fungal diseases. With the increasing threat of pests and diseases to global agricultural systems, CRISPR/Cas has the potential to transform crop breeding practices and contribute to food sustainability. The next step in this research and development will be crucial to unlock the full potential of CRISPR/Cas technology for future food security.

22.4 Engineering Nanocarriers for Gene Delivery

Nanocarriers designed for gene delivery are at the forefront of the future of biotechnology and genetic engineering. These carriers function by transporting nucleic acids (DNA, RNA) into target cells, enabling them to express therapeutic genes or reduce the expression of disease-causing genes. The properties that make nanocarriers unique in terms of size, surface chemistry, and response to specific stimuli put them at the forefront of gene-delivery applications. In this introduction, an overview of various engineered nanocarriers and their mechanisms of action will be provided to create a better understanding of their potential use in gene therapy.

22.4.1 Mechanisms of gene delivery

There are various strategies for improving the transport of nucleic acids with the aid of nanocarriers. For example, cationic lipid nanocarriers are the most studied as they can encapsulate negatively charged nucleic acids and consequently form lipoplexes that can penetrate via endosomes (Dey *et al.*, 2021). They also pointed out that the pH-sensitive charge-reversing nanocarriers, mainly based on N-arginine-N-octyl chitosan, not only improve drug-delivery efficiency but also alleviate the concerns of potential toxicity due to their safety in clinical use and biocompatibility. This pH-sensitive property enables the release of the nucleic acid into endosomes, where the acidic environment changes with the release of the gene from the lipid system.

22.4.2 Types of engineered nanocarriers

Lipid-based nanocarriers (LPNs) are the most commonly used type of nanocarrier for gene transfer. Cell-specific target components can be conjugated to them, enabling the internalization of nucleic acids into specific cell types (Khalil *et al.*, 2020). This success in transferring genetic material has been further underpinned by clinical applications, i.e. mRNA vaccines using LNPs.

Polymer nanocarriers

Like dendrimers and micelles, polymersomes can also be easily designed with different functionalities. Engineered polymersomes can be designed to respond to a variety of different stimuli, including pH changes or the presence of enzymes for controlled release (Cheng *et al.*, 2021). Recently, a polytherapeutic system for the treatment of hepatocellular carcinoma based on the mechanisms described above led to combined immunotherapy and gene regulation using glutathione-specific polymer nanocarriers in cancer therapy (Wang *et al.*, 2020).

INORGANIC NANOCARRIERS. There are a variety of nanocarriers, including gold and iron oxide nanoparticles, which are also being researched for gene delivery. These particles can be functionalized with various biomolecules that increase their targeting accuracy, thereby increasing the efficiency of gene delivery (Lee *et al.*, 2015). For example, magnetic nanoparticles have been used for MRI-guided gene delivery (Wang *et al.*,

2015), enabling the specific localization of therapeutic molecules in the target tissue.

22.4.3 Types of nanocarriers

Nanocarriers have profoundly changed the landscape of gene delivery in biotechnology, providing novel methods to package nucleic acids efficiently into target cells. Including liposomes, nanoparticles (like gold and silica), polymers, and hydrogels etc., these nanocarriers have special characteristics that allows carriers to deliver genetic material efficiently. In this part we discuss different types of nanocarriers, how they work, and where they are employed in gene delivery.

Liposomes

Liposomes are spherical vesicles consisting of lipid bilayers that can encapsulate nucleic acid, protein or drug quantities, which in turn makes them a versatile vector system for efficient gene delivery. Due to their biocompatibility and their ability to effectively bind cell membranes, they are a good cargo for the delivery of drugs into cells. Liposomes can be stabilized by designing their surface with ligands or polymers (Negishi *et al.*, 2010) to improve tumor targeting. In addition, laminin-derived peptides have been reported to significantly increase liposome-mediated gene transfer into tumor cells and thus improve gene therapy for cancer. Furthermore, liposomes can be programmed to be triggered in response to specific stimuli (e.g. pH or temperature) and release their cargo on command (Li and Sun, 2019).

Nanoparticles

Due to their unique physical and chemical properties, nanoparticles, including gold and silica nanoparticles, have gained significant attention for their potential in gene delivery.

Gold nanoparticles (AuNPs)

Gold nanoparticles (AuNPs) are particularly notable for their ease of synthesis, biocompatibility, and ability to be functionalized with various biomolecules. They can effectively deliver nucleic acids, such as siRNA and plasmid DNA, into cells. For example, AuNPs have been shown to enhance siRNA-mediated gene silencing in cancer cell lines, demonstrating their potential as effective gene delivery vehicles (Yildiran, 2022). Furthermore, AuNPs can be conjugated with targeting ligands to improve specificity for certain cell types, thereby increasing the efficiency of gene delivery (Kanu *et al.*, 2022). Their unique optical properties also enable real-time tracking of cellular uptake and distribution, providing valuable insights into the dynamics of gene delivery (Yang *et al.*, 2015).

Among the wide range of nanoparticles, AuNPs are especially interesting due to their simple synthesis in addition to biocompatibility and possibility for conjugation with various biomolecules. In cells, they can deliver nucleic acids such as siRNA and plasmid DNA. For example, the effectiveness of AuNPs as a gene delivery vehicle was shown by Yildiran (2022) and cause higher knockdown in cancer cell lines with siRNA technologies. Additionally, one of the most important aspects that enhance gene delivery using AuNPs is the conjugation with different targeting ligands for specific cell types increasing efficiency in this context (Kanu *et al.*, 2022). Moreover, their exceptional optical properties enable tracking of cellular uptake and distribution in real time, impacting significantly on the mechanisms involving gene delivery (Yang *et al.*, 2015).

Silica nanoparticles

Another class of nanocarriers that have been thoroughly investigated for gene delivery applications is silica nanoparticles. Their surface properties can be readily adjusted for the attachment of targeting and therapeutic agents. Silica nanoparticles have been used to deliver plasmid DNA and siRNA efficiently, with reports providing evidence for their protective ability towards the nucleic acids against enzymatic degradation both *in vivo* and *in vitro* (Gan *et al.*, 2012). Furthermore, they also reported that silica nanoparticles can be designed to respond behaviorally to environmental cues (such as changing pH or temperature), thereby allowing controlled release of the trapped cargo species from their solids fashion.

Polymers and hydrogels

Polymers and hydrogels are another widely used category of nanocarriers for nucleic acid delivery, suitable to encapsulate the drug providing controlled release profiles. These polymers are powerful candidates for the synthesis of various types of polyethylenimine (PEI), chitosan, and polylactic-co-glycolic acid (PLGA). These classes, providing tunable properties, make their unique way in gene therapy applications. PEI-conjugated AuNPs also increased gene transfection efficiency with low toxicity (Hu *et al.*, 2010).

Hydrogels, which consist of 3D networks formed by hydrophilic polymers have been successfully used as gene carriers. This enables them to encapsulate nucleic acids and facilitates a sustained release profile, which essentially makes these suitable for tissue engineering and regenerative medicine applications (Nguyen *et al.*, 2015). Introduction of stimuli-responsive elements into hydrogels permit the on-demand and site-specific release of therapeutic agents based on particular environmental cues, such as pH variation or temperature.

22.4.4 Nanocarrier design for agricultural applications

The development of nanocarriers for agriculture is a complex endeavor that combines various fields such as materials science, biotechnology, and agricultural engineering. Nanocarriers are crucial for the targeted delivery of biomolecules such as genetic material to plants (i.e. DNA and RNA) so that plant improvement, disease resistance, stress tolerance, etc., can be effectively enhanced by them. The use of nanocarriers in agriculture can ensure safer and sustainable crop production that meets the ever-increasing demands for food sustainability.

The first use of nanomaterials for plant cultivation was nanoparticles of mesoporous silica, which were biolistically introduced into tobacco leaves, and mesoporous silica nanoparticles were also used for plant genome insertion into corn embryos. Carbon nanotubes and peptides have also been used to transport RNA and DNA for gene expression in various subcellular locations such as the nucleus and chloroplast (Fig. 22.2).

Selection of nanocarrier materials

The first step in formulating nanocarriers is the selection of materials, which can efficiently encapsulate and shield nucleic acids with biocompatible stability (Fig. 22.3). The typical materials employed in nanocarrier development involve:

LIPIDS. Found in an extensive use of liposomes, which are enclosed by phospholipid bilayers and they have the capability to carry both hydrophilic as well as aerophobic materials. They can be modified to increase stability and targeting properties (Hamada *et al.*, 2018).

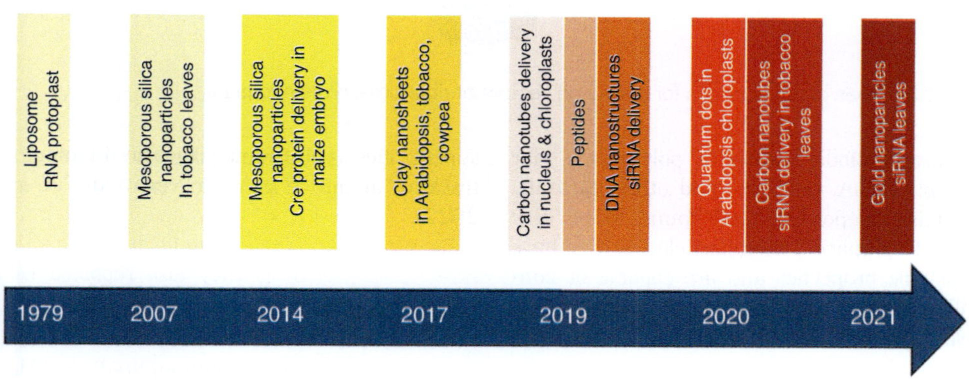

Fig. 22.2. Timeline of nanomaterials developed for plant genetic engineering.

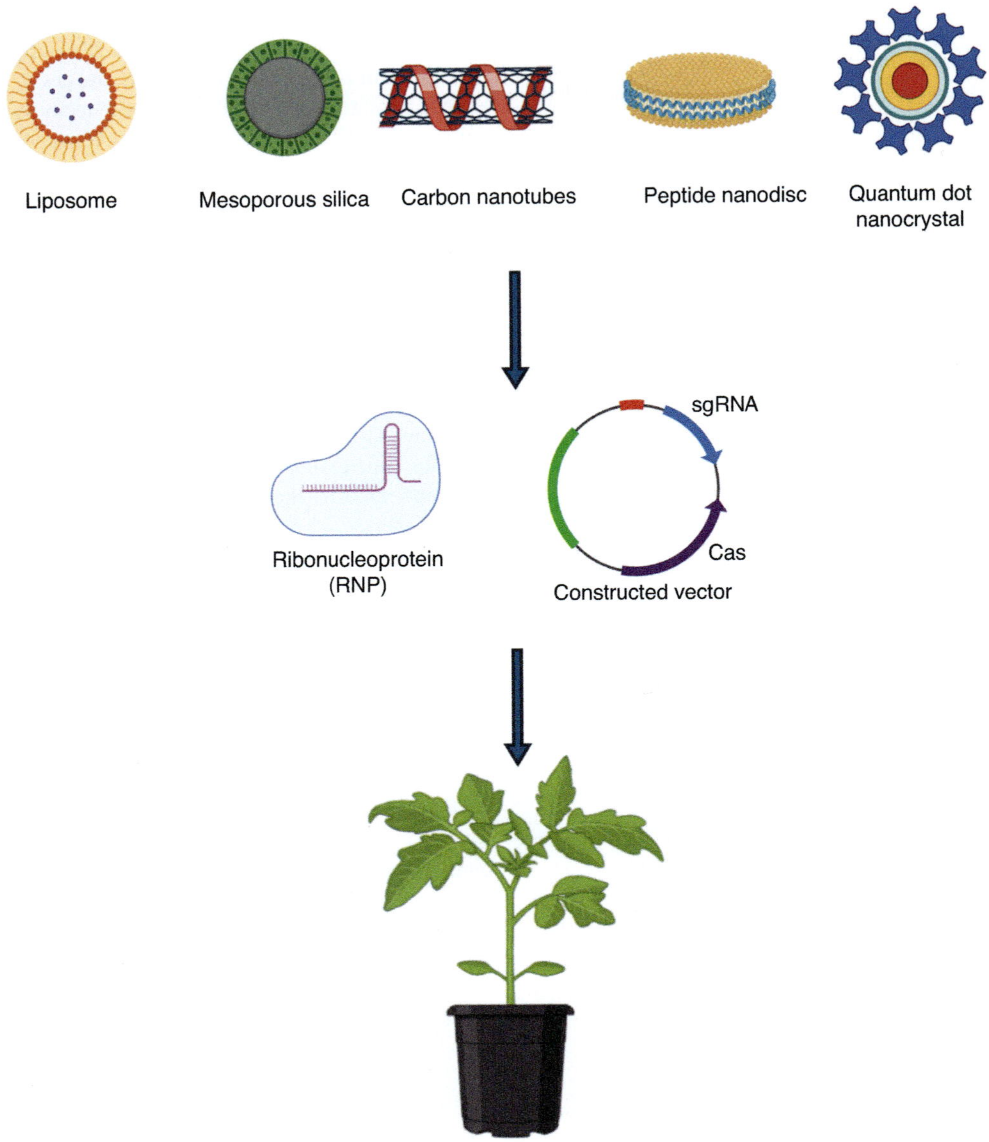

Fig. 22.3. Types of nanocarriers for delivering vectors or ribonucleoprotein to plant.

POLYMERS. Biodegradable polymers, for example PLGA, chitosan, and other naturally occurring biopolymers are found to be very useful in preparing nanoparticles as they have adjustable properties and are capable of controlled release or delivery, which makes them a wise choice to choose as carriers.

INORGANIC NANOPARTICLES. Gold and silica nanoparticles are noteworthy for their optical superiorities as electronic properties in imaging tracking alongside gene delivery (Aliu *et al.*, 2022).

Functionalization of nanocarriers

Once the base materials have been selected, the nanocarriers are functionalized in order to achieve greater target specificity and better uptake into the cells. Functionalization is a surface modification process in which the surface of nanocarriers is modified by different ligands, antibodies, or peptides that specifically bind to receptors on plant target cells. The results show that targeting OSCs in this way increases the efficiency of gene delivery by enhancing endocytosis (Tang *et al.*, 2023).

The next step is coating with biocompatible polymers: coating nanocarriers with biocompatible polymers can increase the stability of the loaded nanoparticles in the biological environment and reduce toxicity. A common practice is to coat nanoparticles with a hydrophilic shell using compounds such as polyethylene glycol (PEG), which serves to increase circulation time and reduce the likelihood of being detected by immune surveillance (Hamada *et al.*, 2018).

Encapsulation of genetic material

The encapsulation of nucleic acids is an important issue in the development of nanocarriers. Efficient encapsulation can be achieved by various methods, such as electrostatic assembly, in which positively charged nanocarriers interact electrostatically with negatively charged nucleic acids to form strong nanocomplexes. This also works well with liposomes and polymeric nanoparticles (Yu *et al.*, 2021). In the formation of polymeric nanoparticles, nucleic acid-polymer complexes can be encapsulated during nanoparticle synthesis either by solvent evaporation or by coacervation without the risk of degradation of the genetic material (Li *et al.*, 2022b).

Characterization of nanocarriers

In the development of nanocarriers for targeted gene delivery, characterization is very important to know whether the developed nanocarriers meet the required criteria. Some of the most important parameters to evaluate are the size and shape of the nanocarriers. Size and morphology (shape) play a crucial role in cell kinetics. Therefore, techniques such as DLS (dynamic light scattering) and TEM are used to characterize the drug-loaded nanoparticles (Kumari *et al.*, 2022). The surface charge zpotentia of nanocarriers is also important for their stability and interaction with cells. A larger positive charge often increases cellular uptake (Banakar *et al.*, 2019). In addition, the release kinetics must be investigated. The release profile of the encapsulated nucleic acids must be investigated to ensure a controlled release over a desired time period leading to efficient gene expression (Lin *et al.*, 2023).

Delivery mechanisms

The final phase of the design strategy relates to the selection of a suitable delivery system for novel nanocarriers (Fig. 22.4). This may include endocytosis, in which the nanocarriers can be internalized into the plant cells by endocytosis, which can facilitate functionalization by targeting ligands (Tang *et al.*, 2023). The second type is nanocarrier delivery or biolistics, which in some cases utilizes biolistics for the direct delivery of nanocarriers to plant tissues (Sturme *et al.*, 2022). The third mechanism is *Agrobacterium*-mediated transformation, in which engineered nanocarriers can be co-administered with *Agrobacter tumefaciens* in another set of crops to promote the transfer of genetic information into plant cells (Calvache *et al.*, 2022).

The development of nanocarriers for agricultural applications is therefore a multi-stage process that includes the choice of material, functionalization, encapsulation strategies, characterization, and transport. With further research progress in this regard, the potential beneficial applications of these nanocarriers to improve plant traits and disease resistance or stress tolerance will provide a way to solve global agricultural problems. The use of nanoparticles for gene delivery not only offers the opportunity to incorporate nanotechnology into crop improvement, but can also strategically realign molecular biology to find more innovative ways to engineer crops and ensure global food security in the face of climate change.

22.4.5 Advantages and challenges of nanocarrier-based delivery

The application of nanocarrier-based delivery systems in crop engineering has been seeing growing interest with the purpose to improve input use efficacy containing genetic material, pesticides, and fertilizers. While these engineered

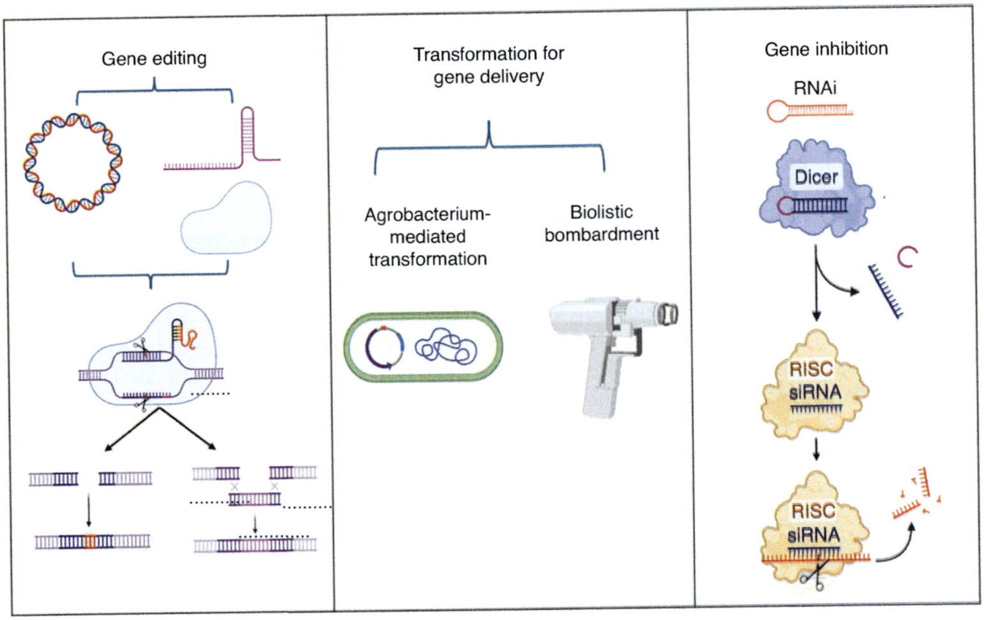

Fig. 22.4. Schematic pathways of gene delivery in plant.

nanocarriers provide myriad advantages including enhanced bioavailability, site-directed delivery, and reduced toxicity, they also pose some limitations which need to be overcome for successful translation. Here we consider the pros and cons of nanocarrier-based delivery for CRISPR applications to crops.

Advantages of nanocarrier-based delivery

Nanocarriers can significantly facilitate the dissolution and metabolic stability of agrochemicals, ultimately leading to increased bioavailability. For example, nanopesticides have been reported to have higher bioavailability than conventional pesticides that contain them and are therefore more effective against pests and pathogens (Galhardi *et al.*, 2022). This higher bioavailability leads to a lower application rate, which ultimately means a lower environmental impact.

One of the most important features for the use of fan nanocarriers is the targeted delivery to a specific location on plants. Functionalization with targeted ligands on the nanocarriers, such as antibodies or peptides, improves the specificity of delivery and allows the active ingredients to be delivered directly to the target cells/tissues (Pérez-de-Luque, 2020). In combination with the surface ligand coating approach, this targeted method should reduce unwanted off-target effects and increase the therapeutic efficiency of the administered drug.

In addition, nanocarriers can be adapted for the controlled release of certain active ingredients (adjuvants) over time. This property is particularly useful in agriculture, where consistent efficacy over time makes the lifespan of crops more predictable. The development of polymeric nanocarriers that selectively release their cargo in response to environmental stimuli, such as changes in pH or temperature, would mean that the active ingredients are only released when needed (Li *et al.*, 2021).

In addition, nanocarriers can reduce the toxicity of conventional agrochemicals. In this context, by using nanocarriers to encapsulate the active ingredients, it is not only possible to reduce their systemic exposure but also (though this is more a form of hypothetical knowledge) to mitigate adverse effects on non-target organisms (including beneficial insects and soil microorganisms) (Worrall *et al.*, 2018). This likely reduces toxicity and makes agriculture more sustainable.

Nonetheless, nanocarriers can stabilize sensitive compounds such as nucleic acids and proteins that are susceptible to degradation under storage conditions or in various applications. Similarly, there has been success in improving RNA interference (RNAi) applications for pest control using lipid-based nanocarriers to protect the sensitive RNA molecules from degradation (Zheng *et al.*, 2019). The analysis is important to promote the shelf life and performance of agricultural inputs.

Challenges of nanocarrier-based delivery

Scale-up for the commercial production of nanoparticles often encounters insurmountable hurdles in research and development, as many nanocarriers require sophisticated and cumbersome processes. However, batch-to-batch variations in production can lead to inconsistent performance in terms of nanocarrier properties and also in the field (Valetti *et al.*, 2013). Given the urgent need to develop reproducible and reliable protocols for the production of nanocarriers, researchers require standardized systems for evaluation.

The application of nanotechnology in the agricultural sector has raised significant regulatory issues related to the safety and environmental impact of the use of nanocarriers. Most regulatory frameworks for nanomaterials are still a work in progress, and compliance can be very complex for researchers and companies moving towards nanocarrier-based products (Pérez-de-Luque, 2020). Nanocarriers in agricultural applications require comprehensive safety assessments to avert potential hazards.

Nanocarriers could reduce the hazards of agrochemicals, but their questionable effects on the environment have not yet been sufficiently researched. Nanomaterials have been suspected to accumulate in soils and water bodies and their ecological safety for non-target organisms has been questioned, so further research is needed (Galhardi *et al.*, 2022). Sustainable agricultural practices therefore depend on understanding the fate and behavior of nanocarriers in the environment.

The use of nanocarrier-based delivery systems to support development and production may incur additional costs compared to the conventional use of agricultural inputs. However, the cost of materials and determination also limits general implementation, especially in smallholder farming systems (Liu *et al.*, 2022a). Solving these costly problems is necessary to promote the technology of nanocarriers in agriculture.

The mechanism of interaction between nanovectors and the penetrating substrates varies depending on the plant species, developmental stage and biophysical properties under different environmental conditions. This ability is crucial for the development of nanocarriers suitable for targeted applications (Lyu *et al.*, 2021). More in-depth studies are needed to explore the pathway and process of nanocarrier uptake, transport, and distribution in plants.

Overall, nanocarrier-based delivery systems offer several advantages for plant breeding, such as increased bioavailability, site-specific action, and release behavior, attenuated toxic effects associated with drugs, and consistent improvement in stability. Nevertheless, concerns regarding scalability of production and technological acceptance (regulatory hurdles) impact on the natural environment and changing costs must be considered to ensure sustainability and complexity of plant biology. Further research and development will be crucial to overcome these challenges and increase adoption, e.g. by introducing nanocarrier-based solutions for sustainable agricultural practices such as solubilization of pesticides or safe delivery systems in isolated/encapsulated nanocarriers.

22.5 Integration of CRISPR/Cas and Nanocarriers in Crop Engineering

22.5.1 Strategies for combining CRISPR/Cas and nanocarriers

Combining CRISPR/Cas technology with engineered nanocarriers is believed to offer an attractive avenue for improving gene delivery in crop engineering. While the precision of CRISPR/Cas systems for targeted genome editing is unique and nanocarriers have inherently efficient gene-delivery properties, this combination makes them ideal partners to exploit enhanced efficiency with great specificity. Here we describe the major strategies to integrate CRISPR/Cas with nanocarriers and discuss how these can be used in agricultural biotechnology.

Design of nanocarriers for CRISPR/Cas delivery

The first approach is the development of a fully engineered nanocarrier system for the targeted delivery of CRISPR/Cas components (i.e. Cas9 protein and gRNAs). In this context, various nanocarriers such as liposomes, polymeric nanoparticles, and inorganic nanoparticles can be used. This has been exemplified by the promotion of the use of nano-liposomal lecithin particles, which provide a system for the delivery of CRISPR/Cas9 complexes (Lin *et al.*, 2023), thereby increasing the efficiency of gene editing for therapeutic targets. Here we describe the design of such nanocarriers considering size efficiency, cellular interaction, and lack of toxicity.

Targeted delivery mechanisms

At the same time, a crucial approach is to functionalize nanocarriers with targeted ligands that can increase their selectivity towards specific plant tissues or cell types. Ligands such as peptides or antibodies attached to the surface of nanocarriers can direct the CRISPR/Cas components into specific cells (Wei *et al.*, 2020). For example, Lin *et al.* (2023) recently reported the use of folate receptor-mediated release in cancer cells and found greater target tissue selectivity among a variety of different protein packaging methods tested with Cas9.RNP complexes. It should be possible to modify this strategy for agricultural purposes by using ligands targeting plant-specific receptors.

Use of stimuli-responsive nanocarriers

Stimuli-responsive nanocarriers, on the other hand, are a new trend to improve CRISPR/Cas delivery. The nanocarriers can be engineered to discharge their cargo only in response to specific environmental stimuli such as changes in pH, temperature, or light. A typical example is the CRISPR/Cas9 system developed by Pan *et al.* (2019), activated by near-infrared up-conversion, which can be used to achieve remote gene editing. This controlled release of CRISPR components at a specific time and location could lead to better results in gene editing in plants through the use of stimulus-dependent systems.

Co-delivery systems

In the case of CRISPR/Cas, co-delivery systems with other therapeutics and signaling molecules are used to improve the results of gene editing. For example, the carboxylated branched poly (β-amino ester) nanoparticles could also efficiently release cytosolic proteins during CRISPR/Cas9 genome editing (Rui *et al.*, 2019). A team of plant scientists has developed a method to improve delivery methods that make the CRISPR/Cas9 gene-editing tool more efficient. This could lead to the improvement of plant traits that are important for global food security.

Evaluation of delivery efficiency and safety

Finally, the imaging approach provides detailed information on the efficiency and safety of different nanocarrier-based systems for CRISPR/Cas delivery. Properties such as the stability of the nanocarriers, their ability to protect nucleic acids from degradation, and their ability to mediate gene editing in the target cells will be investigated. In addition, the toxicity of the nanocarrier itself to non-target organisms and the environment needs to be assessed (Wei *et al.*, 2020). They also pointed out that systemic delivery of CRISPR/Cas9 RNPs via nanoparticles is suitable for tissue-specific gene correction and drew attention to the critical safety considerations associated with this approach.

22.5.2 Successful applications of combining nanoparticles and CRISPR/Cas in crop engineering

Combining CRISPR/Cas tech with nano-homework has successfully applied these systems to the crops, which will increase both efficacy and specificity of gene carriers. These advances have huge potential for enhancing crop traits, from disease resistance through stress tolerance to increased yield. Here are few significant examples of successful applications demonstrating the potential to interface CRISPR/Cas with nanocarrier system.

Liposome-mediated delivery of CRISPR/Cas9

In the study by Sousa *et al.* (2022), CRISPR/Cas9 systems were transferred *in vitro* using multivalent cationic liposome-DNA complexes. In their study, these liposomes were able to encapsulate and shield the CRISPR components to enable gene modification in the desired cellular configurations. In addition to improved transfection efficiency, the use of liposome-based systems also led to a reduction in off-target effects, showing that these carriers are promising for agricultural applications.

Nanoparticle delivery for enhanced gene editing

Lee *et al.* (2017) demonstrated the *in vivo* administration of Cas9 ribonucleoprotein (RNP) complexes using nanoparticles that enabled RNP nucleases to induce homology-directed DNA repair. In this study, a nanoparticle-based delivery system was shown to deliver CRISPR/Cas9 gene-editing material into target cells. The delivery of CRISPR/Cas9 to cell nuclei by nanoparticles is efficient and could lead to unproblematic applications in plant breeding that require precise gene mutation.

Ribonucleoprotein complexes in wheat

Liang *et al.* (2017) successfully used particle bombardment to introduce CRISPR/Cas9 RNP complexes into immature bread wheat embryo cells. This approach enabled efficient and DNA-free genome engineering, a valuable trait for hard-to-regenerate crop species. The strategy, which was successfully applied to wheat for the first time using nanocarrier systems, shows that gene editing is also possible in other crops and gives an insight into how it could work.

Hydrogel nanoparticles for targeted delivery

Zhou *et al.* (2024) has developed a site-specific nanosystem for CRISPR/Cas9 gene therapy based on liposomes and hydrogel nanoparticles. Although this system was developed for cancer therapy, the concepts of targeted delivery and controlled release can also be applied to agrochemicals. Such precision would increase the efficiency of plant breeding, as a high degree of accuracy is required when modifying plant genes using powerful tools such as CRISPR/Cas9.

Gold nanoparticles for gene delivery

In a study by Zhou *et al.* (2024), AuNPs were used as a delivery system for CRISPR/Cas9 components to deliver gene-editing tools directly into target cells. The properties of AuNPs include functionalization, which is well established and optically traceable for monitoring cellular uptake. This study demonstrates the promise of using AuNPs in the co-delivery of CRISPR/Cas9 systems for application in agricultural biotechnology.

22.6 Bioreducible Lipid Nanoparticles

Liu *et al.* (2018) demonstrated *in vivo* gene editing after delivery of CRISPR/Cas9 mRNA and sgRNA using bioreducible lipid nanoparticles. This method validated the rational use of lipid nanoparticles for delivery of CRISPR components with minimal toxicity and suggests that our system may have clinical utility for *in vivo* applications of this innovative modality. Their previous work also demonstrates the utility of these bioreducible lipid nanoparticles for gene editing, highlighting their potential for plant breeding, where effective, safe carriers are essential.

22.6.1 Future directions

CRISPR/Cas nanocarrier integration in plant breeding represents a new opportunity to sustainably improve agricultural productivity. In this rapidly evolving research landscape, several unrealized advances are conceivable to realize the full potential of merging these two powerful tools. Below are some recommendations for future developments in the integration of CRISPR/Cas and nanocarriers with respect to crop fertility.

Development of targeted nanocarriers

Further research is needed to develop and program nanocarriers that can selectively and precisely penetrate the desired tissues or cells to efficiently introduce CRISPR/Cas into plants. One approach for this is the functionalization of nanocarriers with ligands that bind to specific receptors on the surface of plant cells. For example, it has been reported that the uptake efficiency and translocation of polymeric nanocarriers in plants is highly dependent on charge and aspect ratio (Zhang *et al.*, 2023). By tuning these properties, the appropriate and effective manipulation of genes can be controlled to enable more successful CRISPR/Cas applications for crop engineering.

Stimuli-responsive nanocarriers

In the future, it is an intriguing possibility to explore stimuli-sensitive nanocarriers capable of discharging CRISPR/Cas elements in a trigger reliant manner (e.g. pH, temperature, or light). These kinds of systems allow for the timely and spatiotemporal delivery of your CRISPR/Cas components giving an independent delivery on when and where those genetic materials are released in the plant. This method can increase the accuracy during gene editing and decrease off-targets effect; hence, it will overall increase efficiency of this technology.

Multifunctional nanocarriers

The unique ability of nanocarriers to combine more than one functionality in a single CRISPR/Cas carrier system could be the key to highly effective solutions. Further studies could investigate the potential for co-administration of CRISPR/Cas components in combination with other therapeutics (e.g. RNA interference molecules [RNAi] or plant growth regulators) to achieve a synergistic effect on plant productivity. For example, co-administration of CRISPR/Cas and RNAi can enhance host resistance to a disease by affecting different pathways that are part of the plant's own defense mechanisms (Knott and Doudna, 2018).

Enhanced safety and regulatory compliance

Safety and regulatory compliance remain the main concerns in the increasing use of nanocarriers in agriculture. The ecological and toxicological profiles of nanocarrier systems should be explored in future research, coupled with the introduction of appropriate methods for their evaluation. It emphasizes the need for responsible handling and evidence-based use of CRISPR technologies in agriculture (Ahmad *et al.*, 2021). Consequently, the development of such an approach may be an important prerequisite to ensure public confidence and regulatory compliance for nanocarrier-based applications in crop engineering.

Integration with advanced breeding techniques

CRISPR/Cas, in conjunction with nanocarriers, can be coupled with improved breeding technologies such as genomic selection and marker-assisted breeding to achieve faster crop improvement. Combining the precision of CRISPR/Cas technology with nanocarrier systems offers a faster and more effective way to develop crops with advantageous traits. Such integration of different methods for genetic improvement can increase the overall benefit of breeding programs, which will have a positive impact on the adaptability and readiness of crops under changing environmental stress (Waghmode *et al.*, 2024).

Exploration of new nanomaterials

There is also a need to explore various other nanomaterials that can be used as carriers for the CRISPR/Cas delivery system, which is the direction of future work. Novel materials such as carbon-based nanomaterials, biodegradable polymers and metal-organic frameworks (MOFs), which are characterized by special properties, are of great interest for improving gene delivery. Biodegradable nanocarriers are an example of reducing environmental impact without interfering with the delivery of CRISPR/Cas components (Roy *et al.*, 2022). Research into these new materials can pave the way for more environmentally friendly and efficient agricultural delivery systems.

Regulatory and ethical considerations

Regulatory and ethical issues concerning GM crops and nanotechnology are of paramount importance for successful operation of these technologies in agriculture. Considering the progress in application of CRISPR/Cas as well as nanocarriers to engineer crops, it is crucial to warrant more regulatory frameworks and safety assessments and public acceptance. This paper will specifically discuss the current regulations for GM crops and nanotechnology, risk assessments (safety as well as environmental), public perception of such technologies, their ethical issues etc.

Regulatory framework for GM crops and nanotechnology

The regulatory system for genetically modified plants varies greatly around the world. Different countries deal differently with biotechnology and its public health concerns. The regulatory process in the USA is primarily administered by the U.S. Department of Agriculture (USDA), the Environmental Protection Agency (EPA), and the Food and Drug Administration (FDA) (Brookes and Barfoot, 2016). In contrast, the EU has a more stringent regulatory system and conducts full risk assessments and labeling of GM products (Brookes and Barfoot, 2017), which can significantly slow down approval times.

Nanotechnology is clearly still a rapidly developing area for which a regulatory framework is not yet in place. Nanomaterials may fall under the existing regulatory framework for chemicals in many countries, but specific guidance documents for the use of nanotechnology in agriculture are less common. An international, effective, and safe policy regime for the use of nanotechnology in agriculture is imperative (Rzymski and Królczyk, 2016). As nanomaterials differ in some respects from other materials, the potential negative impacts and contamination risks of their use need to be researched, while at the same time they need to be safely integrated into agricultural practices.

Safety and environmental impact assessments

The safety of GM foods has been assessed in terms of risk to human health, non-target ecosystems, and the environment. These assessments are important to identify unintended adverse effects associated with the introduction of GM crops into nature. However, the use of GM crops can alter patterns of pesticide use, with potential consequences for environmental health, as studies have shown (Nakai *et al.*, 2015; Anderson *et al.*, 2021). In many cases, regulators attach great importance to environmental risk assessment (ERA) when determining the potential impact on biodiversity, soil health, and water quality of a GM crop version under study (Cárdenas Pardo *et al.*, 2023).

The same applies to the agricultural sector, so here, too, a comprehensive safety analysis is needed to investigate the risks associated with the introduction of nanocarriers in agriculture. These assessments of nanomaterials should address in detail toxicity, environmental behavior, and interaction with biological systems (Ding and Li, 2023). Previous research has shown that the environmental behavior of nanomaterials depends on composition and surface properties (Brookes and Barfoot, 2018), so a case-by-case assessment was considered appropriate.

Public perception and ethical issues

Interdisciplinary studies show that the public image of genetically modified crops and nanotechnology is of central importance for their acceptance and use. Safety, environment, and ethics are among the case-related factors that shape public attitudes towards GMO technologies. It has also been shown that consumer perceptions of GM foods differ from their attitudes towards GM drugs as a whole, and that consumer scepticism is more pronounced for food (Sun *et al.*, 2015). The difference may be due to ethical concerns related to the intentional spread of GM organisms in the environment rather than the use of GM technology for diagnostics, which is seen as more directly beneficial (Brookes, 2022).

The ethical concerns with both GM crops and nanotechnology are also related to the monopolization of agricultural biotechnology by corporations, the potential impact on small farmers, and issues of operational ethics with regard to the patenting of life. The commercialization of genetically modified crops challenges our food sovereignty, as well as the right of farmers to save and exchange seeds (Van Der Voet and Paoletti, 2019). Furthermore, the risk of unintended consequences such as pest resistance or biodiversity loss with these technologies requires careful analysis to assess the scale and nature of potential long-term impacts (Brookes and Barfoot, 2020).

22.7 Conclusion

Taken together, the combination of CRISPR/Cas technology with engineered nanocarriers is a unique advance in plant breeding and offers new opportunities to address current challenges in modern agriculture. In this chapter, we have discussed the integration of these aspects along with their advanced methods such as intricate mechanisms of gene delivery and controlled design of nanocarriers for efficient transformation leading to better plant traits such as disease resistance, stress tolerance, and higher yields.

The combination of nanocarriers and CRISPR/Cas offers several advantages. These systems offer better bioavailability, site-specific delivery and controlled release, and lower toxicity. In addition, the versatility of nanoparticle design may enable multifunctional nanocarrier systems that allow co-targeting of CRISPR and other therapeutic agents, which could open new horizons in the field of plant improvement through the development of stress-resistant and nutrient-enhanced plants.

However, this chapter also highlighted the regulatory and ethical implications of production with genome and nanotechnology. The regulatory frameworks are very different in different regions, which means that they all need to undergo a thorough safety and environmental impact assessment before they can be applied. To promote consumer and stakeholder acceptance, it is crucial to consider public perception of genetically modified crops as well as moral concerns related to nanotechnology.

The future of CRISPR/Cas integration with nanocarriers in plant breeding seems promising. This will require continued research and innovation to overcome existing challenges – from production scalability to safety issues and regulatory barriers. Advances in novel CRISPR/Cas-based genome-editing tools and methods for site-specific gene regulation, as well as the integration of cutting-edge therapeutic interventions such as Optide or other nano-agents targeted based on stimulus-mediated responsiveness, could provide an additional advantage in delivering the consequences of precision-level applications and drive these developments towards the use of novel nanocarriers as effectively precise biobullets.

As food security is expected to become more important globally in the face of climate change and population growth, the adoption of these new biotechnological solutions will be central to sustainable agriculture. The opportunity to change the game for plant breeding through the use of emerging CRISPR/Cas technology combined with the development of nanocarriers offers a pathway to resilient, high-yielding, and increasingly sustainable crops to feed future generations.

References

Agbenyo, W., Jiang, Y., Jia, X., Wang, J., Ntim-Amo, G. *et al.* (2022) Does the adoption of climate-smart agricultural practices impact farmers' income? Evidence from Ghana. *International Journal of Environmental Research and Public Health* 19(7), 3804. DOI: 10.3390/ijerph19073804.

Ahmad, S., Tang, L., Shahzad, R., Mawia, A.M., Rao, G.S. *et al.* (2021) CRISPR-based crop improvements: A way forward to achieve zero hunger. *Journal of Agricultural and Food Chemistry* 69(30), 8307–8323. DOI: 10.1021/acs.jafc.1c02653.

Aliu, E., Lee, K. and Wang, K. (2022) CRISPR RNA-guided integrase enables high-efficiency targeted genome engineering in *Agrobacterium tumefaciens*. *Plant Biotechnology Journal* 20(10), 1916–1927. DOI: 10.1111/pbi.13872.

Anderson, J., Bachman, P., Burns, A., Chakravarthy, S., Goodwin, L. *et al.* (2021) Streamlining data requirements for the environmental risk assessment of Genetically Modified (GM) crops for cultivation approvals. *Journal of Regulatory Science* 9(1), 26–37. DOI: 10.21423/JRS-V09I1ANDERSON.

Arenas-Calle, L.N., Whitfield, S. and Challinor, A.J. (2019) A Climate Smartness Index (CSI) based on greenhouse gas intensity and water productivity: Application to irrigated rice. *Frontiers in Sustainable Food Systems* 3, 105. DOI: 10.3389/fsufs.2019.00105.

Aryal, J.P., Rahut, D.B., Maharjan, S. and Erenstein, O. (2018) Factors affecting the adoption of multiple climate-smart agricultural practices in the Indo-Gangetic plains of India. *Natural Resources Forum* 42(3), 141–158. DOI: 10.1111/1477-8947.12152.

Avci, U., Sipahi, H. and et al, A. (2023) Genome editing of gene families for crop improvement. In: Ricroch, A., Eriksson, D., Miladinovic, D., Sweet, J. and Laere, K. (eds) *A Roadmap for Plant Genome Editing*. Springer, Cham, Switzerland, pp. 3–15. DOI: 10.1007/978-3-031-46150-7_1.

Balo, S. and Mahata, D. (2022) A review of climate-smart agriculture is a new approach to farming system. *International Journal of Environment and Climate Change* 12(11), 2682–2692. DOI: 10.9734/ijecc/2022/v12i1131256.

Banakar, R., Eggenberger, A.L., Lee, K., Wright, D.A., Murugan, K. *et al.* (2019) High-frequency random DNA insertions upon co-delivery of CRISPR-Cas9 ribonucleoprotein and selectable marker plasmid in rice. *Scientific Reports* 9(1), 19902. DOI: 10.1038/s41598-019-55681-y.

Barman, H.N., Sheng, Z., Fiaz, S., Zhong, M., Wu, Y. *et al.* (2019) Generation of a new thermo-sensitive genic male sterile rice line by targeted mutagenesis of TMS5 gene through CRISPR/Cas9 system. *BMC Plant Biology* 19(1), 9. DOI: 10.1186/s12870-019-1715-0.

Belay, A.D., Kebede, W.M. and Golla, S.Y. (2023) Determinants of climate-smart agricultural practices in smallholder plots: Evidence from Wadla district, northeast Ethiopia. *International Journal of Climate Change Strategies and Management* 15(5), 619–637. DOI: 10.1108/IJCCSM-06-2022-0071.

Brookes, G. (2022) Genetically Modified (GM) crop use 1996–2020: Environmental impacts associated with pesticide use change. *GM Crops & Food* 13(1), 262–289. DOI: 10.1080/21645698.2022.2118497.

Brookes, G. and Barfoot, P. (2016) Environmental impacts of Genetically Modified (GM) crop use 1996–2014: Impacts on pesticide use and carbon emissions. *GM Crops & Food* 7(2), 84–116. DOI: 10.1080/21645698.2016.1192754.

Brookes, G. and Barfoot, P. (2017) Environmental impacts of Genetically Modified (GM) crop use 1996–2015: Impacts on pesticide use and carbon emissions. *GM Crops & Food* 8(2), 117–147. DOI: 10.1080/21645698.2017.1309490.

Brookes, G. and Barfoot, P. (2018) Environmental impacts of Genetically Modified (GM) crop use 1996–2016: Impacts on pesticide use and carbon emissions. *GM Crops & Food* 9(3), 109–139. DOI: 10.1080/21645698.2018.1476792.

Brookes, G. and Barfoot, P. (2020) Environmental impacts of Genetically Modified (GM) crop use 1996–2018: Impacts on pesticide use and carbon emissions. *GM Crops & Food* 11(4), 215–241. DOI: 10.1080/21645698.2020.1773198.

Calvache, C., Vazquez-Vilar, M., Selma, S., Uranga, M., Fernández-Del-Carmen, A. *et al.* (2022) Strong and tunable anti-CRISPR/Cas activities in plants. *Plant Biotechnology Journal* 20(2), 399–408. DOI: 10.1111/pbi.13723.

Cárdenas Pardo, N.J., Rodriguez Robayo, D.E., Fernandez Lizarazo, J.C., Peña-Quemba, D.C. and McGale, E. (2023) Exploring the future of GM technology in sustainable local food systems in Colombia. *Frontiers in Genome Editing* 5, 1181811. DOI: 10.3389/fgeed.2023.1181811.

Chandra, A. and Shmelev, S. (2017) The relevance of political ecology perspectives for smallholder climate-smart agriculture: A review. *Journal of Political Ecology* 24(1), 821–842. DOI: 10.2458/v24i1.20969.

Chandrasekaran, J., Brumin, M., Wolf, D., Leibman, D., Klap, C. *et al.* (2016) Development of broad virus resistance in non-transgenic cucumber using CRISPR/Cas9 technology. *Molecular Plant Pathology* 17(7), 1140–1153. DOI: 10.1111/mpp.12375.

Cheng, X., Li, D., Xu, J., Wei, B., Fang, Q, *et al.* (2021) Self-assembled ternary hybrid nanodrugs for overcoming tumor resistance and metastasis. *Acta Pharmaceutica Sinica B* 11, 3595–3607. DOI: 10.1016/j.apsb.2021.03.041.

De Pinto, A. and Ulimwengu, J.M. (2017) A thriving agricultural sector in a changing climate: Meeting Malabo declaration goals through climate-smart agriculture. In: *ReSAKSS Annual Trends and Outlook Report 2016*. International Food Policy Research Institute, Washington, DC. DOI: 10.2499/9780896292949.

Dey, A.K., Nougarède, A., Clément, F., Fournier, C., Jouvin-Marche, E. *et al.* (2021) Tuning the immunostimulation properties of cationic lipid nanocarriers for nucleic acid delivery. *Frontiers in Immunology* 12, 722411. DOI: 10.3389/fimmu.2021.722411.

Ding, Z. and Li, C. (2023) Can regional environmental quality improve green innovation performance? An empirical analysis from China. *Frontiers in Environmental Science* 22(11), 1276224. DOI: 10.3389/fenvs.2023.127622.

Ekpa, D., Tsado, E.K. and Bodaga, T. (2018) The effect of climate smart agricultural practices on poverty status of maize rural farming household in Sokoto state, Nigeria. *Journal of Agricultural Science and Practice* 3(5), 97–106. DOI: 10.31248/JASP2018.104.

Ekpa, D., Tiri, G. and Mamman, M. (2023) Evaluation of climate smart indicators influencing agricultural practices on sorghum enterprise. *Katsina* and *Sokoto states. Fudma Journal of Agriculture and Agricultural Technology* 9(1), 129–136. DOI: 10.33003/jaat.2023.0901.17.

Fawole, B.E. and Aderinoye-Abdulwahab, S.A. (2021) Farmers' adoption of climate smart practices for increased productivity in Nigeria. In: Oguge, N., Ayal, D., Adeleke, L. and Silva, I. (eds) *African Handbook of Climate Change Adaptation*. Springer, Cham, Switzerland, pp. 495–508. DOI: 10.1007/978-3-030-45106-6_227.

Ferrer, A.J.G., Thanh, L.H., Chuong, P.H., Kiet, N.T., Trang, V.T. *et al.* (2023) Farming household adoption of climate-smart agricultural technologies: Evidence from North-Central Vietnam. *Asia-Pacific Journal of Regional Science* 7(2), 641–663. DOI: 10.1007/s41685-023-00296-5.

Gairhe, J.J., Adhikari, M., Ghimire, D., Khatri-Chhetri, A. and Panday, D. (2021) Intervention of climate-smart practices in wheat under rice-wheat cropping system in Nepal. *Climate* 9(2), 19. DOI: 10.3390/cli9020019.

Galani, Y.J.H., Ligowe, I.S., Kieffer, M., Kamalongo, D., Kambwiri, A.M. *et al.* (2022) Conservation agriculture affects grain and nutrient yields of maize (*Zea mays* L.) and can impact food and nutrition security in Sub-Saharan Africa. *Frontiers in Nutrition* 8, 804663. DOI: 10.3389/fnut.2021.804663.

Galhardi, J.A., Wang, P., Bueno, V., Ghoshal, S., Gravel, V. *et al.* (2022) Field evaluation of the potential effects of polymer and silica-based nanopesticides on strawberries and agricultural soils. *Environmental Science* 9(10), 3833–3843. DOI: 10.1039/D2EN00329E.

Gan, Q., Lu, X., Dong, W., Yuan, Y., Qian, J. *et al.* (2012) Endosomal pH-activatable magnetic nanoparticle-capped mesoporous silica for intracellular controlled release. *Journal of Materials Chemistry* 22(31), 15960–15968. DOI: 10.1039/c2jm32020g.

Hamada, H., Liu, Y., Nagira, Y., Miki, R., Taoka, N. *et al.* (2018) Biolistic-delivery-based transient CRISPR/Cas9 expression enables in planta genome editing in wheat. *Scientific Reports* 8(1), 14422. DOI: 10.1038/s41598-018-32714-6.

Hu, C., Peng, Q., Chen, F., Zhong, Z. and Zhuo, R. (2010) Low molecular weight polyethylenimine conjugated gold nanoparticles as efficient gene vectors. *Bioconjugate Chemistry* 21(5), 836–843. DOI: 10.1021/bc900374d.

Kalaitzandonakes, N., Willig, C. and Zahringer, K. (2023) The economics and policy of genome editing in crop improvement. *The Plant Genome* 16(2), e20248. DOI: 10.1002/tpg2.20248.

Kalinga, A.S., Kangalawe, R.Y. and Lyimo, J.G. (2021) The significance of the diversification of food crops in abating impacts of climate variability: The case of Rungwe district, Tanzania. *Journal of the Geographical Association of Tanzania* 41(2), 1–21. DOI: 10.56279/JGAT.v41i2.1.

Kanu, G.A., Parambath, J.B., Abu Odeh, R.O. and Mohamed, A.A. (2022) Gold nanoparticle-mediated gene therapy. *Cancers* 14(21), 5366. DOI: 10.3390/cancers14215366.

Kaptymer, B.L., Abdulkerim Ute, J. and Negeso Hule, M. (2019) Climate smart agriculture and its implementation challenges in Africa. *Current Journal of Applied Science and Technology* 38(4), 1–13. DOI: 10.9734/cjast/2019/v38i430371.

Khalil, I.A., Younis, M.A., Kimura, S. and Harashima, H. (2020) Lipid nanoparticles for cell-specific *in vivo* targeted delivery of nucleic acids. *Biological & Pharmaceutical Bulletin* 43(4), 584–595. DOI: 10.1248/bpb.b19-00743.

Khan, N., Mujtaba, G., Khalid, M. and Amjad, I. (2023) Effects of global climate change: Adapting agricultural crops for a warmer world. *Biological and Agricultural Sciences Research Journal* 2023(1), 15–15. DOI: 10.54112/basrj.v2023i1.15.

Kim, M.S., Ko, S.R., Jung, Y.J., Kang, K.K., Lee, Y.J. *et al.* (2023) Knockout mutants of OsPUB7 generated using CRISPR/Cas9 revealed abiotic stress tolerance in rice. *International Journal of Molecular Sciences* 24(6), 5338. DOI: 10.3390/ijms24065338.

Knott, G.J. and Doudna, J.A. (2018) CRISPR-Cas guides the future of genetic engineering. *Science* 361(6405), 866–869. DOI: 10.1126/science.aat5011.

Kottegoda, N., Nimanka, S., Fernand, N., Silva, M. and Munaweera, I. (2023) Climate smart agriculture: The role of fertilizer innovations and efficient plant nutrient management. *Vidyodaya Journal of Science* 1(s1). DOI: 10.31357/vjs.v1is1.6709.

Kumari, D., Prasad, B. and Dwivedi, P. (2022) Genome editing platforms in rice (*Oryza sativa* L.). *Basic Methodology and Troubleshooting*. DOI: 10.21203/rs.3.rs-1961833/v1.

Kurgat, B.K., Lamanna, C., Kimaro, A., Namoi, N., Manda, L. *et al.* (2020) Adoption of climate-smart agriculture technologies in Tanzania. *Frontiers in Sustainable Food Systems* 4, 55. DOI: 10.3389/fsufs.2020.00055.

Larriba, E., Yaroshko, O. and Pérez-Pérez, J.M. (2024) Recent advances in tomato gene editing. *International Journal of Molecular Sciences* 25(5), 2606. DOI: 10.3390/ijms25052606.

Lassoued, R., Macall, D.M., Hesseln, H., Phillips, P.W.B. and Smyth, S.J. (2019) Benefits of genome-edited crops: Expert opinion. *Transgenic Research* 28(2), 247–256. DOI: 10.1007/s11248-019-00118-5.

Lee, N., Yoo, D., Ling, D., Cho, M.H., Hyeon, T. *et al.* (2015) Iron oxide based nanoparticles for multimodal imaging and magnetoresponsive therapy. *Chemical Reviews* 115(19), 10637–10689. DOI: 10.1021/acs.chemrev.5b00112.

Lee, K., Conboy, M., Park, H.M., Jiang, F., Kim, H.J. *et al.* (2017) Nanoparticle delivery of Cas9 ribonucleoprotein and donor DNA *in vivo* induces homology-directed DNA repair. *Nature Biomedical Engineering* 1(11), 889–901. DOI: 10.1038/s41551-017-0137-2.

Li, Y.L. and Sun, B.H. (2019) Nano-delivery materials: Review of development and application in drug/gene transport. *Key Engineering Materials* 803, 158–166. DOI: 10.4028/www.scientific.net/KEM.803.158.

Li, X., Wang, Y., Chen, S., Tian, H., Fu, D. *et al.* (2018) Lycopene is enriched in tomato fruit by CRISPR/Cas9-mediated multiplex genome editing. *Frontiers in Plant Science* 9, 559. DOI: 10.3389/fpls.2018.00559.

Li, P., Huang, Y., Fu, C., Jiang, S.X., Peng, W. *et al.* (2021) Eco-friendly biomolecule-nanomaterial hybrids as next-generation agrochemicals for topical delivery. *EcoMat* 3(5), e12132. DOI: 10.1002/eom2.12132.

Li, G., Ma, Y., Wang, X., Cheng, N., Meng, D. *et al.* (2022a) CRISPR/Cas9 gene editing of NtAITRs, a family of transcription repressor genes, leads to enhanced drought tolerance in tobacco. *International Journal of Molecular Sciences* 23(23), 15268. DOI: 10.3390/ijms232315268.

Li, J.-W., Zeng, T., Xu, Z.-Z., Li, J.-J., Hu, H. *et al.* (2022b) Ribozyme-mediated CRISPR/Cas9 gene editing in pyrethrum (*Tanacetum cinerariifolium*) hairy roots using a RNA polymerase II-dependent promoter. *Plant Methods* 18(1), 32. DOI: 10.1186/s13007-022-00863-5.

Liang, Z., Chen, K., Li, T., Zhang, Y., Wang, Y. *et al.* (2017) Efficient DNA-free genome editing of bread wheat using CRISPR/Cas9 ribonucleoprotein complexes. *Nature Communications* 8(1), 14261. DOI: 10.1038/ncomms14261.

Lin, Y., Wilk, U., Pöhmerer, J., Hörterer, E., Höhn, M. *et al.* (2023) Folate receptor-mediated delivery of Cas9 RNP for enhanced immune checkpoint disruption in cancer cells. *Small* 19(2), 2205318. DOI: 10.1002/smll.202205318.

Liu, J., Song, L., Liu, S., Jiang, Q., Liu, Q. *et al.* (2018) A DNA-based nanocarrier for efficient gene delivery and combined cancer therapy. *Nano Letters* 18(6), 3328–3334. DOI: 10.1021/acs.nanolett.7b04812.

Liu, Q., Yang, F., Zhang, J., Liu, H., Rahman, S. *et al.* (2021) Application of CRISPR/Cas9 in crop quality improvement. *International Journal of Molecular Sciences* 22(8), 4206. DOI: 10.3390/ijms22084206.

Liu, W., Ma, X., Kheyr, S.M., Dong, A. and Zhang, J. (2022a) Covalent organic frameworks as nanocarriers for improved delivery of chemotherapeutic agents. *Materials* 15(20), 7215. DOI: 10.3390/ma15207215.

Liu, X., Zhang, S., Jiang, Y., Yan, T., Fang, C. *et al.* (2022b) Use of CRISPR/Cas9-based gene editing to simultaneously mutate multiple homologous genes required for pollen development and male fertility in maize. *Cells* 11(3), 439. DOI: 10.3390/cells11030439.

López-Ridaura, S., Frelat, R., Wijk, M., Valbuena, D., Krupnik, T. *et al.* (2018) Climate smart agriculture, farm household typologies and food security. *Agricultural Systems* 159, 57–68. DOI: 10.1016/j.agsy.2017.09.007.

Lyu, Y., Yang, C., Lyu, X. and Pu, K. (2021) Active delivery of CRISPR system using targetable or controllable nanocarriers. *Small* 17(24), 2005222. DOI: 10.1002/smll.202005222.

Makate, C., Wang, R., Makate, M. and Mango, N. (2016) Crop diversification and livelihoods of smallholder farmers in Zimbabwe: Adaptive management for environmental change. *SpringerPlus* 5, 1–18. DOI: 10.1186/s40064-016-2802-4.

Malnoy, M., Viola, R., Jung, M.H., Koo, O.J., Kim, S. *et al.* (2016) DNA-free genetically edited grapevine and apple protoplast using CRISPR/Cas9 ribonucleoproteins. *Frontiers in Plant Science* 7. DOI: 10.3389/fpls.2016.01904.

Manaye, A., Tesfamariam, B., Tesfaye, M., Worku, A. and Gufi, Y. (2021) Tree diversity and carbon stocks in agroforestry systems in northern Ethiopia. *Carbon Balance and Management* 16(1), 14. DOI: 10.1186/s13021-021-00174-7.

Manda, J., Alene, A., Gardebroek, C., Kassie, M. and Tembo, G. (2015) Adoption and impacts of sustainable agricultural practices on maize yields and incomes: Evidence from rural Zambia. *Journal of Agricultural Economics* 67(1), 130–153.

Manda, J., Alene, A.D., Gardebroek, C., Kassie, M. and Tembo, G. (2016) Adoption and impacts of sustainable agricultural practices on maize yields and incomes: Evidence from rural Zambia. *Journal of Agricultural Economics* 67(1), 130–153. DOI: 10.1111/1477-9552.12127.

Mango, N., Makate, C., Tamene, L., Mponela, P. and Ndengu, G. (2018) Adoption of small-scale irrigation farming as a climate-smart agriculture practice and its influence on household income in the Chinyanja Triangle, southern Africa. *Land* 7(2), 49. DOI: 10.3390/land7020049.

Marshall, K. (2023) Enhancing crop yields through CRISPR technology: A promising approach for sustainable agriculture. DOI: 10.20944/preprints202307.0908.v1.

Mujeyi, A., Mudhara, M. and Mutenje, M. (2021) The impact of climate smart agriculture on household welfare in smallholder integrated crop–livestock farming systems: Evidence from Zimbabwe. *Agriculture & Food Security* 10(1), 1–15. DOI: 10.1186/s40066-020-00277-3.

Muzorewa, W. and Chitakira, M. (2021) Climate-smart livelihood strategies in rural and urban communities in eastern zimbabwe: an in-depth literature study. *South African Geographical Journal* 103(3), 395–410.

Naim, F., Dugdale, B., Kleidon, J., Brinin, A., Shand, K. *et al.* (2018) Gene editing the phytoene desaturase alleles of Cavendish banana using CRISPR/Cas9. *Transgenic Research* 27, 451–460. DOI: 10.1007/s11248-018-0083-0.

Nakai, S., Hoshikawa, K., Shimono, A. and Ohsawa, R. (2015) Transportability of confined field trial data from cultivation to import countries for environmental risk assessment of genetically modified crops. *Transgenic Research* 24(6), 929–944. DOI: 10.1007/s11248-015-9892-6.

Nascimento, F. dos S., Rocha, A. de J., Soares, J.M. da S., Mascarenhas, M.S., Ferreira, M. dos S. *et al.* (2023) Gene editing for plant resistance to abiotic factors: A systematic review. *Plants* 12(2), 305. DOI: 10.3390/plants12020305.

Negishi, Y., Omata, D., Iijima, H., Takabayashi, Y., Suzuki, K. *et al.* (2010) Enhanced laminin-derived peptide AG73-mediated liposomal gene transfer by bubble liposomes and ultrasound. *Molecular Pharmaceutics* 7(1), 217–226. DOI: 10.1021/mp900214s.

Nerkar, G., Devarumath, S., Purankar, M., Kumar, A., Valarmathi, R. *et al.* (2022) Advances in crop breeding through precision genome editing. *Frontiers in Genetics* 13, 880195. DOI: 10.3389/fgene.2022.880195.

Nguyen, M.M., Gianneschi, N.C. and Christman, K.L. (2015) Developing injectable nanomaterials to repair the heart. *Current Opinion in Biotechnology* 34, 225–231. DOI: 10.1016/j.copbio.2015.03.016.

Odipio, J., Alicai, T., Ingelbrecht, I., Nusinow, D.A., Bart, R. *et al.* (2017) Efficient CRISPR/Cas9 genome editing of phytoene desaturase in cassava. *Frontiers in Plant Science* 8, 1780. DOI: 10.3389/fpls.2017.01780.

Ojoko, E.A., Akinwunmi, J.A., Yusuf, S.A. and Oni, O.A. (2017) Factors influencing the level of use of Climate-Smart Agricultural Practices (CSAPs) in Sokoto state, Nigeria. *Journal of Agricultural Sciences, Belgrade* 62(3), 315–327. DOI: 10.2298/JAS1703315O.

Pan, Y., Yang, J., Luan, X., Liu, X., Li, X. *et al.* (2019) Near-infrared upconversion-activated CRISPR-Cas9 system: A remote-controlled gene editing platform. *Science Advances* 5(4), eaav7199. DOI: 10.1126/sciadv.aav7199.

Pérez-de-Luque, A. (2020) Guest edited collection: Nanotechnology in agriculture. *Scientific Reports* 10(1), 15738. DOI: 10.1038/s41598-020-73198-7.

Phiri, A.T., Charimbu, M., Edewor, S.E. and Gaveta, E. (2022) Sustainable scaling of climate-smart agricultural technologies and practices in Sub-Saharan Africa: The case of Kenya, Malawi, and Nigeria. *Sustainability* 14(22), 14709. DOI: 10.3390/su142214709.

Prasad, V. and Cole, D. (2023) Synergizing agroforestry and climate-smart agriculture for climate change adaptation in Australia and Pacific Island countries. DOI: 10.20944/preprints202310.1031.v1.

Razzaq, A., Kaur, P., Akhter, N., Wani, S.H. and Saleem, F. (2021) Next-generation breeding strategies for climate-ready crops. *Frontiers in Plant Science* 12, 620420. DOI: 10.3389/fpls.2021.620420.

Roy, S., Kumar, V., Behera, B.K., Parhi, J., Mohapatra, S. *et al.* (2022) CRISPR/Cas genome editing: Can it become a game changer in future fisheries sector? *Frontiers in Marine Science* 9, 924475. DOI: 10.3389/fmars.2022.924475.

Rui, Y., Wilson, D.R., Choi, J., Varanasi, M., Sanders, K. *et al.* (2019) Carboxylated branched poly (*β-amino ester*) nanoparticles enable robust cytosolic protein delivery and CRISPR-Cas9 gene editing. *Science Advances* 5(12), eaay3255. DOI: 10.10.1126/sciadv.aay3255.

Rzymski, P. and Królczyk, A. (2016) Attitudes toward genetically modified organisms in Poland: To GMO or not to GMO. *Food Security* 8, 689–697. DOI: 10.10.1007/s12571-016-0572-z.

Salava, H., Thula, S., Mohan, V., Kumar, R. and Maghuly, F. (2021) Application of genome editing in tomato breeding: Mechanisms, advances, and prospects. *International Journal of Molecular Sciences* 22(2), 682. DOI: 10.10.3390/ijms22020682.

Schenke, D. and Cai, D. (2020) Applications of CRISPR/Cas to improve crop disease resistance: Beyond inactivation of susceptibility factors. *Iscience* 23(9). DOI: 10.10.1016/j.isci.2020.101478.

Shahzad, M.F., Abdulai, A. and Issahaku, G. (2021) Adaptation implications of climate-smart agriculture in rural Pakistan. *Sustainability* 13(21), 11702. DOI: 10.10.3390/su132111702.

Shi, J., Gao, H., Wang, H., Lafitte, H.R., Archibald, R.L. *et al.* (2017) ARGOS8 variants generated by CRISPR-Cas9 improve maize grain yield under field drought stress conditions. *Plant Biotechnology Journal* 15(2), 207–216. DOI: 10.10.1111/pbi.12603.

Singh, A.K., Jha, S.K., Majumdar, B., Roy, M.L., Sarkar, S. *et al.* (2019) Impacts of climate smart jute farming on resource use efficiency, productivity and economic benefits in rural Eastern India. *Outlook on Agriculture* 48(1), 75–82. DOI: 10.1177/0030727019829488.

Sousa, D.A., Gaspar, R., Ferreira, C.J.O., Baltazar, F., Rodrigues, L.R. *et al.* (2022) *In vitro* CRISPR/Cas9 transfection and gene-editing mediated by multivalent cationic liposome-DNA complexes. *Pharmaceutics* 14(5), 1087. DOI: 10.3390/pharmaceutics14051087.

Sturme, M.H., Berg, J.P., Bouwman, L.M., Schrijver, A., Maagd, R.A. *et al.* (2022) Occurrence and nature of off-target modifications by CRISPR-Cas genome editing in plants. *ACS Agricultural Science & Technology* 2(2), 192–201. DOI: 10.1021/acsagscitech.1c00270.

Sun, J., Wu, W., Tang, H. and Liu, J. (2015) Spatiotemporal patterns of non-genetically modified crops in the era of expansion of genetically modified food. *Scientific Reports* 5(1), 14180. DOI: 10.1038/srep14180.

Tabe-Ojong, M.P.J., Lokossou, J.C., Gebrekidan, B. and Affognon, H.D. (2023) Adoption of climate-resilient groundnut varieties increases agricultural production, consumption, and smallholder commercialization in West Africa. *Nature Communications* 14(1), 5175. DOI: 10.1038/s41467-023-40781-1.

Tadesse, M., Simane, B., Abera, W., Tamene, L., Ambaw, G. *et al.* (2021) The effect of climate-smart agriculture on soil fertility, crop yield, and soil carbon in southern Ethiopia. *Sustainability* 13(8), 4515. DOI: 10.3390/su13084515.

Tang, Q., Wang, X., Jin, X., Peng, J., Zhang, H. *et al.* (2023) CRISPR/Cas technology revolutionizes crop breeding. *Plants* 12(17), 3119. DOI: 10.3390/plants12173119.

Teklewold, H., Mekonnen, A. and Kohlin, G. (2017) Does adoption of multiple climate-smart practices improve farmers' climate resilience? Empirical evidence from the Nile Basin of Ethiopia. *Climate Change Economics* 08(1), 1750001. DOI: 10.1142/S2010007817500014.

Teklewold, H., Mekonnen, A. and Kohlin, G. (2019) Climate change adaptation: A study of multiple climate-smart practices in the Nile Basin of Ethiopia. *Climate and Development* 11(2), 180–192. DOI: 10.1080/17565529.2018.1442801.

Thierfelder, C., Rusinamhodzi, L., Setimela, P., Walker, F. and Eash, N.S. (2016) Conservation agriculture and drought-tolerant germplasm: Reaping the benefits of climate-smart agriculture technologies in central Mozambique. *Renewable Agriculture and Food Systems* 31(5), 414–428. DOI: 10.1017/S1742170515000332.

Tran, M.T., Son, G.H., Song, Y.J., Nguyen, N.T., Park, S. *et al.* (2023) CRISPR-Cas9-based precise engineering of S1HvPRP1 protein towards multi-stress tolerance in tomato. *Frontiers in Plant Science* 14, 1186932. DOI: 10.3389/fpls.2023.1186932.

Tripathi, L., Ntui, V.O., Tripathi, J.N. and Kumar, P.L. (2020) Application of CRISPR/Cas for diagnosis and management of viral diseases of banana. *Frontiers in Microbiology* 11, 609784. DOI: 10.3389/fmicb.2020.609784.

Valetti, S., Mura, S., Stella, B. and Couvreur, P. (2013) Rational design for multifunctional non-liposomal lipid-based nanocarriers for cancer management: Theory to practice. *Journal of Nanobiotechnology* 11, 1–17. DOI: 10.1186/1477-3155-11-S1-S6.

Van Der Voet, H. and Paoletti, C. (2019) Equivalence testing approaches in genetically modified organism risk assessment. *Journal of Agricultural and Food Chemistry* 67(49), 13506–13508. DOI: https://10.1021/acs.jafc.9b05149.

Waghmode, P., Pitha, C.C. and Banka, V. (2024) The role of genome editing technologies in sustainable agriculture. *Journal of Advances in Biology & Biotechnology* 27(8), 543–558. DOI: 10.9734/jabb/2024/v27i81168.

Wang, Q., Li, J., An, S., Chen, Y., Jiang, C. *et al.* (2015) Magnetic resonance-guided regional gene delivery strategy using a tumor stroma-permeable nanocarrier for pancreatic cancer. *International Journal of Nanomedicine* 4479–4490. DOI: 10.2147/IJN.S84930.

Wang, C., Tariq, R., Ji, Z., Wei, Z., Zheng, K. *et al.* (2019) Transcriptome analysis of a rice cultivar reveals the differentially expressed genes in response to wild and mutant strains of *Xanthomonas oryzae* pv. *oryzae*. *Scientific Reports* 9(1), 3757. DOI: 10.1038/s41598-019-39928-2.

Wang, G., Zhu, D., Zhou, Z., Piao, Y., Tang, J. *et al.* (2020) Glutathione-specific and intracellularly labile polymeric nanocarrier for efficient and safe cancer gene delivery. *ACS Applied Materials & Interfaces* 12(13), 14825–14838. DOI: 10.1021/acsami.9b22394.

Wang, T., Xun, H., Wang, W., Ding, X., Tian, H. *et al.* (2021) Mutation of GmAITR genes by CRISPR/Cas9 genome editing results in enhanced salinity stress tolerance in soybean. *Frontiers in Plant Science* 12, 779598. DOI: 10.3389/fpls.2021.779598.

Wang, Y., Tang, Q., Pu, L., Zhang, H. and Li, X. (2022) CRISPR-Cas technology opens a new era for the creation of novel maize germplasms. *Frontiers in Plant Science* 13, 1049803. DOI: 10.3389/fpls.2022.1049803.

Wei, T., Cheng, Q., Min, Y.L., Olson, E.N. and Siegwart, D.J. (2020) Systemic nanoparticle delivery of CRISPR-Cas9 ribonucleoproteins for effective tissue specific genome editing. *Nature Communications* 11(1), 1–12. DOI: 10.1038/s41467-020-17029-3.

Wei, X., Long, Y., Yi, C., Pu, A., Hou, Q. *et al.* (2023) Bibliometric analysis of functional crops and nutritional quality: Identification of gene resources to improve crop nutritional quality through gene editing technology. *Nutrients* 15(2), 373. DOI: 10.3390/nu15020373.

Wekesa, B.M., Ayuya, O.I. and Lagat, J.K. (2018) Effect of climate-smart agricultural practices on household food security in smallholder production systems: Micro-level evidence from Kenya. *Agriculture & Food Security* 7(1), 1–14. DOI: 10.1186/s40066-018-0230-0.

Worrall, E.A., Hamid, A., Mody, K.T., Mitter, N. and Pappu, H.R. (2018) Nanotechnology for plant disease management. *Agronomy* 8(12), 285. DOI: 10.3390/agronomy8120285.

Xie, K. and Yang, Y. (2013) RNA-guided genome editing in plants using a CRISPR-Cas system. *Molecular Plant* 6(6), 1975–1983. DOI: 10.1093/mp/sst119.

Yadav, R.K., Tripathi, M.K., Tiwari, S., Tripathi, N., Asati, R. *et al.* (2023) Genome editing and improvement of abiotic stress tolerance in crop plants. *Life* 13(7), 1456. DOI: 10.3390/life13071456.

Yang, X., Yang, M., Pang, B., Vara, M. and Xia, Y. (2015) Gold nanomaterials at work in biomedicine. *Chemical Reviews* 115(19), 10410–10488. DOI: 10.1021/acs.chemrev.5b00193.

Yildiran, A.B. (2022) SiRNA mediated gene silencing in the pancreatic cancer capan-1 cell line. *Türk Doğa ve Fen Dergisi* 11(4), 98–103. DOI: 10.46810/tdfd.1097432.

Yin, K. and Qiu, J.L. (2019) Genome editing for plant disease resistance: Applications and perspectives. *Philosophical Transactions of the Royal Society B* 374(1767), 20180322. DOI: 10.1098/rstb.2018.0322.

Yu, C., Li, L., Hu, P., Yang, Y., Wei, W. *et al.* (2021) Recent advances in stimulus-responsive nanocarriers for gene therapy. *Advanced Science* 8(14), 2100540. DOI: 10.1002/advs.202100540.

Zenda, T., Liu, S., Dong, A. and Duan, H. (2021) Advances in cereal crop genomics for resilience under climate change. *Life* 11(6), 502. DOI: 10.3390/life11060502.

Zhang, H., Zhang, J., Wei, P., Zhang, B., Gou, F. *et al.* (2014) The CRISPR/Cas9 system produces specific and homozygous targeted gene editing in rice in one generation. *Plant Biotechnology Journal* 12(6), 797–807. DOI: 10.1111/pbi.12200.

Zhang, C., Srivastava, A.K. and Sadanandom, A. (2019) Targeted mutagenesis of the SUMO protease, Overly Tolerant to Salt1 in rice through CRISPR/Cas9-mediated genome editing reveals a major role of this SUMO protease in salt tolerance. *BioRxiv* 555706. DOI: 10.3389/fpls.2021.663849.

Zhang, Y., Martinez, M.R., Sun, H., Sun, M., Yin, R. *et al.* (2023) Charge, aspect ratio, and plant species affect uptake efficiency and translocation of polymeric agrochemical nanocarriers. *Environmental Science & Technology* 57(22), 8269–8279. DOI: 10.1021/acs.est.3c01154.

Zheng, Y., Hu, Y., Yan, S., Zhou, H., Song, D. *et al.* (2019) A polymer/detergent formulation improves dsRNA penetration through the body wall and RNAi-induced mortality in the soybean aphid *Aphis glycines*. *Pest Management Science* 75(7), 1993–1999. DOI: 10.1002/ps.5313.

Zhou, H., Zhang, Y., Xia, L. and Toki, S. (2021) Editorial: Targeted genome editing in crops. *Frontiers in Genome Editing* 3. DOI: 10.3389/fgeed.2021.757916.

Zhou, S., Li, Y., Wu, Q. and Gong, C. (2024) Nanotechnology-based CRISPR/Cas9 delivery system for genome editing in cancer treatment. *MedComm – Biomaterials and Applications* 3(1), e70. DOI: 10.1002/mba2.70.

Zhu, J.K. (2022) The future of gene-edited crops in China. *National Science Review* 9(4), wac063. DOI: 10.1093/nsr/nwac063.

Zhu, M., Chen, G., Dong, T., Wang, L., Zhang, J. *et al.* (2015) SlDEAD31, a putative DEAD-box RNA helicase gene, regulates salt and drought tolerance and stress-related genes in tomato. *PLoS One* 10(8), e0133849. DOI: 10.1371/journal.pone.0133849.

23 Metagenomic Insights Into Plant–probiotic Bacterial Interactions Boosting Crop Growth and Yield

Lamya Muazzeda Medha, Sharmin Islam and Tofazzal Islam*
Institute of Biotechnology and Genetic Engineering, Gazipur Agricultural University, Gazipur, Bangladesh

Abstract
Metagenomics techniques provide profound insights into these interactions by elucidating the structural diversity and functional genes of beneficial microbes associated with plants. This approach reveals the mechanisms by which probiotic bacteria boost plant growth and enhance yield, while also bolstering endurance against stresses, both biotic and abiotic. In-depth understanding of these complex dynamics is necessary for leveraging beneficial microbes in the development of sustainable agricultural practices, reducing reliance on chemical inputs, and engineering microbiomes for improved carbon and nitrogen fixation in soils – ultimately enhancing soil health. This chapter reviews the potential of metagenomics approaches to explore the intricate interactions of plant-associated probiotic bacteria and their significant contributions to crop growth, soil health, and sustainable agriculture.

Keywords: metagenome, next-generation sequencing (NGS), microbial diversity, bioinformatics tools, plant probiotic, climate change

23.1 Introduction

The global challenge of feeding an ever-expanding population is becoming increasingly complex. By 2050, the global population is projected to reach around 9.8 billion people, according to estimates from the United Nations (https://population.un.org/wpp/ [accessed 2/5/25]). This demographic shift is expected to intensify demand for food, particularly for animal-derived products, especially in developing countries. In fact, agricultural productivity must increase by 70% to meet the escalating requirements (Shi *et al.*, 2023). However, intensive farming practices currently employed in agriculture have created significant obstacles to meeting this demand. These methods have led to heightened levels of pests and diseases (Islam *et al.*, 2016c, Islam *et al.*, 2020), which, in turn, necessitate the extensive use of pesticides and chemical fertilizers. Unfortunately, excessive application of these substances causes environmental degradation (Sarker *et al.*, 2023). Furthermore, intensive agriculture contributes to reduced crop diversity, depletion of soil nutrients, loss of genetic variability, and exacerbation of climate change. Consequently, there is an urgent need for innovative agricultural strategies that

*Corresponding author: tofazzalislam@bsmrau.edu.bd

© CAB International 2025. *Stress-Resilient Crops: Coordinated omics-CRISPR-Nanotechnology Strategies* (ed. Jen-Tsung Chen)
DOI: 10.1079/9781800627307.0023

prioritize higher yields and resilience to environmental stresses while ensuring human and environmental health (Menendez and Garcia-Fraile, 2017).

At the heart of these strategies lies the rhizosphere, a vibrant soil region closely connected to the root system of the plants, marked by intense microbial interactions. This microbial community facilitates crucial signaling and material exchanges at the plant–microbe interface (Nwachukwu and Babalola, 2021). The rhizosphere is profoundly influenced by root exudates, which feed a diverse array of microorganisms collectively referred to as the root microbiome (Brader *et al.*, 2017). During plant growth, some of these microbes migrate into the roots, establishing themselves within the plant's internal tissues as part of the root endosphere, a critical subset of the rhizosphere microbiome (Brader *et al.*, 2017). The interactions between plants and these microbes can be neutral, beneficial, or detrimental, significantly impacting both plant health and microbial community structure. Within this microbial consortium, plant growth-promoting probiotics (PGPB) hold particular significance. These beneficial bacteria enhance plant development through various mechanisms, including nitrogen fixation, nutrient enhancement, microbial antagonism, and stimulation of plant stress responses (Shameer and Prasad, 2018). Beyond the rhizosphere, plants also host diverse bacterial communities as epiphytes and endophytes, which parallel the roles of probiotics in animal systems (Dheeman *et al.*, 2024). Research has demonstrated that PGPB can substantially enhance crop yields by improving nutrient availability, supporting vital physiological functions, reducing stress levels, and bolstering plant immunity against pathogens (Gouda *et al.*, 2018). Thus, a deeper exploration of the relationships between beneficial microbes and plant systems is imperative in advancing modern agricultural practices (Khan *et al.*, 2018).

Metagenomics presents a promising avenue for investigating the diversity of microbes across diverse environments excluding the need for traditional culturing methods (Hoque *et al.*, 2019). This cutting-edge method has gained traction in the field of large-scale genomics, enabling the study of the taxonomic classification along with functional characteristics of microbial populations within a wide array of environments, including agricultural contexts (French *et al.*, 2021; Hoque *et al.*, 2022). By deepening the knowledge of microbial roles in plant systems, metagenomics elucidates the intricate relationships of soil microbial communities with plant systems, which are vital for enhancing agricultural productivity (Trivedi *et al.*, 2021). Notably, metagenomic analyses have linked the growth and productivity of rice to specific strains of probiotics, such as *Paraburkholderia fungorum* and *Delftia* species, through modulations in the composition of microbial communities and diversity within the roots and soil surrounding the rhizosphere (Islam *et al.*, 2023). Earlier metagenomic research unlocked the potential for discovering novel antibiotics and enzymes, and it is now increasingly employed to identify unknown pathogens during the onset or resurgence of disease outbreaks (Hoque *et al.*, 2019, 2021). Amplicon-based sequencing provides efficient as well as cost-effective means for analysing microbial communities, which is invaluable for monitoring health and wellbeing (Islam *et al.*, 2021). Moreover, metagenomics serves a crucial role in probiotic development (Xu *et al.*, 2019).

Despite the remarkable potential of metagenomics, the analysis and interpretation of its extensive sequencing data pose significant challenges, often requiring advanced computational expertise (Vincent *et al.*, 2017). Consequently, ongoing research endeavors in metagenomics aim to illuminate the diversity, functions, and metabolic pathways of plant-associated microbes, with the ultimate goal of integrating these insights into sustainable agricultural practices (Adeleke *et al.*, 2023). This chapter explores the metagenomic analysis of plant probiotic bacteria, emphasizing its potential contributions to sustainable agriculture and the ecological services it can provide.

23.2 The Concept of Plant Probiotic Bacteria and their Interaction with Plant

Plant Probiotic Bacteria (PPB) are beneficial microbes that support plant health by effectively colonizing specific ecological niches, inducing systemic resistance (ISR), and directly antagonizing pathogens (Nwachukwu and Babalola, 2022). A notable subset of PPB is referred to as plant growth-promoting rhizobacteria (PGPR) (Hardoim *et al.*, 2015). PGPR are indigenous soil-dwelling bacteria that enhance the productivity of plants as well as bolster plant immunity. They thrive in the rhizosphere, a dynamic zone where plant roots interact with soil microorganisms, influenced by the exudates released by the roots.

PPB can be categorized based on their mode of interaction with plants: (i) free-living rhizobacteria that reside in the soil outside plant cells and promote growth through the secretion of beneficial metabolites into the rhizosphere (Gray and Smith, 2005); and (ii) endophytes that inhabit the internal tissues or cells of plants, facilitating direct metabolic exchanges with the host to enhance its growth (Gray and Smith, 2005). Hardoim *et al.* (2015) noted that most endophytes typically dwell in the intercellular spaces of the plant, although some can establish more intricate mutualistic associations by penetrating plant cells (see Table 23.1).

Examples of mutualistic associations include rhizobia, which forms symbiotic relationships with leguminous plants, enabling them to fix atmospheric nitrogen within specialized root nodules (Brewin, 1991; Suzaki and Kawaguchi, 2014). Similarly, *Frankia* forms nitrogen-fixing nodules in actinorhizic plants such as *Alnus* (Pawlowski and Demchenko, 2012), while cyanobacteria associate with cycads, further illustrating the diversity of mutualistic interactions (Vessey *et al.*, 2005; Menendez and Garcia-Fraile, 2017).

Recent advances in genetic sequencing techniques over the last two decades have enhanced the knowledge of the intricate diversity of the plant microbiome and its contribution to plant growth, development, and survival (Shi *et al.*, 2023). The microbial communities present in soil include a diverse array of beneficial organisms such as nitrogen-fixing cyanobacteria and plant growth-promoting rhizobacteria, bacteria that suppress diseases, actinomycetes, and microorganisms that degrade toxins (Singh *et al.*, 2011). For instance, *Bacillus megaterium* has been shown to inhibit growth in abscisic acid (ABA)-deficient plants (Porcel *et al.*, 2014). Conversely, common PGPB, such as different strains of *Bacilli* and *Pseudomonas*, are widely distributed in soil, playing vital roles in supporting agricultural productivity.

23.3 Underlying Mechanism of Plant Probiotics in Boosting Crop Growth

The rhizosphere is a dynamic environment teeming with millions of microorganisms that significantly influence plant growth through various mechanisms. Among these microorganisms, bacteria are the most abundant and ecologically vital. They can inhabit the soil either as free-living entities or in symbiotic relationships with plants, exhibiting both beneficial and detrimental effects on their hosts.

Beneficial bacteria, including both free-living and endophytic, significantly contribute to promoting crop growth through direct and indirect pathways. Notable genera that have been identified to have beneficial influence on plant development include *Bacillus, Burkholderia, Parabukholderia, Bradyrhizobium, Arthrobacter, Klebsiella, Rhizobium, Chromobacterium, Enterobacter, Erwinia, Flavobacterium, Micrococcus, Frankia, Kocuria, Azospirillum, Microbacterium, Lysobacter, Azotobacter Ochrobactrum, Pseudomonas,* and *Serratia*. While some of these species thrive freely within the soil, others possess endophytic interactions with plants (Sarkar *et al.*, 2018).

The interplay between PPB and plants leads to complex interactions, with root exudates serving as pivotal mediators. These exudates provide essential nutrients that attract and support microbial communities in the rhizosphere. In turn, plants employ regulatory mechanisms to facilitate the colonization of beneficial microbes, thereby enhancing their own health and productivity (Zhalnina *et al.*, 2018).

Multiple aspects of plant biology, including nutrition, growth, and overall health are profoundly influenced by microbial colonization in the rhizosphere (Shi *et al.*, 2023). Plant probiotics have been reported to boost plant growth by producing substances that promote growth and other processes like the reclamation of soil, inhibition of plant pathogens, phosphorus and nitrogen solubilization, nutrient mobilization, and the stimulation of stress tolerance and disease resistance (Fig. 23.1) (Rahman *et al.*, 2018).

Table 23.1. Mutualistic plant-probiotic bacterial interactions.

Bacterial partner	Plant partner	Interaction	Benefit to plant	Benefit to microbe	References
Stenotrophomonas rhizophila SR80	Wheat	Recruited by wheat under pathogen attack (e.g. *Fusarium pseudograminearum*) to promote disease resistance	Enhances plant growth and activates defense mechanisms against pathogens	Gains carbon-rich root exudates from the plant	Liu *et al*., 2021
Rhizobium species	Legumes (e.g. peas, beans)	*Rhizobium* forms nitrogen-fixing nodules on legume roots.	Nitrogen fixation improves growth and soil fertility.	Receives carbohydrates and shelter in plant roots	Dong *et al*., 2021
Bacillus subtilis DHK & *Rhizo SF 48*	Maize	Inoculated maize shows increased biomass and water content under drought conditions.	Enhances plant biomass and drought tolerance	Gains nutrients from root exudates and thrives in drought-prone soil	Gowtham *et al*., 2020; Sood *et al*., 2020
Frankia species	Non-leguminous woody plants (e.g. shrubs, trees)	*Frankia* forms nitrogen-fixing nodules in the roots of actinorhizal plants.	Enhanced nitrogen availability, especially in poor soils	Gains carbon compounds from plant roots	Sprent and Parsons, 2000; Van and Pawlowski, 2017
Anabaena, *Nostoc* species	Mosses, ferns, gymnosperms, angiosperms	Nitrogen-fixing cyanobacteria (e.g. *Anabaena*, *Nostoc*) establish symbiosis with various plants, performing photosynthesis and nitrogen fixation in low-oxygen environments.	Provides nitrogen for plant growth	Access to carbon compounds from plants and protection in low-oxygen conditions	Wei *et al*., 2006; Manyu and Guodong, 2019
Pseudomonas fluorescens DR7	Millet	Recruited under drought conditions to improve millet growth and increase soil-moisture retention	Increases drought tolerance and soil moisture content	Gains carbon-rich exudates from plant roots and thrives in moist soil zones	Niu *et al*., 2017

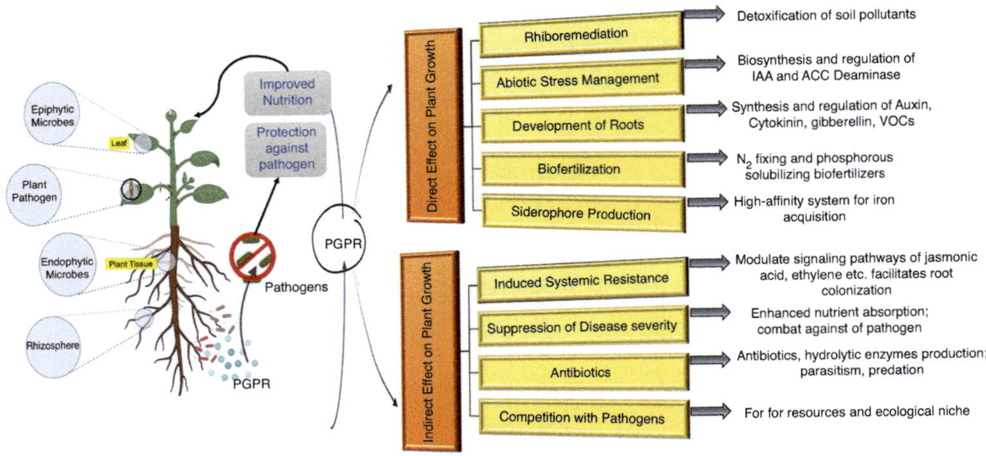

Fig. 23.1. Mechanisms of plant probiotic bacteria in growth improvement.

Research reveals that beneficial microbes enhance nitrate absorption by plant roots (Richardson *et al.*, 2009) and facilitate atmospheric N_2 fixing (Stokstad, 2016), solubilize phosphorus to improve the availability of nutrients (Oteino *et al.*, 2015), and mobilize vital nutrients (Jacoby *et al.*, 2017). In addition, these microbes assist in the rehabilitation of toxic heavy-metal-contaminated soils (Ma *et al.*, 2016) and contribute to overall soil health through diverse activities (Adesemoye *et al.*, 2009). Furthermore, PGPB can induce plant immune responses, boosting resistance to diseases (Chung *et al.*, 2016; Rahman *et al.*, 2018). They are known to release hormones, vitamins, and growth-regulating compounds (Glick, 2014), enhance drought tolerance and stress tolerance due to salinity (Shrivastava and Kumar, 2015), and suppress harmful pathogens and diseases (Saraf *et al.*, 2014; Islam *et al.*, 2016c).

Overall, by improving nutrient availability, enhancing stress resilience, and providing protection against pathogens, PGPB ultimately contribute to increased plant growth, yield, and agricultural productivity (Majeed *et al.*, 2018).

23.4 Role of Plant Probiotic Bacteria in Stress Management and Crop Improvement

Growth, development, and yield of crops are significantly influenced by a number of abiotic factors including poor soil quality, inadequate nutrient availability, heavy-metal contamination, drought, and salinity. The physicochemical properties and nutrient composition of soil are essential for optimal plant development. Multiple investigations have reported the capacity of crops to tolerate abiotic stresses such as extreme temperature or cold, drought, high saline concentration, and heavy metal exposure, while simultaneously improving nutrient absorption and providing defense against a range of parasitic plant pathogens. For example, maize (Touceda-González *et al.*, 2015), rice (*Oryza sativa*) (Sarkar *et al.*, 2018; Verma *et al.*, 2018), wheat (*Triticum aestivum*) (Raheem *et al.*, 2018; Surovy *et al.*, 2024), chickpea (Abd_Allah *et al.*, 2018), potato (*Solanum tuberosum*) (Gururani *et al.*, 2013), and common bean (Remans *et al.*, 2008) crops showed positive effect against such stresses.

Drought, salinity, and heavy-metal stress rank among the major abiotic challenges affecting crops worldwide. Furthermore, biotic stresses due to pathogen exposure – including bacteria, viruses, fungi, and nematodes – pose additional constraints on crop productivity. Current management strategies often rely on high fertilizer application to address nutrient depletion and widespread pesticide use for disease control. Although these practices can boost crop yields, they may also lead to environmental degradation, ecosystem disruption, and health risks.

Plant growth-promoting probiotic bacteria (PGPB) offer a promising alternative for mitigating abiotic stresses and enhancing crop yields, potentially reducing dependence on synthetic fertilizers. Certain PGPB species are recognized for their abilities to fix nitrogen, solubilize phosphate, remediate heavy-metal contamination, and improve crop resilience under drought and saline conditions (refer to Table 23.2).

Endophytic and symbiotic PGPB, such as *Azorhizobium*, *Mesorhizobium*, *Rhizobium*, and *Bradyrhizobium* play vital roles in leguminous plants, while species like *Azospirillum*, *Burkholderia*, *Gluconacetobacter*, and *Herbaspirillum* are influential in non-leguminous crops. Non-symbiotic bacteria of rhizospheric origin, including *Clostridium*, *Azotobacter*, and *Azospirillum*, are also essential for nitrogen fixation and overall plant health (MacLean *et al.*, 2007).

Additionally, plant growth-promoting bacteria (PGPB) species such as *Clostridium*, *Bacillus pumilus*, *Burkholderia phytofirmans*, and *Mesorhizobium* are also critical for nitrogen fixation and overall plant health. *Pseudomonas reactans*, *P. libanensis*, and several *Azotobacter* species have shown promise in the bioremediation of heavy-metal-contaminated soils (Ma *et al.*, 2016; Sobariu *et al.*, 2017; Martins *et al.*, 2018). Other species, including *Bacillus amyloliquefaciens*, *Arthrobacter*, *Brevibacterium*, *Streptomyces*, *Azospirillum brasilense*, *Serratia marcescens*, *Pseudomonas fragi*, and *Sinorhizobium meliloti*, contribute to solubilize phosphate and support the growth of an expanded range of crops (Arif *et al.*, 2017; García *et al.*, 2017; Zheng *et al.*, 2017).

Research has demonstrated that PGPB can enhance seed germination, growth performance and yield, while also improving dry biomass, nutrient absorption, enhanced tolerance to abiotic stress, and safeguarding against pathogens in major crops such as rice, maize, wheat, maize, potato, chickpea, and common bean (Majeed *et al.*, 2018). The integration of PGPB into agricultural practices represents a sustainable strategy to mitigate the challenges posed by abiotic and biotic stresses, ultimately leading to increased food security and environmental preservation (Masenya *et al.*, 2024).

23.5 Metagenomics to Understand Microbial Community

Metagenomics is a pivotal tool in microbial ecology and genomics allowing researchers to explore microbial compositions of a given ecosystem by directly using environmental samples, bypassing the requirement of isolation or culture on media. Metagenomics, based on sequencing followed by DNA extraction of those samples, offers a holistic view of microbial genetic diversity and functional potential, significantly advancing our understanding of microbial interactions within various environments, including agricultural ecosystems (Fig. 23.2) (Pérez-Cobas *et al.*, 2020).

Recent advancements in technologies, such as amplicon-based next-generation sequencing and shotgun metagenomics, have expanded our capacity to profile the microbial composition of samples comprehensively (Jansson and Hofmockel, 2018). Shotgun metagenomics, which sequences all the DNA present in a sample, permits an unbiased exploration of entire genomes of microbes, facilitating the discovery of novel genes and organisms. This approach has become instrumental in analysing microbial diversity and functional roles in environments like soil and plant roots (Zhang *et al.*, 2021). Illumina platforms, including MiSeq and HiSeq, and nanopore (e.g. Oxford nanopore technology), are frequently employed in these studies because of their capability to generate substantial volumes of data. For instance, research using the Illumina HiSeq has uncovered microbial diversity in agricultural soils, pinpointing key microbial phyla which include Actinobacteria, Firmicutes and Proteobacteria (Wang *et al.*, 2019; Babalola *et al.*, 2020).

In addition to Illumina sequencing, emerging third-generation sequencing technologies like MinION and PacBio are gaining prominence for their ability to produce ultra-long reads and enable faster sequencing times. These advancements provide deeper insights into bacterial and archaeal populations, particularly through full-length 16S rRNA gene sequences (Goodwin *et al.*, 2016; Johnson *et al.*, 2019).

The processing of metagenomics data relies heavily on bioinformatics tools. Quality control involves preprocessing tools such as MG-RAST,

Table 23.2. Contributions of various plant probiotic bacteria in enhancing soil quality, nutrient uptake, stimulating crop growth, and managing stress.

Plant probiotic bacteria	Crops	Role in boosting crop growth	Reference
Arthrobacter protophormiae, B. subtilis, Dietzia natronolimnaea	Wheat (*T. aestivum*)	Improved tolerance to salinity and drought, as well as better growth and biomass	Barnawal *et al.*, 2017
Azospirillum brasilense, Herbaspirillum sp., *Seropedicae* sp.	Maize (*Z. mays*)	Enhanced nitrogen acquisition, growth, and grain yield	Martins *et al.*, 2018
Azospirillum spp.	Maize (*Z. mays*)	Promotes drought resistance and improves growth performance	Menendez and Garcia-Fraile, 2017
Bacillus amyloliquefaciens	Rice (*O. sativa*)	Enhanced tolerance to salinity stress and improved growth traits	Shahzad *et al.*, 2017
B. amyloliquefaciens	Mustard (*Brassica juncea*) Tomato (*Lycopersicon esculentum*) Radish (*Raphanus sativus*)	Improved tolerance to salinity elevated solubilization of phosphorus, nutrient uptake, improved growth, and overall physiological performance	Kim *et al.*, 2017
B. phytofirmans	Maize (*Zea mays*)	Mitigates metal stress (Zn, Pb) and supports crop growth	Touceda-González *et al.*, 2015
B. subtilis	Potato (*Solanum tuberosum*)	IAA and cytokinin production, increases zeatin-riboside content in shoots and the mass of roots	Sorokan *et al.*, 2021
B. subtilis (BTS-3) and B. velezensis	Wheat (*T. aestivum*)	Suppression of fungal pathogens (*Magnaporthe oryzae Triticum)*	Surovy *et al.*, 2024
Bacillus sp., *Achromobacter xylosoxidans*	Chickpea (*Cicer arietinum*)	Improved tolerance to salinity and fungal diseases, increased nutrient uptake	Egamberdieva *et al.*, 2017
Bacillus sp., Enterobacter sp., *Moraxella* sp., *Pseudomonas* sp.	Wheat (*T. aestivum*)	Increased drought tolerance, enhanced vegetative growth, and improved yield	Raheem *et al.*, 2018
Bacillus subtilis	Chickpea (*C. arietinum*)	Tolerance to salinity stress, improved nutrient absorption, and increased growth	Abd_Allah *et al.*, 2018
Burkholderia cereus	Mungbean (*Vigna radiate*)	Plant growth and yield enhancement, enhanced tolerance to salt, greater mineral content, and better dry biomass	Islam *et al.*, 2016a, b
Enterobacter sp.	Rice (*O. sativa*)	Phosphorus solubilization, salt tolerance, enhanced germination, and growth stimulation	Sarkar *et al.*, 2018

Continued

Table 23.2. Continued

Plant probiotic bacteria	Crops	Role in boosting crop growth	Reference
Pseudomonas. aurantiaca, P. extremorientalis	Wheat (*T. aestivum*)	Improved germination, seedling development, root elongation, and salinity tolerance	Egamberdieva, 2009
P. fluorescens, Azospirillum brasilense	Wheat (*T. aestivum*)	Increased phosphorus solubilization, tolerance to salinity, and improvements in plant height and dry weight	Kadmiri *et al.*, 2018
Pantoea vagans, Paenibacillus sp., *P. baetica*	Rice (*O. sativa*)	Protects from phytopathogens like *Curvularia* sp., *Alternaria* sp., and *Fusarium* oxysporum), with increased root hair development and improved growth of seedlings	Verma *et al.*, 2018
Paraburkholderia fungorum; Delftia sp.	Rice (*O. sativa*)	Indole-3-acetic acid (IAA) biosynthesis, atmospheric nitrogen fixation; essential mineral, i.e. phosphorus and potassium solubilization	Islam *et al.*, 2023
Planomicrobium chinense, B. cereus, P. fluorescens	Wheat (*T. aestivum*)	Enhanced drought resistance, relative water content, phosphate solubility; improved nutrient uptake, dry biomass as well as growth; suppression of plant pathogens	Khan *et al.*, 2018
Pseudomonas libanensis, P. reactans	Smoothstem turnip (*Brassica oxyrrhina*)	Boosts tolerance to salinity and drought, as well as heavy-metal stress, and enhances leaf pigmentation, ultimately improves plant growth	Ma *et al.*, 2016

UCHIME, and Trimmomatic (Bolger *et al.*, 2014). Data analysis typically utilizes platforms like MOTHUR and QIIME 2, which facilitate the removal of chimeras and the grouping of reads by barcodes (Santamaria *et al.*, 2018; Almeida and De Martinis, 2019). To classify reads into taxonomic and functional categories, software such as MEGAHIT and Meta-IDBA is employed, while tools like SqueezeMeta and DAS Tool assist in the binning and separation of assembled contigs (Sieber *et al.*, 2018; Almeida and De Martinis, 2019; Tamames and Puente-Sánchez, 2019).

De novo assemblers, counting Meta-IDBA, MetaVelvet, and Contig Extender, are utilized for sequencing novel microbial genomes when no reference metagenome is available for alignment (Kumar *et al.*, 2018). The size and continuity of the sequence gaps determined the coverage quality of data, which influences how metagenome sequence reads are assembled into clusters. However, it is particularly essential to note that *de novo* assembly poses computational challenges, often classified as NP-Hard, thus making effective resolution difficult. Challenges associated with *de novo* assemblers include high memory usage and extended assembly times, particularly for paired-end or joined datasets compared to single-end datasets, as illustrated by tools like ABySS (Khan *et al.*, 2018).

For functional annotation and gene prediction, tools such as Metagene, MetaGene Mark, MG-RAST, and IMG/M homology-based databases (e.g. Uniprot, Pfam, SEED) are commonly employed (Almeida and De Martinis, 2019). Overall, metagenomics has emerged as an invaluable approach for unveiling microbial diversity and functionality, fostering advancements in agriculture, medicine, and environmental science. The integration of metagenomics with

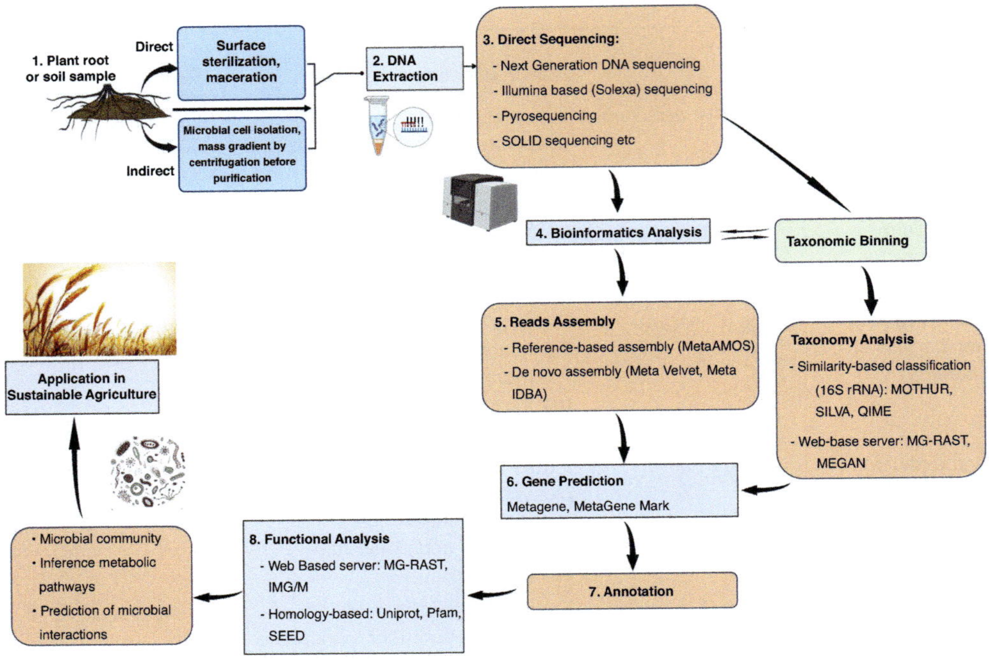

Fig. 23.2. Major steps of metagenomics analysis.

other omics strategies, such as transcriptomics and proteomics, holds tremendous potential for deepening our understanding of complex microbial communities (Trivedi *et al.*, 2021).

23.6 Application of Metagenomics in the Study of Plant Probiotic Bacteria

Metagenomic approaches have revolutionized our understanding of previously unculturable bacteria, particularly those with traits that promote plant growth. This cutting-edge approach enhances crop productivity by revealing new genes, metabolites, and previously unidentified microorganisms that may benefit plants (Leveau, 2007). For instance, an analysis of the taxonomic and functional diversity of microbes in both bulk soil and the rhizosphere of soybean revealed that microbial assembly is driven by niche-based processes, which are significantly influenced by specific plant and environmental factors (Mendes *et al.*, 2014).

Advancements in metagenomic approaches like direct extraction and amplification of bacterial DNA from plant tissues, coupled with sophisticated bioinformatics tools, have deepened the knowledge of bacterial community structure and phylogeny (Priya *et al.*, 2021). For example, an investigation of rice roots using metagenomic techniques showed the dominance of γ-proteobacteria, especially enterobacter-related endophytes, in addition to smaller quantities of V- and IV- proteobacteria (Sun *et al.*, 2008; Priya *et al.*, 2021). Similarly, Tsurumaru *et al.* (2015) reported that Actinobacteria, Alphaproteobacteria, and Betaproteobacteria predominantly compose the bacterial community of the sugar beet (*Beta vulgaris* L.) taproot. A study on the microbiome of sorghum roots and stems using metagenomics showed distinct differences in bacterial composition between these tissues; however, both shared dominance by agriculturally significant genera including *Agrobacterium*, *Sphingobacterium*, *Microbacterium*, *Herbaspirillum*, *Pseudomonas*, *Stenotrophomonas*, and *Erwinia* (Maropola *et al.*, 2015).

Screening of functional attributes of microbial communities present in rhizosphere zones in stressful conditions has led to the discovery of novel metabolic pathways and the bacterial metabolites derived from them (Sasse *et al.*, 2018). Metagenomic analysis of sugar beet taproots identified genes that are responsible for plant growth promotion such as phosphate solubilization, siderophore production, nitrogen fixation, and ACC deaminase activity (Tsurumaru *et al.*, 2015). Notably, genes for β−1,3-glucanase, known for its disease-suppressive properties, were frequently detected, along with those growth-promoting genes. However, indole-3-acetic acid (IAA)-producing genes were largely absent, indicating they may not play any role in the growth and development of sugar beet. A metagenomic study of the rhizospheric soil of rice (*O. sativa*) revealed the presence of a microbial community that helps in atmospheric nitrogen fixation, indole-3-acetic acid (IAA) production, and phosphorus and potassium solubilization (Islam *et al.*, 2023). Such functional analyses provide critical insights for optimizing the management strategies of crops.

Endophytic bacteria play significant roles in plant health by producing antibacterial and antifungal compounds. The abundance of these bacteria suggests a complex interplay within the plant endosphere, where metabolic signaling occurs. The diffusible signal factor system, the N-acyl homoserine lactone (AHL) system, and the autoinducer-2 system are three types of quorum-sensing systems found in rice microbiome (Sessitsch *et al.*, 2012; Miliute *et al.*, 2015).

Furthermore, metagenomic libraries have revealed novel β-glucosidases, crucial for organic matter degradation and nutrient cycling, by screening DNA from winter wheat topsoil (Biver *et al.*, 2014). Tan *et al.* (2014) similarly employed a function-based screening technique to identify two new phytases from wheat soil, one of which belonged to a previously uncharacterized group. Metagenomic library screenings help to document the antibiotic properties of microbial communities present in the rhizosphere, with several studies utilizing pathogen indicators like *Xanthomonas*, *Erwinia*, *Rhizoctonia*, and *Fusarium* (Kim *et al.*, 2006; Priya *et al.*, 2021). Techniques such as high-performance liquid chromatography (HPLC) and colorimetric assays have been employed to screen for IAA production and nitrogen-fixing enzymes, while immunoaffinity chromatography has aided in cytokine detection (Radwan *et al.*, 2002; Leveau and Lindow, 2005; Priya *et al.*, 2021). Additionally, metagenomics has played a critical role in identifying biocontrol agents like chitinase-producing clones and biosurfactant-producing microbes derived from heavy-metal-contaminated soils (Leveau and Preston, 2008; Sachdev and Cameotra, 2013).

Environmental concerns regarding pesticide pollution have spurred metagenomic investigations into rhizoremediation strategies. Radl *et al.* (2019) employed shotgun metagenomics to analyze taxonomic and functional changes in the rhizosphere soil of apple trees affected by replanting disease. Their findings revealed a significant reduction in Actinobacteria, which are vital for organic matter decomposition, juxtaposed with an upregulation of stress-sensing genes in the diseased soils. In contrast, healthy soils exhibited an abundance of genes related to nutrient uptake and sensing, contributing to a clearer understanding of the mechanisms underlying replanting disease.

Moreover, metagenomics has extended its utility to uncovering microbial communities in seeds. Research by Truyens *et al.* (2015) has shown that microbial communities present in the seed endosphere and spermosphere can enhance the germination capacity of seeds along with plant growth, particularly under stress conditions. Epiphytic fungi like *Fusarium* and *Penicillium* have been observed to promote seed germination in tropical climates (Tamura *et al.*, 2008; Priya *et al.*, 2021).

Finally, metagenomic studies have provided insights into antibiotic resistance gene (ARG) profiles across various environments. Li *et al.* (2015) identified resistance genes for aminoglycosides, bacitracin, β-lactams, chloramphenicol, macrolides, quinolones, sulfonamides, and tetracyclines, which are prevalent in medical and veterinary applications. Investigations into the interplay between plant systems and microbial communities in extreme and diverse environments have elucidated adaptive strategies to mitigate abiotic stress. For instance, Fonseca *et al.* (2018) examined soil samples from the Amazon rainforest, revealing that nitrogen and carbon cycling are critical processes within the rhizosphere, underscoring their significance in maintaining soil health. In a related study, Hu *et al.* (2020) established that geological factors significantly influence the composition and diversity of bacterial communities present in soil on the Tibetan Plateau, accounting for 67.9% of variation in plant communities and 35.9% in bacterial communities, thus enhancing our understanding of soil ecosystem functions.

In addition to environmental stress, crop productivity is significantly affected by replanting disease, which is characterized by reduced yields when fields are replanted with the same or related fruit and nut crops. Although this phenomenon is widespread globally, its underlying causes remain poorly understood.

Radl *et al.* (2019) conducted a pioneering study using shotgun metagenomics to investigate the taxonomic as well as functional shifts in the soil present in rhizosphere soil associated with replanting of apples. Their findings revealed a significant decrease in Actinobacteria, which are essential for organic matter decomposition, in soils affected by replanting disease compared to control samples. In the rhizospheres impacted by this disease, genes associated with stress sensing were more actively expressed, while genes involved in sensing and nutrient uptake were upregulated in the control soil. Moreover, genes for aromatic compound degradation were less abundant in the affected soil, causing phenolic compounds to accumulate (Priya *et al.*, 2021).

23.7 Prospects, Challenges, and Future Directions

The declining cost of metagenome sequencing has led to a growing number of studies employing a combination of metabarcoding and metagenomic techniques. This progress has significantly enhanced our understanding of previously unexplored soil microorganisms, bolstering crop productivity and promoting sustainable agriculture through improved phytopathogen resistance and nutrient cycling. Nevertheless, several challenges impede the full potential of metagenomic studies in plant microbiomes.

One prominent issue is the over-representation of abundant microbial species in datasets, which can mask the presence and significance of rare species. Furthermore, distinguishing microbial DNA from the substantial quantities of host DNA in plant samples remains a significant challenge in shotgun metagenomics (Adeleke *et al.*, 2023). Data pooling presents another hurdle, characterizing the complete metagenomes of community members individually and linking the diversity across various environments. However, advances in bioinformatics tools and shotgun metagenomics methodologies are in development to address these challenges (Nwachukwu and Babalola, 2022).

Sample contamination is a common but often under-reported issue, typically representing a small fraction of overall reads. Preprocessing steps, such as PCR amplification, can introduce duplicate reads that necessitate careful filtering. Additionally, uneven genome coverage can complicate analyses, as some genomic regions may be well sequenced while others are not. Continuous improvements in sequencing technologies are gradually addressing these limitations, particularly regarding the accuracy and duplicate reads (Massart *et al.*, 2014).

Interpreting amplicon data can pose difficulties due to its relatively low resolution, resulting in considerable variations in phenotype and genomic content, even among members of the same taxonomic group. To mitigate this, community data from endophytic analyses are often processed prior to chimera removal and denoising, which streamlines interpretation (Fadiji and Babalola, 2020). The *de novo* assembly of metagenomic data remains computationally intensive, demanding substantial memory and processing power (Khan *et al.*, 2018). Meanwhile, the high costs associated with sequencing limit the ability to conduct replicates in metagenomic studies, complicating result validation.

The rapid advancement of sequencing technologies has resulted in an influx of metagenomic data, raising concerns about data storage. Developing a standardized data storage platform potentially utilizing cloud computing will be crucial for managing the extensive metadata associated with plant-associated microbial communities (Jünemann *et al.*, 2017). While computational methods have made strides, the tasks of binning and assembling large datasets continue to challenge researchers (Dröge and

McHardy, 2012). There is an urgent need for robust computational techniques that leverage supercomputing capabilities to accelerate microbial genomic data analysis. Recent advancements are improving the management and analysis of large sequence libraries, with new bioinformatics tools facilitating the efficient processing of extensive databases and expedited genome reconstruction. Affordable gene chips are now capable of distinguishing between active and inactive gene clones, yet further innovation is needed in functional analysis, particularly concerning heterologous gene expression and large library screening (Handelsman, 2004).

Lastly, single-cell sequencing offers notable advantages over shotgun sequencing by working with smaller datasets derived from individual cells, thus simplifying the assembly process (Laudadio *et al.*, 2019)

23.8 Conclusion

The diversity of probiotic bacteria and their intricate interactions with plants present significant implications for growth, homeostasis, and survival, holding great promise for sustainable agriculture. Conventional culture-dependent approaches are severely limited, often identifying only about 1% of the total microbial diversity present in any ecosystem. Consequently, the adoption of omics technologies and metagenomics is critical for discovering yet-to-be-cultured probiotics. These methodologies offer profound insights into the structural, functional, and metabolic pathways of communities of microbes, advancing our understanding of their phylogenetic relationships and taxonomic roles. NGS, including Illumina as well as nanopore technologies, has redefined metagenomics by facilitating exhaustive analyses of microbial diversity and abundance that transcend the capabilities of culture-based techniques. Shotgun metagenomics, in particular, has empowered researchers to delineate the genetic composition of microbial communities, comparing vast amounts of sequence data from uncultivated microbes. Recent advancements in techniques are linking the functional potential of microbial endophyte communities to their metabolic capabilities. Overall, these innovations in metagenomics are paving the way for the discovery of novel genetic traits in probiotic bacteria, which could be harnessed to enhance sustainable agricultural practices and tackle a variety of environmental challenges.

References

Abd_Allah, E.F., Alqarawi, A.A., Hashem, A., Radhakrishnan, R., Al-Huqail, A.A. *et al.* (2018) Endophytic bacterium *Bacillus subtilis* (BERA 71) improves salt tolerance in chickpea plants by regulating the plant defense mechanisms. *Journal of Plant Interactions* 13(1), 37–44. DOI: 10.1080/17429145.2017.1414321.

Adeleke, B.S., Muller, D. and Babalola, O.O. (2023) A metagenomic lens into endosphere microbial communities, promises, and discoveries. *Letters in Applied Microbiology* 76(2), vac030. DOI: 10.1093/lambio/ovac030.

Adesemoye, A.O., Torbert, H.A. and Kloepper, J.W. (2009) Plant growth-promoting rhizobacteria allow reduced application rates of chemical fertilizers. *Microbial Ecology* 58(4), 921–929. DOI: 10.1007/s00248-009-9531-y.

Almeida, O.G. and De Martinis, E.C.P. (2019) Bioinformatics tools to assess metagenomic data for applied microbiology. *Applied Microbiology and Biotechnology* 103(1), 69–82. DOI: 10.1007/s00253-018-9464-9.

Arif, M.S., Shahzad, S.M., Yasmeen, T., Riaz, M., Ashraf, M. *et al.* (2017) Improving plant phosphorus (p) acquisition by phosphate-solubilizing bacteria. In: Naeem, M. and Gill, S.S. (eds) *Essential Plant Nutrients: Uptake, Use Efficiency, and Management*. Springer, pp. 513–556. DOI: 10.1007/978-3-319-58841-4_21.

Babalola, O.O., Alawiye, T.T., Lopez, C.R. and Ayangbenro, A.S. (2020) Shotgun metagenomic sequencing data of sunflower rhizosphere microbial community in south africa. *Data in Brief* 31, 105831. DOI: 10.1016/j.dib.2020.105831.

Barnawal, D., Bharti, N., Pandey, S.S., Pandey, A., Chanotiya, C.S. *et al.* (2017) Plant growth-promoting rhizobacteria enhance wheat salt and drought stress tolerance by altering endogenous phytohormone levels and TaCTR1/TaDREB2 expression. *Physiologia Plantarum* 161(4), 502–514. DOI: 10.1111/ppl.12614.

Biver, S., Stroobants, A., Portetelle, D. and Vandenbol, M. (2014) Two promising alkaline β-glucosidases isolated by functional metagenomics from agricultural soil, including one showing high tolerance towards harsh detergents, oxidants and glucose. *Journal of Industrial Microbiology & Biotechnology* 41(3), 479–488. DOI: 10.1007/s10295-014-1400-0.

Bolger, A.M., Lohse, M. and Usadel, B. (2014) Trimmomatic: a flexible trimmer for illumina sequence data. *Bioinformatics* 30(15), 2114–2120. DOI: 10.1093/bioinformatics/btu170.

Brader, G., Compant, S., Vescio, K., Mitter, B., Trognitz, F. *et al.* (2017) Ecology and genomic insights into plant-pathogenic and plant-nonpathogenic endophytes. *Annual Review of Phytopathology* 55(1), 61–83. DOI: 10.1146/annurev-phyto-080516-035641.

Brewin, N.J. (1991) Development of the legume root nodule. *Annual Review of Cell Biology* 7(1), 191–226. DOI: 10.1146/annurev.cb.07.110191.001203.

Chung, J.H., Song, G.C. and Ryu, C.M. (2016) Sweet scents from good bacteria: Case studies on bacterial volatile compounds for plant growth and immunity. *Plant Molecular Biology* 90, 677–687. DOI: 10.1007/s11103-015-0344-8.

Dheeman, S., Patwal, S., Dobhal, S., Kaur, H. and Rana, D. (2024) Soil bacterial consortia in *rhizobium*–legume interactions. *Soil Bacteria: Biofertilization and Soil Health* 375–390. DOI: 10.1007/978-981-97-3473-3_14.

Dong, W., Zhu, Y., Chang, H., Wang, C., Yang, J, *et al.* (2021) An SHR–SCR module specifies legume cortical cell fate to enable nodulation. *Nature* 589(7843), 586–590. DOI: 10.1038/s41586-020-3016-z.

Dröge, J. and McHardy, A.C. (2012) Taxonomic binning of metagenome samples generated by next-generation sequencing technologies. *Briefings in Bioinformatics* 13(6), 646–655. DOI: 10.1093/bib/bbs031.

Egamberdieva, D. (2009) Alleviation of salt stress by plant growth regulators and IAA producing bacteria in wheat. *Acta Physiologiae Plantarum* 31(4), 861–864. DOI: 10.1007/s11738-009-0297-0.

Egamberdieva, D., Wirth, S.J., Shurigin, V.V., Hashem, A., and Abd Allah, E.F. (2017) Endophytic bacteria improve plant growth, symbiotic performance of chickpea (*Cicer arietinum* L.) and induce suppression of root rot caused by *Fusarium solani* under salt stress. *Frontiers in Microbiology* 8, 1887. DOI: 10.3389/fmicb.2017.01887.

Fadiji, A.E. and Babalola, O.O. (2020) Metagenomics methods for the study of plant-associated microbial communities: a review. *Journal of Microbiological Methods* 170, 105860. DOI: 10.1016/j.mimet.2020.105860.

Fonseca, J.P., Hoffmann, L., Cabral, B.C.A., Dias, V.H.G., Miranda, M.R. *et al.* (2018) Contrasting the microbiomes from forest rhizosphere and deeper bulk soil from an Amazon rainforest reserve. *Gene* 642, 389–397. DOI: 10.1016/j.gene.2017.11.039.

French, E., Kaplan, I., Iyer-Pascuzzi, A., Nakatsu, C.H., and Enders, L. (2021) Emerging strategies for precision microbiome management in diverse agroecosystems. *Nature Plants* 7(3), 256–267. DOI: 10.1038/s41477-020-00830-9.

García, J.E., Maroniche, G., Creus, C., Suárez-Rodríguez, R., Ramirez-Trujillo, J.A. *et al.* (2017) *In vitro* PGPR properties and osmotic tolerance of different *Azospirillum* native strains and their effects on growth of maize under drought stress. *Microbiological Research* 202, 21–29. DOI: 10.1016/j.micres.2017.04.007.

Glick, B.R. (2014) Bacteria with ACC deaminase can promote plant growth and help to feed the world. *Microbiological Research* 169(1), 30–39. DOI: 10.1016/j.micres.2013.09.009.

Goodwin, S., McPherson, J.D. and McCombie, W.R. (2016) Coming of age: Ten years of next-generation sequencing technologies. *Nature Reviews Genetics* 17(6), 333–351. DOI: 10.1038/nrg.2016.49.

Gouda, S., Kerry, R.G., Das, G., Paramithiotis, S., Shin, H.S. *et al.* (2018) Revitalization of plant growth promoting rhizobacteria for sustainable development in agriculture. *Microbiological Research* 206, 131–140. DOI: 10.1016/j.micres.2017.08.016.

Gowtham, H.G., Singh, B., Murali, M., Shilpa, N., Prasad, M. *et al.* (2020) Induction of drought tolerance in tomato upon the application of ACC deaminase producing plant growth promoting rhizobacterium *Bacillus subtilis* Rhizo SF 48. *Microbiological Research* 234, 126422. DOI: 10.1016/j.micres.2020.126422.

Gray, E.J. and Smith, D.L. (2005) Intracellular and extracellular PGPR: commonalities and distinctions in the plant–bacterium signaling processes. *Soil Biology and Biochemistry* 37(3), 395–412. DOI: 10.1016/j.soilbio.2004.08.030.

Gururani, M.A., Upadhyaya, C.P., Baskar, V., Venkatesh, J., Nookaraju, A. *et al.* (2013) Plant growth-promoting rhizobacteria enhance abiotic stress tolerance in *Solanum tuberosum* through inducing changes in the expression of ROS-scavenging enzymes and improved photosynthetic performance. *Journal of Plant Growth Regulation* 32(2), 245–258. DOI: 10.1007/s00344-012-9292-6.

Handelsman, J. (2004) Metagenomics: application of genomics to uncultured microorganisms. *Microbiology and Molecular Biology Reviews* 68(4), 669–685. DOI: 10.1128/MMBR.68.4.669-685.2004.

Hardoim, P.R., van Overbeek, L.S., Berg, G., Pirttilä, A.M., Compant, S, *et al.* (2015) The hidden world within plants: ecological and evolutionary considerations for defining functioning of microbial endophytes. *Microbiology and Molecular Biology Reviews* 79(3), 293–320. DOI: 10.1128/MMBR.00050-14.

Hoque, M.N., Istiaq, A., Clement, R.A., Sultana, M., Crandall, K.A. *et al.* (2019) Metagenomic deep sequencing reveals association of microbiome signature with functional biases in bovine mastitis. *Scientific Reports* 9(1), 13536. DOI: 10.1038/s41598-019-49468-4.

Hoque, M.N., Sarkar, M.M.H., Rahman, M.S., Akter, S., Banu, T.A, *et al.* (2021) SARS-cov-2 infection reduces human nasopharyngeal commensal microbiome with inclusion of pathobionts. *Scientific Reports* 11(1), 24042. DOI: 10.1038/s41598-021-03245-4.

Hoque, M.N., Rahman, M.S., Islam, T., Sultana, M., Crandall, K.A. *et al.* (2022) Induction of mastitis by cow-to-mouse fecal and milk microbiota transplantation causes microbiome dysbiosis and genomic functional perturbation in mice. *Animal Microbiome* 4(1), 43. DOI: 10.1186/s42523-022-00193-w.

Hu, Y., Wang, S., Niu, B., Chen, Q., Wang, J. *et al.* (2020) Effect of increasing precipitation and warming on microbial community in *tibetan* alpine steppe. *Environmental Research* 189, 109917. DOI: 10.1016/j.envres.2020.109917.

Islam, F., Yasmeen, T., Arif, M.S., Ali, S., Ali, B. *et al.* (2016a) Plant growth promoting bacteria confer salt tolerance in *vigna* radiata by up-regulating antioxidant defense and biological soil fertility. *Plant Growth Regulation* 80(1), 23–36. DOI: 10.1007/s10725-015-0142-y.

Islam, F., Yasmeen, T., Arif, M.S., Riaz, M., Shahzad, S.M. *et al.* (2016b) Combined ability of chromium (Cr) tolerant plant growth promoting bacteria (PGPB) and salicylic acid (SA) in attenuation of chromium stress in maize plants. *Plant Physiology and Biochemistry* 108, 456–467. DOI: 10.1016/j.plaphy.2016.08.014.

Islam, M.T., Croll, D., Gladieux, P., Soanes, D.M., Persoons, A. *et al.* (2016c) Emergence of wheat blast in Bangladesh was caused by a South American lineage of *Magnaporthe oryzae*. *BMC Biology* 14(1), 84. DOI: 10.1186/s12915-016-0309-7.

Islam, M.T., Gupta, D.R., Hossain, A., Roy, K.K., He, X. *et al.* (2020) Wheat blast: A new threat to food security. *Phytopathology Research* 2, 1–13. DOI: 10.1186/s42483-020-00067-6.

Islam, S.R., Tanzina, A.Y., Foysal, M.J., Hoque, M.N., Rumi, M.H. *et al.* (2021) Insights into the nutritional properties and microbiome diversity in sweet and sour yogurt manufactured in Bangladesh. *Scientific Reports* 11(1), 22667. DOI: 10.1038/s41598-021-01852-9.

Islam, T., Fatema, H.M.N., Gupta, D.R., Mahmud, N.U., Sakif, T.I. *et al.* (2023) Improvement of growth, yield and associated bacteriome of rice by the application of probiotic Paraburkholderia and Delftia. *Frontiers in Microbiology* 14, 1212505. DOI: 10.3389/fmicb.2023.1212505.

Jacoby, R., Peukert, M., Succurro, A., Koprivova, A. and Kopriva, S. (2017) The role of soil microorganisms in plant mineral nutrition-current knowledge and future directions. *Frontiers in Plant Science* 8, 1617. DOI: 10.3389/fpls.2017.01617.

Jansson, J.K. and Hofmockel, K.S. (2018) The soil microbiome – from metagenomics to metaphenomics. *Current Opinion in Microbiology* 43, 162–168. DOI: 10.1016/j.mib.2018.01.013.

Johnson, J.S., Spakowicz, D.J., Hong, B.-Y., Petersen, L.M., Demkowicz, P. *et al.* (2019) Evaluation of 16S rRNA gene sequencing for species and strain-level microbiome analysis. *Nature Communications* 10(1), 5029. DOI: 10.1038/s41467-019-13036-1.

Jünemann, S., Kleinbölting, N., Jaenicke, S., Henke, C., Hassa, J. *et al.* (2017) Bioinformatics for NGS-based metagenomics and the application to biogas research. *Journal of Biotechnology* 261, 10–23. DOI: 10.1016/j.jbiotec.2017.08.012.

Kadmiri, I.M., Chaouqui, L., Azaroual, S.E., Sijilmassi, B., Yaakoubi, K, *et al.* (2018) Phosphate-solubilizing and auxin-producing rhizobacteria promote plant growth under saline conditions. *Arabian Journal for Science and Engineering* 43(7), 3403–3415. DOI: 10.1007/s13369-017-3042-9.

Khan, A.R., Pervez, M.T., Babar, M.E., Naveed, N. and Shoaib, M. (2018) A comprehensive study of de novo genome assemblers: current challenges and future prospective. *Evolutionary Bioinformatics Online* 14, 1176934318758650. DOI: 10.1177/1176934318758650.

Kim, D.S., Brooks, D.J. and Riedl, H. (2006) Lethal and sublethal effects of abamectin, spinosad, methoxyfenozide and acetamiprid on the predaceous plant bug *Deraeocoris brevis* in the laboratory. *Biocontrol* 51(4), 465–484. DOI: 10.1007/s10526-005-1028-0.

Kim, M.-J., Radhakrishnan, R., Kang, S.-M., You, Y.-H., Jeong, E.-J. *et al.* (2017) Plant growth promoting effect of *Bacillus amyloliquefaciens* h-2-5 on crop plants and influence on physiological changes in soybean under soil salinity. *Physiology and Molecular Biology of Plants* 23(3), 571–580. DOI: 10.1007/s12298-017-0449-4.

Kumar, S., Stecher, G., Li, M., Knyaz, C. and Tamura, K. (2018) MEGA x: molecular evolutionary genetics analysis across computing platforms. *Molecular Biology and Evolution* 35(6), 1547–1549. DOI: 10.1093/molbev/msy096.

Laudadio, I., Fulci, V., Stronati, L. and Carissimi, C. (2019) Next-generation metagenomics: methodological challenges and opportunities. *Omics: A Journal of Integrative Biology* 23(7), 327–333. DOI: 10.1089/omi.2019.0073.

Leveau, J.H. (2007) The magic and menace of metagenomics: prospects for the study of plant growth-promoting rhizobacteria. *European Journal of Plant Pathology* 119(3), 279–300. DOI: 10.1007/s10658-007-9186-9.

Leveau, J.H. and Lindow, S.E. (2005) Utilization of the plant hormone indole-3-acetic acid for growth by *Pseudomonas putida* strain 1290. *Applied and Environmental Microbiology* 71(5), 2365–2371. DOI: 10.1128/AEM.71.5.2365-2371.2005.

Leveau, J.H.J. and Preston, G.M. (2008) Bacterial mycophagy: definition and diagnosis of a unique bacterial–fungal interaction. *New Phytologist* 177(4), 859–876. DOI: 10.1111/j.1469-8137.2007.02325.x.

Li, A.D., Li, L.G. and Zhang, T. (2015) Exploring antibiotic resistance genes and metal resistance genes in plasmid metagenomes from wastewater treatment plants. *Frontiers in Microbiology* 6, 1025. DOI: 10.3389/fmicb.2015.01025.

Liu, H., Li, J., Carvalhais, L.C., Percy, C.D., Prakash Verma, J, *et al.* (2021) Evidence for the plant recruitment of beneficial microbes to suppress soil-borne pathogens. *New Phytologist* 229(5), 2873–2885. DOI: 10.1111/nph.17057.

Ma, Y., Rajkumar, M., Zhang, C. and Freitas, H. (2016) Inoculation of brassica oxyrrhina with plant growth promoting bacteria for the improvement of heavy metal phytoremediation under drought conditions. *Journal of Hazardous Materials* 320, 36–44. DOI: 10.1016/j.jhazmat.2016.08.009.

MacLean, A.M., Finan, T.M. and Sadowsky, M.J. (2007) Genomes of the symbiotic nitrogen-fixing bacteria of legumes. *Plant Physiology* 144(2), 615–622. DOI: 10.1104/pp.107.101634.

Majeed, A., Muhammad, Z. and Ahmad, H. (2018) Plant growth promoting bacteria: role in soil improvement, abiotic and biotic stress management of crops. *Plant Cell Reports* 37(12), 1599–1609. DOI: 10.1007/s00299-018-2341-2.

Manyu, K. and Guodong, J. (2019) A biomarker of diazotrophic cyanobacteria: heterocyst glycolipids and their research progress. *Advances in Earth Science* 34(9), 901.

Maropola, M.K.A., Ramond, J.B. and Trindade, M. (2015) Impact of metagenomic DNA extraction procedures on the identifiable endophytic bacterial diversity in sorghum bicolor (*L. moench*). *Journal of Microbiological Methods* 112, 104–117. DOI: 10.1016/j.mimet.2015.03.012.

Martins, M.R., Jantalia, C.P., Reis, V.M., Döwich, I. and Polidoro, J.C. (2018) Impact of plant growth-promoting bacteria on grain yield, protein content, and urea-15 N recovery by maize in a Cerrado Oxisol. *Plant and Soil* 422, 239–250. DOI: 10.1007/s11104-017-3193-1.

Masenya, K., Manganyi, M.C. and Dikobe, T.B. (2024) Exploring cereal metagenomics: unravelling microbial communities for improved food security. *Microorganisms* 12(3), 510. DOI: 10.3390/microorganisms12030510.

Massart, S., Olmos, A., Jijakli, H. and Candresse, T. (2014) Current impact and future directions of high throughput sequencing in plant virus diagnostics. *Virus Research* 188, 90–96. DOI: 10.1016/j.virusres.2014.03.029.

Mendes, L.W., Kuramae, E.E., Navarrete, A.A., Veen, J.A. and Tsai, S.M. (2014) Taxonomical and functional microbial community selection in soybean rhizosphere. *The ISME Journal* 8(8), 1577–1587. DOI: 10.1038/ismej.2014.17.

Menendez, E. and Garcia-Fraile, P. (2017) Plant probiotic bacteria: solutions to feed the world. *AIMS Microbiology* 3(3), 502–524. DOI: 10.3934/microbiol.2017.3.502.

Miliute, I., Buzaite, O., Baniulis, D. and Stanys, V. (2015) Bacterial endophytes in agricultural crops and their role in stress tolerance: a review. *Zemdirbyste-Agriculture* 102(4), 465–478. DOI: 10.13080/z-a.2015.102.060.

Niu, X., Song, L., Xiao, Y. and Ge, W. (2017) Drought-tolerant plant growth-promoting rhizobacteria associated with foxtail millet in a semi-arid agroecosystem and their potential in alleviating drought stress. *Frontiers in Microbiology* 8, 2580. DOI: 10.3389/fmicb.2017.02580.

Nwachukwu, B.C. and Babalola, O.O. (2021) Perspectives for sustainable agriculture from the microbiome in plant rhizosphere. *Plant Biotechnology Reports* 15(3), 259–278. DOI: 10.1007/s11816-021-00676-3.

Nwachukwu, B.C. and Babalola, O.O. (2022) Metagenomics: a tool for exploring key microbiome with the potentials for improving sustainable agriculture. *Frontiers in Sustainable Food Systems* 6, 886987. DOI: 10.3389/fsufs.2022.886987.

Oteino, N., Lally, R.D., Kiwanuka, S., Lloyd, A., Ryan, D. *et al.* (2015) Plant growth promotion induced by phosphate solubilizing endophytic pseudomonas isolates. *Frontiers in Microbiology* 6, 745. DOI: 10.3389/fmicb.2015.00745.

Pawlowski, K. and Demchenko, K.N. (2012) The diversity of actinorhizal symbiosis. *Protoplasma* 249(4), 967–979. DOI: 10.1007/s00709-012-0388-4.

Pérez-Cobas, A.E., Gomez-Valero, L. and Buchrieser, C. (2020) Metagenomic approaches in microbial ecology: an update on whole-genome and marker gene sequencing analyses. *Microbial Genomics* 6(8), e000409. DOI: 10.1099/mgen.0.000409.

Porcel, R., Zamarreño, Á.M., García-Mina, J.M. and Aroca, R. (2014) Involvement of plant endogenous ABA in *Bacillus megaterium* PGPR activity in tomato plants. *BMC Plant Biology* 14, 1–12. DOI: 10.1186/1471-2229-14-36.

Priya, P., Aneesh, B. and Harikrishnan, K. (2021) Genomics as a potential tool to unravel the rhizosphere microbiome interactions on plant health. *Journal of Microbiological Methods* 185, 106215. DOI: 10.1016/j.mimet.2021.106215.

Radl, V., Winkler, J.B., Kublik, S., Yang, L., Winkelmann, T. *et al.* (2019) Reduced microbial potential for the degradation of phenolic compounds in the rhizosphere of apple plantlets grown in soils affected by replant disease. *Environmental Microbiome* 14, 1–12. DOI: 10.1186/s40793-019-0346-2.

Radwan, S.S., Al-Hasan, R.H., Salamah, S. and Al-Dabbous, S. (2002) Bioremediation of oily sea water by bacteria immobilized in biofilms coating macroalgae. *International Biodeterioration & Biodegradation* 50(1), 55–59. DOI: 10.1016/S0964-8305(02)00067-7.

Raheem, A., Shaposhnikov, A., Belimov, A.A., Dodd, I.C. and Ali, B. (2018) Auxin production by rhizobacteria was associated with improved yield of wheat (*Triticum aestivum L.*) under drought stress. *Archives of Agronomy and Soil Science* 64(4), 574–587. DOI: 10.1080/03650340.2017.1362105.

Rahman, M., Sabir, A.A., Mukta, J.A., Khan, Md., Mohi-Ud-Din, M, *et al.* (2018) Plant probiotic bacteria *Bacillus* and *Paraburkholderia* improve growth, yield and content of antioxidants in strawberry fruit. *Scientific Reports* 8(1), 2504. DOI: 10.1038/s41598-018-20235-1.

Remans, R., Beebe, S., Blair, M., Manrique, G., Tovar, E. *et al.* (2008) Physiological and genetic analysis of root responsiveness to auxin-producing plant growth-promoting bacteria in common bean (*Phaseolus vulgaris L.*). *Plant and Soil* 302, 149–161. DOI: 10.1007/s11104-007-9462-7.

Richardson, A.E., Barea, J.M., McNeill, A.M. and Prigent-Combaret, C. (2009) Acquisition of phosphorus and nitrogen in the rhizosphere and plant growth promotion by microorganisms. *Plant and Soil* 321, 305–339. DOI: 10.1007/s11104-009-9895-2.

Sachdev, D.P. and Cameotra, S.S. (2013) Biosurfactants in agriculture. *Applied Microbiology and Biotechnology* 97(3), 1005–1016. DOI: 10.1007/s00253-012-4641-8.

Santamaria, M., Fosso, B., Licciulli, F., Balech, B., Larini, I. *et al.* (2018) ITSoneDB: a comprehensive collection of eukaryotic ribosomal RNA Internal Transcribed Spacer 1 (ITS1) sequences. *Nucleic Acids Research* 46(D1), D127–D132. DOI: 10.1093/nar/gkx855.

Saraf, M., Pandya, U. and Thakkar, A. (2014) Role of allelochemicals in plant growth promoting rhizobacteria for biocontrol of phytopathogens. *Microbiological Research* 169(1), 18–29. DOI: 10.1016/j.micres.2013.08.009.

Sarkar, A., Ghosh, P.K., Pramanik, K., Mitra, S., Soren, T. *et al.* (2018) A halotolerant *Enterobacter* sp. displaying ACC deaminase activity promotes rice seedling growth under salt stress. *Research in Microbiology* 169(1), 20–32. DOI: 10.1016/j.resmic.2017.08.005.

Sarker, A., Al Masud, M.A., Deepo, D.M., Das, K., Nandi, R, *et al.* (2023) Biological and green remediation of heavy metal contaminated water and soils: a state-of-the-art review. *Chemosphere* 332, 138861. DOI: 10.1016/j.chemosphere.2023.138861.

Sasse, J., Martinoia, E. and Northen, T. (2018) Feed your friends: do plant exudates shape the root microbiome. *Trends in Plant Science* 23(1), 25–41. DOI: 10.1016/j.tplants.2017.09.003.

Sessitsch, A., Hardoim, P., Döring, J., Weilharter, A., Krause, A. *et al.* (2012) Functional characteristics of an endophyte community colonizing rice roots as revealed by metagenomic analysis. *Molecular Plant-Microbe Interactions®* 25(1), 28–36. DOI: 10.1094/MPMI-08-11-0204.

Shahzad, R., Khan, A.L., Bilal, S., Waqas, M., Kang, S.-M. *et al.* (2017) Inoculation of abscisic acid-producing endophytic bacteria enhances salinity stress tolerance in *Oryza sativa*. *Environmental and Experimental Botany* 136, 68–77. DOI: 10.1016/j.envexpbot.2017.01.010.

Shameer, S. and Prasad, T.N.V.K.V. (2018) Plant growth promoting rhizobacteria for sustainable agricultural practices with special reference to biotic and abiotic stresses. *Plant Growth Regulation* 84(3), 603–615. DOI: 10.1007/s10725-017-0365-1.

Shi, X., Zhao, Y., Xu, M., Ma, L., Adams, J.M. *et al.* (2023) Insight into plant-microbe interactions in rhizosphere for sustainable agriculture in the new crops era. *New Crops*. DOI: 10.1016/j.ncrops.2023.11.002.

Shrivastava, P. and Kumar, R. (2015) Soil salinity: a serious environmental issue and plant growth promoting bacteria as one of the tools for its alleviation. *Saudi Journal of Biological Sciences* 22(2), 123–131. DOI: 10.1016/j.sjbs.2014.12.001.

Sieber, C.M., Probst, A.J., Sharrar, A., Thomas, B.C., Hess, M. *et al.* (2018) Recovery of genomes from metagenomes via a dereplication, aggregation and scoring strategy. *Nature Microbiology* 3(7), 836–843. DOI: 10.1038/s41564-018-0171-1.

Singh, J.S., Pandey, V.C. and Singh, D.P. (2011) Efficient soil microorganisms: a new dimension for sustainable agriculture and environmental development. *Agriculture, Ecosystems & Environment* 140(3–4), 339–353. DOI: 10.1016/j.agee.2011.01.017.

Sobariu, D.L., Fertu, D.I.T., Diaconu, M., Pavel, L.V., Hlihor, R.M. *et al.* (2017) Rhizobacteria and plant symbiosis in heavy metal uptake and its implications for soil bioremediation. *New Biotechnology* 39, 125–134. DOI: 10.1016/j.nbt.2016.09.002.

Sood, G., Kaushal, R. and Sharma, M. (2020) Alleviation of drought stress in maize (*Zea mays L.*) by using endogenous endophyte bacillus subtilis in north west himalayas. *Acta Agriculturae Scandinavica, Section B – Soil & Plant Science* 70(5), 361–370. DOI: 10.1080/09064710.2020.1743749.

Sorokan, A., Veselova, S., Benkovskaya, G. and Maksimov, I. (2021) Endophytic strain *Bacillus subtilis* 26D increases levels of phytohormones and repairs growth of potato plants after Colorado potato beetle damage. *Plants* 10(5), 923. DOI: 10.3390/plants10050923.

Sprent, J.I. and Parsons, R. (2000) Nitrogen fixation in legume and non-legume trees. *Field Crops Research* 65(2–3), 183–196. DOI: 10.1016/S0378-4290(99)00086-6.

Stokstad, E. (2016) The nitrogen fix. *Science* 353, 1225–1227. DOI: 10.1126/science.353.6305.1225.

Sun, L., Qiu, F., Zhang, X., Dai, X., Dong, X, et ala. (2008) Endophytic bacterial diversity in rice (Oryza sativa L.) roots estimated by 16S rDNA sequence analysis. *Microbial Ecology* 55(3), 415–424. DOI: 10.1007/s00248-007-9287-1.

Surovy, M.Z., Dutta, S., Mahmud, N.U., Gupta, D.R., Farhana, T. *et al.* (2024) Biological control potential of worrisome wheat blast disease by the seed endophytic bacilli. *Frontiers in Microbiology* 15, 1336515. DOI: 10.3389/fmicb.2024.1336515.

Suzaki, T. and Kawaguchi, M. (2014) Root nodulation: A developmental program involving cell fate conversion triggered by symbiotic bacterial infection. *Current Opinion in Plant Biology* 21, 16–22. DOI: 10.1016/j.pbi.2014.06.002.

Tamames, J. and Puente-Sánchez, F. (2019) SqueezeMeta, a highly portable, fully automatic metagenomic analysis pipeline. *Frontiers in Microbiology* 9, 3349. DOI: 10.3389/fmicb.2018.03349.

Tamura, R., Hashidoko, Y., Ogita, N., Limin, S.H. and Tahara, S. (2008) Requirement for particular seed-borne fungi for seed germination and seedling growth of *Xyris complanata*, a pioneer monocot in topsoil-lost tropical peatland in Central Kalimantan, Indonesia. *Ecological Research* 23(3), 573–579. DOI: 10.1007/s11284-007-0411-y.

Tan, H., Mooij, M.J., Barret, M., Hegarty, P.M., Harington, C. *et al.* (2014) Identification of novel phytase genes from an agricultural soil-derived metagenome. *Journal of Microbiology and Biotechnology* 24(1), 113–118. DOI: 10.4014/jmb.1307.07007.

Touceda-González, M., Brader, G., Antonielli, L., Ravindran, V.B., Waldner, G. *et al.* (2015) Combined amendment of immobilizers and the plant growth-promoting strain *Burkholderia phytofirmans* PsJN favours plant growth and reduces heavy metal uptake. *Soil Biology and Biochemistry* 91, 140–150. DOI: 10.1016/j.soilbio.2015.08.038.

Trivedi, P., Mattupalli, C., Eversole, K. and Leach, J.E. (2021) Enabling sustainable agriculture through understanding and enhancement of microbiomes. *The New Phytologist* 230(6), 2129–2147. DOI: 10.1111/nph.17319.

Truyens, S., Weyens, N., Cuypers, A. and Vangronsveld, J. (2015) Bacterial seed endophytes: genera, vertical transmission and interaction with plants. *Environmental Microbiology Reports* 7(1), 40–50. DOI: 10.1111/1758-2229.12181.

Tsurumaru, H., Okubo, T., Okazaki, K., Hashimoto, M., Kakizaki, K, *et al.* (2015) Metagenomic analysis of the bacterial community associated with the taproot of sugar beet. *Microbes and Environments* 30(1), 63–69. DOI: 10.1264/jsme2.ME14109.

Van, N.T. and Pawlowski, K. (2017) Frankia and actinorhizal plants: symbiotic nitrogen fixation. In: Arora, N. and Mehnaz, S. (eds) *Rhizotrophs: Plant Growth Promotion to Bioremediation*. Springer, pp. 237–261. DOI: 10.1007/978-981-10-4862-3_12.

Verma, S.K., Kingsley, K., Bergen, M., English, C., Elmore, M. *et al.* (2018) Bacterial endophytes from rice cut grass (*Leersia oryzoides L.*) increase growth, promote root gravitropic response, stimulate root hair formation, and protect rice seedlings from disease. *Plant and Soil* 422, 223–238. DOI: 10.1007/s11104-017-3339-1.

Vessey, J.K., Pawlowski, K. and Bergman, B. (2005) Root-based n 2-fixing symbioses: legumes, actinorhizal plants, *Parasponia* sp. and cycads. *Plant and Soil* 274, 51–78. DOI: 10.1007/s11104-005-5881-5.

Vincent, A.T., Derome, N., Boyle, B., Culley, A.I. and Charette, S.J. (2017) Next-generation sequencing (NGS) in the microbiological world: how to make the most of your money. *Journal of Microbiological Methods* 138, 60–71. DOI: 10.1016/j.mimet.2016.02.016.

Wang, C., Zhou, X., Guo, D., Zhao, J., Yan, L, *et al.* (2019) Soil pH is the primary factor driving the distribution and function of microorganisms in farmland soils in northeastern China. *Annals of Microbiology* 69(13), 1461–1473. DOI: 10.1007/s13213-019-01529-9.

Wei, Z., Li-ping, Z., Run-lei, Y. and Zhi-tang, L.U. (2006) Nitrogen fixation activity of *Frankia* sp strains. *Journal of Hebei University (Natural Science Edition)* 26(1), 21.

Xu, H., Zhao, F., Hou, Q., Huang, W., Liu, Y. *et al.* (2019) Metagenomic analysis revealed beneficial effects of probiotics in improving the composition and function of the gut microbiota in dogs with diarrhoea. *Food & Function* 10(5), 2618–2629. DOI: 10.1039/C9FO00087A.

Zhalnina, K., Louie, K.B., Hao, Z., Mansoori, N., Rocha, U.N. *et al.* (2018) Dynamic root exudate chemistry and microbial substrate preferences drive patterns in rhizosphere microbial community assembly. *Nature Microbiology* 3(4), 470–480.

Zhang, L., Chen, F., Zeng, Z., Xu, M., Sun, F. *et al.* (2021) Advances in metagenomics and its application in environmental microorganisms. *Frontiers in Microbiology* 12, 766364. DOI: 10.3389/fmicb.2021.766364.

Zheng, B.-X., Hao, X.-L., Ding, K., Zhou, G.-W., Chen, Q.-L. *et al.* (2017) Long-term nitrogen fertilization decreased the abundance of inorganic phosphate solubilizing bacteria in an alkaline soil. *Scientific Reports* 7(1), 42284. DOI: 10.1038/srep42284.

24 Plant Disease Resistance through CRISPR-Cas Genome Editing: Opportunities and Challenges

Farin Tasnova Hia, Md Omar Kayess and Tofazzal Islam*
Institute of Biotechnology and Genetic Engineering, Gazipur Agricultural University, Gazipur, Bangladesh

Abstract
Plant diseases caused by various microorganisms such as fungi, bacteria, viruses, and nematodes are responsible for 20–40% of global yield losses, posing a significant threat to food security in the face of an ever-growing population. The CRISPR-Cas system has gained significant scientific attention due to its user-friendly application, high efficacy, precise targeting, and versatility in genetic manipulation. Given the pressing need for improved plant disease resistance, genome-editing technologies like CRISPR-Cas have revolutionized plant biology, facilitating the rapid development of disease-resistant cultivars. This approach offers a strategic advantage over traditional breeding methods, as it allows for more predictable, cost-effective, and adaptable modifications to both plant and pathogen genomes across diverse agricultural environments. The efficacy of CRISPR-Cas genome editing has been demonstrated in various key crops, including rice, maize, and soybean, where researchers have successfully enhanced resistance to specific pathogens. This chapter provides a comprehensive overview of recent advancements in plant protection through CRISPR-Cas technologies, highlighting specific applications across different plant communities and their effectiveness against viral, fungal, bacterial, and nematode infections. Additionally, we critically examine the potential benefits and challenges associated with CRISPR-Cas-based genome-editing strategies, offering insights into its implications for future agricultural practices.

Keywords: plant disease, genome editing, CRISPR-Cas, resilience, opportunities, challenges, climate change

24.1 Introduction

Plant diseases caused by fungi, bacteria, viruses, nematodes, and other pathogens present substantial challenges to global food security and agricultural sustainability, with severe implications for crop yields and economic stability (Savary *et al.*, 2019; Rabiey *et al.*, 2019; Hossain *et al.*, 2020; He *et al.*, 2021; Jeger *et al.*, 2021; Ristaino *et al.*, 2021). Climate change exacerbates these challenges, as it undermines the natural defense mechanisms of plants, leading to increased susceptibility to diseases (Hunjan and Lore, 2020; Hossain *et al.*, 2021). This vulnerability results in more frequent disease outbreaks, expanded geographic distribution of pathogens, and altered host specificity, all of which further threaten ecological diversity (Jeger, 2022; Semenza *et al.*, 2022). According to estimates, plant diseases are responsible for 20–40% of yield losses in five major food crops:

*Corresponding author: tofazzalislam@bsmrau.edu.bd

© CAB International 2025. *Stress-Resilient Crops: Coordinated omics-CRISPR-Nanotechnology Strategies* (ed. Jen-Tsung Chen)
DOI: 10.1079/9781800627307.0024

21.5% in wheat, 30.0% in rice, 22.6% in maize, 17.2% in potato, and 21.4% in soybean (Finegold *et al.*, 2019; Baldi and La Porta, 2020; Ayaz *et al.*, 2023). These crops collectively account for approximately half of global human calorie intake (FAOSTAT, 2018), resulting in annual financial losses estimated at $220 billion (FAO, 2019). With the global population projected to rise from 8.2 billion to nearly 10 billion by 2050, crop production must increase by 70–110% to meet future food demands.

Various plant diseases, such as rice blast caused by *Magnaporthe oryzae* pathotype *Oryzae*, late blight of potato and tomato by *Phytophthora infestans*, leaf spot in wheat and barley by *Bipolaris sorokiniana* (Kang *et al.*, 2021; Mahmud *et al.*, 2022; Paul *et al.*, 2022b; Kashyap *et al.*, 2023), Fusarium head blight in wheat by *Fusarium* spp., brown spot in rice (*Bipolaris oryzae*) (Sobanbabu *et al.*, 2018), wheat blast disease by *M. oryzae* pathotype *Triticum* (Islam *et al.*, 2016, 2020, 2024; Islam and Kasfy, 2023; Islam, 2024; Kasfy *et al.*, 2024), and southern corn leaf blight (*B. maydis*) (Manzar *et al.*, 2022), have been linked to significant global crop losses. Soilborne pathogens, including *Rhizoctonia*, *Fusarium*, *Aphanomyces*, and *Pythium* species, can dramatically reduce legume yields, often exacerbating disease severity through synergistic interactions (Panth *et al.*, 2020). Fungal pathogens alone account for an estimated 10% of global yield loss (Mounika and Lal, 2021), with severe cases resulting in losses exceeding 60%, underscoring the urgent need for effective disease management strategies (Różewicz *et al.*, 2021). Bacterial pathogens contribute to an estimated 27% yield loss (Khan and Jairajpuri, 2010) due to over 150 taxa in families such as Xanthomonadaceae, Pseudomonadaceae, and Enterobacteriaceae, causing various diseases including galls, scabs, blights, wilts, and cankers (Joshi *et al.*, 2020). Additionally, plant viruses account for around 47% of total crop losses (Dutta *et al.*, 2022), with significant reductions in yields of economically important crops based on infection incidence and severity (Rao and Reddy, 2020). Plant parasitic nematodes (PPNs) also contribute notably to these losses, causing 5–20% yield reductions in cereal crops and complicating disease dynamics (Khan *et al.*, 2021; Parrado and Quintanilla, 2024).

The rising global population, coupled with stagnant agricultural production, intensifies the strain on the food supply (Islam *et al.*, 2016; Fukase and Martin, 2020). The concurrent increase in plant diseases and their evolution in changing environments poses a formidable obstacle to securing a stable food supply (Chakraborty and Newton, 2011). Traditionally, plant disease management has relied heavily on chemical interventions, particularly pesticides (Singh *et al.*, 2021). However, this reliance has resulted in significant environmental issues, including pollution and biodiversity loss (Damalas and Eleftherohorinos, 2011) and has driven the emergence of pathogen resistance, which limits the effectiveness of chemical controls while squandering financial resources (Lucas *et al.*, 2015; Ahmad *et al.*, 2020). Furthermore, conventional plant breeding methods are often labor-intensive, costly, and inefficient, hampered by difficulties in candidate resistant gene identification and introgression (Gai and Wang, 2024). These strategies struggle to keep pace with the rapidly evolving threats posed by phytopathogens and changing environmental conditions.

In light of these pressing challenges, there is an urgent need to develop innovative, eco-friendly, and sustainable approaches to plant disease management. Recent advancements in genome-editing (GE) technologies have revolutionized our ability to enhance plant resistance to diseases. Techniques such as CRISPR-Cas offer high efficiency and precision in targeting specific DNA sequences, enabling significant advancements in our understanding of plant–pathogen interactions (Islam, 2019; Manzoor *et al.*, 2024). By facilitating the modification of genes associated with disease susceptibility, CRISPR-Cas genome editing can accelerate the development of resistant plant genotypes and improve the efficacy of disease management strategies (Fang and Tyler, 2016; Haque *et al.*, 2018; Islam, 2019; Li *et al.*, 2022a). This chapter provides a comprehensive overview of GE technologies, with a specific focus on CRISPR-Cas and its potential applications in enhancing plant disease resistance. We delve into the underlying mechanisms of GE, examine recent advancements in CRISPR-Cas tools, and explore their implications for sustainable disease management. Additionally, we address the challenges and ethical considerations associated with these innovative approaches.

By highlighting the transformative potential of genome editing, the chapter aims to contribute to the ongoing dialog on developing sustainable solutions for food security in an increasingly complex agricultural landscape.

24.2 Molecular Components and Cellular Mechanisms of CRISPR-Cas Genome Editing

The CRISPR (Clustered Regularly Interspaced Short Palindromic Repeats) system was first identified in *Escherichia coli* by Ishino *et al.* (1987) and later defined by Jansen *et al.* (2002) as repetitive DNA sequences flanked by unique spacer regions. Subsequent phylogenetic analyses revealed an evolutionary relationship between CRISPR loci and exogenous genetic elements such as plasmids and phages (Jaganathan *et al.*, 2018). Further explorations into homology-directed repair mechanisms for genome modification led to the identification of the CRISPR-Cas9 system as a highly efficient approach for targeted genome editing (GE) (Mojica *et al.*, 2005; Liu *et al.*, 2017a).

The CRISPR-Cas9 system employs a guided RNA complex to achieve targeted DNA cleavage. This complex consists of the Cas9 endonuclease and two RNA components: a transactivating CRISPR RNA (tracrRNA) and a CRISPR RNA (crRNA). Together, these RNA components direct the Cas9 enzyme to specific DNA sequences, allowing for precise genome editing (Kumar and Jain, 2015). The combination of tracrRNA and crRNA results in a synthetic construct known as single-guide RNA (sgRNA) (Qi *et al.*, 2013; Mali *et al.*, 2013; Murovec *et al.*, 2017). The 20-base pair sequence within the sgRNA is crucial for the accurate recognition and binding of the target DNA sequence, as elaborated by Hussain *et al.* (2018).

Cas9, as a nuclease enzyme, exhibits catalytic activity capable of inducing double-stranded DNA breaks. When paired with a guide RNA (gRNA), Cas9 effectively cleaves specific DNA sequences, leading to alterations in the genetic sequence that may inhibit the synthesis of targeted proteins (Tang *et al.*, 2019). The repair of double-strand breaks typically occurs through two primary mechanisms: non-homologous end-joining (NHEJ) and homologous recombination (HR) (Islam, 2019; El-Mounadi *et al.*, 2020). NHEJ can result in small insertions or deletions that may disrupt gene function, while HR allows for more precise genetic modifications by utilizing a homologous DNA template to repair the break.

The CRISPR-Cas9 system fundamentally operates through a multi-step process as described by Jiang and Doudna (2017):

1. **Recognition.** Genome editing begins with the formation of a complex between the synthetic guide RNA, which is complementary to the target locus and the Cas9 nuclease. This complex facilitates precise recognition and cleavage of the target DNA.
2. **Binding.** The Cas9-gRNA complex acts as a programmable DNA endonuclease, scanning the genome for sequences that match the gRNA. Upon locating the target site, the complex establishes a stable duplex through Watson-Crick base pairing.
3. **Cutting.** The CRISPR-Cas9 system ensures precise, site-specific DNA cleavage at predetermined genomic locations via guided recognition and binding of the gRNA.
4. **Repair.** Upon detection of a double-strand break (DSB), the cell activates repair mechanisms, primarily through two pathways:

    - **Non-homologous End-joining (NHEJ).** This process can result in small nucleotide changes at the repair site, potentially leading to insertions or deletions that disrupt the gene's functional integrity.
    - **Homology-directed Repair (HDR).** This precise repair mechanism uses a homologous DNA sequence as a template to accurately correct double-strand breaks, enabling the targeted introduction or replacement of specific nucleotide sequences.

Overall, the CRISPR-Cas9 system has emerged as a revolutionary tool in genome editing, favored for its efficiency and versatility, surpassing traditional methods such as ZFNs and TALENs.

24.3 Plant Disease Resistance by Genome Editing

Genome-editing technologies, particularly CRISPR-Cas, hold significant potential for combating plant diseases (Haque *et al.*, 2018; Islam, 2019; Li *et al.*, 2022a). By precisely altering the genomes of plant pathogens, researchers can engineer traits that enhance disease resistance or control (Zhang *et al.*, 2018c). Specifically, CRISPR-Cas9 technology offers a highly efficient and streamlined approach to genome engineering in microbes, often surpassing traditional methods that rely on homologous recombination. This technique not only facilitates targeted modifications but also allows for comprehensive genome-wide functional analyses of plant pathogens, proving to be an invaluable tool for elucidating disease mechanisms (Das *et al.*, 2023).

Since the initial demonstration of effective genome editing in *Escherichia coli* using CRISPR technology (Liu *et al.*, 2020), its applications have broadened significantly to include various bacterial species, such as *Bacillus*, *Pseudomonas*, *Streptomyces*, *Corynebacterium*, and *Yersinia*. In contrast, traditional breeding methods focussed on developing disease resistance can be labor-intensive and time-consuming. Moreover, the genetic markers linked to resistance often affect plant growth and development, potentially leading to unintended consequences (Das *et al.*, 2023). For example, researchers successfully edited the BSR-K1 locus in rice using CRISPR/SpCas9 technology, resulting in enhanced tolerance to *Magnaporthe oryzae* and *Xanthomonas oryzae* pv. *oryzae* (Das *et al.*, 2023). Additionally, another promising strategy for improving crop disease resistance involves the correction of faulty R genes through CRISPR-Cas9-mediated precise base editing (Zhou *et al.*, 2022). Tables 24.1–24.3 provide a summary of genome-editing efforts targeting economically significant crop plants for disease resistance, highlighting the transformative potential of these technologies in agricultural research.

24.3.1 Fungal pathogens resistance by genome editing

Contemporary research into the molecular mechanisms underpinning plant–pathogen interactions has paved the way for innovative strategies aimed at enhancing fungal resistance in crop plants. Certain genes and their products have emerged as promising candidates for improving plant resilience against fungi, making them ideal targets for CRISPR-Cas genome editing (Borrelli *et al.*, 2018; Islam, 2019). Mycotoxins, which are secondary toxic metabolites produced by mycotoxigenic fungi, present a significant threat to both human and animal health when ingested through contaminated food and feed. Fungi, as the predominant agents of plant diseases, impose considerable detrimental effects on agriculture. Their diverse life cycles and remarkable genetic adaptability complicate disease management strategies, as they can swiftly infect new hosts, circumvent resistance conferred by R genes, and develop resistance to fungicides (Doehlemann *et al.*, 2017).

Genetic engineering approaches are being utilized to combat the widespread fungal pathogen powdery mildew by modifying host susceptibility (S) genes. A key focus is on developing wheat cultivars with durable and broad-spectrum resistance. A significant breakthrough in this area was the discovery of barley Mlo mutants, which provide long-lasting immunity to powdery mildew (Lyngkjær *et al.*, 2000). Recent findings by Wang *et al.* (2023) demonstrated that resistance to the powdery mildew pathogen *Blumeria graminis* f. sp. *tritici* (Bgt) in wheat was achieved through the simultaneous mutation of all six *TaMlo* gene sets using TALEN and CRISPR technologies. Extensive research has also been conducted on negative regulators and susceptibility genes associated with defense pathways. For instance, knocking out the wheat homologs of these genes (*TaMLOs*) via the CRISPR/SpCas9 system resulted in enhanced resistance to Bgt (Wang *et al.*, 2014).

Additionally, *Magnaporthe oryzae*, the causative agent of rice blast disease, poses a critical threat to global rice production (Younas *et al.*, 2024). Ethylene responsive factors (ERFs), members of the APETELA2/ERF (AP2/ERF) superfamily, play a vital role in helping rice plants respond to various biotic and abiotic stressors (Park *et al.*, 2023). *M. oryzae*, as well as environmental stressors such as ABA and salt, can induce the expression of *OsERF922* and *OsSEC3A*. The latter gene has been identified as a negative regulator of rice blast resistance; its silencing via RNA interference (RNAi) has been

Table 24.1. CRISPR-Cas-mediated disease resistance to agronomic and model crops.

Crop name	Target gene	Resistance against	Nuclease	References
Rice	*LshCas13a*	Rice stripe mosaic virus (RSMV)	Cas13	Zhang *et al*., 2019b
	OsTFIIAγ1 or OsTFIIAγ5 (TF)	Bacterial blight	Cas9	Xu *et al*., 2021a; b; c; d
	OsSWEET14	Bacterial blight	Cas9	Arulganesh *et al*., 2021; Duy *et al*., 2021
	OsPrx30	Bacterial blight		Liu *et al*., 2021a, b, c
	OsXa13	Bacterial blight		Li *et al*., 2022a, b, c
	OsDjA2 and OsERF104	Rice blast	Cas9	Milazzo *et al*., 2022
	OsDjA2 and OsERF104	Rice blast	Cas9	Távora *et al*., 2022
	OsHRC	Rice blast	Cas9	Ding *et al*., 2023
	OsPita-Fuhui2663	Rice blast	Cas9	He *et al*., 2022
	OsUMP1	Rice blast, bacterial blight, and sheath blight	Cas9	Hu *et al*., 2023
	OsMORE1a	Rice blast and bacterial blight	Cas9	Kim *et al*., 2022
	OsEBEAvrXa23	Bacterial blight	Cas9	Wei *et al*., 2021a;b
	OsCPR5.1	Rice yellow mottle virus (RYMV)	Cas9	Arra *et al*., 2024
	Bsr-d1	*Magnaporthe oryzae*	Cas9	Zhang *et al*., 2024 Li *et al*., 2017
	OsCul3a	*Xanthomonas oryzae/ Magnaporthe oryzae*	Cas9	Gao *et al*., 2020 Liu *et al*., 2017b
	Os8N3	*Xanthomonas oryzae*	Cas9	Kim *et al*., 2019
	OsXa13	*Xanthomonas oryzae*	Cas9	Li *et al*., 2019a
	OsPi21, OsXa13	*Xanthomonas oryzae*	Cas9	Li *et al*., 2019b
	OseIF4G	Rice tungro spherical virus (RTSV)	Cas9	Macovei *et al*., 2018
	OsSec3A	*Magnaporthe oryzae*	Cas9	Ma *et al*., 2018
	OsCYP71A1	Insects: brown planthopper and striped stemborer	Cas9	Lu *et al*., 2018
	OsERF922	*Magnaporthe oryzae (rice blast)*	Cas9	Wang *et al*., 2016
	OsSWEET13	*Xanthomonas oryzae*	Cas9	Zhou *et al*., 2015
	OsSWEET11/14	*Xanthomonas oryzae* (bacterial blight)	Cas9	Jiang *et al*., 2013

Continued

Table 24.1. Continued

Crop name	Target gene	Resistance against	Nuclease	References
	OsMPK5	Fungal (*Magnaporthe grisea*) and bacterial (*Burkholderia glumae*) pathogens	Cas9	Xie and Yang, 2013
Wheat	*TaPDIL5-1*	Wheat yellow mosaic virus (WYMV)	Cas9	Kan *et al*., 2022
	TaNFXL1	*Fusarium graminearum*	Cas 9	Brauer *et al*., 2020
	TaEDR1	Powdery mildew	Cas9	Zhang *et al*., 2017
	TaMlo-A1, TaMlo B1, and TaMlo-D1	*Blumeria graminis* f. sp. *tritici*	TALEN & Cas9	Wang *et al*., 2014
	TaMlo	Powdery mildew	Cas9	Shan *et al*., 2013
	Tamlo-R32	Powdery mildew	TALEN vector	Li *et al*., 2022b
	TaTMT3B	Growth & yield	Cas9	Li *et al*., 2022b
Maize	Cry1F	*Spodoptera frugiperda*	Cas9	Kumari *et al*., 2024
	ZmNLB18 (WAK)	*Setosphaeria turcica* (northern leaf blight)	Cas9HDR	USDA/USDA/APHIS (2017) (DuPontPioneer)
Barley	*eIF4E1*	Barley mild mosaic virus (BaMMV)	Cas9	Hoffie *et al*., 2021
	Viral genome (MP/CP, Rep/RepA, LIR)	Wheat dwarf virus (WDV)	Cas9	Kis *et al*., 2019
Rye	*HvMORC1*	*Fusarium graminearum*	Cas9	Kumar *et al*., 2018
	CpTrpE, Cppyr4	*Claviceps purpurea*	Cas9	Králová *et al*., 2021
Soybean	*GmCDPK38*	Common cutworm	Cas9	Li *et al*., 2022d
	PsSu(z)12	*Phytophthora sojae*	Cas9	Wang *et al*., 2020
	GmF3H1/2,FNSII 1	Soybean mosaic virus (SMV)	Cas9	Zhang *et al*., 2019a
	PsAvr3c	*Phytophthora sojae*	Cas9	Huang *et al*., 2017
	Avr4/6	*Phytophthora sojae*	Cas9	Fang and Tyler, 2016
Cotton	*cpk33/74*	*Spodoptera litura*	Cas9	Wang *et al*., 2024a
	Gh14-3-3d	*Verticillium dahliae*	Cas9	Zhang *et al*., 2018b

Continued

Table 24.1. Continued

Crop name	Target gene	Resistance against	Nuclease	References
Tobacco	*LshCas13a, LwaCas13a*	Turnip mosaic virus (TuMV) Cucumber mosaic virus (CMV)	Cas13	Yu *et al.*, 2022
	LshCas13a, LwaCas13a	Turnip mosaic virus (TuMV) Tobacco mosaic virus (TMV), potato virus X (PVX)	Cas13	Mahas *et al.*, 2019
	UbEF1B and CCR4/ NOT3	*Geminivirus*	Cas9	Li *et al.*, 2024a
	CP, Hc-Pro	Turnip mosaic virus (TuMV)	Cas13	Aman *et al.*, 2018
	IR, CP, RCRII of Rep	Gemini viruses (CLCuKoV, MeMV)	Cas9	Ali *et al.*, 2016
	RBS, IRhairpin,3x Rep	Bean yellow dwarf virus (BeYDV)	Cas9	Baltes *et al.*, 2015
	IR, CP, RCRII of Rep	Tomato yellow leaf curl virus (TYLCV)	Cas9	Ali *et al.*, 2016
Arabidopsis	*At2OGO (DMR6 like)*	*Fusarium graminearum* (head blight)	Cas9	Low *et al.*, 2020
	AtErf019	*Phytophthora parasitica*	Cas9	Lu *et al.*, 2020
	AtIAN9	*Pseudomonas syringae* (bacterial speck)	Cas9	Wang *et al.*, 2018c
	AtEIF(iso)4E	Turnip mosaic virus (TMV)	Cas9	Pyott *et al.*, 2016
Arabidopsis N. benthamiana	*1A, CP,3'UTR-A*	Cucumber mosaic virus (CMV)/Tobacco mosaic virus (TMV)	FnCas9 >RNA	Zhang *et al.*, 2018a
	IR,CP,Rep	Beet severe curly top virus (BSCTV)	Cas9	Ji *et al.*, 2015

shown to improve resistance to *M. oryzae* (Liu *et al.*, 2012).

CRISPR/SpCas9-based genetic engineering technology has been successfully applied across various fungal species, including *Alternaria alternata* (Wenderoth *et al.*, 2017), *Leptosphaeria maculans* (Idnurm *et al.*, 2017), and *Fusarium* spp. like *F. oxysporum* (Wang *et al.*, 2018c), *F. graminearum* (Gardiner and Kazan, 2018), and *F. fujikuroi* (Kumar *et al.*, 2021). In these fungi, both SpCas9 and sgRNA can be introduced stably or transiently through polyethyleneglycol (PEG) transformation, *Agrobacterium*-mediated transformation, electroporation, and biolistic transformation (Schuster and Kahmann, 2019). Alternatively, the SpCas9/sgRNA ribonucleoprotein (RNP) complex can be prepared *in vitro* and applied directly to *M. oryzae* and *F. oxysporum* (Foster *et al.*, 2018; Wang *et al.*, 2018c).

To further enhance resistance against the powdery mildew pathogen *Oidium neolycopersici*, a CRISPR-Cas9-mediated gene-editing strategy targeting the *SlPMR4* gene – a homolog of the *PMR4* gene associated with callose deposition – was employed (Imam *et al.*, 2016). Another study revealed that null mutants of Ossec3a exhibited increased levels of salicylic acid (SA) and upregulated genes related to pathogenesis and SA signaling pathways, enhancing resistance to *M. oryzae* (Tyagi *et al.*, 2020). Additionally, a single nucleotide polymorphism (SNP) within the Taedr1 gene, a wheat homolog of the EDR gene, conferred heightened resistance to *Erysiphe cichoracearum*. Importantly, the genetic modification of Taedr1 minimally affected plant growth while providing a broad resistance profile against various pathogens, including bacteria and oomycetes (Zhang *et al.*,

Table 24.2. CRISPR-Cas-mediated disease resistance to vegetable crops.

Crop name	Target gene	Resistance against	Nuclease	References
Potato	*LshCas13a*	Potato virus Y (PVY)	Cas13	Zhan *et al.*, 2019
	StPM1	*Phytopthora infestans*	Cas9	Bi *et al.*, 2024
	StNRL1	*Phytopthora infestans*	Cas9	Norouzi *et al.*, 2024
	Stdmr6-1	*Phytopthora infestans, Alternaria solani*, Common scab	Cas9	Karlsson *et al.*, 2024
	Viral genome (*P3, CI, Nib, CP*)	Potato virus Y(PVY)	Cas13	Zhan *et al.*, 2019
Tomato	*SlHyPRP1* and *SlDEA1*	Bacterial leaf spot and Bacterial wilt	Cas9	Saikia *et al.*, 2024
	DND1	*Oidium neolycopersici* (Powdery mildew)	Cas9	Li *et al.*, 2024b
	miR482b and *miR482c*	*Phytophthora infestans* (Late blight)	Cas9	Hong *et al.*, 2021
	PECTATELYASE (SlPL)	Gray mold	Cas9	Silva *et al.*, 2021
	MAX1	*Phelipanche aegyptiaca*	Cas9	Bari *et al.*, 2021
	SlPMR4	*Oidium neolycopersici* (powdery mildew)	Cas9	Santillán Martínez *et al.*, 2020
	eIF4E1	Pepper mottle virus	Cas9	Yoon *et al.*, 2020
	SlJAZ2	*Pseudomonas syringae* pv. tomatoDC3000	Cas9	Ortigosa *et al.*, 2019
	DICER-LIKE2b (SlDCL2b)	Tomato mosaic virus	Cas9	Wang *et al.*, 2018a
	SlDCL2a and *SlDCL2b*	Potato virus X		Wang *et al.*, 2018c
	SlMLO1	*Oidium neolycopersici* (powdery mildew)	Cas9	Nekrasov *et al.*, 2017
	SlDMR6	*P. syringae, Phytophthora capsici, Xanthomonas* spp.	Cas9	Toledo Thomazella *et al.*, 2016
Cucumber	*eIF4E*	Cucumber vein yellowing virus; zucchini yellow mosaic virus; papaya ringspot mosaic virus-W	Cas9	Chandrasekaran *et al.*, 2016
Rapeseed	*BnCRT1a*	*Verticillium longisporum* (stem striping)	Cas9	Pröbsting *et al.*, 2020
Cassava	*BnWRKY70*	*Sclerotinia sclerotiorum* (stem rot)	Cas9	Sun *et al.*, 2018
	nCBP-1/2	Cassava brown streak virus (CBSV)	Cas9	Gomez *et al.*, 2019
	CR, AC1, AC2, AV2	African cassava mosaic virus	Cas9	Mehta *et al.*, 2019
Sweet potato	*PspCas13b, RfxCas13d*	Sweet potato chlorotic stunt virus (SPCSV)-RNase3	Cas13	Yu *et al.*, 2022

2017). Furthermore, editing the *VvWRKY52* gene in grapevine conferred resistance against *Botrytis cinerea* (Tyagi *et al.*, 2020).

In a separate initiative, a novel *Phytophthora*-resistant variety of *Theobroma cacao* was generated through a transient leaf transformation

Table 24.3. CRISPR-Cas-mediated disease resistance to fruits crops.

Crop name	Target gene	Resistance against	Nuclease	References
Grapevine	DMR6-1	Downy mildew (*Plasmopara viticola*)	Cas9	Djennane *et al.*, 2024
	VvWRKY52	*Botrytis cinerea* (gray mold)	Cas9	Wang *et al.*, 2018b
	VvMLO-7	*Erysiphe necator* (powdery mildew)	Cas9 RNP	Malnoy *et al.*, 2016
Watermelon	*Clpsk1*	*Fusarium oxysporum*	Cas9	Zhang *et al.*, 2020
Citrus	*CsLOB1*	*X. citri* subsp. *citri*	Cas9	Peng *et al.*, 2017
Duncan grapefruit	*CsLO1*	*X. citri* subsp. *citri*	Cas9	Jia *et al.*, 2017
Litchi	*PlPAE5*	*Peronophythora litchii*	Cas9	Kong *et al.*, 2019
Papaya	*PpalEPICs*	*Phytophthora palmivora*	Cas9	Gumtow *et al.*, 2018
Banana	MusaENODL3	*Xanthomonas campestris* pv. *musacearum* (Xcm)	Cas9	Ntui *et al.*, 2024
	Downy mildew resistance 6 (DMR6)	Xanthomonas Bacterial Wilt (*Xanthomonas campestris* pv. *musacearum* (Xcm)	Cas9	Tripathi *et al.*, 2021
	eBSV	Banana streak virus (eBSV)	Cas9	Tripathi *et al.*, 2019
	RGA2, Ced9	Fusarium wilt caused by *Fusarium oxysporum* f. sp. Cubense tropical race4(TR4)	Cas9	Dale *et al.*, 2017
Cacao	*TcNPR3*	*Phytophthora tropicalis*	Cas9	Fister *et al.*, 2018
Apple	*DIPM-1, DIPM-2, DIPM-4*	*Erwinia amylovora* (fire blight)	Cas9	Malnoy *et al.*, 2016

approach. To bolster this resistance, CRISPR-Cas9 genetic engineering was utilized to target the *NPR3* gene, a crucial regulator of plant defense. This resulted in transgenic embryos that successfully developed into mature plants exhibiting improved resistance to *P. tropicalis* (Fister *et al.*, 2018).

24.3.2 Bacterial pathogens resistance by genome editing

The diversity, rapid reproduction, and various transmission modes of bacterial pathogens present substantial challenges for their control, particularly during outbreaks. Hundreds of bacterial species inflict damage on agricultural systems. Traditional bacteriological management strategies primarily rely on genetic resistance, agronomic practices, and biocontrol agents to prevent and mitigate pathogen impacts on plants (Kerr, 2016).

One prominent example of bacterial disease in agriculture is rice bacterial blight, caused by *Xanthomonas oryzae* pv. *oryzae* (Xoo). This disease can lead to yield reductions of 10–20% (Ou, 1985), with losses sometimes exceeding 50% under unfavorable conditions (Mew *et al.*, 1993). Xoo employs a type III secretion system to introduce TALE effector proteins into rice cells. These effectors specifically target susceptibility (S) genes, leading to their upregulation and facilitating successful infection. A critical component of this process is the effector-binding element (EBE) within the *OsSWEET14* promoter, which binds the TALE protein AvrXa7 from the Philippine strain PXO86. This interaction triggers the expression of *OsSWEET14*, a gene that encodes a sucrose efflux transporter. By leveraging *OsSWEET14*, PXO86 extracts sugars from rice cells, providing essential nutrients for its growth and pathogenicity.

To selectively disrupt AvrXa7's binding to the *OsSWEET14* promoter without inhibiting OsSWEET14's vital role in plant development, a pair of TALENs was designed to target the EBEs within the promoter's region (Li *et al.*, 2014). The successful genetic enhancement (GE) of the *xa23* gene in rice demonstrates this approach's efficacy, significantly boosting resistance to bacterial blight and leaf streak by stacking effector-binding elements in its promoter (Wang *et al.*, 2024b). A similar method has been applied to the Xa27 gene (Xu *et al.*, 2024). Furthermore, epigenome editing has shown promise in cassava, where targeted methylation of the *MeSWEET10a* gene's promoter effectively blocked the pathogen's capacity to cause disease, leading to heightened resistance (Veley *et al.*, 2022).

Citrus canker, caused by *Xanthomonas citri* ssp. *citri*, involves the activation of the *CsLOB1* gene, which is part of the LBD family and was identified by Hu *et al.* (2014). The effector PthA4 from Xcc activates CsLOB1 expression, facilitating canker development. Utilizing CRISPR/Cas9 technology for targeted mutagenesis of the CsLOB1 gene has conferred resistance to Xcc (Jia *et al.*, 2017; Peng *et al.*, 2017). Notably, *CsLOB1* null mutants maintain growth characteristics similar to wild-type plants, highlighting their potential for engineering disease resistance in premium citrus cultivars.

Tomatoes, economically critical but susceptible to bacterial pathogens such as *Pseudomonas syringae* and *Xanthomonas* spp. (Schwartz *et al.*, 2015), have shown promise in genetic resistance as well. The *SlDMR6-1* gene, homologous to the human DMP1 gene, is activated upon infection. CRISPR/Cas9-generated null mutants of this gene exhibit resistance to these bacterial diseases without compromising overall tomato growth and development. These findings suggest that *SlDMR6-1* is a significant contributor to tomato susceptibility to bacterial infections, making it a valuable target for developing resistant varieties (Toledo Thomazella *et al.*, 2016).

Fungal pathogens can degrade the plant cell wall, allowing entry and infection. In response, plants secrete inhibitors and produce callose to fortify their cell walls. One innovative strategy for engineering fungal resistance involves targeting genes associated with callose deposition, critical to maintaining cell-wall integrity. The CRISPR-Cas9 system has been employed to modify the *SlPMR4* gene, a homolog of the *PMR4* gene known to influence callose biosynthesis (Santillán Martínez *et al.*, 2020). This genetic intervention aimed to enhance resistance to the powdery mildew pathogen *Oidium neolycopersici*. The loss of *PMR4* function, due to an inversion mutation, increased salicylic acid levels, activated the hypersensitive response (HR) in plants, and conferred resistance.

CRISPR-Cas9 technology demonstrates remarkable versatility in genome editing, successfully targeting a wide array of organisms, including model species like *Arabidopsis thaliana* and *Nicotiana benthamiana*, as well as economically significant cereal and fruit crops. Its application extends even to complex genomes with varying ploidy levels, underscoring the broad applicability of this technology in advancing plant disease resistance.

24.3.3 Viral resistance by genome editing

The obligate parasitic nature of plant viruses, combined with their rapidly mutating genomes and high infection rates, poses a significant threat to crop yields, resulting in substantial agricultural losses (Das *et al.*, 2018). Plant viruses are categorized into two primary groups based on their genomic structure: DNA viruses and RNA viruses. DNA viruses, such as those from the Geminiviridae family, can be further subdivided into double-stranded (dsDNA) and single-stranded (ssDNA) types. While ssDNA geminiviruses exploit plant cellular machinery for their replication, RNA viruses exhibit a wider range of forms, including dsRNA, negative-strand ssRNA, positive-strand ssRNA, and retroviruses. Despite their ssDNA genome, geminiviruses can inflict considerable yield losses on economically vital crops such as cucumbers, tomatoes, potatoes, wheat, and barley through their unique rolling circle replication mechanism.

Viruses lack autonomous translational machinery and must hijack the host cell's transcription and translation factors to propagate. This dependence underscores potential antiviral strategies focussed on either disrupting viral genes associated with pathogenicity or modifying host susceptibility factors to hinder viral replication (Tyagi *et al.*, 2020). Recent advancements in plant biotechnology have harnessed CRISPR-Cas9 systems, employing both Cas9 and Cas13a nucleases to induce genetic modifications that confer resistance to plant DNA viruses (Hirano *et al.*, 2016). A landmark study in 2015 demonstrated the viability of using CRISPR-Cas9 technology to engineer resistance against DNA viruses, with three independent research teams successfully targeting geminiviruses such as the tomato leaf curl virus, beet severe curly top virus, and bean yellow dwarf virus in model organisms like *Arabidopsis thaliana* and *Nicotiana benthamiana* (Ali *et al.*, 2018).

In their 2016 study, Ali and colleagues effectively introduced genetic changes in the non-coding regions of four distinct begomoviruses – merremia mosaic virus, tomato yellow leaf curl virus, cotton leaf curl kokhran virus, and beet curly top virus – using CRISPR technology. They engineered sgRNA/Cas9 constructs that targeted four genomic loci of wheat dwarf virus (WDV), resulting in plants that exhibited resistance through the absence of viral symptoms and decreased viral activity (Kis *et al.*, 2019). Additionally, CRISPR techniques were employed to inactivate endogenous banana streak virus (eBSV) in banana plants (Tripathi *et al.*, 2019). Mehta *et al.* (2019) designed six guide RNAs targeting overlapping regions of the African cassava mosaic virus (ACMV) genome, focussing on sequences within the AC2/AC3, AC1/AC2, AV1/AV2, AC1, AV1, and intergenic regions (IR).

More recently, Lee *et al.* (2023) utilized the CRISPR/Cas9 approach to perform a precise mutation in the eIF(iso)4E gene of the TuMV-susceptible 'Seoul' Chinese cabbage cultivar, leading to the development of a TuMV-resistant variety. Similarly, Chandrasekaran *et al.* (2016) found that disrupting the eIF4E gene significantly enhanced cucumber immunity against several pathogens, including the zucchini yellow mosaic virus, cucumber vein yellowing virus, and papaya ring spot mosaic virus-W. Pyott *et al.* (2016) demonstrated resistance to turnip mosaic virus (TuMV) infection in *Arabidopsis* by knocking out the eIF(iso)4E gene, achieving this without affecting traits like biomass and flowering time. Furthermore, targeted mutations in the eIF4E gene of the tetraploid potato variety Kruda conferred immunity against potato virus Y (Noureen *et al.*, 2022). Genetic variations of eIF4E, particularly the eIF4E1 isoform, have been shown to influence tomato susceptibility to viruses, as it confers resistance against various pathogens, including PepMoV, CMV, and PVYN. Targeted editing of this gene holds promise for enhancing tomato disease tolerance.

24.3.4 Nematode resistance by genome editing

Plant-parasitic nematodes, such as *Meloidogyne*, *Ditylenchus*, *Heterodera*, and *Globodera* species, pose significant threats to crop health and productivity. Genetic engineering (GE) strategies present promising methods for managing root knot nematodes (RKN). For instance, Gheysen and Vanholme (2007) demonstrated that introducing double-stranded RNA (dsRNA) targeting critical nematode genes into host plants can effectively impart resistance against RKN infection. These nematodes employ a sophisticated array of proteins to establish successful infections. Recent work by Huang *et al.* (2023) utilized precision gene editing to alter the *OsHPP04* gene, leading to the development of genetically stable rice mutants, both with and without exogenous DNA. Notably, these mutants displayed significantly enhanced tolerance to the damaging rice RKN, *Meloidogyne graminicola*. Additionally, the 16D10 gene has emerged as a key target for nematode control; RNA interference (RNAi) aimed at silencing this gene resulted in a remarkable 74–81% reduction in nematode development and reproduction (Huang *et al.*, 2006).

Cyst nematodes, which are highly specialized sedentary endoparasites, also pose a considerable risk to crops. Host-induced RNA interference targeting a set of parasitism genes in *Heterodera schachtii*, the cyst nematode affecting sugar beet, effectively reduced female nematode populations when *Arabidopsis thaliana* was used as a host organism (Sindhu *et al.*, 2009). Similarly, GE-based silencing of the *Mi-cpl-1* gene significantly diminished nematode susceptibility (Dutta *et al.*, 2015). Though complete resistance was not achieved, RNAi-based GE lines led to a notable 23–64% decrease in nematode development.

Studies have consistently indicated that manipulating specific genes can enhance resistance to plant-parasitic nematodes. For example, targeting the *Mi-Rpn7* gene in *Meloidogyne incognita* has shown promise; reducing Mi-Rpn7 activity via RNAi resulted in a significant decrease in nematode infection by 55.2–66.5% (Niu *et al.*, 2012). Moreover, Tian *et al.* (2019) engineered transgenic soybean plants to bolster resistance against the soybean cyst nematode by silencing two essential nematode genes, HgY25 and HgPrp17, which are critical for reproduction and development.

Kaur *et al.* (2017) highlighted the potential of RNA interference for managing RKN infestations in *Arabidopsis thaliana*. By targeting crucial genes involved in RNA processing and integration, they effectively suppressed nematode development and reproduction, leading to a significant reduction in gall formation, female abundance, and egg production. In another study, Tsygankova *et al.* (2019) explored RNAi-based GE as a strategy to enhance bread wheat's resistance to the cereal cyst nematode, *Heterodera avenae*, employing microbially derived bioregulators to facilitate this innovative *in vitro* approach.

24.4 Challenges in Plant Disease Resistance Using CRISPR Technology

CRISPR-Cas technology presents a promising and efficient method for enhancing plant immunity through genetic modification. However, its implementation is accompanied by several challenges. One significant concern is the potential for off-target editing, which arises when the guide RNA inadvertently binds to genomic sequences outside the intended target site (Hsu *et al.*, 2013). Moreover, the evolving nature of pathogens complicates these efforts, as they continuously adapt to overcome existing resistance mechanisms. Additionally, while engineering plant resistance genes, such as the S gene, can provide enhanced immunity, it often leads to trade-offs that may negatively impact plant fitness (Tyagi *et al.*, 2020).

Knocking out susceptible genes presents a seemingly straightforward strategy that can indeed result in durable resistance because pathogens typically exert low selective pressure against plant defense systems (Patel *et al.*, 2024). However, this approach is not without risks. Certain mutations in resistance (R) genes, while nonlethal, may trigger pleiotropic effects such as dwarfism, nutrient deficiencies, and suppression of gene replication processes (Ahmad *et al.*, 2020). For instance, the development of 'triple knockouts' in *TaMLO* wheat led to resistance against powdery mildew but was also associated with leaf chlorosis (Wang *et al.*, 2014). Therefore, comprehensive field-based research, including greenhouse experiments and multi-environmental yield trials, is essential to assess these potential negative side effects (Borrelli *et al.*, 2018).

Although transgenic plants engineered through CRISPR-Cas9 have shown significantly reduced viral loads and maintained their phenotypes across generations, their classification as genetically modified organisms (GMOs) impose adherence to rigorous regulations, which could meet resistance in certain regions. Consequently, the potential for off-target mutations remains a considerable concern (Tyagi *et al.*, 2020).

To mitigate off-target effects, enhancing Cas9's specificity for its intended target sequence is crucial (Majeed *et al.*, 2018; Irfan *et al.*, 2024). Strategies to improve specificity include developing modified enzymes that target various protospacer adjacent motifs (PAMs). Additionally, novel CRISPR-Cas nucleases, such as Cas12a and Cas13a, have been identified, expanding the toolkit for genome editing and RNA interference. Cas12a is specific to DNA, while Cas13a targets single-stranded RNA (Koonin and Makarova, 2019). However, delivering these components – specifically, the Cas9 enzyme and guide RNA – into plant cells remains challenging, especially within recalcitrant species. The intricate interactions between plants and pathogens often complicate the creation of effective resistance, implying that modifying a single gene may insufficiently confer comprehensive immunity and that investigating multiple genes can prove complex (Veillet *et al.*, 2020; Rato *et al.*, 2021). Pathogens' rapid evolution can quickly render previously effective resistance mechanisms obsolete, necessitating a continuous refinement of genome-editing strategies.

While genome editing holds promise for enhancing disease resistance, it may also inadvertently affect other important traits, such as yield and quality (Miladinovic *et al.*, 2021; Rato *et al.*, 2021). To garner public acceptance, the durability of genome-edited resistance genes must be addressed. Strategies for enhancing the persistence of these genes align with those employed for introgressed resistance genes, including gene stacking with varying modes of action and exploring alternative resistance mechanisms beyond the NBS-LRR receptor kinases. Additionally, integrated pest management practices, including crop rotation and the use of biocontrol agents, should be central to these strategies (Borrelli *et al.*, 2018).

Despite their similarities to Cas9, novel genome-editing systems exhibit superior efficacy in certain contexts, particularly in controlling plant diseases. Their unique mechanisms of target nucleotide cleavage and pre-crRNA processing provide advantages over traditional Cas9 (Langner *et al.*, 2018). However, the CRISPR-Cas9 system's potential to induce unwanted chromosomal rearrangements – such as translocations, inversions, large deletions, and insertions – poses a risk due to complex endogenous DNA repair mechanisms activated by Cas nucleases (Kosicki *et al.*, 2018). Additionally, the DSBs induced by Cas9 can trigger detrimental cellular effects, including activation of apoptotic pathways and reduced efficacy in transformation and genetic editing (Roy *et al.*, 2018). Notably, not all edited plant cells are capable of regenerating into whole plants, particularly when substantial genomic alterations occur (Zhang *et al.*, 2020).

Innovatively, the development of nuclease-deficient Cas9 enzymes fused with proteins such as deaminases or recombinases has enabled precise base editing and site-specific recombination without the creation of DSBs, thus addressing some limitations associated with traditional editing methods (Standage-Beier *et al.*, 2019). The original *Streptococcus* pyogenes Cas9 (SpCas9) system's strict requirement for specific PAM sequences adjacent to the target DNA presents challenges in targeting certain genes, particularly those involved in specialized biological pathways (Jinek *et al.*, 2012; Langner *et al.*, 2018). Although CRISPR-Cas9 can recognize a range of PAM sequences, this flexibility may inadvertently increase the risk of off-target modifications.

Regulatory hurdles also present significant obstacles to the agricultural use of genome-edited plants (Entine *et al.*, 2021; Hundleby and Harwood, 2022). Delays in the commercialization of new varieties often stem from complex regulations surrounding targeted mutagenesis and genome editing, which can involve extensive field testing and data collection. These regulations aim to alleviate public concerns about the safety of GMOs. In contrast, the guidelines set forth by the United States Department of Agriculture exempt CRISPR-Cas9-edited crops from certain regulatory scrutiny, accelerating their pathway to commercialization (Waltz, 2018).

In conclusion, while CRISPR-Cas technology provides transformative opportunities for enhancing plant disease resistance, the associated challenges – ranging from off-target effects and regulatory complexities to unintended consequences on plant fitness – underscore the need for ongoing research and innovative approaches to ensure the successful implementation and public acceptance of these technologies in agriculture.

24.5 Future Opportunities of CRISPR Genome Editing for Plant Disease Resistance

The increasing threat of plant diseases to global food security underscores the urgent need for innovative approaches to enhance crop resilience. Although traditional breeding techniques have been effective, they are often slow and limited in their ability to tackle multifaceted disease challenges. The emergence of CRISPR-Cas genome-editing technology offers a transformative method for expediting the development of disease-resistant plants. By allowing precise and targeted modifications to plant genomes, CRISPR technology has the potential to significantly improve plant health and reduce the detrimental effects of plant pathogens.

24.5.1 Tailored approaches to combat disease

CRISPR genome-editing holds considerable promise for agricultural applications, allowing for the precise targeting of specific genes (Afzal *et al.*, 2020; Rao and Wang, 2021). In plants, resistance to diseases is largely governed by resistance (R) genes, which facilitate the recognition and response to pathogens (Gupta *et al.*, 2021; Zhao *et al.*, 2023). Through CRISPR technology, researchers can effectively modify R genes to enhance their functionality, thereby increasing plant resistance to a range of diseases (Zaidi *et al.*, 2020; Gosavi *et al.*, 2020; Zhang *et al.*, 2021). This targeted approach allows for the rapid integration of disease-resistant traits into crops, contrasting sharply with the protracted timeline of traditional breeding methods.

The precision of CRISPR enables researchers to knock out specific susceptibility genes, imparting resistance against diverse pathogens, including viruses, bacteria, and fungi (Schenke, 2020; Tyagi *et al.*, 2021; Talakayala *et al.*, 2022). Noteworthy examples include the targeted mutagenesis of the *OsSWEET* gene family, which has been demonstrated to enhance resistance to bacterial blight in rice (Li *et al.*, 2016; Duy *et al.*, 2021), and the successful editing of the *SlJAZ2* gene in tomato, conferring resistance to bacterial speck disease (Ortigosa *et al.*, 2019). Modifying individual genes for disease resistance is generally more straightforward compared to the multi-gene coordination required for enhancing metabolic pathways or abiotic stress tolerance (Borrelli *et al.*, 2018). Moreover, CRISPR's multiplexing capability enables simultaneous editing of multiple genes, accelerating the development of crops with broad-spectrum disease resistance. Recent advances demonstrate the successful application of CRISPR-Cas technology for modifying susceptibility genes, such as the enhancement of powdery mildew resistance in wheat through the targeted editing of the *MLO* locus (Bui *et al.*, 2023; Aroge *et al.*, 2024). This precision-based gene editing holds significant potential for alleviating the impacts of emerging plant diseases across various agricultural systems.

24.5.2 Genome-wide editing for complex traits

Disease resistance in plants is often dictated by a complex interplay of multiple genes (Li *et al.*, 2021; Tao, 2024). CRISPR-Cas technology facilitates multiplexed genome editing, allowing researchers to simultaneously modify several genes (Abdelrahman *et al.*, 2021; Vazquez-Vilar *et al.*, 2021). This capability is particularly beneficial for enhancing plant disease resistance, as it enables targeted alterations across multiple pathways involved in the plant defense response (Singh *et al.*, 2022; Talakayala *et al.*, 2022). By simultaneously targeting genes involved in pathogen recognition and immune response pathways, CRISPR could significantly bolster plant immunity. This multiplexed editing approach presents the opportunity to develop crops resistant to a wide variety of diseases, which could lessen the reliance on multiple disease management strategies (Rato *et al.*, 2021; Chattopadhyay *et al.*, 2022). A single genome-editing event could yield crops that are resistant to bacterial blight, rusts, and viral infections, offering a streamlined solution to manage diverse disease threats (Hüdig *et al.*, 2022; Yıldırım *et al.*, 2023).

24.5.3 Augmenting host resistance to a diverse pathogen array

CRISPR-based genome-editing holds significant promise for fostering broad-spectrum disease resistance. Unlike conventional methods that target specific pathogens, CRISPR can modify regulatory genes that influence the plant's overall defense mechanisms (Tyagi *et al.*, 2021; Paul *et al.*, 2022a). This strategic approach potentially reduces the risk of pathogen adaptation, a common concern with single-gene resistance strategies. By targeting genes essential for pathogen survival or virulence, such as those involved in pathogen entry or toxin production,

CRISPR can enhance plant resilience against various threats. Additionally, CRISPR may allow for modifications that enable immune cells to resist viral infections, possibly paving the way for innovative treatments (Liu *et al.*, 2021c; Misra *et al.*, 2023; Manzoor *et al.*, 2024). By priming plants to defend against a broad spectrum of pathogens, this strategy could have profound implications for improving crop yields across diverse agricultural landscapes.

24.5.4 CRISPR-Cas and epigenetic interventions

In addition to its core gene-editing capabilities, CRISPR technology offers exciting prospects for manipulating epigenetic regulation, thus indirectly controlling gene expression (Martella and Fisher, 2021; Chakraborty and Islam, 2022). By targeting epigenetic modifications linked to disease resistance genes, CRISPR could revolutionize how we engineer plant–pathogen interactions. Future research could explore the use of CRISPR to adjust epigenetic markers such as DNA methylation and histone acetylation, enabling conditional gene expression. A notable study by Feng *et al.* (2021) demonstrated how CRISPR can be harnessed to modify DNA methylation in *Arabidopsis thaliana*, delaying flowering by increasing methylation at a targeted gene. This method could be instrumental in managing flowering-dependent vectors and controlling plant diseases. By creating libraries of mutants with diverse epigenetic modifications, researchers could identify key genes or epigenetic factors influencing traits like biotic stress tolerance or yield, allowing plants to dynamically regulate immune responses and optimize energy allocation for enhanced disease resistance (Weiss *et al.*, 2022).

24.5.5 Addressing regulatory and ethical hurdles

While the promise of CRISPR-mediated plant breeding is substantial, significant regulatory and ethical challenges remain (Ahmad *et al.*, 2021). Public concerns regarding GMOs and the potential ecological implications of CRISPR-edited crops necessitate careful consideration (Munawar *et al.*, 2024). However, CRISPR-edited plants are often regarded as more acceptable than traditional GMOs due to their reliance on precise modifications of the organism's native genetic material rather than the introduction of foreign DNA (Ahmad *et al.*, 2023). The regulatory framework for CRISPR-edited crops is evolving, with many countries developing policies that distinguish between conventionally modified organisms and those that are gene-edited (Rozas *et al.*, 2022). As CRISPR technology matures, we can expect regulatory processes to become more efficient, facilitating the adoption of CRISPR-based solutions for improving disease resistance in agriculture (Zaidi *et al.*, 2020; Zhu *et al.*, 2021; Mushtaq *et al.*, 2021).

24.6 Concluding Remarks

CRISPR-Cas genome editing represents a transformative opportunity for enhancing plant disease resistance. Its precise and versatile capabilities open unprecedented avenues for engineering crops and pathogens to promote sustainable agricultural practices. By developing crops with tailored immune responses, CRISPR could significantly reduce the negative impact of plant diseases on global food security. A thorough understanding of the intricate dynamics of plant–pathogen interactions requires integrating multiple 'omics' technologies, including transcriptomics, proteomics, and metabolomics, which can provide valuable insights into gene expression, protein abundance, and metabolite profiles. These findings could identify a wide range of cellular targets critical for developing resilient plant cultivars. Anticipated advancements in CRISPR technology, in conjunction with complementary genetic engineering and omics strategies, have the potential to yield crop varieties that are better equipped to face environmental and biotic challenges. These resilient varieties will play a critical role in ensuring sustainable food security for populations around the globe. As regulatory landscapes mature and public perceptions evolve, CRISPR is poised to become a foundational element of sustainable and resilient agricultural practices. We envision that this technology will accelerate research in plant and microbial biology, fostering breakthroughs that contribute both to fundamental research and effective disease management.

References

Abdelrahman, M., Wei, Z., Rohila, J.S. and Zhao, K. (2021) Multiplex genome-editing technologies for revolutionizing plant biology and crop improvement. *Frontiers in Plant Science* 12, 721203. DOI: 10.3389/fpls.2021.721203.

Afzal, S., Sirohi, P. and Singh, N.K. (2020) A review of CRISPR associated genome engineering: Application, advances and future prospects of genome targeting tool for crop improvement. *Biotechnology Letters* 42, 1611–1632. DOI: 10.1007/s10529-020-02950-w.

Ahmad, N., Rahman, M.U., Mukhtar, Z., Zafar, Y. and Zhang, B. (2020) A critical look on CRISPR-based genome editing in plants. *Journal of Cellular Physiology* 235(2), 666–682. DOI: 10.1002/jcp.29052.

Ahmad, A., Ghouri, M.Z., Munawar, N., Ismail, M., Ashraf, S. *et al.* (2021) Regulatory, ethical, and social aspects of CRISPR crops. In: Ahmad, A., Khan, S.H. and Khan, Z. (eds) *CRISPR Crops: The Future of Food Security*. Springer, Singapore, pp. 261–287. DOI: 10.1007/978-981-15-7142-8_9.

Ahmad, A., Jamil, A. and Munawar, N. (2023) GMOs or non-GMOs? The CRISPR conundrum. *Frontiers in Plant Science* 14, 1232938. DOI: 10.3389/fpls.2023.1232938.

Ali, Z., Ali, S., Tashkandi, M., Zaidi, S.S. and Mahfouz, M.M. (2016) CRISPR/Cas9-mediated immunity to geminiviruses: Differential interference and evasion. *Scientific Reports* 6(1), 26912. DOI: 10.1038/srep26912.

Ali, Z., Eid, A., Ali, S. and Mahfouz, M.M. (2018) Pea early-browning virus-mediated genome editing via the CRISPR/Cas9 system in *Nicotiana benthamiana* and *Arabidopsis*. *Virus Research* 244, 333–337. DOI: 10.1016/j.virusres.2017.10.009.

Aman, R., Ali, Z., Butt, H., Mahas, A., Aljedaani, F. *et al.* (2018) RNA virus interference via CRISPR/Cas13a system in plants. *Genome Biology* 19(1), 1. DOI: 10.1186/s13059-017-1381-1.

Aroge, T., Zhu, Y., Jin, D.-N., Dara, M.Z.N., Feng, J. *et al.* (2024) Omics and CRISPR-Cas9 molecular perception: A progressive review approach for powdery mildew disease management. *Physiological and Molecular Plant Pathology* 130, 102217. DOI: 10.1016/j.pmpp.2024.102217.

Arra, Y., Auguy, F., Stiebner, M., Chéron, S., Wudick, M.M. *et al.* (2024) Rice yellow mottle virus resistance by genome editing of the *Oryza sativa* L. ssp. japonica nucleoporin gene OsCPR5.1 but not OsCPR5.2. *Plant Biotechnology Journal* 22(5), 1299–1311. DOI: 10.1111/pbi.14266.

Arulganesh, T., Kumam, Y. and Kumar, K. (2021) Genome editing of elite rice cultivar, co51 for bacterial leaf blight resistance. *Electronic Journal of Plant Breeding* 12(4), 1060–1068. DOI: 10.37992/2021.1204.147.

Ayaz, M., Li, C.H., Ali, Q., Zhao, W., Chi, Y.K. *et al.* (2023) Bacterial and fungal biocontrol agents for plant disease protection: Journey from lab to field, current status, challenges, and global perspectives. *Molecules* 28(18), 6735. DOI: 10.3390/molecules28186735.

Baldi, P. and La Porta, N. (2020) Molecular approaches for low-cost point-of-care pathogen detection in agriculture and forestry. *Frontiers in Plant Science* 11, 570862. DOI: 10.3389/fpls.2020.570862.

Baltes, N.J., Hummel, A.W., Konecna, E., Cegan, R., Bruns, A.N. *et al.* (2015) Conferring resistance to geminiviruses with the CRISPR-Cas prokaryotic immune system. *Nature Plants* 1(10), 15145. DOI: 10.1038/nplants.2015.145.

Bari, V.K., Nassar, J.A. and Aly, R. (2021) CRISPR/Cas9 mediated mutagenesis of MORE AXILLARY GROWTH 1 in tomato confers resistance to root parasitic weed *Phelipanche aegyptiaca*. *Scientific Reports* 11(1), 3905. DOI: 10.1038/s41598-021-82897-8.

Bi, W., Liu, J., Li, Y., He, Z., Chen, Y. *et al.* (2024) CRISPR/Cas9-guided editing of a novel susceptibility gene in potato improves phytophthora resistance without growth penalty. *Plant Biotechnology Journal* 22(1), 4–6. DOI: 10.1111/pbi.14175.

Borrelli, V.M.G., Brambilla, V., Rogowsky, P., Marocco, A. and Lanubile, A. (2018) The enhancement of plant disease resistance using CRISPR/Cas9 technology. *Frontiers in Plant Science* 9, 1245. DOI: 10.3389/fpls.2018.01245.

Brauer, E.K., Balcerzak, M., Rocheleau, H., Leung, W., Schernthaner, J. *et al.* (2020) Genome editing of a deoxynivalenol-induced transcription factor confers resistance to *Fusarium graminearum* in wheat. *Molecular Plant-Microbe Interactions* 33(3), 553–560. DOI: 10.1094/MPMI-11-19-0332-R.

Bui, T.P., Le, H., Ta, D.T., Nguyen, C.X., Le, N.T. *et al.* (2023) Enhancing powdery mildew resistance in soybean by targeted mutation of MLO genes using the CRISPR/Cas9 system. *BMC Plant Biology* 23(1), 533. DOI: 10.1186/s12870-023-04549-5.

Chakraborty, S. and Newton, A.C. (2011) Climate change, plant diseases and food security: An overview. *Plant Pathology* 60(1), 2–14. DOI: 10.1111/j.1365-3059.2010.02411.x.

Chakraborty, M. and Islam, T. (2022) Antifungal secondary metabolites against blast fungus *Magnaporthe oryzae*. In: Sayyed, R.Z., Singh, A. and Ilyas, N. (eds) *Antifungal Metabolites of Rhizobacteria for Sustainable Agriculture*. Springer International Publishing, Cham, Switzerland, pp. 23–51. DOI: 10.1007/978-3-031-04805-0_2.

Chandrasekaran, J., Brumin, M., Wolf, D., Leibman, D., Klap, C. *et al.* (2016) Development of broad virus resistance in non-transgenic cucumber using CRISPR/Cas9 technology. *Molecular Plant Pathology* 17(7), 1140–1153. DOI: 10.1111/mpp.12375.

Chattopadhyay, A., Purohit, J., Mehta, S., Parmar, H., Karippadakam, S. *et al.* (2022) Precision genome editing toolbox: Applications and approaches for improving rice's genetic resistance to pathogens. *Agronomy* 12(3), 565. DOI: 10.3390/agronomy12030565.

Dale, J., James, A., Paul, J.-Y., Khanna, H., Smith, M. *et al.* (2017) Transgenic cavendish bananas with resistance to *Fusarium wilt* tropical race 4. *Nature Communications* 8(1), 1496. DOI: 10.1038/s41467-017-01670-6.

Damalas, C.A. and Eleftherohorinos, I.G. (2011) Pesticide exposure, safety issues, and risk assessment indicators. *International Journal of Environmental Research and Public Health* 8(5), 1402–1419. DOI: 10.3390/ijerph8051402.

Das, A., Sharma, N. and Prasad, M. (2018) CRISPR/Cas9: A novel weapon in the arsenal to combat plant diseases. *Frontiers in Plant Science* 9, 2008. DOI: 10.3389/fpls.2018.02008.

Das, J., Kumar, B., Saha, B., Jaiswal, S., Iquebal, M.A. *et al.* (2023) Genome-wide identification and characterization of tissue specific long non-coding RNAs and circular RNAs in common carp (*Cyprinus carpio* L.). *Frontiers in Genetics* 14, 1239434. DOI: 10.3389/fgene.2023.1239434.

Ding, Y., Zhang, F., Sun, F., Liu, J., Zhu, Z. *et al.* (2023) Loss of OsHRC function confers blast resistance without yield penalty in rice. *Plant Biotechnology Journal* 21(8), 1516–1518. DOI: 10.1111/pbi.14061.

Djennane, S., Gersch, S., Le-Bohec, F., Piron, M.-C., Baltenweck, R. *et al.* (2024) CRISPR/Cas9 editing of downy mildew resistant 6 (DMR6-1) in grapevine leads to reduced susceptibility to *Plasmopara viticola*. *Journal of Experimental Botany* 75(7), 2100–2112. DOI: 10.1093/jxb/erad487.

Doehlemann, G., Ökmen, B., Zhu, W. and Sharon, A. (2017) Plant pathogenic fungi. In: *Microbiology Spectrum*, Vol. 5 (1). DOI: 10.1128/microbiolspec.funk-0023-2016.

Dutta, T.K., Papolu, P.K., Banakar, P., Choudhary, D., Sirohi, A. *et al.* (2015) Tomato transgenic plants expressing hairpin construct of a nematode protease gene conferred enhanced resistance to root-knot nematodes. *Frontiers in Microbiology* 6, 260. DOI: 10.3389/fmicb.2015.00260.

Dutta, P., Kumari, A., Mahanta, M., Biswas, K.K., Dudkiewicz, A. *et al.* (2022) Advances in nanotechnology as a potential alternative for plant viral disease management. *Frontiers in Microbiology* 13, 935193. DOI: 10.3389/fmicb.2022.935193.

Duy, P.N., Lan, D.T., Pham Thu, H., Thi Thu, H.P., Nguyen Thanh, H. *et al.* (2021) Improved bacterial leaf blight disease resistance in the major elite Vietnamese rice cultivar TBR225 via editing of the OsSSWEET14 promoter. *PLoS One* 16(9), e0255470. DOI: 10.1371/journal.pone.0255470.

El-Mounadi, K., Morales-Floriano, M.L. and Garcia-Ruiz, H. (2020) Principles, applications, and biosafety of plant genome editing using CRISPR-Cas9. *Frontiers in Plant Science* 11, 56. DOI: 10.3389/fpls.2020.00056.

Entine, J., Felipe, M.S.S., Groenewald, J.-H., Kershen, D.L., Lema, M. *et al.* (2021) Regulatory approaches for genome edited agricultural plants in select countries and jurisdictions around the world. *Transgenic Research* 30(4), 551–584. DOI: 10.1007/s11248-021-00257-8.

Fang, Y. and Tyler, B.M. (2016) Efficient disruption and replacement of an effector gene in the oomycete *Phytophthora sojae* using CRISPR/Cas9: *Phytophthora sojae* CRISPR/Cas9. *Molecular Plant Pathology* 17(1), 127–139. DOI: 10.1111/mpp.12318.

FAO (2019) *Production Year Book*. Food and Agriculture Organization of the United Nations, Rome.

FAOSTAT (2018) Statistical Division of the UN Food and Agriculture Organization. Available at: http://www.fao.org/faostat/en/#home (accessed 30 March 2018).

Feng, W., Newbigging, A.M., Tao, J., Cao, Y., Peng, H. *et al.* (2021) CRISPR technology incorporating amplification strategies: Molecular assays for nucleic acids, proteins, and small molecules. *Chemical Science* 12(13), 4683–4698. DOI: 10.1039/D0SC06973F.

Finegold, C., Ried, J., Denby, K. and Gurr, S. (2019) Global burden of crop loss. *Gates Open Research* 3(1).

Fister, A.S., Landherr, L., Maximova, S.N. and Guiltinan, M.J. (2018) Transient expression of CRISPR/Cas9 machinery targeting TcNPR3 enhances defense response in *Theobroma cacao*. *Frontiers in Plant Science* 9, 268. DOI: 10.3389/fpls.2018.00268.

Foster, A.J., Martin-Urdiroz, M., Yan, X., Wright, H.S., Soanes, D.M. *et al.* (2018) CRISPR-Cas9 ribonucleoprotein-mediated co-editing and counterselection in the rice blast fungus. *Scientific Reports* 8(1), 14355. DOI: 10.1038/s41598-018-32702-w.

Fukase, E. and Martin, W. (2020) Economic growth, convergence, and world food demand and supply. *World Development* 132, 104954. DOI: 10.1016/j.worlddev.2020.104954.

Gai, Y. and Wang, H. (2024) Plant disease: A growing threat to global food security. *Agronomy* 14(8), 1615. DOI: 10.3390/agronomy14081615.

Gao, Z., Liu, Q., Zhang, Y., Chen, D., Zhan, X. *et al.* (2020) OsCUL3a-associated molecular switches have functions in cell metabolism, cell death, and disease resistance. *Journal of Agricultural and Food Chemistry* 68(19), 5471–5482. DOI: 10.1021/acs.jafc.9b07426.

Gardiner, D.M. and Kazan, K. (2018) Selection is required for efficient cas9-mediated genome editing in *Fusarium graminearum*. *Fungal Biology* 122(2–3), 131–137. DOI: 10.1016/j.funbio.2017.11.006.

Gheysen, G. and Vanholme, B. (2007) RNAi from plants to nematodes. *Trends in Biotechnology* 25(3), 89–92. DOI: 10.1016/j.tibtech.2007.01.007.

Gomez, M.A., Lin, Z.D., Moll, T., Chauhan, R.D., Hayden, L. *et al.* (2019) Simultaneous CRISPR/Cas9-mediated editing of cassava eIF4E isoforms nCBP-1 and nCBP-2 reduces cassava brown streak disease symptom severity and incidence. *Plant Biotechnology Journal* 17(3), 421–434. DOI: 10.1111/pbi.12987.

Gosavi, G., Yan, F., Ren, B., Kuang, Y., Yan, D. *et al.* (2020) Applications of CRISPR technology in studying plant-pathogen interactions: Overview and perspective. *Phytopathology Research* 2(1), 21. DOI: 10.1186/s42483-020-00060-z.

Gumtow, R., Wu, D., Uchida, J., and Tian, M. (2018) A Phytophthora palmivora extracellular cystatin-like protease inhibitor targets papain to contribute to virulence on papaya. *Molecular Plant-Microbe Interactions* 31(3), 363–373. DOI: 10.1094/MPMI-06-17-0131-FI.

Gupta, S., Kumar, A., Patel, R., and Kumar, V. (2021) Genetically modified crop regulations: scope and opportunity using the CRISPR-Cas9 genome editing approach. *Molecular Biology Reports* 48(5), 4851–4863. DOI: 10.1007/s11033-021-06477-9.

Haque, E., Taniguchi, H., Hassan, Md M., Bhowmik, P., Karim, M.R. *et al.* (2018) Application of CRISPR/Cas9 genome editing technology for the improvement of crops cultivated in tropical climates: Recent progress, prospects, and challenges. *Frontiers in Plant Science* 9, 617. DOI: 10.3389/fpls.2018.00617.

He, D.C., He, M.H., Amalin, D.M., Liu, W., Alvindia, D.G. *et al.* (2021) Biological control of plant diseases: An evolutionary and eco-economic consideration. *Pathogens* 10(10), 1311. DOI: 10.3390/pathogens10101311.

He, N., Huang, F., Yu, M., Zhu, Y., Li, Q.Q. *et al.* (2022) Analysis of a rice blast resistance gene *pita-*fuhui2663 and development of selection marker. *Scientific Reports* 12(1), 14917. DOI: 10.1038/s41598-022-19004-y.

Hirano, S., Nishimasu, H., Ishitani, R. and Nureki, O. (2016) Structural basis for the altered PAM specificities of engineered CRISPR-Cas9. *Molecular Cell* 61(6), 886–894. DOI: 10.1016/j.molcel.2016.02.018.

Hoffie, R.E., Otto, I., Perovic, D., Budhagatapalli, N., Habekuß, A. *et al.* (2021) Targeted knockout of eukaryotic translation initiation factor 4E confers bymovirus resistance in winter barley. *Frontiers in Genome Editing* 3, 784233. DOI: 10.3389/fgeed.2021.784233.

Hong, Y., Meng, J., He, X., Zhang, Y., Liu, Y. *et al.* (2021) Editing miR482b and miR482c simultaneously by CRISPR/Cas9 enhanced tomato resistance to *Phytophthora* infestans. *Phytopathology* 111(6), 1008–1016. DOI: 10.1094/PHYTO-08-20-0360-R.

Hossain, A., Islam, M.T. and Islam, M.T. (2020) Wheat (*Triticum aestivum* L.) in the rice-wheat systems of South Asia is influenced by terminal heat stress at late sown condition: A case in Bangladesh. In: Shabala, S. (ed.) *Plant Stress Physiology*. CAB International, Wallingford, UK. DOI: 10.5772/intechopen.91828.

Hossain, A., Islam, T. and Islam, M.T. (2021) Wheat (*Triticum aestivum* L.) in the rice-wheat systems of South Asia is influenced by terminal heat stress at late sown condition: A case in Bangladesh. In: Shabala, S. (ed.) *Plant Stress Physiology*. CAB International, Wallingford, UK. DOI: 10.5772/intechopen.91828.

Hsu, P.D., Scott, D.A., Weinstein, J.A., Ran, F.A., Konermann, S. *et al.* (2013) DNA targeting specificity of RNA-guided cas9 nucleases. *Nature Biotechnology* 31(9), 827–832. DOI: 10.1038/nbt.2647.

Hu, Y., Zhang, J., Jia, H., Sosso, D., Li, T. *et al.* (2014) Lateral organ boundaries 1 is a disease susceptibility gene for citrus bacterial canker disease. *Proceedings of the National Academy of Sciences of the United States of America* 111(4), E521–E529. DOI: 10.1073/pnas.1313271111.

Hu, X.-H., Shen, S., Wu, J.-L., Liu, J., Wang, H. *et al.* (2023) A natural allele of proteasome maturation factor improves rice resistance to multiple pathogens. *Nature Plants* 9(2), 228–237. DOI: 10.1038/s41477-022-01327-3.

Huang, G., Allen, R., Davis, E.L., Baum, T.J. and Hussey, R.S. (2006) Engineering broad root-knot resistance in transgenic plants by RNAi silencing of a conserved and essential root-knot nematode parasitism gene. *Proceedings of the National Academy of Sciences of the United States of America* 103(39), 14302–14306. DOI: 10.1073/pnas.0604698103.

Huang, J., Gu, L., Zhang, Y., Yan, T., Kong, G. *et al.* (2017) An oomycete plant pathogen reprograms host pre-mRNA splicing to subvert immunity. *Nature Communications* 8(1). DOI: 10.1038/s41467-017-02233-5.

Huang, Q., Lin, B., Cao, Y., Zhang, Y., Song, H. *et al.* (2023) CRISPR/Cas9-mediated mutagenesis of the susceptibility gene oshpp04 in rice confers enhanced resistance to rice root-knot nematode. *Frontiers in Plant Science* 14, 1134653. DOI: 10.3389/fpls.2023.1134653.

Hüdig, M., Laibach, N. and Hein, A.C. (2022) Genome editing in crop plant research: Alignment of expectations and current developments. *Plants* 11(2), 212. DOI: 10.3390/plants11020212.

Hundleby, P. and Harwood, W. (2022) Regulatory constraints and differences of genome-edited crops around the globe. In: Wani, S.H. and Hensel, G. (eds) *Genome Editing: Current Technology Advances and Applications for Crop Improvement*. Springer International Publishing, Cham, Switzerland, pp. 319–341. DOI: 10.1007/978-3-031-08072-2_17.

Hunjan, M.S. and Lore, J.S. (2020) Climate change: Impact on plant pathogens, diseases, and their management. In: Jabran, K., Florentine, S. and Chauhan, B.S. (eds) *Crop Protection under Changing Climate*. Springer, pp. 85–100. DOI: 10.1007/978-3-030-46111-9_4.

Hussain, B., Lucas, S.J. and Budak, H. (2018) CRISPR/Cas9 in plants: At play in the genome and at work for crop improvement. *Briefings in Functional Genomics* 17(5), 319–328. DOI: 10.1093/bfgp/ely016.

Idnurm, A., Urquhart, A.S., Vummadi, D.R., Chang, S., Wouw, A.P. *et al.* (2017) Spontaneous and CRISPR/Cas9-induced mutation of the osmosensor histidine kinase of the canola pathogen *Leptosphaeria maculans*. *Fungal Biology and Biotechnology* 4(1), 1–2. DOI: 10.1186/s40694-017-0043-0.

Imam, J., Singh, P.K. and Shukla, P. (2016) Plant microbe interactions in post genomic era: Perspectives and applications. *Frontiers in Microbiology* 7, 1488. DOI: 10.3389/fmicb.2016.01488.

Irfan, M., Majeed, H., Iftikhar, T. and Ravi, P.K. (2024) A review on molecular scissoring with CRISPR/Cas9 genome editing technology. *Toxicology Research* 13(4). DOI: 10.1093/toxres/tfae105.

Ishino, Y., Shinagawa, H., Makino, K., Amemura, M. and Nakata, A. (1987) Nucleotide sequence of the iap gene, responsible for alkaline phosphatase isozyme conversion in *Escherichia coli*, and identification of the gene product. *Journal of Bacteriology* 169(12), 5429–5433. DOI: 10.1128/jb.169.12.5429-5433.1987.

Islam, M.T. (2019) CRISPR-Cas technology in modifying food crops. *CABI Reviews* 2019, 1–16. DOI: 10.1079/PAVSNNR201914050.

Islam, T. and Kasfy, S.H. (2023) CRISPR-based point-of-care plant disease diagnostics. *Trends in Biotechnology* 41(2), 144–146. DOI: 10.1016/j.tibtech.2022.10.002.

Islam, T. (2024) Genomic surveillance for tackling emerging plant diseases, with special reference to wheat blast. *CABI Reviews* 19(1). DOI: 10.1079/cabireviews.2024.0050.

Islam, M.T., Croll, D., Gladieux, P., Soanes, D.M., Persoons, A. *et al.* (2016) Emergence of wheat blast in Bangladesh was caused by a South American lineage of *Magnaporthe oryzae*. *BMC Biology* 14, 1. DOI: 10.1186/s12915-016-0309-7.

Islam, M.T., Gupta, D.R., Hossain, A., Roy, K.K., He, X. *et al.* (2020) Wheat blast: A new threat to food security. *Phytopathology Research* 2(1), 1–3. DOI: 10.1186/s42483-020-00067-6.

Islam, M.T., Nago, C., Yoshioka, M., Vy, T.T.P., Tosa, Y. *et al.* (2024) Identification of *Rmg11* in tetraploid wheat as a new blast resistance gene with tolerance to high temperature. *Phytopathology* 114(8), 1878–1883. DOI: 10.1094/PHYTO-02-24-0074-R.

Jaganathan, D., Ramasamy, K., Sellamuthu, G., Jayabalan, S. and Venkataraman, G. (2018) CRISPR for crop improvement: An update review. *Frontiers in Plant Science* 9, 985. DOI: 10.3389/fpls.2018.00985.

Jansen, R., Embden, J.D.A. van, Gaastra, W. and Schouls, L.M. (2002) Identification of genes that are associated with DNA repeats in prokaryotes. *Molecular Microbiology* 43(6), 1565–1575. DOI: 10.1046/j.1365-2958.2002.02839.x.

Jeger, M.J. (2022) The impact of climate change on disease in wild plant populations and communities. *Plant Pathology* 71(1), 111–130. DOI: 10.1111/ppa.13434.

Jeger, M.J., Beresford, R., Bock, C., Brown, N., Fox, A. *et al.* (2021) Global challenges facing plant pathology: Multidisciplinary approaches to meet the food security and environmental challenges in the mid-twenty-first century. *CABI Agriculture and Bioscience* 2(1), 1–8. DOI: 10.1186/s43170-021-00042-x.

Ji, X., Zhang, H., Zhang, Y., Wang, Y. and Gao, C. (2015) Establishing a CRISPR-Cas-like immune system conferring DNA virus resistance in plants. *Nature Plants* 1, 15144. DOI: 10.1038/nplants.2015.144.

Jia, H., Zhang, Y., Orbović, V., Xu, J., White, F.F. *et al.* (2017) Genome editing of the disease susceptibility gene CsLOB1 in citrus confers resistance to citrus canker. *Plant Biotechnology Journal* 15(7), 817–823. DOI: 10.1111/pbi.12677.

Jiang, F. and Doudna, J.A. (2017) CRISPR-Cas9 structures and mechanisms. *Annual Review of Biophysics* 46(1), 505–529. DOI: 10.1146/annurev-biophys-062215-010822.

Jiang, W., Zhou, H., Bi, H., Fromm, M., Yang, B. *et al.* (2013) Demonstration of CRISPR/Cas9/sgRNA-mediated targeted gene modification in *Arabidopsis*, tobacco, sorghum and rice. *Nucleic Acids Research* 41(20), e188. DOI: 10.1093/nar/gkt780.

Jinek, M., Chylinski, K., Fonfara, I., Hauer, M., Doudna, J.A. *et al.* (2012) A programmable dual-RNA-guided DNA endonuclease in adaptive bacterial immunity. *Science* 337(6096), 816–821. DOI: 10.1126/science.1225829.

Joshi, T., Sharma, P., Joshi, T., Pandey, S., Pande, V. *et al.* (2020) A spotlight on the recent advances in bacterial plant diseases and their footprint on crop production. In: Mandal, S. and Bhatt, P. (eds) *Recent Advancements in Microbial Diverisity*. Academic Press, pp. 37–69. DOI: 10.1016/b978-0-12-821265-3.00003-7.

Kan, J., Cai, Y., Cheng, C., Jiang, C., Jin, Y. *et al.* (2022) Simultaneous editing of host factor gene TaPDILI5-1 homoeoalleles confers wheat yellow mosaic virus resistance in hexaploid wheat. *The New Phytologist* 234(2), 340–344. DOI: 10.1111/nph.18002.

Kang, H., Peng, Y., Hua, K., Deng, Y., Bellizzi, M. *et al.* (2021) Rapid detection of wheat blast pathogen *Magnaporthe oryzae* Triticum pathotype using genome-specific primers and cas12a-mediated technology. *Engineering* 7(9), 1326–1335. DOI: 10.1016/j.eng.2020.07.016.

Karlsson, M., Kieu, N.P., Lenman, M., Marttila, S., Resjö, S. *et al.* (2024) CRISPR/Cas9 genome editing of potato *StDMR6-1* results in plants less affected by different stress conditions. *Horticulture Research* 11(7), uhae130. DOI: 10.1093/hr/uhae130.

Kasfy, S.H., Hia, F.T. and Islam, T. (2024) Do CRISPR-based disease diagnosis methods qualify as point-of-care diagnostics for plant diseases? *The Nucleus* 67(1), 11–24. DOI: 10.1007/s13237-024-00489-z.

Kashyap, A.S., Manzar, N., Maurya, A., Mishra, D.D., Singh, R.P. *et al.* (2023) Development of diagnostic markers and applied for genetic diversity study and population structure of *Bipolaris sorokiniana* associated with leaf blight complex of wheat. *Journal of Fungi* 9(2), 153. DOI: 10.3390/jof9020153.

Kaur, P., Shukla, N., Joshi, G., VijayaKumar, C., Jagannath, A. *et al.* (2017) Genome-wide identification and characterization of mirnaome from tomato (*Solanum lycopersicum*) roots and root-knot nematode (*Meloidogyne incognita*) during susceptible interaction. *PLoS One* 12(4), e0175178. DOI: 10.1371/journal.pone.0175178.

Kerr, A. (2016) Biological control of crown gall. *Australasian Plant Pathology* 45(1), 15–18. DOI: 10.1007/s13313-015-0389-9.

Khan, M.R. and Jairajpuri, M.S. (2010) *Nematode Infestations. Part I: Food Crops; Part II: Industrial Crops*. National Academy of Sciences.

Khan, M.R., Ahamad, I. and Shah, M.H. (2021) Emerging important nematode problems in field crops and their management. In: Singh, K.P., Jahagirdar, S. and Sarma, B.K. (eds) *Emerging Trends in Plant Pathology*. Springer, pp. 33–62. DOI: 10.1007/978-981-15-6275-4_3.

Kim, C.-Y., Park, J.-Y., Choi, G., Kim, S., Vo, K.T.X. *et al.* (2022) A rice gene encoding glycosyl hydrolase plays contrasting roles in immunity depending on the type of pathogens. *Molecular Plant Pathology* 23(3), 400–416. DOI: 10.1111/mpp.13167.

Kim, Y.A., Moon, H. and Park, C.J. (2019) CRISPR/Cas9-targeted mutagenesis of *Os8N3* in rice to confer resistance to *Xanthomonas oryzae* pv. *oryzae*. *Rice* 12(1), 67. DOI: 10.1186/s12284-019-0325-7.

Kis, A., Hamar, É., Tholt, G., Bán, R. and Havelda, Z. (2019) Creating highly efficient resistance against wheat dwarf virus in barley by employing CRISPR/Cas9 system. *Plant Biotechnology Journal* 17(6), 1004–1006. DOI: 10.1111/pbi.13077.

Kong, G., Wan, L., Deng, Y.Z., Yang, W., Li, W. *et al.* (2019) Pectin acetylesterase PAE5 is associated with the virulence of plant pathogenic oomycete *Peronophythora litchii*. *Physiological and Molecular Plant Pathology* 106, 16–22. DOI: 10.1016/j.pmpp.2018.11.006.

Koonin, E.V. and Makarova, KS. (2019) Origins and evolution of CRISPR-Cas systems. *Philosophical Transactions of the Royal Society B* 374(1772), 20180087. DOI: 10.1098/rstb.2018.0087.

Kosicki, M., Tomberg, K. and Bradley, A. (2018) Repair of double-strand breaks induced by CRISPR-Cas9 leads to large deletions and complex rearrangements. *Nature Biotechnology* 36(8), 765–771. DOI: 10.1038/nbt.4192.

Králová, M., Bergougnoux, V. and Frébort, I. (2021) CRISPR/Cas9 genome editing in ergot fungus *Claviceps purpurea*. *Journal of Biotechnology* 325, 341–354. DOI: 10.1016/j.jbiotec.2020.09.028.

Kumar, V. and Jain, M. (2015) The CRISPR-cas system for plant genome editing: Advances and opportunities. *Journal of Experimental Botany* 66(1), 47–57. DOI: 10.1093/jxb/eru42a.

Kumar, N., Galli, M., Ordon, J., Stuttmann, J., Kogel, K.H. *et al.* (2018) Further analysis of barley MORC1 using a highly efficient RNA-guided Cas9 gene-editing system. *Plant Biotechnology Journal* 16(11), 1892–1903. DOI: 10.1111/pbi.12924.

Kumar, A., Chadha, S. and Rath, D. (2021) CRISPR-Cas9 system for functional genomics of filamentous fungi: Applications and challenges. In: Sharma, V.K., Shah, M.P., Parmar, S. and Kumar, A. (eds) *Fungi Bio-Prospects in Sustainable Agriculture, Environment and Nano-Technology*. Academic Press, pp. 541–576. DOI: 10.1016/B978-0-12-821734-4.00016-2.

Kumari, R., Saha, T., Kumar, P. and Singh, A.K. (2024) CRISPR/Cas9-mediated genome editing technique to control fall armyworm (*Spodoptera frugiperda*) in crop plants with special reference to maize. *Physiology and Molecular Biology of Plants* 30(7), 1161–1173. DOI: 10.1007/s12298-024-01486-x.

Langner, T., Kamoun, S. and Belhaj, K. (2018) CRISPR crops: Plant genome editing toward disease resistance. *Annual Review of Phytopathology* 56(1), 479–512. DOI: 10.1146/annurev-phyto-080417-050158.

Lee, Y.R., Siddique, M.I., Kim, D.S., Lee, E.S., Han, K. *et al.* (2023) CRISPR/Cas9-mediated gene editing to confer turnip mosaic virus (TuMV) resistance in Chinese cabbage (*Brassica rapa*). *Horticulture Research* 10(6), uhad078. DOI: 10.1093/hr/uhad078.

Li, T., Liu, B., Chen, C.Y. and Yang, B. (2014) TALEN utilization in rice genome modifications. *Methods* 69(1), 9–16. DOI: 10.1016/j.ymeth.2014.03.019.

Li, J., Meng, X., Zong, Y., Chen, K., Zhang, H. *et al.* (2016) Gene replacements and insertions in rice by intron targeting using CRISPR-Cas9. *Nature Plants* 2(10), 16139. DOI: 10.1038/nplants.2016.139.

Li, W., Zhu, Z., Chern, M., Yin, J., Yang, C. *et al.* (2017) A natural allele of a transcription factor in rice confers broad-spectrum blast resistance. *Cell* 170(1), 114–126. DOI: 10.1016/j.cell.2017.06.008.

Li, C., Li, W., Zhou, Z., Chen, H., Xie, C. *et al.* (2019a) A new rice breeding method: CRISPR/Cas9 system editing of the Xa13 promoter to cultivate transgene-free bacterial blight-resistant rice. *Plant Biotechnology Journal* 18(2), 313–315. DOI: 10.1111/pbi.13217.

Li, S., Shen, L., Hu, P., Liu, Q., Zhu, X. *et al.* (2019b) Developing disease-resistant thermosensitive male sterile rice by multiplex gene editing. *Journal of Integrative Plant Biology* 61(12), 1201–1205. DOI: 10.1111/jipb.12774.

Li, Y., Li, W. and Li, J. (2021) The CRISPR/Cas9 revolution continues: From base editing to prime editing in plant science. *Journal of Genetics and Genomics* 48(8), 661–670. DOI: 10.1016/j.jgg.2021.05.001.

Li, C., Zhou, L., Wu, B., Li, S., Zha, W. *et al.* (2022a) Improvement of bacterial blight resistance in two conventionally cultivated rice varieties by editing the noncoding region. *Cells* 11(16), 2535. DOI: 10.3390/cells11162535.

Li, S., Lin, D., Zhang, Y., Deng, M., Chen, Y. *et al.* (2022b) Genome-edited powdery mildew resistance in wheat without growth penalties. *Nature* 602(7897), 455–460. DOI: 10.1038/s41586-022-04395-9.

Li, Z., Huang, Q., Lin, B., Guo, B., Wang, J. *et al.* (2022c) CRISPR/Cas9-targeted mutagenesis of a representative member of a novel pr10/bet v1-like protein subfamily significantly reduces rice plant height and defense against *Meloidogyne graminicola*. *Phytopathology Research* 4(1), 38. DOI: 10.1186/s42483-022-00143-z.

Li, X., Hu, D., Cai, L., Wang, H., Liu, X. *et al.* (2022d) CALCIUM-DEPENDENT PROTEIN KINASE38 regulates flowering time and common cutworm resistance in soybean. *Plant Physiology* 190(1), 480–499. DOI: 10.1093/plphys/kiac260.

Li, H., Gong, P., Xu, X., Zhou, X. and Li, F. (2024a) Knockout of the virus replication-related genes UbEF1b and CCR4/NOT3 by CRISPR/Cas9 confers high-efficiency and broad-spectrum resistance to geminiviruses in *Nicotiana benthamiana*. *Plant Biotechnology Journal* 22(4), 793. DOI: 10.1111/pbi.14265.

Li, R., Cui, L., Martina, M., Bracuto, V., Meijer-Dekens, F. *et al.* (2024b) Less is more: CRISPR/Cas9-based mutations in DND1 gene enhance tomato resistance to powdery mildew with low fitness costs. *BMC Plant Biology* 24(1), 763. DOI: 10.1186/s12870-024-05428-3.

Liu, D., Chen, X., Liu, J., Ye, J. and Guo, Z. (2012) The rice ERF transcription factor *oserf922* negatively regulates resistance to *Magnaporthe oryzae* and salt tolerance. *Journal of Experimental Botany* 63(10), 3899–3911. DOI: 10.1093/jxb/ers079.

Liu, C., Zhang, L., Liu, H. and Cheng, K. (2017a) Delivery strategies of the CRISPR-Cas9 gene-editing system for therapeutic applications. *Journal of Controlled Release* 266, 17–26. DOI: 10.1016/j.jconrel.2017.09.012.

Liu, Q., Ning, Y., Zhang, Y., Yu, N., Zhao, C. *et al.* (2017b) OsCUL3a negatively regulates cell death and immunity by degrading OsNPR1 in rice. *The Plant Cell* 29(2), 345–359. DOI: 10.1105/tpc.16.00650.

Liu, Z., Dong, H., Cui, Y., Cong, L. and Zhang, D. (2020) Application of different types of CRISPR/Cas-based systems in bacteria. *Microbial Cell Factories* 19(1), 172. DOI: 10.1186/s12934-020-01431-z.

Liu, H., Dong, S., Li, M., Gu, F., Yang, G. *et al.* (2021a) The class III peroxidase gene OsPRX30, transcriptionally modulated by the AT-hook protein OsATH1, mediates rice bacterial blight-induced ROS accumulation. *Journal of Integrative Plant Biology* 63(2), 393–408. DOI: 10.1111/jipb.13040.

Liu, T.Y., Knott, G.J., Smock, D.C.J., Desmarais, J.J., Son, S. *et al.* (2021b) Accelerated RNA detection using tandem CRISPR nucleases. *Nature Chemical Biology* 17(9), 982–988. DOI: 10.1038/s41589-021-00842-2.

Liu, X., Song, L., Zhang, H., Lin, Y., Shen, X. *et al.* (2021c) Rice ubiquitin-conjugating enzyme OsUBC26 is essential for immunity to the blast fungus *Magnaporthe oryzae*. *Molecular Plant Pathology* 22(12), 1613–1623. DOI: 10.1111/mpp.13132.

Low, Y.C., Lawton, M.A. and Di, R. (2020) Validation of barley 2OGO gene as a functional orthologue of *Arabidopsis* DMR6 gene in fusarium head blight susceptibility. *Scientific Reports* 10(1), 9935. DOI: 10.1038/s41598-020-67006-5.

Lu, H., Luo, T., Fu, H., Wang, L., Tan, Y. *et al.* (2018) Resistance of rice to insect pests mediated by suppression of serotonin biosynthesis. *Nature Plants* 4(6), 338–344. DOI: 10.1038/s41477-018-0152-7.

Lu, W., Deng, F., Jia, J., Chen, X., Li, J. *et al.* (2020) The *Arabidopsis thaliana* gene aterf019 negatively regulates plant resistance to *Phytophthora parasitica* by suppressing PAMP-triggered immunity. *Molecular Plant Pathology* 21(9), 1179–1193. DOI: 10.1111/mpp.12971.

Lucas, J.A., Hawkins, N.J. and Fraaije, B.A. (2015) The evolution of fungicide resistance. *Advances in Applied Microbiology* 90, 29–92. DOI: 10.1016/bs.aambs.2014.09.001.

Lyngkjær, M.F., Newton, A.C., Atzema, J.L. and Baker, S.J. (2000) The barley mlo-gene: An important powdery mildew resistance source. *Agronomie* 20(7), 745–756. DOI: 10.1051/agro:2000173.

Ma, J., Chen, J., Wang, M., Ren, Y., Wang, S. *et al.* (2018) Disruption of OsSEC3a increases the content of salicylic acid and induces plant defense responses in rice. *Journal of Experimental Botany* 69(5), 1051–1064. DOI: 10.1093/jxb/erx458.

Macovei, A., Sevilla, N.R., Cantos, C., Jonson, G.B., Slamet-Loedin, I. *et al.* (2018) Novel alleles of rice eF4G generated by CRISPR/Cas9-targeted mutagenesis confer resistance to rice tungro spherical virus. *Plant Biotechnology Journal* 16(11), 1918–1927. DOI: 10.1111/pbi.12927.

Mahas, A., Ali, Z., Tashkandi, M. and Mahfouz, M.M. (2019) Virus-mediated genome editing in plants using the CRISPR/Cas9 system. *Methods in Molecular Biology* 311–326. DOI: 10.1007/978-1-4939-8991-1_23.

Mahmud, N.U., Gupta, D.R., Paul, S.K., Chakraborty, M., Mehebub, M.S. *et al.* (2022) Daylight-driven rechargeable TiO2 nanocatalysts suppress wheat blast caused by Magnaporthe oryzae Triticum. *Bulletin of the Chemical Society of Japan* 95(8), 1263–1271. DOI: 10.1246/bcsj.20220010.

Majeed, M., Soliman, H., Kumar, G., El-Matbouli, M. and Saleh, M. (2018) Editing the genome of *Aphanomyces invadans* using CRISPR/Cas9. *Parasites & Vectors* 11(1). DOI: 10.1186/s13071-018-3134-8.

Mali, P., Esvelt, K.M. and Church, G.M. (2013) Cas9 as a versatile tool for engineering biology. *Nature Methods* 10(10), 957–963. DOI: 10.1038/nmeth.2649.

Malnoy, M., Viola, R., Jung, M.-H., Koo, O.-J., Kim, S. *et al.* (2016) DNA-free genetically edited grapevine and apple protoplast using CRISPR/Cas9 ribonucleoproteins. *Frontiers in Plant Science* 7. DOI: 10.3389/fpls.2016.01904.

Manzar, N., Kashyap, A.S., Maurya, A., Rajawat, M.V.S., Sharma, P.K. *et al.* (2022) Multi-gene phylogenetic approach for identification and diversity analysis of *Bipolaris maydis* and *Curvularia lunata* isolates causing foliar blight of *Zea mays*. *Journal of Fungi* 8(8), 802. DOI: 10.3390/jof8080802.

Manzoor, S., Nabi, S.U., Rather, T.R., Gani, G., Mir, Z.A. *et al.* (2024) Advancing crop disease resistance through genome editing: A promising approach for enhancing agricultural production. *Frontiers in Genome Editing* 6, 1399051. DOI: 10.3389/fgeed.2024.1399051.

Martella, A. and Fisher, D.I. (2021) Regulation of gene expression and the elucidative role of CRISPR-based epigenetic modifiers and CRISPR-induced chromosome conformational changes. *The CRISPR Journal* 4(1), 43–57. DOI: 10.1089/crispr.2020.0108.

Mehta, D., Stürchler, A., Anjanappa, R.B., Zaidi, S.S.-E.-A., Hirsch-Hoffmann, M. *et al.* (2019) Linking CRISPR-Cas9 interference in cassava to the evolution of editing-resistant geminiviruses. *Genome Biology* 20(1), 80. DOI: 10.1186/s13059-019-1678-3.

Mew, T.W., Alvarez, M.A., Leach, E.J. and Swings, J. (1993) Focus on bacterial blight of rice. *Plant Disease* 77(1), 5–12. DOI: 10.1094/PD-77-0005.

Miladinovic, D., Antunes, D., Yildirim, K., Bakhsh, A., Cvejić, S. *et al.* (2021) Targeted plant improvement through genome editing: From laboratory to field. *Plant Cell Reports* 40(6), 935–951. DOI: 10.1007/s00299-020-02655-4.

Milazzo, J., Adre, H., Tharreau, D., Franco, O.L. and et al, A. (2022) CRISPR/Cas9-targeted knockout of rice susceptibility genes OsDJA2 and OsERF104 reveals alternative sources of resistance. *Rice Science* 29(6), 535–544.

Misra, V., Mall, A.K., Pandey, H., Srivastava, S. and Sharma, A. (2023) Advancements and prospects of CRISPR/Cas9 technologies for abiotic and biotic stresses in sugar beet. *Frontiers in Genetics* 14, 1235855. DOI: 10.3389/fgene.2023.1235855.

Mojica, F.J.M., Díez-Villaseñor, C., García-Martínez, J. and Soria, E. (2005) Intervening sequences of regularly spaced prokaryotic repeats derive from foreign genetic elements. *Journal of Molecular Evolution* 60(2), 174–182. DOI: 10.1007/s00239-004-0046-3.

Mounika, K. and Lal, A.A. (2021) Major foliar fungal diseases of annual and perennial crops: Challenging national food and nutritional security. *Journal of Plant Pathology & Microbiology* 12, 563.

Munawar, N., Ahsan, K. and Ahmad, A. (2024) CRISPR-edited plants' social, ethical, policy, and governance issues. In: Abd-Elsalam, K.A. and Ahmad, A. (eds) *Global Regulatory Outlook for CRISPRized Plants*. Academic Press, pp. 367–396. DOI: 10.1016/B978-0-443-18444-4.00011-9.

Murovec, J., Pirc, Ž. and Yang, B. (2017) New variants of CRISPR RNA-guided genome editing enzymes. *Plant Biotechnology Journal* 15(8), 917–926. DOI: 10.1111/pbi.12736.

Mushtaq, M., Dar, A.A., Basu, U., Bhat, B.A., Mir, R.A. *et al.* (2021) Integrating CRISPR-Cas and next generation sequencing in plant virology. *Frontiers in Genetics* 12, 735489. DOI: 10.3389/fgene.2021.735489.

Nekrasov, V., Wang, C., Win, J., Lanz, C., Weigel, D. *et al.* (2017) Rapid generation of a transgene-free powdery mildew resistant tomato by genome deletion. *Scientific Reports* 7(1), 482. DOI: 10.1038/s41598-017-00578-x.

Niu, J., Jian, H., Xu, J., Chen, C., Guo, Q. *et al.* (2012) RNAi silencing of the *Meloidogyne incognita* Rpn7 gene reduces nematode parasitic success. *European Journal of Plant Pathology* 134(1), 131–144. DOI: 10.1007/s10658-012-9971-y.

Norouzi, M., Nazarain-Firouzabadi, F., Ismaili, A., Ahmadvand, R. and Poormazaheri, H. (2024) CRISPR/Cas StNRL1 gene knockout increases resistance to late blight and susceptibility to early blight in potato. *Frontiers in Plant Science* 14, 1278127. DOI: 10.3389/fpls.2023.1278127.

Noureen, A., Khan, M.Z., Amin, I., Zainab, T., and Mansoor, S. (2022) CRISPR/Cas9-mediated targeting of susceptibility factor eIF4E-enhanced resistance against potato virus Y. *Frontiers in Genetics* 13, 922019. DOI: 10.3389/fgene.2022.922019.

Ntui, V.O., Tripathi, J.N., Shah, T., and Tripathi, L. (2024) Targeted knockout of early nodulin-like 3 (MusaENODL3) gene in banana reveals its function in resistance to *Xanthomonas wilt* disease. *Plant Biotechnology Journal* 22(5), 1101–1112. DOI: 10.1111/pbi.14248.

Ortigosa, A., Gimenez-Ibanez, S., Leonhardt, N., and Solano, R. (2019) Design of a bacterial speck resistant tomato by CRISPR/Cas9-mediated editing of S1JAZ2. *Plant Biotechnology Journal* 17(3), 665–673. DOI: 10.1111/pbi.13006.

Ou, S.H. (1985) Bacterial leaf blight. In: *Rice Diseases*, 2nd edn. Mycological Institute, Kew, UK.

Panth, M., Hassler, S.C., and Baysal-Gurel, F. (2020) Methods for management of soilborne diseases in crop production. *Agriculture* 10(1), 16. DOI: 10.3390/agriculture10010016.

Park, S., Shi, A., Meinhardt, L.W. and Mou, B. (2023) Genome-wide characterization and evolutionary analysis of the AP2/ERF gene family in lettuce (*Lactuca sativa*). *Scientific Reports* 13(1), 21990. DOI: 10.1038/s41598-023-49245-4.

Parrado, L.M. and Quintanilla, M. (2024) Plant-parasitic nematode disease complexes as overlooked challenges to crop production. *Frontiers in Plant Science* 15, 1439951.

Patel, S., Sharma, A., Batra, N., Chetani, R., Trivedi, P.C. *et al.* (2024) Perspectives and overview of CRISPR/Cas technology in plant pathogenesis. In: Kumar, A., Arora, S., Ogita, S., Yau, Y.-Y. and Mukherjee, K. (eds) *Gene Editing in Plants: CRISPR-Cas and Its Applications*. Springer Nature Singapore, pp. 571–588.

Paul, B., Chaubet, L., Verver, D.E. and Montoya, G. (2022a) Mechanics of CRISPR-Cas12a and engineered variants on λ-DNA. *Nucleic Acids Research* 50(9), 5208–5225. DOI: 10.1093/nar/gkab1272.

Paul, S.K., Mahmud, N.U., Gupta, D.R., Rani, K., Kang, H. *et al.* (2022b) Oryzae pathotype of *Magnaporthe oryzae* can cause typical blast disease symptoms on both leaves and spikes of wheat under a growth room condition. *Phytopathology Research* 4(1), 1–12. DOI: 10.1186/s42483-022-00114-4.

Peng, A., Chen, S., Lei, T., Xu, L., He, Y. *et al.* (2017) Engineering canker-resistant plants through CRISPR/Cas9-targeted editing of the susceptibility gene CsLOB1 promoter in citrus. *Plant Biotechnology Journal* 15(12), 1509–1519. DOI: 10.1111/pbi.12733.

Pröbsting, M., Schenke, D., Hossain, R., Häder, C., Thurau, T. *et al.* (2020) Loss of function of CRT1a (calreticulin) reduces plant susceptibility to *Verticillium longisporum* in both *Arabidopsis thaliana* and oilseed rape (*Brassica napus*). *Plant Biotechnology Journal* 18(11), 2328–2344. DOI: 10.1111/pbi.13394.

Pyott, D.E., Sheehan, E. and Molnar, A. (2016) Engineering of CRISPR/Cas9-mediated potyvirus resistance in transgene-free *Arabidopsis* plants. *Molecular Plant Pathology* 17(8), 1276–1288. DOI: 10.1111/mpp.12417.

Qi, L.S., Larson, M.H., Gilbert, L.A., Doudna, J.A., Weissman, J.S. *et al.* (2013) Repurposing CRISPR as an RNA-guided platform for sequence-specific control of gene expression. *Cell* 152(5), 1173–1183. DOI: 10.1016/j.cell.2013.02.022.

Rabiey, M., Hailey, L.E., Roy, S.R., Grenz, K., Al-Zadjali, M.A.S. *et al.* (2019) Endophytes vs tree pathogens and pests: Can they be used as biological control agents to improve tree health? *European Journal of Plant Pathology* 155(3), 711–729. DOI: 10.1007/s10658-019-01814-y.

Rao, G.P. and Reddy, M.G. (2020) Overview of yield losses due to plant viruses. In: Awasthi, L.P. (ed.) *Applied Plant Virology*. Elsevier, pp. 531–562. DOI: 10.1016/B978-0-12-818654-1.00038-4.

Rao, M.J. and Wang, L. (2021) CRISPR/Cas9 technology for improving agronomic traits and future prospective in agriculture. *Planta* 254, 1–6. DOI: 10.1007/s00425-021-03716-y.

Rato, C., Carvalho, M.F., Azevedo, C. and Oblessuc, P.R. (2021) Genome editing for resistance against plant pests and pathogens. *Transgenic Research* 30(4), 427–459. DOI: 10.1007/s11248-021-00262-x.

Ristaino, J.B., Anderson, P.K., Bebber, D.P., Brauman, K.A., Cunniffe, N.J. *et al.* (2021) The persistent threat of emerging plant disease pandemics to global food security. *Proceedings of the National Academy of Sciences* 118(23), e2022239118. DOI: 10.1073/pnas.2022239118.

Roy, B., Zhao, J., Yang, C., Luo, W., Xiong, T. *et al.* (2018) CRISPR/Cascade 9-mediated genome editing: Challenges and opportunities. *Frontiers in Genetics* 9, 240. DOI: 10.3389/fgene.2018.00240.

Rozas, P., Kessi-Pérez, E.I. and Martínez, C. (2022) Genetically modified organisms: Adapting regulatory frameworks for evolving genome editing technologies. *Biological Research* 55(1), 31. DOI: 10.1186/s40659-022-00399-x.

Różewicz, M., Wyzińska, M. and Grabiński, J. (2021) The most important fungal diseases of cereals: Problems and possible solutions. *Agronomy* 11(4), 714. DOI: 10.3390/agronomy11040714.

Saikia, B., Remya, S., Debbarma, J., Maharana, J., Sastry, G.N. *et al.* (2024) CRISPR/Cas9-based genome editing and functional analysis of SiHyPRP1 and SIDEA1 genes of *Solanum lycopersicum* L. in imparting genetic tolerance to multiple stress factors. *Frontiers in Plant Science* 15, 1304381. DOI: 10.3389/fpls.2024.1304381.

Santillán Martínez, M.I., Bracuto, V., Koseoglou, E., Appiano, M., Jacobsen, E. *et al.* (2020) CRISPR/Cas9-targeted mutagenesis of the tomato susceptibility gene PMR4 for resistance against powdery mildew. *BMC Plant Biology* 20(1), 284. DOI: 10.1186/s12870-020-02497-y.

Savary, S., Willocquet, L., Pethybridge, S.J., Esker, P., McRoberts, N. *et al.* (2019) The global burden of pathogens and pests on major food crops. *Nature Ecology & Evolution* 3(3), 430–439. DOI: 10.1038/s41559-018-0793-y.

Schenke, D. (2020) CRISPR/Cas9 or prime editing? It depends on .... *Open Life Sciences* 15(1), 868–870. DOI: 10.1515/biol-2020-0109.

Schuster, M. and Kahmann, R. (2019) CRISPR-Cas9 genome editing approaches in filamentous fungi and oomycetes. *Fungal Genetics and Biology* 130, 43–53. DOI: 10.1016/j.fgb.2019.04.016.

Schwartz, A.R., Potnis, N., Timilsina, S., Wilson, M., Patané, J. *et al.* (2015) Phylogenomics of *Xanthomonas* field strains infecting pepper and tomato reveals diversity in effector repertoires and identifies determinants of host specificity. *Frontiers in Microbiology* 6, 535. DOI: 10.3389/fmicb.2015.00535.

Semenza, J.C., Rocklöv, J. and Ebi, K.L. (2022) Climate change and cascading risks from infectious disease. *Infectious Diseases and Therapy* 11(4), 1371–1390. DOI: 10.1007/s40121-022-00647-3.

Shan, Q., Wang, Y., Li, J., Zhang, Y., Chen, K. *et al.* (2013) Targeted genome modification of crop plants using a CRISPR-Cas system. *Nature Biotechnology* 31(8), 686–688. DOI: 10.1038/nbt.2650.

Silva, C.J., Abeele, C., Ortega-Salazar, I., Papin, V., Adaskaveg, J.A. *et al.* (2021) Host susceptibility factors render ripe tomato fruit vulnerable to fungal disease despite active immune responses. *Journal of Experimental Botany* 72(7), 2696–2709. DOI: 10.1093/jxb/eraa601.

Sindhu, A.S., Maier, T.R., Mitchum, M.G., Hussey, R.S., Davis, E.L. *et al.* (2009) Effective and specific in planta RNAi in cyst nematodes: Expression interference of four parasitism genes reduces parasitic success. *Journal of Experimental Botany* 60(1), 315–324. DOI: 10.1093/jxb/ern289.

Singh, V.K., Singh, R., Kumar, A. and Bhadouria, R. (2021) Current status of plant diseases and food security. In: Kumar, A. and Droby, S. (eds) *Food Security and Plant Disease Management*. Woodhead Publishing, pp. 19–35. DOI: 10.1016/B978-0-12-821843-3.00019-2.

Singh, J., Sharma, D., Brar, G.S., Sandhu, K.S., Wani, S.H. *et al.* (2022) CRISPR/Cas tool designs for multiplex genome editing and its applications in developing biotic and abiotic stress-resistant crop plants. *Molecular Biology Reports* 49(12), 11443–11467. DOI: 10.1007/s11033-022-07741-2.

Sobanbabu, G., Sabarinathan, K.G., Parthiban, V.K. and Ramamoorthy, V. (2018) Isolation, screening and identification of virulent isolates of *Bipolaris oryzae* causing rice brown spot and *Sarocladium oryzae* causing sheath rot disease. *International Journal of Current Microbiology and Applied Sciences* 7(9), 930–939. DOI: 10.20546/ijcmas.2018.709.112.

Standage-Beier, K., Brookhouser, N., Balachandran, P., Zhang, Q., Brafman, D.A. *et al.* (2019) RNA-guided recombinase-Cas9 fusion targets genomic DNA deletion and integration. *The CRISPR Journal* 2(4), 209–222. DOI: 10.1089/crispr.2019.0013.

Sun, Q., Lin, L., Liu, D., Wu, D., Fang, Y. *et al.* (2018) CRISPR/Cas9-mediated multiplex genome editing of the bnWRKY11 and bnWRKY70 genes in *Brassica napus* L. *International Journal of Molecular Sciences* 19(9), 2716. DOI: 10.3390/ijms19092716.

Talakayala, A., Mekala, G.K., Reddy, M.K., Ankanagari, S. and Garladinne, M. (2022) Manipulating resistance to mungbean yellow mosaic virus in greengram (*Vigna radiata* L): Through CRISPR/Cas9 mediated editing of the viral genome. *Frontiers in Sustainable Food Systems* 6, 911574. DOI: 10.3389/fsufs.2022.911574.

Tang, X., Ren, Q., Yang, L., Bao, Y., Zhong, Z. *et al.* (2019) Single transcript unit CRISPR 2.0 systems for robust Cas9 and Cas12a mediated plant genome editing. *Plant Biotechnology Journal* 17(7), 1431–1445. DOI: 10.1111/pbi.13068.

Tao, Y. (2024) Progress of construction of gene-modified soybean based on the CRISPR/Cas9 system. *Highlights in Science, Engineering and Technology* 91, 178–185. DOI: 10.54097/1j3ams13.&quot.

Távora, F.T.P.K., Meunier, A.C., Vernet, A., Portefaix, M., Milazzo, J. *et al.* (2022) CRISPR/Cas9-targeted knockout of rice susceptibility genes OsDJA2 and OsERF104 reveals alternative sources of resistance. *Rice Science* 29(6), 535–544. DOI: 10.1016/j.rsci.2022.04.001.

Tian, B., Li, J., Vodkin, L.O., Todd, T.C., Finer, J.J. *et al.* (2019) Host-derived gene silencing of parasite fitness genes improves resistance to soybean cyst nematodes in stable transgenic soybean. *Theoretical and Applied Genetics* 132(9), 2651–2662. DOI: 10.1007/s00122-019-03379-0.

Toledo Thomazella, D.P., Brail, Q., Dahlbeck, D. and Staskawicz, B.J. (2016) CRISPR-Cas9 mediated mutagenesis of a DMR6 ortholog in tomato confers broad-spectrum disease resistance. *Biorxiv*. DOI: 10.1101/064824.

Tripathi, J.N., Ntui, V.O., Shah, T. and Tripathi, L. (2021) CRISPR/Cas9-mediated editing of DMR6 orthologue in banana (*Musa* spp.) confers enhanced resistance to bacterial disease. *Plant Biotechnology Journal* 19(7), 1291–1293. DOI: 10.1111/pbi.13614.

Tripathi, J.N., Ntui, V.O., Ron, M., Muiruri, S.K., Britt, A. *et al.* (2019) CRISPR/Cas9 editing of endogenous banana streak virus in the B genome of *Musa* spp. overcomes a major challenge in banana breeding. *Communications Biology* 2(1), 46. DOI: 10.1038/s42003-019-0288-7.

Tsygankova, V.A., Andrusevich, Y.V., Shysha, E.N., Biliavska, L.O., Galagan, T.O. *et al.* (2019) RNAi-mediated resistance against plant parasitic nematodes of wheat plants obtained *in vitro* using bioregulators of microbiological origin. *Current Chemical Biology* 13(1), 73–89. DOI: 10.2174/2212796812666180507130017.

Tyagi, S., Kumar, R., Das, A., Won, S.Y. and Shukla, P. (2020) CRISPR-Cas9 system: A genome-editing tool with endless possibilities. *Journal of Biotechnology* 319, 36–53. DOI: 10.1016/j.jbiotec.2020.05.008.

Tyagi, S., Kumar, R., Kumar, V., Won, S.Y. and Shukla, P. (2021) Engineering disease resistant plants through CRISPR-Cas9 technology. *GM Crops & Food* 12(1), 125–144. DOI: 10.1080/21645698.2020.1831729.

USDA/APHIS (2017) Biotechnology Regulatory Services. Available at: https://www.aphis.usda.gov/biotechnology/downloads/reg_loi/1707601_air_inquiry_cbidel.pdf (accessed 7 May 2025).

Vazquez-Vilar, M., Garcia-Carpintero, V., Selma, S., Bernabé-Orts, J.M., Sanchez-Vicente, J. *et al.* (2021) The GB4.0 platform, an all-in-one tool for CRISPR/Cas-based multiplex genome engineering in plants. *Frontiers in Plant Science* 12, 689937. DOI: 10.3389/fpls.2021.689937.

Veillet, F., Kermarrec, M.P., Chauvin, L., Chauvin, J.E. and Nogué, F. (2020) CRISPR-induced indels and base editing using the *Staphylococcus aureus* aCs9 in potato. *PLoS One* 15(8), e0235942. DOI: 10.1371/journal.pone.0235942.

Veley, K.M., Elliott, K., Jensen, G., Zhong, Z., Feng, S. *et al.* (2022) Improving disease resistance in plants by editing the epigenome. *BioRxiv*. DOI: 10.1101/2022.01.04.474801.

Waltz, E. (2018) With a free pass, CRISPR-edited plants reach market in record time. *Nature Biotechnology* 36(1), 6–8. DOI: 10.1038/nbt0118-6b.

Wang, Y., Cheng, X., Shan, Q., Zhang, Y., Liu, J. *et al.* (2014) Simultaneous editing of three homoeoalleles in hexaploid bread wheat confers heritable resistance to powdery mildew. *Nature Biotechnology* 32(9), 947–951. DOI: 10.1038/nbt.2969.

Wang, F., Wang, C., Liu, P., Lei, C., Hao, W. *et al.* (2016) Enhanced rice blast resistance by CRISPR/Cas9-targeted mutagenesis of the ERF transcription factor gene OsERF922. *PloS One* 11(4), e0154027. DOI: 10.1371/journal.pone.0154027.

Wang, T., Deng, Z., Zhang, X., Wang, H., Wang, Y. *et al.* (2018a) Tomato DCL2b is required for the biosynthesis of 22-nt small RNAs, the resulting secondary siRNAs, and the host defense against ToMV. *Horticulture Research* 5, 62. DOI: 10.1038/s41438-018-0073-7.

Wang, X., Tu, M., Wang, D., Liu, J., Li, Y. *et al.* (2018b) CRISPR/Cas9-mediated efficient targeted mutagenesis in grape in the first generation. *Plant Biotechnology Journal* 16(4), 844–855. DOI: 10.1111/pbi.12832.

Wang, Q., Cobine, P.A. and Coleman, J.J. (2018c) Efficient genome editing in *Fusarium oxysporum* based on CRISPR/Cas9 ribonucleoprotein complexes. *Fungal Genetics and Biology* 117, 21–29. DOI: 10.1016/j.fgb.2018.05.003.

Wang, L., Chen, H., Li, J., Shu, H., Zhang, X. *et al.* (2020) Effector gene silencing mediated by histone methylation underpins host adaptation in an oomycete plant pathogen. *Nucleic Acids Research* 48(4), 1790–1799. DOI: 10.1093/nar/gkz1160.

Wang, Y., Zhuoma, Q., Xu, Z., Peng, Y. and Wang, M. (2023) Virulence and genetic types of *Blumeria graminis* f. sp. *hordei* in Tibet and surrounding areas. *Journal of Fungi* 9(3), 363. DOI: 10.3390/jof9030363.

Wang, F., Liang, S., Wang, G., Hu, T., Fu, C. *et al.* (2024a) CRISPR-Cas9-mediated construction of a cotton CDPK mutant library for identification of insect-resistance genes. *Plant Communications* 5(11), 101047. DOI: 10.1016/j.xplc.2024.101047.

Wang, M., Li, S., Li, H., Song, C., Xie, W. *et al.* (2024b) Genome editing of a dominant resistance gene for broad-spectrum resistance to bacterial diseases in rice without growth penalty. *Plant Biotechnology Journal* 22(3), 529–531. DOI: 10.1111/pbi.14233.

Wei, H., Wang, X., He, Y., Xu, H. and Wang, L. (2021a) Clock component *osprr73* positively regulates rice salt tolerance by modulating *oshkt2;1*-mediated sodium homeostasis. *The EMBO Journal* 40(3), e105086. DOI: 10.15252/embj.2020105086.

Wei, Z., Abdelrahman, M., Gao, Y., Ji, Z., Mishra, R. *et al.* (2021b) Engineering broad-spectrum resistance to bacterial blight by CRISPR-Cas9-mediated precise homology directed repair in rice. *Molecular Plant* 14(8), 1215–1218. DOI: 10.1016/j.molp.2021.05.012.

Weiss, T., Crisp, P.A., Rai, K.M., Song, M., Springer, N.M. *et al.* (2022) Epigenetic features drastically impact CRISPR–Cas9 efficacy in plants. *Plant Physiology* 190(2), 1153–1164. DOI: 10.1093/plphys/kiac285.

Wenderoth, M., Pinecker, C., Voß, B. and Fischer, R. (2017) Establishment of CRISPR/Cas9 in *Alternaria alternata*. *Fungal Genetics and Biology* 101, 55–60. DOI: 10.1016/j.fgb.2017.03.001.

Xie, K. and Yang, Y. (2013) RNA-guided genome editing in plants using a CRISPR-Cas system. *Molecular Plant* 6(6), 1975–1983. DOI: 10.1093/mp/sst119.

Xu, J., Wang, X., Zu, H., Zeng, X., Baldwin, I.T. *et al.* (2021a) Molecular dissection of rice phytohormone signaling involved in resistance to a piercing-sucking herbivore. *New Phytologist* 230(4), 1639–1652. DOI: 10.1111/nph.17251.

Xu, R., Liu, X., Li, J., Qin, R. and Wei, P. (2021b) Identification of herbicide resistance osacc1 mutations via in planta prime-editing-library screening in rice. *Nature Plants* 7(7), 888–892. DOI: 10.1038/s41477-021-00942-w.

Xu, X., Xu, Z., Li, Z., Zakria, M., Zou, L. *et al.* (2021c) Increasing resistance to bacterial leaf streak in rice by editing the promoter of susceptibility gene *ossulrt3; 6*. *Plant Biotechnology Journal* 19(6), 1101–1103. DOI: 10.1111/pbi.13602.

Xu, X., Xu, Z., Ma, W., Haq, F., Li, Y. *et al.* (2021d) Tale-triggered and itale-suppressed xa1 resistance to bacterial blight is independent of ostfiiaγ1 or ostfiiaγ5 in rice. *Journal of Experimental Botany*. DOI: 10.1093/jxb/erab054.

Xu, Z., Xu, X., Li, Y., Liu, L., Wang, Q. *et al.* (2024) Tal6b/avrXa27a, a hidden TALE targeting the susceptibility gene OsSWEET11a and the resistance gene Xa27 in rice. *Plant Communications* 5(2), 100721. DOI: 10.1016/j.xplc.2023.100721.

Yoon, Y.-J., Venkatesh, J., Lee, J.-H., Kim, J., Lee, H.-E. *et al.* (2020) Genome editing of *eiF4E1* in tomato confers resistance to pepper mottle virus. *Frontiers in Plant Science* 11, 1098. DOI: 10.3389/fpls.2020.01098.

Younas, M.U., Ahmad, I., Qasim, M., Ijaz, Z., Rajput, N. *et al.* (2024) Progress in the management of rice blast disease: The role of avirulence and resistance genes through gene-for-gene interactions. *Agronomy* 14(1), 163. DOI: 10.3390/agronomy14010163.

Yu, Y., Pan, Z., Wang, X., Bian, X., Wang, W. *et al.* (2022) Targeting of SPCSV-RNAse3 via CRISPR-Cas13 confers resistance against sweet potato virus disease. *Molecular Plant Pathology* 23(1), 104–117. DOI: 10.1111/mpp.13146.

Yıldırım, K., Miladinović, D., Sweet, J., Akin, M., Galović, V. *et al.* (2023) Genome editing for healthy crops: Traits, tools and impacts. *Frontiers in Plant Science* 14, 1231013. DOI: 10.3389/fpls.2023.1231013.

Zaidi, S.S.-E.-A., Mahas, A., Vanderschuren, H. and Mahfouz, M.M. (2020) Engineering crops of the future: CRISPR approaches to develop climate-resilient and disease-resistant plants. *Genome Biology* 21(1), 289. DOI: 10.1186/s13059-020-02204-y.

Zhan, X., Zhang, F., Zhong, Z., Chen, R., Wang, Y. *et al.* (2019) Generation of virus-resistant potato plants by RNA genome targeting. *Plant Biotechnology Journal* 17(9), 1814–1822. DOI: 10.1111/pbi.13102.

Zhang, Y., Bai, Y., Wu, G., Zou, S., Chen, Y. *et al.* (2017) Simultaneous modification of three homoeologs of TaEDR1 by genome editing enhances powdery mildew resistance in wheat. *The Plant Journal* 91(4), 714–724. DOI: 10.1111/tpj.13599.

Zhang, T., Zheng, Q., Yi, X., An, H., Zhao, Y. *et al.* (2018a) Establishing RNA virus resistance in plants by harnessing CRISPR immune system. *Plant Biotechnology Journal* 16(8), 1415–1423. DOI: 10.1111/pbi.12881.

Zhang, Z., Ge, X., Luo, X., Wang, P., Fan, Q. *et al.* (2018b) Simultaneous editing of two copies of gh14-3-3d confers enhanced transgene-clean plant defense against *Verticillium dahliae* in allotetraploid upland cotton. *Frontiers in Plant Science* 9, 842. DOI: 10.3389/fpls.2018.00842.

Zhang, Y., Massel, K., Godwin, I.D. and Gao, C. (2018c) Applications and potential of genome editing in crop improvement. *Genome Biology* 19(1), 210. DOI: 10.1186/s13059-018-1586-y.

Zhang, P., Du, H., Wang, J., Pu, Y., Yang, C. *et al.* (2019a) Multiplex CRISPR/Cas9-mediated metabolic engineering increases soya bean isoflavone content and resistance to soya bean mosaic virus. *Plant Biotechnology Journal* 18(6), 1384–1395. DOI: 10.1111/pbi.13302.

Zhang, Z., Hua, L., Gupta, A., Tricoli, D., Edwards, K.J. *et al.* (2019b) Development of an agrobacterium-delivered CRISPR/Cas9 system for wheat genome editing. *Plant Biotechnology Journal* 17(8), 1623–1635. DOI: 10.1111/pbi.13088.

Zhang, M., Liu, Q., Yang, X., Xu, J., Liu, G. *et al.* (2020) CRISPR/Cas9-mediated mutagenesis of ClPSK1 in watermelon to confer resistance to *Fusarium oxysporum* f. sp. Niveum. *Plant Cell Reports* 39(5). DOI: 10.1007/s00299-020-02516-0.

Zhang, S., Zhang, R., Gao, J., Song, G., Li, J. *et al.* (2021) CRISPR/Cas9-mediated genome editing for wheat grain quality improvement. *Plant Biotechnology Journal* 19(9), 1684–1686. DOI: 10.1111/pbi.13647.

Zhang, Y., Lin, X.-F., Li, L., Piao, R.-H., Wu, S. *et al.* (2024) CRISPR/Cas9-mediated knockout of Bsr-d1 enhances the blast resistance of rice in Northeast China. *Plant Cell Reports* 43(4), 100. DOI: 10.1007/s00299-024-03192-0.

Zhao, F., Ding, X., Liu, Z., Yan, X., Chen, Y. *et al.* (2023) Application of CRISPR/Cas9-based genome editing in ecotoxicology. *Environmental Pollution*. DOI: 10.3389/fpls.2018.00842.

Zhou, J., Peng, Z., Long, J., Sosso, D., Liu, B. *et al.* (2015) Gene targeting by the TAL effector pthXo2 reveals cryptic resistance gene for bacterial blight of rice. *Plant Journal* 82(4), 632–643. DOI: 10.1111/tpj.12838.

Zhou, J., Liu, Y., Wei, Y., Zheng, S., Gou, S. *et al.* (2022) Eliminating predictable DNA off-target effects of cytosine base editor by using dual guiders including sgRNA and TALE. *Molecular Therapy* 30(7), 2443–2451. DOI: 10.1016/j.ymthe.2022.04.010.

Zhu, Y., Feng, F., Hu, G., Wang, Y., Yu, Y. *et al.* (2021) A genome-wide CRISPR screen identifies host factors that regulate SARS-CoV-2 entry. *Nature Communications* 12(1), 961. DOI: 10.1038/s41467-021-21213-4.

Index

Locators in *italics* and **bold**, indicate *figure* and **table**, respectively.